The Prokaryotes

Eugene Rosenberg (Editor-in-Chief)
Edward F. DeLong, Stephen Lory, Erko Stackebrandt and Fabiano Thompson (Eds.)

# The Prokaryotes

## Firmicutes and Tenericutes

Fourth Edition

With 83 Figures and 133 Tables

Springer Reference

*Editor-in-Chief*
Eugene Rosenberg
Department of Molecular Microbiology and Biotechnology
Tel Aviv University
Tel Aviv, Israel

*Editors*
Edward F. DeLong
Department of Biological Engineering
Massachusetts Institute of Technology
Cambridge, MA, USA

Center for Microbial Oceanography: Research and Education
University of Hawaii, Manoa
Honolulu, HI, USA

Stephen Lory
Department of Microbiology and Immunobiology
Harvard Medical School
Boston, MA, USA

Erko Stackebrandt
Leibniz Institute DSMZ-German Collection of Microorganisms and Cell Cultures
Braunschweig, Germany

Fabiano Thompson
Laboratory of Microbiology, Institute of Biology, Center for Health Sciences
Federal University of Rio de Janeiro (UFRJ)
Ilha do Fundão, Rio de Janeiro, Brazil

ISBN 978-3-642-30119-3 ISBN 978-3-642-30120-9 (eBook)
ISBN 978-3-642-30121-6 (print and electronic bundle)
DOI 10.1007/978-3-642-30120-9
Springer Heidelberg New York Dordrecht London

Library of Congress Control Number: 2014949495



Printed on acid-free paper

Springer is part of Springer Science+Business Media (www.springer.com)

# Foreword

The purpose of this brief foreword is unchanged from the first edition; it is simply to make you, the reader, hungry for the scientific feast that follows. These 11 volumes on the prokaryotes offer an expanded scientific menu that displays the biochemical depth and remarkable physiological and morphological diversity of prokaryote life. The size of the volumes might initially discourage the unprepared mind from being attracted to the study of prokaryote life, for this landmark assemblage thoroughly documents the wealth of present knowledge. But in confronting the reader with the state of the art, the Handbook also defines where more work needs to be done on well-studied bacteria as well as on unusual or poorly studied organisms.

This edition of *The Prokaryotes* recognizes the almost unbelievable impact that the work of Carl Woese has had in defining a phylogenetic basis for the microbial world. The concept that the ribosome is a highly conserved structure in all cells and that its nucleic acid components may serve as a convenient reference point for relating all living things is now generally accepted. At last, the phylogeny of prokaryotes has a scientific basis, and this is the first serious attempt to present a comprehensive treatise on prokaryotes along recently defined phylogenetic lines. Although evidence is incomplete for many microbial groups, these volumes make a statement that clearly illuminates the path to follow.

There are basically two ways of doing research with microbes. A classical approach is first to define the phenomenon to be studied and then to select the organism accordingly. Another way is to choose a specific organism and go where it leads. The pursuit of an unusual microbe brings out the latent hunter in all of us. The intellectual challenges of the chase frequently test our ingenuity to the limit. Sometimes the quarry repeatedly escapes, but the final capture is indeed a wonderful experience. For many of us, these simple rewards are sufficiently gratifying so that we have chosen to spend our scientific lives studying these unusual creatures. In these endeavors, many of the strategies and tools as well as much of the philosophy may be traced to the Delft School, passed on to us by our teachers, Martinus Beijerinck, A. J. Kluyver, and C. B. van Niel, and in turn passed on by us to our students.

In this school, the principles of the selective, enrichment culture technique have been developed and diversified; they have been a major force in designing and applying new principles for the capture and isolation of microbes from nature. For me, the "organism approach" has provided rewarding adventures. The organism continually challenges and literally drags the investigator into new areas where unfamiliar tools may be needed. I believe that organism-oriented research is an important alternative to problem-oriented research, for new concepts of the future very likely lie in a study of the breadth of microbial life. The physiology, biochemistry, and ecology of the microbe remain the most powerful attractions. Studies based on classical methods as well as modern genetic techniques will result in new insights and concepts.

To some readers, this edition of *The Prokaryotes* may indicate that the field is now mature, that from here on it is a matter of filling in details. I suspect that this is not the case. Perhaps we have assumed prematurely that we fully understand microbial life. Van Niel pointed out to his students that—after a lifetime of study—it was a very humbling experience to view in the microscope a sample of microbes from nature and recognize only a few. Recent evidence suggests that microbes have been evolving for nearly 4 billion years. Most certainly, those microbes now domesticated and kept in captivity in culture collections represent only a minor portion of the species that have evolved in this time span. Sometimes we must remind ourselves that evolution is actively taking place at the present moment. That the eukaryote cell evolved as a chimera of certain prokaryote parts is a generally accepted concept today. Higher as well as lower eukaryotes evolved in contact with prokaryotes, and evidence surrounds us of the complex interactions between eukaryotes and prokaryotes as well as among prokaryotes. We have so far only scratched the surface of these biochemical interrelationships. Perhaps the legume nodule is a pertinent example of nature caught in the act of evolving the "nitrosome," a unique nitrogen-fixing organelle. The study of prokaryotes is proceeding at such a fast pace that major advances are occurring yearly. The increase of this edition to four volumes documents the exciting pace of discoveries.

To prepare a treatise such as *The Prokaryotes* requires dedicated editors and authors; the task has been enormous. I predict that the scientific community of microbiologists will again show its appreciation through use of these volumes—such that the pages will become "dog-eared" and worn as students seek basic information for the hunt. These volumes belong in the laboratory, not in the library. I believe that a most effective way to introduce students to microbiology is for them to isolate microbes from nature, that is, from their habitats in soil, water, clinical specimens, or plants. *The Prokaryotes* enormously simplifies this process and should encourage the construction of courses that contain a wide spectrum of diverse topics. For the student as well as the advanced investigator, these volumes should generate excitement.

Happy hunting!

Ralph S. Wolfe
*Department of Microbiology*
*University of Illinois at Urbana-Champaign*

# Preface

During most of the twentieth century, microbiologists studied pure cultures under defined laboratory conditions in order to uncover the causative agents of disease and subsequently as ideal model systems to discover the fundamental principles of genetics and biochemistry. Microbiology as a discipline onto itself, e.g., microbial ecology, diversity, and evolution-based taxonomy, has only recently been the subject of general interest, partly because of the realization that microorganisms play a key role in the environment. The development and application of powerful culture-independent molecular techniques and bioinformatics tools has made this development possible. The fourth edition of *The Prokaryotes* has been updated and expanded in order to reflect this new era of microbiology.

The first five volumes of the fourth edition contain 34 updated and 43 entirely new chapters. Most of the new chapters are in the two new sections: Prokaryotic Communities and Bacteria in Human Health and Disease. A collection of microorganisms occupying the same physical habitat is called a "community," and several examples of bacterial communities are presented in the Prokaryotic Communities section, organized by Edward F. DeLong. Over the last decade, important advances in molecular biology and bioinformatics have led to the development of innovative culture-independent approaches for describing microbial communities. These new strategies, based on the analysis of DNA directly extracted from environmental samples, circumvent the steps of isolation and culturing of microorganisms, which are known for their selectivity leading to a nonrepresentative view of prokaryotic diversity. Describing bacterial communities is the first step in understanding the complex, interacting microbial systems in the natural world.

The section on Bacteria in Human Health and Disease, organized by Stephen Lory, contains chapters on most of the important bacterial diseases, each written by an expert in the field. In addition, there are separate general chapters on identification of pathogens by classical and non-culturing molecular techniques and virulence mechanisms, such as adhesion and bacterial toxins. In recognition of the recent important research on beneficial bacteria in human health, the section also includes chapters on gut microbiota, prebiotics, and probiotics. Together with the updated and expanded chapter on Bacterial Pharmaceutical Products, this section is a valuable resource to graduate students, teachers, and researchers interested in medical microbiology.

Volumes 6–11, organized by Erko Stackebrandt and Fabiano Thompson, contain 265 chapters in total on each of the ca. 300 known prokaryotic families, in some cases even higher taxa. Each chapter presents both the historical and current taxonomy of these taxa, mostly above the genus level; molecular analyses (e.g., DDH, MLSA, riboprinting, and MALDI-TOF); genomic and phenetic properties of the taxa covered; genome analyses including nonchromosomal genetic elements; phenotypic analyses; methods for the enrichment, isolation, and maintenance of members of the family; ecological studies; clinical relevance; and applications.

As in the third edition, the volumes in the fourth edition are available both as hard copies and as eReferences. The advantages of the online version include no restriction of color illustrations, the possibility of updating chapters continuously and, most importantly, libraries can place their subscribed copies on their servers, making it available to their community in offices and laboratories. The editors thank all the chapter authors and the editorial staff of Springer, especially Hanna Hensler-Fritton, Isabel Ullmann, Daniel Quiñones, Alejandra Kudo, and Audrey Wong, for making this contribution possible.

Eugene Rosenberg
Editor-in-Chief

# About the Editors

**Eugene Rosenberg (Editor-in-Chief)**
Department of Molecular Microbiology and Biotechnology
Tel Aviv University
Tel Aviv
Israel

**Eugene Rosenberg** holds a Ph.D. in biochemistry from Columbia University (1961) where he described the chemical structures of the capsules of *Hemophilus influenzae*, types B, E, and F. His postdoctoral research was performed in organic chemistry under the guidance of Lord Todd in Cambridge University. He was an assistant and associate professor of microbiology at the University of California at Los Angeles from 1962 to 1970, where he worked on the biochemistry of *Myxococcus xanthus*. Since 1970, he has been in the Department of Molecular Microbiology and Biotechnology, Tel Aviv University, as an associate professor (1970–1974), full professor (1975–2005), and professor emeritus (2006–present). He has held the Gol Chair in Applied and Environmental Microbiology since 1989. He is a member of the American Academy of Microbiology and European Academy of Microbiology. He has been awarded a Guggenheim Fellowship, a Fogarty International Scholar of the NIH, the Pan Lab Prize of the Society of Industrial Microbiology, the Proctor & Gamble Prize of the ASM, the Sakov Prize, the Landau Prize, and the Israel Prize for a "Beautiful Israel."

His research has focused on myxobacteriology; hydrocarbon microbiology; surface-active polymers from *Acinetobacter*; bioremediation; coral microbiology; and the role of symbiotic microorganisms in the adaptation, development, behavior, and evolution of animals and plants. He is the author of about 250 research papers and reviews, 9 books, and 16 patents.

**Edward F. DeLong**
Department of Biological Engineering
Massachusetts Institute of Technology
Cambridge, MA
USA
and
Center for Microbial Oceanography: Research and Education
University of Hawaii, Manoa
Honolulu, HI
USA

**Edward DeLong** received his bachelor of science in bacteriology at the University of California, Davis, and his Ph.D. in marine biology at Scripps Institute of Oceanography at the University of California, San Diego. He was a professor at the University of California, Santa Barbara, in the Department of Ecology for 7 years, before moving to the Monterey Bay Aquarium Research Institute where he was a senior scientist and chair of the science department, also for 7 years. He has worked for the past 10 years as a professor at the Massachusetts Institute of Technology in the Department of Biological Engineering, and the Department of Civil and Environmental Engineering, and in August 2014 joined the University of Hawaii as a professor of oceanography. DeLong's scientific interests focus primarily on central questions in marine microbial genomics, biogeochemistry, ecology, and evolution. A large part of DeLong's efforts have been devoted to the study of microbes and microbial processes in the ocean, combining laboratory and field-based approaches. Development and application of genomic, biochemical, and metabolic approaches to study and exploit microbial communities and processes is his other area of interest. DeLong is a fellow in the American Academy of Arts and Science, the U.S. National Academy of Science, and the American Association for the Advancement of Science.

**Stephen Lory**
Department of Microbiology and Immunobiology
Harvard Medical School
Boston, MA
USA

**Stephen Lory** received his Ph.D. degree in microbiology from the University of California in Los Angeles in 1980. The topic of his doctoral thesis was the structure-activity relationships of bacterial exotoxins. He carried out his postdoctoral research on the basic mechanism of protein secretion by Gram-negative bacteria in the Bacterial Physiology Unit at Harvard Medical School. In 1984, he was appointed assistant professor in the Department of Microbiology at the University of Washington in Seattle, becoming full professor in 1995. While at the University of Washington, he developed an active research program in host-pathogen interactions including the role of bacterial adhesion to mammalian cells in virulence and regulation of gene expression by bacterial pathogens. In 2000, he returned to Harvard Medical School where he is currently a professor of microbiology and immunobiology. He is a regular reviewer of research projects on various scientific panels of governmental and private funding agencies and served for four years on the Scientific Council of Institute Pasteur in Paris. His current research interests include evolution of bacterial virulence, studies on post-translational regulation of gene expression in *Pseudomonas*, and the development of novel antibiotics targeting multi-drug-resistant opportunistic pathogens.

**Erko Stackebrandt**
Leibniz Institute DSMZ-German Collection of Microorganisms and Cell Cultures
Braunschweig
Germany

**Erko Stackebrandt** holds a Ph.D. in microbiology from the Ludwig-Maximilians University Munich (1974). During his postdoctoral research, he worked at the German Culture Collection in Munich (1972–1977), 1978 with Carl Woese at the University of Illinois, Urbana Champaign, and from 1979 to 1983 he was a member of Karl Schleifer's research group at the Technical University, Munich. He habilitated in 1983 and was appointed head of the Departments of Microbiology at the University of Kiel (1984–1990), at the University of Queensland, Brisbane, Australia (1990–1993), and at the Technical University Braunschweig, where he also was the director of the DSMZ-German Collection of Microorganisms and Cell Cultures GmbH (1993–2009). He is involved in systematics, and molecular phylogeny and ecology of Archaea and Bacteria for more than 40 years. He has been involved in many research projects funded by the German Science Foundation, German Ministry for Science and Technology, and the European Union, working on pure cultures and microbial communities. His projects include work in soil and peat, Mediterranean coastal waters, North Sea and Baltic Sea, Antarctic Lakes, Australian soil and artesian wells, formation of stromatolites, as well as on giant ants, holothurians, rumen of cows, and the digestive tract of koalas. He has been involved in the description and taxonomic revision of more than 650 bacteria taxa of various ranks. He received a Heisenberg stipend (1982–1983) and his work has been awarded by the Academy of Science at Göttingen, Bergey's Trust (Bergey's Award and Bergey's Medal), the Technical University Munich, the Australian Society for Microbiology, and the American Society for Microbiology. He held teaching positions in Kunming, China; Budapest, Hungary; and Florence, Italy. He has published more than 600 papers in refereed journals and has written more than 80 book chapters. He is the editor of two Springer journals and served as an associate editor of several international journals and books as well as on national and international scientific and review panels of the German Research Council, European Science Foundation, European Space Agency, and the Organisation for Economic Co-Operation and Development.

**Fabiano Thompson**
Laboratory of Microbiology
Institute of Biology
Center for Health Sciences
Federal University of Rio de Janeiro (UFRJ)
Ilha do Fundão
Rio de Janeiro
Brazil

**Fabiano Thompson** became a professor of the Production Engineer Program (COPPE-UFRJ) in 2014 and the director of research at the Institute of Biology, Federal University of Rio de Janeiro (UFRJ), in 2012. He was an oceanographer at the Federal University of Rio Grande (Brazil) in 1997. He received his Ph.D. in biochemistry from Ghent University (Belgium) in 2003, with emphasis on marine microbial taxonomy and biodiversity. Thompson was an associate researcher in the BCCM/LMG Bacteria Collection (Ghent University) in 2004; professor of genetics in 2006 at the Institute of Biology, UFRJ; and professor of marine biology in 2011 at the same university. He has been a representative of UFRJ in the National Institute of Metrology (INMETRO) since 2009. Thompson is the president of the subcommittee on the Systematics of Vibrionaceae–IUMS and an associate editor of *BMC Genomics* and *Microbial Ecology.* The Thompson Lab in Rio currently performs research on marine microbiology in the Blue Amazon, the realm in the southwestern Atlantic that encompasses a variety of systems, including deep sea, Cabo Frio upwelling area, Amazonia river-plume continuum, mesophotic reefs, Abrolhos coral reef bank, and Oceanic Islands (Fernando de Noronha, Saint Peter and Saint Paul, and Trindade).

# Table of Contents

# List of Contributors

**André Antunes**
IBB–Institute for Biotechnology and Bioengineering, Centre of Biological Engineering
Micoteca da Universidade do Minho, University of Minho
Braga
Portugal
and
Rianda Research–Centro de Investigação em Energia, Saúde e Ambiente
Coimbra
Portugal

**Mitchell F. Balish**
Department of Microbiology
Miami University
Oxford, OH
USA

**Wajdi Ben Hania**
CNRS/INSU, IRD, Aix-Marseille Université, Université du Sud Toulon-Var
Marseille
France

**Marcelo Rafael Benítez-Ahrendts**
National University of Jujuy
Jujuy
Argentina

**Johanna Björkroth**
Department of Food Hygiene and Environmental Health, Faculty of Veterinary Medicine
University of Helsinki
Helsinki
Finland

**Alain Blanchard**
INRA, UMR 1332 Biologie du Fruit et Pathologie
University of Bordeaux
Villenave d'Ornon
France

**Amel Bouanane-Darenfed**
Faculté de Biologie, Laboratoire de Biologie Cellulaire et Moléculaire (Equipe de Microbiologie)
Université des Sciences et de la Technologie Houari Boumediene, Bab ezzouar
Alger
Algérie

**Matthew E. Caldwell**
Department of Microbiology and Plant Biology
University of Oklahoma
Norman, OK
USA

**Leonor Carrillo**
National University of Jujuy
Jujuy
Argentina

**Young-Hyo Chang**
Korean Collection for Type Cultures, Biological Resource Centre
Korea Research Institute of Bioscience and Biotechnology
Daejeon
Republic of Korea

**Rafael R. de la Haba**
Department of Microbiology and Parasitology
University of Sevilla
Sevilla
Spain

**Leon Dicks**
Department of Microbiology
University of Stellenbosch
Stellenbosch
South Africa

**Christine E. R. Dodd**
School of Biosciences
University of Nottingham Sutton Bonnington Campus
Loughborough, Leicestershire
UK

**Akihito Endo**
Department of Food and Cosmetic Science
Tokyo University of Agriculture
Hokkaido
Japan

**Marie-Laure Fardeau**
Aix-Marseille Université, Université du Sud
Toulon-Var, CNRS/INSU, IRD, MIO, UM 110
Marseille
France

**Giuseppe Firrao**
Dipartimento di Scienze Agrarie ed Ambientali
Università degli Studi di Udine
Udine
Italy

**Gail E. Gasparich**
Fisher College of Science and Mathematics
Towson University
Towson, MD
USA

**Markus Göker**
Leibniz Institute DSMZ-German Collection of Microorganisms and Cell Cultures GmbH
Braunschweig
Germany

**Per Johansson**
Department of Food Hygiene and Environmental Health, Faculty of Veterinary Medicine
University of Helsinki
Helsinki
Finland

**Estelle Jumas-Bilak**
Equipe Pathogènes et Environnements, UMR5119 ECOSYM
Université Montpellier 1
Montpellier
France
and
Laboratoire d'Hygiène hospitalière
Centre Hospitalier Régional Universitaire de Montpellier
Montpellier
France

**Paul A. Lawson**
Department of Microbiology and Plant Biology
University of Oklahoma
Norman, OK
USA

**Ing-Ming Lee**
Molecular Plant Pathology Laboratory, USDA, ARS
Beltsville, MD
USA

**Niall A. Logan**
Department of Life Sciences
Glasgow Caledonian University
Glasgow
Scotland, UK

**Stephen Lory**
Department of Microbiology and Immunobiology
Harvard Medical School
Boston, MA
USA

**Michael T. Madigan**
Department of Microbiology
Southern Illinois University
Carbondale, IL
USA

**Hélène Marchandin**
Equipe Pathogènes et Environnements, UMR5119 ECOSYM
Université Montpellier 1
Montpellier
France
and
Laboratoire de Bactériologie
Centre Hospitalier Régional Universitaire de Montpellier
Montpellier
France

**Carmine Marcone**
Dipartimento di Farmacia
Università degli Studi di Salerno
Fisciano, Salerno
Italy

**Marta Martini**
Dipartimento di Scienze Agrarie ed Ambientali
Università degli Studi di Udine
Udine
Italy

**Meghan May**
Department of Biological Sciences
Towson University
Towson, MD
USA

**Shanmugam Mayilraj**
Microbial Type Culture and Gene Bank (MTCC)
CSIR-Institute of Microbial Technology
Chandigarh
India

**Jim McLauchlin**
Public Health England
Food Water and Environmental Microbiology Services
London
UK

**Koji Mori**
Biological Resource Center
National Institute of Technology and Evaluation (NBRC)
Kisarazu, Chiba
Japan

**Raúl Muñoz**
Marine Microbiology Group, Department of Ecology and Marine Resources
Institut Mediterráni d'Estudis Avancats (CSIC-UIB)
Esporles, Illes Balears
Spain

**Timo T. Nieminen**
Ruralia Institute
University of Helsinki
Seinäjoki
Finland

**Bernard Ollivier**
Aix-Marseille Université, Université du Sud
Toulon-Var, CNRS/INSU, IRD, MIO, UM 110
Marseille
France

**Ingar Olsen**
Department of Oral Biology, Faculty of Dentistry
University of Oslo
Oslo
Norway

**Aharon Oren**
Department of Plant and Environmental Sciences
The Institute of Life Sciences, The Hebrew University of Jerusalem
Jerusalem
Israel

**G. S. N. Reddy**
CSIR-Centre for Cellular and Molecular Biology
Hyderabad
India

**Catherine E. D. Rees**
School of Biosciences
University of Nottingham Sutton Bonnington Campus
Loughborough, Leicestershire
UK

**Laura B. Regassa**
Department of Biology
Georgia Southern University
Statesboro, GA
USA

**Elina Säde**
Department of Food Hygiene and Environmental Health, Faculty of Veterinary Medicine
University of Helsinki
Helsinki
Finland

**Cristina Sánchez-Porro**
Department of Microbiology and Parasitology
University of Sevilla
Sevilla
Spain

**W. Matthew Sattley**
Department of Biology
Indiana Wesleyan University
Marion, IN
USA

**Carmen Scheuner**
Leibniz Institute DSMZ-German Collection of Microorganisms and Cell Cultures GmbH
Braunschweig
Germany

**Bernhard Schink**
Department of Biology
University of Konstanz
Konstanz
Germany

**Peter Schumann**
Leibniz Institute DSMZ-German Collection of Microorganisms and Cell Cultures GmbH
Braunschweig
Germany

**S. Shivaji**
CSIR-Centre for Cellular and Molecular Biology
Hyderabad
India

**Alexander Slobodkin**
Winogradsky Institute of Microbiology
Russian Academy of Sciences
Moscow
Russia

**Tatyana Sokolova**
Winogradsky Institute of Microbiology
Russian Academy of Sciences
Moscow
Russia

**T. N. R. Srinivas**
Regional Centre
CSIR-National Institute of Oceanography
Visakhapatnam
India

**Erko Stackebrandt**
Leibniz Institute DSMZ-German Collection of Microorganisms and Cell Cultures GmbH
Braunschweig
Germany

**Antonio Ventosa**
Department of Microbiology and Parasitology
University of Sevilla
Sevilla
Spain

**Susanne Verbarg**
Leibniz Institute DSMZ-German Collection of Microorganisms and Cell Cultures GmbH
Braunschweig
Germany

**Juergen Wiegel**
Department of Microbiology
The University of Georgia
Athens, GA
USA

# Firmicutes

# 1 The Family *Aerococcaceae*

*Erko Stackebrandt*
Leibniz Institute DSMZ-German Collection of Microorganisms and Cell Cultures GmbH, Braunschweig, Germany



**Abstract**

The family *Aerococcaceae* is a member of the order *Lactobacillales*, phylum Firmicutes. It comprises the genera *Aerococcus*, *Abiotrophia*, *Dolosicoccus*, *Eremococcus*, *Facklamia*, *Globicatella*, and *Ignavigranum* and has been created on the basis of the phylogenetic position of its members as analyzed by comparative 16S rRNA gene sequence analysis. Members are Gram-positive, nonmotile and non-spore-forming, facultative anaerobic, catalase-negative ovoid cocci or coccibacilli. Lysine has been reported to occur as the diagnostic diamino acid in the peptidoglycan. The habitat includes a wide range of environments, e.g., human specimens, household, schoolrooms, yard, street and hospital environments, and marine sites, and they have also been reported in human infections. This chapter is a short update of species described since 2005, which are not covered in the chapter *Aerococcaceae* in *Bergey's Manual of Systematic Bacteriology*, 2nd edition.

**Table 1.1**
**Some properties of *Aerococcus* species described since 2005. *Aerococcus viridans* is included as a reference**

| Property | *A. urinaeequi*[a, b] | *A. vaginalis* | *A. viridans*[b] | *A. suis* |
|---|---|---|---|---|
| Type strain | ATCC 29723$^T$ | BV2$^T$ | NCTC 8251$^T$ | 1821/02$^T$ |
| Acid from | | | | |
| Lactose | + | – | + | – |
| Mannitol | + | – | + | – |
| Maltose | + | + | + | –(+[b]) |
| Ribose | + | + | + | + (delayed) |
| Sorbitol | – | – | – | – |
| Sucrose | + | – | + | – |
| Tagatose | - | + | – | + (–[b]) |
| Trehalose | + | – | – | – |
| API 20 strep | | | | |
| Arginine | – | – | – | + |
| Hippurate | + | + | + | – |
| Aesculin | – | – | + | – |
| API ZYM | | | | |
| Alkaline phosphatase | + | + | + | – |
| C-4 esterase | – | + | + | + |
| C-8 esterase | – | - | + | + |
| Acid phosphatase | + | + | – | + |
| ß-Galactosidase | - | + | – | + |
| ß-Glucosidase | + | – | – | – |
| Mol% G+C | 39.5 | 44.7 | 41.8[c] | 37 |
| Habitat | | Vaginal mucosa of beef cow | Clinical specimens | Brain from pig with meningitis |
| | Felis et al. 2005 | Tohno et al. 2014 | Collins and Falsen 2009 | Vela et al. 2007 |

[a]Originally described as *Pediococcus urinaeequi*
[b]Emended by Tohno et al. (2014)
[c]Bohacek et al. (1969)
Abbreviation: *d* strain dependent, *nr* not recorded

E. Rosenberg et al. (eds.), *The Prokaryotes – Firmicutes and Tenericutes*, DOI 10.1007/978-3-642-30120-9_349,

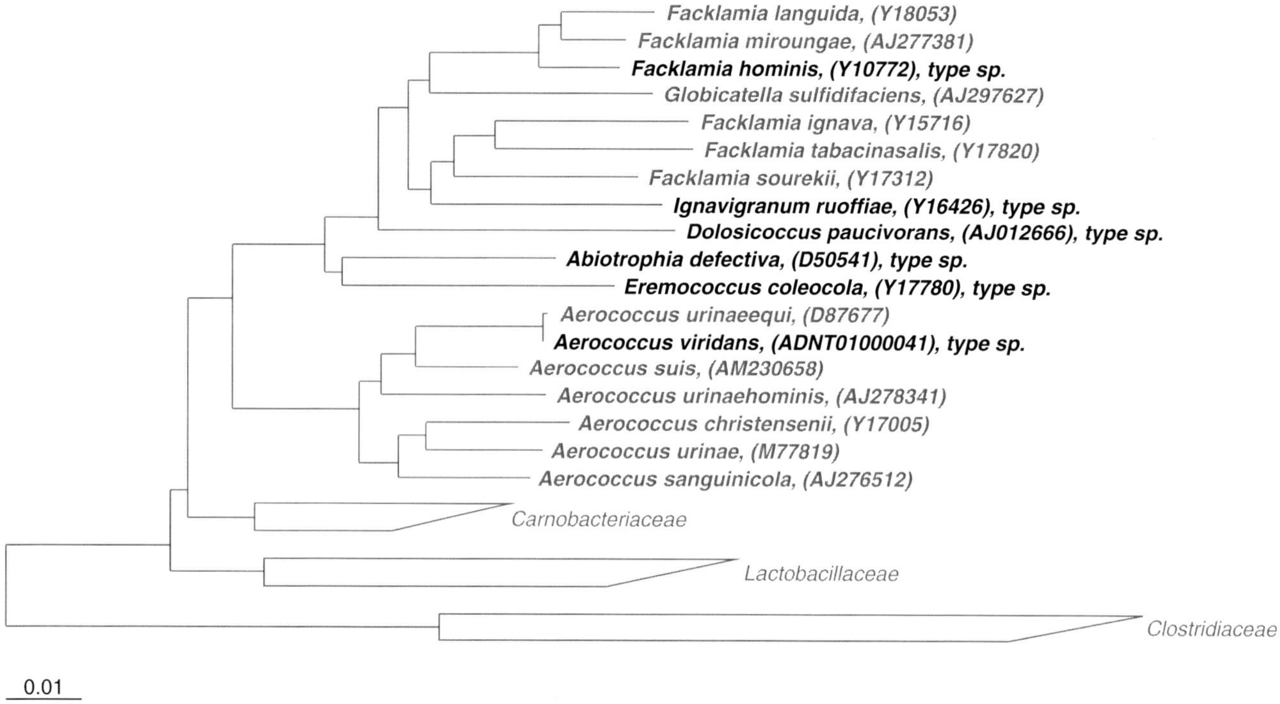


**Fig. 1.1**
**Neighbor-joining genealogy reconstruction based on the RAxML algorithm (Stamatakis 2006) of the sequences of members of the family *Aerococcaceae* and some neighboring taxa present in the LTP_106 (Yarza et al. 2010). The tree was reconstructed by using a subset of sequences. Representative sequences from closely relative genera were used to stabilize the tree topology. In addition, a 40 % maximum frequency filter was applied to remove hypervariable positions from the alignment. Scale bar indicates estimated sequence divergence**

**Table 1.2**
**Examples of 16S rRNA gene sequence of clones retrieved from environmental samples**

| Type species | Example accession numbers | Habitat | Reference |
|---|---|---|---|
| *Facklamia hominis* | FM873983, FM873872 | House dust | Taubel et al. 2009 |
| | AM697040 | Indoor environment | Rintala et al. 2008 |
| | AY958863 | Vaginal epithelium | Hyman et al. 2005 |
| | HQ811680 | Inflammatory bowel | Li et al. 2012 |
| | HM317482, HM272006, JF116170 | Skin microbiome | Kong et al. 2012 |
| *Ignavigranum ruoffiae* | FM872514 | House dust | Taubel et al. 2009 |
| *Abiotrophia defectiva* | AY879307, AY879308 | Bloodstream and endovascular infections | Senn et al. 2006 |
| | AM420170 | Gingivitis | Bolivar et al. 2012 |
| | JQ469612, JQ469565 | Human mouth | Davis et al., unpublished |
| | DQ346440 | Human oral microflora | Diaz et al. 2006 |
| *Eremococcus coleocola* | GQ009094, GQ001484 | Skin microbiome | Grice et al. 2009 |
| *Aerococcus viridans* | JN644571 | Midgut *Culex* spp. | Chandel et al. 2013 |
| | JN713500 | Canine oral microflora | Dewhirst et al. 2012 |
| | AM697119 | Indoor environment | Rintala et al. 2008 |
| | FM875629 | House dust | Taubel et al. 2009 |

**Table 1.3**
**Selection of incomplete or published genomes of members of the family *Aerococcus* according to the GOLD database genomes.org/cgi-bib/Gold/Search.cgi**

| Species | Strain | Accession number | Sequence status |
|---|---|---|---|
| *Aerococcus christensenii* | DSM 15819 | Gi0043392 | Incomplete |
| *Aerococcus viridans* | LL1 | Gi17211 | Qin et al. 2012 |
| *Aerococcus urinaeequi* | DSM 20341 | Gi11098 | Permanent draft |
| *Aerococcus urinae* | ACS-120-V-Col10a | Gc01697 | Complete and deposited |
| *Aerococcus christensenii* | DSM 15819 | Gi03497 | Incomplete |
| *Abiotrophia defectiva* | ATCC $49176^T$ | Gi03551 | Permanent draft |
| *Eremococcus coleocola* | DSM 15696 | Gi11475 | Permanent draft |
| *Facklamia languida* | CCUG 37842 | Gi07365 | Permanent draft |
| *Facklamia hominis* | CCUG 36813 | Gi07363 | Permanent draft |
| *Facklamia tabacinasalis* | FAM208_56 | Gi23620 | Incomplete |
| *Facklamia ignava* | CCUG 37419 | Gi07364 | Permanent draft |
| *Facklamia sourekii* | ATCC 700629 | Gi15305 | Incomplete |

## Taxonomy and Phylogeny

The range of genera included in *Aerococcaceae* (Ludwig et al. 2009a), order *Lactobacillales* (Ludwig et al. 2009b) has not been expanded since 2009, and only a few novel species were described (*Table 1.1*). Differentiation of genera is mostly done on the basis of phenotypic tests, vancomycin resistance, growth in 6.5 % NaCl, or acid and/or gas production from glucose fermentation. The diagnostic amino acid of peptidoglycan is L-lysine (direct cross-linked), and the major fatty acid (>60 %) of *Aerococcus* species is $C_{18:1}\omega9c$ (Tohno et al. 2014).

The neighbor-joining tree (*Fig. 1.1*) and the maximum-likelihood tree (not shown) see the family as a monophyletic clade, branching adjacent to *Carnobacteriaceae* and *Lactobacillaceae*. The genus *Facklamia* appears polyphyletic, as *Globicatella sulfidifaciens* and *Ignavigranum rouffiae* form two independent lineages among *Facklamia* species. Members of these three genera are also phenotypically similar as noted by Collins (2009a, b). Properties recorded to differentiate members of the three genera could as well be used for species differentiation in a redefined genus *Facklamia*. The species *Aerococcus urinaeequi* is highly related to *A. viridans* (99.9 % 16S rRNA gene sequence similarity), but genomically different by DNA-DNA reassociation experiments with 51.0 % similarity between the type strains of the two species which confirms their separate taxonomic status. The description of all *Aerococcus* species has recently been emended (Tohno et al. 2014).

As judged from low number of entries into public sequence databases, the number of clone sequences which are closely related (>99 % 16S rRNA gene sequence similarity) to described family members are rare, and clones related to different genera may occur in the same habitat (*Table 1.2*). *Table 1.3* is a selection of *Aerococcaceae* strains for which genome sequences (different degree of completeness) have been deposited.

## References


Bohacek J, Blazicek G, Kocur M, Solberg O, Clausen OG (1969) Deoxyribonucleic acid base composition of pediococci and aerococci. Arch Microbiol 67:58–61

Bolivar I, Whiteson K, Stadelmann B, Baratti-Mayer D, Gizard Y, Mombelli A, Pittet D, Schrenzel J (2012) CONSRTM The Geneva study group on Noma (GESNOMA) bacterial diversity in oral samples of children in Niger with acute Noma, acute necrotizing gingivitis, and healthy controls. PLoS Negl Trop Dis 6:E1556

Chandel K, Mendki MJ, Parikh RY, Kulkarni G, Tikar SN, Sukumaran D, Prakash S, Parashar BD, Shouche YS, Veer V (2013) Midgut microbial community of *Culex quinquefasciatus* mosquito populations from India. PLoS ONE 8, E80453

Collins MD (2009a) Genus III *Dolosicoccus*. In: De Vos P, Garrity GM, Jones D, Krieg NR, Ludwig W, Rainey FA, Schleifer KH, Whitman WB (eds) Bergey's manual of systematic bacteriology, 2nd edn, Firmicutes. Springer, New York, pp 538–540

Collins MD (2009b) Genus VII *Ignavigranum*. In: De Vos P, Garrity GM, Jones D, Krieg NR, Ludwig W, Rainey FA, Schleifer KH, Whitman WB (eds) Bergey's manual of systematic bacteriology, 2nd edn, Firmicutes. Springer, New York, pp 546–547

Collins MD, Falsen E (2009) Genus I *Aerococcus*. In: VOS PDE, Garrity GM, Jones D, Krieg NR, Ludwig W, Rainey FA, Schleifer KH, Whitman WB (eds) Bergey's manual of systematic bacteriology, vol 3, 2nd edn, The Firmicutes. Springer, Dordrecht/Heidelberg/London/New York, pp 533–536

Dewhirst FE, Klein EA, Thompson EC, Blanton JM, Chen T et al (2012) The canine oral microbiome. PLoS ONE 7, E36067

Diaz PI, Chalmers NI, Rickard AH, Kong C, Milburn CL, Palmer RJ Jr, Kolenbrander PE (2006) Molecular characterization of subject-specific oral microflora during initial colonization of enamel. Appl Environ Microbiol 72:2837–2848

Felis GE, Torriani S, Dellaglio F (2005) Reclassification of *Pediococcus urinaeequi* (*ex* Mees 1934) Garvie 1988 as *Aerococcus urinaeequi* comb. nov. Int J Syst Evol Microbiol 55:1325–1327

Grice EA, Kong HH, Conlan S, Deming CB, Davis J, Young AC, Bouffard GG, Blakesley RW, Murray PR, Green ED, Turner ML, Segre JA (2009) CONSRTM NISC comparative sequencing program topographical and temporal diversity of the human skin microbiome. Science 324:1190–1192

Hyman RW, Fukushima M, Diamond L, Kumm J, Giudice LC, Davis RW (2005) Microbes on the human vaginal epithelium. Proc Natl Acad Sci USA 102:7952–7957

Kong HH, Oh J, Deming C, Conlan S, Grice EA et al (2012) CONSRTM NISC Comparative sequence program temporal shifts in the skin microbiome associated with disease flares and treatment in children with atopic dermatitis. Genome Res 22:850–859

Li E, Hamm CM, Gulati AS, Sartor RB, Chen H, Wu X et al (2012) Inflammatory bowel diseases phenotype, *C. difficile* and NOD2 genotype are associated with shifts in human ileum associated microbial composition. PLoS ONE 7, E26284

Ludwig W, Schleifer KH, Whitman WB (2009a) Order II. *Lactobacillales* ord. nov. In: De Vos P, Garrity GM, Jones D, Krieg NR, Ludwig W, Rainey FA, Schleifer KH, Whitman WB (eds) Bergey's manual of systematic bacteriology, 2nd edn, Firmicutes. Springer, New York, p 464, Validation List no. 132 (2010) Int J Syst Evol Microbiol 60, 469–472

Ludwig W, Schleifer KH, Whitman WB (2009b) Family II. *Aerococcaceae* fam. nov. In: De Vos P, Garrity GM, Jones D, Krieg NR, Ludwig W, Rainey FA, Schleifer KH, Whitman WB (eds) Bergey's manual of systematic bacteriology, 2nd edn, Firmicutes. Springer, New York, p 533, Validation List no. 132 (2010) Int J Syst Evol Microbiol 60, 469–472

Qin N, Zheng B, Yang F, Chen Y, Guo J, Hu X, Li L (2012) Genome Sequence of *Aerococcus viridans* LL1. J Bacteriol 194:4143

Rintala H, Pitkaranta M, Toivola M, Paulin L, Nevalainen A (2008) Diversity and seasonal dynamics of bacterial community in indoor environment. BMC Microbiol 8:56

Senn L, Entenza JM, Greub G, Jaton K, Wenger A, Bille J, Calandra T, Prod'hom G (2006) Bloodstream and endovascular infections due to *Abiotrophia defectiva* and *Granulicatella* species. BMC Infect Dis 6:9. doi:10.1186/1471-2334-6-9

Stamatakis A (2006) RAxML-VI-HPC maximum likelihood-based phylogenetic analyses with thousands of taxa and mixed models. Bioinformatics 22:2688–2690

Taubel M, Rintala H, Pitkaranta M, Paulin L, Laitinen S, Pekkanen J, Hyvarinen A, Nevalainen A (2009) The occupant as a source of house dust bacteria. J Allergy Clin Immunol 124:834–840

Tohno M, Kitahara M, Matsuyama S, Kimura K, Ohkuma M, Tajima K (2014) *Aerococcus vaginalis* sp. nov. isolated from the vaginal mucosa of a beef cow in Japan, and emended descriptions of *A. suis, A. viridans, A. urinaeequi, A. urinaehominis, A. urinae, A. christensenii* and *A. sanguinicola*. Int J Syst Evol Microbiol 64:1229–1236

Vela AI, García N, Latre MV, Casamayor A, Sánchez-Porro C, Briones V, Ventosa A, Domínguez L, Fernández-Garayzábal JF (2007) *Aerococcus suis* sp. nov., isolated from clinical specimens from swine. Int J Syst Evol Microbiol 57:1291–1294

Yarza P, Ludwig W, Euzéby J, Amann R, Schleifer K-H, Glöckner FO, Rosselló-Móra R (2010) Update of the All-species living-tree project based on 16S and 23S rRNA sequence analyses. Syst Appl Microbiol 33:291–299

# 2 The Family *Alicyclobacillaceae*

*Erko Stackebrandt*
Leibniz Institute DSMZ-German Collection of Microorganisms and Cell Cultures GmbH, Braunschweig, Germany

*Introduction* ................................................ *7*

*Taxonomy* ................................................ *7*

**Abstract**
The family *Alicyclobacillaceae* comprises the genera *Alicyclobacillus*, *Kyrpidia*, and *Tumebacillus*. The family is phylogenetically monophyletic, branching next to *Bacillaceae*, *Paenibacillaceae*, *Listeriaceae*, and *Planococcaceae*. In contrast to its genus affiliation, *Alicyclobacillus pohliae* branches outside the radiation of other *Alicyclobacillus* species between members of *Kyrpidia* and *Tumebacillus*. The inclusion of *Kyrpidia* into the family led to the emendation of *Alicyclobacillaceae* in that acid may or may not be produced from carbohydrates (Klenk et al. 2011 Stand Genomic Sci 5:121–134, 2011).

## Introduction

The family *Alicyclobacillaceae* (da Costa and Rainey 2009, Euzéby 2010) comprises the genera *Alicyclobacillus* (Witsotzkey et al. 1992), *Kyrpidia* (Klenk et al. 2011), and *Tumebacillus* (Steven et al. 2008). The family is a member of the order *Bacillales* (Prévot 1953), class *Bacilli* (Ludwig et al. 2009; Euzéby 2010), and phylum Firmicutes (Gibbons and Murray 1978). Since its last comprehensive coverage in *Bergey's Manual of Systematic Bacteriology* (da Costa and Rainey 2009), *Tumebacillus* and *Kyrpidia* as well as several new species of *Alicyclobacillus* have been described and added to the family (see http://www.bacterio.net/) (➋ *Fig. 2.1*). The present communication will be restricted to a short coverage of the novel taxa, and the reader is referred to the history, enrichment, and taxonomy of *Alicyclobacillus* species covered in *Bergey's Manual of Systematic Bacteriology* until 2008.

## Taxonomy

Four new *Alicyclobacillus* species have been validly named since 2008, and their properties are listed in ➋ *Table 2.1*, which follow the format of Table 38 in *Bergey's Manual of Systematic Bacteriology* (da Costa et al. 2009) to facilitate comparison with species described until then. All strains are Gram positive (variable reactions were recorded for *A. aeris*). In none of the strains, the presence of ω-alicyclic fatty acids, hopanoids, or sulfonolipids was recorded. None of the strains grow in 5 % NaCl, and all strains produce acid from glucose.

The genus *Kyrpidia* embraces the former species *Bacillus tusciae*, which has been first shown by 16S rRNA gene sequence analysis to group outside the main bacillus proper (Rainey et al. 1994). *Kyrpidia tusciae* (*Bacillus tusciae*) is defined by chemoautotrophic growth with $H_2$ as electron donor and $CO_2$ (fixed via ribulose-bisphosphate cycle) as carbon source. A few organic acids, amino acids, and alcohol but not sugars serve as carbon sources (Bonjour and Aragno 1984). The complete genome sequence of *K. tusciae* DSM $2912^T$ and its comparison with that of *Alicyclobacillus acidocaldarius* DSM $446^T$ (Mavromatis et al. 2010) and *Bacillus subtilis* strain 168 (Kunst et al. 1997) have been published by Klenk et al. (2011) and reveal that all three species share a core of 1,363 genes, while the former two type strains share additional 387 genes. Unique for the genome of *K. tusciae* are genes coding for enzymes of the Calvin cycle, one of which has already been mentioned in the original species description (Bonjour and Aragno 1984).

The genus *Tumebacillus* contains up to now three species isolated from soil and wastewater. Some properties distinguishing members of the genera *Kyrpidia* and *Tumebacillus* are listed in ➋ *Table 2.2*.

E. Rosenberg et al. (eds.), *The Prokaryotes – Firmicutes and Tenericutes*, DOI 10.1007/978-3-642-30120-9_356,

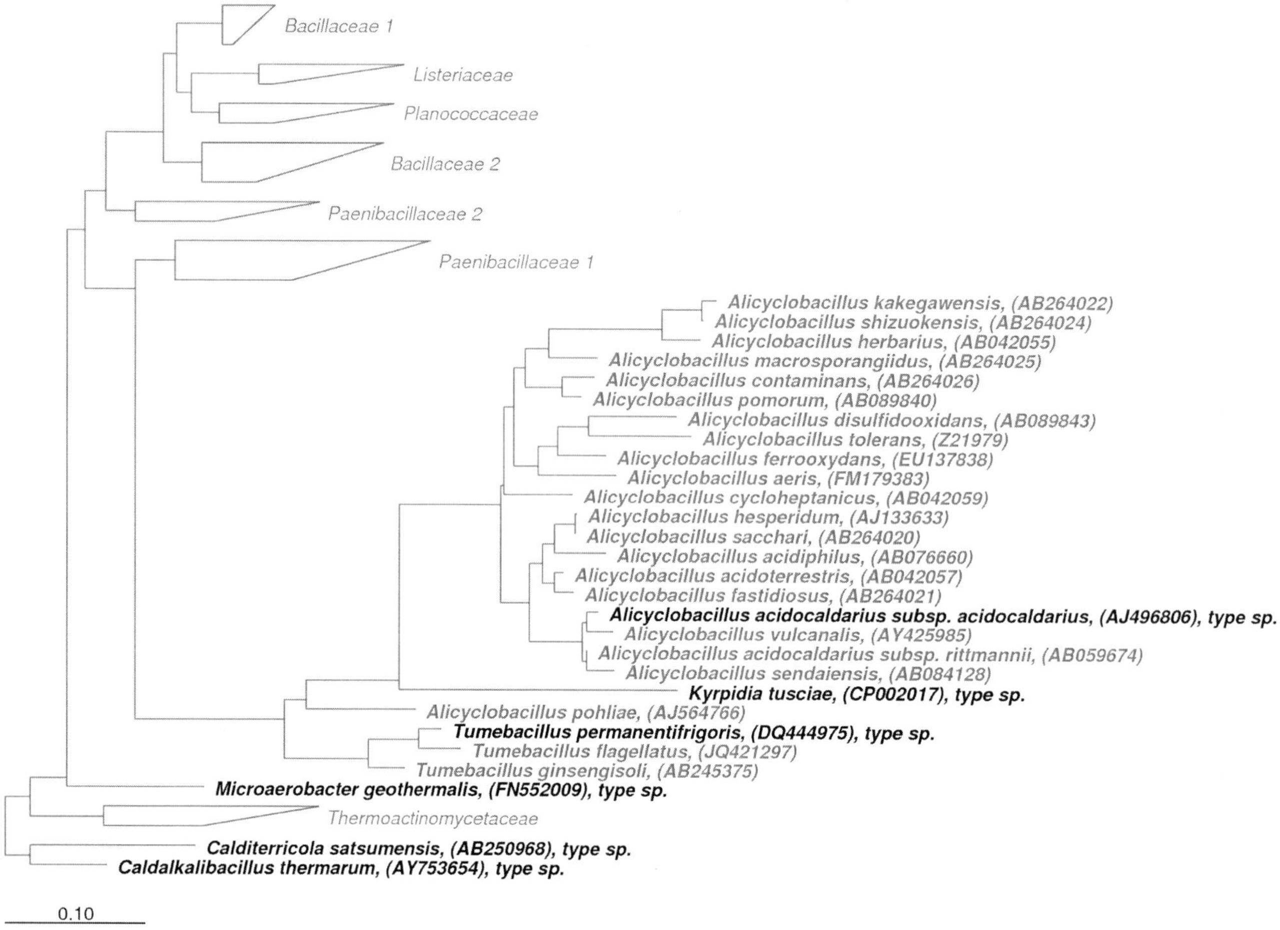


**Fig. 2.1**

Maximum likelihood genealogy reconstruction based on the RAxML algorithm (Stamatakis 2006) of the sequences of all members of the family *Alicyclobacillaceae* present in the LTP_106 (Yarza et al. 2010). Representative sequences from close relative genera were used to stabilize the tree topology. In addition, a 40 % maximum frequency filter was applied to remove hypervariable positions from the alignment. Scale bar indicates estimated sequence divergence

**Table 2.1**

Properties of novel *Alicyclobacillus* species described since 2008

| Characteristics | *A. aeris*[a] | *A. pohliae*[b] | *A. ferrooxydans*[c] | *A. consociatus*[d] |
|---|---|---|---|---|
| Cell size (μm) | 0.4–0.5 x 1.5–2.5 | 0.4–0.6 x 1.5–2.5 | 0.4–0.6 x 1.0–1.5 | 0.8–1.0 x 2.0–5.0 |
| Motility | Peritrichous flagella | nr | – | – |
| Spores | + | Round | nr | Spherical, terminal |
| Sporangia | nr | + | nr | – |
| Anaerobic growth | – | Facultative | – | – |
| Optimum temperature (°C) | 30 | 55 | 28 | 30 |
| Growth temperature range (°C) | 25–30 | 42–60 | 17–40 | 15–45 |
| Optimum pH | 3.5 | 5.5 | 3.0 | 6.5 |
| Growth pH range | 2.0–6.0 | 4.5–7.5 | 2.0–6.0 | 5.5–10.5 |

**Table 2.1 (continued)**

| Characteristics | *A. aeris*[a] | *A. pohliae*[b] | *A. ferrooxydans*[c] | *A. consociatus*[d] |
|---|---|---|---|---|
| *Growth in NaCl* | | | | |
| 1 % | + | + | + | nr |
| 2 % | + | + | + | + |
| 3 % | – | + | + | + |
| 4 % | – | + | – | – |
| Growth factors | Yeast extract | nr | Yeast extract | nr |
| Nitrate reduction | + | nr | – | – |
| Presence of | | | | |
| Oxidase | – | – | + | w |
| Catalase | – | – | + | – |
| Growth on mineral substrates | $Fe^{2+}$, $S^0$, $K_2S_4O_6$ | $Fe^{2+}$ | $Fe^{2+}$, $K_2S_4O_6$, pyrite | nr |
| *Acid production from* | | | | |
| *N*-acetylglucosamine | – | + | nr | nr |
| Adonitol | + | nr | nr | – |
| Amygdalin | – | – | – | nr |
| D-Arabinose | – | + | – | nr |
| L-Arabinose | + | + | + | – |
| D-Arabitol | + | – | – | – |
| Arbutin | – | w | – | nr |
| Cellobiose | – | + | – | – |
| Dulcitol | – | nr | nr | – |
| Esculin | + | + | + | nr |
| D-Fructose | + | + | nr | nr |
| D-Galactose | + | + | – | nr |
| ß-Gentiobiose | – | + | – | nr |
| Gluconate | – | nr | nr | nr |
| Glycerol | + | – | – | nr |
| Glycogen | – | + | – | nr |
| Erythritol | + | – | nr | – |
| Inositol | – | + | – | – |
| Inulin | – | – | – | nr |
| 2-Ketogluconate | – | + | nr | nr |
| 5-Ketogluconate | + | + | + | nr |
| Lactose | – | + | – | – |
| D-Lyxose | – | + | + | nr |
| Maltose | – | + | – | – |
| D-Mannose | + | + | nr | nr |
| Mannitol | + | – | – | – |
| Melibiose | – | + | – | – |
| Melezitose | nr | w | – | nr |
| Methyl-α-D-glucoside | – | w | nr | nr |
| Methyl-α-D-mannoside | – | – | + | nr |
| D-Raffinose | – | w | – | – |
| Rhamnose | – | + | – | – |

**Table 2.1 (continued)**

| Characteristics | *A. aeris*[a] | *A. pohliae*[b] | *A. ferrooxydans*[c] | *A. consociatus*[d] |
|---|---|---|---|---|
| Ribose | + | + | nr | nr |
| Salicin | – | w | – | – |
| Sorbitol | + | – | + | – |
| L-Sorbose | – | + | – | nr |
| Starch | – | nr | nr | nr |
| Sucrose | – | + | – | – |
| D-Tagatose | – | + | + | nr |
| Trehalose | – | w | + | – |
| D-Turanose | – | + | + | nr |
| Xylitol | + | – | – | nr |
| D-Xylose | + | + | – | w |
| L-Xylose | + | + | – | nr |
| Presence of hopanoids | nr | nr | nr | nr |
| Presence of a sulfonolipid | nr | nr | nr | nr |
| Major menaquinone | MK-7 | nr | MK-7 | MK-7 |
| Major fatty acids | Anteiso-C15:0, iso-C16:0, anteiso-C17:0 | Iso-C15:0, iso-C17:0[e] | Anteiso-C15:0, anteiso-C17:0, iso-C16:0, iso-C15:0 | Anteiso-C15:0, iso-C15:0[e] |
| Mol% G+C | 51.2 | 55.1 | 48.6 | 47.0 |
| Habitat | Copper mine, Inner Mongolia, China | Geothermal soil, Mt. Melbourne, Antarctica | Solfataric soil, China | Blood sample |

[a]Guo et al. (2009)
[b]Imperio et al. (2008)
[c]Jiang et al. (2008)
[d]Glaeser et al. (2013)
[e]The fatty acid composition of *A. pohliae* and A. *consociatus* depends upon media composition (Glaeser et al. 2013)
nr, not recorded; w, weak

**Table 2.2**
**Properties of species of recently described typed strains of novel genera of the family *Alicyclobacillaceae***

| Characteristics | *Kyrpidia tusciae*[a] | *Tumebacillus permanentifrigoris*[b] | *Tumebacillus ginsengisoli*[c] | *Tumebacillus flagellatus*[d] |
|---|---|---|---|---|
| Cell size (μm) | 0.8 x 4.0–5.0 | 0.5 x 3.0–3.5 | 0.5–0.8 x 3.0–6.0 | 0.5 x 3.1–4.2 |
| Motility | + | – | – | + |
| Gram stain | Positive | Positive, older cells occasionally negative | Positive | Positive |
| Spores | Ova, subterminal | Oval, terminal | Oval, terminal | Oval, terminal |
| Anaerobic growth | – | – | – | – |
| Optimum temperature (°C) | 55 | 25–30 | nr | 37 |
| Growth temperature range (°C) | No growth at 35 or 65 °C | 5–37 | 20–42 | 20–42 |
| Optimum pH | 4.2–4.8 | nr | nr | 5.5 |
| Growth pH range | Weak growth at 3.5 – no growth at pH 6 | 5.5–8.9 | 5.0–8.5 | 4.5–8.5 |
| Growth on 1 % NaCl (w/v) | – | – | – | + |
| Nitrate reduction | + | | + | + |

**Table 2.2 (continued)**

| Characteristics | *Kyrpidia tusciae*[a] | *Tumebacillus permanentifrigoris*[b] | *Tumebacillus ginsengisoli*[c] | *Tumebacillus flagellatus*[d] |
|---|---|---|---|---|
| Oxidase | + | – | + | + |
| Catalase | w | – | | – |
| Chemolithoautrophy on | $H_2$ | $S^o$, $Na_2SO_3$, $Na_2S_2O_3$ | $NO_3$ | $S^o$, $NO_3$ $NO_2^-$ |
| Assimilation of | | | | |
| Sucrose | – | w | + | – |
| L-Arabinose | – | w | + | – |
| Melibiose | – | w | + | – |
| D-Mannose | – | w | + | – |
| D-Xylose | – | – | + | + |
| D-Sorbitol | – | – | w | + |
| D-Fructose | – | + | – | + |
| Maltose | – | + | – | + |
| *N*-acetylglucosamine | – | w | – | – |
| Acid production from glucose | – | + | + | – |
| Major menaquinone | MK-7[e] | MK-7 | MK-7 | MK-7 |
| Major fatty acids | Iso-C15:0, iso-C17:0[e] | Iso-C15:0 | Iso-C 15:0 anteiso-C 15:0 | Iso-C15:0, anteiso-C15:0 |
| Mol% G+C | 57–58 | 53.1 | 55.6 | 53.7 |
| Habitat | Geothermal solfatara pond, Italy | Permafrost, Canadian high Arctic | Soil of a ginseng field, South Korea | Starch wastewater, China |

[a]Bonjour and Aragno (1984)
[b]Steven et al. (2008)
[c]Baek et al. (2011)
[d]Wang et al. (2013)
[e]As indicated by Klenk et al. (2011)

## References


Baek SH, Cui Y, Kim SC, Cui CH, Yin C, Lee ST, Im WT (2011) *Tumebacillus ginsengisoli* sp. nov., isolated from soil of a ginseng field. Int J Syst Evol Microbiol 61:1715–1719

Bonjour F, Aragno M (1984) *Bacillus tusciae*, a new species of thermoacidophilic, facultatively chemolithoautotrophic, hydrogen oxidizing sporeformer from a geothermal area. Arch Microbiol 139:397–401

da Costa MS, Rainey FA (2009) Family II. *Alicyclobacillaceae* fam. nov. In: De Vos P, Garrity G, Jones D, Krieg NR, Ludwig W, Rainey FA, Schleifer KH, Whitman WB (eds) Bergey's manual of systematic bacteriology, vol 3, 2nd edn. Springer, New York, p 229

da Costa MS, Rainey FA, Albuquerque L (2009) Genus I *Alicyclobacillus*. In: De Vos P, Garrity G, Jones D, Krieg NR, Ludwig W, Rainey FA, Schleifer KH, Whitman WB (eds) Bergey's manual of systematic bacteriology, vol 3, 2nd edn. Springer, New York, pp 229–243

Euzéby J (2010) List of new names and new combinations previously effectively, but not validly, published. Int J Syst Evol Microbiol 60:469–472

Gibbons NE, Murray RGE (1978) Proposals concerning the higher taxa of bacteria. Int J Syst Bacteriol 28:1–6

Glaeser SP, Falsen E, Martin K, Kämpfer P (2013) *Alicyclobacillus consociatus* sp. nov., isolated from a human clinical specimen. Int J Syst Evol Microbiol 63:3623–3627

Guo X, You XY, Liu LJ, Zhang JY, Liu SJ, Jiang CY (2009) *Alicyclobacillus aeris* sp. nov., a novel ferrous- and sulfur-oxidizing bacterium isolated from a copper mine. Int J Syst Evol Microbiol 59:2415–2420

Imperio T, Viti C, Marri L (2008) *Alicyclobacillus pohliae* sp. nov., a thermophilic, endospore-forming bacterium isolated from geothermal soil of the North–West slope of Mount Melbourne (Antarctica). Int J Syst Evol Microbiol 58:221–225

Jiang CY, Liu Y, Liu YY, You XY, Guo X, Liu SJ (2008) *Alicyclobacillus ferrooxydans* sp. nov., a ferrous-oxidizing bacterium from solfataric soil. Int J Syst Evol Microbiol 58:2898–2903

Klenk H-P, Lapidus A, Chertkov O, Copeland A, Glavina Del Rio T, Nolan M, Lucas S et al (2011) Complete genome sequence of the thermophilic, hydrogen-oxidizing *Bacillus tusciae* type strain (T2[T]) and reclassification in the new genus, *Kyrpidia* gen. nov. as *Kyrpidia tusciae* comb. nov. and emendation of the family *Alicyclobacillaceae* da Costa and Rainey, 2010. Stand Genomic Sci 5:121–134

Kunst F, Ogasawara N, Moszer I, Albertini AM, Alloni G, Azevedo V, Bertero MG, Bessières P, Bolotin A, Borchert S (1997) The complete genome sequence of the Gram-positive bacterium *Bacillus subtilis*. Nature 390:249–256

Ludwig W, Schleifer KH, Whitman WB (2009) Class I. *Bacilli* class nov. In: De Vos P, Garrity GM, Jones D, Krieg NR, Ludwig W, Rainey EA, Schleifer KH, Withman WB (eds) Bergey's manual of systematic bacteriology, vol 3, 2nd edn. Springer, Dordrecht/Heidelberg/London/New York, pp 19–20 (The Firmicutes)

Mavromatis K, Sikorski J, Lapidus A, Glavina Del Rio T, Copeland A, Tice H, Cheng J-F, Lucas S et al (2010) Complete genome sequence of *Alicyclobacillus acidocaldarius* type strain (104-IA$^{T}$). Stand Genomic Sci 2:9–18

Prévot AR (1953) In: Hauduroy P, Ehringer G, Guillot G, Magrou J, Prévot AR, Rosset RA, Urbain A (eds) Dictionnaire des Bactéries Pathogènes, 2nd edn. Masson, Paris, pp 1–692

Rainey FA, Fritze D, Stackebrandt E (1994) The phylogenetic diversity of thermophilic members of the genus *Bacillus* as revealed by 16S rDNA analysis. FEMS Microbiol Lett 115:205–211

Stamatakis A (2006) RAxML-VI-HPC: maximum likelihood-based phylogenetic analyses with thousands of taxa and mixed models. Bioinformatics 22:2688–2690

Steven B, Chen MQ, Greer CW, Whyte LG, Niederberger TD (2008) *Tumebacillus permanentifrigoris* gen. nov., sp. nov., an aerobic, spore-forming bacterium isolated from Canadian high Arctic permafrost. Int J Syst Evol Microbiol 58:1497–1501

Wang Q, Xie N, Qin Y, Shen N, Zhu J, Mi H, Hung R (2013) *Tumebacillus flagellatus* sp. nov., an α-amylase/pullulanase-producing bacterium isolated from cassava wastewater. Int J Syst Evol Microbiol 63:3138–3142

Wisotzkey JD, Jurtshuk P Jr, Fox GE, Deinhard G, Poralla K (1992) Comparative sequence analyses on the 16S rRNA (rDNA) of *Bacillus acidocaldarius*, *Bacillus acidoterrestris*, and *Bacillus cycloheptanicus* and proposal for creation of a new genus, *Alicyclobacillus* gen. nov. Int J Syst Bacteriol 42:263–269

Yarza P, Ludwig W, Euzéby J, Amann R, Schleifer K-H, Glöckner FO, Rosselló-Móra R (2010) Update of the All-Species Living-Tree project based on 16S and 23S rRNA sequence analyses. Syst Appl Microbiol 33:291–299

# 3 The Family *Caldicoprobacteraceae*

*Amel Bouanane-Darenfed*[1] · *Marie-Laure Fardeau*[2] · *Bernard Ollivier*[2]
[1]Faculté de Biologie, Laboratoire de Biologie Cellulaire et Moléculaire (Equipe de Microbiologie), Université des Sciences et de la Technologie Houari Boumediene, Bab ezzouar, Alger, Algérie
[2]Aix-Marseille Université, Université du Sud Toulon-Var, CNRS/INSU, IRD, MIO, UM 110, Marseille, France



**Abstract**

The family *Caldicoprobacteraceae* belongs to the order *Clostridiales*, phylum Firmicutes. It embraces the genus *Caldicoprobacter* which contains three recognized species: *C. oshimai*, *C. algeriensis*, and *C. guelmensis*, together with *Acetomicrobium faecale* which should be reclassified within the genus *Caldicoprobacter* as *C. faecale*, comb. nov. Members of the family are defined by a wide range of morphological and chemotaxonomic properties including the cellular fatty acids content. They are all strictly anaerobic thermophilic heterotrophic rod-shaped bacteria using sugars, but not proteinaceous compounds. Members of the family are found in herbivore feces and terrestrial hot springs.

## Taxonomy, Historical, and Current

### Short Description of the Family

Cal'di.co'pro.bac'te.ra'ce.ae. N.L. masc. n. *Caldicoprobacter* type genus of the family; -aceae ending to denote a family; N.L. masc. pl. n. *Caldicoprobacteraceae* the family of *Caldicoprobacter* spp. Members of this family pertain to the order *Clostridiales*, phylum Firmicutes with *Caldicoprobacter* being the type genus of this family (Yokoyama et al. 2010). Gram-staining positive. Includes thermophilic spore-forming and nonspore-forming rod-shaped cells with anaerobic chemoorganotrophic metabolism. Nonmotile. Cellular fatty acids are composed mainly of 17- and 15-carbon-containing saturated, branched fatty acids (iso-$C_{17:0}$, iso-$C_{15:0}$, and anteiso-$C_{17:0}$). The peptidoglycan type is A1$\gamma$. Polar lipids were not analyzed. G+C values of DNA range between 41 and 45 mol%. Isolated from sheep feces and hot springs.

### Phylogenetic Structure of the Family and Its Genera

Phylogenetic analysis of *Caldicoprobacter* spp. using the 16S rRNA gene sequences revealed that they form a separate branch within the order *Clostridiales* and also include *Acetomicrobium faecale* isolated from sewage sludge incubated at 72 °C (Winter et al. 1987). *A. faecale* 16S rRNA gene sequence was established in December 2011 (➋ *Fig. 3.1*) after the description of the family *Caldicoprobacteraceae*. It demonstrates that *A. faecale* should be reclassified within the genus *Caldicoprobacter* as *C. faecale*, comb. nov. (Bouanane-Darenfed et al. 2012). Based mainly on its phenotypical and genetic characteristics, *A. faecale* (Winter et al. 1987) was first recognized as a member of the genus *Acetomicrobium* (Goodfellow et al. 2011), family *Bacteroidaceae*, order *Bacteroidales*, phylum Bacteroidetes, with *A. flavidum* being the type species of this genus (Soustschek et al. 1984).

Moreover, the 16S rRNA gene sequence of *A. flavidum*, established in January 2012, demonstrated that *A. flavidum* and *A. faecale* should not be considered at the same genus level and confirmed the affiliation of *A. faecale* to the genus *Caldicoprobacter* (98 % and 98.9 % similarity with *C. algeriensis* and *C. oshimai*, respectively) as already proposed by Bouanane-Darenfed et al. (2012). Taking into account their 16S rRNA gene sequences, it appears that *A. flavidum*, the three *Anaerobaculum* (*Ab.*) species described so far, *Ab. mobile* (Javier-Menes and Muxi 2002), *Ab. thermoterrenum* (Rees et al. 1997), and *Ab. hydrogeniformans* (Maune and Tanner 2012), and *A. flavidum* belong to the same phylogenetic clade with high levels (>95 %) of similarity (➋ *Fig. 3.2*). In this respect, these three *Anaerobaculum* species should be reclassified within the genus *Acetomicrobium* which has priority over the genus *Anaerobaculum* (Rees et al. 1997) since being validated later after the genus *Acetomicrobium* (validation list,

E. Rosenberg et al. (eds.), *The Prokaryotes – Firmicutes and Tenericutes*, DOI 10.1007/978-3-642-30120-9_395,

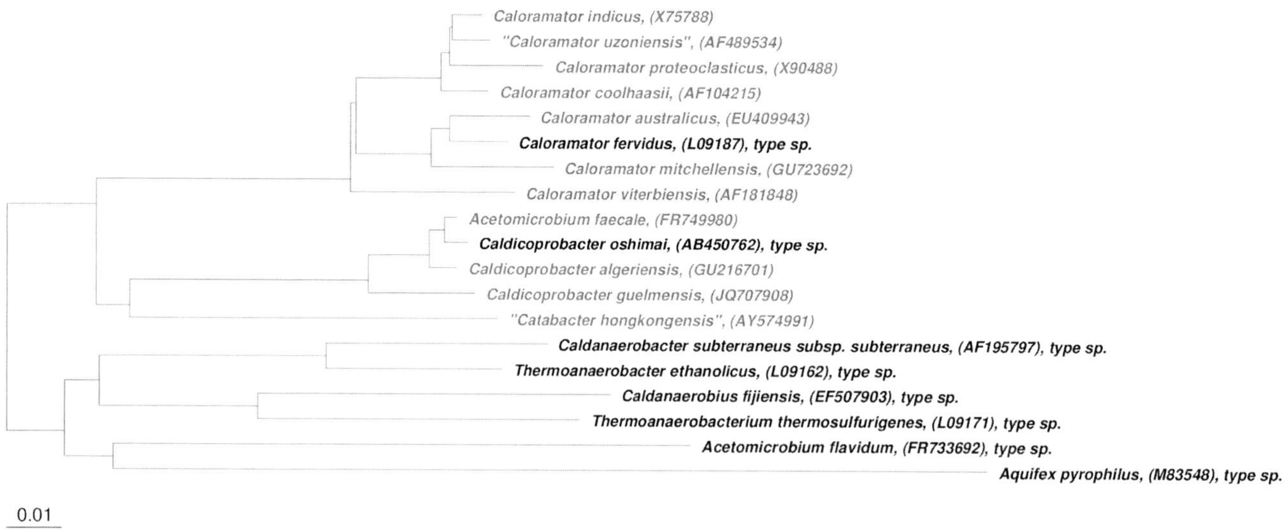


**Fig. 3.1**
Phylogenetic reconstruction of the family *Caldicoprobacteraceae* based on 16S rRNA and created using the neighbor-joining algorithm with the Jukes-Cantor correction. The sequence dataset and alignment were used according to the All-Species Living Tree Project (LTP) database (Yarza et al. 2010; http://www.arb-silva.de/projects/living-tree). The tree topology was stabilized with the use of a representative set of nearly 750 high-quality type strain sequences proportionally distributed among the different bacterial and archaeal phyla. In addition, a 40 % maximum frequency filter was applied in order to remove hypervariable positions and potentially misplaced bases from the alignment. Scale bar indicates estimated sequence divergence

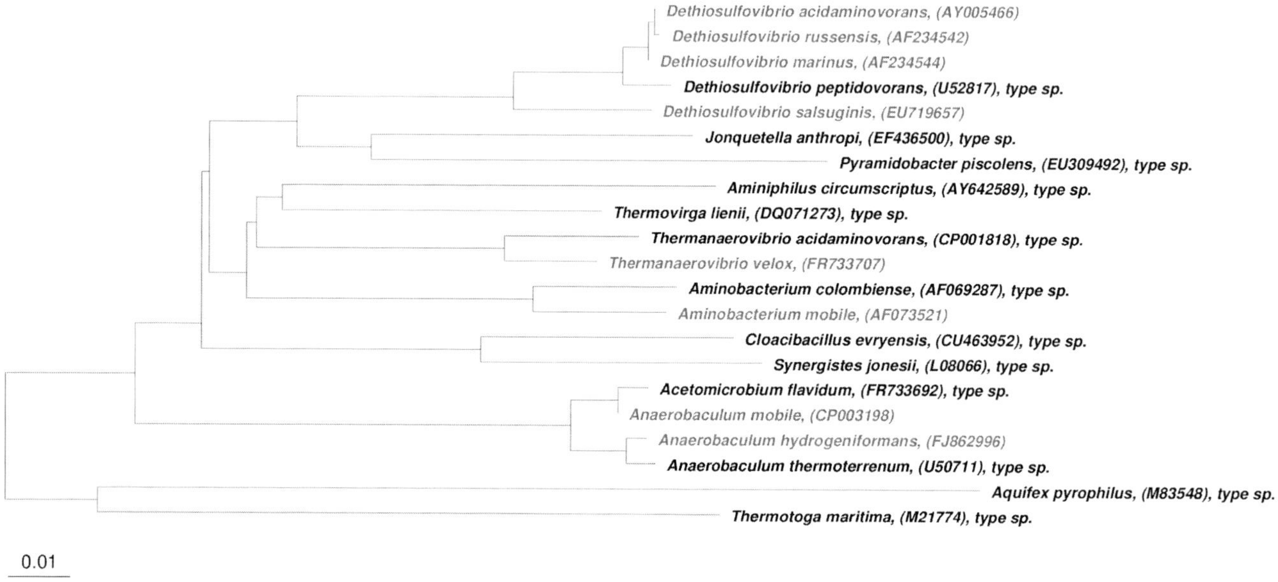


**Fig. 3.2**
Phylogenetic position of the species *Acetomicrobium flavidum* based on 16S rRNA and created using the neighbor-joining algorithm with the Jukes-Cantor correction. The sequence dataset and alignment were used according to the All-Species Living Tree Project (LTP) database (Yarza et al. 2010; http://www.arb-silva.de/projects/living-tree). The tree topology was stabilized with the use of a representative set of nearly 750 high-quality type strain sequences proportionally distributed among the different bacterial and archaeal phyla. In addition, a 40 % maximum frequency filter was applied in order to remove hypervariable positions and potentially misplaced bases from the alignment. Scale bar indicates estimated sequence divergence

**Table 3.1**

**Discriminating characteristics of *Caldicoprobacter guelmensis, C. oshimai, C. algeriensis,* and *Acetobacterium faecale***

| Characteristics | 1 | 2 | 3 | 4 |
|---|---|---|---|---|
| Isolation source | Hot spring | Sludge samples | Sheep feces | Hot spring |
| Motility | – | + | – | - |
| Morphology | Rod | Rod | Spore-forming Rod | Rod |
| Gram | Positive | Negative | Positive | Positive |
| Optimum temperature (°C) | 65 (45–85) | 70–73 | 70 (44–77) | 65 (55–75) |
| Optimum pH (range) | 6.8 (5–9) | 6.5 (5.5–9) | 7.2 (5.9–8.6) | 6.9 (6.2–8.3) |
| Optimum NaCl % | 0–2 | 0–3 | 0–2 | <5 |
| Growth substrate | | | | |
| Xylan | + | – | + | + |
| Raffinose | – | ND | + | + |
| Melibiose | – | + | ND | + |
| Mannitol | – | – | – | + |
| Pyruvate | + | – | ND | + |
| End products of glucose fermentation | Acetate, lactate, $H_2$, $CO_2$ | Acetate, lactate, ethanol, $H_2$, $CO_2$ | Acetate, lactate, ethanol, $H_2$, $CO_2$ | Acetate, lactate, ethanol, $H_2$, $CO_2$ |
| G+C % | 41.6 | 45 | 45.4 | 44.7 |

*1 Caldicoprobacter guelmensis* (Bouanane-Darenfed et al. 2012), *2 Acetomicrobium faecale* (Winter et al. 1987), *3 Caldicoprobacter oshimai* (Yokoyama et al. 2010), *4 Caldicoprobacter algeriensis* (Bouanane-Darenfed et al. 2011) +, positive; –, negative; *ND*, not determined

1985, Int. J. Syst. Bacteriol. 35, 223–225). Consequently, *Acetomicrobium* spp. should belong to family *Synergistaceae*, order *Synergistales*, phylum Synergistetes (see phylum Synergistetes in this book). To date, the genus *Caldicoprobacter* comprises three genera, *C. oshimai*, *C. algeriensis*, and *C. guelmensis*, isolated from mesothermic (Yokoyama et al. 2010) and hot environments (Bouanane-Darenfed et al. 2011, 2012).

## Molecular Analyses

### G+C Content of the Genomic DNA and DNA-DNA Hybridization Studies

Despite *C. algeriensis* differed from *C. oshimai* by the absence of spores, its lower tolerance to NaCl, and the use of mannitol, both bacteria were closely phylogenetically related (98.5 % similarity), thus justifying to perform DNA/DNA hybridization studies. Their DNA/DNA homology was found sufficiently low (45.5 %) to ascertain *C. algeriensis* as a novel species of *Caldicoprobacter* (Bouanane-Darenfed et al. 2011). Because of high similarities in the 16S rRNA gene sequences between *A. faecale* and *C. algeriensis* (98 % similarity), but also *C. oshimai* (98.9 % similarity), DNA/DNA hybridization studies should be performed to reorganize the genus *Caldicoprobacter* and assign *A. faecale* as a novel *Caldicoprobacter* species. In contrast, there was no need to perform DNA/DNA hybridization studies between *C. guelmensis* and the other *Caldicoprobacter* spp. and even *A. faecale* as the former had very low similarity (<97 %) in the 16S rRNA gene sequences with them (Bouanane-Darenfed et al. 2012). The G+C content of *Caldicoprobacter* spp. ranges from 41 to 45 mol% (*Table 3.1*).

## Phenotypic Analyses

The main features of members of *Caldicoprobacteraceae* consisting only of *Caldicoprobacter* spp. and *Acetomicrobium faecale* to be reclassified as a *Caldicoprobacter* sp. are listed in *Tables 3.1* and *3.2.*

*Caldicoprobacter* Yokoyama et al. 2010

Cal'di.co'pro.bac'ter. L. adj. caldus hot; Gr. N. Kopros dung; N. L. masc. n. bacter rod; N.L. masc. n. *Caldicoprobacter* a rod from dung growing at elevated temperatures.

Vegetative cells were straight to curved rods, 0.3–0.6 um in diameter and 2.0–14 um long. They appeared singly or in pairs or occasionally as long chains. They were not motile with the exception of *A. faecale* which has been reported to be weakly motile (Winter et al. 1987). Cells of *Caldicoprobacter* spp. (*C. oshimai*, *C. algeriensis*, and *C. guelmensis*) stained Gram-positive, whereas that of *A. faecale* stained Gram-negative. The cell wall structure of *C. oshimai* and *C. algeriensis* was a single layer (Gram-positive type). In contrast, sections for electron microscopy of *C. guelmensis* cells revealed a cell wall with a structure of three thin layers (Bouanane-Darenfed et al. 2012). The only species reported to form spores was *C. oshimai.* Spores were observed in the stationary growth phase. Single spherical endospores (0.4–0.6 um in diameter) were formed in a swollen terminus.

**Table 3.2**
**Predominant cellular fatty acids (>1 %) in *Caldicoprobacter guelmensis*, *C. oshimai*, and *C. algeriensis***

| Cellular fatty acid (%) | *C. guelmensis* | *C. oshimai* | *C. algeriensis* |
|---|---|---|---|
| Iso-$C_{15:0}$ | 36.7 | 23.7 | 13.3 |
| Anteiso-$C_{15:0}$ | 6.6 | 5.7 | 3.4 |
| $C_{16:0}$ | 8.0 | 6.8 | 14.7 |
| Iso-$C_{17:0}$ | 32.0 | 31.1 | 24.7 |
| Iso-$C_{16:0}$ | 4.0 | 2.4 | 11.1 |
| Iso-$C_{17:0}$ 3OH | 4.4 | 8.1 | 5.9 |
| Anteiso-$C_{17:0}$ | 8.4 | 12.9 | 12.7 |

In the death phase, the swollen termini separated from the rods and the spores were subsequently released from the swollen body (Yokoyama et al. 2010).

*Caldicoprobacter* spp. are thermophilic bacteria which grew at temperature ranging from 44 °C to 85 °C, with *C. guelmensis* being the only species growing over 80 °C. They are neutrophilic, growing optimally at pH around 7.0 and can grow in the absence of NaCl. They all require yeast extract for growth and ferment a wide range of substrates including glucose, fructose, galactose, lactose, mannose, xylose, and cellobiose, but not proteinaceous compounds (e.g., peptone, casein, casamino acids). They can be distinguished by the use of mannitol, xylan, raffinose, and pyruvate (➲ *Table 3.1*). *Caldicoprobacter guelmensis* produced acetate, lactate, $CO_2$, and $H_2$ from sugar fermentation (Bouanane-Darenfed et al. 2012). In addition to these end products of metabolism, ethanol was also detected when *C. oshimai*, *C. algeriensis*, and *A. faecale* grew on sugars (➲ *Table 3.1*). The major membrane fatty acids present in *Caldicoprobacter* spp. were branched, saturated fatty acids with odd numbers of carbon atoms (e.g., iso-$C_{17:0}$, iso-$C_{15:0}$, and anteiso-$C_{17:0}$) (➲ *Table 3.2*). Analyses of the cell wall peptidoglycan revealed the presence of alanine and glutamic acid (Yokoyama et al. 2010) and/or meso-diaminopimelic acid (Yokoyama et al. 2010; Bouanane-Darenfed et al. 2012) indicating the peptidoglycan type A1$\gamma$. The cell wall sugars in *C. oshimai* were galactose and mannose (Yokoyama et al. 2010).

## Isolation, Enrichment, and Maintenance Procedures

*Caldicoprobacter* species can be enriched from herbivore feces or from waters and/or sediments of terrestrial hot springs using culture media and procedures similar to those described by Yokoyama et al. (2010) or Bouanane-Darenfed et al. (2011, 2012) in the presence of complex organic compounds (e.g., yeast extract, tryptone, or biotrypcase) with glucose, xylose, or xylan (birchwood xylan and/or beechwood xylan) as the possible energy sources. Cysteine and/or sulfide should be used as the reductive agents. At least three subcultures in the same growth conditions, at temperature around 70 °C, are needed before isolation.

After several transfers, the enrichment cultures are serially diluted using (1) the Hungate roll-tube method (Hungate 1969) with the addition of Gelrite (0.8 %) in a basal culture medium containing glucose and yeast extract (Bouanane-Darenfed et al. 2012) or (2) the modified Phytagel-shake roll-tube technique (Ljungdhal and Wiegel 1986) with 5 % Phytagel and 0.12 % $MgCl_2.6H_2O$ being added to a basal culture medium containing xylose, xylan, yeast extract, and tryptone (Yokoyama et al. 2010). Colonies, which should appear after 3–8 days incubation at around 70 °C, are picked, and serial dilution in roll tubes is repeated until the cultures are found to be axenic.

Stock cultures can be maintained on the medium described by Yokoyama et al. (2010) or Bouanane-Darenfed et al. (2011, 2012) and transferred at least monthly. Liquid cultures retain viability after several weeks of storage at room temperature, or when lyophilized, or after storage at – 80 °C in the basal culture medium containing 20 % glycerol (v/v). Viability is best maintained from mid-exponential phase cultures.

## Ecology

### Habitat

*Caldicoprobacter algeriensis* and *C. guelmensis* were isolated from Algerian hot springs (Bouanane-Darenfed et al. 2011, 2012) whereas *C. oshimai* was isolated from sheep feces (Yokoyama et al. 2010). Interestingly, this latter bacterium had clone OTU4 (99.5 % similarity) retrieved from cow feces enrichment cultures as its closest phylogenetic relative, thus demonstrating that similar microorganisms, despite being most probably dormant, might prevail in herbivore feces (Hobel et al. 2004; Yokoyama et al. 2010). However, with the isolation of *Caldicoprobacter* spp. from terrestrial hot springs together with the retrieval of unidentified clones phylogenetically related to these species from such ecosystems (clone LNE-5) (Yokoyama et al. 2010) and from a subterrestrial high pH groundwater (clone CVCloAm2Ph136) (Tiago and Verissimo 2012), we may expect

*Caldicoprobacter* spp. as being also potential contributors in organic matter degradation in extreme environments (Bouanane-Darenfed et al. 2011).

Moreover, molecular studies also provided evidence of the presence of unidentified clones having more than 96 % similarity with *Caldicoprobacter* spp. in (a) reactors conducted under thermophilic conditions (clones Hb and EG11) (Hobel et al. 2004; Abreu et al. 2010; Yokoyama et al. 2010), (b) waste activated sludge alkaline fermentation (clone 108) (Zhang et al. 2010), (c) anaerobic bacterial cellulolytic community enriched from coastal marine sediments (clone 21) (Ji et al. 2012), and (d) compost plants (clone PS2456) (Partanen et al. 2010), thus suggesting that *Caldicoprobacter* spp. inhabit various mesothermic and hot environments.

## Application

### Waste Treatment and Removal

With the exception of *A. faecale*, all other *Caldicoprobacter* species displayed a xylanolytic activity at 70 °C that was found extracellular for *C. algeriensis* and *C. guelmensis* (Bouanane-Darenfed et al. 2011, 2012). In this respect, these species may be used to degrade hemicellulosic material, known as the second most abundant component of plant fiber (Yokoyama et al. 2010).

### Added-Value Products

*Acetomicrobium faecale*, which should be reassigned to a *Caldicoprobacter* species (see above), was shown to ferment arabinose, xylose, and ribose while producing equimolar amounts of acetate and ethanol. It was therefore suggested to be a potential candidate for the biotechnological production of ethanol since neither yeasts nor *Zymomonas mobilis* can use these sugars (Winter et al. 1987).

## References


Abreu AA, Alves JI, Pereira MA, Karakashev D, Alves MM, Angelidaki I (2010) Engineered heat treated methanogenic granules: a promising biotechnological approach for extreme thermophilic biohydrogen production. Bioresour Technol 101:9577–9586

Bouanane-Darenfed A, Fardeau ML, Grégoire P, Joseph M, Kebbouche-Gana S, Benayad T, Hacene H, Cayol JL, Ollivier B (2011) *Caldicoprobacter algeriensis* sp. nov. A new thermophilic anaerobic, xylanolytic bacterium isolated from an Algerian hot spring. Curr Microbiol 62:826–832

Bouanane-Darenfed A, Ben Hania W, Hacene H, Cayol JL, Ollivier B, Fardeau ML (2012) *Caldicoprobacter guelmensis* sp. nov., a new thermophilic anaerobic, xylanolytic bacterium isolated from an Algerian hot spring. Int J Syst Evol Microbiol 63:2049–2053

Goodfellow M, Kämpfer P, Chun J, De Vos P, Rainey FA, Whitman WB et al (2011) *Acetomicrobium*. In: Krieg NR, Staley JT, Brown DR, Hedlund BP, Paster BJ, Ward NL, Ludwig W, Whitman WB (eds) Bergey's manual of systematic bacteriology, vol 4, 2nd edn. Springer, New York, pp 42–44

Hobel CFV, Marteinsson VT, Hauksdottir S, Fridjonsson OH, Skirnisdottir S, Hreggvidsson GO, Kristjansson JK (2004) Use of low nutrient enrichments to access novel amylase genes in silent diversity of thermophiles. World J Microbiol Biotechnol 20:801–809

Hungate RE (1969) A roll tube method for the cultivation of strict anaerobes. In: Norris JR, Ribbons DW (eds) Methods in microbiology, vol 3B. Academic, New York, pp 117–132

Javier-Menes R, Muxi L (2002) *Anaerobaculum mobile* sp. nov., a novel anaerobic, moderately thermophilic, peptide-fermenting bacterium that uses crotonate as electron acceptor, and emended description of the genus *Anaerobaculum*. Int J Syst Evol Microbiol 52:157–164

Ji S, Wang S, Tan Y, Chen X, Schwarz W, Li F (2012) An untapped bacterial cellulolytic community enriched from coastal marine sediment under anaerobic and thermophilic conditions. FEMS Microbiol Lett 335:39–46

Ljungdhal LG, Wiegel J (1986) Anaerobic fermentations. In: Demain AL, Solomon NA (eds) Manual of industrial microbiology and biotechnology. American Society for Microbiology, Washington, DC, pp 84–96

Maune MW, Tanner RS (2012) Description of *Anaerobaculum hydrogeniformans* sp. nov., an anaerobe that produces hydrogen from glucose, and emended description of the genus *Anaerobaculum*. Int J Syst Evol Microbiol 62:832–838

Partanen P, Hultman J, Paulin L, Auvinen P, Romantschuk M (2010) Bacterial diversity at different stages of the composting process. Bacterial diversity at different stages of the composting process. BMC Microbiol 10:94

Rees GN, Patel BKC, Grassia GS, Sheehy AJ (1997) *Anaerobaculum thermoterrenum* gen. nov., sp. nov., a novel thermophilic bacterium which ferments citrate. Int J Syst Bacteriol 47:150–154

Soustschek E, Winter J, Schindler F, Kandler O (1984) *Acetomicrobium flavidum*, gen. nov., sp. nov, a thermophilic anaerobic bacterium from sewage sludge, forming acetate, $CO_2$ and $H_2$ from glucose. Syst Appl Microbiol 5:377–390

Tiago I, Verissimo A (2012) Microbial and functional diversity of a subterrestrial high pH groundwater associated to serpentinization. Environ Microbiol 15:1687–1706

Winter J, Braun E, Zabel H-P (1987) *Acetomicrobium faecale* spec. nov., a strictly anaerobic bacterium from sewage sludge, producing ethanol from pentoses. Syst Appl Microbiol 9:71–76

Yarza P, Ludwig W, Euzeby J, Amann R, Schleifer KH, Glöckner FO, Rossello-Mora R (2010) Update of the all-species living tree project based on 16S and 23S rRNA sequence analyses. Syst Appl Microbiol 33:291–299

Yokoyama H, Wagner ID, Wiegel J (2010) *Caldicoprobacter oshimai* gen. nov., sp. nov., an anaerobic, xylanolytic, extremely thermophilic bacterium isolated from sheep faeces, and proposal of *Caldicoprobacteraceae* fam. nov. Int J Syst Evol Microbiol 60:67–71

Zhang P, Chen Y, Zhou Q, Zheng X, Zhu X, Zhao Y (2010) Understanding short-chain fatty acids accumulation enhanced in waste activated sludge alkaline fermentation: kinetics and microbiology. Environ Sci Technol 44:9343–9348

# 4 The Family *Carnobacteriaceae*

*Paul A. Lawson · Matthew E. Caldwell*
Department of Microbiology and Plant Biology, University of Oklahoma, Norman, OK, USA



**Abstract**
*Carnobacteriaceae*, a family of the order *Lactobacillales*, within the phylum Firmicutes, embraces the genera *Carnobacterium*, *Alkalibacterium*, *Allofustis*, *Alloiococcus*, *Atopobacter*, *Atopococcus*, *Atopostipes*, *Bavariicoccus*, *Desemzia*, *Dolosigranulum*, *Granulicatella*, *Isobaculum*, *Jeotgalibaca*, *Lacticigenium*, *Marinilactibacillus*, *Pisciglobus*, and *Trichococcus*. Circumscribed mainly on the basis of 16S rRNA gene sequence analysis and defined by a wide range of morphological and chemotaxonomic properties, such as polar lipids, fatty acids, amino acids of peptidoglycan, and whole-cell sugars which are used for the delineation of genera and species. Members of the family are mainly found associated with food products, human, and cold environmental sources.

## Taxonomy, Historical, and Current

*Car.no.bac.te.ri.a'ce.ae. N.L. neut. n. Carnobacterium* type genus of the family; suff. *–aceae* ending denoting family; *N.L. fem. pl. n. Carnobacteriaceae* the *Carnobacterium* family.

Phylogenetically a member of the order *Lactobacillales* (Ludwig et al. 2009, 2010) within the phylum Firmicutes. The family contains the type genus *Carnobacterium (*Collins et al. 1987), *Alkalibacterium* (Ntougias and Russell 2001), *Allofustis* (Collins et al. 2003), *Alloiococcus* (Aguirre and Collins 1992b), *Atopobacter* (Lawson et al. 2000), *Atopococcus* (Collins et al. 2005), *Atopostipes* (Cotta et al. 2004a), *Bavariicoccus* (Schmidt et al. 2009), *Desemzia* (Stackebrandt et al. 1999), *Dolosigranulum* (Aguirre et al. 1993), *Granulicatella* (Collins and Lawson 2000),

E. Rosenberg et al. (eds.), *The Prokaryotes – Firmicutes and Tenericutes*, DOI 10.1007/978-3-642-30120-9_209,

*Isobaculum* (Collins et al. 2002), *Jeotgalibaca* (Lee et al. 2014), *Lacticigenium* (Iino et al. 2009), *Marinilactibacillus* (Ishikawa et al. 2003), *Pisciglobus*, and *Trichococcus* (Scheff et al. 1984), emended by (Liu et al. 2002). Circumscribed mainly upon 16S rRNA gene sequences, organisms are Gram-stain-positive with morphological forms from rods to coccoid forms. Usually facultatively anaerobic, but some species may grow aerobically or microaerophilically. Do not form endospores; may be motile or nonmotile. Usually catalase-negative. The cell wall may contain the diagnostic diamino acids lysine, ornithine, or *meso*-diaminopimelic acid. The type genus is *Carnobacterium*.

Although workers in the area had recognized the presence of "atypical" lactobacilli and other aberrant strains within the described genera, it was only with the advent of 16S gene sequencing that these discrepancies could be resolved and the true diversity of this group reported (➋ *Fig. 4.1*). Indeed, from the 1980s, these analyses led not only to the creation of the genus *Carnobacterium* but also to a proliferation of novel genera phylogenetically related to *Carnobacterium* before being formally designated as a family (Ludwig et al. 2009, 2010). From ➋ *Fig. 4.1*, it is apparent that the family shares a close relationship with both *Aerocococcaceae* and *Lactobacillaceae* with the majority of genera form a robust group. However, it should be noted that the genera *Desemzia* and *Pisciglobus* fall outside this phylogenetic clade and may share a closer relationship to the *Aerocococcaceae*. The description of novel genera in the future should stabilize the tree and resolve some of the discrepancies seen in the tree.

## Molecular Analysis

### DNA-DNA Hybridization Studies

As mentioned, the *Carnobacteriaceae* are in large part grouped together based on 16S rRNA gene sequence analysis. But while grouping at the genus level separates the family, it is not always the case for species within the genus. There are many examples within the *Carnobacteriaceae* where 16S rRNA gene sequences cannot simply delineate species to species. As an example, within the genus *Trichococcus*, all species are related by 99 % or greater 16S rRNA gene sequence similarity. DNA-DNA hybridization was used to confirm uniqueness between different species, along with varying physiological traits. Of most obvious is the 100 % gene sequence similarity between *T. collinsii* and *T. patagoniensis*. The results of DNA-DNA hybridization showed only a relatedness value of 45 ± 1 % between the two strains and consequently separated the two into separate species (Pikuta et al. 2006). DNA-DNA relatedness also indicated that *T. collinsii* was distinct from *T. pasteurii* (57.9 %), *T. palustris* (34.4 %), and *T. patagoniensis* (45 %) (Liu et al. 2002; Pikuta et al. 2006). Conversely, reassociation values indicated that *T. palustris* was separate from *T. pasteurii* (40.4 %) and *T. patagoniensis* (47 %). While the type species *T. flocculiformis* has not been hybridized versus other type species within *Trichococcus*, hybridization values >70 % with other strains (strains NDP Ben 77, Ben 200, and Ben 201) indicated they were all inclusive to the same species (Liu et al. 2002). This is also demonstrated in other *Carnobacteriaceae* species, namely, *Marinilactibacillus psychrotolerans* and *M. piezotolerans*. 16S rRNA gene sequence analysis indicates a greater than 99 % similarity between the two species, but less than 20 % DNA-DNA relatedness (Toffin et al. 2005). This, along with other differences in phenotypic data, separated the two into different species.

The genus *Carnobacterium* was originally determined in large part on the basis of DNA-DNA hybridization studies. Collins et al. (1987) produced a DNA-DNA homology table from 74 strains of known *Lactobacillus* species and other related lactobacilli that differed in physiological parameters. Homology results indicated that 15 of the strains formed a tight cluster that included both *Lactobacillus piscicola* NCDO 2762$^T$ and *L. carnis* NCDO 2764$^T$, thus indicating that they were the same species. A second group was also determined with high DNA-DNA relatedness values that was separate from the first and included *L. divergens* NCDO 2763$^T$. Two of the tested strains (MT44 and MT45) also formed a separate but distinct group that differed from all other strains tested. In addition, a fourth group of previously studied strains showed high DNA-DNA homology to each other, but were clearly distinct from the other groups (Collins et al. 1987). The results of the DNA-DNA hybridization assays, along with other relevant biochemical tests, validated the inclusion of these strains into the new genus, *Carnobacterium*, namely, as *C. piscicola*, *C. divergens*, *C. gallinarum* NCFB 2766$^T$ (=MT44$^T$), and *C. mobile* NCFB 2765$^T$ (=MT37L$^T$).

*Carnobacterium funditum* strain pf3$^T$ and *Carnobacterium alterfunditum* strain pf4$^T$ were separated originally by differing phenotypes between the two strains, but only *C. funditum* had its 16S rRNA gene sequence analyzed (Franzmann et al. 1991). 16S rRNA gene sequence analysis determined the closest phylogenetic neighbor to *C. funditum* was *C. mobile* strain MT37LT. DNA-DNA analysis showed a relatedness value of only 38 ± 8 % between *C. funditum* and *C. alterfunditum* indicating separate species, which was subsequently confirmed showing a relatedness value of *C. mobile* to *C. funditum* of 26 ± 2 % and 34 ± 2 % to *C. alterfunditum*. The 16S rRNA gene sequence of *C. alterfunditum* was published later and indicated only a 95.8 % similarity to *C. funditum* (Spielmeyer et al. 1993).

DNA-DNA hybridization was also used to differentiate between species of *Carnobacterium* and the type strains of *C. piscicola*, *C. divergens*, *C. mobile*, and *C. gallinarum* (Dicks et al. 1995). Using total soluble protein patterns of all strains, the authors were able to place most of the known strains of *Carnobacterium* into two clusters (I or II) associated with either *C. divergens* or *C. piscicola*, respectively. The type strains of *C. mobile* strain NCFB 2765$^T$ and *C. gallinarum* strain NCFB 2766$^T$ resulted in non-clustering outliers. DNA-DNA reassociation values were obtained when all strains were tested for homology against *C. divergens* LV6 and *C. piscicola* NCFB 2762$^T$. All of the strains that were grouped in Cluster I by total soluble cell protein patterns were also classified as the same species as *C. divergens*, with the identical result observed for

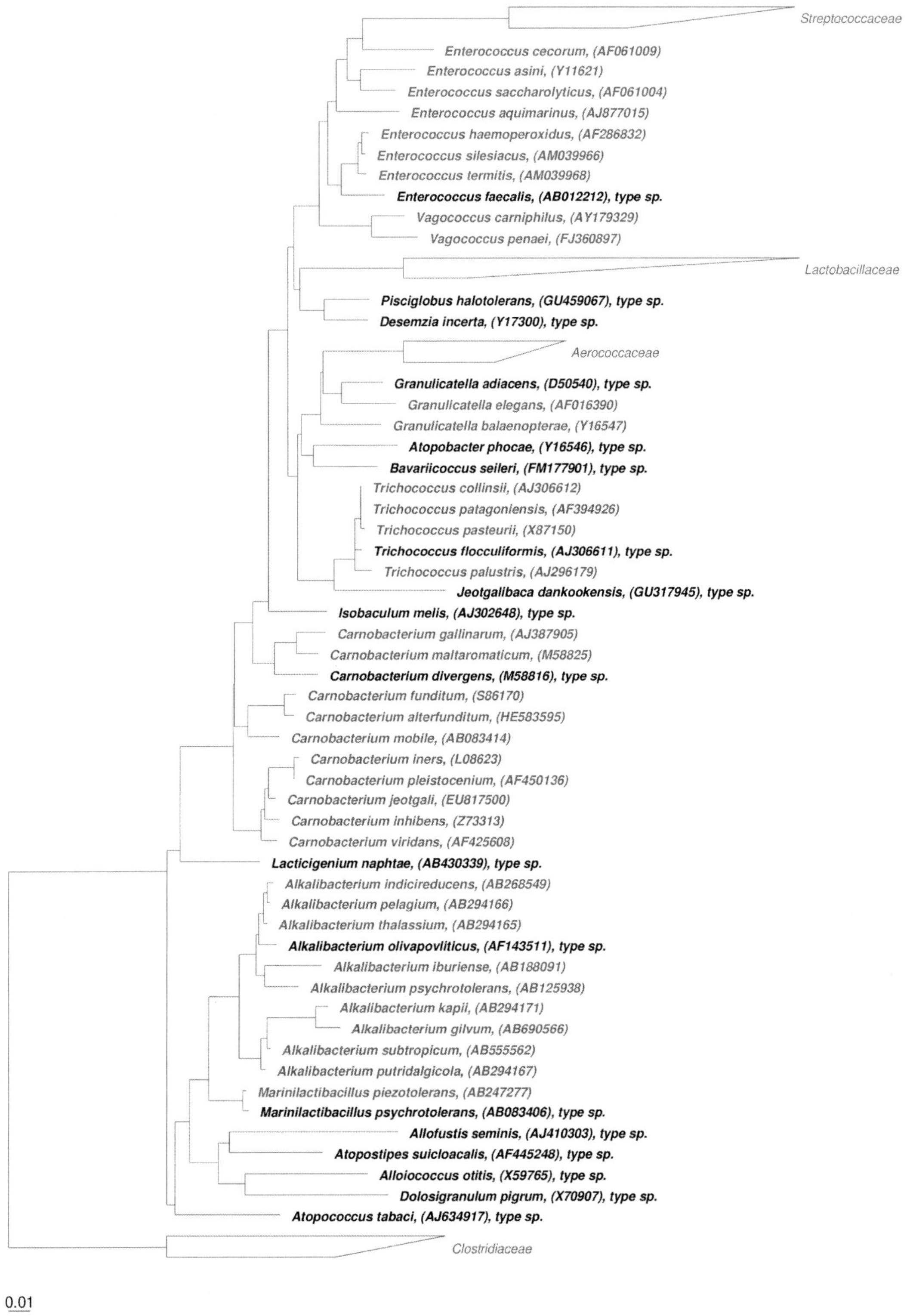


**Fig. 4.1**

Phylogenetic reconstruction of the family *Carnobacteriaceae* based on 16S rRNA and created using the maximum likelihood algorithm RAxML (Stamatakis 2006). The sequence dataset and alignment were used according to the All-Species Living Tree Project (LTP) database (Yarza et al. 2010; http://www.arb-silva.de/projects/living-tree). Representative sequences from closely related taxa were used as out-groups. In addition, a 10 % maximum frequency filter was applied in order to remove hyper variable positions and potentially misplaced bases from the alignment. Scale bar indicates estimated sequence divergence

Cluster II strains with *C. piscicola*. *C. mobile* and *C. gallinarum* were again shown to be only distantly related to either of these strains with homology values only ranging from 7 % to 18 % (Champomier et al. 1989a, b; Dicks et al. 1995).

First suggested by Collins and his colleagues using 16S rRNA gene sequence analysis (Collins et al. 1991), the two species *Lactobacillus maltaromicus* DSM 20342$^T$ and *Carnobacterium piscicola* DSM 20730$^T$ (with 100 % gene sequence similarity between the two) were later confirmed by DNA-DNA hybridization analysis (85 % homology) to belong to the same species (Mora et al. 2003). Further DNA-DNA hybridization events between *L. maltaromicus* DSM 20342$^T$ and *C. gallinarum* SM 4847$^T$ and between *L. maltaromicus* and *C. divergens* DSM 20623$^T$ indicated relatedness values of only 26 % and 29 %, respectively, further indicating a separate species. The combined resulting species name and type strain for *L. maltaromicus* and *C. piscicola* was *Carnobacterium maltaromaticum* strain DSM 20342$^T$ (Mora et al. 2003).

DNA-DNA hybridization was also used to distinguish the proposed new species of *Carnobacterium pleistocenium* strain FTR1$^T$ from its phylogenetically closest relative *Carnobacterium alterfunditum* pf4$^T$. Even though 16S rRNA gene sequence analysis showed 99.8 % similarity between the two species, a DNA-DNA relatedness value of only 39 ± 1.5 % was observed (Pikuta et al. 2005). Similarly, the 16S rRNA gene sequence of *Carnobacterium jeotgali* strain MS3$^T$ was most closely related to *C. pleistocenium* strain FTR1$^T$ at 98.95 %. But only a low value of 16 % DNA-DNA relatedness was observed between the two species (Kim et al. 2009). This is further exemplified in the latest species added to the *Carnobacterium* genus, *Carnobacterium iners*. The closest phylogenetic neighbor of *C. iners* is *C. funditum* at 99.2 % similarity. But DNA-DNA hybridization analysis between *C. iners* strain LMG 26642$^T$ and *C. funditum* LMG 14461$^T$ produced a low relatedness value of 18 %, confirming *Carnobacterium iners* as a novel species (Snauwaert et al. 2013).

Although not formally validated as a new species as yet, Lamosa and colleagues used DNA-DNA hybridization experiments to indicate that their new strain, *Carnobacterium* sp. 17–4, was a new species of *Carnobacterium* (Lamosa et al. 2011). Relatedness values for *Carnobacterium* strain 17–4 against *C. alterfunditum* DSM 5972$^T$ varied between 26.5 % and 28.5 % and varied 17.4–20.3 % against *C. viridans* DSM 14451$^T$ (the two closest phylogenetic neighbors to strain 17–4), thus indicating a new species.

DNA-DNA hybridization was also used to distinguish between the three isolated strains of *Alkalibacterium olivapovliticus*, the type species for the genus *Alkalibacterium* (Ntougias and Russell 2001). The results indicated relatedness values between 88 % and 99 % between all strains indicating they were the same species. When describing the second species of *Alkalibacterium*, *A. psychrotolerans* strain IDR2-2$^T$, Yumoto et al. (2004) used DNA-DNA hybridization to show uniqueness between strain IDR2-2$^T$ and *A. olivapovliticus* (24.3 % relatedness), as well as the phylogenetically close relative *Marinilactibacillus psychrotolerans* (7.6 % relatedness).

Because of the similarity of isolation source for *Alkalibacterium psychrotolerans*, *A. iburiense* and *A. indicireducens*, DNA-DNA hybridizations assays were determined for the different strains as well as differing species. DNA-DNA relatedness values (81.1–100 %) proved that the three strains of *A. iburiense* strain M3$^T$, 41A, and 41C all belong to the same species (Nakajima et al. 2005). Further, that *A. iburiense* strain M3$^T$ differed from *A. psychrotolerans* JCM 12281$^T$, *A. olivapovliticus* NCIMB 13710$^T$, and *M. psychrotolerans* NCIMB 13873$^T$ by relatedness values of 14.1 %, 7.3 %, and 3.9 %, respectively, when DNA from strain M3$^T$ is used as a probe and thus confirming a new species. In a similar fashion, Yumoto et al. (2008) used DNA-DNA homology to indicate that the three strains of *A. indicireducens* strain A11$^T$, F11, and F12 were the same species (87–97 %) and that *A. indicireducens* strain A11$^T$ was indeed unique with relatedness values *for A. psychrotolerans* JCM 12281$^T$, *A. olivapovliticus* NCIM 13710$^T$, and *A. iburiense* JCM 12662$^T$ of 40 %, 21 %, and 34 %, respectively, when DNA from strain A11$^T$ was used as a probe.

Ishikawa and colleagues used DNA-DNA hybridization experiments to show that their ten isolates constitute four separate genomic groups that ultimately represented four species within the genus *Alkalibacterium*, namely, *A. thalassium*, *A. pelagium*, *A. putridalgicola*, and *A. kapii* (Ishikawa et al. 2009). The authors analyzed and presented an extensive DNA-DNA hybridization table examining the four previous known species of the genus *Alkalibacterium* against all ten new strains. The closest phylogenetic neighbor to all four of the new species was *A. indicireducens*. While the 16S rRNA gene sequence similarities between *A. indicireducens* strain JCM 14232$^T$ and both *A. thalassium* strain T117-1-2$^T$ (group 1) and *A. pelagium* strain T143-1-1$^T$ (group 2) are close at 99.7 % and 99.8 %, respectively, their DNA-DNA relatedness is only 32 % and 35 %, respectively. In addition, *A. thalassium* and *A. pelagium* have at most 46 % homology in DNA-DNA analysis, thus indicating separation into individual species. Slightly more 16S rRNA gene sequence divergence is observed between *A. indicireducens* strain JCM 14232$^T$ and both group 3 strains (represented by *A. putridalgicola* strain T1297-2-1$^T$) at 98.1–98.2 % and group 4 strains (represented by *A. kapii* strain T22-1-2$^T$) at 97.1–97.2 %. However, DNA-DNA relatedness values between the *A. indicireducens* and *A. putridalgicola* are low at 8 % and only slightly higher for *A. kapii* at 12 %. In addition, when *A. subtropicum* O24-2$^T$ was hybridized with *A. putridalgicola* T129-2-1$^T$ (its closest phylogenic neighbor at 99.6 % 16S rRNA gene sequence similarity), a relatedness value of only 27 % was obtained, again indicating separate species (Ishikawa et al. 2011). In subsequent work, the nine strains of *Alkalibacterium gilvum* were shown to all have >70 % DNA-DNA relatedness to each other yet had low homology to all other known species of *Alkalibacterium* with relatedness values varying between 6 % and 22 % (Ishikawa et al. 2013).

Two clinical strains now recognized as *Granulicatella adiacens* and *Abiotrophia defectiva* (see Collins and Lawson 2000) were first described in 1961 (Frenkel and Hirsch 1961) and were later separated into two species, *Streptococcus defectivus* and *Streptococcus adjacens*, by DNA-DNA hybridization which showed <10 % relatedness between the two (Bouvet et al. 1989). Kanamoto and colleagues used extensive DNA-DNA hybridization studies (tied with RFLP and PCR probe-based genomic analysis) to separate 45 *Abiotrophia* strains into four groups or genotypes (Kanamoto et al. 2000). *A. defectiva* (formerly *Streptococcus defectivus*) represented group 1, *Granulicatella adiacens* (formerly *Abiotrophia adiacens*, formerly *Streptococcus adjacens*) represents group 2, and *Granulicatella elegans* (formerly *Abiotrophia elegans*) represents group 4. Group 3 was proposed to encompass strains closely associated with group 2 strains, in which Kanamoto et al. proposed the name "*Abiotrophia para-adiacens*" (subsequently "*Granulicatella para-adiacens*"). In addition to the 45 strains used by Kanamoto et al. (2000) in their DNA-DNA analysis, they also included another *Carnobacteriaceae*, *Dolosigranulum pigrum* NCFB 2893$^T$, in their studies. Relatedness values were low when *A. defectiva* was analyzed versus *D. pigrum* (5.2 %), *Granulicatella adiacens* was analyzed versus *D. pigrum* (4.1 %), and "*Granulicatella para-adiacens*" was analyzed against *D. pigrum* (2.5 %). DNA-DNA analysis was not performed with *G. elegans* against *D. pigrum* in this study. *Granulicatella balaenopterae* has not been subjected to DNA-DNA homology analysis.

Schmidt et al. (2009) used DNA-DNA hybridization studies to aid in resolving that the six isolates of their new genus *Bavariicoccus* were actually members of a single species. Relatedness percentages were all >70 %, except for the type species (*Bavariicoccus seileri* strain WCC 4188$^T$) to one other strain (strain WCC 4189), where relatedness values varied between 54 % and 65 %. However, since all other phenotypic, genotypic, and 16S rRNA gene sequence data were identical, it was concluded it also belonged to the same species.

There are several genera within the *Carnobacteriaceae* that are represented by a single species. A number of these microorganisms have not been subjected to DNA-DNA hybridization studies as a result. These include *Allofustis seminis*, *Atopobacter phocae*, *Atopococcus tabaci*, *Atopostipes suicloacalis*, *Isobaculum melis*, and *Lacticigenium naphtae*. *Bavariicoccus* (above) and *Pisciglobus* are also represented by a single species, but DNA-DNA hybridization studies between their isolated strains were performed. *Pisciglobus halotolerans* strains C01$^T$ and C02 indicated 99.8 % DNA-DNA relatedness to each other, but showed <5 % to *Desemzia incerta* DSM 20581$^T$, its phylogenetically closest neighbor at 96.9 % 16S rRNA gene sequence similarity (Tanasupawat et al. 2011). As new species and strains are added to the *Carnobacteriaceae* in the future, DNA-DNA hybridization will certainly become more integral in resolving all the new species and strains.

## PCR Primer-Based Detection

*Carnobacterium* species have special significance because of their prominence as food-associated lactic acid bacteria. Scarpellini and associates devised a species-specific PCR amplification method for the identification of all of the then-known species within *Carnobacterium* (Scarpellini et al. 2002). The PCR-based method used both the *HaeIII*/*HinfI* restriction analysis of the 16S rRNA and the amplification of the internal transcribed spacer (ITS) between the 16S and the 23S rRNA.

Because of the clinical importance of some of the *Carnobacteriaceae*, laboratory studies have tried to distinguish species from species and, in some cases, strain to strain by PCR detection. The importance of *Granulicatella* sp. and other nutritionally variant streptococci (NVS) as opportunistic pathogens has been of particular interest. A PCR assay was designed and tested by Roggenkamp and colleagues with specific 16S rRNA gene sequence primer sets for the detection and identification of *Granulicatella* sp. (formerly *Abiotrophia*) (Roggenkamp et al. 1998a). Using eight strains of *Granulicatella adiacens*, four strains of *Granulicatella elegans*, and three strains of *Abiotrophia defectiva*, in addition to 17 known bacteria and 40 blood-isolated Gram-positive and Gram-negative microorganisms, their method could identify *Granulicatella* sp. and *A. defectiva* from all those tested, as well as singly detect the most fastidious species *G. elegans*.

Most analyses for the identification of *Granulicatella* sp. (or *Abiotrophia defectiva*) were based on the 16S rRNA gene sequence, but other PCR assays for specific genes have also been successfully targeted. Using the *rpoB* gene sequence (encoding a beta-subunit of RNA polymerase) as a target for detection, researchers used a 740-bp amplicon to accurately identify *Granulicatella* species as well as other NVS and *Streptococcus* species (Drancourt et al. 2004). Hung and colleagues developed a multiplexed PCR method that allowed for the simple, rapid, and accurate discrimination of *Granulicatella*, *Abiotrophia*, and *Gemella* species at the genus level using *groESL* gene target (Hung et al. 2010). Molecular probing and diagnosis of *Granulicatella* infections have been recently reviewed and include 16S rRNA gene sequence analysis, analysis of the ribosomal 16S–23S intergenic spacer region (ITS), partial *rpoB* gene sequencing, and array-based methods for detection (Tung and Chang 2011). Detection of the Cha gene sequence (a fibronectin-binding protein) in *Granulicatella adiacens* and *Abiotrophia defectiva* was also applied successfully and could separate these two species from *Granulicatella elegans* and "*G. para-adiacens*" which are Cha-negative (Yamaguchi et al. 2011).

Another of the medically important species within the *Carnobacteriaceae* is *Alloiococcus otitis* (also often referred to "*Alloiococcus otitidis*"). An early PCR-based probe detection was developed specifically for the detection of *A. otitis* within mix cultures (Aguirre and Collins 1992a). Because of the fastidious nature and slow growth of *A. otitis*, it was suggested that it often went undetected in clinical screenings (Faden and Dryja 1989). Using the primer pairs Ao1-Ao11 or Ao2-Ao11, PCR amplification was tested using the DNA from three *A. otitis* strains and 58 other reference strains. Amplification products (900 or 800 bp) were specific only for *A. otitis* DNA, with the Ao1-Ao11 primer pairs showing a tenfold greater sensitivity.

A multiplex-PCR method was described that distinguished between four known ear infection pathogens, namely,

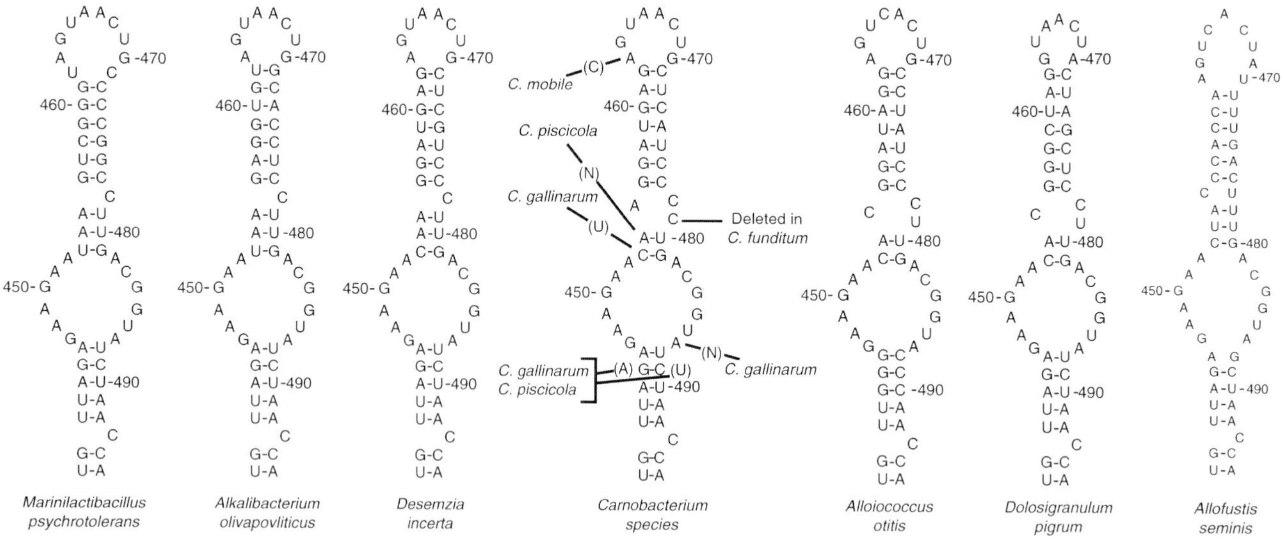


**Fig. 4.2**
**Nucleotide sequences and secondary structures of the V6 region of the 16S rRNA of the new isolates and closely related bacteria. Numbers correspond to positions in the *E. coli* sequence (Data taken from Ishikawa et al. 2003; Ntougias and Russell 2001; Lawson 2009)**

"*Alloiococcus otitidis*," *Haemophilus influenzae*, *Moraxella catarrhalis*, and *Streptococcus pneumonia* in middle ear effusion (MEE) samples (Hendolin et al. 2000). Four species-specific primers are used in the initial multiplex-PCR reaction, followed by a ligase detection reaction (LDR) with fluorescently labeled primers (Cy-5 labeled) specific for "*A. otitidis*." The use of a multiplex-PCR method for the diagnosis of a pathogen was again used to examine middle ear fluids collected from children suffering from otitis media infection and which no cultivable bacterial isolate could be obtained. Detection of "*A. otitidis*" was found in 32 % of the samples from these children and in 9 % of the cases in which no other infecting microorganism was observed (Kaur et al. 2010).

For reference, Woo and colleagues created a 16S rRNA gene sequence database with all medically important microorganisms listed in the *Manual of Clinical Microbiology* (Woo et al. 2011). The database, called 16SpathDB (http://147.8.74.24/16SpathDB), contains 1,014 16S rRNA gene sequences from 1,010 unique bacterial species from the list (as of 2010). It was intended to discriminate against environmental isolates and misnamed gene sequences for clinical sequence searches, which are often encountered when using BLAST searches in the GenBank database. The medically important *Carnobacteriaceae* included in this database are *Alloiococcus otitis*, *Granulicatella adiacens*, *G. elegans*, and *Dolosigranulum pigrum*.

## 16S rRNA Secondary Structure

Ntougias and Russell (2001) when describing *Alkalibacterium* showed that the predicted secondary structure of the V6 region of the terminal ring in the V6 region was different among closely related genera. Subsequently, Ishikawa et.al. (2003) demonstrated such differences that were also present when describing *Marinilactibacillus*, and similar differences have also been observed with *Allofustis seminis* (Lawson 2009) when compared to members of the genera *Alkalibacterium*, *Alloiococcus*, *Desemzia*, *Carnobacterium*, and *Dolosigranulum* (➋ *Fig. 4.2*). However, not all genera and species have been examined, and some genera to date contain only a single species so it remains to be determined if these structural differences are stable.

## FTIR

Quantitative analysis by Fourier transform infrared (FTIR) spectroscopy of microbial species composition has been used as a fast technique for the classification and identification of microorganisms (Oberreuter et al. 2002). Schmidt and colleagues used FTIR to classify six strains of *Bavariicoccus* against phylogenetically related strains *Atopobacter phocae* CCUG 42358$^T$, *Carnobacterium divergens* DSM 20623$^T$, *Trichococcus flocculiformis* DSM 2094$^T$, *Enterococcus hirae* DSM 20160$^T$, and *Vagococcus carniphilus* DSM 17031$^T$. FTIR typing indicated the six *Bavariicoccus* strains formed a separate group from the phylogenetically closely related genera (Schmidt et al. 2009), therefore helping to confirm the new genus and species of *Bavariicoccus*. Because of the fastidious growth requirements of both *Granulicatella* and *Abiotrophia*, their type species were not included in the analysis.

FTIR spectroscopy was also utilized to exam 67 strains of *Carnobacterium* versus other lactic acid bacteria (Lai et al. 2004). The *Carnobacterium* strains clustered into seven of nine distinguishable clusters and were generally in line with previously published phylogenetic data. Only two strains tested, both previously ascribed to *C. divergens* (strains 694 and C749) were the only outliers forming single-member clusters.

## Riboprinting and Ribotyping

Using PCR-RFLP (PCR amplification followed by restriction fragment length polymorphism) to detect genome digested DNA with *Hae*III and *Msp*I, Ohara-Nemoto and colleagues classified RLFP patterns between 92 isolates which included 11 strains of *Abiotrophia defectiva* and 81 strains of *Granulicatella adiacens* (formerly *Abiotrophia adiacens*) from other similar clinically important microorganisms (Ohara-Nemoto et al. 1997). It was projected that since the degree of clinical severity is closely correlated to the delay in bacteriological diagnosis for infective endocarditis, this assay would provide a more rapid detection and treatment option against these opportunistic pathogens.

Using a combination of PCR-RFLP, DNA-DNA hybridization, and 16S rRNA gene sequence analysis, Kanamoto and colleagues classified 45 *Abiotrophia* strains (including *Abiotrophia defectiva* ATCC 49176$^{T}$, *Granulicatella adiacens* ATCC 49175$^{T}$, and *Granulicatella elegans* DSM 11693$^{T}$) into four separate genotype groups (Kanamoto et al. 2000). Each of the type strains were separated into three of the four groups, with the remaining group closely related to *G. adiacens.* Kanamoto et al. (2000) suggested the name "*Abiotrophia para-adiacens*" (subsequently "*G. para-adiacens*") as this new species. *G. balaenopterae* was not included in this study and has not been subjected to DNA-DNA hybridization or RFLP assays against any of the type species.

Using 45 strains representing each of the *Carnobacterium* species and 11 species of related genera (including *Desemzia, Enterococcus, Lactobacillus, Lactococcus, Leuconostoc, Streptococcus,* and *Weissella*), Rachman and colleagues used RFLP assays to distinguish between all eight of the known *Carnobacterium* species (at that time) (Rachman et al. 2004). By examining the 16S–23S rRNA gene intergenic spacer region (ISR), they developed species-specific primers within the large ISR (L-ISR). Priming for this region and utilizing a nested PCR approach, they found that when using HindIII restriction enzyme, individual profiles could be obtained for each of the known strains and was specific for *Carnobacterium* species. At the time of the study, only one strain was known to exist for *C. funditum, C. alterfunditum, C. inhibens,* and *C. viridans.* It was suggested that this approach could prove to be a reliable tool for the quick analysis for new strains, ecotypes, and niches for *Carnobacterium* species. Earlier work presented by Kabadjova et al. (2002) but using ISR-RFLP as a basis for classification were able to separate 42 strains tested into four groups representing the *Carnobacterium* species that are mostly commonly associated with food, namely, *C. divergens, C. gallinarum, C. mobile, and C. piscicola* (*C. piscicola* would be renamed *Carnobacterium maltaromaticum* in 2003) (Kabadjova et al. 2002). In this study, the ISR amplifications were subsequently digested with both HinfI and HindIII restriction endonucleases. These results were essentially duplicated by Laursen and colleagues using 79 strains of Carnobacterium and the same restriction enzymes but also included the strains *C. viridans* and *C. inhibens* (Laursen et al. 2005).

Terminal restriction fragment length polymorphism (T-RFLP) was also successfully used for the identification of *Dolosigranulum pigrum* within bacterial communities of the human anterior nares (nostrils) (Camarinha-Silva et al. 2012). *D. pigrum* could be distinguished by its profile when amplified community DNA was sequentially digested with the AseI, TspRI, and ApekI restriction enzymes and assayed.

## MALDI-TOF

Matrix-assisted laser desorption ionization time-of-flight mass spectrometry (MALDI-TOF MS) can provide a tool for the fast and accurate identification of pathogens. By comparing the mass spectra obtained from bacterial proteins to the mass spectra of many reference strains, rapid identification of unknown microorganisms are achieved. An isolate of *Granulicatella adiacens* was identified by this method in the examination of patients with community-acquired pneumonia (CAP) (Xiao et al. 2012). In addition, from all isolates obtained from throat swab samples, the authors were able to identify 12 genera and 30 species from the 212 original isolates obtained. But while MALDI-TOF MS may prove to be a cost-effective alternative for quickly identifying pathogens, it still has some drawbacks as related to the members of the *Carnobacteriaceae.* In a study comprising 239 isolates, identification by MALDI-TOF MS resulted in six isolates that could not be determined by their protein profiles (Tekippe et al. 2013). Of the six unidentified organisms, *Dolosigranulum pigrum* and a *Granulicatella* sp. were among the unidentifiable by the extraction procedure and system used, although *D. pigrum* was not present in the system software database employed by the investigators. The lack of identification of *Granulicatella* strains to the species level was again observed using the Bruker MALDI-TOF MS system by Schulthess and colleagues (Schulthess et al. 2013). While the system and extraction procedure they utilized could positively identify isolates to the genus level of *Granulicatella* (and *Gemella* strains) 90 % of the time, only 50 % of these isolates could be determined to the species level. Similarly, Ratcliffe and colleagues compared two MALDI-TOF MS systems, the Bruker MS and Vitek MS, for the species identification of 10 species of *Granulicatella adiacens,* one of *G. elegans* and three species of *Abiotrophia defectiva* (Ratcliffe et al. 2013). The Vitek MS system identified all 14 isolates correctly to the species level, while the Bruker MS identified 8 of 10 *G. adiacens* species and all three *A. defectiva* species but required repeated testing to identify the final two *G. adiacens* species and three repeated analyses to identify the *G. elegans.*

**Table 4.1**
**List of members of the *Carnobacteriaceae* that have had their genome sequenced or scheduled to be completed**

| Microorganism | Bioproject # | Size (Mb) | % GC | Genes | Proteins |
|---|---|---|---|---|---|
| *Carnobacterium maltaromaticum* ATCC 35586[a] | PRJNA168324, PRJNA73249 | 3.54 | 34.5 | 3,325 | – |
| *Carnobacterium maltaromaticum* strain LMA28 | PRJNA179370, PRJEB5449 | 3.65 | 34.5 | 3,659 | 3,581 |
| "*Carnobacterium gilichinskyi*" strain WN1359[T] | CP006812[b] | 2.35[c] | 35.3 | 2,234 | 2,152 |
| "*Carnobacterium*" sp. 17-4 | PRJNA65789, PRJNA60607 | 2.69[c] | 35.2 | 2,575 | 2,474 |
| "*Carnobacterium*" sp. AT7 | PRJNA54673, PRJNA19299 | 2.44[c] | 35.3 | 2,489 | 2,388 |
| *Allofustis seminis* DSM15817[T] | PRJNA199283, PRJNA174172 | 1.44 | 37.6 | 1,478 | 1,404 |
| *Alloiococcus otitis* ATCC 51267[T] | PRJNA52161, PRJNA182884 | 1.78 | 44.6 | 1,691 | 1,630 |
| *Atopococcus tabaci* DSM 17538[T] | PRJNA185553, 1012088 | 2.27 | 45.5 | 2,431 | 2,352 |
| *Bavariicoccus seileri* DSM 19936[T] | PRJNA188834, 1013858 | 2.24 | 38.2 | 2,195 | 2,128 |
| *Dolosigranulum pigrum* ATCC 51524[T] | PRJNA83001, PRJNA52171 | 1.85 | 39.5 | 1,723 | 1,692 |
| *Granulicatella adiacens* ATCC 49175[T] | PRJNA55951, PRJNA37271 | 1.92 | 37.6 | 1,949 | 1,920 |
| *Granulicatella elegans* ATCC 700633[T] | PRJNA40873, PRJNA38745 | 1.88 | 33.3 | 1,742 | 1,701 |
| *Lacticigenium naphtae* DSM 19658[T] | PRJNA188919, 1013831 | 2.13 | 37.6 | 2,016 | 1,967 |

(Bioproject # refers to NCBI Genome Database project number or JGI Project ID)
[a] = type strain of *C. piscicola*, see Mora et al. 2003
[b] = GenBank Accession #
[c] = including plasmid
Scheduled to be completed (Bioproject # or JGI Project ID):
*Carnobacterium maltaromaticum* 38b (PRJEB254), *Alkalibacterium kapii* FAM208_38 (PRJEB304), *Bavariicoccus seileri* WCC 4188 (PRJEB359), *Dolosigranulum* sp. KPL1914 (PRJNA169449), *Granulicatella adiacens* CC94D (PRJNA71553), "*Granulicatella para-adiacens*" M562 (PRJNA38785), *Marinilactibacillus psychrotolerans* 42ea (PRJEB266), *Marinilactibacillus psychrotolerans* FAM208_59 (PRJEB307), *Carnobacterium inhibens* K1 (1022086), *Carnobacterium gallinarum* MT44 (1022036), *Carnobacterium mobile* MT37L (1022039), *Carnobacterium viridans* MPL-11 (1022042), *Carnobacterium maltaromaticum* MX5 (1022045), *Carnobacterium divergens* 66 (1022048), *Carnobacterium funditum* pf3 (1022051), *Carnobacterium alterfunditum* pf4 (1022054)

## Genome Analyses

The first of the *Carnobacteriaceae* to have their genome sequenced (and patented) was the medically important organism *Alloiococcus otitis* ("*Alloiococcus otitidis*") (Fletcher et al. 2003). The genome of *A. otitis* was re-sequenced and released as part of the Human Microbiome Project (HMP) Reference Genomes and deposited at NCBI's (National Center for Biotechnology Information) Genome Database (http://www.ncbi.nlm.nih.gov/genome/10857?project_id=52161). Also because of the clinical and medical importance of these microorganisms, *Granulicatella adiacens*, *G. elegans*, and *Dolosigranulum pigrum* ATCC 51524 have also had their genome sequenced and distributed as part of the HMP. ➋ *Table 4.1* indicates their relative size (Mb) and the number of putative genes and proteins based on analysis of the sequence data.

The genomes of other *Carnobacteriaceae* have also been sequenced (or are scheduled to be sequenced) and deposited in conjunction with NCBI, JGI (DOE's Joint Genome Institute-Integrated Microbial Genomes), and GOLD (Genomes Online Database) (➋ *Table 4.1*). Though only one specific case of infection by the species *C. maltaromaticum* has been reported in the literature (see Pathogenicity, Clinical Relevance), virulence factors common to other pathogenic microorganisms have been annotated within the genome of two closely related strain of *C. maltaromaticum*, strains ATCC 35586[T] and LMA 28 (Leisner et al. 2012; Rahman et al. 2013). Analysis of the genome of *C. maltaromaticum* ATCC 35586[T] showed it contains 59 potential virulence gene motifs, including adhesion and survival genes to host organs and extracellular matrices, capsule synthesis, cell wall modification, invasion, and resistance to toxic compounds. More recently, the complete genome of *Carnobacterium maltaromaticum* strain LMA 28 has been completed and published (Cailliez-Grimal et al. 2013; Rahman et al. 2013). Along with several similar virulence genes found in ATCC 35586[T], numerous other genes associated with colonization of the intestinal tract were observed including bile salt hydrolysis, production of exopolysaccharides, and antimicrobial resistance genes. Additionally, analysis of the genomes of strains ATCC 35586[T] and LMA 28 revealed several genes known to be involved in branched-chain amino acid metabolism not present in the genomes of *Carnobacterium* strains 17–4 or AT7. It was suggested that the larger genome size of strains ATCC 35586[T] and LMA 28 might be responsible for these two species to adapt to more diverse environments than other *Carnobacterium* species (Cailliez-Grimal et al. 2013, ➋ *Table 4.1*).

*Carnobacterium* sp. 17–4 which is a psychrotolerant, lactic acid bacterium isolated from seawater (Lamosa et al. 2011) has also had its genome sequenced and published (Voget et al. 2011). The production of a gluconeotrehalose by this species has made it a potentially industrially important microorganism (see Formation of Value Added Products). Strain 17–4 carries one plasmid, pCAR50 (50,105 bp), which putatively harbors 54 protein-encoding gene regions (Voget et al. 2011). Both the complete genome (GenBank accession # CP002563) and plasmid (accession #CP002564) have been deposited in GenBank. *Carnobacterium* sp. AT7 was isolated from the Aleutian Trench at a depth of 2,500 m (Lauro et al. 2007) and has also been sequenced and deposited in GenBank. Both strains have similar genome profiles (➋ *Table 4.1*), and comparison through global sequence alignment indicated the two strains shared 1816 orthologous genes (74.7 %) along with each containing a single plasmid (Voget et al. 2011). Though the genome sizes seem comparable, only replication-associated proteins were shared between the plasmid of strains 17–4 and the putative plasmid of strain AT4 (76,048 bp) indicating different content as well as size. The genome of *Carnobacterium* strain AT7 was also investigated for the presence of clustered regularly interspaced short palindromic repeats (CRISPR) units which are hypervariable loci broadly distributed in both bacteria and archaea (Horvath et al. 2009). These repeated DNA regions have been shown to provide acquired immunity against foreign genetic elements such as viruses or phage predation. Analysis revealed that strain AT7 did not have any CRISPR regions within its genome. It was suggested that since only a single strain of *Carnobacterium* was examined, it may not be representative of the entire genus. This observation may prove valid as the genome of *Carnobacterium maltaromaticum* strains are approximately 1 Mb larger than *Carnobacterium* strain AT7 (➋ *Table 4.1*).

Pikuta et al. (2005) estimated of the genome sizes of *Carnobacterium alterfunditum* strain pf4$^T$ and *Carnobacterium pleistocenium* strain FTR1$^T$ by determining DNA reassociation kinetics utilizing the equation described by Gillis et al. (1970). A genome size of $1.9 \times 10^9$ Da was calculated for *C. alterfunditum* strain pf4$^T$ and $2.1 \times 10^9$ Da for *C. pleistocenium* strain FTR1$^T$. Using standard conversion factors, the genome size of both *C. pleistocenium and C. alterfunditum* equate to approximately 3.2 Mb and 2.9 Mb, respectively, which is in correlation with other *Carnobacterium* genome sizes (➋ *Table 4.1*).

Along with *Allofustis seminis* DSM 15817$^T$, *Atopococcus tabaci* DSM 17538$^T$, *Bavariicoccus seileri* DSM 19936$^T$, and *Lacticigenium naphtae* DSM 19658$^T$, other *Carnobacteriaceae* strains other than the type species have had their genome sequence completed as well (➋ *Table 4.1*). *Carnobacterium maltaromaticum* sp., *Alkalibacterium kapii*, *Bavariicoccus seileri* sp., *Dolosigranulum* sp., *Granulicatella adiacens* sp., "*Granulicatella para-adiacens*" sp., and two species of *Marinilactibacillus psychrotolerans* are all scheduled for genome sequencing and release in the near future.

## Plasmids

Only one genus of the family *Carnobacteriaceae* has been shown to contain plasmids, though future genome analysis of all members may prove otherwise. The genus *Carnobacterium* has several species (and numerous strains) that have been shown to contain a plasmid (➋ *Table 4.8*). The two most prominent species to contain plasmids are *C. maltaromaticum* (formerly *C. piscicola*, though still often referred to as *C. piscicola* in the literature) and *C. divergens*. Laursen and colleagues assayed 111 *Carnobacterium* strains for the presence of a plasmid and showed that 47 % of *C. divergens* strains and 38 % of *C. maltaromaticum* were positive for a plasmid (Laursen et al. 2005). In addition, they showed that the plasmid sizes differed among the differing strains (though more often of a large size, >25–30 kb), as well as the number of plasmids in each strain, with 20 % of the plasmid-positive strains of *C. divergens* and 4 % of *C. maltaromaticum* strains containing two or more plasmids. Three of the tested strains of *Carnobacterium mobile* were shown to contain multiple plasmids that differed from the type species of *C. mobile* strain LMG 9842$^T$. An important function of plasmids within the *Carnobacterium* strains is that they have often been associated with the production of bacteriocins which inhibit other potentially more harmful bacteria from proliferating in packaged food products such as meat, fish, and cheeses (see section "➋ Applications").

## Phenotypic Analyses

Although the family *Carnobacteriaceae* was circumscribed mainly on the basis of 16S rRNA gene sequence analysis, ➋ *Table 4.2* shows morphological, biochemical, and chemotaxonomic properties used for the delineation of genera and species. API data is taken from the Culture Collection University of Göteborg (www.ccug.se) unless where stated in the original descriptions.

### *Carnobacterium* Collins, Farrow, Philips, Ferensu, and Jones 1987, 314$^{VP}$

car.no.bac.teri.um. L. gen. n. *carnis* of flesh; N.L. neut. n. *bacterium* small rod; N.L. neut. n. *Carnobacterium* small rod from flesh.

Cells are Gram-positive staining, non spore-forming straight rods that may or may not be motile. The cells are arranged singly, in pairs, or in short/long irregularly curved chains. The organism is facultatively anaerobic and heterofermentative. Catalase and oxidase are not produced, and nitrate is not reduced to nitrite. Mesophilic and psychrotolerant with growth at 0 °C is observed. Growth is obtained at neutral pH and in alkalitolerant conditions. NaCl is not required for growth with some species tolerating up to 5–6 % NaCl. The majority of species produce L (+) isomer form of lactic acid from glucose. The major fatty acids are of the straight-chain saturated and monounsaturated

**Table 4.2**
**Morphological, biochemical, and chemotaxonomic properties of *Carnobacteriaceae***

| Characteristic | 1 | 2 | 3 | 4 | 5 | 6 | 7 | 8 | 9 |
|---|---|---|---|---|---|---|---|---|---|
| Cell shape | Straight, slender rods | Rods | Rods | Ovoid | Irregular rods | Coccoid | Short rods | Coccoid | Short rods |
| Motility | +/− | + | nd | nd | nd | nd | − | nd | − |
| Gram reaction | + | +/− | + | nd | + | + | + | + | + |
| Metabolism | FA | FA | FA | Aerobic | FA | Aerobic | FA | Aerobic | FA |
| End products from glucose | L (L) | L (DL) | nd | No product | nd | nd | F, A, L | A, E, L (DL) | F, A, L |
| Temperature (°C) | 0–40 | −1.8–45 | nd | No growth at 10 or 45 | 25–40 | 28–32 | 28–32 | 10–40 | nd |
| Major fatty acids | $C_{16:0}$, $C_{16:1}$, $C_{16:1\omega 9c}$, $C_{18:0}$, $C_{18:1\ \omega 9c}$ | $C_{16:0}$, $C_{16:1\ 7c}$, $C_{16:1\ 9c}$, $C_{18:0}$ | $C_{16:0}$, $C_{16:1}$, $C_{16:1}$, $C_{18:0\omega 9c}$ | $C_{16:0}$, $C_{16:1}$, $C_{16:1}$, $C_{18:1\omega 9c}$, $C_{18:2\omega 6,9c}$, *ante* $C_{18:0}$, | nd | $C_{16:1\ \omega 9c}$, $C_{16:0}$, $C_{18:1\ \omega 9c}$ | $C_{14:0}$, $C_{16:0}$, $C_{16:1}$, $C_{18:0}$, $C_{18:1}$ | $C_{16:0}$, $C_{18:1\ \omega 9c}$ | $C_{16:0}$, $C_{16:1\ \omega 9c}$, $C_{18:1\ \omega 9c}$ |
| Cell wall murein | m-Dpm | A4α, Orn-D-Asp or Orn-D-Glu A4β, L-Orn-D-Asp, L-Orn- D-Glu | A1α, L-Lys direct | nd | A4β, L-Orn-D-Asp | A4α, L-Lys-L-Glu | A4α, L-Lys-D-Asp | A4α L-Lys-D-Asp | A4α L-Lys-D-Asp |
| DNA G+C content (mol%) | 32–44 | 39–43 | 39 | 44–45 | nd | 46 | 44 | 38 | 44 |
| Source | Vacuum-packaged meat and related products, cheese, fish, Antarctic lake, permafrost | Wash water from olives, cheese, fermented polygonum indigo, decaying marine algae, decaying sea grass, salted/fermented shrimp paste, salted/raw fish | Swine semen | Human middle ear | Dead seal | Tobacco | Swine manure storage pit | Cheese | Insect ovaries (*Tibicen linnei*), *metalworking fluids* |

| Characteristic | 10 | 11 | 12 | 13 | 14 | 15 | 16 | 17 |
|---|---|---|---|---|---|---|---|---|
| Cell shape | Ovoid | Coccoid | Rods | Coccoid | Oval Rods | Straight Rods | Coccoid | Spherical to ovoid |
| Motility | − | − | − | − | + | + /− | − | nd |
| Gram reaction | + | + | + /− | + | − | + | + | + |
| Metabolism | FA | FA | FA | Aerobic | FA | FA | FA | Aerotolerant |
| End products from glucose | nd | nd | A, L | ND | L (L) | L (L) | L (L) | F, A, L, E |
| Temperature (°C) | No growth at 10 or 45 | No growth at 10 or 45 | No growth at 10 or 45 | 10–37 | 30 | −1.8–45 | 15–40 | −5–40 |
| Major fatty acids | $C_{14:0}$, $C_{16:1\omega 9c}$ $C_{16:0}$, $C_{16:1}$, $C_{18:0}$, $C_{18:1\omega 9c}$ | $C_{16:0}$, $C_{16:1\ \omega 9c}$, $C_{18:1\omega 9c}$, $C_{18:2\ \omega\ 6,9c}$, $C_{18:0}$ | $C_{14:0}$, iso-$C_{15:0}$, anteiso-$C_{15:0}$, $C_{16:0}$ | $C_{14:0}$, $C_{16:0}$, $C_{16:1\omega 9c}$, $C_{18:1\omega 9c}$, | $C_{14:0}$, $C_{16:0}$, $C_{16:1\ \omega 9c}$, $C_{18:1\ \omega 9c}$ | $C_{16:0}$, $C_{16:1}$, $C_{18:0}$, $C_{18:19c}$ | $C_{14:0}$, $C_{16:1\ \omega 9c}$, $C_{16:0}$ | $C_{14:0}$, $C_{16:0}$, $C_{16:1}$, $C_{16:17c}$, $C_{18:1\omega 9c}$ , $C_{18:\ \omega 17c}$ |
| Cell wall murein | A4α L-Lys-D-Asp | A4β, L-Orn-D-Asp A3α, Lys-L-Thr-Gly | A3α, Lys-L-Thr-Gly | A4α (L-Lys-D-Glu) | A4α, L-Lys-L-Glu | A4β, L-Orn-D-Asp | L-Lys | A4α L-Lys-D-Asp |

**Table 4.2 (continued)**

| Characteristic | 10 | 11 | 12 | 13 | 14 | 15 | 16 | 17 |
|---|---|---|---|---|---|---|---|---|
| DNA G+C content (mol%) | 42 | 36–37.5 | 39 | 39.6 | 38 | 34–42 | 38.6 | 45–49 |
| Source | Human sources | Human clinical material, canine oral microbiome, lungs of Minke whale | Intestine of dead badger | Saeujeot (Korean traditional food) | Crude oil | Deep seafloor sediment, living sponge, decaying marine algae, raw Japanese ivory shell | Fish sauce | Sewage bulking sludge, hydrocarbon-contaminated soil, swamp sediment, penguin guano |

1, *Carnobacterium* (Hammes and Hertel 2009) (Kim et al. 2009); 2, *Alkalibacterium* (Ishikawa et al. 2003; Ntougias and Russell 2009; Ishikawa et al. 2013); 3, *Allofustis* (Collins et al. 2003); 4, *Alloiococcus* (Aguirre and Collins 1992b); 5, *Atopobacter* (Lawson et al. 2000); 6, *Atopococcus* (Collins et al. 2005); 7, *Atopostipes* (Cotta et al. 2004a); 8, *Bavariicoccus* (Schmidt et al. 2009); 9, *Desemzia* (Stackebrandt et al. 1999); 10, *Dolosigranulum pigrum* (Aguirre et al. 1993); 11, *Granulicatella* (Collins and Lawson 2000); 12, *Isobaculum* (Collins et al. 2002; 13, *Jeotgalibaca* (Lee et al. 2014); 14, *Lacticigenium* (Iino et al. 2009); 15, *Marinilactibacillus* (Yamasato and Ishikawa 2009); 16, Pisciglobus (Tanasupawat et al. 2011); 17, *Trichococcus* (Rainey 2009)
+ positive, – negative, *nd* no data, *A* acetate, *L* lactate, *FA* facultatively anaerobic, *m-Dpm meso*-diaminopimelic acid

types; cyclopropane ring derivatives are present in some species. The cell wall possesses *meso*-diaminopimelic acid in the peptidoglycan. The G+C content of genomic DNA ranges from 32 to 44 mol %.

Morphological, biochemical, and chemotaxonomic properties useful in the differentiation of members of the genus *Carnobacterium* are given in ➲ *Table 4.3*.

Since the original description of the genus by Collins et al. (1987), there are now 11 validly named species that can be broadly separated into two groups: those isolated from animal and food products (healthy and diseased fish, vacuum-packaged. meat products stored at low temperatures) and those isolated from environmental samples of lacustrian origin (sediments of lake in Antarctica and frozen lake in Pleistocene permafrost, Alaska). The type species is *Carnobacterium divergens.*

*Carnobacterium divergens* (Holzapfel and Gerber 1983) Collins, Farrow, Phillips, Ferusu, and Jones 1987, $315^{VP}$ (*Lactobacillus divergens* Holzapfel and Gerber 1983, p. 530). (Holzapfel and Gerber 1983) (di.ver'gens. L. part. adj. *divergens* deviate, diverge).

Cells are nonmotile arranged as straight, slender, and relatively short rods with rounded ends; generally 0.5–0.7 × 1.0–1.4 μm, occurring singly, in pairs, and short chains. Colonies are cream colored to white, convex, shiny, varying from 0.5 to 1.5 mm on standard-I-agar and MRS agar without acetate. Growth in acetate-containing media is suppressed in the presence of citrate and glucose, but is stimulated when ribose or fructose is added. The surface of the colony is generally affected by aerobiosis. Under certain conditions, the organism is heterofermentative, producing L (+)-lactic acid, $CO_2$, ethanol, and acetate from hexoses. In addition to L (+)-lactic acid, acetate and ethanol are produced from ribose. Growth is observed in 10 % NaCl. Final pH reached in MRS broth (without acetate) is between 5.0 and 5.3 after 4 d. No production of extrapolysaccharide from sucrose or mannitol from fructose. Gas production from gluconate or malate is not observed. Catalase is produced on heme-containing media. Gelatinase, indole, and $H_2O_2$ are not produced. Isolated from vacuum-packaged, refrigerated meat and fish. DNA G+C content (mol% is between 33.7 and 36.4).

For the type species CCUG $30094^T$, using the API rapid ID 32 strep kit, positive reactions for acetoin, *N*-acetyl-β-glucosaminidase, arginine dihydrolase, cyclodextrin, β-glucosidase, glycyl tryptophan arylamidase, maltose, β-mannosidase, melezitose, methyl β-D-glucopyranoside, pyroglutamic acid arylamidase, ribose, saccharose, and trehalose. Negative reactions are observed for alanine phenylalanine proline arylamidase, alkaline phosphatase, L-arabinose, D-arabitol, α-galactosidase, β-glucuronidase, β-galactosidase, glycogen, hippurate, lactose, mannitol, melibiose, pullulan, raffinose, sorbitol, tagatose, and urea.

Type strain: $66^T$, =ATCC $35677^T$, =CCUG $30094^T$, =CIP $101029^T$, =DSM $20623^T$, =NBRC $15683^T$, =JCM $5816^T$, =JCM $9133^T$, =LMG $9199^T$, =NCIMB $11952^T$, and =NRRL B-$14830^T$. GenBank accession numbers *(16S rRNA gene):* AB680940, M58816, and X54270.

*Carnobacterium alterfunditum* Franzmann, Höpfl, Weiss, and Tindall 1993, $188^{VP}$ (Effective publication: Franzmann et al. 1991, p. 261.) (al'ter.fun.di'tum. L. adj. *alter* another; L. adj. *funditus* from the bottom; N.L. neut. adj. *alterfunditum* another [*carnobacterium*] from the [lake] bottom).

Cells are rod-shaped (1.3 × 2.5–12.5 μm) occurring singly, in pairs, or short chains (typically of four cells) and are motile by a single subpolar flagellum. Older cells may be Gram-stain-negative and also lose motility. Anaerobic, with better growth at 20 °C. Carbohydrates are fermented, but gas is not produced. L (+)-lactic acid is the major end product from D (+)-glucose with traces of ethanol, acetic acid, and formic acid. D (−)-ribose is fermented to lactic acid and moderate amounts of ethanol, acetic acid, and formic acid. Glycerol is mainly fermented to acetic acid and formic acid, and traces of ethanol are produced. Growth occurs in media containing 0.1 % yeast extract without

**Table 4.3**

**Morphological, biochemical and chemotaxonomic properties useful in the differentiation of members of the genus *Carnobacterium***

| Characteristic | 1 | 2 | 3 | 4 | 5 | 6 | 7 |
|---|---|---|---|---|---|---|---|
| Growth at 0 °C | + | + | + | + | 4 | + | 4 |
| Growth at 30 °C | – | + | – | nd | nd | + | + |
| Growth at 40 °C | –(+)[b] | + | –(+)[a] | nd | nd | – | nd |
| Motility | + | – | + | – | + | + | – |
| Arginine hydrolysis | + | + | – | + | nd | + | – |
| Hydrolysis of esculin | +/–[c] | nd | – | + | + | + | + |
| Major fatty acid[a] | $C_{16:0}$, $C_{16:1c7}$, $C_{18:19c}$ | $C_{16:0}$, $C_{16:1\ \omega9c}$, $C_{18:1\omega9c}$ | $C_{16:0}$, $C_{16:1c7}$, $C_{18:19c}$ | $C_{14:0}$, $C_{16:0}$, $C_{16:1\omega9c}$, $C_{18:1\omega9c}$ | nd | $C_{16:0}$, $C_{16:1}$, $C_{18:1c9}$, $C_{18:2c9/12}$/$C_{18:0}$ | $C_{16:0}$, $C_{16:1\omega9c}$, $C_{18:1\omega9c}$ |
| Acid from: | | | | | | | |
| Arabinose | – | – | – | – | – | – | – |
| Mannitol | – | – | + | – | – | – | + |
| Ribose | + | + | + | + | – | + | – |
| Trehalose | – | + | + | + | – | + | – |
| DNA G+C content (mol%) | 33–34 | 33–6.4 | 32–34 | 34.3–36.4 | 34 | nd | 43.9 |
| Source | Anoxic water, Ace lake Antarctica, rainbow trout | Vacuum-packaged refrigerated meat, fish | Anaerobic monimolimnion of Antarctic lake | Ice slush surrounding chicken meat, meat products | Cyanobacterial mat of an Antarctic lake | Intestines of healthy fish | Jeotgal |

| Characteristic | 8 | 9 | 10 | 11 |
|---|---|---|---|---|
| Growth at 0 °C | d | + | + | 2+ |
| Growth at 30 °C | + | + | – | + |
| Growth at 40 °C | d | – | – | – |
| Motility | – | + | + | – |
| Arginine hydrolysis | – | + | nd | – |
| Hydrolysis of esculin | + | + | – | |
| Major fatty acid[a] | $C_{14:0}$, $C_{16:0}$, $C_{16:1\omega9c}$, $C_{18:1\omega9c}$ | $C_{16:0}$, $C_{16:1\ \omega9c}$, $C_{18:1\omega9c}$ | $C_{16:0}$, $C_{16:1c7}$ $C_{18:1c9}$ | nd |
| Acid from: | | | | |
| Arabinose | – | – | + | – |
| Mannitol | + | – | + | – |
| Ribose | + | + | + | – |
| Trehalose | + | + | + | + |

**Table 4.3 (continued)**

| Characteristic | 8 | 9 | 10 | 11 |
|---|---|---|---|---|
| DNA G+C content (mol%) | 33.7–36.4 | 35.5–37.2 | 42 | nd |
| Source | Dairy products, meat, fish (healthy and diseased), human plasma and pus | Irradiated chicken meat, shrimp | Alaskan permafrost | Green, discolored vacuum-packaged bologna sausage |

1, *C. alterfunditum* (Franzmann et al. 1991); 2, *C. divergens* (Collins et al. 1987); 3, *C. funditum* (Franzmann et al. 1991); 4, *C. gallinarum* (Collins et al. 1987); 5, *C. iners* (Snauwaert et al. 2013), 6, *C. inhibens*; (Jöborn et al. 1999), 7, *C. jeotgali* (Kim et al. 2009); 8, *C. maltaromaticum* (Mora et al. 2003); 9, *C. mobile* (Collins et al. 1987); 10, *C. pleistocenium*, (Pikuta et al. 2005); 11, *C. viridans* (Holley et al. 2002)
Additional data taken from (Hammes and Hertel 2009)
+ positive, − negative, *d* different results between strains, *nd* no data
[a]taken from original publication or where absent CCUG (www.ccug.se)
[b]according to (Franzmann et al. 1991)
[c]negative in esculin-PY141 but positive in API 20E test

added sodium salts; the optimum concentration of NaCl is 0.1 M. Chopped meat medium and litmus milk remain unchanged. No growth is seen on MRS or SL broths. The spent fermentation broth of PY-amygdalin-2 % NaCl smells similar to the seeds of dried prunes. The optimum initial pH for growth is 7.0–7.4. Pyruvate, lactate, formate, acetate, methanol, betaine, trimethyl ammonium, chloride, glycine, and an atmosphere of $H_2$:$CO_2$ (2 bar, 80:20) do not stimulate growth. Gelatinase-negative. The cell wall contains *meso*-diaminopimelic acid. The predominate fatty acid is $C_{18:1\omega9c}$. Respiratory lipoquinones are not produced. Isolated from the anaerobic monimolimnion of an Antarctic lake (with approximately the salinity of seawater) and rainbow trout. DNA G+C content (mol%) is 32–36 $T_m$.

For the type species CCUG 34643$^T$, using the API rapid ID 32 strep kit, positive reactions for β-glucosidase, maltose, methyl β-D-glucopyranoside (w), saccharose (w), and trehalose. Negative reactions are observed for *N*-acetyl-β-glucosaminidase, acetoin, alanine phenylalanine proline arylamidase, alkaline phosphatase, L-arabinose, D-arabitol, arginine dehydrolase, cyclodextrin, α-galactosidase, β-glucuronidase, β-galactosidase, glycogen, glycyl tryptophan arylamidase, hippurate, lactose, mannitol, β-mannosidase, melezitose, melibiose, pyroglutamic acid arylamidase, pullulan, raffinose, ribose, sorbitol, tagatose, and urea.

According to the API ZYM system, a weak positive reaction is for esterase (C $_4$). Negative for acid phosphatase, *N*-acetyl-β-glucosaminidase, alkaline phosphatase, and α-chymotrypsin, cystine arylamidase, esterase lipase (C $_8$), α-fucosidase, α-glucosidase, β-glucosidase, α-galactosidase, β-galactosidase, β-glucuronidase, leucine arylamidase, lipase (C $_{14}$), α-mannosidase, naphthol-AS-BI-phosphohydrolase, trypsin, and valine arylamidase.

Type strain: pf4$^T$ (=ACAM 313$^T$, =ATCC 49837$^T$, =CCUG 34643$^T$, =CIP 105796$^T$, =DSM 5972$^T$, =NBRC 15548$^T$, =JCM 12498$^T$, =LMG 13520$^T$). GenBank accession number (16S rRNA gene): AB680898.

*Carnobacterium funditum* Franzmann, Höpfl, Weiss, and Tindall 1993, 188$^{VP}$ (Franzmann et al. 1993) (Effective publication: Franzmann et al. 1991, p. 260.) (fun.di'tum. L. neut. adj. *funditum* from the bottom).

Cells are rods (0.8–1.3 × 1.7–0.8 μm) occurring singly, in pairs, or short chains (typically of four cells), motile by a single subpolar flagellum. Old cells usually are Gram-stain negative and are nonmotile. Anaerobic, with better growth at 20 °C. L (+)-lactic acid is the major end product from D (+)-glucose with traces of ethanol, acetic acid, and formic acid. D (−)-ribose is fermented to lactic acid and moderate amounts of ethanol, acetic acid, and formic acid. Glycerol is mainly fermented to acetic acid and formic acid, and traces of ethanol are produced. Gas is not produced. No growth in MRS or SL broths. At least 0.1 % yeast extract is required for good growth. Chopped meat medium and litmus milk remain unchanged. Gelatinase-negative. The optimum initial pH for growth is between 7.0 and 7.4. Sodium is required for growth with optimal growth at 1.7 %. Pyruvate, lactate, formate, acetate, methanol, betaine, trimethyl ammonium, chloride, glycine, and an atmosphere of H2:CO2 (2 bar, 80:20) do not stimulate growth. $C_{18:1\ \omega9c}$ is the predominant fatty acid. Respiratory lipoquinones are not produced. Isolated from the anaerobic monimolimnion of an Antarctic lake of about seawater salinity. DNA G+C content (mol%) ranges between 32 and 35 $T_m$.

For the type species CCUG 34644$^T$, using the API rapid ID 32 strep kit, positive reactions for glycyl tryptophan arylamidase, maltose (w), saccharose (w), and trehalose. Negative reactions are observed for *N*-acetyl-β-glucosaminidase, acetoin, alanine phenylalanine proline arylamidase, alkaline phosphatase, L-arabinose, D-arabitol, arginine dehydrolase, cyclodextrin, α-galactosidase, β-galactosidase, β-glucosidase, β-glucuronidase, glycogen, hippurate, lactose, mannitol, β-mannosidase, melezitose, melibiose, methyl β-D-glucopyranoside, pyroglutamic acid arylamidase, pullulan, raffinose, ribose, sorbitol, tagatose, and urea.

According to the API ZYM system, only weak positive reactions are obtained for esterase ($C_4$), β-glucosidase, leucine arylamidase, and naphthol-AS-BI-phosphohydrolase.

Negative reactions for acid phosphatase, *N*-acetyl-β-glucosaminidase, alkaline phosphatase, α-chymotrypsin, cystine arylamidase, esterase lipase ($C_8$), α-fucosidase, α-galactosidase, β-galactosidase, β-α-glucosidase, glucuronidase, lipase ($C_{14}$), α-mannosidase, trypsin, and valine arylamidase.

Type strain: pf3$^T$, =ACAM 312$^T$, =ATCC 49836$^T$, =CCUG 34644$^T$, =CIP 106503$^T$, =DSM 5970$^T$, =NBRC15549$^T$, =JCM 12499$^T$, and =LMG 14461$^T$. GenBank accession number (16S rRNA gene): AB680899, S86170.

*Carnobacterium gallinarum* Collins, Farrow, Phillips, Ferusu, and Jones 1987, 315$^{VP}$. (gal.li.na'rum. (L. fem. n. *gallina* a hen; L. fem.gen.p.n. *gallinarum* of hens).

Cells stain Gram-positive, are nonmotile and nonspore-forming, straight, slender rods which occur singly or in short chains. Colonies are white, convex, shiny, and circular. Facultatively anaerobic. L (+)-lactic acid is produced glucose with no gas production. Lysine decarboxylase, ornithine decarboxylase, tryptophan deaminase, and urease are not produced. Indole and $H_2S$ are also not produced. The cellular fatty acids are of the straight-chain saturated and monounsaturated types with tetradecanoic, hexadecanoic, and 9,10-octadecenoic acids predominating. Isolated from ice slush from around chicken carcasses and various meat products. DNA G+C content (mol%) is 34.3–35.4 $T_M$.

For the type species CCUG 30095$^T$, using the API rapid ID 32 strep kit, positive reactions for acetoin (w), *N*-acetyl-β-glucosaminidase, arginine dehydrolase, β-glucosidase, glycyl tryptophan arylamidase, lactose, maltose, β-mannosidase, pyroglutamic acid arylamidase, ribose, saccharose, tagatose, trehalose, and urea (w). Negative reactions are observed for alanine phenylalanine proline arylamidase, alkaline phosphatase, L-arabinose, D-arabitol, cyclodextrin, α-galactosidase, β-galactosidase, β-glucuronidase, glycogen, hippurate, mannitol, melezitose, melibiose, methyl β-D-glucopyranoside, pullulan, raffinose, and sorbitol.

The type strain is MT44$^T$ (=ATCC 49517$^T$, =CCUG 30095$^T$, =CIP 103160$^T$, =DSM 4847$^T$, =JCM 12517$^T$, =LMG 9841$^T$, =NCIMB 12848$^T$, =NRRL B-14832$^T$). GenBank accession number (16S rRNA gene): AJ387905, X54269.

*Carnobacterium iners.* Snauwaert, Hoste, De Bruyne, Peeters, De Vuyst, Willems, and Vandamme 2013, 1374$^{VP}$ (in'ers. L. neut. adj. *iners* inactive, lazy).

Cells are psychrophilic, Gram-stain-positive, catalase-negative, facultatively anaerobic, and motile rods, approximately 1.5 mm wide and 3–6 mm long, occurring singly or in pairs or short chains. Colonies grown for 10 days on TSBY salt agar at 4 °C are approximately 0.8 mm in diameter, white, opaque, smooth, and circular with undulate margins and an umbonate elevation. No gas is produced from glucose. Produces D-and L-isomers of lactic acid in a ratio of 1:9. Optimal growth is observed at 4 °C, but grows at 15 °C and 20 °C. Grows with 1–2 % NaCl, but not with 4–10 % NaCl. Using the API 50 CHL *Lactobacillus* identification system (bioMerieux) and GEN III Omnilog ID system (Biolog), the organism does not produce acid from glucose, fructose, mannose, *N*-acetylglucosamine, esculin, cellobiose, maltose, trehalose, gentiobiose, glycerol, erythritol, D- or L-arabinose, ribose, D- or L-xylose, adonitol, methyl β-D-xylopyranoside, galactose, sorbose, rhamnose, dulcitol, inositol, mannitol, sorbitol, methyl a-D-mannopyranoside, methyl a-D-glucopyranoside, amygdalin, arbutin, salicin, lactose, melibiose, sucrose, inulin, melezitose, raffinose, starch, glycogen, xylitol, turanose, D-lyxose, D-tagatose, D- or L-fucose, D- or L-arabitol, potassium gluconate, potassium 2-ketogluonate, or potassium 5-ketogluconate. The cell wall contains meso-diaminopimelic acid. The DNA G+C content of the type strain is 34 mol%. Isolated from a cyanobacterial mat growing in the littoral zone of a continental Antarctic lake (Forlidas Pond, Pensacola Mountains) in December 2003. Type strain, LMG 26642 $^T$ (=CCUG 62000$^T$). GenBank accession number (16S rRNA gene): HE58395.

*Carnobacterium inhibens* Jöborn, Dorsch, Olsson, Westerdahl, and Kjelleberg 1999, 1897$^{VP}$ (in.hi'bens. L. part. adj. *inhibens* inhibiting, referring to the growth-inhibitory activity that the bacterium shows). Cells are motile (monotrichous), nonspore-forming rods occurring singly, in pairs, or as chains of four cells. No growth on MRS medium. Colonies on TSA at 20 °C are circular, entire, convex, and semitranslucent and 1–2 mm in diameter. The color of the colonies is whitish at aerobic growth conditions and buff at anaerobic growth conditions. pH range supporting growth is between 5.5 and 9.0. Catalase is produced on heme-containing media. Hippurate is hydrolyzed, but $H_2S$ is not produced and nitrate not reduced. The most abundant cellular fatty acids are $C_{16:0}$ (31.1 %), $C_{16:1}$ (24.2 %), and $C_{18:1\ \omega 9c}$ (23.4 %). Other fatty acids are $C_{18:2\ \omega 6,9c}$ or $C_{18:0}$ (10.8 %), $C_{14:0}$ (5.4 %), and $C_{18:0}$ (3.5 %). The peptidoglycan type is not known, and DNA G+C content (mol%) has not been determined. Habitat is the intestines of healthy fish.

For the type species CCUG 31728$^T$, using the API rapid ID 32 strep kit, positive reactions for hippurate, β-glucosidase, maltose, mannitol, methyl β-D-glucopyranoside, ribose, saccharose, and trehalose. Negative reactions are observed for acetoin, *N*-acetyl-β-glucosaminidase, alanine phenylalanine proline arylamidase, alkaline phosphatase, L-arabinose, D-arabitol, arginine dehydrolase, cyclodextrin, α-galactosidase, β-galactosidase, β-glucuronidase, glycyl tryptophan arylamidase, glycogen, lactose, β-mannosidase, melezitose, melibiose, pullulan, pyroglutamic acid arylamidase, raffinose, sorbitol, tagatose, and urea.

Type strain: K1$^T$, =CCUG 31728$^T$, =CIP 106863$^T$, and =DSM 13024$^T$. GenBank accession number (16S rRNA gene): Z773313.

*Carnobacterium jeotgali* Kim, Seon, Roh, Nam, Yoon, and Bae, 2009, 3171$^{VP}$. (je.ot.ga'li. N.L. gen. n. *jeotgali* of jeotgal, a traditional Korean fermented seafood).

Cells are Gram-positive staining, 3.5 ± 0.7 mm long and 0.7–0.8 mm wide, and occur more frequently in chains than as single cells. The organism is nonmotile with no flagella. Facultatively anaerobic bacteria with colonies about 1 mm in diameter, irregular in shape, and pale yellow with a rough surface. Grows at 4–37 °C (optimum 30 °C), at pH 5.5–9.0 (optimum pH 8.5), and in the presence of 0–5 % (w/v) NaCl (optimum 2 % NaCl). Oxidase and catalase are not produced. Assimilates erythritol, D-fructose, inositol, D-mannitol and esculin, but not glycerol, D-arabinose, L-arabinose, D-ribose, D-xylose, L-xylose, D-adonitol, methyl β-D-xyloside, D-galactose, D-glucose, D-mannose, L-sorbose, L-rhamnose, dulcitol, D-sorbitol, methyl α-D-mannoside, *N*-acetylglucosamine, amygdalin, arbutin, cellobiose, maltose, D-lactose, melibiose, sucrose, trehalose, inulin, melezitose, raffinose, starch, glycogen, xylitol, gentiobiose, turanose, D-lyxose, D-tagatose, D-fucose, L-fucose, D-arabitol, L-arabitol, gluconate, or 2,5-ketogluconate. In the API 20NE and API ZYM test kits, enzyme activities are negative for nitrate reduction, indole production, D-glucose fermentation, L-arginine dihydrolase, and for hydrolysis of β-galactosidase, gelatin, and urea, but positive for hydrolysis of esculin. Positive reactions for acid phosphatase, naphthol-AS-BI-phosphohydrolase, and β-glucuronidase, but negative for alkaline phosphatase, esterase ($C_4$), esterase lipase ($C_8$), leucine arylamidase, cystine arylamidase, trypsin, α-chymotrypsin, β-glucosidase, lipase ($C_{14}$), valine arylamidase, α-galactosidase, α-glucosidase, β-glucosidase, *N*-acetyl-β-glucosaminidase, α-mannosidase, and α-fucosidase. The predominant fatty acids are $C_{16:\,0}$, $C_{16:\,1v9c}$, and $C_{18:\,1v9c}$. The G+C content of the genomic DNA of the type strain is 43.9 mol%. Isolated from "toha jeotgal," a traditional Korean fermented food. The type strain, MS3$^T$ (=KCTC 13251$^T$, =JCM 15539$^T$). GenBank accession number (16S rRNA gene): EU817500.

*Carnobacterium maltaromaticum* (Miller et al. 1974) Mora, Scarpellini, Franzetti, Colombo, and Galli 2003, 677$^{VP}$ (*Lactobacillus maltaromicus* Miller et al. 1974, p. 352; *Lactobacillus piscicola* Hiu et al. 1984, p. 399; *Lactobacillus carnis* Shaw and Harding 1985, p. 296; *Carnobacterium piscicola* Collins et al. 1987, p. 315.) (malt.a.ro.mat.ic'um. N.L. neut. n. *maltum -i* malt; L. adj. *aromaticus -a -um* aromatic, fragrant; N.L. neut. adj. *maltaromaticum* possessing a malt-like aroma).

Cells are rod-shaped (0.5–0.7 × 3.0 μm) occurring singly or in chains and are nonmotile. Facultatively anaerobic. L (+)-lactic acid, ethanol, and acetate are produced heterofermentatively, but gas production is weak and frequently undetectable. Grows in MRS, TSBY, and brain-heart infusion media. Major cellular fatty acids are straight-chain saturated and monounsaturated acids, with tetradecanoic, hexadecanoic, and 9- and 10-octadecenoic acids predominating. Isolated from dairy products, meat, fish (healthy and diseased), and human plasma and pus. DNA G+C content (mol%) is 33.7–36.4 $T_m$.

For the type species CCUG 30142$^T$, using the API rapid ID 32 strep kit, positive reactions for acetoin, *N*-acetyl-β-glucosaminidase, alkaline phosphatase (w), cyclodextrin, β-galactosidase (w), β-glucosidase, glycyl tryptophan arylamidase (w), hippurate (w), lactose, maltose, β-mannosidase, methyl β-D-glucopyranoside, pyroglutamic acid arylamidase, ribose, saccharose, and trehalose. Negative reactions are observed for alanine phenylalanine proline arylamidase, L-arabinose, D-arabitol, arginine dehydrolase, α-galactosidase, β-glucuronidase, glycogen, mannitol, melezitose, melibiose, pullulan, raffinose, sorbitol, tagatose, and urea.

Type strain is ATCC 27865$^T$ (=CCUG 30142$^T$, =CIP 103135$^T$, =DSM 20342$^T$, =JCM 1154$^T$, =LMG 6903$^T$, =NRRL B-14852$^T$). GenBank accession number (16S rRNA gene): M58825, X54420.

*Carnobacterium mobile* Collins, Farrow, Phillips, Ferusu, and Jones 1987, 315$^{VP}$ (mo'bi.le. L. neut. adj. mobile movable or motile).

Gram-stain-positive, motile, nonspore-forming, straight, slender rods which occur singly or in short chains. Colonies are white, convex, shiny, and circular. Facultatively anaerobic and L (+)-lactic acid is produced heterofermentatively. Gas production from glucose is observed for most strains in arginine-MRS broth. All strains are lysine decarboxylase, ornithine decarboxylase, tryptophan deaminase, and urease-negative. The cellular fatty acids are of the straight-chain saturated and monounsaturated types, with hexadecanoic, hexadecenoic, and 9,10-octadecenoic acids predominating. Isolated from irradiated chicken meat and from shrimp. DNA G+C content (mol%) is 35.5–37.2 $T_M$. Type strain is MT37L$^T$ (=ATCC 49516$^T$, =CCUG 30096$^T$, =CIP 103159$^T$, =DSM 4848$^T$, =JCM 12516$^T$, =LMG 9842$^T$, =NCIMB 12847$^T$, NRRL B-14831$^T$). GenBank accession number (16S rRNA gene): AB083414, X54271.

*Carnobacterium piscicola* (Collins, Farrow, Phillips, Ferusu, and Jones 1987 315$^{VP}$) was reclassified by Mora et al. (2003) as *C. maltaromicus* now considered a junior heterotypic synonym of *Lactobacillus maltaromicus* and a synonym (s) of *Lactobacillus carnis*, *Lactobacillus maltaromicus*, and *Lactobacillus piscicola* (Hiu et al. 1984).

*Carnobacterium pleistocenium* Pikuta, Marsic, Bej, Tang, Krader, and Hoover 2005, 477$^{VP}$. (plei.sto.ce'ni.um. N.L. neut. adj. *pleistocenium* belonging to the Pleistocene, a geological epoch).

Cells stain Gram-positive, are motile, small rods with rounded ends, 0.7–0.8 Å ∼ 1.0–1.5 μm. Facultatively anaerobic. Growth occurs at 22 °C in the pH range of 6.5–9.5. The range of NaCl for growth is 0–5 % (w/v) with the optimum being 0.5 % (w/v) NaCl. Growth occurs using D-glucose, D-fructose, D-mannose, D-maltose, sucrose, lactose, starch, D-mannitol, peptone, Bacto tryptone, Casamino acids, and yeast extract.

End products of growth are acetate, ethanol, and traces of carbon dioxide. Isolated from a sample of permafrost from Fox Tunnel, Alaska. DNA G+C content (mol%) is 42 ± 1.5 $T_m$. Type strain is FTR1$^T$ (=ATCC BAA-754$^T$, =CIP 108033$^T$, =JCM 12174$^T$). GenBank accession number (16S rRNA gene): AF450136.

*Carnobacterium viridans* Holley, Guan, Peirson, and Yost 2002, 1884$^{VP}$. (vi'ri.dans. N.L. adj. *viridans* from L. v. *viridare* to make green, referring to the production of a green color in cured meat by the organism).

Cells stain Gram-positive, are nonmotile, slightly curved rods that occur singly or in pairs, or as straight rods (0.8 × 3.6 ± 0.6 μm). Facultatively anaerobic. Grows satisfactorily in BHI, APT, M5, and CTSI media, but poorly on a variety of media including MRS and SL agar. On blood agar base with 0.8 % sheep blood, it is β-hemolytic. Grows over a range of pH from 5.5 to 9.1. Produces predominantly L (+)-lactic acid from glucose and neither gas nor $H_2S$. Does not grow in 4 % (w/v) NaCl but will tolerate 26.4 % (w/v) NaCl (saturated brine) for long periods at 4 °C. Growth is observed between 2 °C and 30 °C. Ammonia is not produced from arginine. No gas is produced from glucose, nitrate is not reduced, and $H_2S$ is not produced. The Voges-Proskauer reaction is negative. Acid is not produced from amygdalin, inulin, mannitol, methyl α-D-glucoside, ribose, or D-xylose. Acid is produced when grown on galactose, glucose, fructose, mannose, *N*-acetylglucosamine, esculin, cellobiose, maltose, lactose, sucrose, trehalose, and tagatose (API50CHL). The organism also metabolizes *N*-acetyl-D mannosamine, arbutin, dextrin, gentiobiose, glucose 6-phosphate, maltotriose, 3-methyl D-glucose, salicin, α-hydroxybutyric acid, α-ketovaleric acid, pyruvic acid, and uridine (Biolog AN). Negative for all other substrates used in the API 50CHL and Biolog AN panels. The cell wall peptidoglycan contains *meso*-DAP. Isolated from green, discolored vacuum-packaged bologna-type sausage. DNA G+C content (mol%) has not been determined. Type strain: MPL-11$^T$ (=ATCC BAA-336$^T$, =DSM 14451$^T$, =JCM 12222$^T$). GenBank accession number (16S rRNA gene): AF425608.

## *Alkalibacterium* Ntougias and Russell 2001, 1169$^{VP}$

Al.ka.li.bac'te.ri.um. N.L. n. *alkali* (from Ar. Article *al* the; Ar. n. *qaliy* ashes of saltwort) alkali; L. neut. n. *bacterium* a small rod; N.L. neut. n. *Alkalibacterium* living under alkaline conditions.

Gram-positive staining rods that are nonspore-forming. Cells range in size from 0.4–1.2 × 0.7–3.7 μm and occur singly, in pairs, or clusters. Most species are motile by polar or peritrichous flagella but some are nonmotile. Facultatively anaerobic, with all species being catalase and oxidase negative. Lactate is the major product from glucose fermentation, though is variable by pH, and aerobically glucose is metabolized to lactate and acetate (Ishikawa et al. 2009). They are alkaliphilic, growing above pH 8.0, and as high as pH 11. Cells are halotolerant with growth up to 17 % NaCl and mesophilic to psychrotolerant with growth from 4–15 °C to 35–45 °C. Quinones have not been detected. The predominant cellular fatty acids are $C_{16:0}$, $C_{16:1\omega9c}$, and $C_{18:1\omega9c}$. The peptidoglycan type is either 4Aα or 4Aβ with predominantly Orn-D-Asp or Lys-D-Asp configuration. The DNA G+C content (mol %) varies from 36.8 to 47.1. *Alkalibacterium* species can be differentiated by certain phenotypic traits, motility, cell wall type, or % mol G+C content as listed in ➋ *Table 4.4*. The type species is *Alkalibacterium olivapovliticus*.

*Alkalibacterium olivapovliticus* corrig. Ntougias and Russell 2001, 1169$^{VP}$. (o.li.va.pov'lit.i.cus. L. n. *oliva* olives; Gr. n. *apovlito* waste disposal; N.L. gen. n. *olivapovliticus* from the waste of the olives).

Cells are Gram-positive, obligately alkaliphilic, nonspore-forming rods, as are all other species within the genus *Alkalibacterium*. For the type species, growth is optimum at pH 9.0–9.4, 3–5 % NaCl (w/v) and 27–32 °C, though other strains are slightly higher (e.g., pH 9.8–10.2 and 0–10 % NaCl). Substrates utilized for growth are cellobiose, glucose, glucose 6-phosphate, starch, and sucrose. All strains were moderately or weakly positive for glutamate, malate, maltose, mannose, and trehalose. No growth was observed for acetate, arabinose, cellulose, ethanol, fructose, galactose, glycerol, lactose, mannitol, melezitose, melibiose, inositol, raffinose, rhamnose, ribose, sorbitol, succinate, or xylose. Yeast extract but not amino acids could be used as sole carbon and energy source. The major phospholipids within all three strains are phosphatidylglycerol, diphosphatidylglycerol, and phosphatidylserine, plus an unknown phospholipid. Growth is inhibited by ampicillin, carbenicillin, chloramphenicol, penicillin G, kanamycin, streptomycin, and trimethoprim, with some strains are sensitive to amoxicillin, miconazole, and neomycin. The DNA G+C content (mol %) is 39.7. The type strain is WW2-SN4a$^T$ (=DSM 13175$^T$, =CIP 107107 = NCIMB 13710 $^T$) (Ntougias and Russell 2001).

*Alkalibacterium psychrotolerans* Yumoto, Hirota, Nodasaka, Yokota, Hoshino, and Nakajima 2004, 2382$^{VP}$ (psy.chro.to'le. rans. Gr. adj. *psychros* cold; L. part. adj. *tolerans* tolerating; N.L. neut. part. adj. *psychrotolerans* tolerating cold environments).

Facultative anaerobic rod-shaped cells (0.4–0.9 × 0.7–3.1 μm) and peritrichously flagellated. Optimum growth occurs at pH 9.5–10.5, 2–12 % NaCl (w/v) and at 34 °C, but can tolerate pH 9–12, 0–17 % NaCl, and temperatures between 5 °C and 45 °C. Cells utilize arabinose, glucose, maltose, and xylose, but not adonitol, dulcitol, erythritol, galactose, inositol, inulin mannitol, melibiose, sorbitol, sucrose, raffinose, rhamnose, or xylitol. Hydrolyzes starch but not gelatin. Major end product form glucose is lactate. The predominant fatty acid are $C_{14:0}$, $C_{16:0}$, $C_{16:1\omega9c}$, and $C_{18:1\omega9c}$. The cell wall peptidoglycan consists of a type A4α and L-Lys (L-Orn)-D-Asp, and the purified peptidoglycan contains D-Asp, L-Orn + L-Lys, L-Glu, and L-Ala at a molar ratio of 0.4:1.1:1:1.6. The DNA G+C content (mol %) is 40.6. The type strain is IDR2-2$^T$ (=JCM 12281$^T$, =NCIMB 13981$^T$). Isolated from polygonum indigo (*Polygonum tinctorium* Lour.) fermentation liquor (Yumoto et al. 2004).

*Alkalibacterium iburiense* Nakajima, Hirota, Nodasaka, and Yumoto 2005, 1529$^{VP}$. (i.bu.ri.en'e. N.L. neut. adj. *iburiense* from Iburi, the place where the microorganism was isolated).

The physiology of this organism separates it from other species of *Alkalibacterium* (➋ *Table 4.4*). Cells are rods (0.5–0.7 × 1.3–2.7 μm) that form colonies equal in size (2–2.5 mm) both aerobically or anaerobically. The type strain can be differentiated because of a wider range of substrates utilized. Growth can occur on arabinose, fructose, glucose, *N*-acetylglucosamine, glycogen, maltose, mannose, rhamnose, sucrose, trehalose, and xylose, but not on arbutin, galactose, inositol, mannitol, melibiose, raffinose, or sorbitol. With other strains growth is variable on fructose, maltose, mannose, rhamnose, sucrose, trehalose, and xylose. Starch is hydrolyzed, but gelatin is not. Optimum growth occurs at pH 9.5–10.5, 3–13 %

**Table 4.4**

**Comparison of selected characteristics useful in differentiating type species within the genus *Alkalibacterium***

| | 1 | 2 | 3 | 4 | 5 | 6 | 7 | 8 | 9 | 10 |
|---|---|---|---|---|---|---|---|---|---|---|
| *Type strain* | WW2-SN4a$^T$ | IDR2-2$^T$ | M3$^T$ | A11$^T$ | T117-1-2$^T$ | T143-1-1$^T$ | T129-2-1$^T$ | T22-1-2$^T$ | O24-2$^T$ | 3AD-1$^T$ |
| *Isolation source* | Edible olive oil-water, Greece | Fermented polygonum indigo, Japan | Contaminated alkali broth from polygonum indigo, Japan | Fermented polygonum indigo, Japan | Marine algae, Thailand | Decaying sea grass, Thailand | Decaying marine algae, Thailand | Fermented shrimp paste, Thailand | Decaying marine algae, Japan | Brie cheese, France |
| *Mortality* | + | + | + | + | – | + | + | – | – | – |
| *Substrate fermentation* | | | | | | | | | | |
| L-Arabinose | – | + | + | – | – | + | + | – | – | W |
| Lactose | – | – | +* | ND | W | + | W | – | + | – |
| Melibiose | – | – | – | – | – | – | – | – | – | – |
| Raffinose | – | – | – | – | W | W | + | – | + | – |
| Rhamnose | – | – | + | – | – | – | – | – | – | – |
| Sucrose | + | – | + | – | – | – | – | – | – | – |
| Trehalose | + | ND | + | – | + | + | + | + | + | + |
| Xylose | – | + | + | – | – | + | + | + | + | + |
| *Antibiotic sensitivity* | | | | | | | | | | |
| Chloramphenicol | S | WS* | R | R | R | R | R | R | WS | V |
| Kanamycin | S | R | R | R | R | R | R | R | S | V |
| Trimethoprim | S | S* | R | R | R | S | S | R | S | V |
| *Peptidoglycan type* | A4β Orn-D-Asp | A4α Lys(Om)-D-Asp | A4α Lys-D-Asp | A4β Lys(Om)-D-Asp | A4β Om-D-Asp | A4β Om-D-Asp | A4β Om-D-Asp | A4β Om-D-Asp | A4β Om-D-Asp | A4β Om-D-Asp |
| *DNA G+C content (mol%)* | 39.7 | 40.6 | 42.6–43.2 | 47.1 | 41.7 | 42.2 | 42.5 | 39.4 | 43.7 | 36.8 |
| *Culture collections* | = DSM 13175$^T$ | =JCM 12281$^T$ | = JCM12662$^T$ | = JCM14232$^T$ | = DSM19181$^T$ | = DSM19183$^T$ | = DSM 19182$^T$ | = DSM19180$^T$ | = DSM23664$^T$ | = DSM25751$^T$ |
| | = NCIMB 13710$^T$ | = NCIMB 13981$^T$ | = NCIMB 14024$^T$ | = NCIMB 15253$^T$ | = NBRC 103241$^T$ | = NBRC 103242$^T$ | = NBRC 103243$^T$ | = NBRC 103247$^T$ | = NBRC 107172$^T$ | = JCM 18271$^T$ |
| | = CIP 107107$^T$ | = CIP 108641$^T$ | = CIP 108873$^T$ | | = NRIC 0718$^T$ | = NRIC 0719$^T$ | = NRIC 0720$^T$ | = NRIC 0724$^T$ | | |
| | | = BCRC 17792$^T$ | = BCRC 17791$^T$ | | | | | | | |
| *GenBank accession number* | AF143511 | AB125938 | AB188091 | AB268549 | AB294165 | AB294166 | AB294167 | AB294171 | AB555562 | AB690566 |

Taxa: 1, *Alkalibacterium olivoapovliticus* (data from Ntougias and Russell 2001); 2, *A. psychrotolerans* (Yumoto et al. 2004); 3, *A. iburiense* (Nakajima et al. 2005); 4, *A. indicireducens* (Yumoto et al. 2008); 5, *A. thalassium* (Ishikawa et al. 2009); 6, *A. pelagium* (Ishikawa et al. 2009); 7, *A. putridalgicola* (Ishikawa et al. 2009); 8, *A. kapii* (Ishikawa et al. 2009); 9, *A. subtropicum* (Ishikawa et al. 2011); 10, *A. gilvum* (Ishikawa et al. 2013)

+ positive, – negative, *w* weakly positive, *S* susceptible, *R* resistant, *WS* weakly susceptible, *V* variable among strains, *ND* no data available

NaCl (w/v), and at 30–37 °C, but ranges from pH 9 to 12, 0 to 16 % NaCl, and 10 to 45 °C. Lactate is the predominant end product of glucose metabolism. Growth is inhibited by amoxicillin, ampicillin, and penicillin G but not by chloramphenicol, kanamycin, ketoconazole, miconazole, sulfamethoxazole, and trimethoprim. Cell wall peptidoglycan is of type A4α, Lys-D-Asp, and the predominant cellular fatty acids are $C_{16:0}$, $C_{16:1\omega7c}$, and $C_{18:1\omega9c}$. The DNA G+C content (mol %) is 42.6–43.2. The type strain is M3$^T$ (=JCM 12662$^T$, =NCIMB 14024$^T$) and was isolated from a contaminated culture in alkali broth, and strains 41A and 41C were isolated from a polygonum indigo (*Polygonum tinctorium* Lour.) fermentation liquor (Nakajima et al. 2005).

*Alkalibacterium indicireducens* Yumoto, Hirota, Nodasaka, Tokiwa, and Nakajima 2008, 904$^{VP}$ (in.di.ci.re.du'cens. L. n. *indicum* indigo; L. part. adj. *reducens* bringing or leading back, reducing; N.L. part. adj. *indicireducens* indigo-reducing).

Cells are straight rod-shaped (0.4–1.2 × 1.7–3.7 μm) that are motile but distinctively motile depending on the strain. Colonies are equal in size (0.5–2 mm) both when grown aerobically or anaerobically but are smaller than *A. iburiense*. Substrates utilized are limited to glucose, fructose (type strain), and sucrose weakly. Fermentation of fructose is variable for other strains. Arabinose, galactose, inositol, maltose, mannose, melibiose, raffinose, rhamnose, sorbitol, trehalose, and xylose are not utilized. Cellulose, starch, and xylan are hydrolyzed, but casein and gelatin are not. Optimum growth occurs at pH 9.5–11.5, 1–11 % NaCl (w/v), and at 20–30 °C, but ranges from pH 9 to 12.3, 0 to 14 % NaCl, and 15 to 40 °C. Lactate is the predominant end product of glucose metabolism. Like *A. iburiense*, growth is not inhibited by chloramphenicol, kanamycin, ketoconazole, miconazole, sulfamethoxazole, or trimethoprim, but is inhibited by amoxicillin, ampicillin, and penicillin G. The predominant cellular fatty acids are $C_{16:0}$, $C_{16:1\omega7c}$, and $C_{18:1\omega9c}$. The cell wall peptidoglycan is type A4α, L-Lys (L-Orn)-D-Asp. The DNA G+C content (mol %) is 47.0–47.8. The type strain is A11$^T$ (=JCM 14232$^T$, =NCIMB 14253 $^T$) (Yumoto et al. 2008). Isolated from polygonum indigo (*Polygonum tinctorium* Lour.) fermentation liquor.

*Alkalibacterium thalassium* Ishikawa, Tanasupawat, Nakajima, Kanamori, Ishizaki, Kodama, Okamoto-Kainuma, Koizumi, Yamamoto, and Yamasato 2009, 1222$^{VP}$ (tha.las'si.um. N.L. neut. adj. *thalassium* (from Gr. adj. *thalassios -a* –on) of/from the sea).

Cells are nonmotile, long straight rods (0.6–1.4 × 3.6–9.0 μm) that appear singly, in pairs, or short chains. Colony formation varies between surface and sunken; using 2.5 % NaCl GYPF agar medium after 3 days at 30 °C, deep colonies (1.0–2.0 mm) are pale yellow, opaque, and lenticular, while surface colonies (0.5–1.0 mm) are round, convex, entire, and pale yellow to transparent. Optimum growth occurs at pH 9.0, 2.5–11 % NaCl (w/v), and at 37 °C, but ranges from pH 7.0 to 11.0, 0 to 11 % NaCl, and 10 to 42.5 °C. Can utilize cellobiose, galactose, glycerol (weakly), lactose (weakly), raffinose (weakly), salicin, and trehalose for growth. Cannot utilize D- or L-arabinose, inulin, mannitol, melibiose, melezitose, α-methyl-glucoside, rhamnose, or xylose. Does not reduce nitrate, liquefy gelatin, or produce ammonia from arginine. The major fermentation product from glucose is lactate, with lesser amounts of formate, acetate, and ethanol made a molar ratio of approximately 2:1:1. At higher pH values, lactate yield decreases. Aerobically, glucose is metabolized to acetate and lactate. Inhibited by ampicillin but not chloramphenicol, kanamycin, or trimethoprim. The peptidoglycan is type A4β, Orn-D-Asp, and the major cellular fatty acids are $C_{16:0}$, $C_{16:1\omega9c}$, and $C_{18:1\omega9c}$. The DNA G+C content (mol %) is 41.7. The type strain is T117-1-2$^T$ (=DSM 19181$^T$, =NBRC 103241$^T$, =NRIC 0718$^T$), which was isolated from a decaying marine alga.

*Alkalibacterium pelagium* Ishikawa, Tanasupawat, Nakajima, Kanamori, Ishizaki, Kodama, Okamoto-Kainuma, Koizumi, Yamamoto, and Yamasato 2009, 1223$^{VP}$. (pe.la'gi.um. N.L. neut. adj. *pelagium* of the sea, marine).

Cells are rods (0.3–0.9 × 1.8–7.2 μm) that appear singly, in pairs, or short chains and are motile by peritrichous flagella. Similarly, colony formation varies between surface and sunken; using 2.5 % NaCl GYPF agar medium after 3 days at 30 °C, deep colonies (1.0–2.0 mm) are pale yellow, opaque, and lenticular, while surface colonies (0.5–1.0 mm) are round, convex, entire, and pale yellow to transparent. Halotolerant. Optimum growth occurs at pH 9.0–9.5, 0.5–1.5 % NaCl (w/v) and at 37 °C, but ranges from pH 7.0 to 11.0, 0 to 17 % NaCl, and 10 to 47.5 °C. *A. pelagium* can be differentiated by substrate utilization. Cells can utilize L-arabinose, cellobiose, galactose, lactose, raffinose (weakly), salicin, sucrose, trehalose, and xylose for growth, but cannot utilize D-arabinose, glycerol, inulin, mannitol, melibiose, melezitose, α-methyl-glucoside, rhamnose, and sorbitol. Does not reduce nitrate, liquefy gelatin, or produce ammonia from arginine. The major fermentation product from glucose is lactate, with lesser amounts of formate, acetate, and ethanol made a molar ratio of ~2:1:1. At higher pH values, lactate yield decreases. Glucose is metabolized to acetate and lactate under aerobic conditions. Inhibited by ampicillin and trimethoprim but not by chloramphenicol or kanamycin. The peptidoglycan is type A4β, Orn-D-Asp, and the predominant cellular fatty acids are $C_{16:0}$, $C_{16:1\omega9c}$, and $C_{18:1\omega9c}$. The DNA G+C content (mol %) is 42.2. The type strain is T143-1-1$^T$ (=DSM 19183$^T$, =NBRC 103242$^T$, =NRIC 0719$^T$) and was isolated from decaying sea grass.

*Alkalibacterium putridalgicola* Ishikawa, Tanasupawat, Nakajima, Kanamori, Ishizaki, Kodama, Okamoto-Kainuma, Koizumi, Yamamoto, and Yamasato 2009, 1223$^{VP}$. (pu.tri.dal.gi'co.la. L. adj. *putridus* rotten, decayed; L. fem. n. *alga* seaweed; L. suff. *-cola* (from L. n. *incola*) dweller; N.L. n. *putridalgicola* dweller on putrid marine algae).

Again, like *A. pelagium*, *A. putridalgicola* cells are motile, straight but larger rods (0.6–0.9 × 2.7–8.0 μm) that appear singly, in pairs, or short chains. Similar colony formation varies between surface and sunken; using 2.5 % NaCl GYPF agar medium after 3 days at 30 °C, deep colonies (1.0–2.0 mm) are pale yellow, opaque, and lenticular, while surface colonies (0.5–1.0 mm) are round, convex, entire, and pale yellow to transparent. Slightly halophilic and halotolerant. Optimum growth occurs at pH 8.0–9.0, 2.0–4.0 % NaCl (w/v), and at 37–40 °C, but ranges from pH 6.5 to 10.0, 0 to 20 % NaCl, and −1.8 to 45 °C. Cells can utilize L-arabinose, cellobiose, galactose, inulin, lactose (weakly), raffinose, salicin, sucrose, trehalose, and xylose for growth, but cannot utilize D-arabinose, glycerol, mannitol, melibiose, melezitose, α-methyl-glucoside, rhamnose, and sorbitol. Produces ammonia from arginine, but does not reduce nitrate or liquefy gelatin. The major fermentation product from glucose is lactate, with lesser amounts of formate, acetate, and ethanol made a molar ratio of approximately 2:1:1, but decreasing lactate as pH increases. Aerobically, glucose is metabolized to acetate and lactate. Inhibited by ampicillin and trimethoprim but not by chloramphenicol or kanamycin. The peptidoglycan is type A4β, Orn-D-Asp, and the predominant cellular fatty acids are $C_{16:0}$, $C_{16:1\omega9c}$, and $C_{18:1\omega9c}$. The DNA G+C content (mol %) is 41.0–43.0 (type strain 42.5). The type strain is T129-2-1$^T$ (=DSM 19182$^T$, =NBRC 103243$^T$, =NRIC 0720$^T$) and was isolated from decaying marine alga.

*Alkalibacterium kapii* Ishikawa, Tanasupawat, Nakajima, Kanamori, Ishizaki, Kodama, Okamoto-Kainuma, Koizumi, Yamamoto, and Yamasato 2009, 1223$^{VP}$. (ka'pi.i. N.L. n. *kapium* ka-pi (a fermented shrimp paste in Thailand); N.L. gen. n. *kapii* of ka-pi).

Cells are rods (0.6–1.1 × 1.8–3.6 μm) that appear singly, in pairs, or short chains. Most strains are motile, but the type species is nonmotile. Colony formation varies between surface and sunken; using 2.5 % NaCl GYPF agar medium after 3 days at 30 °C, deep colonies (1.0–2.0 mm) are creamy yellow, opaque, and lenticular, while surface colonies (0.5–1.0 mm) are round, convex, entire, and creamy yellow to transparent. Optimum growth occurs at pH 8.5–9.0, 1.5–2.5 % NaCl (w/v), and at 25–37 °C, but ranges from pH 6.0–6.5 to 10.0, 0 to 21 % NaCl, and 5 to 42.5 °C. *A. kapii* can be differentiated by substrate utilization, able to only grow on sorbitol, sucrose, and trehalose. It cannot utilize D- or L-arabinose, cellobiose, galactose, glycerol, inulin, lactose, mannitol, melibiose, melezitose, α-methyl-glucoside, raffinose, rhamnose, salicin, sucrose, trehalose, and xylose. Does not reduce nitrate, liquefy gelatin, or produce ammonia from arginine. The major fermentation product from glucose is lactate, with lesser amounts of formate, acetate, and ethanol made a molar ratio of approximately 2:1:1. At higher pH values, lactate yield decreases. Aerobically, glucose is metabolized to acetate and lactate. Inhibited by ampicillin but not by chloramphenicol, kanamycin, or trimethoprim. The peptidoglycan is type A4β, Orn-D-Glu, and the major cellular fatty acids are $C_{16:0}$, $C_{16:1\omega9c}$, and $C_{18:1\omega9c}$. The DNA G+C content (mol %) is 38.4–39.4 (type strain 39.4). The type strain is T22-1-2$^T$ (=DSM 19180$^T$, =NBRC 103247$^T$, =NRIC 0724$^T$) and was isolated from fermented shrimp paste. Reference strains (T78-1-2, T82-5-1 & T171-1-1) were isolated from "ka-pi," salted fish, and raw fish.

*Alkalibacterium subtropicum* Ishikawa, Nakajima, Ishizaki, Kodama, Okamoto-Kainuma, Koizumi, Yamamoto, and Yamasato 2011, 3001$^{VP}$. (sub.tro'pi.cum. L. pref. *sub* under, below, slightly; L. neut. adj. *tropicum* tropical; N.L. neut. adj. *subtropicum* subtropical, referring to the subtropical region where strains were isolated).

Cells of *Alkalibacterium subtropicum* are motile, long straight rods (0.5–0.9 × 3.0–20.0 μm) that appear singly, in pairs, or short chains become elongated in older cultures. Colony formation varies between surface and sunken; using 2.5 % NaCl GYPF agar medium after 3 days at 30 °C, deep colonies (1.0–2.0 mm) are creamy white, opaque, and lenticular, while surface colonies (0.5–1.0 mm) are round, convex, entire, and creamy white to transparent. Slightly halophilic and halotolerant. Optimum growth occurs at pH 8.0–8.5, 1.0–3.0 % NaCl (w/v), and at 20–30 °C, but ranges from pH 7.5 to 9.5, 0 to 17 % NaCl, and 15 to 40 °C. *A. subtropicum* utilizes cellobiose, dulcitol (weak), fructose, glucose, gluconate, glycerol (weak), inulin, lactose, maltose, mannose, melibiose, sucrose, trehalose, mannitol, raffinose, ribose, salicin, sorbitol, starch, and xylose for growth. D-Arabinose and rhamnose are not fermented. Variable reactions are observed for adonitol, L-arabinose, galactose, inositol, melezitose, and methyl α-glucoside. The major fermentation product from glucose is lactate, with lesser amounts of formate, acetate, and ethanol made a molar ratio of approximately 2:1:1, with lactate decreasing at increasing pH. Glucose is metabolized to acetate and lactate under aerobic conditions. Growth is inhibited by ampicillin, chloramphenicol, kanamycin, and trimethoprim. The peptidoglycan is type A4β, Orn-D-Asp, and the predominant cellular fatty acids are $C_{14:0}$, $C_{16:0}$, and $C_{16:1\omega9c}$. The DNA G+C content (mol %) is 43.7. The type strain is 3 AD-1$^T$ (=DSM 23664$^T$, =NBRC 107172 $^T$) and was isolated from decaying marine alga.

*Alkalibacterium gilvum* Ishikawa, Yamasato, Kodama, Yasuda, Matsuyama, Okamoto-Kainuma, and Koizumi 2013, 1475$^{VP}$. (gil'vum L. adj. *gilvum* pale yellow, referring to the pale yellow color of the colony of the bacterium).

Nine strains of *Alkalibacterium gilvum* were isolated that stained Gram-positive and were microaerophilic, nonsporeforming, and nonmotile rods. The cells occurred singly, in pairs, or chains and ranged from 0.5 to 0.9 × 2.5–10.9 μm in size. Cells are catalase and oxidase negative and do not reduce nitrate. The organisms are slightly halophilic (optimum 2.0–5.0 % NaCl (w/v)) but can tolerate 0–17.5 % NaCl depending on the strain. Optimum pH is 8.5–9.5, and the optimum temperature is 20–30 °C. Glucose is metabolized aerobically to acetate and lactate and anaerobically fermented primarily to lactate with formate, acetate, and ethanol produced. The type strain (strain 3 AD-1$^T$) could utilize arabinose (weakly), cellobiose fructose, galactose, glucose, glycerol, lactose, maltose, mannose, ribose, salicin, sucrose, and trehalose for growth and energy. It could not use adonitol, dulcitol, gluconate, inositol, inulin, mannitol, melibiose, melezitose, α-methyl-glucoside, raffinose, rhamnose, sorbitol, starch, and xylose. The cell wall peptidoglycan consists of a type A4β Orn-D-Asp. The predominant cellular fatty acids are $C_{14:0}$, $C_{16:0}$, $C_{16:1\omega9C}$, and $C_{18:1\omega9C}$. The G+C content of the DNA is 36.0–37.6 mol% (strain 3 AD-1$^T$, 36.8 mol%). The type strain is 3 AD-1$^T$ (=DSM 25751$^T$, =JCM 18271$^T$) and was isolated from mold-ripened, soft Brie cheese.

## *Allofustis* Collins, Higgins, Messier, Fortin, Hutson, Lawson, and Falsen 2003, 813$^{VP}$

Al.lo.fus'tis. Gr. prefix *allos* another, the other; L. masc. n. *fustis* stick; N.L. masc. n. *Allofustis* the other stick or rod.

Cells are Gram-stain-positive, nonspore-forming and rod-shaped. Facultatively anaerobic. Catalase, oxidase, and indole are not produced and the Voges-Proskauer test is negative. Nitrate is not reduced. The enzymes arginine dihydrolase, leucine arylamidase, and pyroglutamic acid arylamidase are produced. The long-chain cellular fatty acids of the organism are of the straight-chain saturated and monounsaturated types, with $C_{16:0}$, $C_{16:1}$, $C_{18:0}$, and $C_{18:1\omega9c}$ predominating. The cell wall murein is type L-lysine-direct (A1α). The DNA G+C content (mol%) is 39.

Using the API rapid ID 32A kit, positive reactions for alanine arylamidase, alkaline phosphatase, arginine arylamidase, arginine dihydrolase, α-fucosidase, β-glucosidase, glycine arylamidase, histidine arylamidase, leucine arylamidase, leucyl glycine arylamidase, *N*-acetyl-β-glucosaminidase, phenylalanine arylamidase, proline arylamidase, pyroglutamic acid arylamidase, serine arylamidase, and tyrosine arylamidase. Negative for α-arabinosidase, α-galactosidase, β-galactosidase, β-galactosidase-6-phosphase, α-glucosidase, β-glucuronidase, glutamic acid decarboxylase, glutamyl glutamic acid arylamidase, indole, mannose, nitrate, raffinose, and urease.

According to the API ZYM system, positive for acid phosphatase, alkaline phosphatase, leucine arylamidase, valine arylamidase, and naphthol-AS-BI-phosphohydrolase is weak. Negative for α-chymotrypsin, cystine arylamidase, esterase ($C_4$), esterase lipase ($C_8$), α-fucosidase, α-glucosidase, β-glucosidase, α-galactosidase, β-galactosidase, β-glucuronidase, lipase (C 14), α-mannosidase, *N*-acetyl-β-glucosaminidase, and trypsin.

It is also pertinent to note that the predicted secondary structure of the V6 region of the 16S rRNA is useful in the assignment of organisms within this suprageneric cluster of organisms. In particular, nucleotides at positions 457–462 and the complementary nucleotide sequence at positions 471–476 appear to be especially informative at the genus level.

The type and only strain isolated to date 01-570-1$^T$ (=CCUG 45438$^T$, =CIP 107425$^T$, =DSM 15817$^T$) was isolated from porcine semen.

## *Alloiococcus* Aguirre and Collins 1992, 83$^{VP}$

Al.loi.o.coc'cus. Gr. adj. *alloios* different; N.L. n. *coccus* coccus; N.L. masc. n. *Alloiococcus* different coccus, referring to the phylogenetic distinctiveness of the organism.

The genus *Alloiococcus* is still only represented by a single species, namely, *Alloiococcus otitis* (o.ti'tis. M.L. n. otitis, inflammation of the ear). Cells stain Gram-positive; are ovoid in shape appearing in pairs, tetrads, and clusters; and do not form spores. They are aerobic and nutritionally fastidious. All strains are oxidase negative but may or may not be catalase positive (type strain 7760$^T$ is catalase positive). Growth is slow and appears as small alpha-hemolytic colonies on blood agar at 37 °C. Because *A. otitis* is so fastidious, acids are not produced from carbohydrates. Fermentation does not occur on arabinose, dulcitol, glucose, glycerol, inulin, lactose, maltose, mannose, mannitol, raffinose, rhamnose, ribose, salicin, sucrose, sorbitol, trehalose, or xylose. Starch and esculin are not hydrolyzed, but hippurate may or may not be hydrolyzed. Can grow in 6.5 % NaCl but not 10 % NaCl. Mesophilic with no growth seen at 10 or 45 °C. Produces pyrrolidonyl arylamidase, β-galactose, and leucine arylamidase and is vancomycin sensitive. The DNA G+C (mol %) is 44–45.

The type strain is NCFB 2890$^T$ (=7760$^T$) (=ATCC 51267$^T$, = DSM 7252$^T$, = CCM 4306$^T$, = CCUG 32997$^T$, = CIP 103508$^T$, = CNCTC 7328$^T$, = IFO 15545$^T$, = LMG 17751$^T$, = NCFB 2890$^T$, = NBRC 15545$^T$, = NCIMB 702890$^T$, = UC12635$^T$) and was isolated from fluid collected from the inner ear of a child suffering from chronic otitis media with effusion.

## *Atopobacter* Lawson, Foster, Falsen, Ohlén, and Collins 2000, 1758$^{VP}$

A.to.po.bac'ter. Gr. adj. *atopos* having no place, strange; L. masc. n. *bacter* rod; M.L. masc. n. *Atopobacter* strange rod.

Cells are Gram-stain-positive, short, nonspore-forming, irregular rods. Facultatively anaerobic and catalase-negative. No growth is observed at 45 °C. Acid is produced from D-glucose and some other sugars, but no gas is produced. The enzymes arginine dihydrolase, pyroglutamic acid arylamidase, and pyrrolidonyl arylamidase are produced. Esculin, gelatin, hippurate, and urea are not hydrolyzed. Nitrate is not reduced and the Voges-Proskauer reaction is negative. The cell wall murein is type L-ornithine-D-aspartic acid (A4β). Type species: *Atopobacter phocae* Lawson, Foster, Falsen, Ohlén, and Collins 2000, 1759$^{VP}$.

The genus to date still only contains a single species, *Atopobacter phocae*, and therefore the characteristics provided below refer to this species. Cells consist of short, irregular rods. On Columbia agar supplemented with 5 % horse blood, small (pin-sized), gray-colored, smooth colonies are formed after 24 h at 37 °C. Nonhemolytic. Growth occurs at 25 °C but not at 45 °C. Acid, but no gas, is produced from D-glucose. Acid is also formed from glycogen, maltose, pullulan, and D-ribose. Acid may or may not be formed from cyclodextrin, lactose, sucrose, and trehalose. Acid is not produced from D-arabitol, L-arabinose, mannitol, melibiose, melezitose, methyl-β-D-glucopyranoside, raffinose, sorbitol, tagatose, or D-xylose. Esculin, gelatin, hippurate, and urea are not hydrolyzed. The cell wall contains an L-ornithine-D-aspartic acid-type murein (variation A4 β).

Using the API ZYM test system, positive for acid phosphatase, alkaline phosphatase, naphthol-AS-BI-phosphohydrolase, arginine dihydrolase, esterase $C_4$, ester lipase $C_8$, pyroglutamic acid arylamidase, pyrazinamidase, and pyrrolidonyl arylamidase. Activity is not detected for chymotrypsin, cysteine arylamidase, α-fucosidase, α-galactosidase, α-glucosidase, β-glucosidase, β-glucuronidase, glycine tryptophan arylamidase, α-mannosidase, β-mannosidase, lipase ($C_{14}$), *N*-acetyl β-glucosaminidase, trypsin, or valine arylamidase. Activity for alanyl phenylalanine proline arylamidase, β-galactosidase, and leucine arylamidase may or may not be detected. Nitrate is not reduced.

*Atopobacter phocae* M1590/94/2$^T$ (=ATCC BAA-285$^T$, = CCUG 42358$^T$, = CIP 106392$^T$) was originally recovered from dead common seals (Lawson et al. 2000). The species has, however, been isolated subsequently in mixed culture from an otter head abscess (Foster, Lawson, and Collins, unpublished results). The habitat of *Atopobacter phocae* is not known.

## *Atopococcus* Collins, Wiernik, Falsen, and Lawson 2005, 1695$^{VP}$

A.to.po.coc'cus. Gr. adj. *atopos* having no place, strange; Gr. n. *coccus* a grain or berry; N.L. masc. n. *Atopococcus* a strange coccus.

Cells are cocci that occur in pairs or short chains and stain Gram-positive. Aerobic, nonmotile, and endospores are not formed. Acid is produced from D-glucose and some other carbohydrates. Pyroglutamic acid arylamidase is produced, but arginine dihydrolase is not. Catalase and urease are not produced. Nitrate is not reduced. Optimum growth temperature is 28–32 °C, but no growth is observed above 32 °C. The major long-chain fatty acids are the straight-chain and monounsaturated types. The cell wall murein contains L-Lys, type A4α (L-Lys-D-Asp). Isolated from moist, powdered tobacco. DNA G+C content (mol%) is 46.0. Type species is *Atopococcus tabaci.*

The genus contains only one species, *Atopococcus tabaci*, and therefore the additional characteristics provided below refer to this species. Grows on Columbia blood agar base supplemented with 5 % horse blood and displays α-hemolysis. Optimal growth temperature on tryptic soy agar is 30 °C. Halotolerant, growing in 8–9 % NaCl.

Nitrate is not reduced, and the Voges-Proskauer test is negative. The fatty acid content is $C_{10:0}$ (0.7 %), $C_{12:0}$ (2.6 %), $C_{14:0}$ (8.3 %), $C_{14:1}$ (4.5 %), $C_{16:1\omega9c}$ (41.9 %), $C_{16:0}$ (15.8 %), iso- $C_{17:1}$ (5.3 %), $C_{18:1\omega9c}$ (9.0 %), and iso-$C_{19:1}$ (1.9 %). No respiratory quinones are detected. The cell wall contains A4α-type murein composed of L-Lys-L-Glu. Amino acids consist of lysine, alanine, and glutamic acid present in molar ratios of 1.0 Lys:1.9 Ala:2.4 Glu. The partial hydrolysate contains the peptides L-Ala-D-Glu and L-Lys-D-Ala. Dinitrophenylation reveals that the *N*-terminus of the interpeptide bridge is glutamate. The only known source from which *Atopococcus tabaci* has been isolated is powdered tobacco (Collins et al. 2005).

Using the API rapid ID 32 strep kit, positive reactions for β-galactosidase, β-glucosidase, hippurate, lactose, *N*-acetyl-β-glucosaminidase, melibiose, melezitose, trehalose, and pyroglutamic acid arylamidase. Negative for acetoin, L-arabinose, D-arabitol, alanine phenylalanine proline arylamidase, arginine dehydrolase, cyclodextrin, α-galactosidase, β-glucuronidase, glycyl tryptophan arylamidase, glycogen, mannitol, β-mannosidase, methyl β-D-glucopyranoside, pullulan, raffinose, sorbitol, tagatose, and urea. Weak reactions are obtained for maltose, ribose, and sucrose.

According to the API ZYM system, positive for acid phosphatase, alkaline phosphatase, α-glucosidase, leucine arylamidase, and valine arylamidase. Negative for α-chymotrypsin, cystine arylamidase, esterase lipase ($C_8$), α-fucosidase, α-galactosidase, β-galactosidase, β-glucosidase, β-glucuronidase, lipase ($C_{14}$), α-mannosidase, *N*-acetyl-β- glucosaminidase naphthol-AS-BI-phosphohydrolase, and trypsin. A weak reaction is obtained for esterase ($C_4$).

The type strain, CCUG 48253$^T$ (=CIP 108502$^T$, DSM 17538$^T$), was isolated from moist powdered tobacco.

## *Atopostipes* Cotta, Whitehead, Collins, and Lawson 2004b, 1425$^{VP}$ (Effective Publication: Cotta, Whitehead, Collins and Lawson 2004a, 193)

A.to.po.sti'pes. Gr. adj. *atopos* having no place, strange; L. masc. n. *stipes* rod; N.L. masc. n. *Atopostipes* a strange rod, referring to its distinct phylogenetic position. *Atopostipes suicloacalis* (su.i.clo.a.ca'lis. L. n. *sus* pig, L. n. adj. *cloacale* pertaining to a sewer (manure canal), N. L. masc. n. *suicloacalis* from pig manure) represents a single species genus (Cotta et al. 2003, 2004a, b).

Cells are short rods that stain Gram-positive that are nonspore-forming and nonmotile. The facultatively anaerobic strain is catalase and urease-negative, nitrate is not reduced, and indole is not formed. Products from glucose are lactate, acetate, and formate. Optimum growth temperature is 28–30 °C, with no growth above 32 °C. Amygdalin, cellobiose, esculin, glucose, lactate (weakly), lactose, maltose, mannose, raffinose, and sucrose are utilized as energy sources. Arabinose, cellulose, inulin, inositol, melibiose, rhamnose, sorbitol, trehalose, and xylose are not utilized. Positive reactions are detected using the API rapid ID 32A for α-arabinosidase, α-galactosidase, β-glucosidase, and *N*-acetyl-β-glucosaminidase when cells were grown in RGM-rumen fluid media with glucose and positive for β-galactosidase and β-galactosidase-6-phosphate when grown in RGM-rumen fluid media with lactose. All other enzymes activities were negative. Cell wall peptidoglycan is of the L-Lys variation, type A4α (L-Lys-D-Asp), and the predominant fatty acids are $C_{14:0}$, $C_{16:0}$, $C_{16:1\omega9c}$, $C_{18:1\omega9c}$, and $C_{18:0}$. The DNA G+C content (mol %) is 43.9. The types strain is PPC79$^T$ (=DSM 15692$^T$, = NRRL 23919$^T$) and was isolated from swine manure slurry.

## *Bavariicoccus* Schmidt, Mayr, Wenning, Glöckner, Busse, and Scherer 2009, 2441$^{VP}$

Ba.va.ri.i.coc'cus. L. fem. *Bavaria* Bavaria, Germany; N.L. masc. n. *coccus* coccus from Gr. masc. n. *kokkos* berry; N.L.

masc. n. *Bavariicoccus* a coccoid-shaped bacterium isolated in Bavaria.

*Bavariicoccus* is comprised of a single species, but six strains were originally isolated (Schmidt et al. 2009). Cells are Gram-positive staining, aerotolerant, catalase-negative, and nonspore-forming cocci. The most predominant fatty acids are $C_{16:0}$ (16.1–29.9 %), $C_{16:1\omega9}$ (1.2–8.0 %), $C_{18:0}$ (4.4–22.4 %), and $C_{18:1\omega9}$ (35.3–72.6 %). Cell wall peptidoglycan consists of alanine, glutamic acid, lysine, and aspartic acid, with a configuration type A4α (L-Lys-D-Asp). Polar lipids contain a large portion of an unknown glycolipid and lesser amounts of diphosphatidylglycerol and phosphatidylglycerol. May or may not use lactose. No cholesterol is found which helps in distinguishing *Bavariicoccus* sp. from *Atopobacter phocae*. The DNA G+C content (mol %) is 38–39.

*Bavariicoccus seileri* (sei'le.ri. N.L. gen. masc. n. *seileri* named in honor of Herbert Seiler, former microbiologist of the Technical University of Munich with great merit in FTIR spectroscopic identification of microorganisms) is the type species. Facultatively anaerobic, *B. seileri* will grow well aerobically at 30 °C on TSA or anaerobically at 34 °C on APT (All-Purpose Tween) agar. Cell diameters are 0.9–1.2 μm. Mesophilic growth occurs between 10 °C and 40 °C and at pH 5.5, but can grow up to 11 % NaCl (w/v). A heterotrophic lactic acid bacterium, *B. seileri*, produces lactate, ethanol, and acetate from glucose. Growth occurs on amygdalin, arbutin, cellobiose, fructose, galactose, β-gentiobiose, glucose, lactose, maltose, pyruvate, salicin, and trehalose. No growth is observed for adonitol, arabinose, arabitol, dulcitol, erythritol, fucose, gluconate, glycerol, glycogen, inositol, inulin, 2- or 5-ketogluconate, lyxose, mannitol, methyl α-mannoside, melibiose, melezitose, pullulan, raffinose, rhamnose, ribose, sorbose, sorbitol, starch, sucrose, tagatose, turanose, xylitol, methyl β-xyloside, or xylose. Hippurate is not hydrolyzed. API results showed positive reactions for β-glucosidase, pyrrolidonyl arylamidase, and leucine arylamidase, but negative for α-galactosidase, β-galactosidase, β-glucuronidase, and alkaline phosphatase. Several unknown phospholipids (polar, glyco-, phospho-, and amino-) are also present. The predominant fatty acids are $C_{16:0}$ (21.3 %), $C_{16:1\omega9}$ (6.3 %), $C_{18:0}$ (9.0 %). and $C_{18:1\omega9}$ (53.0 %). The DNA G+C content (mol %) is 38.

The type strain for *Bavariicoccus seileri* is WCC $4188^T$ (=CCUG $55508^T$ = DSM $19936^T$). The type and reference strains, WCC 4187 (=CCUG 55507) and WCC 4189 (=CCUG 55509), were isolated from the surface and smear water of German smear-ripened soft cheeses.

## *Desemzia* Stackebrandt, Schumann, Swiderski, and Weiss 1999, 187[VP]

De.sem'zi.a. N.L. fem. n. *Desemzia* arbitrary name, derived from the abbreviation DSMZ (Deutsche Sammlung von Mikroorganismen und Zellkulturen).

Gram-stain-positive, short rods, occurring singly and occasionally in pairs. Cells from older cultures tend to stain variably or loss the stain totally and are not acid-fast. The organisms is microaerophilic and nonspore-forming and undergoes a fermentative type of metabolism. Catalase and oxidase are not produced. Peptidoglycan contains lysine as the diagnostic amino acid (peptidoglycan type L-lysine-D-glutamic acid; variation A4α). Mycolic acids and isoprenoid quinones are absent. Straight-chain saturated and monounsaturated fatty acids, hexadecanoic ($C_{16:0}$), hexadecenoic ($C_{16:1}$), and cis-vaccenic ($C_{18:1w7}$) predominate. The DNA G+C content is 40 mol%. The type species is *Desemzia incerta* (Steinhaus 1941) Stackebrandt, Schumann, Swiderski, and Weiss 1999, 187[VP] ("*Bacterium incertum*" Steinhaus 1941; *Brevibacterium incertum* Breed 1953, p. 14.) (Breed 1953).

Using the API rapid ID 32 strep kit, positive reactions for α-galactosidase, β-glucuronidase, β-galactosidase, β-glucosidase, glycyl tryptophan arylamidase, hippurate, lactose, maltose, β-mannosidase, methyl β-D-glucopyranoside, *N*-acetyl-β-glucosaminidase, trehalose, pyroglutamic acid arylamidase, raffinose, and sucrose. Negative reactions for acetoin, alanine phenylalanine proline arylamidase, L-arabinose, D-arabitol, arginine dehydrolase, cyclodextrin, glycogen, mannitol, melibiose, melezitose, pullulan, ribose, sorbitol, tagatose, and urea.

According to the API ZYM system, the only positive reaction is for α-glucosidase. Negative for acid phosphatase, alkaline phosphatase, cystine arylamidase, α-fucosidase, α-galactosidase, β-galactosidase, β-glucuronidase, lipase ($C_{14}$), α-mannosidase, naphthol-AS-BI-phosphohydrolase *N*-acetyl-β-glucosaminidase, trypsin, and valine arylamidase. Weak reaction for α-chymotrypsin, esterase ($C_4$), esterase lipase ($C_8$), β-glucosidase, and leucine arylamidase.

The Biolog identification system GP MicroPlate™ incubated under an atmosphere of 85 % $N_2$ and 15 % $O_2$ revealed that the following substrates were utilized: D-fructose, D-glucose, L-lactic acid, maltose, maltotriose, D-melibiose, D-mannose, D-trehalose, D-psicose, α-methyl D-glucoside, and *N*-acetyl-D-glucosamine (reading after 48 h of incubation). These data complement physiological and nutritional data given by Steinhaus (1941), Breed (1953), and Jones and Keddie (1986).

The history of this organism is somewhat confusing; the organism was first isolated during the course of a study on bacteria isolated from seven orders of the class *Hexapoda*, a Gram-stain-positive bacterium was isolated by Steinhaus (1941) from the ovaries of the lyreman cicada, *Tibicen linnei*. Although the physiological and cultural characteristics resembled those of the genus *Listeria*, the isolate was tentatively classified as *Bacterium incertum* due to the taxonomic uncertainty of the *Listeria* group (Steinhaus 1941). In 1953, Breed transferred this species to the genus *Brevibacterium* as *Brevibacterium incertum*, and the genus became a depository for misclassified strains with superficial morphological and physiological similarities. On the basis of chemotaxonomic and phylogenetic grounds including analysis of the amino acid composition of the peptidoglycan (Schleife and Kandler 1972) and composition of isoprenoid quinones and fatty acids (Kroppenstedt and Kutzner 1978; Collins and Kroppenstedt 1983), it became apparent that the organism was not a member of the genus

*Brevibacterium*. The phylogenetic affiliation and taxonomic position of *Brevibacterium incertum* was finally resolved using 16S rDNA sequencing (Rainey et al. 1996), base composition of DNA (Marmur 1961; Mesbah et al. 1989) and isoprenoid quinones (Tindall 1990; Groth et al. 1996). Stackebrandt et al. (1999) brought together these data and reclassified *incertum* as *Desemzia incerta*. The type strain is ATCC 8363$^T$ (=CCUG 38799$^T$, = CIP 104227$^T$, = CIP 106501$^T$, = DSM 20581$^T$, = NBRC 12145$^T$, = JCM 1969$^T$, = NCIMB 9892$^T$).

## *Dolosigranulum* Aguirre, Morrison, Cookson, Gay, and Collins 1994, 370$^{VP}$ (Effective Publication: Aguirre, Morrison, Cookson, Gay, & Collins 1993, 610)

Do.lo.si.gra'nu.lum. L. adj. *dolosus* crafty, deceitful; L neut. n. *granulum* a small grain; N.L. neut. n. *Dolosigranulum* a deceptive small grain.

*Dolosigranulum pigrum* (Aguirre et al. 1993) is a single strain genus and was originally isolated from a patient's spinal fluid who suffered from acute multiple sclerosis and from the lens and eye swab of a lady with a neurotropic cornea. The type strain, NCFB 2975$^T$ (=R91/1468$^T$), cells are Gram-stain positive, ovoid in shape that occur in pairs or groups. The strain is nonspore-forming, nonmotile, catalase-negative, mesophilic, and facultatively anaerobic though some strains grow better aerobically. Arginine dehydrogenase, pyrrolidonyl arylamidase, and leucine arylamidase are produced. Cell wall murein is type A and is composed of L-lysine (type Lys-D-Asp). The DNA G+C content (mole %) is 40.5.

*Dolosigranulum pigrum* (pi' grum, L. n. adj. *pigrum* lazy). Cells are slow growing, weakly α-hemolytic on blood agars, and does not grow at 6.5 % NaCl, 40 % bile esculin, or at 10 or 45 °C. Grows best aerobically. Acid but not gas is produced from glucose, galactose, fructose, mannose, maltose, and fucose. Acid is not produced from adonitol, lactose, pullulan, or raffinose. Maltose and sucrose are positive in the API ID32 Strep system, and activity is detected for β-galactosidase and pyroglutamic acid arylamidase. Glycyl tryptophan arylamidase has been detected. Esculin and hippurate are not hydrolyzed, pyruvate is not utilized, and urease is not detected. Cell wall murein type is Lys-D-Asp. The DNA G+C content (mol %) is 40.5. The type strain is NCFB 2975$^T$ (=R91/1468$^T$) (=ATCC 51524$^T$, = CCUG 33392$^T$, = CIP 104051$^T$, = IFO 15550$^T$, = KCTC 15002$^T$, = LMG 15126$^T$, = NBRC 15550$^T$, = NCFB 2975$^T$, = NCIMB 702975$^T$).

## *Granulicatella* Collins and Lawson 2000, 367$^{VP}$

Gra.nu.li.ca.tel'la. L. neut. dim. n. *granulum* small grain; L. fem. dim. n. *catella* small chain; N.L. fem. dim. n. *Granulicatella* small chain of small grains.

*Granulicatella* species are Gram-stain-positive cocci that appear singly, in pairs, or chains when grown under optimal conditions. They are all nonmotile, nonspore-forming, facultatively anaerobic microorganisms that are catalase and oxidase negative. Because of similar habitats, phenotypic traits, and characteristics, the *Granulicatella* species were all previously *Abiotrophia* sp. (Bouvet et al. 1989; Kawamura et al. 1995; Roggenkamp et al. 1998b; Lawson et al. 1999; Collins and Lawson 2000; Kanamoto et al. 2000). Identifications and characteristics that differentiate *Granulicatella* species from each other and from *Abiotrophia defectiva* are listed in ➲ Table 4.5. Colony morphology is similar for all species forming small (<0.2 mm diameter) colonies on fresh sheep blood agar plates supplemented 10 mg pyridoxal $l^{-1}$ or 100 mg cysteine $l^{-1}$. Lactic acid is produced from glucose. α- or β-galactosidases are not produced in any species, which helps distinguish it from *Abiotrophia defectiva*. The DNA G+C content (mol %) is 36.3–37.4. The type species for the genus is *Granulicatella adiacens*.

*Granulicatella adiacens* (Bouvet et al. 1989) Collins and Lawson 2000, 367$^{VP}$ (*Streptococcus adjacens* Bouvet et al. 1989, p. 293; *Abiotrophia adiacens* Kawamura et al. 1995, p. 802).

Cell morphology of *G. adiacens* (ad'ia.cens. L. fem. adj. *adjacens* adjacent, indicating that this organism can grow as satellite colonies adjacent to other bacterial growth) is pleomorphic depending on growth conditions and can vary between cocci, chains, coccobacilli, and globular, rod-shaped cells when grown in broth media supplemented with cysteine or pyridoxal. A tendency toward rod shape morphology is observed in stationary phase. *G. adiacens* will grow as a satellite colony to *Staphylococcus epidermidis* on horse blood TSA agar or sheep blood agar. α-hemolysis will occur on sheep blood agar. Produces a red chromophore when boiled at pH 2 for 5 min. Inulin is fermented by some strains, but arabinose, glycogen, lactose, mannitol, raffinose, ribose, sorbitol, starch, and trehalose are not. Arginine and hippurate are not hydrolyzed. Produces pyrrolidonyl arylamidase and leucine aminopeptidase, but not alkaline phosphatase. β-glucosidase and β-glucuronidase are produced by some strains. *G. adiacens* has been isolated from the throat flora, urine, and blood of patients with endocarditis. The DNA G+C content (mol %) is 36.6–37.4. The type strain is GaD$^T$ (=ATCC 49175$^T$, = CCM 4671$^T$, = CCUG 27809$^T$, = CIP 103243$^T$, = DSM 9848$^T$, = KCTC 3661$^T$, = LMG 14496$^T$, = NCTC 13000$^T$).

*Granulicatella elegans* (Roggenkamp et al. 1998b) Collins and Lawson 2000, 367$^{VP}$ (*Abiotrophia elegans* Roggenkamp et al. 1998b, p. 103).

Cells of *G. elegans* (e'le.gans. L. adj. *elegans* choice, elegant, fastidious, referring to the fastidious growth requirements) are Gram-stain-positive but variant in shape dependent on nutritional state. When adequately supplied with nutrients, cells are coccoid in short chains. Lack of adequate nutrients; cells are swollen and elongated. Like *G. adiacens*, *G. elegans* will grow as a satellite colony to *S. epidermidis* on TSA sheep blood agar plates. *G. elegans* has a relatively narrow temperature range, growing regularly from 27 °C to 37 °C, but will not grow at 20 °C or 40 °C. Growth occurs on THB or casein-soy peptone bouillon with 0.01 % cysteine hydrochloride but not with the same media supplemented with 0.001 % pyridoxal

**Table 4.5**

**Comparison of selected characteristics useful to differentiate type species within the genus *Granulicatella* and compared with *Abiotrophia defectiva*, also a nutritionally variant streptococci (NVS) and clinical relative to *G. adiacens***

| Characteristic | *G. adiacens*[a] | *G. elegans*[b] | *G. balaenopterae*[c] | *"G. para-adiacens"*[d] | *A. defectiva*[e] |
|---|---|---|---|---|---|
| *Type strain* | GaD$^T$ | B1333$^T$ | M1975/96/1$^T$ | VPI 6807B* | SC10$^T$ |
| *Isolation source* | Patients with endocarditis, France | Patients with endocarditis, Germany | Sputum, the United States | Patients with endocarditis, France | |
| *Pyridoxal dependence* | + | –[§] | – | + | + |
| *Acid production from* | | | | | |
| Pullulan | – | – | + | – | **V**[§] |
| Sucrose | + | + | – | + | + |
| Tagatose | + | – | – | – | –[§] |
| Trehalose | – | – | + | – | **V**[§] |
| Hydrolysis of hippurate | – | **V** | – | ND | – |
| *Production of* | | | | | |
| Arginine dihydrolase | – | + | + | – | – |
| α-Galactosidase | – | – | – | – | + |
| β-Galactosidase | – | – | – | – | + |
| β-Glucuronidase | **V** | – | – | **V** | – |
| *Murein type* | A3α | ND | A4β | ND | A1α |
| *Culture collection* | = ATCC 49175$^T$ | = ATCC 700633$^T$ | = ATCC 700813$^T$ | = ATCC 27527 | = ATCC 49176$^T$ |
| | = DSM 9848$^T$ | = DSM 11693$^T$ | = DSM15827$^T$ | | = DSM 9849$^T$ |
| | = CCUG 27809$^T$ | = CCUG 38949$^T$ | = CCUG 37380$^T$ | | = CCUG 27804$^T$ |
| | = CIP 103243$^T$ | = CIP 105513$^T$ | = CIP 105938$^T$ | | = CIP 103242$^T$ |
| | = LMG 14496$^T$ | | | | |
| | = NCTC 13000$^T$ | | | | |
| *GenBank accession number* | D50540 | AF016390 | Y16547 | AB022027 | D50541 |

Data taken from: [a](Bouvet et al. 1989; Kawamura et al. 1995; Collins and Lawson 2000); [b](Roggenkamp et al. 1998b; Collins and Lawson 2000); [c](Lawson et al. 1999; Collins and Lawson 2000); [d](Kanamoto et al. 2000); [e](Bouvet et al. 1989; Kawamura et al. 1995)
+ positive, – negative, *v* variable among strains, *ND* no data available

hydrochloride. A red chromophore is produced when the organism is boiled at pH 2 for 5 min. *G. elegans* will utilize raffinose but not glycogen, inulin, lactose, starch, or trehalose. Hippurate is hydrolyzed. Pyrrolidonyl arylamidase and leucine aminopeptidase are produced but not α- or β-galactosidase, β-glucuronidase, alkaline phosphatase, or β-glucosidase. The type strain is B1333$^T$ (=ATCC 700633$^T$, = CCUG 38949$^T$, = CIP 105513$^T$, = CDC 4067-96$^T$, = CCM 4945$^T$, = DSM 11693$^T$, = LMG 19514$^T$).

*Granulicatella balaenopterae* (Lawson et al. 1999) Collins and Lawson 2000, 368$^{VP}$ (*Abiotrophia balaenopterae* Lawson et al. 1999, p. 505).

The third species of *Granulicatella* share all the properties of the genus coupled with these subtle differences. The type species, *G. balaenopterae* (bal.aen.op'ter.ae. N.L. fem. n. *balaenopterae* pertaining to the minke whale, *Balaenoptera acutorostrata*, from which the organism was isolated), form cell colonies that are tiny (<0.2 mm on Columbia agar with 5 % horse blood at 37 °C) and do not require pyridoxal hydrochloride or satellitism for growth. Substrates utilized for growth are glucose, maltose, pullulan, and trehalose. Cells will not produce acid from arabinose, arabitol, cyclodextrin, glycogen, lactose, mannitol, melibiose, melezitose, raffinose, sucrose, sorbitol, tagatose, or xylose. Esculin is hydrolyzed, but gelatin and hippurate are not; nitrate is not reduced. Arginine dihydrolase, pyroglutamic acid arylamidase (weak), *N*-acetyl-glucosaminidase, ester lipase ($C_8$), leucine arylamidase, and urease (weak) activities are detected. All other enzymes tested for in API kits were negative. The cell wall contains an L-Orn-D-Asp directly cross-linked murein (type A4β). The G+C content of DNA (mol %) is 37. It was isolated from the lung of a deceased minke whale, so the habitat remains unknown. The type strain is M1975/96/1$^T$ (=CCUG 37380$^T$, = ATCC 700813$^T$, = DSM 15827$^T$, = CIP 105938$^T$, = NCIMB 13829$^T$).

## Other Organisms

Kanamoto and colleagues delineated the three known species of *Abiotrophia* strains (as they were previously known) with a fourth group that would now belong to the genus *Granulicatella* (Kanamoto et al. 2000). The species name proposed for this fourth group was "*Abiotrophia para-adiacens*." The species name though has not been validly named as per the rules of the Bacteriological Code and therefore lacks standing in the nomenclature. "*A. para-adiacens*" requires pyridoxal for growth and has strain-specific varied growth on tagatose and sucrose. It does not ferment pullulan or trehalose. β-glucosidase and *N*-acetyl-β-glucosaminidase may or may not be produced. α- and β-galactosidase and arginine dihydrolase are not produced. A chromophore is produced. Because the species has not been validly published, there is no type strain (although strain TKT1 would likely be applied); however, the previously deposited microorganism *Gemella morbillorum* strain VPI 6807B (=ATCC 27527; originally deposited as *Streptococcus morbillorum* Prevot) has been renamed "*Granulicatella para-adiacens*" in the ATCC catalog as suggested by Kanamoto et al. (2000).

## *Isobaculum* Collins, Hutson, Foster, Falsen, and Weiss 2002, 209$^{VP}$

(Iso.bac'u.lum. Gr. adj. *isos* alike, similar; L. neut. n. *baculum* small rod; N.L. neut. n. *Isobaculum* the one like a stick or a rod).

*Isobaculum melis* (me'lis. L. fem. n. *meles* badger, L. gen. fem. n. *melis* of the badger). *Isobaculum melis* was isolated from a deceased badger and still stands as a single species that represents the genus (Collins et al. 2002). Cells stain Gram-stain-positive but can readily decolorize to Gram-stain-negative. The strain is nonspore-forming, nonmotile, nonpigmented, and nonhemolytic. It is facultatively anaerobic and is both catalase and oxidase negative, and menaquinones are absent. Growth will occur at 10 °C but will not occur at 45 °C or with 6.5 % NaCl. *I. melis* is a heterofermentative lactic acid bacterium producing lactate and acetate as end products of glucose metabolism. Acid is produced from glucose, glycerol, ribose, and trehalose but not from arabinose, inulin, lactose, maltose, melezitose, melibiose, raffinose, sorbitol, sorbose, or sucrose. Pyruvate is not utilized. Using the API systems, the same sugars are used, but the additional sugars arabitol, cyclodextrin, glycogen, mannitol, pullulan, tagatose, xylose, or methyl β-glucopyranoside are not. Esculin is hydrolyzed, but gelatin, hippurate, and starch are not. Using the API rapid ID32Strep and API CORYNE tests can distinguish it from other *Carnobacterium* species and *D. incerta* and reveal a positive presence for arginine dihydrolase, β-glucosidase, β-mannosidase, phosphoaminidase, and pyroglutamic acid arylamidase, with weak reactions for acid phosphatase, ester lipase ($C_4$), and esterase ($C_8$). All other enzymes tested in the API systems are negative, although alkaline phosphatase and *N*-acetyl-β-glucosaminidase may or may not exist. The strain is sensitive to vancomycin (30 ug), Voges-Proskauer is negative, and nitrate is not reduced. The murein type is Lys-L-Thr-Gly, and the major long-chain fatty acids are $C_{16:0}$, $C_{18:0}$, and $C_{18:1\omega9c}$. The DNA G+C content (mole %) is 39. The type strain is M577-94$^T$ (=CCUG 37660$^T$, = DSM 13760$^T$, = CIP 107375$^T$).

## *Jeotgalibaca* Lee, Trujillo, Kang, and Ahn 2014, 1733$^{VP}$

Je.ot.ga.li.ba'ca. N.L. n. jeotgalum (from Korean n. jeotgal) jeotgal, traditional Korean food; L. fem. n. *baca*, a grain or berry, and in bacteriology a coccus; N.L. fem. n. *Jeotgalibaca*, coccus from jeotgal.

Cells consist of Gram-stain-positive cocci which are arranged in tetrads, sarcinae, or irregular conglomerates. Nonspore-forming and nonmotile cells that are aerobic and chemoheterotrophic. Oxidase and catalase are not produced. The major fatty acids are $C_{16:1\ \omega9c}$, $C_{18:1\ \omega9c}$, $C_{16:0}$, and $C_{14:0}$. Polar lipids include DPG, PG PE, and several unknown glycolipids, amino lipids, and phospholipids. Peptidoglycan type A4α. The type species is *Jeotgalibaca dankookensis.*

*Jeotgalibaca dankookensis* (dan.ko.ok'en.sis. N.L. fem. n. dankookensis, of or belonging to Dankook University). In addition to those given in the genus description: cells are 1.0–1.3 µm in diameter. Good growth is obtained on tryptic soy agar but not on NA, PDA, or R2A; weak growth occurs on one-tenth-strength marine agar. Colonies on TSA agar are circular, convex, and pale orange. Growth occurs at 10–37 °C (optimum, 28 °C), at pH 7.0–9.0, and on TSA agar supplemented with NaCl up to 9 %. Nitrate is not reduced. DNA, casein, chitin, starch, Tween 80, and carboxymethyl cellulose are not degraded. Results based on the commercial systems Biolog GN2, API 20NE, API 50CH, and API ID32 GN are the following: positive for assimilation of D-ribose, D-glucose, D-fructose, D-galactose, D-maltose, D-mannose, arbutin, and salicin. Positive for esculin hydrolysis and β-galactosidase production; negative for indole production, glucose fermentation, and urease. Assimilation (Biolog GN2) of D-fructose, D-galactose, D-glucose, D-psicose, D-trehalose, D-mannose, D-maltose, *N*-acetyl-glucosamine, pyruvic acid, D-gluconic acid, α-keto butyric acid, inosine, uridine, thymidine, and glycerol. The major fatty acids are $C_{16:1\ \omega9c}$ (35.1 %), $C_{18:1\ \omega9c}$ (26.1 %), $C_{16:0}$ (18.5 %), and $C_{14:0}$ (6.9 %).

According to the API ZYM gallery, alkaline phosphatase, esterase ($C_4$), esterase lipase ($C_8$), chymotrypsin, acid phosphatase, and naphthol-AS-BI-phosphohydrolase activities are present; *N*-acetyl-β-glucosaminidase, β-glucosidase, α-fucosidase, leucine arylamidase, lipase ($C_{14}$), valine arylamidase, cystine arylamidase, α-glucosidase, α-galactosidase, β-glucuronidase, β-glucosidase, α-mannosidase, and trypsin activities are absent. The DNA G+C content of the type strain is 39.6 mol%. The type strain, EX-07$^T$ (=KCCM 90229$^T$ = JCM 19215$^T$), was isolated from *saeujeot* (traditional Korean food) in Cheonan Dankook University, South Korea.

## *Lacticigenium* Iino, Suzuki, and Harayama 2009, 779$^{VP}$

(Lac.ti.ci.ge'ni.um. N.L. n. *acidum lacticum* lactic acid; N.L. neut. suff. *-genium* (from Gr. v. *gennao* to produce) that which produces; N.L. neut. n. *Lacticigenium* a bacterium that produces lactic acid).

The Gram stain reaction is negative with conventional Gram stain, but positive with the KOH test. Cells are oval rods, nonsporulating, and motile by peritrichous flagella. Facultatively anaerobic, mesophilic, and neutrophilic. Catalase is not produced. The major cellular fatty acid is $C_{16:\,1\ \omega7c}$. Cell wall murein is type A4α containing L-Lys-L-Glu. The G+C content of the genomic DNA is 38 mol% (as determined by HPLC). The type species is *Lacticigenium naphtae*.

*Lacticigenium naphtae* (naph'tae. L. n. *naphta* crude petroleum; L. gen. n. *naphtae* of crude petroleum). Cells are 0.6–0.761.8–2.5 mm in size. Growth occurs at or below 30 °C, but not at 35 °C, with an optimum at 30 °C. The pH range for growth is 6.5–8.5, with an optimum around pH 7.0. Growth occurs below 17 % (w/v) NaCl, with an optimum at 3 % (w/v). Using the API CHL test system, acids is produced from L-arabinose, ribose, glucose, fructose, mannose, *N*-acetylglucosamine, amygdalin, arbutin, salicin, cellobiose, maltose, sucrose, trehalose, gentiobiose, and 5-ketogluconate. L-lactic acid is the major end product from glucose. Sulfate, sulfite, thiosulfate, elemental sulfur, nitrate, nitrite, and fumarate are not used as electron acceptors. The G+C content of genomic DNA is 37.8 mol% (HPLC). The type strain is MIC1-18$^T$ (=NBRC 101988$^T$, = DSM 19658$^T$), which was isolated from a crude oil sample collected from an oil-water well in Akita, Japan.

## *Marinilactibacillus* Ishikawa, Nakajima, Yanagi, Yamamoto, and Yamasato 2003, 719$^{VP}$

Ma.ri.ni.lac.ti.ba.cil'lus. L. adj. *marinus* marine; L. n. lac, lactis milk; L. n. *bacillus* a small rod; N.L. masc. n. *Marinilactibacillus* marine lactic acid rodlet.

Gram-stain-positive, nonspore-forming, straight rods that occur singly, in pairs, or in short chains. It is motile by the use of peritrichous flagella. The organism is facultative anaerobe. Catalase and oxidase are not produced. Negative for nitrate reduction and gelatin liquefaction but does hydrolyzes casein. Mesophilic and psychrotolerant and is also alkaliphilic, slightly halophilic, and highly halotolerant. L (+) Lactic acid is the major end product from D (+) glucose; trace to small amounts of formate, acetate, and ethanol are produced with a molar ratio of approximately 2:1:1, without gas formation. The peptidoglycan is of the A4β, Orn-D-Glu type. Cellular fatty acids are primarily of the straight-chain saturated and monounsaturated even-carbon-numbered types. The major fatty acids are $C_{16:\,0}$, $C_{16:1\ \omega7}$, $C_{18:0}$, and $C_{18:1\ \omega9}$. Respiratory quinones and cytochromes are absent. DNA G+C content (mol%) is between 34.6 and 36.2. The type species is *Marinilactibacillus psychrotolerans*.

*Marinilactibacillus psychrotolerans* psy.chro.to'le.rans. Gr. adj. *psychros* cold; L. part. adj. *tolerans* tolerating; N.L. adj. *psychrotolerans* tolerating cold temperature.

In addition to the characteristics that define the genus, it has the characteristics described below. Deep colonies in agar medium are pale yellow, opaque, and lenticular with diameters of 2–4 mm. Cells are 0.4–0.5 × 2.3–4.5 μm and elongated in older cultures. Grows evenly throughout a column of semisolid agar medium. Acid is produced from a fairly wide range of carbohydrates, sugar alcohols, and related carbon compounds. Sodium gluconate is fermented without gas production. The G+C content of the DNA of strain M13-2$^T$ is 36.2 mol%. Isolated from a living sponge, raw Japanese ivory shell and decomposing alga. The type strain has been deposited at the IAM Culture Collection; the Institute of Molecular and Cellular Biosciences; the University of Tokyo, Tokyo, Japan; the NITE Biological Resource Center (NBRC); the National Institute of Technology and Evaluation, Kisarazu, Japan; the National Collections of Industrial, Food and Marine Bacteria (NCIMB), Aberdeen, the United Kingdom; and the Nodai Culture Collection Center (NRIC), Tokyo University of Agriculture, Tokyo, Japan, under the accession numbers IAM 14980$^T$, NBRC 100002$^T$, NCIMB 13873$^T$, and NRIC 0510$^T$, respectively.

*Marinilactibacillus piezotolerans* pie.zo.to'le.rans. Gr. v. *piezo* to press; L. part. adj. *tolerans* tolerating, N.L. part. adj. *piezotolerans* tolerating high hydrostatic pressure.

The organism consists of Gram-positive staining cells that are nonspore-forming and nonmotile rods, 2–2.2 × 0.3–0.35 μm in size. It is a facultative anaerobe that grows between 4 °C and 50 °C, with the optimum growth around 37–40 °C; no growth is detected above 50 °C. It grows in NaCl concentrations ranging from 0 to 120 g L$^{-1}$, with the optimum at approximately 10–20 g L$^{-1}$; no growth detected at or above 130 g NaCl L$^{-1}$. Growth occurs at pH values between 5.5 and 10.0, with the optimum at around pH 7.0–8.0. The optimum hydrostatic pressure for growth is 0.1 MPa, with tolerance up to 30 MPa. Under optimal growth conditions (YPG medium, 37 °C, pH 7.0, and 20 g NaCl L$^{-1}$), the doubling time is approximately 57 min. Sulfate, thiosulfate, elemental sulfur, L-cysteine, iron oxide, nitrate, and nitrite are not reduced. The main components of the

**Table 4.6**
**Characteristics useful in the differentiation of *M. piezotolerans* and *M. psychrotolerans***

| Characteristic | *M. piezotolerans* | *M. psychrotolerans* |
|---|---|---|
| pH range | 5.5–10.0 | 8.5–9.0 |
| pH optimum | 7.0–8.0 | 8.8–9.0 |
| NaCl range (%) | 0–12.0 | 0–20.5 |
| NaCl optima (%) | 0.1–0.2 | 2–3.75 |
| Catalase | + | – |
| β-galactosidase | + | – |
| acetoin | + | – |
| *Acid from* | | |
| D-arabinose | w | – |
| L-arabinose | + | – |
| D-cellulose | + | w |
| D-galactose | + | – |
| Lactose | w | – |
| D-maltose | + | – |
| D-melezitose | – | w |
| D-ribose | + | – |
| sucrose | + | – |
| D-sorbitol | – | w |
| Starch | + | – |
| Major fatty acids | $C_{14:0}$, $C_{16:0}$, $C_{16:1}$, | $C_{16:0}$, $C_{16:1}$, $C_{18:1}$, |
| DNA mol % | 42 | 34.6–36.2 |
| Source | Deep seafloor sediment | Living sponge, raw Japanese ivory shell, and decomposing alga |

+ positive reaction, – negative reaction, *w* weak reaction

lipid complex of the cells are phosphatidylglycerols (25 %), diphosphatidylglycerols (34 %), and tentatively identified ammonium-containing phosphatidylserines (32 %); phosphatidylethanolamines are minor compounds, accounting for 9 %. The most abundant fatty acyl side chains (PLFAs) of these phospholipids are $C_{16:0}$ (44.7 %), $C_{14:0}$ (31.5 %), and $C_{16:1}$ (14.1 %).Quinones are not detected. The G+C content of the DNA of the type strain is 42.0 mol% (as determined by HPLC). The type strain is LT20$^T$ (=DSM 16108$^T$, =JCM 12337$^T$), which was isolated from deep sub-seafloor sediment.

Characteristics of the two species are given in *Table 4.6* from information derived from the original publications. However, it is pertinent to note that in the most recent edition of *Bergey's Manual of Systematic Bacteriology*, written by Yamasato and Ishikawa, it is clear that some features such as catalase activity, fatty acid content, fermentation, and G+C mol % provided in the original publication are called into question (Yamasato and Ishikawa 2009).

## *Pisciglobus* (Tanasupawat, Thongsanit, Thawai, Lee, and Lee 2011, 1690$^{VP}$)

Pis.ci.glo'bus. L. n. *piscis* fish; L. masc. n. *globus* ball, sphere, globe; N.L. masc. n. *Pisciglobus* a sphere (coccus) from fish.

*Pisciglobus halotolerans* is another genus within the *Carnobacteriaceae* that is represented by two strains of a single species (Tanasupawat et al. 2011). Cells stain Gram-positive, are nonmotile, and are facultatively anaerobic. The nonspore-forming cocci appear in pairs, tetrads, or packets and have nonpigmented colonies on MRS plates. The strains are homofermentative lactic acid bacteria when grown on glucose. They can tolerate up to 10 % NaCl and grow between pH 5–9 and at 15–40 °C. The cell wall peptidoglycan consists of the L-Lys type, and the major fatty acid is $C_{18:1\omega 9c}$. The DNA G+C content (mol %) is 38.6–38.7.

*Pisciglobus halotolerans* (ha.lo.to'le.rans. Gr. n. *hals* salt; L.pres. part. *tolerans* tolerating; N.L. part. adj. *halotolerans* salt-tolerating). Cells are 0.6–1 μm in diameter. Positive hydrolysis of arginine and positive for the methyl red reaction. Voges-Proskauer reaction is negative, nitrate is not reduced, and hydrogen sulfide is not produced. Negative for the hydrolysis of casein, gelatin, starch, and tributyrin. Catalase positive when grown in the presence of hematin. The strain can utilize and produce acid from esculin, arbutin, cellobiose, galactose, gentiobiose (weakly), glucose, *N*-acetylglucosamine, fructose, lactose, maltose, mannose, mannitol, methyl α-glucoside, methyl α-glucopyranoside (weakly), ribose, sucrose, salicin, and trehalose. No acid is produced from adonitol, arabinose, arabitol, dulcitol, erythritol, fucose, gluconate, glycogen, glycerol, inositol, inulin, 2-ketogluconate, 5-ketogluconate, lyxose, melibiose, melezitose, methyl α-mannopyranoside, methyl β-xyloside, raffinose, rhamnose, sorbose, sorbitol, starch, tagatose, turanose, xylitol, or xylose. The DNA G+C content (mol %) is 38.6. The type strain for *Pisciglobus halotolerans* is strain C01$^T$ (=KCTC 13150$^T$, = TISTR 1958$^T$, = PCU 316$^T$). Two strains (C01$^T$, C02) were isolated from fermented fish sauce, but its habitat is unknown.

## *Trichococcus* Scheff, Salcher, and Lingens 1984b, 356$^{VP}$ (Effective Publication: Scheff, Salcher and Lingens 1984a, 118.) Emend. Liu Tanner, Schumann, Weiss, Mckenzie, Janssen, Seviour, Lawson, Allen, and Seviour, 2002, 1124

(Tri.cho.coc.cus. Gr. n. *thrix* hair; L. masc. n. *coccus* a grain or berry; N.L. masc. n. *Trichococcus* a hair of cocci).

Gram-positive to Gram variable, pleomorphic cocci that appear as cocci, ovals, or olive-shaped with tapered ends, occurring singly, in pairs, chains, elongated chains, clumps, or conglomerates and depends upon the growth conditions of the organisms. They are nonspore-forming, mesophilic, and facultative anaerobes, and they are all catalase and oxidase negative. Characterization data is presented in *Table 4.7.* Mostly nonmotile but *T. patagoniensis* is motile. The *Trichococcus*

**Table 4.7**
**Comparison of selected characteristics useful in differentiating type species within the genus *Trichococcus***

| | *T. flocculiformis*[a] | *T. pasteurii*[b] | *T. palustris*[c] | *T. collinsii*[d] | *T. patagoniensis*[e] |
|---|---|---|---|---|---|
| *Type strain* | Echt[T] | KoTa2[T] | Z-7189[T] | 37AN3[*T] | PmagG1[T] |
| *Isolation source* | Bulking sludge, Germany | Anaerobic digester, Germany | Swamp, Russia | Hydrocarbon-contaminated soil, the United States | Penguin Guano, Chile |
| *Motility* | – | – | – | – | + |
| *Substrate utilization* | | | | | |
| Arabinose | – | – | – | – | + |
| Lactose | + | + | + | – | + |
| Malate | ND | + | – | + | – |
| Maltose | + | + | + | – | + |
| Ribose | + | – | – | – | + |
| Trehalose | + | + | – | + | – |
| *Temperature range (°C)* | | | | | |
| | 4–40 | 0–42 | 0–33 | −5 to 36 | −5 to 35 |
| *pH growth range (%, w/v)* | 5.9–9.0 | 5.5–8.8 | 6.2–8.4 | 6.0–9.0 | 6.0–10.0 |
| *mol % G+C* | 46.8 | 45.2 | 47.5 | 47.0 | 45.8 |
| *Culture collection* | = ATCC 51221[T] | = ATCC 35945[T] | = DSM9172[T] | = ATCC BAA-296[T] | = ATCC BAA-756[T] |
| | = DSM 2094[T] | = DSM 2381[T] | = CIP 105359[T] | = DSM 14526[T] | = DSM 18806[T] |
| | | = CCUG 37395[T] | | | = JCM 12176[T] |
| | | = CIP 104580[T] | | | = CIP 108035[T] |
| | | = NCIMB 13421[T] | | | |
| *GenBank accession number* | YI7301, AJ306611 | X87150 | AJ296179 | AJ306612 | AF394926 |

Data taken from: [a](Scheff et al. 1984); [b](Schink 1984; Janssen et al. 1995); [c](Zhilina et al. 1995; Liu et al. 2002); [d](Liu et al. 2002); [e](Pikuta et al. 2006)
+ positive, – negative, *ND* no data available

species are all psychoactive and are often detected in colder environments, but prefers mesophilic conditions. Optimum temperature range is 25–30 °C, but ranges from −5 °C to 40 °C. Fermentative growth on glucose produces lactate, acetate, formate, and ethanol, but aerobically only lactate and acetate are produced. The results of Biolog substrate testing are varied among the differing type species and strains and can be used to differentiate them (Liu et al. 2002) (*T. patagoniensis* not included). Major fatty acids tend toward $C_{14:0}$, $C_{16:0}$, $C_{18:1\omega9c}$, and/or $C_{16:1}$ or $C_{16:1\omega7c}$. The DNA G+C contents (mol %) range from 45 to 49, and the cell wall peptidoglycan consists of type A4α, L-Lys-D-Asp. Habitats include activated sludge, sediment, and soil. The genus contains five species: *T. flocculiformis* (type species), *T. pasteurii*, *T. collinsii*, *T. palustris*, and *T. patagoniensis*.

*Trichococcus flocculiformis* (Scheff et al. 1984b, 356[VP]) (Effective publication: Scheff et al. 1984a, p. 118.) emend. Liu, Tanner, Schumann, Weiss, McKenzie, Janssen, Seviour, Lawson, Allen, and Seviour 2002, p. 1124. (floc.cu.li'form.is. L. n. *floccus* a flock of wool; N.L. dim. adj. *flocculus* like a small floc of wool; L. n. *forma* shape; L. adj. *flocculiformis* small-floc-shaped).

Cells are spherical to ovoid (1–1.5 μm by 1.0–2.5 μm) and form filaments from twenty to several hundred cells. Nonspore-forming and facultatively anaerobic. Optimum growth occurs at pH 8.0, 0.5 % NaCl (w/v), and at 25–30 °C, but ranges from pH 5.8 to 9.0 and 4–39 °C. Growth occurs on cellobiose, fructose, galactose, glucose, lactose, maltose, mannose, ribose, sucrose, trehalose, and xylose, but not on agar, arabinose, casein, cellulose, chitin, DNA, erythritol, esculin, galactitol, gelatin, pectin, ribitol, starch, tributyrin, Tween 80, and urea. Some strains produce acid from adonitol, inositol, mannitol, sorbitol, and raffinose. Acetylmethylcarbinol, indole, and hydrogen sulfide are not produced. Reduction of nitrate is variable among strains, as is production of urease. Methyl red test is positive. Cell wall peptidoglycan contains the amino acids ala, glu, asp, and lys. Predominant fatty acids are $C_{14:0}$ (8–14 %), $C_{16:0}$ (12–24 %), $C_{16:1}$ (39–50 %), $C_{17:1iso}$ (1–4 %), $C_{18:0}$

(1–3 %), and $C_{18:1\omega9c}$ (14–23 %). The DNA G+C content (mol%) is 46.8. Type strain: Echt$^{T}$ and was isolated from bulking sludge.

*Trichococcus pasteurii* (Schink 1984) Liu, Tanner, Schumann, Weiss, McKenzie, Janssen, Seviour, Lawson, Allen, and Seviour 2002, 1124$^{VP}$ (*Ruminococcus pasteurii* Schink 1984, p. 413; *Lactosphaera pasteurii* Janssen et al. 1995, p. 570). (pas.teu'ri.i. N.L. gen. n. *pasteurii* referring to Louis Pasteur, who probably first enriched and observed this bacterium during studies on tartrate fermentation).

Cells were originally described as cocci (1.0–1.5 um) that are nonmotile and nonspore-forming (Schink 1984). Later studies indicated that cells are pleomorphic depending on growth conditions, appearing spherical, ovoid, or olive-shaped, occurring singly, in pairs, chains, small irregular packets, or as irregular conglomerates. Facultatively anaerobic. Optimum growth occurs at pH 7.0–8.0, 0.5 % NaCl (w/v), and at 25–30 °C, but ranges from pH 5.5 to 8.8, 0 to 1 % NaCl (w/v), and 0 to 42 °C. Biotin (vitamin $B_7$) is required as a growth factor. Cells utilize cellobiose, citrate, fructose, galactose, glucose, lactose, laminarin (weakly), malate, maltose, mannitol, mannose, oxaloacetate, pyruvate, rhamnose, raffinose, sorbitol, sorbose, starch, sucrose, L-tartrate, trehalose, and oat spelt xylan for growth. Arabinose, arabinogalactan, cellulose (amorphous), carboxymethyl cellulose, chitin, glycogen, gumkaraya, gum locust bean, 3-hydroxybutyrate, lactate, lichenan, malonate, mannan, pullulan, ribose, succinate, D-tartrate, meso-tartrate, xylose, amino acids, and alcohols are not utilized. Major fermentation products from glucose are lactate, acetate, formate, and ethanol. The end products of citrate, pyruvate, and L-tartrate are acetate, formate, and $CO_2$. Nitrate, sulfate, sulfite, thiosulfate, sulfur, and fumarate are not reduced, and urease and gelatin are not hydrolyzed. Major fatty acids are $C_{12:0}$ (1 %), $C_{14:0}$ (16 %), $C_{16:0}$ (15 %), $C_{16:1\omega7c}$ (48 %), $C_{18:0}$ (1 %), and $C_{18:1\omega9}$ (17 %). The DNA G+C content (mol%) is 45.0. The type strain is KoTa2$^{T}$ (=ATCC 35945$^{T}$, = CCUG 37395$^{T}$, = CIP 104580$^{T}$, = DSM 2381$^{T}$) and was isolated anoxic digester sludge.

*Trichococcus palustris* Liu, Tanner, Schumann, Weiss, McKenzie, Janssen, Seviour, Lawson, Allen, and Seviour 2002, 1125$^{VP}$ (*Ruminococcus palustris* Zhilina et al. 1995, p. 577). (pa.lu'stris. L. adj. *palustris* swamp-inhabiting *Ruminococcus*). Cells are Gram-stain-positive, cocci, or elongated cocci with slightly tapered ends (0.75–1.3 × 0.7–1.0 μm). They appear singly, paired, or short chains joined by a mucous capsule and do not form spores. Facultatively anaerobic and nonmotile. Optimum growth occurs at pH 7.5, 0.1 % NaCl (w/v), and at 30 °C, but ranges from pH 6.2 to 8.4 and temperatures from 0 °C to 33 °C. Can utilize cellobiose, fructose, glucose, *N*, *N*-acetylglucosamine lactose, maltose, mannitol, mannose, pyruvate, raffinose, sorbitol, and sucrose for growth. Cannot utilize adonite, arabinose, betaine, Casamino acids, cellulose (microcrystalline), choline, dulcitol, erythritol, fucose, fumarate, galactose, glucosamine, glutamate, glutamine, glycerol, glycine, glycogen, histidine, inositol, lactate, malate, melibiose, methionine, mono-, trimethylamine, peptone, rhamnose, ribose, serine, sorbose, starch, succinate, trehalose, xylose, formate, acetate, propionate, butyrate, methanol, ethanol, propanol, butanol, and $H_2 + CO_2$. Urease-negative, and nitrate is not reduced. Predominant fatty acids are $C_{14:0}$ (21 %), $C_{16:0}$ (15 %), $C_{16:1\omega9c}$ (20 %), $C_{18:0}$ (4 %), and $C_{18:1\omega9c}$ (22 %). The DNA G+C content (mol%) is 47.5. The type strain is Z-7189$^{T}$ (CIP 105359$^{T}$, = DSM 9172$^{T}$), and it was isolated from a swamp.

*Trichococcus collinsii* Liu, Tanner, Schumann, Weiss, McKenzie, Janssen, Seviour, Lawson, Allen, and Seviour 2002, 1124$^{VP}$ emend. Pikuta et al. (2006, p. 2060). (*Trichococcus collinsii* (coll.ins'i.i. N.L. gen. n. *collinsii* referring to Matthew D. Collins, a contemporary English microbiologist who contributed significantly to our understanding of the lactic acid bacteria)).

Cells are facultative cocci that appear singly or paired with some tapering of cells. Small chains form in R2A broth. Facultatively anaerobic and nonmotile. Optimum growth occurs at pH 7.5, 0.1 % NaCl (w/v), and at 25–30 °C, but ranges from pH 6.0 to 9.0 and −5 °C to 36 °C. Cells can utilize and grow on allantoin, citrate, malate, and tartrate; produces acid from mannitol. Only strain 37AN3$^{*T}$ produces acid from adonitol, inositol, raffinose, and sorbitol. Urease-negative, and nitrate is not reduced. The DNA G+C content (mol %) is 47.0. Predominant fatty acids consist of $C_{12:0}$ (6–12 %), $C_{14:0}$ (46–57 %) $C_{16:0}$ (14–18 %), and $C_{16:1}$ (18–27 %). The type strain is 37AN3$^{*T}$ (=ATCC BAA-296$^{T}$, = DSM 14526$^{T}$). Two strains (37AN3$^{*T}$ and 45AN2) were isolated from hydrocarbon-contaminated soil.

*Trichococcus patagoniensis* Pikuta, Hoover, Bej, Marsic, Whitman, Krader and Tang 2006, 2060$^{VP}$. (pa.ta.go.ni.en'sis. N.L. masc. adj. *patagoniensis* pertaining to Patagonia, the region of South America where the sample for the type strain was collected).

Cells can be differentiated from other species within *Trichococcus* by motility (➋ *Table 4.7*). Cells stain Gram-stain-variable and are motile, nonspore-forming, cocci (1.3–2.0 μm) that appear from spherical, ovoid to olive-shaped and occur singly, in pairs, chains, or irregular conglomerates. Colonies grown at 4 °C are circular or convex (0.5–4.0 mm) white and mucoid to slimy. Facultatively anaerobic and catalase-negative. Optimum growth occurs at pH 8.5, 0.5 % NaCl (w/v), and at 28–30 °C, but ranges from pH 6.0 to 10.0, 0–6.5 % NaCl, and −5 °C to 35 °C. *T. patagoniensis* can utilize arabinose, citrate, glucose, fructose, maltose, mannitol, mannose, pyruvate, ribose, sucrose, and starch for growth. It could not utilize acetate, acetone, betaine, butyrate, Casamino acids, ethanol, formate, glycerol trimethylamine, lactate, lactose, methanol, pectin, peptone, propionate, trehalose, triethylamine, Bacto tryptone, and yeast extract. Metabolic end products of glucose fermentation are lactate, formate, acetate, ethanol, and $CO_2$. Growth is inhibited by ampicillin, chloramphenicol, gentamicin, kanamycin, rifampicin, and tetracycline. The major fatty acids are $C_{14:0}$ (11 %), $C_{16:0}$ (16 %), $C_{16:1\omega7c}$ (43 %), and $C_{18:1\omega9c}$ (22 %). The DNA G +C content (mol %) is 45.82. The type strain is PmagG1$^{T}$ (ATCC BAA-756$^{T}$, = CIP 108035$^{T}$, = JCM 12176$^{T}$), and it was isolated from guano of Magellanic penguins (*Spheniscus magellanicus*) in south Chilean Patagonia (Pikuta et al. 2006).

## Other Strains

"*Trichococcus* strain ES5" was isolated for its ability to produce 1,3-propanediol from glycerol. As shown by almost all strains of *Trichococcus*, strain ES5 had a high 16S rRNA gene sequence similarity (99 %) with its phylogenetically closest relative *T. flocculiformis* DSM 2094$^T$ (van Gelder et al. 2012). Strain ES5 is coccus-shaped and grew singly, in pairs, or short chains of four cells or less (distinguishing it from *T. flocculiformis* which grows in long chains). Oxidase and catalase-negative like other *Trichococcus* species. Oxygen tolerant (no growth above 1 % $O_2$ in gas phase) but grows best in sulfide-reduced medium. Requires ammonium as nitrogen source and must have vitamin $B_{12}$ (cobalamin) supplemented in media to grow, but has no other vitamin requirement. Does not require yeast extract but grows better when it is supplemented in the medium (0.2 g/l). Optimum growth occurs between pH values 6.5–9.0, 0.03–0.15 % NaCl (w/v), and at 35 °C and ranges from pH 6.5 to 9.0 (does not grow below pH 6.5), 0.03 to 3.4 % NaCl, and 2 to 40 °C. Utilizes cellobiose, fructose, glucose, glycerol, maltose, mannitol, mannose, pyruvate, sucrose, and xylose for growth. Cannot grow on arabinose, citrate, lactate, malate, methanol, sorbitol, xylitol, or olive oil. Does not need yeast extract for growth. Products formed from glucose (or other sugars) are lactate, formate, acetate, and ethanol; products from glycerol are 1,3-propanediol, lactate, formate, and acetate. Utilization of glycerol separates it from all other *Trichococcus* species. The type strain would be strain ES5 (=DSM 23957), and the GenBank accession number is HM773034.

## Isolation, Enrichment, and Maintenance Procedures

*Carnobacteriaceae* are recovered from three main sources, namely, human and animal sources, food products, and aquatic environmental sources (Hammes and Hertel 2006). Carnobacteria and lactobacilli share many phenotypic and nutritional requirement properties and thus may be found in the same habitats and thus are often co-isolated. Due to these similar properties, no true selective media are available. Although a selective medium can be used for the isolation of *C. divergens* and *C. maltaromicus* (formally *C. piscicola*), no selective media exist for *Carnobacterium* as a whole; however, a number of media modifications can be made to increase the probability of isolating these organisms. The type genus of *Carnobacteriaceae*, *Carnobacterium*, was created to encompass two misclassified *Lactobacillus* strains (*L. divergens* and *L. piscicola*) and two novel strains (*C. mobile* and *C. gallinarum*) recovered from poultry meat maintained at low temperatures (Collins et al. 1987). Although *Carnobacterium* species of human origin or associated with foods have been shown to be similar to lactobacilli in their nutritional requirements, they are not aciduric and either do not grow or do not readily grow on acetate. Indeed, by increasing the pH to 8.5, the omission of acetate and the substitution of sucrose for glucose from MRS medium have been shown to favor the growth of carnobacteria while simultaneously suppressing the growth of lactobacilli that are found in association with carnobacteria in food (Hammes et al. 2002). However, it is pertinent to note that *C. funditum* and *C. alterfunditum* do not grow on MRS without acetate. For *C. maltaromicus* (*C. piscicola*) isolates from a diseased fish, a requirement for folic acid, riboflavin, pantothenate, and niacin but not for vitamin $B_{12}$, biotin, thiamine, or pyridoxal (Hiu et al. 1984) has been demonstrated. *C. divergens* and *C. maltaromicus* (*C. piscicola*) isolates from meat do not require thiamine (De Bruyn et al. 1987). Selective Cresol Red Thallium Acetate Sucrose (CTAS) agar is recognized as the only selective medium used for the isolation and enumeration of carnobacteria (WPCM 1989). *C. divergens* and *C. maltaromicus* (*C. piscicola*) [peptone from casein, 10.0 g; yeast extract, 10.0 g; sucrose, 20.0 g; Tween 80, 1.0 g; sodium citrate, 15.0 g; $MnSO_4 \cdot {}_4H_2O$, 4.0 g $K_2HPO_4$, 2.0 g; thallium acetate, 1.0 g; nalidixic acid, 0.04 g; cresol red, 0.004 g; triphenyl tetrazolium chloride, 0.01 g; agar, 15.0 g; water 1 l]. Two percent inulin can be substituted for sucrose to distinguish *C. maltaromicus* (*C. piscicola*) and *enterococci*. Wasney et al. (2001) has shown that with modifications to this medium, a wider variety of *Carnobacterium* spp. can be recovered. For isolation from a diseased fish, brain-heart infusion or tryptic soy agar has been used (Hiu et al. 1984), and for isolation from vacuum-packaged meat, standard-I-agar (Merck) (pH 7.2–7.5), CASO medium (Merck), or tryptic soy agar (Difco) with 0.3 % added yeast extract and aerobic incubation at 25 °C for 3 d or 7 °C for 10 d is recommended (Hammes and Hertel 2006).

*C. funditum* and *C. alterfunditum* were isolated from the anaerobic waters of a meromictic lake in Antarctica. Although isolated on "trypticase 141" which includes trypticase, yeast extract, and fructose of the carbohydrate (Franzmann et al. 1991), the organism can be grown and maintained on a simpler medium TSBY salt medium [DSM 466: trypticase soy broth, 30.00 g; yeast extract, 3.00 g; NaCl, 13.00 g; KCl, 0.34 g; $MgCl_2 \cdot 6H_2O$, 4.00 g; $MgSO_4 \cdot 7H_2O$, 3.45 g; $NH_4Cl$, 0.25 g; $CaCl_2 \cdot 2H_2O$, 0.14 g; distilled water 1 l, pH to 7.2]. The optimum temperature and pH are 22–23 °C and 7.0–7.4, respectively. For optimal growth, *Carnobacterium funditum* and *Carnobacterium alterfunditum* require 1.7 % and 0.6 % NaCl, respectively (Franzmann et al. 1991). *C. pleistocenium* was isolated from permafrost tunnel in Fox, Alaska, and requires mineral salts, yeast extract, and peptone. Although it has not been determined if they are essential, vitamins are normally added to the medium. Enrichment cultures were obtained by using standard anaerobic technique and with a medium containing ($g\ L^{-1}$): NaCl, 10.0; KCl, 0.3; $KH_2PO_4$, 0.3; $MgSO_4 \cdot 7H_2O$, 0.1; $NH_4Cl$, 1.0; $CASO_4 \cdot 7H_2O$, 0.0125; $NaHCO_3$, 0.4; $Na_2S \cdot 9H_2O$, 0.4; resazurin, 0.0001; yeast extract, 0.2; peptone, 3.0; 2 ml vitamin solution (Wolin et al. 1963); and 1 ml trace mineral solution (Whitman et al. 1982). The final pH was 7.14–7.2. Pure cultures were obtained by the dilution method in Hungate tubes on medium with a modified NaCl concentration of 0.5 % (w/v) using the roll-tube method on 3 % (w/v) agar medium (Pikuta et al. 2005). *Carnobacterium iners* is also another species isolated from a cold water aquatic source. Specifically, the organism was recovered from a microbial mat sample originating from the littoral zone (situated under 15 cm of clear ice and 15 cm of water) of a continental Antarctic lake (Forlidas Pond) in the Pensacola Mountains, Antarctic. The organism was also grown aerobically on trypticase soy broth with yeast extract salt medium [TSBY salt; 3 % trypticase soy broth (Oxoid), 0.3 % yeast extract (Oxoid), 1.3 % NaCl (Merck), 0.034 % KCl (UCB), 0.4 % $MgCl_2$ .$6H_2O$ (Sigma-Aldrich), 0.345 %; $MgSO_4.7H_2O$ (UCB), 0.025 % $NH_4$ Cl (Merck), 0.014 % $CaCl_2.2H_2O$ (Merck), 2 % agar no. 2 (LabM), pH 7.2] at 4 °C for 10 days and 20 °C for 4 days, respectively.

*C. inhibens* (Jöborn et al. 1999) was isolated from a 3-year-old Atlantic salmon (*Salmo salar*) during screening for fish intestinal bacteria with the ability to secrete substances that suppressed *Vibrio anguillarum* and *Aeromonas salmonicida* (Jöborn et al. 1999). Intestinal samples were aseptically collected and incubated for 1 h at 15 °C in tryptic soy broth plus 2 % (w/v) NaCl before being exposed to $10^{-7}$ *V. anguillarum* and incubated for a further 18 h. The sample was then plated on tryptic soy agar plus 2 % (w/v) NaCl and incubated for 24 h. Colonies were picked and tested for inhibitory activity again *V. anguillarum* and *A. salmonicida*, inhibitory bacteria were picked and subjected to a preliminary characterization that demonstrated the organisms similarity to members of the *Carnobacterium* genus. A selective medium was developed by the addition of nalidixic acid (80 μg $ml^{-1}$) to TSA plus 2 % (w/v) NaCl.

*C. jeotgali* (Kim et al. 2009) was isolated from a Koren traditional fermented food called "toha jeotgal" made from freshwater shrimp meat and salt. The organism was isolated on TSA with no additional information on selective media.

As mentioned previously, *Carnobacterium maltaromaticum* was described by Mora et al. (2003) that demonstrated that *L. maltaromicus* (Miller et al. 1974) and *C. piscicola* (Collins et al. 1987) were in fact members of the same species and should be considered synonyms. *L. maltaromicus* was recovered from producers' milk criticized as possessing a malty flavor. Samples were subcultured onto trypticase soy agar (TSA, BBL) and the plates incubated at 25 °C for 5 d. Morphological and physiological methods demonstrated that the isolates belonged to a novel species of *Lactobacillus*. *Lactobacillus piscicola* was isolated in 1970 from a diseased adult cutthroat trout (*Salmo clarki*) reared at Bandon Trout Hatchery in Coos County, Oregon. The organism was isolated from the fish by streaking kidney tissue onto brain-heart infusion agar or tryptic soy agar. *Carnobacterium mobile* was originally isolated from minced chicken meat, packed anaerobically, and irradiated at room temperature and in the frozen state with a wide range of doses of 4 MeV cathode rays (Thornley 1957; Thornley and Sharp 1959). The meat samples were streaked onto a wide range of media and colonies picked for further investigation (see Thornley (1957) for a complete list of media used). These authors described these as atypical lactobacilli and were identified as *C. mobile* by Collins et al. (1987) as members of the newly described genus *Carnobacterium* in the same study. *Carnobacterium viridans* strain MPL-11$^T$was found to be responsible for causing green discoloration of the commercially prepared vacuum-packaged bologna sausage after opening. A sample of meat (10 g) from one aseptically opened pack showing green discoloration was homogenized (Stomacher), diluted in 0.1 % peptone, and plated on an all-purpose Tween (APT, Difco), M5 (Zúñiga et al. 1993), and MRS agars (De Man et al. 1960). Plates were incubated anaerobically at 25 °C or 30 °C for 48 h (BBL Gaspak); colonies were picked and subjected to morphological, phenotypic, and phylogenetic analyses.

The genus *Alkalibacterium* was founded on the isolation of three strains of bacteria isolated from the alkaline wash waters (pH 10.9) used to produce edible olives (Ntougias and Russell 2001). Isolation was performed using a dilution agar technique with 50 % diluted edible olive wash water with 2 % agar. Growth and maintenance of the strains utilized a GYEC (glutamate/yeast extract/carbonate) medium containing L-glutamate (0.05 M), yeast extract (0.5 %, w/v), and one of two buffers (0.1 M $Na_2CO_3$/1 mM $K_2HPO_4$, pH 10.5, or 0.1 M $NaHCO_3$/1 mM $K_2HPO_4$, pH 9) also containing $NH_4SO_4$ (0.1 %, w/v) and $MgSO_4$ (0.1 mM) (Quirk et al. 1991).

*Alkalibacterium psychrotolerans*, three strains of *A. iburiense*, and three strains of *A. indicireducens* were all enriched on indigo-containing (0.01 %) PYA (peptone-yeast extract-alkaline) broth at 27 °C as described (Yumoto et al. 2004, 2008; Nakajima et al. 2005). The PYA medium used consists of peptone (8 g), yeast extract (3 g), $K_2HPO_4$ (1 g), EDTA (3.5 mg), $ZnSO_4 \cdot 7H_2O$ (3 mg), $FeSO_4 \cdot 7H_2O$ (10 mg), MnSO4·$H_2O$ (2 mg), $CuSO_4 \cdot 5H_2O$ (1 mg), $Co(NO_3)_2 \cdot 6H_2O$ (2 mg), and $H_3BO_3$ (1 mg) in 1 L of $NaHCO_3/Na_2CO_3$ buffer (100 mM in deionized $H_2O$ at pH 10). Enrichments of 5 % fermentation liquor (fermented polygonum indigo (*Polygonum tinctorium* Lour.)) in 100 ml of medium were incubated at 27 °C and monitored for reduction of indigo. All enrichments went through five transfers with indigo-containing medium before isolation was attempted. Isolation and maintenance was achieved using reinforced clostridial agar (RCA, Sigma) containing $NaHCO_3/Na_2CO_3$ buffer at pH 10 and incubated in argon-flushed jars at 27 °C. Five other species of *Alkalibacterium*, *A. thalassium*, *A. pelagium*, *A. putridalgicola*, *A. kapii*, and *A. subtropicum* were enriched on a 7 % NaCl GYPF (glucose-yeast extract-peptone-fish extract) broth and isolated on 7 % NaCl GYPF agar (1.3 % agar) as described with the following exceptions: pH was adjusted to 9.0 by the addition of 6.1 g $Na_2CO_3$ $l^{-1}$ and 8.9 g $NaHCO_3$ $l^{-1}$ (Ishikawa et al. 2003, 2011). Enrichments and isolation were carried out anaerobically at 30 °C.

Nine strains of "*Alkalibacterium gilvum*" were isolated using a saline (7 % NaCl) and alkaline (pH 9.5) media (Ishikawa et al. 2013). The isolations occurred through either direct pour plates or enrichment, with subsequent pour-plating using GYPF (glucose-yeast extract-peptone-fish extract) at pH 9.5 and 7 % NaCl (Ishikawa et al. 2003). Cultivation and maintenance of the strains were done on 2.5 % NaCl GYPF (pH 9.0) agar or broths at 30 °C.

*Allofustis seminis* (Collins et al. 2003) was isolated from stored pig semen submitted by private artificial insemination centers located in Canada. The organism was isolated from mixed cultures containing *Klebsiella* spp., *Serratia* spp., and *Pseudomonas* spp. using repeated subculturing on Columbia agar with 5 % horse blood at 37 °C under anaerobic conditions. No other details are available.

*Alloiococcus otitis* (Aguirre and Collins 1992b) was isolated from tympanocentesis fluid collected within the inner ear of children suffering from chronic otitis media (Faden and Dryja 1989). Original isolation was performed on a variety of media including trypticase soy agar (TSA) with 5 % sheep blood, phenylethanol agar with 5 % sheep blood, and chocolate blood agar incubated at 37 °C with 5 % $CO_2$. MacConkey agar and brain-heart infusion (BHI) broth plates were also used but incubated at 37 °C with no $CO_2$. Maintenance of the organisms was carried out on blood agar plates or Todd-Hewitt broth with 5 % horse serum at 37 °C (Aguirre and Collins 1992b). An updated version for maintenance and growth of *Alloiococcus* sp. includes BHI broth supplemented with 0.07 % lecithin and 0.5 % Tween 80 (BHIS broth) or with BHI agar with 5 % rabbit blood at 37 °C (Collins 2009). Additionally, it was suggested that Todd-Hewitt broths and TSA with 5 % sheep blood are not sufficient to grow all strains of *Alloiococcus* species.

*Atopobacter* (Lawson et al. 2000) was isolated from deceased common seals (*Phoca vitulina*). One strain was isolated following a postmortem examination of a common seal pup that was described as having a general lymphadenopathy and acutely congested lungs with pulmonary hemorrhaging. The strain was recovered from the small intestine and from the mesenteric, external iliac, prescapular, and pancreatic lymph nodes. A second strain was recovered from the liver, spleen, and blood from an adult animal that had died from gunshot wounds. Both strains were isolated using repeated subculturing on Columbia agar with 5 % horse blood at 37 °C in air with % $CO_2$. No other details were provided on enrichment or selective media for this species provided.

*Atopococcus tabaci* (Collins et al. 2005) was isolated from moist snuff tobacco from an unnamed commercial source. The single strain was recovered using Columbia agar with 5 % horse blood at 37 °C under anaerobic conditions. No other details are available.

*Atopostipes suicloacale* was isolated from a swine manure storage pit by anaerobic serial dilution of the sample and then plating on agar medium containing 40 % clarified rumen fluid (Cotta et al. 2003). General growth/maintenance for *A. suicloacale* used routine growth medium (RGM)-glucose agar plates containing rumen fluid (Hespell et al. 1987) at 24 °C.

Six strains of *Bavariicoccus seileri* were isolated from the surface and smear water of German red smear soft cheese by aerobic incubation on Plate Count Agar with 3 % NaCl for 72 h at 30 °C. While all isolates grew well on common commercial media for lactic acid bacteria (e.g., APT or TSA), fastest growth was achieved using M17 medium (Elliker et al. 1956; Terzaghi and Sandine 1975) with 2 % glucose without shaking at 30 °C.

*Desemzia incerta* (Stackebrandt et al. 1999) was originally isolated by Steinhaus (1941) from the ovaries of lyreman cicada (*Tibicen linnei*) using blood and chocolate agar as it did not grow well on nutrient agar. The ovaries were placed either directly on media in Petri dishes or into tubes of sterile saline to be streaked out on nutrient agar, glucose agar, North's gelatin chocolate agar (defibrinated blood added to medium at 80 °C), and blood agar and cultivated at 37 °C (for detailed information on the isolation of the organism from the insect, see Stackebrandt 2009). After initial cultivation, little growth was seen on nutrient agar, and the temperature was reduced to room temperature to better match the temperature of the insect. Grows well on Columbia blood agar at room temperature and 37 °C after 24 h. For broth cultures, CASO medium (Oxoid, no. CM 129) and in PYG medium after 48 h.

The clinically important organism *Dolosigranulum pigrum* is a slow-growing organism and was cultured in Todd-Hewitt broth (Oxoid Ltd., UK) at 37 °C when it was received in the Collins lab (Aguirre et al. 1993). It can be maintained on rich, blood-based agars which include Columbia agar with 5 % (v/v) horse blood, brain-heart infusion (BHI) with 5 % rabbit blood, or *Brucella* agar with 5 % sheep blood. Maintenance can be kept with BHI broth or Todd-Hewitt broth at 37 °C.

The three valid species that comprise the genus *Granulicatella* can all be isolated on blood agars at 37 °C anaerobically or aerobically with $CO_2$ atmosphere (3–5 %). Originally isolated as nutritionally variant streptococcus (NVS), *G. adiacens* and *Abiotrophia defectiva* were growth dependent on sulfhydryl-supplemented media or grown as satellite colonies around other bacteria (Frenkel and Hirsch 1961). *G. elegans* requires L-cysteine for growth which cannot be substituted by pyridoxal. Maintenance of all strains can be achieved with rich media (e.g., BHI, Columbia, Schaedler) supplemented with 5 % blood at 37 °C. Use of Todd-Hewitt or brain-heart infusion broths supplemented with 10 mg pyridoxal HCl (vitamin $B_6$) or 100 mg cysteine (per L) grown anaerobically at 37 °C overnight is also adequate. However, *G. balaenopterae* does not require the addition of pyridoxal or L-cysteine.

*Isobaculum melis* was isolated from the small intestine of badger (deceased) so its natural habitat remains unknown (Collins et al. 2002). It was isolated aerobically on Columbia agar supplemented with 5 % defibrinated horse blood at 37 °C. Maintenance of this strain can be carried out with peptone-yeast extract-glucose medium (PYG) at 37 °C.

*Lacticigenium naphtae* (Iino et al. 2009) was recovered from crude oil collected from an oil-water separation tank of an oil-water-extracting well in Akita Prefecture, Japan. The sample was kept anaerobic, and 0.5 ml was used to inoculate 20 ml HSm medium in a vial sealed with a tight-fitting butyl rubber stopper. HSm medium was composed of ($l^{-1}$): 0.355 g KCl, 0.14 g $KH_2PO_4$, 0.14 g $CaCl_2.2H_2O$, 0.25 g $NH_4$ Cl, 4.0 g $MgCl_2$ $.6H_2O$, 3.45 g $MgSO_4.7H_2$ O, 18.0 g NaCl, 2.0 mg $Fe(NH_4)_2$ $(SO_4)_2.6H_2O$, 1.0 g sodium acetate, 2.0 g yeast extract (Becton Dickinson), 2.0 g trypticase peptone (BBL), 5.0 g $NaHCO_3$, 10.0 ml trace elements solution (Balch et al. 1979) containing 25.0 mg $NiCl_2.6H_2O$, 2.0 g $(NH4)_2$ $Ni(SO_4)_2.6H_2O$, 0.3 g $Na_2SeO_3.5H_2O$, and 10.0 mg $Na_2WO_4.2H_2O$. Prior to inoculation, the pH of the medium was adjusted to 7.0 with 6 M HCl, dissolved air was removed by flushing with $H_2$ /$CO_2$ (4:1, v/v;

approx. 150 kPa), and 10 ml vitamin solution $l^{-1}$ (Wolin et al. 1963) and 10 ml sterile stock solution $l^{-1}$ containing 0.5 g $Na_2S$/cysteine HCl were added. The enrichment culture was cultivated at 25 °C for 3 weeks and transferred several times to fresh HSm medium. Subsequently, serial decimal dilutions ($10^{-1}$ to $10^{-10}$) of the enrichment culture were made with 2.0 %(w/v) saline, and 0.1 ml of the diluted samples was spread on HSm agar (1.5 %, w/v) plates and cultivated aerobically at 25 °C for 3 weeks. Colonies were picked and grown on HSm agar until pure.

*Marinilactibacillus psychrotolerans* (Ishikawa et al. 2003) was isolated from a variety of samples that included algae, decomposing sponges, crabs, fish, and shellfish. The samples were collected from two separate locations; the first was from Oura Beach, Miura Peninsula, Kanagawa Prefecture, in the middle of the Japanese mainland, a temperate area. 7 % NaCl GYPB isolation broth (10 g glucose, 5 g yeast extract (Oriental Yeast), 5 g polypeptone (Nippon Seiyaku), 5 g beef extract (Difco), 1 g $K_2HPO_4$, 70 g NaCl, 10 g sodium acetate, 0.5 g Tween 80, 10 mg cycloheximide, 10 mg colistin, 15 mg nalidixic acid, 20 mg monofluoroacetic acid, 10 mg sodium azide, 15 g $Na_2CO_3$ (as a buffer), and 10 % (w/v) seawater to 1,000 ml. The final pH was 10.0. The medium was sterilized by filtration through a membrane filter with a 0.2 μm pore size. The plating medium was 7 % NaCl GYPB isolation agar (1.3 % agar) supplemented with 5 g $CaCO_3$. $Na_2CO_3$ and $CaCO_3$ were autoclaved separately at 121 °C for 15 min and added aseptically (final pH, 10.0). The agar medium for overlaying on the pour plates contained 0.1 % (w/v) sodium thioglycolate and the inorganic ingredients of the 7 % NaCl GYPB isolation broth.

A second set of samples were collected from a foreshore site near the Oujima Islet and a fish market in the city of Naha, both in Okinawa in the southern most part of Japan, a subtropical area. Two sets of medium were used for enrichment. One medium, the 7 % NaCl GYPF isolation broth, was the same as the 7 % NaCl GYPB isolation broth but with Extract Bonito instead of beef extract, with sodium acetate and Tween 80 omitted, and with 5 ml of a salts solution added, and prepared with distilled water to 1,000 ml, pH 7.5. The medium was sterilized by filtration. The salt solution was composed of ($ml^{-1}$): 40 mg $MgSO4.7H_2O$, 2 mg $MnSO_4.4H_2O$, and 2 mg $FeSO_4.7H_2O$ (Okada et al. 1992). For plating, the 7 % NaCl GYPF isolation agar (1.3 % agar) supplemented with 5 g $CaCO^3$ $l^{-1}$ was used. The overlaying agar medium contained 0.1 % (w/v) sodium thioglycolate and the inorganic ingredients of the 7 % NaCl GYPF isolation broth. The other media for the enrichment were 12 and 18 % (for a subsequent second enrichment) NaCl GYPFSK isolation broths (10 g glucose, 5 g yeast extract, 5 g polypeptone, 5 g Extract Bonito, 50 ml soy sauce, 10 g $K_2HPO_4$, 110/170 g NaCl, 1 g sodium thioglycolate, 5 ml salts solution, 10 mg cycloheximide, and distilled water to 1,000 ml). The medium was adjusted to pH 7.5 and autoclaved at 110 °C for 10 min. For plating, the 12 % NaCl GYPFSK isolation agar (2.0 % agar, final pH 7.5) supplemented with 5 g $CaCO_3$ $l^{-1}$was used.

For enrichment with the 7 % NaCl GYPB isolation broth (Miura Peninsula samples) or the 7 % NaCl GYPF isolation broth (Okinawa samples), small pieces of the samples (intestinal contents or whole bodies for animal samples) were soaked in 5 ml enrichment medium immediately after collection. After incubation at 30 °C for 3 days, a portion of the enrichment culture broth, whose pH had decreased to below 7.0, was placed into fresh medium and incubated anaerobically at 30 °C for 2 days. A portion of the second enrichment culture broth was poured and the plate overlaid with the overlaying agar medium. In another isolation series of the Okinawa samples, the 12 % NaCl GYPFSK isolation broth was used for a first enrichment incubated for 21 days, and the 18 % NaCl GYPFSK isolation broth was used for a second enrichment incubated for 15 days, both at 25 °C in standing culture. Pour-plated agar media were not overlaid. Lenticular colonies were picked up, and the isolates were purified with repeated plating. Six isolates were recovered, and M13-$2^T$ obtained from a living sponge from the Miura Peninsula was designated as the type strain. One isolate was from a raw Japanese ivory shell (*Babylonia japonica*) cultured with the 7 % NaCl GYPF isolation broth enrichment, and another isolate was from a decomposing alga cultured with the 12–18 % NaCl GYPFSK isolation broth enrichment.

*M. piezotolerans* was recovered by Toffin et al. (2005) from a sediment core collected at 4.15 m below the seafloor from a water depth of 4790.7 m in the Pacific Ocean at Nankai Trough, off the coast of Japan. Samples were enriched anaerobically in 50 ml vials containing 10 ml medium MM, which consisted of the following ($l^{-1}$ distilled water): 23 g NaCl, 3 g $MgCl_2.6H_2O$, 4 g $Na_2$ $SO_4$ , 0.7 g KCl, 0.15 g $CaCl_2$, 0.5 g $NH_4Cl$, 0.27 g $KH_2$ $PO_4$, 15 ml 1 M $NaHCO_3$, 0.1 g yeast extract (Difco), 1 ml trace elements (Widdel and Bak 1992), 1 ml vitamin solution (Widdel and Bak 1992), 1 ml thiamine (0.01 %, w/v; Widdel and Bak 1992), 1 ml vitamin B12 (0.005 %, w/v; Widdel and Bak 1992), 1.0 g sodium acetate, 2.0 g monomethylamine, 5.0 g sodium formate, 0.5 % (v/v) methanol, and 0.5 mg resazurin. The pH of the medium was adjusted to 7.2 at room temperature before autoclaving. Sterile medium was reduced by adding 0.5 g sodium sulfide $l^{-1}$ and then distributed into serum vials before inoculation.

Cultures were incubated at 25 °C in the dark. Positive enrichments were subcultured into the same medium under anaerobic conditions. Subsequent enrichment cultures were grown on agar plates (1 %, w/v; Difco), incubated anaerobically at 25 °C. Colonies were picked subjected to dilution to extinction technique, followed by repeated streaking onto plates until pure. Optimal growth conditions are obtained on YPG medium ($l^{-1}$ distilled water: 25 g NaCl, 3 g $MgCl_2$, 0.5 g KCl, 4 g $Na_2SO_4$, 2 g glucose, 5 g yeast extract (Difco), 5 g peptone (Difco), 34.6 g PIPES, and 0.05 g $KH_2PO_4$, 37 °C, pH 7.0, and 20 g NaCl $l^{-1}$).

Two strains (C01$^T$, C02) of the Gram-positive, catalase-negative cocci of *Pisciglobus halotolerans* were isolated using the pour-plate technique with MRS agar (de Man et al. 1960) containing 5 % NaCl at 30 °C for 5 days. Although *P. halotolerans* can tolerate 10 % NaCl, maintenance of *P. halotolerans* used MRS broth with 0–5 % NaCl at 30–37 °C at circum neutral pH.

Isolation of the first species of *Trichococcus*, *T. flocculiformis*, was performed by adding one drop of bulking sludge material onto UA plate media and incubated at 25 °C (Scheff et al. 1984). UA medium consists of (per liter): 1 g peptone, glucose, 1 g urea 20 g, 2 g $KH_2PO_4$, and 0.012 g phenol red. Maintenance and propagation was performed on M 69 medium: 10 g Bacto tryptone, 3 g glucose (filter-sterilized), 1.1 g $MgCl_2.6\ H_20$, and 9 g $Na_2SO4.10\ H_2O$ at 30 °C. Addition of 12 g agar per liter of medium was used for all platings. *Trichococcus pasteurii* was isolated for its ability to degrade tartrate (2,3-dihydroxybutanedioic acid). Enrichment cultures for the propagation of tartrate utilizing organisms used a basal freshwater mineral media described previously (Widdel and Pfennig 1981; Schink and Pfennig 1982). *T. pasteurii* was isolated utilizing the agar shake culture method (Pfennig 1978) with the same medium, using L-tartrate as the carbon source (*T. pasteurii* cannot utilize D- or *m*-tartrate). The medium was adjusted to a pH of 7.2–7.3, and all incubations were carried out at 28 °C.

*Trichococcus palustris* was isolated from a swamp covered with a 30 cm layer of ice, topped with 50 cm of snow near the vicinity of Moscow, Russia (Zhilina et al. 1995). Enrichment was achieved using Pfennig's medium supplemented with 0.2 % yeast extract and Wolin's vitamins supplemented with glucose (5 g/l) as the carbon source. All work was performed under strictly anaerobic conditions, the media was reduced with sulfide ($Na_2S \times 9H_2O$, 0.5 g/l), the pH adjusted to 7.0, and incubations were carried out at 6 °C. *T. palustris* was isolated by extinction dilution series followed by the roll-tube method (Hungate 1969) in the same medium with glucose as the substrate. Maintenance of the organism was performed with the same media, but will grow in trypticase soy yeast extract medium at pH of 7.0–7.2 at 30 °C.

*T. collinsii* was enriched from soil using half-strength tryptic soy broth (TSB) at 26 °C. Maintenance was maintained on the same or R2A broth or agar (Reasoner and Geldreich 1985). *Trichococcus patagoniensis* was enriched by placing a homogenized sample into a strictly anaerobic, minimal salts medium plus vitamins (pH 7.8) containing glucose as carbon source and incubated at 4 °C (Pikuta et al. 2006). Isolation was achieved using anaerobic roll tubes (3 % agar) with the same medium at 4 °C. *Trichococcus* strain ES5 was enriched from methanogenic sludge in a bicarbonate-buffered medium with glycerol as the carbon source and the addition of bromoethanesulfonic acid (BESA, 20 mM) to inhibit methanogens as described previously (Stams et al. 1993). Isolation was achieved by serial dilution of the enrichment cultures in the same medium at 30 °C.

## Maintenance

Members of the family can be maintained short-term by stab inoculation in the appropriate medium for the species in screw-capped vials and kept at 4 °C. The nonaciduric nature of the carnobacteria should be considered, and selecting APT, CASO, or D-MRS media in the pH range 7.0–8.5 improves viability. Stab cultures in D-MRS agar (pH 8.0–8.5) should be kept at 1–4 °C and transferred every 2–3 weeks. Addition of 5 % calcium carbonate to cultivation broth may serve to protect vitality of stock cultures over several weeks at 1–4 °C. Lyophilization used for lactobacilli gives satisfactory results. Superior results are obtained by cryopreservation at −80 °C using glycerol peptone protective broth for suspending late logarithmic cells, harvested by centrifugation, or by rinsing surface growth from agar media. Borch and Molin (1988) recommended the storage of strains as dense cultures in APT broth at −20 °C.

For medium term maintenance, 20 % (v/v) glycerol suspensions in appropriate medium at −20 °C or at −80 °C is recommended. Long-term preservation is by lyophilization or in liquid nitrogen; cryoprotective agents (milk solids, lactose, or horse serum) should be added to the cell suspension, and ampoules sealed under high vacuum and stored at 8–12 °C (Hammes and Hertel 2006).

## Ecology

### Habitat

As previously stated *Carnobacteriaceae* are recovered from three main sources, namely, human and animal sources, foodstuffs, and aquatic environmental sources (Hammes and Hertel 2006). Members of the *Carnobacteriaceae* that have been most often associated with clinical infections or disease have all been associated as members of the normal human microbiome, with exception to the rare cases of species of *Carnobacterium* (see Pathogenicity, Clinical Relevance). However, many of the genera are circumscribed upon a single strain from a single source, and therefore final conclusions on their habitat and ecological roles must be approached with caution.

*Carnobacterium* species have been recovered from geographically and ecologically diverse sources but appear to have adapted to cold temperatures. Six species (*C. divergens*, *C. gallinarum*, *C. jeotgali*, *C. maltaromaticum*, *C. mobile*, and *C. viridians*) were isolated from refrigerated vacuum-packaged meat or fish and dairy products. However, only *C. divergens* and *C. maltaromaticum* are frequently isolated from fish, meat, and some dairy products (Ringø and Holzapfe 2000; Leisner et al. 2007; Afzal et al. 2010). In a study of the spoilage activity of *Carnobacterium* spp., (Casaburi et al. 2011) reported that although these organisms can often be isolated from food products, evidence suggests that they have little activity in spoilage processes (Casaburi et al. 2011). Indeed reports suggest that *Carnobacterium* spp. may have a protective affect (see section "➋ Food Preservation or Probiotics"). *C. alterfunditum*, *C. funditum*, and *C. iners* were isolated from Antarctica lakes or ponds with *C. pleistocenium* recovered from Alaskan permafrost. Franzmann and coworkers suggested that *C. alterfunditum* and *C. funditum Carnobacterium* spp. play a role in the initial production of a reduced environment (Franzmann et al. 1991). In addition, a number of novel strains have been isolated from tissue samples of a 1-month-old baby mammoth found in Yamal Peninsula in Northwestern Siberia (Pikuta et al. 2011).

*Alkalibacterium* species have isolated from numerous alkaline environments such as soda lakes, edible olive oil-water, indigo production, marine alga, or sea grass, as well as food products like fermented shrimp paste (ka-pi), salted fish, raw fish, and Brie cheese. *Alkalibacterium* species, particularly *A. indicireducens*, *A. olivoapovliticus*, and *A. psychrotolerans* species, were all observed to be in high abundance in wheat straw pulp mill waste (often called black liquor) (Yang et al. 2010). In the spring and winter where pH values are >10 and the temperature is below 45 °C in black liquor storing pools, *Alkalibacterium* species represented >50 % of clonal library assayed from the pools.

*Allofustis seminis* was described in 2003 by Collins et al. and was recovered from pig semen; to date, no other strains have been reported in the literature.

*Atopococcus tabaci* was described in 2005 by Collins et al. and was recovered from tobacco; to date, no other strains have been reported in the literature.

*Alloiococcus otitis* is considered a member of the human microbiome (Frank et al. 2003) and considered an opportunistic pathogen often associated with chronic otitis media infections. Though not physically isolated, the gene sequence of *A. otitis* (or a close species) has been found associated with the sheep scab mite *Psoroptes ovis*, a mite which causes allergic dermatitis in sheep (Hogg and Lehane 1999).

*Atopostipes suicloacalis* strain PPC79$^T$ was isolated from a swine manure storage pit, but *Atopostipes* species have also been observed in black liquors (wheat straw pulp mill waste) (Yang et al. 2010), within aggregates of fed-batch composting reactor (Watanabe et al. 2008) and within water-miscible metalworking fluids (MWF) and water preparation basis (WPB) from industrial plants around Germany (Lodders and Kaempfer 2012). Several *Carnobacteriaceae* were identified by 16S rRNA gene sequence within the MWF including *Desemzia*, *Trichococcus*, and *Atopostipes*, with *Atopostipes* also found with WPB. This is not the only case where *Trichococcus* and *Atopostipes* species were found coinhabiting the same microbial habitat. In examination of the bovine vagina and uterus microbiome, species of both *Trichococcus* and *Atopostipes* (along with *Lactobacillus sp.*, *Aerococcus sp.*, and *Weissella sp.*) were reported as the most prevalent bacterial species within both areas (Peter et al. 2013).

*Bavariicoccus seileri* strains were isolated from smear waters and surfaces of southern Germany-produced red smear soft cheeses (Schmidt et al. 2009). It has also been isolated and detected by DGGE (denaturing gradient gel electrophoresis) profiling from surface samples of some Danish cheeses as well (Gori et al. 2013).

*Dolosigranulum pigrum* is also part of the normal human microbiota but is most prevalent in the human nares (nostrils) and upper respiratory tract (Ling et al. 2013). Also known as the nutritionally variant streptococci (NVS), *Granulicatella adiacens* and *G. elegans* are both part of the normal human oral microbiota and are known to inhabit the respiratory, gastrointestinal, and urogenital tracts (Ruoff 1991; Dewhirst et al. 2010; Diaz et al. 2012). Both *G. adiacens* and *G. elegans* were originally isolated from blood cultures of patients suffering endocarditis (Frenkel and Hirsch 1961; Roggenkamp et al. 1998b). The third recognized species of *Granulicatella*, *G. balaenopterae*, was isolated from mixed cultures of lung fluid and the spleen, as well as the sole isolate from the liver and kidneys of a deceased minke whale (*Balaenoptera acutorostrata*). *Granulicatella* species also appear to be part of the normal oral flora of canines (Dewhirst et al. 2012).

*Isobaculum melis*'s habitat is not precisely known since it was isolated from the gut contents of a deceased badger (Collins et al. 2002). Another reference to *Isobaculum* sp. was the detection of its 16S rRNA gene sequence in a study describing the differences in bacterial community of women with bacterial vaginosis (BV) versus healthy women (Ling et al. 2010).

*Lacticigenium naphtae* was described in 2009 by Iino et al. and was recovered from crude oil; to date, no other strains or have been reported in the literature.

*Marinilactibacillus psychrotolerans* was originally isolated from a living sponge, decaying marine algae, and raw Japanese ivory shell (Ishikawa et al. 2003). Additional strains have been recovered from soft cheeses (Maoz et al. 2003; Feurer et al. 2004) and from spoiled dry-cured hams and mold-ripened soft cheeses (Rastelli et al. 2005). Using molecular methods, the presence of *M. psychrotolerans* was also detected in deep-sea sediments (Inagaki et al. 2003).

In addition, *Pisciglobus halotolerans* was isolated from fermented fish sauce (Tanasupawat et al. 2011) that makes distinguishing its true habitat very difficult.

Most of the *Trichococcus* species have been isolated from strictly anaerobic environments. *T. flocculiformis* was isolated from bulking sludge from several sewage treatment plants located in West Germany. Similarly, *T. pasteurii* (formerly *Lactosphaera pasteurii*, formerly *Ruminococcus pasteurii*) was isolated from anoxic digester sludge of the municipal sewage plant at Konstanz, West Germany, but its habitat was suggested to be anoxic muds in freshwater lakes and creeks (Schink 1984; Janssen et al. 1995). *T. palustris* was isolated from a swamp water/soil slurry covered with ice and snow just outside of Moscow, Russia (Zhilina et al. 1995). *T. patagoniensis* was isolated from guano collected from Magellanic penguins that live in the southern region of Chilean Patagonia. And *Trichococcus* strain ES5 was enriched from methanogenic granular sludge reactor processing the wastewater stream of a paper mill.

*T. collinsii* was originally isolated from anaerobic, gas-condensate-contaminated soil with a redox potential that was predominantly methanogenic, but has also been recovered in other contaminated sites as well (Ballerstedt et al. 2004; Yoshida et al. 2005).

Recent work has also indicated that *Trichococcus* species are one of the three most dominant bacterial species found within municipal sewer systems of Milwaukee, WI, USA (VandeWalle et al. 2012; Newton et al. 2013). Using pyrosequencing of the V6 region of the 16S rRNA gene sequence, pyrotags representing *Trichococcus* species (along with *Acinetobacter* (16.1 %), *Aeromonas* (9.8 %)) accounted for 7.7 % of the overall microbial community. Seasonal variation of the *Trichococcus* population was observed during the 3-year investigation with increases seen in the colder and spring months and sharp declines in late summer. The variations coincided with increases and decreases of *Acinetobacter* and *Aeromonas* species as well, indicating that the population of the microorganisms remained constant, but the concentration levels changed seasonally. While little 16S rRNA gene sequence variation (>98 %) was observed for the population of *Trichococcus* within this study (with one sequence representing 86 % of all *Trichococcus* species), it should be stated that the validly published *Trichococcus* species isolated to date have 99 % and/or greater 16 s rRNA gene sequence similarity. The presence of *Trichococcus* species within activated sludge (AS) (not the influent/out fluent) of sewage plants was also observed to be temperature dependent. *Trichococcus* species represented between 1.55 % and 5.53 % of the population of AS samples that where in cold weather locations, but represented 0–0.96 % in subtropical environments (Zhang et al. 2012).

## Pathogenicity, Clinical Relevance

As mentioned previously, the *Carnobacteriaceae* are a monophyletic group and are therefore wide ranging in both the habitats they colonize as well as their ability to directly cause disease or embody opportunistic pathogens. Some of the genera and species within the *Carnobacteriaceae* do not cause disease nor have been associated with any mammalian illnesses. Namely, half of the *Carnobacterium* species, all of the *Alkalibacterium* species, all of the *Trichococcus* species, *Allofustis seminis*, *Atopobacter phocae*, *Atopococcus tabaci*, *Atopostipes suicloacalis*, *Bavariicoccus seileri*, *Desemzia incerta*, *Isobaculum melis*, *Lacticigenium naphtae*, *Marinilactibacillus psychrotolerans* or *M. piezotolerans*, and *Pisciglobus halotolerans*. But while these bacteria present little to no clinical relevance to humans, other microorganisms within the *Carnobacteriaceae* have a significant impact to human health.

While there is very little evidence for any of the *Alkalibacterium* species to cause disease, there is reference to the isolation of an *Alkalibacterium* isolate within the tissue/blood samples of a patient infected with *Bartonella* (Cadenas et al. 2007). The *Alkalibacterium* species was identified by partial 16S rRNA gene sequence but was not associated with any of the validly published species described herein. This is also the case for an *Isobaculum* sp. which was detected in the vaginal area of women who were suffering from bacterial vaginosis (BV) (Ling et al. 2010). But, *Isobaculum* species were detected in both women who were diagnosed with BV and BV-negative women and could not be strongly associated with diseased versus non-diseased cases, as well as many other genera and species of microorganisms.

Only a few cases of human infection by species of *Carnobacterium* have been reported in the literature, and they have been attributed to a *Carnobacterium* sp. (strain Y6), *Carnobacterium piscicola* (*C. piscicola* was renamed *Carnobacterium maltaromaticum* in 2003), *Carnobacterium divergens* strains BM4489 and BM4490, and a blood-cultured *Carnobacterium* sp. strain 12266/2009. *Carnobacterium* sp. strain Y6, along with 7 other opportunistic pathogenic bacteria, was isolated from a gangrenous lesion of a young girl that started from a small puncture on her finger (Xu et al. 1997). Strain Y6 had typical physiological characteristics of other *Carnobacterium* species, and at that time, 16S rRNA gene sequence analysis revealed that *Carnobacterium alterfunditum* was the nearest phylogenetic neighbor at 97.2 %. Reanalysis of the sequence of strain Y6 now shows it to be more closely related to the validly published *Carnobacterium inhibens* strain $K1^T$ with 98 % gene sequence similarity, although an even closer association (98.7 %) is observed with *Carnobacterium* strain 17–4, whose complete genome has been sequenced. The authors suggested that strain Y6 is an environmental isolate that entered the wound while the child washed clothes in a pool and eventually established itself in an opportunistic pathogenic role. This is partly based on the fact that the strain Y6 could not grow at 37 °C (optimal growth at 30 °C, but could grow at 35 °C).

A strain of *C. maltaromaticum* was isolated from pus of a patient's infection after the traumatic amputation of his hand as a result of an industrial sawmill accident (Chmelař et al. 2002). *C. maltaromaticum* was confirmed by 16S rRNA gene sequence analysis, phenotypic tests, and FAME analysis. While there is little evidence for further human pathogenesis for *C. maltaromaticum* species (opportunistic or otherwise), analysis of the genome of *C. maltaromaticum* ATCC $35586^T$ (type strain of former *Carnobacterium piscicola*) indicated a number of putative virulence factors (Leisner et al. 2012). Though not human-related, *C. maltaromaticum* has been indicated as a pathogen in numerous fish species, with virulence ranging from generally low (occurring mainly in stressed fish being susceptible) (Starliper et al. 1992) to the possible causative agent of meningoencephalitis resulting in death in juvenile salmon sharks (Schaffer et al. 2013).

The two strains of *Carnobacterium divergens*, BM4489 and BM4490, were both clinical isolates obtained from separate infections within the country of Norway (Meziane-Cherif et al. 2008). *C. divergens* strain BM4489 was recovered from the blood of a cesarean-born baby treated with the antibiotic ampicillin. Eight months later, *C. divergens* BM4490 was isolated from a febrile lymphoma patient that had been previously treated with penicillin G. Both strains proved to be resistant to ampicillin, penicillin, and other β-lactam antibiotics. Sequencing and cloning of a 912 bp region from strain BM4489 revealed the coding sequence of a new class A penicillinase (i.e., class A β-lactamase). While prevalent in Gram-negative bacteria, β-lactamases have only been observed in a few Gram-positive bacteria, and this was the first case for a lactic acid bacterium. The authors suggest that the β-lactamase gene was part of a mobile genetic element acquired by the *Carnobacterium* strains since it's genetic position was adjacent to a resolvase and was not observed in the type species of *C. divergens* (CIP $101209^T$).

Another reported human case of a bacteremia by a *Carnobacterium* species has also been reported in the literature

(Hoenigl et al. 2010). Strain 12266/2009 was isolated from blood cultures of a patient suffering from acute pain of the neck and back of the head, fever, and malaise. 16S rRNA gene sequence analysis revealed that this strain was most closely related to the type species of *Carnobacterium mobile* with >99 % sequence identity. The authors however remain cautious of the causal association of this strain to the symptoms exhibited by the patient. While the strain was clearly isolated from the initial blood sample and the patient had extensive exposure to fish/fish products and further suffered increased clinical deterioration upon the initial antibiotic treatment of ceftriaxone (of which strain 12266/2009 was resistant), the strain could only be isolated from the single initial blood sample which does not rule out the possibility of contamination. The authors also speculate on the possibility that this strain may have masked a *Listeria monocytogenes* infection, though no other pathogen or causative agent could be detected.

When discussing the clinical importance of *Granulicatella adiacens*, it is also necessary to relate it to *Abiotrophia defectiva*, in which cases have often been reported of dual infection by both species (Brouqui and Raoult 2001; Christensen and Facklam 2001; Gensheimer et al. 2010; Cargill et al. 2012). These two clinically important microorganisms were first reported in 1961 (Frenkel and Hirsch 1961) and broadly categorized as nutritionally variant streptococci (NVS), before systematically being separated into independent species, *Streptococcus defectivus* and *Streptococcus adjacens* (Bouvet et al. 1989). Using 16S rRNA gene sequence analysis, Kawamura and colleague proposed a new genus, *Abiotrophia*, to classify the NVS strains as *A. adiacens* and *A. defectiva* (Kawamura et al. 1995). Later it was determined that the genus was not monophyletic, and *Abiotrophia adiacens* was transferred to the new genus, *Granulicatella* (as *G. adiacens*), while *Abiotrophia defectiva* remains in good standing in the nomenclature. While both species are typically associated with the same types of infections, that is not always the case. In contrast, since both species are part of the normal oral human flora, *G. adiacens* and *A. defectiva* (along with *Gemella morbillorum*) have been suggested to be having a beneficial role in preventing dental caries (Gross et al. 2012).

The genus *Granulicatella* has only three validly published species (e.g., *G. adiacens*, *G. balaenopterae*, and *G. elegans*) but has been associated with more serious clinical infections than any other *Carnobacteriaceae* species (as reviewed in Cargill et al. 2012; De Luca et al. 2013). An opportunistic pathogen of the normal human oral microbiota (Zaura et al. 2009), *G. adiacens*, has been linked most often with endocarditis (Brouqui and Raoult 2001; Garibyan and Shaw 2013; Shailaja et al. 2013). Because of the fastidious nature of *G. adiacens* and its requirement for pyridoxal and/or L-cysteine, it is often associated with culture-negative infective endocarditis (IE). But, *G. adiacens* has also been associated with numerous afflictions including bacteremia (resulting in sepsis), endovascular, skin (carbuncles), central nervous system, ocular, bone and joint, and genitourinary infections (Ruoff 1991; Heath et al. 1998; Christensen and Facklam 2001; Hepburn et al. 2003; Zheng et al. 2004; Chang et al. 2008; Gensheimer et al. 2010; Bizzarro et al. 2011; Gardenier et al. 2011; Swain and Otta 2012; De Luca et al. 2013; Mougari et al. 2013). With the advent of pyrosequencing of the 16S rRNA gene sequence, the prevalence of *G. adiacens* has been found in a greater number of cases and has been associated with Papillon-Lefèvre syndrome (Albandar et al. 2011), childhood dental caries (Jiang et al. 2013), acute endodontic infections (Siqueira and Rôças 2006; Hsiao et al. 2012), and even colorectal cancer (Chen et al. 2012). In another study looking for potential biomarkers of disease, the authors linked increased levels of *G. adiacens* in a pancreatic cancer patient's oral microbiota (with a consequent decrease in *Streptococcus mitis* and *Neisseria elongata*) when compared with noncancer subjects (Farrell et al. 2012).

*Granulicatella elegans* has been implicated as a causative agent of infectious endocarditis (IE) as well (Roggenkamp et al. 1998b; Wang et al. 2012). Casalta and colleagues used 16S rRNA gene sequence analysis for the detection of *Granulicatella elegans* on a patient's removed cardiac valve (Casalta et al. 2002). The patient showed every sign of infectious endocarditis (IE), but no infecting bacteria could be cultured from blood. Because *G. elegans* is extremely fastidious and requires 0.01 % L-cysteine hydrochloride for growth, it was thought that the blood test vials used for diagnosis (which did not contain supplemented cysteine HCl) were the reasons for a failed detection of infecting bacteria. Another case in which the patient showed signs of acute endocarditis, *G. elegans* was isolated from the blood of the patient that ultimately resulted in cardiac surgery (Ohara-Nemoto et al. 2005). The patient in question had previously had dental procedure 3 months prior to the symptoms occurring. Isolation of a strain of *G. elegans* from oral cavity dental plaque of the patient showed that the two strains were identical in biochemical characterizations and had the same levels of antibiotic susceptibilities, suggesting the pathway in which *G. elegans* was introduced into blood stream. *G. elegans* was also isolated from blood cultures of a newborn baby after an emergency cesarean section was performed on the mother (Quartermain et al. 2013). The mother reported no instances of endocarditis, dental procedures, or infections prior to, during, or after the pregnancy. While neonatal bacteremia has been observed previously for *G. adiacens* and in this case *G. adiacens* was found to be part of the cervical flora (Bizzarro et al. 2011), no direct correlation to how *G. elegans* was introduced to the placental region was found, though premature rupture of membranes may have been the cause.

In addition to *G. adiacens* and *G. elegans*, "*G. para-adiacens*" has also been detected as a causative agent of bacteremia (Abdul-Redha et al. 2007). A report of "*Granulicatella para-adiacens*" as an infective agent of bacteremia was observed in immunosuppressed patients, mainly febrile neutropenic patients (Senn et al. 2006). As noted previously, the epithet of the organism deposited as *Gemella morbillorum* ATCC 27527 has been changed to "*Granulicatella para-adiacens*" by ATCC as suggested by Kanamoto et al. (2000). This particular strain, though unrelated to the type strain of *Gemella morbillorum* (ATCC 27824), may also be associated with other infections previously attributed to *G. morbillorum* as well.

*Alloiococcus otitis* was originally isolated from the fluid within the middle ear of a young child suffering from chronic otitis media (Faden and Dryja 1989). Like most of the other clinically important members of the *Carnobacteriaceae*, it is fastidious in its growth and can often be missed in diagnosis because of this. That notwithstanding, *A. otitis* is still continually isolated from children's ear infections the world over, but it is often identified with the epithet "*Alloiococcus otitidis*." After publication of *Alloiococcus otitis* (Aguirre and Collins 1992b), a proposed name change to "*Alloiococcus otitidis*" was submitted in the *Journal of Clinical Microbiology* (von Graevenitz 1993a, b). Because the name change was not published in the *International Journal of Systematic Bacteriology* (as per Rule 27 of the Bacteriological Code), the name is not validly published so having no standing in the literature. However, medical microbiologists and clinicians accepted the name change and have published extensively with the epithet "*Alloiococcus otitidis*" with almost all cases referring to chronic otitis media with effusion (OME) (Beswick et al. 1999; Hendolin et al. 1999; Leskinen et al. 2004; Ashhurst-Smith et al. 2007; Harimaya et al. 2007; Janapatla et al. 2011; Aydin et al. 2012). But otitis media is not the only type of infection attributed to *A. otitis*, as the first case of acute endophthalmitis (inflammation of the inner eye) caused by "*A. otitidis*" was recently reported after insertion of Ozurdex®, an implanted drug delivery system used to treat macular edema due to central or branch retinal vein occlusions (Marchino et al. 2013). In addition, the first case of endocarditis by "*A. otitidis*" was also reported (Cakar et al. 2013). The patient had a history of chronic otitis, including loss of hearing in his left ear. Still, some contention exists on whether *Alloiococcus otitis* ("*A. otitidis*") is a causative pathogen or an opportunistic pathogen (Aguirre and Collins 1992a; Fraise et al. 2001; Arar et al. 2008; de Miguel and Macias 2008; Matsuo et al. 2011; Marsh et al. 2012).

In contrast to its role as a pathogen, Ling and colleagues observed *Alloiococcus* species in lower abundance in women suffering from bacterial vaginosis (BV) when compared with women who were BV-negative (healthy women) (Ling et al. 2010). Though *Alloiococcus* species were still among the most 11 predominant genera in BV-positive women, the results suggested that *Alloiococcus* strains (in conjunction with *Lactobacillus* species) might be a previously unrecognized member of a healthy vaginal microbiome. Regardless of whether *A. otitis* is reported in the literature as an opportunistic pathogen or not, the epithet "*A. otitidis*" has been firmly established in the medical community.

*Dolosigranulum pigrum* has arisen as another very important clinical microorganism. Originally isolated from the human spinal cord fluid and an eye infection (Aguirre et al. 1993), it has been associated with numerous human afflictions though most often as an opportunistic pathogen. It has been associated with cystic fibrosis, synovitis, nosocomial pneumonia, blepharitis, septicemia, and keratitis (Laclaire and Facklam 2000; Hall et al. 2001; Hoedemaekers et al. 2006; Lécuyer et al. 2007; Bittar et al. 2008; Lopes et al. 2012; Sampo et al. 2013; Venkateswaran et al. 2013). Most studies indicate that *D. pigrum* is an opportunistic pathogen, but it has been suspected of causing disease, in particular acute cholecystitis accompanied by acute pancreatitis (Lin et al. 2006). It has also been associated with the infection of a prosthetic hip joint resulting in arthritis (Johnsen et al. 2011), as well as being detected within a plaque sample extracted from a patient with carotid atherosclerosis (Renko et al. 2013). There is a growing body of scientific work that is now indicating a possible bacterial role in coronary heart disease. Clinical species of *Dolosigranulum* that have been isolated are generally vulnerable to antibiotic treatment though. Antimicrobial susceptibility for 27 strains of *D. pigrum* were tested and indicated that while most of the strains were susceptible to a wide range of antibiotics, 25 strains were intermediate resistant or resistant to erythromycin (Laclaire and Facklam 2000). Though *Dolosigranulum* species have often been allied with differing infections, it has also been purported as a protective organism within the human microflora particularly for acute otitis media (AOM) within small children (Laufer et al. 2011; Pettigrew et al. 2012).

## Applications

### Waste Treatment and Removal

*Atopostipes suicloacalis* was found to be a contaminating bacterial species within the water-miscible metalworking fluids (MWF) and water preparation basis (WPB) from industrial plants around Germany (Lodders and Kaempfer 2012). While it is unclear that *A. suicloacalis* was directly involved in mineral oil/oil degradation, its presence is interesting since it was originally isolated from swine-associated manure pit. In the analysis of black liquor (pulp mill wastewaters), *Atopostipes* species were also found to be one of the top ten genera populating the waste storage pools (Yang et al. 2010). DGGE analysis indicated only a 94.9 % similarity (partial sequence) to an uncultured *Atopostipes* clone and a 93.8 % similarity to *A. suicloacalis*. In this same analysis, *Alkalibacterium* species (*A. indicireducens*, *A. olivoapovliticus*, and *A. psychrotolerans*) were found in numbers >50 % of the clones obtained from DNA extracted from the wastewaters. It was suggested that because of such high abundance and the isolation of *Alkalibacterium* strain Y5 (99.8 % similarity to *A. indicireducens*) *Alkalibacterium* species, along with *Clostridium*, *Halomonas*, and *Bacillus* species, are important members of waste treatment process of black liquors in lowering the pH and reducing chemical oxygen demand (COD). As opposed to being beneficial bacterium, *A. suicloacalis* was also found to become a dominant member of aggregates that formed within a fed-batch composting (FBC) reactor (Watanabe et al. 2008). FBC operations run under optimal conditions do not normally form aggregates. It is only under prolonged operation that the aggregates form, helping to create small anaerobic environments where *A. suicloacalis* and other anaerobes and facultative anaerobes can thrive.

"*Trichococcus* strain R210" was isolated from granular sludge bed reactor material from a biological sulfate reduction installation in Yixing, China. The primary process of the plant is the microbial reduction of sulfate to sulfide (and eventually to sulfur) using a citrate-containing waste stream as the carbon source. Using a serial dilution strategy, "*Trichococcus* strain R210" was enriched and isolated in bicarbonate-buffered mineral medium without the presence of sulfate in the medium (Stams et al. 2009). Examination of the process indicated that sulfate reduction was not correlated directly to citrate metabolism, but due to citrate's end products of metabolism, formate and acetate. The citrate was being primary metabolized by "*Trichococcus* strain R210" and another strain, "*Veillonella* strain S101." "*Trichococcus* strain R210" is most closely related to *Trichococcus pasteurii* with a 16S rRNA gene sequence similarity of 99.5 %.

## Formation of Value-Added Products

*Carnobacterium* strain 17–4 is a psychrotolerant, lactic acid microorganism that produces a unique solute during growth (Lamosa et al. 2011). The production of α-glucopyranosyl-(1–3)-β-glucopyranosyl-(1–1)-α-glucopyranose (abbreviated as gluconeotrehalose) can potentially have industrial and commercial importance due to the stable nature of related neotrehaloses under high thermal and acidic conditions.

While no metabolic or physiological role for gluconeotrehalose has been ascertained for *Carnobacterium* strain 17–4, the unusual and unknown biosynthetic pathways may prove valuable in commercial production of this or other value-added products.

The microbial production of valuable chemicals from the waste products of industrial processes is often the impetus for new biotechnological applications. A new strain of *Trichococcus* (strain ES5) has shown promise in the bacterial conversion of glycerol to 1,3-propanediol (van Gelder et al. 2012). With the advent of "Green" initiatives around the globe to replace fossil fuels, biodiesel production has increased dramatically, but so has its resulting waste product glycerol. 1,3-propanediol is a chemical with growing economic importance because it can be utilized in the synthesis of polyesters and polyurethanes. *Trichococcus* strain ES5 produces 1,3-propanediol as its major end product from glycerol metabolism and thus may prove valuable as a suitable industrial microbial catalyst.

## Food Preservation or Probiotics

*Carnobacterium* species have been widely studied because of their association with both the potential for food spoilage, as well as food protective abilities in the meat and fish industries (as reviewed in Laursen et al. 2005; Leisner et al. 2007; Doyle et al. 2013; Ghanbari and Jami 2013). The protection of food has attained particular focus because of the production of bacteriocins by numerous *Carnobacterium* strains. Bacteriocins are generally classified as small proteins with antimicrobial activity against similarly related bacteria. They typically fall within one of three categories; class I, class II, or class III bacteriocins. Those belonging to class I are commonly called lantibiotics because they often contain the unusual amino acid lanthionine and are classified by the physical nature of the protein, either elongated (type A) or globular (type B). Class II bacteriocins are nonmodified, heat-stable proteins with a molecular mass <10 kDa. They are further divided into three subclasses; class IIa (pediocin-like bacteriocins), class IIb (two peptide bacteriocins), or class IIc (others bacteriocins). The bacteriocins that belong to class III are categorized as proteins with a molecular mass of >30 kDa.

The production of bacteriocins by *Carnobacterium* species, as well as other lactic acid bacteria (LAB), has been extensively reviewed in the literature (Skaugen et al. 2003; Drider et al. 2006; Leisner et al. 2007; Calo-Mata et al. 2008; Rihakova et al. 2009; Afzal et al. 2010). This includes the characterization of the bacteriocins, genes involved, modes of action, as well as the application of *Carnobacterium* species (or their isolated bacteriocins) onto food products for testing the inhibition of known food-borne pathogens (in particular *Listeria monocytogenes*). It is not our intent to rewrite this information but give an awareness of the known bacteriocins produced by *Carnobacterium* species (no other members of the *Carnobacteriaceae* have been shown to produce bacteriocins). ➋ *Table 4.8* lists the known bacteriocins that have been characterized and produced by *Carnobacterium* species. It should be noted that while *C. piscicola* was renamed in 2003 to *C. maltaromaticum* (Mora et al. 2003), the epithet has still often been associated with later published bacteriocin-producing strains. Several other *Carnobacterium*-associated bacteriocins have been described in the literature but have later been shown to be identical to other bacteriocins or have not been completely characterized. These include piscicolin 61 (identical to carnobacteriocin A) (Holck et al. 1994), carnocin KZ213 (formerly BLIS 213) (Khouiti and Simon 1997, 2004), the bacteriocin produced by *C. maltaromaticum* strain A9b (identical to the previously characterized carnobacteriocin B2) (Nilsson et al. 2002), the bacteriocin produced by *C. maltaromaticum* strain C2 (Tulini and De Martinis 2010), piscicolin KH1 (Hashimoto et al. 2011), divercin AS7 (Józefiak et al. 2012), and a potential new circular bacteriocin from *Carnobacterium* sp. 17–4 (Voget et al. 2011). This new circular bacteriocin was elucidated from the whole genome study of strain 17–4, in which the genetic organization is orthologous to that of carnocyclin A, but differs significantly in gene sequence. This is of particular interest since *C. maltaromaticum* strain UAL307 produces carnocyclin A (as well as the bacteriocins piscicolin 126 and carnobacteriocin BM1) and has been approved as a live or attenuated (heat-treated) additive for the preservation of food (e.g., meat, poultry products) in the United States and Canada (http://www.fsis.usda.gov/OPPDE/rdad/FSISDirectives/7120.1.pdf) under the trade name Micocin®. The actual strain of *Carnobacterium maltaromaticum* that has been approved by the US Department of Agriculture (USDA) as a food additive is strain CB1, which is identical to strain UAL307 (personal communication).

**Table 4.8**
**List of characterized bacteriocins produced by *Carnobacterium* species**

| Bacteriocin | Producing strain[a] | Class | Size (Da) | Genetic location | Gene(s) | Accession # | Original reference |
|---|---|---|---|---|---|---|---|
| Carnocin UI49 | *C. maltaromaticum* UI49 | Class I | 4,635 | Chromosomal | ND | P36960 | Stoffels et al. (1992) |
| Carnocin H | *Carnobacterium* sp. 377 | Class II | 8,250–9000[b] | ND | ND | – | Blom et al. (2001) |
| Carnobacteriocin B2 | *C. maltaromaticum* LV17B | Class IIa | 4,970 | Plasmid | cbnB2 (canCP52) | P38580 | Quadri et al. (1994) |
| Carnobacteriocin BM1 | *C. maltaromaticum* LV17B | Class IIa | 4,525 | Chromosomal | cbnBM1 | P38579 | Quadri et al. (1994) |
| Piscicolin 126 | *C. maltaromaticum* JG126 | Class IIa | 4,417 | Chromosomal | pisA | P80569 | Jack et al. (1996) |
| Piscicolin V1a | *C. maltaromaticum* V1[c] | Class IIa | 4,416 | ND | ND | – | Bhugaloo-Vial et al. (1996) |
| Divercin V41 | *C. divergens* V41 | Class IIa | 4,509 | Chromosomal | dvnV41 | Q9Z4J1 | Métivier et al. (1998) |
| Carnocin CP5[d] | *C. maltaromaticum* CP5 | Class IIa | ~3,000 | Plasmid | ND | – | Herbin et al. (1997) |
| Divergicin M35 | *C. divergens* M35 | Class IIa | 4,519 | ND | ND | P84962 | Tahiri et al. (2004) |
| Piscicolin CS526 | *C. maltaromaticum* CS526 | Class IIa | 4,430 | ND | ND | – | Yamazaki et al. (2005) |
| Carnobacteriocin A | *C. maltaromaticum* LV17A | Class IIc | 5,053 | Plasmid | cbnBA (psc61) | P38578 | Worobo et al. (1994) |
| Divergicin A | *C. divergens* LV13 | Class IIc | 4,224 | Plasmid | dvnA | Q3SAX6 | Worobo et al. (1995) |
| Divergicin 750 | *C. divergens* 750 | Class IIc | 3,448 | Plasmid | dvn750 | Q46597 | Holck et al. (1996) |
| Carnocyclin A | *C. maltaromaticum* UAL307[e] | Circular | 5,862 | Chromosomal | cclA | B2MVM5 | Martin-Visscher et al. (2008) |
| Carnobacteriocin X | *C. maltaromaticum* C2[f] | Class IId | 3,587 | chromosomal | cbnX | – | Tulini et al. (2014) |

*ND* not determined
[a]Note the species *Carnobacterium piscicola* was renamed *Carnobacterium maltaromaticum*, Mora et al. 2003
[b]Estimated size by amino acid (aa) composition from Blom et al. (2001), and avg. MW of aa's
[c]Also produces piscicolin V1b which is identical to carnobacteriocin BM1
[d]Represented by CP51 (identical to carnobacteriocin BM1) and CP52 ( identical to carnobacteriocin B2)
[e]Also produces piscicolin 126 and carnobacteriocin BM1
[f]Also produces B1, BM1, and a variant carnobacteriocin B2

Because of the efficacy for the application of *C. maltaromaticum* for the direct protection of meat and poultry, it is also being exploited for the indirect protection of refrigerated packaged food items. Small sachets of *C. maltaromaticum* infused in a nutrient-rich gel and in the presence of acid fuchsin (an acid-base indicator) have been employed as attachable food labels (called time temperature integrators or TTI's) for the monitoring of cold-chain maintenance. When applied to the outer layer of a refrigerated packaged food item, if the item is not kept cold, the subsequent rise in temperature will induce *C. maltaromaticum* to metabolize the nutrient gel thus producing lactic acid and changing the color label from green to red (i.e., indication of improper storage or handling of the cold product). This system has recently passed the European Food Safety Authority's evaluation that it does not raise a safety concern for consumers (EFSA CEF Panel 2013).

*Alkalibacterium kapii* and *Marinilactibacillus psychrotolerans* has also been examined for their protective capacity during cheese production. When the marine lactic acid bacteria *A. kapii* and *M. psychrotolerans* were applied at day 7 as part of "old-new" cheese smear consortia in commercial Raclette-type cheeses, an increase in numbers was observed, with a subsequent decrease in *Listeria* counts (Roth et al. 2010). The prevention of Listeria growth was later tested and indicated a weak growth-inhibitory effect by *M. psychrotolerans* to *Listeria* species, but no inhibitory effect for *A. kapii*. But when applied in situ, both had a $\log^{-1}$ effect on controlling *Listeria* growth (Roth et al. 2011).

## Acknowledgments

We to thank Pablo Yarza of the Marine Microbiology Group, Institut Mediterrani d'Estudis Avanc¸ats (CSIC-UIB), C/ Miquel Marque's 21,E-07190 Esporles, Illes Balears, Mallorca, Spain for the construction of the phylogenetic tree.

## References


Abdul-Redha RJ, Prag J, Sonksen UW, Kemp M, Andresen K, Christensen JJ (2007) *Granulicatella elegans* bacteraemia in patients with abdominal infections. Scand J Infect Dis 39:830–833

Afzal MI, Jacquet T, Delaunay S, Borges F, Millière J-B, Revol-Junelles A-M, Cailliez-Grimal C (2010) *Carnobacterium maltaromaticum*: identification, isolation tools, ecology and technological aspects in dairy products. Food Microbiol 27:573–579

Aguirre M, Collins MD (1992a) Development of a polymerase chain reaction-probe test for identification of *Alloiococcus otitis*. J Clin Microbiol 30:2177–2180

Aguirre M, Collins MD (1992b) Phylogenetic analysis of *Alloiococcus otitis* gen. nov., sp. nov., an organism from human middle ear fluid. Int J Syst Bacteriol 42:79–83

Aguirre M, Morrison D, Cookson BD, Gay FW, Collins MD (1993) Phenotypic and phylogenetic characterization of some Gemella-like organisms from human infections: description of *Dolosigranulum pigrum* gen. nov., sp. nov. J Appl Bacteriol 75:608–612

Aguirre M, Morrison D, Cookson BD, Gay FW, Collins MD (1994) Validation of the publication of new names and new combinations previously effectively published outside the IJSB: list no. 49. Int J Syst Bacteriol 44:370–371

Albandar JM, Khattab R, Monem F, Barbuto SM, Paster BJ (2011) The subgingival microbiota of Papillon-Lefèvre syndrome. J Periodontol 83:902–908

Arar S, Vinogradov E, Shewmaker PL, Monteiro MA (2008) A polysaccharide of *Alloiococcus otitidis*, a new pathogen of otitis media: chemical structure and synthesis of a neoglycoconjugate thereof. Carbohydr Res 343:1079–1090

Ashhurst-Smith C, Hall ST, Walker P, Stuart J, Hansbro PM, Blackwell CC (2007) Isolation of *Alloiococcus otitidis* from indigenous and non-indigenous Australian children with chronic otitis media with effusion. FEMS Immunol Med Microbiol 51:163–170

Aydin E, Taştan E, Yücel M, Aydoğan F, Karakoç E, Arslan N, Kantekin Y, Demirci M (2012) Concurrent assay for four bacterial species including *Alloiococcus otitidis* in middle ear, nasopharynx and tonsils of children with otitis media with effusion: a preliminary report. Clin Exp Otorhinolaryngol 5:81–85

Balch WE, Fox GE, Magrum LJ, Woese CR, Wolfe RS (1979) Methanogens: reevaluation of a unique biological group. Microbiol Rev 43:260–296

Ballerstedt H, Hantke J, Bunge M, Werner B, Gerritse J, Andreesen JR, Lechner U (2004) Properties of a trichlorodibenzo-*p*-dioxin-dechlorinating mixed culture with a *Dehalococcoides* as putative dechlorinating species. FEMS Microbiol Ecol 47:223–234

Beswick AJ, Lawley B, Fraise AP, Pahor AL, Brown NL (1999) Detection of *Alloiococcus otitis* in mixed bacterial populations from middle-ear effusions of patients with otitis media. Lancet 354:386–389

Bhugaloo-Vial P, Dousset X, Metivier A, Sorokine O, Anglade P, Boyaval P, Marion D (1996) Purification and amino acid sequences of piscicocins V1a and V1b, two class IIa bacteriocins secreted by *Carnobacterium piscicola* V1 that display significantly different levels of specific inhibitory activity. Appl Environ Microbiol 62:4410–4416

Bittar F, Richet H, Dubus J-C, Reynaud-Gaubert M, Stremler N, Sarles J, Raoult D, Rolain J-M (2008) Molecular detection of multiple emerging pathogens in sputa from cystic fibrosis patients. PLoS One 3:e2908

Bizzarro MJ, Callan DA, Farrel PA, Dembry L-M, Gallagher PG (2011) *Granulicatella adiacens* and early-onset sepsis in neonate. Emerg Infect Dis 17:1971–1973

Blom H, Katla T, Nissen H, Holo H (2001) Characterization, production, and purification of carnocin H, a bacteriocin produced by *Carnobacterium* 377. Curr Microbiol 43:227–231

Borch E, Molin G (1988) Numerical taxonomy of psychrotrophic lactic acid bacteria from prepacked meat and meat products. Antonie von Leeuwenhoek 54:301–323

Bouvet A, Grimont F, Grimont PAD (1989) *Streptococcus defectivus* sp. nov. and *Streptococcus adjacens* sp. nov., nutritionally variant streptococci from human clinical specimens. Int J Syst Bacteriol 39:290–294

Breed RS (1953) The families developed from *Bacteriaceae* Cohn with a description of the family *Brevibacteriaceae*. Riass Commun VI Congr Int Microbial Roma 1:10–15

Brouqui P, Raoult D (2001) Endocarditis due to rare and fastidious bacteria. Clin Microbiol Rev 14:177–207

Cadenas MB, Maggi RG, Diniz PPVP, Breitschwerdt KT, Sontakke S, Breithschwerdt EB (2007) Identification of bacteria from clinical samples using *Bartonella* alpha-*proteobacteria* growth medium. J Microbiol Methods 71:147–155

Cailliez-Grimal C, Chaillou S, Anba-Mondoloni J et al (2013) Complete chromosome sequence of *Carnobacterium maltaromaticum* LMA 28. Genome Announc 1:1–2

Cakar M, Demirbas S, Yildizoglu U, Arslan E, Balta S, Kara K, Guney M, Demirkol S, Celik T (2013) First report of endocarditis by *Alloiococcus otitidis* spp. in a patient with a history of chronic otitis. J Infect Public Health 6:494–495

Calo-Mata P, Arlindo S, Bochme K, de Miguel T, Pascoal A, Barros-Velazquez J (2008) Current applications and future trends of lactic acid bacteria and their bacteriocins for the biopreservation of aquatic food products. Food Bioprocess Technol 1:43–63

Camarinha-Silva A, Wos-Oxley ML, Jauregui R, Becker K, Pieper DH (2012) Validating T-RFLP as a sensitive and high-throughput approach to assess bacterial diversity patterns in human anterior nares. FEMS Microbiol Ecol 79:98–108

Cargill JS, Scott KS, Gascoyne-Binzi D, Sandoe JAT (2012) *Granulicatella* infection: diagnosis and management. J Med Microbiol 61:755–761

Casaburi A, Nasi A, Ferrocino I, Di Monaco R, Mauriello G, Villani F, Ercolini D (2011) Spoilage related activity of *Carnobacterium maltaromaticum* strains in air-stored and vacuum-packed meat. Appl Environ Microbiol 77:7382–7393

Casalta JP, Habib G, La Scola B, Drancourt M, Caus T, Raoult D (2002) Molecular diagnosis of *Granulicatella elegans* on the cardiac valve of a patient with culture-negative endocarditis. J Clin Microbiol 40:1845–1847

Champomier M-C, Montel M-C, Talon R (1989a) Nucleic acid relatedness studies on the genus *Carnobacterium* and related taxa. J Gen Microbiol 135:1391–1394

Champomier M-C, Montel M-C, Talon R (1989b) Nucleic acid relatedness studies on the genus *Carnobacterium* and related taxa [Erratum]. J Gen Microbiol 135:2345

Chang SH, Lee CC, Chen IC, Hsieh MR, Chen SC (2008) Infectious intracranial aneurysms caused by *Granulicatella adiacens*. Diagn Microbiol Infect Dis 60:201–204

Chen W, Liu F, Ling Z, Tong X, Xiang C (2012) Human intestinal lumen and mucosa-associated microbiota in patients with colorectal cancer. PLoS One 7:e39743

Chmelař D, Matušek A, Korger J, Durnová E, Steffen M, Chmelařová E (2002) Isolation of *Carnobacterium piscicola* from human pus-case report. Folia Microbiol 47:455–457

Christensen JJ, Facklam RR (2001) *Granulicatella* and *Abiotrophia* species from human clinical specimens. J Clin Microbiol 39:3520–3523

Collins MD (2009) Genus IV. *Alloiococcus*. In: de Vos P, Garrity G, Jones D, Krieg NR, Ludwig W, Rainey FA, Schleifer K-H, Whitman WB (eds) The *Firmicutes*, vol 3, 2nd edn, Bergey's manual of systematic bacteriology. Springer, New York, pp 562–563

Collins MD, Kroppenstedt RM (1983) Lipid composition as a guide to the classification of some coryneform bacteria containing an A4$\alpha$ type peptidoglycan (Schleifer and Kandler). Syst Appl Microbiol 4:95–104

Collins MD, Lawson PA (2000) The genus *Abiotrophia* (Kawamura et al) is not monophyletic: proposal of *Granulicatella* gen. nov., *Granulicatella adiacens* comb. nov., *Granulicatella elegans* comb. nov. and *Granulicatella balaenopterae* comb. nov. Int J System Evol Microbiol 50:365–369

Collins MD, Wiernik A, Falsen E, Lawson PA (2005) *Atopococcus tabaci* gen. nov., sp. nov., a novel gram-positive, catalase-negative, coccus-shaped bacterium isolated from tobacco. Int J Syst Evol Microbiol 55:1693–1696

Collins MD, Farrow JAE, Phillips BA, Ferusu S, Jones D (1987) Classification of *Lactobacillus divergens, Lactobacillus piscicola,* and some catalase-negative, asporogenous, rod-shaped bacteria from poultry in a new genus, *Carnobacterium.* Int J Syst Bacteriol 37:310–316

Collins MD, Hutson RA, Foster G, Falsen E, Weiss N (2002) *Isobaculum melis* gen. nov., sp. nov., a *Carnobacterium*-like organism isolated from the intestine of a badger. Int J Syst Evol Microbiol 52:207–210

Collins MD, Higgins R, Messier S, Fortin M, Hutson RA, Lawson PA, Falsen E (2003) *Allofustis seminis* gen. nov., sp. nov., a novel gram-positive, catalase-negative, rod-shaped bacterium from pig semen. Int J Syst Evol Microbiol 53:811–814

Collins MD, Rodrigues U, Ash C, Aguirre M, Farrow JAE, Martinez-Murcia A, Phillips BA, Williams AM, Wallbanks S (1991) Phylogenetic analysis of the genus *Lactobacillus* and related lactic acid bacteria as determined by reverse transcriptase sequencing of 16S rRNA. FEMS Microbiol Lett 77:5–12

Cotta MA, Whitehead TR, Zeltwanger RL (2003) Isolation, characterization and comparison of bacteria from swine faeces and manure storage pits. Environ Microbiol 5:737–745

Cotta MA, Whitehead TR, Collins MD, Lawson PA (2004a) *Atopostipes suicloacale* gen. nov., sp. nov., isolated from an underground swine manure storage pit. Anaerobe 10:191–195

Cotta MA, Whitehead TR, Collins MD, Lawson PA (2004b) Validation of publication of new names and new combinations previously effectively published outside the IJSEM. List no. 99. Int J Syst Evol Microbiol 54:1425–1426

De Bruyn IN, Louw AI, Visser L, Holzapfel WH (1987) *Lactobacillus divergens* is a homofermentative organism. Syst Appl Microbiol 9:173–175

De Luca M, Amodio D, Chiurchiù S, Castelluzzo MA, Rinelli G, Bernaschi P, Calò Carducci FI, D'Argenio P (2013) *Granulicatella* bacteraemia in children: two cases and review of the literature. BMC Pediatr 13:61

De Man JD, Rogosa M, Sharpe ME (1960) A medium for the cultivation of *Lactobacilli.* J Appl Bacteriol 23:130–135

De Miguel MI, Macias AR (2008) Serous otitis media in children: implication of *Alloiococcus otitidis.* Otol Neurotol 29:526–530

Dewhirst FE, Chen T, Izard J, Paster BJ, Tanner ACR, Yu W-H, Lakshmanan A, Wade WG (2010) The human oral microbiome. J Bacteriol 192:5002–5017

Dewhirst FE, Klein EA, Thompson EC, Blanton JM, Chen T, Milella L, Buckley CMF, Davis IJ, Bennett M-L, Marshall-Jones ZV (2012) The canine oral microbiome. PLoS One 7:e36067

Diaz PI, Dupuy AK, Abusleme L, Reese B, Obergfell C, Choquette L, Dongari-Bagtzoglou A, Peterson DE, Terzi E, Strausbaugh LD (2012) Using high throughput sequencing to explore the biodiversity in oral bacterial communities. Mol Oral Microbiol 27:182–201

Dicks LMT, Janssen B, Dellaglio F (1995) Differentiation of *Carnobacterium divergens* and *Carnobacterium piscicola* by numerical analysis of total soluble cell protein patterns and DNA-DNA hybridizations. Curr Microbiol 31:77–79

Doyle MP, Steenson LR, Meng J (2013) Bacteria in food and beverage production. In: Rosenberg E, DeLong E, Lory S, Stackebrandt E, Thompson F (eds) The prokaryotes. Springer, Berlin, pp 241–256

Drancourt M, Roux V, Fournier PE, Raoult D (2004) *rpoB* gene sequence-based identification of aerobic gram-positive cocci of the genera *Streptococcus, Enterococcus, Gemella, Abiotrophia,* and *Granulicatella.* J Clin Microbiol 42:497–504

Drider D, Fimland G, Héchard Y, McMullen LM, Prévost H (2006) The continuing story of class IIa bacteriocins. Microbiol Mol Biol Rev 70:564–582

EFSA CEF Panel (EFSA Panel on Food Contact Materials, Enzymes, Flavourings and Processing Aids) (2013) Scientific opinion on the safety evaluation of a time-temperature indicator system, based on *Carnobacterium maltaromaticum* and acid fuchsin for use in food contact materials. EFSA J 11:3307

Elliker PR, Anderson AW, Hannesson G (1956) An agar medium for lactic acid streptococci and lactobacilli. J Dairy Sci 39:1611–1612

Faden H, Dryja D (1989) Recovery of a unique bacterial organism in human middle ear fluid and its possible role in chronic otitis media. J Clin Microbiol 27:2488–2491

Farrell JJ, Zhang L, Zhou H, Chia D, Elashoff D, Akin D, Paster BJ, Joshipura K, Wong DTW (2012) Variations of oral microbiota are associated with pancreatic diseases including pancreatic cancer. Gut 61:582–588

Feurer C, Irlinger F, Spinnler HE, Glaser P, Vallaeys T (2004) Assessment of the rind microbial diversity in a farmhouse-produced vs. a pasteurized industrially produced soft red-smear cheese using both cultivation and rDNA-based methods. J Appl Microbiol 97:546–556

Fletcher LD, McMichael JC, Russell DP, Zagursky RJ (2003) Complete genome sequence of *Alloiococcus otitidis,* identification of open reading frames encoding polypeptide antigens, and immunogenic compositions and their uses. Wyeth Holdings Corporation. Madison, New Jersey, p 1019

Fraise AP, Pahor AL, Beswick AJ (2001) Otitis media with effusion: the role of *Alloiococcus otitidis.* Ann Med 33:1–3

Frank DN, Spiegelman GB, Davis W, Wagner E, Lyons E, Pace NR (2003) Culture-independent molecular analysis of microbial constituents of the healthy human outer ear. J Clin Microbiol 41:295–303

Franzmann PD, Hoepfl P, Weiss N, Tindall BJ (1991) Psychrotrophic, lactic acid-producing bacteria from anoxic waters in Ace Lake, Antarctica; *Carnobacterium funditum* sp. nov. and *Carnobacterium alterfunditum* sp. nov. Arch Microbiol 156:255–262

Franzmann PD, Höpfl P, Weiss N, Tindall BJ (1993) Validation of the publication of new names and new combinations previously effectively published outside the IJSB: list no. 44. Int J Syst Bacteriol 43:188–189

Frenkel A, Hirsch W (1961) Spontaneous development of *L* form of streptococci requiring secretions of other bacteria or sulphydryl compounds for normal growth. Nature 191:728–730

Gardenier JC, Hranjec T, Sawyer RG, Bonatti H (2011) *Granulicatella adiacens* bacteremia in an elderly trauma patient. Surg Infect (Larchmt) 12:251–253

Garibyan V, Shaw D (2013) Bivalvular endocarditis due to *Granulicatella adiacens.* Am J Case Rep 14:435–438

Gensheimer WG, Reddy SY, Mulconry M, Greves C (2010) *Abiotrophia/Granulicatella* tubo-ovarian abscess in an adolescent virginal female. J Pediatr Adolesc Gynecol 23:e9–e12

Ghanbari M, Jami M (2013) Lactic acid bacteria and their bacteriocins: a promising approach to seafood biopreservation. In: Kongo M (ed) Lactic acid bacteria - R & D for food, health and livestock purposes. pp 381–404. InTech, Rijeka, Croatia, doi:10.5772/50705

Gillis M, De LJ, De CM (1970) The determination of molecular weight of bacterial genome DNA from renaturation rates. Eur J Biochem 12:143–153

Gori K, Ryssel M, Arneborg N, Jespersen L (2013) Isolation and identification of the microbiota of Danish farmhouse and industrially produced surface-ripened cheeses. Microb Ecol 65:602–615

Gross EL, Beall CJ, Kutsch SR, Firestone ND, Leys EJ, Griffen AL (2012) Beyond *Streptococcus mutans*: dental caries onset linked to multiple species by 16S rRNA community analysis. PLoS One 7:e47722

Groth I, Schumann P, Weiss N, Martin K, Rainey FA (1996) *Agrococcus jenensis* gen nov, sp nov, a new genus of actinomycetes with diaminobutyric acid in the cell wall. Int J Syst Bacteriol 46:234–239

Hall GS, Gordon S, Schroeder S, Smith K, Anthony K, Procop GW (2001) Case of synovitis potentially caused by *Dolosigranulum pigrum.* J Clin Microbiol 39:1202–1203

Hammes WP, Hertel C (2006) The genera *Lactobacillus* and *Carnobacterium.* In: Dworkin M, Falkow S, Rosenberg E, Schleifer K-H, Stackebrandt E (eds) The prokaryotes. Springer, New York, pp 320–403

Hammes WP, Hertel C (2009) Genus I. *Carnobacterium.* In: De Vos P, Garrity GM, Jones D, Krieg NR, Ludwig W, Rainey FA, Schleifer K-H, Whitman WB (eds) The *Firmicutes,* vol 3, Bergey's manual of systematic bacteriology. Springer, New York, pp 549–557

Hammes WP, Weiss N, Holzapfe W (2002) The genera *Lactobacillus* and *Carnobacterium.* In: Balows A, Dworkin M, Harder W, Schleifer K-H (eds) The prokaryotes: a handbook on the biology of bacteria: ecophysiology, isolation, identification, applications. Springer-Verlag, New York, pp 1535–1594

Harimaya A, Takada R, Himi T, Yokota S, Fujii N (2007) Evidence of local antibody response against *Alloiococcus otitidis* in the middle ear cavity of children with otitis media. FEMS Immunol Med Microbiol 49:41–45

Hashimoto K, Bari MK, Omatsu Y, Kawamoto S, Shima J (2011) Biopreservation of kamaboko (steamed surimi) using piscicolin KH1 produced by *Carnobacterium maltalomaticum* KH1. Jpn J Food Microbiol 28:193–200

Heath CH, Bowen SF, McCarthy JS, Dwyer B (1998) Vertebral osteomyelitis and discitis associated with *Abiotrophia adiacens* (nutritionally variant *Streptococcus*) infection. Aust N Z J Med 28:663

Hendolin PH, Paulin L, Ylikoski J (2000) Clinically applicable multiplex PCR for four middle ear pathogens. J Clin Microbiol 38:125–132

Hendolin PH, Kärkkäinen U, Himi T, Markkanen A, Ylikoski J (1999) High incidence of *Alloiococcus otitis* in otitis media with effusion. Pediatr Infect Dis J 18:860–865

Hepburn MJ, Fraser SL, Rennie TA, Singleton CM, Delgado B Jr (2003) Septic arthritis caused by *Granulicatella adiacens*: diagnosis by inoculation of synovial fluid into blood culture bottles. Rheumatol Int 23:255–257

Herbin S, Mathieu F, Brulé F, Branlant C, Lefebvre G, Lebrihi A (1997) Characteristics and genetic determinants of bacteriocin activities produced by *Carnobacterium piscicola* CP5 isolated from cheese. Curr Microbiol 35:319–326

Hespell RB, Wolf R, Bothast RJ (1987) Fermentation of xylans by *Butyrivibrio fibrisolvens* and other ruminal bacteria. Appl Environ Microbiol 53:2849–2853

Hiu SF, Holt RA, Sriranganathan N, Seidler RJ, Fryer JL (1984) *Lactobacillus piscicola*, a new species from salmonid fish. Int J Syst Bacteriol 34:393–400

Hoedemaekers A, Schulin T, Tonk B, Melchers WJG, Sturm PDJ (2006) Ventilator-associated pneumonia caused by *Dolosigranulum pigrum*. J Clin Microbiol 44:3461–3462

Hoenigl M, Grisold AJ, Valentin T, Leitner E, Zarfel G, Renner H, Krause R (2010) Isolation of *Carnobacterium* sp. from a human blood culture. J Med Microbiol 59:493–495

Hogg JC, Lehane MJ (1999) Identification of bacterial species associated with the sheep scab mite (*Psoroptes ovis*) by using amplified genes coding for 16S rRNA. Appl Environ Microbiol 65:4227–4229

Holck A, Axelsson L, Schillinger U (1994) Purification and cloning of piscicolin 61, a bacteriocin from *Carnobacterium piscicola* LV61. Curr Microbiol 29:63–68

Holck A, Axelsson L, Schillinger U (1996) Divergicin 750, a novel bacteriocin produced by *Carnobacterium divergens* 750. FEMS Microbiol Lett 136:163–168

Holley RA, Guan TY, Peirson M, Yost CK (2002) *Carnobacterium viridans* sp. nov., an alkaliphilic, facultative anaerobe isolated from refrigerated, vacuum-packed bologna sausage. Int J Syst Evol Microbiol 52:1881–1885

Holzapfel WH, Gerber ES (1983) *Lactobacillus divergens* sp. nov., a new heterofermentative *Lactobacillus* species producing L(+)-Lactate. Syst Appl Microbiol 4:522–534

Horvath P, Coûté-Monvoisin A-C, Romero DA, Boyaval P, Fremaux C, Barrangou R (2009) Comparative analysis of CRISPR loci in lactic acid bacteria genomes. Int J Food Microbiol 131:62–70

Hsiao WWL, Li KL, Liu Z, Jones C, Fraser-Liggett CM, Fouad AF (2012) Microbial transformation from normal oral microbiota to acute endodontic infections. BMC Genomics 13:345–359

Hung W-C, Tseng S-P, Chen H-J, Tsai J-C, Chang C-H, Lee T-F, Hsueh P-R, Teng L-J (2010) Use of *groESL* as a target for identification of *Abiotrophia*, *Granulicatella*, and *Gemella* species. J Clin Microbiol 48:3532–3538

Hungate RE (1969) Roll tube method for cultivation of strict anaerobes. In: Norris JR (ed) Methods in microbiology, vol 3b. Academic, London, pp 117–132

Iino T, K-i S, Harayama S (2009) *Lacticigenium naphtae* gen. nov., sp. nov., a halotolerant and motile lactic acid bacterium isolated from crude oil. Int J Syst Evol Microbiol 59:775–780

Inagaki F, Suzuki M, Takai K, Oida H, Sakamoto T, Aoki K, Nealson KH, Horikoshi K (2003) Microbial communities associated with geological horizons in coastal subseafloor sediments from the Sea of Okhotsk. Appl Environ Microbiol 69:7224–7235

Ishikawa M, Nakajima K, Yanagi M, Yamamoto Y, Yamasato K (2003) *Marinilactibacillus psychrotolerans* gen. nov., sp. nov., a halophilic and alkaliphilic marine lactic acid bacterium isolated from marine organisms in temperate and subtropical areas of Japan. Int J Syst Evol Microbiol 53:711–720

Ishikawa M, Yamasato K, Kodama K, Yasuda H, Matsuyama M, Okamoto-Kainuma A, Koizumi Y (2013) *Alkalibacterium gilvum* sp. nov., slightly halophilic and alkaliphilic lactic acid bacterium isolated from soft- and semi-hard cheeses. Int J Syst Evol Microbiol 63:1471–1478

Ishikawa M, Nakajima K, Ishizaki S, Kodama K, Okamoto-Kainuma A, Koizumi Y, Yamamoto Y, Yamasato K (2011) *Alkalibacterium subtropicum* sp. nov., a slightly halophilic and alkaliphilic marine lactic acid bacterium isolated from decaying marine algae. Int J Syst Evol Microbiol 61:2996–3002

Ishikawa M, Tanasupawat S, Nakajima K, Kanamori H, Ishizaki S, Kodama K, Okamoto-Kainuma A, Koizumi Y, Yamamoto Y, Yamasato K (2009) *Alkalibacterium thalassium* sp. nov., *Alkalibacterium pelagium* sp. nov., *Alkalibacterium putridalgicola* sp. nov. and *Alkalibacterium kapii* sp. nov., slightly halophilic and alkaliphilic marine lactic acid bacteria isolated from marine organisms and salted foods collected in Japan and Thailand. Int J Syst Evol Microbiol 59:1215–1226

Jack RW, Wan J, Gordon J, Harmark K, Davidson BE, Hillier AJ, Wettenhall RE, Hickey MW, Coventry MJ (1996) Characterization of the chemical and antimicrobial properties of piscicolin 126, a bacteriocin produced by *Carnobacterium piscicola* JG126. Appl Environ Microbiol 62:2897–2903

Janapatla R-P, Chang H-J, Hsu M-H, Hsieh Y-C, Lin T-Y, Chiu C-H (2011) Nasopharyngeal carriage of *Streptococcus pneumoniae*, *Haemophilus influenzae*, *Moraxella catarrhalis*, and *Alloiococcus otitidis* in young children in the era of pneumococcal immunization, Taiwan. Scand J Infect Dis 43:937–942

Janssen PH, Evers S, Rainey FA, Weiss N, Ludwig W, Harfoot CG, Schink B (1995) *Lactosphaera* gen. nov., a new genus of lactic acid bacteria, and transfer of *Ruminococcus pasteurii* Schink 1984 to *Lactosphaera pasteurii* comb. nov. Int J Syst Bacteriol 45:565–571

Jiang W, Zhang J, Chen H (2013) Pyrosequencing analysis of oral microbiota in children with severe early childhood dental caries. Curr Microbiol 67:537–542

Jöborn A, Dorsch M, Olsson JC, Westerdahl A, Kjelleberg S (1999) *Carnobacterium inhibens* sp nov., isolated from the intestine of Atlantic salmon (*Salmo salar*). Int J Syst Bacteriol 49:1891–1898

Johnsen BO, Ronning EJ, Onken A, Figved W, Jenum PA (2011) *Dolosigranulum pigrum* causing biomaterial-associated arthritis. APMIS 119:85–87

Jones D, Keddie R (1986) *Brevibacterium*. In: Sneath PHA, Mair NS, Sharp ME, Holt JG (eds) Bergey's manual of systematic bacteriology, vol 2. Williams & Wilkins, Baltimore, pp 1301–1313

Józefiak D, Sip A, Rutkowski A, Rawski M, Kaczmarek S, Wołuń-Cholewa M, Engberg RM, Højberg O (2012) Lyophilized *Carnobacterium divergens* AS7 bacteriocin preparation improves performance of broiler chickens challenged with *Clostridium perfringens*. Poult Sci 91:1899–1907

Kabadjova P, Dousset X, Le CV, Prévost H (2002) Differentiation of closely related *Carnobacterium* food isolates based on 16S-23S ribosomal DNA intergenic spacer region polymorphism. Appl Environ Microbiol 68:5358–5366

Kanamoto T, Sato S, Inoue M (2000) Genetic heterogeneities and phenotypic characteristics of strains of the genus *Abiotrophia* and proposal of *Abiotrophia para-adiacens* sp. nov. J Clin Microbiol 38:492–498

Kaur R, Adlowitz DG, Casey JR, Zeng M, Pichichero ME (2010) Simultaneous assay for four bacterial species including *Alloiococcus otitidis* using multiplex-PCR in children with culture negative acute otitis media. Pediatr Infect Dis J 29:741–745

Kawamura Y, Hou XG, Sultana F, Liu S, Yamamoto H, Ezaki T (1995) Transfer of *Streptococcus adjacens* and *Streptococcus defectivus* to *Abiotrophia* gen. nov. as *Abiotrophia adiacens* comb. nov. and *Abiotrophia defectiva* comb. nov., respectively. Int J Syst Bacteriol 45:798–803

Khouiti Z, Simon J-P (1997) Detection and partial characterization of a bacteriocin produced by *Carnobacterium piscicola* 213. J Ind Microbiol Biotechnol 19:28–33

Khouiti Z, Simon J-P (2004) Carnocin KZ213 produced by *Carnobacterium piscicola* 213 is adsorbed onto cells during growth. Its biosynthesis is regulated by temperature, pH and medium composition. J Ind Microbiol Biotechnol 31:5–10

Kim M-S, Roh SW, Nam Y-D, Yoon J-H, Bae J-W (2009) *Carnobacterium jeotgali* sp. nov., isolated from a Korean traditional fermented food. Int J Syst Evol Microbiol 59:3168–3171

Kroppenstedt RM, Kutzner HJ (1978) Biochemical taxonomy of some problem actinomycetes. Zentralbl Bakteriol Parasitenkd Infektionskr Hyg Abt 6:25–133

Laclaire L, Facklam R (2000) Antimicrobial susceptibility and clinical sources of *Dolosigranulum pigrum* cultures. Antimicrob Agents Chemother 44:2001–2003

Lai S, Goodacre R, Manchester LN (2004) Whole-organism fingerprinting of the genus *Carnobacterium* using Fourier transform infrared spectroscopy (FT-IR). Syst Appl Microbiol 27:186–191

Lamosa P, Mingote AI, Groudieva T, Klippel B, Egorova K, Jabbour D, Santos H, Antranikian G (2011) Gluconeotrehalose is the principal organic solute in the psychrotolerant bacterium *Carnobacterium* strain 17–4. Extremophiles 15:463–472

Laufer AS, Metlay JP, Gent JF, Fennie KP, Kong Y, Pettigrew MM (2011) Microbial communities of the upper respiratory tract and otitis media in children. mBio 2:e00245-10

Lauro FM, Chastain RA, Blankenship LE, Yayanos AA, Bartlett DH (2007) The unique 16S rRNA genes of piezophiles reflect both phylogeny and adaptation. Appl Environ Microbiol 73:838–845

Laursen BG, Bay L, Cleenwerck I, Vancanneyt M, Swings J, Dalgaard P, Leisner JJ (2005) *Carnobacterium divergens* and *Carnobacterium maltaromaticum* as spoilers or protective cultures in meat and seafood: phenotypic and genotypic characterization. Syst Appl Microbiol 28:151–164

Lawson PA (2009) Genus III. *Allofustis*. In: De Vos P, Garrity GM, Jones D, Krieg NR, Ludwig W, Rainey FA, Schleifer K-H, Whitman WB (eds) The *Firmicutes*, vol 3, Bergey's manual of systematic bacteriology. Springer, New York, pp 559–561

Lawson PA, Foster G, Falsen E, Sjödén B, Collins MD (1999) *Abiotrophia balaenopterae* sp. nov., isolated from the minke whale (*Balaenoptera acutorostrata*). Int J Syst Bacteriol 49:503–506

Lawson PA, Foster G, Falsen E, Ohlen M, Collins MD (2000) *Atopobacter phocae* gen. nov., sp nov., a novel bacterium isolated from common seals. Int J Syst Bacteriol 50:1755–1760

Lécuyer H, Audibert J, Bobigny A, Eckert C, Jannière-Nartey C, Buu-Hoï A, Mainardi J-L, Podglajen I (2007) *Dolosigranulum pigrum* causing nosocomial pneumonia and septicemia. J Clin Microbiol 45:3474–3475

Lee D-G, Trujillo ME, Kang H, Ahn T-Y (2014) *Jeotgalibaca dankookensis* gen. nov., sp. nov., a member of the family *Carnobacteriaceae*, isolated from seujeot (Korean traditional food). Int J Syst Bacteriol 64:1729–1735

Leisner JJ, Laursen BG, Prévost H, Drider D, Dalgaard P (2007) *Carnobacterium*: positive and negative effects in the environment and in foods. FEMS Microbiol Rev 31:592–613

Leisner JJ, Hansen MA, Larsen MH, Hansen L, Ingmer H, Sørensen SJ (2012) The genome sequence of the lactic acid bacterium, *Carnobacterium maltaromaticum* ATCC 35586 encodes potential virulence factors. Int J Food Microbiol 152:107–115

Leskinen K, Hendolin P, Virolainen-Julkunen A, Ylikoski J, Jero J (2004) *Alloiococcus otitidis* in acute otitis media. Int J Pediatr Otorhinolaryngol 68:51–56

Lin J-C, Hou S-J, Huang L-U, Sun J-R, Chang W-K, Lu J-J (2006) Acute cholecystitis accompanied by acute pancreatitis potentially caused by *Dolosigranulum pigrum*. J Clin Microbiol 44:2298–2299

Ling Z, Liu X, Luo Y, Yuan L, Nelson KE, Wang Y, Xiang C, Li L (2013) Pyrosequencing analysis of the human microbiota of healthy Chinese undergraduates. BMC Genomics 14:390

Ling Z, Kong J, Liu F, Zhu H, Chen X, Wang Y, Li L, Nelson KE, Xia Y, Xiang C (2010) Molecular analysis of the diversity of vaginal microbiota associated with bacterial vaginosis. BMC Genomics 11:488

Liu JR, Tanner RS, Schumann P, Weiss N, McKenzie CA, Janssen PH, Seviour EM, Lawson PA, Allen TD, Seviour RJ (2002) Emended description of the genus *Trichococcus*, description of *Trichococcus collinsii* sp. nov., and reclassification of *Lactosphaera pasteurii* as *Trichococcus pasteurii* comb. nov. and of *Ruminococcus palustris* as *Trichococcus palustris* comb. nov. in the low-G+C gram-positive bacteria. Int J Syst Evol Microbiol 52:1113–1126

Lodders N, Kaempfer P (2012) A combined cultivation and cultivation-independent approach shows high bacterial diversity in water-miscible metalworking fluids. Syst Appl Microbiol 35:246–252

Lopes SP, Ceri H, Azevedo NF, Pereira MO (2012) Antibiotic resistance of mixed biofilms in cystic fibrosis: impact of emerging microorganisms on treatment of infection. Int J Antimicrob Agents 40:260–263

Ludwig W, Schleife K-H, Whitman WB (2009) Family III. *Carnobacteriaceae* fam. nov. In: De Vos P, Garrity GM, Jones D, Krieg NR, Ludwig W, Rainey FA, Schleifer K-H, Whitman WB (eds) The *Firmicutes*, vol 3, Bergey's manual of systematic bacteriology. Springer, New York, p 549

Ludwig W, Schleife K-H, Whitman WB (2010) Validation list #132. List of new names and new combinations previously effectively, but not validly, published. Int J Syst Bacteriol 60:469–472

Maoz A, Mayr R, Scherer S (2003) Temporal stability and biodiversity of two complex antilisterial cheese-ripening microbial consortia. Appl Environ Microbiol 69:4012–4018

Marchino T, Vela JI, Bassaganyas F, Sánchez S, Buil JA (2013) Acute-onset endophthalmitis caused by *Alloiococcus otitidis* following a dexamethasone intravitreal implant. Case Rep Ophthalmol 4:37–41

Marmur J (1961) A procedure for the isolation of deoxyribonucleic acid from microorganisms. J Mol Biol 3:208–218

Marsh RL, Binks MJ, Beissbarth J, Christensen P, Morris PS, Leach AJ, Smith-Vaughan HC (2012) Quantitative PCR of ear discharge from Indigenous Australian children with acute otitis media with perforation supports a role for *Alloiococcus otitidis* as a secondary pathogen. BMC Ear Nose Throat Disord 12:11

Martin-Visscher LA, van Belkum MJ, Garneau-Tsodikova S, Whittal RM, Zheng J, McMullen LM, Vederas JC (2008) Isolation and characterization of carnocyclin A, a novel circular bacteriocin produced by *Carnobacterium maltaromaticum* UAL307. Appl Environ Microbiol 74:4756–4763

Matsuo J, Harimaya A, Fukumoto T, Nakamura S, Yoshida M, Takahashi K, Iida M, Fujii N, Yamaguchi H (2011) Impact of anaerobic and oligotrophic conditions on survival of *Alloiococcus otitidis*, implicated as a cause of otitis media. J Infect Chemother 17:478–482

Mesbah M, Premachandran U, Whitman WB (1989) Precise measurement of the G+C content of deoxyribonucleic-acid by high-performance liquid-chromatography. Int J Syst Bacteriol 39:159–167

Métivier A, Pilet MF, Dousset X, Sorokine O, Anglade P, Zagorec M, Piard JC, Marion D, Cenatiempo Y, Fremaux C (1998) Divercin V41, a new bacteriocin with two disulfide bonds produced by *Carnobacterium divergens* V41: primary structure and genomic organization. Microbiology 144:2837–2844

Meziane-Cherif D, Decré D, Høiby EA, Courvalin P, Périchon B (2008) Genetic and biochemical characterization of CAD-1, a chromosomally encoded new class A penicillinase from *Carnobacterium divergens*. Antimicrob Agents Chemother 52:551–556

Miller A, Morgan ME, Libbey LM (1974) *Lactobacillus maltaromicus*, a new species producing a malty aroma. Int J Syst Bacteriol 24:346–354

Mora D, Scarpellini M, Franzetti L, Colombo S, Galli A (2003) Reclassification of *Lactobacillus maltaromicus* (Miller et al. 1974) DSM 20342$^T$ and DSM 20344 and *Carnobacterium piscicola* (Collins et al. 1987) DSM 20730$^T$ and DSM 20722 as *Carnobacterium maltaromaticum* comb. nov. Int J Syst Evol Microbiol 53:675–678

Mougari F, Jacquier H, Berçot B, Hannouche D, Nizard R, Cambau E, Zadegan F (2013) Prosthetic knee arthritis due to *Granulicatella adiacens* after dental cares. J Med Microbiol 62:1624–1627

Nakajima K, Hirota K, Nodasaka Y, Yumoto I (2005) *Alkalibacterium iburiense* sp. nov., an obligate alkaliphile that reduces an indigo dye. Int J Syst Evol Microbiol 55:1525–1530

Newton RJ, Bootsma MJ, Morrison HG, Sogin ML, McLellan SL (2013) A microbial signature approach to identify fecal pollution in the waters off an urbanized coast of Lake Michigan. Microb Ecol 65:1011–1023

Nilsson L, Nielsen MK, Ng Y, Gram L (2002) Role of acetate in production of an autoinducible class IIa bacteriocin in *Carnobacterium piscicola* A9b. Appl Environ Microbiol 68:2251–2260

Ntougias S, Russell NJ (2001) *Alkalibacterium olivoapovliticus* gen. nov., sp. nov., a new obligately alkaliphilic bacterium isolated from edible-olive washwaters. Int J Syst Evol Microbiol 51:1161–1170

Ntougias S, Russell NJ (2009) Genus II. *Alkalibacterium*. In: De Vos P, Garrity GM, Jones D, Krieg NR, Ludwig W, Rainey FA, Schleifer K-H, Whitman WB (eds) The *Firmicutes*, vol 3, Bergey's manual of systematic bacteriology. Springer, New York, pp 549–557

Oberreuter H, Seiler H, Scherer S (2002) Identification of coryneform bacteria and related taxa by Fourier-transform infrared (FT-IR) spectroscopy. Int J Syst Evol Microbiol 52:91–100

Ohara-Nemoto Y, Tajika S, Sasaki M, Kaneko M (1997) Identification of *Abiotrophia adiacens* and *Abiotrophia defectiva* by 16S rRNA gene PCR and restriction fragment length polymorphism analysis. J Clin Microbiol 35:2458–2463

Ohara-Nemoto Y, Kishi K, Satho M, Tajika S, Sasaki M, Namioka A, Kimura S (2005) Infective endocarditis caused by *Granulicatella elegans* originating in the oral cavity. J Clin Microbiol 43:1405–1407

Okada S, Uchimura T, Kozaki M (1992) Nyusankin Jikken Manyuaru (Laboratory manual for lactic acid bacteria) (in Japanese). Asakura-shoten, Tokyo

Peter S, Holder C, Braun N, Jung M, Einspanier R, Gabler C (2013) Detection of lactic acid bacteria in the bovine uterus and vagina by real-time PCR approaches. Reprod Biol 13:53–54

Pettigrew MM, Laufer AS, Gent JF, Kong Y, Fennie KP, Metlay JP (2012) Upper respiratory tract microbial communities, acute otitis media pathogens and antibiotic use in healthy and sick children. Appl Environ Microbiol 78:6262–6270

Pfennig N (1978) *Rhodocyclus purpureus* gen. nov. and sp. nov., a ring-shaped, vitamin $B_{12}$-requiring member of the Family *Rhodospirillaceae*. Int J Syst Bacteriol 28:283–288

Pikuta EV, Fisher D, Hoover RB (2011) Anaerobic cultures from preserved tissues of baby mammoth. In: vol 8152. pp 81520U-81520U-81510

Pikuta EV, Marsic D, Bej A, Tang J, Krader P, Hoover RB (2005) *Carnobacterium pleistocenium* sp. nov., a novel psychrotolerant, facultative anaerobe isolated from permafrost of the Fox Tunnel in Alaska. Int J Syst Evol Microbiol 55:473–478

Pikuta EV, Hoover RB, Bej AK, Marsic D, Whitman WB, Krader PE, Tang J (2006) *Trichococcus patagoniensis* sp. nov., a facultative anaerobe that grows at $-5\,^{\circ}C$, isolated from penguin guano in Chilean Patagonia. Int J Syst Evol Microbiol 56:2055–2062

Quadri LEN, Sailer M, Roy KL, Vederas JC, Stiles ME (1994) Chemical and genetic characterization of bacteriocins produced by *Carnobacterium piscicola* LV17B. J Biol Chem 269:12204–12211

Quartermain L, Tailor H, Njenga S, Bhattacharjee P, Rao GG (2013) Neonatal *Granulicatella elegans* bacteremia, London, UK. Emerg Infect Dis 19:1165–1166

Quirk PG, Guffanti AA, Plass RJ, Clejan S, Krulwich TA (1991) Protonophore-resistance and cytochrome expression in mutant strains of the facultative alkaliphile *Bacillus firmus* OF4. Biochim Biophys Acta 1058:131–140

Rachman C, Kabadjova P, Valcheva R, Prévost H, Dousset X (2004) Identification of *Carnobacterium* species by restriction fragment length polymorphism of the 16S-23S rRNA gene intergenic spacer region and species-specific PCR. Appl Environ Microbiol 70:4468–4477

Rahman A, Gleinser M, Lanhers M-C et al (2013) Adaptation of the lactic acid bacterium *Carnobacterium maltaromaticum* LMA 28 to the mammalian gastrointestinal tract: from survival in mice to interaction with human cells. Int Dairy J 34:93–99

Rainey FA (2009) Genus XIII. *Trichococcus*. In: De Vos P, Garrity GM, Jones D, Krieg NR, Ludwig W, Rainey FA, Schleifer K-H, Whitman WB (eds) The *Firmicutes*, vol 3, Bergey's manual of systematic bacteriology. Springer, New York, pp 585–588

Rainey FA, Ward-Rainey N, Kroppenstedt RM, Stackebrandt E (1996) The genus *Nocardiopsis* represents a phylogenetically coherent taxon and a distinct actinomycete lineage: proposal of *Nocardiopsaceae* fam nov. Int J Syst Bacteriol 46:1088–1092

Rastelli E, Giraffa G, Carminati D, Parolari G, Barbuti S (2005) Identification and characterization of halotolerant bacteria in spoiled dry-cured hams. Meat Sci 70:241–246

Ratcliffe P, Fang H, Thidholm E, Boräng S, Westling K, Özenci V (2013) Comparison of MALDI-TOF MS and VITEK 2 system for laboratory diagnosis of *Granulicatella* and *Abiotrophia* species causing invasive infections. Diagn Microbiol Infect Dis 77:216–219

Reasoner DJ, Geldreich EE (1985) A new medium for the enumeration and subculture of bacteria from potable water. Appl Environ Microbiol 49:1–7

Renko J, Koskela KA, Lepp PW, Oksala N, Levula M, Lehtimäki T, Solakivi T, Kunnas T, Nikkari S, Nikkari S (2013) Bacterial DNA signatures in carotid atherosclerosis represent both commensals and pathogens of skin origin. Eur J Dermatol 23:53–58

Rihakova J, Belguesmia Y, Petit VW, Pilet MF, Prevost H, Dousset X, Drider D (2009) Divercin V41 from gene characterization to food applications: 1998–2008, a decade of solved and unsolved questions. Lett Appl Microbiol 48:1–7

Ringø E, Holzapfe W (2000) Identification and characterization of carnobacteria associated with the gills of Atlantic salmon (*Salmo salar* L.). Syst Appl Microbiol 23:523–527

Roggenkamp A, Leitritz L, Baus K, Falsen E, Heesemann J (1998a) PCR for detection and identification of *Abiotrophia* spp. J Clin Microbiol 36:2844–2846

Roggenkamp A, Abele-Horn M, Trebesius KH, Tretter U, Autenrieth IB, Heesemann J (1998b) *Abiotrophia elegans* sp. nov., a possible pathogen in patients with culture-negative endocarditis. J Clin Microbiol 36:100–104

Roth E, Schwenninger SM, Eugster-Meier E, Lacroix C (2011) Facultative anaerobic halophilic and alkaliphilic bacteria isolated from a natural smear ecosystem inhibit *Listeria* growth in early ripening stages. Int J Food Microbiol 147:26–32

Roth E, Miescher Schwenninger S, Hasler M, Eugster-Meier E, Lacroix C (2010) Population dynamics of two antilisterial cheese surface consortia revealed by temporal temperature gradient gel electrophoresis. BMC Microbiol 10:74

Ruoff KL (1991) Nutritionally variant streptococci. Clin Microbiol Rev 4:184–190

Sampo M, Ghazouani O, Cadiou D, Trichet E, Hoffart L, Drancourt M (2013) *Dolosigranulum pigrum* keratitis: a three-case series. BMC Ophthalmol 13:31

Scarpellini M, Mora D, Colombo S, Franzetti L (2002) Development of genus/species-specific PCR analysis for identification of *Carnobacterium* strains. Curr Microbiol 45:24–29

Schaffer PA, Lifland B, Van SS, Casper DR, Davis CR (2013) Meningoencephalitis associated with *Carnobacterium maltaromaticum*-like bacteria in stranded juvenile salmon sharks (*Lamna ditropis*). Vet Pathol 50:412–417

Scheff G, Salcher O, Lingens F (1984a) *Trichococcus flocculiformis* gen. nov. sp. nov - a new gram-positive filamentous bacterium Isolated from bulking sludge. Appl Microbiol Biot 19:114–119

Scheff G, Salcher O, Lingens F (1984b) *Trichococcus flocculiformis* gen. nov., sp. nov. In validation of the publication of new names and new combinations previously effectively published outside the IJSB, list no. 15. Int J Syst Bacteriol 34:355–356

Schink B (1984) Fermentation of tartrate enantiomers by anaerobic bacteria, and description of two new species of strict anaerobes, *Ruminococcus pasteurii* and *Ilyobacter tartaricus*. Arch Microbiol 139:409–414

Schink B, Pfennig N (1982) Fermentation of trihydroxybenzenes by *Pelobacter acidigallici* gen. nov. sp. nov., a new strictly anaerobic, non-sporeforming bacterium. Arch Microbiol 133:195–201

Schleife KH, Kandler O (1972) Peptidoglycan types of bacterial cell-walls and their taxonomic implications. Bacteriol Rev 36:407–477

Schmidt VSJ, Mayr R, Wenning M, Gloeckner J, Busse H-J, Scherer S (2009) *Bavariicoccus seileri* gen. nov., sp. nov., isolated from the surface and smear water of German red smear soft cheese. Int J Syst Evol Microbiol 59:2437–2443

Schulthess B, Brodner K, Bloemberg GV, Zbinden R, Böttger EC, Hombach M (2013) Identification of gram-positive cocci by use of matrix-assisted laser desorption ionization-time of flight mass spectrometry: comparison of different preparation methods and implementation of a practical algorithm for routine diagnostics. J Clin Microbiol 51:1834–1840

Senn L, Entenza JM, Greub G, Jaton K, Wenger A, Bille J, Calandra T, Prod'hom G (2006) Bloodstream and endovascular infections due to *Abiotrophia defectiva* and *Granulicatella* species. BMC Infect Dis 6:9

Shailaja TS, Sathiavathy KA, Unni G (2013) Infective endocarditis caused by *Granulicatella adiacens*. Indian Heart J 65:447–449

Shaw BG, Harding CD (1985) Atypical lactobacilli from vacuum packaged meats: comparison by DNA hybridization, cell composition and biochemical tests with a description of *Lactobacillus carnis* sp. nov. Syst Appl Microbiol 6:291–297

Siqueira JF Jr, Rôças IN (2006) *Catonella morbi* and *Granulicatella adiacens*: new species in endodontic infections. Oral Surg Oral Med Oral Pathol Oral Radiol Endod 102:259–264

Skaugen M, Cintas LM, Nes IF (2003) Genetics of bacteriocin production in lactic acid bacteria. In: Wood JB (ed) Genetics of lactic acid bacteria. Kluwer Academic, New York, pp 225–260

Snauwaert I, Hoste B, De Bruyne K, Peeters K, De Vuyst L, Willems A, Vandamme P (2013) *Carnobacterium iners* sp. nov., a psychrophilic, lactic acid-producing bacterium from the littoral zone of the Forlidas pond (Pensacola mountains), Antarctica. Int J Syst Evol Microbiol 63:1370–1375

Spielmeyer W, McMeekin TA, Miller J, Franzmann PD (1993) Phylogeny of the Antarctic bacterium, *Carnobacterium alterfunditum*. Polar Biol 13:501–503

Stackebrandt E, Schumann P, Swiderski J, Weiss N (1999) Reclassification of *Brevibacterium incertum* (Breed 1953) as *Desemzia incerta* gen. nov., comb. nov. Int J Syst Bacteriol 49:185–188

Stackebrandt E (2009) Genus VIII. *Desemzia*. In: De Vos P, Garrity GM, Jones D, Krieg NR, Ludwig W, Rainey FA, Schleifer KH, Whitman WB (eds) Bergey's manual of systematic bacteriology, 2nd edn., vol 3. The Firmicutes. Springer, New York, pp 568–572

Stamatakis A (2006) RAxML-VI-HPC: maximum likelihood-based phylogenetic analyses with thousands of taxa and mixed models. Bioinformatics 22:2688–2690

Stams AJM, Van Dijk JB, Dijkema C, Plugge CM (1993) Growth of syntrophic propionate-oxidizing bacteria with fumarate in the absence of methanogenic bacteria. Appl Environ Microbiol 59:1114–1119

Stams AJM, Huisman J, Garcia EPA, Muyzer G (2009) Citric acid wastewater as electron donor for biological sulfate reduction. Appl Microbiol Biotechnol 83:957–963

Starliper CE, Shotts EB, Brown J (1992) Isolation of *Carnobacterium piscicola* and an unidentified gram-positive bacillus from sexually mature and post-spawning rainbow trout *Oncorhynchus mykiss*. Dis Aquat Organ 13:181–187

Steinhaus EA (1941) A study of the bacteria associated with thirty species of insects. J Bacteriol 42:757–790

Stoffels G, Nissen-Meyer J, Gudmundsdottir A, Sletten K, Holo H, Nes IF (1992) Purification and characterization of a new bacteriocin isolated from a *Carnobacterium* sp. Appl Environ Microbiol 58:1417–1422

Swain B, Otta S (2012) *Granulicatella adiacens*—an unusual causative agent for carbuncle. Indian J Pathol Microbiol 55:609–610

Tahiri I, Desbiens M, Benech R, Kheadr E, Lacroix C, Thibault S, Ouellet D, Fliss I (2004) Purification, characterization and amino acid sequencing of divergicin M35: a novel class IIa bacteriocin produced by *Carnobacterium divergens* M35. Int J Food Microbiol 97:123–136

Tanasupawat S, Thongsanit J, Thawai C, Lee KC, Lee J-S (2011) *Pisciglobus halotolerans* gen. nov., sp. nov., isolated from fish sauce. Int J Syst Evol Microbiol 61:1688–1692

Tekippe EM, Shuey S, Winkler DW, Butler MA, Burnham CA (2013) Optimizing identification of clinically relevant gram-positive organisms using the Bruker Biotyper MALDI-TOF MS system. J Clin Microbiol 51:1421–1427

Terzaghi BE, Sandine WE (1975) Improved medium for lactic streptococci and their bacteriophages. Appl Microbiol 29:807–813

Thornley MJ (1957) Observations on the microflora of minced chicken meat irradiated with 4 mev cathode rays. J Appl Bacteriol 20:286–298

Thornley MJ, Sharp ME (1959) Microorganisms from chicken meat related to both lactobacilli and aerobic sporeformers. J Appl Bacteriol 22:368–376

Tindall BJ (1990) Lipid-composition of *Halobacterium lacusprofundi*. Fems Microbiol Lett 66:199–202

Toffin L, Zink K, Kato C, Pignet P, Bidault A, Bienvenu N, Birrien J-L, Prieur D (2005) *Marinilactibacillus piezotolerans* sp. nov., a novel marine lactic acid bacterium isolated from deep sub-seafloor sediment of the Nankai Trough. Int J Syst Evol Microbiol 55:345–351

Tulini FL, De Martinis ECP (2010) Improved adsorption-desorption extraction applied to the partial characterization of the antilisterial bacteriocin produced by *Carnobacterium maltaromaticum* C2. Braz J Microbiol 41:493–496

Tulini FL, Lohans CT, Bordon KCF, Zheng J, Arantes EC, Vederas JC, De Martinis ECP (2014) Purification and characterization of antimicrobial peptides from fish isolate *Carnobacterium maltaromaticum* C2: Carnobacteriocin X and carnolysins A1 and A2. Int J Food Microbiol 173:81–88

Tung SK, Chang TC (2011) *Granulicatella*. In: Liu D (ed) Molecular detection of human bacterial pathogens. CRC Press, Boca Raton, pp 249–255

van Gelder AH, Aydin R, Alves MM, Stams AJM (2012) 1,3-Propanediol production from glycerol by a newly isolated *Trichococcus* strain. Microb Biotechnol 5:573–578

VandeWalle JL, Goetz GW, Huse SM, Morrison HG, Sogin ML, Hoffmann RG, Yan K, McLellan SL (2012) *Acinetobacter, Aeromonas* and *Trichococcus* populations dominate the microbial community within urban sewer infrastructure. Environ Microbiol 14:2538–2552

Venkateswaran N, Kalsow CM, Hindman HB (2013) Phlyctenular keratoconjunctivitis associated with *Dolosigranulum pigrum*. Ocul Immunol Inflamm 22:242–245

Voget S, Klippel B, Daniel R, Antranikian G (2011) Complete genome sequence of *Carnobacterium* sp. 17–4. J Bacteriol 193:3403–3404

von Graevenitz A (1993a) Revised nomenclature of *Alloiococcus otitis*. J Clin Microbiol 31:472

von Graevenitz A (1993b) Revised nomenclature of *Alloiococcus otitis*. J Clin Microbiol 31:1402

Wang M, Lin J, Sun S, Yuan X (2012) Identification of *Granulicatella elegans* from a case of infective endocarditis, based on the phenotypic characteristics and 16S rRNA gene sequence. J Cardiovasc Med. doi:10.2459/JCM.0b013e328356a471

Wasney MA, Holley RA, Jayas DS (2001) Cresol red thallium acetate sucrose inulin (CTSI) agar for the selective recovery of *Carnobacterium* spp. Int J Food Microbiol 64:167–174

Watanabe K, Nagao N, Toda T, Kurosawa N (2008) Changes in bacterial communities accompanied by aggregation in a fed-batch composting reactor. Curr Microbiol 56:458–467

Whitman WB, Ankwanda E, Wolfe RS (1982) Nutrition and carbon metabolism of *Methanococcus voltae*. J Bacteriol 149:852–863

Widdel F, Pfennig N (1981) Studies on dissimilatory sulfate-reducing bacteria that decompose fatty acids. I. Isolation of new sulfate-reducing bacteria enriched with acetate from saline environments. Description of *Desulfobacter postgatei* gen. nov., sp. nov. Arch Microbiol 129:395–400

Widdel F, Bak F (1992) Gram-negative mesophilic sulfate-reducing bacteria. In: Balows A, Trüper H, Dworkin M, Harder W, Schleifer K-H (eds) The prokaryotes. Springer, New York, pp 3352–3378

Wolin EA, Wolin MJ, Wolfe RS (1963) Formation of methane by bacterial extracts. J Biol Chem 238:2882–2886

Woo PCY, Teng JLL, Yeung JMY, Tse H, Lau SKP, Yuen K-Y (2011) Automated identification of medically important bacteria by 16S rRNA gene sequencing using a novel comprehensive database, 16SpathDB. J Clin Microbiol 49:1799–1809

Worobo RW, Henkel T, Sailer M, Roy KL, Vederas JC, Stiles ME (1994) Characteristics and genetic determinant of a hydrophobic peptide bacteriocin, carnobacteriocin A, produced by *Carnobacterium piscicola* LV17A. Microbiology 140:517–526

Worobo RW, Van Belkum MJ, Sailer M, Roy KL, Vederas JC, Stiles ME (1995) A signal peptide secretion--dependent bacteriocin from *Carnobacterium divergens*. J Bacteriol 177:3143–3149

WPCM (1989) IUMS/JCFMH working party for cultural media. Cresol red thallium acetate sucrose (CTAS) agar. In: Baird, Corry, Curtis, Mossel, Skovgaard (eds) Pharmacopoeia of cultural media for food microbiology: additional monographs. Elsevier, Amsterdam, pp 129–131

Xiao D, Zhao F, Lv M, Zhang H, Zhang Y, Huang H, Su P, Zhang Z, Zhang J (2012) Rapid identification of microorganisms isolated from throat swab specimens of community-acquired pneumonia patients by two MALDI-TOF MS systems. Diagn Microbiol Infect Dis 73:301–307

Xu J, Yang H, Lai X, Fu X, Wu J, Huang L, Yu X, Wu Y, Wu Y, Liu B (1997) Etiological study for a case of multi-bacterial synergistic gangrene. Chin Sci Bull 42:511–517

Yamaguchi T, Soutome S, Oho T (2011) Identification and characterization of a fibronectin-binding protein from *Granulicatella adiacens*. Mol Oral Microbiol 26:353–364

Yamasato K, Ishikawa M (2009) Genus XII. *Marinilactibacillus*. In: De Vos P, Garrity GM, Jones D, Krieg NR, Ludwig W, Rainey FA, Schleifer K-H, Whitman WB (eds) The *Firmicutes*, vol 3, Bergey's manual of systematic bacteriology. Springer, New York, pp 580–584

Yamazaki K, Suzuki M, Kawai Y, Inoue N, Montville TJ (2005) Purification and characterization of a novel class IIa bacteriocin, piscicocin CS526, from surimi-associated *Carnobacterium piscicola* CS526. Appl Environ Microbiol 71:554–557

Yang C, Niu Y, Su H, Wang Z, Tao F, Wang X, Tang H, Ma C, Xu P (2010) A novel microbial habitat of alkaline black liquor with very high pollution load: microbial diversity and the key members in application potentials. Bioresour Technol 101:1737–1744

Yarza P, Ludwig W, Euzeby J, Amann R, Schleifer KH, Glockner FO, Rossello-Mora R (2010) Update of the All-Species Living Tree Project based on 16S and 23S rRNA sequence analyses. Syst Appl Microbiol 33:291–299

Yoshida N, Takahashi N, Hiraishi A (2005) Phylogenetic characterization of a polychlorinated-dioxin-dechlorinating microbial community by use of microcosm studies. Appl Environ Microbiol 71:4325–4334

Yumoto I, Hirota K, Nodasaka Y, Tokiwa Y, Nakajima K (2008) *Alkalibacterium indicireducens* sp. nov., an obligate alkaliphile that reduces indigo dye. Int J Syst Evol Microbiol 58:901–905

Yumoto I, Hirota K, Nodasaka Y, Yokota Y, Hoshino T, Nakajima K (2004) *Alkalibacterium psychrotolerans* sp., nov., a psychrotolerant obligate alkaliphile that reduces an indigo dye. Int J Syst Evol Microbiol 54:2379–2383

Zaura E, Keijser BJF, Huse SM, Crielaard W (2009) Defining the healthy "core microbiome" of oral microbial communities. BMC Microbiol 9:259

Zhang T, Shao M-F, Ye L (2012) 454 Pyrosequencing reveals bacterial diversity of activated sludge from 14 sewage treatment plants. ISME J 6:1137–1147

Zheng X, Freeman AF, Villafranca J, Shortridge D, Beyer J, Kabat W, Dembkowski K, Shulman ST (2004) Antimicrobial susceptibilities of invasive pediatric *Abiotrophia* and *Granulicatella* isolates. J Clin Microbiol 42:4323–4326

Zhilina TN, Kotsyurbenko OR, Osipov GA, Kostrikina NA, Zavarzin GA (1995) *Ruminococcus palustris* sp. nov.—a psychroactive anaerobic organism from a swamp. Mikrobiologiya 64:674–680

Zúñiga M, Pardo I, Ferrer S (1993) An improved medium for distinguishing between homofermentative and heterofermentative lactic acid bacteria. Int J Food Microbiol 18:37–42

# 5 The Family *Clostridiaceae*, Other Genera

*Erko Stackebrandt*
Leibniz Institute DSMZ-German Collection of Microorganisms and Cell Cultures GmbH, Braunschweig, Germany



**Abstract**
The family *Clostridiaceae*, containing beside the type genus *Clostridium* more than 30 additional genera of the Firmicutes, constitutes a physiologically and phylogenetically heterogeneous taxon, which are generally monospore-forming, anaerobic Gram-positive-staining rods, with meso-diaminopimelic acid as the diagnostic diamino acid in their peptidoglycan. Morphological variations, such as clusters consisting of multiples of four cells (*Sarcina*) and cells with multiple spores (*Anaerobacter*) or no spores, do occur. Many *Clostridium* species which do not belong to the authentic genus defined by the type species *Clostridium butyricum* are scattered among validly named *Clostridiaceae* genera and are taxonomically treated as "incertae sedis." This brief overview concentrates on genera and species described since 2006, which are not covered in the chapter *Clostridiaceae* in *Bergey's Manual of Systematic Bacteriology*, 2nd edition.

## Taxonomy

The pioneering 16S rRNA-gene sequence-based study of Collins et al. (1994) defined 14 *Clostridium* groups (Group I to XIV), restricting Group I as *Clostridium sensu stricto*. Some of the *Clostridium* species were reclassified as new genera (*Caloramator, Filifactor, Moorella, Oxobacter*), and, as a result of the study, authors were encouraged to create novel genera for those new species which had the general appearance of *Clostridium* species but grouped outside Group I. Despite the establishment of a "Taxonomic Subcommittee on Clostridia and *Clostridium*-like Organisms," a coherent solution to the reclassification of the *Clostridium*-like organisms could not be provided because of the lack of genus-specific properties which reliably define related genera. The situation was reversed with the taxonomy of the genus *Sarcina* which clusters within the radiation of authentic *Clostridium* species (Fox et al. 1980). Because of the date of publication *Sarcina* (Goodsir 1842) has priority over *Clostridium* (Prazmowski 1880), the Code of Nomenclature (Lapage et al. 1992) would request all *Clostridium* species, including those of medical importance, to be reclassified as Sarcina spp. To avoid this step with major implications to clinical microbiology, the name *Clostridium* and the type species *Clostridium butyricum* were conserved (nomen periculosum, Rule 56a of the Code).

The RaXML tree depicted in ➲ *Fig. 5.1* clearly shows the phylogenetic heterogeneity of genera listed in the *List of prokaryotic names with standing in nomenclature* (http://www.bacterio.net/) to be members of the family *Clostridiaceae*, as the families *Peptostreptococcaceae* and *Eubacteriaceae* form sister clades with one or groups of *Clostridiaceae* family members. The most deeply branching genus *Lutispora* (Shiratori et al. 2008) has been transferred to the family *Gracilibacteraceae* (Stackebrandt 2015).

The lasts comprehensive coverage of the family *Clostridiaceae* has been presented by Wiegel (2009) in *Bergey's Manual of Systematic Bacteriology*. At the time of that publication, the family comprised 11 genera, while 5 years later the family contains 27 genera. No novel species were described since 2006 for the genera *Anaerobacter, Anoxynatronum, Caloranaerobacter, Caminicella, Oxobacter, Sarcina, Thermohalobacter*, and *Tindallia*. Additional species were described for *Clostridium* (dealt with by in a separate chapter), *Alkaliphilus, Caloramator*, and *Natronincola*. ➲ *Table 5.1* is a list of genera and species described since 2006, together with some of the salient feature of these taxa (➲ *Table 5.2*).

## Genome Sequences

Besides members of the genus *Clostridium*, several strains of the family were subjected to genome sequences, available as finished, permanent drafts, or incomplete sequences. ➲ *Table 5.3* is a summary of the current status of these studies as listed in the Gold database (genomes.org/cgi-bib/Gold/Search.cgi).

E. Rosenberg et al. (eds.), *The Prokaryotes – Firmicutes and Tenericutes*, DOI 10.1007/978-3-642-30120-9_214,

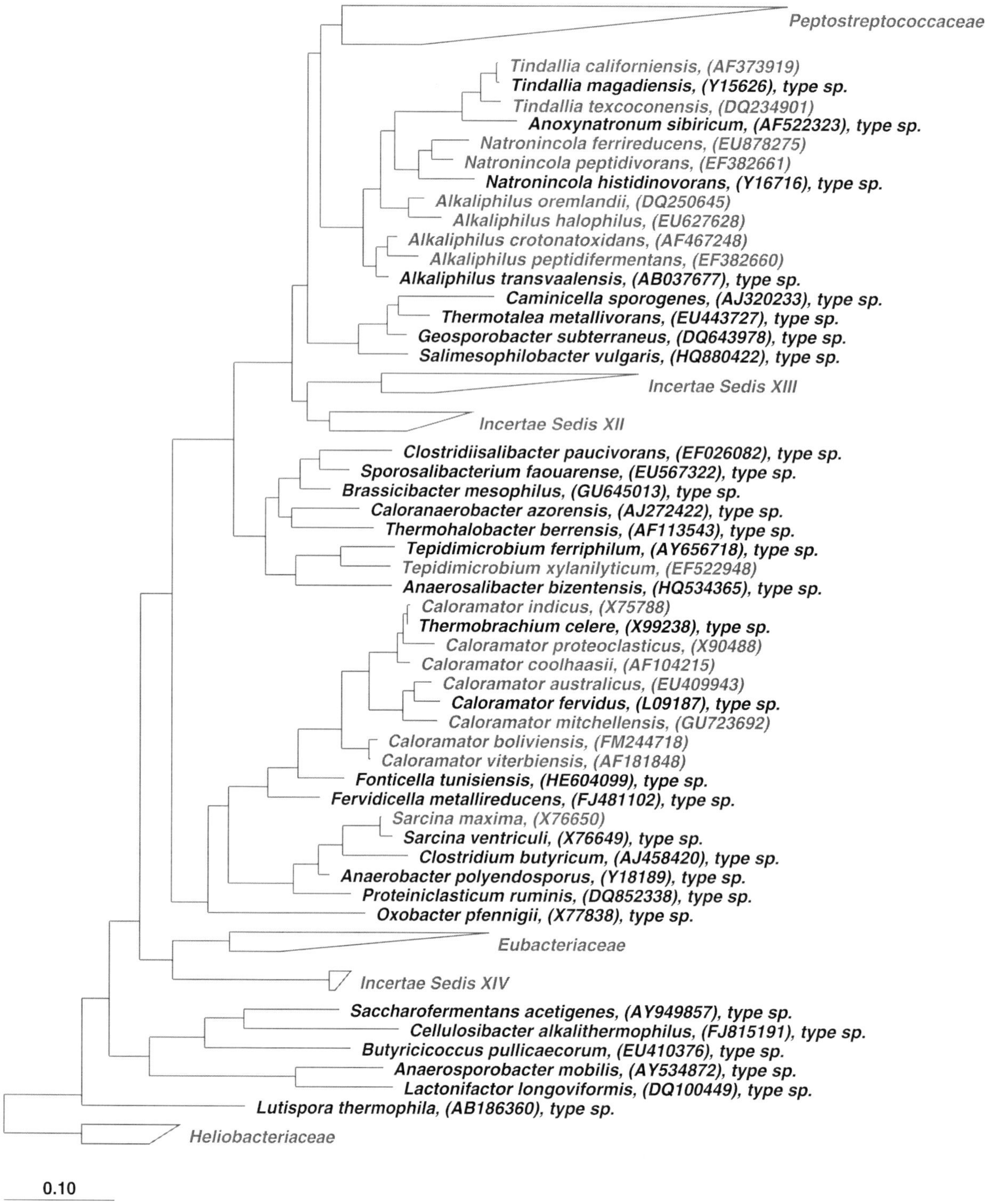


**Fig. 5.1**

Neighbor-joining genealogy reconstruction based on the RAxML algorithm (Stamatakis 2006) of the sequences of members of the family *Clostridiaceae* (except for the genus *Clostridium*) and some neighboring taxa present in the LTP_106 (Yarza et al. 2010). The tree was reconstructed by using a subset of sequences. Representative sequences from closely relative genera were used to stabilize the tree topology. In addition, a 40 % maximum frequency filter was applied to remove hypervariable positions from the alignment. Scale bar indicates estimated sequence divergence

**Table 5.1**
**Species published since 2006, for genera described before 2006**

| Genus | *Alkaliphilus* | *Alkaliphilus* | *Alkaliphilus* | *Natronincola* | *Natronincola* |
|---|---|---|---|---|---|
| Species | *oremlandii* | *peptidofermentans* | *halophilus* | *ferrireducens* | *peptidovorans* |
| Gram stain | Positive | Positive | Positive | Positive | Positive |
| Motility | + | + | + | + | + (young cells) |
| Morphology | Rods | Rods | Slightly curved rods | Slightly curved rods, palisades, capsule | Slightly curved rods, palisades, capsule |
| Spore formation | + | + | + | + | + |
| Growth optimum (° C) | 37 | 35 | 32 | 35–37 | 35–37 |
| Heterotrophyon | Lactate, fumarate, glycerol fermentation. Arsenate and thiosulfate respiration with low molecular weight organic acids as electron acceptors | YE, CA, peptone, tryptone, meat extract, some amino acids<br>Does not ferment carbohydrates, organic acids or alcohols<br>Uses, e.g., Fe(III), crotonate, fumarate, $S_2O_3^{2-}$ as electron acceptors | YE, CA, tryptone, fructose, sucrose, xylose, ribose, lactate, tartrate | YE, proteinaceous substrates, histidine; no sugars | YE, proteinaceous substrates, pyruvate; no sugars |
| Fermentation products from carbohydrates | nd | nd | F, A | A, P, f | A, F, p, ib, $H_2$ |
| Major fatty acids (>10 %) | nd | iso-$C_{15:0}$, $C_{19:1}\omega7c$, $C_{16:0}$, i-$C_{17:0}$, | iso-$C_{15:0}$, iso-$C_{15:1}$ F, iso-$C_{13:0}$ | $C_{16:1}\omega7c$, $C_{16:0}$, $C_{18:1}\omega9$ | $C_{14:0}$, $C_{16:1}\omega7c$, $C_{16:0}$, $C_{18:1}\omega9$ |
| Mol% G+C | 36.1 | 33.8 | 28.5 | 35.3 | 35.5 |
| Habitat | Sediment of Ohio River, USA | Sediment of lake Verkhnee Beloe, Russia | Sediment of salt lake Xiaokule, China | Sediment of lake Khadyn, Russia | Sediment of lake Khadyn, Russia |
| Publication | Fisher et al. 2008 | Zhilina et al. 2009a | Wu et al. 2010 | Zhilina et al. 2009b | Zhilina et al. 2009b |

| Genus | *Caloramator* | *Caloramator* | *Caloramator* | *Caloramator* |
|---|---|---|---|---|
| Species | *mitchellensis* | *boliviensis* | *quimbayensis* | *australicus* |
| Gram stain | Negative with positive wall structure | Variable | Positive | Positive |
| Motility | | Monotrichous | + | Peritrichous |
| Morphology | Slightly curved rods | Slightly curved rods | Slightly curved rods | Slightly curved rods |
| Spore formation | – | + | + | - |
| Growth optimum (° C) | 55 | 60 | 50 | 60 |

**Table 5.1 (continued)**

| | | | | |
|---|---|---|---|---|
| Heterotrophyon | YE and/or tryptone and various carbohydrates.<br>Vanadium(V) reduction | YE and various carbohydrates | YE with sugars, amino acids, starch dextrin | YE plus various carbohydrates<br>Fe(III), Mn (IV), and $S_0$ reduction |
| Carbohydrates | A, E, $CO_2$, $H_2$ | E, a, l, p, trace of $CO_2$, $H_2$ | (with YE) F, A, E, L | E, A |
| Major fatty acids (>10 %) | nd | anteiso-$C_{15:0}$, $C_{16:0}$, iso-$C_{16:0}$, $C_{15:1}$, iso-$C_{14:0}$, $C_{13:0}$, $C_{14:0}$ | | nd |
| Mol% G+C | 34 | 32.6 | 32.6 | 34 |
| Habitat | Thermal water Great Artesian Basin, Australia | Sediment water, hot spring, Bolivia | Hot spring, Andean region Columbia | Thermal water Great Artesian Basin, Australia |
| Publication | Ogg and Patel 2011 | Crespo et al. 2012 | Rubiano-Labrador et al. 2013 | Ogg and Patel 2009a |

*YE* yeast extract, *CA* casamino acids
*F* formate, *A* acetate, *P* propionate, *E* ethanol; *iv* isovalerate, *ib* iso-butyrate, *B* butyrate, *L* lactate. Major product: capital letters, small amounts: lower case letter
*nd* not determined

**Table 5.2**
**New genera and species of *Clostridiaceae* described since 2006**

| Genus | *Anaerosalibacter* | *Anaerosporobacter* | *Brassicibacter* | *Butyricicoccus* | *Cellulosibacter* | *Clostridiisalibacter* | *Fervidicella* |
|---|---|---|---|---|---|---|---|
| Species | *bizertensis* | *mobilis* | *mesophilus* | *pullicaecorum* | *alkalithermophilus* | *paucivorans* | *metallireducens* |
| Gram stain | Positive | Positive | Negative | Positive | Positive | Positive | Negative |
| Motility | Laterally inserted flagella | Peritrichous flagella | + | − | + | + | |
| Morphology | Rods | Rods | Rods | Coccoid, pairs | Rods | Rods | Curved rods |
| Spore formation | + | + | − | − | + | + | + |
| Growth optimum (° C) | 40 | 30 | 37 | 37 | 55 | 40–45 | 50 |
| Heterotrophyon | YE, peptone; some saccharides and organic acids | Saccharides | YE, CA, peptone, tryptone. Some saccharides and amino acids | YE, some saccharides, cellobiose | Many saccharides and polysaccharides | YE, CA, peptone; pyruvate, fumarate, and succinate | tryptone, yeast extract, casamino acids, Fe(III) reduction |
| End products from | | | | | | | |
| Carbohydrates | nd | F, A, $H_2$ | | B, $H_2$, $CO_2$ (in M2GSC broth, Barcenilla et al. 2000) | A, E, P, b | pyruvate fermentation: A, $H_2$, and $CO_2$ | E, A, $CO_2$, $H_2$ |
| Peptone-yeast | nd | | P, F, A, E, iv | | | | |
| Major fatty acids (>10 %) | iso-$C_{15:0}$, iso-$C_{15:0}$DMA, possibly iso-$C_{15:0}$ aldehyde | $C_{16:0}$, $C_{16:0}$3-OH, iso-$C_{17:0}$ I/anteiso-B | iso-$C_{15:0}$, iso-$C_{15:0}$DMA | nd | iso-$C_{14:0}$ 3OH, iso-$C_{15:0}$, iso-$C_{16:0}$, $C_{16:0}$ | iso-$C_{17:1}$ $\omega$3c (or $\omega$10c), $C_{14:0}$ | nd |
| Mol% G+C | 31.1 | 41 | 28.2 | 54.5 | 30 | 33 | 35.4 |
| Habitat | Waste motor oil storage tank, Tunisia | Forest soil, Korea | Food industry waste water, China | Cecum, broiler, Belgium | Soil, coconut garden, Thailand | Olive mill wastewater, Morocco | Thermal water Great Artesian Basin, Australia |
| Publication | Rezgui et al. 2012 | Jeong et al. 2007 | Fang et al. 2012 | Eeckhaut et al. 2008 | Watthanalamloet et al. 2012 | Liebgott et al. 2008 | Ogg and Patel 2010 |

**Table 5.2 (continued)**

| Genus | *Fonticella* | *Geosporobacter* | *Lactonifactor* | *Proteiniclasticum* | *Saccharofermentans* | *Salimesophilobacter* |
|---|---|---|---|---|---|---|
| Species | *tunisiensis* | *subterraneus* | *longoviformis* | *ruminis* | *acetigenes* | *vulgaris* |
| Gram stain | Positive | Positive | Positive to variable | Negative | Positive | positive |
| Motility | – | – | nd | – | – | Lateral flagellum |
| Morphology | Rods | Rods | Rods | Rods | Oval | Rods |
| Spore formation | | + | nd | – | + | |
| Growth optimum (° C) | 55 | 42 | range 25–45 | 38–39 | 37 | 35 |
| Heterotrophy on | Various saccharides | Saccharides and amino acids | Saccharides | Peptone, tryptone, amino acids | Saccharides and alcohols | Peptone, tryptone, xylose, glycerol; Fe(III) reduction |
| End products from | | | | | | |
| Carbohydrates | F, A, E, $CO_2$ | $CO_2$, $H_2$ | nd | | A, L, F, trace of $CO_2$, $H_2$ | A, F, P, L |
| Peptone-yeast | | | | A, P, Ib | | |
| Major fatty acids (>10 %) | $C_{14:0}$, iso-$C_{15:0}$, $C_{17:0}$, $C_{165:0}$ | nd | $C_{16:0}$ | iso-$C_{14:0}$, iso-$C_{13\ :0}$, iso-$C_{15:0}$, $C_{14\ :0}$, anteiso-$C_{13\ :0}$ | iso-$C_{15:0}$, anteiso-$C_{15:0}$, iso-$C_{14:0}$ 3-OH | iso-$C_{15:0}$, $C_{14:0}$, $C_{16:0}$, $C_{16:1}$ *cis*9 |
| Mol% G+C | 37.2 | 42.2 | 48 | 41 | 55.6 | 37 |
| Habitat | Hot spring, Tunisia | Deep aquifer, Paris Basin, France | Feces, human adult | Yak rumen, China | Brewery waste water, China | Paper mill wastewater |
| Publication | Fraj et al. 2013 | Klouche et al. 2007 | Clavel et al. 2007 | Zhang et al. 2010 | Chen et al. 2010 | Zhang et al. 2013 |

| Genus | *Sporosalibacterium* | *Tepidimicrobium* | *Tepidimicrobium* | *Thermotalea* |
|---|---|---|---|---|
| Species | *faouarense* | *ferriphilum* | *xylanilyticum* | *metallivorans* |
| Gram stain | Positive | Positive | Positive | Negative |
| Motility | + | Peritrichous flagella | Peritrichous flagella | Peritrichous flagella |
| Morphology | Rods | Curved rods | Rods | Curved rods |
| Spore formation | + | – | + | – |
| Growth optimum (° C) | 40 | 50 | 60 | 50 |
| Heterotrophy on | YE-various saccharides | Peptone, tryptone, casamino acids, YE, beef extract | Saccharides and proteinaceous compounds | Saccharides and organic acids |
| | | Fe(III) reduction | Fe(III) reduction | Fe(III) reduction |
| End products from | | | | |
| Carbohydrates | Pyruvate fermentation: A, $H_2$, $CO_2$ | nd | A, E, B, $CO_2$, $H_2$ | E, A |
| Peptone-yeast | | nd | | |
| Major fatty acids (>10 %) | iso-$C_{15:0}$, iso-$C_{14:0}$ 3-OH and/or iso-$C_{15}$ DMA | nd | nd | nd |
| Mol% G+C | 30.7 | 33 | 36.2 | 48 |
| Habitat | Hydrocarbon polluted soil, Tunisia | Freshwater hot spring, Russia | Anaerobic thermophilic waste digester, China | Thermal water Great Artesian Basin, Australia |
| Publication | Rezgui et al. 2011 | Slobodkin et al. 2006 | Niu et al. 2009 | Ogg and Patel 2009b |

For abbreviations see footnote of Table 5.1

**Table 5.3**
**Examples for species with genome sequences at various stages of completion**

| Taxon | Strain number | Gold identification | Status |
|---|---|---|---|
| *Alkaliphilus transvaalensis* | ATCC 700919 | Gi 0051614 | Incomplete |
| "Alkaliphilus metalliredigens" | QYMF | Gc 00587 | Finished |
| *Caloramator australicus* | RC3 | Gi 06829 | Permanent draft |
| Caloramator spp. | ALD01 | Gi 0021406 | Permanent draft |
| *Butyricicoccus pullicaecorum* | 1.2 | Gi 7127 | Permanent draft |
| *Clostridiisalibacter paucivorans* | DSM 22131 | Gi 11264 | Incomplete |
| *Fervidicella metallireducens* | AeB | Gi 06858 | Incomplete |
| Proteiniclasticum ruminis | DSM 24773 | Gi 0054515 | Incomplete |
| Thermotalea metallivorans | B2-1 | Gi 06857 | Incomplete |

## References


Barcenilla A, Pryde SE, Martin JC, Duncan SH, Stewart CS, Henderson C, Flint HJ (2000) Phylogenetic relationships of butyrate-producing bacteria from the human gut. Appl Environ Microbiol 66:1654–1661

Chen S, Niu L, Zhang Y (2010) *Saccharofermentans acetigenes* gen. nov., sp. nov., an anaerobic bacterium isolated from sludge treating brewery wastewater. Int J Syst Evol Microbiol 60:2735–2738

Clavel T, Lippman R, Gavini F, Doré J, Blaut M (2007) *Clostridium saccharogumia* sp. nov. and *Lactonifactor longoviformis* gen. nov., sp. nov., two novel human faecal bacteria involved in the conversion of the dietary phytoestrogen secoisolariciresinol diglucoside. Syst Appl Microbiol **30**, 16–26. Validation List no. 115. Int J Syst Evol Microbiol 57:893–897

Collins MD, Lawson PA, Willems A, Cordoba JJ, Fernandez-Garayzabal J, Garcia P, Cai J, Hippe H, Farrow JAE (1994) The phylogeny of the genus *Clostridium*: proposal of five new genera and eleven new species combinations. Int J Syst Bacteriol 44:812–826

Crespo C, Pozzo T, Nordberg Karlsson E, Alvarez MT, Mattiasson B (2012) *Caloramator boliviensis* sp. nov., a thermophilic, ethanol-producing bacterium isolated from a hot spring. Int J Syst Evol Microbiol 62:1679–1686

Eeckhaut V, Van ImmerseeL F, Teirlynck E, Pasmans F, Fievez V, Snauwaert C, Haesebrouck F, Ducatelle R, Louis P, Vandamme P (2008) *Butyricicoccus pullicaecorum* gen. nov., sp. nov., an anaerobic, butyrate-producing bacterium isolated from the caecal content of a broiler chicken. Int J Syst Evol Microbiol 58:2799–2802

Fang MX, Zhang WW, Zhang YZ, Tan HQ, Zhang XQ, Wu M, Zhu XF (2012) *Brassicibacter mesophilus* gen. nov., sp. nov., a strictly anaerobic bacterium isolated from food industry wastewater. Int J Syst Evol Microbiol 62:3018–3023

Fisher E, Dawson AM, Polshyna G, Lisak J, Crable B, Perera E, Ranganathan M, Thangavelu M, Basu P, Stolz JF (2008) Transformation of inorganic and organic arsenic by Alkaliphilus oremlandii sp. nov. strain OhILAs. Ann N Y Acad Sci, 1125, 230–241 Validation List no 127 (2009). Int J Syst Evol Microbiol 59:923–925

Fox GE, Stackebrandt E, Hespell RB, Gibson J, Maniloff J, Dyer T, Wolfe RS, Balch W, Tanner R, Magrum L, Zablen LB, Blakemore R, Gupta R, Luehrsen KR, Bonen L, Lewis BJ, Chen KN, Woese CR (1980) The phylogeny of prokaryotes. Science 209:457–463

Fraj B, Ben Hania W, Postec A, Hamdi M, Ollivier B, Fardeau ML (2013) *Fonticella tunisiensis* gen. nov., sp. nov., isolated from a hot spring. Int J Syst Evol Microbiol 63:1947–1950

Goodsir J (1842) History of a case in which a fluid periodically ejected from the stomach contained vegetable organisms of an undescribed form. Edinb Med Surg J 57:430–443

Jeong H, Lim YW, Yi H, Sekiguchi Y, Kamagata Y, Chun J (2007) *Anaerosporobacter mobilis* gen. nov., sp. nov., isolated from forest soil. Int J Syst Evol Microbiol 57:1784–1787

Klouche N, Fardeau ML, Lascourrèges JF, Cayol JL, Hacene H, Thomas P, Magot M (2007) *Geosporobacter subterraneus* gen. nov., sp. nov., a spore-forming bacterium isolated from a deep subsurface aquifer. Int J Syst Evol Microbiol 57:1757–1761

Lapage SP, Sneath PHA, Lessel EF, Skerman VBD, Seeliger HPR, Clark WA (1992) International code of nomenclature of bacteria: bacteriological code, 1990 revision. ASM Press, Washington, DC

Liebgott PP, Joseph M, Fardeau ML, Cayol JL, Falsen E, Chamkh F, Qatibi AI, Labat M (2008) *Clostridiisalibacter paucivorans* gen. nov., sp. nov., a novel moderately halophilic bacterium isolated from olive mill wastewater. Int J Syst Evol Microbiol 58:61–67

Niu L, Song L, Liu X, Dong X (2009) *Tepidimicrobium xylanilyticum* sp. nov., an anaerobic xylanolytic bacterium, and emended description of the genus *Tepidimicrobium*. Int J Syst Evol Microbiol 59:2698–2701

Ogg CD, Patel BKC (2009a) *Caloramator australicus* sp. nov., a thermophilic, anaerobic bacterium from the Great Artesian Basin of Australia. Int J Syst Evol Microbiol 59:95–101

Ogg CD, Patel BKC (2009b) *Thermotalea metallivorans* gen. nov., sp. nov., a thermophilic, anaerobic bacterium from the Great Artesian Basin of Australia aquifer. Int J Syst Evol Microbiol 59:964–971

Ogg CD, Patel BKC (2010) *Fervidicella metallireducens* gen. nov., sp. nov., a thermophilic, anaerobic bacterium from geothermal waters. Int J Syst Evol Microbiol 60:1394–1400

Ogg CD, Patel BKC (2011) *Caloramator mitchellensis* sp. nov., a thermoanaerobe isolated from the geothermal waters of the Great Artesian Basin of Australia, and emended description of the genus *Caloramator*. Int J Syst Evol Microbiol 61:644–653

Prazmowski A (1880) Untersuchung über die Entwickelungsgeschichte und Fermentwirkung einiger Bacterien-Arten. Inaugural Dissertation. Hugo Voigt, Leipzig pp 1–58

Rezgui R, Ben Ali Gam Z, Ben Hamed S, Fardeau ML, Cayol JL, Maaroufi A, Labat M (2011) *Sporosalibacterium faouarense* gen. nov., sp. nov., a moderately halophilic bacterium isolated from oil-contaminated soil. Int J Syst Evol Microbiol 61:99–104

Rezgui R, Maaroufi A, Fardeau ML, Ben Ali Gam Z, CayolL JL, Ben Hamed S, Labat M (2012) *Anaerosalibacter bizertensis* gen. nov., sp. nov., a halotolerant bacterium isolated from sludge. Int J Syst Evol Microbiol 62:2469–2474

Rubiano-Labrador C, Baena S, Díaz-Cárdenas C, Patel BKC (2013) *Caloramator quimbayensis* sp. nov., an anaerobic, moderately thermophilic bacterium isolated from a terrestrial hot spring. Int J Syst Evol Microbiol 63:1396–1402

Shiratori H, Ohiwa H, Ikeno H, Ayame S, Kataoka N, Miya A, Beppu T, Ueda K (2008) *Lutispora thermophila* gen. nov., sp. nov., a thermophilic, spore-forming bacterium isolated from a thermophilic methanogenic bioreactor digesting municipal solid wastes. Int J Syst Evol Microbiol 58:964–969

Slobodkin AI, Tourova TP, Kostrikina NA, Lysenko AM, German KE, Bonch-Osmolovskaya E, Birkeland NK (2006) *Tepidimicrobium ferriphilum* gen. nov., sp. nov., a novel moderately thermophilic, FeIII-reducing bacterium of the order *Clostridiales*. Int J Syst Evol Microbiol 56:369–372

Stackebrandt E (2015) Family *Gracilibacteraceae* and transfer of the genus *Lutispora* into *Gracilibacteraceae*. In: Rosenberg E, DeLong EF, Lory S, Stackebrandt E, Thompson F (eds) The Prokaryotes, Springer, New York

Stamatakis A (2006) RAxML-VI-HPC: maximum likelihood-based phylogenetic analyses with thousands of taxa and mixed models. Bioinformatics 22:2688–2690

Watthanalamloet A, Tachaapaikoon C, Lee YS, Kosugi A, Mori Y, Tanasupawat S, Kyu KL, Ratanakhanokchai K (2012) *Cellulosibacter alkalithermophilus* gen. nov., sp. nov. an anaerobic alkalithermophilic, cellulolytic-xylanolytic bacterium isolated from soil of a coconut garden. Int J Syst Evol Microbiol 62:2330–2335

Wiegel J (2009) Family I *Clostridiaceae*. In: de Vos P, Garrity GM, Jones D, Krieg NR, Ludwig W, Rainey FA, Schleifer K-H, Whitman WB (eds) Bergey's manual of systematic bacteriology, vol 3, 2nd edn. Springer, New York, pp 736–738

Wu XY, ShiKL XXW, Wu M, Oren A, Zhu XF (2010) *Alkaliphilus halophilus* sp. nov., a strictly anaerobic and halophilic bacterium isolated from a saline lake, and emended description of the genus *Alkaliphilus*. Int J Syst Evol Microbiol 60:2898–2902

Yarza P, Ludwig W, Euzéby J, Amann R, Schleifer K-H, Glöckner FO, Rosselló-Móra R (2010) Update of the All-species living-tree project based on 16S and 23S rRNA sequence analyses. Syst Appl Microbiol 33:291–299

Zhang K, Song L, Dong X (2010) *Proteiniclasticum ruminis* gen. nov., sp. nov., a strictly anaerobic proteolytic bacterium isolated from yak rumen. Int J Syst Evol Microbiol 60:2221–2225

Zhang YZ, Fang MX, Zhang WW, Li TT, Wu M, Zhu XF (2013) *Salimesophilobacter vulgaris* gen. nov., sp. nov., an anaerobic bacterium isolated from paper-mill wastewater. Int J Syst Evol Microbiol 63:1317–1322

Zhilina TN, Zavarzina DG, Kolganova TV, Lysenko AM, Tourova TP (2009a) *Alkaliphilus peptidofermentans* sp. nov., a new alkaliphilic bacterial soda lake isolate capable of peptide fermentation and FeIII reduction. Microbiology 78:445–454, [translated from Mikrobiologiya, 2009, **78**, 496–505]. Validation List no 130 (2009) Int J Syst Evol Microbiol **59**, 2647-2648

Zhilina TN, Zavarzina DG, Osipov GA, Kostrikina NA, Tourova TP (2009b) *Natronincola ferrireducens* sp. nov., and *Natronincola peptidovorans* sp. nov., new anaerobic alkaliphilic peptolytic iron-reducing bacteria isolated from soda lakes. Microbiology 78:455–467, [translated from Mikrobiologiya, 2009, **78**, 506–518]. Validation List no 130 (2009) Int J Syst Evol Microbiol **59**, 2647–2648

# 6 The Family *Enterococcaceae*

*Stephen Lory*
Department of Microbiology and Immunobiology, Harvard Medical School, Boston, MA, USA

**Abstract**
This chapter summarizes the general properties of the genera *Enterococcus*, *Melissococcus*, *Tetragenococcus*, and *Vagococcus* within the family *Enterococcacea*.

The order*Lactobacillae* includes the family *Enterococcaceae* with four genera: *Enterococcus*, *Melissococcus*, *Tetragenococcus*, and *Vagococcus*. They can be found in diverse environments, and many are colonizers of humans. In general, the *Enterococcaceae* are fastidious, and consequently, they thrive in environments where various nutritional needs are provided, usually by other living or dead organisms. In addition to the intestinal tracks of mammals, birds, fish, and insects, they can be recovered from decaying plant and animal material or, in aquatic environments, in close association with living or decomposing algae.

The most extensively studied members of this family belong to the genus *Enterococcus* (referred to as Enterococci) and, based on their 16S rRNA gene sequences, it includes over 40 species (➋ *Fig. 6.1*). Several *Enterococcus* species are found as part of the human normal microbiota primarily in the intestinal tract but also on the skin. A number of *Enterococcus* species have been recovered from wild or domesticated animals including dogs, cattle, rodents, and fowl or found as free-living environmental isolates, but they could conceivably be environmental contaminants of enteric origin (Byappanahalli et al. 2012). They react strongly with the crystal violet component of the Gram stain and, when examined in a light microscope, appear as completely spheroid or slightly elongated cells frequently found in short chains. Enterococci are facultative anaerobes and obtain energy as well as building blocks chemoorganotrophically from a diverse source of organic molecules. Several *Enterococcus* species also produce lactic acid by homofermentative glucose metabolism. Enterococci are routinely cultivated on rich media with occasional pigment production and hemolysis. At least two species (*E. casseliflavus* and *E. gallinarium*) are motile, and in their chromosomes, they contain genes for the synthesis and assembly of functional flagella and determinants for chemotactic responses (Palmer et al. 2012).

Distinct strains of *E. faecium* and *E. faecalis* account for virtually all clinical infections caused by *Enterococcaceae*, particularly in nosocomial settings. These two species are responsible for a significant fraction of hospital-acquired wound infections, bacteremia, and urinary tract infections (Fisher and Phillips 2009). The analyses of complete genomes of multiple isolates of *E. faecium* and *E. faecalis* have provided insights into the evolution dynamics of these organisms (van Schaik and Willems 2010). Genomes of virulent *A. faecalis* have evolved from commensal strains with the acquisition of genomic islands specifying antibiotic resistance and pathogenicity determinants via horizontal gene transfer (Palmer et al. 2012; Gilmore et al. 2013). The rapid evolution of the genomes of lineages responsible for the majority of *A. faecalis* infections can be in part attributed to the absence of a molecular mechanism limiting the acquisition of foreign DNA (the so-called CRISPR-Cas system) present in commensal or less virulent strains (Palmer and Gilmore 2010). Similarly, hypervirulent clonal lineages (clades) of *E. faecium* arise by horizontal gene transfer and recombination favoring adaptation to host defenses and antibiotic challenge (Gilmore et al. 2013). Presence of genes associated with virulence does not entirely define the pathogenic potential of all *Enterococci*, as genomes of many environmental or commensal isolates contain such genes, yet they lack disease-causing capabilities. A number of *Enterococcus* species have been successfully used as probiotics for the treatment of diarrhea in humans and farm animals (Franz et al. 2011).

Other members of the *Enterococcaceae* family have been less well studied beyond their microbiological and biochemical characterization. The genus *Vagococcus* consists of eight species: *V. fluvialis*, *V. salmoninarum*, *V. lutrae V. fessus*, *V. carniphilus*, *V. elongatus*, *V. penaei, and V. acidifermentans*. They are recovered primarily from poultry, fish, or infected aquatic animals or wastewater treatment facilities. One strain of *V. lutrae* was isolated from an infected tooth, and its genome was sequenced. However, the genome provides little information regarding its human adaptability (Lebreton et al. 2013). The genus *Melissococcus* is represented by a single species, *M. plutonius*. This organism is an important insect pathogen; it is responsible for the European foulbrood, an infectious disease of honeybee larvae. Two physiologically distinct but morphologically similar strains of *M. plutonius* with identical 16S rRNA gene sequences have been described (Arai et al. 2012). The so-called typical strains from Europe and Australia are similar in their growth requirements; they are fastidious, grow under microaerophilic or anaerobic conditions with carbon dioxide, and require potassium in their media. In contrast, atypical strains from Japan do not require potassium, utilize a more extended range of carbohydrates, and grow aerobically on a variety of media, and furthermore, these two strains differ in their ability to ferment various sugars. The genomes of both of these strains have been sequenced. While the genetic basis accounting for the various physiological differences between these two strains has not yet been determined, the potassium requirement of the typical strains was shown to be the consequence of mutations in two genes involved in potassium metabolism encoding an $Na^+/H^+$ antiporter and cation-transporting ATPase (Takamatsu et al. 2013).

E. Rosenberg et al. (eds.), *The Prokaryotes – Firmicutes and Tenericutes*, DOI 10.1007/978-3-642-30120-9_346,

**Fig. 6.1**

Phylogenetic reconstruction of the family *Enterococcaceae* based on 16S rRNA and created using the neighbor-joining algorithm with the Jukes-Cantor correction. The sequence datasets and alignments were used according to the All-Species Living Tree Project (LTP) database (Yarza et al. 2010; http://www.arb-silva.de/projects/living-tree). The tree topology was stabilized with the use of a representative set of nearly 750 high quality type strain sequences proportionally distributed among the different bacterial and archaeal phyla. In addition, a 40 % maximum frequency filter was applied in order to remove hypervariable positions and potentially misplaced bases from the alignment. Scale bar indicates estimated sequence divergence

The genus *Tetragenococcus* encompasses five species: *T. halophilus*, *T. muriaticus*, *T. solitarius*, *T. osmophilus*, and *T. koreensis*. These species are salt and alkaline tolerant, with *T. muriaticus* having an absolute growth requirement for NaCl. Genetic variants of *T. halophilus* can also be isolated from osmophilic, sugar-rich food sources. This distinction leads to a subdivision of *T. halophilus* into subspecies: *T. halophilus* subsp. *halophilus* representing isolates from high-salt environments, while a cluster of isolates from contaminated solutions with high sugar content are referred to as *T. halophilus* subsp. *flandriensis* (Justé et al. 2012). Genomes of representative organisms from both subspecies are currently being sequenced. Preliminary annotation of their sequenced genomes provided clues to the basis of salt tolerance of the osmophilic strain. The *T. halophilus* subsp. *halophilus* genome encodes determinants for the uptake and biosynthesis of choline and betaine, providing an ability to synthesize osmoprotectants, which allow these organisms to tolerate high levels of salt in their environment (Rediers et al. 2011).

## References


Arai R, Tominaga K, Wu M, Okura M, Ito K, Okamura N, Onishi H, Osaki M, Sugimura Y, Yoshiyama M, Takamatsu D (2012) Diversity of *Melissococcus plutonius* from honeybee larvae in Japan and experimental reproduction of European foulbrood with cultured atypical isolates. PLoS One 7(3):e33708

Byappanahalli MN, Nevers MB, Korajkic A, Staley ZR, Harwood VJ (2012) Enterococci in the environment. Microbiol Mol Biol Rev 76(4):685–706

Fisher K, Phillips C (2009) The ecology, epidemiology and virulence of *Enterococcus*. Microbiology 155(Pt 6):1749–1757

Franz CM, Huch M, Abriouel H, Holzapfel W, Gálvez A (2011) Enterococci as probiotics and their implications in food safety. Int J Food Microbiol 151(2):125–140

Gilmore MS, Lebreton F, van Schaik W (2013) Genomic transition of enterococci from gut commensals to leading causes of multidrug-resistant hospital infection in the antibiotic era. Curr Opin Microbiol 16(1):10–16

Justé A, Van Trappen S, Verreth C, Cleenwerck I, De Vos P, Lievens B, Willems K (2012) Characterization of *Tetragenococcus* strains from sugar thick juice reveals a novel species, *Tetragenococcus osmophilus* sp. nov., and divides *Tetragenococcus halophilus* into two subspecies, *T. halophilus* subsp. halophilus subsp. nov. and *T. halophilus* subsp. flandriensis subsp. nov. Int J Syst Evol Microbiol 62(1):129–137

Lebreton F, Valentino MD, Duncan LB, Zeng Q, Manson McGuire A, Earl AM, Gilmore MS (2013) High-Quality draft genome sequence of *Vagococcus lutrae* Strain LBD1, isolated from the largemouth bass *Micropterus salmoides*. Genome Announc 1(6). pii: e01087-13

Palmer KL, Gilmore MS (2010) Multidrug-resistant enterococci lack CRISPR-cas. MBio 1(4). pii: e00227-10

Palmer KL, Godfrey P, Griggs A, Kos VN, Zucker J, Desjardins C, Cerqueira G, Gevers D, Walker S, Wortman J, Feldgarden M, Haas B, Birren B, Gilmore MS (2012) Comparative genomics of enterococci: variation in *Enterococcus faecalis*, clade structure in *E. faecium*, and defining characteristics of *E. gallinarum* and *E. casseliflavus*. MBio 3(1):e00318-11

Rediers H, Busschaert P, Crauwels S, Lievens B, Willems KA (2011) Draft genome assemblies of halophilic and osmophilic *Tetragenococcus halophilus* strains. In: Abstract, symposium on lactic acid bacteria edition:10 location: Egmond aan Zee, The Netherlands, 28 Aug–1 Sept 2011

Takamatsu D, Arai R, Miyoshi-Akiyama T, Okumura K, Okura M, Kirikae T, Kojima A, Osaki M (2013) Identification of mutations involved in the requirement of potassium for growth of typical *Melissococcus plutonius* strains. Appl Environ Microbiol 79(12):3882–3886

van Schaik W, Willems RJ (2010) Genome-based insights into the evolution of enterococci. Clin Microbiol Infect 16(6):527–532

Yarza P, Ludwig W, Euzeby J, Amann R, Schleifer KH, Glöckner FO, Rossello-Mora R (2010) Update of the All-Species Living Tree Project based on 16S and 23S rRNA sequence analyses. Syst Appl Microbiol 33:291–299

# 7 The Families *Erysipelotrichaceae* emend., *Coprobacillaceae* fam. nov., and *Turicibacteraceae* fam. nov.

*Susanne Verbarg · Markus Göker · Carmen Scheuner · Peter Schumann · Erko Stackebrandt*
Leibniz Institute DSMZ-German Collection of Microorganisms and Cell Cultures GmbH, Braunschweig, Germany



**Abstract**
The family *Erysipelotrichaceae*, comprising 10 genera and 12 validly named species, is a family of the order *Erysipelotrichales*, class *Erysipelotrichia* within the phylum Firmicutes, remotely related by 16S rRNA gene sequence analysis with some members of Tenericutes (Mollicutes). The phenotype encompasses microaerophilic and anaerobic, spore- and nonsporing organisms, embracing rod-shaped cells to helical and curled rods, appearing singly, in short chains or V-forms. In addition to authentic members of the family, several (misclassified) members of the genera *Streptococcus*, *Eubacterium*, and *Clostridium* are affiliated to the family. Based upon full genome analyses and 16S rRNA gene sequence analyses, the family is polyphyletic and two new families are described on the basis of the 16S rRNA gene tree topology. All members are associated to one or several different hosts, often mammals, but also birds, fish, and marine invertebrates. Besides the obligate pathogen *Erysipelothrix rhusiopathiae*, causing erysipeloid in humans and erysipelas in swine, most of the other members are found as opportunistic pathogens affecting various parts of the body. Cultivation-based and cultivation-independent studies have revealed their presence in diverse environmental samples but rarely in significant numbers.

This contribution contains some paragraphs on taxonomic aspects taken from the 3rd edition of The Prokaryotes (Stackebrandt et al. 2005).

## Taxonomy

The type genus of the family is *Erysipelothrix*, containing the species *Erysipelothrix rhusiopathiae* (Rosenbach 1909), *E. tonsillarum* (Takahashi et al. 1987), and *E. inopinata* (Verbarg et al. 2004). The following paragraphs on the history of *Erysipelothrix* species have been taken from the article published in the 3rd edition of The Prokaryotes (Stackebrandt et al. 2005).

In 1876, Koch first isolated this slender, pleomorphic, Gram-positive bacillus from the blood of mice that had been inoculated subcutaneously with blood from putrefied meat (Koch 1878) and was designated *E. muriseptica*. In 1882, Löffler observed a similar organism in the cutaneous blood vessels of a pig that had died of swine erysipelas and published the first good description of the organism (Loeffler 1886). It is probable that a bacillus observed a few months previously by Pasteur and Dumas in pigs dying of rouget was the same organism as that described by Loeffler (Pasteur and Dumas 1882). Trevisan proposed the name *E. insidiosa* in 1885. Rosenbach was the first to establish *Erysipelothrix* as a human pathogen. In 1909, he reported the isolation of the organism from a patient with localized cutaneous lesions and coined the term "erysipeloid" to distinguish these lesions from those of human erysipelas (Rosenbach 1909). Subsequently, *Erysipelothrix* has been identified as the cause of infection in many animal species.

Rosenbach (1909) distinguished three species, *E. muriseptica*, *E. porci*, and *E. erysipeloides*, based on their murine, porcine, and human origins, respectively. The name *Bacterium rhusiopathiae* (Migula 1900) antedated the name *E. porci*. The combination *Erysipelothrix rhusiopathiae* was first proposed by

E. Rosenberg et al. (eds.), *The Prokaryotes – Firmicutes and Tenericutes*, DOI 10.1007/ 978-3-642-30120-9_205,

Buchanan (1918). At least 36 names have appeared in the literature for species of this genus. With the appreciation that all strains belonged to a single species, the name *E. insidiosa* was proposed for *E. rhusiopathiae*, *E. muriseptica*, and *E. erysipeloides* (Langford and Hansen 1953, 1954). In 1966, Shuman and Wellmann proposed that the name *E. insidiosa* be rejected in favor of *E. rhusiopathiae* which means literally "erysipelas thread of red disease."

With more strains being subjected to taxonomic studies, the great variation in serological, biochemical, chemical, and genomic properties of *E. rhusiopathiae* was noted (Erler 1972; Feist 1972; Flossmann and Erler 1972; White and Mirikitani 1976; Takahashi et al. 1992), and consequently the species *E. tonsillarum* (also named *E. tonsillae* in the older literature (Takahashi et al. 1989)) was described for avirulent *Erysipelothrix* strains of serotype 7, frequently isolated from the tonsils of apparently healthy pigs (Takahashi et al. 1987). The species *E. inopinata* was isolated from sterile-filtered vegetable broth (Verbarg et al. 2004).

Taxonomically, *Erysipelothrix* was once thought to be closely related to *Listeria* (Barber 1939) but results of studies of cell wall peptidoglycan (Schleifer and Kandler 1972; Verbarg et al. 2004), fatty acid patterns (Tadayon and Carroll 1971), DNA hybridization studies (Stuart and Welshimer 1974), and numerical taxonomic studies (Feresu and Jones 1988; Davis and Newton 1969; Jones 1975; Stuart and Pease 1972; Wilkinson and Jones 1977) did not support this relationship. There are no common antigens between strains of *Erysipelothrix* and *Listeria monocytogenes* as detected by immunodiffusion or passive hemagglutination tests (Pleszczynska 1972). Differences between the two genera have been demonstrated in cell wall chemistry by chromatography and infrared spectrophotometry. Paper and thin-layer chromatography of acid hydrolysates of the purified cell wall shows that *Erysipelothrix* is clearly distinguishable from *Listeria*. While *Erysipelothrix rhusiopathiae* has lysine and glycine in the cell wall (Mann 1969), *Listeria* has meso-diaminopimelic acid.

In the 1980s the genus *Erysipelothrix* was classified among the regular nonspore-forming Gram-positive rods (Jones 1986). However, even earlier, enzyme and DNA-base ratio studies revealed a closer relationship of *Erysipelothrix* to the family *Lactobacillaceae* than to *Corynebacteriaceae* (Flossmann and Erler 1972; White and Mirikitani 1976). In a study of more than 200 strains of coryneform bacteria using 173 morphological, physiological, and biochemical tests and computer analysis, eight clusters were identified and *Erysipelothrix* was most closely related to *Streptococcus pyogenes* (Jones 1975). In another study, the most closely related genus to *Erysipelothrix* was *Gemella* (Wilkinson and Jones 1977). In general, results of numerical taxonomic studies indicate that strains of *E. rhusiopathiae* form a distinct cluster that is most similar to the streptococci (Jones 1986)."

16S rRNA gene sequence analyses showed *Erysipelothrix* to be a member of the phylum Firmicutes, branching within *Clostridium* cluster XVI (Collins et al. 1994). The monogeneric family *Erysipelotrichaceae* was established to accommodate the genera *Erysipelothrix*, *Holdemania*, *Bulleidia*, *Solobacterium*, and some non-authentic members of *Clostridium* (*C. innocuum*), *Streptococcus* (*S. pleomorphus*), and *Eubacterium* (*E. biforme*) (Verbarg et al. 2004). The genera *Allobacterium* and the reclassified *Catenibacterium mitsuokai* (formerly *Lachnospiraceae*), *Coprobacillus cateniformis* (formerly *Clostridium catenaforme*), and *Turicibacter* (formerly taxon incertae sedis among *Bacillales*) were later added to the family (Stackebrandt 2009a). For the purpose of providing a more comprehensive phylogenetic structure, the family was placed as a monogeneric family into the order *Erysipelotrichales* (Ludwig et al. 2009a), class *Erysipelotrichia* (Ludwig et al. 2009b), the latter taxa being defined solely on the basis of its phylogenetic position.

The recent 16S rRNA tree of the Living Tree project also adds *Clostridium cocleatum*, *C. spiroforme*, *C. saccharogumia*, *C. ramosum*, *Eubacterium cylindroides*, *E. tortuosum*, and *E. dolichum*, as well as *Sharpea azabuensis* and *Anaerorhabdus furcosa* as members of the family. The latter two genera are not listed as members of the family according to Euzeby's List of genera included in families (http://www.bacterio.cict.fr/). According to the RaXML (➲ *Fig. 7.1*), neighbor-joining (not shown) trees, as well as the LPT tree, the authentic members of the family, form three lineages: The *Erysipelothrix* clade clusters with members of the Tenericutes while the *Coprobacillus* clade branches below this lineage. The not yet reclassified species fall into two clusters: one related to *Allobacterium* of the *Erysipelothrix* clade and the other appearing as a sister group of the *Coprobacillus* clade. This relationship is not obvious from the generalized 16S rRNA dendrograms depicted by Ludwig et al. (2012), showing members of *Erysipelotrichaceae* as a sister clade of Bacilli. Tenericutes, however, were not included in the dendrogram. *Turicibacter sanguinis*, originally placed into *Erysipelotrichaceae* (Stackebrandt 2009a), groups outside this family in all the trees mentioned above, where it forms a separate line of descent within the Firmicutes. The recently published phylogenetic tree of the 23S rRNA and of ribosomal proteins (Davis et al. 2013) discusses in detail the evolutionary origin of the family with respect to its position relative to Tenericutes (Mollicutes) and Firmicutes. The topology of the 23S rRNA gene tree is in accord with that shown in ➲ *Fig. 7.1* in that *Turicibacter* constitutes a separate lineage. They differ, however, in the position of the *Erysipelothrix* and the *Coprobacillus* clusters which form two lineages within a coherent cluster which are not separated by *Acoleplasma laidlawii* and relatives.

This chapter will focus on the 11 validly named genera and 13 validly named species, paying less attention to the as yet not reclassified *Streptococcus*, *Eubacterium*, and *Clostridium* members of this family. Their phenotypic data will be summarized in tables of fatty acid composition (➲ *Tables 7.1*, *7.2*, and *7.7*) and phenotypic analyses (➲ *Tables 7.3*, *7.4*, *7.5*, *7.6*, and *7.8*), but the other properties of these species are not further considered, except for their pathogenic potential and the occurrence of strains and clone sequences in various habitats (➲ *Tables 7.11* and *7.12*). Significant comparative chemotaxonomic and phenotypic characterization is needed in order to provide the data necessary for a proper reclassification.

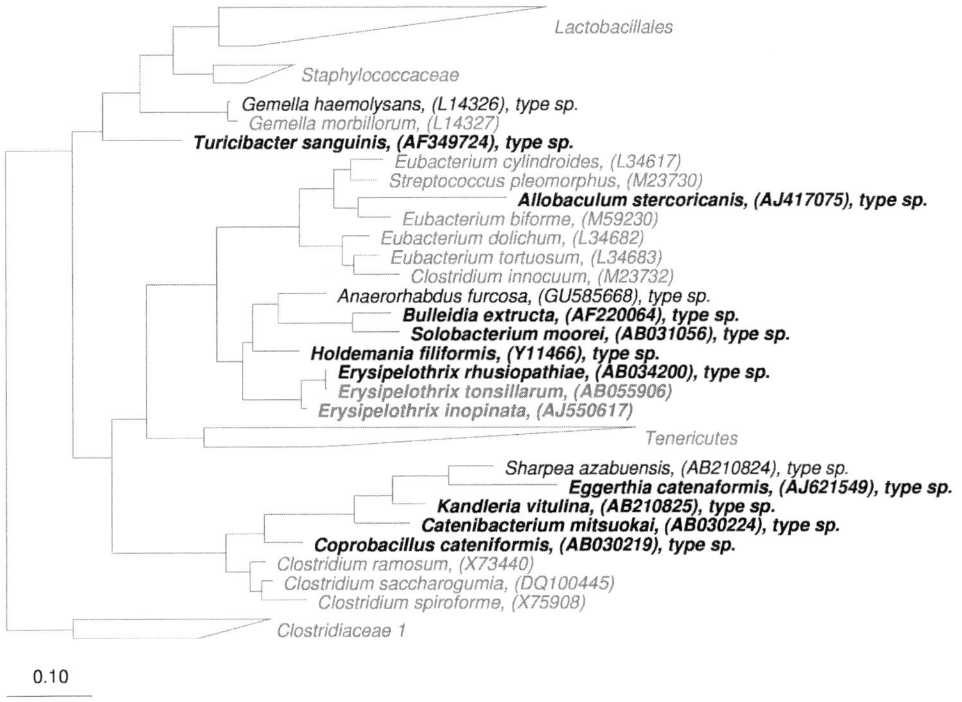


**Fig. 7.1**
**Phylogenetic reconstruction of the family *Erysipelotrichaceae* based on 16S rRNA gene sequences and created using the maximum likelihood algorithm RAxML (Stamatakis 2006). These sequence datasets and alignments were used according to the All-Species Living Tree Project (LTP) database (Yarza et al. 2010; http://www.arb-silva.de/projects/living-tree). Representative sequences from closely related taxa were used as out-groups. In addition, a 40 % maximum frequency filter was applied in order to remove hypervariable positions and potentially misplaced bases from the alignment. Scale bar indicates estimated sequence divergence**

**Table 7.1**
**Fatty acid composition (>1 % of total) of microaerophilically grown cells of *E. rhusiopathiae* DSM 5055$^T$, DSM 5056, DSM 5057, DSM 5058, *E. tonsillarum* DSM 14972$^T$, and *E. inopinata* DSM 15511$^T$**

| Fatty acids | *E. rhusiopathiae* DSM 5055$^T$ and some other strains | *E. tonsillarum* DSM 14972$^T$ | *E. inopinata* DSM 15511$^T$ |
|---|---|---|---|
| $C_{10:0}$ | – | – | 2.96 |
| $C_{14:0}$ | 1.59–1.90 | 2.66 | 2.95 |
| C16:1 cis-8 | 1.0–1.25 | – | 1.24 |
| $C_{16:0}$ | 24.2–31.7 | 28.2 | 34.2 |
| $C_{17:0}$ | 1.0–1.30 | 1.35 | 1.21 |
| $C_{18:2}$ | 6.46–7.37 | 5.04 | 2.77 |
| C18:1 cis-9 | 30.15–39.33 | 32.51 | 33.12 |
| C18:1 cis-11 | 1.03–1.94 | – | – |
| C18:1 cis-12 | 4.83–5.85 | 5.84 | 2.53 |
| $C_{18:0}$ | 10.18–19.95 | 19.64 | 12.88 |
| $C_{20:4}$ | 1.11–1.44 | 1.47 | – |

The phenotypic and phylogenetic heterogeneity of members of the *Erysipelotrichaceae* according to the 16S rRNA gene tree and full genome tree suggest a redefinition of the family may be warranted (see *Figs. 7.1* and *7.3* and *Tables 7.3, 7.4*, and *7.6*).

The separate rooting and isolated position of *Turicibacter sanguinis* and the separation of lineages within the main *Erysipelotrichaceae* cluster justifies the description of two new families, *Turicibacteraceae* and *Coprobacillaceae*, and the description is provided below. The redefined *Erysipelotrichaceae* family also appears phylogenetically heterogeneous but the differences in the 16S rRNA gene tree and the whole genome tree prevent further classification, and future taxonomic changes should be reconsidered only after thorough comparative polyphasic analyses of all members.

## Molecular Analyses

### DNA-DNA Hybridization

As most of the genera of *Erysipelotrichaceae* are monospecific, DNA reassociation experiments (Takahashi et al. 1987, 1992, 2000, 2008) were done on *E. rhusiopathiae* and *E. tonsillarum* strains. Several serotypes were affiliated to either one of the species (Takahashi et al. 1992), while two other serovars with low similarities to either species indicated the presence of two additional genomic species. The 16S rRNA gene sequences of the *E. rhusiopathiae* strain DSM 555$^T$ is highly similar to the sequence of *E. tonsillarum* DSM 14972$^T$ (99.8 % similarity). DNA-DNA reassociation values obtained for these strains

**Table 7.2**

**Fatty acid composition (>1 % of total) of members of the *Erysipelotrichaceae*: 1. *E. rhusiopathiae* DSM 5055$^T$, 2. *E. tonsillarum* DSM 14972$^T$, 3. *E. inopinata* DSM 15511$^T$, 4. *Bulleidia extructa* DSM 13220$^T$ (Downes et al. 2000), 5. *Holdemania filiformis* DSM 12042$^T$ (Willems et al. 1997), 6. *Solobacterium moorei* DSM 22971$^T$ (Kageyama and Benno 2000c, Validation List 75 2000), 7. *Allobaculum stercoricanis* DSM 13633$^T$ (Greetham et al. 2004; Validation List 110 2006), 8. *Eubacterium tortuosum* DSM 3987$^T$ (Debono 1912), 9. *Streptococcus pleomorphus* DSM 20574$^T$ (Barnes et al. 1977), and 10. *Clostridium innocuum* DSM 22910$^T$ (Smith and King 1962). Strains were grown anaerobically on blood agar**

| Fatty acid | 1 | 2 | 3 | 4 | 5 | 6 | 7 | 8 | 9 | 10 |
|---|---|---|---|---|---|---|---|---|---|---|
| $C_{12:0}$ | – | – | – | – | – | – | – | 2.46 | – | 2.34 |
| $C_{14:0}$ | 1.84 | 1.80 | 2.76 | 1.50 | 2.18 | 1.14 | 1.09 | 10.88 | 8.93 | 7.58 |
| $C_{16:0}$ ALDE | – | – | – | – | 2.47 | 3.97 | – | 2.39 | 1.26 | 1.68 |
| $C_{15:0}$ | – | – | – | – | – | – | – | 1.72 | – | – |
| $C_{16:1}$ cis 7 | – | – | – | – | – | – | – | 1.21 | – | 3.11 |
| $C_{16:1}$ cis 9 | 1.57 | 1.17 | 1.26 | 1.19 | 1.34 | – | – | – | 3.31 | 4.15 |
| $C_{16:0}$ | 33.53 | 34.25 | 44.61 | 31.46 | 21.67 | 11.33 | 35.27 | 28.66 | 21.62 | 26.59 |
| Feature 6 | – | – | – | – | – | – | – | 1.03 | – | – |
| $C_{16:0}$ DMA | – | – | – | – | 11.68 | 20.08 | 0.62 | 12.59 | 4.69 | 9.87 |
| $C_{18:0}$ ALDE | – | – | – | – | – | 2.08 | – | – | 2.92 | 1.26 |
| $C_{17:0}$ DMA | – | – | – | – | – | 1.02 | – | – | – | – |
| $C_{17:0}$ | – | 1.00 | – | 1.82 | – | – | – | 1.29 | – | – |
| $C_{18:2}$ cis 9,12 | 22.44 | 22.85 | 12.05 | 19.48 | 21.77 | 8.20 | 18.20 | 5.84 | 7.28 | 11.14 |
| $C_{18:1}$ cis-9 | 19.93 | 21.48 | 26.95 | 20.74 | 17.68 | 18.62 | 23.01 | 7.79 | 21.54 | 14.86 |
| Feature 10 | 4.75 | 4.57 | 2.81 | 6.20 | 3.42 | – | 4.21 | 2.19 | 4.26 | 2.78 |
| $C_{18:1}$ t11? | – | – | – | – | – | 6.62 | – | – | – | – |
| $C_{18:0}$ | 11.42 | 10.46 | 5.36 | 15.13 | 6.22 | 7.76 | 15.06 | 3.2 | 12.47 | 10.33 |
| Feature 11 | – | – | – | – | – | 4.41 | – | 1.65 | – | – |
| $C_{18:1}$ cis 9 DMA | – | – | – | – | 1.08 | 2.78 | – | 4.92 | – | – |
| $C_{18:1}$ cis 11 DMA | 0.24 | 0.22 | – |  | 1.09 | – | – | 0.95 | 0.36 | – |
| $C_{18:0}$ DMA | – | – | – | – | 3.23 | 7.31 | – | 1.74 | 4.55 | 2.67 |
| 19cyc 9,10/:1 | – | – | – | – | – | – | – | 1.59 | – | – |
| 19cyc 9,10 DMA | – | – | – | – | – | – | – | 1.28 | – | – |

Feature 10 ($C_{18:1}$c11/t9/t6 or UN 17.834), Feature 11 ($C_{18:2}$ DMA or $C_{17:0}$ iso 3OH)

range only between 18 % and 36 % which confirm differences at the physiological level, hence their separate species status. The 16S rRNA gene sequence of *E. inopinata* DSM 15511$^T$ is less than 97.5 % similar to that of the other two type strains, while it shares 99.9 % similar to that of strain Pecs 56 (AB055907) listed as unpublished in the EMBL database. 16S rDNA gene sequences of E. *rhusiopathiae* strains serotype 13 (AB019249) and serotype 18 (AB019250) (Takeshi et al. 1999), covering the 3′ half of the molecule (about 790 nucleotides), share 97.5 % and 97.8 % similarity, respectively, with the corresponding fragment of the gene of strain DSM 15511$^T$.

As pointed out by Stackebrandt (2009b), serovar 13 may be affiliated to *E. inopinata* because of high 16S rRNA similarities. In another study Takahashi et al. (2008) hybridized almost 100 strains from a broad variety of sources (e.g., pigs with acute or chronic erysipelas, healthy animals, environmental samples). Most of the strains causing localized or lesions in swine were linked *to E. rhusiopathiae,* while almost all strains of *E. tonsillarum* were avirulent for swine.

## DNA Patterns

Riboprint patterns have been determined for several strains of *Erysipelothrix* and found sufficiently discriminative to differentiate between strains (Verbarg et al. 2004), supporting the presence *of E. inopinata.* Automated ribotyping (Okatani et al. 2004) was found to be more discriminative for *Erysipelothrix* strains than randomly amplified

**Table 7.3**

**Differential characteristics of the genera included in the *Erysipelothrix* branch of the *Erysipelotrichaceae* family (Data are from original genus and species descriptions)**

| Property | *Erysipelothrix rhusiopathiae* DSM 5055$^T$ | *Erysipelothrix tonsillarum* DSM 14972$^T$ | *Erysipelothrix inopinata* DSM 15511$^T$ | *Anaerorhabdus furcosa* | *Bulleidia extructa* DSM 13220$^T$ | *Holdemania filiformis* DSM 12042$^T$ | *Solobacterium moorei* DSM 22971$^T$ | *Allobaculum stercoricanis* DSM 13633$^T$ |
|---|---|---|---|---|---|---|---|---|
| Morphology | Slender curved rods, filaments | | | Pleomorphic rods, in pairs or short chains | Straight or curved rods | Rods, pairs or chains | Rods | Rods, pairs or chains |
| Fermentation end products-check | L, A, F | n.d. | n.d. | L, A | L, A | A, L | A, L, b | L, b |
| Type and diagnostic amino acid of peptidoglycan type | B<br>D-Glu-Gly-L-Lys-L-Lys | n.d. | B<br>Lys-n.d. | no m-$A_2$pm | n.d. | B<br>D-Glu-L-Asp-L-Lys | n.d. | m-$A_2$pm |
| Major fatty acids | $C_{16:0}$, $C_{18:2}$ cis 9,12, $C_{18:1}$ cis-9, $C_{18:0}$ | $C_{16:0}$, $C_{18:2}$ cis 9,12, $C_{18:1}$ cis-9, $C_{18:0}$ | $C_{16:0}$, $C_{18:2}$ cis 9,12, $C_{18:1}$ cis-9, $C_{18:0}$ | n.d. | $C_{16:0}$, $C_{18:2}$ cis 9,12, $C_{18:1}$ cis-9, $C_{18:0}$ | $C_{16:0}$, $C_{16:0}$ DMA, $C_{18:2}$ cis 9,12, $C_{18:1}$ cis-9 | $C_{16:0}$, $C_{16:0}$ DMA, $C_{18:2}$ cis 9,12, $C_{18:1}$ cis-9 | $C_{16:0}$, $C_{18:2}$ cis 9,12, $C_{18:1}$ cis-9, $C_{18:0}$ |
| Isolation source | Animal, human infection | Pigs, tonsils | Vegetable broth | Lung and abdominal abscesses; human and pig feces | Human oral | Human feces | Human feces | Human feces |
| DNA mol% G+C | 36–40 | 36–40 | 37.5 | 34 | 38 | 38 | 37–39 | 37.9 |

Abbreviations: *A* acetic acid, *F* formic acid, *L* lactic acid, *B* butyric acid, capital letter, major amount, small letter, minor amount
Peptidoglycan; *B* B-type, *Asp* aspartic acid, *Lys* lysine, *gly* glycine, *n.d.*, not determined
+ positive, − negative, *w* weak

**Table 7.4**
**Diagnostic properties of members of the *Allobaculum* branch of the emended *Erysipelothricaceae* family (Data from the original descriptions and from Wade (2009) and Rainey et al. (2009))**

| Property | *Allobaculum stercoricanis* DSM 13633$^T$ | *Eubacterium biforme* DSM 3989$^T$ | *Eubacterium cylindroides* DSM 3983$^T$ | *Eubacterium dolichum* DSM 3991$^T$ | *Eubacterium tortuosum* DSM 3987$^T$ | *Clostridium innocuum* DSM 22910$^T$ | *Streptococcus pleomorphus* DSM 20574$^T$ |
|---|---|---|---|---|---|---|---|
| Morphology | Rods, pairs or chains | Long rods to short oval cocci-bacillus | Long rods, singly or in pairs, and long chains | Thin rods in long chains | Thin rods in long chains | Straight rods, singly or in pairs | Medium-dependant pleomorphic cocci |
| Spore formation | – | – | – | – | | +, terminal oval | – |
| Fermentation products of glucose[a] | L, B, e | L, B, c, $CO_2$ | L, B, a, s, f, little or no gas | B, a, l, no gas | L, a, b, s, variable amounts of gas | B, L, A, f, s, abundant $H_2$ | L, no gas |
| Acid from | Glucose, cellobiose, fructose, galactose, maltose, salicin, sucrose | Galactose, glucosamine, glucose, laevulose, mannitol, mannose, trehalose | Glucose, inulin, pectin | Glucose | Weak from glucose, sucrose | Glucose, salicin, mannitol, sucrose, cellobiose, fructose, galactose, inulin, mannose, ribose, trehalose | Glucose, mannose, fructose |
| No acid from | Amygdalin, arabinose, glycogen, inositol, lactose, mannitol, mannose, melezitose, raffinose, rhamnose, ribose, sorbitol, starch, trehalose, xylose | Adonitol, arabinose, dulcitol, erythritol, glycerol, inositol, inulin, melezitose, rhamnose, sorbitol, or xylose | Adonitol, dextrin, dulcitol, galactose, glycerol, sorbose | Adonitol, dextrin, dulcitol, galactose, glycerol, inulin, sorbose | Adonitol, dextrin, dulcitol, galactose, glycerol, inulin, sorbose | Amygdalin, arabinose, glycogen, maltose, melezitose, melibiose, raffinose, rhamnose, xylose, lactose, sorbitol | Arabinose, cellobiose, dextrin, galactose, inositol, lactose, maltose, mannitol, salicin, starch, sucrose, xylose |
| Habitat | Dog feces | Human feces | Human feces | Human feces | Turkey liver granulomas, turkey enteritis, human feces, soil, freshwater | Human infections, abscesses, empyema fluids, normal intestine flora of infants and adults | Intestines of poultry and occasionally human feces |
| Mol% G+C | 37.9 | 32 | 31 | nd | nd | 43–44 | 39.4 |

Abbreviations: [a]*L* lactic acid, *B* butyric acid, *c* caproic acid, *a* acetic acid, *f* formic acid, *s* succinic acid, *e* ethanol, capital letter, major amount; small letter, minor amount

polymorphic DNA (RFLP) and traditional ribotyping. Amplified fragment length polymorphism (AFLP) was determined for over 150 strains of *Erysipelothrix* isolated from swine from Brazil and affiliated to 18 different serotypes. The majority could be linked to E. rhusiopathiae and only 3 % to E. tonsillarum (Coutinho et al. 2011). Random amplified polymorphic DNA analysis was used in comparative strain analysis by Imada et al. (2004) and Okatani et al. (2000).

## 16S rRNA Gene Sequence Approaches

This method is by far the most widely used method to rapidly determine the genomic/phylogenetic novelty of members of the family as all newly described type strains of type species of newly described genera based upon results of this method (Kiuchi et al. 2000; Verbarg et al. 2004). As indicated in ➲ *Tables 7.12*, this method is also widely used to investigate the microbiota in the clinical environment. This method is often accompanied by the

**Table 7.5**

**API32A reactions of members of the *Erysipelotrichaceae* family. Taxa: 1, *Erysipelothrix inopinata* DSM 15511$^T$; 2, *E. rhusiopathiae* DSM 5055$^T$; 3, *E. tonsillarum* DSM 14972$^T$; 4, *Bulleidia extructa* DSM 13220$^T$; 5, *Solobacterium moorei* DSM 22971$^T$; 6, *Holdemania filiformis* DSM 12042$^T$; 7, *Allobaculum stercoricanis* DSM 13633$^T$; 8, *Eubacterium cylindroides* DSM 3983$^T$; 9, *Streptococcus pleomorphus* DSM 20574$^T$; 10, *Eubacterium biforme* DSM 3989$^T$; 11, *Clostridium innocuum* DSM 22910$^T$; 12, *Eubacterium dolichum* DSM 3991$^T$; 13, *Eubacterium tortuosum* DSM 3987$^T$. According to API32A all strains were negative for α-galactosidase, β-galactosidase-6-phosphate, α-arabinosidase, β-glucuronidase, raffinose fermentation, glutamic acid decarboxylase, and production of indole**

| Property | 1 | 2 | 3 | 4 | 5 | 6 | 7 | 8 | 9 | 10 | 11 | 12 | 13 |
|---|---|---|---|---|---|---|---|---|---|---|---|---|---|
| Urease | – | – | – | – | + | – | – | – | – | – | – | – | – |
| ADH | – | – | w | w | + | – | – | – | – | – | – | + | + |
| β-galactosidase | – | + | – | – | – | + | + | – | – | w | – | – | + |
| α-glucosidase | + | – | – | – | + | + | – | – | – | – | – | – | – |
| β-glucosidase | + | – | w | – | + | – | + | – | – | – | w | – | – |
| *N*-acetyl-β-glucosaminidase | + | + | + | – | – | – | + | – | – | – | – | + | – |
| Mannose fermentation | – | w | – | – | + | – | – | – | + | – | w | – | + |
| α-fucosidase | – | – | – | – | – | + | – | – | – | – | – | – | – |
| Nitrate reduction | – | – | – | – | + | – | – | – | – | – | – | – | + |
| Alkaline phosphatase | + | – | – | – | + | – | + | – | – | + | w | – | – |
| Arginine arylamidase | w | + | + | + | + | – | + | – | + | – | – | – | + |
| Proline arylamidase | + | – | + | – | – | – | – | – | – | – | – | – | w |
| Leucyl glycine arylamidase | – | w | + | – | – | – | – | – | w | – | – | – | – |
| Phenylalanine arylamidase | w | + | + | – | – | + | – | – | – | – | – | – | – |
| Leucine arylamidase | – | + | + | + | + | – | – | – | w | – | – | – | – |
| Pyroglutamic acid arylamidase | – | + | + | – | – | + | – | w | + | w | w | – | w |
| Tyrosine arylamidase | + | + | + | w | – | + | – | – | – | – | – | – | – |
| Alanine arylamidase | – | + | + | – | – | – | – | – | – | – | – | – | – |
| Glycine arylamidase | – | w | + | – | – | – | w | – | w | – | – | – | w |
| Histidine arylamidase | – | w | + | – | + | – | w | – | w | – | – | + | – |
| Glutamyl glutamic acid arylamidase | – | – | w | – | – | – | – | – | – | – | – | – | – |
| Serine arylamidase | – | w | + | – | – | – | – | – | – | – | – | – | – |

use of quantitative real-time PCR for increasing the diagnostic sensitivity (To et al. 2009). TGGE of 16S rRNA fragments has been used to identify members of *Erysipelothrix* in mites (Valiente Moro et al. 2009). Automated ribotyping was used by Verbarg et al. (2004) for species differentiation and by Okatani et al. (2004) to discriminate between strains which could not be discriminated by the PFGE technique (Okatani et al. 2001). Discrimination between *E. rhusiopathiae* and *E. tonsillarum* was achieved by a multiplex PCR system, using a DNA- and a 16S rRNA-specific nucleotide sequence stretch (Yamazaki 2006). A similar system, based upon identification of surface protective antigen (*spa*) types (Shen et al. 2010), was used by Bender et al. (2010) to characterize strains of *Erysipelothrix* strains from various samples.

## Other Methods

A set of 96 aerobic Gram-positive bacilli analyzed by the Biotyper MALDI-TOF system included two strains of *E. rhusiopathiae* (Farfour et al. 2012). A more detailed MALDI-TOF analysis of members of *Erysipelotrichaceae* and related species is shown in ➲ *Fig. 7.2*, based upon whole-cell protein extracts. Masses were analyzed by using a Microflex L20 mass spectrometer (Bruker Daltonics) equipped with a $N_2$ laser. Sample preparation for MALDI-TOF MS protein analysis was carried out according to the ethanol/formic acid extraction protocol (Bruker Daltonics) as described by Tóth et al. (2008). The MALDI-TOF mass spectra were analyzed with the Biotyper software (version 3.1, Bruker Daltonics).

As most species are pheno- and genotypically only distantly related, most of the strains, except for the pair *Coprobacillus cateniformis* DSM 15921$^T$ and *Sharpea azabuensis* DSM 18934$^T$, show remote MALDI-TOF pattern similarities. Spectra are available upon request from Peter Schumann by sending requests to psc@dsmz.de. Other though rarely applied methods are pulsed field electrophoresis (Okatani et al. 2001; Opriessnig et al. 2004), comparison of DNA restriction fragments (Makino et al. 1994; Ahrne et al. 1995), protein patterns (Bernath et al. 1997, 2001; Tamura et al. 1993), and multi-locus enzyme electrophoresis

**Table 7.6**

**Some taxonomic properties of members of the family *Coprobacillaceae* fam. nov. Except for fatty acid composition data were taken from the original descriptions: 1. *Coprobacillus cateniformis* RCA1-24[T] (Kageyama and Benno 2000a; Validation List N 74 2000), 2. *Catenibacterium mitsuokai* RCA 14-39[T] (Kageyama and Benno 2000b), 3. *Kandleria vitulina* ATCC 27783[T] (Salvetti et al. 2011), 4. *Eggerthia catenaformis* ATCC 25536[T] (Salvetti et al. 2011), 5. *Sharpea azabuensis* ST18[T] (Morita et al. 2008), 6. *Clostridium cocleatum* I50[T] (Kaneuchi et al. 1979), 7. *C. ramosum* ATCC 25582[T] (Holdeman et al. 1971), 8. *C. saccharogumia* SDG-Mt85-3Db[T] (Clavel et al. 2007; Validation List N 115 2007), 9. *C. spiroforme* ATCC 29900[T] (Kaneuchi et al. 1979)**

| Property | 1 | 2 | 3 | 4 | 5 | 6 | 7 | 8 | 9 |
|---|---|---|---|---|---|---|---|---|---|
| Morphology | Chains of small rods | Chains of rods | Rods, single or in pairs | Chains of short rods | Rods, short chains | Rods, circular or spiral | Straight rods, short chains, V arrangement | Helically coiled rods | Helically coiled rods |
| Spore formation | –[a] | – | – | – | n.d. | + | + | – | + |
| Fermentation end products-check | A, L[b] | n.d., no gas | L | L | L, $CO_2$ | A, F, I | F, A, I | n.d. | A, F, I |
| Diagnostic amino acid of peptidoglycan type | n.d. | m-$A_2$pm | m-$A_2$pm | L-Lys-L-$Ala_3$ | m-$A_2$pm | n.d. | m-$A_2$pm | n.d. | n.d. |
| Isolation source | Human feces | Human feces | Calf and bovine rumen | Human feces, pleural infections | Horse feces | Human feces, mice, rats, chicken cecum | Human feces, various human infections | Human feces | Human feces |
| Major fatty acids | $C_{16:0}$, $C_{18:1}$ cis 9, $C_{18:1}$cis 9 DMA | $C_{16:1}$ cis 9, $C_{16:0}$, $C_{16:0}$ DMA | $C_{16:0}$ | $C_{16:0}$, $C_{18:0}$ | $C_{16:1}$ cis 9, $C_{16:0}$, $C_{16:0}$ DMA | n.d | n.d | n.d | n.d |
| DNA mol% G+C | 32–34 | 36–39 | 34–37 | 31–33 | 37.4 | 28–29 | 26 | 29–32 | 27 |

[a]Terminal spores were detected in DSMZ medium 78 after 24 h (Rüdiger Pukall, personal communication)
[b]Abbreviation: F, A, and L, major amounts of formic, acetic, and lactic acid, respectively; I, minor amounts of lactic acid; m-$A_2$pm, meso-diaminopimelic acid

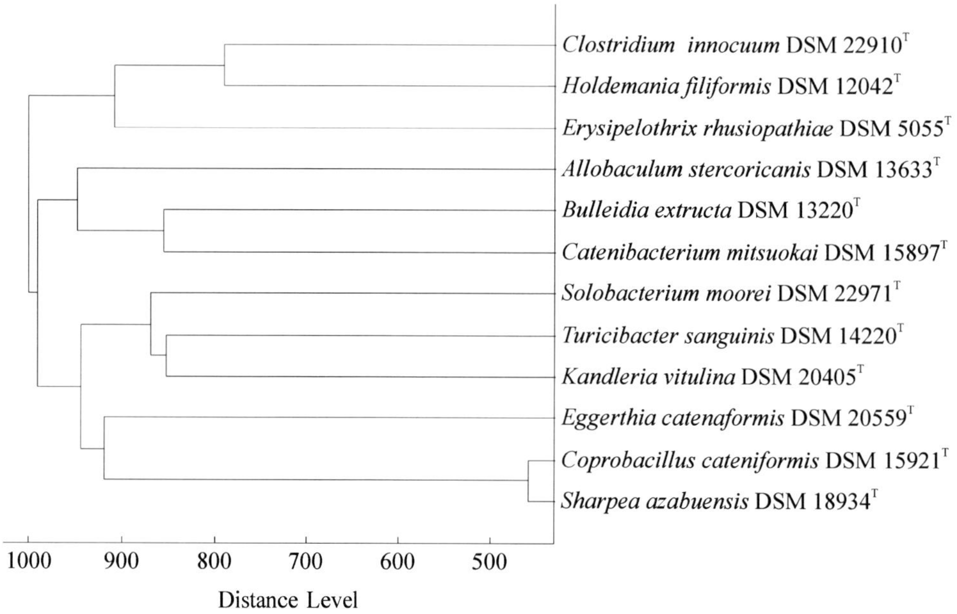


**Fig. 7.2**

**Score-oriented dendrogram generated by the Biotyper software (version 3.1, Bruker Daltonics) showing the similarity of MALDI-TOF mass spectra of cell extracts of some type strains of the family *Erysipelotrichaceae***

(Chooromoney et al. 1994). Serotyping is still widely applied (Imada et al. 2004; Ozawa et al. 2009; Coutinho et al. 2011) especially in the clinical environment. Serotype differentiation by letter and number systems applied to *Erysipelothrix* has been summarized by Stackebrandt et al. (2005). For the differentiation of *E. rhusiopathiae* and *E. tonsillarum* and for discriminating *Erysipelothrix* strains from non-*Erysipelothrix* strains, Yamazaki (2006) designed two PCR systems (one non-16S rRNA, one 16S rRNA system) that resulted in unambiguous analysis.

## Phages and Plasmid

The scientific literature indicates several reports on the presence of *Erysipelothrix* phages, but all studies are written in either Russian, Polish, Bulgarian, or Ukrainian languages. A report on plasmids is available by Noguchi et al. (1993) who found several *E. rhusiopathiae* strains to harbor per strain between 1 and 6 plasmids of unknown function, ranging from 1.4 to 86 kb in size.

## Genome Analysis

Protein sequences of 29 genomes were retrieved from the NCBI or IMG website. *Erysipelotrichaceae* and *Clostridiaceae* of interest were *Allobaculum stercoricanis* DSM 13633$^T$ (IMG taxon ID 2516493028), *Bulleidia extructa* W1219$^T$ (ADFR00000000), *Catenibacterium mitsuokai* DSM 15897$^T$ (ACCK00000000), *Clostridium ramosum* DSM 1402$^T$ (ABFX00000000), *C. spiroforme* DSM 1552$^T$ (ABIK00000000), *Erysipelothrix rhusiopathiae* ATCC 19414$^T$ (ACLK00000000), *E. rhusiopathiae* Fujisawa (AP012027), *E. tonsillarum* DSM 14972$^T$ (IMG taxon ID 2515154211), *Eubacterium biforme* DSM 3989$^T$ (ABYT00000000), *E. cylindroides* T2-87 (FP929041), *E. dolichum* DSM 3991$^T$ (ABAW00000000), *Holdemania filiformis* DSM 12042$^T$ (ACCF00000000), *Solobacterium moorei* F0204 (AECQ00000000), and *Turicibacter sanguinis* PC909 (ADMN00000000). Additionally, one genome sequence per family of *Lactobacillales* was examined, using the type genera of the family (and type strains) where available. The selected genomes were *Aerococcus viridans* ATCC 11563$^T$ (ADNT00000000), *Dolosigranulum pigrum* ATCC 51524$^T$ (AGEF00000000), *Melissococcus plutonius* ATCC 35311$^T$ (AP012200, AP012201), *Lactobacillus delbrueckii* subsp. *bulgaricus* ATCC 11842$^T$ (CR954253), and *Leuconostoc mesenteroides* ATCC 8293$^T$ (CP000414, CP000415). From the order *Bacillales*, *Gemella haemolysans* ATCC 10379$^T$ (ACDZ00000000), *Gemella haemolysans* M341 (ACRO00000000), and *Gemella morbillorum* M424 (ACRX00000000) were analyzed. From the phylum Tenericutes, all currently available type-strain genomes were selected, that is, *Acholeplasma laidlawii* PG-8A$^T$ (CP000896), *Haloplasma contractile* SSD-17B$^T$ (AFNU00000000), *Mesoplasma florum* L1$^T$ (AE017263), *Mycoplasma mycoides* SC PG1$^T$ (BX293980), and *Ureaplasma urealyticum* serovar 8 ATCC 27618$^T$ (AAYN00000000). The *Clostridiaceae* strains *Alkaliphilus oremlandii* OhILAs$^T$ (CP000853) and *Clostridium butyricum* E4 str. BoNT E BL5262 (ACOM00000000) were used as out-group.

The genome sequences were phylogenetically investigated as described in Spring et al. (2010), Anderson et al. (2011), Göker et al. (2011), and Abt et al. (2012). That is, maximum likelihood (ML) and maximum-parsimony (MP) phylogenetic trees were inferred from two distinct supermatrices (concatenated alignments), a "full" matrix using all alignment comprising at least four sequences and a matrix using the "core genes" only, i.e., those alignments containing 29 sequences, as well as from an ortholog-content and a gene-content matrix.

The full supermatrix contained 2,815 genes and 739,145 characters, whereas the core-gene supermatrix comprised 119 genes and 31,829 characters. Both matrices were analyzed under ML with the suggested model PROTGAMMALGF. The gene-content matrix comprised 13,990 characters and the ortholog-content matrix 18,220 characters. For both matrices, the BINGAMMA model was used as implemented in RaxML (Stamatakis 2006).

The resulting full-supermatrix ML topology is shown in ❯ *Fig. 7.3* together with ML and MP bootstrap support values from all analyses. The phylogenetic tree shows all examined *Erysipelotrichaceae* except *Turicibacter sanguinis* PC909 clustered together with the *Clostridiales* species *Clostridium ramosum* DSM 1402$^T$, *C. spiroforme* DSM 1552$^T$, *Eubacterium biforme* DSM 3989$^T$, *E cylindroides* T2-87, and *E. dolichum* DSM 3991$^T$ in a single clade. Its deeper branches are not supported by the gene-content and ortholog-content analyses but maximally supported by the full-supermatrix and core-gene analyses. The monophyly of the group containing *Allobaculum stercoricanis* DSM 13633$^T$, *Eubacterium biforme* DSM 3989$^T$, *E. cylindroides* T2-87, and *E. dolichum* DSM 3991$^T$ and the monophyly of the group comprising *Clostridium ramosum* DSM 1402$^T$, *C. spiroforme* DSM 1552$^T$, and *Catenibacterium mitsuokai* DSM 15897$^T$ are supported by all analyses; only the support for the positioning of *Eubacterium dolichum* DSM 3991$^T$ is missing in the MP analyses of the gene-content and ortholog-content matrices. This confirms the 16S rRNA analysis (see above). Both groups were also found in Davis et al. (2013) (but based on a distinct taxon sampling), who analyzed 16S rRNA and 23S rRNA genes as well as ribosomal proteins.

The remaining *Erysipelotrichaceae* examined (except *Turicibacter sanguinis* PC909) cluster into two distinct clades. These two clades were also found in the 16S rRNA analysis, but there they formed sister clades (see above). The first clade includes *Erysipelothrix rhusiopathiae* ATCC 19414$^T$, *E. rhusiopathiae* Fujisawa, and *E. tonsillarum* DSM 14972$^T$. The second clade comprises *Bulleidia extructa* W1219$^T$, *Holdemania filiformis* DSM 12042$^T$, and *Solobacterium moorei* F0204. This group was also found in the 23S rRNA gene and ribosomal protein analyses conducted by Davis et al. (2013). In contrast to the 16S rRNA gene analysis, when using genome-scale data, *Turicibacter sanguinis* PC909 clusters together with *Haloplasma contractile* SSD-17B$^T$ with maximal bootstrap support in all analyses.

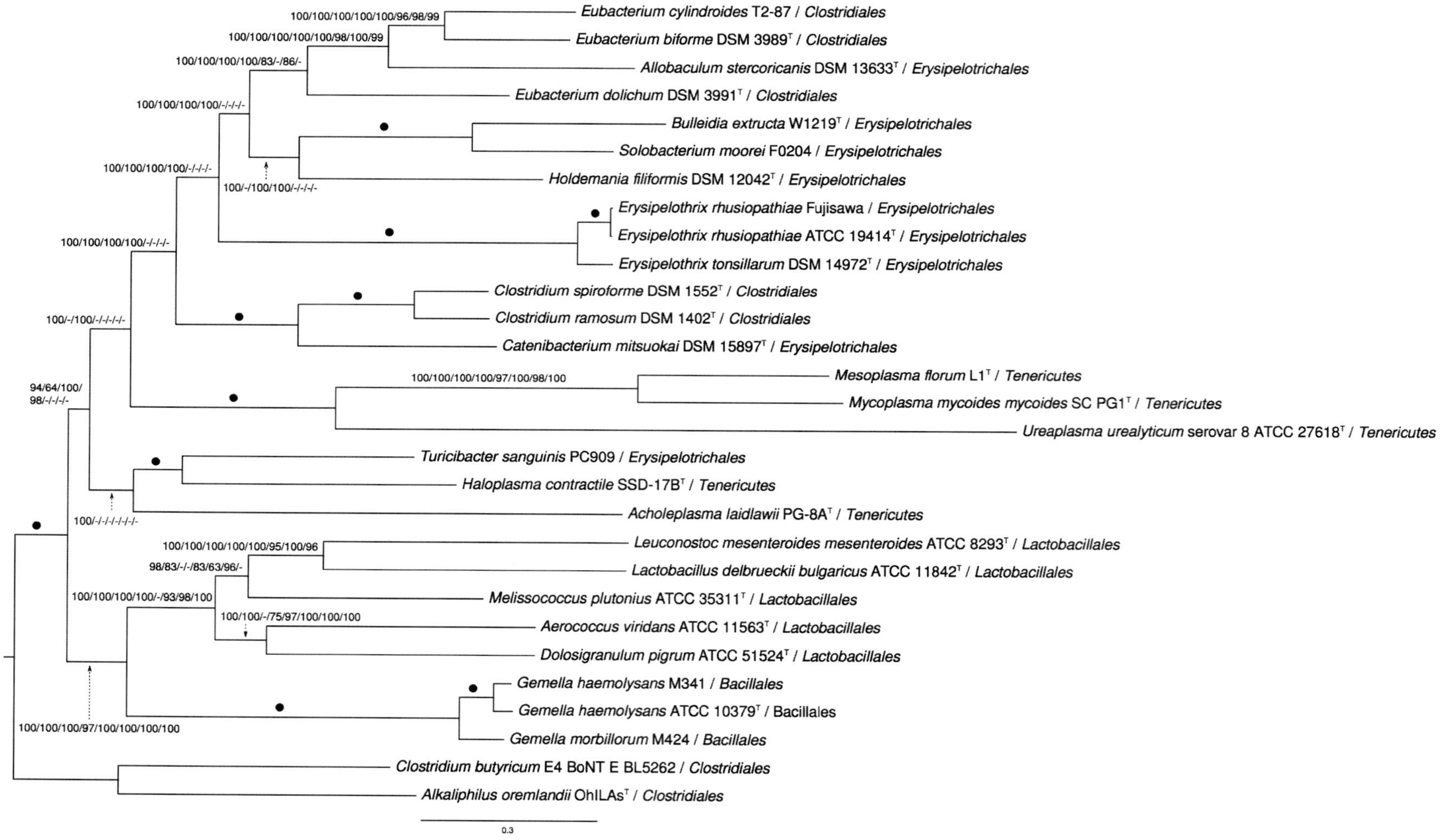

**Fig. 7.3**

Phylogenetic tree inferred from the full supermatrix under the maximum likelihood (ML) criterion and rooted with *Alkaliphilus oremlandii* OhILAs$^T$ and *Clostridium butyricum* E4 str. BoNT E BL5262. The branches are scaled in terms of the expected number of substitutions per site. Numbers on the nodes (from *left to right*) are bootstrapping support values (if larger than 60 %) from (i) ML "full" supermatrix, (ii) maximum-parsimony (MP) "full" supermatrix, (iii) ML core-gene, (iv) MP core-gene, (v) ML gene-content, (vi) MP gene-content, (vii) ML ortholog-content, and (viii) MP ortholog-content analysis. *Dots denote branches with maximal bootstrap support in all analyses*

The other examined taxa (*Bacillales*, *Lactobacillales*, and *Tenericutes*) are grouped according to the current taxonomy with the exception that the *Tenericutes* do not form a monophyletic group. Rather, they cluster into two distinct groups: the first one comprising *Mesoplasma florum* L1$^T$, *Mycoplasma mycoides* SC PG1$^T$, and *Ureaplasma urealyticum* serovar 8 ATCC 27618$^T$ and the second one containing *Acholeplasma laidlawii* PG-8A$^T$ and *Haloplasma contractile* SSD-17B$^T$ and also conflict with the classification *Turicibacter sanguinis* PC909. But this separation of *Tenericutes* is unsupported by the majority of the conducted analyses.

## Chemotaxonomic Analyses

The main feature that distinguishes members of *Erysipelothrix* from most of its phylogenetic neighbors is the presence of a B-cell wall type in which the peptide bridge is formed between amino acids at positions 2 and 4 of adjacent peptide side chains and not, as in the vast majority of bacteria, between amino acids at positions 3 and 4. The rare B-type, which does not contain meso- or LL-diaminopimelic acid (m- or LL-A$_2$pm) as a peptidoglycan constituent, also occurs in some other members of the presently defined family *Erysipelotrichaceae*, in some other members of *Firmicutes*, and in *Microbacteriaceae*, phylum Actinobacteria (Schleifer and Kandler 1972; Schumann 2011). As the peptidoglycan type is a taxonomic marker of weight, we screened other members of *Erysipelotrichaceae* using a sensitive mass spectrometric method, apt for organisms for which low cell masses only can be obtained. Whole-cell hydrolysates (4 N Hall, 100 °C, 16 h) were examined for the presence of 2,6-diaminopimelic acid by gas chromatography/mass spectrometry by using published protocols (Schumann 2011). The agreement in the gas chromatographic retention time with this of the authentic standard and mass spectrometric fragment ions at 380, 324, 306, and 278 m/z was indicative for the presence of the N-heptafluorobutyryl diaminopimelic acid isobutyl ester.

The absence of A$_2$pm was not taken as an indication for the presence of the B-type as the A-type constituent lysine has been found in, i.e., *Eggerthia catenaformis* and *Clostridium innocuum* (Peltier et al. 2011). The A-type diamino acid A$_2$pm was detected in *Turicibacter sanguinis* DSM 14220$^T$, *Allobaculum stercoricanis* DSM 13633$^T$, *Solobacterium moorei* DSM 22971$^T$, and *Bulleidia extructa* DSM 13220$^T$ but not, as reported before, in *E. rhusiopathiae* DSM 2055$^T$, *E. inopinata* DSM 15511$^T$, *E. tonsillarum* DSM 14972$^T$, and *Holdemania filiformis* DSM 12042$^T$. It appears that only certain members of the *Erysipelothrix* lineage possess a B-type peptidoglycan which was analyzed in detail only for *E. rhusiopathiae* and *E. inopinata* (with the interpeptide bridge composed of Gly → L-Lys → L-Lys, type B1δ) (Schubert and Fiedler 2001; Verbarg et al. 2004) and *Holdemania filiformis* ATCC 51649$^T$ (with the interpeptide bridge composed of L-Asp → L-Lys) (Willems et al. 1997). Other members, such as *Solobacterium moorei*, *Bulleidia extructa*, and *Anaerorhabdus furcosa* (investigated as *Bacteroides furcosus* by Hammann and Werner (1981)), lack meso-A$_2$pm, which is present in certain members of the *Coprobacillus* (e.g., *Kandleria vitulina* and *Catenibacterium mitsuokai*) and *Allobaculum* lineages. Members affiliated to the family as defined today vary widely in the peptidoglycan composition, having either the B- or the A-type, the latter with the variations of possessing either A$_2$pm or lysine in position 3 of the peptide subunit. The variations may be greater as not all family members were investigated. As a conclusion, the peptidoglycan type cannot be used for the delineation of higher taxa.

Investigations on cell wall sugars were restricted to *E. rhusiopathiae*. Several sugars occur in the cell wall (arabinose, galactose, glucose, glucose-6-phosphate, galactose-6-phosphate, ribose, and xylose; Feist 1972). Mycolic acids are not present. The cellular fatty acid (FA) composition of *E. rhusiopathiae* and *E. tonsillarum* was reported by Takahashi et al. (1994), von Graevenitz et al. (1991), and Julak et al. (1989), mainly used in the identification of asporogenous, aerobic Gram-positive rods. The FA composition for all type three *Erysipelothrix* type strains was investigated with microaerophilically grown cells (Verbarg et al. 2004), analyzed by Microbial Identification System (MIDI) Sherlock Version 4.5 (method BHILB database) (Sasser 1990). The pattern is dominated by C18:1-cis-9 (>30 %), C16:0 (>24 %), and C18:0 (>10 %) FA methyl esters (❷ *Table 7.1*). This pattern differs from that of *Holdemania filiformis* ATCC 51649$^T$ which contains higher amounts of C18:1-cis-9 (50 %), additional minor components, and significant amounts of dimethyl acetal (C18:1-cis-9 [12 %] and C16:0 [4 %]) (Willems et al. 1997). In order to obtain a comprehensive overview on the FA composition of the majority of *Erysipelothrix* members, all cells were grown under anaerobic conditions (❷ *Table 7.2*) on blood agar for 2 days at 37 °C to promote growth of all strains. *Erysipelothrix* strains grew sparsely only under these conditions. The influence of growth conditions and growth media on qualitative and quantitative FA composition is obvious for the *Erysipelothrix* strains which lack the fatty acid C$_{18:2}$ cis 9,12 and show a much lower C$_{18:0}$ content in *E. inopinata* DSM 15511$^T$ cells grown on blood agar medium. The quantitative composition of C$_{16:0}$ is high in all strains (except for *Solobacterium moorei* DSM 22971$^T$) and, except for *Eubacterium tortuosum* DSM 3987$^T$, also for C$_{18:1}$ cis-9 and C$_{18:2}$ cis 9,12, the latter compound being also low for strain DSM 22971$^T$ which shows higher amounts of C$_{14:0}$ and C$_{16:0}$ DMA fatty acids. The latter compound is also increased in *Holdemania filiformis* DSM 12042$^T$.

## Taxonomic Changes

### Family *Erysipelotrichaceae* emend.

E.ry.si.pe.lo.tri.cha'ce.ae. M.L. fem. n. *Erysipelothrix* type genus of the family; -aceae, ending to denote a family; M.L. fem. pl. n. *Erysipelotrichaceae*, the *Erysipelothrix* family.

Straight, or slightly curved, slender rods; some strains with a tendency to form long filaments or a zigzag formation of individual rods. Nonmotile. Endospores may be produced. Microaerophilic to facultatively anaerobic. Catalase negative. Chemoorganotrophic. Acid but no gas may be produced from glucose and other carbohydrates. If investigated cytochromes and isoprenoid quinones absent. Peptidoglycan type and variation vary: while some members contain peptidoglycan belonging to the B-cross-linking type, other strains express the A-cross-linking type, possessing either lysine or $A_2pm$ as diagnostic diamino acid. In most strains predominant fatty acids are $C_{16:0}$, $C_{18:1}$ cis-9, $C_{18:2}$ cis 9,12, and $C_{18:0}$. Some strains pathogenic for mammals and birds. The mol% G+C of the DNA is 34–40. Found in a wide range of human and animal (fish, birds, mammals) hosts, but also in feces-contaminated soil.

The type genus is *Erysipelothrix* (Migula 1900) Buchanan 1918, 55.

Comments: The emended family comprises validly named genera and species, some of which must be considered misclassified strains of the genera *Eubacterium*, *Streptococcus*, and *Clostridium*. According to the 16S rRNA gene tree, the family consists of two sister branches, i.e., defined by *Erysipelothrix* and *Allobaculum* and their respective relatives. The main phenotypic differences apt to characterize members of the two branches are indicated in ➋ *Tables 7.3* and *7.4*. The type strain of the species *Anaerorhabdus furcosa* was not available for analysis. Extensive comparative polyphasic analyses will be necessary to properly describe the entire family.

Phenotypic properties (API 32A) for all members of the two branches of the *Erysipelotrichaceae* family are indicated in ➋ *Table 7.5*. Reactions were determined for the present study.

## Family *Coprobacillaceae* fam. nov.

Co.pro.ba.cil.la'ce.ae. N.L. fem. *Coprobacillaceae*, type genus of the family, -aceae ending to denote family, N.L. *Coprobacillaceae* the *Coprobacillus* family.

The family is described on the basis of phylogenetic analyses of the 16S rRNA sequences and includes the genus *Coprobacillus*, *Catenibacterium*, *Kandleria*, *Eggerthia*, *and Sharpea* and the misclassified *Clostridium* species *C. cocleatum*, *C. ramosum*, *C. saccharogumia*, and *C. spiroforme*. Both genera, *Catenibacterium* and *Coprobacillus*, were published in 2000: *Catenibacterium mitsuokai* in the International Journal of Systematic Bacteriology (IJSB) (Kageyama and Benno 2000b) and *Coprobacillus cateniformis* in Microbiology and Immunology (Kageyama and Benno 2000a). As the latter name was validated on page 949 of the IJSB, it antedates the description of *C. mitsuokai* (page 1589 of IJSB); subsequently *Coprobacillus* has name priority. Strictly anaerobic, spores may be present. If tested, the peptidoglycan is of the A-type, containing m-$A_2pm$ as diagnostic amino acid. L-Lys has been reported in one species. The fatty acid composition is indicated in ➋ *Table 7.8*. Phylogenetically a sister clade of *Erysipelotrichaceae*, phylum Firmicutes. Inhabitants of intestines of human or rumen or cecum of some vertebrates, rarely from birds (see Table 7.13).

Type genus *Coprobacillaceae* Kageyama and Benno, 2000, $949^{VP}$.

Differential characters of genera included in *Coprobacillaceae* fam.nov.ae indicated in ➋ *Table 7.6*. The fatty acid composition of these members are shown in ➋ *Table 7.7*, while the API 32A reactions are listed in ➋ *Table 7.8*.

## Family *Turicibacteraceae* fam. nov.

Tu.ri.ci.bac ter.a' ce.ae N.L. masc.n. *Turicibacteraceae*, type genus of the family, -aceae ending to denote family; N.L. *Turicibacteraceae*, the *Turicibacter* family.

The family is described on the basis of phylogenetic analyses of the 16S rRNA sequence and presently contains the monospecific-type genus *Turicibacter* Bosshard et al. (2002), as its only member. It is presently composed of anaerobic, Gram-positive, nonspore-forming, and fermenting bacteria.

The type genus of the family is *Turicibacter* Bosshard, Zbinden, and Altwegg 2012, $1266^{VP}$.

Fatty acid composition and phenotypic data are indicated in ➋ *Tables 7.7* and *7.8*, respectively.

## Isolation, Enrichment, and Maintenance Procedures

Except for members of the microaerophilic genus *Erysipelothrix*, all other strains are anaerobes, isolated from a wide range of anaerobic habitats. Several procedures have been devised for the isolation of *E. rhusiopathiae*. These methods may also refer to *E. tonsillarum*, classified as *E. rhusiopathiae* in the past. Most procedures are based on the ability of the organism to grow in the presence of various substances which are bactericidal or bacteriostatic for other organisms, e.g., phenol (0.2 % w/v), potassium tellurite (0.05 % w/v), sodium azide (0.1 % w/v), thallous acetate (0.02 % w/v), 2,3,5-triphenyltetrazolium chloride (0.2 % w/v), and crystal violet (0.001 % w/v) (Sneath et al. 1951; Ewald 1981). Reboli and Farrar (1991) and Stackebrandt et al. (2005) summarized the enrichment and isolation of *E. rhusiopathiae* from pig and human blood, skin, and feces samples and give media recipes (Packer 1943; Wood 1965; Harrington and Hulse 1971). While in previous years detection of the organisms was done by a fluorescent antibody technique (Dacres and Groth 1959; Seidler et al. 1971; Harrington et al. 1974), today identification of *Erysipelothrix* spp. is done quickly and reliably by MALDI-TOF or 16S rRNA gene sequence-based techniques.

*E. inopinata* was enriched in the course of preparation of a vegetable CSB medium (peptone vegetable, 20.0 g; D(+)-glucose, 2.5 g; $K_2HPO_4$, 2.5 g; water, 1,000 ml). The water used for dilution was heated to 80 °C for 1 h and allowed

**Table 7.7**

**Fatty acid composition (>1 % of total) of the family *Coprobacillaceae* fam. nov. 1. *Coprobacillus cateniformis* DSM 15921$^T$, 2. *Sharpea azabuensis* DSM 18934$^T$, 3. *Kandleria vitulina* DSM 20406, 4. *Catenibacterium mitsuokai* DSM 15897$^T$, 5. *Eggerthia catenaformis* DSM 20559$^T$ and *Turicibacteraceae* fam nov., 6. *Turicibacter sanguinis* DSM 14220. Strains were grown anaerobically on blood agar for 2 days at 37 °C, with the exception of strain DSM 15921$^T$, which was grown anaerobically on PYG 1 day at 37 °C. Analysis was conducted using the Microbial Identification System (MIDI) Sherlock Version 4.5 (method BHILB database) as described by Sasser (1990)**

| Fatty acid | 1 | 2 | 3 | 4 | 5 | 6 |
|---|---|---|---|---|---|---|
| $C_{14:0}$ iso | – | – | – | – | 1.20 | 8.93 |
| $C_{14:0}$ | 5.67 | 2.99 | 2.33 | 2.56 | 1.94 | 2.00 |
| Feature 4 | – | 2.91 | 1.40 | 0.59 | – | – |
| $C_{15:0}$ iso | – | – | – | – | 2.49 | 4.60 |
| $C_{15:0}$ anteiso | – | – | – | – | 6.77 | 21.33 |
| $C_{16:0}$ ALDE | 0.47 | 3.10 | 1.80 | 3.68 | 0.58 | – |
| $C_{15:0}$ | – | – | 0.25 | – | 0.85 | 1.29 |
| $C_{16:0}$ iso | – | – | – | – | 0.55 | 1.37 |
| $C_{16:1}$ cis 9 | 6.50 | 17.16 | 9.24 | 14.29 | 2.38 | – |
| $C_{16:0}$ | 24.29 | 29.88 | 31.61 | 23.70 | 24.66 | 10.35 |
| $C_{16:1}$ cis 9 DMA | 1.26 | 9.63 | 6.17 | 2.76 | 0.77 | – |
| $C_{16:0}$ DMA | 2.47 | 16.04 | 9.17 | 22.68 | 2.79 | – |
| Feature 7 | 3.02 | – | 0.59 | 0.44 | 0.72 | – |
| $C_{17:0}$ iso | – | – | – | – | 1.04 | 0.71 |
| $C_{17:0}$ anteiso | – | – | 0.82 | 0.39 | 2.02 | 1.24 |
| $C_{17:0}$ | – | – | 0.80 | – | 1.72 | 2.37 |
| $C_{18:2}$ cis 9,12 | – | 2.38 | 4.49 | 4.19 | 5.22 | 2.07 |
| $C_{18:1}$ cis 9 | 18.30 | 3.00 | 7.99 | 2.63 | 6.30 | 5.76 |
| Feature 10 | 7.62 | 3.06 | 2.78 | 7.79 | 5.75 | 2.52 |
| $C_{18:1}$ t11? | – | – | 6.50 | – | – | – |
| $C_{18:0}$ | 5.46 | 1.92 | 4.64 | 2.03 | 11.21 | 17.83 |
| Feature 11 | – | 1.91 | 3.53 | 2.02 | 3.82 | – |
| $C_{18:1}$ cis 9 DMA | 16.31 | 2.10 | 3.00 | 2.95 | 3.83 | – |
| $C_{18:1}$ cis 11 DMA | 4.74 | 0.6 | 1.05 | 3.14 | 2.45 | – |
| $C_{18:0}$ DMA | 1.77 | 0.89 | 1.09 | 1.68 | 1.10 | – |
| $C_{19:0}$ | – | – | – | – | 2.18 | 3.73 |
| $C_{:20:1}$ cis 11 | – | – | – | – | 1.06 | 4.26 |
| $C_{20:0}$ | – | – | – | – | 4.81 | 7.89 |

Feature 4 ($C_{15:2}$ or $C_{15:1}$ cis 7), Feature 7 ($C_{17:2}$, 16.760 or $C_{17:1}$ cis 8), Feature 10 ($C_{18:1}$ 11/t9/t6 or UN17.834), Feature 11 ($C_{18:2}$ DMA or $C_{17:0}$ iso 3OH)

to cool down to room temperature. The dehydrated medium was then added to the water, and the solution was filtered through a membrane filter (pore width, 0.2 μm). Following incubation of a medium sample at room temperature for 3 days, the medium became turbid. Microscopic analysis and plating in medium TSA (tryptic soy agar: casein peptone, 15 g; soy peptone, 5.0 g; NaCl, 5.0 g; agar, 15.0 g; water, 1,000 ml; pH, 7.3) and TSS (TSA + 5 % sheep blood) indicated the presence of a single contaminant, MF-EP02$^T$ (DSM 15511$^T$), the type strain of *E. inopinata* (Verbarg et al. 2004).

Isolation and identification of other members of the family is indicated in *Tables 7.11*. The isolation media used vary widely.

Media and growth condition used for maintenance and physiological and chemotaxonomic analyses by the German Collection of Microorganisms and Cell Cultures (DSMZ) are indicated in *Table 7.9*

## Maintenance Procedures

The organisms may be preserved for several months by stab inoculation into screw-capped tubes of nutrient agar (pH 7.4). After overnight growth at 30 °C, the tubes are tightly closed and kept at room or refrigerator temperature in the dark.

**Table 7.8**

**Phenotypic properties that differentiate the species of the new families *Coprobacillaceae* and *Turicibacteraceae* according to the API32A test panel. Reactions were determined for the present study. Taxa: 1, *Eggerthia catenaformis* DSM 20559$^T$; 2, *Sharpea azabuensis* DSM 18934$^T$; 3, *Kandleria vitulina* DSM 20406$^T$; 4, *Catenibacterium mitsuokai* DSM 15897$^T$; 5, *Coprobacillus cateniformis* DSM 15921$^T$; 6, *Clostridium ramosum* DSM 1402$^T$; 7, *Clostridium saccharogumia* DSM 17460$^T$; 8, *Clostridium spiroforme* DSM 1552$^T$; 9, *Turicibacter sanguinis* DSM 14220$^T$. All strains were positive for *N*-acetyl-β-glucosaminidase. All strains were negative for urease, alcohol dehydrogenase, α-arabinosidase, β-glucuronidase, glutamic acid decarboxylase, α-fucosidase, nitrate reduction, production of indole, proline arylamidase, leucyl glycine arylamidase, phenylalanine arylamidase, tyrosine arylamidase, alanine arylamidase, glutamyl glutamic acid arylamidase, and serine arylamidase**

| Properties | 1 | 2 | 3 | 4 | 5 | 6 | 7 | 8 | 9 |
|---|---|---|---|---|---|---|---|---|---|
| α-galactosidase | – | + | + | – | – | – | w | – | + |
| β-galactosidase | – | + | + | + | – | + | + | w | + |
| β-galactosidase-6-phosphate | w | – | – | + | + | + | – | – | w |
| α-glucosidase | + | + | + | – | – | w | – | – | + |
| β-glucosidase | + | + | + | + | + | + | + | – | – |
| Mannose fermentation | – | + | + | + | + | + | + | – | – |
| Raffinose fermentation | – | – | – | + | – | – | – | – | – |
| Alkaline phosphatase | – | – | – | – | – | – | – | – | + |
| Arginine arylamidase | + | + | + | + | – | – | – | – | – |
| Leucine arylamidase | w | w | w | – | – | – | – | – | – |
| Pyroglutamic acid arylamidase | + | – | – | + | + | – | – | – | + |
| Glycine arylamidase | w | w | + | + | – | – | – | – | – |
| Histidine arylamidase | – | w | + | + | – | – | – | – | – |

Abbreviation: *w* weak

Longer-term preservation (over 5 years) may be achieved by freezing on glass beads at 70 °C (Feltham et al. 1978). The organisms can also be preserved by freeze-drying or storage in liquid nitrogen.

## Pathogenicity, Clinical Relevance

*Erysipelothrix rhusiopathiae*, identified as a human pathogen at the end of the nineteenth century, is pathogenic for animals and humans, causing localized cutaneous lesion, named erysipeloid in humans and erysipelas in swine. Other symptoms occurring in humans are in most cases in a generalized cutaneous form involving lesions that progress from the initial site of infection or appear in remote areas and a septicemic form often associated with endocarditis (e.g., see *Tables 7.10*). As summarized in the review by Wood (1974b), Fidalgo and Riley (2004), and Wang et al. (2010), disease in swine may be either acute or chronic, resulting in the development of arthritis and endocarditis. Human infection closely resembles that seen in swine, with both acute and chronic forms also. Swine erysipelas caused by E. rhusiopathiae is the disease of greatest prevalence and economic importance. Bacterial infection of traumatized skin of an individual who was in contact with meat or other animal products, waste, or soil contained with E. rhusiopathiae.

Various virulence factors, such as a heat labile capsule, neuraminidase, and hyaluronidase, have been suggested as being involved in the pathogenicity of E. rhusiopathiae. The surface protective antigen (Spa) protein of *Erysipelothrix rhusiopathiae* has been shown to be highly immunogenic (To and Nagai 2007). The spa type of strains of aquatic origin was more variable than those of terrestrial origin (Ingebritson et al. 2010). A recombinant spaA protein has been used for the detection of anti-*Erysipelothrix* spp. IgG antibodies in pigs (Giménez-Lirola et al. 2012).

In the absence of specific antibodies, the organism evades phagocytosis by phagocytic cells, but even if phagocytized, it is able to replicate intracellularly in these cells (Shimoji 2000). Most strains produce hyaluronidase, and it has been speculated that virulence is correlated with hyaluronidase production; good hyaluronidase producers usually belong to serovar 1 (Ewald 1957, 1981). Neuraminidase is produced in differing amounts by all strains, and neuraminidase has been reported to play a significant role in bacterial attachment and subsequent invasion into host cells (Krasemann and Müller 1975; Müller and Seidler 1975; Müller and Krasemann 1976; Nikolov and Abrashev 1976); the level of neuraminidase activity appears to correlate with virulence.

Control of animal disease by sound husbandry, herd management, good sanitation, and immunization procedures is recommended. Vaccines against erysipelas were developed as early as the mid-1950s, and reviews were published by Bairey and Vogel (1973) and Wood (1984). Vaccines against E. rhusiopathiae infection include the use of an attenuated acapsular strain of *Erysipelothrix rhusiopathiae* YS-1, carrying foreign antigens on its surface (Shimoji et al. 2002). *Erysipelothrix rhusiopathiae* Koganei 65–0.15 strain is used as a live swine erysipelas vaccine for subcutaneous injection (Ogawa et al. 2009). As postulated by Neumann et al. (2009), "a live attenuated *E. rhusiopathiae* strain did not appear to become persistently established in pigs post-vaccination, did not cause any local or systemic signs consistent with swine erysipelas, and was therefore unlikely to revert to a virulent state when used in a field setting."

In vitro, most strains of *E. rhusiopathiae* (and probably *E. tonsillarum*) are resistant to sulfonamides, colistin, gentamicin, kanamycin, neomycin, novobiocin, and polymyxin and sensitive to penicillin, streptomycin, chloramphenicol, and tetracycline (Sneath et al. 1951; Füzi 1963; Wood 1965). Reboli and Farrar (1989) report sensitivity against penicillins, cephalosporins, erythromycin, and clindamycin and variable susceptibility to chloramphenicol and tetracycline. A literature search revealed that cefotaxime is consistently active against *E. rhusiopathiae* (Cormican and Jones 1995).

## Habitat

*E. rhusiopathiae* has a wide distribution in nature and has been isolated from a range of mammals (cattle, horses, dogs, cats, rodents [mice, rats]), birds (hen, turkey, ducks, geese, pheasants), fish and marine invertebrates (including fish slime, fish boxes, and cephalopods and crustaceans, including oysters and lobsters (Fidalgo et al. 2000)), and even marsupials (for other examples, see *Table 7.10*). Domesticated as well as wild animals can be affected. The source of infectious organisms is usually asymptomatic animals which are exposed to the agent by surface water runoff, wild mammals, wild birds, pets, and biting insects (mites). Pathogenic strains of *E. rhusiopathiae* have been isolated from the feces of sick animals which contaminate the environment and groundwater, pigpen soil, and abattoir effluent (Wang et al. 2002). The shedding of the organisms by asymptomatic pigs into the soil of pigpens is probably the reason that *E. rhusiopathiae* has been isolated from farms on which no cases of swine erysipelas have

**Table 7.9**

**Growth medium and growth condition for type strains of members of *Erysipelothricaceae*. Numbers refer to DSMZ Catalogue of Microorganisms (http://www.dsmz.de/catalogues/catalogue-microorganisms/culture-technology/list-of-media-for-microorganisms.html)**

| Taxon | Growth medium[a] | Growth condition |
|---|---|---|
| *Erysipelothrix rhusiopathiae* | 693 | Microaerophilic 37 °C |
| *Erysipelothrix tonsillarum* | 693 | Microaerophilic 37 °C |
| *Erysipelothrix inopinata* | 215 | Microaerophilic 30 °C |
| *Holdemania filiformis* | 104 | Anaerobic 37 °C |
| *Solobacterium moorei* | 1,203 or 110 plus hemin + vitamin K1 | Anaerobic 37 °C |
| *Bulleidia obstructa* | 104 or 110 with 0.05 % Tween 80 | Anaerobic 35 °C |
| *Anaerorhabdus furcosa* | ATCC medium 593 = DSMZ medium 110 | Anaerobic 37 °C |
| *Clostridium innocuum* | 78 | Anaerobic 37 °C |
| *Eubacterium tortuosum* | 455 | Anaerobic 37 °C |
| *Eubacterium dolichum* | 78 with vitamin $K_1$, 0.1 mg/l | Anaerobic 37 °C |
| *Eubacterium biforme* | 78 with 0.1 % Tween 80 | Anaerobic 37 °C |
| *Eubacterium cylindroides* | 78 with vitamin $K_1$, 0.1 mg/l | Anaerobic 37 °C |
| *Allobaculum stercoricanis* | 104 or 110 | Anaerobic 37 °C |
| *Streptococcus pleomorphus* | 104 | Anaerobic 37 °C |
| *Coprobacillus cateniformis* | 78 or 104 | Anaerobic 37 °C |
| *Catenibacterium mitsuokai* | 104 or 110 | Anaerobic 37 °C |
| *Kandleria vitulina* | 232, pre-reduced | Anaerobic 37 °C |
| *Eggerthia catenaformis* | 104 | Anaerobic 37 °C |
| *Sharpea azabuensis* | 58 | Anaerobic 37 °C |
| *Clostridium ramosum* | 110 | Anaerobic 37 °C |
| *Clostridium saccharogumia* | 104b | Anaerobic 37 °C |

**Table 7.9 (continued)**

| Taxon | Growth medium[a] | Growth condition |
|---|---|---|
| *Clostridium spiroforme* | 110 | Anaerobic 37 °C |
| *Turicibacter sanguinis* | 110 | Anaerobic 37 °C |

[a]Media composition

**58. Bifidobacterium medium**

Casein peptone, tryptic digest 10.00 g, yeast extract 5.00 g, meat extract 5.00 g, Bacto Soytone 5.00 g, glucose 10.00 g, $K_2HPO_4$ 2.00 g, $MgSO_4 \times 7H_2O$ 0.20 g, $MnSO_4 \times H_2O$ 0.05 g, Tween 80 1.00 ml, NaCl 5.00 g, cysteine-HCl x $H_2O$ 0.50 g, salt solution (see below) 40.00 ml, Resazurin (25 mg/100 ml) 4.00 ml, distilled water 950.00 ml. The cysteine are added after the medium has been boiled and cooled under $CO_2$. Adjust pH to 6.8 using 8 N NaOH. Distribute under $N_2$ and autoclave

Salt solution: $CaCl_2 \times 2H_2O$ 0.25 g, $MgSO_4 \times 7H_2O$ 0.50 g, $K_2HPO_4$ 1.00 g, $KH_2PO_4$ 1.00 g, $NaHCO_3$ 10.00 g, NaCl 2.00 g, distilled water 1000.00 ml

**78. Chopped meat medium**

As medium 119 without carbohydrates: Casitone 30.0 g, yeast extract 5.0 g, $K_2HPO_4$ 5.0 g, Resazurin 1.0 mg

**110. Chopped meat medium with carbohydrates**

Ground beef (fat free) 500 g, distilled water 1,000 ml, NaOH 1 N 25.0 ml. Use lean beef or horse meat. Remove fat and connective tissue before grinding. Mix meat, water, and NaOH, and then boil for 15 min with stirring. Cool to room temperature, skim fat off surface, and filter, retaining both meat particles and filtrate. To the filtrate add water to a final volume of 1,000 ml, and then add:

Casitone 30.0 g, yeast extract 5.0 g, $K_2HPO_4$ 5.0 g, Resazurin 1.0 mg, glucose 4.0 g, cellobiose 1.0 g, maltose 1.0 g, starch (soluble) 1.0 g

To make medium anoxic boil it, cool under nitrogen atmosphere, add 0.5 g/l cysteine hydrochloride, and adjust pH to 7.0. Dispense anoxically 7 ml medium into Hungate tubes (for strains demanding meat particles, put those first into the tube [use 1 part meat particles to 4 or 5 parts fluid]). Autoclave at 121 °C for 30 min. For agar slants use 15 g agar per 1000.0 ml medium

For hemin (10 ml) and Vitamin K1 (0.20 ml) solution, see medium 104

**104. PYG medium (modified)**

Trypticase peptone 5.00 g, peptone 5.00 g, yeast extract 10.0 g, beef extract 5.00 g, glucose 5.00 g, $K_2HPO_4$ 2.00 g, Tween 80 1.00 ml, cysteine-HCl $\times$ $H_2O$ 0.50 g, Resazurin 1.00 mg, salt solution (see below) 40.0 ml, distilled water 950 ml, hemin solution (see below) 10.00 ml, Vitamin K 1 solution (see below) 0.20 ml. The vitamin K 1, hemin solution, and the cysteine are added after the medium has been boiled and cooled under $CO_2$. Adjust pH to 7.2 using 8 N NaOH. Distribute under $N_2$ and autoclave

Salt solution: $CaCl_2 \times 2H_2O$ 0.25 g, $MgSO_4 \times 7H_2O$ 0.50 g, $K_2HPO_4$ 1.00 g, $KH_2PO_4$ 1.00 g, $NaHCO_3$ 10.0 g, NaCl 2.00 g, distilled water 1000.00 ml

Hemin solution: Dissolve 50 mg hemin in 1 ml 1 N NaOH; make up to 100 ml with distilled water. Store refrigerated

Vitamin K 1 solution: Dissolve 0.1 ml of vitamin K 1 in 20 ml 95 % ethanol and filter sterilize. Store refrigerated in a brown bottle

**104b. PYX medium**

Trypticase peptone 5.0 g, peptone from meat (pepsin-digested) 5.0 g, yeast extract 10.0 g, glucose 5.0 g, Resazurin 1.0 mg, salt solution (see medium 104) 40.0 ml, cysteine-HCl $\times$ $H_2O$ 0.5 g, distilled water 1000.0 ml. Dissolve ingredients (except glucose and cysteine), boil medium for 1 min, and then cool to room temperature under $N_2$ gas atmosphere. Add cysteine and adjust pH to 7.0. Thereafter, dispense under 80 % $N_2$ and 20 % $CO_2$ gas atmosphere in culture vessels and autoclave. After autoclaving add glucose or any other substrate from a sterile, anoxic stock solution prepared under $N_2$. Adjust pH of autoclaved medium with an anoxic, sterile stock solution of $NaHCO_3$ (5 % w/v) prepared under 80 % $N_2$ and 20 % $CO_2$

**215. BHI medium**

Brain-heart infusion (Difco) 37.0 g, distilled water 1000.0 ml

**232. MRS medium with cysteine**

Casein peptone, tryptic digest 10.00 g, meat extract 10.00 g, yeast extract 5.00 g, glucose 20.0 g, Tween 80 1.00 g, $K_2HPO_4$ 2.00 g, Na-acetate 5.00 g, $(NH_4)_2$ citrate 2.00 g, $MgSO_4 \times 7H_2O$ 0.20 g, $MnSO_4 \times H_2O$ 0.05 g, distilled water 1000.00 ml. Adjust pH to 6.2–6.5 and add 0.05 % cysteine hydrochloride

**455. AMB medium**

Trypticase peptone 4.00 g, yeast extract 5.00 g, starch, soluble 1.00 g, meat infusion 25.00 g, glucose 5.00 g, cysteine 1.00 g, $CaCl_2$ 0.01 g, $MgSO_4$ 0.20 g, $(NH_4)_2SO_4$ 1.00 g, $K_2HPO_4$ 15.00 g, distilled water 1000.00 ml, pH 6.9

**693. Columbia blood medium**

Columbia agar base supplemented with 5 % defibrinated sheep blood

**1203. Fastidious anaerobe agar**

F.A.A.; LAB 090; LAB M 46.0 g, add deionized water to 1000.0 ml. Allow to soak for 10 min, swirl to mix, and then sterilize by autoclaving at 121 °C for 15 min. Cool to 47 °C and then aseptically add 5–10 % of sterile defibrinated horse blood, mix well, and pour plates. pH: 7.2 ± 0.2

occurred for many years (Wood and Packer 1972). Apparently healthy swine and asymptomatic swine commonly harbor this organism in their tonsils and other lymphoid tissues (Wood 1974a). Some of these isolates are included in *E. tonsillarum* (Takahashi et al. 1987). *E. rhusiopathiae* has food safety implications, because it can survive for several months in animal tissues such as frozen or chilled pork, cured and smoked ham and pickled bacon, and feed by-products such as dried blood. Viable organisms have been recovered from a buried carcass after 9 months. Heat (e.g., 60 °C for 15 min) and direct sunlight diminish the viability of *E. rhusiopathiae* which also does not grow at 10 % (w/v) NaCl. Low temperature, alkaline conditions, and organic matter favor its survival (Woodbine 1950; Grieco and Sheldon 1970; Ewald 1981).

Human skin infection, known as erysipeloid, results from direct handling of contaminated organic matter such as swine carcasses, fish, and poultry. The infections are largely limited to veterinarians, butchers, and fish handlers. Generally, infection is

**Table 7.10**
**Examples of recent findings on the occurrence and clinical symptoms of *Erysipelothrix rhusiopathiae***

| Environment | References |
|---|---|
| Marine environment, fish, seals | Finkelstein and Oren 2011; Sinclair et al. 2013 |
| Fish, seal, marine environment | Opriessnig et al. 2013 |
| Diseased and healthy animals, retail meats, fish | Tlougan et al. 2010 |
| Diseased animals, healthy animals, fish, retail meats, and environmental material | Takahashi et al. 2008 |
| Pig production facility, groundwater | Hong et al. 2013 |
| Subsurface biosphere of metamorphic rocks transformation | Brazelton et al. 2013 |
| Manure, feed, central-line water, oral fluids, and swabs collected from walls, feed lines, air inlets, exhaust fans, and nipple drinkers of facilities of clinically affected pigs | Bender et al. 2010; Cordero et al. 2010 |
| Child, septic arthritis | Mukhopadhyay et al. 2012 |
| Human, endocarditis | Basu and Tewari 2013 |
| Human, endocarditis, presumed osteomyelitis | Romney et al. 2001 |
| Human, knee arthroplasty | Traer et al. 2008 |
| Pigs, endocarditis | Jensen et al. 2010 |
| Pigs, erysipelas | Ozawa et al. 2009 |
| Pigs, poultry, emus, red poultry mite *(Dermanyssus gallinae)* | Eriksson et al. 2009 |
| Dogs, erysipeloid | Foster et al. 2012 |
| Dogs, endocarditis | Takahashi et al. 2000 |
| Bovine tonsils | Hassanein et al. 2001; Hassanein et al. 2003 |
| Racing pigeons | Cousquer 2005 |
| Bottlenose dolphins, pneumonia, retrospective study | Venn-Watson et al. 2012 |
| Laying hens, erysipelas | Eriksson et al. 2013 |
| Chicken | Nakazawa et al. 1998 |
| Numbat (*Myrmecobius fasciatus*), erysipelas | Vaughan-Higgins et al. 2013 |
| Diseased blue penguin, lung and intestine | Boerner et al. 2004 |
| Scorpion fish, hand infection | Veraldi et al. 2009 |
| Red poultry mite (*Dermanyssus gallinae*) | Valiente Moro et al. 2009 |

confined to the skin of the hands and lower arms where the organisms gain entry through cuts and abrasions. Only rarely does the infection become systemic, causing arthritis and endocarditis. The occurrence of early infections by *E. rhusiopathiae* in men has been summarized by Stackebrandt et al. (2005).

*E. inopinata* was isolated in the course of the validation of production processes in aseptic manufacturing of pharmaceuticals when a vegetable-based growth medium was tested for its dilution performance (Verbarg et al. 2004).

Rarely only are 16S rRNA gene clone sequences of E. rhusiopathiae and related species detected in environmental clone libraries (e.g., clone JQ798989, Hernandez et al. unpublished, from the dung beetle *Thorectes lusitanicus*). Many clone sequences are less than 97 % related to the 16S rRNA sequence of the type strain of *E. rhusiopathiae*, and their affiliation to members of *Erysipelothrix* is uncertain (e.g., Snaidr et al. 1997, clone Z94007; Kalmokoff et al. 2011, clone HQ716188; Liu and Yang unpublished clone GQ480105 from sewage).

Other members of the family as defined by Euzeby (➋ *Fig. 7.1*) thrive in mainly anaerobic conditions and are routinely found in feces and the intestine of a wide range of mammals (➋ *Table 7.11*). Besides *E. rhusiopathiae*, most members thrive as commensals but may turn into opportunistic pathogens under appropriate conditions. Reports of their presence in a variety of mostly human infections are increasing.

A high number of 16S rRNA clones of *Turicibacter sanguinis*, *Clostridium innocuum*, and *Eubacterium biforme* have been identified in various, mostly anaerobic habitats (➋ *Table 7.12*). Fewer or no reports are available for *Bulleidia moorei*, *Solobacterium sanguinis*, *Holdemania filiformis*, *Allobaculum*, and *Anaerorhabdus* and for the other misclassified members of *Eubacterium* and *Streptococcus pleomorphus*. In some cases clone sequences from several species were reported in a single habitat (Turnbaugh et al. 2009; Li et al. 2012).

**Table 7.11**
**Examples for the isolation site and identification approach of non-*Erysipelothrix* strains of *Erysipelotrichaceae* and newly described families**

| Taxon | Host | Location | Goal of study | Detection approach | Identification method | Taxon-specific result | References |
|---|---|---|---|---|---|---|---|
| **Other members of *Erysipelothricaceae*** | | | | | | | |
| *Allobaculum stercoricanis* | Mice | Cecum | Shifts by influence of diet | Nonculture | 16S rRNA sequencing | Increase by high-amylose maize | Tachon et al. 2013 |
| *Allobaculum stercoricanis* | Rats | Gut | Shifts induced by carcinogen | Nonculture | 16S rRNA V3 pyrosequencing, DGGE | Increase by 1,2-dimethyl hydrazine treatment | Wei et al. 2010 |
| *Allobaculum stercoricanis* | Rats | Gut | Assessment after berberine treatment | Non culture | 16S rRNA V3 pyrosequencing | Increase after berberine treatment | Zhang et al. 2012b |
| *Allobaculum stercoricanis* | Sheep | Rumen | MA | Nonculture | 16S rRNA sequencing, qPCR | Not cultured but detectable by nonculture approach | Stiverson et al. 2011 |
| *Bulleidia extructa* | Children | Purulent root canal abscess | MA of abscess | Nonculture | 16S rRNA sequencing, DGGE | Minor component of microbiota | Yang et al. 2010 |
| *Bulleidia extructa* | HIV-positive humans | Necrotizing ulcerative periodontitis (NUP) | MA | Nonculture | 16S rRNA sequencing, oligo-probes | Commonly present in NuP | Paster et al. 2002 |
| *Bulleidia extructa* | Human | Periodontal disease | MA, oligo-probe design | Nonculture | 16S rRNA-based information | Rarely present in patients | Booth et al. 2004 |
| *Bulleidia extructa* | Human | Odontogenic infections | Identification of misidentified *Eubacterium* spp. | Culture | 16S rRNA sequencing, phenotypic | 2 of 105 isolates were assigned to *B. extructa* | Downes et al. 2001 |
| *Holdemania filiformis* | Human | Feces | Characterization | Culture | 16S rRNA sequencing, phenotype | Description | Willems et al. 1997 |
| *Holdemania filiformis* | Pigs | Feces | MA, effects of Bt MON810 maize | Culture, nonculture | 16S rRNA, high throughput sequencing | Higher abundance than in isogenic/ Bt treatment | Buzoianu et al. 2012 |
| *Solobacterium moorei* | Human | Femoral thrombophlebitis | MA | Culture | 16S rRNA restriction analysis | Solobacterium with pathogenic potential | Martin et al. 2007 |
| *Solobacterium moorei* | Human | Proctitis, cervix carcinoma | MA | Culture | 16S rRNA sequencing | Molecular analysis superior to phenotypic analysis | Lau et al. 2006 |
| *Solobacterium moorei* | Human | Bacteraemia | MA | Culture | 16S rRNA sequencing | Recovery from blood culture | Detry et al. 2006; Pedersen et al. 2011 |
| *Solobacterium moorei* | Human | Halitosis | MA | Nonculture | 16S rRNA sequencing | Associated with halitosis | Kazor et al. 2003 |

| | | | | | | | |
|---|---|---|---|---|---|---|---|
| *Solobacterium moorei* | Human | Endodontic infections | MA | Nonculture | 16S rRNA sequencing | Associated with refractory cases | Rolph et al. 2001 |
| *Solobacterium moorei* | Human | Mixed wound infection | Solobacterium assessment | Nonculture | 16S rRNA sequencing | Component in some mixed surgical wound infections | Zheng et al. 2010 |
| *Solobacterium moorei* | Human | Peri-implantitis | MA | Nonculture | 16S rRNA sequencing clone libraries | Present at inflammatory sites | Koyanagi et al. 2010 |
| *Solobacterium moorei* | Human | Refractory periodontitis | MA | Nonculture | HOMIM, microarray | Present in refractory periodontitis | Colombo et al. 2009 |
| *Solobacterium moorei* | Human | Root-filled teeth associated with periradicular lesions | MA | Culture | 16S rRNA sequencing, phenotypic | Among the most prevalent species | Schirrmeister et al. 2009 |
| *Solobacterium moorei* | Human | Halitosis | Identification of *S. moorei* | Culture, nonculture | 16S rRNA sequencing, phenotypic | Associated with halitosis | Haraszthy et al. 2008 |
| *Solobacterium moorei* | Pure culture | Halitosis | Characterization of *S. moorei* | Culture | VSC production | Conversion of cysteine into hydrogen sulfide | Tanabe and Grenier 2012 |
| *Solobacterium moorei* | Human | African-American children with aggressive periodontitis | MA | Nonculture | 16S rRNA-based microarrays | Abundant in localized aggressive periodontitis | Shaddox et al. 2012 |
| *Solobacterium moorei* | Human | Infections of root canals | MA | Nonculture | 16S rRNA clone sequencing | Among prevalent species | Zhang et al. 2012a |
| *Solobacterium moorei* | Human | Bisphosphonate-related osteonecrosis of the jaw | MA | Nonculture | 16S rRNA DGGE, sequencing | Higher presence under antibiotics | Ji et al. 2012 |
| *Streptococcus pleomorphus* | Chicken, turkey, dugs | Caeca | Identification | Culture | Phenotypic | Species description | Barnes et al. 1977 |
| *Eubacterium cylindroides* | Human | Intestine | Description | Culture | Phenotype | Species description | Cato et al. 1974 |
| *Eubacterium cylindroides* | Gerbil | Stomach | MA, infected with H. pylori | Nonculture | 16S rRNA, qPCR | Detected only in uninfected stomach | Osaki et al. 2012 |
| *Eubacterium cylindroides* | Human | Feces | MA, continuous culture system | Non culture | 16S rRNA gene oligo FISH probes | Present in high numbers | Child et al. 2006 |
| *Eubacterium cylindroides* | Human | Feces | MA, influence of different carbohydrate energy sources | Non culture | 16S rRNA gene oligo FISH probes | Pronounced increase by dahlia inulin | Duncan et al. 2003 |
| *Eubacterium cylindroides* | Human | Intestine | Isolation of steroid-3-sulfate-desulfating strains | Culture | Phenotypic | 2 isolates | Van Eldere et al. 1988 |
| *Eubacterium biforme* | Bovine | Intestine | MA, identification of ceftiofur degradation | Culture | Phenotypic | Presence of ceftiofur-degrading β-lactamases | Wagner et al. 2011 |
| *Eubacterium biforme* | Chimpanzee | Feces, influence of fiber-diet | MA, influence of fiber-diet | Nonculture | DNA, DGGE | High numbers under high-fiber diets | Kisidayová et al. 2009 |
| *Eubacterium biforme* | Human, animal | Intestine | Identification of anaerobes | Nonculture | 16S rRNA (TRFPs) | Validation of detection | Khan et al. 2001 |

**Table 7.11** (continued)

| Taxon | Host | Location | Goal of study | Detection approach | Identification method | Taxon-specific result | References |
|---|---|---|---|---|---|---|---|
| *Eubacterium tortuosum* | Bobwhite quail | Liver, hepatic granulomas | Identification of pathogen | Culture | Phenotypic, histochemical | Identification | Williams et al. 2007 |
| *Eubacterium tortuosum* | Chicken | Liver, splenic and hepatic granulomas | Verification of *E. tortuosum* as pathogen | Culture | Phenotypic | Pure culture of *E. tortuosum* cause no symptoms | Hafner et al. 1994 |
| *Eubacterium tortuosum* | Sheep, goat | Lung, liver, intestine, udder abscesses | MA | Culture | Phenotypic | Rare presence in 1 goat | Tadayon et al. 1980 |
| *Eubacterium dolichum* | Human | Feces | Description | Culture | Phenotype | Species description | Moore et al. 1976 |
| *Clostridium innocuum* | Human | Colorecteral surgery | MA | Culture | Phenotype | Among the most frequent anaerobes | Goldstein et al. 2009 |
| *Clostridium innocuum* | Human | Feces of healthy Japanese and Canadians | MA | Culture | Phenotype | Lower numbers in Japanese | Benno et al. 1986 |
| ***Coprobacillaceae* fam. nov.** | | | | | | | |
| *Coprobacillus cateniformis* | Human | Intestine | MA, reduction of daidzin | Culture | 16S rRNA sequencing | Tentative new species MRG-1 | Park et al. 2011 |
| *Coprobacillus cateniformis* | Human | Amniotic cavity after membrane rupture | MA of inflammation | Culture, nonculture | 16S rRNA PCR | Presence of *Coprobacillus* spp., usually associated with gastrointestinal tract | DiGiulio et al. 2010 |
| *Coprobacillus cateniformis* | Human | Irritable bowel syndrome, feces | MA | Nonculture | 16S rRNA, qPCR | High presence in all samples | Lyra et al. 2009; Kassinen et al. 2007 |
| *Coprobacillus cateniformis* | Human | Healthy boy, feces | MA, reduction of daidzin | Culture | 16S rRNA sequencing | Tentative new species TM-40 | Tamura et al. 2007 |
| *Sharpea azabuensis* | Thoroughbred horses | Feces | Characterization | Culture | 16S rRNA sequencing, phenotype | Description | Morita et al. 2008 |
| *Eggerthella cateniformis* | Human | Feces, intestinal and pleural infections | Characterization | Culture | 16S rRNA sequencing, phenotype | Description | Salvetti et al. 2011 |
| *Kandleria vitulina* | Calf, bovine | Rumen | Characterization | Culture | 16S rRNA sequencing, phenotype | Description | Salvetti et al. 2011 |
| *Catenibacterium mitsuokai* | Pigs | Gastrointestinal microbiota | Shifts due to *Salmonella* shedding | Nonculture | 16S rRNA sequencing | Intermediate increase in high-shedder pigs | Bearson et al. 2013 |
| *Catenibacterium mitsuokai* | Human | Stool in patients with end-stage renal disease | MA | Nonculture | 16S rRNA sequencing, microarray analysis | Increase as compared to healthy human | Vaziri et al. 2013 |

| | | | | | | | |
|---|---|---|---|---|---|---|---|
| *Clostridium ramosum* | Human | Bacteremia | Identification | | | Unusual cause of infection | Nanda and Voskuhl 2006 |
| *Clostridium ramosum* | Human | Children with and without acute diarrhea | MA | | | Presence in stool from children with and without diarrhea | Ferreira et al. 2004 |
| *Clostridium ramosum* | Human | Bone trophism, spondylodiscitis | Identification | | | Infection cleared using amoxicillin and metronidazole | Lavigne et al. 2003 |
| *Clostridium ramosum* | Human | Children with acute lymphatic leukemia, mucositis | Identification | Culture | Phenotype | Possible pathogenic role | van der Vorm et al. 1999 |
| *Clostridium spiroforme* | Rabbit | Intestine | Review | Culture | Phenotype | Causing enterotoxemia | Borriello 1995 |
| *Clostridium cocleatum* | Mice, rats, hamsters, rabbits | Feces | MA | Culture | Phenotype | Abundant presence, though not in rabbits | Lee et al. 1991 |
| *Clostridium cocleatum* | Mice | Intestine | Identification | Culture | Phenotype | Presence of numerous glucosidase activities | Boureau et al. 1993 |
| *Clostridium cocleatum* | Mice | Intestine, colitis | MA | Nonculture | 16S rRNA, DGGE | Higher numbers than in non-colitic animals | Bibiloni et al. 2005 |
| ***Turicibacteraceae* fam. nov.** | | | | | | | |
| *Turicibacter sanguinis* | Human | Blood culture, acute appendicitis | characterization | Culture | 16S rRNA, phenotype | Description | Bosshard et al. 2002 |
| *Turicibacter sanguinis* | Piglets | Ileum | MA under chlortetracycline | Nonculture | 16S rRNA clone sequencing | Decrease in *Turicibacter* phylotypes | Rettedal et al. 2009 |
| *Turicibacter sanguinis* | Pigs | Swine waste lagoons | MA | Nonculture | 16S rRNA-based FISH analyses | High relative abundance | Goh et al. 2009 |
| *Turicibacter sanguinis* | Dogs | Acute diarrhea and idiopathic inflammatory bowel disease | MA | Nonculture | 16S rRNA, 454-pyrosequencing, qPCR | Decrease in acute hemorrhagic diarrhea | Suchodolski et al. 2012 |
| *Turicibacter sanguinis* | Dairy cows | Subacute ruminal acidosis (SARA): feeding experiments | MA | Nonculture | 16S rRNA, 454-pyrosequencing | High increase after induction of SAID | Mao et al. 2012 |

**Table 7.12**
**Examples for the occurrence of non-*E. rhusiopathiae* 16S rRNA gene clone sequences of members of *Erysipelothricaceae* in environmental samples**

| Taxon | Habitat | Accession number | Author |
|---|---|---|---|
| *Turicibacter sanguinis* | Swine manure | JN173098 | Talbot et al. unpublished |
| | Human intestine | JX543365, JX543369, JX543370 | Ferrer et al. 2013 |
| | Spacecraft assembly clean room | DQ532261 | Moissl et al. 2007 |
| | Hot compost | AM500735, AM500746, and more | Guo et al. unpublished |
| | Human feces | GQ897733, GQ897860 | Waddington et al. unpublished |
| | Human ileum | HQ743880, HQ790438, HQ790448 | Li et al. 2012 |
| *Bulleidia extructa* | Intubated patients colonized by *P. aeruginosa* | EF509096 | Flanagan et al. 2007 |
| | Human, oral sample | AM420268 | Bolivar et al. 2012 |
| *Solobacterium moorei* | Human ileum | HQ819746, HQ815165, and more | Li et al. 2012 |
| *Holdemania filiformis* | Human ileum | HQ761555 | Li et al. 2012 |
| | Human intestine, mucosa | FJ504785, FJ505558, FJ507319 | Walker et al. 2011 |
| | Human intestine | FJ371255 | Turnbaugh et al. 2009 |
| *Allobaculum stercoricanis* | Wolf (*Canis lupus*) | FJ978506, FJ978532, and more | |
| | Human skin with atopic dermatitis | HM330539 | Kong et al. 2012 |
| *Eubacterium cylindroides* | Human feces | GQ897865 | Waddington et al. unpublished |
| | Human ileum | HQ802059, HQ801802, HQ801941 | Li et al. 2012 |
| *Streptococcus pleomorphus* | Indoor environment | AM697427 | Rintala et al. 2008 |
| | Human intestine | FJ371296 | Turnbaugh et al. 2009 |
| *Eubacterium biforme* | Human ileum | HQ792980 | Li et al. 2012 |
| | Human feces | GQ897588, GQ897669, and many more | Waddington et al. unpublished |
| *Clostridium innocuum* | Human intestine | EF398997, EF399419, and more | Li et al. 2008 |
| | Human feces | HQ259734 | Roger et al. 2010 |

## References

Abt B, Han C, Scheuner C, Lu M, Lapidus A, Nolan M, Lucas S, Hammon N, Deshpande S, Cheng J-F, Tapia R, Goodwin L, Pitluck S, Liolios K, Pagani I, Ivanova N, Mavromatis K, Mikhailova N, Huntemann M, Pati A, Chen A, Palaniappan K, Land M, Hauser L, Brambilla E-M, Rohde M, Spring S, Gronow S, Göker M, Woyke T, Bristow J, Eisen JA, Markowitz V, Hugenholtz P, Kyrpides NC, Klenk H-P, Detter JC (2012) Complete genome sequence of the termite hindgut bacterium *Spirochaeta coccoides* type strain (SPN1$^{T}$), reclassification in the genus *Sphaerochaeta* as *Sphaerochaeta coccoides* comb. nov. and emendations of the family *Spirochaetaceae* and the genus *Sphaerochaeta*. Stand Genomic Sci 6:194–209

Ahrne S, Stenstrom IM, Jensen NE, Pettersson B, Uhlen M, Molin G (1995) Classification of *Erysipelothrix* strains on the basis of restriction fragment length polymorphisms. Int J Syst Bacteriol 45:382–385

Anderson IJ, Scheuner C, Göker M, Mavromatis K, Hooper SD, Porat I, Klenk H-P, Ivanova N, Kyrpides NC (2011) Novel insights into the diversity of catabolic metabolism from ten haloarchaeal genomes. PLoS One 6:e20237

Bairey MH, Vogel JH (1973) Erysipelas immunizing product review. Proc Annu Meet U S Anim Health Assoc 77:340–344

Barber M (1939) A comparative study of *Listeria* and *Erysipelothrix*. J Pathol Bacteriol 48:11–23

Barnes EM, Impey CS, Stevens BJ, Peel JL (1977) *Streptococcus pleomorphus* sp. nov.: an anaerobic streptococcus isolated mainly from the caeca of birds. J Gen Microbiol 102:45–53

Basu R, Tewari P (2013) Mitral regurgitation jet around neoannulus: mitral valve replacement in *Erysipelothrix rhusiopathiae* endocarditis. Ann Card Anaesth 16:129–132

Bearson SM, Allen HK, Bearson BL, Looft T, Brunelle BW, Kich JD, Tuggle CK, Bayles DO, Alt D, Levine UY, Stanton TB (2013) Profiling the gastrointestinal microbiota in response to *Salmonella*: low versus high *Salmonella* shedding in the natural porcine host. Infect Genet Evol 16C:330–340

Bender JS, Shen HG, Irwin CK, Schwartz KJ, Opriessnig T (2010) Characterization of *Erysipelothrix* species isolates from clinically affected pigs. Clin Vaccine Immunol 17:1605–1611

Benno Y, Suzuki K, Suzuki K, Narisawa K, Bruce WR, Mitsuoka T (1986) Comparison of the fecal microflora in rural Japanese and urban Canadians. Microbiol Immunol 30:521–532

Bernath S, Kucsera G, Kadar I, Horvath G, Morovjan G (1997) Comparison of the protein patterns of *Erysipelothrix rhusiopathiae* strains by SDS-PAGE and autoradiography. Acta Vet Hung 45:417–425

Bernath S, Nemet L, Toth K, Morovjan G (2001) Computerized comparison of the protein compositions of *Erysipelothrix rhusiopathiae* and *Erysipelothrix tonsillarum* strains. J Vet Med B Infect Dis Vet Public Health 48:73–79

Bibiloni R, Simon MA, Albright C, Sartor B, Tannock GW (2005) Analysis of the large bowel microbiota of colitic mice using PCR/DGGE. Lett Appl Microbiol 41:45–51

Boerner L, Nevis KR, Hinckley LS, Weber ES, Frasca S Jr (2004) *Erysipelothrix* septicemia in a little blue penguin (*Eudyptula minor*). J Vet Diagn Invest 16:145–149

Bolivar I, Whiteson K, Stadelmann B, Baratti-Mayer D, Gizard Y, Mombelli A, Pittet D, Schrenzel J, Consortium: The Geneva Study Group on Noma (GESNOMA) (2012) Bacterial diversity in oral samples of children in Niger with acute Noma, acute necrotizing gingivitis, and healthy controls. PLoS Negl Trop Dis 6:E1556

Booth V, Downes J, Van den Berg J, Wade WG (2004) Gram-positive anaerobic bacilli in human periodontal disease. J Periodontal Res 39:213–220

Borriello SP (1995) Clostridial disease of the gut. Clin Infect Dis 20(Suppl 2): S242–S250

Bosshard PP, Zbinden R, Altwegg M (2002) *Turicibacter sanguinis* gen. nov., sp. nov., a novel anaerobic, Gram-positive bacterium. Int J Syst Evol Microbiol 52:1263–1266

Boureau H, Decré D, Carlier JP, Guichet C, Bourlioux P (1993) Identification of a *Clostridium cocleatum* strain involved in an anti-*Clostridium difficile* barrier effect and determination of its mucin-degrading enzymes. Res Microbiol 144:405–410

Brazelton WJ, Morrill PL, Szponar N, Schrenk MO (2013) Bacterial communities associated with subsurface geochemical processes in continental serpentinite springs. Appl Environ Microbiol 79:3906

Buchanan RE (1918) Studies in the nomenclature and classification of the bacteria. V. Subgroups and genera of the *Bacteriaceae*. J Bacteriol 3:27–61

Buzoianu SG, Walsh MC, Rea MC, O'Sullivan O, Crispie F, Cotter PD, Ross RP, Gardiner GE, Lawlor PG (2012) The effect of feeding Bt MON810 maize to pigs for 110 days on intestinal microbiota. PLoS One 7:e33668

Cato EP, Salmon CNW, Holdeman LV (1974) *Eubacterium cylindroides* (Rocchi) Holdeman and Moore: emended description and designation of neotype strain. Int J Syst Bacteriol 24:256–259

Child MW, Kennedy A, Walker AW, Bahrami B, Macfarlane S, Macfarlane GT (2006) Studies on the effect of system retention time on bacterial populations colonizing a three-stage continuous culture model of the human large gut using FISH techniques. FEMS Microbiol Ecol 55:299–310

Chooromoney KN, Hampson DJ, Eamens GJ, Turner MJ (1994) Analysis of *Erysipelothrix rhusiopathiae* and *Erysipelothrix tonsillarum* by multilocus enzyme electrophoresis. J Clin Microbiol 32:371–376

Clavel T, Lippman R, Gavini F, Doré J, Blaut M (2007) *Clostridium saccharogumia* sp. nov. and *Lactonifactor longoviformis* gen. nov., sp. nov., two novel human faecal bacteria involved in the conversion of the dietary phytoestrogen secoisolariciresinol diglucoside. Syst Appl Microbiol 30:16–26

Collins MD, Lawson PA, Willems A, Cordoba JJ, Fernandez-Garayzabal J, Garcia P, Hippe H, Farrow JAE (1994) The phylogeny of the genus *Clostridium*: proposal of five new genera and eleven new species combinations. Int J Syst Bacteriol 44:812–826

Colombo AP, Boches SK, Cotton SL, Goodson JM, Kent R, Haffajee AD, Socransky SS, Hasturk H, Van Dyke TE, Dewhirst F, Paster BJ (2009) Comparisons of subgingival microbial profiles of refractory periodontitis, severe periodontitis, and periodontal health using the human oral microbe identification microarray. J Periodontol 80:1421–1432

Cordero A, García M, Herradora M, Ramírez G, Martínez R (2010) Bacteriological characterization of wastewater samples obtained from a primary treatment system on a small scale swine farm. Bioresour Technol 101:2938–2944

Cormican MG, Jones RN (1995) Antimicrobial activity of cefotaxime tested against infrequently isolated pathogenic species (unusual pathogens). Diagn Microbiol Infect Dis 22:43–48

Cousquer G (2005) Erysipelas outbreak in racing pigeons following ingestion of compost. Vet Rec 156:656

Coutinho TA, Imada Y, de Barcellos DE, de Oliveira SJ, Moreno AM (2011) Genotyping of Brazilian *Erysipelothrix* spp. strains by amplified fragment length polymorphism. J Microbiol Methods 84:27–32

Dacres WG, Groth AH Jr (1959) Identification of *Erysipelothrix insidiosa* with fluorescent antibody. J Bacteriol 78:298–299

Davis GHG, Newton KG (1969) Numerical taxonomy of some named coryneform bacteria. J Gen Microbiol 56:195–214

Davis JJ, Xia F, Overbeek RA, Olsen GJ (2013) The genomes of the *Erysipelotrichia* clarify the firmicute origin of the Mollicutes. Int J Syst Evol Microbiol. doi:10.1099/ijs.0.048983-0

Debono M (1912) On some anaerobical bacteria of the normal human intestine. Zentralbl Bakteriol Parasitenkd Infektionskr Hyg Abt I 62:229–234

Detry G, Pierard D, Vandoorslaer K, Wauters G, Avesani V, Glupczynski Y (2006) Septicemia due to *Solobacterium moorei* in a patient with multiple myeloma. Anaerobe 123:160–162

DiGiulio DB, Romero R, Kusanovic JP, Gómez R, Kim CJ, Seok KS, Gotsch F, Mazaki-Tovi S, Vaisbuch E, Sanders K, Bik EM, Chaiworapongsa T, Oyarzún E, Relman DA (2010) Prevalence and diversity of microbes in the amniotic fluid, the fetal inflammatory response, and pregnancy outcome in women with preterm pre-labor rupture of membranes. Am J Reprod Immunol 641:38–57

Downes J, Olsvik B, Hiom SJ, Spratt DA, Cheeseman SL, Olsen I, Weightman AJ, Wade WG (2000) *Bulleidia extructa* gen. nov., sp. nov., isolated from the oral cavity. Int J Syst Bacteriol 50:979–983

Downes J, Munson MA, Spratt DA, Kononen E, Tarkka E, Jousimies-Somer H, Wade WG (2001) Characterisation of *Eubacterium*-like strains isolated from oral infections. J Med Microbiol 50:947–951

Duncan SH, Scott KP, Ramsay AG, Harmsen HJ, Welling GW, Stewart CS, Flint HJ (2003) Effects of alternative dietary substrates on competition between human colonic bacteria in an anaerobic fermentor system. Appl Environ Microbiol 69:1136–1142

Eriksson H, Jansson DS, Johansson KE, Baverud V, Chirico J, Aspan A (2009) Characterization of *Erysipelothrix rhusiopathiae* isolates from poultry, pigs, emus, the poultry red mite and other animals. Vet Microbiol 137:98–104

Eriksson H, Nyman AK, Fellström C, Wallgren P (2013) Erysipelas in laying hens is associated with housing system. Vet Rec 173:18

Erler W (1972) Serologisch, chemische und immunochemische Untersuchungen an Rotlaufbakterien. X. Die Differenzierung der Rotlaufbakterien nach chemischen Merkmalen. Arch Exp Veterinarmed 26:809–816

Ewald FW (1957) Das Hyaluronidase-Bildungsvermögen von Rotlaufbakterien. Monatsh Tierheilk 9:333–341

Ewald FW (1981) The genus *Erysipelothrix*. In: Starr MP, Stolp H, Trüper HG, Balows A, Schlegel HG (eds) The prokaryotes: a handbook on habitats, isolation, and identification of bacteria. Springer, New York, pp 1688–1700

Farfour E, Leto J, Barritault M, Barberis C, Meyer J, Dauphin B et al (2012) Evaluation of the Andromas matrix-assisted laser desorption ionization-time of flight mass spectrometry system for identification of aerobically growing Gram-positive bacilli. J Clin Microbiol 50:2702–2707

Feist H (1972) Serologische, chemische und immunchemische Untersuchungen an Rotlaufbakterien. XII. Das Murein der Rotlaufbakterien. Arch Exp Veterinarmed 26:825–834

Feltham RKA, Power AK, Pell PA, Sneath PHA (1978) A simple method for storage of bacteria at −76°C. J Appl Bacteriol 44:313–316

Feresu SB, Jones D (1988) Taxonomic studies on *Brochothrix*, *Erysipelothrix*, *Listeria* and atypical lactobacilli. J Gen Microbiol 134:1165–1183

Ferreira CE, Nakano V, Avila-Campos MJ (2004) Cytotoxicity and antimicrobial susceptibility of *Clostridium difficile* isolated from hospitalized children with acute diarrhea. Anaerobe 10:171–177

Ferrer M, Ruiz A, Lanza F, Haange SB, Oberbach A et al (2013) Microbiota from the distal guts of lean and obese adolescents exhibit partial functional redundancy besides clear differences in community structure. Environ Microbiol 15:211–226

Fidalgo SG, Riley TV (2004) Detection of *Erysipelothrix rhusiopathiae* in clinical and environmental samples. Methods Mol Biol 268:199–205

Fidalgo SG, Wang Q, Riley TV (2000) Comparison of methods for detection of *Erysipelothrix* spp. and their distribution in some Australasian seafoods. Appl Environ Microbiol 66:2066–2070

Finkelstein R, Oren I (2011) Soft tissue infections caused by marine bacterial pathogens: epidemiology, diagnosis, and management. Curr Infect Dis Rep 13:470–477

Flanagan JL, Brodie EL, Weng L, Lynch SV, Garcia O et al (2007) Loss of bacterial diversity during antibiotic treatment of intubated patients colonized with *Pseudomonas aeruginosa*. J Clin Microbiol 45:1954–1962

Flossmann KD, Erler W (1972) Serologische, chemische und immunchemische Untersuchungen an Rothufbakterien. XI. Isolierung und Charakterisierung von Desoxyribonukleinsäuren aus Rotlaufbakterien. Arch Exp Veterinarmed 26:817–824

Foster JD, Hartmann FA, Moriello KA (2012) A case of apparent canine erysipeloid associated with *Erysipelothrix rhusiopathiae* bacteraemia. Vet Dermatol 23:528-e108

Füzi M (1963) A neomycin sensitivity test for the rapid differentiation of *Listeria monocytogenes* and *Erysipelothrix rhusiopathiae*. J Pathol Bacteriol 85:524–525

Giménez-Lirola LG, Xiao CT, Halbur PG, Opriessnig T (2012) Development and evaluation of an enzyme-linked immunosorbent assay based on a recombinant SpaA protein (rSpaA415) for detection of anti- *Erysipelothrix* spp. IgG antibodies in pigs. J Microbiol Methods 91:191–197

Goh SH, Mabbett AN, Welch JP, Hall SJ, McEwan AG (2009) Molecular ecology of a facultative swine waste lagoon. Lett Appl Microbiol 48:486–492

Göker M, Scheuner C, Klenk H-P, Stielow JB, Menzel W (2011) Codivergence of mycoviruses with their hosts. PLoS One 6:e22252

Goldstein EJ, Citron DM, Merriam CV, Abramson MA (2009) Infection after elective colorectal surgery: bacteriological analysis of failures in a randomized trial of cefotetan vs. ertapenem prophylaxis. Surg Infect (Larchmt) 10:111–118

Greetham HL, Gibson GR, Giffard C, Hippe H, Merkhoffer B, Steiner U, Falsen E, Collins MD (2004) *Allobaculum stercoricanis* gen. nov., sp. nov., isolated from canine feces. Anaerobe 10:301–307

Grieco MH, Sheldon C (1970) *Erysipelothrix rhusiopathiae*. Ann N Y Acad Sci 174:523–532

Hafner S, Harmon BG, Thayer SG, Hall SM (1994) Splenic granulomas in broiler chickens produced experimentally by inoculation with *Eubacterium tortuosum*. Avian Dis 38:605–609

Hammann R, Werner H (1981) Presence of diaminopimelic acid in propionate-negative *Bacteroides* species and in some butyric acid-producing strains. J Med Microbiol 14:205–212

Haraszthy VI, Gerber D, Clark B, Moses P, Parker C, Sreenivasan PK, Zambon JJ (2008) Characterization and prevalence of *Solobacterium moorei* associated with oral halitosis. J Breath Res 21:017002

Harrington R Jr, Hulse DC (1971) Comparison of two plating media for the isolation of *Erysipelothrix rhusiopathiae* from enrichment broth culture. Appl Microbiol 22:141–142

Harrington R Jr, Wood RL, Hulse DC (1974) Comparison of a fluorescent antibody technique and cultural method for the detection of *Erysipelothrix rhusiopathiae* in primary broth cultures. Am J Vet Res 35:461–462

Hassanein R, Sawada T, Kataoka Y, Itoh K, Suzuki Y (2001) Serovars *of Erysipelothrix* species isolated from the tonsils of healthy cattle in Japan. Vet Microbiol 82:97–100

Hassanein R, Sawada T, Kataoka Y, Gadallah A, Suzuki Y (2003) Molecular identification of *Erysipelothrix* isolates from the tonsils of healthy cattle by PCR. Vet Microbiol 95:239–245

Holdeman LV, Cato EP, Moore WEC (1971) *Clostridium ramosum* (Vuillemin) comb. nov.: emended description and proposed neotype strain. Int J Syst Bacteriol 21:35–39

Hong PY, Yannarell AC, Dai Q, Ekizoglu M, Mackie RI (2013) Monitoring the perturbation of soil and groundwater microbial communities due to pig production activities. Appl Environ Microbiol 79:2620–2629

Imada Y, Takase A, Kikuma R, Iwamaru Y, Akachi S, Hayakawa Y (2004) Serotyping of 800 strains of *Erysipelothrix* isolated from pigs affected with erysipelas and discrimination of attenuated live vaccine strain by genotyping. J Clin Microbiol 42:2121–2126

Ingebritson AL, Roth JA, Hauer PJ (2010) *Erysipelothrix rhusiopathiae*: association of Spa-type with serotype and role in protective immunity. Vaccine 28:2490–2496

Jensen HE, Gyllensten J, Hofman C, Leifsson PS, Agerholm JS, Boye M, Aalbæk B (2010) Histologic and bacteriologic findings in valvular endocarditis of slaughter-age pigs. J Vet Diagn Invest 22:921–927

Ji X, Pushalkar S, Li Y, Glickman R, Fleisher K, Saxena D (2012) Antibiotic effects on bacterial profile in osteonecrosis of the jaw. Oral Dis 18:85–95

Jones D (1975) A numerical taxonomic study of coryneform and related bacteria. J Gen Microbiol 87:52–96

Jones D (1986) Genus *Erysipelothrix* Rosenbach 367[al]. In: Sneath PH, Mair NS, Sharpe ME (eds) Bergey's manual of systematic bacteriology, vol 2. Williams and Wilkins, Baltimore, pp 1245–1249

Julak J, Ryska M, Koruna I, Mencikova E (1989) Cellular fatty acids and fatty aldehydes of *Listeria* and *Erysipelothrix*. Zentralbl Bakteriol 272:171–180

Kageyama A, Benno Y (2000a) *Coprobacillus catenaformis* gen. nov., sp. nov., a new genus and species isolated from human feces. Microbiol Immunol 44:23–28

Kageyama A, Benno Y (2000b) *Catenibacterium mitsuokai* gen. nov., sp. nov., a Gram-positive anaerobic bacterium isolated from human faeces. Int J Syst Evol Microbiol 50:1595–1599

Kageyama A, Benno Y (2000c) Phylogenic and phenotypic characterization of some *Eubacterium*-like isolates from human feces: description of *Solobacterium moorei* gen. nov., sp. nov. Microbiol Immunol 44:223–227

Kalmokoff M, Waddington LM, Thomas M, Liang KL, Ma C, Topp E, Dandurand UD, Letellier A, Matias F, Brooks SP (2011) Continuous feeding of antimicrobial growth promoters to commercial swine during the growing/finishing phase does not modify faecal community erythromycin resistance or community structure. J Appl Microbiol 110:1414–1425

Kaneuchi C, Mizayato T, Shinjo T, Mitsuoka T (1979) Taxonomic study of helically coiled, sporeforming anaerobes isolated from the intestines of humans and other animals: *Clostridium cocleatum* sp. nov. and *Clostridium spiroforme* sp. nov. Int J Syst Bacteriol 29:1–12

Kassinen A, Krogius-Kurikka L, Mäkivuokko H, Rinttilä T, Paulin L, Corander J, Malinen E, Apajalahti J, Palva A (2007) The fecal microbiota of irritable bowel syndrome patients differs significantly from that of healthy subjects. Gastroenterology 133:24–33

Kazor CE, Mitchell PM, Lee AM, Stokes LN, Loesche WJ, Dewhirst FE, Paster BJ (2003) Diversity of bacterial populations on the tongue dorsa of patients with halitosis and healthy patients. J Clin Microbiol 41:558–563

Khan AA, Nawaz MS, Robertson L, Khan SA, Cerniglia CE (2001) Identification of predominant human and animal anaerobic intestinal bacterial species by terminal restriction fragment patterns (TRFPs): a rapid, PCR-based method. Mol Cell Probes 15:349–355

Kisidayová S, Váradyová Z, Pristas P, Piknová M, Nigutová K, Petrzelková KJ, Profousová I, Schovancová K, Kamler J, Modrý D (2009) Effects of high- and low-fiber diets on fecal fermentation and fecal microbial populations of captive chimpanzees. Am J Primatol 71:548–557

Kiuchi A, Hara M, Pham HS, Takikawa K, Tabuchi K (2000) Phylogenetic analysis of the *Erysipelothrix rhusiopathiae* and *Erysipelothrix tonsillarum* based upon 16S rRNA. DNA Seq 11:257–260

Koch R (1878) Untersuchungen über die Atiologie der Wundinfektionskrankheiten. Vogel, Leipzig

Kong HH, Oh J, Deming C, Conlan S, Grice EA, Beatson MA, Nomicos E, Polley EC, Komarow HD, Murray PR, Turner ML, Segre JA (2012) Consortium: comparative sequence program. 2012. Temporal shifts in the skin microbiome associated with disease flares and treatment in children with atopic dermatitis. Genome Res 22:850–859

Koyanagi T, Sakamoto M, Takeuchi Y, Ohkuma M, Izumi Y (2010) Analysis of microbiota associated with peri-implantitis using 16S rRNA gene clone library. J Oral Microbiol 24:2

Krasemann C, Müller HE (1975) Die Virulenz von *Erysipelothrix-rhusiopathiae*-Stämmen und Neuraminidase-Produktion. Zentralbl Bakteriol Parasitenkd Infektionskr Hyg Abt 1 Orig Reihe A 23:1206–1213

Langford GC, Hansen PA (1953) *Erysipelothrix* insidiosa. Riass Commun VI Congr Int Microbiol Roma 1:18

Langford GC, Hansen PA (1954) The species of *Erysipelothrix*. Antonie Van Leeuwenhoek J Microbiol Serol 20:87–92

Lau SK, Teng JL, Leung KW, Li NK, Ng KH, Chau KY, Que TL, Woo PC, Yuen KY (2006) Bacteremia caused by *Solobacterium moorei* in a patient with acute proctitis and carcinoma of the cervix. J Clin Microbiol 44:3031–3034

Lavigne JP, Bouziges N, Sotto A, Leroux JL, Michaux-Charachon S (2003) Spondylodiscitis due to *Clostridium ramosum* infection in an immunocompetent elderly patient. J Clin Microbiol 41:2223–2226

Lee WK, Fujisawa T, Kawamura S, Itoh K, Mitsuoka T (1991) Isolation and identification of clostridia from the intestine of laboratory animals. Lab Anim 25:9–15

Li M, Wang B, Zhang M, Rantalainen M, Wang S et al (2008) Symbiotic gut microbes modulate human metabolic phenotypes. Proc Natl Acad Sci U S A 105:2117–2122

Li E, Hamm CM, Gulati AS, Sartor RB, Chen H et al (2012) Inflammatory bowel diseases phenotype, C. difficile and NOD2. Genotype are associated with shifts in human ileum associated microbial composition. PLoS One 7(6), E26284

Loeffler FA (1886) Experimentelle Untersuchungen über Schweinerotlauf. Arb Kais Gesundheitsamt 1:46–55

Ludwig W, Schleifer KH, Whitman WB (2009a) Order I. *Erysipelothrichales*. In: De Vos P, Garrity GM, Jones D, Krieg NR, Ludwig W, Rainey FA, Schleifer KH, Whitman WB (eds) Bergey's manual of systematic bacteriology, 2nd edn. Springer, New York, p 1298

Ludwig W, Schleifer KH, Whitman WB (2009b) Class III. *Erysipelotrichia*. In: De Vos P, Garrity GM, Jones D, Krieg NR, Ludwig W, Rainey FA, Schleifer KH, Whitman WB (eds) Bergey's manual of systematic bacteriology, 2nd edn. Springer, New York, p 1298

Ludwig W, Euzéby J, Schumann P, Busse H-J, Trujillo ME, Kämpfer P, Whitman WB (2012) Road map of the Actinobacteria. In: Whitman WB, Goodfellow M, Kämpfer P, Busse H-J, Trujillo ME, Garrity G, Ludwig W, Suzuki K (eds) Bergey's manual of systematic bacteriology, vol 5, 2nd edn. Springer, New York, pp 1–28

Lyra A, Rinttilä T, Nikkilä J, Krogius-Kurikka L, Kajander K, Malinen E, Mättö J, Mäkelä L, Palva A (2009) Diarrhoea-predominant irritable bowel syndrome distinguishable by 16S rRNA gene phylotype quantification. World J Gastroenterol 15:5936–5945

Makino S-I, Okada Y, Maruyama T, Ishikawa K, Takahashi T, Nakamura M, Ezaki T, Morita H (1994) Direct and rapid detection of *Erysipelothrix rhusiopathiae* DNA in animals by PCR. J Clin Microbiol 32:1526–1531

Mann S (1969) Über die Zellwandbausteine *von Listeria monocytogenes und Erysipelothrix rhusiopathiae*. Zentralbl Bakteriol Parasitenkd Infektionskr Hyg Abt 1 Orig Reihe A 209:510–518

Mao S, Zhang R, Wang D, Zhu W (2012) The diversity of the fecal bacterial community and its relationship with the concentration of volatile fatty acids in the feces during subacute rumen acidosis in dairy cows. BMC Vet Res 8:237

Martin CA, Wijesurendra RS, Borland CD, Karas JA (2007) Femoral vein thrombophlebitis and septic pulmonary embolism due to a mixed anaerobic infection including *Solobacterium moorei*: a case report. J Med Case Rep 1:40

Migula W (1900) System der Bakterien. Handbuch der Morphologie, Entwicklungsgeschichte und Systematik der Bacterien. G. Fischer Verlag, Jena

Moissl C, Osman S, La Duc MT, Dekas A, Brodie E, DeSantis T, Venkateswaran K (2007) Molecular bacterial community analysis of clean rooms where spacecraft are assembled. FEMS Microbiol Ecol 61:509–521

Moore WEC, Johnson JL, Holdeman LV (1976) Emendation of *Bacteriodaceae* and *Butyrivibrio* and descriptions of *Desulfomonas* gen. nov. and ten new species in the genera *Desulfomonas*, *Butyrivibrio*, *Eubacterium*, *Clostridium*, and *Ruminococcus*. Int J Syst Bacteriol 26:238–252

Morita H, Shiratori C, Murakami M, Takami H, Toh H, Kato Y, Nakajima F, Takagi M, Akita H, Masaoka T, Hattori M (2008) *Sharpea azabuensis* gen. nov., sp. nov., a Gram-positive, strictly anaerobic bacterium isolated from the faeces of thoroughbred horses. Int J Syst Evol Microbiol 58:2682–2686

Mukhopadhyay C, Shah H, Vandana K, Munim F, Vijayan SA (2012) A child with *Erysipelothrix* arthritis-beware of the little known. Asian Pac J Trop Biomed 2:503–504

Müller HE, Krasemann C (1976) Immunität gegen *Erysipelothrix rhusiopathiae*-Infektion durch aktive Immunizierung mit homologer Neuraminidase. Z Immunitätsforsch 151:237–241

Müller HE, Seidler D (1975) Über das Vorkommen Neuraminidase-neutralizierender Antikörper bei chronisch rotlaufkranken Schweinen. Zentralbl Bakteriol Parasitenkd Infektionskr Hyg Abt 1 Orig Reihe A 230:51–58

Nakazawa H, Hayashidani H, Higashi J, Kaneko K, Takahashi T, Ogawa M (1998) Occurrence of *Erysipelothrix* spp. in broiler chickens at an abattoir. J Food Prot 61:907–909

Nanda N, Voskuhl GW (2006) Lung abscess caused by *Clostridium ramosum*. J Okla State Med Assoc 99:158–160

Neumann EJ, Grinberg A, Bonistalli KN, Mack HJ, Lehrbach PR, Gibson N (2009) Safety of a live attenuated *Erysipelothrix rhusiopathiae* vaccine for swine. Vet Microbiol 135:297–303

Nikolov P, Abrashev I (1976) Comparative studies of the neuraminidase activity of *Erysipelothrix insidiosa*. Activity of virulent strains and avirulent variants of *Erysipelothrix insidiosa*. Acta Microbiol Virol Immunol (Sofia) 3:28–31

Noguchi N, Sasatsu M, Takahashi T, Ohmae K, Terakado N, Kono M (1993) Detection of plasmid DNA in *Erysipelothrix rhusiopathiae* isolated from pigs with chronic swine erysipelas. J Vet Med Sci 55:349–350

Ogawa Y, Oishi E, Muneta Y, Sano A, Hikono H, Shibahara T, Yagi Y, Shimoji Y (2009) Oral vaccination against mycoplasmal pneumonia of swine using a live *Erysipelothrix rhusiopathiae* vaccine strain as a vector. Vaccine 27:4543–4550

Okatani AT, Hayashidani TH, Takahashi T, Taniguchi T, Ogawa M, Kaneko K-I (2000) Randomly amplified polymorphic DNA analysis of *Erysipelothrix* spp. J Clin Microbiol 38:4332–4336

Okatani AT, Uto T, Taniguchi T, Horisaka T, Horikita T, Kaneko K-I, Hayashidani H (2001) Pulsed-field gel electrophoresis in differentiation of *Erysipelothrix* species strains. J Clin Microbiol 39:4032–4036

Okatani TA, Ishikawa M, Yoshida S, Sekiguchi M, Tanno K, Ogawa M, Horikita T, Horisaka T, Taniguchi T, Kato Y, Hayashidani H (2004) Automated ribotyping: a rapid typing method for analysis of *Erysipelothrix* spp. strains. J Vet Med Sci 66:729–733

Opriessnig T, Hoffmann LJ, Harris DL, Gaul SB, Halbur PG (2004) *Erysipelothrix rhusiopathiae*: genetic characterization of midwest US isolates and live commercial vaccines using pulsed-field gel electrophoresis. J Vet Diagn Invest 16:101–107

Opriessnig T, Shen HG, Bender JS, Boehm JR, Halbur PG (2013) *Erysipelothrix rhusiopathiae* isolates recovered from fish, a Harbour Seal (Phoca vitulina) and the marine environment are capable of inducing characteristic cutaneous lesions in pigs. J Comp Pathol 148:365–372

Osaki T, Matsuki T, Asahara T, Zaman C, Hanawa T, Yonezawa H, Kurata S, Woo TD, Nomoto K, Kamiya S (2012) Comparative analysis of gastric bacterial microbiota in Mongolian gerbils after long-term infection with *Helicobacter pylori*. Microb Pathog 53:12–18

Ozawa M, Yamamoto K, Kojima A, Takagi M, Takahashi T (2009) Etiological and biological characteristics of *Erysipelothrix rhusiopathiae* isolated between 1994 and 2001 from pigs with swine erysipelas in Japan. J Vet Med Sci 7:697–702

Packer RA (1943) The use of sodium azide and crystal violet in a selective medium for *Erysipelothrix rhusiopathiae* and streptococci. J Bacteriol 46:343–349

Park HY, Kim M, Han J (2011) Stereospecific microbial production of isoflavanones from isoflavones and isoflavone glucosides. Appl Microbiol Biotechnol 91:1173–1181

Paster BJ, Russell MK, Alpagot T, Lee AM, Boches SK, Galvin JL, Dewhirst FE (2002) Bacterial diversity in necrotizing ulcerative periodontitis in HIV-positive subjects. Ann Periodontol 71:8–16

Pasteur L, Dumas M (1882) Sur le rouget, ou mal rouge des porcs. Extrait d'une Lettre. C R Hebd Seances Acad Sci Paris 95:1120–1121

Pedersen RM, Holt HM, Justesen US (2011) *Solobacterium moorei* bacteremia: identification, antimicrobial susceptibility, and clinical characteristics. J Clin Microbiol 49:2766–2768

Peltier J, Courtin P, El Meouche I, Lemée L, Chapot-Chartier MP, Pons JL (2011) *Clostridium difficile* has an original peptidoglycan structure with a high level of N-acetylglucosamine deacetylation and mainly 3-3 cross-links. J Biol Chem 286:29053–29062

Pleszczynska E (1972) Comparative studies on *Listeria* and *Erysipelothrix*. I. Analysis of whole antigens. II. Analysis of antigen fractions. Pol Arch Weter 15:463–471

Rainey FA, Hollen BJ, Small A (2009) Genus I. *Clostridium*. In: De Vos P, Garrity GM, Jones D, Krieg NR, Ludwig W, Rainey FA, Schleifer KH, Whitman WB (eds) Bergey's manual of systematic bacteriology. pp 738–828

Reboli AC, Farrar WE (1989) *Erysipelothrix rhusiopathiae*: an occupational pathogen. Clin Microbiol Rev 2:354–359

Reboli AC, Farrar WE (1991) The genus *Erysipelothrix*. In: Balows A, Trüper HG, Harder W, Schleifer KH (eds) The prokaryotes. A handbook on the biology of bacteria: ecophysiology, isolation, identification, applications. Springer, New York, pp 1629–1642, 1992

Rettedal E, Vilain S, Lindblom S, Lehnert K, Scofield C, George S, Clay S, Kaushik RS, Rosa AJ, Francis D, Brözel VS (2009) Alteration of the ileal microbiota of weanling piglets by the growth-promoting antibiotic chlortetracycline. Appl Environ Microbiol 75:5489–5495

Rintala H, Pitkaranta M, Toivola M, Paulin L, Nevalainen A (2008) Diversity and seasonal dynamics of bacterial community in indoor environment. BMC Microbiol 8:56

Roger LC, Costabile A, Holland DT, Hoyles L, McCartney AL (2010) Examination of faecal *Bifidobacterium* populations in breast- and formula-fed infants during the first 18 months of life. Microbiology 156:3329–3341

Rolph HJ, Lennon A, Riggio MP, Saunders WP, MacKenzie D, Coldero L, Bagg J (2001) Molecular identification of microorganisms from endodontic infections. J Clin Microbiol 39:3282–3289

Romney M, Cheung S, Montessori V (2001) *Erysipelothrix rhusiopathiae* endocarditis and presumed osteomyelitis. Can J Infect Dis 12:254–256

Rosenbach FJ (1909) Experimentelle, morphologische und klinische Studien über krankheitserregende Mikroorganismen des Schweinerotlaufs, des Erysipeloids und der Mausesepticamie. Z Hyg Infekt 63:343–371

Salvetti E, Felis GE, Dellaglio F, Castioni A, Torriai S, Lawson PA (2011) Reclassification of *Lactobacillus catenaformis* Eggerth 1935 Moore and Holdeman 1970 and *Lactobacillus vitulinus* Sharpe et al. 1973 as *Eggerthia catenaformis* gen. nov., comb. nov. and *Kandleria vitulina* gen. nov., comb. nov., respectively. Int J Syst Evol Microbiol 61:2520–2524

Sasser M (1990) Identification of bacteria by gas chromatography of cellular fatty acids. USFCC Newsl 20:1–6

Schirrmeister JF, Liebenow AL, Pelz K, Wittmer A, Serr A, Hellwig E, Al-Ahmad A (2009) New bacterial compositions in root-filled teeth with periradicular lesions. J Endod 35:169–174

Schleifer KH, Kandler O (1972) Peptidoglycan types of bacterial cell walls and their taxonomic implications. Bacteriol Rev 36:407–477

Schubert K, Fiedler F (2001) Structural investigations on the cell surface of *Erysipelothrix rhusiopathiae*. Syst Appl Microbiol 24:26–30

Schumann P (2011) Peptidoglycan structure. Methods Microbiol 38:101–129

Seidler D, Trautwein G, Bohm KH (1971) Nachweis von *Erysipelothrix insidiosa* mit fluoreszierenden Antikörpern. Zentralbl Veterinarmed B 18:280–292

Shaddox LM, Huang H, Lin T, Hou W, Harrison PL, Aukhil I, Walker CB, Klepac-Ceraj V, Paster BJ (2012) Microbiological characterization in children with aggressive periodontitis. J Dent Res 91:927–933

Shen HG, Bender JS, Opriessnig T (2010) Identification of the surface protective antigen (spa) types in *Erysipelothrix* spp. reference strains and diagnostic samples by spa multiplex real-time and conventional PCR assays. J Appl Microbiol 109:1227–1233

Shimoji Y (2000) Pathogenicity of *Erysipelothrix rhusiopathiae*: virulence factors and protective immunity. Microbes Infect 2:965–972

Shimoji Y, Oishi E, Kitajima T, Muneta Y, Shimizu S, Mori Y (2002) *Erysipelothrix rhusiopathiae* YS-1 as a live vaccine vehicle for heterologous protein expression and intranasal immunization of pigs. Infect Immun 70:226–232

Sinclair M, Hawkins A, Testro A (2013) Something fishy: an unusual *Erysipelothrix rhusiopathiae* infection in an immunocompromised individual. BMJ Case Rep. doi: 10.1136/bcr-2013-008873

Smith LDS, King E (1962) *Clostridium innocuum*, sp. n., a spore-forming anaerobe isolated from human infections. J Bacteriol 83:938–939

Snaidr J, Amann R, Huber I, Ludwig W, Schleifer KH (1997) Phylogenetic analysis and in situ identification of bacteria in activated sludge. Appl Environ Microbiol 63:2884–2896

Sneath PHA, Abbott JD, Cunliffe AC (1951) The bacteriology of erysipeloid. Br Med J 2:1063–1066

Spring S, Scheuner C, Lapidus A, Lucas S, Del Rio TG, Tice H, Copeland A, Cheng J-F, Chen F, Nolan M, Saunders E, Pitluck S, Liolios K, Ivanova N, Mavromatis K, Lykidis A, Pati A, Chen A, Palaniappan K, Land M, Hauser L, Chang Y-J, Jeffries CD, Goodwin L, Detter JC, Brettin T, Rohde M, Göker M, Woyke T, Bristow J, Eisen JA, Markowitz V, Hugenholtz P, Kyrpides NC, Klenk H-P (2010) The genome sequence of *Methanohalophilus mahii* SLP$^T$ reveals differences in the energy metabolism among members of the *Methanosarcinaceae* inhabiting freshwater and saline environments. Archaea 2010:690737

Stackebrandt E (2009a) Genus I. *Erysipelothrix*. In: De Vos P, Garrity GM, Jones D, Krieg NR, Ludwig W, Rainey FA, Schleifer KH, Whitman WB (eds) Bergey's manual of systematic bacteriology, 2nd edn. Springer, New York, pp 1299–1306

Stackebrandt E (2009b) Family I. *Erysipelothricaceae*. In: De Vos P, Garrity GM, Jones D, Krieg NR, Ludwig W, Rainey FA, Schleifer KH, Whitman WB (eds) Bergey's manual of systematic bacteriology, 2nd edn. Springer, New York, p 1299

Stackebrandt E, Reboli AC, Farrar WE (2005) The Genus *Erysipelothrix*. In: Dworkin M, Falkow S, Rosenberg E, Schleifer K-H, Stackebrandt E (eds) The prokaryotes, 3rd edn. Springer, New York (electronic, release 3.1)

Stamatakis A (2006) RAxML-VI-HPC: maximum likelihood-based phylogenetic analyses with thousands of taxa and mixed models. Bioinformatics 22:2688–2690

Stiverson J, Morrison M, Yu Z (2011) Populations of select cultured and uncultured bacteria in the rumen of sheep and the effect of diets and ruminal fractions. Int J Microbiol 2011:750613

Stuart MR, Pease PE (1972) A numerical study on the relationships of *Listeria* and *Erysipelothrix*. J Gen Microbiol 73:551–565

Stuart SE, Welshimer HJ (1974) Taxonomic re-examination of *Listeria* Pirie and transfer of *Listeria grayi* and *Listeria murrayi* to a new genus *Murraya*. Int J Syst Bacteriol 24:177–185

Suchodolski JS, Markel ME, Garcia-Mazcorro JF, Unterer S, Heilmann RM, Dowd SE, Kachroo P, Ivanov I, Minamoto Y, Dillman EM, Steiner JM, Cook AK, Toresson L (2012) The fecal microbiome in dogs with acute diarrhea and idiopathic inflammatory bowel disease. PLoS One 7:e51907

Tachon S, Zhou J, Keenan M, Martin R, Marco ML (2013) The intestinal microbiota in aged mice is modulated by dietary resistant starch and correlated with improvements in host responses. FEMS Microbiol Ecol 83:299–309

Tadayon RA, Carroll KK (1971) Effect of growth conditions on the fatty acid composition of *Listeria monocytogenes* and comparison with the fatty acids of *Erysipelothrix* and *Corynebacterium*. Lipids 6:820–825

Tadayon RA, Cheema AH, Muhammed SI (1980) Microorganisms associated with abscesses of sheep and goats in the south of Iran. Am J Vet Res 41:798–802

Takahashi T, Fujisawa T, Benno Y, Tamura Y, Sawada T, Suzuki S, Muramatsu M, Mitsuoka T (1987) *Erysipelothrix tonsillarum* sp. nov., isolated from tonsils of apparently healthy pigs. Int J Syst Bacteriol 37:166–168

Takahashi T, Tamura Y, Sawada T, Suzuki S, Muramatsu M, Fujisawa T, Benno Y, Mitsuoka T (1989) Enzymatic profiles of *Erysipelothrix rhusiopathiae* and *Erysipelothrix tonsillae*. Res Vet Sci 47:275–276

Takahashi T, Fujisawa T, Tamura Y, Suzuki S, Muramatsu M, Sawada T, Benno Y, Mitsuoka T (1992) DNA relatedness among *Erysipelothrix rhusiopathiae* strains representing all twenty-three serovars and *Erysipelothrix tonsillarum*. Int J Syst Bacteriol 42:469–473

Takahashi T, Tamura Y, Endo YS, Hara N (1994) Cellular fatty acid composition of *Erysipelothrix rhusiopathiae* and *Erysipelothrix tonsillarum*. J Vet Med Sci 56:385–387

Takahashi T, Fujisawa T, Yamamoto K, Kijima M, Takahashi T (2000) Taxonomic evidence that serovar 7 of *Erysipelothrix* strains isolated from dogs with endocarditis are *Erysipelothrix tonsillarum*. J Vet Med B Infect Dis Vet Public Health 47:311–313

Takahashi T, Fujisawa T, Umeno A, Kozasa T, Yamamoto K, Sawada T (2008) A taxonomic study on *Erysipelothrix* by DNA-DNA hybridization experiments with numerous strains isolated from extensive origins. Microbiol Immunol 52:469–478

Takeshi K, Makino S, Ikeda T, Takada N, Nakashiro A, Nakanishi K, Oguma K, Katoh Y, Sunagawa H, Ohyama T (1999) Direct and rapid detection by PCR of *Erysipelothrix* sp. DNAs prepared from bacterial strains and animal tissues. J Clin Microbiol 37:4093–4098

Tamura Y, Takahashi T, Zarkasie K, Nakamura M, Yoshimura H (1993) Differentiation of *Erysipelothrix rhusiopathiae* and *Erysipelothrix tonsillarum* by sodium dodecyl sulfate-polyacrylamide gel electrophoresis of cell proteins. Int J Syst Bacteriol 43:111–114

Tamura M, Tsushida T, Shinohara K (2007) Isolation of an isoflavone-metabolizing, *Clostridium*-like bacterium, strain TM-40, from human faeces. Anaerobe 13:32–35

Tanabe S, Grenier D (2012) Characterization of volatile sulfur compound production by *Solobacterium moorei*. Arch Oral Biol 57:1639–1643

Tlougan BE, Podjasek JO, Adams BB (2010) Aquatic sports dermatoses: part 3. On the water. Int J Dermatol 49:1111–1120

To H, Nagai S (2007) Genetic and antigenic diversity of the surface protective antigen proteins of *Erysipelothrix rhusiopathiae*. Clin Vaccine Immunol 14:813–820

To H, Koyama T, Nagai S, Tuchiya K, Nunoya T (2009) Development of quantitative real-time polymerase chain reaction for detection of and discrimination between *Erysipelothrix rhusiopathiae* and other *Erysipelothrix* species. J Vet Diagn Invest 21:701–706

Tóth EM, Schumann P, Borsodi AK, Kéki Z, Kovács AL, Márialigeti K (2008) *Wohlfahrtiimonas chitiniclastica* gen. nov., sp. nov., a new gammaproteobacterium isolated from *Wohlfahrtia magnifica* (Diptera: Sarcophagidae). Int J Syst Evol Microbiol 58:976–981

Traer EA, Williams MR, Keenan JN (2008) *Erysipelothrix rhusiopathiae* infection of a total knee arthroplasty an occupational hazard. J Arthroplasty 23:609–611

Turnbaugh PJ, Hamady M, Yatsunenko T, Cantarel BL, Duncan A et al (2009) A core gut microbiome in obese and lean twins. Nature 457:480–484

Validation List N 74 (2000) Int J Syst Evol Microbiol 50:949–950

Validation List N 75 (2000) Int J Syst Evol Microbiol 50:1415–1417

Validation List N 110 (2006) Int J Syst Evol Microbiol 56:1459–1460

Validation List N 115 (2007) Int J Syst Evol Microbiol 57:893–897

Valiente Moro C, Thioulouse J, Chauve C, Normand P, Zenner L (2009) Bacterial taxa associated with the hematophagous mite *Dermanyssus gallinae* detected by 16S rRNA PCR amplification and TTGE fingerprinting. Res Microbiol 160:63–70

van der Vorm ER, von Rosenstiel IA, Spanjaard L, Dankert J (1999) Gas gangrene in an immunocompromised girl due to a *Clostridium ramosum* infection. Clin Infect Dis 28:923–924

Van Eldere J, Robben J, De Pauw G, Merckx R, Eyssen H (1988) Isolation and identification of intestinal steroid-desulfating bacteria from rats and humans. Appl Environ Microbiol 54:2112–2117

Vaughan-Higgins RJ, Bradfield K, Friend JA, Riley TV, Vitali SD (2013) Erysipelas in a numbat (*Myrmecobius fasciatus*). J Zoo Wildl Med 44:208–211

Vaziri ND, Wong J, Pahl M, Piceno YM, Yuan J, DeSantis TZ, Ni Z, Nguyen TH, Andersen GL (2013) Chronic kidney disease alters intestinal microbial flora. Kidney Int 83:308–315

Venn-Watson S, Daniels R, Smith C (2012) Thirty year retrospective evaluation of pneumonia in a bottlenose dolphin *Tursiops truncatus* population. Dis Aquat Organ 99:237–242

Veraldi S, Girgenti V, Dassoni F, Gianotti R (2009) Erysipeloid: a review. Clin Exp Dermatol 34:859–862

Verbarg S, Rheims H, Emus S, Frühling A, Kroppenstedt RM, Stackebrandt E, Schumann P (2004) *Erysipelothrix inopinata* sp. nov., isolated in the course of sterile filtration of vegetable peptone broth, and description of *Erysipelotrichaceae* fam. nov. Int J Syst Evol Microbiol 54:221–225

Von Graevenitz A, Osterhout G, Dick J (1991) Grouping of some clinically relevant gram-positive rods by automated fatty acid analysis. Diagnostic implications. APMIS 99:147–154

Wade WG (2009) Genus I. *Eubacterium*. In: De Vos P, Garrity GM, Jones D, Krieg NR, Ludwig W, Rainey FA, Schleifer KH, Whitman WB (eds) Bergey's manual of systematic bacteriology, 2nd edn. Springer, New York, pp 865–881

Wagner RD, Johnson SJ, Cerniglia CE, Erickson BD (2011) Bovine intestinal bacteria inactivate and degrade ceftiofur and ceftriaxone with multiple beta-lactamases. Antimicrob Agents Chemother 55:4990–4998

Walker AW, Sanderson JD, Churcher C, Parkes GC, Hudspith BN, Rayment N, Brostoff J, Parkhill J, Dougan G, Petrovska L (2011) High-throughput clone library analysis of the mucosa-associated microbiota reveals dysbiosis and differences between inflamed and non-inflamed regions of the intestine in inflammatory bowel disease. BMC Microbiol 11:7

Wang Q, Fidalgo S, Chang BJ, Mee BJ, Riley TV (2002) The detection and recovery of *Erysipelothrix* spp. in meat and abattoir samples in Western Australia. J Appl Microbiol 92:844–850

Wang Q, Chang BJ, Riley TV (2010) *Erysipelothrix rhusiopathiae*. Vet Microbiol 140:405–417

Wei H, Dong L, Wang T, Zhang M, Hua W et al (2010) Structural shifts of gut microbiota as surrogate endpoints for monitoring host health changes induced by carcinogen exposure. FEMS Microbiol Ecol 73:577–586

White TG, Mirikitani FK (1976) Some biological and physical chemical properties of *Erysipelothrix rhusiopathiae*. Cornell Vet 66:152–163

Wilkinson BJ, Jones D (1977) A numerical taxonomic survey of *Listeria* and related bacteria. J Gen Microbiol 98:399–421

Willems A, Moore WEC, Weiss N, Collins MD (1997) Phenotypic and phylogenetic characterization of some *Eubacterium*-like isolates containing a novel type B wall murein from human feces: description of *Holdemania filiformis* gen. nov., sp. nov. Int J Syst Bacteriol 47:1201–1204

Williams SM, Hafner S, Sundram Y (2007) Liver granulomas due to *Eubacterium tortuosum* in a seven-week-old Bobwhite quail. Avian Dis 51:797–799

Wood RL (1965) A selective liquid medium utilizing antibiotics for isolation of *Erysipelothrix insidiosa*. Am J Vet Res 26:1303–1308

Wood RL (1974a) Isolation of pathogenic *Erysipelothrix rhusiopathiae* from feces of apparently healthy swine. Am J Vet Res 35:41–43

Wood RL (1974b) *Erysipelothrix* infection. In: Hubbert WT, McCullough WF, Schnurrenberger PR (eds) Diseases transmitted from animals to man, 6th edn. Thomas, Springfield, pp 271–281

Wood RL (1984) Swine erysipelas—a review of prevalence and research. J Am Vet Med Assoc 184:944–949

Wood RL, Packer R (1972) Isolation of *Erysipelothrix rhusiopathiae* from soil and manure of swine-raising premises. Am J Vet Res 33:1611–1620

Woodbine M (1950) *Erysipelothrix rhusiopathiae*. Bacteriology and chemotherapy. Bacteriol Rev 14:161–178

Yamazaki Y (2006) A multiplex polymerase chain reaction for discriminating *Erysipelothrix rhusiopathiae* from *Erysipelothrix tonsillarum*. J Vet Diagn Invest 18:384–387

Yang QB, Fan LN, Shi Q (2010) Polymerase chain reaction-denaturing gradient gel electrophoresis, cloning, and sequence analysis of bacteria associated with acute periapical abscesses in children. J Endod 36:218–223

Yarza P, Ludwig W, Euzeby J, Amann R, Schleifer KH, Glöckner FO, Rossello-Mora R (2010) Update of the All-Species Living Tree Project based on 16S and 23S rRNA sequence analyses. Syst Appl Microbiol 33:291–299

Zhang C, Hou BX, Zhao HY, Sun Z (2012a) Microbial diversity in failed endodontic root-filled teeth. Chin Med J (Engl) 125(6):1163–1168

Zhang X, Zhao Y, Zhang M, Pang X, Xu J, Kang C, Li M, Zhang C, Zhang Z, Zhang Y, Li X, Ning G, Zhao L (2012b) Structural changes of gut microbiota during berberine-mediated prevention of obesity and insulin resistance in high-fat diet-fed rats. PLoS One 7:e42529

Zheng G, Summanen PH, Talan D, Bennion R, Rowlinson MC, Finegold SM (2010) Phenotypic and molecular characterization of *Solobacterium moorei* isolates from patients with wound infection. J Clin Microbiol 48:873–876

# 8 The Family *Eubacteriaceae*

*Erko Stackebrandt*
Leibniz Institute DSMZ-German Collection of Microorganisms and Cell Cultures GmbH, Braunschweig, Germany



**Abstract**
The family *Eubacteriaceae*, defined by the phylogenetic position of its members, comprises the genera *Eubacterium*, *Acetobacterium*, *Alkalibacter*, *Alkalibaculum*, *Anaerofustis*, *Garciella*, and *Pseudoramibacter*. Only those few species are members of the authentic genus *Eubacterium* which are phylogenetically related to the type species *Eubacterium limosum*. Non-authentic *Eubacterium* species are phylogenetically placed in other families with more close relationships to species of other genera. As nearly all authentic and non-authentic members of *Eubacteriaceae* have been extensively covered in the chapter on the genus *Eubacterium* in the latest edition of *Bergey's Manual*, 2nd ed., Firmicutes (Ludwig et al. 2009), this communication will be restricted to taxa described since then.

## Introduction

For many decades the genus *Eubacterium* has been a dumping ground for Gram-stain-positive, nonspore-forming anaerobic uniform or pleomorphic rods, many of which belong to different lineages according to their phylogenetic position based upon 16S rRNA sequence analysis (Yarza et al. 2010). Today, only those few species are considered authentic *Eubacterium* species which cluster around the type species *Eubacterium limosum* (*E. aggregans*, *E. callandri*, *E. barkeri*). Other species, still carrying the generic name *Eubacterium*, are related to other genera, placed in different families of Firmicutes, e.g. *E. biforme* (*Erysipelothrichaceae*), *E. multiforme* (*Costridiaceae* group 1), *E. cellulosolvens* and *E. sabbureum* (*Lachnospiriaceae*), *E. yurii* (*Peptostreptococcaceae*), *E. angustum* (*Clostridiaceae* group 4) or *E. coprostanoligenes* (*Ruminococcaceae*). Their taxonomic status still awaits reclassification which appears difficult on the basis of missing phenotypic properties distinguishing them from their phylogenetic neighbors (Wade 2009). The genus *Garciella*, with *G. nitratireducens* as the type species (Miranda-Tello et al. 2003), clusters outside the *Eubacteriaceae* and together with some genera of Clostridia group 3, (e.g. *Clostridiisalibacter paucivorans* *Sporosalibacterium faouarense*, *Thermohalobacter berrensis*, and *Caloranaerobacter azorensis*) (➲ *Fig. 8.1*). *G. nitratireducens* was tentatively included in the family by Ludwig et al. (2009) because of its phylogenetic relatedness to other family members (Lawson 2009; Willems and Collins 2009).This decision was adopted by Euzeby (http://www.bacterio.net/e/eubacteriaceae.html) but its membership to *Eubacteriaceae* remains to be fully explored.

Since the last comprehensive coverage of the family *Eubacteriaceae* in the 2nd. edition of *Bergey's Manual of Systematic Bacteriology*, only a single genus, *Alkalibaculum* (Allen et al. 2010), with a single species, *A bacchi*, has been affiliated to the family. Therefore, only this taxon will be included in this taxonomic update of the family and the properties are compared to those type species described previously.

## Taxonomy

Three members of the family, *Eubacterium limosum*, *Acetobacterium limosum* and *Alkalibaculum bachii*, have been reported to contain peptidoglycan of the B type (Schleifer and Kandler 1972), with serine at position 1 of the peptide subunit with either D-lysine, D-ornithine or both as the cross-linking amino acid ({L-Ser} [L-Orn] D-Glu-D-Lys(D-Orn). *A. bachii*, in addition, contains aspartic acid at an undetermined position. This peptidoglycan composition has neither been reported in other members of neighboring families nor in other members of *Eubacteriaceae*.

*Alkalibaculum bachii* has a pH optimum of growth at 8.0–8.5 and is motile by peritrichous flagella. It uses $H_2$:$CO_2$, CO:$CO_2$, and various sugars and alcohols as growth substrates. Endproduct of glucose fermentation is acetate, while acetate, $CO_2$ and ethanol is produced from CO:$CO_2$. Vitamins are required for growth. Major fatty acids (<10 %) are $C_{14:0}$ and $C_{16:0}$ DMA. *Alkalibacter saccharofermentans* is its closest phylogenetic neighbor. The type strain of *A. bachii*, strain CP11$^T$, was isolated from livestock-impacted soil.

Complete or draft genome sequences are available for *Eubacterium limosum* KIST612 (Gc01410), *Acetobacterium woodii* WB1, DSM 1030$^T$ (Gc02124), *Anaerofustis stercorihominis* DSM 17244$^T$ (Gi0207) and *Pseudoramibacter alactolyticus* ATCC 23263$^T$ (Gi04127).

E. Rosenberg et al. (eds.), *The Prokaryotes – Firmicutes and Tenericutes*, DOI 10.1007/978-3-642-30120-9_359,

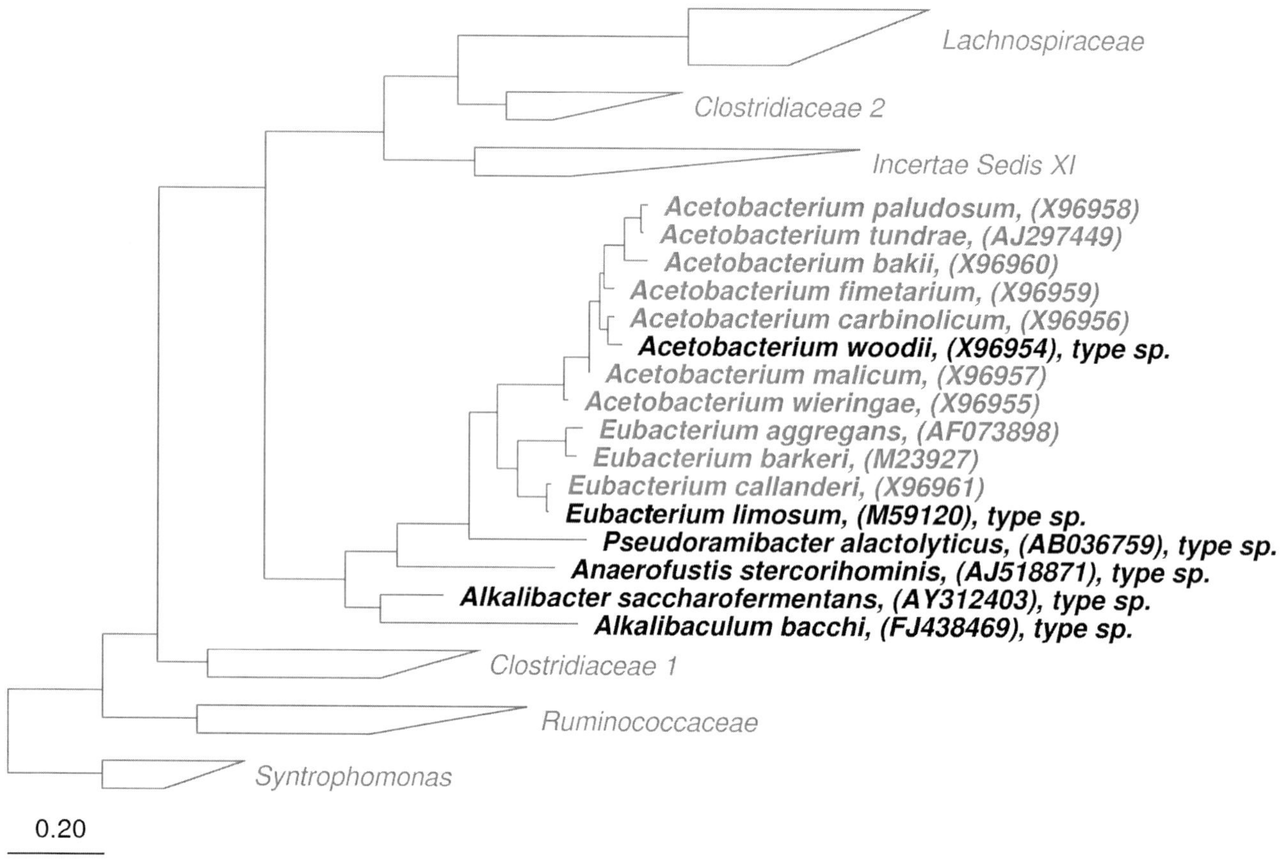


**Fig. 8.1**
**Maximum likelihood genealogy reconstruction based on the RAxML algorithm (Stamatakis 2006) of the sequences of all authentic members of the family *Eubacteriaceae* present in the LTP_106 (Yarza et al. 2010). Representative sequences from close relative genera were used to stabilize the tree topology. In addition, a 40 % maximum frequency filter was applied to remove hypervariable positions from the alignment. Scale bar indicates estimated sequence divergence**

## References


Allen TD, Caldwell ME, Lawson PA, Huhnke RL, Tanner RS (2010) *Alkalibaculum bacchi* gen. nov., sp. nov., a CO-oxidizing, ethanol-producing acetogen isolated from livestock-impacted soil. Int J Syst Evol Microbiol 60:2483–2489

Lawson P (2009) Genus IV. *Anaerofustis*. In: De Vos P, Garrity GM, Jones D, Krieg NR, Ludwig W, Rainey FA, Schleifer K-H, Whitman WB (eds) Bergey's manual of systematic bacteriology, the Firmicutes, vol 3, 2nd edn. Springer, New York, pp 897–900

Ludwig W, Schleifer KH, Whitman WB (2009) Family II. *Eubacteriaceae*. In: De Vos P, Garrity GM, Jones D, Krieg NR, Ludwig W, Rainey FA, Schleifer K-H, Whitman WB (eds) Bergey's manual of systematic bacteriology, the Firmicutes, vol 3, 2nd edn. Springer, New York, p 865

Miranda-Tello E, Fardeau ML, Sepulveda J, Fernandez L, Cayol JL, Thomas P, Ollivier B (2003) *Garciella nitratireducens* gen. nov., sp. nov., an anaerobic, thermophilic, nitrate- and thiosulfate-reducing bacterium isolated from an oilfield separator in the Gulf of Mexico. Int J Syst Evol Microbiol 53:1509–1514

Schleifer KH, Kandler O (1972) Peptidoglycan types of bacterial cell walls and their taxonomic implications. Bacteriol Rev 36:407–477

Stamatakis A (2006) RAxML-VI-HPC: maximum likelihood-based phylogenetic analyses with thousands of taxa and mixed models. Bioinformatics 22:2688–2690

Wade WG (2009) Genus I. *Eubacterium* Prévot 1938. In: De Vos P, Garrity GM, Jones D, Krieg NR, Ludwig W, Rainey FA, Schleifer K-H, Whitman WB (eds) Bergey's manual of systematic bacteriology, the Firmicutes, vol 3, 2nd edn. Springer, New York, pp 865–891

Willems A, Collins MD (2009) Genus VI. *Pseudoramibacter*. In: De Vos P, Garrity GM, Jones D, Krieg NR, Ludwig W, Rainey FA, Schleifer K-H, Whitman WB (eds) Bergey's manual of systematic bacteriology, the Firmicutes, vol 3, 2nd edn. Springer, New York, p 902

Yarza P, Ludwig W, Euzéby J, Amann R, Schleifer K-H, Glöckner FO, Rosselló-Móra R (2010) Update of the All-Species Living-Tree Project based on 16S and 23S rRNA sequence analyses. Syst Appl Microbiol 33:291–299

# 9 The Family *Fusobacteriaceae*

*Ingar Olsen*
Department of Oral Biology, Faculty of Dentistry, University of Oslo, Oslo, Norway



**Abstract**
The family *Fusobacteriaceae*, which falls in the *Fusobacteria* class and the order *Fusobacteriales*, consists of microaerophilic to obligate anaerobic Gram-negative rods. All of them are nonmotile and fermentative. The members ferment carbohydrates or amino acids and peptides producing various organic acids such as acetic, propionic, butyric, formic, or succinic acid depending on the bacterium and the substrate. Habitats are oral and intestinal mucosae of animals including mammals, as well as anaerobic sediments. *Fusobacteriaceae* includes the genera *Cetobacterium, Fusobacterium, Ilyobacter, Propionigenium*, and *Psychrilyobacter*. The type genus is *Fusobacterium* Knorr 1922[AL]. "*F. naviforme*" and "*I. delafieldii*" fall outside the phylogenetic tree established for *Fusobacteriaceae*.

## Taxonomy: Current and Historical

### Short Description of the Family

Fu.so.bact.te.ri.a.ce'a.e. N.L. neut.n. *Fusobacterium* type genus of the family; suff. *-aceae* ending to denote a family; N.L. fem. pl.n. *Fusobacteriaceae* the *Fusobacterium* family. The description of the family is derived from that of Staley and Whitman (2011). The family consists of microaerotolerant to obligate anaerobic, Gram-negative rods. All named species are nonmotile and fermentative. Carbohydrates, amino acids, and peptides are fermented with the production of various organic acids such as acetic, propionic, isobutyric, formic, or succinic acid according to the substrate and species. The organisms are isolated from anoxic environments including sediments, as well as the oral and intestinal habitats of animals comprising mammals.

Following the road map of several phyla (Ludwig et al. 2011), the family *Fusobacteriaceae* consists of five genera: *Fusobacterium, Cetobacterium, Ilyobacter, Propionigenium*, and *Psychrilyobacter*. This definition will be used in the present overview. The genus *Fusobacterium* is paraphyletic including the lineage *Cetobacterium*. The genera *Ilyobacter* and *Propionigenium* are also intermixed. In addition *Psychrilyobacter* forms a deep branch in the family tree. The phylogenetic relationship between *Fusobacteriaceae* and related families is indicated in ➲ *Figs. 9.1* and ➲ *9.2*.

According to Staley and Whitman (2011), the family *Fusobacteriaceae* includes the genera *Cetobacterium, Fusobacterium, Ilyobacter*, and *Propionigenium*. *Psychrilyobacter* was not listed. Previously Garrity et al. (2005) had listed the genera *Fusobacterium, Ilyobacter, Leptotrichia, Propionigenium, Sebaldella, Streptobacillus*, and *Sneathia* under *Fusobacteriaceae*. *Cetobacterium* was listed under Family II *Incertae sedis*[VP] as Genus I. These authors emphasized that the taxonomic scheme was a work-in-progress one, based on data available in October 2003, and that some rearrangement and amendment should be expected to occur as new data became available. The closely related genera *Leptotrichia, Sebaldella, Streptobacillus*, and *Sneathia* will not be dealt with here.

In the genus *Fusobacterium, F. nucleatum* is the type species. Other listed species of the genus are *F. canifelinum, F. equinum, F. gonidiaformans, F. mortiferum, F. naviforme, F. necrogenes, F. necrophorum, F. perfoetens, F. periodonticum, F. russii, F. simiae, F. ulcerans*, and *F. varium* (Gharbia et al. 2011). *F. nucleatum* is divided in five subspecies: *-nucleatum, -animalis, -fusiforme, -polymorphum*, and *-vincentii* while *F. necrophorum* is divided into two subspecies: *-necrophorum* and *-funduliforme*. From the phylogenetic tree for *Fusobacteriaceae* established in ➲ *Figs. 9.1* and ➲ *9.2*, "*F. naviforme*" does not seem to be a member of the genus *Fusobacterium*. "*F. naviforme*" will therefore not be considered as a fusobacterium in the present review. In ➲ *Fig. 9.2*, "*F. naviforme*" is closer to *Firmicutes* than to *Fusobacteria*.

In the genus *Cetobacterium*, there are two species, *C. ceti* which is the type species and *C. somerae* which reflects a deeper branch. In the genus *Ilyobacter, I. polytropus* is the type species. Other species are *I. insuetus* and *I. tartaricus*. "*I. delafieldii*" does

E. Rosenberg et al. (eds.), *The Prokaryotes – Firmicutes and Tenericutes*, DOI 10.1007/978-3-642-30120-9_213,

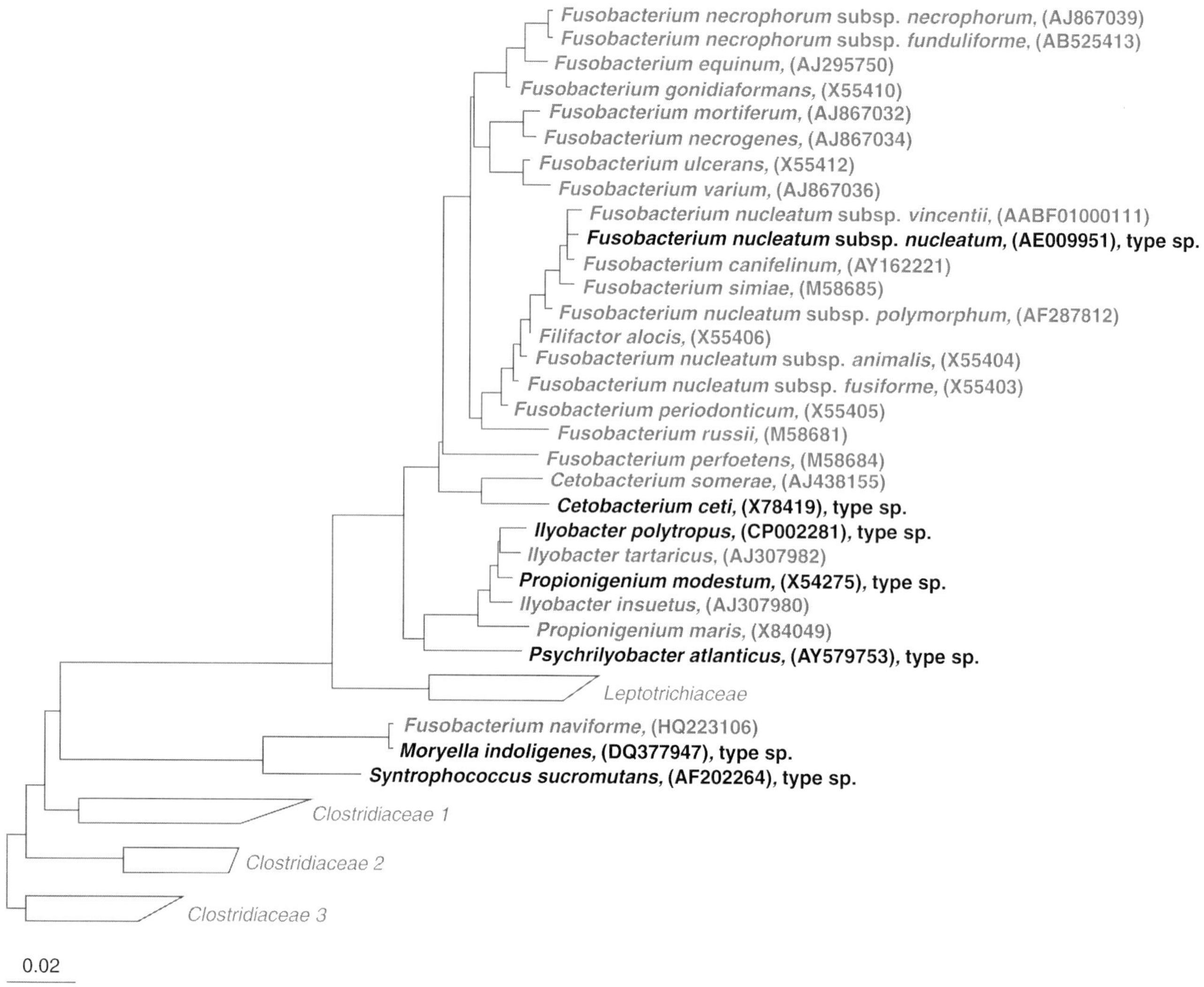


**Fig. 9.1**

**Phylogenetic reconstruction of the family *Fusobacteriaceae* and related families based on 16S rRNA and created using the neighbor-joining algorithm with the Jukes-Cantor correction. The sequence datasets and alignments were used according to the All-Species Living Tree Project (LTP) database (Yarza et al. 2010; http://www.arb-silva.de/projects/living-tree). The tree topology was stabilized with the use of a representative set of nearly 750 high-quality-type strain sequences proportionally distributed among the different bacterial and archaeal phyla. In addition, a 40 % maximum frequency filter was applied in order to remove hypervariable positions and potentially misplaced bases from the alignment. Scale bar indicates estimated sequence divergence**

not fall within the family *Fusobacteriaceae* (Fig. 9.2). It should not be listed in the genus *Ilyobacter* because sequence analysis of its 16S rRNA gene has indicated a place within the genus *Clostridium* (Brune et al. 2002), although spores cannot be detected (Janssen and Harfoot 1990). The cell wall architecture of strain 10crl of "*I. delafieldii*" also resembles that of a Gram-positive bacterium with a complex structure, although it stains Gram-negative. "*I. delafieldii*" will not be included in *Fusobacteriaceae* in the present overview.

*P. modestum* is the type species in the genus *Propionigenium*. Another species is *P. maris*.

*Ps. atlanticus* is the type and so far only species of the new genus *Psychrilyobacter*.

Several reviews have been made on the taxonomy and biology of the genus *Fusobacterium* (e.g., Bolstad et al. 1996; Citron 2002; Jousimies-Somer and Summanen 2002; Hofstad and Olsen 2005; Shah et al. 2009; Gharbia et al. 2011; Liu and Dong 2011). The current subspeciation of *F. nucleatum* (Dzink et al. 1990; Gharbia and Shah 1989, 1990a, 1992) has been challenged (Jousimies-Somer 1997). Also Rogers (1998), using allozyme electrophoresis, found the subspeciation of *F. nucleatum* of doubtful validity. Neither was there any difference in the pathogenicity between the subspecies based on physiology and metabolic properties. However, George et al. (1997), using arbitrarily primed PCR with two primers, found unique profiles for the five subtypes. Also Citron (2002) examining the 16S-23S internal transcribed spacer (ITS) regions, found that the five subspecies of *F. nucleatum* could be distinguished from each other and also from closely related species.

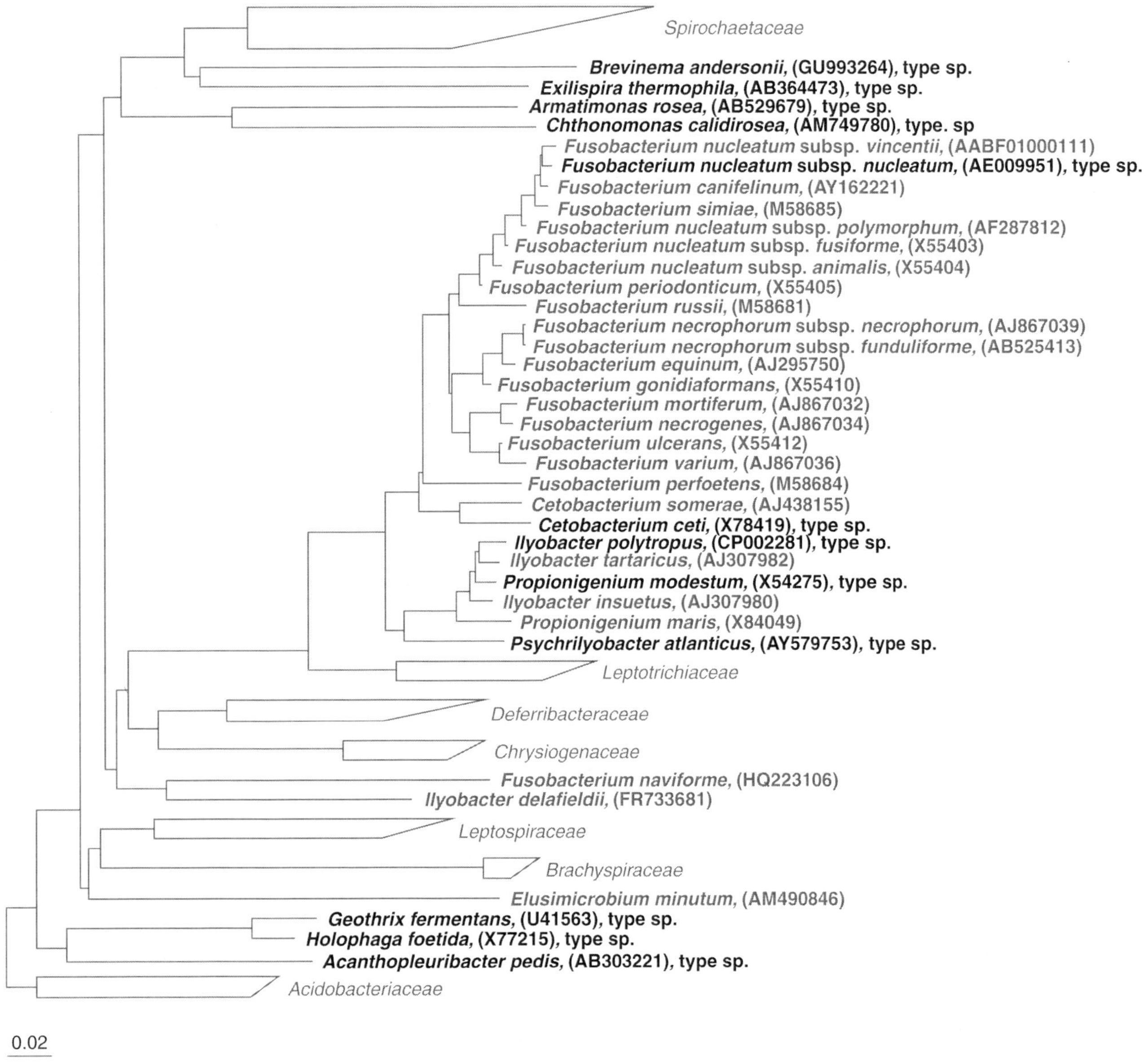


**Fig. 9.2**

**Phylogenetic reconstruction of the genus *Fusobacterium* based on 16S rRNA and created using the neighbor-joining algorithm with the Jukes-Cantor correction. Data suggest that "*F. naviforme*" is closer to *Firmicutes* than to *Fusobacteria.* The sequence datasets and alignments were used according to the All-Species Living Tree Project (LTP) database (Yarza et al. 2010; http://www.arb-silva.de/projects/living-tree). The tree topology was stabilized with the use of a representative set of nearly 750 high-quality-type strain sequences proportionally distributed among the different bacterial and archaeal phyla. In addition, a 40 % maximum frequency filter was applied in order to remove hypervariable positions and potentially misplaced bases from the alignment. Scale bar indicates estimated sequence divergence**

"*F. alocis*" and "*F. sulci*" have been reclassified as *Filifactor alocis* and *Eubacterium sulci*, respectively, because they clustered among Gram-positive genera (Jalava and Eerola 1999). Duncan et al. (2002) transferred "*F. prausnitzii*" to *Faecalibacterium prausnitzii*. "*F. polysaccharolyticum*" was transferred to *C. polysaccharolyticum* (Gylswyk et al. 1983). *F. periodonticum* is closely related to *F. nucleatum* (Jousimies-Somer 1997; Jousimies-Somer and Summanen 2002). Analysis of a 40-kDa outer membrane protein, Fom A, indicated that *F. periodonticum* and *F. nucleatum* are related but on a lower level than the subspecies level (Bolstad et al. 1995). *F. necrophorum* subsp. *necrophorum* includes biovar A of *F. necrophorum*, while *F. necrophorum* subsp. *funduliforme* comprises biovar B of *F. necrophorum*. *F. varium* includes "*F. pseudonecrophorum*" (biovar C of *F. necrophorum*) (Bailey and Love 1993; Citron 2002). According to Citron (2002), "*F. symbiosum*" contains spores, "*F. praecutum*" is motile, and "*F. plauti*" is Gram positive. All have been reclassified into other genera.

The species description of *P. maris* was amended by Watson et al. (2000) who emphasized its ability of reductive dehalogenation of bromophenols found in the linings of some marine infaunal burrows.

## Phylogenetic Structure of the *Fusobacteriaceae* and Its Genera

The phylogenetic structure of the members of the family *Fusobacteriaceae* and related bacteria is given in ➲ *Figs. 9.1* and ➲ *9.2*. All subspecies of *F. nucleatum* affiliate in these trees. "*F. naviforme*" falls out of the family somewhere close to the *Clostridiaceae*, and depending on the calculation method used, it branches from different points. There is also a considerable intermix between *Ilyobacter* and *Propionigenium*. Comparative rRNA gene sequence analyses indicated a monophyletic and separate status of the group constituted by *I. polytropus*, *I. tartaricus*, *P. modestum*, *P. maris*, and strain VenChi2$^{T}$ of *I. insuetus* (Brune et al. 2002). These authors concluded that the members of the different genera are phylogenetically intermixed and taxonomic revision will be necessary. It might be reasonable to unite the species of the *Propionigenium-Ilyobacter* group in a common genus (Ludwig et al. 1998). However, the large phenotypic differences between the existing species of the *Ilyobacter-Propionigenium* group and their unexplored metabolic activity appear counter-indicative and would call for additional genera to be established (Brune et al. 2002). It is also clear from ➲ *Figs. 9.1* and ➲ *9.2* that *Ps. atlanticus* makes a deep branch in the family tree. Phylogenetic analyses by Zhao et al. (2009) demonstrated the affiliation of this species to the family *Fusobacteriaceae* with 87–93 % gene sequence similarity, and they found that the genera *Ilyobacter* and *Propionigenium* were closely related (92.5–93.4 %).

Lawson et al. (1991) studied the phylogenetic interrelationships of 14 members of the genus *Fusobacterium* using 16S rRNA sequences. A considerable intrageneric heterogeneity was demonstrated. *F. nucleatum* together with *F. nucleatum* subsp. *nucleatum*, *-polymorphum*, *-fusiforme*, and *-animalis*, *F. periodonticum*, *F. simiae*, and "*F. alocis*" showed high levels of sequence homology. *F. mortiferum*, *F. varium*, and *F. ulcerans* also constituted a phylogenetically coherent group, as were *F. gonidiaformans* and *F. necrophorum*. *F. russii* and *F. necrogenes* showed no specific relationship with any of the other fusobacteria.

Interestingly, Mira et al. (2004) found the phylogenetic position and evolutionary relationships of fusobacteria uncertain. Particularly intriguing is their relatedness to low G + C Gram-positive bacteria (Firmicutes) by ribosomal molecular phylogenies, although they have a typical Gram-negative outer membrane. The data of Mira et al. (2004) indicated that *Fusobacterium* has a core genome of a very different nature to other bacterial lineages and branches out of Firmicutes. About 35–56 % of *Fusobacterium* genes were estimated to have a xenologous origin from bacteroidetes, proteobacteria, spirochetes, and Firmicutes. The close physical contact in dental plaque may facilitate horizontal gene transfer supporting that this niche is a specific bacterial gene pool.

## Molecular Analyses

The relationship between oral fusobacteria was studied by DNA-DNA hybridization (Potts et al. 1983). The panel included 16 strains of *F. nucleatum* and five strains from other *Fusobacterium* species. *F. nucleatum* represented a heterogeneous group of organisms related to *F. periodonticum* and *F. simiae* but was not related to any other of the *Fusobacterium* species tested.

Conrads et al. (2002) used 16S-23S rDNA ITS sequences to analyze the phylogenetic relationships between species of the genus *Fusobacterium*. After ITS primer amplification, most of the *Fusobacterium* strains together with *L. buccalis* gave one major and two to three weaker distinct bands with lengths 800–830 bp and 1,000–1,100 bp. Six other patterns were also detected within *Fusobacterium* demonstrating the heterogeneity of the genus. The ITS-DNA sequences and ITS relative lengths allowed differentiation of species and subspecies in most cases. There was a striking difference between "*F. prausnitzii*" and the genus *Fusobacterium*. *F. nucleatum* subspecies formed a cluster with *F. simiae*, *F. periodonticum*, and "*F. naviforme*." Other clusters constituted *F. necrophorum* subspecies with *F. gonidiaformans*, and *F. varium* with *F. mortiferum* and *F. ulcerans*. Separate branches were formed by *F. russii* and *F. perfoetens*. Further, high similarity was found between *F. necrophorum* subsp. *necrophorum* and *-funduliforme* on one hand and between *F. varium* and *F. mortiferum* on the other.

The partial *rpoB* gene (approximately 2,419 bp), the *zinc protease* gene (878 bp), and the 16S rRNA genes (approximately 1,500 bp) of the type strains of the five subspecies of *F. nucleatum* were examined together with 28 clinical isolates of *F. nucleatum*, and 10 strains of *F. periodonticum* used as control (Kim et al. 2010). The *rpoB* and *zinc protease* gene sequences separated well the subspecies of *F. nucleatum* and gave better resolution than did the 16S rRNA gene (➲ *Fig. 9.3*). It was suggested that *F. nucleatum* subsp. *vincentii* and *F. nucleatum* subsp. *fusiforme* may form a single subspecies, while five clinical isolates could possibly form a new subspecies of *F. nucleatum*.

## Genomics

Kapatral et al. (2002), analyzing *F. nucleatum* subsp. *nucleatum* strain ATCC 25586$^{T}$, found several features of its core metabolism to be similar to those of species of *Clostridium*, *Enterococcus*, and *Lactococcus*. The genome of this strain contained 2.17 Mbp encoding 2,067 open reading frames (ORFs) and was organized in a single circular chromosome with 27 mol% G+C content. Nine very high molecular weight outer membrane proteins could be predicted from the genome sequence. None of these had been reported previously in the literature.

The genome of *F. nucleatum* subsp. *vincentii* (ATCC 49256$^{T}$) was compared with that of *F. nucleatum* ATCC 25586$^{T}$

(Kapatral et al. 2003). The authors found 441 ORFs in *F. nucleatum* subsp. *vincentii* that had no orthologs in *F. nucleatum*. Among these, 118 ORFs were unique to *F. nucleatum* subsp. *vincentii* but had no known function, while 323 ORFs had functional orthologs in other bacteria. In contrast to *F. nucleatum*, *F. nucleatum* subsp. *vincentii* is unlikely to incorporate galactopyranose, galacturonate, and sialic acid into its O-antigen. Furthermore, genes for eukaryotic type serine/threonine kinase and phosphatase transpeptidase E-transglycosylase Pbp 1A were found in *F. nucleatum* subsp. *vincentii* but not in *F. nucleatum*.

*F. nucleatum* subsp. *polymorphum* ATCC 10953$^T$ has a chromosome of 2,429,698 kb and a plasmid (pFN3) of 11.9 kbp (Karpathy et al. 2007). Compared to the fusobacterial

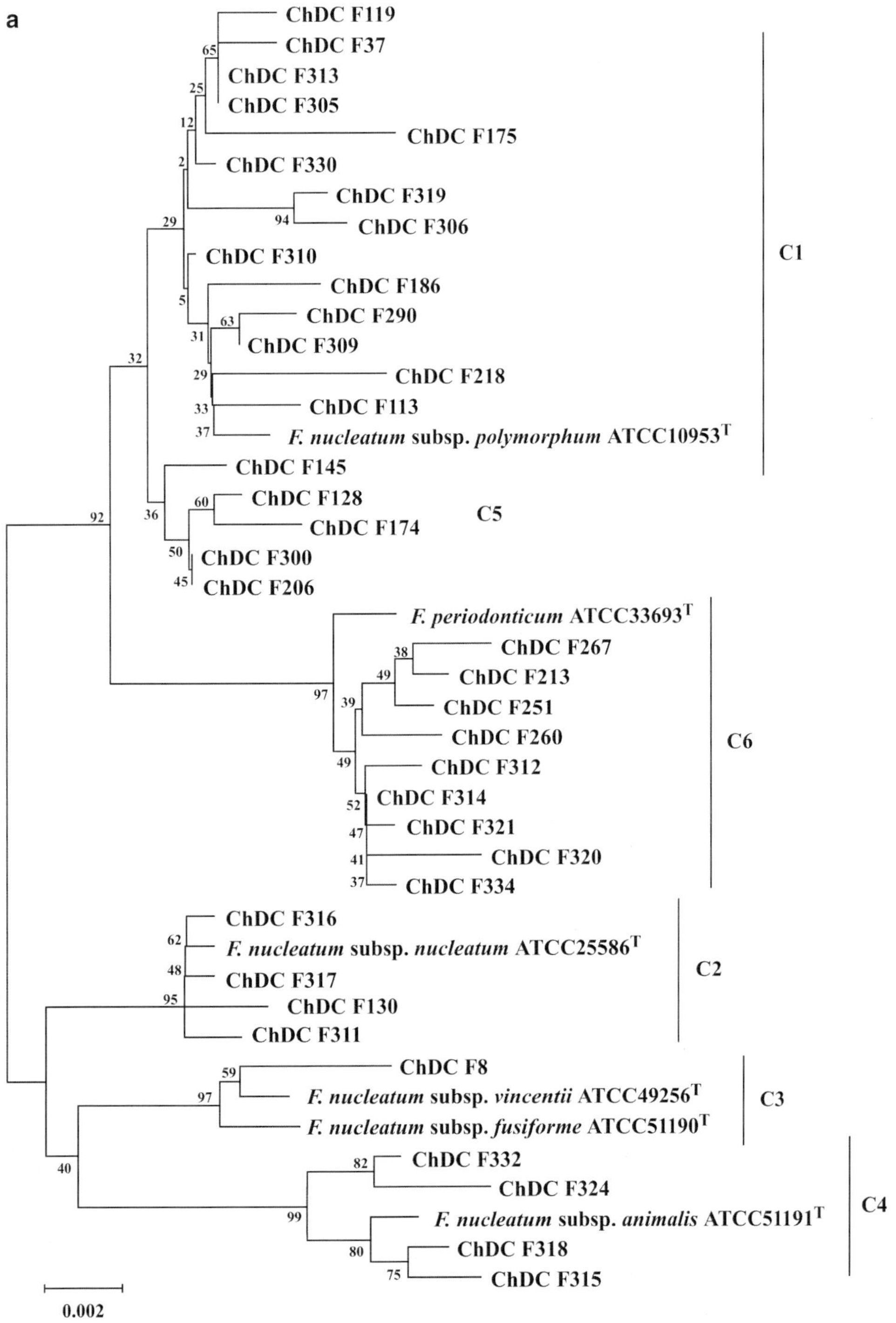


**Fig. 9.3** (continued)

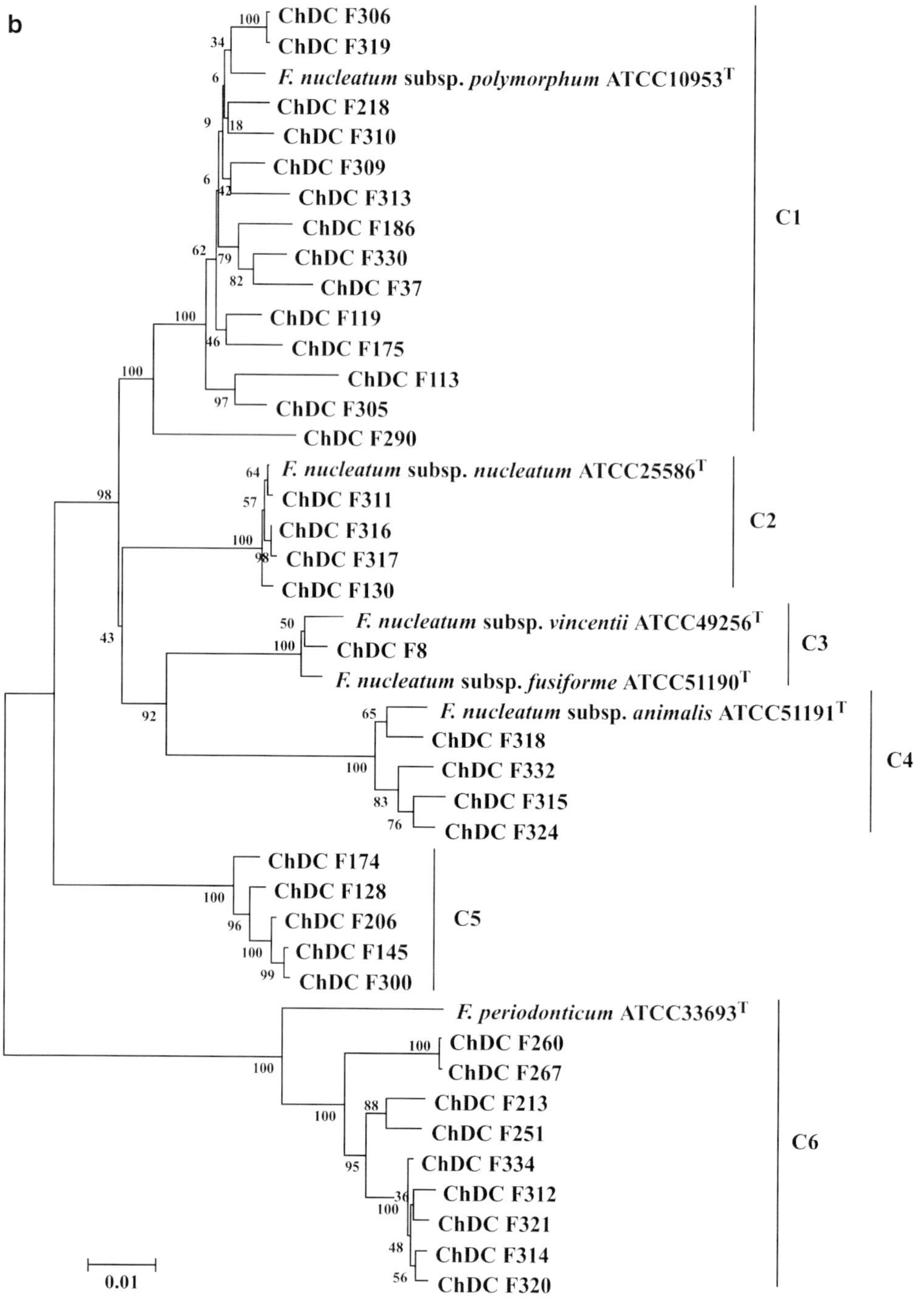


**Fig. 9.3 (continued)**

genomes sequenced from *F. nucleatum* subsp. *nucleatum* and *F. nucleatum* subsp. *vincentii*, 627 ORFs were identified that were unique to *F. nucleatum* subsp. *polymorphum*. A large percentage of these ORFs were located within 1 of 28 regions or islands containing five or more genes. Proteins that showed similarity (17 %) were most similar to proteins from clostridia. Others were most similar to those from *Bacillus* and *Streptococcus*. The genome also contained five composite ribozyme/transposons similar to the Cd*ISt* IStrons found in *Clostridium difficile* but not in other fusobacterial genomes.

The complete genome sequence of *I. polytropus* strain CuHbu1$^T$ was determined by Sikorski et al. (2010). The genome strain CuHbu1$^T$ was 3,132, 314 bp long with 2,934 protein-coding and 108 RNA genes consisting of two chromosomes (2 and 1 Mbp long) (*Fig. 9.4*) and one plasmid.

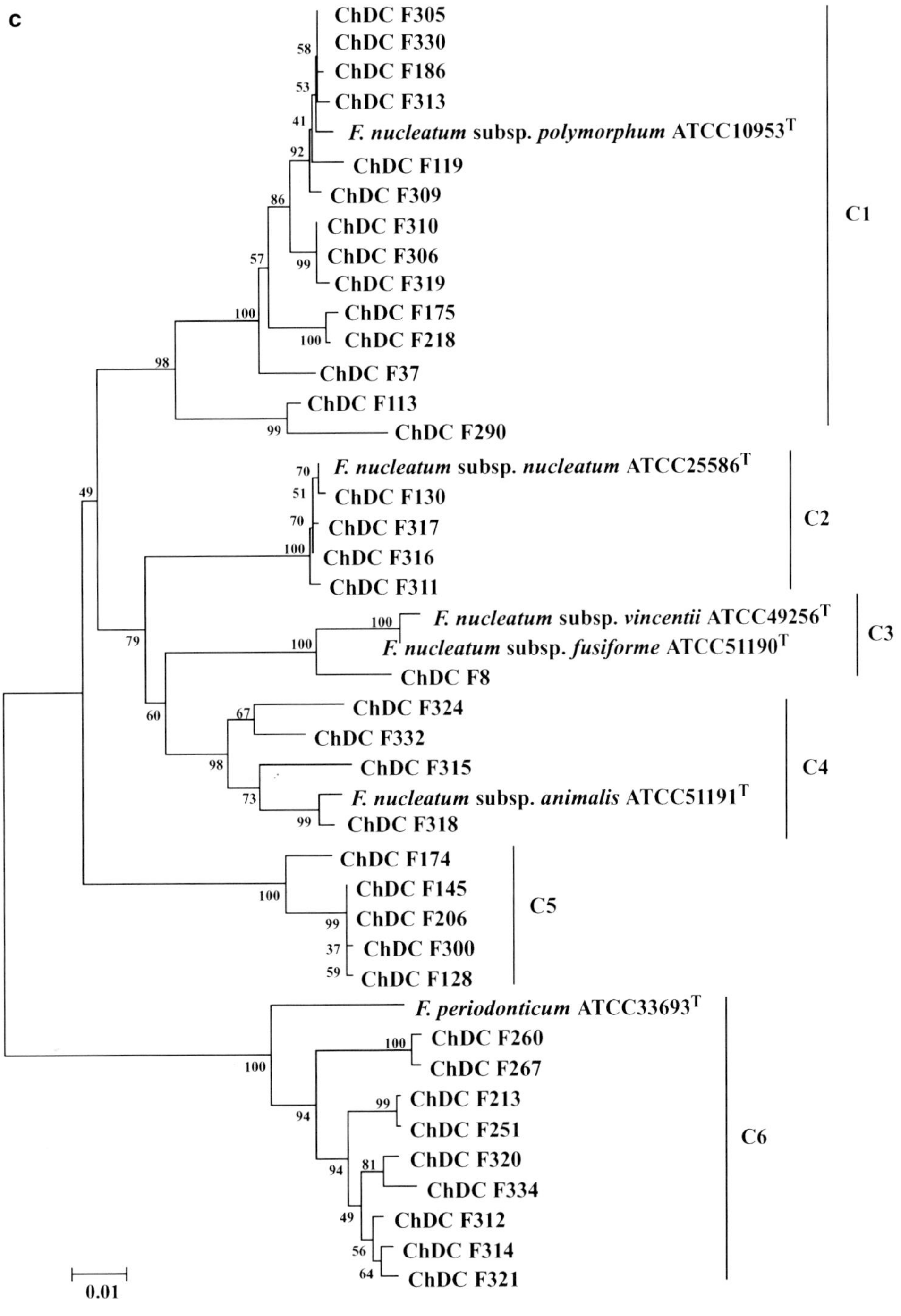


**Fig. 9.3**
**Phylogenetic tree based on the partial nucleotide sequences (a) of 16S rRNA genes (about 1.5 kb), (b) *rpoB* (about 2,419 bp out of 3,335 bp), and (c) the zinc protease gene (878 bp out of 1,227 bp) of type strains and clinical isolates of *F. nucleatum* and *F. periodonticum*. Resulting tree topology evaluated by bootstrap analyses of the neighbor-joining tree based on 1,000 resamplings (Kim et al. 2010) (Courtesy of J Clin Microbiol)**

The *atp* operon of *I. tartaricus* strain DSM 2382 was completely sequenced by conventional technique and inverse polymerase reaction technique (Meier et al. 2003). Nine ORFs were detected that were attributed to eight structural genes of the $F_1F_0$ ATP synthase and *atpI* gene. The genes were arranged in one operon with the sequence *atpIBEFHAGDC* that comprised 6,992 base pairs with a G + C content of 38.1 %. The $F_1F_0$ ATP synthase of *I. tartaricus* had a calculated mass of 510 kDa and included 4,810 amino acids. Significant identities with the *atp* genes of other $Na^{+}$-translocating $F_1F_0$ ATP synthases were recognized.

## Phages

A new bacteriophage Fnpϕ02 was isolated from *F. nucleatum* (Machuca et al. 2010). The virion consisted of an icosahedral head and a segmented tail (➲ *Fig. 9.5*). The size of the phage genome was approximately 59 kbp of double-stranded DNA, and it probably belonged to the *Siphoviridae* family. A small fragment of the phage DNA was cloned and sequenced and showed 93 % nucleotide identity with phage PA6 of *Propionibacterium acnes* and amino acid identity with fragments of two proteins, Gp3 and Gp6, of this phage.

A prophage and integrated conjugal plasmid of a 10 kb region, homologous to the *S. typhimurium* propanediol utilization locus, was detected in *F. nucleatum* subsp. *polymorphum* (Karpathy et al. 2007).

Cryptic phages were isolated in *F. nucleatum* subsp. *vincentii* where six phage contigs encoding 110 ORFs were detected (Kapatral et al. 2003). The average G + C content of the phage DNA was found to be approximately 28 %, and the codon usage was equal to the chromosome DNA (Kapatral et al. 2003). One of the phage regions harbored 66 ORFs and two had 14 ORFs. Each of them showed amino acid sequence similarity to the phage-like element PBSX protein XkdK, XkdM, and XdkT of *Desulfitobacterium hafniense* DCB-2 phage. The fourth region had six ORFs with amino acid sequences equal to the Gram-negative bacteriophage-P2. Regions 5 and 6 had three and seven phage ORFs, respectively. Their sequences were similar to the Gram-positive bacteriophage TP901. In *F. nucleatum*, no phage sequences were found.

Bacteriophages have also been detected in *F. varium* (Andrews et al. 1997) and *F. necrophorum* (Tamada et al. 1985).

Three plasmids from *F. nucleatum* have been sequenced: pFN1, pPA52, and pKH9 (McKay et al. 1995; Haake et al. 2000; Bachrach et al. 2004).

# Phenotypic Analyses

The main phenotypic features of members of *Fusobacteriaceae* are listed in ➲ *Table 9.1*, while phenotypic characters of different *Fusobacterium* species are given in ➲ *Table 9.2*.

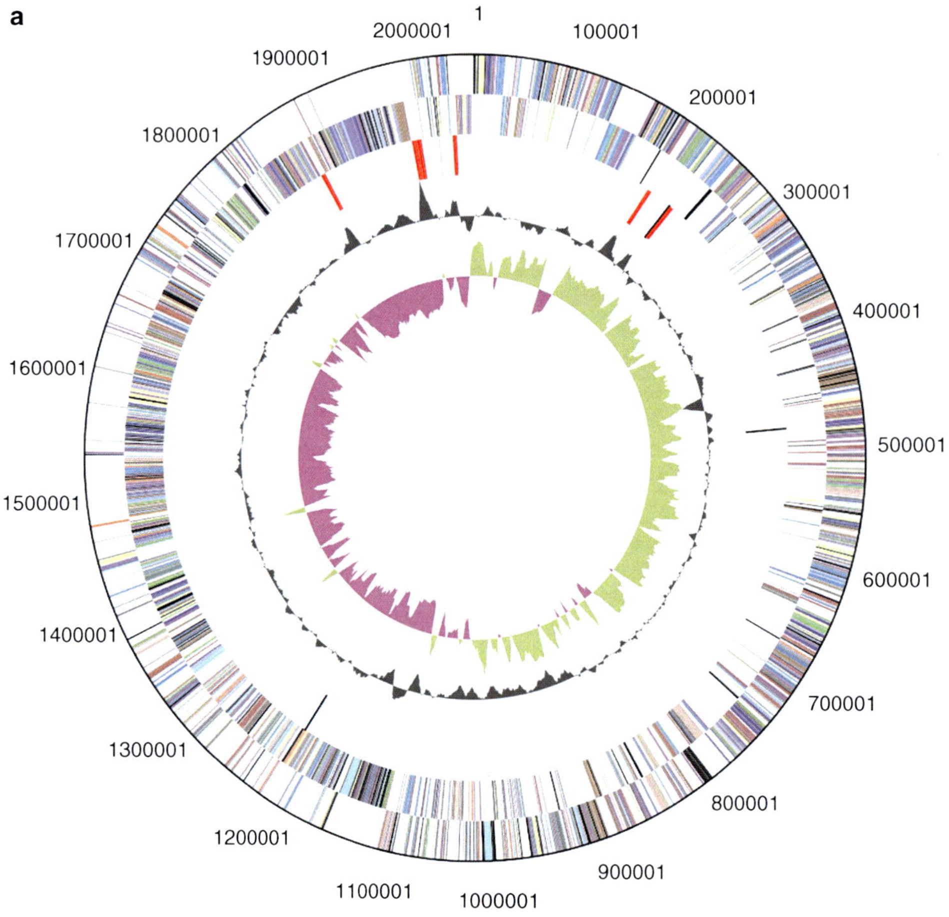


**◘ Fig. 9.4 (continued)**

**b**

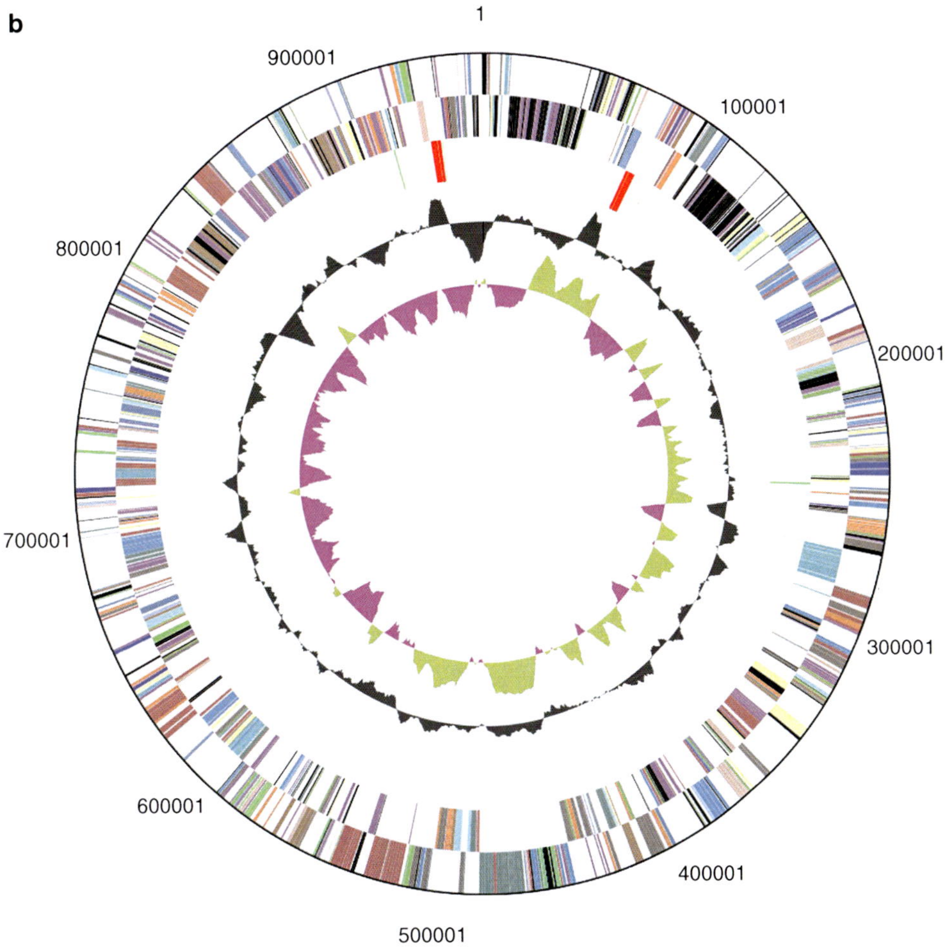


**Fig. 9.4**

**The 3,132,314 bp long genome of *I. polytropus* consists of two chromosomes (1 and 2). (a) shows graphical circular map of chromosome 1. From outside to the center: genes on forward strand (color by COG categories), genes on reverse strand (color by COG categories), RNA genes (tRNAs *green*, rRNAs red, other RNAs *black*), GC content, GC skew. (b) Graphical circular map of chromosome II. From outside to the center: genes on forward strand (color by COG categories), genes on reverse strand (color by COG categories), RNA genes (tRNAs *green*, rRNAs *red*, other RNAs *black*), GC content, GC skew. Chromosome II was identified due to its two 16S rRNA gene copies (Courtesy of Sikorski et al. 2010)**

## *Fusobacterium* Knorr 1922, 4[AL]

Fu.so.bac.te'ri.um. L.n. *fusus* a spindle; L. neut. n. *bacterium* a small rod; N.L. neut. n. *Fusobacterium* a small spindle-shaped rod.

A recent review on *Fusobacterium*, which should be consulted for additional details, is written by Gharbia et al. (2011). *Fusobacterium* is a small Gram-negative rod. However, the cells are pleomorphic. Some appear as filaments that have spindle-shaped pointed ends (fusiforms). Others are coccobacilli. Cell width is variable. The cells may be single, appear in pairs end to end, or can be long coiled filaments. They may show irregular staining.

*F. nucleatum*, which is the type species of the genus, has slender, spindle-shaped cells with tapered or pointed ends, 0.4–0.7 μm thick and 4–10 μm long. Cells appear singly, in tandem pairs, or in bundles. Old cultures of *F. periodonticum* and *F. simiae* often contain filaments. Otherwise their cell morphology is similar to that of *F. nucleatum*. Cells of *F. necrophorum* are pleomorphic. They are often curved with rounded or tapered ends, occasionally with spherical enlargements and vary in length from coccobacilli to long threads. Gonidial cell forms are found in *F. gonidiaformans*. No filaments are seen in *F. varium*. Cells of *F. mortiferum* are extremely pleomorphic. Globular forms, swellings, and threads occur. In *F. equinum*, short rods predominate.

Fusobacteria grow best at 35–37 °C and at pH near 7. They do not require a particularly low redox potential but are killed by exposure to oxygen, probably through production of peroxides. Colonies of *F. nucleatum* are 1–2 mm in diameter after 2 days of anaerobic incubation. They are white-yellow, speckled, smooth, or breadcrumb-like and have a fetid smell. Usually there is no

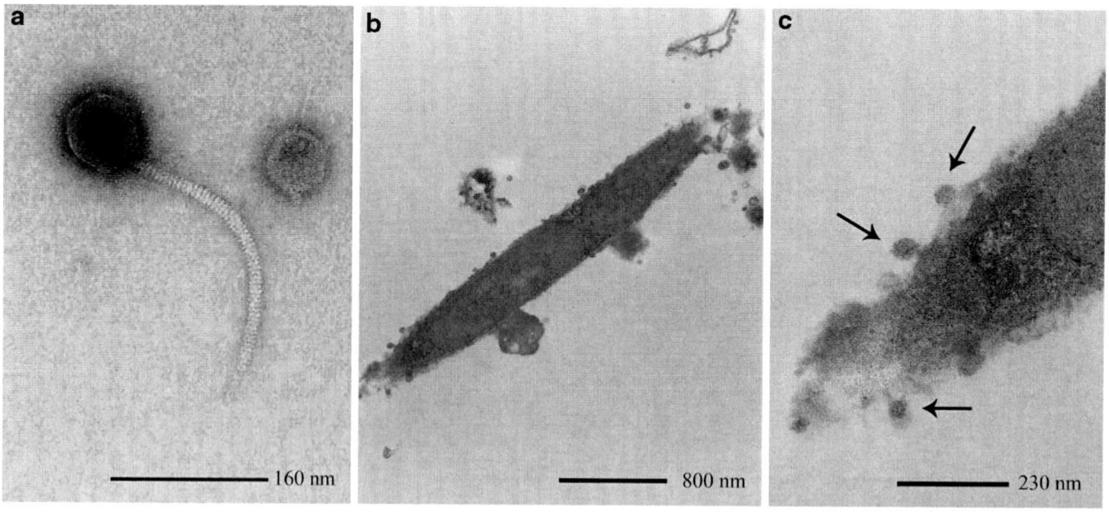


**Fig. 9.5**
Transmission electron micrographs. a. Morphology of phage Fnpϕ02; b. *F. nucleatum* subsp. *polymorphum* (Fnp) cell after inoculation with Fnpϕ02; c. Higher magnification of the Fnp cell lysis with liberation of the phage particles (*arrows*) (From Machuca et al. (2010). Courtesy of Appl Environ Microbiol)

**Table 9.1**
Properties of *Fusobacteriaceae*

| Characteristic | *F. nucleatum* | *C. ceti/C. somerae* | *I. polytropus/ I. tartaricus/I. insuetus* | *P. modestum/ P. maris* | *Ps. psychrilyobacter* |
|---|---|---|---|---|---|
| Habitat | Human mouth, gastrointestinal tract, animal cavity | Mammalian intestinal tract and oral cavity | Anoxic sediment and sludge | Anaerobic mud | Cold deep marine sediment |
| Morphology | Spindle-shaped or pleomorphic rods | Short pleomorphic rods | Round-ended rods or cocci | Short rods or cocci | Short rods |
| Growth at | | | | | |
| 4°C | – | – | – | – | + |
| 37°C | + | + | Unknown | + | |
| Optim. temp. | 37 | 37 | 28–34 | 33–37 | 18.5 |
| NaCl req. | – | Unknown | + | + | + |
| Utilization of | | | | | |
| succinate | Unknown | Unknown | Unknown | + | – |
| 3-hydroxbut., tartrate or quinate | Unknown | Unknown | v | Unknown | – |
| DNA G + C | 26–34 | 29–31 | 31.7–36.7 | 32.9–41 | 28.1 |

hemolysis on blood agar. For *F. necrophorum* colonies, which are 2 mm in diameter, the smell is putrid. Colonies are flat and circular with scalloped or erosive edge. The colony is white to gray in *F. nucleatum* subsp. *necrophorum* and gray or translucent with smooth surface in *F. nucleatum* subsp. *funduliforme*. Colonies of *F. necrophorum* are similar to those of *F. nucleatum* subsp. *funduliforme*. Most other colonies of fusobacteria are also similar to colonies of *F. nucleatum* subsp. *funduliforme*.

Fusobacteria metabolize peptone or carbohydrates in PY-glucose broth producing butyrate, often with acetate and lower levels of lactate, propionate, succinate, and formate and can utilize amino acids and peptides in the absence of proteases. They are catalase and nitrate negative. Sensitive to kanamycin and colistin, but resistant to vancomycin.

Isolation of fusobacteria from clinical samples, e.g., periodontal pockets, requires careful sampling avoiding resident

**Table 9.2**
**Characters differentiating between *Fusobacterium* species** [a]

| Character | *F. nucleatum* subsp. *nucleatum* | *F. nucleatum* subsp. *fusiforme* | *F. nucleatum* subsp. *polymorphum* | *F. nucleatum* subsp. *vincentii* | *F. nucleatum* subsp. *animalis* | *F. canifelinum* | *F. equinum* | *F. gonadiaformans* | *F. mortiferum* | *F. necrogenes* | *F. necrophorum* subsp. *necrophorum* | *F. necrophorum* subsp. *funduliforme* | *F. perfoetens* | *F. periodonticum* | *F. russii* | *F. simiae* | *F. ulcerans* | *F. varium* |
|---|---|---|---|---|---|---|---|---|---|---|---|---|---|---|---|---|---|---|
| Cellobiose | – | – | – | – | – | – | – | – | $w^-$ | $w^-$ | – | – | – | – | – | – | – | – |
| Esculin hydrolysis | – | – | – | – | – | – | – | – | $+^-$ | + | – | – | – | – | – | – | – | – |
| Utilization of | | | | | | | | | | | | | | | | | | |
| Fructose | $w^-$ | $w^-$ | $w^-$ | $w^-$ | $w^-$ | $w^-$ | – | – | $w^a$ | $w^a$ | $-^w$ | $-^w$ | w $w^-$ | $w^-$ | – | w | – | $w^a$ |
| Gelatin | $-^w$ | $-^w$ | $-^w$ | $-^w$ | $-^w$ | $-^w$ | $-^w$ | – | – | – | v | v | – – | – | – | – | – | – |
| Glucose | $-^w$ | $-^w$ | $-^w$ | $-^w$ | $-^w$ | $-^w$ | – | $w^-$ | $a^w$ | $w^a$ | $-^w$ | $-^w$ | w $w^-$ | $w^-$ | – | w | $a^w$ | $w^a$ |
| Lactose | – | – | – | – | – | – | – | – | $w^a$ | – | – | – | – – | – | – | – | – | – |
| Maltose | – | – | – | – | – | – | – | – | $w^-$ | – | – | – | – – | – | – | – | – | – |
| Mannose | – | – | – | – | – | – | – | – | $w^a$ | $w^a$ | – | – | – – | – | – | – | $a^w$ | w |
| Raffinose | – | – | – | – | – | – | – | – | v | $-^w$ | – | – | – – | – | – | – | – | – |
| Sucrose | – | – | – | – | – | – | – | – | v | $-^w$ | – | – | w– | – | – | – | – | – |
| Indole | + | + | + | + | + | + | + | + | – | – | + | + | – + | + | – | + | – | $+^-$ |
| Nitrate | – | – | – | – | – | – | | – | – | – | – | – | – – | – | – | – | + | – |
| Bile growth | – | – | – | – | + | – | + | – | + | + | v | v | – – | – | – | + | + | + |
| Lipase | – | – | – | – | – | – | + | – | – | – | $+^-$ | | – – | – | – | w | – | $-^w$ |
| β-Hemolysis | – | – | – | – | – | – | – | – | – | – | + | – | – – | – | – | – | – | – |
| Gas produced in agar | – | – | – | – | – | – | nt | + | + | + | + | + | + – | – | – | – | + | + |
| Lactate → propionate | – | – | – | – | – | – | + | – | – | – | + | + | – – | – | – | – | – | – |
| Threonine → propionate | + | + | + | + | + | + | + | + | + | + | + | + | + + | + | – | + | + | + |
| Activity of | | | | | | | | | | | | | | | | | | |
| *N*-Acetyl-glucosaminidase | – | – | – | – | – | – | – | – | – | – | – | – | + – | – | – | – | – | – |
| Alkaline phosphatase | – | – | – | – | – | – | – | – | + | + | + | – | – – | – | w | – | – | w |
| Acid-phosphatase | – | – | – | – | – | – | + | w | + | + | + | + | w– | – | + | – | + | w |

**Table 9.2 (continued)**

| Character | *F. nucleatum* subsp. *nucleatum* | *F. nucleatum* subsp. *fusiforme* | *F. nucleatum* subsp. *polymorphum* | *F. nucleatum* subsp. *vincentii* | *F. nucleatum* subsp. *animalis* | *F. canifelinum* | *F. equinum* | *F. gonadiaformans* | *F. mortiferum* | *F. necrogenes* | *F. necrophorum* subsp. *necrophorum* | *F. necrophorum* subsp. *funduliforme* | *F. perfoetens* | *F. periodonticum* | *F. russii* | *F. simiae* | *F. ulcerans* | *F. varium* |
|---|---|---|---|---|---|---|---|---|---|---|---|---|---|---|---|---|---|---|
| α-Galactosidase | – | – | – | – | – | – | – | – | + | + | – | – | – – | – | – | – | – | – |
| β-Galactosidase | – | – | – | – | – | – | – | – | + | + | – | – | + – | – | – | – | – | – |
| β-Glucuronidase | – | – | – | – | – | – | – | – | – | – | – | – | – – | – | – | – | – | – |
| α-Glucosidase | – | – | – | – | – | – | – | – | – | – | – | – | – – | – | – | – | – | – |
| β-Glucosidase | – | – | – | – | – | – | – | – | + | – | – | – | – | – | – | – | – | – |
| Esterase (C4) | – | – | – | – | – | – | + | – | – | – | w | w | – – | – | – | w | – | – |
| Esterase (C8) | – | – | – | – | – | – | + | – | – | – | w | w | – – | – | – | – | – | – |

[a]Adopted from Gharbia et al. (2011). Symbols: w most strains weakly positive, some negative; $-^{w}$ most strains negative, some weakly positive $a^{w}$ strongly acid with some weakly acid reactions, $w^{a}$ usually weakly acid with occasional strong acid reactions, *nt* not tested

microflora, transport of samples in gas-tight glass tubes with a prereduced anaerobically sterilized (PRAS) medium, anaerobic culture (80 % $N_2$, 10 % $H_2$ and 10 % $CO_2$), and a rich medium for growth, e.g., Fastidious Anaerobe Agar (FAA, Lab M) supplemented with 5 % sheep or horse blood. All fusobacteria grow on blood agar based on proteose peptone, tryptone, or trypticase. Also brain-heart infusion broth supplemented with yeast extract provides good growth. A selective medium helps enumeration of fusobacteria in the resident microflora. Supplements such as josamycin, vancomycin, and norfloxacin plus 5 % defibrinated horse blood are useful for stimulation of growth and preventing growth of other anaerobic and facultative organisms. Fusobacteria can be stored frozen in PRAS medium with a cryoprotective substance such as DMSO or as a lyophilized preparation. For *F. varium* and *F. mortiferum*, rifampin blood agar is useful for growth. For preliminary diagnosis tablets for analysis of bile resistance, alkaline phosphatase and ortho-nitrophenyl-β-D-galactopyranoside (ONPG) can be used together with commercial kits containing preformed enzymes for identification of anaerobic bacteria. The DNA G + C content is 26–34 mol%.

## *Cetobacterium*

Ce.to.bac.te'ri.um. Gr. N. *kêtos* whale; L. neut. n. *bacterium* a rod; N.L. neut. N. *Cetobacterium* a bacterium found in association with whales.

According to Edwards et al. (2011), whose overview should be consulted for additional details, *Cetobacterium* has short, pleomorphic nonspore-forming, rod-shaped cells which may have central swellings and appear as filaments. They stain Gram negative and are nonmotile, microaerotolerant, and catalase negative. Acetic acid is the main end product from fermentation of peptones or carbohydrates. Small amounts of butyric, propionic, lactic, and succinic acid may or may not be produced. *Cetobacterium* is indole positive and ONPG is hydrolyzed. Alkaline phosphatase, acid phosphatase, and phosphohydrolase are generated. α- or β-Galactosidase, α-glucosidase, and urease may or may not be produced, and nitrate may or may not be reduced to nitrite. The organisms are resistant to 20 % bile and to vancomycin, but sensitive to kanamycin, colistin sulfate, cefoxitin, clindamycin, imipenem, and metronidazole. The DNA G + C content is 29–31 mol%. *Cetobacterium* was first isolated from the intestinal contents of a porpoise and from a lesion in the mouth of a minke whale. Later *C. somerae* was isolated from human feces.

Colonies are gray or waxy with circular or scalloped to erose edges. They are slightly raised, smooth, dull, and opaque. Diameter is 2–4 mm. Weak hemolysis on sheep and horse blood agar. *C. somerae* grows in 2 % but not 6 % oxygen, while *C. ceti* cannot grow in 10 % $CO_2$ or in air.

*C. ceti* colonies on blood agar are 2–4 mm in diameter after 48 h at 37 °C, waxy, gray, and circular with scalloped to erose edges. Weak hemolysis on sheep and horse blood agar. There is no growth at 25 °C or 45 °C. Catalase is negative but indole, ONPG, and phosphatase are positive. Resistant to 20 % bile. Sensitive to colistin sulfate and kanamycin. Sensitivity to penicillin is variable. Major metabolic fatty acids are acetic, propionic, lactic, and succinic acid.

*C. somerae* has rod-shaped, Gram-negative, microaerotolerant cells. Colonies on brucella agar are 2–3 mm in diameter, smooth, circular, entire, and gray after 48 h. Catalase negative and indole positive. Major end product in peptone yeast broth is acetic acid. Nitrate is reduced to nitrite. Esculin may or may not be hydrolyzed; ONPG is hydrolyzed. Urease may or may not be detected.

Resistant to bile. Susceptible to kanamycin, colistin sulfate, cefoxitin, clindamycin, imipenem, and metronidazole.

## *Ilyobacter*

I.ly.o.bac'ter. Gr. fem. n. *ilys* mud; N.L. masc. n. *bacter* rod; N.L. masc. n. *Ilyobacter* a mud-inhabiting rod.

For a more comprehensive review on *Ilyobacter* than the present one, Schink et al. (2011) should be consulted. These organisms are strictly anaerobic, chemoorganotrophic, and nonsporulating with fermentative metabolism. They are nonphotosynthetic, and inorganic electron acceptors are not used. They have no cytochromes and use unusual substrates for growth giving unusual products. It is necessary to include a reductant in the medium for growth. Anoxic marine sediments are typical habitats for these bacteria. Acetate, butyrate, formate (on some substrates), and ethanol are fermentation products. Cells are short to coccoid rods that often occur in pairs or short chains. DNA G + C content is 31.7–36.7 mol%. *I. polytropus*, which is the type species, was enriched and isolated from marine sediments with 3-hydroxybutyrate which is fermented to acetate and butyrate. 1,3-Propanediol and 3-hydroxypropionate are produced from glycerol, while acetate and formate are the only products of pyruvate or citrate fermentation.

Highly selective for *Ilyobacter* is a strictly anoxic, sulfide-reduced mineral medium with 10 mM of either 3-hydroxybutyrate, shikimate, or L-tartrate as the sole organic carbon and energy sources. Selective for the enrichment of *Ilyobacter* is also culture at 27–30 °C. A carbon-buffered standard medium for enrichment and isolation has been described (Widdel and Pfennig 1981; Schink and Pfennig 1982). Roll tubes and anoxic agar deep dilution series can also be used to isolate the bacteria.

Cells are maintained by repeated transfer at 2–3 months' intervals or by freezing in liquid nitrogen.

*I. polytropus* which is the type species of the genus *Ilyobacter* was enriched and cultured from marine sediment with 3-hydroxybutyrate and crotonate that were fermented to acetate and butyrate. Glycerol is metabolized to 1,3-propanediol and 3-hydroxypropionate. Pyruvate and citrate fermentation produce acetate and formate, while glucose and fructose are fermented to acetate, formate, and ethanol.

*I. insuetus* was isolated from marine sediments with quinic acid as the sole source of carbon and energy and is restricted to fermentation of hydroaromatic substrates. Of 30 substrates tested, only quinic and shikimic acid were utilized. This species represents an extreme case of specialization in substrate utilization since neither sugars, alcohols, other carboxylic acids, amino acids, nor aromatic compounds are fermented.

*I. tartaricus* was enriched and isolated from a marine sediment with L-tartaric acid as carbon and energy source. Acetate, formate, and $CO_2$ are fermentation products of glucose and fructose. Sodium ions are required as coupling ions in energy conservation for this species. Selective enrichment from marine sediments occurs with L-tartrate as the sole carbon and energy source.

## *Propionigenium*

Pro.pi.o.ni.ge'ni.um. N.L. n. *acidum propionicum* propionic acid; L. v. *genere* to make, produce; N.L. neut. n. *Propionigenium* propionic acid maker.

This genus has been described recently in a review by Schink and Janssen (2011) which should be consulted for more information. Cells are strictly anaerobic, chemoorganotrophic with fermentative metabolism, nonphotosynthetic, and nonspore-forming. Inorganic electron acceptors are not used. The organisms preferably use decarboxylic acids as substrates in their fermentative metabolism. A reductant is necessary in the medium for growth. Catalase negative. *Propionigenium* has been isolated from anoxic marine and freshwater sediments. The genus was created to comprise strictly anaerobic bacteria able to grow by decarboxylation of succinate to propionate (Schink 1984). Pure cultures are obtained with enrichment cultures from marine sources. Freshwater enrichments grow much slower, pure cultures being isolated when the sodium chloride concentration is increased to 100–150 mM. Contrary to *P. modestum* which is the type species of the genus, *P. maris* is able to ferment carbohydrates, amino acids, and other organic acids in addition to $C_4$ dicarboxylic acids.

*P. modestum* was isolated from a black, anoxic, marine sediment from the Canal Grande of Venice. Also *P. maris* was isolated from marine sediments. Accordingly, marine sediments are the typical environments for these bacteria. Sodium ions serve in their metabolism as coupling ions for energy conservation. *P. maris*-like bacteria were detected in burrows of bromophenol-producing marine infauna. Here they probably are involved in reductive debromination of bromophenols and use organic excretions of the infauna as electron donors for the reductive reaction.

Highly selective for the enrichment of *P. modestum* is a strictly anoxic, sulfide-reduced mineral medium with 20 mM succinate as the sole organic carbon and energy source. Incubation is best at 27–30 °C. A standard carbonate-buffered medium has been described by Widdel and Pfennig (1981) and Schink and Pfennig (1982). After multiple transfers in liquid medium, the dominant population consists of short, coccoid rods which can be isolated in anoxic agar deep dilution series or in roll tubes. *P. maris* needs yeast extract for growth in pure culture.

Cultures are maintained by repeated transfers at 2–3 months' interval or by freezing in liquid nitrogen.

*P. modestum* has rod-shaped to coccoid Gram-negative cells with rounded ends, 0.5–0.6 µm in diameter, 0.5–2.0 µm long. They appear single, in pairs, or in chains. Nonmotile, nonsporing, strictly anaerobic, and chemoorganotrophic. Succinate, fumarate, malate, aspartate, oxaloacetate, and pyruvate are utilized for growth and are fermented to propionate and $CO_2$. For growth, mineral media with a reductant and 1 % NaCl are needed. Selective enrichment can be made in NaCl-containing media with succinate as substrate. pH optimum is 7.1–7.7 and temperature optimum 33 °C. Habitats are anoxic marine or brackish water sediments.

*P. maris* has coccoid to oval, short, Gram-negative, nonsporing rods with rounded ends. They are 1.0 µm × 1.2–2.5 µm and up to 50 µm long under certain culture conditions. Strictly anaerobic and chemoorganotrophic. The metabolism is fermentative with electron acceptors not being used, although bromophenols can be reductively dehalogenated. With yeast extract in the medium, several carbohydrates and amino and organic acids are fermented. Propionate, acetate, and formate are typically produced depending on the substrate. Hydrogen is generated from carbohydrates and yeast extract, and indole is formed from L-tryptophan. Growth can occur in saltwater media with 5–55 g/l NaCl, but anoxic growth conditions are required. pH optimum 6.9–7.7. Temperature optimum 34–37 °C. Anoxic marine or brackish water sediment is the habitat.

## *Psychrilyobacter*

Psy.chri.lyo'bac.ter. Gr. adj. *psychros* cold; N.L. masc. n. *Ilyobacter* a bacterial genus name; N.L. masc. n. *Psychrilyobacter* a psychrotrophic bacterium related to the genus *Ilyobacter.*

*Psychrilyobacter* was described by Zhao et al. (2009). For more details, this paper should be consulted. Cells are obligate, anaerobic, Gram-negative short rods without spores. They grow at low temperatures and require NaCl for growth being slightly halophilic. Utilize sugars, organic and amino acids, and peptone. The metabolism is fermentative with $H_2$ and acetate as major fermentation products. There is no utilization of succinate, L-tartrate, 3-hydroxybutyrate, or quinate and shikimate.

*Ps. atlanticus* is the type and only species of the genus. Cells are 0.5 µm in diameter, 0.5–1 µm in length, and nonmotile. Colonies are round and colorless. Catalase and oxidase negative. Optimum growth temperature is 18.5°C. Slightly halophilic growing at 2 % NaCl. No growth at NaCl $<$0.5 % w/v.

Cells are chemoorganotrophic and fermentative. Utilize glucose, fructose, maltose, *N*-acetyl-D-glucosamine, citrate, pyruvate, casitone, fumarate, lysine, threonine, and aspartate. *Ps. atlanticus* ferments threonine to propionate and $H_2$. The type strain HAW-EB21$^T$ was isolated from a site 215 m deep and 50 nautic miles off the Halifax Harbour in the Atlantic Ocean. The DNA G + C content is 28.1 mol%.

## Identification

A selective medium (JVN) has been developed to isolate *Fusobacterium* species from clinical samples (Brazier et al. 1991). It consists of josamycin, vancomycin, and norfloxacin at 3, 4, and 1 μg/ml, respectively, together with 5 % defibrinated horse blood in Fastidious Anaerobe Agar Base (Lab M). The JVN medium isolated successfully *F. nucleatum*, *F. necrophorum*, and "*F. naviforme*" from the gingivae of 9/16 healthy volunteers and *F. varium* and *F. mortiferum* from fecal suspensions inoculated with these organisms. Another selective medium was devised by Sutter et al. (1971) to allow detection of relatively small amounts of *F. nucleatum* in fecal specimens. Blood agar with 50 mg of rifampicin/ml inhibited the growth of many species of *Bacteroides* and *F. nucleatum* subsp. *fusiforme/-nucleatum* but allowed good growth of *F. varium* and most strains of *F. mortiferum*. The medium was supposed to be of value in studies on the fecal microbiota.

The lack of constitutive β-glucosidase in the genus *Fusobacterium* was used to distinguish it from *Bacteroides* organisms in a 30-min one-tube test (Edberg and Bell 1985). Beighton et al. (1997) found that under standard conditions, hydrolytic (endopeptidase) abilities of fusobacteria could be used to differentiate organisms at the species level. However, further investigation with a larger number of strains was warranted.

Spaulding and Rettger (1937) classified the genus *Fusobacterium* by biochemical and serologic methods, and Jenkins et al. (1992) did numerical taxonomy on 141 different strains of species of *Fusobacterium, Bacteroides, Porphyromonas*, and *Prevotella*. The isolates were tested for 111 phenotypic characters including constitutive enzymes, fermentation of carbohydrates, gas chromatographic analysis of metabolic end products, and cellular carboxylic acid composition. At 94 % similarity level, nine different groups were delineated, one of which consisted of *F. mortiferum, F. necrogenes, F. necrophorum, F. nucleatum*, and *F. varium*.

Classification of fusobacteria based on pyrolysis mass spectrometry resolved groups largely corresponding to *F. necrogenes, F. necrophorum, F. nucleatum, F. mortiferum, F. varium, F. gonidiaformans, F. russii*, and "*F. naviforme*," although *F. nucleatum* strains were divided between two groups (Magee et al. 1989).

Cellular fatty acids of *Fusobacterium* species were examined with gas-liquid chromatography (Jantzen and Hofstad 1981). *F. nucleatum, F. necrophorum, F. mortiferum, F. gonidiaformans*, and *F. varium* had all 3-hydroxytetradecanoate, n-tetradecanoate, hexadecenoate, n-hexadecanoate, octadecanoate, n-octadecanoate, and a component with the properties of octadecadienoate. Distinctive for *F. nucleatum* was 3-hydroxyhexadecanoate. Simpler patterns, characterized by the absence of 3-hydroxytetradecanoate and other fatty acids, were seen in "*F. plauti*," "*F. prausnitzii*," and in most *F. russii* and "*F. naviforme*" strains. All species except *F. mortiferum, F. gonidiaformans*, and "*F. naviforme*" contained substantial amounts of fatty aldehyde dimethyl acetals of chain length C14–C18. Calhoon et al. (1983) examined cellular fatty acid and protein contents of 43 strains of oral fusobacteria from human periodontal and stump-tailed macaque samples. The n16:0 to 3-OH-16:0 ratio differentiated between *F. nucleatum* and the non-oral species *F. varium, F. necrophorum, F. russii, F. necrogenes, F. mortiferum*, and "*F. naviforme*." The soluble protein content, determined by polyacrylamide gel electrophoresis, varied considerably between the species.

The fatty acid composition of lipid A in *F. nucleatum* Fev 1 consisted of β-1′, 6-linked D-glucosamine disaccharides which have two phosphate groups, one in glycosidic and one in ester linkage (Hase et al. 1977). In *F. varium* and *F. mortiferum*, 3-hydroxydecanoic acid was absent. Here only-3-hydroxytetradecanoic acid was found, particularly ester and partly amide bound emphasizing the heterogeneous nature of these organisms.

In lipopolysaccharides (LPSs) of two *F. necrophorum* strains, tetra- and hexadecanoic acids were detected while 3-hydroxy fatty acids were absent (Meisel-Mikolajczyk and Dobrowolska 1974).

*F. nucleatum* strains were also classified into six chemotypes based on the polysaccharide composition of their LPSs (Fredriksen and Hofstad 1978).

Hermansson et al. (1993) found that the O-specific polysaccharide component of LPS in *F. necrophorum* was of the teichoic acid type with repeating units connected by phosphoric diester linkages. The LPSs of *F. nucleatum* subsp. *fusiforme* and *F. nucleatum* subsp. *funduliforme* were clearly different with distinct levels of polysaccharides being demonstrated (Garcia et al. 1999).

Lanthionine is a natural component of the peptidoglycan of *F. nucleatum* FeV 1 (Vasstrand et al. 1979). It replaces *meso*-diaminopimelic acid which is a normal component of the peptidoglycan layer of Gram-negative bacteria. Vasstrand et al. (1982) found that lanthionine of peptidoglycan could serve as a taxonomic marker for *F. nucleatum, F. necrophorum, F. russii*, and *F. gonidiaformans*. Peptidoglycans from these bacteria were suggested to belong to a new chemotype, A 1δ (Vasstrand 1981; Vasstrand et al. 1982). "*F. plauti*" had a peptidoglycan composition atypic of Gram-negative cells. Gharbia and Shah (1990b) found a unique dibasic amino acid linkage of *meso*-lanthionine in fusobacteria except for *F. varium, F. ulcerans*, and "*F. naviforme*" which contained *meso*-diaminopimelic acid. The genus *Fusobacterium* could be divided in two major groups which had either lanthionine or diaminopimelic acid in their peptidoglycan. *F. mortiferum* had both of these substances.

Gharbia and Shah (1990b) differentiated *Fusobacterium* species using electrophoretic migration of glutamate dehydrogenase (GDH). When all recognized members were included, except *F. perfoetens* and "*F. prausnitzii*," they clustered into three broad electrophoretic groups. *F. periodonticum, F. simiae*, and *F. necrophorum* could not be distinguished. However, 2-oxoglutarate reductase (OGR) distinguished between them. Accordingly, neither of the techniques differentiated all species, but used together, they gave unambiguous discrimination of all species except *F. varium* and *F. mortiferum*.

Representative strains of *F. nucleatum* subgroups Fn-1, Fn-2, and Fn-3 were distinguished by ribosomal RNA gene restriction patterns (Lawson et al. 1989). The patterns from DNA digested with EcoR1 or TaqI allowed clustering of the strains in three subgroups. TaqI gave a particularly wide distribution of taxonomically important bands, and the pattern generated was characteristic of each subgroup.

A PCR probe with 100 % sequence identity to 120 deoxynucleotides of *F. nucleatum* Fev1 was prepared by Bolstad and Jensen (1993). It coded for a part of the 40-kDa major outer membrane protein and was labeled with the steroid hapten digoxigenin. The probe distinguished *F. nucleatum* from other Gram-negative bacteria in the periodontal pocket and from other fusobacterial species and different strains of *F. nucleatum*.

Zhou et al. (2009a) identified a 407-bp DNA sequence amplified from the type strain of *F. equinum* (NCTC 13176$^T$) using PCR primers that were based on published *Fusobacterium* leukotoxin (Lkt)A sequences. This made it possible to distinguish *F. equinum* from *F. necrophorum* to which it is quite similar phenotypically.

Bank et al. (2010) developed a selective agar for *F. necrophorum* containing vancomycin and nalidixin which inhibits the growth of most Gram-positive and many Gram-negative bacteria.

Narayanan et al. (1997) who used ribotyping found genetic similarity between isolates of *F. necrophorum* from liver abscesses and the ruminal wall supporting the hypothesis that *F. necrophorum* isolates of liver abscesses originate from the rumen. Interestingly, *F. necrophorum* may interact synergistically with the facultative anaerobic *Arcanobacterium pyogenes* to cause liver abscesses in cattle (Tadepalli et al. 2009).

Phylogenetic analysis of *F. necrophorum* subspecies *necrophorum*, *F. varium*, *F. nucleatum* subsp. *nucleatum*, and *F. nucleatum* subsp. *vincentii* based on *gyrB* gene sequences suggested that gyrB is an accurate genealogical marker for the classification of these species (Jin et al. 2004). The strains formed distinct clusters corresponding to each species, with deep sublines.

The two subspecies of *F. necrophorum*, originally biovars A and B, were established by Shinjo et al. (1991) according to biological and biochemical properties, DNA base compositions and levels of DNA-DNA homology. Ribotyping appeared to be a useful technique to genetically differentiate the two subspecies of *F. necrophorum* (Okwumabua et al. 1996). While only *F. nucleatum* subsp. *necrophorum* had an approximate 2.6-kb band, the less virulent *F. necrophorum* subsp. *funduliforme* had a 4.3-kb band.

*F. necrophorum* subsp. *necrophorum* can be distinguished from *F. necrophorum* subsp. *funduliforme* by colony and cellular morphology, hemagglutination, hemolytic and extracellular enzymatic activities, and virulence in mice (Holdeman et al. 1977; Shinjo et al. 1991; Amoako et al. 1993; Smith and Thornton 1993; Holst et al. 1994). Minimum requirements for a rapid and reliable identification of *F. necrophorum* based on phenotypic characters were proposed by Jensen et al. (2008). All but one (*F. necrophorum* subsp. *necrophorum*) of 357 isolates were identified as *F. necrophorum* subsp. *funduliforme* in all phenotypic tests. It was suggested that *F. necrophorum* can be differentiated from other *Fusobacterium* species by its unique but subspecies-specific colony morphology, susceptibility to kanamycin and metronidazole, the smell of butyric acid, chartreuse color fluorescence, and β-hemolysis on horse blood agar.

Narongwanichgarn et al. (2001) found that RAPD analysis with primer WIL-2 revealed a 2.4 kb band in all *F. necrophorum* subsp. *necrophorum* strains but not in strains of *F. necrophorum* subsp. *funduliforme*. Other useful techniques for distinction between the two subspecies are 16S rRNA sequencing (Nicholson et al. 1994), restriction fragment length polymorphism analysis (Hodgson et al. 1993), and ribotyping (Okwumabua et al. 1996). *F. necrophorum* subsp. *necrophorum* contains the isoleucine and alanine tRNA gene, while *F. necrophorum* subsp. *funduliforme* carries only the isoleucine tRNA gene (Jin et al. 2002). Specific primers made for the hemagglutinin-related protein gene amplified a 250 bp fragment of the genome of *F. necrophorum* subsp. *necrophorum* strains but not from strains of *F. necrophorum* subsp. *funduliforme*, suggesting that this gene is unique to *F. necrophorum* subsp. *necrophorum* (Narongwanichgarn et al. 2003). A one-step duplex PCR technique in a single tube was developed by the authors for rapid detection and differentiation of the *F. necrophorum* subspecies. DNase is produced by *F. nucleatum* subsp. *necrophorum* but not by *F. nucleatum* subsp. *funduliforme* and may be used for distinction between them (Amoako et al. 1993).

Strain-specific PCR primers have been developed for identification of *F. nucleatum* subsp. *fusiforme* (ATCC 51190$^T$ and subsp. *vincentii* ATCC 49256$^T$ (Shin et al. 2010). Two pairs of primers Fs17-F14/Fs17-R14 and Fv35-F1/Fv35-R1 produced strain-specific amplicons from subsp. *fusiforme* and -*vincentii*, respectively. As little as 0.4 or 4 pg, respectively, of the genomic DNA of each target strain could be detected.

Two sets of PCR primers, Fp-F3/Fp-R2 and Fp-F1/Fp-R2, gave amplicons from all *F. periodonticum* isolates but not from 12 other *Fusobacterium* species or subspecies tested as well as representative oral bacteria (Park et al. 2010). The sensitivity of the primer sets was quite good, 4 or 40 pg of the chromosome DNA from *F. periodonticum* ATCC 33693$^T$.

*Cetobacterium* can be distinguished from *Fusobacterium* by its production of acetic and propionic acid since members of the genus *Fusobacterium* produce butyric acid (Foster et al. 1995).

16S rRNA gene sequencing of *Cetobacterium* showed a 91–94 % similarity with *Fusobacterium* and 92 % with *P. nodestum*. *Cetobacterium* exhibited 86 % similarity with *Leptotrichia* and *Sebaldella* (Foster et al. 1995). It can be distinguished from *P. modestum* through its production of large amounts of acetic acid with lesser amounts of propionic, lactic, and succinic acid. *P. modestum* does not ferment carbohydrates but generates large amounts of propionic and lesser amounts of acetic acid exclusively from succinate and other substrates (Schink and Pfennig 1982). Distinctive from *P. modestum* is also a positive indole reaction.

The selective medium for *I. polytropus* is a NaCl-containing mineral medium with 3-hydroxybutyrate as the sole carbon and energy source (Stieb and Schink 1984).

As for *Ilyobacter*, the three species are differentiated from most other strict anaerobic bacteria through their unusual pattern of substrate utilization and product formation, as well as low concentration of G + C. 16S rRNA sequencing separated *Ilyobacter* from all other genera except *Propionigenium*. Brune et al. (2002) showed through 16S rRNA gene sequencing that *Propionigenium* and *Ilyobacter* (except "*I. delafieldii*") belong to the phylum Fusobacteria. Both these genera were found in a distinct cluster separated from the *Sebaldella-Streptobacillus-Leptotrichia* cluster and from *Fusobacterium*. Even if the 16S rRNA sequencing did not separate the *Ilyobacter-Propionigenium* branch, 23S rRNA sequencing demonstrated a monophyletic status at least for the genus *Ilyobacter* (Brune et al. 2002). The 16S rRNA gene sequence of *I. polytropus* showed high similarity to *I. insuetus* (97.3 %) and to *I. tartaricus* (98.3 %) (Sikorski et al. 2010). The similarity with other members of the family *Fusobacteriaceae* varied from 89.5 % to 97.8 %, with *P. modestum* as the closest species. The metabolic properties of *Ilyobacter* and *Propionigenium* are quite distinct. In *I. tartaricus*, e.g., energy metabolism is related to sodium ions which are coupling ions in energy conservation. $Na^+$-ATPase of *I. tartaricus* has together with a similar enzyme in *P. modestum* become a model system for studying the architecture of $F_1F_0$-ATPases and the connection between $Na^+$-ion transport and ATP synthesis (Neumann et al. 1998). *Propionigenium* species are specialists in utilizing $C_4$-dicarboxylic acids. *P. modestum* utilizes only a few of these compounds, while *P. maris* also uses sugars and amino acids and requires yeast extract in the culture medium. Phylogenetic analysis has demonstrated a close relationship between *P. maris* and *P. modestum* (Janssen and Liesack 1995). Both organisms formed a distinct lineage within a phylogenetically coherent group formed by *L. buccalis*, *S. termitidis*, and *Fusobacterium* species as well as *C. rectum* (Both et al. 1991; Janssen and Liesack 1995).

*Psychrilyobacter* represents a new genus in the phylum Fusobacteria (Zhao et al. 2009). *Ps. atlanticus* (strain HAW-EB21$^T$) shows optimal growth at 18.5 °C making it a psychrotrophic bacterium. Members of the genera *Ilyobacter*, *Propionigenium*, and *Fusobacterium* are all mesophilic bacteria with optimum growth temperatures of 28–37 °C. Phylogenetic analyses of the 16S rRNA gene sequences demonstrated the affiliation of *Ps. atlanticus* to the phylum Fusobacteria with 87–93 % similarity. Most closely related were the genera *Ilyobacter* and *Propionigenium* (92.5–93.4 % similarity). Also 23S rRNA sequencing showed a high dissimilarity of *Ps. atlanticus* to members of the genera *Propionigenium* and *Ilyobacter* (9.1–11.5 %) and from those of the genus *Fusobacterium* (14–16 %). The DNA G + C content of 28.1 mol% in *Ps. atlanticus* is lower than that in the genus *Ilyobacter* (31.7–36.7 mol%) and in *Propionigenium* (32.9–41 mol%). *Ps. atlanticus* strain HAW-EB21$^T$ also differed from species in the genera *Propionigenium* and *Ilyobacter* through its utilization of carbon sources. HAW-EB21$^T$ did not grow on succinate which is the carbon source for isolation of members of *Propionigenium* or on L-tartrate, 3-hydroxybutyrate, or quinate (and shikimate) used for isolation of *I. tartaricus*, *I. polytropus*, and *I. insuetus*, respectively. HAW-EB21$^T$ utilized glucose, fructose, and maltose which are not used by *I. insuetus* and most *Fusobacterium* species. The strain did not ferment pyruvate to propionate, unlike *P. modestum*. Neither did it ferment fructose and pyruvate to formate in contrast to *I. tartaricus* and *I. polytropus*. The major membrane fatty acid in HAW-EB21$^T$ is $C_{15-1}$ (30 %) which is absent from the cell membrane of *Fusobacterium*.

## Ecology

The mucosae of humans and animals are the most common colonization sites for fusobacteria (Gharbia et al. 2011). *F. nucleatum* and *F. periodonticum* (Slots et al. 1983) are usually found in the gingival sulcus/periodontal pocket of man, but not exclusively as also the gut can be a source (Strauss et al. 2008). Gharbia et al. (1990) suggested that *F. nucleatum* subsp. *nucleatum* is the most frequent fusobacterium in subgingival sites, while *F. nucleatum* from the gut tends to be similar with the *animalis* subspecies (Strauss et al. 2008). The vagina is the primary site for *F. gonidiaformans*. *F. mortiferum* and *F. varium* are usually located in the human gastrointestinal tract. *F. necrogenes* is rarely isolated from man and the habitat of *F. ulcerans* is unknown. *F. necrophorum* is a normal inhabitant of the gastrointestinal tract of cattle. Important human habitats are the respiratory tract and the oral cavity of man. *F. russii* is found in the normal canine and feline oral flora but has also been recovered from human feces. Isolation sites for *F. equinum* and *F. simiae* are the oral cavity of horses and of the stump-tailed macaque, respectively, while *F. canifelinum* has been isolated from bite wounds in humans inflicted by cats and dogs.

Of the species comprising the genus *Fusobacterium*, only *F. mortiferum* has the capacity to utilize an extraordinary wide variety of sugars (monosaccharides, disaccharides, and both α-and β-glycosides) as energy sources (Bouma et al. 1997). It has been speculated that the genetic information for transport and dissimilation of these carbohydrates comes from other bacteria or that other genes responsible for this have been deleted from other fusobacteria. Anyhow, such events may be responsible for the unique ecological niches habited by different fusobacteria.

*Cetobacterium* species are recovered from the mammalian intestinal tract and the oral cavity (Edwards et al. 2011). The organisms were first isolated from the intestinal content of a porpoise and from a mouth lesion of a minke whale (Foster et al. 1995). Finegold et al. (2003) isolated *C. somerae* from human feces, and Tsuchiya et al. (2008) defined intestinal tracts of freshwater fish as another ecological niche for *C. somerae*.

Anoxic marine sediments are the typical habitats of *Ilyobacter* (Schink et al. 2011). This also relates to species of *Propionigenium* (Schink and Janssen 2011).

Fusobacteria are associated with a wide spectrum of diseases. In children, e.g., fusobacteria have been recovered from abscesses, aspiration pneumonia, paronychia, bites, chronic otitis media, and osteomyelitis (Brook 1994). An association between *F. nucleatum* and otitis media was also reported by Könönen et al. (1999). Fusobacteria can adhere to and invade epithelial cells (HaCaT keratinocytes). They multiply intracellularly for hours and survive intracellularly under aerobic conditions. Newly formed actin filaments could be seen in epithelial cells associated with invasion (Gursoy et al. 2008). *F. nucleatum* and *F. nucleatum* subsp. *polymorphum* and -*vincentii* also entered and located in the cytoplasm of gingival fibroblasts and periodontal ligament fibroblasts in vitro (Dabija-Wolter et al. 2009). *F. nucleatum* subsp. *polymorphum* showed the greatest bacterial mass in fibroblasts. *F. nucleatum* is able to adhere to all oral microorganisms so far tested (Kolenbrander et al. 1989). Realization of its adhesive properties is important to comprehend its pathogenesis. A novel surface adhesin, FadA (Fusobacterium adhesin A) binds to the surface proteins of oral mucosal KB cells (Han et al. 2005). It is highly conserved among *F. nucleatum*, *F. periodonticum*, and *F. simiae* which are the most closely related oral species. In non-oral species, it is absent, including *F. gonidiaformans*, *F. mortiferum*, *F. russii*, *F. ulcerans*, and "*F. naviforme*." A double-cross over *fadA*-deficient deletion mutant that was constructed showed reduced binding to KB and Chinese hamster cells by 70–80 % compared to the wild type. This indicated that FadA has an important role in fusobacterial colonization of the host and due to its uniqueness in oral fusobacteria, may indicate translocation of fusobacteria from the oral cavity when present elsewhere in the body. Cell-to-cell contact may facilitate metabolic communication and enhance the proliferation of cells in severe stages of periodontitis. Also intergeneric coaggregation of oral *Treponema* species with *Fusobacterium* species and intrageneric coaggregation among *Fusobacterium* species have been demonstrated (Kolenbrander et al. 1995). Among 79 bacterial strains representing 16 genera, *Helicobacter pylori* (strains ATCC $43504^T$ and ATCC 43629) were tested for their ability to coaggregate with all these bacteria, of which only two were of nonhuman origin and most were oral (Andersen et al. 1998). There was strong coaggregation between the *Helicobacter* strains and four subspecies strains of *F. nucleatum* and of *F. periodonticum* (ATCC $33693^T$) which all were of human dental plaque origin. On the other hand, the helicobacters did not coaggregate with non-plaque isolates such as *F. mortiferum* ATCC $25557^T$ and *F. ulcerans* ATCC $49185^T$. This indicated that *H. pylori* has a potential for colonizing the oral cavity.

Fusobacteria produce an LPS with strong biologic activity, as well as an Lkt and hemolysin that cause destruction of white and red blood cells. Hemagglutinin production causes platelet aggregation and septic thrombus formation, as seen in Lemierre's syndrome (Shanghai and Kerstein 2001).

*F. nucleatum* isolated from healthy children exhibited a great portion of β-lactamase producers (Könönen et al. 1999) which may deteriorate treatment with β-lactam antibiotics.

Desvaux et al. (2005) gave new insight into protein secretion and virulence mechanisms of *F. nucleatum*. Using the genomic analyses of *F. nucleatum* subsp. *nucleatum* ATCC $25586^T$ and *F. nucleatum* subsp. *vincentii* ATCC $49256^T$, they confirmed absence of a twin-arginine translocation system, a type III secretion system, a type IV secretion system, and a chaperone/usher pathway. Their studies also indicated absence of a type I protein secretion system. This is in contrast to Gram-negative bacteria in general where six major protein secretion systems are recognized: types I, II, III, IV, and V and the chaperone/usher pathway. They also found a type 4 pilus locus and genes encoding type V secretion systems. The type V secretion pathway, representing the largest protein secretion family in Gram-negative bacteria, is believed to be the most important source of virulence factors.

The effect of *F. nucleatum* and *P. gingivalis* on the production of interleukin-1 (IL-1) and IL-inhibitors by human plastic-adherent mononuclear cells from normal donors were examined by Walsh et al. (1989). Unstimulated adherent cells secreted an IL-1 inhibitor spontaneously, while stimulation with *P. gingivalis* reduced synthesis and secretion of IL-1. Bacteria are obviously capable of modulating cytokine production by monocytes thereby altering the local immune response. Similar results were achieved with human polymorphonuclear neutrophils (Yamazaki et al. 1989).

Genome sequence analysis of *F. nucleatum* subspecies *nucleatum* ATCC $25586^T$ indicated that a principal source of energy was the fermentation of glutamate to butyrate (Kapatral et al. 2002). In addition desulfuration of cysteine and methionine yielded ammonia, $H_2S$, methyl mercaptan, and butyrate capable of arresting fibroblast growth. This may prevent wound healing and promote penetration of the gingival epithelium by *F. nucleatum*.

It has been claimed that the nature and magnitude of periodontitis are determined by the host's immune response. Chaushu et al. (2012) demonstrated in mice that the NK killer receptor NKp46 (NCR 1 in mice) binds directly to *F. nucleatum* and that this interaction affects the outcome of *F. nucleatum*-mediated periodontitis by causing fast and transient TNF-α secretion.

The host response to GroEL of *F. nucleatum* may be involved in atherosclerosis and supports the association between periodontitis and systemic disease. Thus, GroEL of *F. nucleatum* upregulated the expression of cytokines, cell adhesion molecules, and procoagulant factors in ApoE $^{-/-}$ mice (Lee et al. 2012). They also induced monocyte adhesion to and transmigration into endothelial cells together with increased uptake of lipids in atherosclerotic lesions.

An asymptomatic fusobacterium-induced dental abscess caused spinal abscess and mitral valve endocarditis (Goolamali et al. 2006) indicating that oral fusobacteria can spread through the blood. *F. nucleatum* was detected in all oral and seven chorionic tissue samples from 24 high-risk pregnant women (Tateishi et al. 2012). *F. nucleatum* LPS stimulated IL-6 and corticotrophin (CRH)-releasing hormone from chorion-derived cells. However, *F. nucleatum* LPS-induced IL-6 and CRH were significantly reduced in TLR-2 or TLR-4 gene-silenced chorion cells. Chorionic tissue from normal pregnant women did not contain *F. nucleatum*. This organism may have an influence on chorionic tissues of abnormal pregnancy outcomes. *F. nucleatum* has been reported to be the most frequently isolated species from amniotic fluid cultures in women with preterm labor and intact membranes (Hill 1998). *F. nucleatum* associated with vaginosis was clearly linked with preterm delivery (Holst et al. 1994). Further, *F. nucleatum* and *Mobiluncus curtisii* were found in amniotic fluid from women with preterm delivery. This *Fusobacterium* species and subspecies isolated from amniotic fluid often match those reported from healthy and diseased subgingival sites, i.e., *F. nucleatum* subsp. *vincentii* and *F. nucleatum* subsp. *nucleatum* (Hill 1998). Mikamo et al. (1998) reported that phospholipase A from *F. nucleatum* stimulates release of arachidonic acid from endometrial cells, which is accompanied by lysophospholipid formation. This increases production and release of prostaglandin $E_2$ which may start labor associated with intra-amniotic infection. These findings support further studies on the relationship between oral fusobacteria and preterm birth.

*F. equinum* is associated with necrotic infections of the respiratory tract in horses such as necrotizing pneumonia, pleuritis, and paraoral infections. The species is closely related to *F. necrophorum*. Using PCR amplification of the *F. equinum* LktA sequence, Zhou et al. (2009a, b) found *F. equinum* on footrot-infected hooves of sheep and cattle. They suggested that *F. equinum* can be involved in footrot infection and that it may occasionally have been mistyped as the closely related *F. necrophorum*. *F. equinum* has an LktA gene and its product is toxic to equine leukocytes (Tadepalli et al. 2008c). Lkt may be an important virulence factor in *F. equinum*.

*F. gonidiaformans* has been reported to cause peritonsillar infection, bacteremia, septicopyemia, and septic trochanteric bursitis (Rubinstein et al. 1974; Lefebvre et al. 1985; Gouby et al. 1987). It has been recovered from urogenital and intestinal tracts in humans and sometimes from infections in other sites (Citron 2002).

The non-oral *F. mortiferum* was found to stimulate IL-1 production by PMNs in vitro. None of periodontopathic bacteria (*F. nucleatum*, *P. gingivalis*, *Tannerella forsythia*, *Aggregatibacter actinomycetemcomitans*) did this although they had an IL-1 enhancing effect by themselves. However, they produced an IL-1 inhibitory fraction suggesting that these bacteria may evade the protective effects of PMNs and a regulatory role for PMNs in chronic periodontitis (Yamazaki et al. 1989).

LPS of *F. mortiferum* had the ability to activate Hageman factor (HF) when preincubated with purified HF factor, followed by addition of purified human prekallikrein (Bjornson 1984). LPS of this organism, as well as that from other selective Gram-negative organisms may initiate the intrinsic pathway of coagulation.

*F. mortiferum* exhibited the highest rate of β-lactamase production (76.9 %) among 129 fusobacteria from 28 US centers (Appelbaum et al. 1990). For comparison it was 64.7 % in *Bacteroides* species (not *B. fragilis*) and 41.1 % in fusobacteria as such. Although β-lactamase production is increasing in non-*B. fragilis Bacteroides* and fusobacteria, it was concluded that amoxicillin-clavulanate, ticarcillin, cefoxitin, imipenem, and metronidazole would still be suitable for treatment of infections caused by these bacteria.

At pH 5.2, *F. mortiferum* showed an intense phosphatase activity and could be differentiated from other *Fusobacterium* and *Bacteroides* species (Porschen and Spaulding 1974). There was an association between hydrolysis and several important pathogens.

*F. necrophorum* is often found in liver abscesses of cattle (Langworth 1977; Tadepalli et al. 2009). It is a normal inhabitant in the rumen but can be opportunistic and cause a variety of necrotic infections (necrobacillosis) in humans and animals. Of these bovine liver abscesses and footrot infection are particularly important to the cattle industry. In man *F. necrophorum* may cause liver and lung abscess, infection of the female genital tract, intra-abdominal infections, and skin infections. Half of the infections appear as Lemierre's syndrome of postanginal sepsis in young previously healthy persons (Citron 2002). It is frequently encountered in mixed infections. Therefore, synergisms between *F. necrophorum* and other pathogens may be important for an infection to develop (Tan et al. 1996). There is a strong association between *F. necrophorum* with clinically distinctive, severe septic infections known as necrobacillosis, postanginal sepsis, or Lemierre's syndrome (Riordan 2007).

Scanning electron microscopy showed that *F. necrophorum* subsp. *necrophorum* cells penetrated into murine and rabbit cheek cell membranes to which they had shown strong adherence. Pretreatment of the cells with hemagglutinin antiserum reduced the degree of attachment (Okada et al. 1999). The hemagglutinin is probably a cell surface component of *F. necrophorum* subsp. *necrophorum* (Kanoe et al. 1998). Further, *F. necrophorum* subsp. *necrophorum* cells may contribute to the degradation of cellular actin during the initial stage of the infection caused by this bacterium (Yamaguchi et al. 1999).

Both *F. necrophorum* subsp. *necrophorum* and *F. necrophorum* subsp. *funduliforme* produce Lkt. *F. necrophorum* subsp. *funduliforme* makes lower concentrations of Lkt, has lower DNAse and phosphatase activity, and is less virulent than *F. necrophorum* subsp. *necrophorum* in mice (Tan et al. 1994). The lower toxicity of *F. necrophorum* subsp. *funduliforme* may in part be due to difference in the LktA gene and reduced transcription (Tadepalli et al. 2008b). Also the hemolytic activity (horse erythrocytes) was higher in *F. necrophorum* subsp. *necrophorum* than in *F. nucleatum* subsp. *funduliforme* (Amoako et al. 1994). According to Tadepalli et al. (2009), hemagglutinin, endotoxin, and Lkt are major virulence factors contributing to colonization and invasion into the liver of cattle causing abscesses. Hemolysins, proteases, and adhesins may also be involved (Nagaraja et al. 2005). Also human *F. necrophorum* isolates have the LktA gene and leukotoxic activity (Tadepalli et al. 2008a). The Lkt is a high molecular weight protein, encoded by a tricistronic leukotoxic operon. The Lkt ORF consists of 9,726 bp and encodes a protein of 3,241 amino acids with an overall molecular weight of 335,956 and does not share sequence similarity with any other Lkt (Narayanan et al. 2001).

The lipid A ratio of *F. nucleatum* subsp. *necrophorum* and *F. nucleatum* subsp. *funduliforme* was 4:1 which was suggested as another reason for their different virulence (Garcia et al. 1999).

*F. necrophorum* subsp. *funduliforme* has factor H binding as a complement evasion mechanism. The binding of the C inhibitor factor H (fH) to evade C attack correlated with the severity of disease suggesting that it contributes to the virulence and survival of *F. necrophorum* subsp. *funduliforme* in the host, e.g., in Lemierre's syndrome (Friberg et al. 2008). Collagenolytic capacity has been demonstrated in the cell wall of *F. nucleatum* subsp. *necrophorum* (Okamoto et al. 2001, 2007) and the same bacterium caused degradation of bovine collagen type 1 in tissue-cultured bovine kidney cells, which may promote infection in vivo (Okada et al. 2000). Also rabbit tissue-cultured cells were affected by a collagenolytic cell wall component of *F. nucleatum* subsp. *necrophorum* (Okamoto et al. 2006). Another study demonstrated a dermotoxic effect of a collagenolytic cell wall component of *F. necrophorum* subsp. *necrophorum* probably contributing to the establishment of necrotic lesions during infection with this bacterium (Okamoto et al. 2005).

*F. russii*, found in the feline and canine oral microbiota, has been isolated from infected dog and cat bites in humans (Jang and Hirsch 1994; Talan et al. 1999).

*F. simiae*, first isolated from the monkey oral microbiota, is rarely associated with human infection (Citron 2002).

*F. ulcerans* has been recovered from tropical ulcers (Adriaans and Garelick 1989).

*C. somerae* was observed in stools from children with autism. The hypothesis of an interaction between the intestinal microbiota and the brain, based on possible interactions of children with autism, has been launched (Martirosian 2004). It may be that alterations in bacterial toxins and other metabolites are involved.

## Application

Deposited material in landfills is degraded under strict anaerobic conditions. Production of new biodegradable polymers should take this into consideration. However, the anaerobic degradation of hydroaromatic compounds by fermentative bacteria has been demonstrated only in the past decade. Several bacterial strains have been enriched and recovered from marine and freshwater sediments with quinic acid as the only source of carbon and energy (Brune et al. 2002). The marine strain Vent Chi2$^T$ and the freshwater strain GolChiI of *I. insuetus* were found to degrade hydroaromatic compounds via novel, fermentative pathways that do not include aromatic intermediates (Brune and Schink 1992). Poly-beta-hydroxybutyrate is depolymerized by the anaerobic bacterium "*I. delafieldii*" by means of an extracellular hydrolase (Schink et al. 1992). Classical beta-oxidation is used by the cells to degrade monomers. Synthetic biodegradable polymers should be designed in the future to contain linkages that can be cleaved by extracellular hydrolytic enzymes.

Changes in the total bacterial community of live Pacific oysters were related to storage temperature with *Psychrilyobacter* dominating at 4 °C (43.8 % of clones) (Fernandez-Piquer et al. 2012). *Psychrilyobacter* was one of three groups of bacteria considered as possible indicators for postharvest temperature control of oysters.

## Acknowledgement

The author wants to thank Raúl Munoz for preparing the phylogenetic trees.

## References

Adriaans B, Garelick H (1989) Cytotoxicity of *Fusobacterium ulcerans*. J Med Microbiol 29:177–180

Amoako KK, Goto Y, Shinjo T (1993) Comparison of extracellular enzymes of *Fusobacterium necrophorum* subsp. *necrophorum* and *Fusobacterium necrophorum* subsp. *funduliforme*. J Clin Microbiol 31:2244–2247

Amoako KK, Goto Y, Shinjo T (1994) Studies on the factors affecting the hemolytic activity of *Fusobacterium necrophorum*. Vet Microbiol 41:11–18

Andersen RN, Ganeshkumar N, Kolenbrander PE (1998) *Helicobacter pylori* adheres selectively to *Fusobacterium* spp. Oral Microbiol Immunol 13:51–54

Andrews DMA, Gharbia SE, Shah HN (1997) Characterization of a novel bacteriophage in *Fusobacterium varium*. Clin Infect Dis 25(Suppl 2):S287–S288

Appelbaum PC, Spangler SK, Jacobs MR (1990) Beta-lactamase production and susceptibilities to amoxicillin, amoxicillin-clavulanate, cefoxitin, imipenem, and metronidazole of 320 non-Bacteroides fragilis Bacteroides isolates and 129 fusobacteria from 28 U.S. centers. Antimicrob Agents Chemother 34:1546–1550

Bachrach G, Haake SK, Glick A, Hazan R, Naor R, Andersen RN, Kolenbrander PE (2004) Characterization of the novel *Fusobacterium nucleatum* plasmid pKH9 and evidence of an addiction system. Appl Environ Microbiol 70:6957–6962

Bailey GD, Love DN (1993) *Fusobacterium necrophorum* is a synonym of *Fusobacterium varium*. Int J Syst Bacteriol 43:819–821

Bank S, Nielsen HM, Mathiasen BH, Leth DC, Hagelskjaer Kristensen L, Prag J (2010) *Fusobacterium necrophorum* – detection and identification on a selective agar. APMIS 118:994–999

Beighton D, Homer KA, De Graaff J (1997) Endopeptidase activities of selected *Porphyromonas* spp., *Prevotella* spp. and *Fusobacterium* spp. of oral and non-oral origin. Archs Oral Biol 42:827–834

Bjornson HS (1984) Activation of Hageman factor by lipopolysaccharides of *Bacteroides fragilis*, *Bacteroides vulgatus*, and *Fusobacterium mortiferum*. Rev Infect Dis 6(Suppl 1):S30–S33

Bolstad AI, Jensen HB (1993) Polymerase chain reaction-amplified nonradioactive probes for identification of *Fusobacterium nucleatum*. J Clin Microbiol 31:528–532

Bolstad AI, Høgh BT, Jensen HB (1995) Molecular characterization of a 40-kDA outer membrane protein, FomA, of *Fusobacterium periodonticum* and comparison with *Fusobacterium nucleatum*. Oral Microbiol Immunol 10:257–264

Bolstad AI, Jensen HB, Bakken V (1996) Taxonomy, biology, and periodontal aspects of *Fusobacterium nucleatum*. Clin Microbiol Rev 9:55–71

Both B, Kaim G, Wolters J, Schleifer KH, Stackebrandt E, Ludwig W (1991) Propionigenium modestum: a separate line of descent within the eubacteria. FEMS Microbiol Lett 62:53–58

Bouma CL, Reizer J, Reizer A, Robrish SA, Thompson J (1997) 6-Phospho-α-D-glucosidase from *Fusobacterium mortiferum*: cloning, expression, and assignment to family 4 of the glycosylhydrolases. J Bacteriol 179:4129–4137

Brazier JS, Citron DM, Goldstein EJ (1991) A selective medium for Fusobacterium spp. J Appl Bacteriol 71:343–346

Brook I (1994) Fusobacterial infections in children. J Infect 28:155–165

Brune A, Schink B (1992) Anaerobic degradation of hydroaromatic compounds by newly isolated fermenting bacteria. Arch Microbiol 158:320–327

Brune A, Evers S, Kaim G, Ludwig W, Schink B (2002) *Ilyobacter insuetus* sp. nov., a fermentative bacterium specialized in the degradation of hydroaromatic compounds. Int J Syst Evol Microbiol 52:429–432

Calhoon DA, Mayberry WR, Slots J (1983) Cellular fatty acids and soluble protein profiles of oral fusobacteria. J Dent Res 62:1181–1185

Chaushu S, Wilensky A, Gur C, Shapira L, Elboim M, Halftek G, Polak D, Achdout H, Bachrach G, Mandelboim O (2012) Direct recognition of *Fusobacterium nucleatum* by the NK cell natural cytotoxicity receptor NKp46 aggravates periodontal disease. PLoS Pathog 8:e1002601. doi:10.1371/journal. ppat. 1002601

Citron DM (2002) Update on the taxonomy and clinical aspects of the genus *Fusobacterium*. Clin Infect Dis 35(Suppl 1):S22–S27

Conrads G, Claros MC, Citron DM, Tyrrell KL, Merriam V, Goldstein EJ (2002) 16S-23S rDNA internal transcribed spacer sequences for analysis of the phylogenetic relationships among species of the genus *Fusobacterium*. Int J Syst Evol Microbiol 52:493–499

Dabija-Wolter G, Cimpan MR, Costea DE, Johannessen AC, Sørnes S, Neppelberg E, Al-Haroni M, Skaug N, Bakken V (2009) *Fusobacterium nucleatum* enters normal human oral fibroblasts in vitro. J Periodontol 80:1174–1183

Desvaux M, Khan A, Beatson SA, Scott-Tucker A, Henderson IR (2005) Protein secretion systems in *Fusobacterium nucleatum*: genomic identification of Type 4 piliation and complete Type V pathways brings new insight into mechanisms of pathogenesis. Biochim Biophys Acta 1713:92–112

Duncan SH, Hold GL, Harmsen HJ, Stewart CS, Flint HJ (2002) Growth requirements and fermentation products of *Fusobacterium prausnitzii*, and a proposal to reclassify it as *Faecalibacterium prausnitzii* gen. nov., comb. nov. Int J Syst Evol Microbiol 52:2141–2146

Dzink JL, Sheenan MT, Socransky SS (1990) Proposal of three subspecies of *Fusobacterium nucleatum* Knorr 1922: *Fusobacterium nucleatum* subsp. *nucleatum* subsp. nov., comb. nov.; *Fusobacterium nucleatum* subsp. *polymorphum* subsp. nov., nom. rev., comb. nov.; and *Fusobacterium nucleatum* subsp. *vincentii* subsp. nov., nom. rev., comb. nov. Int J Syst Bacteriol 40:74–78

Edberg SC, Bell SR (1985) Lack of constitutive beta-glucosidase (esculinase) in the genus *Fusobacterium*. J Clin Microbiol 22:435–437

Edwards KJ, Logan JMJ, Gharbia SE (2011) Genus II *Cetobacterium* Foster, Ross, Naylor, Collins, Ramos, Fernández-Garayzábal and Reid 1996, 362$^{VP}$. In: Krieg NR, Staley JT, Brown DR, Hedlund BP, Paster BJ, Ward NL, Ludwig W, Whitman WB (eds) Bergey's manual of systematic bacteriology, vol 4, 2nd edn, The *Bacteroidetes, Spirochaetes, Tenericutes (Mollicutes), Acidobacteria, Fibrobacteres, Fusobacteria, Dictyoglomi, Gemmatimonadetes, Lentisphaerae, Verrucomicrobia, Chlamydiae*, and *Planctomycetes*. Springer, New York, pp 758–759

Fernandez-Piquer J, Bowman JP, Ross T, Tamplin ML (2012) Molecular analysis of the bacterial communities in the live Pacific oyster (*Crassostrea gigas*) and the influence of postharvest temperature on its structure. J Appl Microbiol 112:1134–1143

Finegold SM, Vaisanen ML, Molitoris DR, Tomzynski TJ, Song Y, Liu C, Collins MD, Lawson PA (2003) *Cetobacterium somerae* sp. nov. from human feces and emended description of the genus *Cetobacterium*. Syst Appl Microbiol 26:177–181

Foster G, Ross HM, Naylor RD, Collins MD, Ramos CP, Fernandez Garayzabal F, Reid RJ (1995) *Cetobacterium ceti* gen. nov., sp. nov., a new Gram-negative obligate anaerobe from sea mammals. Lett Appl Microbiol 21:202–206

Fredriksen G, Hofstad T (1978) Chemotypes of *Fusobacterium nucleatum* lipopolysaccharides. Acta Pathol Microbiol Scand Sect B 86:41–45

Friberg N, Carlson P, Kentala E, Mattila PS, Kuusela P, Meri S, Jarva H (2008) Factor H binding as a complement evasion mechanism for an anaerobic pathogen, *Fusobacterium necrophorum*. J Immunol 181:8624–8632

Garcia GG, Amoako KK, Xu DL, Inoue T, Goto Y, Shinjo T (1999) Chemical composition of endotoxins produced by *Fusobacterium necrophorum* subsp. *necrophorum* and *F. necrophorum* subsp. *funduliforme*. Microbios 100:175–179

Garrity GM, Bell JA, Lilburn T (2005) Appendix 2. Taxonomic outline of *Archaea* and *Bacteria*. In: Brenner DJ, Krieg NR, Staley JT (eds) Bergey's manual of systematic bacteriology, vol 2, 2nd edn. The *Proteobacteria*. Part A Introductory essays. Springer, New York, pp 207–220

George KS, Reynolds MA, Falkler WA Jr (1997) Arbitrarily primed polymerase chain reaction fingerprinting and clonal analysis of oral *Fusobacterium nucleatum* isolates. Oral Microbiol Immunol 12:219–226

Gharbia SE, Shah HN (1989) Glutamate dehydrogenase and 2-oxoglutarate reductase electrophoretic patterns and deoxyribonucleic acid-deoxyribonucleic acid hybridization among human oral isolates of *Fusobacterium nucleatum*. Int J Syst Bacteriol 39:467–470

Gharbia SE, Shah HN (1990a) Heterogeneity within *Fusobacterium nucleatum*, proposal of four subspecies. Lett Appl Microbiol 10:105–108

Gharbia SE, Shah HN (1990b) Identification of *Fusobacterium* species by the electrophoretic migration of glutamate dehydrogenase and 2-oxoglutarate reductase in relation to their DNA base composition and peptidoglycan dibasic amino acids. J Med Microbiol 33:183–188

Gharbia SE, Shah HN (1992) *Fusobacterium nucleatum* subsp. *fusiforme* subsp. nov. and *Fusobacterium nucleatum* subsp. *animalis* subsp. nov. as additional subspecies within *Fusobacterium nucleatum*. Int J Syst Bacteriol 42:296–298

Gharbia SE, Shah HN, Lawson PA, Haapasalo M (1990) Distribution and frequency of *Fusobacterium nucleatum* subspecies in the human oral cavity. Oral Microbiol Immunol 5:324–327

Gharbia SE, Shah HN, Edwards KJ (2011) Genus I. *Fusobacterium* Knorr 1922, 4$^{AL}$. In: Krieg NR, Staley JT, Brown DR, Hedlund BP, Paster BJ, Ward NL, Ludwig W, Whitman WB (eds) Bergey's manual of systematic bacteriology, vol 4, 2nd edn, The *Bacteroidetes, Spirochaetes, Tenericutes (Mollicutes), Acidobacteria, Fibrobacteres, Fusobacteria, Dictyoglomi, Gemmatimonadetes, Lentisphaerae, Verrucomicrobia, Chlamydiae*, and *Planctomycetes*. Springer, New York, pp 748–758

Goolamali SJ, Carulli MT, Davies UM (2006) Spinal abscess and mitral valve endocarditis secondary to asymptomatic fusobacterium-induced dental abscess. J R Soc Med 99:368–369

Gouby A, Dubois A, Bouziges N, Saissi G, Ramuz M (1987) Septicopyemia caused by *Fusobacterium gonidiaformans*. Presse Med 16:34.(Article in French)

Gursoy UK, Könönen E, Uitto VJ (2008) Intracellular replication of fusobacteria requires new actin filament formation of epithelial cells. APMIS 116:1063–1070

Haake SK, Yoder SC, Attarian G, Podkaminer K (2000) Native plasmids of *Fusobacterium nucleatum*: characterization and use in development of genetic systems. J Bacteriol 182:1176–1180

Han YW, Ikegami A, Rajanna C, Kawsar HI, Zhou Y, Li M, Sojar HT, Genco RJ, Kuramitsu HK, Deng CX (2005) Identification and characterization of a novel adhesin unique to oral fusobacteria. J Bacteriol 187:5330–5340

Hase S, Hofstad T, Rietschel ET (1977) Chemical structure of the lipid A component of lipopolysaccharides from *Fusobacterium nucleatum*. J Bacteriol 129:9–14

Hermansson K, Perry MB, Altman E, Brisson JR, Garcia MM (1993) Structural studies of the O-antigenic polysaccharide of *Fusobacterium necrophorum*. Eur J Biochem 212:801–809

Hill GB (1998) Preterm birth: associations with genital and possibly oral microflora. Ann Periodontol 3:222–232

Hodgson AL, Nicholson LA, Doran TJ, Corner LA (1993) Restriction fragment length polymorphism analysis of *Fusobacterium necrophorum* using a novel repeat DNA sequence and 16S rRNA gene probe. FEMS Microbiol Lett 107:205–210

Hofstad T, Olsen I (2005) *Fusobacterium* and *Leptotrichia*. In: Borriello SP, Murray PR, Funke G (eds) Topley & Wilson's microbiology & microbial infections, vol 2, 10th edn, Bacteriology. Hodder Arnold, ASM Press, Washington, DC, pp 1945–1956

Holdeman LV, Cato EP, Moore WEC (1977) Anaerobe laboratory manual, 4th edn. Southern Printing, Blacksburg, pp 23–28

Holst E, Goffeng AR, Andersch B (1994) Bacterial vaginosis and vaginal microorganisms in idiopathic premature labor and association with pregnancy outcome. J Clin Microbiol 32:176–186

Jalava J, Eerola E (1999) Phylogenetic analysis of *Fusobacterium alocis* and *Fusobacterium sulci* based on 16S rRNA gene sequences: proposal of *Filifactor alocis* (Cato, Moore and Moore) comb. nov. and *Eubacterium sulci* (Cato, Moore and Moore) comb. nov. Int J Syst Bacteriol 49:1375–1379

Jang SS, Hirsch DC (1994) Characterization, distribution and microbiological associations of *Fusobacterium* spp. in clinical specimens of animal origin. J Clin Microbiol 32:384–387

Janssen PH, Harfoot CG (1990) *Ilyobacter delafieldii* sp. nov., a metabolically restricted anaerobic bacterium fermenting PHB. Arch Microbiol 154:253–259

Janssen PH, Liesack W (1995) Succinate decarboxylation by *Propionigenium maris* sp. nov., a new anaerobic bacterium from an estuarine sediment. Arch Microbiol 164:29–35

Jantzen E, Hofstad T (1981) Fatty acids of *Fusobacterium* species: taxonomic implications. J Gen Microbiol 123:163–171

Jenkins SA, Drucker DB, Hillier VF, Ganguli LA (1992) Numerical taxonomy of Bacteroides and other genera of Gram-negative anaerobic rods. Microbios 69:139–154

Jensen A, Hagelskjaer Kristensen L, Nielsen H, Prag J (2008) Minimum requirements for a rapid and reliable routine identification and antibiogram of *Fusobacterium necrophorum*. Eur J Clin Microbiol Infect Dis 27:557–563

Jin J, Xu D, Narongwanichgarn W, Goto Y, Haga T, Shinjo T (2002) Characterization of the 16S-23S rRNA intergenic spacer regions among strains of the *Fusobacterium necrophorum* cluster. J Vet Med Sci 64:273–276

Jin J, Haga T, Shinjo T, Goto Y (2004) Phylogenetic analysis of *Fusobacterium necrophorum, Fusobacterium varium* and *Fusobacterium nucleatum* based on *gyrB* gene sequences. J Vet Med Sci 66:1243–1245

Jousimies-Somer HR (1997) Recently described clinically important anaerobic bacteria: taxonomic aspect and update. Clin Infect Dis 25:S78–S87

Jousimies-Somer H, Summanen P (2002) Recent taxonomic changes and terminology update of clinically significant anaerobic gram-negative bacteria (excluding spirochetes). Clin Infect Dis 35(Suppl 1):S17–S21

Kanoe M, Koyanagi Y, Kondo C, Mamba K, Makita T, Kai K (1998) Location of haemagglutinin in bacterial cells of *Fusobacterium nucleatum* subsp. *necrophorum*. Microbios 96:33–38

Kapatral V, Anderson I, Ivanova N, Reznik G, Los T, Lykidis A, Bhattacharyya A, Bartman A, Gardner W, Grechkin G, Zhu L, Vasieva O, Chu L, Kogan Y, Chaga O, Goltsman E, Bernal A, Larsen N, D'Souza M, Walunas T, Pusch G, Haselkorn R, Fonstein M, Kyrpides N, Overbeek R (2002) Genome sequence and analysis of the oral bacterium *Fusobacterium nucleatum* strain ATCC 25586. J Bacteriol 184:2005–2018

Kapatral V, Ivanova N, Anderson I, Reznik G, Bhattacharyya A, Gardner WL, Mikhailova N, Lapidus A, Larsen N, D'Souza M, Walunas T, Haselkorn R, Overbeek R, Kyrpides N (2003) Genome analysis of *F. nucleatum sub spp. vincentii* and its comparison with the genome of *F. nucleatum* ATCC 25586. Genome Res 13:1180–1189

Karpathy SE, Qin X, Giola J, Jiang H, Liu Y, Petrosino JF, Yerrapragada S, Fox GE, Haake SK, Weinstock GM, Highlander SK (2007) Genome sequence of *Fusobacterium nucleatum* subspecies *polymorphum* – a genetically tractable fusobacterium. PLoS One 8:e659. doi:10.1371/journal.pone. 0000659

Kim H-S, Lee D-S, Chang Y-H, Kim MJ, Koh S, Kim J, Seong J-H, Song SK, Shin HS, Son J-B, Jung MJ, Park S-N, Yoo SY, Cho KW, Kim D-K, Moon S, Kim D, Choi Y, Kim B-O, Jang H-S, Kim CS, Kim C, Choe S-J, Kook J-K (2010) Application of *rpoB* and zinc protease gene for use in molecular discrimination of *Fusobacterium nucleatum* subspecies. J Clin Microbiol 48:545–553

Kolenbrander PE, Andersen RN, Moore LVH (1989) Coaggregation of *Fusobacterium nucleatum, Selenomonas fluggei, Selenomonas infelix, Selenomonas noxia*, and *Selenomonas sputigena* with strains from 11 genera of oral bacteria. Infect Immun 57:3194–3203

Kolenbrander PE, Parrish KD, Andersen RN, Greenberg EP (1995) Intergeneric coaggregation of oral *Treponema* spp. with *Fusobacterium* spp. and intrageneric coaggregation among *Fusobacterium* spp. Infect Immun 63:4584–4588

Könönen E, Kanervo A, Salminen K, Jousimies-Somer H (1999) β-Lactamase production and antimicrobial susceptibility of oral heterogenous *Fusobacterium nucleatum* populations in young children. Antimicrob Agents Chemother 43:1270–1273

Langworth BF (1977) *Fusobacterium necrophorum*: its characteristics and role as an animal pathogen. Bacteriol Rev 41:373–390

Lawson FA, Gharbia SE, Shah HN, Clark DR (1989) Recognition of *Fusobacterium nucleatum* subgroups Fn-1, Fn-2 and Fn-3 by ribosomal RNA gene restriction patterns. FEMS Microbiol Lett 53:41–45

Lawson PA, Gharbia SE, Shah HN, Clark DR, Collins MD (1991) Intrageneric relationships of members of the genus *Fusobacterium* as determined by reverse transcriptase sequencing of small-subunit rRNA. Int J Syst Bacteriol 41:347–354

Lee H-R, Jun H-K, Kim H-D, Lee S-H, Choi B-K (2012) *Fusobacterium nucleatum* GroEL induces risk factors of atherosclerosis in human microvascular endothelial cells and ApoE$^{-/-}$ mice. Mol Oral Microbiol 27:109–123

Lefebvre C, Lambert M, Bastien P, Wauters G, Nagant de Deuxchaisnes C (1985) Septic trochanteric bursitis caused by *Fusobacterium gonidiaformans*. J Rheumatol 12:391–392

Liu D, Dong X (2011) Chapter 48. *Fusobacterium*. In: Liu D (ed) Molecular detection of human bacterial pathogens. CRC Press, Boca Raton, pp 543–553

Ludwig W, Strunk O, Klugbauer S, Klugbauer N, Weizenegger M, Neumaier J, Bachleitner M, Schleifer KH (1998) Bacterial phylogeny based on comparative sequence analysis. Electrophoresis 19:554–568

Ludwig W, Euzéby J, Whitman W (2011) Road map of the phyla *Bacteroidetes, Spirochaetes, Tenericutes* (*Mollicutes*), *Acidobacteria, Fibrobacteres, Fusobacteria, Dictyoglomi, Gemmatimonadetes, Lentisphaerae, Verrucomicrobia, Chlamydiae*, and *Planctomycetes*. In: Krieg NR, Staley JT, Brown DR, Hedlund BP, Paster BJ, Ward NL, Ludwig W, Whitman WB (eds) Bergey's manual of systematic bacteriology, vol 4, 2nd edn, The *Bacteroidetes, Spirochaetes, Tenericutes* (*Mollicutes*), *Acidobacteria, Fibrobacteres, Fusobacteria, Dictyoglomi, Gemmatimonadetes, Lentisphaerae, Verrucomicrobia, Chlamydiae*, and *Planctomycetes*. Springer, New York, pp 1–19

Machuca P, Daille L, Vinés E, Berrocal L, Bittner M (2010) Isolation of novel bacteriophage specific for the periodontal pathogen *Fusobacterium nucleatum*. Appl Environ Microbiol 76:7243–7250

Magee JT, Hindmarch JM, Bennett KW, Duerden BI, Aries RE (1989) A pyrolysis mass spectrometry study of fusobacteria. J Med Microbiol 28:227–236

Martirosian G (2004) Anaerobic intestinal microflora in pathogenesis of autism? Postepy Hig Med Dosw 58:349–351.(Article in Polish)

McKay TL, Ko J, Bilalis Y, DiRienzo JM (1995) Mobile genetic elements of *Fusobacterium nucleatum*. Plasmid 33:15–25

Meier T, von Ballmoos C, Neumann S, Kaim G (2003) Complete DNA sequence of the *atp* operon of the sodium-dependent $F_1F_0$ ATP synthase from *Ilyobacter tartaricus* and identification of the encoded subunits. Biochim Biophys Acta 1625:221–226

Meisel-Mikolajczyk F, Dobrowolska T (1974) Comparative immunochemical studies on the endotoxins of two *Fusobacterium necrophorum* strains. Bull Acad Pol Sci Ser Sci Biol 22:555–562

Mikamo H, Kawazoe K, Sato Y, Imai A, Tamaya T (1998) Preterm labor and bacterial intraamniotic infection: arachidonic acid liberation by phospholipase $A_2$ of *Fusobacterium nucleatum*. Am J Obstet Gynecol 179:1579–1582

Mira A, Pushker R, Legault BA, Moreira D, Rodriguez-Valera F (2004) Evolutionary relationships of *Fusobacterium nucleatum* based on phylogenetic analysis and comparative genomics. BMC Evol Biol 4:50. doi:10.1186/1471-2148-4-50

Nagaraja TG, Narayanan SK, Stewart GC, Chengappa MM (2005) *Fusobacterium necrophorum* infections in animals: pathogenesis and pathogenic mechanisms. Anaerobe 11:239–246

Narayanan S, Nagaraja TG, Okwumabua O, Staats J, Chengappa MM, Oberst RD (1997) Ribotyping to compare *Fusobacterium necrophorum* isolates from bovine liver abscesses, ruminal walls, and ruminal contents. Appl Environ Microbiol 63:4671–4678

Narayanan SK, Nagaraja TG, Chengappa MM, Stewart GC (2001) Cloning, sequencing, and expression of the leukotoxin gene from *Fusobacterium necrophorum*. Infect Immun 69:5447–5455

Narongwanichgarn W, Kawaguchi E, Misawa N, Goto Y, Haga T (2001) Differentiation of *Fusobacterium necrophorum* subspecies from bovine pathological lesions by RAPD-PCR. Vet Microbiol 1:383–388

Narongwanichgarn W, Misawa N, Jin JH, Amoako KK, Kawaguchi E, Shinjo T, Haga T, Goto Y (2003) Specific detection and differentiation of two subspecies of *Fusobacterium necrophorum* by PCR. Vet Microbiol 91:183–195

Neumann S, Matthey U, Kaim G, Dimroth P (1998) Purification and properties of the $F_1F_0$ ATPase of *Ilyobacter tartaricus*, a sodium ion pump. J Bacteriol 180:3312–3316

Nicholson LA, Morrow CJ, Corner LA, Hodgson ALM (1994) Phylogenetic relationship of *Fusobacterium necrophorum* A, AB and B biotypes based upon 16S rRNA gene sequence analysis. Int J Syst Bacteriol 44:315–319

Okada Y, Kanoe M, Yaguchi Y, Watanabe T, Ohmi H, Okamoto K (1999) Adherence of *Fusobacterium necrophorum* subspecies *necrophorum* to different animal cells. Microbios 99:95–104

Okada Y, Kanoe M, Okamoto K, Sakamoto K, Yaguchi Y, Watanabe T (2000) Effects of *Fusobacterium necrophorum* subspecies *necrophorum* on extracellular matrix of tissue-cultured bovine kidney cells. Microbios 101:147–156

Okamoto K, Kanoe K, Watanabe T (2001) Collagenolytic activity of a cell wall preparation from *Fusobacterium necrophorum* subsp. *necrophorum*. Microbios 106(Suppl 2):89–95

Okamoto K, Kanoe M, Inoue M, Watanabe T, Inoue T (2005) Dermotoxic activity of a collagenolytic cell wall component from *Fusobacterium necrophorum* subsp. *necrophorum*. Vet J 169:308–310

Okamoto K, Kanoe M, Yaguchi Y, Inoue T, Watanabe T (2006) Effects of a collagenolytic cell wall component from *Fusobacterium necrophorum* subsp. *necrophorum* on rabbit tissue-cultured cells. Vet J 171:380–382

Okamoto K, Kanoe M, Yaguchi Y, Watanabe T, Inoue T (2007) Effects of the collagenolytic cell wall component of *Fusobacterium necrophorum* subsp. *necrophorum* on bovine hepatocytes. Res Vet Sci 82:166–168

Okwumabua O, Tan Z, Staats J, Oberst RD, Chengappa MM, Nagaraja TG (1996) Ribotyping to differentiate *Fusobacterium necrophorum* subsp. *necrophorum* and *F. necrophorum* subsp. *funduliforme* isolated from bovine ruminal contents and liver abscesses. Appl Envron Microbiol 62:469–472

Park SN, Park JY, Kook JK (2010) Development of species-specific polymerase chain reaction primers for detection of *Fusobacterium periodonticum*. Microbiol Immunol 54:750–753

Porschen RK, Spaulding EH (1974) Phosphatase activity of anaerobic organisms. Appl Microbiol 27:744–747

Potts TV, Holdeman LV, Slots J (1983) Relationship among the oral fusobacteria assessed by DNA-DNA hybridization. J Dent Res 62:702–705

Riordan T (2007) Human infection with *Fusobacterium necrophorum* (necrobacillosis), with a focus on Lemierre's syndrome. Clin Microbiol Rev 20:622–659

Rogers AH (1998) Studies on fusobacteria associated with periodontal diseases. Aust Dent J 43:105–109

Rubinstein E, Onderdonk AB, Rahal JJ Jr (1974) Peritonsillar infection and bacteremia caused by *Fusobacterium gonidiaformans*. J Pediatr 85:673–675

Schink B (1984) Fermentation of tartrate enantiomers by anaerobic bacteria, and description of two new species of strict anaerobes, *Ruminococcus pasteurii* and *Ilyobacter tartaricus*. Arch Microbiol 139:409–414

Schink B, Janssen PH (2011) Genus IV. *Propionigenium* Schink and Pfennig 1983, $896^{VP}$ (Effective publication Schink and Pfennig 1982, 215). In: Krieg NR, Staley JT, Brown DR, Hedlund BP, Paster BJ, Ward NL, Ludwig W, Whitman WB (eds) Bergey's manual of systematic bacteriology, vol 4, 2nd edn. The *Bacteroidetes, Spirochaetes, Tenericutes (Mollicutes), Acidobacteria, Fibrobacteres, Fusobacteria, Dictyoglomi, Gemmatimonadetes, Lentisphaerae, Verrucomicrobia, Chlamydiae*, and *Planctomycetes*. Springer, New York, pp 761–765

Schink B, Pfennig N (1982) *Propionigenium modestum* gen. nov. sp. nov., a new strictly anaerobic nonsporeforming bacterium growing on succinate. Arch Microbiol 133:209–216

Schink B, Janssen PH, Frings J (1992) Microbial degradation of natural and synthetic polymers. FEMS Microbiol Rev 9:311–316

Schink B, Janssen PH, Brune A (2011) Genus III. *Ilyobacter* Stieb and Schink 1985, $375^{VP}$ (Effective publication: Stieb and Schink 1984, 145). In: Krieg NR, Staley JT, Brown DR, Hedlund BP, Paster BJ, Ward NL, Ludwig W, Whitman WB (eds) Bergey's manual of systematic bacteriology, vol 4, 2nd edn. The *Bacteroidetes, Spirochaetes, Tenericutes (Mollicutes), Acidobacteria, Fibrobacteres, Fusobacteria, Dictyoglomi, Gemmatimonadetes, Lentisphaerae, Verrucomicrobia, Chlamydiae*, and *Planctomycetes*. Springer, New York, pp 759–761

Shah HN, Olsen I, Bernard K, Finegold SM, Gharbia S, Gupta RS (2009) Approaches to the study of the systematics of anaerobic, Gram-negative, non-sporeforming rods: current status and perspectives. Anaerobe 15:179–194

Shanghai A, Kerstein M (2001) Lemierre's syndrome. South Med J 94:886–887

Shin HS, Kim M-J, Kim H-S, Park S-N, Kim DK, Baek D-H, Kim C, Kook J-K (2010) Development of strain-specific PCR primers for the identification of *Fusobacterium nucleatum* subsp. *fusiforme* ATCC $51190^T$ and subsp. *vincentii* ATCC $49256^T$. Anaerobe 16:43–46

Shinjo T, Fujisawa T, Mitsuoka T (1991) Proposal of two subspecies of *Fusobacterium necrophorum* (Flügge) Moore and Holdeman: *Fusobacterium necrophorum* subsp. *necrophorum* subsp. nov., nom. rev. (ex Flügge 1886), and *Fusobacterium necrophorum* subsp. *funduliforme* subsp. nov., nom. rev. (ex Hallé 1898). Int J Syst Bacteriol 41:395–397

Sikorski J, Chertkov O, Lapidus A, Nolan M, Lucas S, Del Rio TG, Tice H, Cheng J-F, Tapia R, Han C, Goodwin L, Pitluck S, Liolios K, Ivanova N, Mavromatis K, Mikhailova N, Pati A, Chen A, Palaniappan K, Land M, Hauser L, Chang Y-J, Jeffries CD, Brambilla E, Yasawong M, Rohde M, Pukall R, Spring S, Göker M, Woyke T, Bristow J, Eisen JA, Markowitz V, Hugenholtz P, Kyrpides NC, Klenk H-P (2010) Complete genome sequence of *Ilyobacter polytropus* type strain (CuHbu1). Stand Genom Sci 3:304–314

Slots J, Potts TV, Mashimo PA (1983) *Fusobacterium periodonticum*, a new species from the human oral cavity. J Dent Res 62:960–963

Smith GR, Thornton EA (1993) Pathogenicity of *Fusobacterium necrophorum* strains from man and animals. Epidemiol Infect 110:499–506

Spaulding EH, Rettger LF (1937) The Fusobacterium genus: I. Biochemical and serological classification. J Bacteriol 34:535–563

Staley JT, Whitman WB (2011) Family I. *Fusobacteriaceae* fam. nov. In: Krieg NR, Staley JT, Brown DR, Hedlund BP, Paster BJ, Ward NL, Ludwig W, Whitman WB (eds) Bergey's manual of systematic bacteriology, vol 4, 2nd edn, The *Bacteroidetes, Spirochaetes, Tenericutes (Mollicutes), Acidobacteria, Fibrobacteres, Fusobacteria, Dictyoglomi, Gemmatimonadetes, Lentisphaerae, Verrucomicrobia, Chlamydiae*, and *Planctomycetes*. Springer, New York, p 748

Stieb M, Schink B (1984) A new 3-hydroxybutyrate fermenting anaerobe, *Ilyobacter polytropus*, gen. nov. sp. nov., possessing various fermentation pathways. Arch Microbiol 140:139–146

Strauss J, White A, Ambrose C, McDonald J, Allen-Vercoe E (2008) Phenotypic and genotypic analyses of clinical *Fusobacterium nucleatum* and *Fusobacterium periodonticum* isolates from the human gut. Anaerobe 14:301–309

Sutter VL, Sugihara PT, Finegold SM (1971) Rifampicin-blood-agar as a selective medium for the isolation of certain anaerobic bacteria. Appl Microbiol 22:777–780

Tadepalli S, Stewart GC, Nagaraja TG, Narayanan SK (2008a) Human *Fusobacterium necrophorum* strains have a leukotoxin gene and exhibit leukotoxic activity. J Med Microbiol 57:225–231

Tadepalli S, Stewart GC, Nagaraja TG, Narayanan SK (2008b) Leukotoxin operon and differential expressions of the leukotoxin gene in bovine *Fusobacterium necrophorum* subspecies. Anaerobe 14:13–18

Tadepalli S, Stewart GC, Nagaraja TG, Jang SS, Narayanan SK (2008c) *Fusobacterium equinum* possesses a leukotoxin gene and exhibits leukotoxin activity. Vet Microbiol 127:89–96

Tadepalli S, Narayanan SK, Stewart GC, Chengappa MM, Nagaraja TG (2009) *Fusobacterium necrophorum*: a ruminal bacterium that invades liver to cause abscesses in cattle. Anaerobe 15:36–43

Talan DA, Citron DM, Abrahamian FM, Moran GJ, Goldstein EJ (1999) Bacteriologic analysis of infected dog and cat bites. N Engl J Med 340:85–92

Tamada H, Harasawa R, Shinjo T (1985) Isolation of bacteriophage in *Fusobacterium necrophorum*. Nihon Juigaku Zasshi 47:483–486

Tan ZL, Nagaraia TG, Chengappa MM (1994) Biochemical and biological characterization of ruminal *Fusobacterium necrophorum*. FEMS Microbiol Lett 120:81–86

Tan ZL, Nagaraja TG, Chengappa MM (1996) *Fusobacterium necrophorum* infections: virulence factors, pathogenic mechanisms and control measures. Vet Res Commun 20:113–140

Tateishi F, Hasegawa-Nakamura K, Nakamura T, Oogai Y, Komatsuzawa H, Kawamata K, Douchi T, Hatae M, Naguchi K (2012) Detection of

*Fusobacterium nucleatum* in chronic tissues of high-risk pregnant women. J Clin Periodontol 39:417–424

Tsuchiya C, Sakata T, Sugita H (2008) Novel ecological niche of *Cetobacterium somerae*, an anaerobic bacterium in the intestinal tracts of freshwater fish. Lett Appl Microbiol 46:43–48

Van Gylswyk NO, Morris EJ, Els HJ (1983) The sporulation and cell wall structure of *Clostridium polysaccharolyticum* comb. nov. (formerly *Fusobacterium polysaccharolyticum*). J Gen Microbiol 121:491–493

Vasstrand EN (1981) Lysozyme digestion and chemical characterization of the peptidoglycan of *Fusobacterium nucleatum* Fev 1. Infect Immun 33:75–82

Vasstrand EN, Hofstad T, Endresen C, Jensen HB (1979) Demonstration of lanthionine as a natural constituent of the peptidoglycan of *Fusobacterium nucleatum*. Infect Immun 25:775–780

Vasstrand EN, Jensen HB, Miron T, Hofstad T (1982) Composition of peptidoglycans in *Bacteroidaceae*: determination and distribution of lanthionine. Infect Immun 36:114–122

Walsh LJ, Stritzel F, Yamazaki K, Bird PS, Gemmell E, Seymour GJ (1989) Interleukin-1 and interleukin-1 inhibitor production by human adherent cells stimulated with periodontopathic bacteria. Arch Oral Biol 34:679–683

Watson J, Matsui GY, Leaphart A, Wiegel J, Rainey FA, Lovell CR (2000) Reductively debrominating strains of *Propionigenium maris* from burrows of bromophenol-producing marine infauna. Int J Syst Evol Microbiol 50:1035–1042

Widdel F, Pfennig N (1981) Studies on dissimilatory sulfate-reducing bacteria that decompose fatty acids. I. Isolation of new sulfate-reducing bacteria enriched with acetate from saline environments: description of *Desulfobacter postgatei* gen. nov., sp. nov. Arch Microbiol 129:395–400

Yamaguchi M, Kanoe M, Kai K, Okada Y (1999) Actin degradation concomitant with *Fusobacterium necrophorum* subsp. *necrophorum* adhesion to bovine portal cells. Microbios 98:87–94

Yamazaki K, Polak B, Bird PS, Gemmell E, Hara K, Seymour GJ (1989) Effects of periodontopathic bacteria on IL-1 and IL-1 inhibitor production by human polymorphonuclear neutrophils. Oral Microbiol Immunol 4:193–198

Yarza P, Ludwig W, Euzéby J, Amann R, Schleifer KH, Glöckner FO, Rosselló-Móra R (2010) Update of the All-Species Living Tree Project based on 16S and 23S rRNA sequence analyses. Syst Appl Microbiol 33:291–299. doi:10.1016/j.syapm.2010.08.001

Zhao J-S, Manno D, Hawari J (2009) *Psychrilyobacter atlanticus* gen. nov., sp. nov., a marine member of the phylum Fusobacteria that produces $H_2$ and degrades nitramine explosives under low temperature conditions. Int J Syst Evol Microbiol 59:491–497

Zhou H, Bennett G, Kennan RM, Rood JL, Hickford JGH (2009a) Identification of a leukotoxin sequence from *Fusobacterium equinum*. Vet Microbiol 133:394–395

Zhou H, Bennett G, Hickford JGH (2009b) Detection of *Fusobacterium equinum* on footrot infected hooves of sheep and cattle. Vet Microbiol 134:400–401

# 10 The Genus *Geobacillus*

*Niall A. Logan*
Department of Life Sciences, Glasgow Caledonian University, Glasgow, Scotland, UK



**Abstract**
*Bacillus stearothermophilus* was established in 1920, and many isolates of thermophilic, aerobic endosporeformers were subsequently allocated to it, so that the species became heterogeneous. Between the 1960s and the 1980s various phenotypic techniques demonstrated this heterogeneity, and new thermophilic species were proposed, but as late as the first edition of *Bergey's Manual of Systematic Bacteriology*, the authors were unable to take the taxonomy of the *B. stearothermophilus* group any further, in the absence of sufficient data. With the increasing availability of molecular analyses, several novel species were described, and in 2001 the genus *Geobacillus* was proposed to accommodate *B. stearothermophilus* and its relatives. Some other thermophilic *Bacillus* species were subsequently transferred to the new genus. However, this expansion of *Geobacillus*, to 17 species, left the type species, *G. stearothermophilus*, without a modern description based upon a polyphasic taxonomic study. Also, the taxonomic positions of several other species were unclear and other taxa awaited validation. Polyphasic taxonomic studies published in 2011 and 2012 countered the continuing expansion of the genus by showing that a substantial number of species were synonymous and by transferring some other species to *Anoxybacillus* and the new genus *Caldibacillus*. The genus *Geobacillus* now comprises 11 species: *G. stearothermophilus*, *G. caldoxylosilyticus*, *G. jurassicus*, *G. subterraneus*, *G. thermantarcticus*, *G. thermocatenulatus*, *G. thermodenitrificans*, *G. thermoglucosidans*, *G. thermoleovorans*, *G. toebii*, and *G. uzenensis*. This article summarizes the taxonomic history of the genus and outlines the habitats, isolation, and properties of its species.

## Taxonomy, Historical and Current

### Thermophilic *Bacillus* Species

Claus and Berkeley (1986) listed only three *Bacillus* strict thermophiles (that is to say growing at 65 °C and above) in the first edition of *Bergey's Manual of Systematic Bacteriology*: *B. acidocaldarius* (Darland and Brock 1971), *B. schlegelii* (Schenk and Aragano 1979), and *B. stearothermophilus* (Donk 1920). The last-named of these species, having been established for many years, became something of a dumping-ground for any thermophilic, aerobic endosporeformers, and many isolates were allocated to it. This is understandable because, as later studies have shown, there are rather few routine phenotypic differences between the species now accommodated in *Geobacillus*. Indeed, although the original strain of the species was thought to have been lost, Gordon and Smith (1949) considered Donk's description to match most of the obligately thermophilic *Bacillus* strains that they studied. In addition, Walker and Wolf (1971) believed that two of their cultures (NRRL 1170 and NCA 26 [= ATCC 12980$^T$]) represented Donk's original strain.

Although Mishustin (1950) had renamed as *B. thermodenitrificans* the *Denitrobacterium thermophilum* of Ambroz (1913), and Heinen and Heinen (1972) had proposed *B. caldolyticus*, *B. caldotenax*, and *B. caldovelox*, and Golovacheva et al. (1975) had described *B. thermocatenulatus*, all of these species were listed as *species incertae sedis* by Claus and Berkeley.

The heterogeneity of *B. stearothermophilus* was becoming widely appreciated, however. Walker and Wolf (1971) found that their strains of *B. stearothermophilus* formed three groups on the basis of biochemical tests (Walker and Wolf 1961, 1971) and serology (Walker and Wolf 1971; Wolf and Chowhury 1971), and their division of the species was further supported by studies of esterase patterns (Baillie and Walker 1968), further studies with routine phenotypic characters (Logan and Berkeley 1981), and polar lipids (Minnikin et al. 1977). Klaushofer and Hollaus (1970) also recognized these three major subdivisions within the thermophiles. Walker and Wolf (1971) regarded the recognition of only one thermophilic species of *Bacillus* in the seventh edition of *Bergey's Manual of Determinative Bacteriology* (Breed et al. 1957) as a "dramatic restriction," and in the eighth edition of *Bergey's Manual of Determinative Bacteriology*, Gibson

E. Rosenberg et al. (eds.), *The Prokaryotes – Firmicutes and Tenericutes*, DOI 10.1007/978-3-642-30120-9_212,

and Gordon (1974) commented that the species was "markedly heterogeneous" and that "the emphasis on ability to grow at 65 °C has the effect of excluding organisms that have temperature maxima between 55 °C and 65 °C although they have not so far been distinguished from *B. stearothermophilus* by any other property" "As yet," they concluded, "there has been no agreement on how classification in this part of the genus might be improved." Early studies on *Bacillus* thermophiles were comprehensively reviewed by Wolf and Sharp (1981), who also used the scheme of Walker and Wolf (1971) to allocate several thermophilic species to the three previously established groups:

Group 1 was the most heterogeneous assemblage. It comprised strains that produced gas from nitrate, hydrolyzed starch only weakly, and which produced slightly to definitely swollen sporangia with cylindrical to oval spores; it was divided into five subgroups on the basis of growth temperature maxima and minima. This group accommodated the majority of strains received as *B. stearothermophilus*, as well as *B. caldotenax, B. caldovelox, B. kaustophilus* (Prickett 1928), and *B. thermodenitrificans.*

Group 2 contained strains of *B. stearothermophilus* that were described as "relatively inert" and which showed lower temperature ranges than members of the other groups, but which had greater salt tolerance. They produced definitely swollen sporangia with oval spores.

Group 3 strains hydrolyzed starch strongly and produced definitely swollen sporangia with cylindrical to oval spores. They were divided into four subgroups on the basis of certain biochemical characters and growth temperatures. Group 3 included the type strain of *B. stearothermophilus* and *B. calidolactis* (Grinstead and Clegg 1955; Galesloot and Labots 1959) and *B. thermoliquefaciens* (Galesloot and Labots 1959).

Having considered a wide range of evidence, Wolf and Sharp (1981) concluded that the earlier "restrictive attitude" that regarded *B. stearothermophilus* as the only obligate thermophile in the genus was "no longer tenable" and regretted that differences in sporangial morphologies among thermophiles were disregarded by Gibson and Gordon (1974). Wolf and Sharp (1981) also showed the wide range of G+C mol% of 44–69 among the *Bacillus* thermophiles, but they did not emphasize its taxonomic significance. Claus and Berkeley (1986) were unable to take the taxonomy of the *B. stearothermophilus* group any further, in the absence of further data, but noted that the heterogeneity of the species was indicated by the wide range of DNA base composition.

Following the pioneering work of the late 1960s and early 1970s, several new thermophilic species were described, but the overall taxonomy of the group languished for some years, despite the continuing considerable interest in the biology of the thermophiles and potential applications of their enzymes. Of the new species described in the decade before 1980, when the Approved Lists of Bacterial Names were published (Skerman et al. 1980), only *B. acidocaldarius* (Darland and Brock 1971) was included, while *B. caldolyticus, B. caldotenax, B. caldovelox, B. schlegelii, B. thermocatenulatus*, and *B. thermodenitrificans* were excluded. Subsequent proposals of thermophilic taxa included *B. flavothermus* (Heinen et al. 1982), *B. thermoglucosidasius* (Suzuki et al. 1983), *Bacillus tusciae* (Bonjour and Aragno 1984), *B. acidoterrestris* (Deinhard et al. 1987a), *B. cycloheptanicus* (Deinhard et al. 1987b), *B. pallidus* (Scholz et al. 1987), *B. thermoleovorans* (Zarilla and Perry 1987), *B. thermocloacae* (Demharter and Hensel 1989), *B. thermoaerophilus* (Meier-Stauffer et al. 1996), *B. thermoamylovorans* (Combet-Blanc et al. 1995), *Bacillus thermosphaericus* (Andersson et al. 1995), *B. thermantarcticus* (corrig., Nicolaus et al. 2002; *B. thermoantarcticus* [*sic*], Nicolaus et al. 1996), and *B. vulcani* (Caccamo et al. 2000). Also, some species that had been excluded from the Approved Lists were revived: *B. thermoruber* (ex Guicciardi et al. 1968; Manachini et al. 1985), *B. kaustophilus* (ex Prickett 1928; Priest et al. 1988), and *B. thermodenitrificans* (ex Klaushofer and Hollaus 1970; Manachini et al. 2000). Several of these species were subsequently allocated to new genera as follows: *Alicyclobacillus acidocaldarius, A. acidoterrestris, A. cycloheptanicus* (Wisotzkey et al. 1992), *Brevibacillus thermoruber* (Shida et al. 1996), *Aneurinibacillus thermoaerophilus* (Heyndrickx et al. 1997), *Anoxybacillus flavithermus* (Pikuta et al. 2000), and *Ureibacillus thermosphaericus* (Fortina et al. 2001a).

De Bartolemeo et al. (1991) subjected moderately and obligately thermophilic species of *Bacillus* to numerical taxonomic analysis and found four groups within *B. stearothermophilus.* Three of their groups corresponded with those previously recognized by Walker and Wolf (1971) and other authors, while the fourth group comprised biochemically inert strains of high G+C content which were incapable of growing above 65 °C. White et al. (1993) carried out a polyphasic, numerical taxonomic study on a large number of thermophilic *Bacillus* strains and recommended the revival of *B. caldotenax* and *B. thermodenitrificans* and proposed an emended description of *B. kaustophilus.* However, the clusters they found in their numerical analysis revealed considerable heterogeneity within the species and species groups, and these clusters were often only separated from each other by small margins, indicating that separation of some of the species by routine tests would probably be difficult.

Ash et al. (1991) included strains of *B. stearothermophilus, B. acidoterrestris, B. kaustophilus*, and *B. thermoglucosidasius* in their comparison of the 16S rRNA sequences of the type strains of 51 *Bacillus* species. Their strain of *B. acidoterrestris* was later found to have been a contaminant or misnamed culture (Wisotzkey et al. 1992), but their strains of *B. stearothermophilus, B kaustophilus*, and *B. thermoglucosidasius* grouped together in an evolutionary line (called group 5) distinct from other *Bacillus* species, implying that these thermophiles might represent a separate genus. Rainey et al. (1994) compared the 16S rDNA sequences of 16 strains of 14 thermophilic *Bacillus* species and found that strains of *B. caldolyticus, B. caldotenax, B. caldovelox, B. kaustophilus, B. thermocatenulatus, B. thermodenitrificans*, and *B. thermoleovorans* grouped with *B. stearothermophilus* at similarities of greater than 98 %, while *B. thermoglucosidasius* joined the group at 97 % similarity. This group thus constituted group 5 *sensu* Ash et al. (1991), a coherent and phylogenetically distinct group of thermophilic *Bacillus* species that did not, however, include all the obligate thermophiles in the genus. Studholme et al. (1999) examined whether transformability is a trait associated with a particular phylogenetic group of thermophilic *Bacillus*. Two of their three transformable strains, all received as *B. stearothermophilus*, were more closely related to *B. thermodenitrificans* and *B. thermoglucosidasius* when their 16S rDNA sequences were compared; it was concluded therefore that although transformability might be strain-specific, it is not limited to a single thermophilic *Bacillus* species.

## *Geobacillus* and Its Expansion

Following the discovery of two novel thermophilic, aerobic endosporeformers in petroleum reservoirs, Nazina et al. (2001) proposed that the valid species of Ash et al. group 5 should be accommodated in a new genus, *Geobacillus*, along with their new species *G. subterraneus* and *G. uzenensis*. The new genus thus contained eight species: *G. stearothermophilus* (type species), *G. kaustophilus*, *G. subterraneus*, *G. thermocatenulatus*, *G. thermodenitrificans*, *G. thermoglucosidasius*, *G. thermoleovorans*, and *G. uzenensis*. The other thermophilic species *B. pallidus*, *B. schlegelii*, *B. thermantarcticus*, *B. thermoamylovorans*, *B. thermocloacae*, *B. tusciae*, and *B. vulcani* remained in *Bacillus*.

It was clear from the phylogenetic analyses accompanying the proposals of *B. thermantarcticus* (Nicolaus et al. 1996) and *B. vulcani* (Caccamo et al. 2000) that these species belong to *Geobacillus*, and Nazina et al. (2004) formally proposed the transfer of the latter species. Zeigler (2005) recommended the transfer of *Bacillus thermantarcticus* to the genus *Geobacillus* on the basis of full-length *rec*N and 16S rRNA gene sequences, but did not make a formal proposal. Following the proposal of *Geobacillus*, *Saccharococcus caldoxylosilyticus* (Ahmad et al. 2000) was transferred to it as *Geobacillus caldoxylosilyticus* (Fortina et al. 2001b), *B. pallidus* was transferred as *Geobacillus pallidus* (Banat et al. 2004), and six new species, *G. toebii* (Sung et al. 2002), *G. gargensis* (Nazina et al. 2004), *G. debilis* (Banat et al. 2004), *G. lituanicus* (Kuisiene et al. 2004), *G. tepidamans* (Schäffer et al. 2004), and *G. jurassicus* (Nazina et al. 2005), were described. *G. pallidus* was subsequently allocated to a new genus, *Aeribacillus* (Miñana-Galbis et al. 2010). Although *B. schlegelii* and *B. tusciae* remained in *Bacillus*, they lay at some distance from other members of the genus, and the latter species has recently been allocated to a new genus, which the authors described as a sister group of *Alicyclobacillus*, as *Kyrpidia tusciae* (Klenk et al. 2011, 2012).

The expansion of *Geobacillus*, however, left the long-established species and type species of the genus, *G. stearothermophilus*, without a modern description based upon a polyphasic taxonomic study. The description given by Nazina et al. (2001) for *G. stearothermophilus* was largely based upon the one given by Claus and Berkeley (1986) at a time when the species was essentially all-embracing for thermophilic *Bacillus* strains, despite being widely recognized as heterogeneous. Also, strains of several revived taxa, such as *G. kaustophilus* and *G. thermodenitrificans*, might formerly have been classified within "*B. stearothermophilus*" *sensu lato*, yet no emended description of *G. stearothermophilus* had been published following those proposed revivals.

Several further proposals for new taxa, based upon single isolates, also awaited valid publication. "*Geobacillus caldoproteolyticus*" (Chen et al. 2004) was isolated from sewage sludge in Singapore and deposited as DSM 15730 and ATCC BAA-818; "*Geobacillus thermoleovorans* subsp. *stromboliensis*" (Romano et al. 2005) was isolated from a geothermal environment in the Aeolian Islands in Italy and deposited as DSM 15393 and ATCC BAA-979; "*Geobacillus toebii* subsp. *decanicus*" (Poli et al. 2006) was found in hot compost; and "*Geobacillus zalihae*" (Rahman *et al.* 2007) came from Malaysian palm oil mill effluent.

At this time, there were, therefore, 17 valid species of *Geobacillus*, and a number of taxa awaiting validation or transfer to the genus, or whose taxonomic positions required clarification. No emended description of *G. stearothermophilus* has been published following these proposed revivals, and the thermophilic, aerobic endosporeformers have not been investigated by polyphasic taxonomic study since. It is clear, therefore, that some *Geobacillus* species are without useful definitions at present. A further complication, already mentioned, is that the members of this genus yield rather similar profiles upon phenotypic characterization, so that their routine differentiation is difficult. Therefore, *G. stearothermophilus* was without a practically useful definition at the time that the treatment in the second edition of *Bergey's Manual of Systematic Bacteriology* (Logan et al. 2009) was prepared.

## Revision of *Geobacillus*

For many years, the taxonomic positions of *G. kaustophilus* and *G. thermocatenulatus*, along with "*Bacillus caldolyticus*," "*B. caldotenax*," and "*B. caldovelox*" had been unclear. Metabolic studies and phage typing by Sharp et al. (1980) also revealed close relationships between these species. White et al. (1993) considered whether "*B. caldolyticus*," "*B. caldotenax*," and "*B. caldovelox*" (Heinen and Heinen 1972) were synonymous, and they recommended the revival of "*B. caldotenax*," but their DNA relatedness data were inconclusive. They also proposed an emended description of *B. kaustophilus*, but did not validate this proposal. The 16S rRNA gene sequence studies of Rainey et al. (1994) showed that *B. kaustophilus*, *B. thermoleovorans*, "*B. caldolyticus*," "B. *caldotenax*," and "*B. caldovelox*" were closely related and that this group was related to *B. thermocatenulatus*. Sunna et al. (1997b) identified *B. kaustophilus* and *B. thermocatenulatus*, as well as "*B. caldolyticus*," "*B. caldotenax*," and "*B. caldovelox*," as members of *B. thermoleovorans*, on the basis of DNA homologies ranging from 72 % to 88 % between the type and reference strains of all these species. They proposed the merger of these species and gave an emended description of *B. thermoleovorans*, but the proposal was not validated. Nazina et al. (2004) did not support such a merger; however, as they found only 47–54 % DNA relatedness between *G. kaustophilus*, *G. thermoleovorans*, and *G. thermocatenulatus*.

Nazina et al. (2004) had proposed *G. gargensis* and the transfer of *B. vulcani* (Caccamo et al. 2000) to *Geobacillus* on the basis of 99.4 % 16S rRNA gene sequence similarity and 55 % DNA relatedness with *G. kaustophilus*. Other taxa showing high 16S rRNA gene sequence similarities with *G. kaustophilus*, *G. thermocatenulatus*, and *G. thermoleovorans* were *G. lituanicus* (Kuisiene et al. 2004), "*Geobacillus thermoleovorans* subsp. *stromboliensis*" (Romano et al. 2005) and "*Geobacillus zalihae*" (Rahman et al. 2007). In order to resolve the taxonomic confusion, Dinsdale et al. (2011) subjected strains of *G. gargensis*, *G. kaustophilus*, *G. lituanicus*, *G. stearothermophilus*, *G. thermocatenulatus*, *G. thermoleovorans*, "*G. thermoleovorans* subsp. *stromboliensis*," *G. vulcani*, "*Bacillus caldolyticus*," "*B. caldotenax*," and "*B. caldovelox*" to a polyphasic taxonomic study. Their study countered the continuing expansion of the number of species in *Geobacillus* by showing that *G. kaustophilus*, *G. lituanicus*, *G. vulcani*, "*G. thermoleovorans* subsp. *stromboliensis*," *G. vulcani*, "*Bacillus caldolyticus*," "*B. caldotenax*," and "*B. caldovelox*" were all synonyms of *G. thermoleovorans* and that *G. gargensis* was a synonym of *G. thermocatenulatus*. These mergers left 12 validly published species in *Geobacillus*: *G. stearothermophilus*, *G. caldoxylosilyticus*, *G. debilis*, *G. jurassicus*, *G. subterraneus*, *G. thermocatenulatus*, *G. thermodenitrificans*, *G. thermoglucosidasius*, *G. thermoleovorans*, *G. tepidamans*, *G. toebii*, and *G. uzenensis*.

Coorevits et al. (2012) subjected 62 strains of thermophilic aerobic endosporeforming bacteria to polyphasic taxonomic study, including 16S rRNA gene sequence analysis, polar lipid and fatty acid analysis, phenotypic characterization, and DNA-DNA hybridization experiments. Distinct clusters of the species *Geobacillus stearothermophilus*, *Geobacillus thermodenitrificans*, *Geobacillus toebii*, and *Geobacillus thermoglucosidasius* were formed, allowing their descriptions to be emended, and the distinctnesses of the poorly represented species *Geobacillus jurassicus*, *Geobacillus subterraneus*, and *Geobacillus caldoxylosilyticus* were confirmed. As *Bacillus thermantarcticus* clustered between *Geobacillus* species on the basis of 16S rRNA gene sequence analysis, it was transferred to *Geobacillus*. It was also proposed that the name *Geobacillus thermoglucosidasius* should be corrected to *G. thermoglucosidans*. The above-mentioned species, together with *Geobacillus thermoleovorans* and *Geobacillus thermocatenulatus*, formed a monophyletic cluster representing the genus *Geobacillus*.

Cihan et al. (2011) proposed a subspecies of *Geobacillus thermodenitrificans*, *G. thermodenitrificans* subsp. *calidus*. However, their study included only one reference strain, the type strain, of *G. themodenitrificans* and so the diversity of this species was not represented; the appropriateness of creating a subspecies in such circumstances is doubtful.

Furthermore, Coorevits et al. (2012) found that the strain deposited as the type of *G. uzenensis* in the DSMZ (DSM 13551$^T$), and studied by them, was not the authentic strain U$^T$, but actually a strain of *G. subterraneus*. This was confirmed by 100 % $S_P$ and complete DNA relatedness between strain R-35640$^T$ (= DSM 13551$^T$) and their type strain of *B. subterraneus* R-32641$^T$ (= DSM 13552$^T$). Zeigler (2005) had analyzed the full-length *rec*N and 16S rRNA gene sequences of the type strains of *G. subterraneus* and *G. uzenensis*, along with two other isolates described as belonging to *G. subterraneus*, and found that they clustered within the same similarity group. It was not, however, clear whether the close relationship shown by these methods was due to sequencing errors in one or both GenBank entries or that Zeigler's strain for *G. uzenensis* used was not in fact the same as the type strain studied by Nazina et al. (2001). Attempts by Coorevits et al. (2012) to obtain an authentic strain elsewhere were unsuccessful, and it is currently uncertain whether or not *G. uzenensis* is represented by authentic type cultures in two publicly accessible culture collections; the validity of the species therefore remains in some doubt.

The type strain of *Geobacillus tepidamans* and the strain received as "*Geobacillus caldoproteolyticus*" clustered together in 16S rRNA gene sequence analysis and were recovered within a group harboring *Anoxybacillus* species. Their transfers to *Anoxybacillus* (the latter as *A. caldiproteolyticus*) were therefore proposed. The type strain of *Geobacillus debilis* was not closely related to any members of *Anoxybacillus* or *Geobacillus*, however, and it was proposed that this species be placed in a new genus *Caldibacillus* (Coorevits et al. 2012).

In conclusion, therefore, the genus *Geobacillus* comprises 11 species at the time of writing: *G. stearothermophilus*, *G. caldoxylosilyticus*, *G. jurassicus*, *G. subterraneus*, *G. thermantarcticus*, *G. thermocatenulatus*, *G. thermodenitrificans*, *G. thermoglucosidans*, *G. thermoleovorans*, *G. toebii*, and *G. uzenensis* (➋ *Fig. 10.1*).

## Molecular Analyses

Given the industrial interest in thermophilic organisms and their products over many years, it is perhaps surprising that rather few full genome sequences have been published: *G. kaustophilus* (now *G. thermoleovorans*) HTA426 from deep-sea sediment of the Mariana Trench (Takami et al. 2004); *G. thermodenitrificans* NG80-2 from a deep-subsurface oil reservoir (Feng et al. 2007); *G. thermoleovorans* CCB_US3_UF5 from a hot spring in Malaysia (Sakaff et al. 2012); and *G. stearothermophilus* in draft at University of Oklahoma. Comparisons with genomes of closely related mesophilic endosporeformers have not revealed salient features of the thermophile sequences that clearly correlate with adaptations to thermophily; indeed, it has been asked why can these thermophiles not grow at

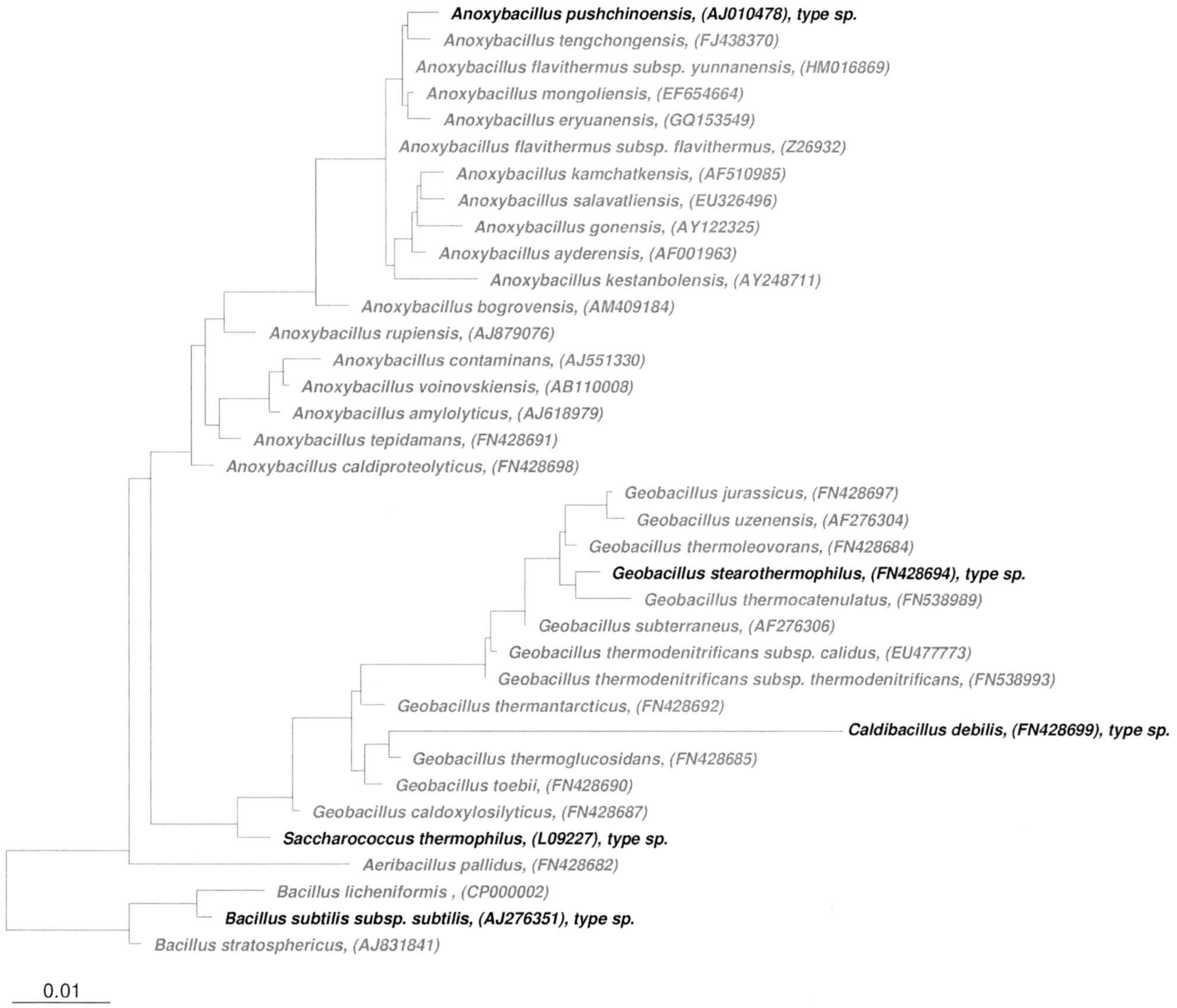


**Fig. 10.1**
**Phylogenetic reconstruction of the genus *Geobacillus* based on 16S rRNA and created using the maximum likelihood algorithm RAxML (Stamatakis 2006). The sequence dataset and alignment were used according to the All-Species Living Tree Project (LTP) database (Yarza et al. 2010; http://www.arb-silva.de/projects/living-tree). Representative sequences from closely related taxa were used as outgroups. In addition, a 40 % maximum frequency filter was applied in order to remove hypervariable positions and potentially misplaced bases from the alignment. Scale bar indicates estimated sequence divergence**

the same temperatures as their mesophilic relatives (Banat and Marchant 2011)? The mol% G+C range for the DNA of the genus is 48.4–54.5 ($T_m$).

## Phenotypic Analyses

### Cultural Properties

Vegetative cells are rod-shaped and occur either singly or in short chains and are motile by means of peritrichous flagella or they are non-motile. The cell wall structure is Gram-positive, but the Gram-stain reaction may vary between positive and negative. One ellipsoidal or cylindrical endospore is produced per cell, and spores are located terminally or subterminally in slightly swollen or non-swollen sporangia (*Fig. 10.2*). Colony morphologies and sizes are variable, and pigments may be produced on certain media.

### Nutrition and Growth Conditions

All species of *Geobacillus* are obligately thermophilic chemo-organotrophs. They are aerobic or facultatively anaerobic, and oxygen is the terminal electron acceptor, replaceable in some species by nitrate. Temperature ranges for growth generally lie

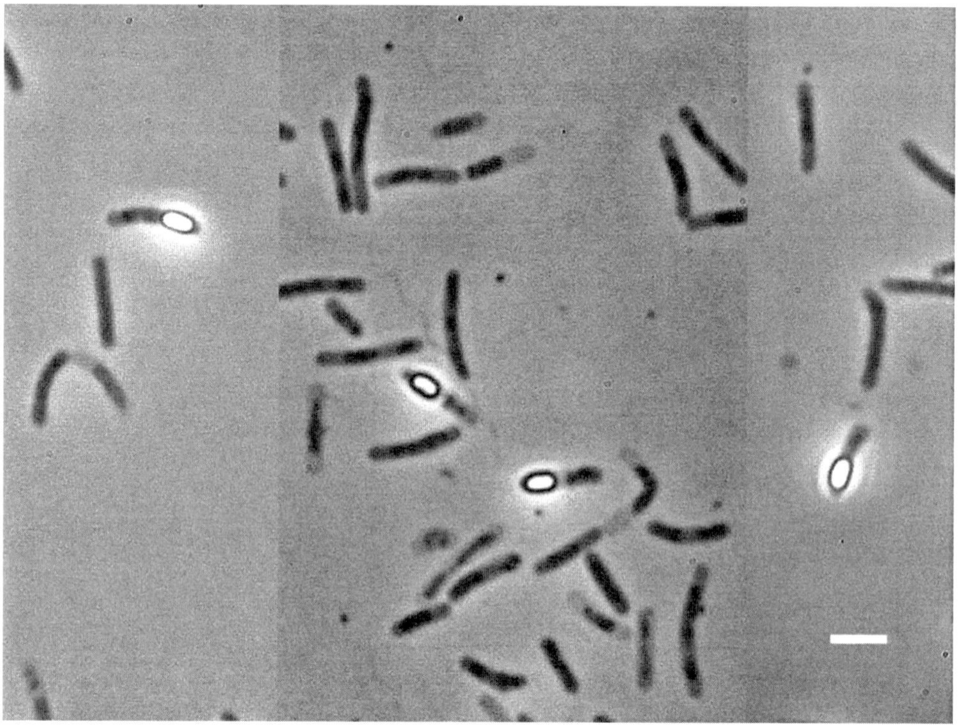

**Fig. 10.2**
**Photomicrograph of type strain of *Geobacillus stearothermophilus* viewed by phase contrast microscopy, showing ellipsoidal, subterminal spores that slightly swell the sporangia. Bar marker represents 2 μm**

between 30 °C and 80 °C, with optima between 50 °C and 60 °C. They are neutrophilic and grow within a relatively narrow pH range of 5.0 to 9.0, and their optima lie within the range pH 6.2–7.5. For the species tested, growth factors, vitamins, NaCl, and KCl are not required, and most strains will grow on routine media such as nutrient agar. A wide range of substrates is utilized, including carbohydrates, organic acids, peptone, tryptone, and yeast extract; the ability to utilize hydrocarbons as carbon and energy sources is a widely distributed property in the genus (Nazina et al. 2001). Most species produce acid but no gas from fructose, glucose, maltose, mannose, and sucrose. Most species produce catalase. Oxidase reaction varies. Phenylalanine is not deaminated, tyrosine is not degraded, and indole is not produced. A strain of *G. thermoleovorans* has been found to have extracellular lipase activity and high growth rates on lipid substrates such as olive oil, soybean oil, mineral oil, tributyrin, triolein, and Tweens 20 and 40 (Lee et al. 1999).

## Cell Wall Composition and Fine Structure

The vegetative cells of the majority of *Bacillus* species that have been studied, and of the examined representatives of several genera, such as *Geobacillus*, whose species were previously accommodated in *Bacillus*, have the most common type of cross linkage in which a peptide bond is formed between the diamino acid in position 3 of one subunit and the D-Ala in position 4 of the neighboring peptide subunit, so that no interpeptide bridge is involved. The diamino acid in the two *Geobacillus* species for which it has been determined, *G. stearothermophilus* (Schleifer and Kandler 1972) and *G. thermoleovorans* (Zarilla and Perry 1987), is diaminopimelic acid, and the configuration has been determined for the former as *meso*-diaminopimelic acid (*meso*-DAP); this cross-linkage is now usually known as DAP-direct (A1g in the classification of Schleifer and Kandler 1972).

Organisms growing at high temperatures need enzyme adaptations to give molecular stability as well as structural flexibility (Kawamura et al. 1998; Alvarez et al. 1999; Perl et al. 2000), heat-stable protein-synthesizing machinery, and adaptations of membrane phospholipid composition. They differ from their mesophilic counterparts in the fatty acid and polar headgroup compositions of their phospholipids.

The effect of temperature on the membrane composition of *G. stearothermophilus* has been intensively studied. Phosphatidylglycerol (PG) and cardiolipin (CL) comprise about 90 % of the phospholipids, but as the growth temperature rises the PG content increases at the expense of the CL content. The acyl-chain composition of all the membrane lipids also alters; the longer, saturated-linear, and *iso* fatty acids with relatively high melting points increase in abundance while *anteiso* fatty acids and unsaturated components with lower melting points decrease. As a result, the organism is able to maintain nearly constant membrane fluidity across its whole growth temperature range; this has been termed homeoviscous adaptation (Martins et al. 1990; Tolner et al. 1997); and alternative theory, homeophasic adaptation, considers that maintenance of the liquid-crystalline phase is more important than an absolute value of membrane fluidity in *Bacteria* (Tolner et al. 1997).

The main menaquinone type is MK-7. The major cellular fatty acid components of *Geobacillus* species following incubation at 55 °C are (with ranges as percent of total given in parentheses) iso-$C_{15:0}$ (20–40 %; mean 29 %), iso-$C_{16:0}$ (6–39 %; mean 25 %), and iso-$C_{17:0}$ (7–37 %; mean 19.5 %), which account for 60–80 % of the total. As minor components, anteiso-$C_{15:0}$ (0.6–6.4 %; mean 2.3 %), $C_{16:0}$ (1.7–11.2 %; mean 5.8 %), and anteiso-$C_{17:0}$ (3.1–18.7 %; mean 7.3 %) are detected (Nazina et al. 2001). The figures given by Fortina et al. (2001b) for *G. caldoxylosilyticus* and Sung et al. (2002) for *G. toebii* generally lie within these ranges, with the exception that strains of the former species showed 45–57 % of iso-$C_{15:0}$.

Direct comparison of profiles between the obligately thermophilic *Geobacillus* species and mesophilic aerobic endosporeformers is not normally possible, as the assays of members of the two groups have not usually been done at the same temperature. For many species of *Bacillus sensu stricto*, fatty acid profiles obtained following incubation at 30 °C were anteiso $C_{15:0}$ (25–66 %), iso $C_{15:0}$ (22–47 %), and anteiso $C_{17:0}$ (2–12 %). For the *B. cereus*, group levels of anteiso $C_{15:0}$ were lower (3–7 %), and amounts of unsaturated fatty acids were generally higher (>10 %) (Kämpfer 1994). The thermotolerant species *B. coagulans* and *B. smithii* show higher amounts of anteiso $C_{17:0}$ (means of 28 % and 42 % respectively) and generally lower amounts of anteiso $C_{15:0}$ (means of 55 % and 12 % respectively) and iso $C_{15:0}$ (means of 9 % and 19 % respectively), but for these data the former was incubated at 30 °C and the latter at 57 °C.

Llarch et al. (1997) compared the fatty acid profiles of aerobic endosporeformers isolated from Antarctic geothermal environments; their six isolates had temperature ranges with minima between 17 °C and 45 °C and maxima between 62 °C and 73 °C, with optima of 60–70 °C. Two strains (temperature ranges 37–70 °C and 45–73 °C) were found to lie nearest to *B. stearothermophilus* in a phenotypic analysis, while two other isolates could be identified as strains of *B. licheniformis* (temperature range 17–68 °C) and *B. megaterium* (temperature range 17–63 °C) whose maximum growth temperatures were extended beyond those seen in strains from temperate environments. The fatty acid profiles for all of these strains were compared following incubation at 45 °C; the two *B. stearothermophilus*-like strains showed profiles of iso-$C_{15:0}$ (19 % and 40 %), iso-$C_{16:0}$ (47 % and 5 %), and iso-$C_{17:0}$ (7.5 % and 23 %), which accounted for 55–73 % of the total, while for minor components the patterns were anteiso-$C_{15:0}$ (2.6 % and 9.6 %), $C_{16:0}$ (4 % and 5.8 %), and anteiso-$C_{17:0}$ (4.6 % and 8.7 %); these profiles are consistent with those reported for *Geobacillus*. The profiles for the *B. licheniformis* and *B. megaterium* strains were iso-$C_{15:0}$ (38.4 % and 20.5 % [means for mesophilic strains of these species, from Kämpfer (1994) were: 33–38 %; 15–48 % respectively]), iso-$C_{16:0}$ (5.2 % and 1.9 % [mesophiles 2 %; 0.9–2.4 %]), and iso-$C_{17:0}$ (24.4 % and 2.3 % [mesophiles 10 %; 0.5–1.7 %]), which accounted for 68–25 % of the total, while for minor components the patterns were anteiso-$C_{15:0}$ (10.2 % and 50 % [mesophiles 30 %; 32–67 %]), $C_{16:0}$ (12.4 % and 3.3 % [mesophiles 2 %; 1.5–2.8 %]), and anteiso-$C_{17:0}$ (6 % and 7.7 % [mesophiles 10 %, 1.7–3 %]). These profiles suggest that any potential distinctions between the rather variable fatty acid profiles of *Geobacillus* species and *Bacillus* species are largely lost when strains of each group are incubated at the same temperature.

Amino acid transport in *G. stearothermophilus* is $Na^+$-dependent, which is unusual for neutrophilic organisms such as these, but common among marine bacteria and alkalophiles; however, the possession of primary and secondary $Na^+$-transport systems may be advantageous to the organism by allowing energy conversion via $Na^+$-cycling when the phospholipid adaptations needed to give optimal membrane fluidity at the organism's growth temperature also result in membrane leakiness (de Vrij et al. 1990; Tolner et al. 1997).

## Enrichment and Isolation Procedures

Thermophiles may easily be obtained by incubating environmental or other samples in routine cultivation media at 65 °C and above. As for other aerobic endosporeformers, it is useful to heat-treat the specimens to select for endospores and encourage their germination: heat treatment may vary from 60 °C to 80 °C for 10 min or longer; 80 °C for 10 min is widely used; the given time allows for a period of heat penetration of the sample followed by a sufficient holding period at temperature; this assumes that the specimen is in an aqueous suspension in a water bath and is adequately immersed. Solid samples may be emulsified in sterile, deionized water, 1:2 [w/v] prior to heating; the unheated control is prepared in the same way, but is unheated, or else the suspension intended for heating may be sampled for cultivation prior to heating. Direct plate cultures are made on appropriate solid media by spreading up to 250 ml volumes from undiluted, and 10-, 100- and 1,000-fold dilutions of the treated sample.

Allen (1953) described enrichment methods for strains belonging to particular physiological groups. A selective procedure for the isolation of flat-sour organisms from food was described by Shapton and Hindes (1963); Yeast-Dextrose-Tryptone Agar contains peptone, 5 g; beef extract, 3 g; tryptone, 2.5 g; yeast extract, 1 g; glucose, 1 g; distilled water, 1,000 ml; dissolve by heating, adjust to pH 8.4 and simmer for 10 min then pass through coarse filter paper if necessary; cool, make back up to 1,000 ml and adjust to pH 7.4; add sufficient agar to solidify and 2.5 ml of 1 % aqueous solution of bromcresol purple; and then sterilize by autoclaving. Prepare a suspension of the food sample in 1/4 strength Ringer's solution and pasteurize it with the molten medium at 108 °C (8 psi) for 10 min, then reduce the temperature to 100 °C and maintain for 20 min; cool to 50 °C; and pour plates and allow to set. Incubate at 55 °C for 48 h and observe for yellow colonies. Donk (1920) reported finding *B. stearothermophilus* in spoiled cans of corn and string beans, but the method of isolation was not described; the reader is referred to the methods described for the isolation of flat-sour organisms.

The following procedures were those used in the isolation of strains of *Geobacillus* species, but do not necessarily represent methods especially designed to enrich or select for those species. *G. caldoxylosilyticus* was isolated from Australian soil by adding 0.1–0.2 g of sample to minimal medium and incubating at 65 °C for up to 24 h (Ahmad et al. 2000). Minimal medium contained xylose, 10 g; $K_2HPO_4$, 4 g; $KH_2PO_4$, 1 g; $NH_4NO_3$, 1 g; NaCl, 1 g; $MgSO_4$, 0.25 g; trace mineral solution, 10 ml; water to 1,000 ml, pH, 6.8; 1.5 % agar was added when a solid medium was desired. Trace mineral solution contained EDTA, 5.0 g; $CaCl_2.2H_2O$, 6.0 g; $FeSO_4.7H_2O$, 6.0 g; $MnCl_2.4H_2O$, 1.15 g; $CoCl_2.6H_2O$, 0.8 g; $ZnSO_4.7H_2O$, 0.7 g; $CuCl_2.2H_2O$, 0.3 g; $HBO_3$, 0.3 g; $(NH_4)_6Mo_7O_{24}.4H_2O$, 0.25 g; and water, 1,000 ml. After two transfers of 1 ml of culture into fresh medium, enrichments were placed on solidified minimal medium and incubated at 65 °C for 24 h. Further isolations from soils taken from China, Egypt, Italy, and Turkey were made by heating samples at 90 °C for 10 min and plating on CESP agar and incubating at 65 °C for 24 h (Fortina et al. 2001b); CESP agar contained casitone, 15 g; yeast extract, 5 g; soytone, 3 g; peptone, 2 g; $MgSO_4$, 0.015 g; $FeCl_3$, 0.007 g; $MnCl_2.H_2O$, 0.002 g; and water, 1,000 ml, pH, 7.2.

*G. jurassicus* was isolated from oilfield formation water by diluting enrichment cultures grown in a modification of the medium of Adkins et al. (1992) ($NH_4Cl$, 1 g; KCl, 0.1 g; $KH_2PO_4$; 0.75 g; $K_2HPO_4$, 1.4 g; $MgSO_4.7H_2O$, 0.2 g; $CaCl_2.2H_2O$, 0.02 g; NaCl, 1.0 g; water, 1,000 ml; pH 7.0) supplemented with 4 $\%^v/_v$ crude oil, incubated at 60 °C, and plating on the same medium solidified with 2 % agar.

*G. subterraneus* and *G. uzenensis* were isolated from serial dilutions of thermophilic hydrocarbon-oxidizing enrichments taken from oilfields; the enrichments were inoculated onto the medium described by Zarilla and Perry (1987) supplemented with 0.1 % *n*-hexadecane and incubated at 55–60 °C (Nazina et al. 2001).

*G. thermantarcticus* was isolated from geothermally heated soil from Cryptogam Ridge, Mount Melbourne, Northern Victoria Land, Antarctica; the medium contained 1 % w/v yeast extract and 0.3 % NaCl at pH 6.0; it was incubated aerobically at 65 °C and strains were isolated by repeated serial dilution and inoculation onto plates of the same basal medium solidified with agar (Nicolaus et al. 1991).

*G. thermocatenulatus* was isolated from a slimy bloom at about 60 °C on the inside surface of a pipe in a steam and gas thermal borehole in thermal zone of Mount Yangan-Tau in the South Urals, using potato-peptone and meat-peptone media (Golovacheva et al. 1965, 1975). "*G. gargensis*," shown to be a synonym of *G. thermocatenulatus*, was isolated from the upper layer of a microbial mat from the Garga spring, Eastern Siberia, by serial dilutions and inoculation onto the agar medium described by Adkins et al. (1992) supplemented with 15 mM sucrose: TES (*N*-tris[hydroxymethyl]methyl-2-aminoethanesulfonic acid), 10 g; $NH_4Cl$, 1 g; NaCl, 0.8 g; $MgSO_4.7H_2O$, 0.2 g; $CaCO_3$ (precipitated chalk), 0.2 g; KCl, 0.1 g; $K_2HPO_4$, 0.1 g; $CaCl_2.2H_2O$, 0.02 g; yeast extract, 0.2 g; trace metal solution, 5 ml; vitamin solution, 10 ml; water to 1,000 ml, pH, 7.0; agar was added to solidify. Trace metal solution (Tanner 1989) contained nitrilotriacetic acid (2 g, pH adjusted to 6 with KOH); $MnSO_4.H_2O$, 1 g; Fe $(NH_4)_2(SO_4)_2.6H_2O$, 0.8 g; $CoCl_2.6H_2O$, 0.2 g; $ZnSO_4.7H_2O$, 0.2 g; $CuCl_2.2H_2O$, 0.02 g; $NiCl_2.6H_2O$, 0.02 g; $Na_2MoO_4.2H_2O$, 0.02 g; $Na_2SeO_4$; 0.02; $Na_2WO_4$, 0.02; and water, 1,000 ml. Vitamin solution (Tanner 1989) contained pyridoxine HCl, 10 mg; thiamine HCl, 5 mg; riboflavin, 5; calcium pantothenate, 5 mg; thioctic acid, 5 mg; *p*-aminobenzoic acid, 5 mg; nicotinic acid, 5 mg; vitamin $B_{12}$, 5 mg; biotin, 2 mg; folic acid, 2 mg; and water, 1,000 ml. Plates were incubated at 60 °C.

Mora et al. (1998) isolated new strains of *G. thermodenitrificans* from soil by suspending 1 g of soil sample in 5 ml sterile distilled water and heat treating at 90 °C for 10 min, then plating 1 ml on nutrient agar and incubating at 65 °C for 24 h.

*G. thermoglucosidans* (formerly *G. thermoglucosidasius*) was isolated from Japanese soil by adding 0.1 g samples to 5 ml of medium I in large (1.8 cm by 19 cm) test tubes and incubating at 65 °C for 18 h with the tubes leaning at an angle of about 10°, followed by further enrichments in tubes of the same medium and then purification of plates of medium I solidified with 3 % agar (Suzuki et al. 1976); medium I contained: peptone, 5 g; meat extract, 3 g; yeast extract, 3 g; $K_2HPO_4$, 3 g; $KH_2PO_4$, 1 g; water, 1,000 ml, pH 7.0. *G. thermoleovorans* was isolated by adding soil, mud, and water samples to L-salts basal medium supplemented with (0.1 % (v/v) *n*-heptadecane and incubating at 60 °C for 1–2 weeks, followed by transfer from turbid cultures to fresh medium of the same composition; after several such transfers, pure cultures were obtained by streaking on plates of L-salts basal medium supplemented with 0.2 % (v/v) *n*-heptadecane and solidified with 2 % agar (Merkel et al. 1978; Zarilla and Perry 1987). L-salts (Leadbetter and Foster 1958) contained $NaNO_3$, 2.0 g; $MgSO_4.7H_2O$, 0.2 g; $NaH_2PO_4$, 0.09 g; KCl, 0.04 g; $CaCl_2$, 0.015 g; $FeSO_4.7H_2O$, 1.0 mg; $ZnSO_4.7H_2O$, 70.0 mg; $H_3BO_3$, 10.0 mg; $MnSO_4.5H_2O$, 10.0 mg; $MoO_3$, 10.0 mg; $CuSO_4.5H_2O$, 5.0 mg; and deionized water, 1,000 ml. Of species recently found to be synonyms of *G. thermoleovorans*, "*G. kaustophilus*" was isolated by Prickett (1928) from uncooled pasteurized milk by plating on a peptonized milk agar, followed by subculturing on the same medium or on nutrient agar supplemented with 1 % yeast extract, 0.25 % tryptophan broth, and 0.05 % glucose; "*G. lituanicus*" was isolated using tenfold serial dilutions of crude oil. The dilutions were inoculated onto Czapek agar and plates were incubated aerobically at 60 °C for 48 h; "*G. vulcani*" was isolated from a marine sediment sample by inoculation into Bacto Marine Broth (Difco) and Medium D (Castenholst 1969; Degryse et al. 1978) and incubating aerobically for 3 days at 65 °C, followed by plating positive cultures onto Bacto Marine Agar (Difco).

*G. toebii* was isolated from a suspension of hay compost plated onto a solid modified basal medium and incubated at 60 °C for 3 days (Sung et al. 2002). The medium contained polypeptone, 5 g; $K_2HPO_4$, 6 g; $KH_2PO_4$, 2 g; yeast extract, 1 g; $MgSO_4.7H_2O$, 0.5 g, L-tyrosine, 0.5 g; agar to solidify; and deionized water, 1,000 ml.

There are rather few routine phenotypic characters that can be used reliably to distinguish between the members of *Geobacillus* (➋ *Table 10.1*). Characters testable by the API System (bioMérieux), especially acid production from a range of carbohydrates, that are valuable for differentiating between *Bacillus* species, show relatively little variation in pattern between several *Geobacillus* species. Most species show 16S rDNA sequence similarities higher than 96.5 %, and so they cluster together quite closely in trees based on such data. They may also show high similarities in other, phenotypic, analyses. The distinction of six species by Nazina et al. (2001) was mainly supported by DNA-DNA homology data, and their differentiation table for eight species was compiled from the literature for all of the six previously established species, so that the characterization methods used were not strictly comparable; furthermore, the data were incomplete for these species. The same is true of the differentiation table that accompanied the description of *G. toebii* (Sung et al. 2002). The species *G. kaustophilus*, *G. stearothermophilus*, *G. thermocatenulatus*, *G. thermoglucosidasius*, and *G. thermoleovorans*, especially, need to be characterized alongside the recently described and other, revived, species in order to allow their descriptions to be emended where necessary. 16S rDNA sequencing is not reliable as a stand-alone tool for identification, and a polyphasic taxonomic approach is advisable for the identification of *Geobacillus* species and the confident recognition of suspected new taxa. Species descriptions accompanying proposals of new species will be based on differing test methods, and reference strains of established taxa are often not included for comparison, so original descriptions should never be relied upon entirely. Nomenclatural types exist for a good reason and are usually easily available; there is no substitute for direct laboratory comparisons with authentic reference strains.

### Maintenance Procedures

*Geobacillus* strains may be preserved on slopes of a suitable growth medium that encourages sporulation, such as nutrient agar or trypticase soy agar containing 5 mg $L^{-1}$ of $MnSO_4.7H_2O$. Slopes should be checked microscopically for spores before sealing, to prevent drying out, and storage in a refrigerator; on such sealed slopes the spores should remain viable for many years. For longer-term preservation, lyophilization and liquid nitrogen may be used, as long as cryoprotectants are added.

## Ecology

Although thermophilic aerobic endosporeformers, and other thermophilic bacteria, might be expected to be restricted to hot environments, they are also very widespread in cold environments and appear to be ubiquitously distributed in soils worldwide. Strains with growth temperature ranges of 40–80 °C can be isolated from soils whose temperatures never exceed 25 °C (Marchant et al. 2002); indeed, Weigel (1986) described how easy it is to isolate such organisms from cold soils and even from Arctic ice. That the spores of endosporeformers may survive in such cool environments without any metabolic activity is understandable, but their wide distribution and contribution of up to 10 % of the cultivable flora suggests that they do not merely represent contamination from hot environments (Marchant et al. 2002). It has been suggested that the direct heating action of the sun on the upper layers of the soil, and local heating from the fermentative and putrefactive activities of mesophiles, might be sufficient to allow the multiplication of thermophiles (Norris et al. 1981). The first described strains of the species now called *G. stearothermophilus* were isolated from spoiled canned corn and string beans. Such organisms, and other *Bacillus* species, have long been important in the canned food and dairy industries and are responsible for "flat sour" spoilage of canned foods and products such as evaporated milk (Kalogridou-Vassilliadu 1992). The organisms may thrive in parts of the food processing plant, and their contaminating spores survive the canning or dairy process and then outgrow in the product if it is held for any time at an incubating temperature. This is a particular problem for foods such as military rations that may need to be stored in tropical climates (Llaudes et al. 2001). This species may represent up to a third of thermophilic isolates from foods (Deák and Temár 1988) and approaching two thirds of the thermophiles in milk (Chopra and Mathur 1984). Other sources of this species include soils, geothermal soil, rice soils (Garcia et al. 1982), desert sand, composts (Blanc et al. 1997), and water, ocean sediments, and shallow marine hydrothermal vents (Caccamo et al. 2001); *G. caldoxylosilyticus* was found in Australian soils and subsequently in uncultivated soils from China, Egypt, Italy, and Turkey and in central heating system water (Obojska et al. 2002). *G. thermodenitrificans* has been isolated from soils from Australia, Asia, and Europe, from shallow marine hydrothermal vents (Caccamo et al. 2001), from sugar beet juice, and, along with *G. thermoglucosidans*, in other soils (Mora et al. 1998) and hot composts (Blanc et al. 1997). *G. toebii* was found in hot hay compost. *G. thermoleovorans* was first cultivated from soil, muds, and activated sludge collected in the USA, and further isolations have been made from shallow marine hydrothermal vents (Maugeri et al. 2001), deep subterranean petroleum reservoirs (Kato et al. 2001), crude oil in Lithuania ("*G. lituanicus*"; Kuisiene et al. 2004), from a shallow hydrothermal vent ("*G. vulcani*"; Caccamo et al. 2000), and from Japanese, Indonesian, and Icelandic hot springs (Sunna et al. 1997a; Lee et al. 1999; Markossian et al. 2000). "*G. kaustophilus,*" now merged into *G. thermoleovorans*, was first isolated from pasteurized milk, and other strains have been found in spoiled canned food, and in geothermal and temperate soils from Iceland, New Zealand, Europe, and Asia (White et al. 1993). *G. jurassicus*, *G. subterraneus*, and *G. uzenensis* were all isolated from the formation waters of high-temperature oilfields in China, Kazakhstan, and Russia, and G. *thermocatenulatus* was isolated from a slimy bloom on the inside surface of a pipe in a steam and gas thermal borehole in thermal zone of Yangan-Tau mountain in the South Urals, and "*G. gargensis,*" now merged with

**Table 10.1**
Characters useful for differentiating between *Geobacillus* species

| Character | *G. stearothermophilus* | *G. caldoxylosilyticus* | *G. jurassicus* | *G. subterraneus* | *G. thermantarcticus* | *G. thermocatenulatus* | *G. thermodenitrificans* | *G. thermoglucosidans* | *G. thermoleovorans* | *G. toebii* | *G. uzenesis* |
|---|---|---|---|---|---|---|---|---|---|---|---|
| | *n* = 15 | *n* = 1 | *n* = 2 | *n* = 1 | *n* = 1 | *n* = 2 | *n* = 9 | *n* = 8 | *n* = 11 | *n* = 6 | *n* = 2 |
| **Sporangia** | | | | | | | | | | | |
| Cylindrical spores | – | v | – | – | – | d | – | d | – | – | – |
| Sporangia swollen | v | + | v | – | + | – | – | d | – | + | d |
| Spores subterminal | + | – | + | + | + | d | + | + | + | – | – |
| Spores terminal | + | + | – | v | + | + | + | + | + | + | + |
| Spores central/ paracentral | – | + | – | – | – | d | – | – | – | – | – |
| **Hydrolysis of** | | | | | | | | | | | |
| Aesculin | +/w | + | + | + | + | + | + | + | + | – | + |
| Casein | +/w | v | + | – | – | – | – | – | d | – | – |
| Gelatin | + | + | + | w | + | + | d | + | + | + | + |
| ONPG | – | – | – | – | – | – | – | – | d | – | – |
| Starch | + | + | + | + | – | – | +/w | d (w) | + | +/w | + |
| Catalase | + | + | + | + | + | + | + | + | + | + | |
| Oxidase | – | + | + | – | +/w | + | – | w | + | + | |
| Nitrate reduction | d | + | + | + | – | d | +/d | d | + | + | + |
| Voges-Proskauer | + | – | – | + | w | + | +/d | d | d (w) | v | – |
| **Acid from** | | | | | | | | | | | |
| *N*-Acetylglucosamine | – | – | – | – | – | d | – | + | d | – | – |
| Amygdalin | – | + | – | – | – | + | – | d | – | – | – |
| L-Arabinose | – | + | + | – | – | – | d (w) | d | – | – | + |
| Arbutin | – | + | – | – | – | + | – | d | – | – | w |
| D-Cellobiose | – | + | + | + | + | + | + | + | d | – | + |
| Galactose | w | + | + | + | w | + | – | – | d | + | + |
| Gentiobiose | – | + | – | – | – | – | – | d | – | – | – |
| Glycerol | w | w | + | + | w | + | – | w | + | – | + |
| Glycogen | + | + | – | + | – | – | – | – | d | – | + |
| Lactose | – | + | – | – | – | – | – | w | d (w) | – | – |
| *meso*-inositol | – | – | – | + | – | – | – | – | d | – | – |
| Mannitol | – | – | + | + | – | + | w | + | d | w | + |
| D-Melezitose | + | – | + | + | – | + | d (w) | – | d (w) | – | – |

| | | | | | | | | | | | |
|---|---|---|---|---|---|---|---|---|---|---|---|
| D-Melibiose | + | + | – | – | – | + | d | – | d | d | – |
| Methyl-D-glucoside | + | w | + | – | + | d | – | + | d | – | + |
| D-Raffinose | + | + | – | – | – | + | – | – | d | – | – |
| Ribose | – | + | + | – | – | + | w | + | + | d | – |
| Salicin | – | + | – | + | + | + | d | + | w | – | + |
| Sucrose | + | + | + | + | + | + | d | + | d | d | + |
| D-Trehalose | + | + | + | + | + | + | d | + | d | + | + |
| D-Turanose | + | + | + | – | + | – | – | + | – | – | – |
| D-Xylose | – | + | + | + | + | – | w | + | d | – | – |
| Anaerobic growth | w/– | + | w | + | – | w | + | + | w | + | + |
| Growth at pH 5 | – | – | + | – | – | + | – | – | w | – | – |
| Growth at pH 9 | + | + | + | – | + | + | + | w | w | + | – |
| Minimum growth temperature (°C) | 30–45 | 50 | 45 | 37 | 37 | 37 | 50 | 40 | 37–40 | 37 | 45 |
| Maximum growth temperature (°C) | 60–70 | 70 | 65 | 60 | 80 | 80 | 70 | <60 | 70 | 70 | 65 |
| Optimum temperature (°C) | 40–60 | 50–65 | 60 | 55–60 | 60 | 60 | 50 | 50 | 60 | 60 | 55–60 |
| Growth in 1 % NaCl (w/v) | + | – | + | + | + | + | – | – | d | + | + |
| Growth in 5 % NaCl (w/v) | – | – | + | – | – | + | – | – | – | – | – |
| Mol % G+C of type strain | 52.8 | 45.8 | 54.5 | 52.3 | 53.7 | 55.2 | 48.4 | 43.4 | 51.2 | 43.9 | 50.4 |

Data for *G. uzenensis* were taken from Logan et al. (2009), because an authentic strain of this species was not available; all other data were taken from Dinsdale et al. (2011) and Coorevits et al. (2012). No entry indicates the data are not available. All strains studied were motile, formed ellipsoidal spores, and produced acid from fructose, glucose, maltose, and mannose

Symbols: +, >85 % strains give positive reaction; –, <15 % of the strains give positive reaction; +/w, positive or weakly positive reaction; w, weak reaction; w/–, weak or negative reaction; +/d, usually positive, but different strains give different reactions; d, different strains give different reactions, *d (w)* different strains give different reactions, but positive reactions are weak, v, results vary

*G. thermocatenulatus*, was isolated from a microbial mat that formed in the Garga spring, in the Transbaikal region of Russia (Nazina et al. 2004). *G. thermantarcticus* was isolated from geothermally heated volcanic soil in Antarctica (Nicolaus et al. 1991, 1996). Unidentified strains belonging to *Geobacillus* have been reported from deep-sea hydrothermal vents lying at 2,000–3,500 m (Marteinsson et al. 1996) and from sea mud of the Mariana Trench at 10,897 m below the surface (Takami et al. 1997).

## Pathogenicity

The body temperatures of humans and other animals lie at or near the minimum temperatures for growth for species of *Geobacillus*, and there have been no reports of infections with these organisms.

## Applications

There has long been great interest in the enzymes and other products of thermophilic aerobic endosporeformers on account of their thermostabilities. *G. stearothermophilus* spores are widely used as bioindicators for sterilization control. A more recent potential application is in the remediation of hydrocarbon-contaminated sites (Feng et al. 2007); as the spores of such organisms appear to be widely distributed, even in temperate sites, it may not be necessary to introduce the organisms (bioaugmentation) other than to enhance initial activity, and biostimulation with raised temperatures and the addition of limiting nutrients may be effective (Banat and Marchant 2011).

## References


Adkins JP, Cornell LA, Tanner RA (1992) Microbial composition of carbonate petroleum reservoir fluids. Geomicrobiol J 10:87–97

Ahmad S, Scopes RK, Rees GN, Patel BKC (2000) *Saccharococcus caldoxylosilyticus* sp. nov., an obligately thermophilic, xylose-utilizing, endospore-forming bacterium. Int J Syst Evol Microbiol 50:517–523

Allen MB (1953) The thermophilic aerobic sporeforming bacteria. Bacteriol Rev 17:125–173

Alvarez M, Wouters J, Maes D, Mainfroid V, Rentier-Delrue F, Wyns L, Depiereux E, Martial JA (1999) Lys13 plays a crucial role in the functional adaptation of the thermophilic triose-phosphate isomerase from *Bacillus stearothermophilus* to high temperatures. J Biol Chem 274:19181–19187

Ambroz A (1913) *Denitrobacterium thermophilum* spec nova, ein Beitrag zur Biologie der thermophilen Bakterien. Zentralbl Bakteriol Parasitenkd Infektionskr Hyg Abt II 37:3–16

Andersson M, Laukkanen M, Nurmiaho-Lassila E-L, Rainey FA, Niemelä SI, Salkinoja-Salonen M (1995) *Bacillus thermosphaericus* sp. nov., a new thermophilic ureolytic *Bacillus* isolated from air. Syst Appl Microbiol 18:203–220

Ash C, Farrow JAE, Wallbanks S, Collins MD (1991) Phylogenetic heterogeneity of the genus *Bacillus* revealed by comparative analyses of small subunit-ribosomal RNA sequences. Lett Appl Microbiol 13:202–206

Baillie A, Walker PD (1968) Enzymes of thermophilic aerobic spore-forming bacteria. J Appl Bacteriol 31:114–119

Banat I, Marchant R (2011) Geobacillus activities in soil and oil contamination remediation. In: Logan NA, De Vos P (eds) Aerobic, endospore-forming soil bacteria. Springer, Berlin, pp 259–270

Banat IM, Marchant R, Rahman TJ (2004) *Geobacillus debilis* sp. nov., a novel obligately thermophilic bacterium isolated from a cool soil environment, and reassignment of *Bacillus pallidus* to *Geobacillus pallidus* comb. nov. Int J Syst Evol Microbiol 54:2197–2201

Blanc M, Marilley L, Beffa T, Aragno M (1997) Thermophilic bacterial communities in hot composts as revealed by most probable number counts and molecular (16S rDNA) methods. FEMS Microbiol Ecol 28:141–149

Bonjour F, Aragno M (1984) *Bacillus tusciae*, a new species of thermoacidophilic, facultatively chemolithotrophic, hydrogen oxidizing sporeformer from a geothermal area. Arch Microbiol 139:397–401

Breed RS, Murray EGD, Smith NR (1957) Bergey's manual of determinative bacteriology, 7th ed. Williams and Wilkins, Baltimore

Caccamo D, Gugliandolo C, Stackebrandt E, Maugeri TL (2000) *Bacillus vulcani* sp. nov., a novel thermophilic species isolated from a shallow marine hydrothermal vent. Int J Syst Evol Microbiol 50:2009–2012

Caccamo D, Maugeri TL, Gugliandolo C (2001) Identification of thermophilic and marine bacilli from shallow thermal vents by restriction analysis of their amplified 16S rDNA. J Appl Microbiol 91:520–524

Castenholst RW (1969) Thermophilic blue-green algae and the thermal environment. Bacteriol Rev 33:476–504

Chen XG, Stabnikova O, Tay JH, Wang JY, Tay ST (2004) Thermoactive extracellular proteases of *Geobacillus caldoproteolyticus*, sp. nov., from sewage sludge. Extremophiles 8:489–498

Chopra AK, Mathur DK (1984) Isolation, screening and characterisation of thermophilic *Bacillus* species isolated from dairy products. J Appl Bacteriol 57:263–271

Cihan AC, Ozcan B, Tekin N, Cokmus C (2011) *Geobacillus themodenitrificans* subsp. *calidus*, subsp. nov., a thermophilic and a-glucosidase producing bacterium isolated from Kizilcahamam, Turkey. J Gen Appl Microbiol 57:83–92

Claus D, Berkeley RCW (1986) Genus *Bacillus* Cohn 1872. In: Sneath PHA, Mair NS, Sharpe ME, Holt JG (eds) Bergey's manual of systematic bacteriology, vol 2. The Williams and Wilkins, Baltimore, pp 1105–1139

Combet-Blanc Y, Ollivier B, Streicher C, Patel BKC, Dwivedi PP, Pot B, Prensier G, Garcia J-L (1995) *Bacillus thermoamylovorans* sp. nov., a moderately thermophilic and amylolytic bacterium. Int J Syst Bacteriol 45:9–16

Coorevits A, Dinsdale AE, Halket G, Lebbe L, De Vos P, Van Landschoot A, Logan NA (2012) Taxonomic revision of the genus *Geobacillus*: emendation of *Geobacillus*, *G. stearothermophilus*, *G. jurassicus*, *G. toebii*, *G. thermodenitrificans* and *G. thermoglucosidans* (nom. corrig., formerly "*thermoglucosidasius*"); transfer of *Bacillus thermantarcticus* to the genus as *G. thermantarcticus*; proposal of *Caldibacillus debilis* gen. nov., comb. nov.; transfer of *G. tepidamans* to *Anoxybacillus* as *A. tepidamans* and proposal of *Anoxybacillus caldiproteolyticus* sp. nov. Int J Syst Evol Microbiol 62:1470–1485

Darland G, Brock TD (1971) *Bacillus acidocaldarius* sp. nov., an acidophilic, thermophilic spore-forming bacterium. J Gen Microbiol 67:9–15

De Bartolemeo A, Trotta F, La Rosa F, Saltalamacchia G, Mastrandrea V (1991) Numerical analysis and DNA base compositions of some thermophilic *Bacillus* species. Int J Syst Bacteriol 41:502–509

de Vrij W, Speelmans G, Heyne RIR, Konings WN (1990) Energy transduction and amino acid transport in thermophilic aerobic and fermentative bacteria. FEMS Microbiol Rev 75:183–200

Deák T, Temár É (1988) Simplified identification of aerobic spore-formers in the investigation of foods. Int J Food Microbiol 6:115–125

Degryse E, Glansdorff N, Piérard A (1978) A comparative analysis of extreme thermophilic bacteria belonging to the genus *Thermus*. Arch Microbiol 117:189–196

Deinhard G, Blanz P, Poralla K, Altan E (1987a) *Bacillus acidoterrestris* sp. nov., a new thermotolerant acidophile isolated from different soils. Syst Appl Microbiol 10:47–53

Deinhard G, Saar J, Krischke W, Poralla K (1987b) *Bacillus cycloheptanicus* sp. nov., a new thermoacidophile containing omega-cycloheptane fatty acids. Syst Appl Microbiol 10:68–73

Demharter W, Hensel R (1989) *Bacillus thermocloacae* sp. nov., a new thermophilic species from sewage sludge. Syst Appl Microbiol 11:272–276

Dinsdale AE, Halket G, Coorevits A, Van Landschoot A, Busse H-J, De Vos P, Logan NA (2011) Emended descriptions of *Geobacillus thermoleovorans* and *Geobacillus thermocatenulatus*. Int J Syst Evol Microbiol 61:1802–1810

Donk PJ (1920) A highly resistant thermophilic organism. J Bacteriol 5:373–374

Feng L, Wang W, Cheng J, Ren Y, Zhao G, Gao C, Tang Y, Liu X, Han W, Peng X, Liu R, Wang L (2007) Genome and proteome of long-chain alkane degrading *Geobacillus thermodenitrificans* NG80-2 isolated from a deep-subsurface oil reservoir. Proc Natl Acad Sci U S A 104:5602–5607

Fortina MG, Pukall R, Schumann P, Mora D, Parini C, Manachini PL, Stackebrandt E (2001a) *Ureibacillus* gen. nov., a new genus to accommodate *Bacillus thermosphaericus* (Andersson et al. 1995), emendation of *Ureibacillus thermosphaericus* and description of *Ureibacillus terrenus* sp. nov. Int J Syst Evol Microbiol 51:447–455

Fortina MG, Mora D, Schumann P, Parini C, Manachini PL, Stackebrandt E (2001b) Reclassification of *Saccharococcus caldoxylosilyticus* as *Geobacillus caldoxylosilyticus* (Ahmad et al. 2000) comb. nov. Int J Syst Evol Microbiol 51:2063–2071

Galesloot TE, Labots H (1959) Thermofiele sporevormers in melk, vooral met betrekking tot de bereiding van gesteriliseerde melk en chocolademelk. Ned Melk Zuiveltijd 13:155–179

Garcia JL, Roussos S, Bensoussan M, Bianchi A, Mandel M (1982) Numerical taxonomy of a thermophilic "*Bacillus*" species isolated from West African rice soils. Ann Microbiol (Paris) 133:471–488

Gibson T, Gordon RE (1974) *Bacillus* Cohn 1872. In: Buchanan RE, Gibbons NE (eds) Bergey's manual of determinative bacteriology, 8th edn. The Williams and Wilkins, Baltimore, pp 529–550

Golovacheva RS, Egorova LA, Loginova LG (1965) Ecology and systematics of aerobic obligate-thermophilic bacteria isolated from thermal localities on Mount Yangan-Tau and Kunashir Isle of the Kuril chain. Microbiology (English translation of Mikrobiologiya) 34:693–698

Golovacheva RS, Loginova LG, Salikhov TA, Kolesnikov AA, Zaitseva GN (1975) A new thermophilic species *B. thermocatenulatus* sp. nov. Microbiology (English translation of Mikrobiologiya) 44:230–233

Gordon RE, Smith NR (1949) Aerobic sporeforming bacteria capable of growth at high temperatures. J Bacteriol 58:327–341

Grinstead E, Clegg LFL (1955) Spore-forming organisms in commercial sterilized milk. J Dairy Res 22:178–190

Guicciardi A, Biffi MR, Manachini PL, Craveri A, Scolastico C, Rindone B, Craveri C (1968) Ricerche preliminary su un nuovo schizomicete termofilo del genere *Bacillus* e caratterizzazione del pigmento rosso prodotto. Ann Microbiol (Milan) 18:191–205

Heinen UJ, Heinen W (1972) Characteristics and properties of a caldoactive bacterium producing extracellular enzymes and two related strains. Arch Mikrobiol 82:1–23

Heinen W, Lauwers AM, Mulders JWM (1982) *Bacillus flavothermus*, a newly isolated facultative thermophile. Antonie Van Leeuwenhoek J Microbiol Serol 48:265–272

Heyndrickx M, Lebbe L, Vancanneyt M, Kersters K, De Vos P, Logan NA, Forsyth G, Nazli S, Ali N, Berkeley RCW (1997) A polyphasic reassessment of the genus *Aneurinibacillus*, reclassification of *Bacillus thermoaerophilus* (Meier-Stauffer et al. 1996) as *Aneurinibacillus thermoaerophilus* comb. nov. and emended descriptions of *A. aneurinilyticus*, of *A. migulanus* and of *A. thermoaerophilus*. Int J Syst Bacteriol 47:808–817

Kalogridou-Vassilliadu D (1992) Biochemical activities of *Bacillus* species isolated from flat sour evaporated milk. J Dairy Sci 75:2681–2686

Kämpfer P (1994) Limits and possibilities of total fatty acid analysis for classification and identification of *Bacillus* species. Syst Appl Microbiol 17:86–96

Kato T, Haruki M, Imanaka T, Morikawa M, Kanaya S (2001) Isolation and characterization of long-chain-alkane degrading *Bacillus thermoleovorans* from deep subterranean petroleum reservoirs. J Biosci Bioeng 91:64–70

Kawamura S, Abe Y, Ueda T, Masumoto K, Imoto T, Yamasaki N, Kimura M (1998) Investigation of the structural basis for thermostability of DNA-binding protein HU from *Bacillus stearothermophilus*. J Biol Chem 273:19982–19987

Klaushofer H, Hollaus F (1970) Zur Taxonomie der hoch-thermophilen, in Zukerfabriksäften vorkommenden aeroben sporenbildner. Z Zuckerrind 20:465–470

Klenk H-P, Lapidus A, Chertkov O, Copeland A, Del Rio TG, Nolan M, Lucas S, Chen F, Tice H, Cheng J-F, Han C, Bruce D, Goodwin L, Pitluck S, Pati A, Ivanova N, Mavromatis K, Daum C, Chen A, Palaniappan K, Chang Y-J, Land M, Hauser L, Jeffries CD, Detter JC, Rohde M, Abt B, Pukall R, Göker M, Bristow J, Markowitz V, Hugenholtz P, Eisen JA (2012) Complete genome sequence of the thermophilic, hydrogen-oxidizing *Bacillus tusciae* type strain (T2T) and reclassification in the new genus, *Kyrpidia* gen. nov. as *Kyrpidia tusciae* comb. nov. and emendation of the family *Alicyclobacillaceae* da Costa and Rainey, 2010. Stand Genomic Sci 5:121–134

Kuisiene N, Raugalas J, Chitavichius D (2004) *Geobacillus lituanicus* sp. nov. Int J Syst Evol Microbiol 54:1991–1995

Leadbetter ER, Foster JW (1958) Studies on some methane-utilizing bacteria. Arch Microbiol 30:91–118

Lee D-W, Koh Y-S, Kim K-J, Kim B-C, Choi H-J, Kim D-S, Suhatono MT, Pyun Y-R (1999) Isolation and characterization of a thermophilic lipase from *Bacillus thermoleovorans* ID-1. FEMS Microbiol Lett 179:393–400

Llarch À, Logan NA, Castellví J, Prieto MJ, Guinea J (1997) Isolation and characterization of thermophilic *Bacillus* species from Deception Island, South Shetland archipelago. Microb Ecol 34:58–65

Llaudes MK, Zhao L, Duffy S, Schaffner DW (2001) Simulation and modelling of the effect of small inoculum size on time to spoilage by *Bacillus stearothermophilus*. Food Microbiol 18:395–405

Logan NA, Berkeley RCW (1981) Classification and identification of members of the genus *Bacillus*. In: Berkeley RCW, Goodfellow M (eds) The aerobic endospore-forming bacteria. Academic, London, pp 105–140

Logan NA, De Vos P, Dinsdale A (2009) Genus *Geobacillus* Nazina et al. 2001. In: De Vos P, Garrity G, Jones D, Krieg NR, Ludwig W, Rainey FA, Schleifer K-H, Whitman WB (eds) Bergey's manual of systematic bacteriology, vol 3, 2nd edn. Springer, New York, pp 144–160

Manachini PL, Fortina MG, Parini C, Craveri R (1985) *Bacillus thermoruber* sp. nov., nom. rev., a red-pigmented thermophilic bacterium. Int J Syst Bacteriol 35:493–496

Manachini PL, Mora D, Nicastro G, Parini C, Stackebrandt E, Pukall R, Fortina MG (2000) *Bacillus thermodenitrificans* sp. nov., nom. rev. Int J Syst Evol Microbiol 50:1331–1337

Marchant R, Banat IM, Rahman TJS, Berzano M (2002) What are high-temperature bacteria doing in cold environments? Trends Microbiol 10:120–121

Markossian S, Becker P, Markl H, Antranikian G (2000) Isolation and characterization of lipid-degrading *Bacillus thermoleovorans* IHI-91 from an Icelandic hot spring. Extremophiles 4:365–371

Marteinsson VG, Birrien J-L, Jeanthon C, Prieur D (1996) Numerical taxonomic study of thermophilic *Bacillus* isolated from three geographically separated deep-sea hydrothermal vents. FEMS Microbiol Ecol 21:255–266

Martins LO, Jurado AS, Madiera VMC (1990) Composition of polar lipid acyl chains of *Bacillus stearothermophilus* as affected by temperature and calcium. Biochim Biophys Acta 1045:17–20

Maugeri TL, Gugliandolo C, Caccamo D, Stackebrandt E (2001) A polyphasic taxonomic study of thermophilic bacilli from shallow, marine vents. Syst Appl Microbiol 24:572–587

Meier-Stauffer K, Busse H-J, Rainey FA, Burghardt J, Scheberl A, Hollaus F, Kuen B, Makristathis A, Sleytr UB, Messner P (1996) Description of *Bacillus thermoaerophilus* sp. nov., to include sugar beet isolates and *Bacillus brevis* ATCC 12990. Int J Syst Bacteriol 46:532–541

Merkel GJ, Underwood WH, Perry JJ (1978) Isolation of thermophilic bacteria capable of growth solely in long-chain hydrocarbons. FEMS Microbiol Lett 3:81–83

Miñana-Galbis D, Pinzón DL, Lorén JG, Manresa Á, Oliart-Ros RM (2010) Reclassification of *Geobacillus pallidus* (Scholz et al. 1988) Banat et al. 2004 as *Aeribacillus pallidus* gen. nov., comb. nov. Int J Syst Evol Microbiol 60:433–446

Minnikin DE, Abdolrahimzadeh H, Wolf J (1977) Taxonomic significance of polar lipids in some thermophilic members of *Bacillus*. In Barker AN, Wolf J, Ellar DJ, Dring GJ, Gould GW (eds). Academic, London, pp 879–893

Mishustin EN (1950) Termofilnie mikroorganiszmi w prirode I praktike. Akademi Nauk SSSR, Moskwa

Mora D, Fortina MG, Nicastro G, Parini C, Manachini PL (1998) Genotypic characterization of thermophilic bacilli: a study on new soil isolates and several reference strains. Res Microbiol 149:711–722

Nazina TN, Tourova TP, Poltaraus AB, Novikova EV, Grigoryan AA, Ivanova AE, Lysenko AM, Petrunyaka VV, Osipov GA, Belyaev SS, Ivanov MV (2001) Taxonomic study of aerobic thermophilic bacilli: descriptions of *Geobacillus subterraneus* gen nov., sp. nov. and *Geobacillus uzenensis* sp. nov. from petroleum reservoirs and transfer of *Bacillus stearothermophilus, Bacillus thermocatenulatus, Bacillus thermoleovorans Bacillus kaustophilus, Bacillus thermoglucosidasius, Bacillus thermodenitrificans* to *Geobacillus* as *Geobacillus stearothermophilus, Geobacillus thermocatenulatus, Geobacillus thermoleovorans Geobacillus kaustophilus, Geobacillus thermoglucosidasius, Geobacillus thermodenitrifi.* Int J Syst Evol Microbiol 51:433–446

Nazina TN, Lebedeva EV, Poltaraus AB, Tourova TP, Grigoryan AA, Sokolova DS, Lysenko AM, Osipov GA (2004) *Geobacillus gargensis* sp. nov., a novel thermophile from a hot spring, and the reclassification of *Bacillus vulcani* as *Geobacillus vulcani* (Caccamo et al. 2000) comb. nov. Int J Syst Evol Microbiol 54:2019–2024

Nazina TN, Sokolova DS, Grigoryan AA, Shestakova NM, Mikhailova EM, Poltaraus AB, Tourova TP, Lysenko AM, Osipov GA, Belyaev SS (2005) *Geobacillus jurassicus* sp. nov., a new thermophilic bacterium isolated from a high-temperature petroleum reservoir, and the validation of the *Geobacillus* species. Syst Appl Microbiol 28:43–53

Nicolaus B, Marsiglia F, Esposito E, Trincone A, Lama L, Sharp R, di Prisco G, Gambacorta A (1991) Isolation of five strains of thermophilic eubacteria in Antarctica. Polar Biol 11:425–429

Nicolaus B, Lama L, Esposito E, Manca MC, di Prisco G, Gambacorta A (1996) *Bacillus thermoantarcticus* sp. nov. from Mount Melbourne, Antarctica: a novel thermophilic species. Polar Biol 16:101–104

Nicolaus B, Lama L, Esposito E, Manca MC, di Prisco G, Gambacorta A (2002) Validation list no. 84. Int J Syst Evol Microbiol 52:3–4

Norris JR, Berkeley RCW, Logan NA, O'Donnell AG (1981) The genera *Bacillus* and *Sporolactobacillus*. In: Starr MP, Stolp H, Truper HG, Balows A, Schlegel HG (eds) The prokaryotes: a handbook on habitats, isolation and identification of bacteria, vol 2. Springer, Berlin/Heidelberg, pp 1711–1742

Obojska A, Ternan NG, Lejczak B, Kafarski P, McMullan G (2002) Organophosphate utilization by the thermophile *Geobacillus caldoxylosilyticus* T20. Appl Environ Microbiol 68:2081–2084

Perl D, Mueller U, Heinemann U, Schmid FX (2000) Two exposed amino acid residues confer thermostability on a cold shock protein. Nat Struct Biol 7:380–383

Pikuta E, Lysenko A, Chuvilskaya N, Mendrock U, Hippe H, Suzina N, Nikitin D, Osipov G, Laurinavichius K (2000) *Anoxybacillus pushchinensis* gen. nov., sp. nov., a novel anaerobic, alkaliphilic, moderately thermophilic bacterium from manure, and description of *Anoxybacillus flavithermus* comb. nov. Int J Syst Evol Microbiol 50:2109–2117

Poli A, Romano I, Caliendo G, Nicolaus G, Orlando P, Falco A, Lama L, Gambacorta A, Nicolaus B (2006) *Geobacillus toebii* subsp. *decanicus* subsp. nov., a hydrocarbon-degrading, heavy metal resistant bacterium from hot compost. J Gen Appl Microbiol 52:223–234

Prickett PS (1928) Thermophilic and thermoduric microorganisms with special reference to species isolated from milk. New York Agric Exp Stat Tech Bull 147:58

Priest FG, Goodfellow M, Todd C (1988) A numerical classification of the genus *Bacillus*. J Gen Microbiol 134:1847–1882

Rahman RNZRA, Leow TC, Salleh AB, Basri M (2007) *Geobacillus zalihae* sp. nov., a thermophilic lipolytic bacterium isolated from palm oil mill effluent in Malaysia. BMC Microbiol 7:77

Rainey FA, Fritze D, Stackebrandt E (1994) The phylogenetic diversity of thermophilic members of the genus *Bacillus* as revealed by 16S rDNA analysis. FEMS Microbiol Lett 115:205–212

Romano I, Poli A, Lama L, Gambacorta A, Nicolaus B (2005) *Geobacillus thermoleovorans* subsp. *stromboliensis* subsp. nov., isolated from the geothermal volcanic environment. J Gen Appl Microbiol 51:183–189

Sakaff MKLM, Rahman AYA, Saito JA, Hou S, Alama M (2012) Complete genome sequence of the thermophilic bacterium *Geobacillus thermoleovorans* CCB_US3_UF5. J Bacteriol 194:1239

Schäffer C, Franck WL, Scheberl A, Kosma P, McDermott TR, Messner P (2004) Classification of isolates from locations in Austria and Yellowstone National Park as *Geobacillus tepidamans* sp. nov. Int J Syst Evol Microbiol 54:2361–2368

Schenk A, Aragano M (1979) *Bacillus schlegelii* a new species of thermophilic, facultatively chemoli-thoautotrophic bacterium oxidizing molecular hydrogen. J Gen Microbiol 115:333–341

Schleifer KH, Kandler O (1972) Peptidoglycan types of bacterial cell walls and their taxonomic implications. Bacteriol Rev 36:407–477

Scholz T, Demharter W, Hensel R, Kandler O (1987) *Bacillus pallidus* sp. nov., a new thermophilic species from sewage. Syst Appl Microbiol 9:91–96

Shapton DA, Hindes WR (1963) The standardization of a spore count technique. Chem Indust 41:230–234

Sharp RJ, Bown KJ, Atkinson A (1980) Phenotypic and genotypic characterization of some thermophilic species of *Bacillus*. J Gen Microbiol 117:201–210

Shida O, Takagi H, Kadowaki K, Komagata K (1996) Proposal for two new genera, *Brevibacillus* gen. nov. and *Aneurinibacillus* gen. nov. Int J Syst Bacteriol 46:939–946

Skerman VBD, McGowan V, Sneath PHA (1980) Approved lists of bacterial names. Int J Syst Bacteriol 30:225–420

Stamatakis A (2006) RAxML-VI-HPC: maximum likelihood-based phylogenetic analyses with thousands of taxa and mixed models. Bioinformatics 22:2688–2690

Studholme DJ, Jackson RA, Leak DJ (1999) Phylogenetic analysis of transformable strains of thermophilic *Bacillus* species. FEMS Microbiol Lett 172:85–90

Sung M-H, Kim H, Bae J-W, Rhee S-K, Jeon CO, Kim K, Kim J-J, Hong S-P, Lee S-G, Yoon J-H, Park Y-H, Baek D-H (2002) *Geobacillus toebii* sp. nov., a novel thermophilic bacterium isolated from hay compost. Int J Syst Evol Microbiol 52:2251–2255

Sunna A, Prowe SG, Stroffregen T, Antranikian G (1997a) Characterization of the xylanases from the new isolated thermophilic xylan-degrading *Bacillus thermoleovorans* strain K-3d and *Bacillus flavothermus* strain LB3A. FEMS Microbiol Lett 148:209–216

Sunna A, Tokajian S, Burghardt J, Rainey F, Antranikian G, Hashwa F (1997b) Identification of *Bacillus kaustophilus, Bacillus thermocatenulatus* and *Bacillus* strain HSR as members of *Bacillus thermoleovorans*. Syst Appl Microbiol 20:232–237

Suzuki Y, Kishigami T, Abe S (1976) Production of extracellular α-glucosidase by a thermophilic *Bacillus* species. Appl Environ Microbiol 31:807–812

Suzuki Y, Kishigami T, Inoue K, Mizoguchi Y, Eto N, Takagi M, Abe S (1983) *Bacillus thermoglucosidasius* sp. nov., a new species of obligately thermophilic bacilli. Syst Appl Microbiol 4:487–495

Takami H, Inoue A, Fuji F, Horikoshi K (1997) Microbial flora in the deepest sea mud of the Mariana Trench. FEMS Microbiol Lett 152:279–285

Takami H, Takaki Y, Chee G-J, Nishi S, Shimamura S, Suzuki H, Matsui S, Uchiyama I (2004) Thermoadaptation trait revealed by the genome sequence of thermophilic *Geobacillus kaustophilus*. Nucleic Acids Res 32:6292–6303

Tanner RS (1989) Monitoring sulfate-reducing bacteria: comparison of enumeration media. J Microbiol Methods 10:83–90

Tolner B, Poolman B, Konings WN (1997) Adaptation of microorganisms and their transport systems to high temperatures. Comp Biochem Physiol 118A:423–428

Walker PD, Wolf J (1961) Some properties of aerobic thermophiles growing at 65°. J Appl Bacteriol 24:iv–v

Walker PD, Wolf J (1971) The taxonomy of *Bacillus stearothermophilus*. In: Barker AN, Gould GW, Wolf J (eds) Spore research 1971. Academic, London, pp 247–262

Weigel J (1986) Methods for isolation and study of thermophiles. In: Brock TD (ed) Thermophiles: general, molecular and applied microbiology. Wiley, New York, pp 17–37

White D, Sharp RJ, Priest FG (1993) A polyphasic taxonomic study of thermophilic bacilli from a wide geographical area. Antonie Van Leeuwenhoek J Microbiol Serol 64:357–386

Wisotzkey JD, Jr Jurtshuk P, Fox GE, Deinhard G, Poralla K (1992) Comparative sequences analyses on the 16S rRNA (rDNA) of *Bacillus acidocaldarius, Bacillus acidoterrestris*, and *Bacillus cycloheptanicus* and proposal for creation of a new genus, *Alicyclobacillus* gen. nov. Int J Syst Bacteriol 42:263–269

Wolf J, Chowhury MSU (1971) Taxonomy of *B. circulans* and *B. stearothermophilus*. In: Barker AN, Gould GW, Wolf J (eds) Spore research 1971. Academic, London, pp 349–350

Wolf J, Sharp RJ (1981) Taxonomic and related aspects of thermophiles within the genus *Bacillus*. In: Berkeley RCW, Goodfellow M (eds) The aerobic endospore-forming bacteria. Academic, London, pp 251–296

Yarza P, Ludwig W, Euzéby J, Amann R, Schleifer K-H, Glöckner FO, Rosselló-Móra R (2010) Update of the all-species living tree project based on 16S and 23S rRNA sequence analyses. Syst Appl Microbiol 33:291–299

Zarilla KA, Perry JJ (1987) *Bacillus thermoleovorans*, sp. nov., a species of obligately thermophilic hydrocarbon utilizing endospore-forming bacteria. Syst Appl Microbiol 9:258–264

Zeigler DR (2005) Application of a recN sequence similarity analysis to the identification of species within the bacterial genus *Geobacillus*. Int J Syst Evol Microbiol 55:1171–1179

# 11 The Family *Gracilibacteraceae* and Transfer of the Genus *Lutispora* into *Gracilibacteraceae*

*Erko Stackebrandt*
Leibniz Institute DSMZ-German Collection of Microorganisms and Cell Cultures GmbH,
Braunschweig, Germany



**Abstract**

*Gracilibacter thermotolerans*, with strain JW/YJL-S1$^T$ as the type strain, is the only species of the genus *Gracilibacter*. The type strain, isolated from sediment of an acid sulfate-containing water of a constructed wetland system (Savannah River Site near Aiken, SC, USA), is obligatorily anaerobic, chemoorganotrophic, asporogenic, and thermotolerant. Its closest phylogenetic neighbor is *Lutispora thermophila*, with which it shares 85.3 % 16S rRNA gene sequence similarity. As these two genera form an individual clade among neighboring lineages with family status, *Lutispora* is transferred into the family *Gracilibacteraceae* on the basis of its phylogenetic position.

## Taxonomic Position

With the increased attempt to create higher taxa on the basis of 16S rRNA gene sequence similarities and the phylogenetic position of *Gracilibacter thermotolerans* within the phylum Firmicutes, the monogeneric family *Gracilibacteraceae* was described. At the time of its description (Lee et al. 2006), *G. thermotolerans* was found to be remotely related to members of *Clostridium* clusters I/II and III (Collins et al. 1994) with the highest gene sequence similarity with *Clostridium thermosuccinogenes* DSM 5807$^T$. The inclusion of additional 16S rRNA sequences to public databases refined the phylogenetic position and *Gracilibacteraceae* is today neighboring *Lutispora thermophila* (Shiratori et al. 2008), and this clade is branching intermediate to genera of *Lachnospiraceae* and *Caldicoprobacteraceae* on the one side and to members of *Clostridiaceae* group I on the other side (➋ *Fig. 11.1*).

## Phenotype

Cells of strain JW/YJL-S1$^T$ were Gram staining negative straight to curved rods, 0.2–0.4 x 2.0–7.0 μm in length. Cells were either single or formed chains. Cells up to 45 mm in length were occasionally detected. Spores were not detected. Retarded peritrichous flagella (1–5 per cell) were detected under the electron microscope, and motility was observed on SIM agar medium (Cappuccin and Sherman 1987). Colonies were less than 1 mm in diameter, circular to irregular, and mostly translucent and filamentous. Growth occurred at 25–54 °C, but not at or below 20 or at or above 58 °C, with an optimum of 42.5–46.5 °C. The pH range for growth was 6.0–8.3 with an optimum at pH 6.8–7.8. The salinity (NaCl) growth range was 0–1.5 % (w/v), with an optimum at 0–0.5 %. Yeast extract was required for growth. Main end products of glucose fermentation were acetate, lactate, and ethanol. Casamino acids, tryptone, peptone, maltose, sucrose, arabinose, fructose, galactose, glucose, mannose, xylose, mannitol, and sorbitol were used as carbon and energy sources, while no growth was observed with cellobiose, lactose, raffinose, ribose, trehalose, inositol, xylitol, acetate, lactate, pyruvate, methanol, or carboxymethyl cellulose as carbon and energy sources.

Fumarate, nitrate, sulfate, sulfite (2 mM), thiosulfate, elemental sulfur, iron(III), anthraquinone-2,6-disulfonate, or manganese(IV) at concentrations of 20 mM were not used as electron acceptors, tested in media containing 20 mM lactate or 0.1 % yeast. No growth occurred in the presence of $H_2/CO_2$ (80:20, v/v). Positive growth occurred on peptone-yeast extract (PY), peptone-yeast extract glucose (PYG), reinforced clostridial medium (RCM; Difco), and thioglycolate broth (Difco). According to the API ZYM system (bioMérieux), strain JW/YJL-S1$^T$ was positive for esterase, leucine arylamidase, acid phosphatase, naphthol-AS-BI-phosphohydrolase, β-galactosidase, α-glucosidase, and β-glucosidase.

The fatty acid composition was dominated by branched-chain compounds: iso-$C_{15:0}$, anteiso-$C_{15:0}$, iso-$C_{16:0}$, and iso-$C_{17:0}$. The mol% G+C content of DNA was 42.8 mol% (HPLC). The 16S rRNA gene sequences of the type strain JW/YJL-S1$^T$ show sequence polymorphism of 2 % divergence.

Of the antibiotics ampicillin, chloramphenicol, erythromycin, rifampicin, streptomycin, and tetracycline tested at concentrations of 10 and 100 mM, strain JW/YJL-S1T was resistant only to 10 mM streptomycin.

E. Rosenberg et al. (eds.), *The Prokaryotes – Firmicutes and Tenericutes*, DOI 10.1007/978-3-642-30120-9_358,

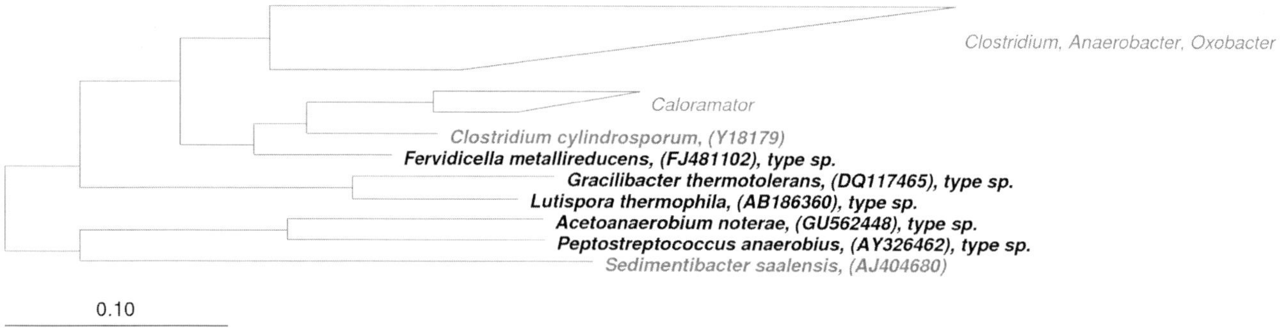


**Fig. 11.1**
**Neighbor-joining genealogy reconstruction based on the RAxML algorithm (Stamatakis 2006) of the sequences of *Gracilibacter* and *Lutispora* species and some neighboring taxa present in the LTP_106 (Yarza et al. 2010). The tree was reconstructed by using a subset of sequences. Representative sequences from close relative genera were used to stabilize the tree topology. In addition, a 40 % maximum frequency filter was applied to remove hypervariable positions from the alignment. Scale bar indicates estimated sequence divergence**

## Proposal to Transfer the Genus *Lutispora* into the Family *Gracilibacteraceae*

The genus *Lutispora* was described by Shiratori et al. (2008) who already mentioned in the original description the phylogenetic relationship to *Gracilibacter*. The type strains of the two monospecific genera are anaerobic, moderately thermophilic, rod shaped, stain Gram-negatively, require yeast extract for growth, and share high amounts of iso-$C_{15:0}$ fatty acids. They differ from each other significantly in that *Lutispora* forms clearly motile spores, has a lower G+C content of DNA of about 6 mol%, and has higher amounts of $C_{14:0}$ and $C_{16:0}$ DMA fatty acids but lower amounts of $C_{16:0}$ and iso-$C_{17:0}$ (*Table 11.1*).

*Lutispora thermophila* EBR46[T] does not utilize carbohydrates but peptone, tryptone, casamino acids, casein hydrolysates, pyruvate, methionine, threonine, tryptophan, cysteine, lysine, and serine for growth. Fermentation products from tryptone are acetate, isobutyrate, propionate, and isovalerate. Hydrogen sulfide is produced from cysteine. Electron acceptors such as fumarate, sulfate, nitrate, elemental sulfur, or iron (III) are not used.

**Table 11.1**
**Properties distinguishing *Gracilibacter thermotolerans* and *Lutispora thermophila* (Data from Shiratori et al. 2008)**

| Property | *Gracilibacter thermotolerans* | *Lutispora thermophila* |
|---|---|---|
| Spore formation | – | + |
| Motility | Rarely | + |
| Mol% G+C of DNA | 42.8 | 36.2 |
| Temperature range (optimum) | 25–54 (43–47) | 40–60 (55–58) |
| pH optimum | 6.8–7.8 | 7.5–8.0 |
| Fatty acids | | |
| $C_{14:0}$ | 2.3 | 21.4 |
| $C_{16:0}$ | 29.0 | 3.9 |
| iso-$C_{17:0}$ | 15.4 | 0.3 |
| $C_{16:0}$ DMA | – | 10.7 |

## Isolation and Maintenance

Strain JW/YJL-S1[T] was isolated from a most probable number tube inoculated with sediment from the upper layer of a wetland system, receiving water from an acid sulfate runoff pond from a coal pile located at the Department of Energy's Savannah River Site near Aiken, SC, USA (Lee 2005). According to the authors, the specific habitat is unknown. The sample site was dominated by iron oxyhydroxide precipitate coating. Single colonies were obtained from dilution rows in 1.5 % (w/v agar) shake roll-tubes. Purity was verified by additional five rounds of single colony isolation using the agar-shake roll-tube method (Ljungdahl and Wiegel 1986). The isolate was routinely cultured in a carbonate-buffered basal medium (Widdel and Bak 1992) supplemented with 20 mM acetate and 0.1 mM ferric citrate at pH 25 °C (Wiegel 1998) and 37 °C under anaerobic conditions (100 % $N_2$) using a modified Hungate technique (Ljungdahl and Wiegel 1986). In the German Collection of Microorganisms and Cell Cultures, strain JW/YJL-S1[T] is routinely cultured anaerobically on CM3 medium (medium 520, http://www.dsmz.de/?id=441) at 45 °C.

*Lutispora thermophila* was isolated from a methanogenic bioreactor in a yeast extract-mineral-vitamin medium, supplemented with various cellulosic substrates as described by Shiratori et al. (2006). The addition of neomycin (50 μg $ml^{-1}$) was used to remove contaminating organisms. Modified PYG medium (medium 104, http://www.dsmz.de/?id=441) is used for routine maintenance.

Cells of both species are long term stored in liquid nitrogen, or lyophilized.

## Ecology

BLASTN search against sequences in GenBank at the time of the original description indicated that strain JW/YJL-S1$^T$ was closely related to uncultured clones mostly obtained from methanogenic environments and to consortia including those from rice paddy-field microcosms, methanogenic fermenter cultures degrading acetate propionate or butyrate (GenBank accession numbers AB221361, AB232817, AB248637, AB232818, AB248624, and AB248638), and uranium reduction enrichment plant (DQ125504and DQ125852). In 2013, this picture did not change, as the clone sequences indicated above are still among those which are most closely (though distantly) related to strain JW/YJL-S1$^T$ and the range of habitats of relative sequences is still narrow but expanded by methanogenic corn stalk-degrading microbial systems (AB742117), potential gas hydrate area, Taiwan (JQ816709), or a hydrocarbon-contaminated aquifer (Tischer et al. 2013).

Similar environments have been described to contain clone sequences highly identical to those of *Lutispora thermophila*, such as packed-bed reactors run under different conditions (Sasaki et al. 2006, 2007) and thermophilic bicompost (Sizova et al. 2011).

## References

Cappuccin JG, Sherman N (1987) Microbiology: a laboratory manual, 2nd edn. Benjamin & Cummins, Menlo Park

Collins MD, Lawson PA, Willems A, Cordoba JJ, Fernandez-Garayzabal J, Garcia P, Cai J, Hippe H, Farrow JAE (1994) The phylogeny of the genus *Clostridium*: proposal of five new genera and eleven new species combinations. Int J Syst Bacteriol 44:812–826

Lee YJ (2005) Microbial diversity in a constructed wetland system for treatment of acid sulfate water. PhD thesis, University of Georgia, Athens

Lee Y-J, Romanek CS, Mills GL, Davis RC, Whitman WB, Wiegel J (2006) *Gracilibacter thermotolerans* gen. nov., sp. nov., an anaerobic, thermotolerant bacterium from a constructed wetland receiving acid sulfate water. Int J Syst Evol Microbiol 56:2089–2093

Ljungdahl LG, Wiegel J (1986) Anaerobic fermentations. In: Demain AL, Solomon NA (eds) Manual of industrial microbiology and biotechnology. American Society for Microbiology, Washington, DC, pp 84–96

Sasaki K, Haruta S, Tatara M, Yamazawa A, Ueno Y, Ishii M, Igarashi Y (2006) Microbial community in a methanogenic packed-bed reactor successfully operating at short hydraulic retention time. J Biosci Bioeng 101:271–273

Sasaki K, Haruta S, Ueno Y, Ishii M, Igarashi Y (2007) Microbial populations in the biomass adhering to supporting material in a packed-bed reactor degrading organic solid waste. Appl Microbiol Biotechnol 75:941–952

Shiratori H, Ikeno H, Ayame S, Kataoka N, Miya A, Hosono K, Beppu T, Ueda K (2006) Isolation and characterization of a new *Clostridium* sp. that performs effective cellulosic waste digestion in a thermophilic methanogenic bioreactor. Appl Environ Microbiol 72:3702–3709

Shiratori H, Ohiwa H, Ikeno H, Ayame S, Kataoka N, Miya A, Beppu T, Ueda K (2008) Lutispora thermophila gen.nov, sp. nov., a thermophilic, spore-forming bacterium isolated from a thermophilic methanogenic bioreactor digesting municipal solid wastes. Int J Syst Evol Microbiol 58:964–969

Sizova MV, Izquierdo JA, Panikov NS, Lynd LR (2011) Cellulose and xylan-degrading thermophilic anaerobic bacteria from biocompost. Appl Environ Microbiol 77:2282–2291

Stamatakis A (2006) RAxML-VI-HPC: maximum likelihood-based phylogenetic analyses with thousands of taxa and mixed models. Bioinformatics 22:2688–2690

Tischer K, Kleinsteuber S, Schleinitz KM, Fetzer I, Spott O, Stange F, Lohse U, Franz J, Neumann F, Gerling S, Schmidt C, Hasselwander E, Harms H, Wendeberg A (2013) Microbial communities along biogeochemical gradients in a hydrocarbon-contaminated aquifer. Environ Microbiol 15:2603–2615

Widdel F, Bak F (1992) Gram-negative mesophilic sulfate-reducing bacteria. In: Balows A, Trüper HG, Dworkin M, Harder W, Schleifer KH (eds) The prokaryotes, vol 4, 2nd edn. Springer, New York

Wiegel J (1998) Anaerobic alkali-thermophiles, a novel group of extremophiles. Extremophiles 2:257–267

Yarza P, Ludwig W, Euzéby J, Amann R, Schleifer K-H, Glöckner FO, Rosselló-Móra R (2010) Update of the All-Species Living-Tree Project based on 16S and 23S rRNA sequence analyses. Syst Appl Microbiol 33:291–299

# 12 The Order *Halanaerobiales*, and the Families *Halanaerobiaceae* and *Halobacteroidaceae*

*Aharon Oren*
Department of Plant and Environmental Sciences, The Institute of Life Sciences, The Hebrew University of Jerusalem, Jerusalem, Israel



**Abstract**
The order *Halanaerobiales*, families *Halanaerobiaceae* and *Halobacteroidaceae*, consists of obligatory anaerobic, moderately halophilic bacteria that require NaCl concentrations between 0.5 and 3.4 M for optimal growth. Representatives have been isolated from anaerobic sediments of salt lakes worldwide, from brines associated with oil reservoirs, and also from fermented salted foods. Some species are thermophilic or alkaliphilic. Although phylogenetically affiliated with the low G+C branch of the *Firmicutes*, the cells show a Gram-negative wall structure, and most species stain Gram-negative. Some representatives of the *Halobacteroidaceae* produce endospores. Most species ferment carbohydrates to acetate, ethanol, $H_2$, $CO_2$, and other fermentation products. Within the *Halobacteroidaceae*, a greater metabolic diversity is found, with some species displaying a homoacetogenic metabolism; growth by anaerobic respiration using different electron acceptors including nitrate, trimethylamine *N*-oxide, selenate, arsenate, or Fe(III); or chemolithoautotrophic growth on hydrogen and elemental sulfur.

## Taxonomy, Historical and Current

Bottom sediments of hypersaline lakes and lagoons may support a rich community of anaerobic halophilic bacteria, as the solubility of oxygen in hypersaline brines is low and the amounts of organic matter available are often high (Oren 1988). It is therefore surprising that the first records of the isolation of obligatory anaerobic fermentative bacteria growing at salt concentrations of 10–20 % and higher were published only in the early 1980s, when *Halanaerobium praevalens* was isolated from the bottom sediments of Great Salt Lake, Utah (Zeikus 1983; Zeikus et al. 1983) and *Halobacteroides halobius* and *Sporohalobacter lortetii* were discovered in Dead Sea sediments (Oren 1983; Oren et al. 1984b). *Halanaerobium praevalens* probably resembles "Bacteroides halosmophilus," isolated by Baumgartner (1937) from solar salt and from salted anchovies. Unfortunately no cultures of that isolate have been preserved.

Order *Halanaerobiales* corrig. Rainey and Zhilina 1995, $879^{VP}$ (Validation List no. 55); Effective Publication: Rainey, Zhilina, Boulygina, Stackebrandt, Tourova and Zavarzin 1995, 193.

Hal.an.ae.ro.bi.a'les. N.L. neut. n. *Halanaerobium*, type genus of the order; suff. *–ales*, ending denoting an order; N.L. fem. pl. n. *Halanaerobiales*, the *Halanaerobiaceae* order.

Cells are rod-shaped and generally stain Gram-negative. Endospores are produced by some species. Strictly anaerobic. Oxidase negative and generally catalase negative. Most species ferment carbohydrates to products including acetate, ethanol, $H_2$, and $CO_2$. Some species may grow fermentatively on amino acids, and others have a homoacetogenic metabolism or may grow by anaerobic respiration on nitrate, trimethylamine *N*-oxide, selenate, arsenate, or Fe(III). Chemolithoautotrophic growth on $H_2$ and elemental sulfur may also occur. Moderately

Dedicated to the memory of George A. Zavarzin (1933–2011), a pioneer of research on anaerobic halophilic microorganisms.

E. Rosenberg et al. (eds.), *The Prokaryotes – Firmicutes and Tenericutes*, DOI 10.1007/978-3-642-30120-9_218,

halophilic. NaCl concentrations between 0.5 and 3.4 M are required for optimal growth, and no growth is observed below 0.3–1.7 M NaCl, depending on the species.

The mol% G+C of the DNA varies between 27 and 45.

Type Genus: *Halanaerobium*

The order *Haloanaerobiales* was created in 1995, based on 16S rRNA sequence comparisons. These resulted in a reclassification of the species of the former family *Haloanaerobiaceae* over two families: the *Haloanaerobiaceae* and the newly created family *Halobacteroidaceae* (Rainey et al. 1995). Physiologically the group is coherent, to the extent that, as yet, no aerobes or non-halophiles are known to cluster phylogenetically within the order.

The genus *Halanaerobium* (originally named *Haloanaerobium* and corrected in accordance with Rule 61 of the Bacteriological Code) (Oren 2000) is now the largest genus within the order (nine species and two subspecies). Based on 16S rRNA sequence comparisons (Rainey et al. 1995), a number of species formerly classified in other genera were transferred to this genus: the former *Halobacteroides acetoethylicus* (Rengpipat et al. 1988a) was reclassified as *Halanaerobium acetethylicum* (Oren 2000; Patel et al. 1995; Rainey et al. 1995), and the former *Haloincola saccharolyticus* (originally described under the name *Haloincola saccharolytica*) (Zhilina et al. 1992b) was renamed as *Halanaerobium saccharolyticum*, with two subspecies, *saccharolyticum* and *senegalense* (Cayol et al. 1994a; Oren 2000; Rainey et al. 1995). The genera *Halobacteroides*, *Acetohalobium*, *Halanaerobacter*, and *Sporohalobacter*, previously classified within the family *Halanaerobiaceae*, were transferred to the *Halobacteroidaceae* (Rainey et al. 1995).

At the time of writing (March 2012), 30 species had been described. The family *Halanaerobiaceae* currently has 4 genera with 12 species; the family *Halobacteroidaceae* contains 11 genera with 18 species (see ➋ *Figs. 12.1* and ➋ *12.2* and ➋ *Tables 12.1*, ➋ *12.2*, ➋ *12.3*, ➋ *12.4*, ➋ *12.5*, ➋ *12.6*, and ➋ *12.7*).

The group was earlier reviewed by Kivistö and Karp (2011); Lowe et al. (1993); Ollivier et al. (1994), and Oren (1986a, 1990, 1993a, b, 2006).

## Phylogenetic Structure of the Family and Its Genera

➋ *Figure 12.3* shows a neighbor-joining phylogenetic tree of the type strains of the 31 species and subspecies of the order *Halanaerobiales*. It may be noted that *Halobacteroides elegans* does not cluster with *Halobacteroides halobius*, the type species of the genus, but with the species of the genus *Halanaerobacter*, suggesting that reclassification of *H. elegans* may be recommended. The family is associated with the low-G+C branch of the *Firmicutes*. The group forms a coherent cluster close to the bifurcation point that separates the *Actinobacteria* and the *Bacillus*/*Clostridium* group (Rainey et al. 1995; Tourova et al. 1995). The deep branching justifies classification in a separate order (Rainey et al. 1995). The order *Halanaerobiales* has been used as a paradigm to demonstrate the application of 16S rRNA gene sequencing and DNA-DNA hybridization in bacterial taxonomy (Tourova 2000). Two families were described: the *Halanaerobiaceae* (Oren et al. 1984a) and the *Halobacteroidaceae* (Rainey et al. 1995).

The species share a low content of G+C in their DNA, generally between 29 mol% and 34 mol%. Exceptions are the thermophilic *Halothermothrix orenii* with a G+C content of 37.9 mol% and the atypical, non-fermentative anaerobic respirer *Halarsenatibacter silvermanii* with 45 mol%.

## Genome Analysis

At the time of writing (March 2012), three complete genome sequences of members of the *Halanaerobiales* had been published: the type strain of *Halanaerobium praevalens* (Ivanova et al. 2011), the thermophilic *Halothermothrix orenii* (Mijts and Patel 2001; Mavromatis et al. 2009), and a haloalkaliphilic hydrogen-producing strain known as "*Halanaerobium hydrogenoformans*," earlier designated as "*Halanaerobium sapolanicus*" (Brown et al. 2011). This organism is not currently available from culture collections. Except for the genome sequence, little information is available about it beyond the fact that it was isolated from the alkaline hypersaline and sulfide-rich Soap Lake, Washington, USA, that it grows optimally at pH 11, 7 % NaCl, and 33 °C and that it produces acetate, formate, and $H_2$ (➋ *Table 12.8*).

The three genomes are 2.3–2.6 Mbp in length and each contains four identical or nearly identical copies of the 16S rRNA gene. Analysis of the *H. orenii* gene showed a few features characteristic for Gram-negative bacteria such as a pathway for lipid A biosynthesis, outer membrane secretion proteins, and two copies of the chaperone OmpH, a periplasmic protein that helps to transport proteins to the outer membrane. There also are a number of sporulation-related genes. The main sporulation regulator Spo0A of bacilli and clostridia is present, but sporulation was never shown in this organism. Genes coding for the biosynthesis of organic osmotic solutes were not detected except for the finding of a gene for sucrose phosphate synthase, suggesting that sucrose can be formed and may possibly act as an osmotic solute (Chua et al. 2008; Mavromatis et al. 2009).

Comparative analysis of the three *Halanaerobiales* genomes did not show an unusually high content of acidic amino acids or a low content of basic amino acids in the encoded proteins. The apparent excess of acidic amino acids in the bulk protein of *Halanaerobium praevalens*, *H. saccharolyticum*, *Halobacteroides halobius*, *Sporohalobacter lortetii*, and *Natroniella acetigena* reported earlier (Detkova and Boltyanskaya 2006; Oren 1986b) is therefore due to the high content of glutamine and asparagine in their proteins, which yield glutamate and aspartate upon acid hydrolysis. The proteins of the *Halanaerobiales*, which are active in the presence of high intracellular KCl concentrations, do thus not possess the typical acidic signature of the

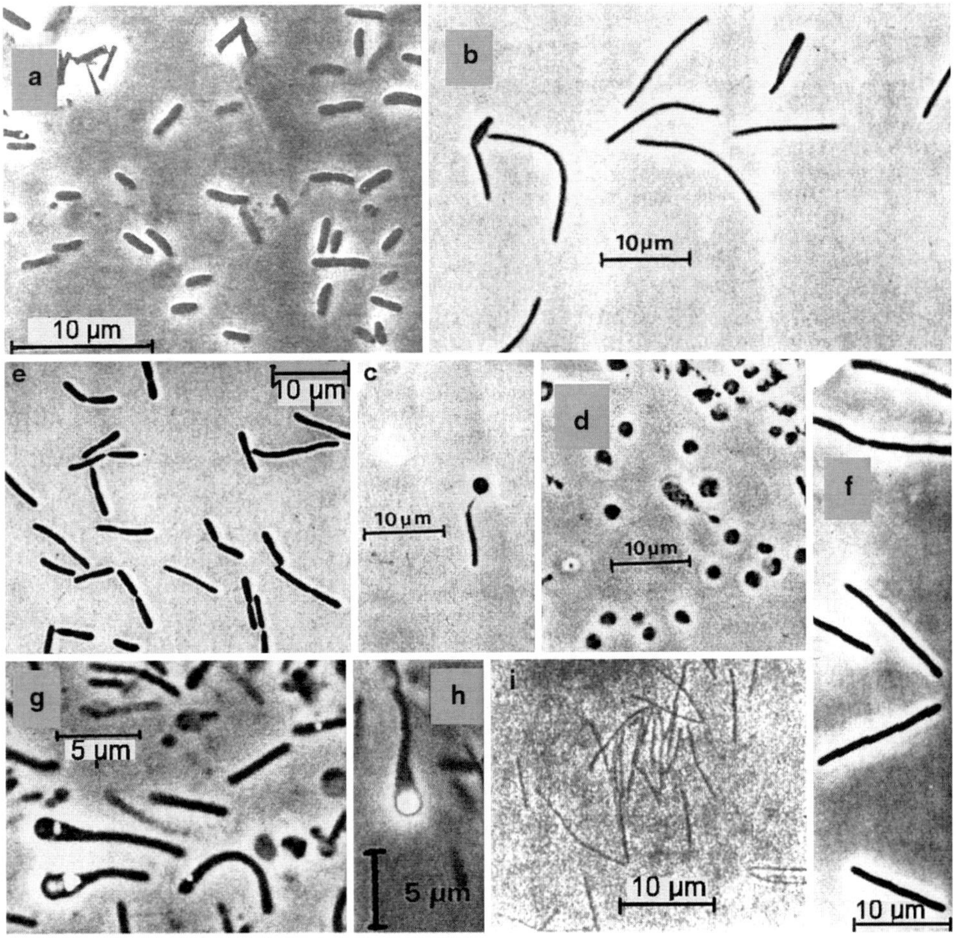


**Fig. 12.1**
**Phase-contrast micrographs of members of the *Halanaerobiales*: (a) *Halanaerobium alcaliphilum*; (b–d) young, senescent, and old cells of *Halobacteroides halobius*; (e) *Acetohalobium arabaticum*; (f) *Natroniella acetigena*; (g) *Sporohalobacter lortetii*; (h) *Orenia marismortui*; (i) *Halothermothrix orenii*. Figures were derived from Tsai et al. (1995), Oren et al. (1984b), Zavarzin et al. (1994), Zhilina et al. (1996), Oren (1983), and Cayol et al. (1994b), respectively; reproduced with permission**

"halophilic" proteins of the Archaea of the order *Halobacteriales* or of the extremely halophilic bacterium *Salinibacter* (Elevi Bardavid and Oren 2012).

### Phages

No phages active on strains of *Halanaerobiales* have yet been described.

## Phenotypic Analyses

### General Comments

Members of the *Halanaerobiales* display a Gram-negative type of cell wall with an outer membrane and periplasmic space (➔ *Fig. 12.2*). *Meso*-diaminopimelic acid was detected in the peptidoglycan of *Halanaerobium saccharolyticum* subsp. *saccharolyticum* (Zhilina et al. 1992b). Most species also show a negative Gram-stain reaction; however, *Halanaerobium tunisiense* and *Halanaerocella petrolearia* stain Gram-positive (Gales et al. 2011; Hedi et al. 2009).

Heat-resistant endospores are produced by a number of species of *Halobacteroidaceae*, including *Sporohalobacter lortetii* (Oren 1983), the three *Orenia* species (Mouné et al. 2000; Oren et al. 1987; Zhilina et al. 1999), and *Natroniella acetigena* (Zhilina et al. 1996). When initially isolated, *Acetohalobium arabaticum* produced spores, but sporulation was not observed during subsequent transfers (Zavarzin et al. 1994). Special conditions may be required for induction of endospore formation. Growth on solid media or in nutrient-poor liquid media enhances sporulation in certain species (Oren 1983; Oren et al. 1987). A phenotypic test which may be correlated with the phylogenetic position of the *Halanaerobiaceae* within the *Firmicutes* and with the ability to form endospores is the

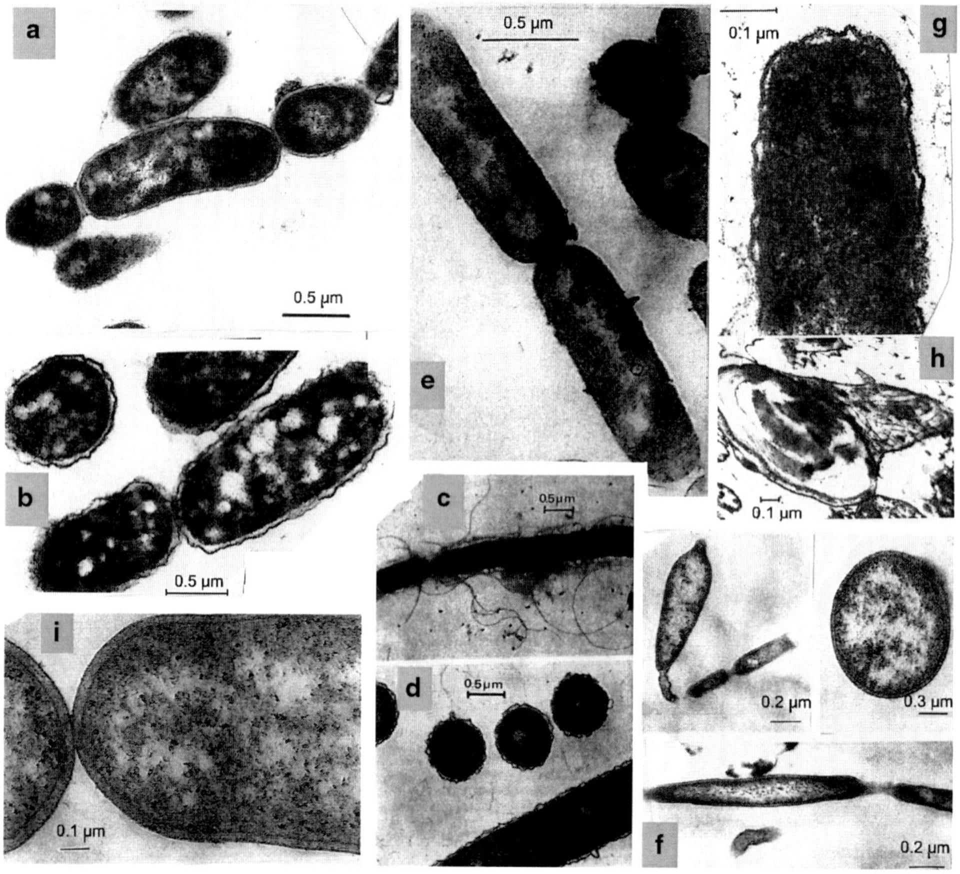


**Fig. 12.2**
**Electron micrographs of members of the *Halanaerobiales*: (a) *Halanaerobium lacusrosei*; (b) *Halanaerobium saccharolyticum*; (c,d) *Halobacteroides halobius*; (e) *Acetohalobium arabaticum*; (f) *Halothermothrix orenii*; (g,h) *Sporohalobacter lortetii*; (i) *Halanaerobium saccharolyticum* subsp. *senegalense*. Figures were derived from Cayol et al. (1995), Zhilina et al. (1992b), Oren et al. (1984b), Zavarzin et al. (1994), Cayol et al. (1994b), Oren (1983), and Cayol et al. (1994a), respectively; reproduced with permission**

hydrolysis of the D-isomer of *N′*-benzoyl-arginine-*p*-nitroanilide (BAPA). Three representatives of the *Halanaerobiales* (*Halobacteroides halobius, Halanaerobium praevalens, Orenia marismortui*) were found to hydrolyze D-BAPA, while L-BAPA was not hydrolyzed. *Sporohalobacter lortetii* degraded neither of the BAPA stereoisomers (Oren et al. 1989).

All members of the *Halanaerobiales* are strict anaerobes. They are oxidase negative, and most species lack catalase, *Halarsenatibacter silvermanii* being the only known exception. All members of the *Halanaerobiaceae* and most members of the *Halobacteroidaceae* obtain their energy by fermenting simple sugars (➲ *Tables 12.9* and ➲ *12.10*). *Halanaerobacter chitinivorans* uses chitin, and *Halocella cellulosilytica* degrades cellulose. Fermentation products typically include acetate, $H_2$, and $CO_2$. Some strains produce in addition butyrate, lactate, propionate, and/or formate. *Halarsenatibacter silvermanii* lives by dissimilatory reduction of arsenate to arsenite, Fe(III) to Fe(II), and elemental sulfur to sulfide. Chemoautotrophic growth occurs with sulfide as the electron donor and arsenate as the electron acceptor (Switzer Blum et al. 2009). Within the family *Halobacteroidaceae*, the metabolic diversity is much greater than within the *Halanaerobiaceae*. Thus, there are species that ferment amino acids, either alone or by using the Stickland reaction. For example, *Halanaerobacter salinarius* and *Halanaerobacter chitinivorans* can use serine as an electron donor using the Stickland reaction while reducing glycine betaine, with the formation of acetate, trimethylamine, $CO_2$, and $NH_3$ (Mouné et al. 1999). *Sporohalobacter lortetii* is primarily an amino acid fermenter, and sugars are poorly used (Oren 1983). Anaerobic respiration also occurs, using different electron acceptors: *Selenihalanaerobacter shriftii* oxidizes glycerol or glucose by anaerobic respiration with nitrate, trimethylamine *N*-oxide, or selenate as electron acceptor (Switzer Blum et al. 2001a).

*Acetohalobium arabaticum* (neutrophilic), *Natroniella acetigena* (alkaliphilic), and *Fuchsiella alkaliacetigena* (alkaliphilic) have a homoacetogenic metabolism, producing acetate as the main end product of their energy metabolism. *Acetohalobium arabaticum* grows on $H_2$ + $CO_2$ or on carbon monoxide as a lithoautotroph, on trimethylamine as a methylotroph, and on other substrates (formate, glycine betaine, lactate, pyruvate, histidine, aspartate, glutamate, and

**Table 12.1**
**Comparison of selected characteristics of members of the genus *Halanaerobium***

| Character | *Halanaerobium praevalens*[a] | *Halanaerobium alcaliphilum*[b] | *Halanaerobium acetethylicum*[c,d,e] | *Halanaerobium salsuginis*[f] | *Halanaerobium saccharolyticum subsp. saccharolyticum*[e,g] |
|---|---|---|---|---|---|
| Earlier name/ basonym | *Haloanaerobium praevalens* | *Haloanaerobium alcaliphilum* | *Halobacteroides acetoethylicus* | *Haloanaerobium salsugo* | *Haloincola saccharolytica*; *Haloanaerobium saccharolyticum* subsp. *saccharolyticum* |
| Type strain | DSM 2228 | DSM 8275 | DSM 3532 | ATCC 51327 | DSM 6643 |
| Cell size | 0.9–1.1 × 2.0–2.6 μm | 0.8 × 3.3–5 μm | 0.4–0.7 × 1–1.6 μm | 0.3–0.4 × 2.6–4 μm | 0.5–0.7 × 1–1.5 μm |
| Morphology | Rods | Rods | Rods | Rods | Rods |
| Motility | – | +, peritrichous | +, peritrichous | – | +, peritrichous |
| | | flagella | flagella | | flagella |
| Endospores | – | – | – | – | – |
| Spheroplasts | – | NR | – | NR | – |
| Gas vesicles | NR | NR | NR | NR | NR |
| NaCl range | 2–30 % | 2.5–25 % | 5–22 % | 6–24 % | 3–30 % |
| NaCl optimum | 13 % | 10 % | 10 % | 9 % | 10 % |
| pH range | 6.0–9.0 | 5.8–10.0 | 5.4–8.0 | 5.6–8.0 | 6.0–8.0 |
| pH optimum | 7.0–7.4 | 6.7–7.0 | 6.3–7.4 | 6.1 | 7.5 |
| Temperature range | 5–50 °C | 25–50 °C | 15–45 °C | 22–51 °C | 15–47 °C |
| Temperature optimum | 37 °C | 37–40 °C | 34 °C | 40 °C | 37–40 °C |
| Doubling time | 4 h | 3.3 h | 7.5 h | 9 h | 3.9 h |
| Carbohydrates utilized | + | + | + | + | + |
| End products of fermentation | Acetate, butyrate, propionate, $H_2$, $CO_2$ | Acetate, butyrate, lactate, $H_2$, $CO_2$ | Acetate, ethanol, $H_2$, $CO_2$ | Acetate, ethanol, $H_2$, $CO_2$ | Acetate, $H_2$, $CO_2$ |
| Major fatty acids | 14:0 16:0 16:1 | NR | 14:0 16:0 16:1 | 14:0 16:0 16:1 $17{:}0_{cyc}$ | 15:1 16:0 16:1 |
| G+C content of DNA (mol%) | 30.3[m] | 31.0 | 32.0 | 34.0 | 31.3 |
| Sample source and site | Sediment, Great Salt Lake, Utah, USA | Sediment, Great Salt Lake, Utah, USA | Filter material, offshore oil rig, Gulf of Mexico | Petroleum reservoir fluid, Oklahoma, USA | Sediment, Lake Sivash, Crimea |

| Character | *Halanaerobium saccharolyticum* subsp. *senegalense*[e,h] | *Halanaerobium congolense*[i] | *Halanaerobium lacusrosei*[j] | *Halanaerobium kushneri*[k] | *Halanaerobium fermentans*[l] |
|---|---|---|---|---|---|
| Earlier name/ basonym | *Haloincola saccharolytica* subsp. *senegalensis* | *Haloanaerobium congolense* | *Haloanaerobium lacusroseus* | *Haloanaerobium kushneri* | *Haloanaerobium fermentans* |
| Type strain | DSM 7379 | DSM 11287 | DSM 10165 | ATCC 700103 | JCM 10494 |
| Cell size | 0.4–0.6 × 2–5 μm | 0.5–1 × 2–4 μm | 0.4–0.6 × 2–3 μm | 0.5–0.8 × 0.7–3.3 μm | 1.0–1.2 × 2.7–3.3 μm |
| Morphology | Rods | Rods | Rods | Rods | Rods |
| Motility | +, peritrichous flagella | – | +, peritrichous flagella | +, peritrichous flagella | +, peritrichous flagella |
| Endospores | – | – | NR | – | – |
| Spheroplasts | NR | NR | NR | NR | NR |

**Table 12.1 (continued)**

| Character | *Halanaerobium saccharolyticum* subsp. *senegalense*[e,h] | *Halanaerobium congolense*[i] | *Halanaerobium lacusrosei*[j] | *Halanaerobium kushneri*[k] | *Halanaerobium fermentans*[l] |
|---|---|---|---|---|---|
| Gas vesicles | NR | NR | NR | NR | NR |
| NaCl range | 5–25 % | 4–24 % | 7.5–34 % | 9–18 % | 7–25 % |
| NaCl optimum | 7.5–12.5 % | 10 % | 18–20 % | 12 % | 10 % |
| pH range | 6.3–8.7 | 6.3–8.5 | NR | 6.0–8.0 | 6–9 |
| pH optimum | 7.0 | 7.0 | 7.0 | 6.5–7.5 | 7.5 |
| Temperature range | 20–47 °C | 20–45 °C | 20–50 °C | 20–45 °C | 15–45 °C |
| Optimum temperature | 40 °C | 42 °C | 40 °C | 35–40 °C | 35 °C |
| Doubling time | 4.2 h | 2.5 h | 2.4 h | 7.3 h | NR |
| Carbohydrates utilized | + | + | + | + | + |
| End products of fermentation | Acetate, $H_2$, $CO_2$ | Acetate, $H_2$, $CO_2$ | Acetate, ethanol, $H_2$, $CO_2$ | Acetate, ethanol, $H_2$, $CO_2$ | Acetate, ethanol, formate, lactate, $H_2$, $CO_2$ |
| Major fatty acids[1] | 14:0 16:0 15:1 16:1 | NR | NR | 14:0 16:0 16:1 | NR |
| G+C content of DNA (mol%) | 31.7 | 34 | 32 | 32.4–36.9 | 33.3 |
| Sample source and site | Sediment, Lake Retba, Senegal | Offshore field, Congo | Sediment, Lake Retba, Senegal | Petroleum reservoir fluid, Oklahoma | Fermented puffer fish ovaries, Japan |

Data taken from:
[a]Zeikus et al. (1983)
[b]Tsai et al. (1995)
[c]Rengpipat et al. (1988a)
[d]Patel et al. (1995)
[e]Rainey et al. (1995)
[f]Bhupathiraju et al. (1994)
[g]Zhilina et al. (1992b)
[h]Cayol et al. (1994a)
[i]Ravot et al. (1997)
[j]Cayol et al. (1995)
[k]Bhupathiraju et al. (1999)
[l]Kobayashi et al. (2000a)
[m]Based on the genome sequence
*NR* not reported

asparagine) as an organotroph. *Fuchsiella* can also grow chemolithoautotrophically (Kevbrin et al. 1995; Zavarzin et al. 1994; Zhilina and Zavarzin 1990a, b; Zhilina et al. 1996, 2012).

Several species (*Halanaerobium saccharolyticum*, *Halanaerobacter lacunarum*, *Halobacteroides halobius*, *Halobacteroides elegans*) can use methanethiol as the sole source of assimilatory sulfur for growth and reduce elemental sulfur to sulfide (Kevbrin and Zavarzin 1992a; Zhilina et al. 1992a, b, 1997). *Acetohalobium arabaticum* slowly reduces sulfur to sulfide, but this was not accompanied by growth enhancement (Kevbrin and Zavarzin 1992b; Zavarzin et al. 1994). *Natroniella acetigena* can grow chemolithoautotrophically by oxidizing $H_2$, using elemental sulfur as electron acceptor (Sorokin et al. 2011). *Halanaerobium congolense* uses thiosulfate and elemental sulfur as electron acceptors. Addition of thiosulfate or sulfur increased the growth yield sixfold and threefold, respectively, and growth rates were enhanced (Ravot et al. 1997). Thiosulfate reduction was also observed in *Orenia marismortui* and in *Halanaerobium congolense* (Oren et al. 1987; Ravot et al. 2005).

High concentrations of $Na^+$, $K^+$, and $Cl^-$, high enough to be at least isotonic with the medium, were measured inside the cells of *Halanaerobium praevalens*, *Halanaerobium acetethylicum*, *Halobacteroides halobius*, and *Natroniella acetigena* (Detkova and Pusheva 2006; Oren 1986b; Oren et al. 1997; Rengpipat et al. 1988b). No organic osmotic solutes have been detected in the anaerobic halophilic bacteria (Oren 1986b; Oren et al. 1997; Rengpipat et al. 1988b), except in the case of *Orenia salinaria*, found to accumulate glycine betaine when grown in medium containing yeast extract (Mouné et al. 2000). The intracellular enzymatic machinery appears to be well adapted to function in the presence of high salt concentrations. The enzymes tested (including glyceraldehyde-3-phosphate dehydrogenase,

**Table 12.2**
**Comparison of selected characteristics of members of the monospecific genera *Halocella, Halothermothrix,* and *Halarsenatibacter* (family *Halanaerobiaceae*)**

| Character | *Halocella cellulosilytica*[a] | *Halothermothrix orenii*[b] | *Halarsenatibacter silvermanii*[c] |
|---|---|---|---|
| Earlier name | *Halocella cellulolytica* | | |
| Type strain | DSM 7362 | OCM 544 | ATCC BAA-1651 |
| Cell size | 0.4–0.6 × 3.8–12 μm | 0.4–0.6 × 10–20 μm | 0.5 × 3 μm |
| Morphology | Rods | Rods | Curved rods |
| Motility | +, peritrichous flagella | +, peritrichous flagella | +, paired subpolar flagella |
| Endospores | – | – | – |
| Spheroplasts | + | NR | – |
| Gas vesicles | NR | NR | – |
| NaCl range | 5–20 % | 4–20 % | 20–35 % |
| NaCl optimum | 15 % | 10 % | 35 % |
| pH range | 5.5–8.5 | 5.5–8.2 | 8.7–9.8 |
| pH optimum | 7.0 | 6.5–7.0 | 9.4 |
| Temperature range | 20–50 °C | 45–68 °C | 28–55 °C |
| Optimum temperature | 39 °C | 60 °C | 44 °C |
| Carbohydrates utilized | + | + | – |
| End products of fermentation | Acetate, ethanol, lactate, $H_2$, $CO_2$ | Acetate, ethanol, $H_2$, $CO_2$ | Not fermentative; reduces arsenate, Fe(III), and sulfur |
| Major fatty acids[1] | 14:0 16:0 15:0$_{anteiso}$ | 14:0 15:0$_{iso}$ 16:0 | 15:0$_{iso}$ 18:0 17:0$_{iso}$ 16:0 |
| G+C content of DNA (mol%) | 29.0 | 38 | 45.2 |
| Sample source and site | Sediment, Lake Sivash, Crimea | Sediment, hypersaline lake, Tunisia | Sediment, Searles Lake, California, USA |

Data taken from:
[a]Simankova et al. (1993)
[b]Cayol et al. (1994b)
[c]Switzer Blum et al. (2009)
*NR* not reported

NAD-linked alcohol dehydrogenase, pyruvate dehydrogenase, and methyl viologen-linked hydrogenase from *Halanaerobium acetethylicum*, the fatty acid synthetase complex of *Halanaerobium praevalens*, hydrogenase and CO dehydrogenase of *Acetohalobium arabaticum*, CO dehydrogenase of *Natroniella acetigena*) function better in the presence of molar concentrations of salts than in salt-free medium (Detkova and Boltyanskaya, 2006; Oren and Gurevich 1993; Pusheva and Detkova 1996; Pusheva et al. 1992; Rengpipat et al. 1988b; Zavarzin et al. 1994).

## The Properties of the Genera and Species of *Halanaerobiales*

Information on the phenotypic properties of the genera and species of the *Halanaerobiales*, as summarized below, was derived from Cayol et al. 2009; Mesbah 2009; Oren 2009b, c, d, e, f; Oren et al. 2009; Rainey 2009; Zavarzin 2009; Zavarzin and Zhilina 2009a, b; and Zhilina et al. 2009 and from the original species descriptions.

### Family *Halanaerobiaceae* corrig.

Oren, Paster and Woese 1984, 503$^{VP}$ (Validation List no. 16) (Effective Publication: Oren, Paster and Woese 1984a, 79).

*Hal.an.ae.ro.bi.a.ce'ae* N.L. neut. n. *Halanaerobium*, type genus of the family; suff. *-aceae*, ending to denote a family; N.L. fem. pl. n. *Halanaerobiaceae*, the *Halanaerobium* family.

Cells are rod-shaped and stain Gram-negative. Endospore formation never observed. Strictly anaerobic. Oxidase and catalase negative. Carbohydrates are fermented to products including acetate, ethanol, $H_2$, and $CO_2$. Moderately halophilic. NaCl concentrations between 1.7 and 2.6 M are required for optimal growth, and no growth is observed below 0.3–1.7 M NaCl, depending on the species.

Type Genus: *Halanaerobium*

**Table 12.3**
**Comparison of selected characteristics of members of the genus *Halobacteroides***

| Character | *H. halobius*[a] | *H. elegans*[b,c] |
|---|---|---|
| Earlier name | | *Halobacteroides halobius* |
| Type strain | ATCC 35273 | DSM 6639 |
| Cell size | 0.5–0.6 × 10–20 μm | 0.3–0.5 × 2–10 μm |
| Morphology | Flexible rods | Curved rods |
| Motility | +, peritrichous flagella | +, peritrichous flagella |
| Endospores | –[d] | + |
| Spheroplasts | + | + |
| Gas vesicles | NR | NR |
| NaCl range | 7–19 % | 10–30 % |
| NaCl optimum | 9–15 % | 10–15 % |
| pH range | NR | 6.5–8.0 |
| pH optimum | NR | 7.0 |
| Temperature range | 30–47 °C | 28–47 °C |
| Optimum temperature | 37–42 °C | 40 °C |
| Doubling time | 1 h | 2 h |
| Carbohydrates utilized | + | + |
| End products of fermentation | Acetate, ethanol, $H_2$, $CO_2$ | Acetate, ethanol, $H_2$, $CO_2$ |
| Major fatty acids[1] | 14:0 16:0 16:1 | 14:0 16:0 16:1 |
| G+C content of DNA (mol%) | 30.7 | 30.5 |
| Sample source and site | Sediment, Dead Sea | Cyanobacterial mat, Lake Sivash, Crimea |

Data taken from:
[a]Oren et al. (1984b)
[b]Zhilina et al. (1997)
[c]Based on 16S rRNA sequence comparison, *Halobacteroides elegans* does not cluster with *Halobacteroides halobius* but with the species of the genus *Halanaerobacter*, suggesting that reclassification of *H. elegans* may be recommended
[d]Similar isolates were recovered from anaerobic sediments following pasteurization, suggesting that heat-stable endospores may be formed (Oren 1987)
*NR* not reported

**Genus *Halanaerobium* corrig.**

Zeikus, Hegge, Thompson, Phelps and Langworthy 1984, 503[VP] (Validation List no. 16), Emend. Rainey, Zhilina, Boulygina, Stackebrandt, Tourova and Zavarzin 1995, 197 (Effective publication: Zeikus, Hegge, Thompson, Phelps and Langworthy 1983, 232).

*Hal.an.ae.ro'bium.* Gr. n. *hals halos*, salt; Gr. pref. *an*, not; Gr. n. *aer*, air; Gr. n. *bios*, life; N.L. neut. n. *Halanaerobium*, salt organism which grows in the absence of air.

Cells rod-shaped, nonmotile or motile by peritrichous flagella, generally staining Gram-negative. Strictly anaerobic, chemoorganotrophic with fermentative metabolism. Carbohydrates are fermented with production of acetate, $H_2$, and $CO_2$; in some species, ethanol, formate, propionate, butyrate, and lactate are found in addition. Thiosulfate and elemental sulfur may be used as electron acceptors in certain species. Halophilic, growing optimally at NaCl concentrations around 1.7–2.5 M and requiring a minimum of 0.3–1.7 M NaCl for growth. Neutral or slightly alkaline pH values are preferred. Endospore formation never observed.

Type Species: *Halanaerobium praevalens*

The main features of members of the genus *Halanaerobium*, updated for March 2012, are listed in ➜ *Table 12.1*.

**Genus *Halocella***

Simankova, Chernych, Osipov and Zavarzin 1994, 182[VP] (Validation List no. 48) (Effective publication: Simankova, Chernych, Osipov and Zavarzin 1993, 389).

Ha.lo.cel'la. Gr. n. *hals halos*, salt; L. fem. n. *cella*, a storeroom and in biology a cell; N.L. fem. n. *Halocella*, salt cell.

Cells are straight or slightly curved rods, non-sporulating, and motile by means of peritrichous flagella. Cell wall of Gram-negative structure. Obligately anaerobic. Moderately halophilic. Ferment carbohydrates, including cellulose, producing acetate, ethanol, lactate, $H_2$, and $CO_2$. Peptides and amino acids are not utilized.

Type Species: *Halocella cellulosilytica*

**Genus *Halothermothrix***

Cayol, Ollivier, Prensier, Guezennec and Garcia 1994b, 538[VP]

Ha.lo.ther'mo.thrix. Gr. n. *hals halos*, salt; Gr. adj. *thermos*, hot; Gr. fem. n. *thrix*, hair; N.L. fem. n. *Halothermothrix*, a thermophilic (fermentative) hair-shaped halophile.

Long rod-shaped bacteria with cells that are 0.4–0.6 × 10–20 μm, occurring mainly singly. Motile by peritrichous flagella. Non-sporulating. Gram stain-negative. Strictly anaerobic. Chemoorganotrophic; ferment carbohydrates to acetate, ethanol, $H_2$, and $CO_2$. NaCl and yeast extract are required for growth. Thermophilic.

Type Species: *Halothermothrix orenii*

**Genus *Halarsenatibacter***

Switzer Blum, Han, Lanoil, Saltikov, White, Tabita, Langley, Beveridge, Jahnke and Oremland 2010, 1985[VP] (Validation List no. 135) (Effective publication: Switzer Blum, Han, Lanoil, Saltikov, White, Tabita, Langley, Beveridge, Jahnke and Oremland 2009, 1958).

Hal.ar.se.na.ti.bac'ter. Gr. n. *hals halos*, salt; N.L. n. *arsenas -atis*, arsenate; N.L. masc. n. *bacter*, rod; N.L. masc. n. *Halarsenatibacter*, halophilic arsenate-utilizing rod.

Gram-negative, motile, strictly anaerobic, slightly curved rods (3.0 by 0.5 μm). Motility achieved by a pair of flagella located along the side of the organism. Extremely halophilic, growing between 20 % and 35 % salt with an optimum at salt saturation. Alkaliphilic. A limited number of organic substrates support growth, including a few sugars and organic acids but not fatty acids or amino acids. Fermentative growth or

**Table 12.4**
**Comparison of selected characteristics of members of the genus *Halanaerobacter***

| Character | *H. chitinivorans*[a] | *H. lacunarum*[b,c] | *H. salinarius*[d] | *H. jeridensis*[e] |
|---|---|---|---|---|
| Earlier name | *Haloanaerobacter chitinovorans* | *Halobacteroides lacunaris* | | |
| Type strain | OCG 229 | DSM 6640 | DSM 12146 | DSM 23230 |
| Cell size | 0.5 × 1.4–8 μm | 0.5–0.6 × 0.7–1 μm | 0.3–0.4 × 5–8 μm | 1.2 × 2.5–6 μm |
| Morphology | Flexible rods | Slightly curved rods | Flexible rods | Rods |
| Motility | +, peritrichous flagella | +, peritrichous flagella | +, peritrichous flagella | + |
| Endospores | – | – | – | – |
| Spheroplasts | + | + | + | – |
| Gas vesicles | NR | NR | NR | – |
| NaCl range | 3–30 % | 10–30 % | 5–30 % | 6–30 % |
| NaCl optimum | 12–18 % | 15–18 % | 14–15 % | 15 % |
| pH range | NR | 6.0–8.0 | 5.5–8.5 | 6–9.6 |
| pH optimum | 7.0 | 6.5–7.0 | 7.4–7.8 | 8.3 |
| Temperature range | 23–50 °C | 25–52 °C | 10–50 °C | 30–60 °C |
| Optimum temperature | 30–45 °C | 35–40 °C | 45 °C | 45 °C |
| Doubling time | 2.5 h | 2.9 h | 2.3 h | NR |
| Carbohydrates utilized | + | + | + | + |
| End products of fermentation | Acetate, isobutyrate, $H_2$, $CO_2$; trimethylamine from glycine betaine in the Stickland reaction | Acetate, ethanol, $H_2$, $CO_2$ | Acetate, ethanol, propionate, formate, $H_2$, $CO_2$; trimethylamine from glycine betaine in the Stickland reaction | Lactate, ethanol, acetate, $H_2$, $CO_2$ |
| Major fatty acids | 16:0 16:1 | 16:0 16:1 | NR | $16{:}1_{cis9}$ 16:0 |
| G+C content of DNA (mol%) | 34.8 | 32.4 | 31.6 | 33.3 |
| Sample source and site | Sediment, saltern pond, California, USA | Silt, Lake Chokrak, Kerch Peninsula | Sediment, saltern pond, France | Sediment, Chott el Djerid, Tunisia |

Data taken from:
[a]Liaw and Mah (1992)
[b]Zhilina et al. (1992a)
[c]Rainey et al. (1995)
[d]Mouné et al. (1999)
[e]Mezghani et al. (2012)
*NR* not reported

microaerophilic growth not observed. Growth is by dissimilatory (respiratory) reduction of arsenate to arsenite, Fe(III) to Fe(II), and elemental sulfur to sulfide. Chemoautotrophic growth occurs with sulfide as the electron donor and arsenate as the electron acceptor. Catalase positive.

Type Species: *Halarsenatibacter silvermanii*

The main features of members of the monospecific genera *Halocella*, *Halothermothrix*, and *Halarsenatibacter*, updated for March 2012, are listed in *Table 12.2.*

## Family *Halobacteroidaceae*

Zhilina and Rainey 1995, 879[VP] (Validation List no. 55) (Effective publication: Rainey, Zhilina, Boulygina, Stackebrandt, Tourova and Zavarzin 1995, 193).

Ha.lo.bac.te.ro.i.da.ce'ae. N.L. masc. n. *Halobacteroides*, type genus of the family; suff. *–aceae*, ending to denote a family; N.L. fem. pl. n. *Halobacteroidaceae*, the *Halobacteroides* family.

**Table 12.5**
**Comparison of selected characteristics of members of the genus *Orenia***

| Character | *O. marismortui*[a,b] | *O. salinaria*[c] | *O. sivashensis*[d] |
|---|---|---|---|
| Basonym | *Sporohalobacter marismortui* | | |
| Type strain | ATCC 35420 | ATCC 700911 | DSM 12596 |
| Cell size | 0.6 × 3–13 μm | 1 × 6–10 μm | 0.5–0.75 × 2.5–10 μm |
| Morphology | Rods | Rods | Flexible rods |
| Motility | +, peritrichous flagella | +, peritrichous flagella | +, peritrichous flagella |
| Endospores | + | + | + |
| Spheroplasts | + | + | + |
| Gas vesicles | – | – | + |
| NaCl range | 3–18 % | 2–25 % | 5–25 % |
| NaCl optimum | 3–12 % | 5–10 % | 7–10 % |
| pH range | NR | 5.5–8.5 | 5.5–7.8 |
| pH optimum | NR | 7.2–7.4 | 6.3–6.6 |
| Temperature range | 25–50 °C | 10–50 °C | Up to 50 °C |
| Optimum temperature | 36–45 °C | 40–45 °C | 40–45 °C |
| Doubling time | 40 min | NR | 3.5 h |
| Carbohydrates utilized | + | + | + |
| End products of fermentation | Acetate, ethanol, butyrate, formate, $H_2$, $CO_2$ | Acetate, ethanol, formate, lactate, $H_2$, $CO_2$ | Acetate, ethanol, formate, butyrate, $H_2$, $CO_2$ |
| Major fatty acids | 14:0 16:0 16:1 18:0 | NR | NR |
| G+C content of DNA (mol%) | 29.6 | 33.7 | 28.6 |
| Sample source and site | Sediment, Dead Sea | Sediment, saltern pond, France | Cyanobacterial mat, hypersaline lagoon, Lake Sivash, Crimea |

Data taken from:
[a]Oren et al. (1987)
[b]Rainey et al. (1995)
[c]Mouné et al. (2000)
[d]Zhilina et al. (1999)
*NR* not reported

Cells are rod-shaped and stain Gram-negative. Endospores produced by some species. Strictly anaerobic. Oxidase and generally catalase negative. Most species ferment carbohydrates to products including acetate, ethanol, $H_2$, and $CO_2$. Some species may grow fermentatively on amino acids; others have a homoacetogenic metabolism or grow by anaerobic respiration while reducing nitrate, trimethylamine *N*-oxide, selenate, or arsenate or chemolithoautotrophically on $H_2$ and elemental sulfur. Moderately halophilic. NaCl concentrations between 1.7 and 2.5 M are required for optimal growth, and no growth is observed below 0.3–1.7 M NaCl, depending on the species.

Type Genus: *Halobacteroides*

### Genus *Halobacteroides*

Oren, Weisburg, Kessel and Woese 1984, 355[VP] (Effective publication: Oren, Weisburg, Kessel and Woese 1984, 68).

*Ha.lo.bac.te.ro'i.des.* Greek n. *hals halos*, salt; N.L. masc. n. *bacter*, a staff or rod; L. suff. –oides (from Gr. suff. *eides*, from Gr. n. *eidos*, that which is seen, form, shape, figure; N.L. masc. n. *Halobacteroides*, rod-like salt organism.

Cells are long, thin, often flexible rods and motile by peritrichous flagella, staining Gram-negative. Endospores may be formed. Strictly anaerobic, chemoorganotrophic with fermentative metabolism. Carbohydrates are fermented with production of acetate, ethanol, $H_2$, and $CO_2$. Halophilic, growing optimally at NaCl concentrations around 1.7–2.6 M and requiring a minimum of 1.2–1.7 M NaCl for growth.

Type Species: *Halobacteroides halobius*

The main features of members of the genus *Halobacteroides*, updated for March 2012, are listed in Table 12.3.

### Genus *Halanaerobacter*

Liaw and Mah 1996, 362[VP] (Validation List no. 56), Emend. Rainey, Zhilina, Boulygina, Stackebrandt, Tourova and Zavarzin 1995, 197; Emend. Mouné, Manac'h, Hirschler, Caumette, Willison and Matheron 1999, 109 (Effective publication: Liaw and Mah 1992, 265).

**Table 12.6**
**Comparison of selected characteristics of members of the genus *Natroniella***

| Character | *N. acetigena*[a] | *N. sulfidigena*[b] |
|---|---|---|
| Type strain | DSM 9952 | DSM 22104 |
| Cell size | 1–1.2 × 6–15 μm | 0.3–0.5 × 3–30 μm |
| Morphology | Rods | Flexible rods |
| Motility | +, peritrichous flagella | +, peritrichous flagella |
| Endospores | + | – |
| Spheroplasts | + | + |
| Gas vesicles | – | – |
| NaCl range | 10–26 % | 1.4–4 M $Na^+$ |
| NaCl optimum | 12–15 % | 3 M $Na^+$ |
| pH range | 8.1–10.7 | 8.1–10.6 |
| pH optimum | 9.7–10.0 | 10.0 |
| Temperature range | 28–42 °C | Up to 41 °C |
| Optimum temperature | 37 °C | 35 °C |
| Carbohydrates utilized | – | + |
| End products of fermentation | Acetate | No fermentative growth observed; chemolithoautotroph or acetate-dependent sulfur respiration |
| Major fatty acids | 14:0 16:1 | 14:0 16:0 16:1 $16:1_{ald}$ |
| G+C content of DNA (mol%) | 31.9 | 31.3–32.0 |
| Sample source and site | Sediment, Lake Magadi, Kenya | Sediment, soda lakes, Wadi Natrun, Egypt, and Kulunda Steppe, Russia |

Data taken from
[a]Zhilina et al. (1996)
[b]Sorokin et al. (2011)
*NR* not reported

Hal.an.ae.ro.bac'ter. Gr. n. *hals halos*, salt; Gr. pref. *an*, not; Gr. n. *aer aeros*, air; N.L. masc. n. *bacter*, rod; N.L. masc. n. *Halanaerobacter*, salt rod which grows in the absence of air.

Cells are rod-shaped or slightly curved, flexible, and motile by means of peritrichous flagella. Gram-stain-negative. Strictly anaerobic. Chemoorganotrophic with fermentative metabolism; some strains can utilize amino acids in the Stickland reaction or with hydrogen as electron donor. Carbohydrates are fermented with production of acetate, $H_2$, and $CO_2$. In some species, ethanol, propionate, formate, and isobutyrate are also formed. Elemental sulfur can be used as electron acceptor in certain species. Halophilic; optimal growth occurs at NaCl concentrations around 2.0–3.0 M. Cells require a minimum of 0.5–1.6 M NaCl for growth. Neutral to slightly alkaline pH values required for optimal growth. Mesophilic to slightly thermotolerant. Endospores not observed. Short degenerate cells and spheroplasts occur in stationary phase.

Type Species: *Halanaerobacter chitinivorans*

The main features of members of the genus *Halanaerobacter*, updated for March 2012, are listed in Table 12.4.

## Genus *Orenia*

Rainey, Zhilina, Boulygina, Stackebrandt, Tourova and Zavarzin 1995, 880$^{VP}$ (Validation List no. 55) (Effective publication: Rainey, Zhilina, Boulygina, Stackebrandt, Tourova and Zavarzin 1995, 197).

O.re'ni.a. N.L. fem. n. *Orenia*, named after Aharon Oren, an Israeli microbiologist.

Rods, 2.5–13 μm in length with rounded ends. Gram-stain-negative. Motile by peritrichous flagella. Spores are round, terminal, or subterminal. Gas vesicles detected in some species. Forms spheroplasts. Strictly anaerobic. Halophilic; optimum NaCl concentration for growth 3–12 %; no growth below 2 % or above 25 %. Mesophilic to slightly thermophilic. Chemoorganotrophic. End products of glucose fermentation include $H_2$, $CO_2$, lactate, acetate, butyrate and ethanol.

Type Species: *Orenia marismortui*

The main features of members of the genus *Orenia*, updated for March 2012, are listed in Table 12.5.

## Genus *Natroniella*

Zhilina, Zavarzin, Detkova and Rainey 1996, 1189$^{VP}$ (Validation List no. 59); Emend. Sorokin, Detkova and Muyzer 2011, 94 (Effective publication: Zhilina, Zavarzin, Detkova and Rainey 1996b, 324).

Na.tro.ni.el'la. N.L. n. *natron* (arbitrarily derived from the Arabic n. *natrun* or *natron*) soda, sodium carbonate; N.L. fem. n. *Natroniella*, organism growing in soda deposits.

Flexible rods, motile by peritrichous flagella. Spores may be formed. Cell wall has Gram-negative structure. Strictly anaerobic. Possesses a respiratory type of homoacetogenic metabolism. Extremely alkaliphilic, developing in soda brines at pH 9–10. Halophilic, growing at 1.7–4.4 M NaCl. Obligately dependent on $Na^+$, $Cl^-$, and $CO_3^{2-}$ ions. Mesophilic. Chemoorganotrophic: some organic acids, amino acids, and alcohols are fermented. Acetate is the product of fermentation. Some representatives have obligate sulfur-dependent respiratory metabolism and are able to grow autotrophically or with acetate as an electron donor with sulfur serving as an electron acceptor.

Type Species: *Natroniella acetigena*

The main features of members of the genus *Natroniella*, updated for March 2012, are listed in Table 12.6.

## Genus *Acetohalobium*

Zhilina and Zavarzin 1990, 470$^{VP}$ (Validation List no. 35) (Effective publication: Zhilina and Zavarzin 1990b, 747).

A.ce.to.ha.lo'bi.um. L. n. *acetum*, vinegar; Gr. n. *hals halos*, salt; Gr. n. *bios*, life; N.L. neut. n. *Acetohalobium*, acetate-producing organism living in salt.

**Table 12.7**

**Comparison of selected characteristics of members of the monospecific genera *Acetohalobium*, *Sporohalobacter*, *Fuchsiella*, *Halarsenatibacter*, *Halanaerobaculum*, *Halonatronum*, *Selenihalanaerobacter*, and *Halanaerocella* (family *Halobacteroidaceae*)**

| Character | *Acetohalobium arabaticum*[a] | *Sporohalobacter lortetii*[b,c] | *Fuchsiella alkaliacetigena*[d] | *Halanaerobaculum tunisiense*[e] |
|---|---|---|---|---|
| Basonym | | *Clostridium lortetii* | – | |
| Type strain | DSM 5501 | ATCC 35059 | VKM B-2667 | DSM 19997 |
| Cell size | 0.7–1 × 2–5 μm | 0.5–0.6 × 2.5–10 μm | 0.2–0.5 × 10–30 μm | 0.7–1 × 4–13 μm |
| Morphology | Curved rods | Rods | Flexible rods | Rods |
| Motility | +, 1–2 subterminal flagella | +, peritrichous flagella | +, peritrichous flagella | – |
| Endospores | Rare | + | + | – |
| Spheroplasts | NR | – | + | – |
| Gas vesicles | NR | + | – | – |
| NaCl range | 10–25 % | 4–15 % | 0–14 % | 14–30 % |
| NaCl optimum | 15–18 % | 8–9 % | 7–8.5 % | 20–22 % |
| pH range | 5.8–8.4 | NR | 8.5–10.7 | 5.9–8.4 |
| pH optimum | 7.4–8.0 | NR | 8.8–9.3 | 7.2–7.4 |
| Temperature range | NR–47 °C | 25–52 °C | 25–45 °C | 30–50 °C |
| Optimum temperature | 38–40 °C | 37–45 °C | 40 °C | 42 °C |
| Doubling time | NR | 8 h | 85 h | 2.1 h |
| Carbohydrates utilized | – | Weak | Only uses lactate, pyruvate, glutamate, ethanol, and propanol | + |
| End products of fermentation | Acetate | Acetate, propionate, isobutyrate, isovalerate, $H_2$, $CO_2$ | Acetate (from $H_2$ + $CO_2$) | Acetate, lactate, butyrate, $H_2$, $CO_2$ |
| Major fatty acids | 16:0 16:1 | 16:0 16:1 | 14:0 $15:0_{anteiso}$ 16:0 | 16:1 16:0 14:0 $12:0_{3-OH}$ 10:0 |
| G+C content of DNA (mol%) | 33.6 | 31.5 | 32.0 | 34.3 |
| Sample source and site | Sediment, Lake Sivash, Crimea | Sediment, Dead Sea | Sediment, soda lake, Altai, Russia | Sediment, Chott el Djerid, Tunisia |

| Character | *Halonatronum saccharophilum*[f] | *Selenihalanaerobacter shriftii*[g] | *Halanaerocella petrolearia*[h,i] |
|---|---|---|---|
| Type strain | DSM 13868 | ATCC BAA-73 | DSM 22693 |
| Cell size | 0.4–0.6 × 3.5–10 μm | 0.6 × 2–6 μm | 0.8–1.2 × 8–15 μm |
| Morphology | Flexible rods | Rods | Flexible rods |
| Motility | +, peritrichous flagella | – | – |
| Endospores | + | – | – |
| Spheroplasts | + | – | + |
| Gas vesicles | | – | – |
| NaCl range | 3–17 % | 10–24 % | 6–26 % |
| NaCl optimum | 7–12 % | 21 % | 15 % |
| pH range | 7.7–10.3 | 6–8.5 | 6.2–8.8 |
| pH optimum | 8–8.5 | 7.2 | 7.3 |
| Temperature range | 18–60 °C | 16–42 °C | 25–47 °C |
| Optimum temperature | 36–55 °C | 38 °C | 40–45 °C |
| Doubling time | 2.5 h | 4.3 h (nitrate); 8.9 h (selenate) | 3.5 h |
| Carbohydrates utilized | + | + | + |

**Table 12.7 (continued)**

| Character | *Halonatronum saccharophilum*[f] | *Selenihalanaerobacter shriftii*[g] | *Halanaerocella petrolearia*[h,i] |
|---|---|---|---|
| End products of fermentation | Acetate, ethanol, formate, $H_2$, $CO_2$ | Acetate, $CO_2$ (with reduction of selenate or nitrate) | Acetate, ethanol, formate, lactate, $H_2$, $CO_2$ |
| Major fatty acids | NR | NR | 16:1 16:0 14:0 |
| G+C content of DNA (mol%) | 34.4 | 31.2 | 32.7 |
| Sample source and site | Sediment, Lake Magadi, Kenya | Sediment, Dead Sea | Hypersaline oil reservoir, Gabon |

Data taken from
[a]Zhilina and Zavarzin (1990b)
[b]Oren (1983)
[c]Oren et al. (1987)
[d]Zhilina et al. (2012)
[e]Hedi et al. (2009)
[f]Zhilina et al. (2001)
[g]Switzer Blum et al. (2001)
[h]Gales et al. (2011)
[i]Name not yet validly published
*NR* not reported

Rod-shaped cells. Motile with 1–2 subterminal flagella. Multiplication by binary fission is by constriction rather than septation. Gram-negative wall structure. Thermoresistant endospores formed by some strains. Strictly anaerobic. Possess a respiratory type of homoacetogenic metabolism. Extremely halophilic, growing at 1.7–4 M NaCl. Neutrophilic. Mesophilic-Metabolism variable; lithoheterotrophic, utilizing $H_2$, formate, and carbon monoxide; methylotrophic, utilizing methylamines and betaine; or chemoorganotrophic, fermenting some amino acids and organic acids. Acetate is the end product with all substrates utilized.

Type Species: *Acetohalobium arabaticum*

#### Genus *Sporohalobacter*

Oren, Pohla and Stackebrandt 1988, 136[VP] (Validation List no. 24) (Effective publication: Oren, Pohla and Stackebrandt 1987, 239).

Spo.ro.ha.lo.bac'ter. Gr. n. *spora*, seed; Greek n. *hals halos*, salt; N.L. n. *bacter* a staff or rod; N.L. masc. n. *Sporohalobacter* spore-producing salt rod.

Gram-negative rod-shaped cells, motile by peritrichous flagella. Halophilic, growing optimally at 1.4–1.5 M NaCl and requiring minimum 0.7 M NaCl for growth. Temperature optimum about 40 °C. Strictly anaerobic. Ferments amino acids with production of acetate, propionate and other acids, $H_2$, and $CO_2$. Sugars poorly used. Endospores produced. Gas vesicles are attached to the endospores in the single species described.

Type Species: *Sporohalobacter lortetii*

#### Genus *Fuchsiella*

Zhilina, Zavarzina, Panteleeva, Osipov, Kostrikina, Tourova and Zavarzin 2012, 1671[VP].

Fuch.si.el'la. N.L. gem. dim. n. *Fuchsiella*, named in the honor of Prof. Georg Fuchs (Freiburg, Germany), who made a most serious contribution to our understanding of multiple pathways of $CO_2$ assimilation by microorganisms.

Gram-negative, spore-forming rods, motile by peritrichous flagella. Obligatory anaerobic. Obligately alkaliphilic and natronophilic. Performing homoacetogenic metabolism of a restricted number of compounds. Able to grow chemolithoautotrophically with $H_2$ + $CO_2$. Few organic compounds are metabolized with external electron acceptors.

Type Species: *Fuchsiella alkaliacetigena*

#### Genus *Halanaerobaculum*

Hedi, Fardeau, Sadfi, Boudabous, Ollivier and Cayol 2009, 923[VP] (Effective publication: Hedi, Fardeau, Sadfi, Boudabous, Ollivier and Cayol 2009, 317).

Hal.an.ae.ro.ba'cu.lum. Gr. n. *hals halos*, salt; Gr. pref. *an*-, not; Gr. n. *aer aeros*, air; L. neut. n. *baculum*, stick; N.L. neut. n. *Halanaerobaculum*, salt stick not living in air.

Cells are Gram-negative, nonmotile, non-sporulating rods appearing singly, in pairs, or occasionally as long chains, halophilic, obligate anaerobes. Metabolize only carbohydrates. Grow at NaCl concentrations ranging from 14 to 30 %. The end products from glucose fermentation are butyrate, lactate, acetate, $H_2$, and $CO_2$.

Type Species: *Halanaerobaculum tunisiense*

#### Genus *Halonatronum*

Zhilina, Garnova, Tourova, Kostrikina and Zavarzin 2001, 263[VP] (Validation List no. 79) (Effective publication: Zhilina, Garnova, Tourova, Kostrikina and Zavarzin 2001a, 70).

*Ha.lo.na.tro'num.* Gr. n. *hals halos* salt; N.L. n. *natron* (arbitrarily derived from the Arabic n. *natrun* or *natron*) soda,

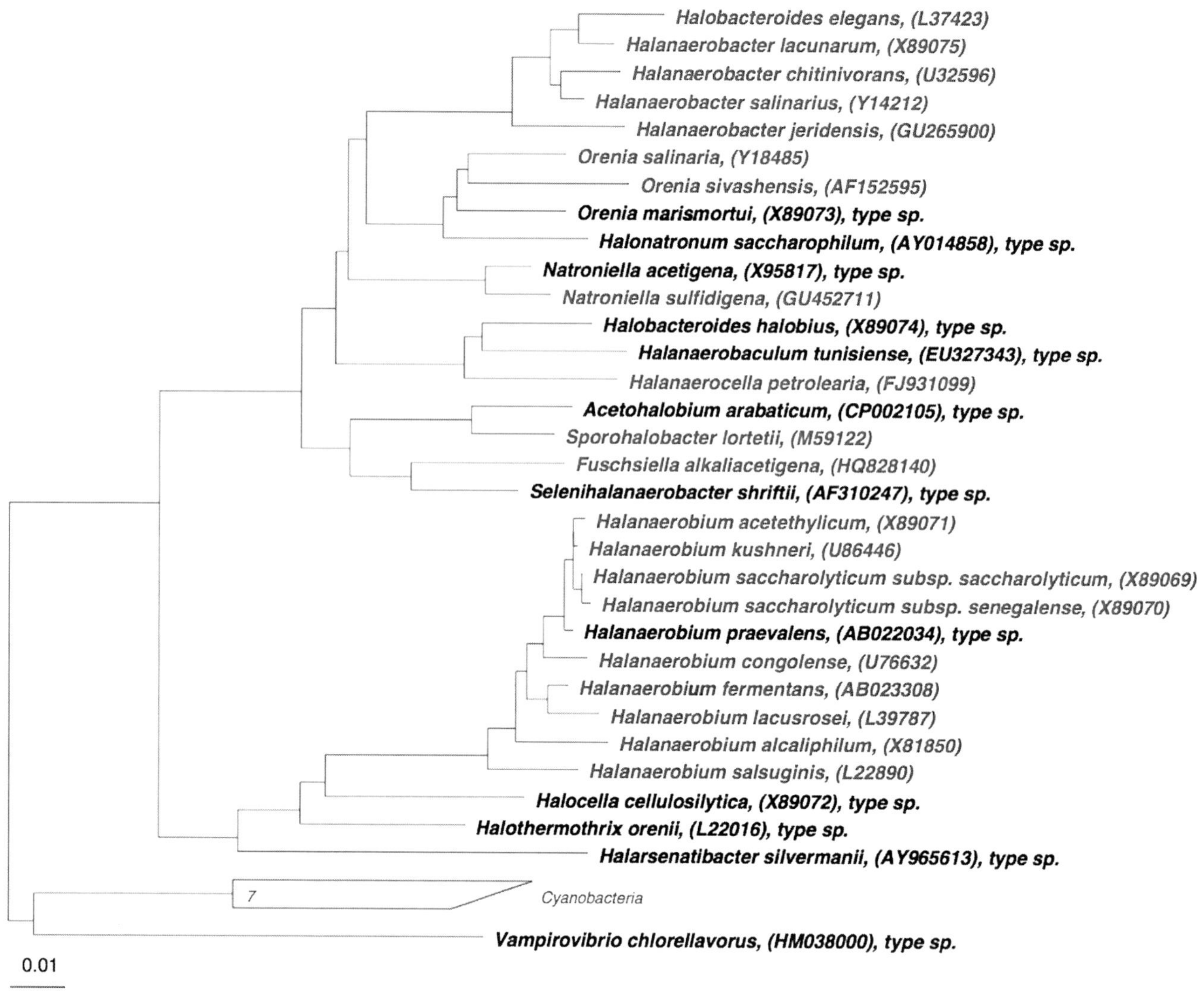


**Fig. 12.3**

**Neighbor-joining genealogy reconstruction of the 31 species and subspecies of the order *Halanaerobiales* present in the LTP_106 (Yarza et al. 2010). The tree was reconstructed by using a subset of sequences 767 type strains of Bacteria and Archaea to stabilize the tree topology. In addition, a 40 % conservational filter for the whole bacterial domain was used to remove hypervariable positions. Numbers in triangles denote number of taxa included. The bar indicates 1 % sequence divergence**

sodium carbonate; N.L. neut. n. *Halonatronum*, an organism growing with salt and soda.

Cells are rod-shaped, flexible, and motile by peritrichous flagella. The cell wall has a Gram-negative structure. Strictly anaerobic, chemoorganotrophic with fermentative metabolism. Carbohydrates, including soluble polysaccharides, are fermented to acetate, ethanol, formate, $H_2$, and $CO_2$. Halophilic and alkaliphilic. Endospores produced.

Type Species: *Halonatronum saccharophilum*

### Genus *Selenihalanaerobacter*

Switzer Blum, Stolz, Oren and Oremland 2001, 1229[VP] (Validation List no. 81) (Effective publication: Switzer Blum, Stolz, Oren and Oremland 2001, 217).

Se.le.ni.hal.an.ae.ro.bac'ter. N.L. n. *selenium* (from Gr. n. *selênê*, the moon), selenium, element 34; Gr. n. *hals halos*, salt; Gr. pref. *an*, not; Gr. n. *aer aeros*, air; N.L. masc. n. *bacter*, a staff or rod; N.L. masc. n. *Selenihalanaerobacter*, the salty anaerobic selenium rod.

Gram-negative rod-shaped cells, nonmotile. Halophilic, growing optimally at 3.6 M NaCl and requiring minimum 1.7 M NaCl for growth. Temperature optimum about 38 °C. Strictly anaerobic. Grows by anaerobic respiration on organic electron donors, using selenate and other electron acceptors. Fermentative growth not observed. Endospores not produced.

Type Species: *Selenihalanaerobacter shriftii*

### Genus *Halanaerocella*

(Effective publication: Gales, Chehider, Joulian, Battaglia-Brunet, Cayol, Postec, Borgomano, Neria-Gonzalez, Lomans, Ollivier and Alazard 2011, 570; the name is yet to be validated).

**Table 12.8**
**Properties of the sequenced genomes of members of the *Halanaerobiales***

| Property | *Halanaerobium praevalens* DSM 2228[Ta] | "*Halanaerobium hydrogenoformans*"[b] | *Halothermothrix orenii* DSM 9562[Tc] |
|---|---|---|---|
| Accession number | CP002175 | CP002304 | CP001098 |
| Genome length (bp) | 2,309,262 | 2,613,116 | 2,578,146 |
| G+C content | 30.3 mol% | 33.1 mol% | 37.9 mol% |
| Extrachromosomal elements | 0 | 0 | 0 |
| % Coding bases | 89.2 % | NR | 88.6 % |
| Number of predicted genes | 2,180 | NR | 2,451 |
| Predicted protein-coding genes | 2,110 | 2,295 | 2,366 |
| % of proteins with putative function | 77.7 % | NR | 80.6 % |
| % assigned to COGs | 80.7 % | NR | 76.6 % |
| Number of 16S rRNA genes | 4 | 4 | 4 |

Data taken from
[a]Ivanova et al. (2011), [b]Brown et al. (2011), [c]Mavromatis et al. (2009), *NR* not reported

Hal.an.ae.ro.cel'la. Gr. h. *hals halos*, salt; Gr. pref. *an*, not; Gr. n. *aer*, air; L. fem. n. *cella*, a store-room and in biology a cell; N.L. fem. n. *Halanaerocella*, salt cell not living in air.

Cells stain Gram-positive, nonmotile, non-sporulating rods occurring singly, in pairs, or occasionally as long chains. Obligate anaerobe metabolizing only carbohydrates. The end products from glucose fermentation are lactate, ethanol, acetate, formate, $H_2$, and $CO_2$.

The Type Species is *Halanaerocella petrolearia.*

The main features of members of the monospecific genera *Acetohalobium*, *Halanaerobacter*, *Sporohalobacter*, *Fuchsiella*, *Halanaerobaculum*, *Halonatronum*, *Selenihalanaerobacter*, and *Halanaerocella* (updated for March 2012), are listed in ➲ *Table 12.7*.

## Isolation, Enrichment, and Maintenance Procedures

Any anoxic reducing medium containing high salt concentrations (5–25 %) and containing a suitable carbon source is a potential enrichment and growth medium for members of the *Halanaerobiales*. A variety of such media have been used for isolation and cultivation. ➲ *Table 12.11* presents a selection. Most species grow as fermenters on simple sugars. Although most species are not extremely sensitive to molecular oxygen, strict anaerobic techniques should be used, including boiling the media under nitrogen or nitrogen-$CO_2$ (80:20) and adding reducing agents such as cysteine, dithionite, or ascorbate + thioglycollate to the boiled media. Protocols for the preparation of media were compiled by Oren (2006); details can be found in the original species descriptions. For the enrichment of thermophiles such as *Halothermothrix*, the incubation temperature should be adjusted to that of the natural environment. More specialized media have been designed for the cultivation of amino acid fermenting, homoacetogenic, selenate- and arsenate-respiring members, and other atypical organisms belonging to the order. For the isolation of *Selenihalanaerobacter*, selenate is the preferred electron acceptor, because nitrate and trimethylamine *N*-oxide also enable anaerobic growth of a variety of facultative anaerobes belonging to other orders.

The formation of heat-resistant endospores has been exploited in a selective enrichment procedure for *Halobacteroides halobius*-like bacteria, based on negative selection by pasteurization of the inoculum for 10–20 min at 80–100 °C (Oren 1987). In view of the number of endospore-forming genera within the family (*Halobacteroides*, *Orenia*, *Sporohalobacter*, *Acetohalobium*, *Natroniella*), such an enrichment strategy could be useful for the isolation of other novel members.

### Maintenance

Many species of *Halobacteroidaceae*, notably the species of the genera *Halobacteroides* (Oren et al. 1984b; Zhilina et al. 1997), *Orenia* (Oren et al. 1987; Mouné et al. 2000; Zhilina et al. 1999), *Haloanaerobacter* (Liaw and Mah 1992; Zhilina et al. 1992; Mouné et al. 1999), *Halonatronum* (Zhilina et al. 2001), and *Natroniella* (Zhilina et al. 1996), easily undergo autolysis, generating spherical degeneration forms (➲ *Fig. 12.1 c, d*). Lysis starts at the end of the exponential growth phase, especially at relatively high growth temperatures. One possibility to avoid death of such cultures is the use of media with a reduced nutrient content and lower growth temperatures (15–25 °C). Weekly transfers may then suffice to maintain viable cultures. Long-term preservation is by freezing anaerobic suspensions in 20 % glycerol at −80 °C (Rengpipat et al. 1988a), by lyophilization, or by storage in liquid nitrogen.

**Table 12.9**
**Substrates used by the type strains of carbohydrate-fermenting species of *Halanaerobiaceae***

| Substrate | *Halanaerobium praevalens* | *Halanaerobium alcaliphilum* | *Halanaerobium acetethylicum* | *Halanaerobium salsuginis* | *Halanaerobium saccharolyticum* subsp. *saccharolyticum* | *Halanaerobium saccharolyticum* subsp. *senegalense* | *Halanaerobium congolense* | *Halanaerobium lacusrosei* | *Halanaerobium kushneri* | *Halanaerobium fermentans* | *Halothermothrix orenii* | *Halocella cellulosilytica* |
|---|---|---|---|---|---|---|---|---|---|---|---|---|
| L-Arabinose | NR | – | NR | + | + | – | – | – | + | – | + | – |
| Cellobiose | – | – | + | – | + | + | NR | + | + | + | + | + |
| Chitin | – | NR | – | – | – | NR | NR | NR | – | NR | NR | NR |
| Erythritol | NR | – | NR | NR | + | NR | NR | NR | NR | NR | NR | NR |
| Fructose | + | + | + | + | + | + | + | + | + | + | + | – |
| Galactose | – | – | – | + | + | – | + | + | + | + | + | + |
| Glucose | + | + | + | + | + | + | + | + | + | + | + | + |
| *N*-acetyl- glucosamine | + | + | + | + | + | NR | NR | NR | NR | + | NR | NR |
| Glycerol | – | – | – | – | + | + | NR | + | + | - | – | NR |
| Lactose | – | – | + | + | + | + | – | + | – | + | – | NR |
| Maltose | NR | + |  | + | + | + | + | + | + | + | – | NR |
| Mannitol | NR | – | NR | – | + | + | NR | + | + | NR | – | NR |
| D-Mannose | + | + | + | + | + | + | + | + | + | + | + | + |
| Melibiose | NR | NR | NR | NR | + | NR | NR | – | NR | NR | + | NR |
| Pectin | + | – | – | – | – | – | NR | NR | – | – | NR | NR |
| Pyruvate | – | + | + | + | + | NR | NR | NR | + | – | NR | NR |
| Raffinose | NR | Slight | NR | + | – | + | NR | NR | NR | + | NR | NR |
| Rhamnose | NR | – | NR | + | – | – | – | – | NR | – | – | NR |
| D-Ribose | NR | – | NR | + | + | + | + | + | NR | + | + | – |
| L-Sorbose | – | NR | – | + | – | – | NR | NR | – | – | – | NR |
| Starch | – | – | – | – | – | NR | NR | + | – | – | + | + |
| Sucrose | – | + | + | + | + | + | + | + | + | + | – | + |
| Trehalose | NR | – | NR | + | + | NR | + | NR | NR | NR | NR | NR |
| D-Xylose | – | – | + | + | + | – | – | + | – | – | + | – |

*NR* not reported

**Table 12.10**
**Substrates used by the type strains of carbohydrate-fermenting species of *Halobacteroidaceae***

| Substrate | *Halobacteroides halobius* | *Halobacteroides elegans* | *Halanaerobacter chitinivorans* | *Halanaerobacter lacunarum* | *Haloanaerobacter salinarius* | *Halanaerobacter jeridensis* | *Orenia marismortui* | *Orenia salinaria* | *Orenia sivashensis* | *Halanaerobaculumtunisiense* | *Halonatronum saccharophilum* | *Halanaerocella petrolearia* |
|---|---|---|---|---|---|---|---|---|---|---|---|---|
| L-Arabinose | – | NR | NR | – | – | – | – | NR | – | NR | NR | NR |
| Cellobiose | – | Weak | + | Weak | NR | v | – | + | + | + | – | + |
| Chitin | NR | – | + | – | – | NR | NR | NR | – | NR | NR | NR |
| Erythritol | NR | NR | NR | – | NR | NR | NR | NR | – | NR | NR | NR |
| Fructose | + | + | + | + | + | + | + | + | – | – | + | + |
| Galactose | + | Weak | NR | NR | + | + | – | – | – | + | NR | + |
| Glucose | + | + | + | + | + | + | + | + | + | + | + | + |
| *N*-acetyl-glucosamine | NR | – | + | – | + | NR | – | – | + | NR | + | NR |
| Glycerol | – | NR | – | – | NR | + | – | NR | – | NR | NR | – |
| Lactose | – | NR | NR | – | – | – | – | NR | – | NR | NR | + |
| Maltose | + | Weak | + | + | + | + | + | + | + | + | + | + |
| Mannitol | – | Weak | NR | + | + | + | NR | + | + | – | – | + |
| D-Mannose | + | + | + | + | + | + | + | – | + | + | – | + |
| Melibiose | – | NR | NR | – | NR | NR | NR | NR | – | NR | NR | NR |
| Pectin | NR | NR | – | – | – | NR | NR | NR | – | NR | NR | NR |
| Pyruvate | + | Weak | – | Weak | – | + | – | – | + | + | – | + |
| Raffinose | + | – | – | – | + | + | – | – | – | NR | NR | – |
| Rhamnose | + | NR | NR | – | – | – | – | NR | – | – | NR | – |
| D-Ribose | – | – | NR | Weak | NR | + | + | NR | + | – | – | – |
| L-Sorbose | – | NR | NR | NR | – | – | NR | NR | – | + | NR | – |
| Starch | + | + | Weak | + | – | + | + | – | + | + | + | NR |
| Sucrose | + | + | + | + | + | + | + | + | + | + | + | + |
| Trehalose | NR | + | NR | + | + | + | NR | + | + | NR | – | NR |
| D-Xylose | – | NR | NR | – | – | + | – | NR | – | – | NR | + |

*NR* not reported

## Ecology

Species belonging to the *Halanaerobiales* can probably be found in any hypersaline anaerobic environment where simple sugars are available or other substrates metabolized by the members of the order. Representatives have been isolated from Great Salt Lake, Utah (Tsai et al. 1995; Zeikus et al. 1983); Salton Sea, California (Shiba 1991; Shiba and Horikoshi 1988; Shiba et al. 1989); Searles Lake, California (Switzer Blum et al. 2009), the Dead Sea (Oren 1983; Oren et al. 1984b, 1987; Switzer Blum et al. 2001); and a hypersaline sulfur spring on the shore of the Dead Sea (Oren 1989); from the alkaline (pH 10.2) hypersaline lake in Magadi, Kenya—shown to harbor a varied anaerobic community, including cellulolytic, proteolytic, saccharolytic, and homoacetogenic bacteria (Shiba and Horikoshi 1988; Zhilina and Zavarzin 1994; Zhilina et al. 1996, 2001)—Big Soda Lake, Nevada (Shiba and Horikoshi 1988; Shiba et al. 1989), soda lakes in Russia (Sorokin et al. 2011; Zhilina et al. 2012), and hypersaline lakes and lagoons in the Crimea (Simankova et al. 1993; Zhilina and Zavarzin 1990b; Zhilina et al. 1991, 1992b) and Senegal (Cayol et al. 1994a, 1995); and from the hypersaline lakes in Tunisia (Cayol et al. 1994b; Hedi et al. 2009; Mezghani et al. 2012) and saltern evaporation ponds in California (Liaw and Mah 1992) and France (Mouné et al. 1999, 2000). Brines associated with oil wells and petroleum reservoirs also yielded a number of interesting species (Bhupathiraju et al. 1991, 1993, 1994, 1999; Gales et al. 2011; Ravot et al. 1997; Rengpipat et al. 1988a). They may also be present in salted fermented foods (Kobayashi et al. 2000a, b). 16S rRNA sequences of yet uncultured organisms affiliated with the *Halanaerobiales* are often recovered in clone libraries prepared from DNA extracted from anaerobic hypersaline environments such as sediments of saltern evaporation ponds (Mouné et al. 2003; Sørensen et al. 2005) and also from anaerobic brines in the depths of the Red Sea (Eder et al. 2001).

The ability to use glycerol, glucosylglycerol, trehalose, cellulose, and chitin may be of particular ecological importance. The first three compounds are accumulated at high concentrations as organic osmotic solutes by aerobic photosynthetic halophilic microorganisms inhabiting salt lakes: glycerol in the green unicellular alga *Dunaliella* and glucosylglycerol and trehalose in a variety of cyanobacteria. Such compounds may then be available to the anaerobic bacterial community in the bottom sediments of these lakes. *Halanaerobium saccharolyticum* and *Halanaerobium lacusrosei* ferment glycerol (Cayol et al. 1994a, 1995; Zhilina et al. 1992). Glycerol oxidation by anaerobic halophiles may be markedly improved through interspecies hydrogen transfer when grown in coculture with $H_2$-consuming sulfate-reducing bacteria (Cayol et al. 1995). *Halanaerobium saccharolyticum* was isolated from a cyanobacterial mat dominated by *Coleofasciculus* (*Microcoleus*) *chthonoplastes*, covering the bottom of a hypersaline lagoon in the Crimea; its ability to use glucosylglycerol, the osmotic solute produced by the cyanobacteria, may be of great ecological importance (Zhilina and Zavarzin 1991). The same organism also degrades trehalose, produced by other cyanobacteria for similar purposes (Zhilina et al. 1992b).

The hypersaline lagoons of the Crimea also contain large masses of dead macroalgae (*Cladophora*). Such environments show high cellulolytic activity. The optimum salt concentration for cellulose decomposition was 15 %, and decomposition was possible up to 25 % salt (Siman'kova and Zavarzin 1992). The cellulose-degrading *Halocella cellulosilytica* was isolated from this habitat (Simankova et al. 1993), and its cellulase complex was characterized in part (Bolobova et al. 1992). Another biopolymer that may be available in large quantities in hypersaline lakes is chitin, derived from the brine shrimp *Artemia* and from larvae of the brine fly which are often abundant in such environments. Evolution of gas bubbles was observed from the sediment of a Californian saltern containing massive amounts of dead brine shrimp. Two strains of *Halanaerobacter chitinivorans* were isolated from this saltern, of which only one grew on chitin (Liaw and Mah 1992). Another substrate that may be available abundantly in hypersaline environments is glycine betaine. This compound is produced as an osmotic solute by the most halophilic among the cyanobacteria and by halophilic anoxygenic photosynthetic bacteria such as *Halorhodospira* species. Glycine betaine is fermented to acetate and trimethylamine by *Halanaerobium alcaliphilum* isolated from Great Salt Lake (Tsai et al. 1995), and by the (non-saccharolytic) *Acetohalobium arabaticum*. The latter species produces only minor amounts of trimethylamine as most is converted to acetate (Zhilina and Zavarzin 1990b). Glycine betaine can also be used as an electron acceptor in the Stickland reaction by *Halanaerobacter salinarius* and *Halanaerobacter chitinivorans*, with $H_2$ or serine as electron donor (Mouné et al. 1999).

Quantitative data on the occurrence of members of the *Halanaerobiales* in hypersaline anoxic environments are scarce. *Halanaerobium praevalens* was reported to be present in Great Salt Lake surface sediment in numbers of up to $10^8$ per mL sediment (Zeikus 1983; Zeikus et al. 1983), while $10^3$–$10^5$ *Halobacteroides* cells were counted per mL of Dead Sea sediment (Oren et al. 1984b). Up to $10^7$–$10^9$ anaerobic halophilic cellulolytic bacteria were enumerated per mL sediment in lagoons of the Arabat strait (Siman'kova and Zavarzin 1992), and up to $4.6 \times 10^3$ anaerobic halophiles were counted in anaerobic brines associated with an oil reservoir in Oklahoma (Bhupathiraju et al. 1991, 1993). The few data available prove that these anaerobic halophiles may form a significant component of the ecosystem in anaerobic hypersaline sediments.

## Pathogenicity, Clinical Relevance

All members of the *Halanaerobiales* are moderately halophilic and do not grow at low salt concentrations. Accordingly, no pathogens are found within the group.

Sensitivity to antibiotics was tested in some species. Within the family *Halanaerobiaceae*, *Halanaerobium salsuginis*, *H. kushneri*, and *H. lacusrosei* were reported to be sensitive to penicillin,

**Table 12.11**

**Media for the growth of members of selected members of the *Halanaerobiales* (all values in g/L, unless stated otherwise). Additional information can be found in the original species description papers and in the website of the Deutsche Sammlung von Mikroorganismen und Zellkulturen: http://www.dsmz.de**

| Compound | *Halanaerobium species* | *Halanaerobium salsuginis* | *Halanaerobium acetethylicum* | *Halobacteroides* spp., *Halanaerobium saccharolyticum* | *Halothermothrix orenii* | *Halocella cellulosilytica* |
|---|---|---|---|---|---|---|
| NaCl | 130 | 120 | 100 | 100–150 | 100 | 150 |
| $MgSO_4 \cdot 7H_2O$ | 8.8 | 0.2 | | | | |
| $MgCl_2 \cdot 6H_2O$ | | | 0.4 | 0.33 | 2.0 | 3.3 |
| KCl | 1.0 | 0.1 | | 0.33 | 4.0 | 0.33 |
| $NH_4Cl$ | | 1.0 | 0.9 | 0.33 | 1.0 | 0.33 |
| $CaCl_2 \cdot 2H_2O$ | | 0.2 | | 0.33 | 0.2 | 0.33 |
| $KH_2PO_4$ | | 0.1 | 0.75 | 0.33 | 0.3 | 0.33 |
| $K_2HPO_4$ | | | 1.5 | | | |
| $FeSO_4 \cdot 7H_2O$ | | | 3 mg | | | |
| $NaHCO_3$ | | | | 1.5[a] | 5.0[a] | 2.5 |
| $Na_2CO_3$ | | | | | | |
| $Na_2S \cdot 9H_2O$ | 0.5[a] | | 1.0[a] | 0.5[a] | 0.2[a] | 0.5[a] |
| Glucose | 5.0[a] | 2.5[a] | 5.0[a] | 5.0[a] | 10[a] | |
| Chitin | | | | | | |
| Microcrystalline cellulose or cellobiose | | | | | | 5.0 or 5.0[a] |
| Trimethylamine HCl or glycine betaine | | | | | | |
| Na-acetate | | | | | 1.0 | |
| Ethanol | | | | | | |
| Yeast extract | 10 | 1.0 | 3.0[a] | | | 2.0 |
| Trypticase | 10 | | 10 | | 0.5 | |
| Peptone | | | | 5.0 | | |
| Casamino acids | 1.0 | | 1.0 | | | |
| Nutrient broth | | | | | | |
| L-Glutamic acid | | | | | | |
| Vitamin solution[b] | 10 mL[a] | 10 mL | 5 mL | 10 mL | | 10 mL |
| Trace element solution | 5 mL[c] | 10 mL[c] | 9 mL[d] | 1 mL[e] | 1 mL[c] | 1 mL[e] |
| Thioglycolate-ascorbate solution[g] | | 25 mL[a] | | | | |
| Cysteine HCl | 0.5 | | | | | |
| Na dithionite | | | | | 10 mg[a] | |
| Resazurin | 0.5 mg | | 1 mg | 2 mg | 1 mg[a] | 2 mg |
| NaOH 2 N | | 10 mL[a] | | | | |
| PIPES-di-K[h] | 1.5 | | | | | |
| Final pH[i] | 7.1–7.3 | 9.0 | 7.2–7.4 | 7.5 | 7.0 | 7.0 |

**Table 12.11 (continued)**

| | *Halobacteroides halobius, Orenia marismortui* | *Acetohalobium arabaticum* | *Halanaerobacter chitinivorans* | *Sporohalobacter lortetii* | *Natroniella acetigena* |
|---|---|---|---|---|---|
| NaCl | 140 | 150 | 100 | 105 | 15.7 |
| $MgSO_4 \cdot 7H_2O$ | | | 9.6 | | |
| $MgCl_2 \cdot 6H_2O$ | 20.3 | 4.0 | 7.0 | 10[c] | 0.1 |
| KCl | 3.7 | 0.33 | 3.8 | 0.75 | 0.2 |
| $NH_4Cl$ | 7.35 | 0.33 | 1.0 | | 1.0 |
| $CaCl_2 \cdot 2H_2O$ | | 0.33 | 0.5 | 3.7 | |
| $KH_2PO_4$ | | 0.33 | 0.4 | | 0.2 |
| $K_2HPO_4$ | | | | | |
| $FeSO_4 \cdot 7H_2O$ | | | | 2 mg | |
| $NaHCO_3$ | 5.0[a] | 4.5[a] | 3.0[a] | | 38.3[a] |
| $Na_2CO_3$ | | | 1.0[a] | | 68.3 |
| $Na_2S \cdot 9H_2O$ | | 0.5[a] | 0.5[a] | | 1.0[a] |
| Glucose | | | 5.0[a] or | | |
| Chitin | | | 5.0 | | |
| Microcrystalline cellulose or cellobiose | | | | | |
| Trimethylamine HCl or glycine betaine | | 2.4[a] or 4.5[a] | | | |
| Na acetate | | | | | |
| Ethanol | | | | | 5 mL[a] |
| Yeast extract | 5.0 | 0.05[a] | 1.0 | 2.0 | 0.2 |
| Trypticase | | | | | |
| Peptone | | | | | |
| Casamino acids | | | | 2.0 | |
| Nutrient broth | | | | 2.0 | |
| L-Glutamic acid | | | | 4.0 | |
| Vitamin solution[b] | | 10 mL[a] | | 10 mL | 10 mL |
| Trace element solution | | 10 mL[c] | 1 mL[c] | 10 mL[c] | 1 mL[f] |
| Thioglycolate-ascorbate solution[g] | | | | | |
| Cysteine HCl | | | 0.5[c] | 0.5[c] | |
| Na dithionite | | | | | |
| Resazurin | 1 mg | 1 mg | 1 mg | 1 mg | 0.5 mg |
| PIPES-di-K[h] | 40 mM | | | | |
| Final pH[i] | 6.5–7.0 | 7.6–8.0 | 7.2 | 6.5 | 9.7-10.0 |

[a]Add separately from sterile anoxic solutions

[b]Vitamin solution containing per liter: biotin, 2 mg; folic acid, 2 mg; pyridoxine HCl, 10 mg; thiamine.HCl·$2H_2O$, 5 mg; riboflavin, 5 mg; nicotinic acid, 5 mg; D-Ca pantothenate, 5 mg; vitamin $B_{12}$, 0.1 mg; *p*-aminobenzoic acid, 5 mg; lipoic acid, 5 mg

[c]Trace element solution containing per liter: nitrilotriacetic acid, 1.5 g; $MgSO_4 \cdot 7H_2O$, 3 g; $MnSO_4 \cdot 2H_2O$, 0.5 g; NaCl, 1 g; $FeSO_4 \cdot 7H_2O$, 100 mg; $CoSO_4 \cdot 7H_2O$, 180 mg; $CaCl_2 \cdot 2H_2O$, 100 mg; $ZnSO_4 \cdot 7H_2O$, 180 mg; $CuSO_4 \cdot 5H_2O$, 10 mg; $KAl(SO_4)_2 \cdot 12H_2O$, 20 mg; $H_3BO_3$, 10 mg; $Na_2MoO_4 \cdot 2H_2O$, 10 mg; $NiCl_2 \cdot 6H_2O$, 25 mg; $Na_2SeO_3 \cdot 5H_2O$, 0.3 mg. First dissolve the nitrilotriacetic acid and adjust to pH 6.5, then add the other minerals. Adjust the final pH to 7.0 with KOH

[d]Trace element solution containing per liter: nitrilotriacetic acid, 12.8 g; $FeCl_2 \cdot 4H_2O$, 200 mg; $MnCl_2 \cdot 4H_2O$, 100 mg; $CoCl_2 \cdot 6H_2O$, 170 mg; $CaCl_2 \cdot 2H_2O$, 100 mg; $ZnCl_2$, 100 mg; $CuCl_2$, 20 mg; $H_3BO_3$, 10 mg; $Na_2MoO_4 \cdot 2H_2O$, 10 mg; $NiCl_2 \cdot 6H_2O$, 26 mg; NaCl, 1 g; $Na_2SeO_3 \cdot 5H_2O$, 20 mg. First dissolve the nitrilotriacetic acid and adjust to pH 6.5 with KOH

[e]Trace element solution containing per liter: 25 % HCl, 10 mL; $FeCl_2 \cdot 4H_2O$, 1.5 g; $ZnCl_2$, 70 mg; $MnCl_2 \cdot 4H_2O$, 100 mg; $H_3BO_3$, 6 mg; $CoCl_2 \cdot 6H_2O$, 190 mg; $CuCl_2 \cdot 2H_2O$, 2 mg; $NiCl_2 \cdot 6H_2O$, 24 mg; $Na_2MoO_4 \cdot 2H_2O$, 36 mg; pH 6.0. First dissolve the $FeCl_2$ in the HCl, then dilute in water, add and dissolve the other salts, and adjust the volume to 1 L

[f]Trace element solution containing per liter: $Na_2EDTA$, 5.2 g; $FeCl_2 \cdot 4H_2O$, 1.5 g; $ZnCl_2$, 70 mg; $MnCl_2 \cdot 4H_2O$, 100 mg; $H_3BO_3$, 6 mg; $CoCl_2 \cdot 6H_2O$, 190 mg; $CuCl_2 \cdot 2H_2O$, 2 mg; $NiCl_2 \cdot 6H_2O$, 24 mg; $Na_2MoO_4 \cdot 2H_2O$, 36 mg; pH 6.0

[g]0.5 g of Na-thioglycolate and 0.5 g of Na-ascorbate in 25 mL $H_2O$, sterilized by filtration

[h]PIPES = piperazine-*N,N'*-*bis*-ethane-sulfonic acid (sesquisodium salt or dipotassium salt have been used in different protocols)

[i]To be adjusted with sterile anoxic HCl, NaOH, or $Na_2CO_3$ (recommended for *Acetohalobium arabaticum*)

chloramphenicol, and tetracycline. An alkaliphilic member of the genus *H. alcaliphilum* resists low concentrations of antibiotics but is inhibited by 200 μg/mL penicillin, 400 μg/mL cycloserine, and 1,000 μg/mL streptomycin. *Halocella cellulosilytica* is inhibited by streptomycin, penicillin, vancomycin, rifampicin, and bacitracin; *Halarsenatibacter silvermanii* is sensitive to vancomycin, kanamycin, penicillin, and tetracycline.

Members of the *Halobacteroidaceae* tested for antibiotics sensitivity include *Halobacteroides halobius* (forming large spheres in the presence of penicillin, also sensitive to chloramphenicol and bacitracin), *Halanaerobacter chitinivorans* (inhibited by chloramphenicol, but not by 100 μg/mL cycloserine, penicillin, streptomycin, or tetracycline), *H. salinarius* (sensitive to chloramphenicol, erythromycin, kanamycin, and tetracycline), *Orenia marismortui* (sensitive to penicillin, bacitracin, novobiocin, erythromycin, polymyxin, and chloramphenicol, but not to streptomycin), *O. salinaria* (sensitive to chloramphenicol, erythromycin, and tetracycline but resistant to kanamycin), and *Fuchsiella alkaliacetigena* (sensitive to vancomycin, novobiocin, and rifampicin).

## Application

### Use in Food Fermentations

*Halanaerobium fermentans* was isolated from "fugunoko nukaduke," a traditional Japanese food prepared from fermented salted puffer fish ovaries. Puffer fish ovaries are salted for at least 6 months, and the ovaries are then fermented naturally with rice bran, fish sauce, and koji for several years. *H. fermentans* may be one of the main bacteria involved in the fermentation process (Kobayashi et al. 2000a). Halophilic anaerobes identified as *Halanaerobium praevalens* (based on 16S rRNA sequence and DNA-DNA hybridization) or *H. alcaliphilum*, producing acetate, butyrate, and propionate, were isolated from canned Swedish fermented herrings ("surströmming") (Kobayashi et al. 2000b). Members of the genus *Halanaerobium* may thus be involved in the manufacturing of traditional fermented food products.

### Industrial Fermentation for Hydrogen and Acetate

The use of anaerobic halophilic bacteria in the industrial fermentation of complex organic matter and the production of organic solvents has been proposed (Lowe et al. 1993; Wise 1987), but any such applications are still in an experimental stage. Recently it was proposed to use *Halanaerobium saccharolyticum* subsp. *saccharolyticum* and subsp. *senegalense* for the industrial production of hydrogen from glycerol formed as by-product of the biodiesel industry. The highest $H_2$ yield (1.6 mol $H_2$/mol glycerol) was obtained with *H. saccharolyticum* subsp. *senegalense* grown at 15 % salt. *H. saccharolyticum* subsp. *saccharolyticum* produced less $H_2$ (0.6 mol/mol glycerol) but also yielded 1,3-propanediol (up to 0.49 mol/mol glycerol) as a valuable by-product (Kivistö et al. 2010). *Halocella cellulosilytica* is a cellulose degrader (Simankova et al. 1993), but its biotechnological potential for cellulose degradation at high salt concentrations has not yet been exploited.

### Enhanced Oil Recovery

Several species of *Halanaerobium* (*H. salsuginis, H. acetethylicum, H. kushneri, H. congolense*) and *Halanaerocella petrolearia* were isolated from brines associated with oil reservoirs (Bhupathiraju et al. 1994, 1999; Gales et al. 2011; Ravot et al. 1997; Rengpipat et al. 1988a). Such bacteria may be applied for microbially enhanced oil recovery from oil reservoirs by plugging of porous reservoirs and by anaerobically metabolizing nutrients with the production of useful products such as gases, biosurfactants, and polymers under the environmental conditions that exist in the reservoirs (Bhupathiraju et al. 1991).

### Treatment of Saline Wastewater

Treatment of saline wastewater in an anaerobic packed bed reactor inoculated with *Halanaerobium lacusrosei* was explored, using model wastewaters with glucose as carbon source applying a gradual increase in salinity from 0 to 5 % or from 3 to 10%. Glucose removal at 70 % efficiency was claimed at 3 % salt (Kapdan and Erten 2007; Kapdan and Boylan 2009). As *H. lacusrosei* does not grow below 6 % salt and has its optimum at 20 % salt (Cayol et al. 1995), it is not clear to what extent the glucose degradation observed was indeed effected by *Halanaerobium*.

Nitrosubstituted aromatic compounds such as nitrobenzene, nitrophenols, 2,4-dinitrophenol, and 2,4-dinitroaniline are reduced to the amino derivatives by *Halanaerobium praevalens* and by *Orenia marismortui* (Oren et al. 1991).

### Enzymes

Several enzymes from members of the *Halanaerobiales* have been cloned, purified, and characterized. One such enzyme is the rhodanese-like protein (thiosulfate: cyanide sulfurtransferase; EC 2.8.1.1) of *Halanaerobium congolense* (Ravot et al. 2005). *Halothermothrix orenii* has become a popular object of such studies because of the prospect of enzymes that function both at high salinity and at high temperature. A few such enzymes have been crystallized to study their structure: α-amylase AmyA (EC 3.2.1.1) (optimum activity at 65 °C in 5 % NaCl, with significant activity at 25 % NaCl) (Li et al. 2002; Mijts and Patel 2002), α-amylase AmyB (Tan et al. 2008), ribokinase (EC 2.7.1.15) (Kori et al. 2012), sucrose phosphate synthase (EC 2.4.1.14) (Chua et al. 2008; Huynh et al. 2005), fructokinase (EC 2.7.1.4) (Chua et al. 2010), and class II 5-enopyruvylshikimate-3-phosphate synthase (EC 2.5.1.19). The latter protein, a key enzyme in the synthesis of aromatic amino acids, when expressed in *Arabidopsis* plants bestowed resistance to glyphosate herbicides (Tian et al. 2012).

## References


Baumgartner JG (1937) The salt limits and thermal stability of a new species of anaerobic halophile. Food Res 2:321–329

Bhupathiraju VK, Sharma PK, McInerney MJ, Knapp RM, Fowler K, Jenkins W (1991) Isolation and characterization of novel halophilic anaerobic bacteria from oil field brines. In: Donaldson EC (ed) Microbial enhancement of oil recovery—recent advances. Elsevier Scientific Publishers, Amsterdam, pp 132–143

Bhupathiraju VK, McInerney MJ, Knapp RM (1993) Pretest studies for a microbially enhanced oil recovery field pilot in a hypersaline oil reservoir. Geomicrobiol J 11:19–34

Bhupathiraju VK, Oren A, Sharma PK, Tanner RS, Woese CR, McInerney MJ (1994) *Haloanaerobium salsugo* sp. nov., a moderately halophilic, anaerobic bacterium from a subterranean brine. Int J Syst Bacteriol 44:565–572

Bhupathiraju VK, McInerney MJ, Woese CR, Tanner RS (1999) *Haloanaerobium kushneri* sp. nov., an obligately halophilic, anaerobic bacterium from an oil brine. Int J Syst Bacteriol 49:953–960

Bolobova AV, Simankova MC, Markovich NA (1992) Cellulase complex of a new halophilic bacterium *Halocella cellulolytica*. Mikrobiologiya 61:557–562

Brown SD, Begemann MB, Mormile MR, Wall JD, Han CS, Goodwin LA, Pitluck S, Land ML, Hauser LJ, Elias DA (2011) Complete genome sequence of the haloalkaliphilic, hydrogen-producing bacterium *Halanaerobium hydrogenoformans*. J Bacteriol 193:3682–3683

Cayol JL, Ollivier B, Lawson Anani Soh A, Fardeau M-L, Ageron E, Grimont PAD, Prensier G, Guezennec J, Magot M, Garcia J-L (1994a) *Haloincola saccharolytica* subsp. senegalensis subsp. nov., isolated from the sediments of a hypersaline lake, and emended description of *Haloincola saccharolytica*. Int J Syst Bacteriol 44:805–811

Cayol J-L, Ollivier B, Patel BKC, Prensier G, Guezennec J, Garcia J-L (1994b) Isolation and characterization of *Halothermothrix orenii* gen. nov., sp. nov., a halophilic, thermophilic, fermentative, strictly anaerobic bacterium. Int J Syst Bacteriol 44:534–540

Cayol J-L, Ollivier B, Patel BKC, Ageron E, Grimont PAD, Prensier G, Garcia J-L (1995) *Haloanaerobium lacusroseus* sp. nov., an extremely halophilic fermentative bacterium from the sediments of a hypersaline lake. Int J Syst Bacteriol 45:790–797

Cayol J-L, Ollivier B, Garcia J-L (2009) Genus III. *Halothermothrix* Cayol, Ollivier, Patel, Prensier, Guezennec and Garcia 1994a, 538[VP]. In: De Vos P, Garrity GM, Jones D, Krieg NR, Ludwig W, Rainey FA, Schleifer K-H, Whitman WB (eds) Bergey's manual of systematic bacteriology, vol 3, 2nd edn, The Firmicutes. Springer, Dordrecht, pp 1202–1206

Chua TK, Bujnicki JM, Tan T-C, Huynh F, Patel BK, Sivaraman J (2008) The structure of sucrose phosphate synthase from *Halothermothrix orenii* reveals its mechanism of action and binding mode. Plant Cell 20:1059–1072

Chua TK, Seetharaman J, Kasprzak JM, Ng C, Patel BKC, Love C, Bujnicki JM, Sivaraman J (2010) Crystal structure of a fructokinase homolog from *Halothermothrix orenii*. J Struct Biol 171:397–401

Detkova EN, Boltyanskaya YV (2006) Relationships between the osmoadaptation strategy, amino acid composition of bulk protein, and properties of certain enzymes of haloalkaliphilic bacteria. Mikrobiologiya 75:259–265

Detkova EN, Pusheva MA (2006) Energy metabolism in halophilic and alkaliphilic acetogenic bacteria. Mikrobiologiya 75:1–11

Eder W, Jahnke LL, Schmidt M, Huber R (2001) Microbial diversity of the brine-seawater interface of the Kebrit Deep, Red Sea, studied via 16S rRNA gene sequences and cultivation methods. Appl Environ Microbiol 67:3077–3085

Elevi Bardavid R, Oren A (2012) The amino acid composition of proteins from anaerobic halophilic bacteria of the order *Halanaerobiales*. Extremophiles 16:567–572

Gales G, Chehider N, Joulian C, Battaglia-Brunet F, Cayol J-L, Postec A, Borgomano J, Neria-Gonzalez I, Lomans BP, Ollivier B, Alazard D (2011) Characterization of *Halanaerocella petrolearia* gen. nov., sp. nov., a new anaerobic moderately halophilic fermentative bacterium isolated from a deep subsurface hypersaline oil reservoir. Extremophiles 15:565–571

Hedi A, Fardeau M-L, Sadfi N, Boudabous A, Ollivier B, Cayol J-L (2009) Characterization of *Haloanaerobaculum tunisiense* gen. nov., sp. nov., a new halophilic fermentative, strictly anaerobic bacterium isolated from a hypersaline lake in Tunisia. Extremophiles 13:313–319

Huynh F, Tan T-C, Swaminathan K, Patel BKC (2005) Expression, purification and preliminary crystallographic analysis of sucrose phosphate synthase (SPS) from *Halothermothrix orenii*. Acta Crystallogr F Struct Biol Cryst Commun 61:116–117

Ivanova N, Sikorski J, Chertkov O, Nolan M, Lucas S, Hammon N, Deshpande S, Cheng J-F, Tapia R, Han C, Goodwin L, Pitluck S, Huntemann M, Liolios K, Pagani I, Mavromatis K, Ovchinikova G, Pati A, Chen A, Palaniappan K, Land M, Hauser L, Brambilla E, Palaniappan K, Kannan KP, Rohde M, Tindall BJ, Göker M, Detter JC, Woyke T, Bristow J, Eisen JA, Markowitz V, Hugenholtz P, Kyrpides NC, Klenk H-P, Lapidus A (2011) Complete genome sequence of *Halanaerobium praevalens* type strain ($GSL^T$). Stand Genom Sci 4:312–321

Kapdan IK, Boylan B (2009) Batch treatment of saline wastewater by *Halanaerobium lacusrosei* in an anaerobic packed bed reactor. J Chem Technol Biotechnol 84:34–38

Kapdan IK, Erten B (2007) Anaerobic treatment of saline wastewater by *Halanaerobium lacusrosei*. Process Biochem 42:449–453

Kevbrin VV, Zavarzin GA (1992a) Methanethiol utilization and sulfur reduction by anaerobic halophilic saccharolytic bacteria. Curr Microbiol 24:247–250

Kevbrin VV, Zavarzin GA (1992b) Effect of sulfur compounds on the growth of the halophilic homoacetogenic bacterium *Acetohalobium arabaticum*. Mikrobiologiya 61:563–567

Kevbrin VV, Zhilina TN, Zavarzin GA (1995) Physiology of the halophilic homoacetic bacterium *Acetohalobium arabaticum*. Mikrobiologiya 64:134–138

Kivistö AT, Karp MT (2011) Halophilic anaerobic fermentative bacteria. J Biotechnol 152:114–124

Kivistö AT, Santala V, Karp MT (2010) Hydrogen production from glycerol using halophilic fermentative bacteria. Bioresour Technol 101:8671–8677

Kobayashi T, Kimura B, Fujii T (2000a) *Haloanaerobium fermentans* sp. nov., a strictly anaerobic, fermentative halophile isolated from fermented puffer fish ovaries. Int J Syst Evol Microbiol 50:1621–1627

Kobayashi T, Kimura B, Fujii T (2000b) Strictly anaerobic halophiles isolated from canned Swedish fermented herrings (Surströmming). Int J Food Microbiol 54:81–89

Kori LD, Hofmann A, Patel BKC (2012) Expression, purification, crystallization and preliminary X-ray diffraction analysis of a ribokinase from the thermohalophile *Halothermothrix orenii*. Acta Cryst F68:240–241

Li N, Patel BKC, Mijts BN, Swaminathan K (2002) Crystallization of an α-amylase, AmyA, from the thermophilic halophile *Halothermothrix orenii*. Acta Cryst D58:2125–2126

Liaw HJ, Mah RA (1992) Isolation and characterization of *Haloanaerobacter chitinovorans* gen. nov., sp. nov., a halophilic, anaerobic, chitinolytic bacterium from a solar saltern. Appl Environ Microbiol 58:260–266

Lowe SE, Jain MK, Zeikus JG (1993) Biology, ecology, and biotechnological applications of anaerobic bacteria adapted to environmental stresses in temperature, pH, salinity, or substrates. Microbiol Rev 57:451–509

Mavromatis K, Ivanova N, Anderson I, Lykidis A, Hooper SD, Sun H, Kunin V, Lapidus A, Hugenholtz P, Patel B, Kyrpides NC (2009) Genome analysis of the anaerobic thermohalophilic bacterium *Halothermothrix orenii*. PLoS One 4:e4192

Mesbah NM (2009) Genus III. *Halanaerobacter* Liaw and Mah 1996, 362[VP] (Effective publication: Liaw and Mah 1992, 265). In: De Vos P, Garrity GM, Jones D, Krieg NR, Ludwig W, Rainey FA, Schleifer K-H, Whitman

WB (eds) Bergey's manual of systematic bacteriology, vol 3, 2nd edn, The Firmicutes. Springer, Dordrecht, pp 1212–1215

Mezghani M, Alazard D, Karray F, Cayol JL, Joseph M, Postec A, Fardeau ML, Tholozan JL, Sayadi S (2012) *Halanaerobacter jeridensis* sp. nov., a novel fermentative, anaerobic bacterium from a hypersaline lake in Tunisia. Int J Syst Evol Microbiol 62:1970–1973

Mijts BM, Patel BKC (2001) Random sequence analysis of genomic DNA of an anaerobic, thermophilic, halophilic bacterium, *Halothermothrix orenii*. Extremophiles 5:61–69

Mijts BM, Patel BKC (2002) Cloning, sequencing and expression of an α-amylase gene, *amyA*, from the thermophilic halophile *Halothermothrix orenii* and purification and biochemical characterization of the recombinant enzyme. Microbiology 148:2343–2349

Mouné S, Manac'h M, Hirschler A, Caumette P, Willison JC, Matheron R (1999) *Haloanaerobacter salinarius* sp. nov., a novel halophilic fermentative bacterium that reduces glycine-betaine to trimethylamine with hydrogen or serine as electron donors; emendation of the genus *Haloanaerobacter*. Int J Syst Bacteriol 49:103–112

Mouné S, Eatock C, Matheron R, Willison JC, Hirschler A, Herbert R, Caumette P (2000) *Orenia salinaria* sp. nov., a fermentative bacterium isolated from anaerobic sediments of Mediterranean salterns. Int J Syst Evol Microbiol 50:721–729

Mouné S, Caumette P, Matheron R, Willison JC (2003) Molecular sequence analysis of prokaryotic diversity in the anoxic sediments underlying cyanobacterial mats of two hypersaline ponds in Mediterranean salterns. FEMS Microbiol Ecol 44:117–130

Ollivier B, Caumette P, Garcia J-L, Mah RA (1994) Anaerobic bacteria from hypersaline environments. Microbiol Rev 58:27–38

Oren A (1983) *Clostridium lortetii* sp. nov., a halophilic obligatory anaerobic bacterium producing endospores with attached gas vacuoles. Arch Microbiol 136:42–48

Oren A (1986a) The ecology and taxonomy of anaerobic halophilic eubacteria. FEMS Microbiol Rev 39:23–29

Oren A (1986b) Intracellular salt concentrations of the anaerobic halophilic eubacteria *Haloanaerobium praevalens* and *Halobacteroides halobius*. Can J Microbiol 32:4–9

Oren A (1987) A procedure for the selective enrichment of *Halobacteroides halobius* and related bacteria from anaerobic hypersaline sediments. FEMS Microbiol Lett 42:201–204

Oren A (1988) Anaerobic degradation of organic compounds at high salt concentrations. Antonie Van Leeuwenhoek 54:267–277

Oren A (1989) Photosynthetic and heterotrophic benthic bacterial communities of a hypersaline sulfur spring on the shore of the Dead Sea. In: Cohen Y, Rosenberg E (eds) Microbial mats. Physiological ecology of benthic microbial communities. American Society for Microbiology, Washington, DC, pp 64–76

Oren A (1990) Anaerobic degradation of organic compounds in hypersaline environments: possibilities and limitations. In: Wise DL (ed) Bioprocessing and biotreatment of coal. Marcel Dekker, New York, pp 155–175

Oren A (1993a) The genera *Haloanaerobium*, *Halobacteroides*, and *Sporohalobacter*. In: Balows A, Trüper HG, Dworkin M, Harder W, Schleifer K-H (eds) The prokaryotes. A handbook on the biology of bacteria: ecophysiology, isolation, identification, applications, vol II, 2nd edn. Springer, New York, pp 1893–1900

Oren A (1993b) Anaerobic heterotrophic bacteria growing at extremely high salt concentrations. In: Guerrero R, Pedrós-Alió C (eds) Trends in microbial ecology. Spanish Society for Microbiology, Madrid, pp 539–542

Oren A (2000) Change of the names *Haloanaerobiales*, *Haloanaerobiaceae* and *Haloanaerobium* to *Halanaerobiales*, *Halanaerobiaceae* and *Halanaerobium*, respectively, and further nomenclatural changes within the order *Halanaerobiales*. Int J Syst Evol Microbiol 50:2229–2230

Oren A (2006) The order *Haloanaerobiales*. In: Dworkin M, Falkow S, Rosenberg E, Schleifer K-H, Stackebrandt E (eds) The prokaryotes. A handbook on the biology of bacteria: ecophysiology and biochemistry, vol 4. Springer, New York, pp 804–817

Oren A (2009a) Order II. *Halanaerobiales* corrig. Rainey and Zhilina 1995a, 879$^{VP}$ (Effective publication: Rainey, Zhilina, Boulygina, Stackebrandt, Tourova and Zavarzin 199b, 193). In: De Vos P, Garrity GM, Jones D, Krieg NR, Ludwig W, Rainey FA, Schleifer K-H, Whitman WB (eds) Bergey's manual of systematic bacteriology, vol 3, 2nd edn, The Firmicutes. Springer, Dordrecht, pp 1191–1194

Oren A (2009b) Family I. *Halanaerobiaceae* corrig. Oren, Paster and Woese 1984b, 503$^{VP}$ (Effective publication: Oren, Paster and Woese, 1984a, 79). In: De Vos P, Garrity GM, Jones D, Krieg NR, Ludwig W, Rainey FA, Schleifer K-H, Whitman WB (eds) Bergey's manual of systematic bacteriology, vol 3, 2nd edn, The Firmicutes. Springer, Dordrecht, pp 1195–1196

Oren A (2009c) Genus I. *Halanaerobium* corrig. Zeikus, Hegge, Thompson, Phelps and Langworthy 1984, 503$^{VP}$ (Effective publication: Zeikus, Hegge, Thompson, Phelps and Langworthy 1983, 232). In: De Vos P, Garrity GM, Jones D, Krieg NR, Ludwig W, Rainey FA, Schleifer K-H, Whitman WB (eds) Bergey's manual of systematic bacteriology, vol 3, 2nd edn, The Firmicutes. Springer, Dordrecht, pp 1196–1201

Oren A (2009d) Family II. *Halobacteroidaceae* Zhilina and Rainey 1995, 879$^{VP}$ (Effective publication: Rainey, Zhilina, Boulygina, Stackebrandt, Tourova and Zavarzin 1995a, 193). In: De Vos P, Garrity GM, Jones D, Krieg NR, Ludwig W, Rainey FA, Schleifer K-H, Whitman WB (eds) Bergey's manual of systematic bacteriology, vol 3, 2nd edn, The Firmicutes. Springer, Dordrecht, pp 1207–1208

Oren A (2009e) Genus I. *Halobacteroides* Oren, Weisburg, Kessel and Woese 1984a, 355$^{VP}$ (Effective publication: Oren, Weisburg, Kessel and Woese 1984b, 68). In: De Vos P, Garrity GM, Jones D, Krieg NR, Ludwig W, Rainey FA, Schleifer K-H, Whitman WB (eds) Bergey's manual of systematic bacteriology, vol 3, 2nd edn, The Firmicutes. Springer, Dordrecht, pp 1208–1209

Oren A (2009f) Genus VIII. *Sporohalobacter* Oren, Pohla and Stackebrandt 1988, 136$^{VP}$ (Effective publication: Oren, Pohla and Stackebrandt 1987, 245). In: De Vos P, Garrity GM, Jones D, Krieg NR, Ludwig W, Rainey FA, Schleifer K-H, Whitman WB (eds) Bergey's manual of systematic bacteriology, vol 3, 2nd edn, The Firmicutes. Springer, Dordrecht, pp 1222–1224

Oren A, Gurevich P (1993) The fatty acid synthetase complex of *Haloanaerobium praevalens* is not inhibited by salt. FEMS Microbiol Lett 108:287–290

Oren A, Paster BJ, Woese CR (1984a) *Haloanaerobiaceae*: a new family of moderately halophilic, obligatory anaerobic bacteria. Syst Appl Microbiol 5:71–80

Oren A, Weisburg WG, Kessel M, Woese CR (1984b) *Halobacteroides halobius* gen. nov., sp. nov., a moderately halophilic anaerobic bacterium from the bottom sediments of the Dead Sea. Syst Appl Microbiol 5:58–70

Oren A, Pohla H, Stackebrandt E (1987) Transfer of *Clostridium lortetii* to a new genus *Sporohalobacter* gen. nov. as *Sporohalobacter lortetii* comb. nov., and description of *Sporohalobacter marismortui* sp. nov. Syst Appl Microbiol 9:239–246

Oren A, Gofshtein-Gandman LV, Keynan A (1989) Hydrolysis of *N*′-benzoyl-D-arginine-*p*-nitroanilide by members of the *Haloanaerobiaceae*: additional evidence that *Haloanaerobium praevalens* is related to endospore-forming bacteria. FEMS Microbiol Lett 58:5–10

Oren A, Gurevich P, Henis Y (1991) Reduction of nitrosubstituted aromatic compounds by the halophilic eubacteria *Haloanaerobium praevalens* and *Sporohalobacter marismortui*. Appl Environ Microbiol 57:3367–3370

Oren A, Gurevich P, Azachi M, Henis Y (1992) Microbial degradation of pollutants at high salt concentrations. Biodegradation 3:387–398

Oren A, Heldal M, Norland S (1997) X-ray microanalysis of intracellular ions in the anaerobic halophilic eubacterium *Haloanaerobium praevalens*. Can J Microbiol 43:588–592

Oren A, Oremland RS, Stolz JF (2009) Genus VII. *Selenihalanaerobacter* Switzer Blum, Stolz, Oren and Oremland 2001, 1229$^{VP}$ (Effective publication: Switzer Blum, Stolz, Oren and Oremland 2001, 217). In: De Vos P, Garrity GM, Jones D, Krieg NR, Ludwig W, Rainey FA, Schleifer K-H, Whitman WB (eds) Bergey's manual of systematic bacteriology, vol 3, 2nd edn, The Firmicutes. Springer, Dordrecht, p 1121

Patel BKC, Andrews KT, Ollivier B, Mah RA, Garcia J-L (1995) Reevaluating the classification of *Halobacteroides* and *Haloanaerobacter* species based on sequence comparisons of the 16S ribosomal RNA gene. FEMS Microbiol Lett 134:115–119

Pusheva MA, Detkova EN (1996) Bioenergetic aspects of acetogenesis on various substrates by the extremely halophilic acetogenic bacterium *Acetohalobium arabaticum*. Mikrobiologiya 65:516–520

Pusheva MA, Detkova EN, Bolotina NP, Zhilina TN (1992) Properties of periplasmatic hydrogenase of *Acetohalobium arabaticum*, an extremely halophilic homoacetogenic bacterium. Mikrobiologiya 61:653–657

Rainey FA (2009) Genus VI. *Orenia* Rainey, Zhilina, Boulygina and Zavarzin 1995b, 880$^{VP}$ (Effective publication: Rainey, Zhilina, Boulygina, Stackebrandt, Tourova and Zavarzin 1995a, 197). In: De Vos P, Garrity GM, Jones D, Krieg NR, Ludwig W, Rainey FA, Schleifer K-H, Whitman WB (eds) Bergey's manual of systematic bacteriology, vol 3, 2nd edn, The Firmicutes. Springer, Dordrecht, pp 1217–1220

Rainey FA, Zhilina TN, Boulygina ES, Stackebrandt E, Tourova TP, Zavarzin GA (1995) The taxonomic status of the fermentative halophilic anaerobic bacteria: description of Halobacteriales ord. nov., *Halobacteroidaceae* fam. nov., *Orenia* gen. nov. and further taxonomic rearrangements at the genus and species level. Anaerobe 1:185–199

Ravot G, Magot M, Ollivier B, Patel BKC, Ageron E, Grimont PAD, Thomas P, Garcia J-L (1997) *Haloanaerobium congolense* sp. nov., an anaerobic, moderately halophilic, thiosulfate- and sulfur-reducing bacterium from an African oil field. FEMS Microbiol Lett 147:81–88

Ravot G, Casalot L, Ollivier B, Loison G, Magot M (2005) *rdlA*, a new gene encoding a rhodanese-like protein in *Halanaerobium congolense* and other thiosulfate-reducing anaerobes. Res Microbiol 156:1031–1038

Rengpipat S, Langworthy TA, Zeikus JG (1988a) *Halobacteroides acetoethylicus* sp. nov., a new obligately anaerobic halophile isolated from deep subsurface hypersaline environments. Syst Appl Microbiol 11:28–35

Rengpipat S, Lowe SE, Zeikus JG (1988b) Effect of extreme salt concentrations on the physiology and biochemistry of *Halobacteroides acetoethylicus*. J Bacteriol 170:3065–3071

Shiba H (1991) Anaerobic halophiles. In: Horikoshi K, Grant WD (eds) Superbugs. Microorganisms in extreme environments. Japan Scientific Societies Press/Springer, Tokyo/Berlin, pp 191–211

Shiba H, Horikoshi K (1988) Isolation and characterization of novel anaerobic, halophilic eubacteria from hypersaline environments of western America and Kenya. In: Proceedings of the FEMS symposium on the microbiology of extreme environments and its biotechnological potential, Estoril, pp 371–374

Shiba H, Yamamoto H, Horikoshi K (1989) Isolation of strictly anaerobic halophiles from the aerobic surface sediments of hypersaline environments in California and Nevada. FEMS Microbiol Lett 57:191–196

Siman'kova MV, Zavarzin GA (1992) Anaerobic decomposition of cellulose in lake Sivash and hypersaline lagoons of Arabat spit. Mikrobiologiya 61:193–197

Simankova MV, Chernych NA, Osipov GA, Zavarzin GA (1993) *Halocella cellulolytica* gen. nov., sp. nov., a new obligately anaerobic, halophilic, cellulolytic bacterium. Syst Appl Microbiol 16:385–389

Sørensen KB, Canfield DE, Teske AP, Oren A (2005) Community composition of a hypersaline endoevaporitic microbial mat. Appl Environ Microbiol 71:7352–7365

Sorokin DY, Detkova EN, Muyzer G (2011) Sulfur-dependent respiration under extremely haloalkaline conditions in soda lake 'acetogens' and the description of *Natroniella sulfidigena* sp. nov. FEMS Microbiol Lett 319:88–95

Switzer Blum J, Stolz JF, Oren A, Oremland RS (2001) *Selenihalanaerobacter shriftii* gen. nov., sp. nov., a halophilic anaerobe from Dead Sea sediments that respires selenate. Arch Microbiol 175:208–219

Switzer Blum J, Han S, Lanoil B, Saltikov C, White B, Tabita FR, Langley S, Beveridge TJ, Jahnke L, Oremland RS (2009) Ecophysiology of "*Halarsenatibacter silvermanii*" strain SLAS-1$^{T}$, gen. nov., sp. nov., a facultative chemoautotrophic arsenate-respirer from salt-saturated Searles Lake, California. Appl Environ Microbiol 75:1950–1960

Tan T-C, Mijts BN, Swaminathan K, Patel BKC, Divne C (2008) Crystal structure of the polyextremophilic α-amylase AmyB from *Halothermothrix orenii*: details of a productive enzyme-substrate complex and an *N* domain with a role in binding raw starch. J Mol Biol 378:850–868

Tian Y-S, Xu J, Xiong A-S, Zhao W, Gao F, Fu X-Y, Peng R-H, Yao Q-H (2012) Functional characterization of Class II 5-enopyruvylshikimate-3-phosphate synthase from *Halothermothrix orenii* H168 in *Escherichia coli* and transgenic *Arabidopsis*. Appl Microbiol Biotechnol 93:241–250

Tourova TP (2000) The role of DNA-DNA hybridization and 16S rRNA gene sequencing in solving taxonomic problems by the example of the order *Haloanaerobiales*. Mikrobiologiya 69:623–634

Tourova TP, Boulygina ES, Zhilina TN, Hanson RS, Zavarzin GA (1995) Phylogenetic study of haloanaerobic bacteria by 16S ribosomal RNA sequences analysis. Syst Appl Microbiol 18:189–195

Tsai C-R, Garcia J-L, Patel BKC, Cayol J-L, Baresi L, Mah RA (1995) *Haloanaerobium alcaliphilum* sp. nov., an anaerobic moderate halophile from the sediments of Great Salt Lake. Utah Int J Syst Bacteriol 45:301–307

Wise DL (1987) Meeting report—first international workshop on biogasification and biorefining of Texas lignite. Res Conserv 15:229–247

Yarza P, Ludwig W, Euzéby J, Amann R, Schleifer K-H, Glöckner FO, Rosseló-Móra R (2010) Update of the all-species living tree project based on 16S and 23S rRNA sequence analyses. Syst Appl Microbiol 33:291–299

Zavarzin GA (2009) Genus II. *Halocella* Simankova, Osipov and Zavarzin 1994, 182$^{VP}$ (Effective publication: Simankova, Chernych, Osipov and Zavarzin 1993, 389). In: De Vos P, Garrity GM, Jones D, Krieg NR, Ludwig W, Rainey FA, Schleifer K-H, Whitman WB (eds) Bergey's manual of systematic bacteriology, vol 3, 2nd edn, The Firmicutes. Springer, Dordrecht, pp 1201–1202

Zavarzin GA, Zhilina TN (2009a) Genus II. *Acetohalobium* Zhilina and Zavarzin 1990b, 470$^{VP}$ (Effective publication: Zhilina and Zavarzin 1990c, 747). In: De Vos P, Garrity GM, Jones D, Krieg NR, Ludwig W, Rainey FA, Schleifer K-H, Whitman WB (eds) Bergey's manual of systematic bacteriology, vol 3, 2nd edn, The Firmicutes. Springer, Dordrecht, pp 1209–1212

Zavarzin GA, Zhilina TN (2009b) Genus V. *Natroniella* Zhilina, Zavarzin, Detkova and Rainey 1996a, 1189$^{VP}$ (Effective publication: Zhilina, Zavarzin, Detkova and Rainey 1996b, 324). In: De Vos P, Garrity GM, Jones D, Krieg NR, Ludwig W, Rainey FA, Schleifer K-H, Whitman WB (eds) Bergey's manual of systematic bacteriology, vol 3, 2nd edn, The Firmicutes. Springer, Dordrecht, pp 1216–1217

Zavarzin GA, Zhilina TN, Pusheva MA (1994) Halophilic acetogenic bacteria. In: Drake HL (ed) Acetogenesis. Chapman and Hall, New York, pp 432–444

Zeikus JG (1983) Metabolic communication between biodegradative populations in nature. In: Slater JH, Whittenbury R, Wimpenny JWT (eds) Microbes in their natural environments. 34th symposium, Society for General Microbiology. Cambridge University Press, Cambridge, pp 423–462

Zeikus JG, Hegge PW, Thompson TE, Phelps TJ, Langworthy TA (1983) Isolation and description of *Haloanaerobium praevalens* gen. nov. and sp. nov., an obligately anaerobic halophile common to Great Salt Lake sediments. Curr Microbiol 9:225–234

Zhilina TN (2009) Genus IV. *Halonatronum* Zhilina, Garnova, Tourova, Kostrikina and Zavarzin 2001a, 263$^{VP}$ (Effective publication: Zhilina, Garnova, Tourova, Kostrikina and Zavarzin 2001b, 70). In: Zavarzin GA, Oren A, De Vos P, Garrity GM, Jones D, Krieg NR, Ludwig W, Rainey FA, Schleifer K-H, Whitman WB (eds) Bergey's manual of systematic bacteriology, vol 3, 2nd edn, The Firmicutes. Springer, Dordrecht, p 1215

Zhilina TN, Zavarzin GA (1990a) Extremely halophilic, methylotrophic, anaerobic bacteria. FEMS Microbiol Rev 87:315–322

Zhilina TN, Zavarzin GA (1990b) A new extremely halophilic homoacetogenic bacterium *Acetohalobium arabaticum* gen. nov., sp. nov. Dokl Akad Nauk SSSR 311:745–747 (in Russian)

Zhilina TN, Zavarzin GA (1991) Anaerobic bacteria participating in organic matter destruction in halophilic cyanobacterial community. J Obshey Biol 52:302–318 (in Russian)

Zhilina TA, Zavarzin GA (1994) Alkaliphilic anaerobic community at pH 10. Curr Microbiol 29:109–112

Zhilina TN, Kevbrin VV, Lysenko AM, Zavarzin GA (1991) Isolation of saccharolytic anaerobes from a halophilic cyanobacterial mat. Microbiology 60:101–107

Zhilina TN, Miroshnikova LV, Osipov GA, Zavarzin GA (1992a) *Halobacteroides lacunaris* sp. nov., new saccharolytic, anaerobic, extremely halophilic organism from the lagoon-like hypersaline lake Chokrak. Microbiology 60:495–503

Zhilina TN, Zavarzin GA, Bulygina ES, Kevbrin VV, Osipov GA, Chumakov KM (1992b) Ecology, physiology and taxonomy studies on a new taxon of *Haloanaerobiaceae, Haloincola saccharolytica* gen. nov., sp. nov. Syst Appl Microbiol 15:275–284

Zhilina TN, Zavarzin GA, Detkova EN, Rainey FA (1996) *Natroniella acetigena* gen. nov. sp. nov., an extremely haloalkaliphilic, homoacetogenic bacterium: a new member of *Haloanaerobiales*. Curr Microbiol 32:320–326

Zhilina TN, Turova TP, Lysenko AM, Kevbrin VV (1997) Reclassification of *Halobacteroides halobius* Z-7287 on the basis of phylogenetic analysis as a new species *Halobacteroides elegans* sp. nov. Microbiology 66:97–103

Zhilina TN, Tourova TP, Kuznetsov BB, Kostrikina NA, Lysenko AM (1999) *Orenia sivashensis* sp. nov., a new moderately halophilic anaerobic bacterium from lake Sivash lagoons. Microbiology 68:452–459

Zhilina TN, Garnova ES, Tourova TP, Kostrikina NA, Zavarzin GA (2001) *Halonatronum saccharophilum* gen. nov. sp. nov.: a new haloalkaliphilic bacterium of the order *Haloanaerobiales* from Lake Magadi. Microbiology 70:64–72

Zhilina TN, Zavarzina DG, Panteleeva AN, Osipov GA, Kostrikina NA, Tourova TP, Zavarzin GA (2012) *Fuchsiella alkaliacetigena* gen. nov., sp. nov., the first alkaliphilic, lithoautotrophic homoacetogenic bacterium. Int J Syst Evol Microbiol 62:1666–1673

# 13 The Family *Haloplasmataceae*

*André Antunes*
IBB–Institute for Biotechnology and Bioengineering, Centre of Biological Engineering, Micoteca da Universidade do Minho, University of Minho, Braga, Portugal
Rianda Research–Centro de Investigação em Energia, Saúde e Ambiente, Coimbra, Portugal



**Abstract**
*Haloplasmataceae* is a family within the order *Haloplasmatales*, which currently includes one single genus and species: *Haloplasma contractile*. This family has unusual phenotypic features –the most noticeable being a unique morphology and cellular contractility cycle– and a distinct phylogenetic position between the *Firmicutes* and the *Tenericutes* (*Mollicutes*).

Members of the *Haloplasmataceae* have been isolated from the upper sediments of a deep-sea anoxic brine in the Red Sea, but cultivation-independent studies have found related sequences in a wide range of biotopes including other extreme environments, contaminated soils and marine sediments, as well as intestinal samples. The isolation and description of new representatives of this family might therefore result in significant changes to the current description.

## Taxonomy: Historical and Current

The family *Haloplasmataceae* currently comprises one single genus and species: *Haloplasma contractile*. Phylogenetically, a member of the order *Haloplasmatales* – a novel lineage within the Bacteria with branching position between the *Firmicutes* and *Tenericutes* (*Mollicutes*) (Antunes et al. 2008; ➔ *Fig. 13.1.*). The closest relatives of the *Haloplasmataceae* include the genera *Gemella*, and *Turicibacter* (family *Erysipelotrichaceae*).

Both the family and order were established based on unusual phenotypic features and distinct phylogenetic position, although the lack of additional isolates precluded the possible establishment of higher-ranked taxa.

The most striking feature of the *Haloplasmataceae* is their unusual morphology and cellular contractility cycle. Cells have central round bodies, with one or two cellular projections that alternate between a coiled and linear state (➔ *Figs. 13.2* and ➔ *13.3*).

### *Haloplasmataceae* Rainey, da Costa, Antunes and Huber 2008

*Haloplasmataceae* (Ha.lo.plas. ma.ta'ce.ae. N.L. n. *Haloplasma*, type genus of the family; suff. *-aceae*, ending to denote a family; N.L. fem. pl. n. *Haloplasmataceae*, the *Haloplasma* family).

The family contains the type genus *Haloplasma*. The description is the same as for this genus.

## Molecular Analyses

### Genome Analyses

Only the draft genome sequence of the *Haloplasma contractile* is currently available (Antunes et al. 2011a). The approach used in this study provided annotation for 84 % of all 3,984 predicted genes. The distribution of genes into COGs functional categories indicates that the highest number of genes are involved in amino acid transport and metabolism (170; 6.7 %), carbohydrate transport and metabolism (165; 6.5 %), and translation, ribosomal structure, and biogenesis (156; 6.2 %), followed by inorganic ion transport and metabolism (139; 5.5 %). The draft genome has a G+C content of 33.0 mol%, only slightly lower than the 33.8 mol% determined for the species by the HPLC method (Antunes et al. 2008).

Data from the draft genome provided some insights on the genetic basis for the unusual morphology and cellular dynamic of *H. contractile*. Antunes et al. (2011a) reported on the split of the *dcw* gene cluster and the disruption of the *murD-ftsW-murG* gene sequences (involved in cellular morphology), and, most importantly, the presence of seven MreB/Mbl homologs, which appears to be the highest copy number ever reported. MreB/Mbl are cytoskeletal elements with a typical helical placement, and have been previously implicated in cellular contraction (Kürner et al. 2005).

E. Rosenberg et al. (eds.), *The Prokaryotes – Firmicutes and Tenericutes*, DOI 10.1007/978-3-642-30120-9_206,

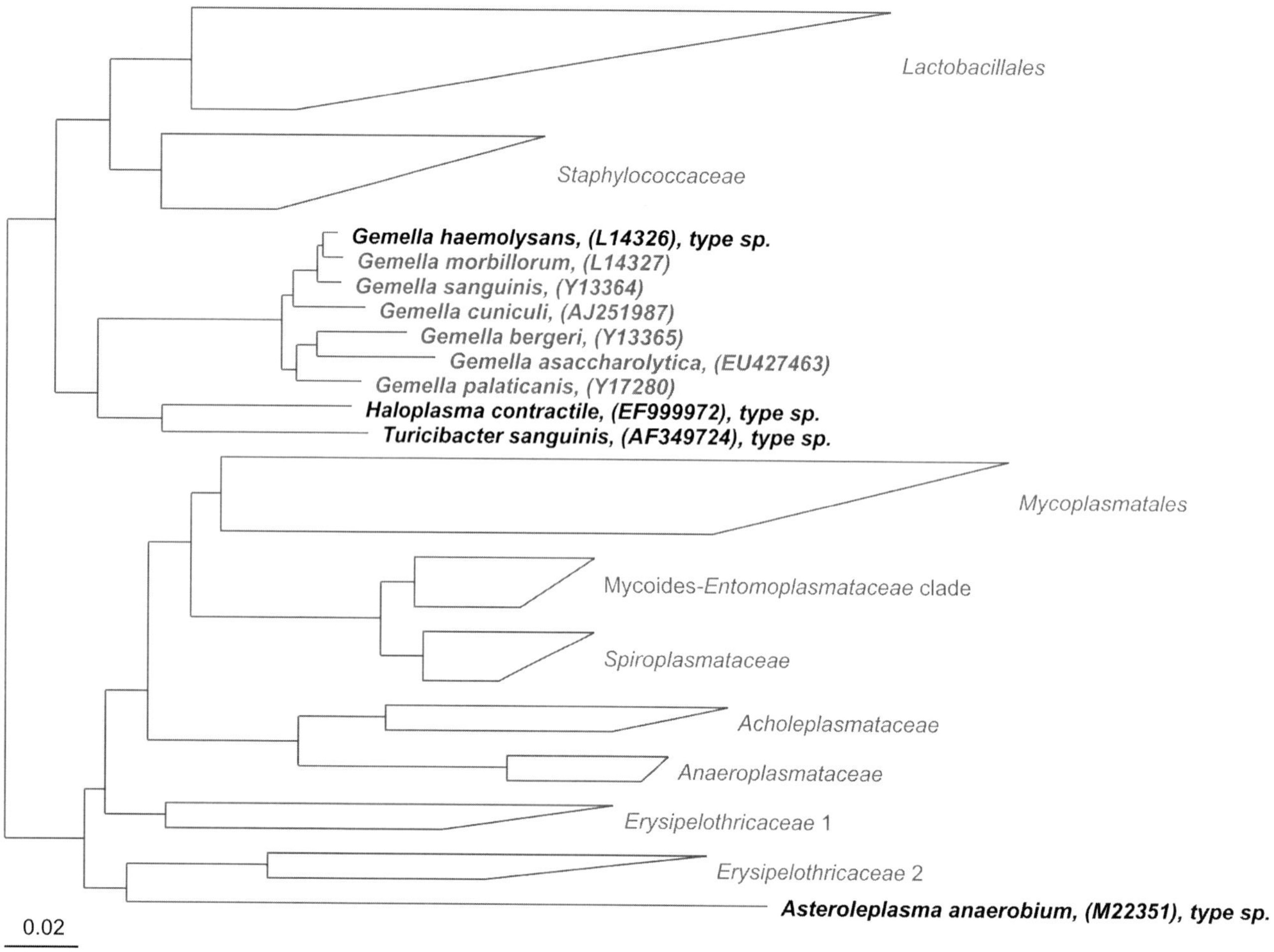


**Fig. 13.1**

**Phylogenetic reconstruction of the family *Haloplasmataceae* based on 16S rRNA and created using the neighbor-joining algorithm with the Jukes-Cantor correction. The sequence datasets and alignments were used according to the All-Species Living Tree Project (LTP) database (Yarza et al. 2010; http://www.arb-silva.de/projects/living-tree). The tree topology was stabilized with the use of a representative set of nearly 750 high quality type strain sequences proportionally distributed among the different bacterial and archaeal phyla. In addition, a 40 % maximum frequency filter was applied in order to remove hypervariable positions and potentially misplaced bases from the alignment. Scale bar indicates estimated sequence divergence**

Furthermore, genomic data for *H. contractile* (Antunes et al. 2011a) led to detection of genes for rhodopsins, later classified as xenorhodopsins – an unusual new class of rhodopsins (Ugalde et al. 2011).

## Phenotypic Analyses

### *Haloplasma* Antunes, Rainey, da Costa and Huber 2008

Ha.lo.plas'ma Gr. n. *hals halos*, salt; Gr. neut. n. *plasma*, something formed or molded, a form; N.L. neut. n. *Haloplasma*, a salt-loving form.

Highly pleomorphic cells generally with a central coccoid body (0.5–1.8 μm) exhibiting one or two "tentacle-like" contractile cellular projections (up to 20 μm in length and 0.1–0.25 μm in diameter). Chains of micrococci often observed in scanning electron micrographs of the cellular projections (➋ *Fig. 13.3e*). Gram-stain-negative. Jerky motility by cellular contraction of the cellular projections (➋ *Fig. 13.2*). Colonies on MM-X agar are very small (0.03–0.05 mm in diameter after 3 days incubation at 22 °C), gold-yellow, round, and frequently with a "fried-egg"-appearance (➋ *Fig. 13.4*). Growth between 1.5 % and 18 % (w/v) NaCl (optimum: 8 %), 10 °C and 44 °C (optimum: 30–37 °C), and pH 6.0 and 8.0 (optimum: 7.0).

Strictly anaerobic, denitrifying, and fermentative, with lactate produced as an end-product of fermentation of the medium components. Peptone and yeast extract are required for growth.

*H. contractile* utilizes only a limited range of Biolog GN2 substrates with clear positives for L-arabinose, and D-psicose,

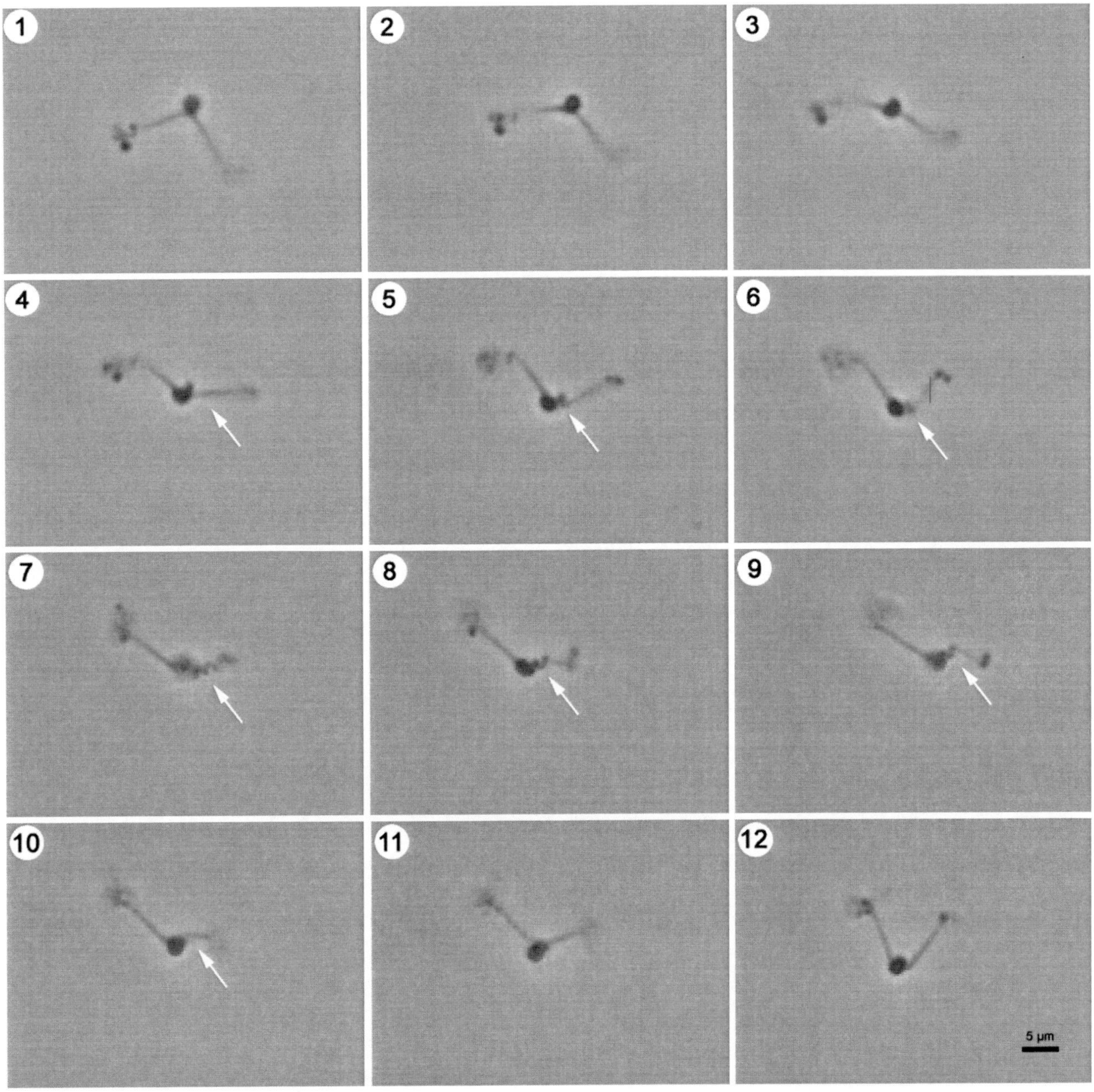


**◘ Fig. 13.2**
**Series of consecutive light micrographs (phase contrast) of *Haloplasma contractile*, illustrating the contraction of the "tentacle-like" cellular projections. The time span between the first and last micrographs was about 8 s (Copyright © American Society for Microbiology, J Bacteriol 190, 2008, 3580–3587)**

and weak positives for α-ketobutyric acid, α-ketoglutaric acid, and α-ketovaleric acid. The following substrates were not utilized: α-cyclodextrin, dextrin, glycogen, Tweens 40 and 80, *N*-acetyl-D-galactosamine, *N*-acetyl-D-glucosamine, adonitol, D-arabitol, D-cellobiose, i-erythritol, D-fructose, L-fucose, D-galactose, gentiobiose, α-D-glucose, myoinositol, α-D-lactose, lactulose, maltose, D-mannitol, D-mannose, D-melibiose, β-methyl D-glucoside, D-raffinose, L-rhamnose, D-sorbitol, sucrose, D-trehalose, turanose, xylitol, methylpyruvate, monomethyl succinate, acetic acid, cis-aconitic acid, citric acid, formic acid, D-galactonic acid lactone, D-galacturonic acid, D-gluconic acid, D-glucosaminic acid, D-glucuronic acid, α-, β- and γ-hydroxybutyric acids, p-hydroxyphenylacetic acid, itaconic acid, D,L-lactic acid, malonic acid, propionic acid, quinic acid, D-saccharic acid, sebacic acid, succinic acid, bromosuccinic acid, succinamic acid, glucuronamide, L-alaninamide, D-alanine, L-alanine, L-alanylglycine, L-asparagine, L-aspartic acid, L-glutamic acid, glycyl L-aspartic acid, glycyl L-glutamic acid, L-histidine, hydroxy-L-proline, L-leucine, L-ornithine, L-phenylalanine, L-proline, L-pyroglutamic acid, D-serine, L-serine, DL-carnitine, γ-aminobutyric acid, urocanic acid, inosine, uridine, thymidine, phenylethylamine, putrescine,

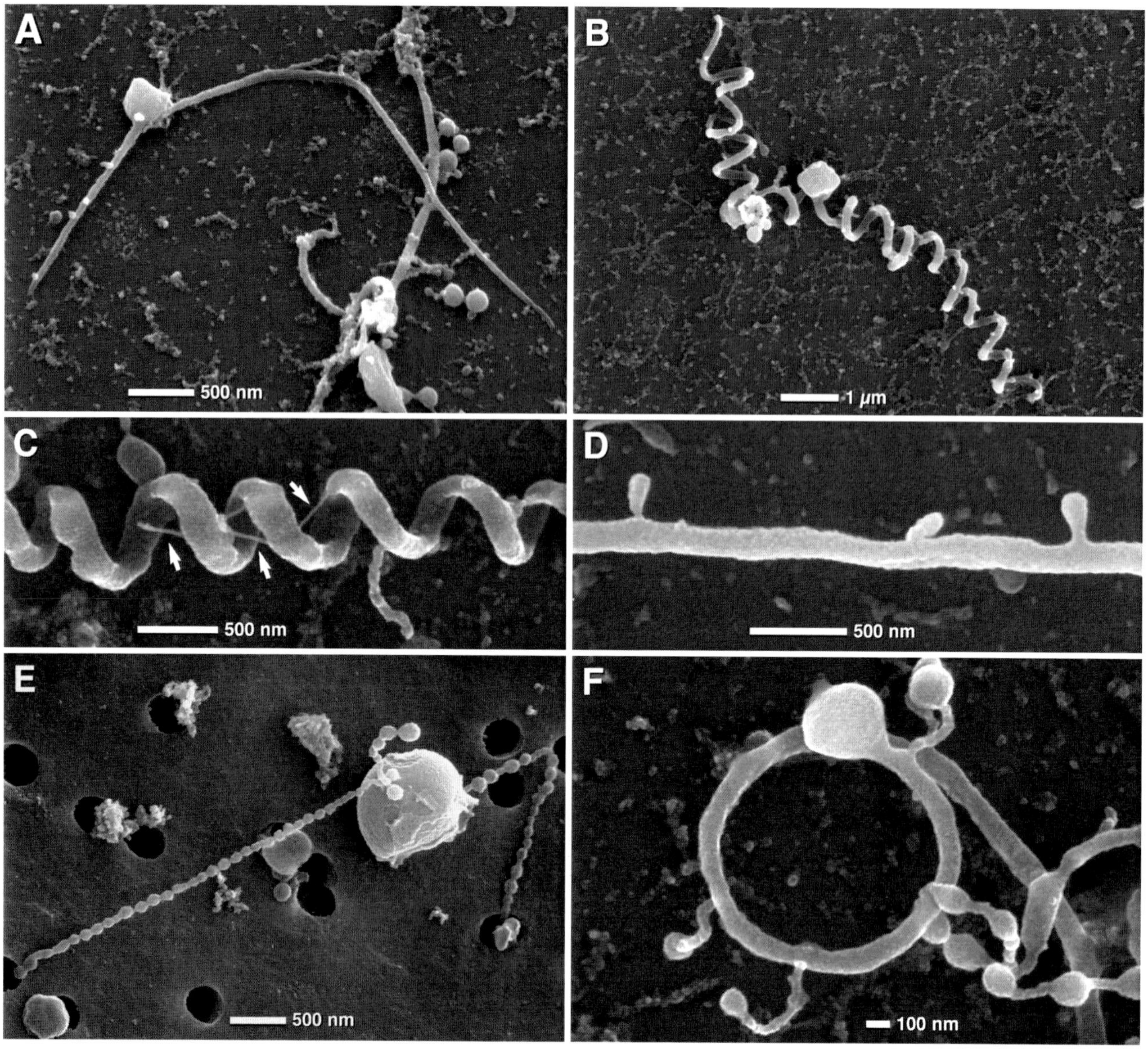


**Fig. 13.3**
**Scanning electron micrographs illustrating the pleomorphic nature of *Haloplasma contractile* and its cellular projections (Copyright © American Society for Microbiology, J Bacteriol 190, 2008, 3580–3587)**

2-aminoethanol, 2,3-butanediol, glycerol, DL-α-glycerol phosphate, glucose 1-phospate, and glucose 6-phosphate.

Cell wall not detected, although genes for peptidoglycan synthesis have been detected from genome analysis. Fatty acid profile dominated by unbranched saturated and unsaturated compounds, with C16:0, C18:0 and C18:1 ω9c as major components. Phosphatidyl glycerol and bisphosphatidyl glycerol are the major polar lipids, and MK-4 is the major respiratory quinone.

Unable to hydrolyze urea or arginine; no phosphatase activity detected. Resistant to penicillin G, ampicillin, and cephalosporin but susceptible to bacitracin and chloramphenicol. Cells filterable through 450-nm pore membranes.

The G+C content of the DNA of the type species is 33.6 mol%.

## Isolation, Enrichment, and Maintenance Procedures

*Haloplasma contractile* was enriched at 22 °C in MM-X medium (Antunes et al. 2008) containing the following (g $L^{-1}$): NaCl (120), $KH_2PO_4$ (0.5), yeast extract (0.2), peptone (0.2), starch (5.0), and artificial seawater (250 ml; Huber et al. 1998).

The medium was prepared under strictly anaerobic conditions according to Balch et al. (1979). After normal preparation procedures, resazurine (5 μg $L^{-1}$) was added as redox indicator and the medium was transferred to Schott flasks, properly stoppered, and bubbled with $N_2$ for at least 45 min and, after that, reduced with $Na_2S.9H_2O$ (0.05 %). 10-ml aliquots of the medium were then dispensed into 28-ml serum tubes, inside

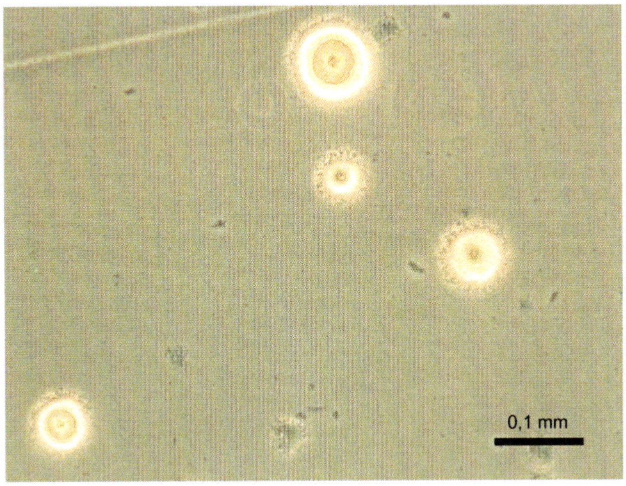


**Fig. 13.4**
**"Fried-egg" colonies of *Haloplasma contractile* grown on MM-X agar after 3-day incubation at 22 °C**

an anaerobic chamber [Gas phase $N_2/H_2 = 95/5$ (v/v)], where they were rubber-stoppered and sealed. The final step consisted in exchanging the gas phase of the tubes with $N_2$ (250 kPa) prior to autoclaving.

Isolation was achieved by visual separation of one single cell of the desired morphology using the "optical tweezers" technique (Huber et al. 1990).

Pure cultures were routinely grown at 30 °C in a variant of MM-X medium containing 60 g $L^{-1}$ of NaCl; growth in solidified MM-X medium was achieved by addition of 1.5 % (w/v) agar.

## Preservation

Long-term storage is best achieved by freezing sterile glass capillaries previously filled with cells suspended in culture media containing 5 % dimethyl sulfoxide (as reviewed by Tindall (2007)). Capillaries are kept at −80 °C or, optimally, at around −140 °C (in the gas phase over liquid nitrogen). No loss of cell viability is usually observed after several years of storage under optimal conditions (A. Antunes, personal observation).

## Ecology

The only current representative of the *Haloplasmataceae* originated from the upper sediments of Shaban Deep – one of the approximately 25 deep-sea brine pools identified in the Red Sea. Formed during the same tectonic splitting of the Arabian and African plates that generated the Red Sea, these brines combine multiple environmental extremes (e.g., absence of oxygen, high pressure, and increased levels of salinity and heavy metals when compared with overlying seawater), and are one of the most unique and extreme environments on Earth (e.g., Antunes et al. 2011a; Hartmann et al. 1998).

Shaban Deep has an approximately rhombic shape comprised of four different basins and extends over a total area of 106 km (Pautot et al. 1984). It reaches a maximum depth of 1,540 m, while the brine-seawater interface occurs near 1,325 m. The $H_2S$-free anoxic brine water is slightly warmer (23 °C) and more acidic (pH 6.0) than overlying seawater, while salinity is close to saturation (25.6–26.1 %; Michaelis et al. 1990; Hartmann et al. 1998).

Sediment pore water analysis revealed high values for sodium, chloride, and potassium, similar to values in the brine, associated with significant increases in the concentration of iron and manganese, and moderate increase in calcium (Antunes et al. 2008; Eder et al. 2002). Analysis of the sediment fraction revealed a predominance of carbonates together with significant amounts of muscovite and quartz (Antunes et al. 2008), with additional reports on high organic content (Botz et al. 2007), and retrieval of petroleum-impregnated sediment samples and Ni-rich massive sulfides (Michaelis et al. 1990; Blum and Puchelt 1991).

*H. contractile* was isolated from a sample collected at the brine-sediment interface of the eastern basin of Shaban Deep (26°13.9′N, 35°21.3′E) at a depth of 1,447 m. The sample had a pH of 6.0, a salinity of 24.4 %, and a temperature of 24.1 °C (Antunes et al. 2008).

Despite the current absence of further isolates, several *Haloplasma*-related sequences have been reported in a variety of culture-independent studies from several different environments. Indeed, such sequences have been detected in a wide range of conditions from hypersaline and other extreme environments (e.g., Isenbarger et al. 2008; López-García et al. 2005), to marine sediments (e.g., Mills et al. 2008), petroleum and poly-aromatic contaminated soils (e.g., Guazzaroni et al. 2013), to gut-associated samples (e.g., Salzman et al. 2002).

## Pathogenicity, Clinical Relevance

The *Haloplasmataceae* currently include no known pathogenic or clinically relevant members. *H. contractile* cells are resistant to penicillin G, ampicillin, and cephalosporin but susceptible to bacitracin, and chloramphenicol (all tested at a 25 μg/ml concentration; Antunes et al. 2008).

## Application

No current application of members of this family has been reported. There is however potential for some applications, as suggested by previous studies.

The unique properties of nanostructured surfaces and structures of microbes have received considerable interest and might lead to interesting new applications (e.g., Moissl et al. 2005). The molecular mechanism and machinery involved in the cellular contractility of *H. contractile* might provide a new molecular tool in biophysics and nanobiotechnology.

Also, genomic data from *H. contractile* (Antunes et al. 2011a) led to detection of genes for rhodopsins that were later classified as xenorhodopsins – an unusual new class of rhodopsins (Ugalde et al. 2011). Microbial rhodopsins are a well-known attractive source for innumerous applications, ranging from holography, to spatial light modulators, artificial retinas, neural network, optical computing, and volumetric and associative optical memories (Hampp 2000; Margesin and Schinner 2001; Miyake and Stingl 2011).

## References


Antunes A, Alam I, Bajic VB, Stingl U (2011b) Genome sequence of *Haloplasma contractile*, an unusual contractile bacterium from a deep-sea anoxic brine lake. J Bacteriol 193:4551–4552

Antunes A, Ngugi DK, Stingl U (2011a) Microbiology of the Red Sea (and other) deep-sea anoxic brine lakes. Environ Microbiol Rep 3:416–433. doi:10.1111/j.1758-2229.2011.00264.x

Antunes A, Rainey F, Wanner G, Taborda M, Pätzold J, Nobre MF, da Costa MS, Huber R (2008) A new lineage of halophilic, wall-less, contractile bacteria from a brine-filled Deep of the Red Sea. J Bacteriol 190:3580–3587

Balch WE, Fox GE, Magrum LJ, Woese CR, Wolfe RS (1979) Methanogens: reevaluation of a unique biological group. Microbiol Rev 43:260–296

Blum N, Puchelt H (1991) Sedimentary-hosted polymetallic massive sulfide deposits of the Kebrit and Shaban Deeps, Red Sea. Miner Deposita 26:217–227

Botz R, Schmidt M, Wehner H, Hufnagel H, Stoffers P (2007) Organic-rich sediments in brine-filled Shaban- and Kebrit Deeps, northern Red Sea. Chem Geol 244:520–553

Eder W, Schmidt M, Koch M, Garbe-Schönberg D, Huber R (2002) Prokaryotic phylogenetic diversity and corresponding geochemical data of the brine-seawater interface of the Shaban Deep, Red Sea. Environ Microbiol 4:758–763

Guazzaroni ME, Herbst FA, Lores I, Tamames J, Pelaez AI, Lopez-Cortes N, Alcaide M, Del Pozo MV, Vieites JM, von Bergen M, Gallego JL, Bargiela R, Lopez-Lopez A, Pieper DH, Rossello-Mora R, Sanchez J, Seifert J, Ferrer M (2013) Metaproteogenomic insights beyond bacterial response to naphthalene exposure and bio-stimulation. ISME J 7:122–136

Hampp N (2000) Bacteriorhodopsin as a photochromic retinal protein for optical memories. Chem Rev 100:1755–1776

Hartmann M, Scholten JC, Stoffers P, Whener F (1998) Hydrographic structure of brine-filled Deeps in the Red Sea – new results from the Shaban, Kebrit, Atlantis II, and Discovery Deep. Mar Geol 144:311–330

Huber R, Woese CR, Langworthy TA, Kristjansson JK, Stetter KO (1990) *Fervidobacterium islandicum* sp. nov., a new extremely thermophilic eubacterium belonging to the "*Thermotogales*". Arch Microbiol 154:105–111

Huber R, Wolfgang E, Heldwein S, Wanner G, Huber H, Rachel R, Stetter KO (1998) *Thermocrinis ruber* gen. nov., sp. nov., a pink-filament-forming hyperthermophilic bacterium isolated from Yellowstone National Park. Appl Environ Microbiol 64:3576–3583

Isenbarger TA, Finney M, Rios-Velazquez C, Handelsman J, Ruvkun G (2008) Miniprimer PCR, a new lens for viewing the microbial world. Appl Environ Microbiol 74:840–849

Kürner J, Frangakis AS, Baumeister W (2005) Cryo-electron tomography reveals the cytoskeletal structure of *Spiroplasma melliferum*. Science 307:436–438

López-García P, Kazmierczak J, Benzerara K, Kempe S, Guyot F, Moreira D (2005) Bacterial diversity and carbonate precipitation in the giant microbialites from the highly alkaline Lake Van, Turkey. Extremophiles 9:263–274

Margesin R, Schinner F (2001) Potential of halotolerant and halophilic microorganisms for biotechnology. Extremophiles 5:73–83

Michaelis W, Jenisch A, Richnow HH (1990) Hydrothermal petroleum generation in Red Sea sediments from the Kebrit and Shaban Deeps. Appl Geochem 5:103–114

Mills HJ, Hunter E, Humphrys M, Kerkhof L, McGuinness L, Huettel M, Kostka JE (2008) Characterization of nitrifying, denitrifying, and overall bacterial communities in permeable marine sediments of the northeastern Gulf of Mexico. Appl Environ Microbiol 74:4440–4453

Miyake S, Stingl U (2011) Proteorhodopsin. In: Encyclopedia of life sciences (eLS). Wiley, Chichester. http://www.els.net [doi:10.1002/9780470015902.a0022837]

Moissl C, Rachel R, Briegel A, Engelhardt H, Huber R (2005) The unique structure of archaeal 'hami', highly complex cell appendages with nano-grappling hooks. Mol Microbiol 56:361–370. doi:10.1111/j.1365-2958.2005.04294.x

Pautot G, Guennoc P, Coutelle A, Lyberis N (1984) Discovery of a large brine deep in the northern Red Sea. Nature 310:133–136

Salzman NH, de Jong H, Paterson Y, Harmsen HJM, Welling GW, Bos NA (2002) Analysis of 16S libraries of mouse gastrointestinal microflora reveals a large new group of mouse intestinal bacteria. Microbiology 148:3651–3660

Tindall BJ (2007) Vacuum drying and cryopreservation of prokaryotes. In: Day JG, Stacey GN (eds) Cryopreservation and freeze-drying protocols, 2nd edn. Humana Press, New Jersey, pp 73–97

Ugalde JA, Podell S, Narasingarao P, Allen EE (2011) Xenorhodopsins, an enigmatic new class of microbial rhodopsins horizontally transferred between archaea and bacteria. Biol Direct 6:52

Yarza P, Ludwig W, Euzéby J, Amann R, Schleifer K-H, Glöckner FO, Rosselló-Móra R (2010) Update of the all-species living tree project based on 16S and 23S rRNA sequence analyses. Syst Appl Microbiol 33:291–299.

# 14 The Family *Heliobacteriaceae*

*W. Matthew Sattley*[1] · *Michael T. Madigan*[2]
[1]Department of Biology, Indiana Wesleyan University, Marion, IN, USA
[2]Department of Microbiology, Southern Illinois University, Carbondale, IL, USA



**Abstract**

Heliobacteria are anoxygenic phototrophic bacteria of the phylum Firmicutes and are distinct from all other anoxygenic phototrophs in many ways. These include their phylogeny, synthesis of the unique photopigment bacteriochlorophyll *g*, production of heat-resistant endospores, and their primarily soil habitat. Five genera of heliobacteria have been described, including a total of 11 species. Heliobacteria are obligate anaerobes, and most species are capable of both phototrophic and chemotrophic growth. Two distinct phylogenetic clades of heliobacteria exist, including a group that inhabits neutral pH soils and a group that inhabits alkaline soils and soda lake ecosystems. As a group, heliobacteria are distant relatives of endospore-forming bacteria of the *Bacillaceae* and *Clostridiaceae*. The genome of the thermophile *Heliobacterium modesticaldum* lacks genes for autotrophy but contains genes encoding key endospore-specific proteins and nitrogenase; the heliobacterial photosynthesis gene cluster encodes the most streamlined photosystem of any known anoxygenic phototroph. Heliobacteria are widespread in paddy soils where their strong nitrogen-fixing capacities may benefit rice plants. The photoheterotrophic lifestyle of the heliobacteria may also benefit from such associations by receiving organic carbon from plant exudates.

## Taxonomy, Historical and Current

### Description of the Family

He.li.o.bac.te.ri.a.cé.a.e. N.L. neut. n. *Heliobacterium* type genus of the family; *-aceae* ending to denote a family; N.L. fem. pl. n. *Heliobacteriaceae* the *Heliobacterium* family.

The family *Heliobacteriaceae* (Madigan et al. 2010) is accommodated within the order *Clostridiales* of the phylum Firmicutes, and is the only group of phototrophic bacteria within the *Firmicutes.* The nearest neighboring family to that of the heliobacteria is the *Peptococcaceae*, a family that contains, among other genera, species of *Desulfotomaculum*, *Pelotomaculum*, and *Desulfitobacterium*, with the latter genus containing species most closely related to members of the *Heliobacteriaceae.* All heliobacteria have genus names with the prefix "Helio," a combining form from the Greek word "helios," meaning "sun." In addition to the type genus *Heliobacterium*, the *Heliobacteriaceae* includes the genera *Heliobacillus*, *Heliophilum*, and *Heliorestis* (Asao and Madigan 2009). A fifth genus within the family was recently proposed for a novel heliobacterial strain, designated "*Candidatus* Heliomonas lunata," which has not yet been isolated in pure culture (Asao et al. 2012).

A total of 11 species within the *Heliobacteriaceae* has been isolated, described, and effectively published (Asao and Madigan 2009; Asao et al. 2012). *Heliobacterium* (*Hbt.*) *chlorum* was the first of the heliobacteria to be discovered (Gest and Favinger 1983). However, because cells of *Hbt. chlorum* are prone to lysis upon reaching late exponential phase (Gest and Favinger 1983), few laboratories routinely culture this species for research purposes. A significant amount of genetic and molecular characterization has been accomplished over the past 30 years using various species of heliobacteria, but it is the thermophilic *Heliobacterium modesticaldum* that has emerged as a model organism for biochemical and genomic studies of heliobacterial physiology and photosynthesis (Sattley et al. 2008; Sattley and Swingley 2013).

The family *Heliobacteriaceae* encompasses all anoxygenic phototrophic bacteria that synthesize bacteriochlorophyll (BChl) *g* as the major photosynthetic pigment and sole bacteriochlorophyll (Brockmann and Lipinski 1983; Michalski et al. 1987). In addition, heliobacteria synthesize small amounts of $8^1$-OH-Chlorophyll (Chl) *a*; both BChl *g* and $8^1$-OH-Chl *a* are esterified with the $C_{15}$ alcohol farnesol (Michalski et al. 1987; van de Meent et al. 1991; ➲ *Fig. 14.1*). Unlike the $C_{40}$ carotenoids found in all other phototrophic bacteria, heliobacteria

E. Rosenberg et al. (eds.), *The Prokaryotes – Firmicutes and Tenericutes*, DOI 10.1007/978-3-642-30120-9_362,

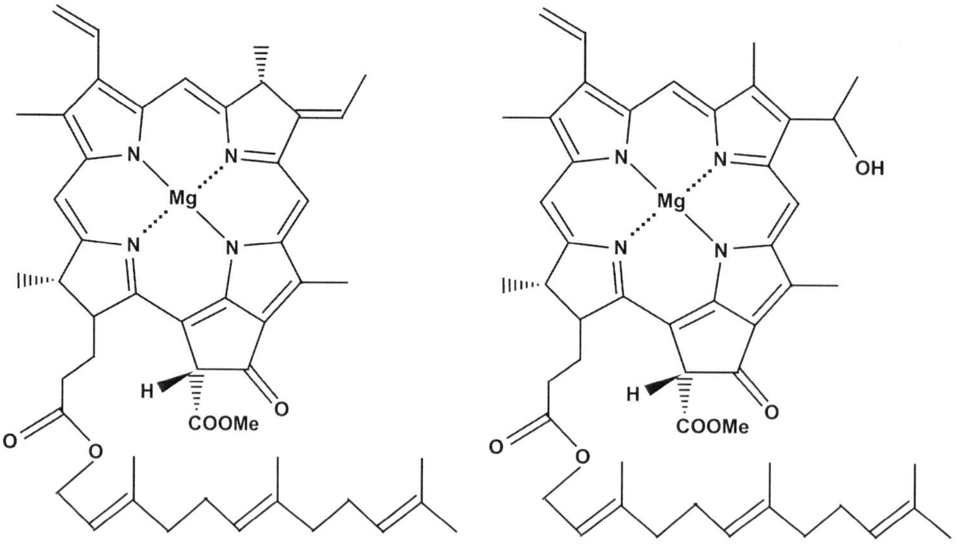

**Bacteriochlorophyll *g*** **8¹-OH-Chlorophyll *a***

**Fig. 14.1**

**Structures of pigments synthesized by species of *Heliobacteriaceae*. Bacteriochlorophyll *g* is the major pigment in heliobacteria, while small amounts of 8¹-OH-Chlorophyll *a* are also synthesized and packaged into the heliobacterial reaction center. In both pigments, the esterifying alcohol is farnesol**

synthesize $C_{30}$ carotenoids—either 4,4′-diaponeurosporene or, in alkaliphilic species, OH-diaponeurosporene glucoside esters (Takaichi et al. 1997; Takaichi et al. 2003). Cells of heliobacteria contain no differentiated, internal photosynthetic membrane systems (e.g., chlorosomes, chromatophores, lamellae, or thylakoids; ➲ *Fig. 14.2*), and therefore, heliobacterial pigments are confined to cytoplasmic membrane-bound photosynthetic reaction centers (Miller et al. 1986).

Growth of heliobacteria is strictly anaerobic, occurring photoheterotrophically or, in non-alkaliphilic species, chemotrophically (in darkness) by fermentation of pyruvate (Kimble et al. 1994). Photoautotrophic growth ($CO_2$ plus $H_2$ or $H_2S$) has not been observed in any species of heliobacteria (Asao and Madigan 2010). Cells of heliobacteria are either straight, curved, or coiled rods, varying in length from a few micrometers to filaments of 20 μm or more (➲ *Fig. 14.2a*; Asao and Madigan 2009). Cells of *Hbt. gestii* are spiral-shaped (Ormerod et al. 1996). Most heliobacteria exhibit flagellar motility, with at least one species, *Hbt. chlorum*, capable of gliding (Asao and Madigan 2009; Asao et al. 2012). Like other *Firmicutes* but unlike all other phototrophic bacteria, heliobacteria have a gram-positive cell wall structure, and like their close relatives, the clostridia, cells of heliobacteria can differentiate into heat-resistant endospores (Ormerod et al. 1996; Kimble-Long and Madigan 2001). With the exception of "*Candidatus* Heliomonas lunata," in which nitrogenase activity was not detected under nitrogen-starved conditions, all heliobacteria that have been tested are strong nitrogen-fixers (Asao and Madigan 2009; Asao et al. 2012). A summary of the major properties of described species of heliobacteria is presented in ➲ *Table 14.1*.

## Phylogenetic Structure of the Family and Its Genera

Phylogenetic studies based on 16S ribosomal RNA gene sequence analyses reveal that species of the family *Heliobacteriaceae* form a monophyletic group within the *Firmicutes* that bifurcates into two distinct clades in neighbor-joining phylogenetic trees. The major clade contains the genera *Heliobacterium*, *Heliobacillus*, and *Heliophilum*, while the minor clade contains all described species of *Heliorestis* (➲ *Fig. 14.3*). Before the identification of "*Candidatus* Heliomonas lunata," an unusual alkaliphilic heliobacterium cultured from the sediment-water interface of Soap Lake, Washington, USA (Asao et al. 2012), the two clades of heliobacteria presented a clear phylogenetic differentiation showing distinct lineages for the neutrophilic heliobacteria and the alkaliphilic *Heliorestis* species. However, the grouping of "*Candidatus* Heliomonas lunata" within the clade containing neutrophilic heliobacteria blurs the phylogenetic delineation of these physiologically and ecologically distinct heliobacteria. Although not closely related to neutrophilic species of heliobacteria (<94 % 16S rRNA gene sequence identity), the placement of "*Candidatus Heliomonas lunata*" near the base of the major clade suggests a possible alkaliphilic origin for the heliobacteria.

The four described species of *Heliorestis* (*Heliorestis* [*Hrs.*] *daurensis*, *Hrs. baculata*, *Hrs. convoluta*, and *Hrs. acidaminivorans*) form a tightly packed clade, with 16S rRNA gene sequence identities ranging from about 94 % to 98 % (➲ *Fig. 14.3*). Species of the major clade made up of the other three established genera of heliobacteria show

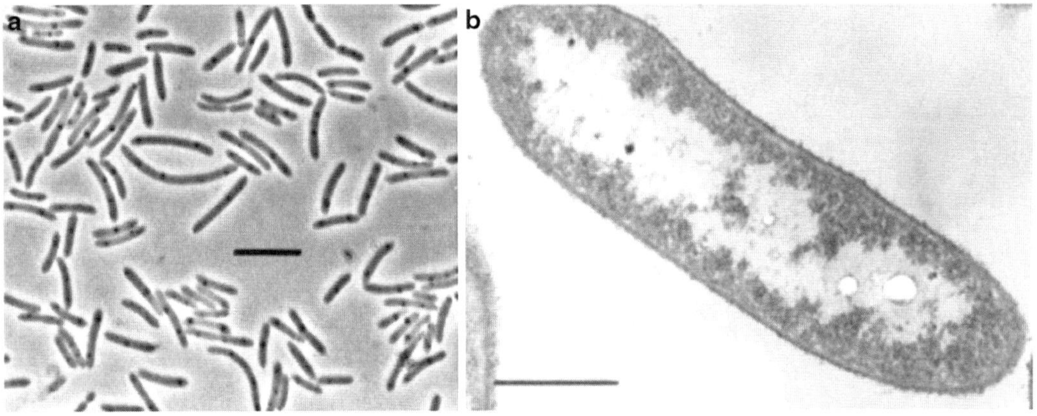


**Fig. 14.2**
**Heliobacteria. (a) Phase-contrast micrograph of cells of the thermophilic heliobacterium, *Heliobacterium modesticaldum*, strain Ice1. The nature of the dark granules in cells is unknown, but is likely to be polyphosphate. Marker bar, 3 μm. (b) Transmission electron micrograph of a longitudinal section of a cell of *Heliobacterium modesticaldum* strain Ice1. Note the absence of intracytoplasmic photosynthetic membranes and chlorosomes**

a greater degree of divergence than within the alkaliphilic clade. *Heliophilum* (*Hph.*) *fasciatum* (the only species of the genus) is the most evolutionarily diverged of the neutrophilic species, having <93 % 16S rRNA gene sequence identity to any other heliobacterium. This species is also distinguished from other heliobacteria based on several phenotypic distinctions (to be discussed later), and therefore, its existence as a novel genus is well supported (Asao and Madigan 2010).

*Heliobacterium chlorum* groups closely with *Heliobacillus* (*Hba.*) *mobilis*, the sole representative of the genus (➲ *Fig. 14.3*). Despite their taxonomic distinction, the two species have 16S rRNA gene sequences that are nearly 98 % identical. By this standard, *Hbt. chlorum* is in fact more closely related to *Hba. mobilis* than it is to other species of *Heliobacterium* (i.e., *Hbt. gestii*, *Hbt. modesticaldum*, *Hbt. sulfidophilum*, or *Hbt. undosum*), to which it shows <96 % 16S rRNA gene sequence identity. Therefore, this is a point of taxonomic uncertainty that warrants further investigation. Because of their close phylogenetic association, as well as several shared phenotypic traits (see later discussion), it seems likely that *Hbt. chlorum* and *Hba. mobilis* are actually two species of the same genus. If true, the genus designation *Heliobacillus* should be abandoned, with *Hba. mobilis* being absorbed into the genus *Heliobacterium*; this was suggested by Bryantseva et al. (2000b) and Asao and Madigan (2010). However, a second possibility is that *Heliobacillus* is a valid genus and should be retained as such, thereby likely resulting in the reclassification of *Hbt. chlorum* as a species of *Heliobacillus*. Genomic DNA-DNA hybridization studies have already been employed for the proper classification of several species of heliobacteria (discussed below), and the use of such an analysis in this case would resolve this issue.

## Molecular Analyses

### DNA-DNA Hybridization Studies

Genomic DNA-DNA hybridization (DDH) studies support taxonomic conclusions for several, but not all, type strains of heliobacteria. The initial descriptions for most species of *Heliorestis* included DDH analysis. Bryantseva et al. (2000a) determined *Hrs. baculata* strain OS-H1$^T$ to have a genomic DNA identity of 43 % to *Hrs. daurensis*, the only described species of the genus at the time. Later, Asao et al. (2006) determined a 48 % DNA identity between *Hrs. convoluta* strain HH$^T$ and *Hrs. daurensis*, its closest phylogenetic relative based on a 16S rRNA gene sequence identity of 97.8 %. DDH analyses for other heliobacteria have been limited to a single study, in which two novel species of *Heliobacterium*, *Hbt. sulfidophilum* and *Hbt. undosum*, were described (Bryantseva et al. 2000b). The authors determined *Hbt. sulfidophilum* strain BR4$^T$ to have a 25 % and 38 % DNA identity to *Hba. mobilis* and *Hbt. chlorum*, respectively. In addition, *Hbt. sulfidophilum* strain BR4$^T$ was found to have a 25 % DNA identity to *Hbt. undosum* strain BG29$^T$, an isolate with which it shared ~97.5 % 16S rRNA gene sequence identity (Bryantseva et al. 2000b). DDH studies have not been conducted for the remaining described species of heliobacteria, including *Hph. fasciatum*, *Hbt. chlorum*, *Hbt. modesticaldum*, and *Hrs. acidaminivorans*.

### Genome Analyses

Thus far, *Heliobacterium modesticaldum* strain Ice1$^T$, a moderate thermophile isolated from volcanic hot spring soils of Reykjanes, Iceland (Kimble et al. 1995), is the only heliobacterium for

**Table 14.1**
Summary of major properties of heliobacteria[a]

| | Neutrophilic heliobacteria | | | | | |
|---|---|---|---|---|---|---|
| **Species** | *Heliobacterium chlorum* | *Heliobacterium gestii* | *Heliobacterium modesticaldum* | *Heliobacterium sulfidophilum* | *Heliobacterium undosum* | *Heliobacillus mobilis* |
| **Morphology** | Rod | Spirillum | Rod/curved rod | Rod | Rod/slightly twisted spirillum | Rod |
| **Dimensions (μm)** | 1 × 7–9 | 1 × 7–10 | 0.8–1 × 2.5–9 | 0.6–1 × 4–7 | 0.8–1.2 × 7–20 | 1 × 7–10 |
| **Motility** | Gliding | Multiple subpolar flagella | Flagella or none | Peritrichous flagella | Peritrichous flagella | Peritrichous flagella |
| **Major carotenoids** | 4, 4′-diaponeuro-sporene | 4, 4′-diaponeuro-sporene | 4, 4′-diaponeuro-sporene | Neurosporene[c] | Neurosporene[c] | 4, 4′-diaponeuro-sporene |
| **Carbon sources photo-metabolized**[d] | P, L, YE | P, L, fructose, glucose, ribose; A, B, ethanol (+ $CO_2$); YE | P, L, A, fructose, glucose, ribose, YE | P, L, A; B (+ $CO_2$); C, malate, YE | P, L, A, C, PR, YE | P, L; A, B (+ $CO_2$); YE |
| **Nitrogen fixation** | + | + | + | + | ND | + |
| **Pyruvate fermentation** | + | + | + | + | + | + |
| **Growth Factor Requirement** | Biotin | Biotin | Biotin | Biotin | Biotin | Biotin |
| **Endospores produced** | N/O | + | + | + | N/O | N/O |
| **Optimum temperature (°C)** | 37–42 | 37–42 | 50–52 | 32 | 31–36 | 38–42 |
| **Optimum pH** | 6.2–7 | 6.2–7 | 6–7 | 7–8 | 7–8 | 6.2–7 |
| **Optimum NaCl (%)**[f] | ND | 0 | 0 | 0 | 0 | ND |
| **Habitat of type strain** | Garden soil at Indiana University Bloomington (USA) | Rice paddy in Thailand | Soil near alkaline hot springs at Reykjanes, Iceland | Hot spring near Bol'shaya River, Russia | Garginskii sulfidic hot spring | Dry soil from Thailand |
| **G+C content (Mol%)** | 52 | 54.8 | 56 | 51.3 | 57.2–57.7 | 50.3 |

[a]Data obtained from Gest and Favinger (1983), Beer-Romero and Gest (1987), Beer-Romero et al. (1988), Kimble and Madigan (1992), Kimble et al. (1995), Ormerod et al. (1996), Stevenson et al. (1997), Takaichi et al. (1997), Bryantseva et al. (1999), Bryantseva et al. (2000a, b), Takaichi et al. (2003), Asao et al. (2006), Tang et al. (2010), and Asao et al. (2012) (Adapted from Asao and Madigan (2010))
[b]All physiological studies were conducted in coculture
[c]HPLC analysis was not conducted in determining the type of carotenoid. Bryantseva et al. (2000b) indicated the presence of neurosporene based on the in vivo absorption spectrum peak at 412 nm. However, it is likely that this is 4, 4′-diaponeuro sporene, as for other neutrophilic heliobacteria
[d]*ND* not determined, *N/O* none observed, *A* acetate, *B* butyrate, *C* casein hydrolysate, *L* lactate, *P* pyruvate, *PR* propionate, *YE* yeast extract
[e]Phase-bright structures resembling endospores were observed in cells present in early enrichment cultures (Bryantseva et al. 1999; Asao et al. 2012); whether these structures were indeed endospores is unknown. Kimble-Long and Madigan (2001) present molecular evidence that endospore formation is a universal property of species of *Heliobacteriaceae*
[f]No cultured heliobacteria have a NaCl requirement

| | Alkaliphilic heliobacteria | | | | |
|---|---|---|---|---|---|
| *Heliophilum fasciatum* | *Heliorestis baculata* | *Heliorestis convoluta* | *Heliorestis daurensis* | *Heliorestis acidaminivorans* | "*Candidatus* Heliomonas lunata"[b] |
| Straight rods with tapered ends grouped in bundles of two to many cells | Rod/curved rod | Coil | Coil/bent filament | Rod | Curved rod |
| 0.8–1 × 5–8 | 0.6–1 × 6–10 | 0.6 × variable | 0.8–1.2 × <20 | 0.6–0.9 × 3–12 | 0.6 × 2–7 |
| Polar to subpolar flagella; cell bundles move as a unit | Peritrichous flagella | Unknown | Peritrichous flagella | Flagella | Flagella |
| 4, 4′-diaponeuro-sporene | OH-diaponeuro-sporene glucoside esters | OH-diaponeuro-sporene glucoside esters | OH-diaponeuro-sporene glucoside esters | OH-diaponeuro-sporene glucoside esters | OH-diaponeuro-sporene glucoside esters |
| P, L; A, B (+ $CO_2$) | P, L, A (+ $CO_3^{2-}$) | P, A, B, PR (+ $HCO_3^-$/$CO_3^{2-}$) | P, A, PR (+ $HCO_3^-$/ $CO_3^{2-}$) | P, A, casamino acids, PR, YE, amino acids (Ala, Arg, Glu, Gln, His, Lys, Ser) | P, B, malate, YE, amino acids (Ala, Lys, Ser, Thr) |
| + | ND | + | ND | + | – |
| + | – | – | – | – | – |
| Biotin | Biotin | None | Biotin | None | YE |
| + | + | N/O | +[e] | +[e] | N/O |
| 37–40 | 30 | 30–35 | 25–30 | 30–37 | 25–30 |
| 7 | 8.5–9 | 8.5 | 9 | 8–9 | 8–9.5 |
| 0 | 0.5–1 | 0–1 | 0 | 0.5–4 | 1.5–3 |
| Rice soil from Tanzania | Shoreline soil of Lake Ostozhe, a soda lake in Russia | Shoreline soil of Lake El Hamra, a soda lake in Egypt | Shoreline soil of Lake Barun Torey, a soda lake in Russia | Shoreline sediment of Lake El Hamra, a soda lake in Egypt | Water/benthic sediment of Soap Lake, Washington State (USA) |
| 51.8 | 45 | ND | 44.9 | ND | ND |

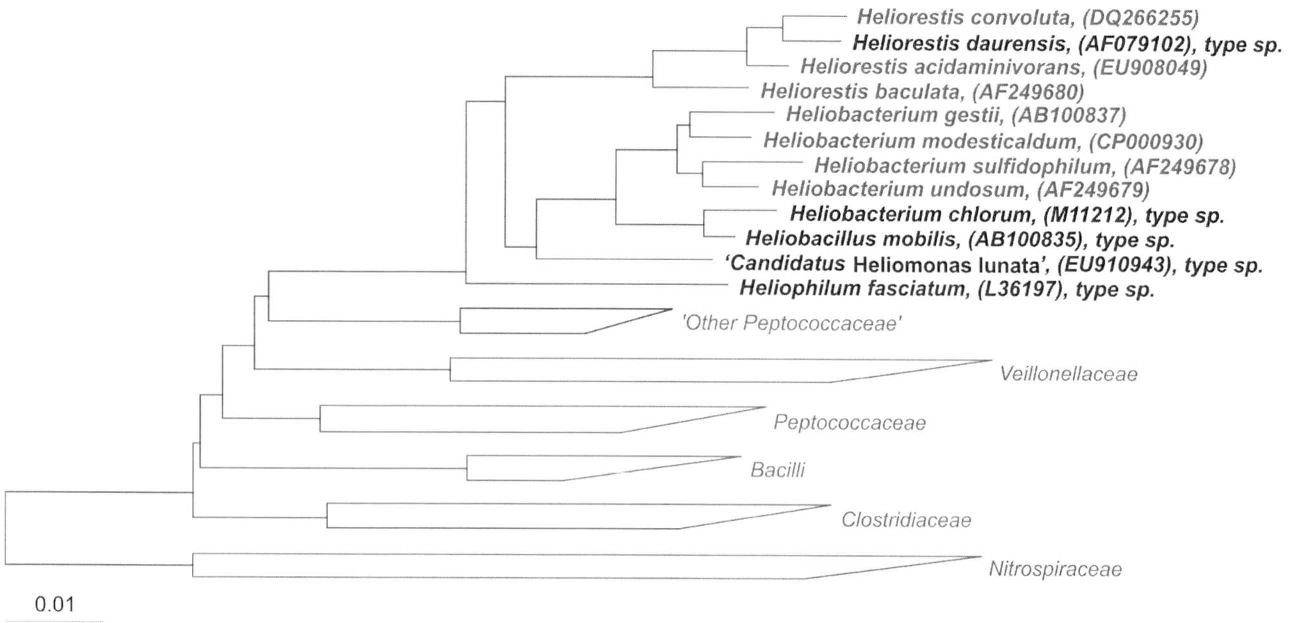


**Fig. 14.3**

**Neighbor-joining phylogenetic tree of heliobacteria and related *Firmicutes* based on comparative 16S rRNA gene sequences. All cultured species of heliobacteria are represented, including the provisional taxon "*Candidatus* Heliomonas lunata." Species in *bold* represent type species of their corresponding genus. Sequences from genera of closely related bacterial families were included to stabilize the tree topology, and a 20 % conservation filter was used to minimize the effect of hypervariable positions**

which a complete genome sequence is available (Sattley et al. 2008). The genome of *Hbt. modesticaldum* strain Ice1$^T$ consists of a single circular chromosome of 3,075,407 bp with a G+C content of 56.98 mol% (Sattley et al. 2008; Sattley and Swingley 2013), a value somewhat higher than that determined by thermal denaturation ($T_m$) of purified DNA (54.6 mol% G+C; Kimble et al. 1995). No plasmids are present in *H. modesticaldum*, and approximately 87 % of the genome encoded gene products. A total of 3,138 open reading frames were predicted, with 3,000 protein-encoding genes, 104 tRNA genes, 24 rRNA genes (in 8 operons), one structural RNA gene, one tmRNA gene, and eight pseudogenes (Sattley et al. 2008). Most of the open reading frames (65.1 %) were assigned a putative function, while the remaining genes were designated as conserved hypothetical (11.1 %) or hypothetical (23.8 %) proteins (Sattley et al. 2008).

An analysis of functional gene roles from the *Hbt. modesticaldum* strain Ice1$^T$ genome showed that the functional category containing the largest number of genes (389; 13 % of the total genome content) was that of energy and central intermediary metabolism. Other functional categories containing a large number of genes included cellular processes (cell division, motility, sporulation, etc.) (273; 9.1 %); protein synthesis, modification, and degradation (247; 8.2 %); and regulatory functions and signal transduction (177; 5.9 %). Sattley et al. (2008) provide a more detailed list of functional role categories and the number of genes they contain.

In addition to the formal genome paper, Sattley and Blankenship (2010) provided a summary of protein coding sequences predicted to play important physiological roles in *Hbt. modesticaldum*. Of particular relevance to the obligately heterotrophic metabolism of heliobacteria was the absence of genes encoding key enzymes for $CO_2$ fixation via any of the known autotrophic pathways, including the Calvin cycle, the reductive tricarboxylic acid cycle, and the hydroxypropionate cycle (Sattley et al. 2008). Curiously, however, if not for the lack of a single gene encoding citrate lyase, heliobacteria could presumably carry out $CO_2$ fixation in a manner similar to that of green sulfur bacteria, in which the reductive tricarboxylic acid cycle supports autotrophic growth (Sattley et al. 2008; Sirevåg and Ormerod 1970).

Genome sequencing for a second heliobacterium, the alkaliphilic *Heliorestis convoluta* strain HH$^T$, is underway, and a complete genome sequence should become available in the near future. The completion of this work will allow the first comparative genomic analyses to be performed on heliobacteria. Indeed, such analyses should provide interesting new insights into the genomics behind essential cellular processes (e.g., carbon and nitrogen metabolism, photophosphorylation, and endosporulation) in a species of each of the phylogenetic clades of known heliobacteria (*Hbt. modesticaldum*, a neutrophilic thermophile, and *Hrs. convoluta*, an alkaliphilic mesophile, *Fig. 14.3*).

## Phenotypic Analyses

### Cultural and Morphological Properties

Healthy cultures of heliobacteria are a brownish-green color, whether in liquid media or as colonies on solid media. Exposure of these strict anaerobes to oxygen in the presence of light causes the spontaneous and presumably irreversible oxidation of the primary photosynthetic pigment, BChl *g*, to a spectroscopic equivalent of Chl *a* (➲ *Fig. 14.4*). After approximately 1 h in such conditions, cells in both liquid and plate cultures transition from their original olive drab to an emerald green color, and this phenomenon is accompanied with significant loss of cell viability.

Positive primary enrichments for heliobacteria often contain endospores that are located either centrally or subterminally (Stevenson et al. 1997); however, for unknown reasons, the appearance of endospores diminishes with subsequent culture transfers and is almost never observed in highly enriched and pure cultures of heliobacteria. Nevertheless, genetic evidence for the capacity to produce endospores exists in every heliobacterium examined (Kimble-Long and Madigan 2001). On Petri plates, heliobacteria produce flat or centrally raised colonies with either entire margins or spreading, irregular margins, depending on the strain (Stevenson et al. 1997). Well-isolated colonies can be relatively large, having a diameter of up to 1 cm.

Although the Gram stain result for heliobacteria is usually negative, ultrathin sections of cells of heliobacteria reveal a Gram-positive cell wall structure, with cells often having a "studded" appearance but lacking an outer membrane (➲ *Fig. 14.2b*). Cells of most species of heliobacteria are medium to large rods with diameters from 0.6 to 1.2 μm and lengths ranging from 2 μm to about 20 μm in species that form short filaments, such as *Hrs. daurensis* and *Hbt. undosum* (➲ *Table 14.1*). Other notable heliobacterial cell morphologies include spirilla (*Hbt. gestii*) and coiled rods (*Hrs. convoluta*), and cells of *Hph. fasciatum* are tapered rods that aggregate into bundles that move as a unit (Ormerod et al. 1996). Motility in most species of heliobacteria is accomplished by flagella in either a polar, subpolar, or peritrichous arrangement. The exception to this is *Hbt. chlorum*, in which flagella are absent and gliding motility is observed (Gest and Favinger 1983).

### Physiology and Metabolism

The defining characteristic of the heliobacteria is their unique pigment BChl *g* (Brockmann and Lipinski 1983). This is present in a small light-harvesting core complex that contains the heliobacteria reaction center. BChl *g* absorbs maximally in the near infrared (785–790 nm, ➲ *Fig. 14.4*) at wavelengths that are not utilized by other phototrophs; presumably this allows the heliobacteria to occupy a unique ecological niche (Madigan 2006). The structure of BChl *g* contains elements that are

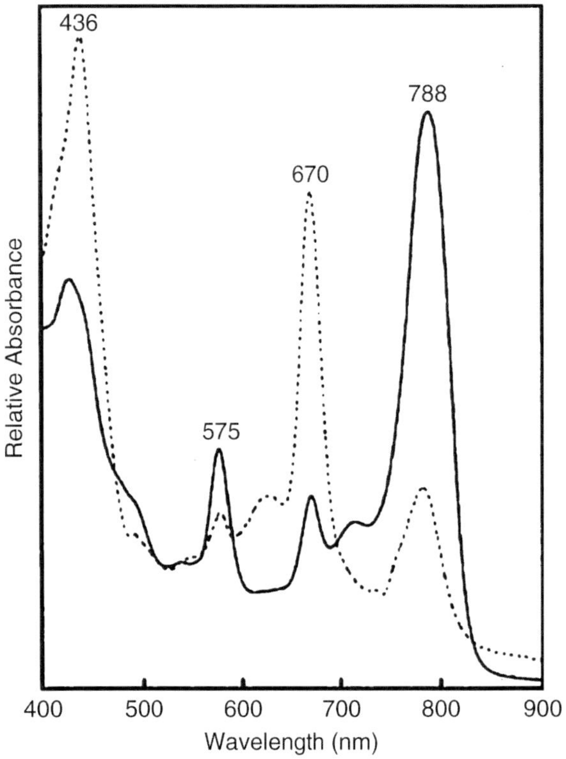


**Fig. 14.4**
**Absorption spectrum of intact cells of *Heliobacterium modesticaldum*. *Solid line*, cells suspended in 30 % bovine serum albumin that was prereduced by the addition of 0.05 % sodium ascorbate. *Dashed line*, spectrum of the same cell suspension exposed to light and air for 1 h. Note major peak of bacteriochlorophyll *g* at 788 nm and of chlorophyll *a* at 670 nm**

reminiscent of both Chl *a* and BChl *b*. Specifically, ring A of BChl *g*, which contains a C-3 vinyl group, is identical to ring A of Chl *a*, while ring B of BChl *g* has a C-8 ethylidene group and is identical to ring B of BChl *b* (Blankenship 2002; ➲ *Fig. 14.1*). In contrast to both Chl *a* and BChl *b*, however, the esterifying alcohol of BChl *g* is farnesol rather than phytol (Michalski et al. 1987; ➲ *Fig. 14.1*).

In addition to BChl *g*, smaller amounts of $8^1$-OH-Chl *a* are also synthesized by heliobacteria. This pigment, also esterified with farnesol (➲ *Fig. 14.1*), is inserted into the heliobacterial reaction center (RC) and participates in RC electron transfer kinetics (Heinnickel and Golbeck 2007; Oh-Oka 2007; van de Meent et al. 1991). Photosynthesis genes in heliobacteria, including those that facilitate pigment biosynthesis and photosystem assembly, are apparently expressed constitutively since cultures of heliobacteria are fully pigmented regardless of whether the cells are grown phototrophically (anoxic/light) or chemotrophically under anaerobic conditions in darkness (Kimble et al. 1994; Tang et al. 2010). For a discussion of putative biosynthetic pathways for BChl *g* and $8^1$-OH-Chl *a* in heliobacteria, see Sattley and Swingley (2013).

The photosynthetic pigments of heliobacteria are contained within an iron–sulfur type (type-I), homodimeric RC that is the simplest known photosynthetic apparatus (Heinnickel and Golbeck 2007; Sattley et al. 2008). Analyses of purified RCs from *Hbt. modesticaldum* indicate that each RC homodimer binds 20 BChl *g*, two BChl *g*′ (the special pair primary electron donor, P798), two $8^1$-OH-Chl *a* (the primary electron acceptor, $A_0$), one 4-4′-diaponeurosporene, and ~1.6 menaquinone, primarily consisting of MQ-9 (Kobayashi et al. 1991; Sarrou et al. 2012; Trost and Blankenship 1989). Protons pumped through a cytochrome (cyt) *bc* complex during electron transfer establish a proton motive force across the cytoplasmic membrane and drive photophosphorylation through ATP synthase (Heinnickel and Golbeck 2007; Sattley et al. 2008). The membrane-bound, diheme cyt $c_{553}$ shuttles electrons from cyt *bc* to the RC (Oh-Oka et al. 2002), and electrons transferred from the RC to a cytoplasmic ferredoxin provide reducing equivalents for nitrogen fixation via a molybdenum-dependent group I nitrogenase (Heinnickel and Golbeck 2007; Sattley et al. 2008). *Heliobacterium gestii* and possibly some other heliobacteria contain a non-molybdenum nitrogenase, as well (Kimble and Madigan 1992b).

Although heliobacteria grow optimally under photoheterotrophic conditions using a limited number of carbon sources, with pyruvate supporting the best growth, neutrophilic species are also able to grow chemotrophically in darkness by pyruvate fermentation (Kimble et al. 1994). Cultures of heliobacteria grown chemotrophically experience a drop in pH as pyruvate is oxidized to acetate to generate ATP by substrate-level phosphorylation (Kimble et al. 1994; Pickett et al. 1994; Sattley et al. 2008). Such cultures must be established in highly buffered media to achieve significant growth yields and retain viability. A complete list of carbon sources that can be photoassimilated by the different species of heliobacteria, as well as several other phenotypic properties, is presented in ➔ *Table 14.1.*

## Enrichment, Isolation, and Maintenance Procedures

### Enrichment

Enrichment of heliobacteria begins with soil as inoculum. Dry soil is suitable, and may even be desirable. The key to successful enrichments is to pasteurize the inoculum (80 °C for 10–15 min). This eliminates any purple or green bacteria that may be present and activates heliobacterial endospores.

Observations of enrichment cultures have shown that samples that contain both purple bacteria and heliobacteria invariably yield purple bacteria if enrichments are established from an unpasteurized inoculum. Using the same inoculum following pasteurization reverses this outcome (Stevenson 1993). It is thus possible that purple bacteria in some way outcompete or even inhibit heliobacteria. In fact, this could be a major reason why heliobacteria were not isolated until the 1980s, decades after the classic enrichment studies of Norbert Pfennig and others. Moreover, because cultures of heliobacteria are green, it would not be surprising if heliobacteria had actually appeared in earlier enrichments for anoxygenic phototrophic bacteria but were simply discarded as algal or cyanobacterial contaminants.

Dilute complex media at neutral pH or mineral salts media containing lactate, pyruvate, or acetate are suitable for enrichment of heliobacteria (see section "➔ Culture Media for Heliobacteria," below). Yeast extract at 0.1–0.2 % final concentration works well for enrichment of neutrophilic species (some yeast extract should be present in all heliobacteria growth media as virtually all species have at least a biotin requirement). Alkaline, sulfidic mineral media work best for the enrichment of soda lake heliobacteria. By contrast, sulfide is unnecessary for the enrichment of neutrophilic heliobacteria, and in some cases, for example, in *Heliophilum fasciatum*, sulfide is strongly growth inhibitory (Kimble et al. 1995; Ormerod et al. 1996).

Heliobacteria are strict anaerobes, and thus, media prereduced by storage for several days in contact with the atmosphere of an anoxic glove box or, alternatively, boiled for several minutes and sealed under $N_2$:$CO_2$ (95:5) before autoclave sterilization (Hungate method) are beneficial but not necessarily required for enrichment of heliobacteria. However, once crude enrichments of heliobacteria are obtained, strictly anaerobic conditions must be maintained in all subsequent isolation and purification procedures.

Light and temperature can also be selective enrichment factors for heliobacteria. Because they lack a peripheral antenna complex, heliobacteria cannot grow at as low a light intensity as purple and green bacteria. However, heliobacteria tolerate high light intensities well and grow best in pure culture at significantly higher light intensities than those typically used for the culture of purple and green bacteria (Beer-Romero 1986; Kimble-Long and Madigan 2002). Incandescent illumination at 40–60 $\mu E \bullet m^{-2} \bullet s^{-1}$ is recommended for both primary isolations and for pure cultures. Incubation temperatures between 35 °C and 40 °C are also useful for enriching heliobacteria (all neutrophilic heliobacteria isolated thus far grow up to at least 42 °C) and tend to discourage growth of purple bacteria. Enrichment cultures for thermophilic heliobacteria should be at 48–50 °C.

### Isolation

Positive enrichments for heliobacteria typically appear as slimy green clumps of cells near the soil/liquid or glass/liquid interface, with only weak suspended growth observed. One must therefore carefully inspect liquid enrichments before concluding that an enrichment culture is negative. Spectral evidence for Bchl *g* (in vivo absorption maxima near 788 nm) of any green or green-brown cell material maintained anoxic during spectral analyses (➔ *Fig. 14.4*) confirms the presence of heliobacteria. Microscopically, pasteurized enrichments typically contain a variety of sporulating organisms, including heliobacteria and clostridia. Heliobacteria are usually the largest sporulating cells in such enrichments (Stevenson et al. 1997).

If heliobacteria are suspected in an enrichment culture, agar plates of the enrichment medium should be streaked within an anoxic glove box and then incubated phototrophically in anoxic jars (streaking plates in an oxygenated atmosphere is rarely successful). For alkaliphilic species, standard agar dilution cultures can be established. For plates or dilution cultures, the agar should be washed three times in distilled water to remove sugars and other soluble organic substances that tend to promote the growth of contaminating organisms. The presence of heliobacteria on plates is signaled within 2–3 days by the formation of green to brownish-green colonies, often showing irregular edges. These can be picked and restreaked within the anoxic glove box to obtain pure cultures. Further details of enrichment methods for isolating heliobacteria and recipes for a variety of media for growing pure cultures of heliobacteria are given in Madigan (2006).

Alkaliphilic heliobacteria are easily enriched in alkaline (pH 9) mineral salts media containing an organic acid, such as pyruvate or lactate, low levels of yeast extract, and sulfide at a concentration of 2–5 mM. Except for the final pH, the composition and preparation of such media are similar to media prepared for the growth of purple and green sulfur bacteria in completely filled screw-capped tubes or bottles.

Alkaliphilic heliobacteria have been isolated from soda lakes of variable salinity, and the addition of NaCl to culture media, typically at marine levels (3 % NaCl), increases the likelihood of successful enrichment. As is the case for neutrophilic species, pasteurization of alkaline samples ensures that purple bacteria (for example, *Ectothiorhodospira* species) will not interfere with enrichment of any heliobacteria that may be present. All heliobacteria characterized to date, including alkaliphilic species, produce endospores, and this is likely a universal property of the *Heliobacteriaceae* (Kimble-Long and Madigan 2001).

## Culture Media for Heliobacteria

Both mineral salts and complex media for the routine culture of neutrophilic heliobacteria are described. Half-strength medium PYE is useful as a basic enrichment medium. Culture media for alkaliphilic heliobacteria require the preparation of separate solutions that are assembled aseptically following autoclave sterilization.

See ❯ *Table 14.2*

Other organic or fatty acids can be added in place of pyruvate, in particular, lactate. If fatty acids are used, sodium bicarbonate (1 g/l) should be added. Adjust the pH to 6.8 with NaOH or HCl, boil the medium to de-gas, and transfer to stoppered tubes or bottles under a stream of $N_2$:$CO_2$ (95:5). Half-filled, neoprene rubber-stoppered culture tubes (for example, 18 × 142 mm Bellco® culture tubes) work well and can be sterilized by autoclaving in a tube press for 20 min. Sterile tubes of media can be stored and inoculated inside an anoxic glove box or stored in normal atmosphere and inoculated using standard "Hungate-type" techniques.

**Table 14.2**
**Pyruvate mineral salts (PMS) medium (per liter of double-distilled water)**

| | |
|---|---|
| MOPS buffer (Sigma, St. Louis) | 10 mM (final concentration) |
| EDTA | 10 mg |
| $MgSO_4 \bullet 7H_2O$ | 200 mg |
| $CaCl_2 \bullet 2H_2O$ | 75 mg |
| $NH_4Cl$ | 1 g |
| $K_2HPO_4$ | 0.9 g |
| $KH_2PO_4$ | 0.6 g |
| Sodium pyruvate | 2 g |
| $Na_2S_2O_3 \bullet 5\ H_2O$ | 0.1 g |
| Trace element solution (see below) | 1 ml |
| Yeast extract | 0.1 g |
| Biotin | 15 μg |

**Table 14.3**
**Pyruvate-yeast extract (PYE) medium (per liter of double-distilled water)**

| | |
|---|---|
| $K_2HPO_4$ | 1 g |
| $MgSO_4 \bullet 7H_2O$ | 200 mg |
| $CaCl_2 \bullet 2H_2O$ | 20 mg |
| $Na_2S_2O_3 \bullet 5H_2O$ | 100 mg |
| Sodium pyruvate | 2 g |
| Yeast extract | 4 g |

See ❯ *Table 14.3*

Adjust to pH7, distribute to culture vessels, and sterilize as described for medium PMS.

See ❯ *Table 14.4*

The carbonate and bicarbonate are autoclaved dry in a 0.5-l bottle and then dissolved in 300-ml sterile, double-distilled water; this solution is then added to the main nutrient solution with gentle stirring. Sulfide is prepared by washing and drying crystals of sodium sulfide, weighing the necessary amount, and then dissolving the crystals in 100 ml of boiling distilled water. This solution is transferred to a screw-cap bottle and sterilized in an autoclave. The salts solution is prepared in a 2-l jar containing a spigot to dispense the sterile medium into tubes or bottles.

After autoclave sterilization, all solutions are mixed (the sulfide solution should be added last), and the pH is adjusted with sterile HCl or NaOH. The final assembled medium should be dispensed immediately into completely filled, sterile, screw-capped tubes or bottles and capped tightly. Medium Hr forms a slight grayish-black precipitate upon storage. Better buffering

**Table 14.4**
**Medium Hr for enrichment and growth of alkaliphilic heliobacteria (per liter of double-distilled water)**

| | |
|---|---|
| Distilled water | 600 ml |
| EDTA | 5 mg |
| $KH_2PO_4$ | 0.5 g |
| $NH_4Cl$ | 0.5 g |
| $MgSO_4\bullet 7H_2O$ | 0.2 g |
| $CaCl_2\bullet 7H_2O$ | 75 mg |
| $Na_2CO_3$ | 2.5 g |
| $NaHCO_3$ | 2.5 g |
| Na acetate | 1 g |
| Na pyruvate | 2 g |
| Yeast extract | 0.1 g |
| Trace elements | 1 ml |
| $Na_2S\bullet 9H_2O$ | 0.5 g |
| Final pH | 9 |

**Table 14.5**
**Trace element solution for all heliobacteria media (per liter of double-distilled water)**

| | |
|---|---|
| Conc. HCl | 1 ml |
| $Na_2$-EDTA | 5.2 g |
| $FeCl_2\bullet 4H_2O$ | 1.5 g |
| $ZnCl_2$ | 70 mg |
| $MnCl_2\bullet 4H_2O$ | 100 mg |
| $H_3BO_3$ | 6 mg |
| $CoCl_2\bullet 6H_2O$ | 190 mg |
| $CuCl_2\bullet 2H_2O$ | 17 mg |
| $NiCl_2\bullet 6H_2O$ | 25 mg |
| $Na_2MoO_4\bullet 2H_2O$ | 188 mg |
| $VSO_4\bullet 2H_2O$ | 30 mg |
| $NaWO_4\bullet 2H_2O$ | 2 mg |

capacity of alkaline media can be achieved by adding 10-mM (final) BICINE buffer (Sigma, St. Louis).

See *Table 14.5*

Add compounds in the above order; the EDTA should be fully dissolved before adding remaining components. Store at 4 °C.

## Maintenance

For short-term (weeks to months) storage of heliobacteria, "stab" cultures remain viable if stored in the light in anoxic jars. Once fully grown, stab cultures can be stored for several months in anoxic jars or in an anoxic chamber exposed to low light. For unknown reasons, stab cultures of heliobacteria retain viability much longer than do liquid cultures.

Growing cultures of heliobacteria should be transferred weekly. Older cultures transition from a brown-green to a bright green color, and this transition signals pigment degradation (*Fig. 14.4*) and is accompanied by reduced viability. Long-term storage of heliobacteria can be facilitated by adding sterile dimethyl sulfoxide (DMSO; pure DMSO can be autoclaved) to a final concentration of 10 % (v/v) to 2 ml of a mid-exponential phase culture in a sterile cryotube. The prepared tube should be chilled on ice for 10 min in darkness and then frozen at −80 °C or lower. Properly prepared cryotubes of heliobacteria retain their viability for years.

## Ecology

Our understanding of the distribution and ecology of heliobacteria has primarily emerged from successful isolations. Based on available data, it appears that soil is the major habitat for heliobacteria (Madigan and Ormerod 1995), and curiously, dry soils seem to be the most reliable substrate for enriching these phototrophs (Stevenson et al. 1997). Isolation studies indicate that rice (paddy) soils are excellent sources of heliobacteria, as all paddy soils tested have been positive (unpublished results). Other agricultural soils are less reliable, and disturbed soils rarely yield heliobacteria (Stevenson et al. 1997). This observation makes the original discovery of heliobacteria from manicured garden soil (Gest 1994; Gest and Favinger 1983) all the more remarkable. In contrast to soils, aquatic habitats, such as sewage, pond and lake waters, and freshwater lake sediments, were uniformly negative in yielding enrichment cultures for heliobacteria (Stevenson et al. 1997).

Highly desiccated paddy soils have proven particularly good sources of heliobacteria (Ormerod et al. 1996; Madigan 2006). This may be due to reduction or elimination of competing phototrophs (e.g., purple bacteria) when paddy soils become desiccated and highly oxic in the dry season. It is possible that these very conditions are what induce endosporulation in heliobacteria, a phenomenon that is generally not observed in highly enriched and pure cultures. The ecology of the heliobacteria is likely to be closely linked to sporulation and germination events, and thus, the ability of these phototrophs to produce endospores—unprecedented among anoxygenic phototrophic bacteria—is likely a major factor in their ecological success in nature.

Alkaliphilic heliobacteria have been isolated from shoreline soils of several soda lakes. Whether these species develop to any significant number in the actual water column of soda lakes is an unanswered question. Only one heliobacterium has ever been cultured from the water of a soda lake, and this was the phylogenetically unique "*Candidatus* Heliomonas lunata" enriched from Soap Lake, Washington (USA) (Asao et al. 2012). Although cultured from a sediment sample, it is possible that this organism was actually enriched from endospores that entered the lake from surrounding soils and became attached to particulate matter that settled to the sediments. Interestingly, however, and despite several attempts, additional strains of "*Candidatus* Heliomonas" could not be enriched from soils surrounding Soap Lake (Asao et al. 2012). So, whether heliobacteria inhabit Soap Lake, per se, or other aquatic habitats (other than hot spring microbial mats) remains an open question. A molecular ecology study of the distribution of heliobacteria in nature using FISH or PCR methods should be straightforward and would likely reveal the true breadth of the habitats of these phototrophs.

Based on isolations, then, it is safe to say that heliobacteria are ecologically distinct from green and purple bacteria, organisms which typically inhabit aquatic environments (Madigan 1988; Pfennig 1989). Although purple nonsulfur bacteria are present in some soils (Gest et al. 1985), like the purple and green sulfur bacteria, their primary habitats are also aquatic (Madigan 1988). An unexplained enigma surrounding the habitats of heliobacteria is why phototrophic organisms, especially obligately anaerobic phototrophs that grow best at high light intensities (Kimble-Long and Madigan 2002), would inhabit soil in the first place. On the one hand, it seems that soils would greatly restrict light penetration, and in addition, anoxic conditions in soils are likely to be variable and highly dependent on moisture. On the other hand, however, soils can be quite heterogeneous in their physicochemical properties, and suitable conditions for the growth of heliobacteria may be more common than it was thought. This is especially so in the rice soil habitat, where growth conditions in the rhizosphere may be seasonally ideal.

## Application

The major application of heliobacteria thus far has been as new experimental tools for biophysical and biochemical studies of photosynthesis (Heinnickel and Golbeck 2007; Sattley and Swingley 2013). The reaction center photocomplex of heliobacteria is a model for green plant photosystem I, and because of this, heliobacteria have been widely studied in terms of basic events in photosynthesis. But their unique ecology and likely association with rice (if not other) plants suggest that heliobacteria may have even more significant applications in agriculture.

The apparent lack of autotrophic capacities in the heliobacteria (Sattley et al. 2008) means that heliobacteria are likely to be most active and abundant in the rhizosphere of soils, regions where organic compounds excreted from plant roots could support their photoheterotrophic lifestyle. The rhizosphere is typically rich in organic plant exudates, and it is likely that heliobacteria exploit these as their major carbon sources.

It may also be more than a coincidence that most heliobacteria fix molecular nitrogen, some at very high rates (Kimble and Madigan 1992a, b). Because of their connection to rice soils, it is thus possible that heliobacteria have developed a loose or even quasi-specific symbiotic relationship with rice plants. In such an association, heliobacteria might flourish in the rhizosphere at the expense of organic carbon excreted from plant roots. In exchange, the heliobacteria might supply the rice plants with fixed nitrogen.

Fixed nitrogen is often limiting in paddy soils, yet rice fields typically remain highly productive without the input of added nitrogen fertilizers. This is almost certainly due to the fact that nitrogen fixation is a major process in paddy soils, and significant nitrogen fixation has previously been linked to anoxygenic phototrophic bacteria in these environments (Buresh et al. 1980; Habte and Alexander 1980). If a specific or even casual rice–heliobacteria relationship exists in paddy soils, the association might be commercially exploited by, for example, coating rice plant roots with heliobacterial endospores before planting, similar to how seeds of leguminous crop plants such as soybeans are coated with cultures of their specific rhizobial symbiont to ensure root nodule development. Inoculation of rice plants with heliobacteria might ensure that a strong association developed quickly, thus promoting the growth of rice plants in marginal soils, not otherwise considered suitable for agricultural purposes.

## Acknowledgments

This chapter was supported in part by grant EF0950550 from the US National Science Foundation to MTM.

## References

Asao M, Madigan MT (2009) Family IV *Heliobacteriaceae* Madigan 2001, 625. In: De Vos P, Garrity GM, Jones D, Krieg NR, Ludwig W, Rainey FA, Schleifer K-H, Whitman WB (eds) Bergey's manual of systematic bacteriology, (the *Firmicutes*), 2nd edn, vol 3. Springer, New York, pp 923–931

Asao M, Madigan MT (2010) Taxonomy, phylogeny, and ecology of the heliobacteria. Photosynth Res 104:103–111

Asao M, Jung DO, Achenbach LA, Madigan MT (2006) *Heliorestis convoluta* sp. nov., a coiled, alkaliphilic heliobacterium from the Wadi El Natrun, Egypt. Extremophiles 10:403–410

Asao M, Takaichi S, Madigan MT (2012) Amino acid-assimilating phototrophic heliobacteria from soda lake environments: *Heliorestis acidaminivorans* sp. nov. and "*Candidatus* Heliomonas lunata". Extremophiles 16:585–595

Beer-Romero P (1986) Comparative studies on *Heliobacterium chlorum*, *Heliospirillum gestii* and *Heliobacillus mobilis*. MA Thesis, Department of Biology, Indiana University, Bloomington

Beer-Romero P, Gest H (1987) *Heliobacillus mobilis*, a peritrichously flagellated anoxyphototroph containing bacteriochlorophyll *g*. FEMS Microbiol Lett 41:109–114

Beer-Romero P, Favinger JL, Gest H (1988) Distinctive properties of bacilliform photosynthetic heliobacteria. FEMS Microbiol Lett 49:451–454

Blankenship RE (2002) Molecular mechanisms of photosynthesis. Blackwell Science, Oxford, UK

Brockmann H, Lipinski A (1983) Bacteriochlorophyll *g*. A new bacteriochlorophyll from *Heliobacterium chlorum*. Arch Microbiol 136:17–19

Bryantseva IA, Gorlenko VM, Kompantseva EI, Achenbach LA, Madigan MT (1999) *Heliorestis daurensis* gen. nov. sp. nov., an alkaliphilic rod to coiled-shaped phototrophic heliobacterium from a Siberian soda lake. Arch Microbiol 172:167–174

Bryantseva IA, Gorlenko VM, Kompantseva EI, Tourova TP, Kuznetsov BB, Osipov GA (2000a) Alkaliphilic heliobacterium *Heliorestis baculata* sp. nov.

and emended description of the genus *Heliorestis*. Arch Microbiol 174:283–291

Bryantseva IA, Gorlenko VM, Tourova TP, Kuznetsov BB, Lysenko AM, Bykova SA, Gal'chenko VF, Mityushina LL, Osipov GA (2000b) *Heliobacterium sulfidophilum* sp. nov. and *Heliobacterium undosum* sp. nov.: sulfide-oxidizing heliobacteria from thermal sulfidic springs. Microbiology (En transl from Mikrobiologiya) 69:325–334

Buresh RJ, Casselman ME, Patrick WH Jr (1980) Nitrogen fixation in flooded soil systems, a review. Adv Agron 33:149–192

Gest H (1994) Discovery of the heliobacteria. Photosynth Res 41:17–21

Gest H, Favinger JL (1983) *Heliobacterium chlorum*, an anoxygenic brownish-green photosynthetic bacterium containing a "new" form of bacteriochlorophyll. Arch Microbiol 136:11–16

Gest H, Favinger JL, Madigan MT (1985) Exploitation of $N_2$ fixation capacity for enrichment of anoxygenic photosynthetic bacteria in ecological studies. FEMS Microbiol Ecol 31:317–322

Habte M, Alexander M (1980) Nitrogen fixation by photosynthetic bacteria in lowland rice culture. Appl Environ Microbiol 39:342–347

Heinnickel M, Golbeck JH (2007) Heliobacterial photosynthesis. Photosynth Res 92:35–53

Kimble LK, Madigan MT (1992a) Nitrogen fixation and nitrogen metabolism in heliobacteria. Arch Microbiol 158:155–161

Kimble LK, Madigan MT (1992b) Evidence for an alternative nitrogenase system in *Heliobacterium gestii*. FEMS Microbiol Lett 100:255–260

Kimble LK, Stevenson AK, Madigan MT (1994) Chemotrophic growth of heliobacteria in darkness. FEMS Microbiol Lett 115:51–55

Kimble LK, Mandelco L, Woese CR, Madigan MT (1995) *Heliobacterium modesticaldum*, sp. nov., a thermophilic heliobacterium of hot springs and volcanic soils. Arch Microbiol 163:259–267

Kimble-Long LK, Madigan MT (2001) Molecular evidence that the capacity for endosporulation is universal among phototrophic heliobacteria. FEMS Microbiol Lett 199:191–195

Kimble-Long LK, Madigan MT (2002) Irradiance effects on growth and bacteriochlorophyll content of phototrophic heliobacteria, purple and green photosynthetic bacteria. Photosynthetica 40:629–632

Kobayashi M, Watanabe T, Ikegami I, van de Meent EJ, Amesz J (1991) Enrichment of bacteriochlorophyll *g'* in membranes of *Heliobacterium chlorum* by ether extraction: unequivocal evidence for its existence in vivo. FEBS Lett 284:129–131

Madigan MT (1988) Microbiology, physiology, and ecology of phototrophic bacteria. In: Zehnder AZB (ed) Biology of anaerobic microorganisms. Wiley, New York, pp 39–111

Madigan MT (1992) The family *Heliobacteriaceae*. In: Balows A, Trüper HG, Dworkin M, Schleifer K-H (eds) The prokaryotes, 2nd edn. Springer, New York, pp 1981–1992

Madigan MT (2006). The family *Heliobacteriaceae*. In: Dworkin M, Falkow S, Rosenberg E, Schleifer K-H, Stackebrandt E (eds) Prokaryotes, vol 4. Springer, New York, pp 951–964

Madigan MT, Ormerod JG (1995) Taxonomy, physiology, and ecology of heliobacteria. In: Blankenship RE, Madigan MT, Bauer CE (eds) Anoxygenic photosynthetic bacteria. Kluwer, Dordrecht, pp 17–30

Madigan MT, Euzéby JP, Asao M (2010) Proposal of *Heliobacteriaceae* fam. nov. Int J Syst Evol Microbiol 60:1709–1710

Michalski TJ, Hunt JE, Bowman MK, Smith U, Bardeen K, Gest H, Norris JR, Katz JJ (1987) Bacteriopheophytin *g*: properties and some speculations on a possible primary role for bacteriochlorophylls *b* and *g* in the biosynthesis of chlorophylls. Proc Natl Acad Sci USA 84:2570–2574

Miller KR, Jacob JS, Smith U, Kolaczkowski S, Bowman MK (1986) *Heliobacterium chlorum*: cell organization and structure. Arch Microbiol 146:111–114

Oh-Oka H (2007) Type 1 reaction center of photosynthetic heliobacteria. Photochem Photobiol 83:177–186

Oh-Oka H, Iwaki M, Itoh S (2002) Electron donation from membrane-bound cytochrome *c* to the photosynthetic reaction center in whole cells and isolated membranes of *Heliobacterium gestii*. Photosynth Res 71:137–147

Ormerod JG, Kimble LK, Nesbakken T, Torgersen YA, Woese CR, Madigan MT (1996) *Heliophilum fasciatum* gen. nov. sp. nov. and *Heliobacterium gestii* sp. nov.: endopore-forming heliobacteria from rice field soils. Arch Microbiol 165:226–234

Pfennig N (1989) Ecology of phototrophic purple and green sulfur bacteria. In: Schlegel HG, Bowien B (eds) Autotrophic bacteria. Springer, New York, pp 97–116

Pickett MW, Williamson MP, Kelly DJ (1994) An enzyme and $^{13}$C-NMR study of carbon metabolism in heliobacteria. Photosynth Res 41:75–88

Sarrou I, Khan Z, Cowgill J, Lin S, Brune D, Romberger S, Golbeck JH, Redding KE (2012) Purification of the photosynthetic reaction center from *Heliobacterium modesticaldum*. Photosynth Res 111:291–302

Sattley WM, Blankenship RE (2010) Insights into heliobacterial photosynthesis and physiology from the genome of *Heliobacterium modesticaldum*. Photosynth Res 104:113–122

Sattley WM, Swingley WD (2013) Properties and evolutionary implications of the heliobacterial genome. In: Beatty JT (ed) Genome evolution of photosynthetic bacteria, vol 66, Advances in botanical research. Academic Press, Elsevier, San Diego, pp 67–97

Sattley WM, Madigan MT, Swingley WD, Cheung PC, Clocksin KM, Conrad AL, Dejesa LC, Honchak BM, Jung DO, Karbach LE, Kurdoglu A, Lahiri S, Mastrian SD, Page LE, Taylor HL, Wang ZT, Raymond J, Chen M, Blankenship RE, Touchman JW (2008) The genome of *Heliobacterium modesticaldum*, a phototrophic representative of the *Firmicutes* containing the simplest photosynthetic apparatus. J Bacteriol 190:4687–4696

Sirevåg R, Ormerod JG (1970) Carbon dioxide—fixation in photosynthetic green sulfur bacteria. Science 169:186–188

Stevenson AK (1993) Isolation and characterization of heliobacteria from soil habitats worldwide. MA Thesis, Department of Microbiology, Southern Illinois University, Carbondale

Stevenson AK, Kimble LK, Woese CR, Madigan MT (1997) Characterization of new heliobacteria and their habitats. Photosynth Res 53:1–12

Takaichi S, Inoue K, Akaike M, Kobayashi M, Oh-Oka H, Madigan MT (1997) The major carotenoid in all species of heliobacteria is the $C_{30}$ carotenoid 4,4′-diaponeurosporene, not neurosporene. Arch Microbiol 168:277–281

Takaichi S, Oh-Oka H, Maoka T, Jung DO, Madigan MT (2003) Novel carotenoid glucoside esters from alkaliphilic heliobacteria. Arch Microbiol 179:95–100

Tang KH, Yue H, Blankenship RE (2010) Energy metabolism of *Heliobacterium modesticaldum* during phototrophic and chemotrophic growth. BMC Microbiol 10:150

Trost JT, Blankenship RE (1989) Isolation of a photoactive photosynthetic reaction center-core antenna complex from *Heliobacillus mobilis*. Biochemistry 28:9898–9904

van de Meent EJ, Kobayashi M, Erkelens C, van Veelen PA, Amesz J, Watanabe T (1991) Identification of $8^1$-hydroxychlorophyll *a* as a functional reaction center pigment in heliobacteria. Biochim Biophys Acta 1058:356–362

# 15 The Family *Lachnospiraceae*

*Erko Stackebrandt*
Leibniz Institute DSMZ-German Collection of Microorganisms and Cell Cultures GmbH,
Braunschweig, Germany



**Abstract**

The family *Lachnospiraceae* is a phylogenetically and morphologically heterogeneous taxon of the class *Clostridia*, phylum Firmicutes. The family, described on the basis of 16S rRNA gene sequence analysis, contains a high number of as yet not reclassified species of other genera which will significantly expand the physiological and chemotaxonomic diversity of the family once these species will be assigned to new genera. All members are anaerobic, fermentative, and chemoorganotrophic, some with strong hydrolyzing activities, e.g., pectin methylesterase, pectate lyase, xylanase, α-L-arabinofuranosidase, ß-xylosidase α- and ß-galactosidase, α- and ß-glucosidase, *N*-acetyl-ß-glucosaminidase, or α-amylase. The human or animal digestive tract is the main habitat for most members; other were isolated from the oral cavity but rarely from soil.

## Introduction

The family *Lachnospiraceae* (Rainey 2009a) is a member of the class *Clostridia* (Rainey 2009b), order *Clostridiales* (Rainey 2009c). According to *Bergey's Manual of Systematic Bacteriology* (De Vos et al. 2009), the family encompasses 19 genera. The higher taxa were added to the Validation List (Euzeby 2010). The family has been described on phylogenetic grounds, i.e., the position of authentic members of *Lachnospiraceae*, branching within the radiation of the order *Clostridiales*, especially associated to members of *Clostridium* rRNA cluster XIVa (Collins et al. 1994; Stackebrandt et al. 1999). The LTP tree (Yarza et al. 2010) groups members of the genera *Clostridium*, *Eubacterium*, and *Ruminococcus* with various representatives of *Lachnospiraceae*, also mentioned in some of the original genus descriptions. Members of the family are strict anaerobes but differ widely in morphology by the presence of straight to curved and short and long rods to cocci (*Coprococcus*, *Synthrophococcus*). Rarely, spores are formed (*Sporobacterium*). The description of family members have been extensively covered in the volume Firmicutes of *Bergey's Manual of Systematic Bacteriology* (De Vos et al. 2009) and this short assay will only cover differences found by comparing members of the family listed in volume 3 Firmicutes to the list of genera in *Lachnospiraceae* as indicated in the List of Prokaryotic Names with Standing in Nomenclature (LPSN, http://www.bacterio.net/) and recent description of novel family members.

## Recent Additions to the Family

The phylogenetic tree (➋ *Fig. 15.1*) includes the genera *Howardella* (Cook et al. 2007), Lactonifactor (Clavel et al. 2007), *Anaerosporobacter* (Jeong et al. 2007), and *Robinsoniella* (Cotta et al. 2009). These genera are absent in *Bergey's Manual* as they were described too late to be considered for this handbook. Except for *Robinsoniella*, the other three genera are also not included in the LPSN List. *Acetatifactor* (Pfeiffer et al. 2012), *Lachnoanaerobaculum* (Hedberg et al. 2012), and *Stomatobaculum* (Sizova et al. 2013), described recently, are included in the family composition of the LPSN List but are missing in ➋ *Fig. 15.1*.

Members of all these genera are strictly anaerobic and have C16:0 as major fatty acid. Individual-type species can be differentiated by a combination of morphology, spore formation, motility, glucose metabolism, and mol% G+C (➋ *Table 15.1*). *Howardella ureilytica* appears a moderately related neighbor of *Lachnospira multipara* and *Anaerosporobacter mobilis* groups with *Butyrivibrio crossotus* while *Lactonifactor longoviformis* form an individual lineage next to *Hespellia* and *Dorea* species (➋ *Fig. 15.1*). As displayed in the original descriptions, *Lachnoanaerobaculum* species are phylogenetic neighbors of *Catonella* and *Johnsonella* species (Hedberg et al. 2012), *Robinsoniella* appears to moderately related to the *Ruminococcus* species of *Lachnospiraceae* (Cotta et al. 2009), *Stomatobaculum* is a distant relative of *Moryella indoligenes* (Sizova et al. 2013), while *Acetatifactor* appears as a most distantly related member of the family (Pfeiffer et al. 2012).

Draft or complete genomes are available for a large number of *Lachnospiraceae* members, such as several unnamed *Lachnospiraceae* bacteria (6_1_37FAA, 8_1_57FAA, 5_1_63FAA, oral taxon 107 str. F0167, or 1_1_57FAA), numerous *Lachnospira* isolates, *Roseburia intestinalis* XB6B4 and M50/1, *Stomatobaculum longum* DSM 24645$^{T}$, *Oribacterium sinus* strain F0268, *Shuttleworthia satelles* DSM 14600, *Anaerostipes hadrus* DSM 3319, *Catonella morbi* VPI D154F-12, several strains of *Roseburia intestinalis including* DSM 14610, and many strains of *Dorea longicatena* including strain DSM 13814. A complete list

E. Rosenberg et al. (eds.), *The Prokaryotes – Firmicutes and Tenericutes*, DOI 10.1007/978-3-642-30120-9_363,

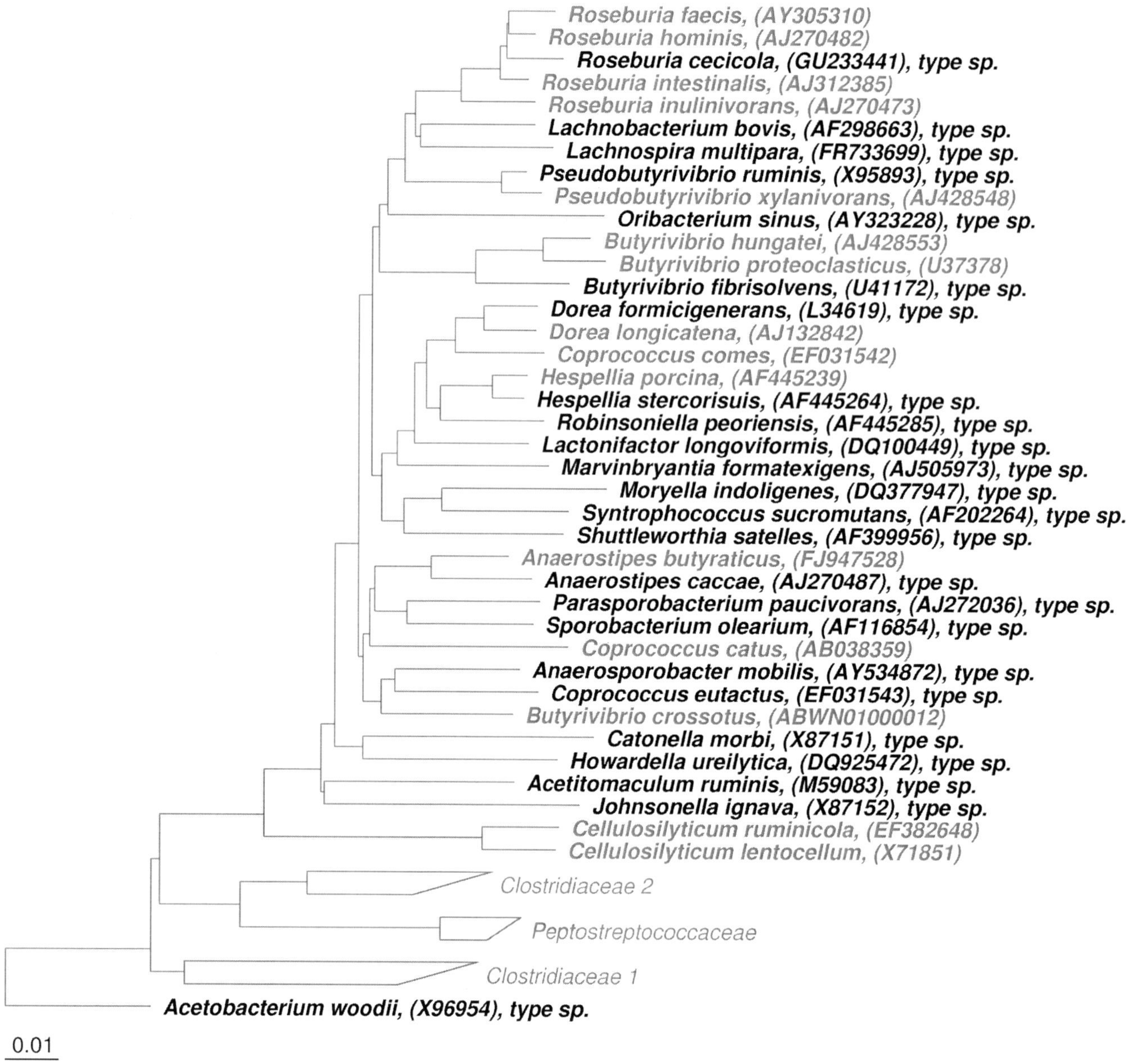


**Fig. 15.1**
Neighbor-joining genealogy reconstruction based on the RAxML algorithm (Stamatakis 2006) of the sequences of validly named members of the family *Lachnospiraceae* and some neighboring taxa present in the LTP_106 (Yarza et al. 2010). The tree was reconstructed by using a subset of sequences. Representative sequences from close relative genera were used to stabilize the tree topology. In addition, a 40 % maximum frequency filter was applied to remove hypervariable positions from the alignment. Scale bar indicates estimated sequence divergence

can be retrieved from http://www.genomesonline.org/cgi-bin/GOLD/index.cgi.

## Ecology

As indicated in *Table 15.1*, most novel family members were found in the gut or rumen of mammals, hence in the same range of members described previously. Analysis of public 16S rRNA gene sequence database by screening entries with the homologous sequence of type strains confirms the niches and expands them slightly, but the occurrence of highly related sequences is rare. For each strain only a few isolates or clones with high BLAST similarities (98–99 %) were found. In some cases (Ley et al. 2006; Li et al. 2012; Ziemer unpublished), the same study revealed the simultaneous presence of some of the members covered in this chapter:

*Howardella urealytica* clones were detected in guts of mammals (Eckburg et al. 2005; Ley et al. 2006, 2008) and human ileum with inflammatory bowel diseases phenotype (Li et al. 2012).

**Table 15.1**
**Properties of family type strains**

| Genus | *Howardella* | *Lactonifactor* | *Acetatifactor* | *Robinsoniella* | *Anaerosporobacter* | *Lachnoanaerobaculum* | *Stomatobaculum* |
|---|---|---|---|---|---|---|---|
| Species | *H. urealytica* | *L. longoviformis* | *A. muris* | *R. peoriensis* | *A. mobilis* | *L. umeaense*[1], *L. orale*[2], *L. sabburreum*[3] | *S. longum* |
| Morphology | Coccoid, short chains | Regular rods, often in pairs | Thin straight rods, single, pairs or short chains | Oval to rod shaped | Rods | Filamentous rods, sometimes curved, aggregates may be formed | Long rods, chains or curved filaments |
| Motility | – | – | – | – | +, peritrichous flagella | – | – |
| Spore formation | – | – | – | +, subterminal | +, terminal | +, terminal | – |
| Carbohydrates | Not fermented | Used as carbon source | Arabinose and xylose fermentation | Wide range fermented | Wide range fermented | Species-specific differences, saccharolytic | Acid produced from glucose, maltose and sucrose |
| End product of glucose fermentation | | | *nd* | Acetate, succinate | Formate, acetate, $H_2$ | Butyrate, acetate, lactate, and small amounts of succinate may be formed | Butyrate, lactate, isovalerate, acetate |
| Major fatty acids (>5 %), major compound underlined | $C_{16:0}$, $C_{18:0}$, unidentified FA | $C_{16:0}$, $C_{14:0}$, iso-$C_{17:1}$,$C_{18:2}$ ω6,9c and anteiso-$C_{18:0}$, $C_{18:0}$, $C_{18:1}$ω9c DMA, iso-$C_{19:1}$ | *C14:0*, $C_{14:0}$ DMA, $C_{16:0}$, $C_{18:0}$, $C_{18:1}$ω9c | $C_{16:0}$, $C_{13:0}$ 3-OH, $C_{14:0}$, $C_{16:1}$ω7c, iso-$C_{17:0}$, $C_{18:1}$ω7c | $C_{16:0}$, $C_{16:0}$ 3-OH, iso-$C_{17:1}$/anteiso B $C_{18:1}$ ω 7c, $C_{16:1}$ω7c and/or iso-$C_{15:0}$ 2-OH | $C_{16:0}$, $C_{14:0}$, $C_{18:1}$ ω7c, | $C_{14:0}$,$C_{14:0}$ DMA, $C_{16:0}$, $C_{16:1}$ω7c DMA, $_{16:1}$ ω7c, $C_{16:0}$ DMA |
| Whole cell sugars | nd | nd | *glu, rib, gal* | gal, glu, rham rib | nd | nd | nd |
| Polar lipids | nd | nd | DPG, PG, PE, PGL,AL, AGL | DGP, PG, PE, GL, PL, AGL, APL | nd | nd | nd |
| G+C | 34 | 48 | 48.5 | 48.7 | 41 | 35–38 | 55–55.3 |
| Isolation source | Rumen fluid, sheep | Faeces of healthy human male adult | Cecum of mouse fed a high calorie diet | Swine manure storage pit, fresh swine manure, deep human wound | Forest soil, Korea | Oral cavity, small intestine, blood, amniotic fluid | Subgingival plaque |
| Type strain | GPC 589$^T$ | ED-Mt61/PYG-s6$^T$ | CT-m2$^T$ | PPC31$^T$ | HY-37-4$^T$ | [1]CD3:22$^T$<br>[2]N1$^T$<br>[3]CCUG 28089$^T$ | DSM 24645$^T$ |

*glu* glucose, *rib* ribose, *gal* galactose, *rham* rhamnose
*DPG* diphosphatidylglycerol, *PG* phosphatidylglycerol, *PE* phosphatidylethanolamine, *PGL* phosphoglycolipid, *AL* aminolipid, *AGL* aminoglycolipid, *APL* aminophospholipid

Clones of *Anaerosporobacter mobilis*, originally described for a strain from forest soil, were found in the gastrointestinal tract of a wood-eating catfish (Watts et al. unpublished; accession number KC000052), among a population of fecal bacteria fed with cellulose or xylan/pectin (Ziemer, unpublished; KC000052), and in the gut of the giant panda (Zhu and Wei, unpublished; JF920374), suggesting that members of this bacterium are involved in the hydrolysis of polysaccharides. Additional strains of *Robinsoniella peoriensis* were isolated from swine manure (Cotta et al. 2003; Whitehead et al. unpublished, DQ681227) but also from feces of a porpoise (McLaughlin et al. 2013). *Acetatifactor muris*, originally isolated from mouse cecum, were reported as strains or clones from the same habitat by Turnbaugh et al. (2008), Smith et al. (2012), and Nozu et al. (unpublished, AB702929), while Grice et al. (2010) detected a clone in a diabetic wound. Strains and clones from

*Lactonifactor longoviformis*, originally described from feces, were reported in human guts (Ley et al. 2006; Jin et al. 2007; Ziemer, unpublished; JQ608233) but also in humanized gnotobiotic mice (Turnbaugh et al. 2009). *Lachnoanaerobaculum* strains from various human sources were reported infected lungs (van der Gast et al. 2011), feces (Kassinen et al. 2007) human ileum (Li et al. 2012), and oral cavity (Kroes et al. 1999; Dewhirst et al. 2010; Sizova et al. 2012). *Stomatobaculum longum* is related to uncultured oral taxa (see Sizova et al. 2013, AF385510; Munson et al. 2004; Perkins et al. 2010; Zhang et al. unpublished, FJ470429; Schulze-Schweifing et al. unpublished, JQ406544) but also from human ileum (Li et al. 2012).

## Medical Relevance

Though the novel members of Lachnospiraceae have been found in samples isolated from inflamed samples such as diabetic wounds (Grice et al. 2010), irritable bowel syndrome (Kassinen et al. 2007; Li et al. 2012), subgingival crevice (Kroes et al. 1999), and cystic fibrosis (van der Gast et al. 2011), their role as causative agent remains obscure and as yet they have to be considered opportunistic pathogens.

Antibiotic susceptibility data are available for *Lactonifactor longoviformis*. The type strain ED-Mt61/PYG-s6 did not grow in the presence of penicillin or rifampicin at concentration as low as 0.5 mg $mol^{-1}$ and for the type strain of *Stomatobaculum longum* DSM $24645^T$ which is susceptible to kanamycin, colistin, vancomycin, metronidazole, penicillin, rifampicin, erythromycin, ampicillin, tetracycline, and bile but is resistant to sulfamethoxazole/trimethoprim.

## References

Clavel T, Lippman R, Gavini F, Doré J, Blaut M (2007) *Clostridium saccharogumia* sp. nov. and *Lactonifactor longoviformis* gen. nov., sp. nov., two novel human faecal bacteria involved in the conversion of the dietary phytoestrogen secoisolariciresinol diglucoside. Syst Appl Microbiol 30:16–26

Collins MD, Lawson PA, Willems A, Cordoba JJ, Fernandez-Garayzabal J, Garcia P, Cai J, Hippe H, Farrow JAE (1994) The phylogeny of the genus *Clostridium*: proposal of five new genera and eleven new species combinations. Int J Syst Bacteriol 44:812–826

Cook AR, Riley PW, Murdoch H, Evans PN, McDonald IR (2007) *Howardella ureilytica* gen. nov., sp. nov., a Gram-positive, coccoid-shaped bacterium from a sheep rumen. Int J Syst Evol Microbiol 57:2940–2945

Cotta MA, Whitehead TR, Zeltwanger RL (2003) Isolation, characterization and comparison of bacteria from swine faeces and manure storage pits. Environ Microbiol 5:737–745

Cotta MA, Whitehead TR, Falsen E, Moore E, Lawson PA (2009) *Robinsoniella peoriensis* gen. nov., sp. nov., isolated from a swine-manure storage pit and a human clinical source. Int J Syst Evol Microbiol 59:150–155

De Vos P, Garrity GM, Jones D, Krieg NR, Ludwig W, Rainey FA, Schleifer KH, Whitman WB (eds) (2009) Bergey's manual of systematic bacteriology, vol 3, 2nd edn. Springer, New York

Dewhirst FE, Chen T, Izard J, Paster BJ, Tanner AC, Yu WH, Lakshmanan A, Wade G (2010) The human oral microbiome. J Bacteriol 192:5002–5017

Eckburg PB, Bi EM, Bernstein CN, Purdom E, Dethlefsen L, Sargent M, Gill SR, Nelson KE, Relman DA (2005) Diversity of the human intestinal microbial flora. Science 308:1635–1638

Euzeby J (2010) List of new names and new combinations previously effectively, but not validly, published. Int J Syst Evol Microbiol 60:469–472

Grice EA, Snitkin ES, Yockey LJ, Bermudez DM, Liechty KW, Segre JA (2010) CONSRTM NISC Comparative sequencing program longitudinal shift in diabetic wound microbiota correlates with prolonged skin defense response. Proc Natl Acad Sci U S A 107:14799–14804

Hedberg ME, Moore ER, Svensson-Stadler L, Hörstedt P, Baranov V, Hernell O, Wai SN, Hammarström S, Hammarström ML (2012) *Lachnoanaerobaculum* gen. nov., a new genus in the *Lachnospiraceae*: characterization of *Lachnoanaerobaculum umeaense* gen. nov., sp. nov., isolated from the human small intestine, and *Lachnoanaerobaculum orale* sp. nov., isolated from saliva, and reclassification of *Eubacterium saburreum* (Prevot 1966) Holdeman and Moore 1970 as *Lachnoanaerobaculum saburreum* comb. nov. Int J Syst Evol Microbiol 62:2685–2690

Jeong H, Lim YW, Yi H, Sekiguchi Y, Kamagata Y, Chun J (2007) *Anaerosporobacter mobilis* gen. nov., sp. nov., isolated from forest soil. Int J Syst Evol Microbiol 57:1784–1787

Jin JS, Kakiuchi N, Hattori M (2007) Enantioselective oxidation of enterodiol to enterolactone by human intestinal bacteria. Biol Pharm Bull 30:2204–2206

Kassinen A, Krogius-Kurikka L, Makivuokko H, Rinttila T, Paulin L, Corander J, Malinen E, Apajalahti J, Palva A (2007) The fecal microbiota of irritable bowel syndrome patients differs significantly from that of healthy subjects. Gastroenterology 133:24–33

Kroes I, Lepp PW, Relman DA (1999) Bacterial diversity within the human subgingival crevice. Proc Natl Acad Sci U S A 96:14547–14552

Ley RE, Turnbaugh PJ, Klein S, GordonJ I (2006) Microbial ecology: human gut microbes associated with obesity. Nature 444:1022–1023

Ley RE, Hamady M, Lozupone C, Turnbaugh PJ, Ramey RR, Bircher JS, Schlegel ML, Tucker TA, Schrenzel MD, Knight R, Gordon JI (2008) Evolution of mammals and their gut microbes. Science 320:1647–1651

Li E, Hamm CM, Gulati AS, Sartor RB, Chen H, Wu X, Zhang T, Rohlf FJ, Zhu W, Gu C, Robertson CE, Pace NR, Boedeker EC, Harpaz N, Yuan J, Weinstock GM, Sodergren E, Frank DN (2012) Inflammatory bowel diseases phenotype, C. difficile and NOD2 genotype are associated with shifts in human ileum associated microbial composition. PLoS One 7:e26284

McLaughlin RW, Zheng J, Ruan R, Wang C, Zhao Q, Wang D (2013) Isolation of *Robinsoniella peoriensis* from the fecal material of the endangered Yangtze finless porpoise, *Neophocaena asiaeorientalis* asiaeorientalis. Anaerobe 20:79

Munson MA, Banerjee A, Watson TF, Wade WG (2004) Molecular analysis of the microflora associated with dental caries. J Clin Microbiol 42:3023–3029

Perkins SD, Woeltje KF, Angenent LT (2010) Endotracheal tube biofilm inoculation of oral flora and subsequent colonization of opportunistic pathogens. Int J Med Microbiol 300:503–511

Pfeiffer N, Desmarchelier C, Blaut M, Daniel H, Haller D, Clavel T (2012) *Acetatifactor muris* gen. nov., sp. nov., a novel bacterium isolated from the intestine of an obese mouse. Arch Microbiol 194:901–907

Rainey FA (2009a) Family V. *Lachnospiraceae*. In: De Vos P, Garrity GM, Jones D, Krieg NR, Ludwig W, Rainey FA, Schleifer KH, Whitman WB (eds) Bergey's manual of systematic bacteriology, vol 3, 2nd edn. Springer, New York, p 921

Rainey FA (2009b) Class II *Clostridia* class. nov. In: De Vos P, Garrity GM, Jones D, Krieg NR, Ludwig W, Rainey FA, Schleifer KH, Whitman WB (eds) Bergey's manual of systematic bacteriology, vol 3, 2nd edn. Springer, New York, p 736

Rainey FA (2009c) Order I *Clostridiales* prevot 1953. In: De Vos P, Garrity GM, Jones D, Krieg NR, Ludwig W, Rainey FA, Schleifer KH, Whitman WB (eds) Bergey's manual of systematic bacteriology, vol 3, 2nd edn. Springer, New York, p 736

Sizova MV, Hohmann T, Hazen A, Paster BJ, Halem SR, Murphy CM, Panikov NS, Epstein SS (2012) New approaches for isolation of previously uncultivated oral bacteria. Appl Environ Microbiol 78:194–203

Sizova M, Muller P, Panikov N, Mandalakis M, Hohmann T, Hazen A, Fowle W, Prozorov T, Bazylinski DA, Epstein SS (2013) *Stomatobaculum longum* gen.

nov., sp. nov., an obligately anaerobic bacterium from the human oral cavity. Int J Syst Evol Microbiol 63:1450–1456

Smith P, Siddharth J, Pearson R, Holway N, Shaxted M, Butler M, Clark N, Jamontt J, Watson RP, Sanmugalingam D, Parkinson SJ (2012) Host genetics and environmental factors regulate ecological succession of the mouse colon issue-associated microbiota. PLoS One 7, E30273

Stackebrandt E, Kramer I, Swiderski J, Hippe H (1999) Phylogenetic basis for a taxonomic dissection of the genus *Clostridium*. FEMS Immunol Med Microbiol 24:253–258

Stamatakis A (2006) RAxML-VI-HPC: maximum likelihood-based phylogenetic analyses with thousands of taxa and mixed models. Bioinformatics 429(22):2688–2690

Turnbaugh PJ, Backhed F, Fulto L, Gordon JI (2008) Diet-induced obesity is linked to marked but reversible alterations in the mouse distal gut microbiome. Cell Host Microbe 3:213–223

Turnbaugh PJ, Ridaura VK, Faith JJ, Rey FE, Knight R, Gordon JI (2009) The effect of diet on the human gut microbiome: a metagenomic analysis in humanized gnotobiotic mice. Sci Transl Med 1:6RA14

van der Gast CJ, Walker AW, Stressmann FA, Rogers GB, Scott P, Daniels TW, Carroll MP, Parkhill J, Bruce KD (2011) Partitioning core and satellite taxa from within cystic fibrosis lung bacterial communities. ISME J 5:780–791

Yarza P, Ludwig W, Euzéby J, Amann R, Schleifer K-H, Glöckner FO, Rosselló-Móra R (2010) Update of the all-species living-tree project based on 16S 445 and 23S rRNA sequence analyses. Syst Appl Microbiol 33:291–299

# 16 The Family *Lactobacillaceae*: Genera Other than *Lactobacillus*

*Leon Dicks*[1] · *Akihito Endo*[2]
[1]Department of Microbiology, University of Stellenbosch, Stellenbosch, South Africa
[2]Department of Food and Cosmetic Science, Tokyo University of Agriculture, Hokkaido, Japan



**Abstract**
Members of the genus *Pediococcus* are Gram-positive, catalase negative, oxidase negative, lactic acid producing bacteria that prefer facultatively aerobic to microaerophilic growth conditions. Glucose and gluconate are fermented to lactic acid, but without the production of $CO_2$, which classifies the genus as homofermentative. Cells are spherical and arranged in a tetrad formation. The genus *Tetragenococcus* has similar cell morphology. *Pediococcus* is genotypically closer related to the *Lactobacillus casei/Lactobacillus paracasei* group within the genus *Lactobacillus* than to *Tetragenococcus*. Pediococci are normally found in the same habitats as *Lactobacillus*, *Leuconostoc* and *Weissella* spp. Most *Pediococcus* spp. have been isolated from plants and fermented plant material. Some strains have been isolated from the gastrointestinal tract (GIT) of humans and animals. *Pediococcus acidilactici* has been isolated from the GIT of carp (*Cyprinus carpio*) and freshwater prawns (*Macrobrachium rosenbergii*). *Pediococcus pentosaceus* has been isolated from tonsils and the nasal cavity of piglets. Only a few rare strains are resistant to vancomycin and they are generally not regarded as pathogens. The genus *Paralactobacillus* was differentiated from homofermentative *Lactobacillus* spp. by its phylogenetic position based on 16S rRNA gene sequences and sugar fermentation profiles. However, the differentiations were denied later and *Paralactobacillus selangorensis*, sole species in the genus *Paralactobacillus*, was reclassified as a species in the genus *Lactobacillus*. The only species in the genus *Sharpea* is *Sharpea azabuensis*. The species is phylogenetically related to *Lactobacillus catenaformis*, but is differentiated based on 16S rRNA gene sequences. This chapter describes the history of the genera *Paralactobacillus*, *Pediococcus* and *Sharpea*.

## *Pediococcus*

### Taxonomy, Historical and Current

Members of the genus *Pediococcus* are described as cocci growing in a plane (Pe.di.o.coc'cus Gr. n. *pedium* a plane surface; Gr. n. *coccus* a grain or berry; n, *Pediococcus* coccus growing in one plane). The species are genetically diverse, as evident from the DNA base composition (mol% G+C) that ranges from 35 to 44 ($T_m$), or 36 to 42 as determined by HPLC (Holzapfel et al. 2009). The type species is *Pediococcus damnosus* (Claussen 1903, p. 68). A full description of the species is found in volume three of the second edition of *Bergey's Manual of Systematic Bacteriology*, pp. 517–520 (Holzapfel et al. 2009). The species not included in this edition of *Bergey's Manual of Systematic Bacteriology* are *Pediococcus argentinicus* (De Bruyne et al. 2008), *Pediococcus ethanolidurans* (Liu et al. 2006), *Pediococcus lolii* (Doi et al. 2009), and *Pediococcus siamensis* (Tanasupawat et al. 2007). A full description of these species can be found in the reference listed for each.

The first description of cocci in tetrads was as early as 1884 (Balcke 1884). Four years later, Lindner (1888) described cocci isolated from beer and named them *Pediococcus acidilactici* and *Pediococcus cerevisiae*. In 1903 the name *P. cerevisiae* was rejected and replaced by *Pediococcus damnosus* (Claussen 1903). The name *P. damnosus* was used in later publications (Mees 1934), although Claussen was of the opinion that the strains could have been the same as the strains described by Balcke in 1884 and that they could have been members of *P. cerevisiae*. The name *P. cerevisiae* was therefore used in subsequent publications (Pedersen 1949; Nakagawa and Kitahara 1959). The confusion in nomenclature of pediococci was discussed by Garvie (1974). The Judicial Commission published Opinion 52, declaring that *P. damnosus* Claussen is the type species of the genus. The generic name was thus conserved over *Pediococcus* (Balcke 1884) and all earlier objective synonyms. However, in the Approved Lists of Bacterial Names (Skerman et al. 1980) and in the Amended Edition of the Approved Lists of Bacterial Names (Skerman et al. 1989; Euzéby 1998), *Pediococcus* (Balcke 1884) was cited as the genus name.

E. Rosenberg et al. (eds.), *The Prokaryotes – Firmicutes and Tenericutes*, DOI 10.1007/978-3-642-30120-9_207,

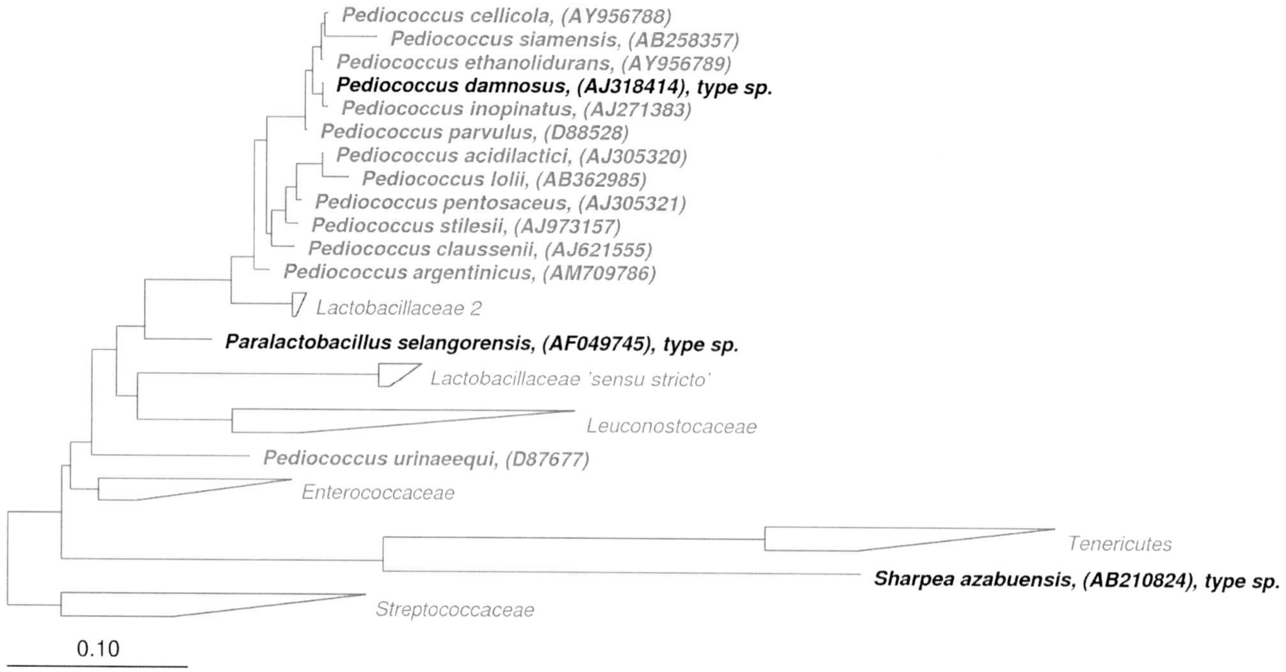


**Fig. 16.1**
**Phylogenetic relationship of *Pediococcus* spp. and related taxa based on 16 S rRNA gene sequences. The tree was constructed by using the maximum likelihood method. *Pediococcus urinaeequi* has been reclassified to *Aerococcus urinaeequi* (Felis et al. 2005) and was used as an out-group**

The genus consists of 12 species, i.e., *Pediococcus acidilactici* (Lindner 1887; Skerman et al. 1980), *Pediococcus argentinicus* (De Bruyne et al. 2008), *Pediococcus cellicola* (Zhang et al. 2005), *Pediococcus claussenii* (Dobson et al. 2002), *Pediococcus damnosus* (Balcke 1884; Claussen 1903; Skerman et al. 1980), *Pediococcus ethanolidurans* (Liu et al. 2006), *Pediococcus inopinatus* (Back 1978a, 1988), *Pediococcus lolii* (Doi et al. 2009), *Pediococcus parvulus* (Günther and White 1961; Skerman et al. 1980), *Pediococcus pentosaceus* (Mees 1934; Skerman et al. 1980), *Pediococcus siamensis* (Tanasupawat et al. 2007), and *Pediococcus stilesii* (Franz et al. 2006).

*Pediococcus urinaeequi ex* (Mees 1934; Garvie 1988) was reclassified as *Aerococcus urinaeequi* (Felis et al. 2005) and *Pediococcus halophilus* as *Tetragenococcus halophilus* (Collins et al. 1990), based on 16S rRNA gene sequences and genomic DNA–DNA hybridizations. *Pediococcus soyae* was reclassified as *Tetragenococcus halophilus* (Weiss 1992). *Pediococcus dextrinicus* (Coster and White 1964) Back 1978 (Approved List 1980) is not closely related to the other pediococci, as shown by 16S rRNA sequence similarity, and was reclassified as *Lactobacillus dextrinicus*. The distant relatedness of "*Pediococcus dextrinicus*" to the other pediococci was confirmed by multilocus sequence analyses of the 16S rRNA gene and Cpn60, PheS, RecA, and RpoA proteins. "*P. dextrinicus*" is further distinguished from the true pediococci by the production of L(+)-lactic acid from glucose via a fructose-1,6-diphosphate FDP inducible L-lactate dehydrogenase (L-LDH; Back 1978a). *Lactobacillus concavus* sp. nov. is closely related (97.9 % 16S rRNA similarity) to the type strain of "*P. dextrinicus*" JCM 5887$^T$. The grouping of "*P. dextrinicus*," a coccus-shaped cell, with rod-shaped *Lactobacillus* cells, necessitates an amendment of the description of the genus *Lactobacillus*, as proposed by Haakensen et al. (2009).

## Molecular Analyses

Eleven of the species within the genus *Pediococcus* form a close group (91.3–99.3 % 16S rRNA sequence similarity) within the *Clostridium* branch (Collins et al. 1991). *Pediococcus damnosus* and *P. inopinatus* share the highest similarity (➲ *Fig. 16.1*). The lowest similarity is shared between *P. lolii* and *P. siamensis* (➲ *Fig. 16.1*). *Pediococcus* spp. are phylogenetically closer related to the *Lactobacillus* branch of lactic acid bacteria (LAB), especially the *Lactobacillus casei/paracasei* group, with *Tetragenococcus* as the closest neighbor at 88–91 % 16S rDNA similarity (Schleifer and Ludwig 1995; Felis and Dellaglio 2007). *Pediococcus* may thus be considered part of the genus *Lactobacillus*.

Species within the genus *Pediococcus* are differentiated by 16S rRNA gene sequencing (Collins et al. 1990; Kurzak et al. 1998; Omar et al. 2000; Barney et al. 2001), ribotyping (Jager and Harlander 1992; Satokari et al. 2000; Barney et al. 2001; Santos et al. 2005), hybridization with species-specific DNA probes (Lonvaud-Funel et al. 1993; Rodriguez et al. 1997; Mora et al. 1997, 1998), comparison of randomly amplified polymorphic DNA (RAPD) PCR profiles (Kurzak et al. 1998; Nigatu et al. 1998; Mora et al. 2000; Simpson et al. 2002; Fujii et al. 2005;

Perez Pulido et al. 2005; Tamang et al. 2005), real-time PCR with strain- and species-specific primers (Delaherche et al. 2004; Lan et al. 2004; Stevenson et al. 2005), pulsed-field gel electrophoresis (PFGE) (Dellaglio et al. 1981; Luchansky et al. 1992; Barros et al. 2001; Simpson et al. 2006), PCR-ELISA (Waters et al. 2005), and genomic DNA–DNA hybridization (Back and Stackebrandt 1978; Dellaglio et al. 1981; Franz et al. 2006). De Bruyne and co-workers (2008) differentiated *Pediococcus* spp. by comparing sequence analyses of the genes encoding the alpha subunits of phenylalanyl-tRNA synthase (*pheS*), RNA polymerase (*rpoA*), and ATP synthase (*atpA*) and described the species *P. argentinicus*. The authors concluded that multilocus sequence analysis based on *pheS*, *rpoA*, and *atpA* genes is one of the most suitable methods to identify *Pediococcus* spp. Of the three subunits, *pheS* sequences provided the best separation. Doi and co-workers (2009) used 16S rRNA gene sequence analyses, DNA–DNA hybridizations, RAPD-PCR, and enterobacterial repetitive intergenic consensus (ERIC)-PCR fingerprinting to distinguish between *P. acidilactici, P. pentosaceus* and *P. lolii*. The genomic diversity within the genus *Pediococcus* has been investigated by Simpson et al. (2002), using RAPD-PCR and PFGE. These techniques proved useful for the rapid classification of *Pediococcus* strains isolated from sources such as food, feed, silage, beer and human clinical samples. Franz and co-workers (2006) used a combination of 16S rRNA gene sequence analysis, fluorescent amplified-fragment length polymorphism (FAFLP) of genomic DNA, and repetitive extragenic palindromic sequence-based PCR (rep-PCR) to differentiate *Pediococcus* spp. *Pediococcus ethanolidurans*, new member of the genus, was described based on DNA–DNA hybridization and *ftsZ* gene sequence similarity (Liu et al. 2006). *Pediococcus cellicola* was described based on 16S rRNA gene sequence analyses (Zhang et al. 2005).

Total genomic DNA–DNA hybridization remains the most reliable method to differentiate among *Pediococcus* spp. (Back and Stackebrandt 1978; Dellaglio et al. 1981; Franz et al. 2006) and is considered the gold standard for genotypic delineation (Mehlen et al. 2004).

The mol% G+C content of the DNA of *Pediococcus* spp. ranges from 35 to 44 ($T_m$) or 36–42, determined by HPLC (Holzapfel et al. 2009), while that of *Aerococcus* spp. ranges from 37 % to 40 % (Schultes and Evans 1971; Wiik et al. 1986) and that of *Tetragenococcus* spp. from 34 % to 38 % (Lee et al. 2005). Differentiation of the genus *Pediococcus* from *Tetragenococcus* or *Aerococcus* based on G+C content alone is thus not possible. Real-time PCR is increasingly being used for quantitative detection and identification of *Pediococcus* spp. among other lactic acid bacteria (Delaherche et al. 2004; Lan et al. 2004; Stevenson et al. 2005).

Banding patterns of PCR-amplified ISR-*Dde*I differentiated *Pediococcus, Lactobacillus, Lactococcus,* and *Enterococcus* spp. (Belgacem et al. 2009). This technique, combined with RFLP obtained from α-*Taq*I digests, proofed valuable in differentiating the species (Belgacem et al. 2009). Other techniques include ribotyping using the automated RiboPrinter® (Satokari et al. 2000; Barney et al. 2001); ARDRA genotyping with *Bfa*I, *Mse*I, and *Alu*I restriction enzymes (Rodas et al. 2003); and rep-PCR (Kostinek et al. 2008).

The genome of *P. pentosaceus* ATCC 25745 (1.83 Mbp) has been sequenced by Makarova et al. (2006). The genome (sequence deposited as NCBI accession number NC_008525) encodes 1757 proteins and has 19 pseudogenes, five rRNA operons, and 55 tRNAs. The genome and plasmids of the type strain of *P. claussenii* (ATCC BAA-344), a common beer contaminant, has been sequenced by Pittet et al. (2012). The sequences were deposited in GenBank under accession numbers CP003137, CP003138, CP003139, CP003140, CP003141, CP003142, CP003143, CP003144, and CP003145 for the chromosome and plasmids 1–8, respectively.

Limited information is available on bacteriophages of pediococci. Temperate bacteriophages of *P. acidilactici*, studied by Caldwell et al. (1999), could be induced with mitomycin C and were classified into two genetic groups.

A number of bacteriocins, pediocins, have been described for *Pediococcus* spp. Most of the pediocins belong to class IIa bacteriocins, i.e., small (less than 10 kDa), non-lanthionine-containing, and *Listeria*-active peptides with a YGNGV (tyrosine-glycine-asparagine-glycine-valine) consensus sequence in the N terminus (Nes et al. 1996; Bauer et al. 2005). The pediocins share many sequence similarities with bacteriocins produced by *Lactobacillus* species, e.g., curvacin A, sakacin P, bavaricin A, and bavaricin MN; *Leuconostoc* species, e.g., leukocin A and mesentericin Y105; *Streptococcus* species, e.g., mundticin; *Enterococcus* species, e.g., enterocin A; and *Carnobacterium* species, e.g., carnobacteriocin B2 and piscicolin 126 (Holzapfel et al. 2006).

## Phenotypic Analyses

Members of the genus are Gram-positive cocci (0.6–1.0 μm in diameter), mostly spherical, but in rare cases ovoid shaped, and are arranged in pairs or tetrads (Axelsson 1998). *Pediococcus dextrinicus* produce non-perpendicular cells without separation of the tetrads, resulting in the formation of clusters (Haakensen et al. 2009). Single cells may be found during early or midexponential growth. Endospores are not produced. Growth is facultatively anaerobic and all species are oxidase and catalase negative. Cytochromes are not present. However, some strains of *P. pentosaceus*, especially strains isolated from goat milk, Feta cheese, and Kaseri cheese, produce catalase or pseudocatalase (Simpson and Taguchi 1995). Pseudocatalase is produced when cells are grown in media with a low carbohydrate concentration (Weiss 1992). The growth temperature range is normally 25–35 °C. However, *P. acidilactici, P. ethanolidurans, P. Siamensis, P. stilesii*, and some strains of *P. pentosaceus* grow at 45 °C, while all other species do not grow at this temperature. Nitrate is not reduced. The interpeptide bridge in peptidoglycan is usually the Lys-D-Asp type (Holzapfel et al. 2006). However, *P. lolii* contains diaminopimelic acid (DAP) in their cell walls. Indole is not formed from tryptophan (Nakagawa and Kitahara 1959). Hippurate is not hydrolysed (Tanasupawat and Daengsubha 1983).

Arginine is usually not hydrolysed, except for a few strains of *P. acidilactici* and *P. pentosaceus* (Simpson and Taguchi 1995). Sugars are fermented using the glycolytic pathway. All species produce DL-lactic acid from glucose, except *P. claussenii*. No $CO_2$ is produced from the fermentation of glucose or gluconate. *Pediococcus dextrinicus* is an exception, but the species is closer related to *Lactobacillus* (see Taxonomy, Historical and Current). Optimum growth occurs at pH 6.0, but not at pH 9 (except for *P. stilesii*). All species grow at pH 4.5 (no data available for growth of *P. siamensis* at low pH). Most species grow at pH 7.5, except *P. damnosus* (Simpson and Taguchi 1995). *P. siamensis* and some strains of *P. acidilactici* and *P. pentosaceus* grow at pH 8.5, while *P. stilesii* grows at pH 9.6. All species grow in the presence of 4–5 % (w/v) NaCl, with *P. acidilactici* and *P. pentosaceus* being able to withstand 10 % NaCl (Simpson and Taguchi 1995). Sensitivity to NaCl varies with composition of growth medium and incubation conditions (Nakagawa and Kitahara 1959; Coster and White 1964). The $\beta$- and $\alpha$-D-glucopyranosidase activities of pediococci are affected by pH, temperature, ethanol, and sugars. Fructose, mannose, and cellobiose are fermented by all species, except *P. lolii* which ferments mannose and cellobiose, but not mannose. Maltose is fermented by all species, except *P. acidilactici, P. claussenii, P. lolii, P. parvulus*, and *P. siamensis* (Simpson and Taguchi 1995; Dobson et al. 2002). Rhamnose, melibiose, melezitose, raffinose, inulin, and $\alpha$-methyl glucoside-D are usually not fermented. Sucrose is not fermented by *P. claussenii, P. lolii, P. parvulus P. inopinatus, P. siamensis* and *P. stilesii* (Simpson and Taguchi 1995; Dobson et al. 2002). Starch and sorbitol are not fermented. Mannitol is fermented by only *P. acidilactici, P. argentinicus*, and *P. claussenii* (Simpson and Taguchi 1995; Dobson et al. 2002; De Bruyne et al. 2008). Variable results were recorded for the fermentation of trehalose, lactose, maltotriose, dextrin, glycerol, salicin, and amygdalin (Simpson and Taguchi 1995; Zhang et al. 2005; Liu et al. 2006; De Bruyne et al. 2008; Doi et al. 2009). The genes encoding the fermentation of raffinose, melibiose, and sucrose are often plasmid encoded (Gonzalez and Kunka 1986; Hoover et al. 1988). Glycerol is fermented by *P. pentosaceus* to lactate, acetate, acetoin, and $CO_2$ in the presence of oxygen (Dobrogosz and Stone 1962). Arabinose, ribose, and xylose are not fermented by *P. damnosus, P. ethanolidurans, P. inopinatus, P. parvulus*, and *P. siamensis*. Strains of *P. pentosaceus* and *P. damnosus* isolated from wine (Manca de Nandra and Strasser de Saad 1995; Lonvaud-Funel and Joyeux 1998) and *P. pentosaceus* isolated from Thai fermented pork sausage (Smitinont et al. 1999) produce exopolysaccharides. Certain strains of *P. cellicola, P. damnosus, P. ethanolidurans, P. inopinatus*, and *P. parvulus* are resistant to ethanol (Davis et al. 1988).

Besides sugar fermentation, genes encoding antibiotic resistance may also be plasmid encoded. For example, an erythromycin resistance is encoded on a 40-MDa plasmid in *P. acidilactici* (Torriani et al. 1987). According to Gonzalez and Kunka (1983), plasmids are frequently transferred between *Pediococcus* and *Enterococcus, Streptococcus*, and *Lactococcus* species.

## Isolation, Enrichment, and Maintenance Procedures

*Pediococcus* spp. grow on MRS (De Man, Rogosa and Sharpe) medium, as most other lactic acid bacteria. *Pediococcus acidilactici* and *P. pentosaceus* are selected by plating onto MDS medium, which is MRS supplemented with cysteine hydrochloride, novobiocin, vancomycin, and nystatin (Simpson et al. 2006). Other semi-selective growth media that have been used to isolate pediococci are SL Medium (Rogosa et al. 1951) and Acetate Agar (Whittenbury 1965). However, Acetate Agar also supports the growth of *Leuconostoc* and *Pediococcus* spp. (Whittenbury 1965). Growth of *Lactobacillus plantarum* and *Lactobacillus casei* is inhibited by the addition of ampicillin.

Growth of *P. damnosus* and *P. claussenii* is stimulated in MRS adjusted to pH 5.5. The addition of beer to MRS (1:1, v/v) also serves as a selective growth medium, especially when plates are incubated at 22 °C in the presence of 90 % $N_2$ and 10 % $CO_2$ (Back 1978b). The addition of cycloheximide prevents the growth of yeasts and molds (Holzapfel et al. 2006). The addition of polymyxin B, acetic acid (at low pH), and thallous acetate prevents the growth of Gram-negative bacteria (Schillinger and Holzapfel 2003). Apart from beer, malt extract, liver concentrate, maltose, L-malic acid, cytidine, thymidine, actidione, and sodium azide have also been used in selective growth media (Taguchi et al. 1990).

Cultures may be stored at 4 °C for 2–3 months as stabs in appropriate growth media that has been supplemented with 0.3 % $CaCl_2$. For long-term storage, active-growing cultures may be mixed with an equal volume of 80 % (v/v) sterile glycerol and stored in cryotubes at −20 °C or −80 °C. Cultures may also be freeze-dried in the presence of skim milk or stored in liquid nitrogen. Vancomycin is sometimes added (Björkroth and Holzapfel 2003).

## Ecology

Pediococci are usually isolated from plant material, fermented beverages, meat, and dairy products. *P. acidilactici, P. parvulus, P. inopinatus, P. stilesii, P. pentosaceus* and *P. lolii* are typically isolated from plants, fruits, and vegetables (Mundt et al. 1969; Back 1978b; Dellaglio et al. 1981; Wilderdyke et al. 2004; Doi et al. 2009) and are thus present in many fermented products. In some of the fermented products, especially cereals, pediococci are closely associated with filamentous fungi and yeasts. *Pediococcus damnosus, P. dextrinicus, P. inopinatus, P. cellicola, P. claussenii*, and *P. ethanolidurans* are usually associated with beer and other alcoholic beverages (Back 1978b; Simpson and Taguchi 1995; Zhang et al. 2005; Liu et al. 2006; Pittet et al. 2012). *Pediococcus stilesii* has been isolated from maize grains (Franz et al. 2006) and *P. siamensis* from fermented tea leaves (*miang*) produced in the northern part of Thailand (Tanasupawat et al. 2007). Pediococci also play a role in the ripening of certain cheese (Olson 1990) and meat fermentations. A few pediococci have been associated with humans

(Sims 1986), the gastrointestinal tract of humans and animals (Ruoff et al. 1988; Walter et al. 2001; Heilig et al. 2002), birds (Juven et al. 1991; Kurzak et al. 1998), and freshwater prawns (Cai et al. 1999a). *Pediococcus pentosaceus* has been isolated from the tonsils and nasal cavity of piglets (Baele et al. 2001; Martel et al. 2003). Some pediococci produce polysaccharides in wine. In one such study the ropy texture produced by *P. parvulus* has been characterized as β-glucan which the cell produces to protect itself against lysozyme (Coulon et al. 2012). Growth of the strain in a white wine-based media was controlled by treating the cells with a combination of lysozyme and β-glucanase (Coulon et al. 2012).

## Pathogenicity, Clinical Relevance

A few pediococci have been isolated from infectious tissue in humans. However, they are considered opportunistic pathogens (Golledge et al. 1990; Mastro et al. 1990; Riebel and Washington 1990; Green et al. 1991; Sarma and Mohanty 1998; Von Witzingerode et al. 2000; Barros et al. 2001; Barton et al. 2001). In most cases infections are only observed in immune compromised individuals (Mastro et al. 1990; Facklam and Elliot 1995; Sarma and Mohanty 1998; Barton et al. 2001). Vancomycin resistance is rare, but has been reported for some strains (Swenson et al. 1990; Tankovic et al. 1993; Ammor et al. 2007; Haakensen et al. 2009). A few cases of resistance to teicoplanin have also been reported (Swenson et al. 1990; Tankovic et al. 1993; Ammor et al. 2007; Haakensen et al. 2009). Most pediococci are resistant to metronidazole, cephalosporin, and cefoxitin (Ammor et al. 2007). Some pediococci are also resistant to quinolone antibiotics and tetracycline (Tankovic et al. 1993; Sarma and Mohanty 1998), but they are generally sensitive to antibiotics such as penicillin, ampicillin, gentamicin, and netilmicin (Mastro et al. 1990; Swenson et al. 1990; Tankovic et al. 1993; Barton et al. 2001; Ammor et al. 2007). Tetracycline resistance in *P. parvulus* is encoded by the *tet* (L) gene (Rojo-Bezares et al. 2006). The *erm*(B) gene, encoding resistance toward erythromycin, has also been reported present in *P. acidilactici* (Rojo-Bezares et al. 2006). The aminoglycoside resistance genes *aac*(6')*Ie-aph*(2")*Ia*, *aac*(6')-*aph*(2") and *ant*(6) have been detected in strains of *P. acidilactici* and *P. parvulus* (Tenorio et al. 2001; Rojo-Bezares et al. 2006; Ammor et al. 2007).

## Application

Since pediococci are usually isolated from plant material, they are considered important in the preservation of plant-derived products such as silage (Cai et al. 1999b; Zhang et al. 2000), sauerkraut (Back 1978a), fermented beans, cucumbers (Stamer 1983), olives, and cereals (Lin et al. 1992; Wilderdyke et al. 2004). *Pediococcus acidilactici*, *P. pentosaceus*, *P. parvulus*, and *P. inopinatus* play an important role as starter cultures in the production of silage (Cai et al. 1999b; Zhang et al. 2000). *Pediococcus pentosaceus* is closely associated with the fermentation of koko, togwa (fermented sorghum, maize, and millet), khamir (bread produced from sorghum), and Hussuwa (fermented sorghum) (Gassem 1999; Mugula et al. 2003a, b; Lei and Jakobsen 2004; Yousif et al. 2010). Pediococci have also been isolated from marcha (murcha), produced in the Himalayan regions of India, Nepal, Tibet, and Bhutan; ragi, produced in Indonesia; nuruk, a product from Korea; bubo in the Philippines; chiu-yueh (chiu nang or lao chao) in China; loog-pang in Thailand (Tamang 1998); and banh men in Vietnam (Lee and Fujio 1999).

Pediococci also occur in fresh and cured meat, raw and fermented sausages (Gevers et al. 2000; Parente et al. 2001), and fresh and marinated fish (Tanasupawat and Daengsubha 1983; Paludan-Muller et al. 2002). Starter cultures of *Pediococcus* spp. have been used in the United States since the late 1950s for the production of fermented sausage (Deibel et al. 1961), dry sausages (Raccach 1987) and Spanish dry-cured ham (Molina et al. 1989). Excess production of diacetyl from citrate may lead to the formation of off-flavors in meat (Holzapfel 1998). Some strains of *P. pentosaceus* produce lipases and have strong leucine and valine arylamidase activities (Nieto et al. 1989; Molina and Toldra 1992).

*Pediococcus pentosaceus* may play a role in the fermentation and maturation of cheese (Callon et al. 2004), especially in blends where a strong aroma is required (Ogier et al. 2002). Strains of *P. pentosaceus* with high β-casein hydrolase, proteinase, aminopeptidase, and dipeptidyl aminopeptidase activities have been isolated from Comté (Bouton et al. 1998), Salers cheese (Callon et al. 2004), and Parmigiano Reggiano (Coppola et al. 1997). Caldwell et al. (1998) genetically modified *Pediococcus* spp. to be used as starter culture in the fermentation of Mozzarella, aimed at finding an alternative for *Streptococcus thermophilus* and solving the problem of bacteriophage contamination.

*Pediococcus damnosus* is considered a major spoilage organism in beer, as it is associated with cloudiness, ropiness, and an off-taste caused by increased diacetyl production (Weiss 1992; Donhauser 1993; Sakamoto and Konings 2003). *Pediococcus inopinatus* has also been isolated occasionally from beer (Back 1978a, 1994, 2000). *Pediococcus acidilactici*, *P. pentosaceus*, and *P. parvulus* are mainly found on malt and in wort at temperatures <50 °C (Barney et al. 2001). These strains and *P. claussenii* are resistant to low alcohol concentrations and hops and grow at higher pH values than most LAB (Back 1978a, 1994; Dobson et al. 2002). *Pediococcus claussenii* also produces an exopolysaccharide in beer (Dobson et al. 2002). In rare cases, the production of tyramine by pediococci has been reported in beer (Izquierdo-Pulido et al. 1997), but at lower levels as observed for lactobacilli (Kalač et al. 2002).

*Pediococcus damnosus* is also considered a spoilage organism in wine, especially at higher pH (Weiller and Radler 1970; Back 1978a, 1994, 2000; Beneduce et al. 2004). Pediococci have also been associated with higher levels of biogenic amines in wine (Weiller and Radler 1976). Production of tyramine is commonly associated with the fermentation of traditional fermented sausages and related to *P. cerevisiae* (and probably *P. pentosaceus*; Rice and Koehler 1976). On the other hand, the use of starter cultures such as selected strains of *P. pentosaceus* resulted in lower amounts of tyramine, putrescine, and cadaverine (Hernandez-Jover et al. 1997).

Two lytic enzymes, both classified as peptidoglycan hydrolases (PGHs), have been described for *P. acidilactici* ATCC 8042 (García-Cano et al. 2011). The PGHs (110 and 99 kDa in size, according to SDS–PAGE) inhibited the growth of *Micrococcus lysodeikticus*, *Staphylococcus aureus*, *Bacillus cereus*, *Listeria monocytogenes*, and *Salmonella typhimurium*. According to mass spectrometry analysis (LC/ESI-MS/MS), the 110-kDa protein is novel. The 99-kDa protein is a *N*-acetylmuramidase that has catalytic sites with *N*-acetylmuramoyl-L-alanine amidase and *N*-acetylglucosaminidase activities. The genes encoding these proteins are located on the genome of strain ATCC 8042. Since the PGH's has a broad growth inhibition spectrum, they may be used to control the growth of pathogenic bacteria in food.

## *Paralactobacillus*

### Taxonomy, Historical and Current

The genus *Paralactobacillus* was proposed in 2000 for the novel species, *Paralactobacillus selangorensis*, isolated from Chili bo, a Malaysian fermented food ingredient (Leisner et al. 2000). The genus was composed of the single species, and isolation of the species has not been reported from any other sources. When the genus was proposed, *P. selangorensis* was placed outside of the *Lactobacillus* phylogenetic group based on 16S rRNA gene sequences (Leisner et al. 2000). In addition, *P. selangorensis*, which is a homofermentative species, was differentiated from homofermentative lactobacilli based on a combination of a few phenotypic characteristics (Leisner et al. 2000). The species was not included in the genus *Lactobacillus*, but proposed as a novel species in the novel genus *Paralactobacillus*. However, this conclusion was questionable, as the authors used only a few species in the phylogenetic analyses and comparison of phenotypic characteristics. A number of articles suggested to place *P. selangorensis* in the *Lactobacillus* phylogenetic group (Ennahar et al. 2003; Hammes and Hertel 2006). Haakensen et al. (2011) studied phylogenetic position of the species by using multilocus sequencing analysis with three housekeeping genes and indicated that *P. selangorensis* is phylogenetically located in the *Lactobacillus* cluster. Moreover, phenotypic characteristics of the species were consistent with those of *Lactobacillus* spp. Thereafter, *P. selangorensis* was reclassified as the species in the genus *Lactobacillus*, *Lactobacillus selangorensis* (Haakensen et al. 2011). The phylogenetic relationship of *P. selangorensis* with other taxa, based on 16S rRNA gene sequences, is shown in the chapter on the genus *Pediococcus*.

## *Sharpea*

The genus *Sharpea* was proposed by Morita et al. (2008). Thus far, only one species has been described, *Sharpea azabuensis*. All four strains, isolated from horse feces, are genetically closely related and group in one cluster based on 16S rRNA gene sequence analysis, phenotypic characteristics, and DNA–DNA relatedness. The phenotypic characteristics of *S. azabuensis* are typical to those described for many species within the family of lactic acid bacteria, i.e., Gram positive, anaerobic, catalase negative, asporogenous, and non-motile. The G+C content of the four strains ranged from 36 to 38 mol%.

Phylogenetically, the species is most closely related to *Lactobacillus catenaformis*. However, 16S rRNA gene sequences of the four strains shared only 89.9 % similarity with *L. catenaformis*. Unlike *L. catenaformis*, *Lactobacillus vitulinus*, and *Catenibacterium mitsuokai*, *S. azabuensis* is heterofermentative and produces $CO_2$ from glucose. The cell-wall peptidoglycan type of *S. azabuensis* is the same as that of *C. mitsuokai* [A1c (L-Ala–D-Glu–m-Dpm)], but different from that of *L. catenaformis* and *L. vitulinus* (L-Lys–L-Ala3).

Based on phenotypic differences and 16S rRNA gene sequence divergence of more than 10 % from *L. catenaformis*, *S. azabuensis* is a novel species within the *Clostridium* subphylum cluster XVII. The type strain is DSM 18934$^{T}$.

The cells are 2–10 μm × 0.7–1.0 μm and arranged in short chains. Colonies on BL agar are 1.0–2.5 mm in diameter, smooth, circular, or slightly irregular and may appear brown, transparent, or butyrous. The optimum growth temperature is 37 °C. No growth was recorded in the presence of 4.5 % (w/v) NaCl at 15 °C. However, the cells grow in the presence of 3 % NaCl at 45 °C. Galactose, fructose, mannose, cellobiose, lactose, melibiose, and starch are fermented. D(−)-lactic acid is produced from the fermentation of glucose. Some strains ferment salicin, trehalose, and raffinose, but rhamnose, mannitol, sorbitol, and melezitose are not fermented.

## References


Ammor MS, Florez AB, Mayo B (2007) Antibiotic resistance in non-enterococcal lactic acid bacteria and bifidobacteria. Food Microbiol 24:559–570

Axelsson L (1998) Lactic acid bacteria: classification and physiology. In: Salminen S, von Wright A (eds) Lactic acid bacteria. Microbiology and functional aspects. Marcel Dekker, New York, pp 1–72

Back W (1978a) Zur Taxonomie der Gattung *Pediococcus*. Phänotypische und genotypische Abgrenzung der bisher bekannten Arten sowie Beschreibung einer neuen bierschädlichen Art: *Pediococcus inopinatus*. Brauwissenschaft 31:237–250, 312–320, 336–343

Back W (1978b) Elevation of *Pediococcus cerevisiae* subsp. *dextrinicus* Coster and White to species status *Pediococcus dextrinicus* (Coster and White) comb. nov. Int J Syst Bacteriol 28:523–527

Back W (1988) Validation list number 25. Int J Syst Bacteriol 38:220–222

Back W (1994) Farbatlas und Handbuch der Getränkebiologie Teil l. Verlag Hans Carl, Nürnberg

Back W (2000) Farbatlas und Handbuch der Getränkebiologie Teil 2. Verlag Hans Carl, Nürnberg

Back W, Stackebrandt E (1978) DNS/DNS-homologiestudien innerhalb der Gattung *Pediococcus*. Arch Microbiol 118:79–85

Baele M, Chiers K, Devriese LA, Smith HE, Wisselink HJ, Vaneechoutte M, Haesebrouck F (2001) The Gram-positive tonsillar and nasal flora of piglets before and after weaning. J Appl Microbiol 91:997–1003

Balcke J (1884) Über häufig vorkommende Fehler in der Bierbereitung. Wochenschrift für Brauerei 1:181–184

Barney M, Volgyi A, Navarro A, Ryder D (2001) Riboprinting and 16S rRNA sequencing for identification of brewery *Pediococcus* isolates. Appl Environ Microbiol 67:553–560

Barros RR, Carvalho MDS, Peralta JM, Facklam RR, Teixeira LM (2001) Phenotypic and genotypic characterization of *Pediococcus* strains isolated from human clinical sources. J Clin Microbiol 39:1241–1246

Barton LL, Rider ED, Coen RW (2001) Bacteremic infection with *Pediococcus*: vancomycin-resistant opportunist. Pediatrics 107:775–776

Bauer R, Chikindas ML, Dicks LMT (2005) Purification, partial amino acid sequence and mode of action of pediocin PD-1, a bacteriocin produced by *Pediococcus damnosus* NCFB 1832. Int J Food Microbiol 101:17–27

Belgacem B, Dousset X, Prevost H, Mania M (2009) Polyphasic taxonomic studies of lactic acid bacteria associated with Tunisian fermented meat based on the heterogeneity of the 16S-23S rRNA intergenic spacer region. Arch Microbiol 191:711–720

Beneduce L, Spano G, Vernile A, Tarantino D, Massa S (2004) Molecular characterization of lactic acid populations associated with wine spoilage. J Basic Microbiol 44:10–16

Björkroth J, Holzapfel WH (2003) Genera *Leuconostoc*, *Oenococcus* and *Weissella*. In: Dworkin M (ed) The prokaryotes, 3rd edn. Springer, New York (electronic version)

Bouton Y, Guyot P, Grappin R (1998) Preliminary characterization of microflora of Comté cheese. J Appl Microbiol 85:123–131

Cai Y, Kumai S, Ogawa M, Benno Y, Nakase T (1999b) Characterization and identification of *Pediococcus* species isolated from forage crops and their application for silage preparation. Appl Environ Microbiol 65:2901–2906

Cai Y, Suyanandana P, Benno Y (1999a) Classification and characterization of lactic acid bacteria isolated from the intestines of common carp and freshwater prawns. J Gen Appl Microbiol 45:177–184

Caldwell SL, Hutkins RW, McMahon DJ, Oberg CJ, Broadbent JR (1998) Lactose and galactose uptake by genetically engineered *Pediococcus* species. Appl Microbiol Biotechnol 49:315–320

Caldwell SL, McMahon DJ, Oberg CJ, Broadbent JR (1999) Induction and characterization of *Pediococcus acidilactici* temperate bacteriophage. Syst Appl Microbiol 22:514–519

Callon C, Millet C, Montel MC (2004) Diversity of lactic acid bacteria isolated from AOC Salers cheese. J Dairy Res 71:231–244

Claussen NH (1903) Etudes sur les bacteries dites sarcines et sur les maladies quelles provoquent dans le biere. C R Trav Lab Carlsberg 6:64–83

Collins MD, Williams AM, Wallbanks S (1990) The phylogeny of *Aerococcus* and *Pediococcus* as determined by 16S rRNA sequence analysis: description of *Tetragenococcus* gen. nov. FEMS Microbiol Lett 70:255–262

Collins MD, Rodrigues U, Ash C, Aguirre M, Farrow JAE, Martinez-Murcia A, Phillips BA, Williams AM, Wallbanks S (1991) Phylogenetic analysis of the genus *Lactobacillus* and related lactic acid bacteria as determined by reverse transcriptase sequencing of 16S rRNA. FEMS Microbiol Lett 77:5–12

Coppola R, Nanni M, Iorizzo M, Sorrentino A, Sorrentino E, Grazia L (1997) Survey of lactic acid bacteria isolated during the advanced stages of the ripening of Parmigiano Reggiano cheese. J Dairy Res 64:305–310

Coster E, White HR (1964) Further studies of the genus *Pediococcus*. J Gen Microbiol 26:185–197

Coulon J, Houlès A, Dimopoulou M, Maupeu J, Dols-Lafargue M (2012) Lysozyme resistance of the ropy strain *Pediococcus parvulus* IOEB 8801 is correlated with beta-glucan accumulation around the cell. Int J Food Microbiol 159:25–29

Davis CR, Wibowo D, Fleet GH, Lee TH (1988) Properties of wine lactic acid bacteria: their potential enological significance. Am J Enol Vitic 39:137–142

De Bruyne K, Franz CMAP, Vancanneyt M, Schillinger U, Mozzi F, de Valdez GF, De Vuyst L, Vandamme P (2008) *Pediococcus argentinicus* sp. nov. from Argentinean fermented wheat flour and identification of *Pediococcus* species by *pheS*, *rpoA* and *atpA* sequence analysis. Int J Syst Evol Microbiol 58:2909–2916

Deibel RH, Wilson GD, Niven CF Jr (1961) Microbiology of meat curing. IV. A lyophilized *Pediococcus cerevisiae* starter culture for fermented sausage. Appl Microbiol 9:239–243

Delaherche A, Claisse O, Lonvaud-Funel A (2004) Detection and quantification of *Brettanomyces bruxellensis* and 'ropy' *Pediococcus damnosus* strains in wine by real-time polymerase chain reaction. J Appl Microbiol 97:910–915

Dellaglio F, Trovatelli LG, Sarra PG (1981) DNA-DNA homology among representative strains of the genus *Pediococcus*. Zbl Bakt Mikrobiol Hyg I Abt Orig C 2:140–150

Dobrogosz WJ, Stone RW (1962) Oxygen metabolism in *Pediococcus pentosaceus*. I. Role of oxygen and catalase. J Bacteriol 84:716–723

Dobson CM, Deneer H, Lee S, Hemmingsen S, Glaze S, Ziola B (2002) Phylogenetic analysis of the genus *Pediococcus*, including *Pediococcus claussenii* sp. nov., a novel lactic acid bacterium isolated from beer. Int J Syst Evol Microbiol 52:2003–2010

Doi K, Nishizaki Y, Fujino Y, Ohshima T, Ohmomo S, Ogata S (2009) *Pediococcus lolii* sp. nov., isolated from ryegrass silage. Int J Syst Evol Microbiol 59:1007–1010

Donhauser S (1993) Mikrobiologie des Bieres. In: Dittrich HH (ed) Mikrobiologie der Lebensmittel. Getränke. Behr's Verlag, Hamburg, pp 109–182

Ennahar S, Cai Y, Fujita Y (2003) Phylogenetic diversity of lactic acid bacteria associated with paddy rice silage as determined by 16S ribosomal DNA analysis. Appl Environ Microbiol 69:444–451

Euzéby JP (1998) Necessary corrections according to the judicial opinions 16, 48 and 52. Int J Syst Evol Microbiol 48:613

Facklam RR, Elliot JA (1995) Identification, classification, and clinical relevance of catalase-negative, Gram-positive cocci, excluding the streptococci and enterococci. Clin Microbiol Rev 8:479–495

Felis GE, Dellaglio F (2007) Taxonomy of lactobacilli and bifidobacteria. Curr Issues Intest Microbiol 8:44–61

Felis GE, Torriani S, Dellaglio F (2005) Reclassification of *Pediococcus urinaeequi* (*ex* Mees 1934) Garvie 1988 as *Aerococcus urinaeequi* comb. nov. Int J Syst Evol Microbiol 55:1325–1327

Franz CMAP, Vancanneyt M, Vandemeulebroecke K, De Wachter M, Cleenwerck I, Hoste B, Schillinger U, Holzapfel WH, Swings J (2006) *Pediococcus stilesii* sp. nov., isolated from maize grains. Int J Syst Evol Microbiol 56:329–333

Fujii T, Nakashima K, Hayashi N (2005) Random amplified polymorphic DNA-PCR based cloning of markers to identify the beer-spoilage strains of *Lactobacillus brevis*, *Pediococcus damnosus*, *Lactobacillus collinoides* and *Lactobacillus coryniformis*. J Appl Microbiol 98:1209–1220

García-Cano I, Velasco-Pérez L, Rodríguez-Sanoja R, Sánchez S, Mendoza-Hernández G, Llorente-Bousquets A, Farrés A (2011) Detection, cellular localization and antibacterial activity of two lytic enzymes of *Pediococcus acidilactici* ATCC 8042. J Appl Microbiol 111:607–615

Garvie EI (1974) Nomenclatural problems of the pediococci. Int J Syst Microbiol 24:301–306

Garvie EI (1988) Validation list number 25. Int J Syst Bacteriol 38:220–222

Gassem MA (1999) Study of the micro-organisms associated with the fermented bread (khamir) produced from sorghum in Gizan region Saudi Arabia. J Appl Microbiol 86:221–225

Gevers D, Huys G, Devlieghere F, Uyttendaele M, Debevere J, Swings J (2000) Isolation and identification of tetracycline resistant lactic acid bacteria from pre-packed sliced meat products. Syst Appl Microbiol 23:279–284

Golledge CL, Stingemore N, Aravena M, Joske D (1990) Septicemia caused by vancomycin-resistant *Pediococcus acidilactici*. J Clin Microbiol 28: 1678–1679

Gonzalez C, Kunka BS (1983) Plasmid transfer in *Pediococcus* spp.: Intergeneric and intrageneric transfer of pIP501. Appl Environ Microbiol 46:81–89

Gonzalez C, Kunka BS (1986) Evidence for plasmid linkage of raffinose utilization and associated galactose and sucrose hydrolase activity in *Pediococcus pentosaceus*. Appl Environ Microbiol 51:105–109

Green M, Barbadora K, Michaels M (1991) Recovery of vancomycin-resistant Gram-positive cocci from pediatric liver transplant recipients. J Clin Microbiol 29:2503–2506

Günther HL, White HR (1961) The cultural and physiological characters of the pediococci. J Gen Microbiol 26:185–197

Haakensen M, Dobson CM, Hill JE, Ziola B (2009) Reclassification of *Pediococcus dextrinicus* (Coster and White 1964) Back 1978 (Approved list 1980) as *Lactobacillus dextrinicus* comb. nov., and emended description of the genus *Lactobacillus*. Int J Syst Evol Microbiol 59:615–621

Haakensen M, Pittet V, Ziola B (2011) Reclassification of *Paralactobacillus selangorensis* Leisner et al. 2000 as *Lactobacillus selangorensis* comb. nov. Int J Syst Evol Microbiol 61:2979–2983

Hammes WP, Hertel C (2006) The genera *Lactobacillus* and *Carnobacterium*. In: Dworkin M, Falkow S, Rosenberg E, Schleifer K-H, Stackebrandt E (eds) The prokaryotes, vol 4, 3rd edn. Springer, New York, pp 320–403

Heilig HG, Zoetendal EG, Vaughan EE, Marteau P, Akkermans AD, de Vos WM (2002) Molecular diversity of *Lactobacillus* spp. and other lactic acid bacteria in the human intestine as determined by specific amplification of 16S rDNA. Appl Environ Microbiol 68:114–123

Hernandez-Jover T, Izquierdo-Pulido M, Veciana-Nogues MT, Marine-Font A, Vidal-Carou MC (1997) Effect of starter cultures on biogenic amine formation during fermented sausage production. J Food Prot 60:825–830

Holzapfel W (1998) The Gram-positive bacteria associated with meat and meat products. In: Davies A, Board R (eds) The microbiology of meat and poultry. Blackie Academic and Professional, London, pp 35–74

Holzapfel WH, Franz CMAP, Ludwig W, Back W, Dicks LMT (2006) Genera *Pediococcus* and *Tetragenococcus*. In: Dworkin M, Falkow S, Rosenberg E, Schleifer K-H, Stackebrandt E (eds) The prokaryotes, 3rd edn, An evolving electronic resource for the microbiological community. Springer, New York. http://www.prokaryotes.com

Holzapfel WH, Franz CMAP, Ludwig W, Dicks LMT (2009) Family I, Genus III. *Pediococcus* Claussen 1903, 68$^{AL}$. In: De Vos P, Garrity G, Jones D, Krieg NR, Ludwig W, Rainy FA, Schleifer K-H, Whitman WB (eds) Bergey's manual of systematic bacteriology, 2nd edn, vol 3 (The *Firmicutes*). Springer, New York, pp 513–532

Hoover DG, Walsh PM, Kolactis KM, Daly MM (1988) A bacteriocin produced by *Pediococcus* species associated with a 5 · 5 MDa plasmid. J Food Prot 51:29–31

Izquierdo-Pulido M, Carceller-Rosa JM, Marine-Font A, Vidal-Carou MC (1997) Tyramine formation by *Pediococcus* spp. during beer fermentation. J Food Prot 60:831–836

Jager K, Harlander S (1992) Characterization of a bacteriocin from *Pediococcus acidilactici* PC and comparison of bacteriocin-producing strains using molecular typing procedures. Appl Microbiol Biotechnol 37:631–637

Juven BJ, Meinersmann RJ, Stern NJ (1991) Antagonistic effects of lactobacilli and pediococci to control intestinal colonization by human enteropathogens in live poultry. J Appl Bacteriol 70:95–103

Kalač P, Šavel J, Křížek M, Pelikánová T, Prokopová M (2002) Biogenic amine formation in bottled beer. Food Chem 79:431–434

Kostinek M, Ban-Koffi L, Ottah-Atikpo M, Teniola D, Schillinger U, Holzapfel WH, Franz CMAP (2008) Diversity of predominant lactic acid bacteria associated with cocoa fermentation in Nigeria. Curr Microbiol 56:306–314

Kurzak P, Ehrmann MA, Vogel RF (1998) Diversity of lactic acid bacteria associated with ducks. Syst Appl Microbiol 21:588–592

Lan Y, Sun S, Tamminga S, Williams BA, Verstegen MW, Erdi G (2004) Real-time PCR detection of lactic acid bacteria in cecal contents of *Eimeria tenella*-Infected broilers fed soybean oligosaccharides and soluble soybean polysaccharides. Poult Sci 83:1696–1702

Lee AC, Fujio Y (1999) Microflora of banh men, a fermentation starter from Vietnam. World J Microbiol Biotechnol 15:57–62

Lee M, Kim MK, Vancanneyt M, Swings J, Kim S-H, Kang MS, Lee S-T (2005) *Tetragenococcus koreensis* sp. nov., a novel rhamnolipid-producing bacterium. Int J Syst Evol Microbiol 55:1409–1413

Lei V, Jakobsen M (2004) Microbiological characterization and probiotic potential of koko and koko sour water, African spontaneously fermented millet porridge and drink. J Appl Microbiol 96:384–397

Leisner JJ, Vancanneyt M, Goris J, Christensen H, Rusul G (2000) Description of *Paralactobacillus selangorensis* gen. nov., sp. nov., a new lactic acid bacterium isolated from chili bo, a Malaysian food ingredient. Int J Syst Evol Microbiol 50:19–24

Lin CL, Bolsen KK, Fung DYC (1992) Epiphytic lactic acid bacteria succession during the pre-ensiling periods of alfalfa and maize. J Appl Bacteriol 73:375–387

Lindner P (1887) Über ein neues in Malzmaischen vorkommendes, milchsäurebildendes Ferment. Wschr Brauerei 4:437–440

Lindner P (1888) Die Sarcina-Organismen der Gährungsgewerbe. Inaug. Diss. der Friedrich Wilhelms Universität 1–59. Zentralbl Bakteriol Parasitenk Infektionskr Hyg (II) 4: 427–429

Liu L, Zhang B, Tong H, Dong X (2006) *Pediococcus ethanolidurans* sp. nov., isolated from the walls of a distilled-spirit-fermenting cellar. Int J Syst Evol Microbiol 56:2405–2408

Lonvaud-Funel A, Joyeux A (1998) A bacterial disease causing ropiness of wine. Sciences des Aliments 8:33–49

Lonvaud-Funel A, Guilloux Y, Joyeux A (1993) Isolation of a DNA probe for identification of glucan-producing *Pediococcus damnosus* in wines. J Appl Bacteriol 74:41–47

Luchansky JB, Glass KA, Harsono KD, Degnan AJ, Faith NG, Cauvin B, Baccus-Taylor G, Arihara K, Bater B, Maurer AJ, Cassens RB (1992) Genomic analysis of *Pediococcus* starter cultures used to control *Listeria monocytogenes* in turkey summer sausage. Appl Environ Microbiol 58:3053–3059

Makarova K, Slesarev A, Wolf Y, Sorokin A, Koonin E, Pavlov A, Pavlova N, Karamychev V, Polouchin N, Shakhova V, Grigoriev I, Lou Y, Rohksar D, Lucas S, Huang K, Goodstein DM, Hawkins T, Plengvidhya V, Welker D, Hughes J, Goh Y, Benson A, Baldwin K, Lee J-H, Diaz-Muniz I, Dosti B, Smeianov V, Wechter W, Barabote R, Lorca G, Altermann E, Barrangou R, Ganesan B, Xie Y, Rawsthorne H, Tamir D, Parker C, Breidt F, Broadbent J, Hutkins R, O'Sullivan D, Steele J, Unlu G, Saier M, Klaenhammer T, Richardson P, Kozyavkin S, Weimer B, Mills D (2006) Comparative genomics of the lactic acid bacteria. Proc Natl Acad Sci USA 103:15611–15616

Manca de Nandra MC, Strasser de Saad AM (1995) Polysaccharide production by *Pediococcus pentosaceus* from wine. Int J Food Microbiol 27:101–106

Martel A, Meulenaere V, Devriese LA, Decostere A, Haesebrouck F (2003) Macrolide and lincosamide resistance in the Gram-positive nasal and tonsillar flora of pigs. Microb Drug Resist 9:293–297

Mastro TD, Spika JS, Lozano P, Appel J, Facklam RR (1990) Vancomycin-resistant *Pediococcus acidilactici*: nine cases of bacteremia. J Infect Dis 161:956–960

Mees RH (1934) Onderzoekingen over de Biersarcina. Thesis. Technical University, Delft, pp 1–110

Mehlen A, Goeldner M, Ried S, Stindl S, Ludwig W, Schleifer K-H (2004) Development of a fast DNA–DNA hybridization method based on melting profiles in microplates. Syst Appl Microbiol 27:689–695

Molina I, Toldra F (1992) Detection of proteolytic activity in microorganisms isolated from dry-cured ham. J Food Sci 57:1308–1310

Molina I, Silla H, Flores J (1989) Studie über die Keimflora trocken gepökelter Schinken. 3. Milchsäurebakterien. Fleischwirtschaft 69:1754–1756

Mora D, Fortina MG, Parini C, Manachini PL (1997) Identification of *Pediococcus acidilactici* and *Pediococcus pentosaceus* based on 16S rRNA and *ldhD* gene-targeted multiplex PCR analysis. FEMS Microbiol Lett 151:231–236

Mora D, Parini C, Fortina MG, Manachini PL (1998) Discrimination among pediocin AcH/PA-1 producer strains by comparison of pedB and pedD amplified genes and multiplex PCR assay. Syst Appl Microbiol 21:454–460

Mora D, Parini C, Fortina MG, Manachini PL (2000) Development of molecular RAPD marker for the identification of *Pediococcus acidilactici* strains. Syst Appl Microbiol 23:400–408

Morita H, Shiratori C, Murakami M, Takami H, Toh H, Kato Y, Nakajima F, Takagi M, Akita H, Masaoka T, Hattori M (2008) *Sharpea azabuensis* gen. nov., sp. nov., a Gram-positive, strictly anaerobic bacterium isolated from the faeces of thoroughbred horses. Int J Syst Evolut Microbiol 58:2682–2686

Mugula JK, Nnko SA, Narvhus JA, Sorhaug T (2003a) Microbiological and fermentation characteristics of togwa, a Tanzanian fermented food. Int J Food Microbiol 80:187–199

Mugula JK, Sorhaug T, Stepaniak L (2003b) Proteolytic activities in togwa, a Tanzanian fermented food. Int J Food Microbiol 84:1–12

Mundt JO, Beattie WG, Wieland FR (1969) Pediococci residing on plants. J Bacteriol 98:938–942

Nakagawa A, Kitahara K (1959) Taxonomic studies on the genus *Pediococcus*. J Gen Appl Microbiol 5:95–126

Nes IF, Diep DB, Håvarstein LS, Brurberg MB, Eijsink V, Holo H (1996) Biosynthesis of bacteriocins in lactic acid bacteria. Antonie van Leeuwenhoek 70:113–128

Nieto P, Molina I, Flores J, Silla MH, Bermell S (1989) Lipolytic activity of microorganisms isolated from dry-cured ham. In: Proceedings of 35th

International Congress of Meat Science and Technology, vol II. Copenhagen, pp 323–329

Nigatu A, Ahrne S, Gashe BA, Molin G (1998) Randomly amplified polymorphic DNA (RAPD) for discrimination of *Pediococcus pentosaceus* and *Ped. acidilactici* and rapid grouping of *Pediococcus* isolates. Lett Appl Microbiol 26:412–416

Ogier J-C, Son O, Gruss A, Tailliez P, Delacroix-Buchet A (2002) Identification of the bacterial microflora in dairy products by temporal temperature gradient gel electrophoresis. Appl Environ Microbiol 68:3691–3701

Olson NF (1990) The impact of lactic acid bacteria on cheese flavor. FEMS Microbiol Rev 87:131–147

Omar NB, Ampe F, Raimbault M, Guyot J-P, Tailliez P (2000) Molecular diversity of lactic acid bacteria from cassava sour starch (Colombia). Syst Appl Microbiol 23:285–291

Paludan-Muller C, Madsen M, Sophanodora P, Gram L, Moller PL (2002) Fermentation and microflora of plaa-som, a thai fermented fish product prepared with different salt concentrations. Int J Food Microbiol 73:61–70

Parente E, Grieco S, Crudele MA (2001) Phenotypic diversity of lactic acid bacteria isolated from fermented sausages produced in Basilicata (Southern Italy). J Appl Microbiol 90:943–952

Pedersen CS (1949) The genus *Pediococcus*. Bacteriol Rev 13:225–232

Perez Pulido R, Ben Omar N, Abriouel H, Lucas Lopez R, Canamero M, Galvez A (2005) Microbiological study of lactic acid fermentation of Caper berries by molecular and culture-dependent methods. Appl Environ Microbiol 71:7872–7879

Pittet V, Abegunde T, Marfleet T, Haakensen M, Morrow K, Jayaprakash T, Schroeder K, Trost B, Byrns S, Bergsveinson J, Kusalik A, Ziola B (2012) Complete genome sequence of the beer spoilage organism *Pediococcus claussenii* ATCC BAA-344$^T$. J Bacteriol 194(5):1271–1272

Raccach M (1987) Pediococci and biotechnology. Crit Rev Microbiol 14:291–309

Rice SL, Koehler PE (1976) Tyrosine and histidine decarboxylase activities of *Pediococcus cerevisiae* and *Lactobacillus* species and the production of tyramine in fermented sausages. J Milk Food Technol 39:166–169

Riebel WJ, Washington JA (1990) Clinical and microbiologic characteristics of pediococci. J Clin Microbiol 28:1348–1355

Rodas AM, Ferrer S, Pardo I (2003) 16S-ARDRA, a tool for identification of lactic acid bacteria isolated from grape must and wine. Syst Appl Microbiol 26:412–422

Rodriguez JM, Cintas LM, Casaus P, Suarez MI, Hernandez PE (1997) Detection of pediocin PA-1-producing pediococci by rapid molecular biology techniques. Food Microbiol 14:363–371

Rogosa J, Mitchell JA, Wiseman RF (1951) A selective medium for isolation and enumeration of oral and fecal lactobacilli. J Bacteriol 62:132–133

Rojo-Bezares B, Saenz Y, Poeta P, Zarazaga M, Ruiz-Larrea F, Torres C (2006) Assessment of antibiotic susceptibility within lactic acid bacteria strains isolated from wine. Int J Food Microbiol 111:234–240

Ruoff KL, Kuritzkes DR, Wolfson JS, Ferraro MJ (1988) Vancomycin-resistant Gram-positive bacteria isolated from human sources. J Clin Microbiol 26:2064–2068

Sakamoto K, Konings WN (2003) Beer spoilage bacteria and hop resistance. Int J Food Microbiol 89:105–124

Santos EM, Jaime I, Rovira J, Lyhs U, Korkeala H, Björkroth J (2005) Characterization and identification of lactic acid bacteria in "morcilla de Burgos". Int J Food Microbiol 97:285–296

Sarma P, Mohanty S (1998) *Pediococcus acidilactici* pneumonitis and bacteremia in a pregnant woman. J Clin Microbiol 36:2392–2392

Satokari R, Mattila-Sandholm T, Suihko M-L (2000) Identification of pediococci by ribotyping. J Appl Bact 88:260–265

Schillinger U, Holzapfel WH (2003) Culture media for lactic acid bacteria. In: Corry JEL, Curtis GDW, Baird RM (eds) Handbook of culture media for food microbiology, vol 37. Boston, Amsterdam, pp 127–140, Chapter 8

Schleifer KH, Ludwig W (1995) Phylogeny of the genus *Lactobacillus* and related genera. Syst Appl Microbiol 18:461–467

Schultes LM, Evans JB (1971) Deoxyribonucleic acid homology of *Aerococcus viridans*. Int J Syst Bacteriol 21:207–209

Simpson WJ, Taguchi H (1995) The genus *Pediococcus*, with notes on the genera *Tetragenococcus* and *Aerococcus*. In: Wood JB, Holzapfel WH (eds) B. The genera of lactic acid bacteria, Academic & Professional, pp 125–172

Simpson PJ, Stanton C, Fitzgerald GF, Ross RP (2002) Genomic diversity within the genus *Pediococcus* as revealed by randomly amplified polymorphic DNA PCR and pulsed-field gel electrophoresis. Appl Environ Microbiol 68:765–771

Simpson PJ, Fitzgerald GF, Stanton C, Ross RP (2006) Enumeration and identification of pediococci in powder-based products using selective media and rapid PFGE. J Microbiol Methods 64:120–125

Sims W (1986) The isolation of pediococci from human saliva. Arch Oral Biol 11:967–972

Skerman VBD, McGowan V, Sneath PHA (1980) Approved list of bacterial names. Int J Syst Bacteriol 30:225–420

Skerman VBD, McGowan V, Sneath PHA (eds) (1989) Approved list of bacterial names, emended edition. American Society for Microbiology, Washington, DC

Smitinont T, Tansakul C, Tanasupawat S, Keeratipibul S, Nacarini L, Bosco M, Cescutti P (1999) Exopolysaccharide-producing lactic acid bacteria strains from traditional Thai fermented foods: isolation, identification and exopolysaccharide characterization. Int J Food Microbiol 51:105–111

Stamer JR (1983) Lactic acid fermentation of cabbage and cucumbers. In: Reed G (ed) Biotechnology, vol 5. Verlag Chemie, Weinheim, pp 365–378

Stevenson DM, Muck RE, Shinners KJ, Weimer PJ (2005) Use of real-time PCR to determine population profiles of individual species of lactic acid bacteria in alfalfa silage and stored corn stover. Appl Microbiol Biotechnol 71:329–338

Swenson JM, Facklam RR, Thornsberry C (1990) Antimicrobial susceptibility of vancomycin-resistant *Leuconostoc*, *Pediococcus*, and *Lactobacillus* species. Antimicrob Agents Chemother 34:543–549

Taguchi H, Ohkochi M, Uehara H, Kojima K, Mawatari M (1990) KOT medium, a new medium for the detection of beer spoilage lactic acid bacteria. J Am Soc Brew Chem 48:72–75

Tamang JP (1998) Role of microorganisms in traditional fermented foods. Indian Food Ind 17:162–167

Tamang JP, Tamang B, Schillinger U, Franz CM, Gores M, Holzapfel WH (2005) Identification of predominant lactic acid bacteria isolated from traditionally fermented vegetable products of the Eastern Himalayas. Int J Food Microbiol 105:347–356

Tanasupawat S, Daengsubha W (1983) *Pediococcus* species and related bacteria found in fermented foods and related materials in Thailand. J Gen Appl Microbiol 29:487–506

Tanasupawat S, Pakdeeto A, Thawai C, Yukphan P, Okada S (2007) Identification of lactic acid bacteria from fermented tea leaves (*miang*) in Thailand and proposals of *Lactobacillus thailandensis* sp. nov., *Lactobacillus camelliae* sp. nov., and *Pediococcus siamensis* sp. nov. J Gen Appl Microbiol 53:7–15

Tankovic J, Leclercq R, Duval J (1993) Antimicrobial susceptibility of *Pediococcus* spp. and genetic basis of macrolide resistance in *Pediococcus acidilactici* HM3020. Antimicrob Agents Chemother 37:789–792

Tenorio C, Zarazaga M, Martinez C, Torres C (2001) Bifunctional enzyme 6'-*N*-aminoglycoside acetyltransferase-2"-*O*-aminoglycoside phosphotransferase in *Lactobacillus* and *Pediococcus* isolates of animal origin. J Clin Microbiol 39:824–825

Torriani S, Vescovo M, Dellaglio F (1987) Tracing *Pediococcus acidilactici* in ensiled maize by plasmid-encoded erythromycin resistance. J Appl Bacteriol 63:543–553

Von Witzingerode HM, Moter A, Halle E, Lohnbrunner H, Kaisers U, Neuhaus P, Halle E (2000) A case of septicemia with *Pediococcus acidilactici* after long term antibiotic treatment. Eur J Clin Microbiol Infect Dis 19:946–948

Walter J, Hertel C, Tannock GW, Lis CM, Munro K, Hammes WP (2001) Detection of *Lactobacillus*, *Pediococcus*, *Leuconostoc*, and *Weissella* species in human feces by using group-specific PCR primers and denaturing gradient gel electrophoresis. Appl Environ Microbiol 67:2578–2585

Waters SM, Doyle S, Murphy RA, Power RF (2005) Development of solution phase hybridization PCR-ELISA for the detection and quantification of *Enterococcus faecalis* and *Pediococcus pentosaceus* in Nurmi-type cultures. J Microbiol Methods 63:264–275

Weiller HG, Radler F (1970) Milchsäurebakterien aus Wein und von Rebenblättern. Zbl Bakt II Abt 124:707–732

Weiller HG, Radler F (1976) Über den Aminosäurestofwechsel von Milchsäurebakterien aus Wein. Zschr Lebensm Unters Forsch 161:259–266

Weiss N (1992) The genera *Pediococcus* and *Aerococcus*. In: Balows A, Trüper HG, Dworkin M, Harder W, Schleifer K-H (eds) The prokaryotes, vol 2, 2nd edn. Springer, New York

Whittenbury R (1965) A study of some pediococci and their relationship to *Aerococcus viridans* and the enterococci. J Gen Microbiol 40:97–106

Wiik R, Torsvik V, Egidius E (1986) Phenotypic and genotypic comparisons among strains of the lobster pathogen *Aerococcus viridans* and other marine *Aerococcus viridans*-like cocci. Int J Syst Bacteriol 36:431–434

Wilderdyke MR, Smith DA, Brashears MM (2004) Isolation, identification and selection of lactic acid bacteria from alfalfa sprouts for competitive inhibition of foodborne pathogens. J Food Prot 67:947–951

Yousif NMK, Huch M, Schuster T, Cho G, Dirar HA, Holzapfel WH, Franz CMAP (2010) Diversity of lactic acid bacteria from Hussuwa, a traditional African fermented sorghum food. Food Microbiol 27:757–768

Zhang JG, Cai Y, Kobayashi R, Mukai S (2000) Characteristics of lactic acid bacteria isolated from forage crops and their effects on silage fermentation. J Sci Food Agric 80:1455–1460

Zhang B, Tong H, Dong X (2005) *Pediococcus cellicola* sp. nov., a novel lactic acid coccus from a distilled-spirit fermenting cellar. Int J Syst Evol Microbiol 55:2167–2170

# 17 The Family *Leptotrichiaceae*

*Stephen Lory*
Department of Microbiology and Immunobiology, Harvard Medical School, Boston, MA, USA

**Abstract**
This chapter discusses briefly the family *Leptotrichiaceae* and its four genera, *Streptobacillus*, *Sneathia*, *Sebaldella*, and *Leptotrichia*.

The family *Leptotrichiaceae* is within the phylum Fusobacteria and includes four genera. Three (*Streptobacillus*, *Sneathia*, and *Sebaldella*) are represented by a single species, while *Leptotrichia* includes six closely related species and a more distal *Leptotrichia goodfellowii*. They have been recovered in association with insects and mammals, including humans. Comparative genome sequence analysis suggests that some species have undergone genome reduction, increasing their host dependence. The phylogenetic relationship of *Leptotrichiaceae* is shown in ➋ *Fig. 17.1*.

The various *Leptotrichia* species are commonly found in the human oral cavity, in the female genitourinary tract, and occasionally, in the intestinal tract. Most members of this genus can be propagated in the laboratory, although several species were identified only by molecular methods, and they could not be cultured. *Leptotrichia* display a Gram-negative spindle-like (fusiform) morphology, occasionally arranged in pairs or chains. They are aerotolerant anaerobes, fastidious, forming colonies over several days in rich media supplemented with various vitamins or blood. They are all capable of fermenting carbohydrates, including disaccharides; however, there is considerable variation among different species (Thompson and Pikis 2012).

Although *Leptotrichiaceae* are common inhabitants of the oral cavity, including dental plaques, the association with tooth decay has not been firmly established. It is able to infect patients undergoing radiation or treatment with bone marrow transplantation for hematological malignances, where it can be readily recovered from blood (Bhally et al. 2005; Couturier et al. 2012; Schrimsher et al. 2013; Cooreman et al. 2011). Common species include *L. trevisanii* and less frequently *Leptotrichia wadei*, *Leptotrichia goodfellowii*, and *Leptotrichia hongkongensis* (Eribe et al. 2004). Blood infections of immunocompetent individuals by *Leptotrichia buccalis*, with unrelated health problems, have been reported; these organisms very likely originated in the patient's oral cavity (Ulstrup and Hartzen 2006).

*Sebaldella termitidis* was found in the intestinal tract of insects, but it was also recovered from a sediment in a wastewater treatment plant. They are Gram-negative, relatively long straight rods, often in pairs or short chains. Its relatively large genome (4.48 mega bases) and few pseudogenes suggest a potential for occupying a variety of niches, yet it has been isolated only from limited environments. Some of its metabolic reactions, such as supplying organic nitrogen, could potentially benefit the insect host. A range of sugars can be catabolized by *S. termitidis* with organic acids as end products (Harmon-Smith et al. 2010).

The urogenital tract of women is commonly colonized with *Sneathia* (Harwich et al. 2012). It is less frequently found in the male urogenital tract, and it appears to be sexually transmitted. In women, the two species *Sneathia sanguinegens* and *Leptotrichia amnionii* (closely related, or the same species as *Sneathia amnii* (Harwich et al. 2012) can cause bacterial vaginosis due to outgrowth of these normal microbiota (Thilesen et al. 2007). However, they are also important pathogens of the female reproductive tract, capable of inducing preterm labor and causing spontaneous abortions. The genome of *S. amnii* (1.34 mega bases) is the smallest of all *Leptotrichiaceae* genomes sequenced with a rather limited metabolic capacity (Harwich et al. 2012). The annotated genome showed that *S. amnii* can utilize glycogen as a carbon and energy source, and judging by a relatively large number of genes expressing small molecule transporters, it has a capacity for the uptake of nutrients, vitamins, and cofactors from its environment. Since vaginal epithelial cells produce glycogen, this metabolic capacity could explain the tissue tropism of the organism. The main method of energy generation is glycolysis, consistent with the life style of *Sneathia* as anaerobes. The requirements for complex media for growth were also confirmed by the analysis of the genome for determinants of biosynthetic pathways, with a notable absence of genes encoding enzymes for biosynthesis of most amino acids as well as purines and pyrimidines.

The genus *Streptobacillus* is represented by a single species, *Streptobacillus moniliformis*. It is found in most wild, laboratory and pet rats, usually part of the normal microbiota in their oropharynx. A bite by an infected rat can result in the transmission of *S. moniliformis* to people, causing rat-bite fever, and a bite or ingestion of food (e.g., unpasteurized milk) contaminated with rat fees or urine can lead to Haverhill fever (Elliott 2007; Gaastra et al. 2009). The symptoms of both of these infections are similar (fever, headaches, chills, and vomiting) followed by the development of skin rash; the different designations refer to the route of acquisition of the organisms. Failure to treat the infection, usually with penicillin or tetracycline, can lead to bacteremia with a mortality rate of 15–20 % (Elliott 2007).

E. Rosenberg et al. (eds.), *The Prokaryotes – Firmicutes and Tenericutes*, DOI 10.1007/978-3-642-30120-9_357,

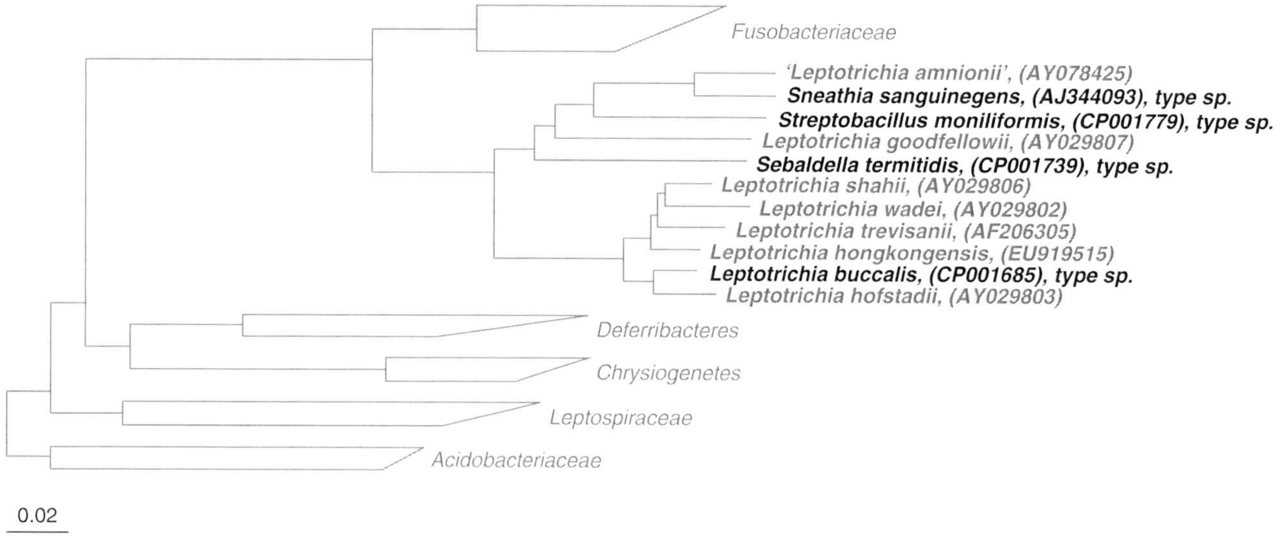


**Fig. 17.1**

**Phylogenetic reconstruction of the family *Leptotrichiaceae* based on 16S rRNA and created using the neighbor-joining algorithm with the Jukes-Cantor correction. The sequence datasets and alignments were used according to the All-Species Living Tree Project (*LTP*) database (Yarza et al. 2010; http://www.arb-silva.de/projects/living-tree). The tree topology was stabilized with the use of a representative set of nearly 750 high-quality type strain sequences proportionally distributed among the different bacterial and archaeal phyla. In addition, a 40 % maximum frequency filter was applied in order to remove hypervariable positions and potentially misplaced bases from the alignment. Scale bar indicates estimated sequence divergence**

## References


Bhally HS, Lema C, Romagnoli M, Borek A, Wakefield T, Carroll KC (2005) *Leptotrichia buccalis* bacteremia in two patients with acute myelogenous leukemia. Anaerobe 11(6):350–353

Cooreman S, Schuermans C, Van Schaeren J, Olive N, Wauters G, Verhaegen J, Jeurissen A (2011) Bacteraemia caused by *Leptotrichia trevisanii* in a neutropenic patient. Anaerobe 17(1):1–3

Couturier MR, Slechta ES, Goulston C, Fisher MA, Hanson KE (2012) Leptotrichia bacteremia in patients receiving high-dose chemotherapy. J Clin Microbiol 50(4):1228–1232

Elliott SP (2007) Rat bite fever and *Streptobacillus moniliformis*. Clin Microbiol Rev 20(1):13–22

Eribe ER, Paster BJ, Caugant DA, Dewhirst FE, Stromberg VK, Lacy GH, Olsen I (2004) Genetic diversity of *Leptotrichia* and description of *Leptotrichia goodfellowii* sp. nov., *Leptotrichia hofstadii* sp. nov., *Leptotrichia shahii* sp. nov. and *Leptotrichia wadei* sp. nov. Int J Syst Evol Microbiol 54(Pt 2):583–592

Gaastra W, Boot R, Ho HTK, Lipman LJA (2009) Rat bite fever. Vet Microbiol 133:211–228

Harwich MD Jr, Serrano MG, Fettweis JM, Alves JM, Reimers MA, Vaginal Microbiome Consortium (additional members), Buck GA, Jefferson KK (2012) Genomic sequence analysis and characterization of *Sneathia amnii* sp. nov. BMC Genomics 13(Suppl 8):S4

Harmon-Smith M, Celia L, Chertkov O, Lapidus A, Copeland A, Glavina Del Rio T, Nolan M, Lucas S, Tice H, Cheng JF, Han C, Detter JC, Bruce D, Goodwin L, Pitluck S, Pati A, Liolios K, Ivanova N, Mavromatis K, Mikhailova N, Chen A, Palaniappan K, Land M, Hauser L, Chang YJ, Jeffries CD, Brettin T, Göker M, Beck B, Bristow J, Eisen JA, Markowitz V, Hugenholtz P, Kyrpides NC, Klenk HP, Chen F (2010) Complete genome sequence of *Sebaldella termitidis* type strain (NCTC 11300). Stand Genomic Sci 2(2):220–227

Schrimsher JM, McGuirk JP, Hinthorn DR (2013) *Leptotrichia trevisanii* sepsis after bone marrow transplantation. Emerg Infect Dis 19(10):1690–1691

Thilesen CM, Nicolaidis M, Lökebö JE, Falsen E, Jorde AT, Müller F (2007) J Clin Microbiol 45(7):2344–2347

Thompson J, Pikis A (2012) Metabolism of sugars by genetically diverse species of oral *Leptotrichia*. Mol Oral Microbiol 27(1):34–44

Ulstrup AK, Hartzen SH (2006) *Leptotrichia buccalis*: a rare cause of bacteraemia in non-neutropenic patients. Scand J Infect Dis 38(8):712–716

Yarza P, Ludwig W, Euzeby J, Amann R, Schleifer KH, Glöckner FO, Rossello-Mora R (2010) Update of the All-Species Living Tree Project based on 16S and 23S rRNA sequence analyses. Syst Appl Microbiol 33:291–299. doi:10.1016/j.syapm.2010.08.001

# 18 The Family *Leuconostocaceae*

*Timo T. Nieminen*[1] · *Elina Säde*[2] · *Akihito Endo*[3] · *Per Johansson*[2] · *Johanna Björkroth*[2]
[1]Ruralia Institute, University of Helsinki, Seinäjoki, Finland
[2]Department of Food Hygiene and Environmental Health, Faculty of Veterinary Medicine, University of Helsinki, Helsinki, Finland
[3]Department of Food and Cosmetic Science, Tokyo University of Agriculture, Hokkaido, Japan



**Abstract**
*Leuconostocaceae* are lactic acid bacteria (LAB) belonging to order *Lactobacillales*. The family consists of genera *Leuconostoc*, *Weissella*, *Oenococcus*, and *Fructobacillus*. The genus *Leuconostoc* was described already in 1878 by van Tieghem. The oldest described species belonging to *Oenococcus* and *Fructobacillus* were originally described as *Leuconostoc* spp. but were later reclassified based on phenotypic and phylogenetic studies. Genus *Weissella* contains species originally classified as *Leuconostoc* or *Lactobacillus* spp.

Like other LAB, *Leuconostocaceae* are Gram positive, catalase negative, and chemoorganotrophic. They grow in rich media supplemented with growth factors and amino acids and generate energy by substrate-level phosphorylation. *Leuconostocaceae* ferment glucose heterofermentatively yielding lactic acid, $CO_2$, ethanol, and/or acetate.

*Leuconostocaceae* are found in environments with high nutrient content, e.g., on green vegetation, roots, and food. Within LAB, *Leuconostocaceae* are characterized by their adaptable fermentation patterns that enable efficient generation of ATP from carbohydrates and, consequently, enhanced growth. Due to their ability to grow rapidly in rich media under elevated $CO_2$ concentration at moderate temperatures, *Leuconostocaceae* are competitive in various food environments and contribute to a number of fermentation processes. The diverse fermentation substrates and products of *Leuconostocaceae* may cause desired or undesired effects on the organoleptic quality of foods.

This contribution is a modified and updated version of previous descriptions of the family (Schleifer, 2009) and the included genera (Björkroth et al., 2009; Björkroth and Holzapfel, 2006; Dicks and Holzapfel, 2009; Holzapfel et al., 2009).

## Taxonomy, Historical and Current

The family *Leuconostocaceae* belongs to the order *Lactobacillales*. Since the late 2000s, this family has contained the genera of *Fructobacillus*, *Leuconostoc*, *Oenococcus*, and *Weissella*. The genus *Leuconostoc* has been the hub of taxonomic reclassifications leading to description of the three other genera.

Historically, *Leuconostoc mesenteroides* was first mentioned by Van Tieghem in 1878 (Van Tieghem 1878) in an article called "Sur la Gomme de Sucrerie (*Leuconostoc mesenteroides*)." The description of the genus *Leuconostoc* is following today the lines published by Garvie (1986). The taxonomic revisions affecting leuconostocs have mainly been due to implementation of phylogenetic analyses and the studies utilizing polyphasic taxonomy approaches. The first phylogenetic analyses of the 16S rRNA gene sequences (Martinez-Murcia and Collins 1990; Martinez-Murcia et al. 1993) resulted in recognition of three distinct lineages within leuconostocs. They were referred as the genus

E. Rosenberg et al. (eds.), *The Prokaryotes – Firmicutes and Tenericutes*, DOI 10.1007/978-3-642-30120-9_208,

*Leuconostoc sensu stricto*, the *Leuconostoc paramesenteroides* group, and *Leuconostoc oenos*. A new genus *Weissella* (Collins et al. 1993) was described to accommodate members of the so-called *L. paramesenteroides* group (including *L. paramesenteroides* and some atypical, heterofermentative lactobacilli). In addition, *L. oenos* has been reclassified as *Oenococcus oeni* (Dicks et al. 1995). More recently, some atypical leuconostocs of plant origin including *Leuconostoc durionis*, *Leuconostoc ficulneum*, *Leuconostoc fructosum*, and *Leuconostoc pseudoficulneum* have been assigned to the new genus *Fructobacillus* (Endo and Okada 2008). After these reclassifications, the genus *Leuconostoc* includes 13 validly published species names (➋ *Table 18.1*) with *L. mesenteroides* as the type species. *L. mesenteroides* is the only species divided into subspecies which have not been established based on phylogenetic or genomic boarders. According to Vancanneyt et al. (2006), *Leuconostoc argentinum* (Dicks et al. 1993) is a later synonym of *Leuconostoc lactis*.

With the exception of *Leuconostoc fallax*, 16S rRNA gene sequence similarities among the type strains of *Leuconostoc* spp. are high, varying from 97.3 % to 99.5 % (Björkroth and Holzapfel 2006). 16S rRNA gene sequence analysis further divides leuconostocs into three evolutionary branches including *Leuconostoc citreum*, *Leuconostoc holzapfelii*, *Leuconostoc lactis*, and *Leuconostoc palmae* in the first branch; *L. mesenteroides* and *L. pseudomesenteroides* in the second; and *Leuconostoc carnosum*, *Leuconostoc gasicomitatum*, *Leuconostoc gelidum*, *Leuconostoc inhae*, and *Leuconostoc kimchii* in the third branch, whereas *L. fallax* is genetically more distinct from the other *Leuconostoc* species (➋ *Fig. 18.1*).

In addition to the 16S rRNA gene, the loci of housekeeping genes *atpA*, *dnaK*, *pheS*, *recN*, and *rpoA* in leuconostocs have been analyzed. The phylogenetic trees constructed on analyses of *pheS*, *rpoA*, and *atpA* loci offered discriminatory power for differentiation of species within the genus *Leuconostoc* and were roughly in agreement with 16S rRNA gene-based phylogeny (Ehrmann et al. 2009; De Bruyne et al. 2007). Comparative sequencing of the additional phylogenetic markers *dnaK* and *recA* confirmed the 16S rRNA gene tree topology in the study describing *L. palmae* (Ehrmann et al. 2009). Arahal et al. (2008) studied the usefulness of *recN* locus and concluded that also *recN* can serve as a phylogenetic marker as well as a tool for species identification. Congruence of evolutionary analyses inside the *Leuconostoc–Oenococcus–Weissella* clade has been assessed by comparative phylogenetic analyses of 16S rRNA, *dnaA*, *gyrB*, *rpoC*, and *dnaK* housekeeping genes (Chelo et al. 2007). Phylogenies obtained with the different genes were in overall good agreement, and a well-supported almost fully resolved phylogenetic tree was obtained when the combined sequence data were analyzed using a Bayesian approach.

Within the genus *Weissella*, several new species have been characterized during the last 5 years, and the genus currently comprises 17 species (➋ *Table 18.2*). The description for the genus is as published by Collins et al. (1993). The type species is *Weissella viridescens* (Niven and Evans 1957) which is synonymous to *Lactobacillus viridescens*.

The genus *Weissella* was proposed by Collins et al. (1993), and the first species included in this genus comprises species previously classified as *Leuconostoc* or *Lactobacillus*. *L. paramesenteroides* (Garvie 1967a), *L. viridescens* (Niven and Evans 1957; Kandler and Abo-Elnaga 1966), *Lactobacillus confusus* (Holzapfel and Van Wyk 1982; Holzapfel and Kandler 1969), *Lactobacillus kandleri* (Holzapfel and Van Wyk 1982), *Lactobacillus minor* (Kandler et al. 1983), and *Lactobacillus halotolerans* (Kandler et al. 1983) kept their specific epithets and were reclassified as *Weissella paramesenteroides*, *W. viridescens*, *Weissella confusa*, *W. kandleri*, *Weissella minor*, and *Weissella halotolerans*, respectively. These species were followed by inclusion of *Weissella hellenica* (Collins et al. 1993), *Weissella thailandensis* (Tanasupawat et al. 2000), *Weissella cibaria* (Björkroth et al. 2002), *Weissella soli* (Magnusson et al. 2002), and *Weissella koreensis* (Lee et al. 2002). In addition, *Weissella kimchii* was proposed by Choi et al. (2002), but it was found as a later heterotypic synonym of *Weissella cibaria* (Ennahar and Cai 2004). *Weissella ghanensis* (De Bruyne et al. 2008), *Weissella beninensis* (Padonou et al. 2010) and *Weissella fabaria* (De Bruyne et al. 2010), *Weissella ceti* (Vela et al. 2011), *Weissella fabalis* (Snauwaert et al. 2013), and *Weissella oryzae* (Tohno et al. 2012) are the latest species suggested to the genus *Weissella*.

*W. confusa*, *W. cibaria*, *W. halotolerans*, *W. hellenica*, *W. kandleri*, *W. koreensis*, *W. minor*, *W. paramesenteroides*, *W. soli*, *W. thailandensis*, and *W. viridescens* share 93.9–99.2 % 16S rRNA encoding gene sequence similarity (Björkroth et al. 2009). Among the recently described species, sequence similarity analyses (Snauwaert et al. 2013) indicated that *W. fabalis* type strain shares the highest sequence similarities with the type strains of *W. fabaria* (97.7 %), *W. ghanensis* (93.3 %), and *W. beninensis* (93.4 %). Five main phylogenetic branches exist based on the 16S rRNA encoding gene analyses. *W. hellenica*, *W. paramesenteroides*, and *W. thailandensis* branch together, as do *W. cibaria* and *W. confusa*. Two other branches are formed by *W. ceti*, *W. minor*, *W. halotolerans*, and *W. viridescens* in one branch and *W. kandleri*, *W. koreensis*, *W. oryzae*, and *W. soli* in another. *W. fabali* s (Snauwaert et al. 2013), *W. fabaria* (De Bruyne et al. 2010), and *W. ghanensis* (De Bruyne et al. 2008) form the fifth branch distinct from the other species within the genus.

*Oenococcus oeni*, type species of the genus *Oenococcus*, had been formerly classified as *Leuconostoc oenos* (Garvie 1967b). The genus *Oenococcus* currently includes two species, which are *Oenococcus kitaharae* and *O. oeni*. *O. oeni* was formerly classified as *Leuconostoc oenos* and reclassified as a member of the novel genus *Oenococcus* (Dicks et al. 1995). A candidate of novel *Oenococcus* species might have been isolated from bioethanol fermenting tank (Lucena et al. 2010), which has not been characterized at a time of writing (September, 2012). Originally, the species *O. oeni* was considered as a genetically homogeneous organism based on the sequencing of rRNA operon (Jeune and Lonvaud-Funel 1997; Zavaleta et al. 1996). However, recent study by Bridier et al. (2010) found diverse genetic groups in the species by multilocus sequence typing (MLST) with sequences

**Table 18.1**
**Phenotypic characteristics of *Leuconostoc* spp.**

| Characteristics | *L. mesenteroides* ssp. *mesenteroides* | *L. mesenteroides* ssp. *dextranicm* | *L. mesenteroides* ssp. *cremoris* | *L. mesenteroides* ssp. *suionicum* | *L. lactis* | *L. pseudomesenteroides* | *L. citreum* | *L. carnosum* | *L. gelidum* | *L. fallax* | *L. kimchii* | *L. gasicomitatum* | *L. inhae* | *L. holzapfelii* | *L. palmae* | *L. miyukkimchii* |
|---|---|---|---|---|---|---|---|---|---|---|---|---|---|---|---|---|
| Acid from | | | | | | | | | | | | | | | | |
| Amygdalin | d | ND | – | + | – | d | d | – | + | ND | + | – | d | – | ND | + |
| L-Arabinose | + | – | – | + | d | d | + | – | + | – | – | + | + | + | – | – |
| Arbutin | d | –[a] | – | + | – | d | + | – | + | – | ND | – | – | – | ND | + |
| Cellobiose | d | d | – | + | – | d | + | d | + | – | + | + | + | – | – | + |
| Galactose | + | d | + | + | + | d | d | – | – | – | + | d | d | + | ND | – |
| Lactose | d | d | d | | + | d | – | – | – | – | + | – | – | – | ND | – |
| Maltose | + | + | – | + | + | + | + | – | d | + | + | + | + | + | – | + |
| Mannitol | d | (d) | – | d | – | – | d | – | – | (d) | + | – | + | – | – | + |
| Mannose | d | d[a] | – | + | + | + | + | d | + | + | + | + | + | + | | + |
| Melibiose | d | d | – | + | d | d | – | d | + | – | ND | + | – | + | – | – |
| Raffinose | d | d[a] | – | – | d | d | – | – | + | – | – | + | – | + | – | – |
| Ribose | d | +[a] | – | + | – | + | – | d | d | + | + | + | + | – | | + |
| Salicin | d | d | – | + | d | d | + | d | + | – | + | – | d | – | – | + |
| Starch | – | –[a] | | – | – | d | – | ND | ND | ND | – | – | – | – | ND | + |
| Sucrose | + | + | – | + | + | d | + | + | + | + | + | + | + | – | + | + |
| Trehalose | + | + | – | + | d | + | + | + | + | (d) | + | + | + | + | – | + |
| D-Xylose | d | d | – | + | – | + | d | – | + | – | – | – | – | – | ND | + |
| Hydrolysis of aesculin | + | d | – | + | – | d | + | d | + | ND | ND | + | + | – | ND | + |
| Dextran production | +[a] | +[a] | –[a] | ND | –[a] | ND | ND | d[b] | d | ND | +[a] | + | ND | + | ND | ND |
| Growth at pH 4.8 | –[a] | –[a] | –[a] | ND | –[a] | ND | ND | ND | ND | ND | ND | ND | + | + | ND | – |
| Growth at 37 °C | d[a] | + | –[a] | + | +[a] | + | d | – | – | + | + | – | – | + | ND | – |
| Peptidoglycan type | Lys-Ser-$Ala_2$[a] | Lys-Ser-$Ala_2$[a] | Lys-Ser-$Ala_2$[a] | ND | Lys-$Ala_2$[a] | Lys-Ser-$Ala_2$[a] | Lys-$Ala_2$[a] | Lys-$Ala_2$[a] | Lys-$Ala_2$[a] | Lys-$Ala_2$ | ND | Lys-$Ala_2$ | ND | ND | ND | ND |
| References and the number of strains examined | Farrow et al. (1989) ($n = 30$) | Garvie (1976) ($n = 21$) | Farrow et al. 1989 ($n = 14$) | Gu et al. (2012) ($n = 2$) | Farrow et al. (1989) ($n = 4$) | Farrow et al. (1989) ($n = 7$) | Farrow et al. (1989) ($n = 11$) | Shaw and Harding (1989) ($n = 15$) | Shaw and Harding (1989) ($n = 30$) | Holzapfel et al. (2009) | Kim et al. (2000a) ($n = 1$) | Björkroth et al. (2000) ($n = 4$) | Kim et al. (2003) ($n = 6$) | De Bruyne et al. (2007) ($n = 1$) | Ehrmann et al. (2009) ($n = 1$) | Lee et al. (2012b) ($n = 1$) |

+, 90 % or more of strains positive; –, 90 % or more of strains negative; d, 11–98 % of strains positive; (), delayed reaction

*ND* no data

[a]Holzapfel et al. (2009)

[b]Björkroth et al. (1998)

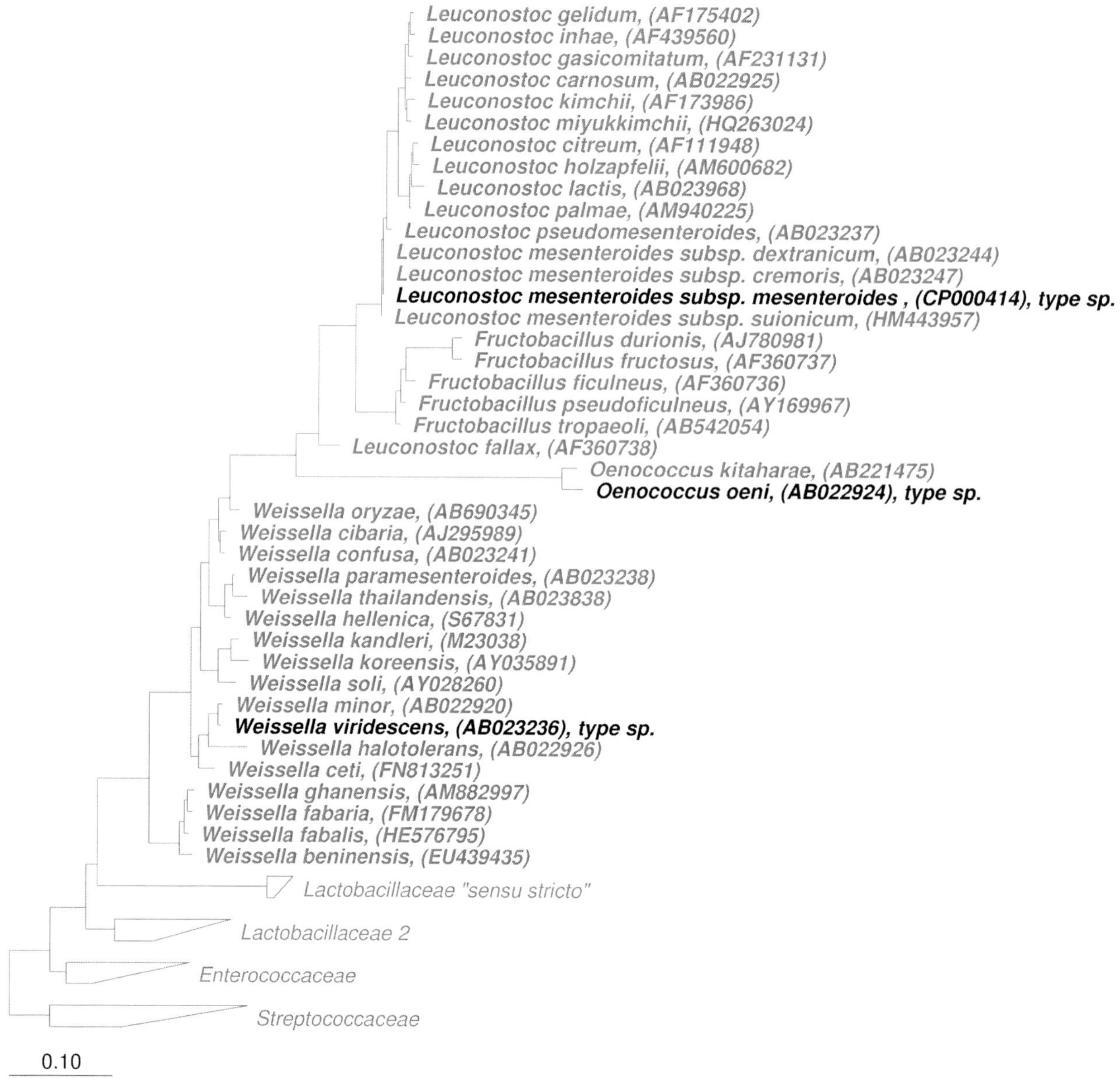


**Fig. 18.1**

**Phylogenetic reconstruction of the family *Leuconostocaceae* based on 16S rRNA and created using the maximum likelihood algorithm RAxML (Stamatakis 2006).The sequence datasets and alignments were used according to the All-Species Living Tree Project (LTP) database (Yarza et al. 2010; http://www.arb-silva.de/projects/living-tree). Representative sequences from closely related taxa were used as outgroups. In addition, a 40 % maximum frequency filter was applied in order to remove hypervariable positions and potentially misplaced bases from the alignment. Scale bar indicates estimated sequence divergence**

of several housekeeping genes. Reclassification of *L. oenos* into the genus *Oenococcus* was carried out based on its unique phylogenetic position, physiological characteristics, total soluble cell protein analysis, and several biochemical characteristics by Dicks et al. (1995). Already in 1993 Martinez-Murcia et al. (1993) showed by comparison of both 16S and 23SrRNA sequences that *L. oenos* does not belong to the same line of descent with the *L. sensu stricto* organisms or *L. paramesenteroides* group of species (the current genus *Weissella*). The second species, *O. kitaharae*, was described from compost of distilled *shochu* residue in Japan (Endo and Okada 2006). These two species share 96.0 % similarity based on 16S rRNA gene sequence. Sequence similarities with other members of family *Leuconostocaceae* are less than 85 %. The high level of phylogenetic divergence of the genus *Oenococcus* compared to that of the other lactic acid bacteria might be explained by the absence of the mismatch mutation repair system in oenococci, which causes a high mutation rate, an excess of recombination, and a rapid genetic evolution (Marcobal et al. 2008). Based on *pheS* sequences, *Oenococcus* spp. still belong to the family

**Table 18.2**
**Phenotypic characteristics of *Weissella* spp.**

| Characteristics | *W. viridescens* | *W. paramesenteroides* | *W. confusa* | *W. halotolerans* | *W. kandleri* | *W. minor* | *W. hellenica* | *W. thailandensis* | *W. cibaria* | *W. koreensis* | *W. soli* | *W. ghanensis* | *W. beniensis* | *W. fabaria* | *W. ceti* | *W. fabalis* | *W. oryzae* |
|---|---|---|---|---|---|---|---|---|---|---|---|---|---|---|---|---|---|
| Acid from | | | | | | | | | | | | | | | | | |
| Amygdalin | ND | ND | ND | ND | ND | ND | ND | – | + | | – | + | + | – | – | – | – |
| L-Arabinose | – | d | – | – | – | – | + | + | + | + | + | – | + | – | – | – | + |
| Arbutin | ND | ND | ND | ND | ND | ND | ND | ND | + | ND | – | + | + | – | – | – | – |
| Cellobiose | – | (d) | + | – | – | + | – | – | + | – | – | + | d | + | – | + | – |
| Fructose | ND | ND | ND | ND | ND | ND | ND | + | + | ND | – | + | + | + | – | + | + |
| Galactose | – | + | + | – | + | – | – | + | – | – | – | – | + | – | – | – | (+) |
| Lactose | ND | ND | ND | ND | ND | ND | ND | – | – | ND | – | – | + | – | – | – | – |
| Maltose | + | + | + | + | – | + | + | + | + | – | + | + | + | – | + | + | + |
| Mannitol | ND | ND | ND | ND | ND | ND | ND | – | – | ND | – | – | + | + | – | – | – |
| Mannose | ND | ND | ND | ND | ND | ND | ND | + | + | ND | + | + | + | + | – | + | + |
| Melibiose | – | + | – | – | – | – | – | + | – | – | + | – | + | – | – | – | (+) |
| Raffinose | – | d | – | – | – | – | – | + | – | – | + | – | + | – | – | – | – |
| Ribose | – | ND | + | + | + | + | – | + | – | + | + | – | d | – | + | – | + |
| Salicin | ND | ND | ND | ND | ND | ND | ND | – | + | ND | + | + | d | – | – | – | – |
| Sucrose | d | + | + | – | – | + | + | + | + | – | + | d | + | – | – | – | – |
| Trehalose | d | + | – | – | – | + | + | + | – | – | + | + | d | + | + | + | + |
| D- Xylose | – | d | + | – | – | – | – | – | + | + | + | – | – | – | – | – | + |
| Ammonia from arginine | – | – | + | + | + | + | – | – | + | + | + | + | + | + | d | + | + |
| Hydrolysis of aesculin | – | ND | + | – | – | + | ND | – | + | – | ND | + | + | + | + | + | – |
| Gas from glucose | ND | ND | ND | ND | ND | ND | ND | + | + | + | ND | d | + | + | – | + | + |
| Lactic acid configuration | DL | D | DL | DL | DL | DL | D | D | DL | D | D | D or DL | DL | DL | DL | D | D |

**Table 18.2 (continued)**

| Characteristics | *W. viridescens* | *W. paramesenteroides* | *W. confusa* | *W. halotolerans* | *W. kandleri* | *W. minor* | *W. hellenica* | *W. thailandensis* | *W. cibaria* | *W. koreeensis* | *W. soli* | *W. ghanensis* | *W. beniensis* | *W. fabaria* | *W. ceti* | *W. fabalis* | *W. oryzae* |
|---|---|---|---|---|---|---|---|---|---|---|---|---|---|---|---|---|---|
| Dextran production | ND | − | + | ND | + | − | − | − | + | + | − | + | d | + | − | + | − |
| Growth at 37°C | ND | ND | ND | ND | ND | ND | ND | + | + | + | + | + | ND | + | + | + | + |
| Peptidoglycan type | Lys-Ala-Ser | Lys-$Ala_2$ or Lys-Ser-$Ala_2$ | Lys-Ala | Lys-Ala-Ser | Lys-Ala-Gly-$Ala_2$ | Lys-Ser-$Ala_2$ | Lys-Ala-Ser | Lys-$Ala_2$ | Lys-Ala (Ser)-Ala | Lys-Ala-Ser | ND | ND | ND | Lys-Ala-Ser | ND | Lys-Ala-Ser | ND |
| Cell morphology | Short rods | Spherical or lenticular | Short rods thickened at one end | Irregular short or coccoid rods | Irregular rods | Irregular short or coccoid rods | Large spherical or lenticular cells | Coccoid | Short rods | Short rods or coccoid | Short rods | Short rods | Short rods or coccoid | Coccoid | Short rods or coccoid | Coccoid | Short rods or coccoid |
| References and the number of strains examined | Collins et al. (1993) | Collins et al. (1993) | Collins et al. (1993) | Collins et al. (1993) | Collins et al. (1993) | Collins et al. (1993) | Collins et al. (1993) | Tanasupawat et al. (2000) (*n* = 5) | Björkroth et al. (2002) (*n* = 18) | Lee et al. (2002) (*n* = 2) | Magnusson et al. (2002) (*n* = 4) | De Bruyne et al. (2008) (*n* = 2) | Padonou et al. (2010) (*n* = 4) | De Bruyne et al. (2010) (*n* = 2) | Vela et al. (2011) (*n* = 9) | Snauwaert et al. (2013) (*n* = 1 | Tohno et al. (2012) (*n* = 2) |

+, 90 % or more of strains positive; −, 90 % or more of strains negative; d, 11–98 % of strains positive; (), delayed reaction

*ND* no data

D, 90 % or more of the lactic acid is D(−), DL more than 25 % of the total lactic acid is L(+)

*Leuconostocaceae* but share different relationships with the other genera when compared to 16S rRNA gene sequence analyses. Sequence similarity of partial *pheS* gene between the two *Oenococcus* spp. is approximately 75 % and less than 70 % between *Oenococcus* spp. and other members in the family *Leuconostocaceae*. Related to the phylogeny, an interesting debate over its evolution speed has occurred. Because of a long branch in the 16S rRNA phylogenetic tree, *O. oeni* is regarded as "rapidly evolving" species (Yang and Woese 1989). This hypothesis was at first questioned based on data generated by *rpoB* gene sequences (Morse et al. 1996), but supported by comparative genome analyses of different species of lactic acid bacteria (LAB), including *O. oeni* (Makarova et al. 2006).

In addition to the 16S rRNA gene phylogeny, analysis with *pheS* (De Bruyne et al. 2010) and *recN* (Arahal et al. 2008) loci has been done. Congruence of evolutionary relationships inside the *Leuconostoc*–*Oenococcus*–*Weissella* clade has been assessed by phylogenetic analyses of 16SrRNA, *dnaA*, *gyrB*, *rpoC*, and *dnaK* (Chelo et al. 2007) housekeeping genes. Phylogenies obtained with the different genes were in overall good agreement, and a well-supported, almost fully resolved phylogenetic tree was obtained when the combined data were analyzed in a Bayesian approach.

The genus *Fructobacillus* currently includes five species. They are *F. durionis* (Leisner et al. 2005), *F. ficulneus* (Antunes et al. 2002), *F. fructosus* (Kodama 1956), *F. pseudoficulneus* (Chambel et al. 2006), and *F. tropaeoli* (Endo et al. 2011). With the exception of *F. tropaeoli*, these species were formerly classified as *Leuconostoc* species (Endo and Okada 2008). *Fructobacillus fructosus*, type species of the genus *Fructobacillus*, had been firstly classified as *Lactobacillus fructosus* based on morphological and physiological characteristics and later reclassified to *Leuconostoc fructosum* based on its phylogenetic position (Kodama 1956; Antunes et al. 2002). *Leuconostoc fructosum* was re-reclassified to *F. fructosus* based on physiological and morphological characteristics and its phylogenetic position (Endo and Okada 2008). Based on the 16S rRNA gene sequences, *Fructobacillus* species are phylogenetically separated into two subclusters. The first subcluster contains *F. fructosus* and *F. durionis* (97.9 % sequence similarity), and the second contains *Fructobacillus ficulneus*, *F. pseudoficulneus*, and *F. tropaeoli* (98.0–99.2 % sequence similarities). The sequence similarity between the two groups ranges from 94.2 % to 99.4 %. *Fructobacillus* species has been also genetically characterized based on sequences of 16S–23S rRNA gene intergenic spacer regions (ISR), *rpoC* and *recA*. Phylogenetic analysis based on the ISR and *rpoC* gene shows similar clustering to that based on 16S rRNA gene, but phylogenetic analysis based on *recA* gene shows different clustering (Endo and Okada 2008; Endo et al. 2011).

## Molecular Analyses

Classification of the members of the family *Leuconostocaceae* using adequate molecular methods gives faster and more consistent and reliable results than schemes based on phenotypic characters. The molecular analyses have provided deeper insights into the phylogeny of the already assigned taxons within *Leuconostocaceae* and led to reclassification of species. In addition to the taxonomy and phylogeny, the motivation of many molecular studies has been more practical: to distinguish and identify relevant strains among closely related isolates. For the molecular characterization of *Leuconostocaceae*, various methods with differing resolving capacity have been reported and proposed; some have proven applicable for species identification, while others provide high discriminatory power and detail strain characterization. The choice of method depends on the scope and purpose of the study as well as on the availability of laboratory facilities. A summary of common molecular methods and their relative performances in differentiation of *Leuconostocaceae* is discussed below. In many studies cited, the results of two or more molecular methods have been combined to achieve better discrimination and more accurate clustering of the given set of isolates. However, only few studies report systematic comparison of different molecular methods and discuss their limitations for characterizing *Leuconostocaceae*.

### DNA–DNA Hybridization Studies

DNA–DNA hybridization assays have been included in many studies to determine interspecies relationships among *Leuconostoc* and *Weissella* species and to reveal whether two isolates should be classified in the same species. Since many closely related species of *Fructobacillus*, *Leuconostoc*, or *Weissella* share high 16S rRNA gene sequence similarity, DNA–DNA hybridization experiments have been necessary to support a proposal for a novel species status.

### DNA Fingerprinting

DNA fingerprinting using pulsed field gel electrophoresis (PFGE) and an appropriate restriction endonuclease provides high level of discrimination, allowing differentiation of closely related strains that are indistinguishable by other methods. Several investigators have used PFGE typing for characterizing *Leuconostocaceae* from dairy, meat, vegetable, and wine-related sources. These studies have demonstrated the success of PFGE typing in differentiating strains in a specific ecosystem or monitoring the presence of particular strains in a mixed population. For instance, PFGE typing has been used in several studies to study strain heterogeneity of *O. oeni* population during malolactic fermentation of wine (Sato et al. 2001; Vigentini et al. 2009; Zapparoli et al. 2012) as well as during an in-plant investigation of a ham spoilage problem to pinpoint potential sources of harmful *L. carnosum* contamination (Björkroth et al. 1998).

Another commonly used DNA fingerprinting technique is ribotyping or restriction fragment length polymorphism (RFLP) analysis of 16S and 23S rRNA genes where the detection of ribotyping fingerprint is accomplished by hybridization with probes. Numerical analysis of ribotyping has been included in polyphasic taxonomy studies on *Leuconostoc* (Björkroth et al. 2000), *Fructobacillus* (Chambel et al. 2006) and *Weissella* (Björkroth et al. 2002) and found to provide species level identification with some intraspecies variation. Subsequently, ribotyping has been applied to detect and identify individual species or strains of *Leuconostocaceae* from various food, animal, and environmental sources. Although ribotyping provides discriminatory capacity for species identification, PFGE appears to be superior for strain differentiation (Björkroth et al. 1998; Vihavainen and Björkroth 2009).

## PCR-Based DNA Fingerprinting Methods

Analysis of (fluorescent) amplified fragment length polymorphism (FAFLP or AFLP) fingerprints is another highly discriminatory characterization tool which has proven useful in the differentiation of *Leuconostoc* (De Bruyne et al. 2007) and *Weissella* species (De Bruyne et al. 2008, 2010). Furthermore, AFLP has been found valuable in typing of *O. oeni* strains (Cappello et al. 2008, 2010).

Amplified ribosomal DNA restriction analysis (ARDRA) is a technical variation of ribotyping comprising of restriction enzyme analysis of PCR amplicons from the rrn operon. Several ARDRA procedures targeting to different regions of the rrn operon have been reported; some give limited resolution being mainly applicable for rapid first-stage screening of isolates, while others provide discriminatory power allowing reliable species identification of *Leuconostocaceae*. For instance, 16S-ARDRA has been used for identification of species of *Leuconostocaceae* from grape must and wine (Rodas et al. 2003) and fermented sausages (Bonomo et al. 2008). Protocols for 16S-ARDRA employing genus-specific primers for *Weissella* (Jang et al. 2002) and *Leuconostoc* (Jang et al. 2003) have been developed to allow identification of *Weissella* and *Leuconostoc* species among other phylogenetically related lactic acid bacteria in food. Furthermore, a 16S–23S rRNA spacer ARDRA method has been developed for identification of lactic acid bacteria and proved useful in identifying *Leuconostoc* species from meat (Chenoll et al. 2003, 2007).

Fingerprinting using randomly amplified polymorphic DNA (RAPD) is another PCR-based tool applied for molecular typing of *Leuconostoc*, *Weissella*, and *O. oeni*. Various studies have demonstrated the success of RADP in monitoring *O. oeni* strains during winemaking (Bartowsky et al. 2003; Reguant and Bordons 2003; Zapparoli et al. 2000). Other workers have analyzed RAPD fingerprints to differentiate species and strain of *Leuconostoc* and *Weissella* from various sources (Aznar and Chenoll 2006; Cibik et al. 2000; De Bruyne et al. 2008; Ehrmann et al. 2009; Nieto-Arribas et al. 2010; Padonou et al. 2010).

Repetitive element palindromic PCR (REP-PCR) with the (GTG)5 primer has been applied for high-throughput screening of large collections of lactic acid bacteria isolates in numerous studies. Numerical analysis of REP-PCR patterns has been reported to be suitable for species identification and for genotypic characterization of *Leuconostoc* (Bounaix et al. 2010a; Vancanneyt et al. 2006) and *Weissella* (Bounaix et al. 2010b; Padonou et al. 2010).

## DNA Sequencing-Based Analysis

Sequence analysis of 16S rRNA gene or its variable regions are widely applied strategies for classification of lactic acid bacteria and have been used for identification of *Leuconostocaceae* from various sources. In addition to 16S rRNA gene sequence analysis, phylogenetic analysis of partial sequences of several protein-coding genes such as dnaA, dnaK, gyrB, pheS, recN, rpoA, or rpoC has been reported to be highly discriminatory, allowing differentiation of species and strains within the family *Leuconostocaceae* (Arahal et al. 2008; Chelo et al. 2007; Ehrmann et al. 2009; De Bruyne et al. 2007, 2010). Furthermore, multilocus sequence typing (MLST) schemes have been proposed and applied for *O. oeni* (de las Rivas et al. 2004; Bilhere et al. 2009; Bridier et al. 2010). These studies have demonstrated that MLST is a powerful method for typing of *O. oeni* strains and provides data that can be used for studying genetic diversity, population structure, and evolutionary mechanism of this organism.

## Protein Profiling

In addition to various DNA-based molecular techniques, analysis of whole-cell protein pattern by sodium dodecyl sulfate-polyacrylamide gel electrophoresis (SDS-PAGE) has proven useful in the differentiation of closely related *Leuconostoc* and *Weissella* and has been widely applied for identification of *Leuconostocaceae* (Dicks et al. 1990; Björkroth et al. 2002; De Bruyne et al. 2007, 2008, 2010). In addition, matrix-assisted laser desorption/ionization time-of-flight mass spectrometry (MALDI-TOF MS) has been increasingly studied and applied for the identification and typing of lactic acid bacteria. This method is based on the analysis of the structural differences of microbial cells; the mass spectra mainly reflect the heterogeneity of ribosomal proteins and, thus, give a specific profile for each organism. A MALDI-TOF MS profiling method has also been reported for the family *Leuconostocaceae* (De Bruyne et al. 2011). The results have demonstrated that MALDI-TOF MS profiling is a rapid, cost-effective, and reliable method, allowing classification of most species of *Fructobacillus*, *Leuconostoc*, and *Weissella* (De Bruyne et al. 2011; Snauwaert et al. 2013).

## Genomes

Within the family *Leuconostocaceae*, seven *Leuconostoc* genomes, one *Oenococcus* genome, and one *Weissella* genome have been completed (➋ *Table 18.3*). In addition, 11 *Leuconostoc* genomes,

**Table 18.3**
*Leuconostocaceae* genomes

| Genome status (September 2012) | Species | Genome size | %GC | Genes | Proteins | Chromosome INSDC | Plasmid INSDC | References |
|---|---|---|---|---|---|---|---|---|
| Complete | *Leuconostoc carnosum* JB16 | 1.77 | 37.09 | 1769 | 1691 | CP003851 | CP003854 CP003852 CP003855 CP003853 | Jung et al. (2012a) |
| Complete | *Leuconostoc citreum* KM20 | 1.9 | 38.88 | 1903 | 1820 | DQ489736 | DQ489738 DQ489739 DQ489740 DQ489737 | Kim et al. (2008) |
| Complete | *Leuconostoc gasicomitatum* LMG 18811 | 1.95 | 36.7 | 1993 | 1912 | FN822744 | None | Johansson et al. (2011) |
| Complete | *Leuconostoc gelidum* JB7 | 1.89 | 36.7 | 1875 | 1796 | CP003839 | None | Jung et al. (2012b) |
| Complete | *Leuconostoc kimchii* IMSNU 11154 | 2.1 | 37.91 | 2209 | 2129 | CP001758 | CP001754 CP001757 CP001756 CP001753 CP001755 | Oh et al. (2010) |
| Complete | *Leuconostoc mesenteroides* subsp. *mesenteroides* ATCC 8293 | 2.08 | 37.66 | 2108 | 2005 | CP000414 | CP000415 | Makarova et al. (2006) |
| Complete | *Leuconostoc mesenteroides* subsp. *mesenteroides* J18 | 2.02 | 37.68 | 2020 | 1937 | CP003101 | CP003104 CP003102 CP003103 CP003105 CP003106 | Jung et al. (2012c) |
| Complete | *Leuconostoc* sp. C2 | 1.88 | 37.9 | 1935 | 1855 | CP002898 | None | Lee et al. (2011c) |
| Complete | *Weissella koreensis* KACC 15510 | 1.44 | 35.52 | 1428 | 1357 | CP002899 | CP002900 | Lee et al. (2011b) |
| Complete | *Oenococcus oeni* PSU-1 | 1.78 | 37.9 | 1864 | 1691 | CP000411 | None | Makarova et al. (2006) |
| Scaffolds or contigs | *Leuconostoc argentinum* KCTC 3773 | 1.72 | 42.9 | 1810 | 1759 | AEGQ00000000 | ND | Nam et al. (2010b) |
| Scaffolds or contigs | *Leuconostoc carnosum* KCTC 3525 | 3.23 | 40.9 | ND | ND | BACM00000000 | ND | Nam et al. (2011) |
| Scaffolds or contigs | *Leuconostoc citreum* LBAE C10 | 1.93 | 38.7 | 2024 | 1971 | CAGE00000000 | ND | Laguerre et al. (2012) |
| Scaffolds or contigs | *Leuconostoc citreum* LBAE C11 | 1.97 | 38.6 | 2089 | 2036 | CAGF00000000 | ND | Laguerre et al. (2012) |
| Scaffolds or contigs | *Leuconostoc citreum* LBAE E16 | 1.8 | 38.9 | 1908 | 1854 | CAGG00000000 | ND | Laguerre et al. (2012) |
| Scaffolds or contigs | *Leuconostoc fallax* KCTC 3537 | 1.64 | 37.5 | 1604 | 1551 | AEIZ00000000 | ND | Nam et al. (2010a) |
| Scaffolds or contigs | *Leuconostoc gelidum* KCTC 3527 | 1.96 | 36.6 | 1978 | 1928 | AEMI00000000 | ND | Kim et al. (2011b) |
| Scaffolds or contigs | *Leuconostoc lactis* KCTC 3528 | 2.01 | 42.6 | 2776 | 2727 | AEOR00000000 | ND | |
| Scaffolds or contigs | *Leuconostoc mesenteroides* subsp. *cremoris* ATCC 19254 | 1.64 | 37.9 | 1903 | 1847 | ACKV00000000 | ND | |
| Scaffolds or contigs | *Leuconostoc pseudomesenteroides* 4882 | 2.01 | 39.1 | 2152 | 2086 | CAKV00000000 | ND | Meslier et al. (2012) |

**Table 18.3 (continued)**

| Genome status (September 2012) | Species | Genome size | %GC | Genes | Proteins | Chromosome INSDC | Plasmid INSDC | References |
|---|---|---|---|---|---|---|---|---|
| Scaffolds or contigs | *Leuconostoc pseudomesenteroides* KCTC 3652 | 3.24 | 38.3 | 3888 | 3832 | AEOQ00000000 | ND | Kim et al. (2011a) |
| Scaffolds or contigs | *Weissella cibaria* KACC 11862 | 2.32 | ND | 2234 | 2154 | AEKT00000000 | ND | Kim et al. (2011c) |
| Scaffolds or contigs | *Weissella confusa* LBAE C39-2 | 2.28 | ND | 2237 | 2156 | CAGH00000000 | ND | Amari et al. (2012b) |
| Scaffolds or contigs | *Weissella koreensis* KCTC 3621 | 1.73 | 35.5 | 1750 | 1672 | AKGG00000000 | ND | Lee et al. (2012a) |
| Scaffolds or contigs | *Weissella paramesenteroides* ATCC 33313 | 1.96 | 37.9 | 2020 | 1952 | ACKU00000000 | ND | |
| Scaffolds or contigs | *Weissella thailandensis* fsh4-2 | ND | 40.0 | 1651 | 1437 | | | |
| | HE575133NDHE575182 | ND | | | | Benomar et al. (2011) | Scaffolds or contigs | |
| *Oenococcus kitaharae* DSM 17330 | 1.84 | 42.7 | 1878 | 1825 | | CM001398 | CM001399 | Borneman et al. (2012a) |
| Scaffolds or contigs | *Oenococcus oeni* ATCC BAA-1163 | 1.75 | 37.9 | 1678 | 1398 | AAUV00000000 | ND | |
| Scaffolds or contigs | *Oenococcus oeni* AWRIB202 | ND | ND | 1831 | 1732 | AJTO00000000 | ND | Borneman et al. (2012b) |
| Scaffolds or contigs | *Oenococcus oeni* AWRIB304 | 1.85 | 37.9 | 1844 | 1743 | AJIJ00000000 | ND | Borneman et al. (2012b) |
| Scaffolds or contigs | *Oenococcus oeni* AWRIB318 | 1.81 | 37.9 | 1798 | 1698 | ALAD00000000 | ND | Borneman et al. (2012b) |
| Scaffolds or contigs | *Oenococcus oeni* AWRIB418 | 1.84 | 37.8 | 1817 | 1739 | ALAE00000000 | ND | Borneman et al. (2012b) |
| Scaffolds or contigs | *Oenococcus oeni* AWRIB419 | 1.79 | 37.8 | 1780 | 1685 | ALAF00000000 | ND | Borneman et al. (2012b) |
| Scaffolds or contigs | *Oenococcus oeni* AWRIB422 | 1.81 | 37.9 | 1812 | 1696 | ALAG00000000 | ND | Borneman et al. (2012b) |
| Scaffolds or contigs | *Oenococcus oeni* AWRIB429 | 1.93 | 37.9 | 2161 | 2161 | ACSE00000000 | ND | Borneman et al. (2010) |
| Scaffolds or contigs | *Oenococcus oeni* AWRIB548 | 1.84 | 37.9 | 1831 | 1713 | ALAH00000000 | ND | Borneman et al. (2012b) |
| Scaffolds or contigs | *Oenococcus oeni* AWRIB553 | 1.76 | 37.7 | 1733 | 1645 | ALAI00000000 | ND | Borneman et al. (2012b) |
| Scaffolds or contigs | *Oenococcus oeni* AWRIB568 | 1.87 | 38.0 | 1879 | 1778 | ALAJ00000000 | ND | Borneman et al. (2012b) |

**Table 18.3 (continued)**

| Genome status (September 2012) | Species | Genome size | %GC | Genes | Proteins | Chromosome INSDC | Plasmid INSDC | References |
|---|---|---|---|---|---|---|---|---|
| Scaffolds or contigs | *Oenococcus oeni* AWRIB576 | 1.88 | 38.0 | 1873 | 1774 | ALAK00000000 | ND | Borneman et al. (2012b) |
| Scaffolds or contigs | *Oenococcus oeni* DSM 20252 | ND | ND | 1705 | 1616 | AJTP00000000 | ND | Borneman et al. (2012b) |
| Scaffolds or contigs | *Fructobacillus fructosus* KCTC 3544 | 1.47 | 44.6 | 1600 | 1550 | AEOP00000000 | ND | |

INSDC International Nucleotide Sequence Database Collaboration
*ND* no data

14 *Oenococcus* genomes, five *Weissella* genomes, and one *Fructobacillus* genome are available as draft genomes made up of a few to many contigs. From genome mapping, the genome size of *Leuconostocaceae* genomes has been estimated to range in size from 1.4 to 2.2 Mb (Chelo et al. 2010), and all completely sequenced genomes also fall within that size range. As noted previously, most *Leuconostocaceae* strains do contain plasmids, although spontaneous curing of plasmids frequently occurs when these strains are maintained in laboratory conditions (Brito and Paveia 1999).

The analysis of the pan genomes within the homogeneous genera *Fructobacillus*, *Leuconostoc*, *Oenococcus*, and *Weissella* has shown that the core genome within one species comprises between 67 % and 80 % of a genome (Borneman et al. 2012b; Johansson et al. 2011). This is in agreement with the core genome proportion of 60 % in the highly divergent species *Lactobacillus casei* (Broadbent et al. 2012). The size of the supragenome for a species is directly proportional to number of sequenced strains until a saturation level is reached. The saturation level corresponds to the size of the complete supragenome and can be calculated when sufficient number of strains have been sequenced (Boissy et al. 2011). The size of the supragenomes characterized for LAB species is two to three times of the size of any individual genome (Boissy et al. 2011; Borneman et al. 2012b; Broadbent et al. 2012).

It has been shown that there is a good correlation between experimentally determined DNA–DNA hybridization (DDH) and digital DDH, calculated from sequence alignment of the genome sequences (Konstantinidis and Tiedje 2005; Auch et al. 2010). This is also the case for the genomes of *Leuconostocaceae*, although the genomes and experimental DDH are not obtained from the same strains in all cases.

All fully sequenced *Leuconostocaceae* genomes contain complete or partial prophages. The prophages of *Oenococcus* have been well characterized (São-José et al. 2004), and they all use tRNA genes as attachment sites in the genome (Borneman et al. 2012b). Genomes from *Fructobacillus* and *Leuconostoc* all have four rrn operons, while genomes from *Oenococcus* have two rrn operons. *Weissella* have previously been shown to have between six and eight rrn operons (Chelo et al. 2010), but the only completed *Weissella* genome, *W. koreensis*, actually have five rrn operons. The rrn operons are usually distributed around the chromosome, except for *L. gasicomitatum* and *L. gelidum*, where all four rrn operons are concentrated on the last quarter of the chromosome.

## Phenotypic Analyses

*Leuconostocaceae* are Gram positive, asporogenous, nonmotile (with the exception of *Weissella beninensis*), chemoorganotrophic, facultative anaerobic, and catalase negative. They are unable to reduce nitrate and grow in rich media supplemented with growth factors and amino acids. *Leuconostocaceae* generate energy by substrate-level phosphorylation. Glucose is fermented heterofermentatively via 6-phosphogluconate/phosphoketolase pathway yielding lactic acid, $CO_2$, ethanol, and/or acetate. Glucose-6-phosphate dehydrogenase and xylulose-5-phosphoketolase are the key enzymes of the pathway (Garvie 1986). Earlier it was thought that *Leuconostocaceae* do not have enzyme fructose 1,6-biphosphate aldolase required for homolactic fermentation, but the genomic analyses have shown that the genes encoding this enzyme are relatively common within the family. The main morphological, metabolic, and chemotaxonomic characters of the genera of *Leuconostocaceae* are shown in *Table 18.4.*

## *Leuconostoc*[AL] van Tieghem (1878), 198[AL] emend. mut. char. Hucker and Pederson (1930), 66[AL]

*Leuconostoc* cells are spherical to ellipsoidal but may also resemble short rods, especially when grown in glucose medium or on solid medium. Cells are often seen in pairs or short chains.

**Table 18.4**
**Morphological, metabolic, and chemotaxonomic characters of genera of *Leuconostocaceae***

| | *Leuconostoc* | *Weissella* | *Oenococcus* | *Fructobacillus* |
|---|---|---|---|---|
| Morphology | Spherical to ellipsoidal | Ellipsoidal to short rods | Spherical to ellipsoidal | Rods |
| Lactic acid enantiomer from glucose | D(–) | D(–) or DL | D(–) | D(–) |
| Hydrolysis of arginine | – | +/– | – | – |
| Dextran from sucrose | –/+ | –/+ | – | – |
| Growth in 10 % ethanol | – | – | +/– | – |
| Peptidoglycan | L-Lys-L-Ser-L-$Ala_2$ or L-Lys- L-$Ala_2$ | L-Lys-L-$Ala_2$ or L-Lys-L-Ala or L-Lys-L-Ala-Gly-L-$Ala_2$ or L-Lys-L-Ala-L-Ser | L-Lys-L-Ala-L-Ser or L-Lys-L-$Ser_2$ | L-Lys-L-Ala |
| Major fatty acids | C14:0, C16:0, C16:1(9), C18:1(9), C19cycl(9) | C14:0, C16:0, C16:1(9), C17:0, C18:0, C18:1(9), C19cycl(9), C19cycl(11) | C16:0, C16:1(9), C18:1(9), C18:1(11), C19cycl(9), C19cycl (11) | C16:0, C16:1(9), C18:1(9), C18:1(11) |
| G+C content of DNA (Mol%) | 36–45 | 37–47 | 37–43 | 42–45 |

*Abbreviations*: *Lys* lysine, *Ala* alanine, *Ser* serine, *Gly* glycine, *ND* no data
Symbols: + positive reaction, – negative reaction, +/– mostly positive, only some strains negative, and –/+ mostly negative, only some strains positive

True cellular capsules are not formed. Some strains produce extracellular dextran, which forms an electron-dense coat on the cell surface.

Leuconostocs develop visible colonies usually only after three to five days of incubation at 25–30 °C. Colonies on commonly used LAB media are smooth, round, grayish white, and less than 1 mm in diameter. Unlike other leuconostocs, most of the *Leuconostoc citreum* strains are able to form yellow-pigmented colonies (Farrow et al. 1989).

The optimal growth temperature is between 20 °C and 30 °C, although most species are able grow at 37 °C. Growth at 4 °C or below has been reported for *L. gelidum*, *L. carnosum*, and *L. gasicomitatum* (Holzapfel et al. 2009). Some psychrotrophic strains grow poorly at 30 °C (Björkroth et al. 2000). Leuconostocs are non-acidophilic and prefer an initial medium pH of 6.5. Most of the species are unable to grow at pH 4.8. Growth is uniform, except when cells in long chains sediment. In stab cultures, growth is concentrated in the lower two thirds. Growth on surface plates is poor under aerobic conditions, but is stimulated when incubated anaerobically.

All leuconostocs produce predominantly D(–) enantiomer lactic acid from glucose and are unable to hydrolyze arginine. *Leuconostoc* species are difficult, sometimes impossible, to distinguish by phenotypic routine testing. Many reactions are strain dependent or are, on the other hand, shared between the different species (*Table 18.1*). Only *L. mesenteroides* subsp. *cremoris* can be easily distinguished from the other leuconostocs owing to its poor carbohydrate fermentation capability. Sugars most helpful for the differentiation of *Leuconostoc* species are L-arabinose, melibiose, and D-xylose.

*Leuconostoc* spp. metabolize glucose heterofermentatively via 6-phosphogluconate/phosphoketolase pathway, yielding lactic acid, $CO_2$, ethanol, and/or acetate. Characteristics to the pathway are that hexoses are initially oxidized to pentoses resulting in generation of NAD(P)H. Under anaerobic conditions, NAD+ is regenerated by reduction of acetyl-CoA to ethanol in a process that does not produce ATP. However, if other means to oxidize NAD(P)H are available, acetyl-CoA can be converted to acetate which doubles the amount ATP produced per unit of hexose consumed. In the presence of oxygen, strains of *L. mesenteroides* use NADH oxidases and NADH peroxidases as alternative mechanisms to regenerate NAD+ (Condon 1987). Leuconostocs are also able to re-oxidize NAD(P)H by using pyruvate, fructose, or citrate as electron acceptors. The cofermentation of several metabolites increases the production of ATP and, subsequently, the growth rate (Zaunmüller et al. 2006). Citrate metabolism was also reported to form proton motive force across the cell membrane in *L. mesenteroides* (Marty-Teysset et al. 1996) which may contribute to the enhanced growth.

Most *Leuconostoc* species have genes encoding *bd*-type cytochrome oxidase, and they do respire in the presence of heme and oxygen (Brooijmans et al. 2009; Johansson et al. 2011; Sijpesteijn 1970). Respiration enables higher biomass production than fermentation (Brooijmans et al. 2009).

Under reducing conditions, leuconostocs may ferment citrate and hexose to diacetyl and acetoin which are important flavor compounds in dairy products. The amount of diacetyl produced is strain dependent (Walker and Gilliland 1987). In a study by Schmitt et al. (1997), *Leuconostoc mesenteroides* subsp. *mesenteroides* produced diacetyl as a result of cofermentation of xylose and citrate but not from glucose and citrate. Xylose reduced the activity of lactate dehydrogenase in comparison to glucose, meaning that less pyruvate was converted to lactate in the presence of xylose. Instead, pyruvate was converted to diacetyl/acetoin. In comparison to glucose, xylose may reduce lactate dehydrogenase activity because generation of pyruvate from xylose generates less NAD(P)H, meaning that less reducing power is available for the formation of lactate from pyruvate, a reaction catalyzed by lactate dehydrogenase. Instead of diacetyl/acetoin, surplus pyruvate formed from citrate could be also converted to acetic acid with a coupled generation of ATP, but this pathway seems not to be beneficial under acidic conditions (Schmitt et al. 1997).

Fermentation of pentoses via phosphogluconate/phosphoketolase pathway generates less NAD(P)H than fermentation of hexoses. Thus, acetyl-CoA produced from pentoses can be converted to acetate without a need of an external electron acceptor for the regeneration of NAD+. Despite the supposed benefits of pentose fermentation, many *Leuconostoc* species seem to be unable to ferment the common pentoses L-arabinose, ribose, or D-xylose, when provided as the sole carbon source (➋ *Table 18.1*). The reason for this is not known. Some leuconostocs are able to co-metabolize pentoses together with other carbon sources, e.g., xylose together with citrate (Schmitt et al. 1997).

Fructose is fermented by all *Leuconostoc* spp., except by some strains of *L. mesenteroides* subsp. *cremoris*. If fructose is used as an electron acceptor, mannitol is formed. The regeneration of NAD(P)H by fructose enables the production of acetate instead of ethanol which results in gain of ATP and enhanced growth. Interestingly, this process has been investigated as a means to produce D-mannitol from fructose by leuconostocs at industrial scale (Kiviharju and Nyyssölä 2008; von Weymarn et al. 2003).

Citrate and malate are the organic acids most frequently fermented by *Leuconostoc* spp. Acetate and tartrate are not utilized. Malate is converted into L(+)-lactate and $CO_2$ by *L. mesenteroides* subsp. *mesenteroides*. Leuconostocs do not metabolize sugar alcohols other than mannitol. Glycogen and starch are generally not degraded with the exception of *L. miyukkimchii* that is able to metabolize starch (Lee et al. 2012b).

Many leuconostocs are able to form dextran from sucrose, and this property has been used as one criterion differentiating the species. However, dextran production among *L. gelidum* and *L. carnosum* is strains dependent. The ability to form dextran is often lost when serial transfers are made in media of increasing salt concentrations (Pederson and Albury 1955). Dextran production from sucrose is dependent on the growth medium (Pederson and Albury 1955).

Little is known about the production of biogenic amines by leuconostocs. No tyramine formation was detected in strains of *Leuconostoc* isolated from fresh- and vacuum-packaged meat (Edwards et al. 1987). Some strains of *L. mesenteroides* subsp. *mesenteroides*, subsp. *cremoris*, and *Leuconostoc paramesenteroides* are known to produce tyramine and tryptamine (Bover-Cid and Holzapfel 1999; de Llano et al. 1998; Moreno-Arribas et al. 2003).

The major fatty acids recorded for *Leuconostoc* spp. are myristic (C14:0), palmitic (C16:0), palmitoleic [C16:1(9)], oleic [C18:1(9)], and dihydrosterculic acid [C19-cyc(9)] (Schmitt et al. 1989; Shaw and Harding 1989; Tracey and Britz 1989). *Leuconostoc* spp. differ from *Oenococcus* spp. and *Fructobacillus* spp. in containing oleic acid, and not vaccenic [C18-1(11)] acid, as the dominant C18:1 fatty acid (Tracey and Britz 1989). *L. carnosum* and *L. gelidum* are clearly differentiated based on their fatty acid profiles (Shaw and Harding 1989).

The interpeptide bridge of the peptidoglycan in leuconostocs consists either Lys-Ser-$Ala_2$ or Lys-$Ala_2$.

## *Weissella*[VP] Collins et al. (1993, 595); emend. Padonou et al. (2010)

The genus *Weissella* harbors two different morphological types: the short rods and the ovoid-shaped cocci. Some strains, e.g., in *W. minor*, are pleomorphic. *Weissella* colonies are 1–2 mm in diameter, white to creamish white, smooth, circular, and convex after 3–4 days of anaerobic growth. Weissellas are nonmotile with the exception of *W. beninensis*, the only motile species belonging to *Leuconostocaceae*. *W. beninensis* has peritrichous flagella (Padonou et al. 2010).

Weissellas are heterofermentative lactic acid bacteria and share most of the metabolic properties with leuconostocs. Unlike leuconostocs, some *Weissella* species produce DL lactic acid from glucose (➋ *Table 18.2*). Most weissellas are able to hydrolyze arginine. Growth occurs at 15 °C, with some species growing at 42–45 °C. All species are able to grow at 37 °C and most species are able to grow at pH 4.8.

Phenotypic tests have been traditionally used to identify *Weissella* species. Cell morphology has some diagnostic value. Hydrolysis of arginine is a simple biochemical test for differentiation. A battery of ten sugars was recommended by Collins et al. (1993) to be used in combination with other phenotypic tests for identification. Among some weissellas, and particularly *W. confusa*, dextran production appears to be a common and a widespread feature.

Similar to leuconostocs, some weissellas have genes encoding *bd*-type cytochrome oxidase required for heme-dependent respiration (Kim et al. 2011c), but functional respiration chain is yet to be reported for weissellas.

Literature describing the production of biogenic amines by *Weissella* spp. is scarce. *Weissella halotolerans* W22 combines an arginine deaminase pathway and an ornithine decarboxylation pathway, which results in generation of biogenic amine putrescine and proton motive force (Pereira et al. 2009).

The cell wall peptidoglycan in weissellas is based on lysine as dipeptide, and, with the exception of *W. kandleri*, all contain alanine or alanine and serine in the interpeptide bridge. In addition, the interpeptide bridge of *W. kandleri* (Lys-L-Ala-Gly-L-$Ala_2$) contains glycine (Holzapfel and Van Wyk 1982).

Fatty acid profiles can be used to differentiate weissellas. Applying a rapid gas chromatographic method, Samelis et al. (1998) could differentiate between *W. viridescens*, *W. paramesenteroides*, *W. hellenica*, and some typical arginine-negative *Weissella* isolates from meats on the basis of their cellular fatty acid

**Table 18.5**
**Phenotypic characteristics of *Fructobacillus* spp. and *Oenococcus* spp.**

| Characteristics | *F. fructosus* | *F. durionis* | *F. ficulneus* | *F. pseudoficulneus* | *F. tropaeoli* | *O. oeni* | *O. kitaharae* |
|---|---|---|---|---|---|---|---|
| Acid from | | | | | | | |
| Galactose | – | – | – | – | – | d | + |
| Maltose | – | (+) | w | – | – | – | + |
| Mannose | – | – | – | – | – | d | + |
| Mannitol | (+) | (+) | (+) | (+) | (+) | – | – |
| Melibiose | – | – | – | – | – | d | + |
| Sucrose | – | + | w | – | – | – | – |
| Trehalose | – | (+) | (+) | – | – | + | + |
| Turanose | – | + | w | – | – | ND | ND |
| Ammonia from arginine | – | – | – | – | ND | d | ND |
| Hydrolysis of aesculin | – | – | – | – | – | + | ND |
| Peptidoglycan type | Lys-Ala | ND | Lys-Ala | ND | ND | Lys-Ala-Ser or Lys-$Ser_2$ | ND |
| Cell morphology | Rods | Rods | Rods | Rods | Rods | Coccoid to elongated cocci | Small ellipsoidal cocci |
| References | Endo et al. (2011) | Endo et al. (2011) | Endo et al. (2011) | Endo et al. (2011) | Endo et al. (2011) | Dicks et al. (1995) | Endo and Okada, (2006) |

+, 90 % or more of strains positive; –, 90 % or more of strains negative; d, 11–98 % of strains positive; (), delayed reaction; w, weakly positive
*ND* no data

composition. *W. viridescens* synthesized eicosenoic (C20:1) acid, while the other two species did not. Unlike *W. paramesenteroides*, *W. hellenica* and *W. viridescens* contained zero to low amounts of cyclopropane fatty acids with 19 carbon atoms, i.e., dihydrosterculic [C19cycl(9)], or lactobacillic acid [C19cycl(11)].

## *Oenococcus*[VP] Dicks et al. (1995) emend. Endo and Okada (2006)

*Oenococcus* species are Gram positive and nonmotile, ellipsoidal to spherical in shape. Growth in broth is slow and usually uniform. Colonies usually develop only after 5 d and are less than 1 mm in diameter.

The optimal growth temperature is between 20 °C and 30 °C. Oenococci prefer anaerobic conditions for growth. They produce D-(–)-lactate, $CO_2$, and ethanol or acetate from glucose (➔ *Table 18.5*) via a pathway not yet fully elucidated. In most species, both NAD and NADP may serve as coenzymes of the glucose-6-phosphate dehydrogenase, but in *O. oeni*, only NADP is required (Garvie 1975). Fermentation profiles of the different *O. oeni* strains vary greatly despite the genetically homogeneous nature of this species.

*O. oeni* is an important organism for malolactic fermentation (MLF) in wine and has several specific characteristics to inhabit in wine, e.g., acidophile and the ability to grow in medium containing 10 % of ethanol. These characteristics differentiate *O. oeni* from other *Leuconostocaceae*, including *O. kitaharae*. *O. kitaharae* is not acidophilic, cannot tolerate 10 % ethanol, and does not perform MLF (Endo and Okada 2006).

The citrate metabolism in *O. oeni* is conducted only when fermentable carbohydrates (e.g., glucose) are available. The cofermentation of citrate and glucose in *O. oeni* is physiologically important for the organism, as co-metabolism of citrate–glucose enhances the ATP synthesis and, consequently, increases the growth rate and biomass yield (Ramos and Santos 1996; Liu 2002).

*O. kitaharae* does not perform MLF. A stop codon has been found in the gene encoding malolactic enzyme in *O. kitaharae* (Borneman et al. 2012a; Endo and Okada 2006).

Some *O. oeni* strains may produce biogenic amines in wine (Bonnin-Jusserand et al. 2011; Izquierdo Cañas et al. 2009; Lucas et al. 2008). Gardini et al. (2005) reported tyramine formation by a strain of *O. oeni* isolated from Italian red wine. The formation of putrescine from arginine by some strains could be demonstrated (Guerini et al. 2002). However, e.g., Moreno-Arribas et al. (2003) could not detect any potential among *O. oeni* strains to form biogenic amines. Production of histamine by *O. oeni* has been extensively analyzed with contradictory results (Garcia-Moruno and Muñoz 2012).

Eighteen fatty acids are associated with *O. oeni* (Tracey and Britz 1987, 1989). The numerical analysis of the fatty acids showed four clusters defined at $r = 0.920$, with five strains unassigned. On the basis of the amounts of oleic acid [C18-1(9)] and C19-cyclopropane fatty acids, the strains of *O. oeni* could also be distinguished from each other. For the majority of *O. oeni* strains, the result obtained with the cellular fatty acid analysis confirmed the phenotypic relationships.

### *Fructobacillus*[VP] Endo and Okada (2008)

Fructobacilli are Gram-positive and nonmotile rods. They produce lactate, acetate, $CO_2$, and trace amounts of ethanol from glucose (➔ *Table 18.5*). Produced lactate is mainly D-isomer. *Fructobacillus* species prefer fructose over glucose as a carbon source. Aerobic culturing or the presence of pyruvate enhances their growth on glucose (Endo and Okada 2008). Because of the characteristics, they are classified as fructophilic LAB (Endo et al. 2009, 2011). They are usually osmotolerant and grow with 30 % (w/v) fructose, except *F. tropaeoli*. *Fructobacillus* spp. are usually poor sugar fermenters, and some of them metabolize only fructose, glucose, and mannitol. On the agar medium, they do not grow on glucose under anaerobic conditions if external electron acceptors are not supplied.

The cell wall peptidoglycan type of *F. ficulneus* is A3α. The predominant fatty acids in *F. ficulneus* and *F. fructosus* are C16:1(9), C16:0, C18:1(9), and C18:1(11) (Antunes et al. 2002).

## Isolation, Enrichment, and Maintenance Procedures

### *Leuconostoc* and *Weissella*

*Leuconostoc* and *Weissella* are isolated using rich media such those routinely used for culturing lactic acid bacteria, including All-Purpose Tween (Evans and Niven 1951), MRS (De Man et al. 1960), and Rogosa SL (Rogosa et al. 1951). A review by Schillinger and Holzapfel (2011) discusses in detail the selective and semi-selective media available and applied for isolation of lactic acid bacteria from different habitat such as meat or dairy products. If psychrotrophic species, such as *L. carnosum*, *L. gasicomitatum*, *L. gelidum*, and *L. inhae*, are expected to occur in the sample, an incubation temperature of 25 °C is recommended. For cultures on solid medium, an anaerobic atmosphere is recommended, while liquid cultures can be maintained in aerobic conditions.

Overall, neither selective agents nor growth conditions have been identified that allow growth and selective isolation of *Leuconostoc* or *Weissella* while inhibiting other lactic acid bacteria. Although selective and differential media for detection and enumeration of *Leuconostoc* have been proposed, they may give unreliable results in cases of samples with large numbers of *Pediococcus* and *Lactobacillus* which share many physiological and metabolic properties with *Leuconostoc* spp. Inclusion of vancomycin (30 μg/mL) in a growth medium may assist the selective isolation of *Leuconostoc* and *Weissella* from mixed bacterial populations. However, as some *Pediococcus* and *Lactobacillus* spp. are also resistant to vancomycin, this strategy is not entirely selective, and the identities of the isolates recovered need to be confirmed.

### *Oenococcus*

*O. oeni* is well known to need a specific growth factor. Tomato juice or grape juice is usually added to the medium to supply the growth factor. The pH of the medium is set at 4.8, as the species has a unique acidophilic characteristic. The species hardly grow under aerobic conditions and prefer anaerobic conditions. Several media have been developed to isolate because of the importance of the species in industry, and acidic tomato broth (ATB) might be one of the most well-used medium for isolation and culture of *O. oeni* (Garvie 1967b; Garvie and Mabbitt 1967). Björkroth and Holzapfel (2006) have summarized the several media for isolation of *O. oeni* from wine.

*O. kitaharae* has growth characteristics different from those of *O. oeni*. Tomato juice or grape juice does not favor the growth of *O. kitaharae*, and low pH prevents its growth. The organism needs a medium rich in nutrients and anaerobic conditions for maximum growth (Endo and Okada 2006). It was originally isolated using MRS agar containing inhibitors of aerobic fungi (sodium azide and cycloheximide). The growth was very slow and weak in MRS broth and MRS agar. Additional nutrients, e.g., half-strength brain heart infusion (BHI) broth, and anaerobic conditions are required to enhance the growth rate and biomass yield of this bacterium.

### *Fructobacillus*

As *Fructobacillus* species possess very unique physiological characteristics, selective enrichment isolation can be conducted (Endo et al. 2009). *Fructobacillus* species prefer fructose over glucose and grow very slowly on glucose under static conditions. They cannot metabolize glucose under anaerobic conditions. However, the presence of external electron acceptors, e.g., pyruvate or oxygen, enhances the growth of *Fructobacillus* species. Thus, enrichment culturing on fructose, e.g., FYP broth (Endo et al. 2009), under aerobic conditions favors their growth, as other LAB usually prefer anaerobic conditions. To inhibit the growth of aerobic bacteria and fungi in enrichment broth, sodium azide and cycloheximide are very useful. The enrichment can be streaked onto the FYP agar and incubated under aerobic conditions for further selection. Certain oxygen-tolerant LAB, e.g., *Lactobacillus plantarum*, *L. brevis*, and *Leuconostoc* spp., may grow as well, but they can be easily differentiated from *Fructobacillus* species based on the poor glucose utilization of *Fructobacillus* species. Because of their unique characteristics, *Fructobacillus* species are regarded as fructophilic LAB.

Differentiation of *Fructobacillus* species from *Lactobacillus kunkeei*, which is also a fructophilic species, requires carbohydrate fermentation patterns or molecular approaches.

## Maintenance Procedures

Most cultures on liquid or solid media remain viable for at least two to three weeks at 4–6 °C. Longer maintenance is in glycerol (10–20 % v/v) or dimethyl sulfoxide (10 % v/v) suspension at −20° (for months) or preferably at −70 °C or lower (for several years). Cultures are also well preserved in liquid nitrogen or by lyophilization (freeze-drying).

# Ecology

Leuconostocs are associated with plants and decaying plant material. They have been detected in green vegetation and roots (Hemme and Foucaud-Scheunemann 2004; Mundt 1967) and in various fermented vegetable products, such as cucumber, kimchi, cabbage, and olives (Kim and Chun 2005; Mäki 2004). In addition to plant-originated material, leuconostocs are frequent in foods of animal origin, including raw milk and dairy products, meat, poultry, and fish (Kim and Chun 2005; Björkroth and Holzapfel 2006). However, healthy warm-blooded animals, including humans, are rarely reported to carry *Leuconostoc* in the microbiota of their gut or mucous membranes, whereas leuconostocs have been recovered from the intestines of fish (Williams and Collins 1990).

*L. carnosum*, *L. gasicomitatum*, and *L. gelidum* have often been associated with food spoilage (Schillinger et al. 2006). Some modified atmosphere packaged meat- and vegetable-based foods have been prone to leuconostoc spoilage manifesting as bulging of the packages, off-odors and smells, and color changes. In addition to the publications cited in this paragraph, leuconostocs have been frequently reported to belong to microbiota of various fermented foods (see section ➋ Application).

*O. oeni* usually predominates at the end and after alcoholic fermentation in fermenting wine and plays a key role in the MLF. This is because of high resistance to $SO_2$ and ethanol in the organism as compared to other bacteria. $SO_2$ is added to wine as an antioxidant and to prevent the growth of undesirable microorganisms (Amerine et al. 1980). In the work by Carreté et al. (2002), the presence of 2 mM of $SO_2$ had no impact on MLF by *O. oeni*, but 5 mM of $SO_2$ caused considerable delay on MLF. Cell growth is not necessary to conduct MLF (Carreté et al. 2002). *O. oeni* is also a responsible organism for MLF in ciders (Sánchez et al. 2012). The cider isolates were separated from wine isolates based on the results of MANOVA analysis of PFGE (Bridier et al. 2010). This is generally supported by MLST (Bridier et al. 2010), suggesting that *O. oeni* strains have had habitat-specific evolution. Quite recently, an interesting study which found DNA of *O. oeni* in cocoa bean fermentation by metagenomic approach was reported (Illeghems et al. 2012).

*O. kitaharae* was originally isolated from a compost of distilled *shochu* residue in Japan (Endo and Okada 2006). The species was also isolated from the wastewater of a starch factory in Japan (Dr. Tomohiro Irisawa, personal communication). The preferred habitat of *O. kitaharae* is still uncertain, but compost, wastewater, sludge, and sewage are possible niches.

The habitats of *Weissella* species are variable and the sources of isolation suggest environmental (soil, vegetation) origin. *W. viridescens*, *W. halotolerans*, and *W. hellenica* have been associated with meat and meat products. *W. viridescens* may cause spoilage of cured meat due to green discoloration (Niven and Evans 1957), and it also is a prevailing spoilage LAB in Spanish blood sausage called Morcilla de Burgos (Koort et al. 2006; Diez et al. 2009; Santos et al. 2005). *W. viridescens* is considered somewhat heat resistant (Niven et al. 1954) which is not a common property for a LAB.

*W. cibaria*, *W. confusa*, *W. koreensis*, and *W. oryzae* have been detected in fermented foods of vegetable origin (Björkroth et al. 2002; Lee et al. 2002), whereas *W. confusa* has been associated with Greek salami (Samelis et al. 1994), Mexican pozol (Ampe et al. 1999), and Malaysian chili bo (Leisner et al. 1999). *Weissella cibaria* and *W. confusa* have also been associated with various types of sour doughs (Galle et al. 2010; Katina et al. 2009; Scheirlinck et al. 2007; De Vuyst et al. 2002). *W. soli* (Magnusson et al. 2002) is the only species known to originate in soil, but *W. paramesenteroides* has also been detected in soil (Chen et al. 2005). In addition, weissellas have been isolated from sediments of a coastal marsh (Zamudio-Maya et al. 2008) and lake water (Yanagida et al. 2007).

*W. ghanensis*, *W. fabaria*, and *W. fabalis* were detected in traditional heap fermentations of Ghanaian cocoa bean (De Bruyne et al. 2008, 2010; Snauwaert et al. 2013). *W. beninensis* (Padonou et al. 2010) originates from submerged fermenting cassava. Weissellas in food fermentations are further discussed in section "➋ Application" of this chapter.

*W. ceti* was isolated from beaked whales (Mesoplodon bidens); nine isolates were obtained from different organs of four animals (Vela et al. 2011).

*Fructobacillus* species can be found in several fructose-rich niches, e.g., fresh flowers and fruits. *F. fructosus* and *F. tropaeoli* were originally isolated from fresh flowers (Kodama 1956; Endo et al. 2011), and *F. ficulneus* and *F. pseudoficulneus* were originally found in ripe figs (Antunes et al. 2002; Chambel et al. 2006). Endo et al. (2009) also isolated a *F. fructosus* strain from a flower and *F. pseudoficulneus* strains from a banana peel and a fig. Moreover, *Fructobacillus* species have been found in several fermented foods produced from fruits. *F. durionis* was originally isolated from *tempoyak*, a Malaysian acid-fermented condiment made from the pulp of the durian fruit (Leisner et al. 2005). Several *Fructobacillus* species have been found in cocoa bean fermentation (Nielsen et al. 2007; Papalexandratou et al. 2011a, b) and wine (Mesas et al. 2011). Moreover, *F. fructosus* has been found from guts of several fructose-related insects, i.e., bumblebees, fruit flies, and giant ants (He et al. 2011; Koch and Schmid-Hempel 2011; Thaochan et al. 2010). This is highly interesting as *Fructobacillus* species do not grow on glucose under anaerobic conditions. They can grow well on fructose under anaerobic conditions.

## Pathogenicity and Clinical Significance

Some *Leuconostoc* species have caused infections, but most of the patients had received vancomycin, had an underlying disease, or were premature babies. These bacteria are not a risk for healthy individuals, and leuconostocs are considered as GRAS organisms (Schillinger et al. 2006). All leuconostocs are intrinsically resistant to vancomycin and other glycopeptide antibiotics; the first clinical reports were published in 1984–1985 (Buu-Hoi et al. 1985; Huygens 1993; Orberg and Sandine 1984: Elisha and Courvalin 1995).

*W. confusa* has been detected in the normal human intestinal microbiota (Stiles and Holzapfel 1997; Walter et al. 2001; Tannock et al. 1999). *W. cibaria* and *W. confusa* have been detected in clinical samples of humans and animals (Björkroth et al. 2002). *W. confusa* has been associated with bacteremia (Olano et al. 2001; Harlan et al. 2011; Salimnia et al. 2011; Lee et al. 2011a) and endocarditis (Flaherty et al. 2003) in humans. As in the case of *Leuconostoc* infection, the infection is mainly due to the natural resistance of these species to vancomycin and an underlying disease or immunosuppression of the host. In addition to human cases, *W. confusa* has been documented as a cause for a systemic infection in a non-immunocompromised primate (*Cercopitheus mona*) (Vela et al. 2003), and unknown *Weissella* strains were isolated from a diseased rainbow trout in China (Liu et al. 2009).

*Oenococcus* and *Fructobacillus* species have not been associated with disease in humans or animals.

## Application

### Meat

As commercial starter organisms for meat fermentations, leuconostocs are not as important as some *Lactobacillus* and *Pediococcus* spp. (Holzapfel 1998). However, leuconostocs and weissellas are repeatedly found in fermented meat products (Albano et al. 2009; Aymerich et al. 2006; Babic et al. 2011; Ben Belgacem et al. 2009; Benito et al. 2007; Danilovic et al. 2011; Kesmen et al. 2012; Papamanoli et al. 2003; Parente et al. 2001; Samelis et al. 1994; Tu et al. 2010), although at lower levels than lactobacilli. *L. mesenteroides* and *W. viridescens* are the species most often encountered in fermented meats, but *L. carnosum*, *L. gelidum*, *L. pseudomesenteroides*, *W. confusa*, and *W. paramesenteroides* are also reported. Weissellas and leuconostocs are associated with the production of bacteriocins (Hastings et al. 1994) which could be of importance in the fermentation process and may contribute to the microbiological safety of the final product.

### Dairy

In contrast to the lactococci, leuconostocs are not competitive growers or important producers of lactic acid in milk. The ability of certain strains to produce the flavor compound diacetyl, however, has led to their frequent incorporation into mixed strain starter cultures in products like buttermilk, butter, and quarg (cream cheese). Leuconostocs form functional associations with lactococci that ferment lactose efficiently to lactate. The subsequent acidification creates favorable conditions for the production of diacetyl from citrate by citrate-lyase-positive *Leuconostoc* strains (Vedamuthu 1994). Strain 91404 of *L. mesenteroides* subsp. *cremoris* was selected by Levata-Jovanovic and Sandine (1997) as an aroma producer in the preparation of experimental cultured buttermilk on the basis of its low diacetyl reductase activity, citrate utilization, and high diacetyl production under acidic conditions, and also because of its growth characteristics and its compatibility with *Lactococcus* strains. Fortification of ripened buttermilk with sodium citrate resulted in a significant increase of diacetyl and acetoin production during buttermilk storage at 5 °C for 2 weeks. Surplus of citrate, low pH of 4.5–4.7, a sufficient number of active, non-growing aroma producers, air incorporation during curd breaking, and low storage temperatures stimulated citrate metabolism and enhanced flavor during the 2 weeks of storage. Optimal development of *L. mesenteroides* subsp. *cremoris* appears to be dependent on the manganese content of the milk, and with values <15 μg/L, it may be outcompeted in a mixed strain starter culture. The ratio of *L. mesenteroides* subsp. *cremoris* to *Lactococcus lactis* in mixed culture is also dependent on the incubation temperature: warmer than 25 °C favors *L. lactis* (Hemme and Foucaud-Scheunemann 2004).

*L. mesenteroides* subsp. *cremoris* plays an important role in the desired $CO_2$ formation in the cheeses such as Gouda and Edam where it comprises ca. 5 % of a typical starter culture, as compared to 2–3 % for Tilsiter (Zickrick 1996). Cogan et al. (1997) studied 4,379 isolates from 35 artisanal dairy products, including 24 artisanal cheeses, and identified 10 % of the LAB strains as *Leuconostoc* spp. The reported proportions of *Leuconostoc* spp. in LAB communities found in artisanal cheeses typically vary between 1 % and 10 % (Campos et al. 2011; Fontana et al. 2010; Menendez et al. 2001; Samelis et al. 2010). Nieto-Arribas et al. (2010) characterized technical properties of 27 *Leuconostoc* isolates from Manchego cheese in order to test their potential as dairy starter cultures. Majority of the isolates belonged to *L. mesenteroides*, although *W. paramesenteroides* and *Leuconostoc lactis* were also found. All isolates grew at high concentrations of NaCl (4.0–4.5 %). They had poor acidifying capacity, no lipolytic activity, and poor capacity to produce diacetyl from citrate. Several isolates showed proteolytic activity. Most of the isolates were considered unsuitable as starter cultures because they grew poorly at pH 4.3.

Weissellas are rarely isolated from cheeses. *W. thailandensis* was a minor part of the halotolerant lactic acid bacteria community in two types of Mexican cheeses that contained 5–6 % of NaCl (Morales et al. 2011). *W. paramesenteroides* was found to be the dominant species of LAB in "dadih," a traditional fermented milk in Indonesia (Hosono et al. 1989). Zakaria et al. (1998) reported *W. paramesenteroides* as one of three predominating LAB species in dadih with different strains of *W. paramesenteroides* having different influences on its viscosity and curd syneresis.

Kefir is milk drink fermented with kefir grains that consist of bacteria and yeasts. *L. mesenteroides* has been reported to be part of the predominating microbiota in kefir strains together with lactobacilli and yeasts (Hsieh et al. 2012; Kowalczyk et al. 2011; Lin et al. 1999). The use of *L. mesenteroides* in formulated starter cultures for kefir production has also been reported (Duitschaever et al. 1987; Marshall and Cole 1985).

It is known that leuconostocs play a minor role in most traditional milk fermentations. Beukes et al. (2001) collected 15 samples of conventionally fermented milk from households in South Africa and Namibia and found that genera *Leuconostoc*, *Lactococcus*, and *Lactobacillus* predominated the microbial communities. Of the leuconostoc isolates, 83 % were identified as *L. mesenteroides* subsp. *dextranicum*. *L. citreum* was a minor group. In traditional Chinese yak milk products investigated by Bao et al. (2012), *L. mesenteroides* subsp. *mesenteroides* predominated. Yu et al. (2011) identified LAB isolated from several traditional fermented dairy products in Mongolia. Of the 668 isolates, 43 (6.4 %) were identified as *Leuconostoc lactis* or *L. mesenteroides*.

## Foods and Beverages of Plant Origin

*L. mesenteroides* subsp. *mesenteroides* plays an important role in the fermentation of vegetables such as sauerkraut and cucumbers. Although not the dominant species on cabbage at the time of shredding, *L. mesenteroides* subsp. *mesenteroides* initiates the fermentation of sauerkraut and is then succeeded by the more acid-tolerant lactobacilli (Pederson 1930; Stamer 1975). The same microbial succession was observed during fermentation of cucumbers or other pickles as well as olives (Vaughn 1985). Kimchi, a traditional Korean food, is produced by the lactic fermentation of vegetables such as Chinese cabbage, radishes, and cucumbers. Like in sauerkraut fermentation, *Leuconostocs* such as *L. citreum*, *L. gelidum*, *L. kimchii*, and *L. mesenteroides* dominate the early stages of fermentation, followed by lactobacilli (Choi et al. 2003; Kim et al. 2000a, b; Lee et al. 1997), while some *Weissella*-like strains were reported for the midstage of fermentation (Choi et al. 2003).

The sequence of LAB in vegetable fermentations is mainly dependent upon the initial load, growth rates, and salt and acid tolerances (Daeschel et al. 1987). Leuconostocs are apparently better adapted to plant materials and initiate growth more rapidly than most of the other LAB. Some leuconostocs, e.g., *L. mesenteroides* subsp. *mesenteroides*, *L. citreum*, *L. gelidum*, and *L. kimchii*, may be favored by their ability to utilize a wide selection of plant carbohydrates, such as L-arabinose, D-xylose, and sucrose (➋ *Table 18.1*). Furthermore, vegetables contain citrate and fructose, which can be utilized by leuconostocs as electron acceptors for faster growth (Zaunmüller et al. 2006). Carbon dioxide produced by leuconostocs replaces the air and creates an anaerobic atmosphere that inhibits aerobic bacteria (Steinkraus 1983).

The concentration of NaCl added to vegetables in the fermentation process affects the composition of bacterial community. *L. mesenteroides* subsp. *mesenteroides* is less salt tolerant than the other LAB involved in vegetable fermentation (Vaughn 1985). In salt stock pickles, the initial salt concentration is two- to threefold higher than that employed in sauerkraut, and *L. mesenteroides* subsp. *mesenteroides* therefore plays a less-active role in pickle fermentations (Stamer 1988).

Another important factor determining the composition of the bacterial community is the fermentation temperature. Kimchi is often fermented at chilled temperatures (−1 °C to 10 °C) which favors psychrotrophic bacteria (Eom et al. 2007), like *L. gasicomitatum* and *L. gelidum*. *W. koreensis* was identified as the species best adapted at kimchi fermentation at −1 °C (Cho et al. 2006).

Although most of the vegetable fermentations are "spontaneous," the inclusion of *Leuconostoc* strains into starter cultures appears beneficial for the fermentation process and for the development of desirable sensory traits. Using a vegetable juice medium (VJM), Gardner et al. (2001) selected LAB strains for mixed starter cultures to be used in lactic acid fermentation of carrot, beet, and cabbage. Compared to spontaneous fermentation, the inoculation of the vegetables with selected mixed starter cultures accelerated acidification and produced a more stable product. Starter cultures consisting of psychrotrophic *L. mesenteroides* have been successfully applied to accelerate the fermentation of kimchi at +4 °C (Jung et al. 2012d). According to Eom et al. (2007), *L. mesenteroides* and *L. citreum* starter cultures can be used to enhance the production of prebiotic oligosaccharides in kimchi-like foods fermented at low temperatures.

*L. mesenteroides* and *L. citreum* may be part of predominating LAB community in artisanal wheat sourdough (Corsetti et al. 2001; Robert et al. 2009) and distinctively influences the bread taste (Lönner and Prove-Akesson, 1989). *W. cibaria* and *W. confusa* are also found, although at lesser proportions (Minervini et al. 2012; Robert et al. 2009). Several leuconostocs and weissellas have been introduced to wheat sourdough for the production of exopolysaccharides from sucrose. This is considered as a means to improve the shelf life, volume, and nutritional value of bread without additives. *W. cibaria* and *W. confusa* strains are potential starter cultures for wheat and sorghum sourdoughs due to their high capacity for the production and exopolysaccharides without strong acidification (Galle et al. 2010; Katina et al. 2009).

*L. mesenteroides* subsp. *mesenteroides* is also predominant and responsible for initiating the fermentation of many traditional lactic acid-fermented foods in the tropics. High numbers of *L. mesenteroides* subsp. *mesenteroides* were isolated from starchy products like cassava (Okafor 1977) or kocho, an African acidic fermented product from false banana (*Ensete ventricosum*; Gashe 1987). Strains of *L. mesenteroides* subsp. *mesenteroides* have been found to produce a highly active linamarase, which hydrolyzes the cyanogenic glucoside linamarin present in cassava (Okafor and Ejiofor 1985). Gueguen et al. (1997) purified and characterized an intracellular β-glucosidase from a strain of *L. mesenteroides* isolated from cassava. When grown on an arbutin-containing medium, it was found to produce an intracellular β-glucosidase. Its cyanogenic activity was suggested to be of potential interest in cassava detoxification, by hydrolyzing the cyanogenic glucosides present in cassava pulp. *W. confusa* was identified as one of the LAB predominating in highly complex microbial communities in Lafun, an African traditional cassava food (Padonou et al. 2009).

Hancioglu and Karapinar (1997) studied the microflora of Boza, a traditional fermented Turkish beverage, prepared by yeast and lactic acid fermentation of cooked maize, wheat, and rice flours. Among the 77 LAB strains isolated during the fermentation, *W. paramesenteroides* (25.6 %), *L. mesenteroides* subsp. *mesenteroides* (18.6 %), *W. confusa* (7.8 %), *L. mesenteroides* subsp. *dextranicum* (7.3 %), and *O. oeni* (3.7 %) were found. *L. mesenteroides* and *Fructobacillus durionis* were part of a complex microbial community in palm wine made of *Borassus akeassii* (Ouoba et al. 2012). Palm wine was fermented at 21–30 °C and had pH of 3.5–4.1 and ethanol content of 0.3–2.7 %. *L. palmae* was originally isolated from palm wine by Ehrmann et al. (2009).

*L. mesenteroides* subsp. *mesenteroides* is also involved in the fermentation of seeds of the African oil bean tree (Antai and Ibrahim 1986) and of cocoa (Ostovar and Keeney 1973; Passos et al. 1984). Lefeber et al. (2011) tested metabolic activities of various cocoa-specific *Lactobacillus*, *Leuconostoc*, *Weissella*, and *Fructobacillus* strains in cocoa pulp simulation medium and concluded that citric acid converting, mannitol-producing, heterofermentative, and/or fructose-loving LAB strains are particularly adapted to cocoa pulp matrix. Of the investigated strains, those belonging to *Lactobacillus fermentum* were considered to be the most suitable for the process. Illeghems et al (2012) considered *Leuconostoc mesenteroides* to be only an opportunistic member of the fermentation process wherein a succession of microbial activities of yeasts, LAB, and acetic acid bacteria takes place. Several *Fructobacillus* species have been commonly seen in spontaneous cocoa bean fermentation carried out in different countries (Ecuador, Brazil and Ghana) (Camu et al. 2007; Papalexandratou et al. 2011a, b), suggesting that they play certain key roles for the fermentation. Possible roles might be fructose fermentation and oxygen consumption (Papalexandratou et al. 2011a, b).

*L. mesenteroides* subsp. *mesenteroides* is also involved in the submerged fermentation of coffee berries, practiced in some highland regions, and by which the oligosaccharide concentration decreases and monosaccharides increase, with a concomitant improvement in coffee quality (Frank and Dela Cruz 1964; Jones and Jones 1984; Müller 1996). Avallone et al. (2001) found that LAB, predominated by *L. mesenteroides*, and yeasts were the microbes mainly responsible for the coffee fermentation. *Leuconostoc holzapfelii* was originally isolated from Ethiopian coffee fermentation (De Bruyne et al. 2007).

Some leuconostocs, lactobacilli, and pediococci are associated with the early stages of fermenting grape must (juice). *Oenococcus oeni*, however, has been reported as the most important and desirable species among the LAB involved in winemaking thanks to its key role in the secondary fermentation of wine, also referred to as the "malolactic fermentation" (MLF). By their high resistance to $SO_2$ and ethanol, *O. oeni* may be present in relatively high numbers at the end of the alcoholic fermentation. At this stage, they play the major role in the production of microbiologically stable wines by converting L-malic acid to L(+)-lactic acid and $CO_2$, decreasing wine acidity by 0.1–0.3 units (Davis et al. 1985; Wibowo et al. 1985). This deacidification is particularly desirable for high-acid wine produced in cool-climate regions (Liu 2002). *Lactobacillus* spp. and *Pediococcus* spp. found in wine can also conduct MLF, but, however, these organisms sometimes cause spoilage problems by production of several undesirable volatile compounds (Bartowsky 2009). Some strains of *O. oeni* are also unsuitable for the MLF. Edwards et al. (1998) identified two *O. oeni* strains that were associated with sluggish and/or stuck fermentations and that were found to slow down some alcoholic fermentations. Better control over the MLF can be achieved by inoculating wines with a selected *O. oeni* strain (Nielsen et al. 1996; Rodríguez-Nogales et al. 2012) commercially available in the major wine-growing areas of industrialized countries.

Besides the MLF, citrate metabolism by *O. oeni* is also regarded as important for quality of wine because of the large quantity of citrate in grape juice. Citrate is generally transformed to lactate, acetate, diacetyl, acetoin, and 2,3-butanediol. These chemicals have an impact on quality of wine both positively and negatively (Bartowsky and Borneman 2011).

In addition to wine, MLF by *O. oeni* is important in fermentation of apple cider. Herrero et al. (2001) used *O. oeni* immobilized in alginate beads for controlled malolactic fermentation of cider. The rates of malic acid consumption were similar to conventional fermentation, but a lower acetic acid content and higher concentration of alcohols were detected with immobilized cells. These features were considered to have beneficial effects on the sensory properties of cider (Herrero et al. 2001). Nedovic et al. (2000) succeeded in improving cider quality and to accelerate the process by continuous fermentation with coimmobilized yeast and *O. oeni* cells.

## Dextran Production

Dextran is a glucose polymer that has many applications in medicine, separation technology, and biotechnology. The ability of *L. mesenteroides* subsp. *mesenteroides* to produce dextrans from sucrose has been exploited for the production of commercially valuable dextran on an industrial scale. In addition to dextran, leuconostocs are able to produce different types of glucose polymers (glucans) such as alternans and levans from sucrose (Cote and Ahlgren 1995). Glucans are synthesized from sucrose by large extracellular glucosyltransferase enzymes, commonly named glucansucrases. Glucosidic bond synthesis occurs without the mediation of nucleotide-activated sugars and cofactors are not necessary (Monchois et al. 1999). Glucansucrases differ in their ability to synthesize glucans with different types of glucosidic linkages (Kralj et al. 2004).

Dextransucrase is economically the most important glucansucrase. It is mainly produced by *L. mesenteroides* subsp. *mesenteroides*. To develop strategies for improved dextransucrase production, Dols et al. (1997) studied dextran production in relation to the growth and energetics of *L. mesenteroides* NRRL B-1299 during metabolism of various sugars. For sucrose-grown cultures, they found that a large fraction of sucrose is converted outside the cell by dextransucrase into dextran and fructose without supporting growth. The fraction entering the cell is phosphorylated by an inducible sucrose phosphorylase and converted to glucose-6-phosphate (G-6-P) by a constitutive phosphoglucomutase and to heterofermentative metabolites (lactate, acetate, and ethanol). Sucrose was found to support a higher growth rate than the monosaccharides.

In the presence of efficient monomer acceptors, like maltose or isomaltose, dextransucrase catalyzes the synthesis of low molecular weight oligosaccharides instead of high molecular weight dextran (Monchois et al. 1999). Some gluco-oligosaccharides have prebiotic properties, meaning that their industrial production is of interest. The structure and chain length of oligosaccharides can be tailored by changing the concentrations of sucrose and acceptor carbohydrate in the medium (Lee et al. 2008).

Maina et al. (2008) studied the production of gluco-oligosaccharides and linear dextran by *W. confusa* E392 and *L. citreum* E497. The gluco-oligosaccharides were characterized by $\alpha$-(1$\rightarrow$2) linked branches that are associated with probiotic properties. In addition, *W. confusa* E392 was found to be a good alternative to widely used *L. mesenteroides* B-512F in the production of linear dextran. Interestingly, dextransucrases of *Weissella* form a distinct phylogenetic group within glucansucrases of other lactic acid bacteria (Amari et al. 2012a).

## References


Albano H, van Reenen CA, Todorov SD, Cruz D, Fraga L, Hogg T, Dicks LMT, Teixeira P (2009) Phenotypic and genetic heterogeneity of lactic acid bacteria isolated from "Alheira", a traditional fermented sausage produced in Portugal. Meat Sci 82:389–398

Amari M, Arango LFG, Gabriel V, Robert H, Morel S, Moulis C, Gabriel B, Remaud-Siméon M, Fontagné-Faucher C (2012a) Characterization of a novel dextransucrase from *Weissella confusa* isolated from sourdough. Appl Microbiol Biotechnol 97(12):5413–5422

Amari M, Laguerre S, Vuillemin M, Robert H, Loux V, Klopp C, Morel S, Gabriel B, Remaud-Siméon M, Gabriel V, Moulis C, Fontagné-Faucher C (2012b) Genome sequence of *Weissella confusa* LBAE C39-2, isolated from a wheat sourdough. J Bacteriol 194:1608–1609

Amerine MA, Berg HW, Kunkee RE, Ough CS, Singleton VL, Webb AD (1980) The technology of wine making, 4th edn. AVI, Westport

Ampe F, Ben Omar N, Moizan C, Wacher C, Guyot J-P (1999) Polyphasic study of the spatial distribution of microorganisms in Mexican pozol, a fermented maize dough, demonstrates the need for cultivation-independent methods to investigate traditional fermentations. Appl Environ Microbiol 65:5464–5473

Antai SP, Ibrahim MH (1986) Microorganisms associated with African locust bean (*Parkia-Filicoidea*) fermentation for "dawadawa" production. J Appl Bacteriol 6l:145–148

Antunes A, Rainey FA, Nobre MF, Schumann P, Ferreira AM, Ramos A, Santos H, da Costa MS (2002) *Leuconostoc ficulneum* sp. nov., a novel lactic acid bacterium isolated from a ripe fig, and reclassification of *Lactobacillus fructosus* as *Leuconostoc fructosum* comb. nov. Int J Syst Evol Microbiol 52:647–655

Arahal DR, Sanchez E, Macian MC, Garay E (2008) Value of *recN* sequences for species identification and as a phylogenetic marker within the family "*Leuconostocaceae*". Int Microbiol 11:33–39

Auch AF, von Jan M, Klenk H-P, Göker M (2010) Digital DNA-DNA hybridization for microbial species delineation by means of genome-to-genome sequence comparison. Stand Genomic Sci 2:117–134

Avallone S, Guyot B, Brillouet JM, Olguin E, Guiraud JP (2001) Microbiological and biochemical study of coffee fermentation. Curr Microbiol 42:252–256

Aymerich T, Martín B, Garriga M, Vidal-Carou MC, Bover-Cid S, Hugas M (2006) Safety properties and molecular strain typing of lactic acid bacteria from slightly fermented sausages. J Appl Microbiol 100:40–49

Aznar R, Chenoll E (2006) Intraspecific diversity of *Lactobacillus curvatus*, *Lactobacillus plantarum*, *Lactobacillus sakei*, and *Leuconostoc mesenteroides* associated with vacuum-packed meat product spoilage analyzed by randomly amplified polymorphic DNA PCR. J Food Prot 69:2403–2410

Babic I, Markov K, Kovacevic D, Trontel A, Slavica A, Dugum J, Cvek D, Svetec IK, Posavec S, Frece J (2011) Identification and characterization of potential autochthonous starter cultures from a Croatian "brand" product "Slavonski kulen". Meat Sci 88:517–524

Bao Q, Liu W, Yu J, Wang W, Qing M, Chen X, Wang F, Zhang J, Zhang W, Qiao J, Sun T, Zhang H (2012) Isolation and identification of cultivable lactic acid bacteria in traditional yak milk products of Gansu Province in China. J Gen Appl Microbiol 58:95–105

Bartowsky EJ (2009) Bacterial spoilage of wine and approaches to minimize it. Lett Appl Microbiol 48:149–156

Bartowsky EJ, Borneman AR (2011) Genomic variations of *Oenococcus oeni* strains and the potential to impact on malolactic fermentation and aroma compounds in wine. Appl Microbiol Biotechnol 92:441–447

Bartowsky EJ, McCarthy JM, Henschke P (2003) Differentiation of Australian wine isolates of *Oenococcus oeni* using random amplified polymorphic DNA (RAPD). Aust J Grape Wine Res 9:122–126

Ben Belgacem Z, Dousset X, Prévost H, Manai M (2009) Polyphasic taxonomic studies of lactic acid bacteria associated with Tunisian fermented meat based on the heterogeneity of the 16S-23S rRNA gene intergenic spacer region. Arch Microbiol 191:711–720

Benito MJ, Martín A, Aranda E, Pérez-Nevado F, Ruiz-Moyano S, Córdoba MG (2007) Characterization and selection of autochthonous lactic acid bacteria isolated from traditional Iberian dry-fermented salchichón and chorizo sausages. J Food Sci 72:M193–M201

Benomar N, Abriouel H, Lee H, Cho G-S, Huch M, Pulido RP, Holzapfel WH, Gálvez A, Franz CMAP (2011) Genome sequence of *Weissella thailandensis* fsh4-2. J Bacteriol 193:5868

Beukes EM, Bester BH, Mostert JF (2001) The microbiology of South African traditional fermented milks. Int J Food Microbiol 63:189–197

Bilhere E, Lucas PM, Claisse O, Lonvaud-Funel A (2009) Multilocus sequence typing of *Oenococcus oeni*: detection of two subpopulations shaped by intergenic recombination. Appl Environ Microbiol 75:1291–1300

Björkroth J, Holzapfel WH (2006) Genera *Leuconostoc*, *Oenococcus* and *Weissella*. In: Dworkin M, Falkow S, Rosenberg E, Schleifer K-H, Stackebrandt E (eds) The prokaryotes: a handbook on the biology of bacteria: *Firmicutes*, *Cyanobacteria*, vol 4, 3rd edn. Springer, Dordrecht/Heidelberg/London/New York, pp 267–319

Björkroth KJ, Vandamme P, Korkeala HJ (1998) Identification and characterization of *Leuconostoc carnosum*, associated with production and spoilage of vacuum-packaged, sliced, cooked ham. Appl Environ Microbiol 64:3313–3319

Björkroth J, Geisen R, Schillinger U, Weiss N, De Vos P, Holzapfel WH, Korkeala HJ, Vandamme P (2000) Characterization of *Leuconostoc gasicomitatum* sp. nov. associated with spoiled raw tomato-marinated broiler meat strips packaged under modified atmosphere. Appl Environ Microbiol 66:3764–3772

Björkroth J, Schillinger U, Geisen R, Weiss N, Holzapfel WH, Korkeala HJ, Vandamme P (2002) Taxonomic study of *Weissella confusa* and description of *Weissella cibaria* sp. nov., a novel species detected in food and clinical samples. Int J Syst Evol Microbiol 52:141–148

Björkroth J, Dicks LMT, Holzapfel WH (2009) Genus III. *Weissella* Collins, Samelis, Metaxopoulos and Wallbanks 1994, $370^{VP}$ (Effective publication: Collins, Samelis Metaxopoulos and Wallbanks 1993, 597). In: De Vos P, Garrity GM, Jones D, Krieg NR, Ludwig W, Rainey FA, Schleifer KH, Whitman WB (eds) Bergey's manual of systematic bacteriology (The Firmicutes), vol 3, 2nd edn. Springer, Dordrecht/Heidelberg/London/New York, pp 643–653

Boissy R, Ahmed A, Janto B, Earl J, Hall BG, Hogg JS, Pusch GD, Hiller LN, Powell E, Hayes J, Yu S, Kathju S, Stoodley P, Post JC, Ehrlich GD, Hu FZ (2011) Comparative supragenomic analyses among the pathogens *Staphylococcus aureus*, *Streptococcus pneumoniae*, and *Haemophilus influenzae* using a modification of the finite supragenome model. BMC Genomics 12:187

Bonnin-Jusserand M, Grandvalet C, David V, Alexandre H (2011) Molecular cloning, heterologous expression, and characterization of ornithine decarboxylase from *Oenococcus oeni*. J Food Prot 74:1309–1314

Bonomo MG, Ricciardi A, Zotta T, Parente E, Salzano G (2008) Molecular and technological characterization of lactic acid bacteria from traditional fermented sausages of Basilicata region (Southern Italy). Meat Sci 80:1238–1248

Borneman AR, Bartowsky EJ, McCarthy J, Chambers PJ (2010) Genotypic diversity in *Oenococcus oeni* by high-density microarray comparative genome hybridization and whole genome sequencing. Appl Microbiol Biotechnol 862:681–691

Borneman AR, McCarthy JM, Chambers PJ, Bartowsky EJ (2012a) Functional divergence in the genus *Oenococcus* as predicted by genome sequencing of the newly-described species, *Oenococcus kitaharae*. PLoS One 7:e29626

Borneman AR, McCarthy JM, Chambers PJ, Bartowsky EJ (2012b) Comparative analysis of the *Oenococcus oeni* pan genome reveals genetic diversity in industrially-relevant pathways. BMC Genomics 13:373

Bounaix MS, Gabriel V, Robert H, Morel S, Remaud-Siméon M, Gabriel B, Fontagné-Faucher C (2010a) Characterization of glucan-producing *Leuconostoc* strains isolated from sourdough. Int J Food Microbiol 144:1–9

Bounaix MS, Robert H, Gabriel V, Morel S, Remaud-Siméon M, Gabriel B, Fontagné-Faucher C (2010b) Characterization of dextran-producing *Weissella* strains isolated from sourdoughs and evidence of constitutive dextransucrase expression. FEMS Microbiol Lett 311:18–26

Bover-Cid S, Holzapfel WH (1999) Improved screening procedure for biogenic amine production by lactic acid bacteria. Int J Food Microbiol 53:33–41

Bridier J, Claisse O, Coton M, Coton E, Lonvaud-Funel A (2010) Evidence of distinct populations and specific subpopulations within the species *Oenococcus oeni*. Appl Environ Microbiol 76:7754–7764

Brito L, Paveia H (1999) Presence and analysis of large plasmids in *Oenococcus oeni*. Plasmid 413:260–267

Broadbent JR, Neeno-Eckwall EC, Stahl B, Tandee K, Cai H, Morovic W, Horvath P, Heidenreich J, Perna NT, Barrangou R, Steele JL (2012) Analysis of the *Lactobacillus casei* supragenome and its influence in species evolution and lifestyle adaptation. BMC Genomics 131:533

Brooijmans R, Smit B, Santos F, van Riel J, de Vos WM, Hugenholtz J (2009) Heme and menaquinone induced electron transport in lactic acid bacteria. Microb Cell Fact 8:28

Buu-Hoi A, Branger C, Acar JF (1985) Vancomycin-resistant streptococci or *Leuconostoc* sp. Antimicrob Agents Chemother 28:458–460

Campos G, Robles L, Alonso R, Nuñez M, Picon A (2011) Microbial dynamics during the ripening of a mixed cow and goat milk cheese manufactured using frozen goat milk curd. J Dairy Sci 94:4766–4776

Camu N, De Winter T, Verbrugghe K, Cleenwerck I, Vandamme P, Takrama JS, Vancanneyt M, De Vuyst L (2007) Dynamics and biodiversity of populations of lactic acid bacteria and acetic acid bacteria involved in spontaneous heap fermentation of cocoa beans in Ghana. Appl Environ Microbiol 73:1809–1824

Cappello M, Stefani D, Grieco F, Logrieco A, Zapparoli G (2008) Genotyping by amplified fragment length polymorphism and malate metabolism performances of indigenous *Oenococcus oeni* strains isolated from Primitivo wine. Int J Food Microbiol 127:241–245

Cappello M, Zapparoli G, Stefani D, Logrieco A (2010) Molecular and biochemical diversity of *Oenococcus oeni* strains isolated during spontaneous malolactic fermentation of Malvasia Nera wine. Syst Appl Microbiol 33:461–467

Carreté R, Vidal MT, Bordons A, Constantí M (2002) Inhibitory effect of sulfur dioxide and other stress compounds in wine on the ATPase activity of *Oenococcus oeni*. FEMS Microbiol Lett 211:155–159

Chambel L, Chelo IM, Zé-Zé L, Pedro LG, Santos MA, Tenreiro R (2006) *Leuconostoc pseudoficulneum* sp. nov., isolated from a ripe fig. Int J Syst Evol Microbiol 56:1375–1381

Chelo IM, Ze-Ze L, Tenreiro R (2007) Congruence of evolutionary relationships inside the *Leuconostoc-Oenococcus-Weissella* clade assessed by phylogenetic analysis of the 16S rRNA gene, dnaA, gyrB, rpoC and dnaK. Int J Syst Evol Microbiol 57:276–286

Chelo IM, Zé-Zé L, Tenreiro R (2010) Genome diversity in the genera *Fructobacillus*, *Leuconostoc* and *Weissella* determined by physical and genetic mapping. Microbiology 156:420–430

Chen Y-S, Yanagida F, Shinohara T (2005) Isolation and identification of lactic acid bacteria from soil using an enrichment procedure. Lett Appl Microbiol 40:195–200

Chenoll E, Macián MC, Aznar R (2003) Identification of *Carnobacterium*, *Lactobacillus*, *Leuconostoc* and *Pediococcus* by rDNA-based techniques. Syst Appli Microbiol 26(4):546–556

Chenoll E, Macián MC, Elizaquível P, Aznar R (2007) Lactic acid bacteria associated with vacuum-packed cooked meat product spoilage: population analysis by rDNA-based methods. J Appl Microbiol 102(2):498–508. doi:10.1111/j.1365-2672.2006.03081.x

Cho J, Lee D, Yang C, Jeon J, Kim J, Han H (2006) Microbial population dynamics of kimchi, a fermented cabbage product. FEMS Microbiol Lett 257:262–267

Choi H-J, Cheigh C-I, Kim S-B, Lee J-C, Lee D-W, Choi S-W, Park J-M, Pyun Y-R (2002) *Weissella kimchii* sp. nov., a novel lactic acid bacterium isolated from kimchi. Int J Syst Appl Microbiol 52:507–511

Choi I-K, Jung S-H, Kim B-J, Park S-Y, Kim J, Han H-U (2003) Novel *Leuconostoc citreum* starter culture system for fermentation of kimchi, a fermented cabbage product. Antonie Van Leeuwenhoek 84(4):247–253

Cibik R, Lepage E, Talliez P (2000) Molecular diversity of *Leuconostoc* mesenteroides and *Leuconostoc* citreum isolated from traditional French cheeses as revealed by RAPD fingerprinting, 16S rDNA sequencing and 16S rDNA fragment amplification. Syst Appl Microbiol 23:267–278

Cogan TM, Barbosa M, Beuvier E, Bianchi-Salvadori B, Cocconcelli PS, Fernandes I, Gomez J, Gomez R, Kalantzopoulos G, Ledda A, Medina M, Rea MC, Rodriguez E (1997) Characterization of the lactic acid bacteria in artisanal dairy products. J Dairy Res 64:409–421

Collins MD, Samelis J, Metaxopoulus J, Wallbanks S (1993) Taxonomic studies on some leuconostoc-like organisms from fermented sausages: Description of a new genus *Weissella* for the *Leuconostoc paramesenteroides* group of species. J Appl Bacteriol 75:595–603

Condon S (1987) Responses of lactic acid bacteria to oxygen. FEMS Microbiol Lett 46:269–280

Corsetti A, Lavermicocca P, Morea M, Baruzzi F, Tosti N, Gobbetti M (2001) Phenotypic and molecular identification and clustering of lactic acid bacteria and yeasts from wheat (species *Triticum durum* and *Triticum aestivum*) sourdoughs of Southern Italy. Int J Food Microbiol 64:95–104

Cote GL, Ahlgren JA (1995) Microbial polysaccharides. In: Kroschwitz JI, Howe-Grant M (eds) Kirk-Othmer encyclopedia of chemical technology, vol 16, 4th edn. Wiley, New York, pp 578–611

Daeschel MA, Andersson RE, Fleming HP (1987) Microbial ecology of fermenting plant materials. FEMS Microbiol Rev 46:357–367

Danilovic B, Jokovic N, Petrovic L, Veljovic K, Tolinacki M, Savic D (2011) The characterisation of lactic acid bacteria during the fermentation of an artisan Serbian sausage (Petrovská Klobása). Meat Sci 88:668–674

Davis G, Silveira NFA, Fleet GH (1985) Occurrence and properties of bacteriophages of *Leuconostoc oenos* in Australian wines. Appl Environ Microbiol 50:872–876

De Bruyne K, Schillinger U, Caroline L, Boehringer B, Cleenwerck I, Vancanneyt M, De Vuyst L, Franz CMAP, Vandamme P (2007) *Leuconostoc holzapfelii* sp. nov., isolated from Ethiopian coffee fermentation and assessment of sequence analysis of housekeeping genes for delineation of *Leuconostoc* species. Int J Syst Evol Microbiol 57:2952–2959

De Bruyne K, Camu N, Lefebvre K, De Vuyst L, Vandamme P (2008) *Weissella ghanensis* sp. nov., isolated from a Ghanaian cocoa fermentation. Int J Syst Evol Microbiol 58:2721–2725

De Bruyne K, Camu N, De Vuyst L, Vandamme P (2010) *Weissella fabaria* sp. nov., from a Ghanaian cocoa fermentation. Int J Syst Evol Microbiol 60:1999–2005

De Bruyne K, Slabbinck B, Waegeman W, Vauterin P, De Baets B, Vandamme P (2011) Bacterial species identification from MALDI-TOF mass spectra through data analysis and machine learning. Syst Appl Microbiol 34(1):20–29

de Las Rivas B, Marcobal A, Munoz R (2004) Allelic diversity and population structure in *Oenococcus oeni* as determined from sequence analysis of housekeeping genes. Appl Environ Microbiol 70:7210–7219

de Llano DG, Cuesta P, Rodríguez A (1998) Biogenic amine production by wild lactococcal and leuconostoc strains. Lett Appl Microbiol 26:270–274

De Man JC, Rogosa M, Sharpe EM (1960) A medium for the cultivation of lactobacilli. J Appl Microbiol 23:130–135

De Vuyst L, Schrijvers V, Paramithiotis S, Hoste B, Vancanneyt M, Swings J, Kalantzopoulos G, Tsakalidou E, Messens W (2002) The biodiversity of lactic acid bacteria in Greek traditional wheat sourdoughs is reflected in both composition and metabolite formation. Appl Environ Microbiol 68:6059–6069

Dicks LMT, Holzapfel WH (2009) Genus II. *Oenococcus* Dicks, Dellaglio and Collins 1995a, 396[VP]. In: De Vos P, Garrity GM, Jones D, Krieg NR, Ludwig W, Rainey FA, Schleifer KH, Whitman WB (eds) Bergey's manual of systematic bacteriology (The Firmicutes), vol 3, 2nd edn. Springer, Dordrecht/Heidelberg/London/New York, pp 635–642

Dicks LM, van Vuuren HJ, Dellaglio F (1990) Taxonomy of *Leuconostoc* species, particularly *Leuconostoc oenos*, as revealed by numerical analysis of total soluble cell protein patterns, DNA base compositions, and DNA-DNA hybridizations. Int J Syst Bacteriol 40:83–91

Dicks LMT, Dellaglio F, Collins MD (1995) Proposal to reclassify *Leuconostoc oenos* as *Oenococcus oeni* corrig. gen. nov., comb. nov. Int J Syst Bacteriol 45:395–397

Diez AM, Bjorkroth J, Jaime I, Rovira J (2009) Microbial, sensory and volatile changes during the anaerobic cold storage of morcilla de Burgos previously inoculated with *Weissella viridescens* and *Leuconostoc mesenteroides*. Int J Food Microbiol 131:168–177

Dols M, Chraibi W, Remaud-Simeon M, Lindley ND, Monsan PF (1997) Growth and energetics of *Leuconostoc mesenteroides* NRRL B-1299 during metabolism of various sugars and their consequences for dextransucrase production. Appl Environ Microbiol 63:2159–2165

Duitschaever CL, Kemp N, Emmons E (1987) Pure culture formulation and procedure for the production of kefir. Milchwissenschaft 42:80–82

Edwards RA, Dainty RH, Hibbard CM, Ramantanis SV (1987) Amines in fresh beef of normal pH and the role of bacteria in changes in concentration observed during storage in vacuum packs at chill temperatures. J Appl Bacteriol 63:427–434

Edwards CG, Haag KM, Collins MD (1998) Identification and characterization of two lactic acid bacteria associated with sluggish/stuck fermentations. Am J Enol Viticult 49:445–448

Ehrmann MA, Freiding S, Vogel RF (2009) *Leuconostoc palmae* sp. nov., a novel lactic acid bacterium isolated from palm wine. Int J Syst Evol Microbiol 59:943–947

Elisha BG, Courvalin P (1995) Analysis of genes encoding D-alanine: D-alanine ligase-related enzymes in *Leuconostoc mesenteroides* and *Lactobacillus* spp. Gene 152:79–83

Endo A, Okada S (2006) *Oenococcus kitaharae* sp. nov., a non-acidophilic and non-malolactic-fermenting oenococcus isolated from a composting distilled shochu residue. Int J Syst Evol Microbiol 56:2345–2348

Endo A, Okada S (2008) Reclassification of the genus *Leuconostoc*, and proposals of *Fructobacillus fructosus* gen. nov., comb. nov., *Fructobacillus durionis* comb. nov., *Fructobacillus ficulneus* comb. nov. and *Fructobacillus pseudoficulneus* comb. nov. Int J Syst Evol Microbiol 58:2195–2205

Endo A, Futagawa-Endo Y, Dicks LMT (2009) Isolation and characterization of fructophilic lactic acid bacteria from fructose-rich niches. Syst Appl Microbiol 32:593–600

Endo A, Irisawa T, Futagawa-Endo Y, Sonomoto K, Itoh K, Takano K, Okada S, Dicks LMT (2011) *Fructobacillus tropaeoli* sp. nov., a novel fructophilic lactic acid bacterium isolated from a flower. Int J Syst Evol Microbiol 61:898–902

Ennahar S, Cai Y (2004) Genetic evidence that *Weissella kimchii* Choi et al. 2002 is a later heterotypic synonym of *Weissella cibaria* Björkroth et al. 2002. Int J Syst Evol Microbiol 54:463–465

Eom H-J, Seo DM, Han NS (2007) Selection of psychrotrophic *Leuconostoc* spp. producing highly active dextransucrase from lactate fermented vegetables. Int J Food Microbiol 117:61–67

Evans JB, Niven CF Jr (1951) Nutrition of the heterofermentative lactobacilli that cause greening of cured meat products. J Bacteriol 62:599–603

Farrow JAE, Facklam RR, Collins MD (1989) Nucleic acid homologies of some vancomycin-resistant leuconostocs and description of *Leuconostoc citreum* sp. nov. and *Leuconostoc pseudomesenteroides*. Int J Syst Bacteriol 39:279–283

Flaherty JD, Levett PN, Dewhirst FE, Troe TE, Warren JR, Johnson S (2003) Fatal case of endocarditis due to *Weissella confusa*. J Clin Microbiol 41:2237–2239

Fontana C, Cappa F, Rebecchi A, Cocconcelli PS (2010) Surface microbiota analysis of Taleggio, Gorgonzola, Casera, Scimudin and Formaggio di Fossa Italian cheeses. Int J Food Microbiol 138:205–211

Frank HA, Dela Cruz AS (1964) Role of incidental microflora in natural decomposition of mucilage layer in Kona coffee. J Food Sci 29:850–853

Galle S, Schwab C, Arendt E, Gänzle M (2010) Exopolysaccharide-forming *Weissella* strains as starter cultures for sorghum and wheat sourdoughs. J Agric Food Chem 58:5834–5841

Garcia-Moruno E, Muñoz R (2012) Does *Oenococcus oeni* produce histamine? Int J Food Microbiol 157:121–129

Gardini F, Zaccarelli A, Belletti N, Faustini F, Cavazza A, Martuscelli M, Mastrocola D, Suzzi G (2005) Factors influencing biogenic amine production by a strain of *Oenococcus oeni* in a model system. Food Control 16:609–616

Gardner NJ, Savard T, Obermeier P, Caldwell G, Champagne CP (2001) Selection and characterisation of mixed starter cultures for lactic acid fermentation of carrot, cabbage, beet and onion vegetable mixtures. Int J Food Microbiol 64:261–275

Garvie EI (1967a) The growth factor and amino acid requirements of species of the genus *Leuconostoc*, including *Leuconostoc paramesenteroides* (sp. nov.) and *Leuconostoc oenos*. J Gen Microbiol 48:439–447

Garvie EI (1967b) *Leuconostoc oenos* sp. nov. J Gen Microbiol 48:431–438

Garvie EI (1975) Some properties of gas forming lactic acid bacteria and their significance in classification. In: Carr JG, Cutting DV, Whiting GC (eds) Lactic acid bacteria in beverages and food. Academic, London

Garvie EI (1976) Hybridization between the deoxyribonucleic acids of some strains of heterofermentative lactic acid bacteria. Int J Syst Bacteriol 26:116–122

Garvie EI (1986) Genus *Leuconostoc* van Tieghem 1878, 198[AL] emend. mut. char. Hucker and Pederson 1930, 66[AL]. In: Sneath PHA, Mair NS, Sharpe MS, Holt JG (eds) Bergey's manual of systematic bacteriology. Williams and Wilkins, Baltimore, pp 1071–1075

Garvie EI, Mabbitt LA (1967) Stimulation of the growth of *Leuconostoc oenos* by tomato juice. Arch Mikrobiol 55:398–407

Gashe BA (1987) Kocho fermentation. J Appl Bacteriol 62:473–478

Gu CT, Wang F, Li CY, Liu F, Huo GC (2012) *Leuconostoc mesenteroides* subsp. *suionicum* subsp. nov. Int J Syst Evol Microbiol 62:1548–1551

Gueguen Y, Chemardin P, Labrot P, Arnaud A, Galzy P (1997) Purification and characterization of an intracellular beta-glucosidase from a new strain of *Leuconostoc mesenteroides* isolated from cassava. J Appl Microbiol 82:469–476

Guerini S, Mangani S, Granchi L, Vincenzini M (2002) Biogenic amine production by *Oenococcus oeni*. Curr Microbiol 44:374–378

Hancioglu O, Karapinar M (1997) Microflora of Boza, a traditional fermented Turkish beverage. Int J Food Microbiol 35:271–274

Harlan NP, Kempker RR, Parekh SM, Burd EM, Kuhar DT (2011) *Weissella confusa* bacteremia in a liver transplant patient with hepatic artery thrombosis. Transpl Infect Dis 13:290–293

Hastings JW, Stiles ME, von Holy A (1994) Bacteriocins of leuconostocs isolated from meat. Int J Food Microbiol 24:75–81

He H, Chen Y, Zhang Y, Wei C (2011) Bacteria associated with gut lumen of *Camponotus japonicus* Mayr. Environ Entomol 40:1405–1409

Hemme D, Foucaud-Scheunemann C (2004) *Leuconostoc*, characteristics, use in dairy technology and prospects in functional foods. Int Dairy J 14:467–494

Herrero M, Laca A, Garcia LA, Diaz M (2001) Controlled malolactic fermentation in cider using *Oenococcus oeni* immobilized in alginate beads and comparison with free cell fermentation. Enzyme Microb Technol 28:35–41

Holzapfel W (1998) The Gram-positive bacteria associated with meat and meat products. In: Davies A, Board R (eds) The microbiology of meat and poultry. Blackie Academic and Professional, London, pp 35–74

Holzapfel WH, Kandler O (1969) Zur Taxonomie der Gattung *Lactobacillus* Beijernick. VI.*Lactobacillus coprophilus* subsp. *confusus* nov. subsp., eine neue Unterart der Untergattung *Betabacterium*. Zentbl Bakteriol Parasitenkd Infektionskr Hyg 123:657–666

Holzapfel WH, van Wyk EP (1982) *Lactobacillus kandleri* sp. nov., a new species of the subgenus *Betabacterium* with glycine in the peptidoglycan. Zentbl Bakteriol Parasitenkd Infektionskr Hyg C3:495–502

Holzapfel WH, Björkroth J, Dicks LMT (2009) Genus I. *Leuconostoc* van Tieghem 1878, $198^{AL}$ emend. mut. char. (Hucker and Pederson 1930), $66^{AL}$. In: De Vos P, Garrity GM, Jones D, Krieg NR, Ludwig W, Rainey FA, Schleifer KH, Whitman WB (eds) Bergey's manual of systematic bacteriology (The Firmicutes), vol 3, 2nd edn. Springer, Dordrecht/Heidelberg/London/New York, pp 624–634

Hosono A, Wardojo R, Otani H (1989) Microbial flora in dadih, a traditional fermented milk in Indonesia. Lebensm Wiss Technol 22:20–24

Hsieh H-H, Wang S-Y, Chen T-L, Huang Y-L, Chen M-J (2012) Effects of cow's and goat's milk as fermentation media on the microbial ecology of sugary kefir grains. Int J Food Microbiol 157:73–81

Hucker GJ, Pederson CS (1930) Studies on the Coccoceae XVI. The Genus *Leuconostoc*. N.Y. Agric Exp Sta Bull 167:3–80

Huygens F (1993) Vancomycin binding to cell walls of non-streptococcal vancomycin-resistant bacteria. J Antimicrob Chemother 32:551–558

Illeghems K, De Vuyst L, Papalexandratou Z, Weckx S (2012) Phylogenetic analysis of a spontaneous cocoa bean fermentation metagenome reveals new insights into its bacterial and fungal community diversity. PLoS One 7:e38040

Izquierdo Cañas PM, Gómez Alonso S, Ruiz Pérez P, Seseña Prieto S, García Romero E, Palop Herreros ML (2009) Biogenic amine production by *Oenococcus oeni* isolates from malolactic fermentation of Tempranillo wine. J Food Prot 72:907–910

Jang J, Kim B, Lee J, Kim J, Jeong G, Han H (2002) Identification of *Weissella* species by the genus-specific amplified ribosomal DNA restriction analysis. FEMS Microbiol Lett 212:29–34

Jang J, Kim B, Lee J, Han H (2003) A rapid method for identification of typical *Leuconostoc* species by 16S rDNA PCR-RFLP analysis. J Microbiol Methods 55:295–302

Johansson P, Paulin L, Vihavainen E, Salovuori N, Alatalo ER, Björkroth KJ (2011) Genome sequence of a food spoilage lactic acid bacterium *Leuconostoc gasicomitatum* LMG 18811T in association with specific spoilage reactions. Appl Environ Microbiol 77:4344–4351

Jones KL, Jones SE (1984) Fermentations involved in the production of cocoa, coffee and tea. Prog Ind Microbiol 19:411–456

Jung JY, Lee SH, Jeon CO (2012a) Complete genome sequence of *Leuconostoc carnosum* strain JB16, isolated from kimchi. J Bacteriol 194:6672–6673

Jung JY, Lee SH, Jeon CO (2012b) Complete genome sequence of *Leuconostoc gelidum* strain JB7, isolated from kimchi. J Bacteriol 194:6665

Jung JY, Lee SH, Lee SH, Jeon CO (2012c) Complete genome sequence of *Leuconostoc mesenteroides* subsp. *mesenteroides* strain J18, isolated from kimchi. J Bacteriol 194:730–731

Jung JY, Lee SH, Lee HJ, Seo H-Y, Park W-S, Jeon CO (2012d) Effects of *Leuconostoc mesenteroides* starter cultures on microbial communities and metabolites during kimchi fermentation. Int J Food Microbiol 153:378–387

Kandler O, Abo-Elnaga IG (1966) Zur Taxonomie der Gattung *Lactobacillus* Beijerinck. IV. *L. corynoides* ein Synonym von *L. viridescens*. Zentrbl Bakteriol Parasitenkd Infektionskr Hyg 120:753–759

Kandler O, Schillinger U, Weiss N (1983) *Lactobacillus halotolerans* sp. nov, nom. rev. and *Lactobacillus minor* sp. nov., nom. rev. System Appl Microbiol 4:280–285

Katina K, Maina NH, Juvonen R, Flander L, Johansson L, Virkki L, Tenkanen M, Laitila A (2009) *In situ* production and analysis of *Weissella confusa* dextran in wheat sourdough. Food Microbiol 26:734–743

Kesmen Z, Yetiman AE, Gulluce A, Kacmaz N, Sagdic O, Cetin B, Adiguzel A, Sahin F, Yetim H (2012) Combination of culture-dependent and culture-independent molecular methods for the determination of lactic microbiota in sucuk. Int J Food Microbiol 153:428–435

Kim M, Chun J (2005) Bacterial community structure in Kimchi, a Korean fermented vegetable food, as revealed by 16S rRNA gene analysis. Int J Food Microbiol 103:91–96

Kim J, Chun J, Han H-U (2000a) *Leuconostoc kimchii* sp. nov., a new species from kimchi. Int J Syst Evol Microbiol 50:1915–1919

Kim B-J, Lee H-J, Park S-Y, Kim J, Han H-U (2000b) Identification and characterisation of *Leuconostoc gelidum* isolated from kimchi, a fermented cabbage product. J Microbiol 38:132–135

Kim B, Lee J, Jang J, Kim J, Han H (2003) *Leuconostoc inhae* sp. nov., a lactic acid bacterium isolated from kimchi. Int J Syst Evol Microbiol 53:1123–1126

Kim JF, Jeong H, Lee J-S, Choi S-H, Ha M, Hur C-G, Kim J-S, Lee S, Park H-S, Park Y-H, Oh TK (2008) The complete genome sequence of *Leuconostoc citreum* KM20. J Bacteriol 190:3093–3094

Kim D-W, Choi S-H, Kang A, Nam S-H, Kim RN, Kim A, Kim D-S, Park H-S (2011a) Genome sequence of *Leuconostoc pseudomesenteroides* KCTC 3652. J Bacteriol 193:4299

Kim D-S, Choi S-H, Kim D-W, Kim RN, Nam S-H, Kang A, Kim A, Park H-S (2011b) Genome sequence of *Leuconostoc gelidum* KCTC 3527, isolated from kimchi. J Bacteriol 193:799–800

Kim D-S, Choi S-H, Kim D-W, Nam S-H, Kim RN, Kang A, Kim A, Park H-S (2011c) Genome sequence of *Weissella cibaria* KACC 11862. J Bacteriol 193:797–798

Kiviharju K, Nyyssölä A (2008) Contributions of biotechnology to the production of mannitol. Recent Pat Biotechnol 2:73–78

Koch H, Schmid-Hempel P (2011) Bacterial communities in central European bumblebees: low diversity and high specificity. Microb Ecol 62:121–133

Kodama R (1956) Studies on the nutrition of lactic acid bacteria. Part IV. *Lactobacillus fructosus* nov. sp., a new species of lactic acid bacteria. J Agr Chem Soc Jpn 30:05–708

Konstantinidis KT, Tiedje JM (2005) Towards a genome-based taxonomy for prokaryotes. J Bacteriol 187:6258–6264

Koort J, Coenye T, Santos EM, Molinero C, Jaime I, Rovira J, Vandamme P, Bjorkroth J (2006) Diversity of *Weissella viridescens* strains associated with "Morcilla De Burgos". Int J Food Microbiol 109:164–168

Kowalczyk M, Kolakowski P, Radziwill-Bienkowska JM, Szmytkowska A, Bardowski J (2011) Cascade cell lyses and DNA extraction for identification of genes and microorganisms in kefir grains. J Dairy Res 79:26–32

Kralj S, van Geel-Schutten GH, Dondorff MMG, Kirsanovs S, van der Maarel MJEC, Dijkhuizen L (2004) Glucan synthesis in the genus *Lactobacillus*: isolation and characterization of glucansucrase genes, enzymes and glucan products from six different strains. Microbiology 150:3681–3690

Laguerre S, Amari M, Vuillemin M, Robert H, Loux V, Klopp C, Morel S, Gabriel B, Remaud-Siméon M, Gabriel V, Moulis C, Fontagné-Faucher C (2012) Genome sequences of three *Leuconostoc citreum* strains, LBAE C10, LBAE C11, and LBAE E16, isolated from wheat sourdoughs. J Bacteriol 194:1610–1611

Le Jeune C, Lonvaud-Funel A (1997) Sequence of DNA 16S/23S spacer region of *Leuconostoc oenos* (*Oenococcus oeni*): application to strain differentiation. Res Microbiol 148:79–86

Lee J-S, Chun CO, Hector M, Kim S-B, Kim H-J, Park B-K, Joo Y-J, Lee H-J, Park C-S, Ahn J-S, Park Y-H, Mheen T-L (1997) Identification of *Leuconostoc* strains isolated from kimchi using carbon-source utilisation patterns. J Microbiol 35:10–14

Lee J-S, Lee KC, Ahn J-S, Mheen T-I, Pyun Y-R, Park Y-H (2002) *Weissella koreensis* sp. nov., isolated from kimchi. Int J Syst Evol Microbiol 52:1257–1261

Lee MS, Cho SK, Eom H-J, Kim S-Y, Kim T-J, Han NS (2008) Optimized substrate concentrations for production of long-chain isomaltooligosaccharides using dextransucrase of *Leuconostoc mesenteroides* B-512F. J Microbiol Biotechnol 18:1141–1145

Lee MR, Huang YT, Liao CH, Lai CC, Lee PI, Hsueh PR (2011a) Bacteraemia caused by *Weissella confusa* at a university hospital in Taiwan, 1997–2007. Clin Microbiol Infect 17:1226–1231

Lee SH, Jung JY, Lee SH, Jeon CO (2011b) Complete genome sequence of *Weissella koreensis* KACC 15510, isolated from kimchi. J Bacteriol 193:5534

Lee SH, Jung JY, Lee SH, Jeon CO (2011c) Complete genome sequence of *Leuconostoc kimchii* strain C2, isolated from kimchi. J Bacteriol 193:5548

Lee JH, Bae J-W, Chun J (2012a) Draft genome sequence of *Weissella koreensis* KCTC 3621T. J Bacteriol 194:5711–5712

Lee SH, Park MS, Jung JY, Jeon CO (2012b) *Leuconostoc miyukkimchii* sp. nov., isolated from brown algae (Undaria pinnatifida) kimchi. Int J Syst Evol Microbiol 62:1098–1103

Lefeber T, Janssens M, Moens F, Gobert W, De Vuyst L (2011) Interesting starter culture strains for controlled cocoa bean fermentation revealed by simulated cocoa pulp fermentations of cocoa-specific lactic acid bacteria. Appl Environ Microbiol 77:6694–6698

Leisner JJ, Pot B, Christensen H, Rusul G, Olsen JE, Wee BW, Muhamad K, Ghazali HM (1999) Identification of lactic acid bacteria from Chili Bo, a Malaysian food ingredient. Appl Environ Microbiol 65:599–605

Leisner JJ, Vancanneyt M, van der Meulen R, Lefebvre K, Engelbeen K, Hoste B, Laursen BG, Bay L, Rusul G, de Vuyst L, Swings J (2005) *Leuconostoc durionis* sp. nov., a heterofermenter with no detectable gas production from glucose. Int J Syst Evol Microbiol 55:1267–1270

Levata-Jovanovic M, Sandine WE (1997) A method to use *Leuconostoc mesenteroides* ssp. cremoris 91404 to improve milk fermentations. J Dairy Sci 80:11–18

Lin CW, Chen HL, Liu JR (1999) Identification and characterisation of lactic acid bacteria and yeasts isolated from kefir grains in Taiwan. Aust J Dairy Technol 54:14–18

Liu SQ (2002) A review: malolactic fermentation in wine – beyond deacidification. J Appl Microbiol 92:589–601

Liu JY, Li AH, Ji C, Yang WM (2009) First description of a novel *Weissella* species as an opportunistic pathogen for rainbow trout *Oncorhynchus mykiss* (Walbaum) in China. Vet Microbiol 136:314–320

Lönner C, Prove-Akesson K (1989) Effects of lactic acid bacteria on the properties of sour dough bread. Food Microbiol 6:19–35

Lucas PM, Claisse O, Lonvaud-Funel A (2008) High frequency of histamine-producing bacteria in the enological environment and instability of the histidine decarboxylase production phenotype. Appl Environ Microbiol 74:811–817

Lucena BT, dos Santos BM, Moreira JL, Moreira AP, Nunes AC, Azevedo V, Miyoshi A, Thompson FL, de Morais MA Jr (2010) Diversity of lactic acid bacteria of the bioethanol process. BMC Microbiol 23:298

Magnusson J, Jonsson H, Schnürer J, Roos S (2002) *Weissella soli* sp. nov., a lactic acid bacterium isolated from soil. Int J Syst Evol Microbiol 52:831–834

Maina NH, Tenkanen M, Maaheimo H, Juvonen R, Virkki L (2008) NMR spectroscopic analysis of exopolysaccharides produced by *Leuconostoc citreum* and *Weissella confusa*. Carbohydr Res 343:1446–1455

Makarova K, Slesarev A, Wolf Y, Sorokin A, Mirkin B, Koonin E, Pavlov A, Pavlova N, Karamychev V, Polouchine N, Shakhova V, Grigoriev I, Lou Y, Rohksar D, Lucas S, Huang K, Goodstein DM, Hawkins T, Plengvidhya V, Welker D, Hughes J, Goh Y, Benson A, Baldwin K, Lee JH, Díaz-Muñiz I, Dosti B, Smeianov V, Wechter W, Barabote R, Lorca G, Altermann E, Barrangou R, Ganesan B, Xie Y, Rawsthorne H, Tamir D, Parker C, Breidt F, Broadbent J, Hutkins R, O'Sullivan D, Steele J, Unlu G, Saier M, Klaenhammer T, Richardson P, Kozyavkin S, Weimer B, Mills D (2006) Comparative genomics of the lactic acid bacteria. Proc Natl Acad Sci U S A 103:15611–15616

Mäki M (2004) Lactic acid bacteria in vegetable fermentations. In: Salminen S, von Wright A, Ouwehand A (eds) Lactic acid bacteria. Microbiological and functional aspects, 3rd edn. Marcel Dekker, New York/Basel, pp 419–430

Marcobal A, Sela DA, Wolf YI, Makarova KS, Mills DA (2008) Role of hypermutability in the evolution of the genus *Oenococcus*. J Bacteriol 190:564–570

Marshall VM, Cole WM (1985) Methods for making kefir and fermented milks based on kefir. J Dairy Res 52:451–456

Martinez-Murcia AJ, Collins MD (1990) A phylogenetic analysis of the genus *Leuconostoc* based on reverse transcriptase sequencing of 16S rRNA. FEMS Microbiol Lett 70:73–84

Martinez-Murcia AJ, Harland NM, Collins MD (1993) Phylogenetic analysis of some leuconostocs and related organisms as determined from large-subunit rRNA gene sequences: assessment of congruence of small- and large-subunit rRNA derived trees. J Appl Bacteriol 74:532–541

Marty-Teysset C, Posthuma C, Lolkema JS, Schmitt P, Divies C, Konings WN (1996) Proton motive force generation by citrolactic fermentation in Leuconostoc mesenteroides. J Bacteriol 178:2178–2185

Menendez S, Godínez R, Centeno JA, Rodríguez-Otero JL (2001) Microbiological, chemical and biochemical characteristics of "Tetilla" raw cows-milk cheese. Food Microbiol 18:151–158

Mesas JM, Rodríguez MC, Alegre MT (2011) Characterization of lactic acid bacteria from musts and wines of three consecutive vintages of Ribeira Sacra. Lett Appl Microbiol 52:258–268

Meslier V, Loux V, Renault P (2012) Genome sequence of *Leuconostoc pseudomesenteroides* strain 4882, Isolated from a dairy starter culture. J Bacteriol 194:6637

Minervini F, Lattanzi A, De Angelis M, Di Cagno R, Gobbetti M (2012) Influence of artisan bakery- or laboratory-propagated sourdoughs on the diversity of lactic acid bacterium and yeast microbiotas. Appl Environ Microbiol 78:5328–5340

Monchois V, Willemot RM, Monsan P (1999) Glucansucrases: mechanism of action and structure-function relationships. FEMS Microbiol Rev 23:131–151

Morales F, Morales JI, Hernández CH, Hernández-Sánchez H (2011) Isolation and partial characterization of halotolerant lactic acid bacteria from two Mexican cheeses. Appl Biochem Biotechnol 164:889–905

Moreno-Arribas MV, Polo MC, Jorganes F, Muñoz R (2003) Screening of biogenic amine production by lactic acid bacteria isolated from grape must and wine. Int J Food Microbiol 84:117–123

Morse R, Collins MD, O'Hanlon K, Wallbanks S, Richardson PT (1996) Analysis of the beta' subunit of DNA-dependent RNA polymerase does not support the hypothesis inferred from 16S rRNA analysis that *Oenococcus oeni* (formerly *Leuconostoc oenos*) is a tachytelic (fast-evolving) bacterium. Int J Syst Bacteriol 46:1004–1009

Müller G (1996) Kaffee, Kakao, Tee, Vanile, Tabak. In: Müller G, Holzapfel WH, Weber H (eds) Mikrobiologie der Lebensmittel: Lebensmittel pflanzlicher Herkunft. Behr's Verlag, Hamburg, pp 431–450

Mundt JO (1967) Spherical lactic acid-producing bacteria of southern-grown raw and processed vegetables. Appl Environ Microbiol 15:1303–1308

Nam S-H, Choi S-H, Kang A, Kim D-W, Kim D-S, Kim RN, Kim A, Park H-S (2010a) Genome sequence of *Leuconostoc fallax* KCTC 3537. J Bacteriol 193:588–589

Nam S-H, Choi S-H, Kang A, Kim D-W, Kim RN, Kim A, Park H-S (2010b) Genome sequence of *Leuconostoc argentinum* KCTC 3773. J Bacteriol 192:6490–6491

Nam S-H, Kim A, Choi S-H, Kang A, Kim D-W, Kim RN, Kim D-S, Park H-S (2011) Genome sequence of *Leuconostoc carnosum* KCTC 3525. J Bacteriol 193:6100–6101

Nedovic VA, Durieuxb A, Van Nedervelde L, Rosseels P, Vandegans J, Plaisant A, Simon J (2000) Continuous cider fermentation with co-immobilized yeast and *Leuconostoc oenos* cells. Enzyme Microb Technol 26:834–839

Nielsen J, Prahl C, Lonvaud-Funel A (1996) Malolactic fermentation in wine by direct inoculation with freeze-dried *Leuconostoc oenos* cultures. Am J Enol Vitic 47:42–48

Nielsen DS, Teniola OD, Ban-Koffi L, Owusu M, Andersson TS, Holzapfel WH (2007) The microbiology of Ghanaian cocoa fermentations analysed using culture-dependent and culture-independent methods. Int J Food Microbiol 114:168–186

Nieto-Arribas P, Seseña S, Poveda JM, Palop L, Cabezas L (2010) Genotypic and technological characterization of Leuconostoc isolates to be used as adjunct starters in Manchego cheese manufacture. Food Microbiol 27:85–93

Niven CF Jr, Evans JB (1957) *Lactobacillus viridescens* nov. spec., a heterofermentative species that produces a green discoloration of cured meat pigments. J Bacteriol 73:758–759

Niven CF Jr, Buettner LG, Evans JB (1954) Thermal tolerance studies on the heterofermentative lactobacilli that cause greening of cured meat products. Appl Microbiol 2:26–29

Oh H-M, Cho Y-J, Kim BK, Roe J-H, Kang S-O, Nahm BH, Jeong G, Han H-U, Chun J (2010) Complete genome sequence analysis of *Leuconostoc kimchii* IMSNU 11154. J Bacteriol 192:3844–3845

Okafor N (1977) Microorganisms associated with cassava fermentation for gari production. J Appl Bacteriol 42:279–284

Okafor N, Ejiofor MAN (1985) The linamarase of *Leuconostoc mesenteroides*, production, isolation and some properties. J Sci Food Agric 36:669–678

Olano A, Chua J, Schroeder S, Minari A, La Salvia M, Hall G (2001) *Weissella confusa* (basonym: *Lactobacillus confusus*) bacteremia: a case report. J Clin Microbiol 39:1604–1607

Orberg PK, Sandine WE (1984) Common occurrence of plasmid DNA and vancomycin resistance in *Leuconostoc* spp. Appl Environ Microbiol 48:1129–1133

Ostovar K, Keeney PG (1973) Isolation and characterization of microorganisms involved in the fermentation of Trinidad's cacao beans. J Food Sci 38:611–617

Ouoba LII, Kando C, Parkouda C, Sawadogo-Lingani H, Diawara B, Sutherland JP (2012) The microbiology of Bandji, palm wine of *Borassus akeassii* from Burkina Faso: identification and genotypic diversity of yeasts, lactic acid and acetic acid bacteria. J Appl Microbiol 113(6):1428–1441

Padonou SW, Nielsen DS, Hounhouigan JD, Thorsen L, Nago MC, Jakobsen M (2009) The microbiota of Lafun, an African traditional cassava food product. Int J Food Microbiol 133:22–30

Padonou SW, Schillinger U, Nielsen DS, Franz CMAP, Hansen M, Hounhouigan JD, Nago MC, Jakobsen M (2010) *Weissella beninensis* sp. nov., a motile lactic acid bacterium from submerged cassava fermentations, and emended description of the genus *Weissella*. Int J Syst Evol Microbiol 60:2193–2198

Papalexandratou Z, Falony G, Romanens E, Jimenez JC, Amores F, Daniel HM, De Vuyst L (2011a) Species diversity, community dynamics, and metabolite kinetics of the microbiota associated with traditional ecuadorian spontaneous cocoa bean fermentations. Appl Environ Microbiol 77:7698–7714

Papalexandratou Z, Vrancken G, De Bruyne K, Vandamme P, De Vuyst L (2011b) Spontaneous organic cocoa bean box fermentations in Brazil are characterized by a restricted species diversity of lactic acid bacteria and acetic acid bacteria. Food Microbiol 28:1326–1338

Papamanoli E, Tzanetakis N, Litopoulou-Tzanetaki E, Kotzekidou P (2003) Characterization of lactic acid bacteria isolated from a Greek dry-fermented sausage in respect of their technological and probiotic properties. Meat Sci 65:859–867

Parente E, Grieco S, Crudele MA (2001) Phenotypic diversity of lactic acid bacteria isolated from fermented sausages produced in Basilicata (Southern Italy). J Appl Microbiol 90:943–952

Passos FML, Silva DO, Lopez A, Ferreira CLLF, Guimaraes WV (1984) Characterization and distribution of lactic-acid bacteria from traditional cocoa bean fermentations in Bahia, Brazil. J Food Sci 49:205–208

Pederson CS (1930) Floral changes in the fermentation of sauerkraut. N Y Agric Exp Sta Techn Bull 168:137

Pederson CS, Albury MN (1955) Variation among the heterofermentative lactic acid bacteria. J Bacteriol 70:702–708

Pereira CI, Romão MVS, Lolkema JS, Crespo MTB (2009) Weissella halotolerans W22 combines arginine deiminase and ornithine decarboxylation pathways and converts arginine to putrescine. J Appl Microbiol 107:1894–1902

Ramos A, Santos H (1996) Citrate and sugar cofermentation in *Leuconostoc oenos*, a $13^{C}$ nuclear magnetic resonance study. Appl Environ Microbiol 62:2577–2585

Reguant C, Bordons A (2003) Typification of *Oenococcus oeni* strains by multiplex RAPD-PCR and study of population dynamics during malolactic fermentation. J Appl Microbiol 95:344–353

Robert H, Gabriel V, Fontagné-Faucher C (2009) Biodiversity of lactic acid bacteria in French wheat sourdough as determined by molecular characterization using species-specific PCR. Int J Food Microbiol 135:53–59

Rodas AM, Ferrer S, Pardo I (2003) 16S-ARDRA, a tool for identification of lactic acid bacteria isolated from grape must and wine. Syst Appl Microbiol 26:412–422

Rodríguez-Nogales JM, Vila-Crespo J, Fernández-Fernández E (2012) Immobilization of *Oenococcus oeni* in Lentikats® to develop malolactic fermentation in wines. Biotechnol Prog 29(1):60–65

Rogosa M, Mitchell JA, Wiseman RF (1951) A selective medium for the isolation and enumeration of oral and faecal lactobacilli. J Bacteriol 62:132–133

Salimnia H, Alangaden GJ, Bharadwaj R, Painter TM, Chandrasekar PH, Fairfax MR (2011) *Weissella confusa*: an unexpected cause of vancomycin-resistant Gram-positive bacteremia in immunocompromised hosts. Transpl Infect Dis 13:94–98

Samelis J, Maurogenakis F, Metaxopoulos J (1994) Characterisation of lactic acid bacteria isolated from naturally fermented Greek dry salami. Int J Food Microbiol 23:179–196

Samelis J, Rementzis J, Tsakalidou E, Metaxopoulos J (1998) Usefulness of rapid GC analysis of cellular fatty acids for distinguishing *Weissella viridescens*, *Weissella paramesenteroides*, *Weissella hellenica* and some nonidentifiable, arginine negative *Weissella* strains of meat origin. Syst Appl Microbiol 21:260–265

Samelis J, Kakouri A, Pappa EC, Matijasic BB, Georgalaki MD, Tsakalidou E, Rogelj A (2010) Microbial stability and safety of traditional Greek Graviera cheese: characterization of the lactic acid bacterial flora and culture-independent detection of bacteriocin genes in the ripened cheeses and their microbial consortia. J Food Prot 73:1294–1303

Sánchez A, Coton M, Coton E, Herrero M, García LA, Díaz M (2012) Prevalent lactic acid bacteria in cider cellars and efficiency of *Oenococcus oeni* strains. Food Microbiol 32:32–37

Santos EM, Diez AM, González-Fernández C, Jaime I, Rovira J (2005) Microbiological and sensory changes in "Morcilla de Burgos" preserved in air, vacuum and modified atmosphere packaging. Meat Sci 71:249–255

São-José C, Santos S, Nascimento J, Brito-Madurro AG, Parreira R, Santos MA (2004) Diversity in the lysis-integration region of oenophage genomes and evidence for multiple tRNA loci, as targets for prophage integration in *Oenococcus oeni*. Virology 325:82–95

Sato H, Yanagida F, Shinohara T, Suzuki M, Suzuki K, Yokotsuka K (2001) Intraspecific diversity of *Oenococcus oeni* isolated during red wine-making in Japan. FEMS Microbiol Lett 202:109–114

Scheirlinck I, Van der Meulen R, Van Schoor A, Vancanneyt M, De Vuyst L, Vandamme P, Huys G (2007) Influence of geographical origin and flour type on diversity of lactic acid bacteria in traditional Belgian sourdoughs. Appl Environ Microbiol 73:6262–6269

Schillinger U, Holzapfel WH (2011) Culture media for lactic acid bacteria. In: Corry J, Curtis G, Baird R (eds) Handbook of culture media for food and water microbiology. Royal Society of Chemistry, Cambridge, UK, pp 174–192

Schillinger U, Björkroth KJ, Holzapfel WH (2006) Lactic acid bacteria. In: de Blackburn CW (ed) Food spoilage microorganisms. Woodhead, Cambridge, UK, pp 541–578 (Chap 20)

Schleifer KH (2009) Family V. *Leuconostocaceae* fam. nov. In: De Vos P, Garrity GM, Jones D, Krieg NR, Ludwig W, Rainey FA, Schleifer KH, Whitman WB (eds) Bergey's manual of systematic bacteriology (The Firmicutes), vol 3, 2nd edn. Springer, Dordrecht/Heidelberg/London/New York, p 624

Schmitt P, Mathot AG, Divies C (1989) Fatty acid composition of the genus *Leuconostoc*. Milchwissenschaft-Milk Sci Int 44:556–559

Schmitt P, Vasseur C, Phalip V, Huang DQ, Diviès C, Prevost H (1997) Diacetyl and acetoin production from the co-metabolism of citrate and xylose by *Leuconostoc mesenteroides* subsp. *mesenteroides*. Appl Microbiol Biotechnol 47:715–718

Shaw BG, Harding CD (1989) *Leuconostoc gelidum* sp. nov. and *Leuconostoc carnosum* sp. nov. from chill-stored meats. Int J Syst Bacteriol 39:217–223

Sijpesteijn AK (1970) Induction of cytochrome formation and stimulation of oxidative dissimilation by hemin in *Streptococcus lactis* and *Leuconostoc mesenteroides*. Antonie Van Leeuwenhoek 36:335–348

Snauwaert I, Papalexandratou Z, De Vuyst L, Vandamme P (2013) Characterization of *Weissella fabalis* sp. nov. and *Fructobacillus tropaeoli* from spontaneous cocoa bean fermentations. Int J Syst Evol Microbiol 63(Pt 5):1709–1716. doi:10.1099/ijs.0.040311-0

Stamatakis A (2006) RAxML-VI-HPC: maximum likelihood-based phylogenetic analyses with thousands of taxa and mixed models. Bioinformatics 22:2688–2690

Stamer JR (1975) Recent developments in the fermentation of sauerkraut. In: Carr JG, Cutting CV, Whiting GC (eds) Lactic acid bacteria in beverages and food. Academic, London, pp 267–280

Stamer JR (1988) Lactic acid bacteria in fermented vegetables. In: Robinson RK (ed) Developments in food microbiology, vol 3. Elsevier, London, pp 67–85

Steinkraus KH (1983) Lactic acid fermentation in the production of foods from vegetables, cereals and legumes. Antonie Van Leeuwenhoek 49:337–348

Stiles ME, Holzapfel WH (1997) Lactic acid bacteria of foods and their current taxonomy. Int J Food Microbiol 36:1–29

Tanasupawat S, Shida O, Okada S, Komagata K (2000) *Lactobacillus acidipiscis* sp. nov. and *Weissella thailandensis* sp. nov., isolated from fermented fish in Thailand. Int J Syst Evol Microbiol 50(Pt 4):1479–1485

Tannock GW, Tilsala-Timisjarvi A, Rodtong S, Ng J, Munro K, Alatossava T (1999) Identification of Lactobacillus isolates from the gastrointestinal tract, silage, and yoghurt by 16S-23S rRNA gene intergenic spacer region sequence comparisons. Appl Environ Microbiol 65:4264–4267

Thaochan N, Drew RA, Hughes JM, Vijaysegaran S, Chinajariyawong A (2010) Alimentary tract bacteria isolated and identified with API-20E and molecular cloning techniques from Australian tropical fruit flies *Bactrocera cacuminata* and *B. tryoni*. J Insect Sci 10:1–16

Tohno M, Kitahara M, Inoue H, Uegaki R, Irisawa T, Ohkuma M, Tajima K (2012) *Weissella oryzae* sp. nov., isolated from fermented rice grain (*Oryza sativa* L. subsp. *japonica*). Int J Syst Evol Microbiol 63(Pt 4):1417–1420

Tracey RP, Britz TJ (1987) A numerical taxonomic study of *Leuconostoc oenos* strains from wine. J Appl Bacteriol 63:523–532

Tracey RP, Britz TJ (1989) Cellular fatty acid composition of *Leuconostoc oenos*. J Appl Bacteriol 66:445–456

Tu R-J, Wu H-Y, Lock Y-S, Chen M-J (2010) Evaluation of microbial dynamics during the ripening of a traditional Taiwanese naturally fermented ham. Food Microbiol 27:460–467

Van Tieghem PH (1878) Sur La Gomme De Sucrerie (*Leuconostoc mesenteroides*). Annal de Sci Nat Bot Ser 7:180–203

Vancanneyt M, Zamfir M, De Wachter M, Cleenwerck I, Hoste B, Rossi F, Dellaglio F, De Vuyst L, Swings J (2006) Reclassification of *Leuconostoc argentinum* as a later synonym of *Leuconostoc lactis*. Int J Syst Evol Microbiol 56:213–216

Vaughn RH (1985) The microbiology of vegetable fermentations. In: Wood BJB (ed) Microbiology of fermented foods, 2nd edn. Elsevier, New York, pp 49–109

Vedamuthu ER (1994) The dairy *Leuconostoc*: use in dairy products. J Dairy Sci Am Dairy Sci Assoc 77(9):2725–2737

Vela AI, Porrero C, Goyache J, Nieto A, Sanchez B, Briones V, Moreno MA, Dominguez L, Fernandez-Garayzabal JF (2003) *Weissella confusa* infection in primate (*Cercopithecus mona*). Emerg Infect Dis 9:1307–1309

Vela AI, Fernández A, de Quirós YB, Herráez P, Domínguez L, Fernández-Garayzábal JF (2011) *Weissella ceti* sp. nov., isolated from beaked whales (*Mesoplodon bidens*). Int J Syst Evol Microbiol 61:2758–2762

Vigentini I, Picozzi C, Tirelli A, Giugni A, Foschino R (2009) Survey on indigenous *Oenococcus oeni* strains isolated from red wines of Valtellina, a cold climate wine-growing Italian area. Int J Food Microbiol 136(1):123–128

Vihavainen EJ, Johanna Björkroth K (2009) Diversity of *Leuconostoc gasicomitatum* associated with meat spoilage. Int J Food Microbiol 136(1):32–36

von Weymarn FNW, Kiviharju KJ, Jääskeläinen ST, Leisola MSA (2003) Scale-up of a new bacterial mannitol production process. Biotechnol Prog 19:815–821

Walker DK, Gilliland SE (1987) Buttermilk manufacture using a combination of direct acidification and citrate fermentation by *Leuconostoc cremoris*. J Dairy Sci 70:2055–2062

Walter J, Hertel C, Tannock GW, Lis CM, Munro K, Hammes WP (2001) Detection of *Lactobacillus*, *Pediococcus*, *Leuconostoc*, and *Weissella* species in human feces by using group-specific PCR primers and denaturing gradient gel electrophoresis. Appl Environ Microbiol 67:2578–2585

Wibowo D, Eschenbruch R, Davis CR, Fleet GH, Lee TH (1985) Occurrence and growth of lactic-acid bacteria in wine: a review. Am J Enol Viticult 36:302–313

Williams AM, Collins MD (1990) Molecular taxonomic studies on *Streptococcus uberis* types I and II. Description of *Streptococcus parauberis* sp. nov. J Appl Bacteriol 68:485–490

Yanagida F, Yi-Sheng C, Masatoshi Y (2007) Isolation and characterization of lactic acid bacteria from lakes. J Basic Microbiol 47:184–190

Yang D, Woese CR (1989) Phylogenetic structure of the "Leuconostocs": an interesting case of a rapidly evolving organism. Syst Appl Microbiol 12(2):145–149

Yarza P, Ludwig W, Euzéby J, Schleifer K-H, Amann R, Amann R, Glöckner FO, osselló-Móra R (2010) Update of the all-species living tree project based on 16S and 23S rRNA sequence analyses. Syst Appl Microbiol 33:291–299

Yu J, Wang WH, Menghe BLG, Jiri MT, Wang HM, Liu WJ, Bao QH, Lu Q, Zhang JC, Wang F, Xu HY, Sun TS, Zhang HP (2011) Diversity of lactic acid bacteria associated with traditional fermented dairy products in Mongolia. J Dairy Sci 94:3229–3241

Zakaria Y, Ariga H, Urashima T, Toba T (1998) Microbiological and rheological properties of the Indonesian traditional fermented milk Dadih. Milchwissenschaft 53:30–33

Zamudio-Maya M, Narváez-Zapata J, Rojas-Herrera R (2008) Isolation and identification of lactic acid bacteria from sediments of a coastal marsh using a differential selective medium. Lett Appl Microbiol 46:402–407

Zapparoli G, Reguant C, Bordons A, Torriani S, Dellaglio F (2000) Genomic DNA fingerprinting of *Oenococcus oeni* strains by pulsed-field gel electrophoresis and randomly amplified polymorphic DNA-PCR. Curr Microbiol 40:351–355

Zapparoli G, Fracchetti F, Stefanelli E, Torriani S (2012) Genetic and phenotypic strain heterogeneity within a natural population of *Oenococcus oeni* from Amarone wine. J Appl Microbiol 113:1365–2672

Zaunmüller T, Eichert M, Richter H, Unden G (2006) Variations in the energy metabolism of biotechnologically relevant heterofermentative lactic acid bacteria during growth on sugars and organic acids. Appl Microbiol Biotechnol 72:421–429

Zavaleta AI, Martinez-Murcia AJ, Rodriguez-Valera F (1996) 16S–23S rDNA intergenic sequences indicate that *Leuconostoc oenos* is phylogenetically homogeneous. Microbiology 142:2105–2114

Zickrick K (1996) Mikrobiologie der Käse. In: Weber H (ed) Mikrobiologie der Lebensmittel: Milch und Milchprodukte. Behr's Verlag, Hamburg, pp 255–351

# 19 *Listeria monocytogenes* and the Genus *Listeria*

*Jim McLauchlin*[1] · *Catherine E. D. Rees*[2] · *Christine E. R. Dodd*[2]
[1]Public Health England, Food Water and Environmental Microbiology Services, London, UK
[2]School of Biosciences, University of Nottingham Sutton Bonnington Campus, Loughborough, Leicestershire, UK



**Abstract**

The genus *Listeria* contains ten species of Gram-positive bacteria, *L. monocytogenes, L. fleischmannii, L. grayi, L. innocua, L. ivanovii, L. marthii, L. rocourtiae, L. seeligeri, L. weihenstephanensis*, and *L. welshimeri*, and has been classified (along with members of the genus *Brochothrix: B. thermosphacta* and *B. campestris*) within the family *Listeriaceae*. Members of this family produce short rods that may form filaments. Cells stain Gram-positive, and the cell walls contain meso-diaminopimelic acid. The major lipid components include saturated straight-chain and methyl-branched fatty acids. Endospores are not produced; menaquinones are the sole respiratory quinones. Growth is aerobic and facultatively anaerobic; glucose is fermented to lactate and other products.

*L. monocytogenes* (and to a lesser extent *L. ivanovii*) which are pathogenic to humans and a range of other animals, and the disease is primarily transmitted by consumption of contaminated food or feed. Human listeriosis is an opportunistic infection which most often affects those with severe underlying illness, the elderly, pregnant women, and both unborn and newly delivered infants. The reported incidence of human listeriosis varies between countries from <1 to >10 cases per million of the total population. Because of the severity of infection, listeriosis is one of the major causes of death from a preventable foodborne illness. Studies of the molecular biology of *L. monocytogenes* have identified a number of virulence factors that promote uptake into nonprofessional phagocytic cells and the process of movement from cell-to-cell by recruiting host cell proteins and remodeling the host cell cytoskeleton. This has made *L. monocytogenes* also of interest both as a tool to help understand eukaryotic cell biology and as a potential therapeutic agent for intracellular delivery of drugs and as a cancer vaccine. The presence of *L. monocytogenes* remains a major challenge for the food industry. Its psychrotrophic nature means that it can grow at or below refrigeration temperatures and it is also relatively tolerant of high solute concentrations, resists desiccation, and therefore can overcome mild food preservation techniques. *L. monocytogenes* is able to form biofilms and can colonize food processing equipment and environments, leading to cross-contamination of processed foods. Hence it is of particular concern in ready-to-eat foods.

## Taxonomy, Historical and Current

Short description of the families and their genera.

Lis.ter.ri.a'ce.ae. N.L. fem. n. *Listeria* type genus of the family; suff. -aceae ending denoting family; N.L. fem. Pl. n. *Listeriaceae* the *Listeria* family.

The family *Listeriaceae* is circumscribed for this volume on the basis of phylogenetic analyses and the 16S rDNA sequences and includes all of the genus *Listeria* and *Brochothrix*. Cells are short rods that may form filaments. Cells stain Gram-positive, and the cell walls contain meso-diaminopimelic acid. The major

E. Rosenberg et al. (eds.), *The Prokaryotes – Firmicutes and Tenericutes*, DOI 10.1007/978-3-642-30120-9_210,

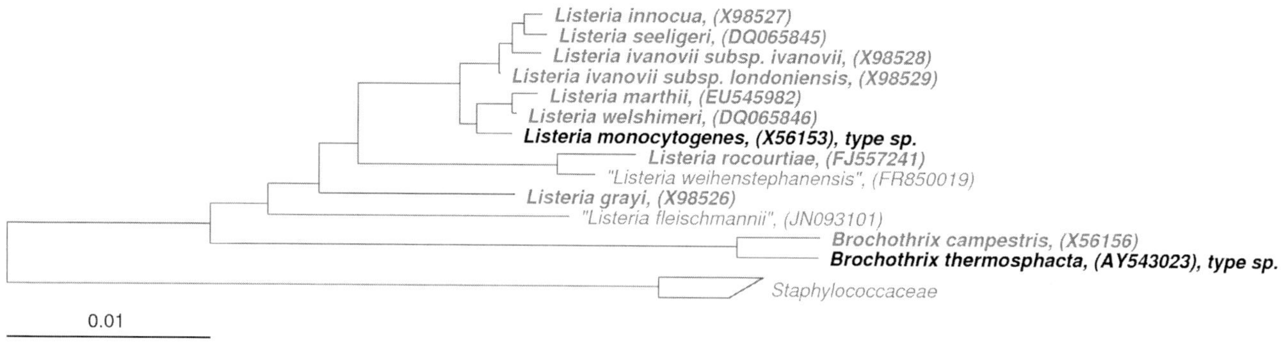


**Fig. 19.1**
**Phylogenetic reconstruction of the family *Listeriaceae* based on 16S rDNA and created using the neighbor-joining algorithm with the Jukes-Cantor correction. The sequence datasets and alignments were used according to the All-Species Living Tree Project (LTP) database (Yarza et al. 2010; http://www.arb-silva.de/projects/living-tree). Candidate species in quotation marks were added to the dataset for the tree calculation. The tree topology was stabilized with the use of a representative set of nearly 750 high quality type strain sequences proportionally distributed among the different bacterial and archaeal phyla. In addition, a 20 % maximum frequency filter was applied in order to remove hypervariable positions and potentially misplaced bases from the alignment. Scale bar indicates estimated sequence divergence**

lipid components include saturated straight-chain and methyl-branched fatty acids. Endospores are not produced; menaquinones are the sole respiratory quinones. Growth is aerobic and facultatively anaerobic; glucose is fermented to lactate and other products.

Type genus: *Listeria* Pirie 1940b.

## Genus *Listeria* Pirie 1940

Lis.te'ri.a. M.L. fem. n. *Listeria* named after Lord Lister, English surgeon and pioneer of antisepsis.

The genus *Listeria* contains ten species, *L. monocytogenes, L. fleischmannii, L. grayi, L. innocua, L. ivanovii, L. marthii, L. rocourtiae, L. seeligeri, L. weihenstephanensis*, and *L. welshimeri*. On the basis of 16S rRNA gene sequencing, members of the genus *Brochothrix* (*B. thermosphacta* and *B. campestris*) are the closest relatives to *Listeria*, which is consistent with chemical and numerical taxonomic approaches (Wilkinson and Jones 1977; Collins et al. 1979, 1991; Talon et al. 1988): these two genera justify family status as *Listeriaceae*.

The family *Listeriaceae* shows relatedness to other low G+C Gram-positive bacteria, especially to species of *Bacillus*. Comparison of the genome organizations of different *Listeria* species shows a high degree of synteny as well as a high degree of conservation in genome organization as compared to other low G+C Gram-positive genera, e.g., *Bacillus, Lactococcus, Staphylococcus*, and *Streptococcus* (Glaser et al. 2001; Buchrieser et al. 2003; Hain et al. 2006; Buchrieser 2007). A phylogenetic reconstruction of the family *Listeriaceae* based on 16S rDNA is presented in ➋ *Fig. 19.1*. This analysis indicates that *L. marthii* and *L. welshimeri* are most closely related at the rDNA level to *L. monocytogenes*. In addition, *L. innocua* is found to cluster together with *L. ivanovii* and *L. seeligeri*. This is in contrast to the relationships generally reported in the literature, when *L. innocua* is often considered to be most closely related to *L. monocytogenes*.

The analysis presented here includes the species *L. fleischmannii, L. rocourtiae*, and *L. weihenstephanensis* which have more recently been described, and the inclusion of additional sequences will influence the results gained. The phylogenetic analysis of the genus performed when *L. marthii* was first described produced a tree that placed this species close to *L. monocytogenes* but, unlike this tree, also included *L. innocua* in the group (Graves et al. 2010). Even when the same species are compared, the deduced relationships can vary depending on the choice of gene sequence included in the analysis, as seen when a tree was generated based on the analysis of sequences of 100 core genes rather than on 16S rDNA sequence alone (den Bakker et al. 2010). Interestingly this produced similar relationships to those reported here, with the exception of *L. innocua*. This analysis also identified three clades within the genus: (i) *L. monocytogenes, L. marthii, L. innocua*, and *L. welshimeri;* (ii) *L. ivanovii* and *L. seeligeri;* and (iii) *L. grayi* but *L. fleischmannii, L. rocourtiae*, and *L. weihenstephanensis* were not included in the analysis. This illustrates that phylogenic analyses produce varying relationships dependent on the characteristics or sequences chosen, and evolutionary relatedness cannot necessarily always be inferred from them.

## Type Species: *Listeria monocytogenes* (Murray, Webb and Swann) Pirie, 1940, 383AL

### Historical Considerations

Following the original isolation of *L. monocytogenes* by Murray and colleagues (Murray et al. 1926), this bacterium was named

*Bacterium monocytogenes*. The name was changed to *Listerella monocytogenes* following the recognition of the same bacterium by Pirie, which had been originally named as *Listerella hepatolytica* (Pirie 1927, 1940a). The genus was renamed *Listeria* (Pirie 1940b) since *Listerella* had already been used for another genus. The conservation of the name *Listeria* was approved by the Judicial Commission on Bacteriological Nomenclature and Taxonomy (Anon 1954).

*Listeria* was originally described as monotypic containing only *Listeria monocytogenes*. *L. monocytogenes sensu lato* was reclassified into *L. monocytogenes* (*sensu stricto;* Rocourt et al. 1982), *L. innocua* (Seeliger 1981), *L. welshimeri* (Rocourt and Grimont 1983), *L. seeligeri* (Rocourt and Grimont 1983), and *L. ivanovii* (Seeliger et al. 1984). It is not therefore possible to be certain of the species designations cited in the older literature.

The genus also includes *L. grayi* (Errebo Larsen and Seeliger 1966). The species *L. murrayi* (Welshimer and Meredith 1971) was combined with *L. grayi* to form two subspecies of *L. grayi* (Rocourt et al. 1992): *L. grayi* subsp. *grayi* and *L. grayi* subsp. *murrayi*.

*L. fleischmannii* (Bertsch et al. 2013), *L. marthii* (Graves et al. 2010), *L. rocourtiae* (Leclercq et al. 2010), and *L. weihenstephanensis* (Lang Halter et al. 2013) were subsequently described.

The species previously known as *Listeria denitrificans* (Prévot 1961) is not a member of the family *Listeriaceae* and has been reclassified into the genus *Jonesia* (Rocourt et al. 1987) which is a monogeneric family and contains two species, *J. denitrificans* and *J. quinhaiensis* (see Chap. 25, "The Families *Jonesiaceae, Ruaniaceae*, and *Borgoriellaceae*," in Volume 6, *The Prokaryotes – Actinobacteria*).

## Short Description and Main Phenotypic Properties

All members of the genus *Listeria* are widely distributed in nature and have been isolated from soil, vegetation, water, sewage, animal feed, fresh and frozen poultry, slaughter house wastes, and in the feces of healthy animals including humans. *L. monocytogenes* is pathogenic to humans and to a wide range of animals, especially sheep and goats. In contrast *L. ivanovii* is primarily a disease of animals with only very rare cases of human infection being reported (Vázquez-Boland et al. 2001). For both organisms, transmission to humans and livestock is predominantly by the consumption of contaminated food or feed. The severity of the disease caused in humans has lead to *L. monocytogenes* being recognized as one of the major foodborne bacterial pathogens. The persistence of *L. monocytogenes* at specific sites within food manufacturing environments for long periods, together with its ability to grow in a wide range of foods at low temperatures (and in foods containing sodium chloride or sodium nitrate as preservative), makes this bacterium of particular concern as a contaminant of refrigerated ready-to-eat foods (Luber et al. 2011).

All members of the genus grow in regular, short rods, 0.4–0.5 μm by 1–2 μm with parallel sides and blunt ends, usually occurring singly or in short chains. In older or rough cultures or when exposed to environmental stress conditions such as low pH and high osmotic pressure, low oxygen, and low temperatures, filaments of $\geq$6 μm in length may develop which can result in an underestimation of cell number by viable count agars as they can divide into individual cells once the stress is removed (Vail et al. 2012).

The cells are Gram-positive with even staining, but some cells, especially in older cultures, lose their ability to retain the Gram stain. The cells are not acid fast; endospores and capsules are not formed, although exopolymeric substances have been reported in biofilms (Renier et al. 2011). All species are motile with peritrichous flagella when cultured <30 °C. All members of the genus exhibit aerobic and facultatively anaerobic metabolism.

Optimal growth temperature is between 30 °C and 37 °C. Temperature limits of growth are <0–45 °C. The genus does not survive heating at 60 °C for 30 min. Growth occurs between pH 6 and pH 9 and in nutrient broth supplemented with up to 10 % (w/v) NaCl. The catalase test is positive and oxidase test negative. Cytochromes are produced. Menaquinones are the sole respiratory quinones; the major quinone contains seven isoprene units (MK-7). Acid but no gas is produced from sugars. The methyl red and Voges-Proskauer tests are positive. Exogenous citrate is not utilized. Organic growth factors are required. Indole is not produced. Aesculin and sodium hippurate are hydrolyzed. Urea, gelatin, casein, and milk are not hydrolyzed.

The cell wall contains a directly cross-linked peptidoglycan based on meso-diaminopimelic acid (meso-DAP) (variation A1γ of Schleifer and Kandler 1972): the cell wall does not contain arabinose. Mycolic acids are not present. The long-chain fatty acids consist of predominantly straight-chain saturated, anteiso-methyl branched-chain types. When grown at 37 °C, the major fatty acids are 14-methylhexadecanoic (*anteiso*-$C_{17:0}$) and 12-methyltetradecanoic (*anteiso*-$C_{15:0}$). However, when *L. monocytogenes* is exposed to cold temperatures, the membrane fatty acid profile changes so that fluidity of the membrane is retained. This leads to reduced levels of isoform fatty acids and an increase in anteiso-form fatty acids, with anteiso-C15:0 becoming the dominant fatty acid followed by anteiso-C17:0 fatty acid (Annous et al. 1997). This change is achieved due to the characteristics of the *L. monocytogenes* FabH (beta-ketoacyl-acyl carrier protein synthase III) which displays a change in its preference for 2-methylbutyryl-CoA, the precursor of odd-numbered anteiso fatty acids, as environmental temperatures drop below 30 °C (Singh et al. 2009).

## Molecular Analyses

Comparative genomics of *Listeria* species shows a high degree of synteny with very few large-scale inversions or rearrangements. This is attributed to the low number of repeat sequences and IS elements present in these genomes. The genome sizes range from 2.8 to 2.9 million bps (G+C content of 36–38 mol%), with 88–90 % being assigned to coding regions, encoding between 2,900 and 3,200 proteins. There are 6 rRNA operons and 66–67 tRNA genes (Barbuddhe et al. 2008; Buchrieser et al. 2011). *L. welshimeri* has one of the smallest genomes within the genus (Hain et al. 2006) and is also distinct in having the lowest G+C content (36.4 %, *c.f. L. monocytogenes, L. innocua, L. seeligeri*, and *L. ivanovii* = 37.0–38.0 % G+C).

Genome comparisons suggest a process of genome acquisition leading to the development of pathogenic species and also subsequent reduction resulting in the generation of "nonpathogenic" species from pathogenic progenitors. This hypothesis also explains the existence of a natural atypical *Listeria innocua* strain which could be a relic of the common ancestor of *L. monocytogenes* and *L. innocua* and is hemolytic due to the presence of the virulence gene cluster LIPI-1, but avirulent due to the absence of the *inl* operon containing the surface invasion genes (Johnson et al. 2004). Similarly divergence in the virulence gene complement of *L. monocytogenes* and *L. ivanovii* is proposed to account for both the differences in host range and pathology of these two organisms (Domínguez-Bernal et al. 2006). Comparative genomics has led to the identification of a range of unique sequences that can be used for the identification of *L. monocytogenes* by PCR amplification. Many of the targets chosen are virulence genes such as hemolysin (*hly*), the phospholipase C gene (*plcA* and *plcB* genes), and the internalin genes *inlA* and *inlB* (Jadhav et al. 2012). Less common factors such as the fibronectin-binding protein (*fbp;* Gilot and Content 2002) and the delayed-type hypersensitivity protein (*dth*-18; Wernars et al. 1991) have been chosen as diagnostic PCR targets. However, not all of these sequences are unique to *L. monocytogenes*, and care is required when interpreting results. Other targets have been chosen that are present in more members of the genus, and in this case identification is based on the amplification of PCR products of a particular size. Examples of this are the invasion-associated protein p60 (*iap*; Hein et al. 2001) and the 16S and 23S rDNA intergenic spacer regions of the rRNA operon (Graham et al. 1997; Somer and Kashi 2003).

Molecular analysis using different methods has consistently identified three distinct lineages within *L. monocytogenes* which have also been named as Genomic Divisions I, II, and III (Wiedmann et al. 1997; Call et al. 2003; Doumith et al. 2004a). The three lineages corresponded to groupings identified by serotyping (Seeliger and Höhne 1979) which was subsequently shown to be based on both teichoic acid substitutions and flagella antigens. The lineages corresponded to serovars 1/2b, 3b, 4b, 4d, and 4e (Lineage I); serovars 1/2a, 1/2c, 3a, and 3c (Lineage II); and serovars 4a, 4c, and the minority of serovar 4b (Lineage III). There are distinct differences in cell surface proteins amongst the different lineages which are independent of the antigens used in serotyping (Nelson et al. 2004). The majority of human infections are due to serovar 4b, 1/2a, and 1/2b, and almost all of the larger outbreaks are due to Lineage I serovar 4b strains. Because of difficulties in production of serotyping reagents, this method has been largely superceded by a PCR-based procedure which identifies the same groupings and is based on polymorphisms amongst surrogate genes (Doumith et al. 2004b). Since (molecular) "serotyping" provides insufficient discrimination for epidemiological purposes, various additional molecular typing methods based on detection of variation in gene sequence in localized regions have been employed including RAPD (random amplification of polymorphic DNA), AFLP (amplified fragment length polymorphism), and RFLP (restriction fragment length polymorphism) including ribotyping (Chen and Knabel 2008). However, pulse-field gel electrophoresis (PFGE) is generally accepted as a robust and reproducible method for subtyping *Listeria* isolates which can be reproducibly employed in multiple sites employing electronic comparisons (Graves and Swaminathan 2001; Gerner-Smidt et al. 2006). Whole genomic comparisons are likely to supercede PFGE in the near future for molecular typing (Lusk et al. 2012; Heger 2012).

Plasmid profiling has also been used to subtype *Listeria* isolates but has not proved to be a generally useful method (Lebrun et al. 1992). Plasmid DNA has been detected in a variety of *Listeria* species, most of which are larger than 20 MDa (Péréz-Díaz et al. 1982; Fistrovici and Collins-Thompson 1990). Most of the larger plasmids in *L. monocytogenes* have an origin that uses rolling circle replication (Kuenne et al. 2010), and a large number of these encode resistance to cadmium, and the *cadAC* operon has been found to be present on a wide range of plasmids isolated from *Listeria* (Lebrun et al. 1994a, b; Kuenne et al. 2010). A plasmid of 54 MDa was detected in the complete sequence of the *L. innocua* genome and was shown to contain 79 genes (Glaser et al. 2001). Smaller plasmids encoding for resistance to tetracycline alone (3 MDa; Poyart-Salmeron et al. 1992) or for multiresistance to the antibiotics chloramphenicol, erythromycin, streptomycin, and tetracycline (25 MDa; Poyart-Salmeron et al. 1990; Quentin et al. 1990; Hadorn et al. 1993; Tsakris et al. 1997) have been detected in *L. monocytogenes*, although these are rare. Resistance of *L. innocua* to trimethoprim has been attributed to the presence of the gene *dfrD* encoded on a 2.5 MDa plasmid (Charpentier and Courvalin 1999). A pediocin-like type II bacteriocin (LisA) has been detected on a 1.9 kDa plasmid of *L. innocua* which is closely related to the pediocin family of bacteriocins produced by lactic acid bacteria (Kalmokoff et al. 2001).

Generally, there are very few transposons found in the genomes of *Listeria* spp. A transposon, similar to *Tn*917 (designated *Tn*5422), has been detected in plasmid DNA of *L. monocytogenes* that encodes the cadmium efflux pump associated with the cadmium resistance phenotype associated with carriage of this plasmid (Lebrun et al. 1994a, b). Although transposons *Tn*1545, *Tn*916, and *Tn*917 (and their derivatives) have been introduced into *L. monocytogenes* as genetic tools (Vázquez-Boland et al. 2001), the delivery systems for these are not ideal and the transposons themselves are quite large. More recently, mariner-based transposon systems have been developed for *Listeria* which are smaller (approx. 1.5 kbp), encoding only an antibiotic resistance gene, and have a better delivery system (Cao et al. 2007; Zemansky et al. 2009). These have now been used to identify a number of novel genes involved in different aspects of *Listeria* biology including virulence (Zemansky et al. 2009); biofilm formation (Chang et al. 2012);

resistance to phage, bile, and antimicrobials (Collins et al. 2010; Dowd et al. 2011; Kim et al. 2012); and low temperature growth (Azizoglu and Kathariou 2010a, b).

Lysogenic phage are commonly carried by *Listeria* (Audurier et al. 1977; Loessner et al. 1994). They are generally morphologically similar with isometric heads and long noncontractile tails and correspond to the *Myoviridae* or *Siphoviridae* families (Rocourt et al. 1986; Loessner et al. 1994). The complete genome sequence of one lysogenic phage, A118, has been reported (Loessner et al. 2000), and phage integration was shown to occur in a homologue of the *Bacillus comK* gene, although no known function has yet been determined for this gene in *Listeria*. To date, no phage conversion has been reported due to toxin genes associated with prophage sequences. Two large, broad host range lytic phage have been sequenced (A511, 134.5 kbp; Klumpp et al. 2008 and P100, 131.4 kbp; Carlton et al. 2005) which both belong to the *Myoviridae* family and share significant DNA homology; however, neither phage is able to bind to or infect serovar 3 strains which lack the rhamnose residue in the polyribitol phosphate backbone of the cell wall teichoic acids. Recently a novel broad host range phage of the *Siphoviridae* family (P70, 67.2 kbp; Schmuki et al. 2012) has been sequenced, and despite the fact that it has very little genetic relatedness to the other two broad host range listeriaphage that have been sequenced, this is also unable to infect serovar 3 strains.

## Phenotypic Analyses

Members of the genus *Listeria* are remarkably similar in their phenotypic characteristics (Wilkinson and Jones 1977; Rocourt et al. 1983; Rocourt and Catimel 1985; Feresu and Jones 1988; Kämpfer et al. 1991; Kämpfer 1992), and a more comprehensive description of the phenotypic properties is given in McLauchlin and Rees (2009).

Colonies of all *Listeria* species growing on media containing agar show limited visual variation and after 24–48 h are 0.5–1.5 mm in diameter, round, translucent, low convex with a smooth surface and entire margin, and nonpigmented with a crystalline semitransparent central appearance. Growth may be sticky when removed from agar surfaces but usually emulsifies easily and may leave a slight impression on the agar surface after removal. Older cultures (3–7 days) are larger, 3–5 mm in diameter, and have a more opaque appearance, sometimes with a sunken center. In semisolid medium composed of 0.25 % (w/v) agar, 8 % (w/v) gelatin, and 1.0 % (w/v) glucose, growth along the stab after 24 h at 37 °C is followed by irregular, cloudy extensions into the medium. Growth spreads slowly through the entire medium, and an umbrella-like zone of maximal growth occurs 3–5 mm below the surface. Rough forms, where individual bacteria do not septate and are unusually long, occur where colonies have an undulating rough surface and an uneven edge: bacteria from these colonies usually autoagglutinate. Some of these long forms in *L. monocytogenes* are due to a defective P60 protein which has murein hydrolase activity and is required for normal septum formation (Wuenscher et al. 1993). The growth of all *Listeria* species generates a characteristic sweet caramel or buttery smell due to the generation of butyric acid. Stable L-forms of *L. monocytogenes* (including their colonial morphology) have been well described (Dell'Era et al. 2009).

Carbohydrate is essential for growth of *Listeria* strains and glucose is the usual choice. All *Listeria* species grow well on most nonselective bacteriological media including blood agar base, nutrient, tryptose, tryptose soy, or brain heart infusion agars. Growth is enhanced by the addition of a suitable fermentable carbohydrate (0.2–1 % (w/v) glucose is suitable for all species), blood, or serum. Fully chemically defined media that successfully support the growth of *L. monocytogenes* in both batch (Friedman and Roessler 1961; Trivett and Meyer 1971; Phan-Thanh and Gormon 1997) and continuous culture (Jones et al. 1995) have been described. Catabolism of glucose proceeds by the Embden-Meyerhof pathway both aerobically and anaerobically. Jones et al. (1979) reported cytochrome $abb_1$ in *L. monocytogenes* NCTC 7973, but in a later study, cytochromes $a_1bdo$ were demonstrated to be present in *L. monocytogenes*, *L. innocua*, and *L. ivanovii* (Feresu, Jones and Collins, unpublished). Under anaerobic conditions, the catabolism of glucose by *Listeria* species is homofermentative, i.e., lactate is produced exclusively (Pine et al. 1989). Under aerobic conditions, cell yields are considerably increased, and all species produce lactic, acetic, isobutyric, and isovaleric acids: there are differences between strains in the relative amounts of lactic and acetic acids produced (Pine et al. 1989). Friedman and Alm (1962) and Daneshvar et al. (1989) also reported the production of acetoin and pyruvate by *L. monocytogenes* under aerobic conditions. There is no evidence for the Entner-Doudoroff pathway, but glucose-6-phosphate dehydrogenase and 6-phosphogluconate dehydrogenase have been reported to be present (Miller and Silverman 1959). Under anaerobic conditions, only hexoses and pentoses support growth, but under aerobic conditions, maltose and lactose support growth of some species, but sucrose does not (Pine et al. 1989). No growth occurred with lactose under anaerobic conditions, but all species tested grow under anaerobic conditions and *L. grayi* utilized both the galactose and glucose moieties but *L. monocytogenes* and *L. innocua* only the glucose (Pine et al. 1989). Analysis of cultures grown at 5 °C in sterile milk suggested that glucose was the major and limiting substrate (Pine et al. 1989).

The cell wall of *L. monocytogenes* has the appearance of a thick multilayered structure, typical for Gram-positive bacteria (Ghosh and Murray 1967). The cell wall peptidoglycan contains meso-diaminopimelic acid as the diamino acid (variation A1γ of Schleifer and Kandler 1972). Alanine and glutamic acid are also present (Schleifer and Kandler 1972; Kamisango et al. 1982, 1983; Fiedler and Ruhland 1987). In addition to *N*-acetylmuramic acid and *N*-acetylglucosamine, glucosamine also occurs as a component of the cell wall polysaccharide (Ullmann and Cameron 1969; Hether et al. 1983; Fiedler and Seger 1983). Ribitol and lipoteichoic acids are present in *L. monocytogenes* (Fiedler 1988): these, together with flagella antigens, are responsible for the serological types (Fiedler et al. 1984; Wendlinger et al. 1996). *Listeria* is naturally resistant to lysozyme and this is due to peptidoglycan modification. In *L. monocytogenes* peptidoglycan is deacetylated by the action of *N*-acetylglucosamine deacetylase (Pgd; Boneca et al. 2007) and acetylated by O-acetylmuramic acid transferase (Oat; Rae et al. 2011). Mutants in these genes are attenuated, and this modification is believed to help the bacterium evade the host immune system.

Mycolic acids are not present (Jones et al. 1979), and MK-7 is the major menaquinone with MK-6 and MK-5 present as minor components for all species examined (Collins and Jones 1981). The polar lipid composition is similar in *L. monocytogenes*, *L. innocua*, and *L. welshimeri* and comprises phosphatidylglycerol, diphosphatidylglycerol, galactosylglucosyldiacylglycerol, and L -lysylcardiolipin (Kosaric and Carroll 1971; Shaw 1974; Fischer and Leopold 1999). Other more polar phospholipids were suggested to be polyprenol phosphate and glycero-1-phospholipid plus a D-ananyl derivative (Fischer and Leopold 1999). The fatty acid composition of *L. monocytogenes*, *L. innocua*, and *L. ivanovii* is very similar. All contain predominantly straight-chain saturated *anteiso*- and *iso*- methyl branched-chain types. The major fatty acids are 14-methylhexadecanoic (*anteiso* -$C_{17:0}$) and 12-methyltetradecanoic (*anteiso*- $C_{15:0}$) (Kosaric and Carroll 1971; Feresu and Jones 1988; Ninet et al. 1992; Nichols et al. 2002). The composition of the fatty acids changes under different growth conditions (Puttmann et al. 1993; Nichols et al. 2002), with a shortening of the fatty acid chain length and alteration of branched chain from iso- to anteiso-forms with decreasing temperature to maintain membrane fluidity and function and permit continued growth at lower temperatures (Annous et al. 1997).

Two to six peritrichous flagella are produced by *Listeria* species (Peel et al. 1988a, b) grown below 30 °C. Tumbling motility is observed for cultures grown between 20 °C and 30 °C, and expression of the structural gene for the flagellin protein (*flaA*) has been shown to be temperature regulated and expressed most strongly below 30 °C (Dons et al. 1992). Although motility is not evident, flagella biosynthesis occurs in strains grown below 20 °C (Mattila et al. 2011). Generally at 37 °C, flagella are repressed and motility is not evident (Peel et al. 1988a, b). A number of regulators have been shown to be required to achieve this temperature-dependent regulation of *L. monocytogenes* flagellin expression, such as MogR (Gründling et al. 2004), DegU (Gueriri et al. 2008), and GmaR (an antirepressor of MogR; Kamp and Higgins 2009). It is believed that this complex regulation is required to allow coordinated expression of the flagellin and virulence genes.

## Isolation, Enrichment, and Maintenance Procedures

A range of selective enrichment and isolation media have been described for *Listeria*. Most rely on a combination of resistance characteristics of the organism: tolerance of acriflavine hydrochloride, nalidixic acid, cycloheximide, and lithium chloride. Commonly used selective diagnostic plating media also rely on acriflavine and lithium chloride resistance for selectivity. A common diagnostic feature is aesculin hydrolysis; in combination with ferrous ions, this produces black iron phenolic compounds derived from aglucon which gives black zones around the colonies. This is the basis of the two commonly used media: Oxford and PALCAM agars. The latter also uses mannitol and phenol red to detect lack of mannitol fermentation characteristic of *Listeria*.

The international standard ISO 112090-1 describes the method for isolation of *L. monocytogenes*. Designed for the detection of low numbers of cells in food samples (typically one cell in 25 g), this method includes the use of a two-stage enrichment using two versions of Fraser broth (half Fraser with lower concentrations of acriflavine hydrochloride and nalidixic acid followed by Fraser broth) and then subcultured onto a selective, diagnostic agar for growth of *Listeria* colonies (➲ *Fig. 19.2*). Originally, Oxford and/or PALCAM agar (Curtis and Lee 1995) was recommended, but more recently, the selective and diagnostic medium ALOA (Agar Listeria Ottaviani Agosti; Ottaviani et al. 1997) has gained popularity. This uses a chromogenic substrate utilized by β-D-glucosidase which all *Listeria* species have and gives blue-turquoise colonies. L-α-phosphatidylinositol is a second diagnostic component; this is hydrolyzed by phospholipase C, found only in *L. monocytogenes* and *L. ivanovii*, and gives a white zone of precipitation around the colonies. There are also commercial chromogenic versions of this agar that allow differentiation between *L. monocytogenes* and *L. ivanovii*.

After inoculation, all these different agars are incubated at 30–37 °C for up to 48 h, either aerobically or under microaerophilic conditions, depending on the specification of the manufacturer. On Oxford and PALCAM agar, *L. monocytogenes* grows as grey-green-colored colonies with a black zone which can extend to the whole surface of the agar if large numbers of colonies are present in which case the whole medium is colored black-brown. On PALCAM, the medium surrounding individual colonies should be cherry-red as mannitol utilizers turn the pH indicator yellow. On ALOA colonies of *Listeria* spp. appear blue in color, with *L. monocytogenes* and *L. ivanovii* producing a halo around the colony. While both PALCAM and ALOA are highly selective, other mannitol-positive enterococci or staphylococci will grow, albeit showing different colony morphology. However, after isolation of presumptive colonies using either type of agar, confirmatory tests are recommended.

Confirmatory testing: presumptive colonies purified on tryptone soya agar are confirmed as *Listeria* as short Gram-positive, catalase-positive, oxidase-negative, non-sporing rods motile at <30 °C but nonmotile at 37 °C. *L. monocytogenes*, *L. ivanovii*, and *L. seeligeri* are hemolytic on agars containing sheep, cow, horse, rabbit, or human blood: the remaining species are nonhemolytic. For *L. monocytogenes* demonstration of β-hemolysis on blood agar is often confirmatory, but this does

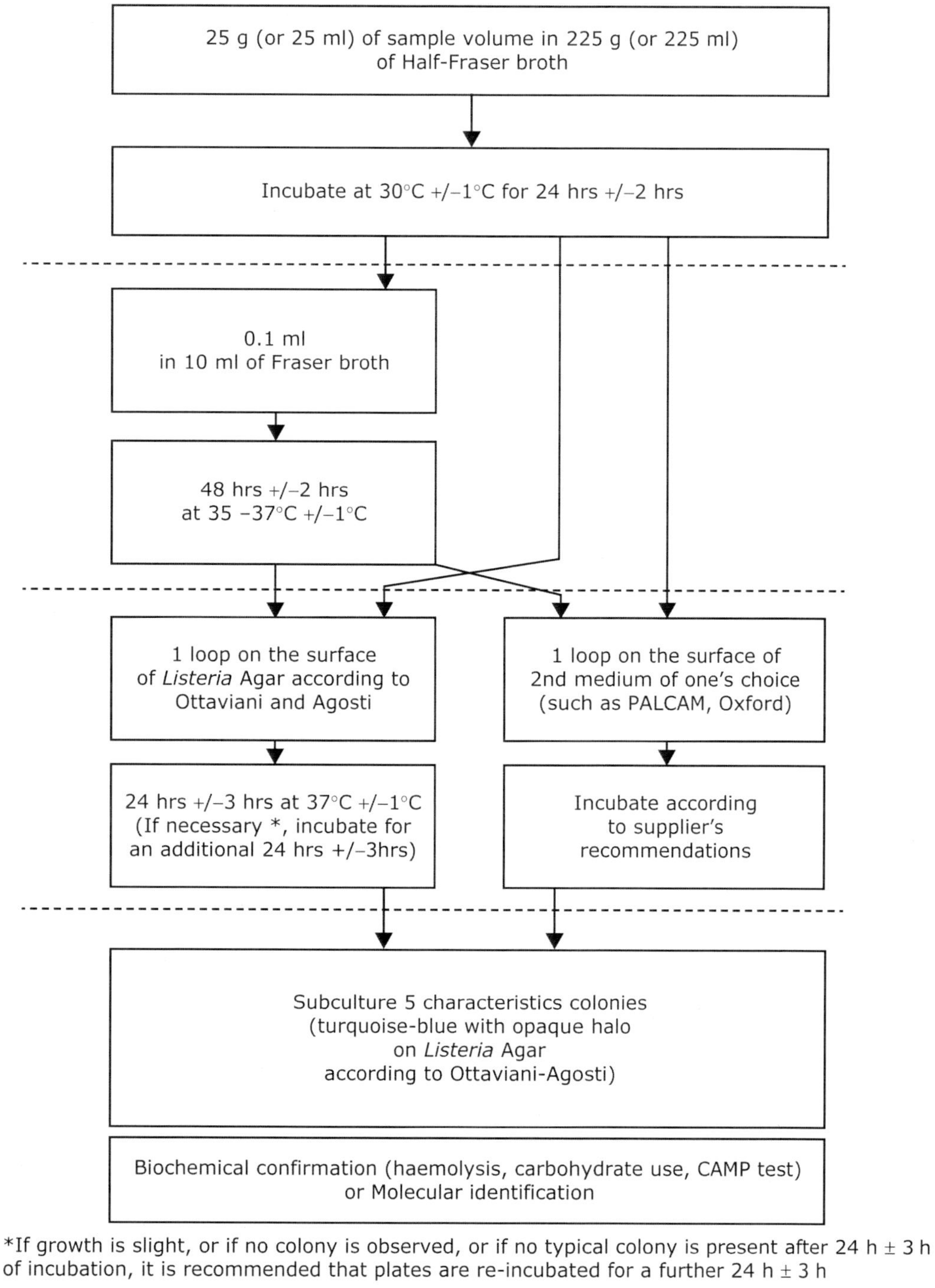
25 g (or 25 ml) of sample volume in 225 g (or 225 ml) of Half-Fraser broth
Incubate at 30°C +/−1°C for 24 hrs +/−2 hrs
0.1 ml in 10 ml of Fraser broth
48 hrs +/−2 hrs at 35 −37°C +/−1°C
1 loop on the surface of Listeria Agar according to Ottaviani and Agosti
1 loop on the surface of 2nd medium of one's choice (such as PALCAM, Oxford)
24 hrs +/−3 hrs at 37°C +/−1°C (If necessary *, incubate for an additional 24 hrs +/−3hrs)
Incubate according to supplier's recommendations
Subculture 5 characteristics colonies (turquoise-blue with opaque halo on Listeria Agar according to Ottaviani-Agosti)
Biochemical confirmation (haemolysis, carbohydrate use, CAMP test) or Molecular identification
*If growth is slight, or if no colony is observed, or if no typical colony is present after 24 h ± 3 h of incubation, it is recommended that plates are re-incubated for a further 24 h ± 3 h

**Fig. 19.2**
**Outline of International Standard for the isolation of *L. monocytogenes* (ISO 112090-1)**

not distinguish the organism from *L. ivanovii* or *L. seeligeri.* The CAMP test or enhancement of hemolysis reactions (Christie et al. 1944) has been traditionally used; *L. monocytogenes, L. seeligeri,* and *L. ivanovii* all show positive CAMP test: *L. monocytogenes* and *L. seeligeri* are CAMP test positive with *Staphylococcus aureus* and CAMP test negative with *Rhodococcus equi*; *L. ivanovii* is CAMP test positive with *R. equi* and CAMP test negative with *S. aureus.* All other *Listeria* species are CAMP test negative.

The colonial growth of *L. monocytogenes* and *L. innocua* is often faster than that of the other species; the latter may only form small colonies after 18–24 h. Similar effects are observed in liquid cultures, and metabolic and sugar fermentation tests should be recorded after >24-h incubation.

Speciation is generally by biochemical testing with commercial kits readily available. Acid production from xylose and rhamnose is a common inclusion. Phenotypic tests to distinguish between the species of *Listeria* are shown in *Table 19.1*.

Members of this genus do not require special procedures for maintenance and medium- and long-term storage. Facultative anaerobic strains may be preserved for some months by stab inoculation in nutrient agar (tryptose agar or other similar media) in screw-capped vials. These should be stored at room or preferably refrigerator temperature. Generally strains are maintained on isolation medium or agar slants at ambient or refrigeration temperatures for a few days. Medium-term maintenance is in 20 % (v/v) glycerol suspensions at −20 °C or at −70 °C. Long-term preservation is by lyophilization or in liquid nitrogen.

## Ecology

The genus *Listeria* is described as having a ubiquitous distribution, and this reflects the recovery of these bacteria from a wide range of natural and artificial environments such as food processing facilities. In addition, they occur in the feces of mammals, including humans, without causing infection and have been isolated from the GI tract of birds and fish. It has been shown that both flies and beetles can become contaminated with *Listeria* and therefore act as a vector for transmission in farm and human environments (Domenichini et al. 1992). More recently, a survey of flies collected from garbage cans outside urban restaurants found *L. monocytogenes* in the gut contents of 3 % of samples (Pava-Ripoll et al. 2012) and suggests that insects are effective vectors of this bacterium. Although present in the gut of these insects, no replication has yet been demonstrated to confirm that these insects act as biological vectors rather than simple mechanical vectors (Putt et al. 1988). Despite the fact that both fruit flies (*Drosophila melanogaster;* Mansfield et al. 2003) and the greater wax moth (*Galleria mellonella*; Mukherjee et al. 2010) have been successfully used as infection models for *L. monocytogenes*, no natural infections of the insects have been described to date. Other researchers have investigated whether amoebae can act as a reservoir for this organism; although evidence has been found to show that *L. monocytogenes* growth is promoted in co-culture with *Acanthamoeba* spp., there are conflicting reports of whether or not *Listeria* can survive and replicate following ingestion by this protozoa (Huws et al. 2008; Akya et al. 2009; Anacarso et al. 2012). Hence whether or not protozoa do act as significant biological vectors in natural environments is yet to be clearly demonstrated. There has been limited attention paid to the survival and invasion of multicellular lower eukaryotic organisms by *Listeria* in the natural environment; however, there are conflicting reports of survival of *L. monocytogenes* in the nematode *Caenorhabditis elegans* using in vitro models (Forrester et al. 2007).

## Survival in the Food Processing Environment

One factor that is believed to contribute to the persistence of *Listeria* in food processing environments is its ability to form biofilms. Biofilms are defined as microbial communities attached to a surface enclosed in a hydrated matrix formed from extracellular polymeric substances, collectively known as glycocalyx (Sandasi et al. 2008). Biofilms are highly organized and specific adaptations of the cells growing in the biofilm are often detected. The expression of at least 30 proteins has been shown to be significantly affected when *Listeria* cells adapt from planktonic to biofilm growth, including flagellin, superoxide dismutase, and 30S ribosomal protein S2 (Trémoulet et al. 2002). Flagella motility is in some way critical in the attachment and biofilm development by *L. monocytogenes*, but the specific role of such motility on attachment and biofilm formation is still unclear since there are conflicting reports in the literature (Vatanyoopaisarn et al. 2000; Lemon et al. 2007; Caly et al. 2009). *L. monocytogenes* is able to colonize and build biofilms on a wide range of surfaces including both hydrophilic and hydrophobic materials, and both the nature of the attachment surface and the growth temperature have an influence on biofilm formation (Chavant et al. 2002). It has been shown that extracellular DNA (eDNA) is present in the *L. monocytogenes* biofilm matrix and that this is needed for both initial attachment and early biofilm formation. However, the origin of this eDNA is still unclear (Harmsen et al. 2010). *L. monocytogenes* produces different biofilm structures in the presence and absence of nutrient flow. Under static conditions, biofilms are less organized and formed only a few layers, whereas under flow conditions, highly organized ball-shaped microcolonies are surrounded by a network of knitted chains (Rieu et al. 2008). Genes involved in the SOS response of *Listeria* (*recA* and *yneA*) are linked to the formation of these knitted biofilm structures (van Der Veen and Abee 2010), and therefore a stress response may trigger the adaption to biofilm growth. The SOS response in bacteria is normally considered to be a stress response induced in response to DNA damage and results in error-prone repair (Michel 2005). However, in *L. monocytogenes*, there is evidence for a wider role since induction of the operon results in a higher heat, $H_2O_2$, and acid resistance, indicating that the SOS response of *L. monocytogenes* can contribute to its survival in the environment (van der Veen et al. 2010).

Once formed, biofilms act as physical barriers to provide protection for the cells against a wide range of treatments, such as surfactants or detergents, and this protection is generally attributed to the formation of the hydrated matrix that surrounds the cells. Belessi et al. (2011) found that the efficiency of different sanitation methods for treatment of *L. monocytogenes* biofilms is variable. The formation of persistent biofilms in food production facilities has been recognized as a contributing factor to outbreaks of listeriosis (Carpentier and Cerf 2011). However, the ability of *Listeria* to form biofilms on plant surfaces may also result in entry of this organism into the food chain. While its saprophytic characteristic means that *Listeria* can be found associated with both the

**Table 19.1**
**Phenotypic properties of *Listeria* species**

| Phenotypic test | *Listeria* species | | | | | | | | | |
|---|---|---|---|---|---|---|---|---|---|---|
| | *L. monocytogenes* | *L. grayi* | *L. fleischmannii* | *L. innocua* | *L. ivanovii* | *L. marthii* | *L. rocourtiae* | *L. seeligeri* | *L. welshimeri* | *L. weihenstephanensis* |
| β-hemolysis on blood agar | + | – | – | – | + | – | – | (+) | – | – |
| Lipase production | + | – | – | – | + | – | – | + | – | – |
| Acid production from: | | | | | | | | | | |
| D-mannitol | – | + | + | – | – | – | + | – | – | + |
| L-rhamnose | + | V | + | + | – | – | + | – | V | + |
| D-xylose | – | – | + | – | + | – | + | + | + | + |
| CAMP (enhancement of hemolysis) against:† | | | | | | | | | | |
| *Staphylococcus aureus* | + | – | – | – | – | NK | – | (+) | – | – |
| *Rhodococcus equi* | – | – | – | – | + | NK | – | – | – | – |
| Amino acid peptidase | – | + | – | + | + | + | + | + | + | – |

+ positive reaction, – negative reaction, (+) weak reaction, *v* variable reaction, *NK* not known
† Non-specific reactions can occur in the CAMP test: unless zone of enhancement has classic diagnostic shape, these non-specific reactions should be disregarded

roots and leaves of plants, there is also a concern that it also has the ability to internalize within plant structures. This too would promote entry into the food chain as it would not be removed from fresh vegetables by simple decontamination steps; however, the evidence for this is still controversial with some publications demonstrating internalization, while others report that this did not occur (Girardin et al. 2005; Jablasone et al. 2005; Kutter et al. 2006; Milillo et al. 2008).

The other factors that contribute to transmission of *Listeria* through the food chain are the natural biological properties of the cell. Its ability to grow at low temperature has long been recognized as an issue for food manufacturers using cold storage as a means of controlling microbial growth (Tasara and Stephan 2006; Chan and Wiedmann 2009). However, the fact that it can tolerate relatively high salt concentrations (Sleator and Hill 2010) and low pH (Cotter and Hill 2003) also allows it to survive these mild food preservation treatments and poses a particular problem in ready-to-eat foods which have been associated with many of the large outbreaks of listeriosis worldwide (Lianou and Sofos 2007).

## Transmission of the Disease in Humans

Consumption of contaminated foods is recognized as the principal route of transmission for this disease in humans, and microbiological and epidemiological evidence supports an association with many food types in both sporadic and epidemic listeriosis, e.g., dairy, meat, vegetable, fish and shellfish, as well as complex foods such as sandwiches (➋ *Table 19.2*). Although diverse in their constituents and manufacturing processes, foods associated with transmission often show the common features of the following: the capability of supporting the multiplication of *L. monocytogenes* (relatively high water activity and near-neutral pH), relatively heavy ($>10^3$/g) contamination with the implicated strain, processed with an extended (refrigerated) shelf life, and consumed without further cooking (Farber and Peterkin 1991).

Outbreaks of human listeriosis involving >100 individuals involving consumption of a specific food from a single source are described; however, most cases appear to be sporadic. Outbreaks have occurred over durations of 6 months to 5 years, and this is likely to represent long-term colonization of a specific site in food manufacturing environments (Carpentier and Cerf 2011) as well as the long incubation periods shown by some patients. *L. monocytogenes* has been shown to survive well in a variety of environments where food is manufactured, processed, or grown, particularly those that are moist with organic material, and it is from such sites that contamination of food occurs (Gandhi and Chikindas 2007). Given the properties and distribution of *L. monocytogenes*, cases related by a common source may be very widely distributed both temporally and geographically.

**Table 19.2**
**Food types associated with human listeriosis**

| **Dairy products** | **Vegetable products** |
|---|---|
| Soft cheese | Coleslaw salad |
| Milk | Vegetable rennet |
| Ice cream/soft cream | Salted mushrooms |
| Butter | Alfalfa tables |
| Chocolate milk | Raw vegetables |
| | Pickled olives |
| | Rice salad |
| | Cut fruit |
| | Melon |
| | Celery |
| **Meat products** | **Fishery products** |
| Cooked chicken | Fish |
| Turkey frankfurters | Shellfish |
| Sausages | Shrimps |
| Pâté and rillettes | Smoked fish and shellfish |
| Pork tongue in aspic | Cod roe |
| Sliced meats | **Complex foods** |
| Meat pies | Pre-prepared sandwiches |
| Hot dogs | |
| Cooked turkey | |

## Transmission of Listeriosis in Animals

Abortion and septicemia in newborn animals result from intrauterine infection acquired from the dam and parallel abortion and early-onset neonatal infection in humans. Septicemia and meningitis in lambs in the first few weeks of life have been attributed to umbilical infection acquired in lambing pens, possibly from contaminated soil or feed: this corresponds to late-onset neonatal infection in humans. Iritis and keratoconjunctivitis usually occur during the winter in silage-fed sheep and cattle. These may occur by direct introduction of contaminated feed into the eye and have been particularly associated where feed is provided in holders or racks at eye level (Low and Donachie 1997).

As in human infection, the majority of cases of animal listeriosis are assumed to be acquired via the oral route. There is a strong association between the feeding of poor-quality silage and all manifestations of listeriosis in sheep and cattle, although cases do occur in the absence of consumption of this type of feed. However, the exact mechanism by which silage feeding leads or predisposes to listeriosis is not clear. Under normal conditions it is impossible to produce silage free of *Listeria*: the organism has been isolated from silage with a pH of <4, albeit in very low numbers. However, where poor-quality silage has been produced and a low pH and anaerobic conditions are not achieved, proliferation of *Listeria* takes place and very high numbers can be found. Poor quality is often also due to insufficient herbage quality or to contamination by soil or feces. The change to production of silage in polythene bales ("big bale" silage) corresponded to increases in ovine listeriosis in the UK. Although the big bale method is more economical than the traditional use of clamps, these are more prone to spoilage and growth of *Listeria*: high numbers of the organism are often associated with sites where the damage to the bags has occurred or at the tied end. The peak in the numbers of animal listeriosis in the spring may reflect a decrease in the quality of silage used for feed (Low and Donachie 1997).

## Pathogenicity, Clinical Relevance

### Clinical Relevance for Humans and Other Animals

Human listeriosis most often affects those with severe underlying illness, the elderly, pregnant women, and both unborn and newly delivered infants. However, patients without these risk factors can also become infected. Individuals at greatest risk of contracting listeriosis include those with malignant neoplasms or undergoing immunosuppressive therapy. Other predisposing conditions include those with agents to reduce stomach acid, AIDS, alcoholism, alcoholic liver disease, diabetes, and recipients of prosthetic heart valves or articulation joints. The disease most often affects the bloodstream, the pregnant uterus, or the central nervous system. In nonpregnant individuals, listeriosis most frequently presents as septicemia without involvement of the central nervous system or, to a lesser extent, as meningitis (with or without septicemia). Patients >60 years of age with concurrent pathologies are now the most common group affected in Europe particularly presenting with invasion of the blood but not the central nervous system. Listeric meningoencephalitis and encephalitis occur less commonly.

In pregnant woman, listeriosis is most often recognized with one or more self-limiting influenza-like episodes (pyrexia and other nonspecific symptoms, although some individuals may be asymptomatic) during or after the latter half of the second trimester, although infection can occur throughout gestation. Maternal listeric meningitis during pregnancy is very rare. During pregnancy, infection spreads from the maternal circulatory system to the fetus, probably via the placenta, although this is not inevitable. Fetal infection developing before the third trimester usually results in an *intrauterine* death. The fetus has severe and overwhelming multisystem infection involving internal organs, with the widespread formation of granulomatous lesions, especially in the liver and placenta: the condition is named *granulomatosis infantiseptica*. Infection of the infant during the third trimester results in either intrauterine death or the delivery of a severely ill neonate (early-onset infection). Early-onset sepsis in the neonate is characterized by nonspecific signs of infection and prematurity. Cutaneous lesions may be present (sometimes with granulomas) and the neonate may be convulsive. Most early-onset cases are septicemic, some with meningitis; however, infants may appear infected only at superficial sites. The degree of severity may be partially dependent on the gestational age at infection. Surviving infants can exhibit long-term sequelae, especially those delivered prematurely or with involvement of the central nervous system. Late-onset neonatal sepsis typically occurs after uncomplicated full-term pregnancies and usually presents as meningitis about 10 days after delivery. *L. monocytogenes* is acquired either from maternal sites during or shortly after delivery (possibly during passage through the birth canal) or from the postnatal environment, including from direct or indirect contact with an early-onset case of neonatal listeriosis.

Deep-seated focal infections caused by *L. monocytogenes* with or without abscess formation occur in a wide variety of sites and are largely confined to immunocompromised adults. Listeric endocarditis occurs, usually in patients with underlying cardiac lesions or with prosthetic heart valves. Diarrheal disease has also been described, although this does not appear in all cases and may be specific to some *L. monocytogenes* strains. Cutaneous and ocular infection occurs. Mortality rates for all forms of adult and juvenile listeriosis vary from 10 % to >50 %, with poor prognostic indicators which include age (>50 years), preexisting disease, early convulsions (in cases of meningitis), and the needs for cardiovascular, renal, or ventilatory support. Residual disabilities may occur. The incubation period between consumption of contaminated food and onset of symptoms varies from <1 to >90 days. Relapses of infection, some greater than 2 years after the original episodes, have been described.

Amongst domestic animals, sheep and goats are most susceptible, although infection also takes place in cattle (Low and Donachie 1997). Infection has been recognized in >40 other species of feral and domesticated animals. Listeriosis presents as a wide range of disorders which parallel much of what has already been outlined for humans, although there are some differences. *L. monocytogenes* is the major pathogen; however, some of the cases of abortion or septicemia in sheep (but less commonly in cattle) are due to *L. ivanovii*. The disease manifests as abortion, septicemia, encephalitis, diarrhea, mastitis, or ocular infections, and the presentation may vary depending on the species involved. Listeriosis in primates manifests similarly to that in humans. Although less commonly associated with predisposing factors (as described for humans), listeriosis in animals shows characteristics of an opportunistic pathogen. The unborn and newly delivered are more susceptible to infection, and encephalitis occurs most often in the adult pregnant animals during the later stages of gestation or shortly after delivery. Outbreaks have been associated with climatic stress (sudden drops in temperature, snow falls, drought, and shortage of food), and cases most often occur in the spring when animals may be in poor condition and exposed to poor-quality feed. Increases in susceptibility of animals to experimental infection have been demonstrated by malnutrition, immunosuppression, viral infection, reducing stomach acid, and other uncharacterized stress factors.

In sheep, goats, and cattle, abortion is recognized late in pregnancy and is rarely accompanied by severe systemic disease in the dam. Aborting animals may excrete the organism in the milk without evidence of mastitis. Septicemia in young animals occurs in the first few weeks of life, some with diarrhea, but without specific symptomatology. Diarrhea and septicemia also occur in older animals (principally ewes).

Unlike human listeriosis, the most common form of listeriosis in animals is as an encephalitis. In ruminants, this takes the form of a unilateral (or less commonly bilateral) cranial nerve paralysis which is often accompanied with ataxia, and moving in circles: hence the name "circling disease." Abortion, septicemia, and encephalitis are usually sporadic in cattle, but can occur as outbreaks amongst flocks of sheep where losses may be heavy. During outbreaks, septicemia and abortion may occur together with cases of encephalitis, but is unusual.

Experimental infection indicates that septicemia can develop in a few days after consumption of contaminated feed, but the incubation period for encephalitis is likely to be much longer (20–30 days). *L. monocytogenes* causes mastitis in cows, sheep, and goats where large numbers of the bacterium can be shed into milk. Keratoconjunctivitis together with iritis occurs in both sheep and cattle. These conditions are usually unilateral. In cases of conjunctivitis, other bacterial or viral pathogens may also be present on the conjunctiva. Listeric abortion, septicemia, and encephalitis have been recognized in pigs, horses, dogs, and cats, but are rare. Listeriosis occurs in rodents and >20 species of birds and is probably rare. Infection is most often recognized in farmed birds, i.e., chickens, turkeys, and ducks. Septicemia and myocardial necrosis are the most common manifestation, and these may be secondary to other infections. Because of the disease severity and rapid onset of clinical symptoms, treatment of infected sheep or cattle is rarely attempted: infected animals are usually destroyed on humanitarian grounds. During outbreaks, mortality rates are often 100 %, and those surviving can exhibit permanent central nervous system disorders. As is found with humans, the pregnant dam with an intrauterine infection is rarely accompanied by severe systemic disease, so it is not necessary to attempt treatment. A listeric abortion does not seem to affect the possibility of subsequent conceptions. The response to antibiotic treatment in cows with listeric mastitis has been poor, and the organism can be excreted for extended periods of time. Hence it is recommended that such animals should not be used for milk production and culled.

## Importance of Listeriosis

The reported incidence of human listeriosis varies between countries from <1 to >10 cases per million of the total population. Although these, in part, may reflect differences in surveillance systems, they probably represent true differences in incidence. Because of the severity of infection, listeriosis is one of the major causes of death from a preventable foodborne illness. The incidence in animals is influenced by feed quality, particularly in the ability to produce good-quality silage.

## Transmission of Listeriosis in Humans

The widespread distribution of *L. monocytogenes* provides numerous potential ways in which the disease may be transmitted to both animals and humans. Although there has been much current interest in infection via the oral route, this is not the only mode of transmission.

Listeriosis may be transmitted by direct contact with the environment, although this has been rarely reported. Contact with infected animals or animal material is a well-recognized risk factor, and in such cases, the disease occurs as papular or pustular cutaneous lesions usually on the upper arms or wrists of farmers or veterinarians 1–4 days after attending bovine abortions (McLauchlin and Low 1994). Although cutaneous listeriosis in adults is invariably mild with a successful resolution, severe systemic infection may follow. Conjunctivitis in poultry workers has also been reported.

Hospital cross-infection between newborn infants occurs. These show a common pattern of an infant born with congenital listeriosis (onset within 1 day of birth). In the same hospital, and within a short period of time, an apparently healthy (or more rarely more than one) neonate is born who typically develops late-onset listeriosis between the 5th and 12th day after delivery. In most of the episodes, the cases are either delivered or nursed in the same or adjacent rooms, and consequently staff and equipment are common to both. Two larger series have been described occurring in Sweden and Costa Rica, where four and seven cases, respectively, resulted from single early-onset cases (Larsson et al. 1978; Schuchat et al. 1991). The likely routes of transmission here involved a contaminated rectal thermometer or a mineral oil bath. In one episode, true person-to-person transmission occurred where, 3 days after delivery, the mother of an early-onset case was nursed in an open ward and handled a neonate from an adjacent bed who subsequently developed late-onset listeriosis (Isaacs and Liberman 1981).

## Pathogenicity

It is a characteristic of the natural disease in both humans and animals that there is usually a low attack rate. The susceptibility to infection may be increased by external factors, some of which have already been mentioned. However, other factors (such as other infectious agents, the nature of the food matrix, or products of the metabolism of other microorganisms) may be of importance. *L. monocytogenes* is a somewhat marginal pathogen. Hence experimental models reflecting the natural infection may work poorly, and relatively large numbers of animals may be needed for a small proportion of these to produce clinical symptoms of the disease.

There is evidence supporting the role of antacid therapy in increasing susceptibility of some patients, and in experimental animal infection. The buffering capacity of some food types may also be of importance in facilitating the survival of the organism, which may then invade at sites further along the gastrointestinal tract, although other routes of infection may occur. Experimental septicemia in animals can be achieved via the respiratory route, and further evidence supporting this possibility comes from one of the cases (aspiration pneumonia and septicemia) which developed after eating contaminated coleslaw salad in the 1981 Canadian outbreak.

Histopathological analysis suggests that the intestinal tract can act as the site of invasion and *L. monocytogenes* (as well as *L. ivanovii*) will readily invade various epithelial and fibroblast cell types growing in vitro suggesting that there may be multiple routes by which this bacterium initially invades the host's cells. In the caecum and colon of animals following oral inoculation, the bacteria can be observed together with an inflammatory reaction in phagocytic cells present in the underlying lamina propria. Following this phase, invasion of the uterine contents or central nervous system (for patients with shorter incubation periods) may occur probably via the circulatory system.

In the liver, the organism is cleared from the blood by the phagocytic Kupffer cells. In their non-activated state, some bacteria will survive, escape to the cell cytoplasm, and subsequently spread to hepatocytes using the process described in the following section. Formation of localized lesions occurs in the liver and also in the spleen.

Intrauterine infection of the fetus results from hematogenous spread from the mother. Abscess formation takes place in the placenta, and this may spread via the umbilical vein or the amniotic fluid to the fetal internal organs. The series of pyrexial episodes observed in the mother may result from reinvasion of maternal bloodstream from placental sites. *L. monocytogenes* is unusual in that it is able to survive and grow in amniotic fluid, and aspiration of this leads to the pathological changes in the fetal respiratory tracts. The presence of high numbers of the organism in amniotic fluid results in widespread contamination of neonatal and maternal surface sites at delivery as well as the postnatal environment and may result in cases of neonatal cross-infection.

Experimental and field studies suggest that encephalitis in sheep and cattle results from *L. monocytogenes* reaching the base of the brain along cranial nerves, particularly the trigeminal nerve. It is assumed that animals eat contaminated feed, particularly silage, and the organism enters the nerves after penetrating the oral mucus membrane or through preexisting areas of trauma such as tooth root scars (which are prominent in sheep during the spring). The mechanism for travel along the nerves is not understood.

## Molecular Pathogenicity

*L. monocytogenes* is an intracellular parasite, and it is in this environment that the pathogen gains protection and evades some of the host's defenses. Molecular biological techniques together with models using experimental animal infection and invasion of mammalian cells grown in vitro have revolutionized the understanding of the process of *L. monocytogenes* pathogenicity. At the cellular level, three distinct processes occur which are cellular invasion, escape from the intracellular vacuole, and actin-based motility including cell-to-cell spreading (Seveau et al. 2007; Cossart 2011).

### Cellular Invasion

*L. monocytogenes* is able to invade a number of non-phagocytic cells. Invasion is mediated by two leucine-rich repeat internalin proteins, InlA and InlB, which are expressed at the surface of *L. monocytogenes.* InlA and InlB specifically interact with host cell proteins and mediate adherence and internalization. InlA is involved with crossing the intestinal and maternofetal barriers and interacts with the mammalian cell surface adhesion protein E-cadherin (E-cad), which, for intestinal tissue, takes place at the tip of the intestinal villus. InlB is a soluble extracellular protein which promotes entry of *L. monocytogenes* into epithelial, endothelial, hepatocyte, and fibroblast cells. InlB interacts with the hepatocyte growth factor receptor (Met) which is a transmembrane signalling receptor involved with mammalian cell growth, migration, and differentiation. The results of these interactions (InlA/E-cad or InlB/Met) subverts the molecular cytoskeletal structures and F-actin dynamics to generate contractile forces that result in engulfment of the bacterium into an intracellular membrane-bound compartment.

### Escape from the Intracellular Vacuole

*L. monocytogenes* is able to escape from the membrane-bound compartment which dissolves by the action of a thiol-activated hemolysin (*hly* gene). This is achieved by interaction with host membrane cholesterol. *L. monocytogenes* also produces two phospholipases (PlcA and PlcB), one of which is activated by a metalloprotease which is also produced by the bacterium. The combined action of the hemolysin and a phospholipase also contribute to the process of escaping from membrane-bound compartments, especially when spreading from cell-to-cell (see next section).

### Actin-Based Motility Including Cell-to-Cell Spreading

*L. monocytogenes* enters the host cell cytoplasm where it multiplies with a doubling time of about one hour. Once in the cytoplasm, the bacterium becomes surrounded by polymerized host cell actin, which becomes preferentially polymerized at the older pole of the bacterial cell by the ActA cell surface protein. Actin polymerization confers intracellular motility to the bacterium which allows invasion of an adjacent mammalian cell. The bacterium is then encapsulated in a double-membrane-bound compartment, and the whole process is repeated without the bacterium entering an extracellular environment.

### Molecular Organization of Virulence Genes

The genes involved with pathogenicity are all located together in a single operon (*plc* A, *hly*, metalloprotease, *act* A, and *plc* B) which are regulated by the positive regulation factor (*prfA* gene) which is located in the same operon. PrfA also regulates the *inl* A and *inl* B genes which are located quite closely on the bacterial chromosome. Other proteins involved in virulence include BSH, a bile salt hydrolase; the surface proteins Auto and VIP; the phosphatases STP and LipA; and the superoxide dismutase (SOD) MnSOD. The peptidoglycan-modifying enzyme PgdA deacetylates the N-acetylglucosamine residues of the *L. monocytogenes* cell wall and confers resistance to lysozyme, and mutants in this gene are highly attenuated. Another peptidoglycan modification, acetylation of muramic acid residues, is induced by gene *oat* A and critical for the survival of *L. monocytogenes* in infected hosts. Listeriolysin S is a toxin similar to the modified peptide streptolysin S, a hemolytic and cytotoxic virulence factor that plays a key role in the virulence of group A streptococci. Interestingly it is only present in a subset of Lineage I strains of *L. monocytogenes*, those responsible for most listeriosis outbreaks (Cossart 2011).

### Antimicrobial Resistance

Various antibiotics have been recommended for treatment of listeriosis; however, *L. monocytogenes* is universally sensitive to ampicillin/penicillin and aminoglycosides which are the treatments of choice: this bacterium is uniformly constituently resistant to cephalosporins. Up to 10 % of isolates are resistant to tetracycline or fluoroquinolones. Plasmids encoding resistance to tetracycline alone and for multiresistance to chloramphenicol, erythromycin, streptomycin, and tetracycline have been observed although these are rare (Threlfall et al. 1998; Charpentier and Courvalin 1999; Morvan et al. 2010; Granier et al. 2011; Lungu et al. 2011).

## Applications

While the intracellular lifestyle of *Listeria* contributes to the high mortality rate associated with infections, this characteristic has been exploited for the development of anticancer agents. One of the limitations discovered when developing antitumor vaccines was that delivery of the tumor-associated antigen (TAA) into the antigen-presenting cells (APCs) was inefficient. This is required for processing and presentation of the TAA to the immune system. Since *L. monocytogenes* is able to infect APCs, including monocytes, macrophages, and dendritic cells, a recombinant bacterium expressing the TAA that can be delivered directly into these cells with high efficiency is highly desirable (Kolb-Maurer et al. 2000). In addition, the wide tissue tropism of this bacterium means that it can induce its own uptake into other nonprofessional cells including epithelial cells, fibroblasts, hepatocytes, endothelial cells, and neurons (Seveau et al. 2007) using its array of surface-associated virulence proteins and cell surface-binding factors (Gravekamp and Paterson 2010). This extends the potential sites that can be targeted by a *Listeria* TAA delivery system.

To facilitate the delivery of the TAA, antigens are usually fused to the hemolysin protein (LLO) or to just the signal sequence of this protein (tLLO; Gunn et al. 2001) so that they are secreted from the cell following internalization (Gravekamp and Paterson 2010). Secretion of hemolysin and the phospholipases can result in the escape of the recombinant vaccine strain from an intracellular host cell vacuole. Once in the cytoplasm, antigens secreted by the bacterium are naturally targeted for both MHC class I and II presentation to stimulate both $CD4^+$ and $CD8^+$ T cells (Hiltbold and Ziegler 1993). It appears that this natural immune response of the host is sufficient to overcome the immune-suppressive activity of regulatory T cells (Tregs) and myeloid-derived suppressor cells (MDCSs) that are normally found in the tumor microenvironment and prevent immune cells from targeting and killing tumors. Hence there appears to be a natural advantage for choosing *L. monocytogenes* as the TAA delivery system. However, some caution has to be applied as the bacterium itself is a very efficient pathogen once infection becomes established. Therefore a range of mutant strains have been developed that are viable but attenuated for clinical applications (Singh and Wallecha 2011). Different recombinant *L. monocytogenes* strains expressing TAAs associated with different cancers have now been evaluated both in a preclinical and, more recently, in a clinical setting (Maciag et al. 2009; Radulovic et al. 2009; Le et al. 2012) and continue to show promise as therapeutic agents.

In the area of food microbiology, bacteriophage specific for *Listeria* are being exploited as a biocontrol agent. For this application, virulent, broad host range bacteriophage such as A511 (Klumpp et al. 2008) or P100 (Carlton et al. 2005) have been chosen that are likely to be able to infect and lyse most environmental isolates encountered in the food. The bacteriophage are applied to the surface of foods during chilled storage (Monk et al. 2010), and studies have shown that these phage can infect and lyse *Listeria* in situ in different food types (Holck and Berg 2009; Guenther et al. 2009; Soni and Nannapaneni 2010; Soni et al. 2010; Bigot et al. 2011; Rossi et al. 2011). A commercial product containing phage P100 (Listex™; Soni et al. 2010) has been approved for use by the FDA for application on RTE meat and poultry (Peek and Reddy 2006). In 2012 Listex™ was also approved for use in Australia and New Zealand by FSANZ and has been accepted as an allowed processing aid in some European countries, with organic status being granted in the Netherlands. While the use of bacteriophage as a biocontrol agent is becoming more widespread, there is still some concern about the safety of adding viruses to food (EFSA 2012).

## References


Akya A, Pointon A, Thomas C (2009) Viability of *Listeria monocytogenes* in co-culture with *Acanthamoeba* spp. FEMS Microbiol Ecol 70:20–29

Anacarso I, de Niederhausern S, Messi P, Guerrieri E, Iseppi R, Sabia C, Bondi M (2012) *Acanthamoeba polyphaga*, a potential environmental vector for the transmission of food-borne and opportunistic pathogens. J Basic Microbiol 52:261–268

Annous BA, Becker LA, Bayles DO, Labeda DP, Wilkinson BJ (1997) Critical role of anteiso-C15:0 fatty acid in the growth of *Listeria monocytogenes* at low temperatures. Appl Environ Microbiol 63:3887–3894

Anon (1954) Opinion 12, conservation of *Listeria* Pirie 1940 as a generic name in bacteriology. Int Bull Bacteriol Nomencl Taxon 4:150–151

Audurier A, Rocourt J, Courtieu AL (1977) Isolement et characterisation de bacteriophages de *Listeria monocytogenes*. Ann Microbiol 128A:185–198

Azizoglu RO, Kathariou S (2010a) Inactivation of a cold-induced putative RNA helicase gene of *Listeria monocytogenes* is accompanied by failure to grow at low temperatures but does not affect freeze-thaw tolerance. J Food Prot 73:1474–1479

Azizoglu RO, Kathariou S (2010b) Temperature-dependent requirement for catalase in aerobic growth of *Listeria monocytogenes* F2365. Appl Environ Microbiol 76:6998–7003

Barbuddhe S, Hain T, Chakraborty T (2008) Comparative genomics and evolution of virulence. In: Lio D (ed) Handbook of *Listeria monocytogenes*. CRC Press, Baco Raton, pp 311–335

Belessi C-EA, Gounadaki AS, Psomas AN, Skandamis PN (2011) Efficiency of different sanitation methods on *Listeria monocytogenes* biofilms formed under various environmental conditions. Int J Food Microbiol 145:S46–S52

Bertsch D, Rau J, Eugster MR, Haug MC, Lawson PA, Lacroix C, Meile L (2013) *Listeria fleischmannii* sp. nov., isolated from cheese. Int J Syst Evol Microbiol 63:526–532

Bigot B, Lee WJ, McIntyre L, Wilson T, Hudson JA, Billington C, Heinemann JA (2011) *Listeria fleischmannii* sp. nov., isolated from cheese. Control of *Listeria monocytogenes* growth in a ready-to-eat poultry product using a bacteriophage. Food Microbiol 28:1448–1452

Boneca IG, Dussurget O, Cabanes D, Nahori MA, Sousa S, Lecuit M, Psylinakis E, Bouriotis V, Hugot JP, Giovannini M, Coyle A, Bertin J, Namane A, Rousselle JC, Cayet N, Prévost MC, Balloy V, Chignard M, Philpott DJ, Cossart P, Girardin SE (2007) A critical role for peptidoglycan N-deacetylation in *Listeria* evasion from the host innate immune system. Proc Natl Acad Sci USA 104:997–1002

Buchrieser C (2007) Biodiversity of the species *Listeria monocytogenes* and the genus *Listeria*. Microbes Infect 9:1147–1155

Buchrieser C, Rusniok C, Kunst F, Cossart P, Glaser P, The Listeria Consortium (2003) Comparison of the genome sequences of *Listeria monocytogenes* and *Listeria innocua*: clues for evolution and pathogenicity. FEMS Immunol Med Microbiol 35:207–213

Buchrieser C, Rusniok C, Garrido P, Hain T, Scortti M, Lampidis R, Karst U, Chakraborty T, Cossart P, Kreft J, Vazquez-Boland JA, Goebel W, Glaser P (2011) Complete genome sequence of the animal pathogen *Listeria ivanovii*, which provides insights into host specificities and evolution of the Genus *Listeria*. J Bacteriol 193:6787–6788

Call DR, Borucki MK, Besser TE (2003) Mixed-genome microarrays reveal multiple serotype and lineage-specific differences among strains of *Listeria monocytogenes*. J Clin Microbiol 41:632–639

Caly D, Takilt D, Lebret V, Tresse O (2009) Sodium chloride affects *Listeria monocytogenes* adhesion to polystyrene and stainless steel by regulating flagella expression. Lett Appl Microbiol 49:751–756

Cao M, Pavinski Bitar A, Marquis H (2007) A mariner-based transposition system for *Listeria monocytogenes*. Appl Environ Microbiol 73:2758–2761

Carlton RM, Noordman WH, Biswas B, de Meester ED, Loessner MJ (2005) Bacteriophage P100 for control of *Listeria monocytogenes* in foods: genome sequence, bioinformatic analyses, oral toxicity study, and application. Regul Toxicol Pharmacol 43:301–312

Carpentier B, Cerf O (2011) Review – persistence of *Listeria monocytogenes* in food industry equipment and premises. Int J Food Microbiol 145:1–8

Chan YC, Wiedmann M (2009) Physiology and genetics of *Listeria monocytogenes* survival and growth at cold temperatures. Crit Rev Food Sci Nutr 49:237–253

Chang Y, Gu W, Fischer N, McLandsborough L (2012) Identification of genes involved in *Listeria monocytogenes* biofilm formation by mariner-based transposon mutagenesis. Appl Microbiol Biotechnol 93:2051–2062

Charpentier E, Courvalin P (1999) Antibiotic resistance in *Listeria* spp. Antimicrob Agents Chemother 43:2103–2108

Chavant P, Martinie B, Meylheuc T, Bellon-Fontaine M-N, Hebraud M (2002) *Listeria monocytogenes* LO28: surface physicochemical properties and ability to form biofilms at different temperatures and growth phases. Appl Environ Microbiol 68:728–737

Chen Y, Knabel S (2008) Strain typing. In: Liu D (ed) Handbook of *Listeria monocytogenes*. CRC Press, Boca Raton, pp 203–240

Christie R, Atkins NE, Munch-Peterson E (1944) A note on a lytic phenomenon shown by group B streptococci. Aust J Exp Biol Med Sci 22:197–200

Collins MD, Jones D (1981) The distribution of isoprenoid quinone structural types in bacteria and their taxonomic implications. Microbiol Rev 45:316–354

Collins MD, Jones D, Goodfellow M, Minnikin DE (1979) Isoprenoid quinone composition as a guide to the classification of *Listeria*, *Brochothrix*, *Erysipelothrix* and *Caryophanon*. J Gen Microbiol 111:453–457

Collins MD, Wallbanks S, Lane DJ, Shah J, Nietupski R, Smida J, Dorsch M, Stackebrandt E (1991) Phylogenetic analysis of the genus *Listeria* based on reverse transcriptase sequencing of 16S rRNA. Int J Syst Bacteriol 41:240–246

Collins B, Curtis N, Cotter PD, Hill C, Ross RP (2010) The ABC transporter AnrAB contributes to the innate resistance of *Listeria monocytogenes* to nisin, bacitracin, and various beta-lactam antibiotics. Antimicrob Agents Chemother 54:4416–4423

Cossart P (2011) Illuminating the landscape of host-pathogen interactions with the bacterium *Listeria monocytogenes*. Proc Natl Acad Sci USA 108:19484–19491

Cotter PD, Hill C (2003) Surviving the acid test: responses of Gram-positive bacteria to low pH. Microbiol Mol Biol Rev 67:429–453

Curtis GD, Lee WH (1995) Culture media and methods for the isolation of *Listeria monocytogenes*. Int J Food Microbiol 26:1–13

Daneshvar MI, Brooks JB, Malcolm GB, Pine L (1989) Analyses of fermentation products of *Listeria* species by frequency-pulsed electron-capture gas–liquid chromatography. Can J Microbiol 35:786–793

Dell'Era S, Buchrieser C, Couvé E, Schnell B, Briers Y, Schuppler M, Loessner MJ (2009) *Listeria monocytogenes* L-forms respond to cell wall deficiency by modifying gene expression and the mode of division. Mol Microbiol 73:306–322

den Bakker HC, Cummings CA, Ferreira V, Vatta P, Orsi RH, Degoricija L, Barker M, Petrauskene O, Furtado MR, Wiedmann M (2010) Comparative genomics of the bacterial genus *Listeria*: genome evolution is characterized by limited gene acquisition and limited gene loss. BMC Genomics 11:688

Domenichini G, Fogliazza D, Pagani M (1992) Studies on the transmission of *Listeria* by means of arthropods. Ital J Food Sci 4:269–278

Domínguez-Bernal G, Müller-Altrock S, González-Zorn B, Scortti M, Herrmann P, Monzó HJ, Lacharme L, Kreft J, Vázquez-Boland JA (2006) A spontaneous genomic deletion in *Listeria ivanovii* identifies LIPI-2, a species-specific pathogenicity island encoding sphingomyelinase and numerous internalins. Mol Microbiol 59:415–432

Dons L, Rasmussen OF, Olsen JE (1992) Cloning and characterization of a gene encoding flagellin of *Listeria monocytogenes*. Mol Microbiol 6:2919–2929

Doumith M, Cazalet C, Simoes N, Frangeul L, Jacquet C, Kunst F, Martin P, Cossart P, Glaser P, Buchrieser C (2004a) New aspects regarding evolution and virulence of *Listeria monocytogenes* revealed by comparative genomics and DNA arrays. Infect Immun 72:1072–1083

Doumith M, Buchrieser C, Glaser P, Jacquet C, Martin P (2004b) Differentiation of the major *Listeria monocytogenes* serovars by multiplex PCR. J Clin Microbiol 42:3819–3822

Dowd GC, Joyce SA, Hill C, Gahan CG (2011) Investigation of the mechanisms by which *Listeria monocytogenes* grows in porcine gallbladder bile. Infect Immun 79:369–379

EFSA Panel on Biological Hazards (BIOHAZ) (2012) Scientific Opinion on the evaluation of the safety and efficacy of Listex$^{TM}$ P100 for the removal of *Listeria monocytogenes* surface contamination of raw fish. EFSA J 10:2615. [43 pp]

Errebo Larsen H, Seeliger HPR (1966) A mannitol fermenting *Listeria*: *Listeria grayi* sp.n. In: Proceedings of the 3rd international symposium on Listeriosis, Biltoven, pp 35–39

Farber JM, Peterkin PI (1991) *Listeria monocytogenes*, a food-borne pathogen. Microbiol Rev 55:476–511

Feresu SB, Jones D (1988) Taxonomic studies on *Brochothrix*, *Erysipelothrix*, *Listeria* and atypical lactobacilli. J Gen Microbiol 134:1165–1183

Fiedler F (1988) Biochemistry of the cell surface of *Listeria* strains: a locating general view. Infection 16:S92–S97

Fiedler F, Ruhland GJ (1987) Structure of *Listeria monocytogenes* cell walls. Bull Inst Pasteur 85:287–300

Fiedler F, Seger J (1983) The murein types of *Listeria grayi*, *Listeria murrayi* and *Listeria denitrificans*. Syst Appl Microbiol 4:444–450

Fiedler F, Seger J, Shrettenbrunner A, Seeliger HPR (1984) The biochemistry of murein and cell wall teichoic acids in the genus *Listeria*. Syst Appl Micobiol 5:360–376

Fischer W, Leopold K (1999) Polar lipids of four *Listeria* species containing L-lysylcardiolipin, a novel lipid structure, and other unique phospholipids. Int J Syst Bacteriol 49:653–662

Fistrovici E, Collins-Thompson CL (1990) Use of plasmid profiles and restriction endonuclease digests in environmental studies of *Listeria* spp. from raw milk. Int J Food Microbiol 10:43–50

Forrester S, Milillo SR, Hoose WA, Wiedmann M, Schwab U (2007) Evaluation of the pathogenicity of *Listeria* spp. in *Caenorhabditis elegans*. Foodborne Pathog Dis 4:67–73

Friedman ME, Alm WL (1962) Effect of glucose concentration in the growth medium on some metabolic activities of *Listeria monocytogenes*. J Bacteriol 84:375–376

Friedman ME, Roessler WG (1961) Growth of *Listeria monocytogenes* in defined media. J Bacteriol 82:528–537

Gandhi M, Chikindas ML (2007) *Listeria*: a foodborne pathogen that knows how to survive. Int J Food Microbiol 113:1–15

Gerner-Smidt P, Hise K, Kincaid J, Hunter S, Rolando S, Hyytiä-Trees E, Ribot EM, Swaminathan B (2006) PulseNet USA: a five-year update. PulseNet taskforce. Foodborne Pathog Dis 3:9–19

Ghosh BK, Murray RG (1967) Fine structure of *Listeria monocytogenes* in relation to protoplast formation. J Bacteriol 93:411–426

Gilot P, Content J (2002) Specific identification of *Listeria welshimeri* and *Listeria monocytogenes* by PCR assays targeting a gene encoding a Fibronectin-Binding Protein. J Clin Microbiol 40:698–703

Girardin H, Morris CE, Albagnac C, Dreux N, Glaux C, Nguyen-The C (2005) Behaviour of the pathogen surrogates *Listeria innocua* and *Clostridium sporogenes* during production of parsley in fields fertilized with contaminated amendments. FEMS Microbiol Ecol 54:287–295

Glaser P, Frangeul L, Buchrieser C, Rusniok C, Amend A, Baquero F, Berche P, Bloecker H, Brandt P, Chakraborty T, Charbit A, Chetouani F, Couvé E, de Daruvar A, Dehoux P, Domann E, Domínguez-Bernal G, Duchaud E, Durant L, Dussurget O, Entian KD, Fsihi H, García-del Portillo F, Garrido P, Gautier L, Goebel W, Gómez-López N, Hain T, Hauf J, Jackson D, Jones LM, Kaerst U, Kreft J, Kuhn M, Kunst F, Kurapkat G, MaduenoE, Maitournam A, Vicente JM, Ng E, Nedjari H, Nordsiek G, Novella S, de Pablos B, Pérez-Diaz JC, Purcell R, Remmel B, Rose M, Schlueter T, Simoes N, Tierrez A, Vázquez-Boland JA, Voss H, Wehland J, Cossart P (2001) Comparative genomics of *Listeria* species. Science 294:849–852

Graham TA, Golsteyn-Thomas EJ, Thomas JE, Gannon VP (1997) Inter- and intraspecies comparison of the 16S-23S rRNA operon intergenic spacer regions of six *Listeria* spp. Int J Syst Bacteriol 4(7):863–869

Granier SA, Moubareck C, Colaneri C, Lemire A, Roussel S, Dao TT, Courvalin P, Brisabois A (2011) Antimicrobial resistance of *Listeria monocytogenes* isolates from food and the environment in France over a 10-year period. Appl Environ Microbiol 77:2788–2790

Gravekamp C, Paterson Y (2010) Harnessing *Listeria monocytogenes* to target tumors. Cancer Biol Ther 9:257–265

Graves LM, Swaminathan B (2001) PulseNet standardized protocol for subtyping *Listeria monocytogenes* by macrorestriction and pulsed-field gel electrophoresis. Int J Food Microbiol 65:55–62

Graves LM, Helsel LO, Steigerwalt AG, Morey RE, Daneshvar MI, Roof SE, Orsi RH, Fortes ED, Milillo SR, den Bakker HC, Wiedmann M, Swaminathan B, Sauders BD (2010) *Listeria marthii* sp. nov., isolated from the natural environment, Finger Lakes National Forest. Int J Syst Evol Microbiol 60:1280–1288

Gründling A, Burrack LS, Bouwer HG, Higgins DE (2004) *Listeria monocytogenes* regulates flagellar motility gene expression through MogR, a transcriptional repressor required for virulence. Proc Natl Acad Sci USA 101:12318–12323

Guenther S, Huwyler D, Richard S, Loessner MJ (2009) Virulent bacteriophage for efficient biocontrol of *Listeria monocytogenes* in ready-to-eat foods. Appl Environ Microbiol 75:93–100

Gueriri I, Cyncynatus C, Dubrac S, Arana AT, Dussurget O, Msadek T (2008) The DegU orphan response regulator of *Listeria monocytogenes* autorepresses its own synthesis and is required for bacterial motility, virulence and biofilm formation. Microbiology 154:2251–2264

Gunn GR, Zubair A, Peters C, Pan ZK, Wu TC, Paterson Y (2001) Two *Listeria monocytogenes* vaccine vectors that express different molecular forms of human papilloma virus-16 (HPV-16) E7 induce qualitatively different T cell immunity that correlates with their ability to induce regression of established tumors immortalized by HPV-16. J Immunol 167:6471–6479

Hadorn K, Hächler H, Schaffner A, Kayser FH (1993) Genetic characterization of plasmid-encoded multiple antibiotic resistance in a strain of *Listeria monocytogenes* causing endocarditis. Eur J Clin Microbiol 12:928–937

Hain T, Steinweg C, Chakraborty T (2006) Comparative and functional genomics of *Listeria* spp. J Biotechnol 126:37–51

Harmsen M, Lappann M, Knochel S, Molin S (2010) Role of extracellular DNA during biofilm formation by *Listeria monocytogenes*. Appl Environ Microbiol 76:2271–2279

Heger M (2012) In pilot project, FDA places MiSeqs in state and federal labs to track food-borne pathogens. Genomeweb Oct 17 available from http://www.genomeweb.com/sequencing/pilot-project-fda-places-miseqs-state-and-federal-labs-track-food-borne-pathogen. Accessed Jan 2013

Hein I, Klein D, Lehnera A, Bubert A, Brandl E, Wagner M (2001) Detection and quantification of the *iap* gene of *Listeria monocytogenes* and *Listeria innocua* by a new real-time quantitative PCR assay. Res Microbiol 15:237–246

Hether NW, Campbell PA, Baker LA, Jackson LL (1983) Chemical composition and biological functions of *Listeria monocytogenes* cell wall preparations. Infect Immun 39:1114–1121

Hiltbold EM, Ziegler HK (1993) Mechanisms of processing and presentation of the antigens of *Listeria monocytogenes*. Infect Agents Dis 2:314–323

Holck A, Berg J (2009) Inhibition of *Listeria monocytogenes* in cooked ham by virulent bacteriophages and protective cultures. Appl Environ Microbiol 75:6944–6946

Huws SA, Morley RJ, Jones MV, Brown MR, Smith AW (2008) Interactions of some common pathogenic bacteria with *Acanthamoeba polyphaga*. FEMS Microbiol Lett 282:258–265

Isaacs D, Liberman MM (1981) Babies cross-infected with *Listeria monocytogenes*. Lancet ii:940

Jablasone J, Warriner K, Griffiths M (2005) Interactions of *Escherichia coli* O157: H7. *Salmonella typhimurium* and *Listeria monocytogenes* plants cultivated in a gnotobiotic system. Int J Food Microbiol 99:7–18

Jadhav S, Bhave M, Palombo EA (2012) Methods used for the detection and subtyping of *Listeria monocytogenes*. J Microbiol Methods 88:327–341

Johnson J, Jinneman K, Stelma G, Smith BG, Lye D, Messer J, Ulaszek J, Evsen L, Gendel S, Bennett RW, Swaminathan B, Pruckler J, Steigerwalt A, Kathariou S, Yildirim S, Volokhov D, Rasooly A, Chizhikov V, Wiedmann M, Fortes E, Duvall RE, Hitchins AD (2004) Natural atypical *Listeria innocua* strains with *Listeria monocytogenes* pathogenicity island 1 genes. Appl Environ Microbiol 70:4256–4266

Jones D, Collins MD, Goodfellow M, Minnikin DE (1979) Chemical studies in the classification of the genus *Listeria* and possibly related bacteria.

In: Ivanov I (ed) Problems of Listeriosis. National Agroindustrial Union, Centre for Scientific Studies, Sofia, pp 17–24

Jones CE, Shama G, Andrews PW, Roberts IS, Jones D (1995) Comparative study of the growth of *Listeria monocytogenes* in defined media and demonstration of growth in continuous culture. J Appl Bacteriol 78:66–70

Kalmokoff ML, Banerjee SK, Cyr T, Hefford MA, Gleeson T (2001) Identification of a new plasmid-encoded sec-dependent bacteriocin produced by *Listeria innocua* 743. Appl Environ Microbiol 67:4041–4047

Kamisango K, Saiki I, Tanio Y, Okumura H, Araki Y, Sekikawa I, Azuma I, Yamamura Y (1982) Structures and biological activities of peptidoglycans of *Listeria monocytogenes* and *Propionibacterium acnes*. J Biochem (Tokyo) 92:23–33

Kamisango K, Fujii H, Okumura H, Saiki I, Araki Y, Yamamura Y, Azuma I (1983) Structural and immunochemical studies of teichoic acid of *Listeria monocytogenes*. J Biochem (Tokyo) 93:1401–1409

Kamp HD, Higgins DE (2009) Transcriptional and post-transcriptional regulation of the GmaR antirepressor governs temperature-dependent control of flagellar motility in *Listeria monocytogenes*. Mol Microbiol 74:421–435

Kämpfer P (1992) Differentiation of *Corynebacterium* spp., *Listeria* spp., and related organism by using fluorogenic substrates. J Clin Microbiol 30:1067–1071

Kämpfer P, Böttcher S, Dott W, Rüden H (1991) Physiological characterization and identification of *Listeria* species. Zbl Bakt Int J Med Microbiol 275:423–435

Kim JW, Dutta V, Elhanafi D, Lee S, Osborne JA, Kathariou S (2012) A novel restriction-modification system is responsible for temperature-dependent phage resistance in *Listeria monocytogenes* ECII. Appl Environ Microbiol 78:1995–2004

Klumpp J, Dorscht J, Lurz R, Bielmann R, Wieland M, Zimmer M, Calendar R, Loessner MJ (2008) The terminally redundant, nonpermuted genome of *Listeria* bacteriophage A511: a model for the SPO1-like myoviruses of Gram-positive bacteria. J Bacteriol 190:5753–5765

Kolb-Maurer A, Gentschev I, Fries HW, Fiedler F, Brocker EB, Kampgen E, Goebel W (2000) *Listeria monocytogenes*-infected human dendritic cells: uptake and host cell response. Infect Immun 68:3680–3688

Kosaric N, Carroll KK (1971) Phospholipids of *Listeria monocytogenes*. Biochim Biophys Acta 239:428–442

Kuenne C, Voget S, Pischimarov J, Oehm S, Goesmann A, Daniel R, Hain T, Chakraborty T (2010) Comparative analysis of plasmids in the genus *Listeria*. PLoS One 5(9):e12511

Kutter S, Hartmann A, Schmid M (2006) Colonization of barley (*Hordeumvulgare*) with *Salmonella enterica* and *Listeria* spp. FEMS Microbiol Ecol 56:262–271

Lang Halter E, Neuhaus K, Scherer S (2013) *Listeria weihenstephanensis* sp. nov., isolated from the water plant *Lemna trisulca* of a German fresh water pond. Int J Syst Evol Microbiol 63:641–647

Larsson S, Cederberg A, Ivarsson S, Svanberg L, Cronberg S (1978) *Listeria monocytogenes* causing hospital-acquired enterocolitis and meningitis in newborn infants. BMJ 2:473–474

Le DT, Brockstedt DG, Nir-Paz R, Hampl J, Mathur S, Nemunaitis J, Sterman DH, Hassan R, Lutz E, Moyer B, Giedlin M, Louis JL, Sugar EA, Pons A, Cox AL, Levine J, Murphy AL, Illei P, Dubensky TW Jr, Eiden JE, Jaffee EM, Laheru DA (2012) A live-attenuated *Listeria* vaccine (ANZ-100) and a live-attenuated *Listeria* vaccine expressing mesothelin (CRS-207) for advanced cancers: phase I studies of safety and immune induction. Clin Cancer Res 18:858–868

Lebrun M, Loulergue J, Chaslus-Dancla E, Audurier A (1992) Plasmids in *Listeria monocytogenes* in relation to cadmium resistance. Appl Environ Microbiol 58:3183–3186

Lebrun M, Audurier A, Cossart P (1994a) Plasmid borne cadmium resistance genes in *Listeria monocytogenes* are similar to *cadA* and *cadC* of *Staphylococcus aureus* and are induced by cadmium. J Bacteriol 176:3040–3048

Lebrun M, Audurier A, Cossart P (1994b) Plasmid borne cadmium resistance genes in *Listeria monocytogenes* are present on Tn5422, a novel transposon closely related to Tn917. J Bacteriol 176:3049–3061

Leclercq A, Clermont D, Bizet C, Grimont PA, Le Flèche-Matéos A, Roche SM, Buchrieser C, Cadet-Daniel V, Le Monnier A, Lecuit M, Allerberger F (2010) *Listeria rocourtiae* sp. nov. Int J Syst Evol Microbiol 60:2210–2214

Lemon KP, Higgins DE, Kolter R (2007) Flagellar motility is critical for *Listeria monocytogenes* biofilm formation. J Bacteriol 189:4418–4424

Lianou A, Sofos JN (2007) A review of the incidence and transmission of *Listeria monocytogenes* in ready-to-eat products in retail and food service environments. J Food Prot 70:2172–2198

Loessner MJ, Krause IB, Henle T, Scherer S (1994) Structural proteins and DNA characteristics of 14 *Listeria* typing bacteriophages. J Gen Virol 75:701–710

Loessner MJ, Inman RB, Lauer P, Calendar R (2000) Complete nucleotide sequence, molecular analysis and genome structure of bacteriophage A118 of *Listeria monocytogenes*: implications for phage evolution. Mol Microbiol 35:324–340

Low JC, Donachie W (1997) A review of *Listeria monocytogenes* and listeriosis. Vet J 153:9–29

Luber P, Crerar S, Dufour C, Farber J, Datta A, Todd ECD (2011) Controlling *Listeria monocytogenes* in ready-to-eat foods: working towards global scientific consensus and harmonization – recommendations for improved prevention and control. Food Control 22:1535–1549

Lungu B, O'Bryan CA, Muthaiyan A, Milillo SR, Johnson MG, Crandall PG, Ricke SC (2011) *Listeria monocytogenes*: antibiotic resistance in food production. Foodborne Pathog Dis 8:569–578

Lusk TS, Ottesen AR, White JR, Allard MW, Brown EW, Kase JA (2012) Characterization of microflora in Latin-style cheeses by next-generation sequencing technology. BMC Microbiol 12:254. doi:10.1186/1471-2180-12-254

Maciag PC, Radulovic S, Rothman J (2009) The first clinical use of a live-attenuated *Listeria monocytogenes* vaccine: a phase I safety study of Lm-LLO-E7 in patients with advanced carcinoma of the cervix. Vaccine 27:3975–3983

Mansfield BE, Dionne MS, Schneider DS, Freitag NE (2003) Exploration of host–pathogen interactions using *Listeria monocytogenes* and *Drosophila melanogaster*. Cell Microbiol 5:901–911

Mattila M, Lindström M, Somervuo P, Markkula A, Korkeala H (2011) Role of *flhA* and *motA* in growth of *Listeria monocytogenes* at low temperatures. Int J Food Microbiol 148:177–183

McLauchlin J, Low JC (1994) Primary cutaneous listeriosis in adults: an occupational disease of veterinarians and farmers. Vet Rec 135:615–617

McLauchlin J, Rees CED (2009) Genus *Listeria*. In: De Vos P, Garrity G, Jones D, Krieg NR, Ludwig W, Rainey FA, Schleifer KH, Whitman WB (eds) Bergey's manual of systematic bacteriology, vol 3, 2nd edn, The low G + C Gram-positive bacteria. Springer, Dordrecht, pp 244–257

Michel B (2005) After 30 years of study, the bacterial SOS response still surprises us. PLoS Biol 3:e255

Milillo SR, Badamo JM, Boor KJ, Wiedmann M (2008) Growth and persistence of *Listeria monocytogenes* isolates on the plant model *Arabidopsis thaliana*. Food Microbiol 25:698–704

Miller IL, Silverman SJ (1959) Glucose metabolism of *Listeria monocytogenes*. Bacteriolo Proc 103

Monk AB, Rees CD, Barrow P, Hagens S, Harper DR (2010) Bacteriophage applications: where are we now? Lett Appl Microbiol 51:363–369

Morvan A, Moubareck C, Leclercq A, Hervé-Bazin M, Bremont S, Lecuit M, Courvalin P, Le Monnier A (2010) Antimicrobial resistance of *Listeria monocytogenes* strains isolated from humans in France. Antimicrob Agents Chemother 54:2728–2731

Mukherjee K, Altincicek B, Hain T, Domann E, Vilcinskas A, Chakraborty T (2010) *Galleria mellonella* as a model system for studying *Listeria* pathogenesis. Appl Environ Microbiol 76:310–317

Murray EGD, Webb RA, Swann MBR (1926) A disease of rabbits characterised by a large mononuclear leucocytosis, caused by a hitherto undescribed bacillus *Bacterium monocytogenes* (n.sp.). J Pathol Bacteriol 29:407–439

Nelson KE, Fouts DE, Mongodin EF, Ravel J, DeBoy RT, Kolonay JF, Rasko DA, Angiuoli SV, Gill SR, Paulsen IT, Peterson J, White O, Nelson WC, Nierman W, Beanan MJ, Brinkac LM, Daugherty SC, Dodson RJ, Durkin AS, Madupu R, Haft DH, Selengut J, Van Aken S, Khouri H, Fedorova N, Forberger H, Tran B, Kathariou S, Wonderling LD, Uhlich GA, Bayles DO, Luchansky JB, Fraser CM (2004) Whole genome comparisons of serotype 4b and 1/2a strains of the food-borne pathogen *Listeria monocytogenes* reveal new insights into the core genome components of this species. Nucleic Acids Res 32:2386–2395

Nichols DS, Presser KA, Olley J, Ross T, McMeekin TA (2002) Variation of branched-chain fatty acids marks the normal physiological range for growth in *Listeria monocytogenes*. Appl Environ Microbiol 68:2809–2813

Ninet B, Traitler H, Aeschlimann JM, Hormna I, Hartman D, Bille J (1992) Quantitative analysis of cellular fatty acids (CFAs) composition of the seven species of *Listeria*. Syst Appl Microbiol 15:76–81

Ottaviani F, Ottaviani M, Agosti M (1997) Differential agar medium for *Listeria monocytogenes*. Ind Aliment 36:888

Pava-Ripoll M, Goeriz Pearson RE, Miller AK, Ziobro GC (2012) Prevalence and relative risk of *Cronobacter* spp., *Salmonella* spp., and *Listeria monocytogenes* associated with the body surfaces and guts of individual filth flies. Appl Environ Microbiol 78:7891–7902

Peek R, Reddy KR (2006) FDA approves use of bacteriophages to be added to meat and poultry products. Gastroenterology 131:1370–1372

Peel M, Donachie W, Shaw A (1988a) Temperature-dependent expression of flagella of *Listeria monocytogenes* studied by electron microscopy, SDS-PAGE and Western blotting. J Gen Microbiol 134:2171–2178

Peel M, Donachie W, Shaw A (1988b) Physical and antigenic heterogeneity in the flagellins of *Listeria monocytogenes* and *L.ivanovii*. J Gen Microbiol 134:2593–2598

Pérez-Díaz JC, Vicente MF, Baquero F (1982) Plasmids in *Listeria*. Plasmid 8:112–118

Phan-Thanh L, Gormon T (1997) A chemically defined minimal medium for the optimal culture of *Listeria*. Int J Food Microbiol 35:91–95

Pine L, Malcolm GB, Brooks JB, Daneshvar MI (1989) Physiological studies on the growth and utilization of sugars by *Listeria* species. Can J Microbiol 35:245–254

Pirie JHH (1927) A new disease of veld rodents, 'Tiger River Disease.'. Publ S Afr Inst Med Res 3:63–186

Pirie JHH (1940a) The genus *Listerella*, Pirie. Science 91:383

Pirie JHH (1940b) *Listeria*: change of name for a genus of bacteria. Nature 145:264

Poyart-Salmeron C, Carlier C, Trieu-Cuot P, Courtieu AL, Courvalin P (1990) Transferable plasmid-mediated antibiotic resistance in *Listeria monocytogenes*. Lancet 335:1422–1426

Poyart-Salmeron C, Trieu-Cuot P, Carlier C, MacGowan A, McLauchlin J, Courvalin P (1992) Genetic basis of tetracycline resistance in clinical isolates of *Listeria monocytogenes*. Antimicrob Agents Chemother 36:463–466

Prévot AR (1961) Listeria Traité de Systématique Bactérienne, vol 2. Dunod, Paris, pp 511–512

Putt SNH, Shaw APM, Woods AJ, Tyler L, James AD (1988) Veterinary epidemiology and economics in Africa – a manual for use in the design and appraisal of livestock health policy. ILCA, University of Reading, Reading

Puttmann M, Ade N, Hof H (1993) Dependence of fatty acid composition of *Listeria* spp. on growth temperature. Res Microbiol 144:279–283

Quentin C, Thibaut MC, Horovitz J, Bebear C (1990) Multiresistant strain of *Listeria monocytogenes* in septic abortion. Lancet 336:375

Radulovic S, Brankovic-Magic M, Malisic E, Jankovic R, Dobricic J, Plesinac-Karapandzic V, Maciag PC, Rothman J (2009) Therapeutic cancer vaccines in cervical cancer: phase I study of Lovaxin-C. J BUON 14:S165–S168

Rae CS, Geissler A, Adamson PC, Portnoy DA (2011) Mutations of the *Listeria monocytogenes* peptidoglycan *N*-Deacetylase and *O*-Acetylase result in enhanced lysozyme sensitivity, bacteriolysis, and hyperinduction of innate immune pathways. Infect Immun 79:3596–3606

Renier S, Hébraud M, Desvaux M (2011) Molecular biology of surface colonization by *Listeria monocytogenes*: an additional facet of an opportunistic Gram-positive foodborne pathogen. Environ Microbiol 13:835–850

Rieu A, Briandet R, Habimana O, Garmyn D, Guzzo J, Piveteau P (2008) *Listeria monocytogenes* EGD-e biofilms: no mushrooms but a network of knitted chains. Appl Environ Microbiol 74:4491–4497

Rocourt J, Catimel B (1985) Charactérisation biochemique des espèces du genre *Listeria*. Zbl Bakt Hyg A 260:221–231

Rocourt J, Grimont PAD (1983) *Listeria welshimeri* sp. nov. and *Listeria seeligeri* sp. nov. Int J Syst Bacteriol 33:866–869

Rocourt J, Grimont F, Grimont PAD, Seeliger HPR (1982) DNA relatedness among serovars of *Listeria monocytogenes sensu lato*. Curr Microbiol 7:383–388

Rocourt J, Schrettenbrunner A, Seeliger HPR (1983) Différenciation biochemique des groupes génomiques de *Listeria monocytogenes* (*sensu lato*). Ann Microbiol (Paris) 134A:65–71

Rocourt J, Gilmore M, Goebel W, Seeliger HPR (1986) DNA relatedness among *Listeria monocytogenes* and *Listeria innocua* bacteriophages. Syst Appl Microbiol 8:42–47

Rocourt J, Wehmeyer U, Stackebrandt E (1987) Transfer of *Listeria denitrificans* to a new genus *Jonesia* gen.nov. as *Jonesia denitrificans* comb.nov. Int J Syst Bacteriol 37:266–270

Rocourt J, Boerlin P, Grimont F, Jacquet C, Piffaretti JC (1992) Assignment of *Listeria grayi* and *Listeria murrayi* to a single species, *Listeria grayi*, with a revised description of *Listeria grayi*. Int J Syst Bacteriol 42:171–174

Rossi LPR, Almeida RCC, Lopes LS, Figueiredo ACL, Ramos MPP, Almeida PF (2011) Occurrence of *Listeria* spp. in Brazilian fresh sausage and control of *Listeria monocytogenes* using bacteriophage P100. Food Control 22:954–958

Sandasi M, Leonard CM, Viljoen AM (2008) The effect of five common essential oil components on *Listeria monocytogenes* biofilms. Food Control 19:1070–1075

Schleifer KH, Kandler O (1972) Peptidoglycan types of bacterial cell walls and their taxonomic implications. Bacteriol Rev 36:407–477

Schmuki MM, Eme D, Loessner MJ, Klumpp J (2012) Bacteriophage P70: unique morphology and unrelatedness to other *Listeria* bacteriophages. J Virol 86:13099

Schuchat A, Lizano C, Broome CV, Swaminathan B, Kim C, Winn K (1991) Outbreak of neonatal listeriosis associated with mineral oil. Pediatr Infect Dis J 10:183–189

Seeliger HP (1981) Nonpathogenic listeriae: *L. innocua* sp. n. (Seeliger et Schoofs, 1977). Zentralbl Bakteriol Mikrobiol Hyg [A] 249:487–493

Seeliger HPR, Höhne K (1979) Serotyping of *Listeria monocytogenes* and related species. In: Bergan T, Norris JR (eds) Methods in microbiology, vol 13. Academic, London, pp 31–49

Seeliger HPR, Rocourt J, Schrettenbrunner A, Grimont PAD, Jones D (1984) *Listeria ivanovii* sp. nov. Int J Syst Bacteriol 34:336–337

Seveau S, Pizarro-Cerda J, Cossart P (2007) Molecular mechanisms exploited by *Listeria monocytogenes* during host cell invasion. Microbes Infect 9:1167–1175

Shaw N (1974) Lipid composition as a guide to the classification of bacteria. Adv Appl Microbiol 17:63–108

Singh R, Wallecha A (2011) Cancer immunotherapy using recombinant *Listeria monocytogenes*: transition from bench to clinic. Hum Vaccin 7:497–505

Singh AK, Zhang Zhu K, Subramanian C, Li Z, Jayaswal RK, Gatto C, Rock CO, Wilkinson BJ (2009) FabH selectivity for anteiso branched-chain fatty acid precursors in low-temperature adaptation in *Listeria monocytogenes*. FEMS Microbiol Lett 301:188–192

Sleator RD, Hill C (2010) Compatible solutes: a listeria passé-partout? Gut Microbes 1:77–79

Somer L, Kashi Y (2003) A PCR method based on 16S rRNA sequence for simultaneous detection of the genus *Listeria* and the species *Listeria monocytogenes* in food products. J Food Prot 66:1658–1665

Soni KA, Nannapaneni RJ (2010) Bacteriophage significantly reduces *Listeria monocytogenes* on raw salmon fillet tissue. J Food Prot 73:32–38

Soni KA, Nannapaneni R, Hagens S (2010) Reduction of *Listeria monocytogenes* on the surface of fresh channel catfish fillets by bacteriophage Listex P100. Foodborne Pathog Dis 7:427–434

Talon R, Grimont PAD, Grimont F, Gasser F, Boeufgras JM (1988) Brochothrix campestris sp. nov. Int J Syst Bacteriol 38:99–102

Tasara T, Stephan R (2006) Cold stress tolerance of *Listeria monocytogenes*: a review of molecular adaptive mechanisms and food safety implications. J Food Prot 69:1473–1484

Threlfall EJ, Skinner JA, McLauchlin J (1998) Antimicrobial resistance in *Listeria monocytogenes* from humans and food in the UK, 1967–96. Clin Microbiol Infect 4:410–412

Trémoulet F, Duche O, Namane A, Martinie B, The European Listeria Genome Consortium, Labadie JC (2002) Comparison of protein patterns of *Listeria monocytogenes* grown in biofilm or in planktonic mode by proteomic analysis. FEMS Microbiol Lett 210:25–31

Trivett TL, Meyer EA (1971) Citrate cycle and related metabolism of *Listeria monocytogenes*. J Bacteriol 107:770–779

Tsakris A, Papa A, Douboyas J, Antoniadis A (1997) Neonatal meningitis due to multi-resistant *Listeria monocytogenes*. J Antimicrob Chemother 39:553–554

Ullmann WW, Cameron JA (1969) Immunochemistry of the cell walls of *Listeria monocytogenes*. J Bacteriol 98:486–493

Vail KM, McMullen LM, Jones TH (2012) Growth and filamentation of cold-adapted, log-phase *Listeria monocytogenes* exposed to salt, acid, or alkali stress at 3 °C. J Food Prot 75:2142–2150

van Der Veen S, Abee T (2010) Dependence of continuous-flow biofilm formation by *Listeria monocytogenes* EGD-e on SOS response factor YneA. Appl Environ Microbiol 76:1992–1995

van der Veen S, van Schalkwijk S, Molenaar D, de Vos WM, Abee T, Wells-Bennik MH (2010) The SOS response of *Listeria monocytogenes* is involved in stress resistance and mutagenesis. Microbiology 156:374–384

Vatanyoopaisarn S, Nazli A, Dodd CE, Rees CE, Waites WM (2000) Effect of flagella on initial attachment of *Listeria monocytogenes* to stainless steel. Appl Environ Microbiol 66:860–863

Vázquez-Boland JA, Kuhn M, Berche P, Chakraborty P, Domínguez-Bernal G, Goebel W et al (2001) *Listeria* pathogenesis and molecular virulence determinants. Clin Microbiol Rev 14:584–640

Welshimer HJ, Meredith AL (1971) *Listeria murrayi* sp. n.: a nitrate-reducing mannitol-fermenting *Listeria*. Int J Syst Bacteriol 21:3–7

Wendlinger G, Loessner MJ, Scherer S (1996) Bacteriophage receptors on *Listeria monocytogenes* cells are the *N*-acetylglucosamine and rhamnose substituents of teichoic acids or the peptidoglycan itself. Microbiology 142:985–992

Wernars K, Heuvelman CJ, Chakraborty T, Notermans SH (1991) Use of the polymerase chain reaction for direct detection of *Listeria monocytogenes* in soft cheese. J Appl Bacteriol 70:121–126

Wiedmann M, Bruce JL, Keating C, Johnson AE, McDonough PL, Batt CA (1997) Ribotypes and virulence gene polymorphisms suggest three distinct *Listeria monocytogenes* lineages with differences in pathogenic potential. Infect Immun 65:2707–2716

Wilkinson BJ, Jones D (1977) A numerical taxonomic survey of *Listeria* and related bacteria. J Gen Microbiol 98:399–421

Wuenscher MD, Kohler S, Bubert A, Gerike U, Goebel W (1993) The *iap* gene of *Listeria monocytogenes* is essential for cell viability, and its gene product, p60, has bacteriolytic activity. J Bacteriol 175:3491–3501

Yarza P, Ludwig W, Euzéby J, Amann R, Schleifer KH, Glöckner FO, Rosselló-Móra R (2010) Update of the all-species living-tree project based on 16S and 23S rRNA sequence analyses. Syst Appl Microbiol 33:291–299

Zemansky J, Kline BC, Woodward JJ, Leber JH, Marquis H, Portnoy DA (2009) Development of a mariner-based transposon and identification of *Listeria monocytogenes* determinants, including the peptidyl-prolyl isomerase PrsA2, that contribute to its hemolytic phenotype. J Bacteriol 191:3950–3964

# 20 The Family *Natranaerobiaceae*

*Aharon Oren*
Department of Plant and Environmental Sciences, The Institute of Life Sciences, The Hebrew University of Jerusalem, Jerusalem, Israel



**Abstract**
The family *Natranaerobiaceae*, first proposed in 2007, is the only described family within the order *Natranaerobiales*, affiliated with the class *Clostridia* in the phylum Firmicutes. Currently (October 2013) it encompasses two genera: *Natranaerobium* (type genus) with two species, and *Natronovirga* with one species. The description of a third genus, *Natranaerobaculum*, with one species is in press. All members of the family are Gram-positive, fermentative polyextremophiles that require high salinity and high pH for growth, and are markedly thermotolerant. They were found in the anaerobic sediments of hypersaline soda lakes of the Wadi El Natrun, Egypt, and Lake Magadi, Kenya.

## Taxonomy

### Order *Natranaerobiales* Mesbah, Hedrick, Peacock, Rohde and Wiegel, 2007, 2511[VP]

Natr.an.ae.ro.bi.a'les. N.L. masc. n. *Natranaerobius*, type genus of the order; suff.—*ales*, ending denoting an order; N.L. fem. pl. n. *Natranaerobiales*, the *Natranaerobius* order.

The order *Natranaerobiales* was proposed both on the basis of the unique phenotypic properties, especially the polyextremophilic nature of its members which all require high salinity and high pH for growth and are markedly thermotolerant, and on the basis of their phylogenetic position within the class *Clostridia* (Mesbah et al. 2007b). The order currently contains one family, the *Natranaerobiaceae*, and the description of the order is therefore the same as for the family.

Family *Natranaerobiaceae* Mesbah et al. 2007b, 2511[VP]

Natr.an.ae.ro.bi.a.ce'ae. L. masc. n. *Natranaerobius*, type genus of the family; *-aceae* ending to denote a family; N.L. fem. n. *Natranaerobiaceae* the family of *Natranaerobius*.

The cells have a Gram-positive-type cell wall and stain Gram-positive. Slender straight or slightly curved rods, generally non-motile. Endospores may be present. Strictly anaerobic halophilic organoheterotrophs, obtaining energy by fermentation or by anaerobic respiration.

Type genus: *Natranaerobius*.

The mol% G+C of the DNA varies between 35.6 and 42.

The family *Natranaerobiaceae* is phenotypically, metabolically, and ecologically homogeneous. The family includes organisms that live by fermentation of sugars or peptides. All members also have a potential for anaerobic respiration using, e.g., thiosulfate, nitrate, or fumarate as the electron acceptor.

At the time of writing (October 2013), the family contained two genera with a total of three species whose names have standing in the nomenclature (➲ *Table 20.2*): *Natranaerobius* (two species) and *Natronovirga* (one species). The description of a third genus, *Natranaerobaculum*, is in press (Zavarzina et al. 2013). The general properties of the three genera are given in ➲ *Table 20.1*. ➲ *Table 20.2* lists the differential morphological, metabolic, and chemotaxonomic characteristics of the type strains of the four described species. 16S rRNA gene sequences of additional, not fully characterized and classified strains, can be found in the GenBank.

## Phylogenetic Structure of the Family and Its Genera

The order *Natranaerobiales* was proposed on the basis of the phylogenetic position of its representatives within the class *Clostridia* in the phylum Firmicutes (Mesbah et al. 2007b). ➲ *Figure 20.1* presents a neighbor-joining tree based on 16S

E. Rosenberg et al. (eds.), *The Prokaryotes – Firmicutes and Tenericutes*, DOI 10.1007/ 978-3-642-30120-9_360,

**Table 20.1**
**The genera classified within the family *Natranaerobiaceae*, as of October 2013**

| Genus | Number of species | Type species | General properties |
|---|---|---|---|
| *Natranaerobius* | 2 | *Natranaerobius thermophilus* | Obligately anaerobic, Gram-positive rods, endospores not observed. Obligately alkaliphilic and thermophilic, halotolerant chemoorganotrophs. The cell wall lacks significant amounts of murein and *meso*-diaminopimelic acid. The fatty acid profile is dominated by branched fatty acids with odd numbers of carbons; dimethylacetals are also present |
| *Natronovirga* | 1 | *Natronovirga wadinatrunensis* | Obligately anaerobic, Gram-positive rods, endospores not observed. Obligately alkaliphilic and thermophilic, halotolerant chemoorganotrophs. The fatty acid profile is dominated by branched fatty acids with 15 carbons. Cell-wall peptidoglycan is of the $\alpha$-4-$\beta$ Orn-Gly-Asp type |
| *Natranaerobaculum* | 1 | *Natranaerobaculum magadiense* | Obligately anaerobic, endospore-forming Gram-positive rods, obligately alkaliphilic and thermophilic, halotolerant chemoorganotrophs |

rRNA sequences. A similar topology is indicated in a maximum likelihood tree (not shown). The three genera form a monophyletic cluster that can be distinguished from others not only on the basis of 16S rRNA sequence comparison but also by their physiological and ecological properties, notably their polyextremophilic nature. Members of the families *Peptococcaceae* (*Clostridiales*), *Thermoanaerobacteraceae* (*Thermoanaerobacterales*), and *Thermolithobacteraceae* (*Thermolithobacterales*) are among the closest neighbors to the *Natranaerobiaceae*. The closest neighbor to the *Natranaerobiaceae* is the species *Dethiobacter alkaliphilus*, a species currently assigned to the *Syntrophomonadaceae* (Clostridiales) (Sorokin et al. 2008).

## Genome Analysis

The genome sequence of the type strain of *Natranaerobius thermophilus* is available (Zhao et al. 2011). This genome consists of one 3,165,557 bp circular chromosome and two plasmids (17,207 bp and 8.689 bp). The sequences were deposited in GenBank as CP001034 (chromosome) and CP001035 and CP001036 (plasmids). The G+C percentages are 36.4 mol% for the chromosome (notably lower than the value of 40.4 mol% determined by HPLC (Mesbah et al. 2007b)), and 34.1 and 35.7 mol% for the two plasmids, respectively. Three rRNA operons are present with nearly identical 16S rRNA gene sequences.

### Phages

No bacteriophages infecting members of the *Natranaerobiaceae* were yet described.

## Phenotypic Analyses

The properties of the genera and species of *Natranaerobiaceae*.

### Genus *Natranaerobius* Mesbah, Hedrick, Peacock, Rohde and Wiegel, 2007, 2511[VP]

Natr.an.ae.ro'bi.us. N.L. n. *natron* derived from Arabic *natrun*, soda (sodium carbonate); Gr. pref. *an*, not; Gr. n. *aer aeros*, air; Gr. masc. n. *bios*, life; N.L. masc. n. *Natranaerobius*, a soda-requiring anaerobe.

Cells are Gram-positive, non-motile, strictly anaerobic, catalase- and oxidase-negative rods. Endospores are not observed. They are alkaliphilic, halophilic, moderately themophilic chemoorganotrophs that obtain energy by fermentation or by anaerobic respiration. The fatty acid profile is dominated by branched fatty acids with odd numbers of carbons.

The mol% G+C of the DNA is 40.4–41, as determined by HPLC.

The type species is *Natranaerobius thermophilus* with type strain JW/NM-WN-LF$^T$ (=DSM 18059$^T$ = ATCC BAA-1301$^T$) (Mesbah et al. 2007b). However, strain BAA-1301$^T$ is no longer available from the ATCC.

The genus *Natranaerobius* currently contains two species: *N. thermophilus* and *N. trueperi*.

Further Comments:

- *Natranaerobius thermophilus* and *Natranaerobius trueperi* lack significant amounts of murein and *meso*-diaminopimelic acid in their cell wall.
- Two additional putative species have been isolated from sediment samples of Lake Magadi in the Kenyan Rift Valley: "Natranaerobius jonesii" (growing optimally at 55 °C, pH 8.5–10.5 and 3.7–3.9 M $Na^+$) and "Natranaerobius grantii"

**Table 20.2**

**Differential morphological, metabolic, and chemotaxonomic characteristics of the type strains of *Natranaerobiaceae* species, including *Natranaerobium magadiense*, the description of which was in press at the time of writing)**

| Genus | *Natranaerobius* | | *Natronovirga* | *Natranaerobaculum* |
|---|---|---|---|---|
| Species | *Natranaerobius thermophilus*[a] | *Natranaerobius trueperi*[b] | *Natronovirga wadinatrunensis*[b] | *Natranaerobaculum magadiense*[c] |
| Type strain | DSM 18059, ATCC BAA-1301[d] | DSM 18760, ATCC BAA-1443[d] | DSM 18770, ATCC BAA-1444[d] | DSM 24923, VKM B-2666 |
| Cell size (μm) | 0.2–0.4 × 3–5 | 0.6 × 2–3 | 0.3–0.4 × 4–5 | 0.2–0.5 × 3–7 |
| Endospore formation | – | – | – | + |
| Motility | – | – | – | + (slow) |
| pH range for growth and optimum[e] | 8.3–10.6 (9.5) | 8.0–10.8 (9.5) | 8.5–11.5 (9.9) | 7.5–10.7 (9.25–9.5) |
| Temperature range for growth and optimum (°C) | 35–56 (53–55) | 26–55 (52) | 26–55 (51) | 20–57 (40–50) |
| Total $Na^+$ range and optimum[f] | 3.0–5.0 (3.3–3.9) | 3.5–4.5 (3.7) | 3.1–5.3 (3.9) | 0.5–2.7 (0.9) |
| Utilization of carbohydrates | + | + | + | – |
| Use of | | | | |
| Cellobiose | + | + | – | – |
| Fructose | + | – | + | – |
| Galactose | NR | NR | + | – |
| Glucose | – | + | + | – |
| Lactose | NR | NR | +[g] | – |
| Mannose | – | – | + | – |
| Pyruvate | + | –/+[h] | + | – |
| Ribose | + | + | + | – |
| Sucrose | + | + | + | – |
| Trehalose | + | – | + | – |
| Xylose | + | NR | – | – |
| Main fermentation products | Acetate, formate | Acetate, lactate | Acetate, lactate | Acetate, succinate, formate, lactate |
| Thiosulfate reduction | + | – | – | + |
| Nitrate reduction | + | + | + | + |
| Cell-wall structure | No significant amounts of murein and *meso*-diaminopimelic acid detected | No significant amounts of murein and *meso*-diaminopimelic acid detected | α-4-β Orn-Gly-Asp type | NR |
| Major fatty acids | $C_{i15:0}$; $C_{i17:0}$; $C_{16:0}$ | $C_{i15:0}$; $C_{a15:0}$ | $C_{i15:0}$; $C_{a15:0}$ | $C_{16:0}$; $C_{16:1\omega7c}$, $C_{18:0}$, $C_{18:1\omega9c}$ |
| DNA G+C content (mol %) | 36.4 (Genome sequence); 40.4 (HPLC) | 41 (HPLC) | 42 (HPLC) | 35.6 (Thermal denaturation) |

+ positive, – negative, *NR* not reported

Data taken from: [a]Mesbah et al. 2007b

[b] Mesbah and Wiegel 2009

[c]Zavarzina et al. 2013

[d]Strain BAA-1301 can no longer be obtained, and at the time of writing strains BAA-1443 and BAA-1444 were not available from the ATCC

[e]The pH values quoted refer to the values measured at or near the optimum growth temperature: 55 °C for *Natranaerobius* and *Natronovirga* species, 45 °C for *Natranaerobaculum*

[f]This row refers to the total medium $Na^+$ concentration, including $Na_2CO_3$, $NaHCO_3$, and NaCl

[g]Lactose was reported to be used, but no β-galactosidase activity was detected

[h]According to the protologue pyruvate is used, but *Table 20.1* in Mesbah and Wiegel 2009 lists utilization of pyruvate as negative

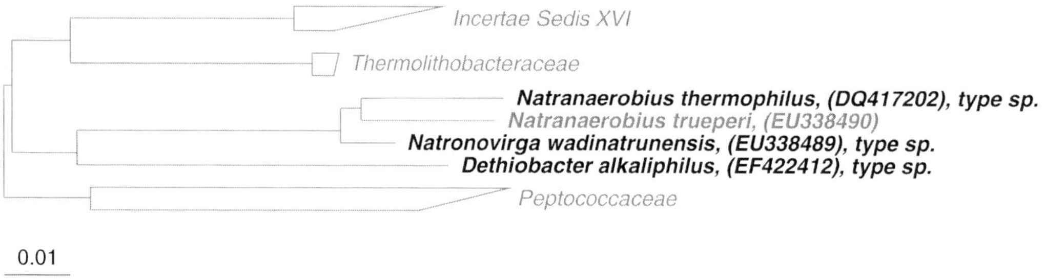


**Fig. 20.1**
**Phylogenetic reconstruction of the order *Natranaerobiales* based on 16S rRNA and created using the neighbor-joining algorithm with the Jukes-Cantor correction. The sequence dataset and alignment were used according to the All-Species Living Tree Project (*LTP*) database (Yarza et al. 2010; http://www.arb-silva.de/projects/living-tree). The tree topology was stabilized with the use of a representative set of nearly 750 high quality type strain sequences proportionally distributed among the different bacterial and archaeal phyla. In addition, a 40 % maximum frequency filter was applied in order to remove hypervariable positions and potentially misplaced bases from the alignment. Scale bar indicates estimated sequence divergence**

(optimal growth at 46 °C, pH 9.5 and 4.3 M $Na^+$) (Mesbah and Wiegel 2008; Bowers et al. 2008, 2009). These isolates are still awaiting formal description.

## Genus *Natronovirga* Mesbah and Wiegel, 2009, 2047$^{VP}$

Na.tro.no.vir'ga. N.L. n. *natron* (arbitrarily derived from the Arabic n. *natrun* or *natron*) soda, sodium carbonate; L. fem. n. *virga*, rod; N.L. fem. n. *Natronovirga*, a soda-requiring rod.

Cells are rod-shaped, non-motile, Gram-positive rods. Endospores were never observed. They are strictly anaerobic and oxidase and catalase negative. Extremely halophilic, obligately alkaliphilic and thermophilic, and chemoorganotrophic. The fatty acid profile is dominated by branched fatty acids with 15 carbons, and the cell-wall peptidoglycan is of the α-4-β Orn-Gly-Asp type.

The mol% G+C of the DNA is 42, as determined by HPLC.

The type species, and currently single species of the genus, is *Natronovirga wadinatrunensis* with type strain JW/NM-WN-LH1$^T$ (=DSM 18870 $^T$ = ATCC BAA-1444 $^T$) (Mesbah and Wiegel 2009). Note that at the time of writing strain BAA-1444 $^T$ was not available from the ATCC.

The description of a third genus, *Natranaerobaculum*, is in press:

Genus *Natranaerobaculum* Zavarzina et al. (2013).

Natr.an.ae.ro.ba'cu.lum. N.L. n. *natron* derived from Arabic *natrun*, soda (sodium carbonate); Gr. pref. *an*, not; Gr. n. *aer aeros*, air; L. neut. n. *baculum*, small stick; N.L. masc. n. *Natranaerobaculum*, a soda-requiring anaerobic rod.

Cells are Gram-positive rods. Slight tumbling motility was observed, but no flagella were visualized in the electron microscope. Endospores are produced. Strictly anaerobic, oxidase- and catalase-negative chemoorganotrophs, using peptides as substrate for fermentation. Sugars are not used as growth substrates. Obligately alkaliphilic and thermophilic, halotolerant chemoorganotrophs. The major fatty acids are $C_{16:0}$; $C_{16:1\omega7c}$, $C_{18:0}$, and $C_{18:1\omega9c}$.

The mol% G+C of the DNA is 35.6, as determined by HPLC.

The type species, and currently single species of the genus, is *Natranaerobaculum magadiense* with type strain Z-1001$^T$ (=DSM 24923$^T$ = VKM B-2666$^T$) (Zavarzina et al. 2013).

Further comments:

- In addition to its fermentative mode of growth, *Natranaerobaculum magadiense* can grow by anaerobic respiration, using thiosulfate, nitrate, arsenate, selenite, Fe(III) citrate, and anthraquinone-2,6-disulfonate as the electron acceptors.
- The polar lipids of *Natranaerobaculum magadiense* are two not further characterized aminophospholipids, four phospholipids, and three other unknown polar lipids.
- The type strain of *Natranaerobaculum magadiense* contains three slightly different 16S rRNA genes.

## Isolation, Enrichment, and Maintenance Procedures

All isolated members of the *Natranaerobiaceae* were recovered from anaerobic sediments of hypersaline extremely alkaline lakes in Africa, located in the Wadi An Natrun, Egypt, or in Kenya. All strains were obtained from enrichment cultures, using complex media with yeast extract, tryptone, and simple sugars such as sucrose as carbon and energy sources. Strictly anaerobic handling techniques are necessary, including boiling media under nitrogen to remove molecular oxygen and addition of cysteine as a reducing agent. The media must contain molar concentrations of $Na^+$ as NaCl, $Na_2CO_3$ and $NaHCO_3$ to maintain the pH in the range of 9–10. Following incubation at 40–55 °C growth was obtained, and colonies were isolated from anaerobic shake-roll tubes (1 % agar)

(Mesbah and Wiegel 2012; Mesbah et al. 2007b; Mesbah and Wiegel 2009). When adjusting the pH to the required alkaline values, it is essential to do so at the proper temperature at which cells will be grown. For the growth media employed, $pH^{55oC}$ values of 8.3, 9.5 and 10.6 were found to correspond to $pH^{25oC}$ of 9.3, 10.5, and 11.2 (Mesbah et al. 2009).

Cultures can be maintained at room temperature for short-time preservation (a few weeks). For long-term preservation, cultures can be stored at −80 °C in prereduced medium mixed with 50 % (v/v) glycerol (Mesbah and Wiegel 2009).

## Physiological and Biochemical Features

Under optimal conditions the species of *Natranaerobius* and *Natronovirga* have doubling times of around 3 h (Mesbah and Wiegel 2009). In addition to their fermentative metabolism on sugars (*Natranaerobius* and *Natronovirga*) and/or peptides (*Natranaerobaculum*) with the formation of products including acetate, formate, and lactate, members of the *Natranaerobiaceae* can grow by anaerobic respiration. Among the electron acceptors used are fumarate, nitrate, Fe(III), thiosulfate, and others. The potential for anaerobic respiration is also reflected in the genome of *Natranaerobius thermophilus* (Zhao et al. 2011): genes annotated include a nitrate reductase, a fumarate reductase, and cytochrome-related proteins.

Fermentative metabolism yields only little energy, and therefore the *Natranaerobiaceae* face interesting challenges with respect to survival under extremes of salinity, high pH, and temperature. They are the best model organisms for the study of the limits of life under multiple environmental extremes (Mesbah and Wiegel 2008, 2012). To cope with the osmotic pressure of their hypersaline environment, they probably use a combination of organic osmotic solutes and intracellular ions. Among the genes annotated in the genome of *Natranaerobius thermophilus* are systems for the de novo synthesis of the compatible solute glycine betaine, 15 genes for betaine ABC transporters, as well as genes for the transport of proline and choline (Zhao et al. 2011). Its proteome is markedly acidic (median pI of the proteins encoded by the genome: 6.27), which may indicate the presence of salt-adapted proteins (Elevi Bardavid and Oren 2012).

To survive at highly alkaline pH values, *Natranaerobius thermophilus* maintains its intracellular pH at about 1 unit below the medium pH over the entire pH range in which growth is possible. At least eight electrogenic $Na^+(K^+)/H^+$ antiporters were identified in this organism to contribute to the acidification of the cytoplasm and to expel cytoplasmic $Na^+$ that tends to accumulate inside the cells during alkaline stress (Mesbah and Wiegel 2011; Mesbah et al. 2009).

Sensitivity tests to different antibiotics were only reported for *Natranaerobaculum magadiense*. The type strain is inhibited by streptomycin, vancomycin, and rifampicin, but is resistant to kanamycin, novobiocin, and penicillin G (Zavarzina et al. 2013).

## Ecology

Members of the family *Natranaerobiaceae* have thus far only been isolated from the anaerobic sediments of the alkaline hypersaline lakes of the Wadi An Natrun, Egypt, and from Lake Magadi, Kenya, where salinity and pH are optimal for these organisms and temperatures are often high due to solar heating. There are more such environments at different locations worldwide, but these have not yet been explored for the presence of this group of organisms. Based on their type of metabolism the members of the *Natranaerobiaceae* are expected to participate in the anaerobic degradation processes in the sediments, fermenting sugars, and amino acids to lactate and to other products that can subsequently be used as electron donors by sulfate reducing bacteria.

Nothing is known yet about the abundance of the members of the family in the anaerobic sediments from which they have been recovered. All extant cultures were derived from enrichment cultures, which provide no quantitative information about the numbers at which the organisms are present. However, no sequences related to the *Natranaerobiales* were found during the analysis of 16S rRNA gene clone libraries prepared from sediments of three of the Wadi An Natrun lakes (Mesbah et al. 2007b) and from the Kenyan soda lakes including Lake Magadi (Rees et al. 2004). This, together with the fact that GenBank currently does not contain other environmental sequences with a high degree of similarity to the *Natranaerobiaceae*, suggests that the members of the family are not among the numerically dominant prokaryotes in the anaerobic alkaline hypersaline environments explored thus far.

## Pathogenicity, Clinical Relevance

No members of the *Natranaerobiaceae* are known to be pathogenic to humans, animals, or plants, as expected for organisms that can only grow under extremes of salinity, temperature, and pH.

## Application

No applications have yet been proposed for any of the species of the three genera currently classified within the *Natranaerobiaceae*.

## Acknowledgment

The author's current studies on anaerobic halophilic bacteria are supported by the Israel Science Foundation (grant no. 343/13).

## References

Bowers KJ, Mesbah NM, Wiegel J (2008) *Natranaerobius 'grantii'* and *Natranaerobius 'jonesii'*, spp. nov., two anaerobic halophilic alkaliphiles isolated from the Kenyan-Tanzanian Rift. Abstract I-007, General Meeting of the American Society for Microbiology, Boston

Bowers KJ, Mesbah NM, Wiegel J (2009) Biodiversity of poly-extremophilic *Bacteria*: does combining the extremes of high salt, alkaline pH and elevated temperature approach a physico-chemial boundary for life? Saline Syst 5:9

Elevi Bardavid R, Oren A (2012) Acid-shifted isoelectric point profiles of the proteins in a hypersaline microbial mat: an adaptation to life at high salt concentrations? Extemophiles 16:787–792

Mesbah NM, Wiegel J (2008) Life at extreme limits. The anaerobic halophilic alkalithermophiles. Ann N Y Acad Sci 1125:44–57

Mesbah NM, Wiegel J (2009) *Natronovirga wadinatrunensis* gen. nov., sp. nov. and *Natranaerobius trueperi* sp. nov., halophilic alkalithermophilic micro-organisms from soda lakes of the Wadi An Natrun, Egypt. Int J Syst Evol Microbiol 59:2042–2048

Mesbah NM, Wiegel J (2011) The $Na^+$-translocating $F_1F_o$-ATPase from the halophilic, alkalithermophile *Natranaerobius thermophilus*. Biochim Biophys Acta 1807:1133–1142

Mesbah NM, Wiegel J (2012) Life under multiple extreme conditions: diversity and physiology of the halophilic alkalithermophiles. Appl Environ Microbiol 78:4074–4082

Mesbah NM, Abou-El-Ela SH, Wiegel J (2007a) Novel and unexpected prokaryotic diversity in water and sediments of the alkaline, hypersaline lakes of the Wadi An Natrun, Egypt. Microb Ecol 54:598–616

Mesbah NM, Hedrick DB, Peacock AD, Rohde M, Wiegel J (2007b) *Natranaerobius thermophilus* gen. nov., sp. nov., a halophilic, alkalithermophilic bacterium from soda lakes of the Wadi An Natrun, Egypt, and proposal of *Natranaerobiaceae* fam. nov. and *Natranaerobiales* ord. nov. Int J Syst Evol Microbiol 57:2507–2512

Mesbah NM, Cook GM, Wiegel J (2009) The halophilic alkalithermophile *Natranaerobius thermophilus* adapts to multiple environmental extremes using a large repertoire of $Na^+(K^+)/H^+$ antiporters. Mol Microbiol 74:270–281

Rees HC, Grant WD, Jones BE, Heaphy S (2004) Diversity of Kenyan soda lake alkaliphiles assessed by molecular methods. Extremophiles 8:63–71

Sorokin DY, Tourova TP, Mußmann M, Muyzer G (2008) *Dethiobacter alkaliphilus* gen. nov. sp. nov., and *Desulfurivibrio alkaliphilus* gen. nov. sp. nov.: two novel representatives of reductive sulfur cycle from soda lakes. Extremophiles 12:431–439

Yarza P, Ludwig W, Euzéby J, Amann R, Schleifer KH, Glöckner FO, Rosselló-Móra R (2010) Update of the all-species living tree project based on 16S and 23S rRNA sequence analyses. Syst Appl Microbiol 33:291–299

Zavarzina DG, Zhilina TN, Kuznetsov BB, Kolganova TV, Osipov GA, Kotelev MS, Zavarzin GA (2013) *Natranaerobaculum magadiense* gen. nov., sp. nov., a new anaerobic alkalithermophilic bacterium from soda lake Magadi. Int J Syst Evol Microbiol (in press). doi:10.1099/ijs/0.054536-0)

Zhao B, Mesbah NM, Dalin E, Goodwin L, Nolan M, Pitluck S, Chertkov O, Brettin TS, Han J, Larimer FW, Land ML, Hauser L, Kyrpides N, Wiegel J (2011) Complete genome sequence of the anaerobic, halophilic alkalithermophilic *Natranaerobius thermophilus*. J Bacteriol 193:4023–4024

# 21 The Family *Paenibacillaceae*

*Shanmugam Mayilraj*[1] · *Erko Stackebrandt*[2]
[1]Microbial Type Culture and Gene Bank (MTCC), CSIR-Institute of Microbial Technology, Chandigarh, India
[2]Leibniz Institute DSMZ-German Collection of Microorganisms and Cell Cultures GmbH, Braunschweig, Germany



**Abstract**
The family *Paenibacillaceae* has been created on the basis of the analysis of 16S rRNA gene sequences. It embraces the species-rich type genera *Paenibacillus*, *Ammoniphilus*, *Aneurinibacillus*, *Brevibacillus*, *Cohnella*, *Oxalophagus*, and *Thermobacillus* and the recently described genera *Fontibacillus* and *Saccharibacillus*. Oval to ellipsoid spores are formed, most species are Gram staining positive, and some stain Gram negative. Other characteristics of taxonomic values are varying such as motility, relationship to oxygen, and catalase formation. The major menaquinone is either MK-7 or MK-6; anteiso-$C_{15:0}$, iso-$C_{15:0}$, iso-$C_{16:0}$, and $C_{16:0}$ are the major fatty acids; and the mol% G+C ranges between 36 and 59. Members of the family are frequently isolated from various soil habitats, compost, and various plant materials but also from freshwater, blood, and feces. The biology of the genus *Cohnella* is described here in greater detail.

## Taxonomy

Most genera of the family *Paenibacillus* (De Vos et al. 2009a) has been extensively covered in the 2nd edition of *Bergey's Manual of Systematic Bacteriology* (De Vos et al. 2009b). Since then, many new species and two new genera (*Fontibacillus*, *Saccharibacillus*) were described. The chapter on *Paenibacillus* covered descriptions of 73 species and additional 13 species, published after submission deadline of the handbook, and was briefly characterized. Since then, 57 additional species (➲ *Table 21.1*) were described as new members of *Paenibacillus*, which indicates the ease at which new organisms are isolated from environmental samples. Five new *Brevibacillus* species as well as two species of *Saccharibacillus* and one new species of each *Fontibacillus and Thermobacillus* were published since then (➲ *Table 21.1*). The genus *Cohnella* is dealt with in a separate chapter.

As mentioned by De Vos et al. (2009a), the family comprises two lineages. The two newly described genera *Saccharibacillus* and *Fontibacillus* group with *Cohnella* and *Paenibacillus*, while the other genera cluster distantly to members of *Brevibacillus*. *Thermobacillus* does not appear as the most deeply branching lineage (Touzel and Prensier 2009) but as a rapidly evolving lineage within the genus *Paenibacillus* (➲ *Fig. 21.2b*). The two species of *Fontibacillus* are remotely related to each other, clustering with different *Paenibacillus* species (➲ *Fig. 21.2a*).

Several genome sequences of family members have been completed (published or unpublished according to the Genomes Online Database (http://www.genomesonline.org/cgi-bin/GOLD/index.cgi?page_requested=Complete+Genome+Projects) such as *Alicyclobacillus* (e.g., strains DSM 446$^T$, DSM 13609$^T$, DSM 22757$^T$), *Aneurinibacillus terranovensis* (DSM 18919$^T$), *Brevibacillus* (e.g., strains NBRC 100599, DSM 25$^T$, phR$^T$), *Cohnella* (e.g., strains DSM 21336$^T$, DSM 17683$^T$), various *Paenibacillus* species (e.g., strains DSM 5050$^T$, DSM 29$^T$, DSM 18201$^T$, YK9$^T$, DSM 15491$^T$, 3016), *Saccharibacillus kuerlensis* (DSM 22868$^T$), and *Thermobacillus composti* (DSM 18247$^T$).

## The Genus *Cohnella*

The genus *Cohnella* was proposed as a member of the family *Paenibacillaceae*, distinguished from the genera *Paenibacillus* and *Bacillus* on the basis of 16S rRNA gene sequence analysis and chemotaxonomic markers (Kämpfer et al. 2006). Members of the genus *Cohnella* are Gram-positive, endospore-forming, aerobic, rod-shaped organisms which are distributed in a wide variety of environments, including volcanic pond, industrial samples, and root nodules. The 19 species of the genus *Cohnella* possess a DNA mol% G+C between 47.6 and 65.1 mol% and contained meso-diaminopimelic acid in the cell-wall peptidoglycan, MK-7 as the predominant menaquinone; diphosphatidylglycerol, phosphatidylglycerol, phosphatidylethanolamine,

E. Rosenberg et al. (eds.), *The Prokaryotes – Firmicutes and Tenericutes*, DOI 10.1007/978-3-642-30120-9_354,

**Table 21.1**
**Species described to be members of genera affiliated to the family *Paenibacillaceae* (except *Cohnella*) since the deadline for submission to the 2nd edition of *Bergey's Manual of Systematic Bacteriology* (De Vos et al. 2009a)**

| Species | Type strain | 16S rRNA gene sequence accession number | Author, effective publication |
|---|---|---|---|
| *Paenibacillus* | | | |
| *P. aestuarii* | CJ25 | EU570250 | Bae et al. 2010 |
| *P. algorifonticola* | XJ259 | GQ383922 | Tang et al. 2011 |
| *P. camelliae* | b11s-2 | EU400621 | Oh et al. 2008 |
| *P. castaneae* | Ch-32 | EU099594 | Valvaerde et al. 2008 |
| *P. catalpae* | D75 | HQ657320 | Zhang et al. 2013 |
| *P. cellulositrophicus* | P2-1 | FJ178001 | Akaracharanya et al. 2009 |
| *P. chartarius* | CCM 7759 | FN689718 | Kämpfer et al. 2012 |
| *P. chungangensis* | CAU 9038 | GU187432 | Park et al. 2011 |
| *P. contaminans* | CKOBP-6 | EF626690 | Chou et al. 2009 |
| *P. edaphicus* | T7 | AF006076 | Hu et al. 2010 |
| *P. filicis* | S4 | GQ423055 | Kim et al. 2009b |
| *P. frigoriresistens* | YIM 016 | JQ314346 | Ming et al. 2012 |
| *P. ginsengihumi* | DCY16 | EF452662 | Kim et al. 2008 |
| *P. glacialis* | KFC91 | EU815294 | Kishore et al. 2010 |
| *P. harenae* | B519 | AY839867 | Jeon et al. 2009 |
| *P. hordei* | RH-N24 | HQ833590 | Kim et al. 2013 |
| *P. hunanensis* | FeL05 | EU741036 | Liu et al. 2010 |
| *P. jilunlii* | Be17 | GQ985393 | Jin et al. 2011b |
| *P. macquariensis subsp. defensor* | M4-2 | AB360546 | Hoshino et al. 2009 |
| *P. macquariensis subsp. macquariensis* | ATCC 23464 | X60625 | Hoshino et al. 2009 |
| *P. montaniterrae* | | | Khianngam et al. 2009a |
| *P. mucilaginosus* | 1480D | AF006077 | Hu et al. 2010 |
| *P. nanensis* | MX2-3 | AB265206 | Khianngam et al. 2009b |
| *P. oceanisediminis* | L10 | JF811909 | Lee et al. 2013 |
| *P. pectinilyticus* | RCB-08 | EU391157 | Park et al. 2009 |
| *P. phoenicis* | 3PO2SA | EU977789 | Benardini et al. 2011 |
| *P. pini* | S22 | GQ423056 | Kim et al. 2009d |
| *P. pinihumi* | S23 | GQ423057 | Kim et al. 2009c |
| *P. pocheonensis* | Gsoil 1138 | AB245386 | Baek et al. 2010 |
| *P. profundus* | Sl 79 | | Romanenko et al. 2013 |
| *P. prosopidis* | PW21 | FJ820995 | Valverde et al. 2010 |
| *P. pueri* | b09i-3 | EU391156 | Kim et al. 2009a |
| *P. puldeungensis* | CAU 9324 | GU187433 | Traiwan et al. 2011 |
| *P. purispatii* | ES_MS17 | EU888513 | Benardini et al. 2011 |
| *P. residui* | MC-246 | FN293173 | Vaz-Moreira et al. 2010 |
| *P. rigui* | WPCB173 | EU939688 | Baik et al. 2011b |
| *P. riograndensis* | SBR5 | EU257201 | Beneduzi et al. 2010 |
| *P. sacheonensis* | SY01 | GU124597 | Moon et al. 2011 |
| *P. sediminis* | GTH-3 | GQ355277 | Wang et al. 2012 |
| *P. septentrionalis* | X13-1 | AB295647 | Khianngam et al. 2009a |
| *P. siamensis* | S5-3 | AB295645 | Khianngam et al. 2009a |
| *P. sonchi* | X19-5 | DQ358736 | Hong et al. 2009 |
| *P. sophorae* | S27 | GQ985395 | Jin et al. 2011a |
| *P. sputi* | KIT 00200-70066-1 | FN394513 | Kim et al. 2010a |

**Table 21.1 (continued)**

| Species | Type strain | 16S rRNA gene sequence accession number | Author, effective publication |
|---|---|---|---|
| *P. taichungensis* | V10537 | EU179327 | Lee et al. 2008 |
| *P. taihuensis* | THMBG22 | JQ398861 | Wu et al. 2013 |
| *P. tarimensis* | SA-7-6 | EF125184 | Wang et al. 2008 |
| *P. telluris* | PS38 | HQ257247 | Lee et al. 2012 |
| *P. thailandensis* | S3-4A | AB265205 | Khianngam et al. 2009b |
| *P. thermoaerophilus* | TC22-2b | AB738878 | Ueda et al. 2013 |
| *P. thermophilus* | WP-1 | JQ824133 | Zhou et al. 2012 |
| *P. tianmuensis* | B27 | FJ719490 | Wu et al. 2011 |
| *P. tundrae* | A10b | EU558284 | Nelson et al. 2009 |
| *P. typhae* | xj7 | JN256679 | Kong et al. 2013 |
| *P. uliginis* | N3/975 | FN556467 | Behrendt et al. 2010 |
| *P. vulneris* | CCUG 53270 | HE649498 | Glaeser et al. 2013 |
| *P. wooponensis* | WPCB018 | EU939687 | Baik et al. 2011a |
| *P. xylanexedens* | B22a | EU558281 | Nelson et al. 2009 |
| *Brevibacillus* | | | |
| *B. aydinogluensis* | PDF25 | HQ419073 | Inan et al. 2012 |
| *B. fluminis* | CJ71 | EU375457 | Choi et al. 2010 |
| *B. massiliensis* | phR | JN837488 | Hugon et al. 2013 |
| *B. nitrificans* | DA2 | AB507254 | Takebe et al. 2012 |
| *B. panacihumi* | DCY35 | EU383033 | Kim et al. 2009e |
| *Fontibacillus* | | | |
| *F. aquaticus* | GPTSA 19 | DQ023221 | Saha et al. 2010 |
| *F. panacisegetis* | P11-6 | GQ303568 | Lee et al. 2011b |
| *Saccharibacillus* | | | |
| *S. sacchari* | GR21 | EU014873 | Rivas et al. 2008 |
| *S. kuerlensis* | HR1 | EU046270 | Yang et al. 2009 |
| *Thermobacillus* | | | |
| *T. composti* | KWC4 | AB254031 | Watanabe et al. 2007 |

and lysyl-phosphatidylglycerol as major polar lipids; and straight-chain saturated (C16:0) and iso (iso-C16:0) and anteiso (anteiso-C15:0) branched fatty acids as the major fatty acids.

## Taxonomy, Historical and Current

The genus *Cohnella* was first proposed by Kämpfer et al. (2006) and later emended by García-Fraile et al. (2008) and Khianngam et al. (2010a). The genus *Cohnella* comprises Gram-positive or Gram-negative, nonmotile or motile strains, most of which are thermotolerant, and aerobic or facultatively anaerobic. The taxonomic status of the genus *Cohnella* began with a detailed study on the molecular and chemical composition analysis of the type species of the genus *Bacillus* and *Paenibacillus* (Kämpfer et al. 2006). The type species of the genus *Cohnella*, *Cohnella thermotolerans*, was clearly moderately related only to species of the genus *Paenibacillus* at 94.4 % 16S rRNA gene sequence similarity level. Further comparative analysis of chemotaxonomic markers, specifically fatty acids (presence of large amounts of iso-C16:0) and the polar lipid composition, indicated that members of the genus *Paenibacillus* were different from the type species of *Cohnella* (presence of lysyl-phosphatidylglycerol, unknown phospholipids, and aminophospholipids). At present, the number of validly published species within *Cohnella* is nineteen, isolated from different ecological niches.

## Molecular Analyses

Molecular intraspecies similarities were determined for most species by comparative analysis of the 16S rRNA gene sequence (Weisburg et al. 1991; Rivas et al. 2002; Lane 1991) (➋ *Fig. 21.1*). The species showing higher than 98 % 16S rRNA gene sequence similarity were defined by DNA-DNA hybridization [DDH] (Ezaki et al. 1989), whereas the strains which showed less than 98 % 16S rRNA gene

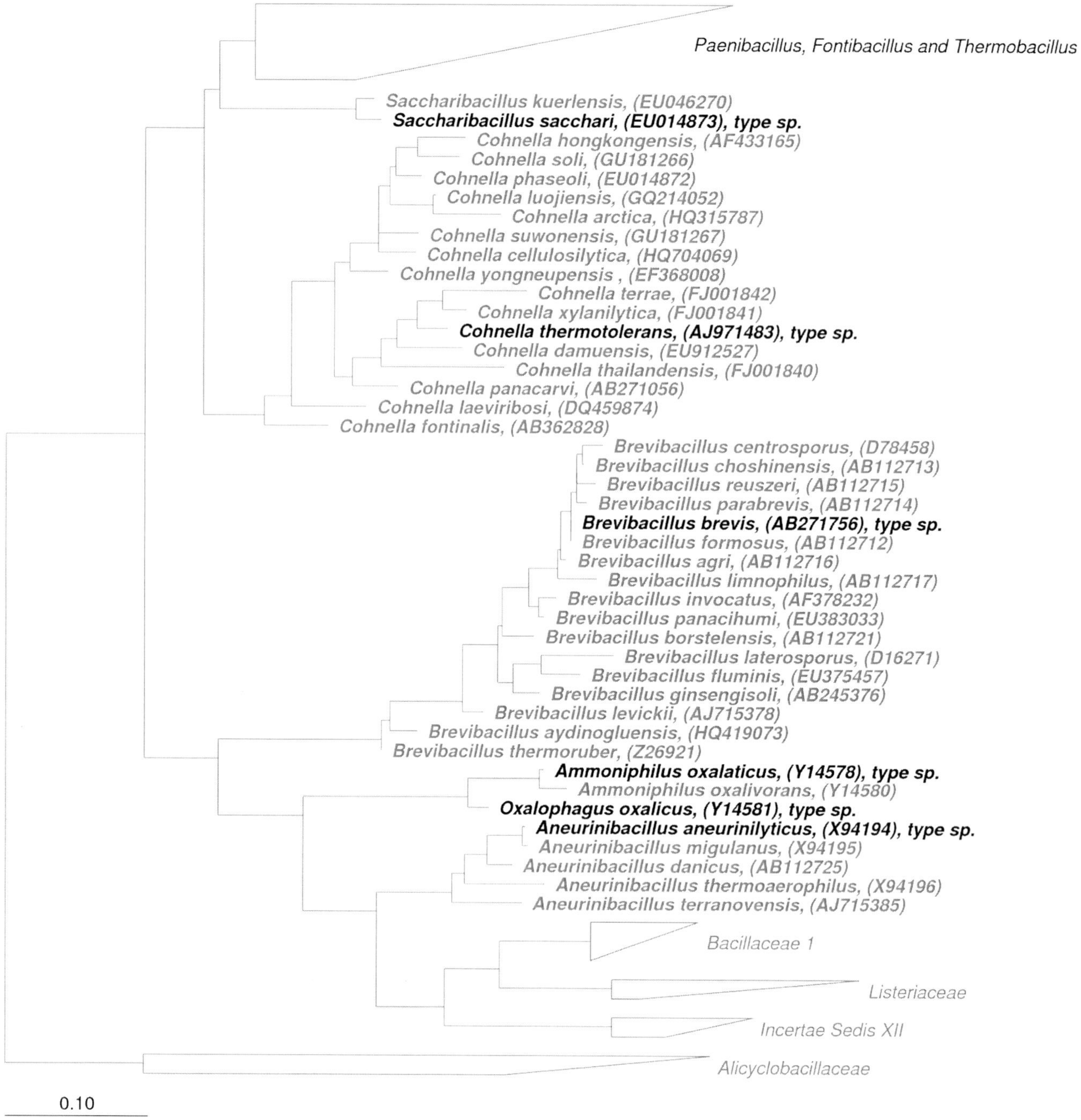


**Fig. 21.1**
**Maximum likelihood genealogy reconstruction based on the RAxML algorithm (Stamatakis 2006) of the sequences of all members of the family *Paenibacillaceae* present in the LTP_106 (Yarza et al. 2010). Representative sequences from close relative genera were used to stabilize the tree topology. In addition, a 40 % maximum frequency filter was applied to remove hypervariable positions from the alignment. Scale bar indicates estimated sequence divergence**

sequence similarities among other type strains of the genus were not included in DDH (Stackebrandt and Ebers 2006). So far none of the type strains of the genus *Cohnella* were analyzed for MALDI-TOF or ribotyping. Whole genome sequences are available for *C. laeviribosi* DSM 21336$^{T}$ (Gi11322) and *C. thermotolerans* DSM 17683$^{T}$ (Gi11323).

## Phenotypic Analyses

*Cohnella* (Coh.nel'la. N. L. fem. dim. n. *Cohnella* named after Ferdinand Cohn, the German microbiologist who first described the bacterial genus *Bacillus* in 1872).

At present, the genus *Cohnella* contains 19 species. The cells are Gram positive or negative, endospore forming, aerobic or facultatively anaerobic, motile or nonmotile, and rod shaped, and most of the species are thermotolerant. Good growth occurs at 25–30 °C; some species grow at 10 or 60 °C and grow in the presence of 3 % NaCl. Species possess a DNA mol% G+C between 47.6 and 65.1 mol% and contain meso-diaminopimelic acid in the cell-wall peptidoglycan; the predominant menaquinone is MK-7, and the major polar lipids are diphosphatidylglycerol, phosphatidylglycerol, phosphatidylethanolamine, and lysyl-phosphatidylglycerol; several unknown phospholipids, unknown aminophospholipids and unknown glycolipids also present;

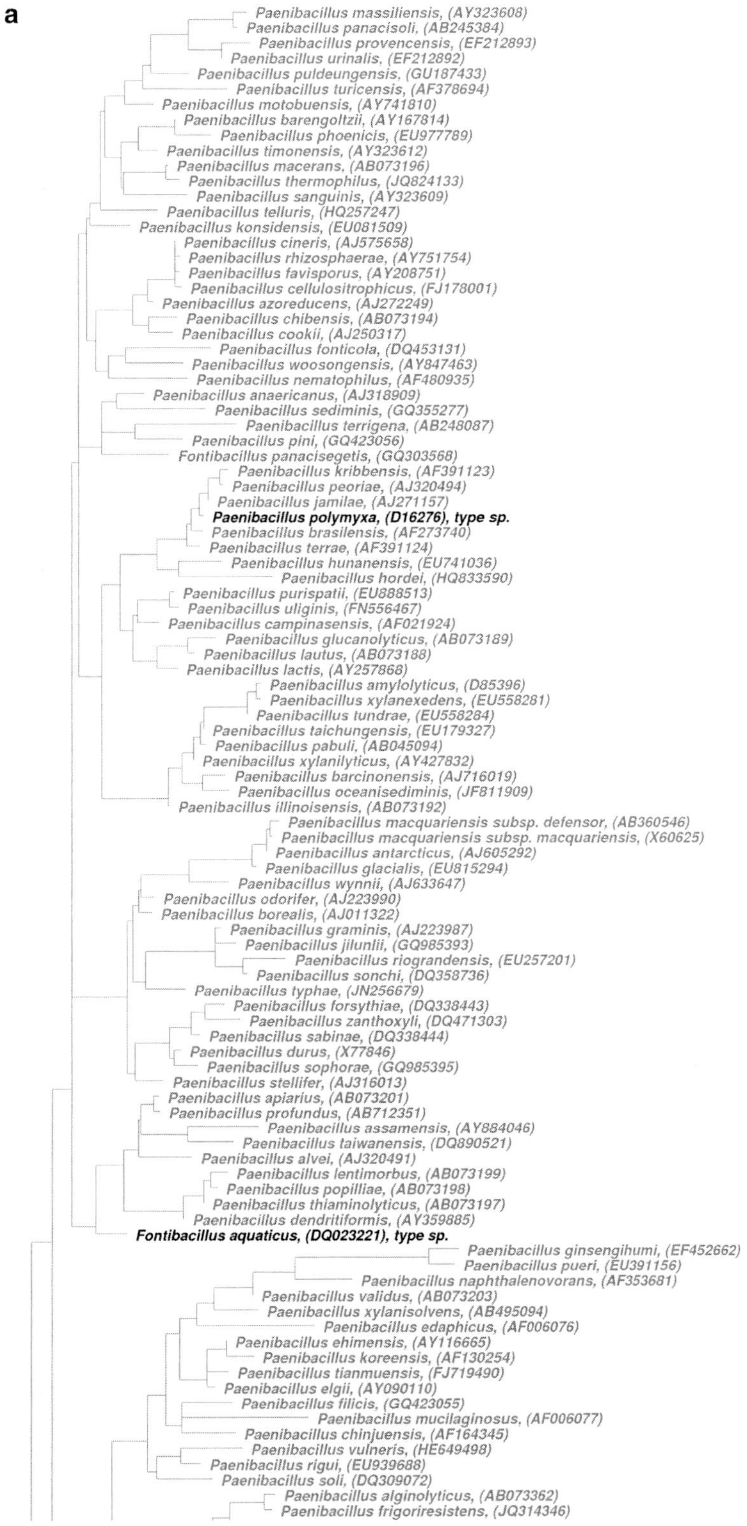


**Fig. 21.2 (continued)**

**Fig. 21.2**

**(a, b) Maximum likelihood genealogy reconstruction based on the RAxML algorithm (Stamatakis 2006) of the sequences of members of the genera *Paenibacillus*, *Fontibacillus*, and *Thermobacillus* present in the LTP_106 (Yarza et al. 2010). Representative sequences from close relative genera were used to stabilize the tree topology. In addition, a 40 % maximum frequency filter was applied to remove hypervariable positions from the alignment. Scale bar indicates estimated sequence divergence**

straight-chain saturated ($C_{16:0}$) as well as iso-$C_{16:0}$ and anteiso branched fatty acids. *Tables 21.2* and *21.3* show differences in phenotype and fatty acid profiles, respectively, between the type strains of the genus *Cohnella*.

The species description concentrates on salient, mainly morphological, cultural, and chemotaxonomic features, omitting genus-specific properties, and the reader is referred to the original species description *Cohnella* in order to obtain a more complete picture of properties.

*Cohnella arctica* (arc′ti.ca. L. fem. adj. *arctica*, northern, from the Arctic, referring to the site where the type strain was isolated).

Cells are aerobic, Gram reaction negative, rod shaped (0.2–0.3 × 1.3–2.3 μm), and motile by means of peritrichous flagella. Oval subterminal spores are formed. Growth occurs on R2A agar, 0.3 × R2A agar, and NA and 0.3 × MB agar, but not on MacConkey agar and TSB agar. Colonies grown on 0.3 × MB agar are orange, circular, convex, and smooth. Growth occurs

**Table 21.2**

Phenotypic differences between the type strains of the genus *Cohnella*. 1. *C. thermotolerans* (data compiled from Kämpfer et al. 2006), 2. *C. soli*, 3. *C. suwonensis* (data compiled from Kim et al. 2011), 4. *C. yongneupensis*, 5. *C. ginsengisoli* (data compiled from Kim et al. 2010b), 6. *C. hongkongensis* (data compiled from Kämpfer et al. 2006), 7. *C. laeviribosi* (data compiled from Cho et al. 2007), 8. *C. phaseoli* (data compiled from García-Fraile et al. 2008), 9. *C. cellulosilytica* (data compiled from Khianngam et al. 2012), 10. *C. arctica* (data compiled from Jiang et al. 2012), 11. *C. damensis* (data compiled from Luo et al. 2010a), 12. *C. fontinalis* (data compiled from Shiratori et al. 2010), 13. *C. luojiensis* (data compiled from Cai et al. 2010), 14. *C. xylanilytica*, 15. *C. terrae* (data compiled from Khianngam et al. 2010b), 16. *C. thailandensis* (data compiled from Khianngam et al. 2010a), 17. *C. panacarvi* (data compiled from Yoon et al. 2007), 18. *C. boryungensis* (data compiled from Yoon and Jung 2012), 19. *C. ferri* (data compiled from Mayilraj et al. 2011)

| Characteristics | 1 | 2 | 3 | 4 | 5 | 6 | 7 | 8 | 9 | 10 | 11 | 12 | 13 | 14 | 15 | 16 | 17 | 18 | 19 |
|---|---|---|---|---|---|---|---|---|---|---|---|---|---|---|---|---|---|---|---|
| Catalase/oxidase | +/+ | –/+ | –/+ | (+)/+ | +/+ | (+)/+ | +/– | (+)/+ | +/+ | +/– | +/+ | +/+ | +/+ | +/+ | +/+ | w/+ | w/+ | +/+ | + |
| Nitrate reduction | – | – | – | – | + | + | – | – | + | – | w | + | – | – | – | – | + | – | + |
| Indole production | – | – | + | – | – | – | – | – | – | – | – | – | – | – | – | – | – | – | – |
| Urease | – | – | – | – | – | – | + | – | + | – | – | – | – | – | – | – | – | – | + |
| Assimilation of | | | | | | | | | | | | | | | | | | | |
| D-Glucose | + | + | + | – | + | + | + | + | + | – | + | + | + | + | – | + | + | + | + |
| L-Arabinose | + | + | + | – | + | + | + | + | + | – | + | + | + | + | – | + | + | + | + |
| D-Mannose | + | – | + | – | + | + | + | + | + | – | + | + | + | + | – | + | + | + | + |
| D-Mannitol | + | – | – | – | + | + | + | – | + | – | – | + | + | + | – | + | – | + | + |
| *N*-acetylglucosamine | – | + | – | – | – | + | – | + | – | – | – | – | – | – | – | – | – | + | + |
| D-Maltose | + | + | + | – | + | + | + | + | + | – | + | + | + | + | – | – | + | + | + |
| Potassium gluconate | + | – | – | – | – | – | – | – | + | – | – | + | – | – | + | – | – | + | + |
| L-Rhamnose | + | + | + | – | – | + | + | – | + | – | ND | – | – | + | + | + | – | + | + |
| D-Ribose | + | + | – | – | – | + | + | + | + | – | ND | – | – | – | – | – | – | + | + |
| Inositol | + | – | – | – | – | + | – | – | – | – | ND | – | – | – | – | – | – | – | + |
| D-Saccharose | + | + | + | – | – | – | + | + | + | – | ND | – | + | + | + | + | – | + | + |
| Glycogen | + | + | + | – | – | – | + | + | + | – | ND | – | – | + | + | + | – | – | – |
| Salicin | + | + | + | – | + | + | – | + | – | – | ND | – | + | + | + | + | + | + | + |
| D-Melibiose | + | + | + | – | + | + | + | + | + | – | ND | - | + | + | + | + | - | + | + |
| L-Fucose | + | - | + | - | - | + | + | + | + | - | ND | - | + | - | - | – | - | + | – |
| D-Sorbitol | + | – | – | – | – | + | + | – | + | – | ND | – | – | – | – | – | – | – | + |
| Potassium 2-ketogluconate | + | + | + | – | – | – | – | – | ND | – | – | – | – | – | – | – | – | – | + |
| Gelatin hydrolysis | – | – | – | – | – | – | (+) | (+) | – | – | – | – | – | w | + | + | – | – | – |
| DNA G+C content (mol%) | 59 | 52,2 | 55,6 | 58,8 | 61,3 | 60,9 | 51 | 60,3 | 58 | 50,3 | 54,3 | 58,6 | 49,6 | 63 | 65,1 | 53,3 | 53,4 | 54,9 | 59,3 |

**Table 21.3**

Cellular fatty acid profiles of the type strains of the genus *Cohnella*. 1. *C. thermotolerans* (data compiled from Kämpfer et al. 2006), 2. *C. soli*, 3. *C. suwonensis* (data compiled from Kim et al. 2011), 4. *C. yongneupensis*, 5. *C. ginsengisoli* (data compiled from Kim et al. 2010b), 6. *C. hongkongensis* (data compiled from Kämpfer et al. 2006), 7. *C. laeviribosi* (data compiled from Cho et al. 2007), 8. *C. phaseoli* (data compiled from García-Fraile et al. 2008), 9. *C. cellulosilytica* (data compiled from Khianngam et al. 2012), 10. *C. arctica* (data compiled from Jiang et al. 2012), 11. *C. damensis* (data compiled from Luo et al. 2010a), 12. *C. fontinalis* (data compiled from Shiratori et al. 2010), 13. *C. luojiensis* (data compiled from Cai et al. 2010), 14. *C. xylanilytica*, 15. *C. terrae* (data compiled from Khianngam et al. 2010b), 16. *C. thailandensis* (data compiled from Khianngam et al. 2010a), 17. *C. panacarvi* (data compiled from Yoon et al. 2007), 18. *C. boryungensis* (data compiled from Yoon and Jung 2012), 19. *C. ferri* (data compiled from Mayilraj et al. 2011)

| Fatty acids | 1 | 2 | 3 | 4 | 5 | 6 | 7 | 8 | 9 | 10 | 11 | 12 | 13 | 14 | 15 | 16 | 17 | 18 | 19 |
|---|---|---|---|---|---|---|---|---|---|---|---|---|---|---|---|---|---|---|---|
| ai-13:0 | ND | 1,7 | 1,2 | 2,2 | ND | 0,8 | tr | 1,4 | 0,5 | 2,5 | ND | ND | ND | ND | ND | ND | 0,8 | tr | 1,6 |
| i-14:0 | 2,1 | 4,6 | 4,1 | 2,4 | 8,2 | 2,3 | 3,9 | 2,6 | 3,4 | 5,9 | 8,2 | 4 | 1,4 | 3 | 3,2 | 4,4 | 3,6 | 4 | 3,1 |
| C14:0 | 1 | 4,4 | 2 | 3,2 | 3,1 | 5 | 1,9 | 1,8 | 2 | 2,8 | ND | 1 | 1,2 | tr | tr | 1,8 | 2,5 | 3,8 | 2,1 |
| i-15:0 | 3,2 | 3,5 | 5,5 | ND | 12,2 | 8,1 | 11,7 | 14,3 | 6,5 | 1,8 | 2,9 | 14,3 | 9,1 | 7,5 | 8,3 | 7,5 | 6 | 3,1 | 5,3 |
| ai-15:0 | 28,4 | 44,4 | 46,5 | 51,1 | 48,9 | 31,2 | 22 | 44,5 | 52,5 | 51,1 | 30,1 | 33,2 | 57,4 | 31,6 | 31,6 | 26,8 | 53,7 | 46,5 | 50,1 |
| C15:0 | 1,4 | ND | ND | ND | ND | 8 | 1,3 | 5,3 | ND | ND | 3,3 | 2,3 | ND | 2 | 4,5 | 5,2 | 2,4 | 2,4 | ND |
| i-16:0 | 45,5 | 16,8 | 21,8 | 18,5 | 15 | 11,9 | 40,5 | 14,1 | 18,9 | 7,5 | 18,9 | 20,6 | 9,1 | 39,2 | 36,1 | 39,5 | 16,8 | 16,5 | 17,1 |
| C16:0 | 6,6 | 19,2 | 11,2 | 13,2 | 6,7 | 25,3 | 9,2 | 8,9 | 9,1 | 6,1 | 12,5 | 12,5 | 6,2 | 5,4 | 5,4 | 7,7 | 9,4 | 18,6 | 11,6 |
| C16:1w11c | ND | ND | ND | ND | ND | 0,9 | tr | 1,9 | ND | 4,2 | ND | ND | ND | ND | ND | ND | 0,5 | ND | 2,3 |
| ai-17:0 | 6,7 | 2,9 | 4,7 | 4,2 | 1,8 | 2,6 | 5,78 | 2,3 | 3 | 1,5 | 2 | 4,3 | 4,8 | 6,4 | 5,6 | 3,8 | 2,6 | 2,8 | ND |
| ai-17:1 | 1,1 | ND | ND | ND | ND | 1,9 | ND | ND | ND | ND | ND | ND | ND | ND | ND | ND | ND | ND | ND |
| C17:0 | ND | ND | ND | ND | ND | 1,2 | tr | 0,5 | 0,5 | 1,6 | ND | tr | ND | tr | tr | tr | ND | ND | ND |
| C17:1w6c | 1 | ND | ND | ND | ND | ND | tr | ND | ND | ND | ND | ND | ND | tr | tr | ND | ND | ND | ND |
| C18:1w7c | 4 | ND | ND | ND | ND | ND | ND | ND | ND | ND | 1,9 | ND | ND | ND | ND | ND | ND | ND | ND |

between 4 and 30 °C with an optimum at 25 °C. The pH range for growth is pH 5.0–8.0, with an optimum growth at pH 6.0–7.0. Growth occurs in the presence of 0.5 % (w/v) NaCl, but no growth occurred in the presence of 1.0 % (w/v) NaCl or higher. The DNA G+C content of the type strain is 50.3 mol%.

The type strain M9-62$^T$ was isolated from a tundra soil near Ny-Ålesund, Svalbard Islands, Norway.

*Cohnella boryungensis* (bo.ryung.en'sis. N.L. Fem. adj. *boryungensis* pertaining to Boryung, from where the type strain was isolated).

Cells are aerobic, Gram staining positive, and rods (0.3–0.5 × 1.0–3.5 μm). Motile by means of peritrichous flagella. Central ellipsoidal endospores are observed in swollen sporangia. Colonies on TSA are circular to slightly irregular, slightly convex, smooth, glistening, pale yellow in color, and 1.0–2.0 mm in diameter after incubation for 3 days at 30 °C. Growth occurs at 10 and 40 °C, but not at 4 and 45 °C. Optimal pH for growth is around 7.5. Growth occurs at pH 5.5 and 9.0, but not at pH 5.0 and 9.5. Growth occurs in the presence of 0–3.0 % (w/v) NaCl with an optimum in the presence of 0.5 % (w/v) NaCl. Susceptible to carbenicillin, cephalothin, chloramphenicol, gentamicin, kanamycin, lincomycin, neomycin, novobiocin, oleandomycin, polymyxin B, streptomycin, and tetracycline, but not to ampicillin and penicillin G. In addition to major polar lipids indicated in the genus description, two unidentified phospholipids and minor amounts of phosphatidylglycerol are present. The DNA G+C content is 54.9 mol% (determined by HPLC).

The type strain BR-29$^T$ was isolated from soil around a coast at Boryung, Korea.

*Cohnella cellulosilytica* (cel.lu.lo.si.ly'ti.ca. N.L. n. *cellulosum*, cellulose; N.L. adj. *lyticus*, able to loose, able to dissolve; N.L. fem. ddj. *cellulosilytica* cellulose dissolving).

Cells of strain FCN3-3$^T$ are Gram-positive, aerobic, and motile rods (0.3–0.4 × 1.4–2.8 μm). Central and subterminal ellipsoidal endospores are observed in swollen sporangia. Colonies are 0.1–0.45 mm in diameter, circular, raise, smooth, translucent, and white yellow colored after 2 days incubation on TSA agar medium. Grows in 3 % NaCl (weakly), at pH 7, 8 and 9 (optimally at pH 7) and at 15 and 30 °C (optimally at 30 °C). Does not grow in 5 % NaCl, at pH 5 and 6 and at 10 and 40 °C. The DNA G+C content is 58.0 mol%.

The type strain is FCN3-3$^T$ was isolated from buffalo feces.

*Cohnella damensis* (dam'ensis. N.L. masc. adj. dam'ensis. pertaining to Damu a village in Tibet, China, where the type strain was isolated).

Cells are Gram staining variable, rod shaped (0.5–0.7 × 1.5–2.5 μm), and motile by means of peritrichous flagella. Colonies on tryptone soybean agar (TSA; Difco) are circular, flat, white cream, opaque, and usually 2–3 mm in diameter within 48 h at 28 °C. Growth occurs from 10 to 40 °C (optimal 28 °C) and from pH 5.5–7.5 (optimal 7.0). Cells grow in the presence of 1 % NaCl. In addition to major polar lipids indicated in the genus description, several unknown phospholipids, unknown aminophospholipids, and unknown glycolipids are present. DNA G+C content is 54.3 mol%.

The type strain 13-25$^T$ was isolated from Damu village in Tibet, China.

*Cohnella ferri* (fer'ri. L. gen. n. ferri, of iron).

Cells are facultative anaerobe, Gram-positive, and motile rods (0.3–0.6 × 0.8–2.4 μm). Ellipsoidal spores develop subterminally in the cells and sporangia are swollen. Colonies are circular, convex and smooth, and creamish yellow pigmented. Growth occurs within temperature range of 15–42 °C (optimum temperature 37 °C), pH 7.0–11.0 (optimum pH 8.0), and up to 2 % NaCl. The DNA G+C content is 59.3 %.

The type strain HIO-4$^T$ was isolated from a hematite ore sample collected from Barbil mining area, District Keonjhar, state of Odisha, India.

*Cohnella fontinalis* (fon.ti.na'lis. L. fem. adj. fontinalis of or from a fountain, referring to the isolation of the type strain from freshwater from a fountain).

Cells are Gram-positive, aerobic, and endospore-forming rods (0.5–0.7 × 1.5–6.5 μm). Motile by means of peritrichous flagella. Colonies are irregular, translucent, cream-colored, and usually 1.0–1.5 mm in diameter within 48 h at 40 °C on TSA. Growth occurs at 25–55 °C (optimum 40 °C) and pH 5.5–8.5 (optimum pH 6.0–7.0). Growth occurs at NaCl concentrations of up to 2.0 % (w/v). The DNA G+C content of the type strain is 58.6 mol%.

The type strain YT-1101$^T$ was isolated from freshwater of a fountain in Japan.

*Cohnella ginsengisoli* (gin.sen.gi.so'li. N.L. n. *ginsengum ginseng*; L. n. solum soil; N.L. gen. n. ginsengisoli of the soil of a ginseng field, the source of the type strain).

Cells are motile, Gram-positive rods (1.6 × 3.0 μm) with ellipsoidal or oval spores positioned centrally or paracentrally in swollen sporangia. Growth occurs at 10–40 °C (optimum 30 °C), at pH 5.0–9.0 (optimum pH 7.0) and in the presence of 0–2 % (w/v) NaCl. The DNA G+C content of the type strain is 61.3 mol% (HPLC).

The type strain GR21-5$^T$ was isolated from ginseng soil in the Youngju region of the Republic of Korea.

*Cohnella hongkongensis* (hong.kong.en'sis. N.L. fem. adj. *hongkongensis* pertaining to Hong Kong).

Cells are aerobic nonmotile, sporulating, Gram-negative straight or slight curved rods. Growth occurs on horse blood agar, cells are nonhemolytic, and colonies are gray with 1 mm in diameter after 24 h of incubation at 37 °C. No enhancement of growth in 5 % $CO_2$. Colonies grew at 50 °C as pinpoint colonies after 72 h of incubation. No growth at 65 °C or on MacConkey agar. The DNA G+C content is 60.9 mol%.

The type strain is HKU3$^T$ was isolated in a patient with neutropenic fever.

*Cohnella laeviribosi* (lae.vi.ri'bo.si. L. adj. laevus left, on the left side; N.L. n. ribosum ribose; N.L. gen. n. laeviribosi referring to L-ribose [isomerase], because the type strain exhibits L-ribose isomerization ability).

Aerobic, nonmotile, and Gram positive. Cells are rod shaped (about 0.5–0.7 × 2.0–7.0 μm). In old cultures, cells become shorter rods or spherical elements. Colonies are circular, flat, smooth, opaque, and white. No growth in the presence of

1 % (w/v) NaCl, with 0.001 % (w/v) lysozyme or under anaerobic conditions on TSA. Grows at 30–60 °C and at pH 5.5–8.0, with optimal growth at 45 °C and pH 6.5. Optimal growth occurs in the presence of 0.2–0.5 % (w/v) NaCl. The DNA G+C content of the type strain is 51 mol%.

The type strain RI-39$^{T}$ isolated from Likupang, a volcanic area in Indonesia.

*Cohnella luojiensis* (lu.o.ji.en'sis. N.L. fem. adj. *Luojiensis* pertaining to Luojia hill, the site of the campus of Wuhan University, where the type strain was characterized).

Cells are strictly aerobic, Gram-positive-staining, rod shaped (0.4–0.6 × 1.2–3.5 μm), and motile by means of peritrichous flagella. Oxidase and catalase positive. Oval subterminal spores are formed. Colonies on TSA are opaque, white, convex, and about 1 mm in diameter after growth at 30 °C for 48 h. Grows at 10–37 °C (optimum 30 °C), at pH 6.0–8.0 (optimum pH 7.0) and with 0–1 % (w/v) NaCl. In addition to major polar lipids indicated in the genus description, two unknown phospholipids and three unknown aminophospholipids are also detected. The DNA G+C content of the type strain is 49.6 mol%.

The type strain HY-22R$^{T}$ was isolated from a soil sample from Xinjiang, China.

*Cohnella panacarvi* (pa.na.car.vi. N.L. n. *Panax-acis*, scientific name of ginseng; arvum, a field; N.L. gen. n. panacarvi, of a ginseng field).

Cells are Gram-positive, aerobic, nonmotile, spore-forming, and thin rod shaped (0.2–0.4 × 1.5–3.5 μm). Spores are oval, central, occurring in swollen sporangia. After two days on R2A, colonies are 0.5–1.0 mm in diameter, circular, convex, nonglossy, and white colored. Grows between 18 °C and 45 °C; the optimum temperature for growth is 30 °C. The bacterium grows within pH values of between 5.5 and 8.0; the optimum pH is 6.5–7.0. The strain tolerates 1 % (w/v) NaCl, but not 2 %. Growth occurs on TSA and nutrient agar but not on MacConkey agar. The G+C content of the genomic DNA is 53.4 mol%.

The type strain Gsoil 349$^{T}$ was isolated from soil of a ginseng field of Pocheon Province, South Korea.

*Cohnella phaseoli* (pha.se.o'li. N.L. masc. n. *Phaseolus* botanical genus name; N.L. gen. n. phaseoli of *Phaseolus*, referring to the isolation source of the type strain, nodules of *Phaseolus coccineus*).

Aerobic, spore-forming rods (0.7 × 2.5 μm). Gram positive. Motile by means of peritrichous flagella. Round or ovoid spores are formed in slightly swollen sporangia, and they are in a central or subterminal position within cells. Colonies on YED are circular, flat, white cream, opaque, and usually 1–3 mm in diameter after 48 h growth at 28 °C. Growth occurs from 10 °C to 45 °C (optimal growth at 28 °C) and pH 6–8 (optimal pH 7). The DNA G+C content of the type strain is 60.3 mol%.

The type strain GSPC1$^{T}$ was isolated from root nodules of *Phaseolus coccineus* in Segovia (Spain).

*Cohnella soli* (so'li. L. gen. n. *soli* of the soil)

Cells are strictly aerobic, Gram positive, motile with peritrichous flagella, and rod shaped (0.6–0.7 × 1.8–3.5 μm). Ellipsoidal bulging positioned subterminal spores are formed. Growth on R2A and NA, but not on TSA, LB, or MacConkey agar. Colonies are white colored and circular. Growth occurs at temperatures in the range of 15–37 °C (optimum 30 °C) and pH 5.0–7.0 (optimum pH 7.0). Salt concentrations above 1.5 % are not tolerated. The DNA G+C content of the type strain is 52.2 mol%.

The type strain YM2-7$^{T}$ was isolated from soil on Yeogi Mountain, Republic of Korea.

*Cohnella suwonensis* (su.won.en'sis. N.L. masc. adj. *Suwonensis* referring to Suwon region, Republic of Korea, where the type strain was first identified).

Cells are strictly aerobic, Gram positive, motile with peritrichous flagella, and rod shaped (0.6–0.7 × 2.0–4.9 μm). Ellipsoidal bulging positioned subterminal spores are formed. Growth occurs on R2A and NA, but not TSA, LB, or MacConkey agar. Colonies are white colored and circular. The strain grows at temperatures in the range of 10–35 °C (optimum 30 °C) and pH 5.0–8.0 (optimum pH 7.0) but not above 1 % NaCl. The DNA G+C content of the type strain is 55.6 mol%.

The type strain WD2-19$^{T}$ was isolated from field soil in the Republic of Korea.

*Cohnella terrae* (ter'rae. L. gen. n. *terrae* of the earth).

Cells are Gram reaction positive, rod shaped (0.3–0.5 × 1.5–4.0 μm), facultatively anaerobic, and motile by means of peritrichous flagella. Central ellipsoidal endospores are observed in swollen sporangia. After 2 days of incubation on C agar medium, colonies are 1–3.5 mm in diameter, circular, flat, and white. Grows at pH 5–9, at 20–45 °C, and under anaerobic conditions. No growth in 3–5 % (w/v) NaCl or at 10, 15, 50, 55, or 60 °C. In addition to major polar lipids indicated in the genus description, unknown phospholipids and aminophospholipids are present. The genomic DNA G+C content of the type strain is 65.1 mol%.

The type strain is MX21-2$^{T}$ was isolated from a soil sample collected in Muang district, Nan province, Thailand.

*Cohnella thailandensis* (thai.lan.den'sis. N.L. fem. ddj. thailandensis pertaining to Thailand, where the type strain was isolated).

Cells of strain S1-3$^{T}$ are Gram-stain-positive, facultatively anaerobic, motile rods (0.2–0.5 × 1.2–2.5 μm). Subterminal ellipsoidal endospores are observed in swollen sporangia. Colonies are 0.5–1.0 mm in diameter, circular, flat, and white after 2 days of incubation on C agar medium. Grows at pH 5 (weakly), pH 6–9 (optimally at 7), and 20–50 °C (optimally at 37 °C), in 3 % NaCl, and under anaerobic conditions. Does not grow in 5 % NaCl and at 10, 15, 55, and 60 °C. The DNA G+C content of the type strain is 53.3 mol%.

The type strain is S1-3$^{T}$ isolated from a soil sample collected in Muang district, Nan province, Thailand.

*Cohnella thermotolerans* (ther.mo.tol'er.ans. Gr. n. *therme* heat; L. pres. part. tolerans tolerating; N.L. part. adj. thermotolerans able to tolerate high temperatures).

Cells are Gram positive, spore forming, aerobic, nonmotile, rod shaped, and thermotolerant. Good growth occurs after 24 h incubation on TS and nutrient agars at 25–30 °C; good growth also occurs at 55 °C. In addition to the major polar lipids given in the genus description, two unknown phospholipids and four unknown aminophospholipids are present. The DNA G+C content is 59 mol%.

The type strain CCUG 47242$^T$ was isolated from a sample of industrial starch production in Sweden.

*Cohnella xylanilytica* (xy.la.ni.ly'ti.ca. N.L. neut. n. *xylanum* xylan; N.L. fem. adj. lytica from Gr. masc. adj. lytikos able to loose, dissolving; N.L. fem. adj. *xylanilytica* xylan dissolving).

Cells are Gram reaction positive, rod shaped (0.3–0.5 × 1.4–3.5 μm), facultatively anaerobic, and motile by means of peritrichous flagella. Central ellipsoidal endospores are observed in swollen sporangia. After 2 days of incubation on C agar medium, colonies are 1–3 mm in diameter, circular, flat, and white. Grows in 3 % (w/v) NaCl (weakly), at pH 6–9, at 20–45 °C, and at 50 °C (weakly) and under anaerobic conditions. Does not grow in 5 % (w/v) NaCl, at pH 5, or at 10, 15, 55, or 60 °C. In addition to the major polar lipids given in the genus description, unknown phospholipids and aminophospholipids are present. The genomic DNA G+C content of the type strain is 63.0 mol%.

The type strain is MX15-2$^T$ was isolated from a soil sample collected in Muang district, Nan province, Thailand.

*Cohnella yongneupensis* (yong.neup.en'sis. N.L. fem. adj. yongneupensis pertaining to Yongneup, an upland wetland of the Republic of Korea, from where the type strain was isolated).

Cells are motile, Gram-positive rods (0.762.5–3.5 μm) with ellipsoidal or oval spores positioned centrally or paracentrally in swollen sporangia. Growth occurs at 10–40 °C (optimum 30 °C), at pH 4.0–9.0 (optimum pH 7.0), and in the presence of 0–1 % (w/v) NaCl. The DNA G+C content of the type strain is 58.8 mol% (HPLC).

The type strain 5YN10-14$^T$ was isolated from the Yongneup wetland in the Republic of Korea.

## Ecology, Isolation, Enrichment, and Maintenance Procedures

The type species of the genus *Cohnella*, *Cohnella thermotolerans*, was isolated from a sample of industrial starch production by using blood agar. Most other type strains of the genus *Cohnella* were isolated from different ecological niches like soil samples using R2A agar (*C. soli*, *C. suwonensis* (Kim et al. 2011), *C. yongneupensis*, *C. ginsengisoli* (Kim et al. 2010b), *C. laeviribosi* (Cho et al. 2007), *C. luojiensis* (Cai et al. 2010), *C. panacarvi* (Yoon et al. 2007), and *C. boryungensis* (Yoon and Jung 2012); soil, marine broth, marine agar (*C. arctica*, Jiang et al. 2012; *C. damensis*, Luo et al. 2010a, b); soil, XC agar containing 10 g of oat spelt xylan, 5 g peptone, 1 g yeast extract, 4 g $K_2HPO_4$, 1 g $MgSO_47H_2O$, 0.2 g KCL, 0.02 g $FeSO_4 \times 7H_2O$, 15 g agar, pH7.0 (*C. xylanilytica* and *C. terrae* Khianngam et al. 2010b; *C. thailandensis*, Khianngam et al. 2010a); water, tryptic soy agar (*C. fontinalis*, Shiratori et al. 2010); root nodules, modified yeast extract-mannitol agar (*C. phaseoli*, García-Fraile et al. 2008); and feces, CMC basal medium containing 5 g of carboxymethyl cellulose (Sigma), 1 g yeast extract (Difco), 1 g $(NH_4)_2SO_4$, 15 g agar, pH 7.0 (*C. cellulosilytica*, Khianngam et al. 2012), and incubated at 28–30 °C for 1–5 days. All the type strains were available from any one of the culture collection centers where they are deposited and preserved in glycerol (10 % v/v) at −80 °C and, for long term, preserved in liquid nitrogen or as freeze-dried cultures.

## Pathogenicity, Clinical Relevance

Since most of the type species were isolated from soil, none of the established type strains belonging to the genus *Cohnella* were related to pathogenicity or clinical relevance. Only the type strain *C. hongkongensis* isolated from a patient with neutropenic fever reported to produce pseudobacteremia (Teng et al. 2003) but was considered a contaminant as the cultures was obtained only from one of four parallel patient's blood samples.

## Application

The type strain *C. laeviribosi* (Cho et al. 2007) reported to be capable of assimilating and isomerizing L-ribose; *C. cellulosilytica* (Khianngam et al. 2012) reported for degradation of cellulose; and *C. xylanilytica*, *C. terrae*, and *C. thailandensis* reported for xylanase production (Khianngam et al. 2010a, b).

## Acknowledgments

I wish to thank Dr. Girish Sahni, Director, CSIR-Institute of Microbial Technology for the encouragement and Ms. Ishwinder Kaur and Ms. Chandandeep Kaur, Microbial Type Culture Collection and Gene Bank (MTCC), Chandigarh, India, for their technical support. Financial assistance provided by DBT and CSIR, Government of India, is greatly acknowledged.

## References

Akaracharanya A, Lorliam W, Tanasupawat S, Lee KC, Lee JS (2009) *Paenibacillus cellulositrophicus* sp. nov., a cellulolytic bacterium from Thai soil. Int J Syst Evol Microbiol 59:2680–2684

Bae JY, Kim KY, Kim JH, Lee K, Cho JC, Cha CJ (2010) *Paenibacillus aestuarii* sp. nov., isolated from an estuarine wetland. Int J Syst Evol Microbiol 60:644–647

Baek SH, Yi TH, Lee ST, Im WT (2010) *Paenibacillus pocheonensis* sp. nov., a facultative anaerobe isolated from soil of a ginseng field. Int J Syst Evol Microbiol 60:1163–1167

Baik KS, Choe HN, Park SC, Kim EM, Seong CN (2011a) *Paenibacillus wooponensis* sp. nov., isolated from wetland freshwater. Int J Syst Evol Microbiol 61:2763–2768

Baik KS, Lim CH, Choe HN, Kim EM, Seong CN (2011b) *Paenibacillus rigui* sp. nov., isolated from a freshwater wetland. Int J Syst Evol Microbiol 61:529–534

Behrendt U, Schumann P, Stieglmeier M, Pukall R, Augustin J, Spröer C, Schwendner P, Moissl-Eichinger C, Ulrich A (2010) Characterization of heterotrophic nitrifying bacteria with respiratory ammonification and denitrification activity - description of *Paenibacillus uliginis* sp. nov., an inhabitant of fen peat soil and *Paenibacillus purispatii* sp. nov., isolated from a spacecraft assembly clean room. Syst Appl Microbiol 33:328–336

Benardini JN, Vaishampayan PA, Schwendner P, Swanner E, Fukui Y, Osman S, Satomi M, Venkateswaran K (2011) *Paenibacillus phoenicis* sp. nov., isolated from the Phoenix Lander assembly facility and a subsurface molybdenum mine. Int J Syst Evol Microbiol 61:1338–1343

Beneduzi A, Costa PB, Parma M, Melo IS, Bodanese-Zanettini MH, Passaglia LMP (2010) *Paenibacillus riograndensis* sp. nov., a nitrogen-fixing species isolated from the rhizosphere of *Triticum aestivum*. Int J Syst Evol Microbiol 60:128–133

Cai F, Wang Y, Qi H, Dai J, Yu B, An H, Rahman E, Fang C (2010) *Cohnella luojiensis* sp. nov., isolated from soil of a Euphrates poplar forest. Int J Syst Evol Microbiol 60:1605–1608

Cho EA, Lee JS, Lee KC, Jung HC, Pan JG, Pyun YR (2007) *Cohnella laeviribosi* sp. nov., isolated from a volcanic pond. Int J Syst Evol Microbiol 57:2902–2907

Choi MJ, Bae JY, Kim KY, Kang H, Cha CJ (2010) *Brevibacillus fluminis* sp. nov., isolated from sediment of estuarine wetland. Int J Syst Evol Microbiol 60:1595–1599

Chou JH, Lee JH, Lin MC, Chang PS, Arun AB, Young CC, Chen WM (2009) *Paenibacillus contaminans* sp. nov., isolated from a contaminated laboratory plate. Int J Syst Evol Microbiol 59:125–129

De Vos P, Ludwig W, Schleifer KH, Whitman WB (2009a) Family IV *Paenibacillaceae*. In: De Vos P, Garrity GM, Jones D, Krieg NR, Ludwig W, Rainey EA, Schleifer KH, Withman WB (eds) Bergey's manual of systematic bacteriology, vol 3, 2nd edn. Springer, Dordrecht/Heidelberg/London/New York, p 269 (The Firmicutes)

De Vos P, Garrity GM, Jones D, Krieg NR, Ludwig W, Rainey EA, Schleifer KH, Withman WB (eds) (2009b) Bergey's manual of Systematic bacteriology, vol 3, 2nd edn. Springer, Dordrecht/Heidelberg/London/New York (The Firmicutes)

Ezaki T, Hashimoto Y, Yabuuchi E (1989) Fluorometric deoxyribonucleic acid-deoxyribonucleic acid hybridization in microdilution wells as an alternative to membrane filter hybridization in which radioisotopes are used to determine genetic relatedness among bacterial strains. Int J Syst Bacteriol 39:224–229

García-Fraile P, Velázquez E, Mateos PF, Martínez-Molina E, Rivas R (2008) *Cohnella phaseoli* sp. nov., isolated from root nodules of *Phaseolus coccineus* in Spain, and emended description of the genus *Cohnella*. Int J Syst Evol Microbiol 58:1855–1859

Glaeser SP, Falsen E, Busse HJ, Kämpfer P (2013) *Paenibacillus vulneris* sp. nov., isolated from a necrotic wound. Int J Syst Evol Microbiol 63:777–782

Hong YY, Ma YC, Zhou YG, Gao F, Liu HC, Chen SF (2009) *Paenibacillus sonchi* sp. nov., a nitrogen-fixing species isolated from the rhizosphere of *Sonchus oleraceus*. Int J Syst Evol Microbiol 59:2656–2661

Hoshino T, Nakabayashi T, Hirota K, Matsuno T, Koiwa R, Fujiu S, Saito I, Tkachenko OB, Matsuyama H, Yumoto I (2009) *Paenibacillus macquariensis* subsp *defensor* subsp. nov., isolated from boreal soil. Int J Syst Evol Microbiol 59:2074–2079

Hu XF, Li SX, Wu JG, Wang JF, Fang QL, Chen JS (2010) Transfer of *Bacillus mucilaginosus* and *Bacillus edaphicus* to the genus *Paenibacillus* as *Paenibacillus mucilaginosus* comb. Nov. and *Paenibacillus edaphicus* comb nov. Int J Syst Evol Microbiol 60:8–14

Hugon P, Mishra AK, Nguyen T-T, Raoult D, Fournier P-E (2013) Non-contiguous finished genome sequence and description of *Brevibacillus massiliensis* sp. nov. Stand Genomic Sci 8:1–14, Validation List no 153 (2013) Int J Syst Evol Microbiol 63: 3131–3134

Inan K, Canakci S, Belduz AO, Sahin F (2012) *Brevibacillus aydinogluensis* sp. nov., a moderately thermophilic bacterium isolated from Karakoc hot spring. Int J Syst Evol Microbiol 62:849–855

Jeon CO, Lim JM, Lee SS, Chung BS, Park DJ XULH, Jiang CL, Kim CJ (2009) *Paenibacillus harenae* sp. nov., isolated from desert sand in China. Int J Syst Evol Microbiol 59:13–17

Jiang F, Dai J, Wang Y, Xue X, Xu M, Li W, Fang C, Peng F (2012) *Cohnella arctica* sp. nov., isolated from Arctic tundra soil. Int J Syst Evol Microbiol 62:817–821

Jin HJ, LV J, Chen SF (2011a) *Paenibacillus sophorae* sp. nov., a nitrogen-fixing species isolated from the rhizosphere of *Sophora japonica*. Int J Syst Evol Microbiol 61:767–771

Jin HJ, Zhou YG, Liu HC, Chen SF (2011b) *Paenibacillus jilunlii* sp. nov., a nitrogen-fixing species isolated from the rhizosphere of *Begonia semperflorens*. Int J Syst Evol Microbiol 61:1350–1355

Kämpfer P, RossellóMora R, Falsen E, Busse HJ, Tindall BJ (2006) *Cohnella thermotolerans* gen. nov., sp. nov., and classification of '*Paenibacillus hongkongensis*' as *Cohnella hongkongensis* sp. nov. Int J Syst Evol Microbiol 56:781–786

Kämpfer P, Falsen E, Lodders N, Martin K, Kassmannhuber J, Busse HJ (2012) *Paenibacillus chartarius* sp. nov., isolated from a paper mill. Int J Syst Evol Microbiol 62:1342–1347

Khianngam S, Tanasupawat S, Lee JS, Lee KC, Akaracharanya A (2009a) *Paenibacillus siamensis* sp. nov., *Paenibacillus septentrionalis* sp. nov. and *Paenibacillus montaniterrae* sp. nov., xylanase-producing bacteria from Thai soils. Int J Syst Evol Microbiol 59:130–134

Khianngam S, Akaracharanya A, Tanasupawat S, Lee KC, Lee JS (2009b) *Paenibacillus thailandensis* sp. nov. and *Paenibacillus nanensis* sp. nov., xylanase-producing bacteria isolated from soil. Int J Syst Evol Microbiol 59:564–568

Khianngam S, Tanasupawat S, Akaracharanya A, Kim KK, Lee KC, Lee JS (2010a) *Cohnella thailandensis* sp. nov., a xylanolytic bacterium from Thai soil. Int J Syst Evol Microbiol 60:2284–2287

Khianngam S, Tanasupawat S, Akaracharanya A, Kim KK, Lee KC, Lee JS (2010b) *Cohnella xylanilytica* sp. nov. and *Cohnella terrae* sp. nov., xylanolytic bacteria from soil. Int J Syst Evol Microbiol 60:2913–2917

Khianngam S, Tanasupawat S, Akaracharanya A, Kim KK, Lee KC, Lee JS (2012) *Cohnella cellulosilytica* sp. nov., isolated from buffalo feces. Int J Syst Evol Microbiol 62:1921–1925

Kim MK, Kim YA, Park MJ, Yang DC (2008) *Paenibacillus ginsengihumi* sp. nov., a bacterium isolated from soil in a ginseng field. Int J Syst Evol Microbiol 58:1164–1168

Kim BC, Jeong WJ, Kim DY, Oh HW, Kim H, Park DS, Park HM, Bae KS (2009a) *Paenibacillus pueri* sp. nov., isolated from Pu'er tea. Int J Syst Evol Microbiol 59:1002–1006

Kim BC, Kim MN, Lee KH, Kwon SB, Bae KS, Shin KS (2009b) *Paenibacillus filicis* sp. nov., isolated from the rhizosphere of the fern. J Microbiol 47:524–529, Validation List no 131 (2010) Int J Syst Evol Microbiol 60: 1–2

Kim BC, Lee KH, Kim MN, Kim EM, Rhee MS, Kwon OY, Shin KS (2009c) *Paenibacillus pinihumi* sp. nov., a cellulolytic bacterium isolated from the rhizosphere of *Pinus densiflora*. J Microbiol 47:530–535, Validation List no 131 (2010) Int J Syst Evol Microbiol 60: 1–2

Kim BC, Lee KH, Kim MN, Kim EM, Min SR, Kim HS, Shin KS (2009d) *Paenibacillus pini* sp. nov., a cellulolytic bacterium isolated from the rhizosphere of pine tree. J Microbiol 47:699–704, Validation List no 137 (2011) Int J Syst Evol Microbiol 61: 1–3

Kim MK, Sathiyaraj S, Pulla RK, Yang DC (2009e) *Brevibacillus panacihumi* sp. nov., a *β*-glucosidase-producing bacterium. Int J Syst Evol Microbiol 59:1227–1231

Kim KK, Lee KC, Yu H, Ryoo S, Park Y, Lee JS (2010a) *Paenibacillus sputi* sp. nov., isolated from the sputum of a patient with pulmonary disease. Int J Syst Evol Microbiol 60:2371–2376

Kim SJ, Weon HY, Kim YS, Anandham R, Jeon YA, Hong SB, Kwon SW (2010b) *Cohnella yongneupensis* sp. nov. and *Cohnella ginsengisoli* sp. nov., isolated from two different soils. Int J Syst Evol Microbiol 60:526–530

Kim SJ, Weon HY, Kim YS, Kwon SW (2011) *Cohnella soli* sp. nov. and *Cohnella suwonensis* sp. nov. isolated from soil samples in Korea. J Microbiol 49:1033–1038

Kim JM, Lee SH, Lee SH, Choi EJ, Jeon CO (2013) *Paenibacillus hordei* sp. nov., isolated from naked barley in Korea. Antonie van Leeuwenhoek 103:3–9, Validation List no 150 (2013) Int J Syst Evol Microbiol 63: 797–798

Kishore KH, Begum Z, Pathan AAK, Shivaji S (2010) *Paenibacillus glacialis* sp. nov., isolated from the Kafni glacier of the Himalayas, India. Int J Syst Evol Microbiol 60:1909–1913

Kong BH, Liu QF, Liu M, Liu Y, Liu L, Li CL, Yu R, Li YH (2013) *Paenibacillus typhae* sp. nov., isolated from roots of *Typha angustifolia L.* Int J Syst Evol Microbiol 63:1037–1044

Lane DJ (1991) 16S/23S sequencing. In: Stackebrandt E, Goodfellow M (eds) Nucleic acid techniques in bacterial systematics. Wiley, Chichester, pp 115–175

Lee FL, Tien CJ, Tai CJ, Wang LT, Liu YC, Chern LL (2008) *Paenibacillus taichungensis* sp. nov., from soil in Taiwan. Int J Syst Evol Microbiol 58:2640–2645

Lee JC, Kim CJ, Yoon KH (2011a) *Paenibacillus telluris* sp. nov., a novel phosphate-solubilizing bacterium isolated from soil. J Microbiol 49:617–621, Validation List no 148 (2012) Int J Syst Evol Microbiol 62: 2549–2554

Lee KC, Kim KK, Eom MK, Kim MJ, Lee JS (2011b) *Fontibacillus panacisegetis* sp. nov., isolated from soil of a ginseng field. Int J Syst Evol Microbiol 61:369–374

Lee J, Shin NR, Jung MJ, Roh SW, Kim MS, Lee JS, Lee KC, Kim YO, Bae JW (2013) *Paenibacillus oceanisediminis* sp. nov. isolated from marine sediment. Int J Syst Evol Microbiol 63:428–434

Liu Y, Liu L, Qiu F, Schumann P, Shi Y, Zou Y, Zhang X, Song W (2010) *Paenibacillus hunanensis* sp. nov., isolated from rice seeds. Int J Syst Evol Microbiol 60:1266–1270

Luo X, Wang Z, Dai J, Zhang L, Fang C (2010a) *Cohnella damensis* sp. nov., a motile xylanolytic bacteria isolated from a low altitude area in Tibet. J Microbiol Biotechnol 20:410–414

Luo X, Wang Z, Dai J, Zhang L, Fang C (2010b) *Cohnella damensis* sp. nov. in list of new names and new combinations previously effectively, but not validly, published, Validation List no. 134. Int J Syst Evol Microbiol 60:1477–1479

Mayilraj S, Ruckmani A, Kaur C, Kaur I, Klenk HP (2011) *Cohnella ferri* sp. nov. a novel member of the genus *Cohnella* isolated from haematite ore. Curr Microbiol 62:1704–1709

Ming H, Nie GX, Jiang HC, Yu TT, Zhou EM, Feng HG, Tang SK, Li WJ (2012) *Paenibacillus frigoriresistens* sp. nov., a novel psychrotroph isolated from a peat bog in Heilongjiang, Northern China. Antonie van Leeuwenhoek 102:297–305, Validation List no 148 (2012) Int J Syst Evol Microbiol 62: 2549–2554

Moon JC, Jung YJ, Jung JH, Jung HS, Cheong YR, Jeon CO, Lee KO, Lee SY (2011) *Paenibacillus sacheonensis* sp. nov., a xylanolytic and cellulolytic bacterium isolated from tidal flat sediment. Int J Syst Evol Microbiol 61:2753–2757

Nelson DM, Glawe AJ, Labeda DP, Cann IKO, Mackie RI (2009) *Paenibacillus tundrae* sp. nov. and *Paenibacillus xylanexedens* sp. nov., psychrotolerant, xylan-degrading bacteria from Alaskan tundra. Int J Syst Evol Microbiol 59:1708–1714

Oh HW, Kim BC, Lee KH, Kim DY, Park DS, Park HM, Bae KS (2008) *Paenibacillus camelliae* sp. nov., isolated from fermented leaves of *Camellia sinensis*. J Microbiol 46:530–534, Validation List no 136 (2010) Int J Syst Evol Microbiol 60: 2509–2510

Park DS, Jeong WJ, Lee KH, Oh HW, Kim BC, Bae KS, Park HY (2009) *Paenibacillus pectinilyticus* sp. nov., isolated from the gut of *Diestrammena apicalis*. Int J Syst Evol Microbiol 59:1342–1347

Park MH, Traiwan J, Jung MY, Nam YS, Jeong JH, Kim W (2011) *Paenibacillus chungangensis* sp. nov., isolated from a tidal-flat sediment. Int J Syst Evol Microbiol 61:281–285

Rivas R, Velázquez E, Willems A, Vizcaíno N, Subba-Rao NS, Mateos PF, Gillis M, Dazzo FB, Martínez-Molina E (2002) A new species of *Devosia* that forms a unique nitrogen-fixing root-nodule symbiosis with the aquatic legume *Neptunia natans* (L.f.) Druce. Appl Environ Microbiol 68:5217–5222

Rivas R, Garcia-Fraile P, Zurdo-Pineiro JL, Mateos PF, Martinez-Molina E, Bedmar EJ, Sanchez-Raya J, Velazquez E (2008) *Saccharibacillus sacchari* gen nov, sp. nov., isolated from sugar cane. Int J Syst Evol Microbiol 58:1850–1854

Romanenko LA, Tanaka N, Svetashev VI, Kalinovskaya NI (2013) *Paenibacillus profundus* sp. nov., a deep sediment bacterium that produces isocoumarin and peptide antibiotics. Arch Microbiol 195:247–254, Validation List no 152 (2013) Int J Syst Evol Microbiol 63: 2365–2367

Saha P, Krishnamurthi S, Bhattacharya A, Sharma R, Chakrabarti T (2010) *Fontibacillus aquaticus* gen nov, sp. nov., isolated from a warm spring. Int J Syst Evol Microbiol 60:422–428

Shiratori H, Tagami Y, Beppu T, Ueda K (2010) *Cohnella fontinalis* sp. nov., a xylanolytic bacterium isolated from fresh water. Int J Syst Evol Microbiol 60:1344–1348

Stackebrandt E, Ebers J (2006) Taxonomic parameters revisited: tarnished gold standards. Microbiol Today 33:152–155

Stamatakis A (2006) RAxML-VI-HPC: maximum likelihood-based phylogenetic analyses with thousands of taxa and mixed models. Bioinformatics 22:2688–2690

Takebe F, Hirota K, Nodasaka Y, Yumoto I (2012) *Brevibacillus nitrificans* sp. nov., a nitrifying bacterium isolated from a microbiological agent for enhancing microbial digestion in sewage treatment tanks. Int J Syst Evol Microbiol 62:2121–2126

Tang QY, Yang N, Wang J, Xie YQ, Ren B, Zhou YG, Gu MY, Mao J, Li WJ, Shi YH, Zhang LX (2011) *Paenibacillus algorifonticola* sp. nov., isolated from a cold spring. Int J Syst Evol Microbiol 61:2167–2172

Teng JLL, Woo PCY, Leung KW, Lau SKP, Wong MKM, Yuen KY (2003) Pseudobacteraemia in a patient with neutropenic fever caused by a novel paenibacillus species: *Paenibacillus hongkongensis* sp. nov. Mol Pathol 56:29–35

Touzel JP, Prensier G (2009) Genus VII *Thermobacillus*. In: De Vos P, Garrity GM, Jones D, Krieg NR, Ludwig W, Rainey EA, Schleifer KH, Withman WB (eds) Bergey's manual of systematic bacteriology, vol 3, 2nd edn. Springer, Dordrecht/Heidelberg/London/New York, pp 321–322 (The Firmicutes)

Traiwan J, Park MH, Kim W (2011) *Paenibacillus puldeungensis* sp. nov., isolated from a grassy sandbank. Int J Syst Evol Microbiol 61:670–673

Ueda J, Yamamoto S, Kurosawa N (2013) *Paenibacillus thermoaerophilus* sp. nov., a moderately thermophilic bacterium isolated from compost. Int J Syst Evol Microbiol 63:3330–3335

Valvaerde A, Peix A, Rivas R, Velazquez E, Salazar S, Santa-Regina I, Rodriguez-Barrueco C, Igual JM (2008) *Paenibacillus castaneae* sp. nov., isolated from the phyllosphere of *Castanea sativa*. Miller. Int J Syst Evol Microbiol 58:2560–2564

Valverde A, Fterich A, Mahdhi M, Ramirez-Bahena MH, Caviedes MA, Mars M, Velazquez E, Rodriguez-Llorente ID (2010) *Paenibacillus prosopidis* sp. nov., isolated from the nodules of *Prosopis farcta*. Int J Syst Evol Microbiol 60:2182–2186

Vaz-Moreira I, Figueira V, Lopes AR, Pukall R, Spröer C, Schumann P, Nunes OC, Manaia CM (2010) *Paenibacillus residui* sp. nov., isolated from urban waste compost. Int J Syst Evol Microbiol 60:2415–2419

Wang M, Yang M, Zhou G, Luo X, Zhang L, Tang Y, Fang C (2008) *Paenibacillus tarimensis* sp. nov., isolated from sand in Xinjiang, China. Int J Syst Evol Microbiol 58:2081–2085

Wang L, Baek SH, Cui Y, Lee HG, Lee ST (2012) *Paenibacillus sediminis* sp. nov., a xylanolytic bacterium isolated from a tidal flat. Int J Syst Evol Microbiol 62:1284–1288

Watanabe K, Nagao N, Yamamoto S, Toda T, Kurosawa N (2007) *Thermobacillus composti* sp. nov., a moderately thermophilic bacterium isolated from a composting reactor. Int J Syst Evol Microbiol 57:1473–1477

Weisburg WG, Barns SM, Pelletier DA, Lane DJ (1991) 16S ribosomal DNA amplification for phylogenetic study. J Bacteriol 173:697–703

Wu X, Fang H, Qian C, Wen Y, Shen X, Li O, Gao H (2011) *Paenibacillus tianmuensis* sp. nov., isolated from soil. Int J Syst Evol Microbiol 61:1133–1137

Wu YF, Wu QL, Liu SJ (2013) *Paenibacillus taihuensis* sp. nov., isolated from an eutrophic lake. Int J Syst Evol Microbiol 63:3652–3658

Yang SY, Liu H, Liu R, Zhang KY, Lai R (2009) *Saccharibacillus kuerlensis* sp. nov., isolated from a desert soil. Int J Syst Evol Microbiol 59:953–957

Yarza P, Ludwig W, Euzéby J, Amann R, Schleifer K-H, Glöckner FO, Rosselló-Móra R (2010) Update of the All-Species Living-Tree Project based on 16S and 23S rRNA sequence analyses. Syst Appl Microbiol 33:291–299

Yoon JH, Jung YT (2012) *Cohnella boryungensis* sp. nov., isolated from soil. Antonie Van Leeuwenhoek 101:769–775

Yoon MH, Ten LN, Im WT (2007) *Cohnella panacarvi* sp. nov., a xylanolytic bacterium isolated from ginseng cultivating soil. J Microbiol Biotechnol 17:913–918

Zhang J, Wang ZT, Yu HM, Ma Y (2013) *Paenibacillus catalpae* sp. nov., isolated from the rhizosphere soil of *Catalpa speciosa*. Int J Syst Evol Microbiol 63:1776–1781

Zhou Y, Gao S, Wei DQ, Yang LL, Huang X, HE J, Zhang YJ, Tang SK, Li WJ (2012) *Paenibacillus thermophilus* sp. nov., a novel bacterium isolated from a sediment of hot spring in Fujian province, China. Antonie van Leeuwenhoek 102:601–609, Validation List no 149 (2013) Int J Syst Evol Microbiol 63: 1–5

# 22 The Family *Pasteuriaceae*

*Erko Stackebrandt*
Leibniz Institute DSMZ-German Collection of Microorganisms and Cell Cultures GmbH, Braunschweig, Germany



**Abstract**
The species of the genus *Pasteuria* are rare examples in bacteriological systematics as their description is solely based on morphology, ultrastructure, and host relationships. As none of them can be grown axenically, the type strains of species have not been deposited in public service collections; most of them can be grown in the laboratory together with its host. Today, after the establishment of the *Candidatus* category, novel taxa are not described as species but receive the *Candidatus* status; e.g., Phylogenetically, *Pasteuria* forms a monophyletic clade within the Firmicutes, branching next the members of the family *Thermoactinosporaceae*. Recent literature on *Pasteuria* concentrates on bacterium-host relationships and ecology, and this contribution adds some of this information to the excellent contribution of Sayre and Starr, revised by Dickson et al. (2009) in *Bergey's Manual of Systematic Bacteriology*, 2nd edition.

## Taxonomy and Biology

The checkered history of the first members of the genus *Pasteuria* has been dealt with in detail and depth by Sayre and Starr (Sayre et al. 2009). Ebert (2008) summarized the current state of knowledge about the coevolution of the Daphnia-parasite system until 2007.

The family *Pasteuriaceae* (Laurent 1890) comprises Gram-positive, nonmotile, dichotomously branching firmicutes, which form a septate mycelium in which the terminal hyphae are enlarged to form sporangia, each of them containing a single refractile endospore. The sporangia and microcolonies are endoparasitic in some invertebrates thriving in freshwater, plants, and soil. Transmission occurs via soil or waterborne spores. Infected hosts fail to reproduce. As shown for the *Daphnia magna-Pasteuria ramosa* system, both host and parasite were affected by food quality, especially polyunsaturated fatty acids (Schlotz et al. 2013). *Pasteuria* cells that pass through both susceptible and resistant *Daphnia* generally remain viable and infectious (King et al. 2013). When parasite spores of *P. ramosa* are challenging *Daphnia magna*, only hosts from susceptible host-parasite genetic combinations show a cellular response (measured as hemocyte density). As described by Auld et al. (2012), this reaction is compatible with the hypothesis that genetic specificity is attributable to barrier defenses at the site of infection (the gut). The immune response is general, reporting the number of parasite spores entering the hemocoel (Auld et al. 2012). Ben-Ami and Routtu (2013) found that *Pasteuria ramosa* isolates killed their host faster than individual *P. ramosa* clones which may point towards a greater genetic heterogeneity, hence pathogenic potential, of isolates. A range of different genotypes of *Daphnia* host attachment of four *P. ramosa* genotypes are not host specific but host genotype specific. As Pasteuria genotypes were never able to reproduce in nonnative host species, the authors (Luijckx et al. 2013b) suggested that genotypes infecting different host species are of different varieties, each with a narrow host range. Resistance of *Daphnia magna* against the *P. ramosa* follows a matching-allele model (Luijckx et al. 2013a).

Clonal genotypes of *P. ramosa* were first investigated by Luijckx et al. (2011) who reported that clones showed more specific interactions with host genotypes than previous studies using isolates suggested. As the presence of multiple genotypes within an isolate may influence the outcome and interpretation of some experiments, the authors recommend caution when studying *P. ramosa* isolates.

The population structure of *P. ramosa* was studied in two segregated ponds based on geography, host resistance phenotype, and host genotype by experimentally infecting D. magna host clones with known resistance phenotypes. The genetic diversity of the parasite isolates was high but strongly differentiated by pond, indicating spatially restricted gene flow. Nearly all infected D. magna hosted more than one parasite haplotype. On the basis of the observation of recombinant haplotypes and relatively low levels of linkage disequilibrium, Andras and Ebert (2013) concluded that *P. ramosa* engages in substantial recombination.

Until now, four species have been described, i.e., *Pasteuria ramosa* (Metchnikoff 1888) (host: Cladocerans such as *Daphnia* and *Moina*), *P. nishizawae* (Sayre et al. 1991) (host: cyst nematode *Heterodera glycines*), *P. penetrans* (Sayre and Starr 1985) (host: Nematodes such as *Meloidogyne* spp), and *P. thornei* (Starr and Sayre 1988) (host: Nematodes such as *Pratylenchus brachyurus*). In addition, two *Candidatus* species were described, i.e., *Candidatus* Pasteuria usgae (Giblin-Davis et al. 2003) (host: Nematodes such as *Belonolaimus longicaudatus*) and *Candidatus* Pasteuria aldrichii (Giblin-Davis et al. 2011) (host: Nematodes of the genus *Bursilla*).

As depicted in ➋ *Fig. 22.1*, the family *Pasteuriaceae* branches next to members of the family *Thermoactinomycetaceae* with which they share a mycelium-like proliferation during

E. Rosenberg et al. (eds.), *The Prokaryotes – Firmicutes and Tenericutes*, DOI 10.1007/978-3-642-30120-9_347,

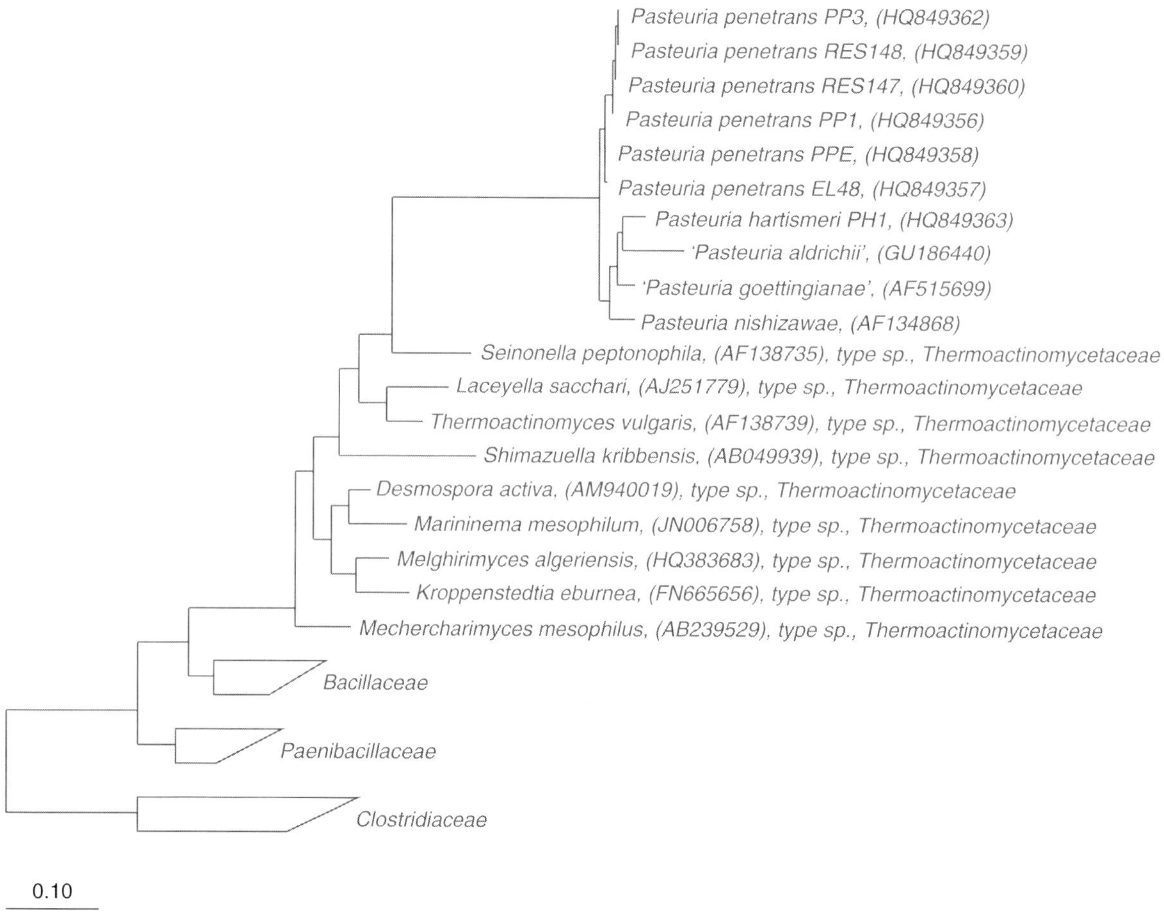


**Fig. 22. 1**

**Neighbor-joining genealogy reconstruction based on the RAxML algorithm (Stamatakis 2006) of the sequences of members of the family *Pasteuriaceae* and some neighboring taxa present in the LTP_106 (Yarza et al. 2010). The tree was reconstructed by using a subset of sequences. Representative sequences from closely relative genera were used to stabilize the tree topology. In addition, a 40 % maximum frequency filter was applied to remove hypervariable positions from the alignment. Scale bar indicates estimated sequence divergence**

development of the vegetative colony towards spore formation (Sayre et al. 2009). In contrast, the tree shown by Sayre and Starr (2009) and Schmidt et al. (2010) sees members of *Alicyclobacillus* to branch between these two families. In the phylogenetic tree shown by the latter authors, *Pasteuria* strains isolated from plant-parasitic nematodes formed a monophyletic clade apart from *Pasteuria* strains associated with bacteriophagous nematodes.

In addition to the described species and *Candidatus* taxa, the latter tree also included nonvalid names such as "Pasteuria hartismerei" (host: *Meloidogyne ardenensis*) and "Pasteuria goettingianae" (host: cyst nematode *Heterodera goettingiana*). Additional sequences of undescribed *Pasteuria* from various hosts (e.g., *Rotylenchulus reniformis*, *Belonolaimus longicaudatus*, and *Hoplolaimus galeatus*) were also included in phylogenetic analysis. 16S rRNA gene sequence comparison of a *Pasteuria* isolate from *Heterodera cajani*, which also infects the potato cyst nematode *Globodera pallida*, revealed 98.6 % similarity to the *Pasteuria nishizawae* (Mohan et al. 2012).

## Molecular Analysis

A multilocus sequence analysis (Charles et al. 2005), using the amino acid sequence alignments of more than 25 housekeeping genes of *P. penetrans*, either individually or concatenated, led to the conclusion that this species is ancestral to members of *Bacillus*, branching between *Staphylococcus aureus* and the *Bacillus* clade. Due to the restricted size of the database of sequenced genomes at that time, the information lacks the precision of todays' 16S rRNA gene sequence-based phylogeny.

DNA for rRNA gene amplification has been extracted from spores (Atibalentja et al. 2004; Schmidt et al. 2010; Mauchline et al. 2010) as performed for the characterization of *Pasteuria nishizawae* (Atibalentja et al. 2000; Noel et al. 2005) or from vegetative cells, e.g., as described for a comparison of *P. ramosa* and *P. penetrans* on the basis of selected sporulation genes and an epitope associated with the spore envelope (Schmidt et al. 2008). This study also developed a monoclonal antibody probe directed against an endospore adhesin epitope, with which spores of different *Pasteuria* spp. were detected and discriminated. A rapid method for isolating *Pasteuria penetrans* endospores was described by Waterman et al (2006).

Polymerase chain reaction analysis with primers specific for *Pasteuria* 16S ribosomal DNA sequences yielded a 549-bp band (Duan et al. 2003). The phylogenetic assessment of *Candidatus* Pasteuria aldrichii (Giblin-Davis et al. 2011) is an example of a complex molecular identification of clones related to Pasteuria spp. as it involved a two-step amplification, cloning, and sequencing strategy in which first the 5′ terminus and then the 3′ terminal fragment of the sequence were obtained. In other report, two separate PCRs were performed using primer pairs 39F and 1166R as well as the primers PsppF4 and PsppR5 (Mauchline et al. 2011).

A single-nucleotide polymorphism (SNP) study was performed on seven populations of *P. penetrans* isolated from a wide range of geographic locations (Mauchline et al. 2011). Based upon the microheterogeneity of 16S rRNA clones, an intraspecies diversity of the species could be detected which was not obvious from published gene sequences. Not all clones could be discriminated which was the case when several protein-encoding genes were analyzed in addition. When the species *P. penetrans*, *P. ramosa*, and "Pasteuria hartismerei" were analyzed with respect to 16S rRNA gene and protein-encoding gene sequences, the latter information provided greater discrimination than the 16S rRNA gene.

Tracing Pasteuria spp. in mixed infection was done (Ben-Ami and Routtu 2013) using variable number of tandem repeat (VNTR) markers which have been described for defining *P. ramosa* strains (Mouton et al. 2007; Mouton and Ebert 2008; Ben-Ami et al. 2008).

A collagen-like protein named Pcl1a (Pasteuria collagen-like protein 1a) was identified in spores of two *P. ramosa* isolates that were selected for their differences in infectivity. The protein contained a 75-amino-acid amino-terminal domain with a potential transmembrane helix domain, a central collagen-like region (CLR) containing Gly-Xaa-Yaa repeats, and a 7-amino-acid carboxy-terminal domain. Distinct differences were found to occur in the CLR region among the two isolates (Mouton et al. 2009). Investigating a higher number of parasite strains, additional 37 novel putative P. ramosa collagen-like protein genes (PCLs) were identified (McElroy et al. 2011).

The gold genome database (http://genomesonline.org/cgi-bin/GOLD/index.cgi) contains three entries for incomplete whole genome sequences, i.e., for *P. ramosa* (Gi08593) and two for *P. penetrans* (Gi00435 and Gi00436).

## References


Andras JP, Ebert D (2013) A novel approach to parasite population genetics: experimental infection reveals geographic differentiation, recombination and host-mediated population structure in *Pasteuria ramosa*, a bacterial parasite of Daphnia. Mol Ecol 22:972–986

Atibalentja N, Noel GR, Domier LL (2000) Phylogenetic position of the North American isolate of *Pasteuria* that parasitizes the soybean cyst nematode, *Heterodera glycines*, as inferred from the 16S rDNA sequence. Int J Syst Evol Microbiol 50:605–613

Atibalentja N, Noel GR, Ciancio A (2004) A simple method for PCR-amplification, cloning, and sequencing of Pasteuria 16SrDNA from small numbers of endospores. J Nematol 36:100–105

Auld SKJR, Edel KH, Little TJ (2012) The cellular immune response of *Daphnia magna* under host-parasite genetic variation and variation in initial dose. Evolution 66:3287–3293

Ben-Ami F, Routtu J (2013) The expression and evolution of virulence in multiple infections: the role of specificity, relative virulence and relative dose. BMC Evol Biol 13:97

Ben-Ami F, Mouton L, Ebert D (2008) The effects of multiple infections on the expression and evolution of virulence in a *Daphnia*-endoparasite system. Evolution 62:1700–1711

Charles L, Carbone I, Davies KG, Bird D, Burke M, Kerry BR, Opperman CH (2005) Phylogenetic analysis of *Pasteuria penetrans* by use of multiple genetic loci. J Bacteriol 187:5700–5708

Duan YP, Castro HF, Hewlett TE, White JH, Ogram AV (2003) Detection and characterization of *Pasteuria* 16S rRNA gene sequences from nematodes and soils. Int J Syst Evol Microbiol 53:105–112, Check for fragment size

Ebert D (2008) Host-parasite coevolution: Insights from the *Daphnia*-parasite model system. Curr Opin Microbiol 11:290–301

Giblin-Davis RM, Williams DS, Bekal S, Dickson DW, Brito JA, Becker JO, Preston JF (2003) '*Candidatus* Pasteuria usgae' sp. nov., an obligate endoparasite of the phytoparasitic nematode *Belonolaimus longicaudatus*. J Syst Evol Microbiol 53:197–200

Giblin-Davis RM, Nong G, Preston JF, Williams DS, Center BJ, Brito JA, Dickson DW (2011) '*Candidatus* Pasteuria aldrichii', an obligate endoparasite of the bacterivorous nematode *Bursilla*. Int J Syst Evol Microbiol 61:2073–2080

King KC, Auld SKJR, Wilson PJ, James J, Little TJ (2013) The bacterial parasite *Pasteuria ramosa* is not killed if it fails to infect: implications for coevolution. Ecol Evol 3:197–203

Laurent E (1890) Sur le microbe des nodosités des légumineuses. Comptes Rendus de l'Académie des Sciences Paris 111:754–756

Luijckx P, Ben-Ami F, Mouton L, Du Pasquier L, Ebert D (2011) Cloning of the unculturable parasite *Pasteuria ramosa* and its *Daphnia* host reveals extreme genotype-genotype interactions. Ecol Lett 14:125–131

Luijckx P, Duneau D, Andras JP, Ebert D (2013a) Cross-species infection trials reveal cryptic parasite varieties and a putative polymorphism shared among host species. Evolution. doi:10.1111/evo.12289

Luijckx P, Fienberg H, Duneau D, Ebert D (2013b) A matching-allele model explains host resistance to parasites. Curr Biol 23:1085–1088

Mauchline TH, Mohan S, Davies KG, Schaff JE, Opperman CH, Kerry BR, Hirsch PR (2010) A method for release and multiple strand amplification of small quantities of DNA from endospores of the fastidious bacterium *Pasteuria penetrans*. Lett Appl Microbiol 50:515–521

Mauchline TH, Knox R, Mohan S, Powers SJ, Kerry BR, Davies KG, Hirsch PR (2011) Identification of new single nucleotide polymorphism-based markers for inter- and intraspecies discrimination of obligate bacterial parasites (Pasteuria spp.) of invertebrates. Appl Environ Microbiol 77:6388–6394

McElroy K, Mouton L, Du Pasquier L, Qi W, Ebert D (2011) Characterisation of a large family of polymorphic collagen-like proteins in the endospore-forming bacterium *Pasteuria ramosa*. Res Microbiol 162:701–714

Metchnikoff E (1888) *Pasteuria ramosa* un représentant des bactéries à division longitudinale. Annales de l'Institut Pasteur (Paris) 2:165–170

Mohan S, Mauchline TH, Rowe J, Hirsch PR, Davies KG (2012) *Pasteuria* endospores from *Heterodera cajani* (Nematoda: *Heteroderidae*) exhibit inverted attachment and altered germination in cross-infection studies with *Globodera pallida* (Nematoda: *Heteroderidae*). FEMS Microbiol Ecol 79:675–684

Mouton L, Ebert D (2008) Variable-number-of-tandem-repeats analysis of genetic diversity in *Pasteuria ramosa*. Curr Microbiol 56:447–552

Mouton L, Nong G, Preston JF, Ebert D (2007) Variable-number tandem repeats as molecular markers for biotypes of *Pasteuria ramosa* in *Daphnia* spp. Appl Environ Microbiol 73:3715–3718

Mouton L, Traunecker E, McElroy K, Du Pasquier L, Ebert D (2009) Identification of a polymorphic collagen-like protein in the crustacean bacteria *Pasteuria ramosa*. Res Microbiol 160:792–799

Noel GR, Atibalentja N, Domier LL (2005) Emended description of *Pasteuria nishizawae*. Int J Syst Evol Microbiol 55:1681–1685

Sayre RM, Starr MP (1985) *Pasteuria penetrans* (*ex* Thorne, 1940) nom. rev., comb. n., sp. n., a mycelial and endospore-forming bacterium parasitic in plant-parasitic nematodes. Proc Helminthol Soc Wash 52:149–165, Validation List no. 20. (1986) Int J Syst Bacteriol 36:354–356

Sayre RM, Wergin WP, Schmidt JM, Starr MP (1991) *Pasteuria nishizawae* sp. nov., a mycelial and endosporeforming bacterium parasitic on cyst nematodes of genera *Heterodera* and *Globodera*. Res Microbiol 142:551–564, Validation List no. 41 (1992) Int J Syst Bacteriol 42:327–328

Sayre RM, Starr MP, revised by Dickson DW, Preston JF III, Giblin-Davis RM, Noel GR, Ebert D, Bird GW (2009) Genus I. *Pasteuria*. In: de Vos P, Garrity GM, Jones D, Krieg NR, Ludwig W, Rainey FA, Schleifer K-H, Whitman WB (eds) Bergey's manual of systematic bacteriology, vol 3, 2nd edn. Springer, New York, pp 328–347

Schlotz N, Ebert D, Martin-Creuzburg D (2013) Dietary supply with polyunsaturated fatty acids and resulting maternal effects influence host -parasite interactions. BMC Ecol 13:41

Schmidt LM, Mouton L, Nong G, Ebert D, Preston JF (2008) Genetic and immunological comparison of the cladoceran parasite *Pasteuria ramosa* with the nematode parasite *Pasteuria penetrans*. Appl Env Microbiol 74:259–264

Schmidt LM, Hewlett TE, Green A, Simmons LJ, Kelley K, Doroh M, Stetina SR (2010) Molecular and morphological characterization and biological control capabilities of a *Pasteuria* ssp. parasitizing *Rotylenchulus reniformis*, the reniform nematode. J Nematol 42:207–217

Stamatakis A (2006) RAxML-VI-HPC: maximum likelihood-based phylogenetic analyses with thousands of taxa and mixed models. Bioinformatics 22:2688–2690

Starr MP, Sayre RM (1988) *Pasteuria thornei* sp. nov. and *Pasteuria penetrans sensu stricto* emend., mycelial and endospore-forming bacteria parasitic, respectively, on plant parasitic nematodes of the genera *Pratylenchus* and *Melodogyne*. Ann Inst Pasteur 139:11–31, Validation List no. 26. (1988) Int J Syst Bacteriol 38:328–329

Waterman JT, McK BD, Opperman CH (2006) A method for Isolation of *Pasteuria penetrans* endospores for bioassay and genomic studies. J Nematol 38:165–167

Yarza P, Ludwig W, Euzéby J, Amann R, Schleifer K-H, Glöckner FO, Rosselló-Móra R (2010) Update of the All-Species Living-Tree Project based on 16S and 23S rRNA sequence analyses. Syst Appl Microbiol 33:291–299

# 23 The Emended Family *Peptococcaceae* and Description of the Families *Desulfitobacteriaceae*, *Desulfotomaculaceae*, and *Thermincolaceae*

*Erko Stackebrandt*
Leibniz Institute DSMZ-German Collection of Microorganisms and Cell Cultures GmbH, Braunschweig, Germany



**Abstract**
The family *Peptococcaceae* is one of several families of the order *Clostridiales*, class *Clostridia*. Besides the type genus *Peptococcus*, the family encompasses the genera *Cryptanaerobacter*, *Dehalobacter*, *Desulfitibacter*, *Desulfitispora*, *Desulfitobacterium*, *Desulfonispora*, *Desulfosporosinus*, *Desulfotomaculum*, *Desulfurispora*, *Pelotomaculum*, *Sporotomaculum*, *Syntrophobotulus*, *and Thermincola*. The family is physiologically (chemoorganotroph, chemolithoheterotroph, chemolithoautotroph, syntrophy with hydrogenotrophs) and phylogenetically heterogeneous. Many of its members were isolated from human material, while others occur in soil, marine, and freshwater sediments or sewage. All members of the family are obligate anaerobes and Gram positive, though some members stain Gram negative. The morphology ranges from spherical to rod-shaped cells while spore formation is genus specific. This brief overview concentrates on genera and species described since 2006, and which are not covered in the chapter *Peptococcaceae* in *Bergey's Manual of Systematic Bacteriology*, 2nd edition.

## Taxonomy

As indicated by Ezaki (2009), the family is phylogenetically heterogeneous, comprising at least two major clades, one defined by *Peptococcus* and related genera and the other by Desulfotomaculum and related genera. Thermincola stands isolated, branching intermediate. The tree depicted in ➲ *Fig. 23.1* includes recently described family and genus members and deviated slightly from the situation given by Ezaki (2009). *Peptococcus* together with *Desulfonispora*, *Desulfitispora*, and *Desulfitibacter* branch separately from other family members as defined in the List of Bacterial Names with Standing in Nomenclature (http://www.bacterio.net/), showing a slightly closer relationship to *Veillonellaceae*, *Halobacteroidaceae*, *Haloanaerobiaceae*, and *Cyanobacteria*. The clade with *Desulfosporosinus*, *Desulfitobacterium*, *Dehalobacter*, and *Syntrophobotulus* branches adjacent to the family *Heliobacteriaceae*, while the species-rich genus *Desulfotomaculum*, together with *Cryptanaerobacter*, Pelotomaculum, *Desulfurispora*, and *Sporotomaculum*, constitutes a second large clade. The family is in need of reclassification, restricting it to the cluster around *Peptococcus*, while the other clades deserve family status. These three additional families are described below on the basis of comparative 16S rRNA gene sequence analysis.

It should be noted that both ML and NJ (not shown) trees see *Cryptanaerobacter phenolicus* to be closely related to *Pelotomaculum schinkii* and other members of *Pelotomaculum*, while *Desulfurispora thermophila* and *Sporotomaculum hydroxybenzoicum* cluster with different subgroups of *Desulfotomaculum*. The latter genus appears phylogenetically heterogeneous and a dissection and reclassification of at least two new genera appear indicated.

The last comprehensive coverage of the family *Peptococcaceae* has been presented by Ezaki (2009) in *Bergey's Manual of Systematic Bacteriology*, 2nd ed, covering descriptions since 2006. This communication concentrates on recently described genera and species and the reader should consult the chapter of Ezaki (2009) for obtaining a more comprehensive overview of the biology of family members. ➲ *Table 23.1* is a list of species belonging to genera described until 2006, together with some of the salient feature of these taxa. ➲ *Table 23.2* compiles recently described genera of *Peptococcaceae* as defined by Euzeby (http://www.bacterio.net/).

E. Rosenberg et al. (eds.), *The Prokaryotes – Firmicutes and Tenericutes*, DOI 10.1007/978-3-642-30120-9_364,

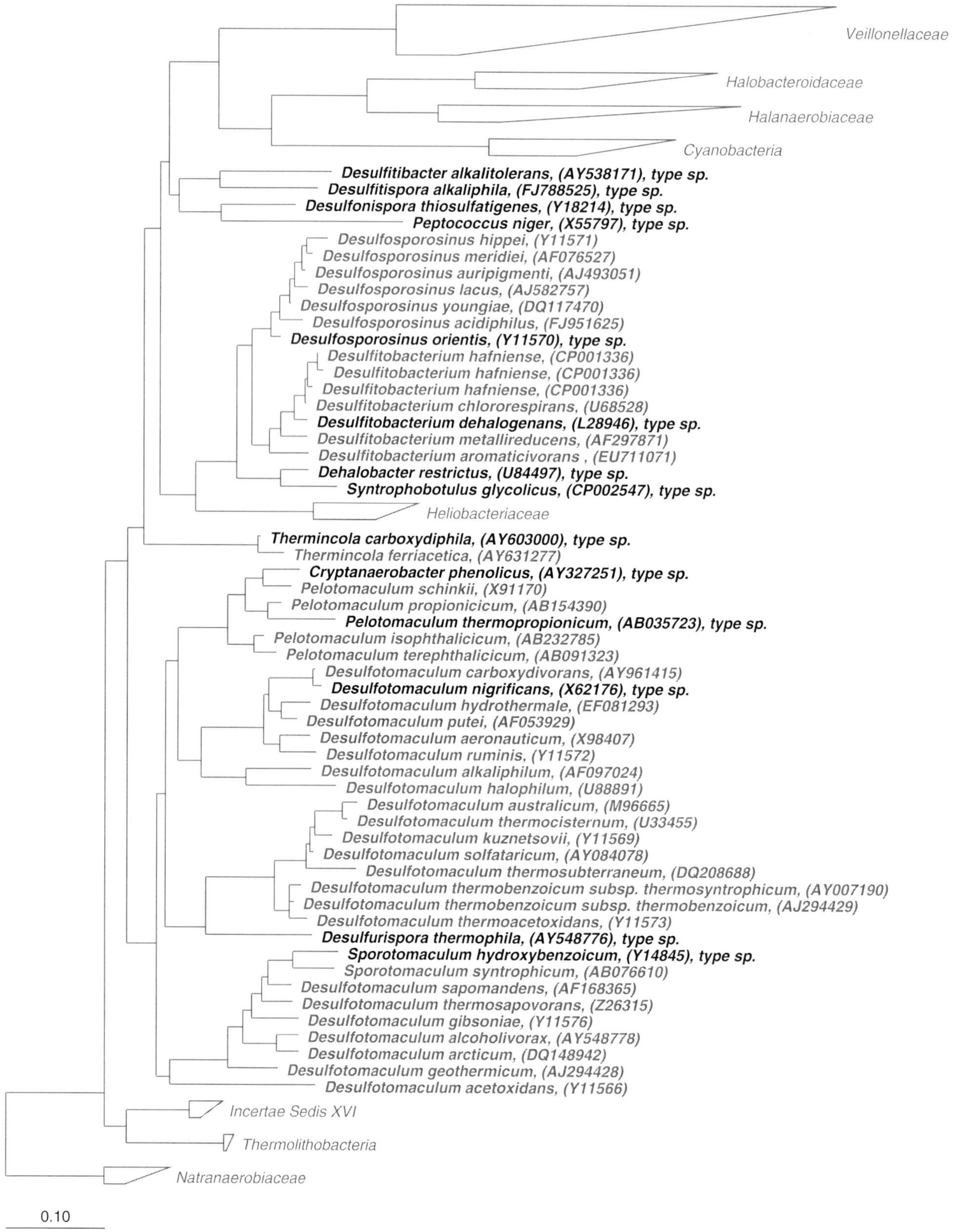


**◘ Fig. 23.1**

Neighbor-joining genealogy reconstruction based on the RAxML algorithm (Stamatakis 2006) of the sequences of members of the family *Peptococcaceae* and some neighboring taxa present in the LTP_106 (Yarza et al. 2010). The tree was reconstructed by using a subset of sequences. Representative sequences from closely relative genera were used to stabilize the tree topology. In addition, a 40 % maximum frequency filter was applied to remove hypervariable positions from the alignment. Scale bar indicates estimated sequence divergence

**Table 23.1**
**Species published since 2006, for genera described before 2006**

| Genus | *Desulfosporosinus* | | | | *Desulfitobacterium* |
|---|---|---|---|---|---|
| Species | *acidiphilus* | *hippei* | *lacus* | *youngiae* | *aromaticivorans* |
| Gram stain | Negative | Negative | Negative | Negative | Positive |
| Motility | – | Subpolar flagella | Peritrichous flagella | n.r. | + |
| Morphology | Curved rods | Curved rods | Rods | Curved rods | Slightly curved rods |
| Spore formation | +, oval, subterminal | +, round, terminal | +, oval, subterminal | +, oval, central to subterminal | +, oval, terminal |
| Growth optimum (°C) | 20–40 (35) | 5.0–37.0 | 4–32 | 8–39 (32–35) | (30) |
| pH range (optimum) | 3.6–5.5 (5.2) | 6.5–7.5. | 6.5–7.5 | 5.7–8.2 (7.0–7.3) | 6.5–7.5 (6.6–7.0) |
| Electron donors | $H_2$, lactate, pyruvate, glycerol, glucose, and fructose as electron donors | Lactate | Lactate | Lactate | Acetate, toluene, phenol, *o*-xylene, and *p*-cresol as carbon and energy sources |
| Electron acceptor | Sulfate | Sulfate, thiosulfate | Sulfate, sulfite, thiosulfate Dissimilatory Fe(III) reduction | Fumarate, sulfate, sulfite, and thiosulfate | Ferrihydrite, ferric citrate |
| Major fatty acids (>10 %) | $C_{14:0}$, iso-$C_{15:0}$, $C_{16:0}$, $C_{16:0}$ DMA | $C_{16:1}$cis9, $C_{18:1}$cis11 | $C_{16:1}$cis9, $C_{16:0}$, $C_{18:1}$cis11 | $C_{16:1}$, $C_{16:0}$ | iso-$C_{15:0}$ |
| Mol% G+C | 42.3 | 42.1 | 42.7 | 36.6 | 47.7 |
| Type strain | SJ4$^T$ | 343$^T$ | STP12$^T$ | JW/YJL-B18$^T$ | UKTL$^T$ |
| Habitat | Pond sediment | Ancient permafrost deposits in Siberia, Russia | Sediments of Lake Stechlin, Germany | Sediment of constructed wetland | Coal-gasification site, Gliwice, Poland |
| Publication | Alazard et al. 2010 | Vatsurina et al. 2008 | Ramamoorthy et al. 2006 | Lee et al. 2009 | Kunapuli et al. 2010 |

| Genus | *Desulfotomaculum* | | | | | |
|---|---|---|---|---|---|---|
| Species | *hydrothermale* | *intricatum* | *peckii* | *varum* | *defluvii* | *alcoholivorax* |
| Gram stain | Negative, positive cell wall structure | Negative | Positive | Positive, negative cell wall structure | Negative | Positive |
| Motility | Peritrichous flagella | Weak | – | n.r | – | + |
| Morphology | Slightly curved rods | Rods | Slightly curved rods | Straight to slightly curved rods | Rods | Rods |
| Spore formation | + | + | +, central to subterminal | + | +, subterminal | +, spherical, central |
| Growth range (° C) (optimum) | 40–60 (55) | 6.4–6.8 (42–45) | 50–65 (55–60) | 37–55 (50) | 25–42 (37) | 33–51 (44–46) |
| pH range (optimum) | 5.8–8.2 (7.1) | 6–7.3 (6.4–6.8) | 5.9–9.2 (6.0–6.8) | 5.0–8.5 (7) | 6.5–8.5 (7.5) | 6.0–7.5 (6.4–7.3) |
| Electron donors | Lactate, pyruvate, formate, ethanol, butanol, glycerol, propanol, and $H_2$ (plus acetate) | Acetate | $H_2$/$CO_2$, propanol, butanol, ethanol | Fructose, mannose, glycerol, lactate, pyruvate, and $H_2$ as electron donors | Acetate, fumarate | Various alcohols and carboxylic as electron donors |
| Electron acceptors | Sulfate, sulfite, thiosulfate, As(V),Fe(III) | Sulfate, sulfite, thiosulfate, sulfur | Sulfate (sulfite, thiosulfate) | Sulfate, sulfite, thiosulfate, sulfur | Sulfate, sulfite, thiosulfate | Sulfate, sulfite, thiosulfate |

**Table 23.1 (continued)**

| Genus | *Desulfotomaculum* | | | | | |
|---|---|---|---|---|---|---|
| Species | *hydrothermale* | *intricatum* | *peckii* | *varum* | *defluvii* | *alcoholivorax* |
| Major fatty acids (>10 %) | iso-$C_{15:0}$, $C_{16:0}$, iso-$C_{17:0}$ | iso-$C_{15:0}$, $C_{16:1}\omega7c$, $C_{16:0}$ | $C_{16:0}$, $C_{18:0}$, iso-$C_{15:0}$, iso-$C_{17:1}$ I/anteiso-$C_{17:1}$ B | iso-$C_{15:0}$, $C_{16:0}$, iso-$C_{17:0}$ | $C_{16:1}$ c9, $C_{16:0}$, $C_{16:1}$ $\omega7c$, and/or iso-$C_{15:0}$ 2-OH | iso-$C_{15:0}$, iso-$C_{17:1}\omega10c$, iso-$C_{17:0}$ |
| Mol% G+C | 46.8 | 41.1 | 44.4 | 52.4 | 45.4 | 48 |
| Type strain | Lam5$^T$ | SR45$^T$ | LINDBHT1$^T$ | RH04-3$^T$ | A5LFS102$^T$ | RE35E1$^T$ |
| Habitat | Hot spring, northeast Tunisia | Lake Mizugaki, Japan | Abattoir wastewaters digester, Tunisia | Microbial mat, bore well, Great Artesian Basin, Australia | Subsurface landfill, Chandigarh, India | Metal and sulfate fluidized-bed reactor |
| Publication | Haouari et al. 2008 | Watanabe et al. 2013 | Jabari et al. 2013 | Ogg and Patel 2011 | Krishnamurthi et al. 2013 | Kaksonen et al. 2008 |

**Table 23.2**
**New genera and species of *Peptococcaceae*, described since 2006**

| Genus | *Desulfitibacter* | *Desulfitispora* | *Desulfurispora* |
|---|---|---|---|
| Species | *alkalitolerans* | *alkaliphila* | *thermophila* |
| Gram stain | Positive | Positive | Positive |
| Motility | +, polar flagellum | +, subterminal flagellum | + |
| Morphology | Rods | Thin long rods | Rods |
| Spore formation | +, terminal, round | +, terminal, swollen sporangia | + |
| Growth optimum (° C) | 23–44 (35–37) | n.r. | 40–67(59–61) |
| pH range (optimum) | 7.6–10.5 (8.0–9.5) | 8.5–10.3 (9.5) | 6.4–7.9 (7.0–7.3) |
| Electron donors | Betaine, formate, lactate, methanol, choline, pyruvate | Lactate, pyruvate | $H_2/CO_2$ (80:20, v/v), alcohols, various carboxylic acids, some sugars |
| Electron acceptors | Sulfur, sulfite, thiosulfate, nitrate, nitrite | Sulfur, sulfite, thiosulfate | Sulfur, sulfite, sulfate, thiosulfate |
| Major fatty acids (>10 %) | n.r. | $C_{16:1}\omega5c$, $C_{16:1}\omega7c$. | iso-$C_{15:0}$, iso-$C_{17:0}$ |
| Mol% G+C | 41.6 | 34.3 | 53.5 |
| Habitat | Heating plant, Denmark | Soda lakes, Kulunda Steppe, Altai, Russia | Sulfidogenic fluidized-bed reactor |
| Type strain | sk.kt5$^T$ | AHT17$^T$ | RA50E1$^T$ |
| Publication | Nielsen et al. 2006 | Sorokin and Muyzer 2010 | Kaksonen et al. 2007 |

*NR* not recorded

**Table 23.3**
**Examples of published and unpublished genome sequences**

| *Taxon* | Strain number | GOLD identification | Status |
|---|---|---|---|
| *Desulfosporosinus orientis* | DSM 765 | Gc02026 | Pester et al. 2012 |
| *Desulfosporosinus acidiphilus* | SJ4 | Gc02346 | Pester et al. 2012 |
| *Desulfosporosinus meridiei* | DSM 13257 | Gc02327 | Pester et al. 2012 |
| *Desulfitobacterium metallireducens* | DSM 15288 | Gi08580 | Lucas et al., unpublished Locus NZ_AGJB01000000 |
| *Desulfitobacterium hafniense* | DCB-2 | Gc00918 | Kim et al. 2012 |
| *Syntrophobotulus glycolicus* | DSM 8271 | Gc01670 | Han et al. 2011 |
| *Pelotomaculum thermopropionicum* | SI | Gc00556 | Kosaka et al. 2008 |
| *Desulfotomaculum ruminis* | DSM 2154 | Gc01775 | Spring et al. 2012 |
| *Desulfotomaculum kuznetsovii* | DSM 6115 | Gc01781 | Visser et al. 2013 |
| *Desulfotomaculum acetoxidans* | DSM 771 | Gc01106 | Spring et al. 2009 |
| *Desulfotomaculum reducens* | MI-1 | Gc00530 | Junier et al. 2010 |

## Genome Sequences

A high number of strains have been subjected to the analysis of genome sequences. Only a few examples of published or deposited sequences of strains of the various genera are given in *Table 23.3*. More information is available in the GOLD database (genomes.org/cgi-bib/Gold/Search.cgi).

## Emendation of the Family *Peptococcaceae*

The family as described by Rogosa (1971) is restricted to the genera *Peptococcus, Desulfonispora, Desulfitispora,* and *Desulfitibacter.*

The family is proposed on the basis of the isolated phylogenetic position of its members among other families of the order *Clostridiales* (*Fig. 23.1*).

The type genus is *Peptococcus* (Kluyver and Van 1936).

## *Desulfitobacteriaceae* fam. nov.

De.sul.fi.to.bac.te.ria' ce.ae. *Desulfitobacterium* type genus of the family, -*aceae* ending to denote a family. N.L.fem. pl. n. *Desulfitobacteriaceae*, the family of *Desulfitobacterium.*

The family *Desulfitobacteriaceae* is described on the basis of phylogenetic analyses of 16S rRNA gene sequences (*Fig. 23.1*). The family consists of the genera *Desulfitobacterium, Dehalobacter, Desulfosporosinus,* and *Syntrophobotulus.*

The type genus is *Desulfitobacterium* (Utkin et al. 1994).

## *Desulfotomaculaceae* fam. nov.

De.sul.fo.to.ma.cu.la' ce.ae. *Desulfotomaculum* type genus of the family, -*aceae* ending to denote a family. N.L.fem. pl. n. *Desulfotomaculaceae*, the family of *Desulfotomaculum.*

The family is proposed on the basis of the isolated phylogenetic position of its members among other families of the order *Clostridiales* (*Fig. 23.1*). The family consists of the genera *Desulfotomaculum, Cryptanaerobacter, Pelotomaculum, Desulfurispora,* and *Sporotomaculum.*

The type genus is *Desulfotomaculum* (Campbell and Postgate 1965).

## *Thermincolaceae* fam. nov.

Therm.in.co.la' ce.ae. *Thermincola* type genus of the family, -*aceae* ending to denote a family. N.L.fem. pl. n. *Thermincolaceae*, the family of *Thermincola.*

The family is proposed on the basis of the isolated phylogenetic position of its members among other families of the order *Clostridiales* (*Fig. 23.1*).

The type genus is *Thermincola* (Sokolova et al. 2005).

## References


Alazard D, Joseph M, Battaglia-Brunet F, Cayol JL, Ollivier B (2010) *Desulfosporosinus acidiphilus* sp nov: a moderately acidophilic sulfate-reducing bacterium isolated from acid mining drainage sediments. Extremophiles 14:305–312, Validation List no 145 (2012) Int J Syst Evol Microbiol 62, 1017–1019

Campbell LL, Postgate JR (1965) Classification of the spore-forming sulfate-reducing bacteria. Bacteriol Rev 29:359–363

Ezaki (2009) *Peptococcaceae.* In:De Vos P, Garrity GM, Jones D, Krieg NR, Ludwig W, Rainey FA, Schleifer K-H, Whitman WB (eds) Bergey's manual of systematic bacteriology, 2nd ed. Vol 3. Springer, New York, pp 969-971

Han C, Mwirichia R, Chertkov O, Held B, Lapidus A et al (2011) Complete genome sequence of *Syntrophobotulus glycolicus* type strain (FlGlyR). Stand Genomic Sci 4:371–380

Haouari O, Fardeau ML, Cayol JL, Casiot C, Elbaz-Poulichet F, Hamdi M, Joseph M, Ollivier B (2008) *Desulfotomaculum hydrothermale* sp. nov., a thermophilic sulfate-reducing bacterium isolated from a terrestrial Tunisian hot spring. Int J Syst Evol Microbiol 58:2529–2535

Jabari L, Gannoun H, Cayol JL, Hamdi M, Ollivier B, Fauque G, Fardeau ML (2013) *Desulfotomaculum peckii* sp. nov., a moderately thermophilic member of the genus *Desulfotomaculum*, isolated from an upflow anaerobic filter treating abattoir wastewaters. Int J Syst Evol Microbiol 63:2082–2087

Junier P, Junier T, Podell S, Sims DR, Detter JC, Lykidis A et al (2010) The genome of the Gram-positive metal- and sulfate-reducing bacterium *Desulfotomaculum reducens* strain MI-1. Environ Microbiol 12:2738–2754

Kaksonen AH, Spring S, Schumann P, Kroppenstedt RM, Puhakka JA (2007) *Desulfurispora thermophila* gen. nov., sp. nov., a thermophilic, spore-forming sulfate-reducer isolated from a sulfidogenic fluidized-bed reactor. Int J Syst Evol Microbiol 57:1089–1094

Kaksonen AH, Spring S, Schumann P, Kroppenstedt RM, Puhakka JA (2008) *Desulfotomaculum alcoholivorax* sp. nov., a moderately thermophilic, spore-forming, sulfate-reducer isolated from a fluidized-bed reactor treating acidic metal- and sulfate-containing wastewater. Int J Syst Evol Microbiol 58:833–838

Kim SH, Harzman C, Davis JK, Hutcheson R, Broderick JB, Marsh TL, Tiedje JM (2012) Genome sequence of *Desulfitobacterium hafniense* DCB-2, a Gram-positive anaerobe capable of dehalogenation and metal reduction. BMC Microbiol 12:21

Kluyver AJ, Van NIELCB (1936) Prospects for a natural system of classification of bacteria. Zentralbl Bakt Parasitenk Infekt und Hygiene. Abt II 94:369–403

Kosaka T, Kato S, Shimoyama T, Ishii S, Abe T, Watanabe K (2008) The genome of *Pelotomaculum thermopropionicum* reveals niche-associated evolution in anaerobic microbiota. Genome Res 18:442–448

Krishnamurthi S, Spring S, Anil Kumar P, Mayilraj S, Klenk HP, Suresh K (2013) *Desulfotomaculum defluvii* sp. nov., a sulfate-reducing bacterium isolated from the subsurface environment of a landfill. Int J Syst Evol Microbiol 63:2290–2295

Kunapuli U, Jahn MK, Lueders T, Geyer R, Heipieper HJ, Meckenstock RU (2010) *Desulfitobacterium aromaticivorans* sp. nov. and *Geobacter toluenoxydans* sp. nov., iron-reducing bacteria capable of anaerobic degradation of monoaromatic hydrocarbons. Int J Syst Evol Microbiol 60:686–695

Lee YJ, Romanek CS, Wiegel J (2009) *Desulfosporosinus youngiae* sp. nov., a spore-forming, sulfate-reducing bacterium isolated from a constructed wetland treating acid mine drainage. Int J Syst Evol Microbiol 59:2743–2746

Nielsen MB, Kjeldsen KU, Ingvorsen K (2006) *Desulfitibacter alkalitolerans* gen. nov., sp. nov., an anaerobic, alkalitolerant, sulfite-reducing bacterium isolated from a district heating plant. Int J Syst Evol Microbiol 56:2831–2836

Ogg CD, Patel BKC (2011) *Desulfotomaculum varum* sp. nov., a moderately thermophilic sulfate-reducing bacterium isolated from a microbial mat colonizing a Great Artesian Basin bore well runoff channel. 3 Biotech 1:139–149, Validation List no 142 (2011) Int J Syst Evol Microbiol 61, 2563–2565

Pester M, Brambilla E, Alazard D, Rattei T, Weinmaier T et al (2012) Complete genome sequences of *Desulfosporosinus orientis* DSM765$^T$, *Desulfosporosinus youngiae* DSM17734$^T$, *Desulfosporosinus meridiei* DSM13257$^T$, and *Desulfosporosinus acidiphilus* DSM22704$^T$. J Bacteriol 194:6300–6301

Ramamoorthy S, Sass H, Langner H, Schumann P, Kroppenstedt RM, Spring S, Overmann J, Rosenzweig RF (2006) *Desulfosporosinus lacus* sp. nov., a sulfate-reducing bacterium isolated from pristine freshwater lake sediments. Int J Syst Evol Microbiol 56:2729–2736

Rogosa M (1971) *Peptococcaceae*, a new family to include the Gram-positive, anaerobic cocci of the genera *Peptococcus*, *Peptostreptococcus* and *Ruminococcus*. Int J Syst Bacteriol 21:234–237

Sokolova TG, Kostrikina NA, Chernyh NA, Kolganova TV, Tourova TP, Bonch-Osmolovskaya EA (2005) *Thermincola carboxydiphila* gen. nov., sp. nov., a novel anaerobic, carboxydotrophic, hydrogenogenic bacterium from a hot spring of the Lake Baikal area. Int J Syst Evol Microbiol 55:2069–2073

Sorokin DY, Muyzer G (2010) Haloalkaliphilic spore-forming sulfidogens from soda lake sediments and description of *Desulfitispora alkaliphila* gen. nov., sp. nov. Extremophiles 14:313–320, Validation List no 134 (2010) Int J Syst Evol Microbiol 60, 1477–1479

Spring S, Lapidus A, Schröder M, Gleim D, Sims D et al (2009) Complete genome sequence *of Desulfotomaculum acetoxidans* type strain (5575). Stand Genomic Sci 21:242–253

Spring S, Visser M, Lu M, Copeland A, Lapidus A, Lucas S et al (2012) Complete genome sequence of the sulfate-reducing firmicute *Desulfotomaculum ruminis* type strain (DL$^T$). Stand Genomic Sci 7:304–319

Stamatakis A (2006) RAxML-VI-HPC: maximum likelihood-based phylogenetic analyses with thousands of taxa and mixed models. Bioinformatics 22:2688–2690

Utkin I, Woese C, Wiegel J (1994) Isolation and characterization of *Desulfitobacterium dehalogenans* gen. nov., sp. nov., an anaerobic bacterium which reductively dechlorinates chlorophenolic compounds. Int J Syst Bacteriol 44:612–619

Vatsurina A, Badrutdinova D, Schumann P, Spring S, Vainshtein M (2008) *Desulfosporosinus hippei* sp. nov., a mesophilic sulfate-reducing bacterium isolated from permafrost. Int J Syst Evol Microbiol 58:1228–1232

Visser M, Worm P, Muyzer G, Pereira IAC, Schaap P J et al (2013) Genome analysis of *Desulfotomaculum kuznetsovii* DOI:10.1601/nm.4341#xrefwindow strain 17$^T$ reveals a physiological similarity with *Pelotomaculum thermopropionicum* DOI:10.1601/nm.4353#xrefwindow strain SI$^T$. Stand Genomic Sci 8(1)

Watanabe M, Kojima H, Fukui M (2013) *Desulfotomaculum intricatum* sp. nov., a sulfate reducer isolated from freshwater lake sediment. Int J Syst Evol Microbiol 63:3574–3578

Yarza P, Ludwig W, Euzéby J, Amann R, Schleifer K-H, Glöckner FO, Rosselló-Móra R (2010) Update of the All-Species Living-Tree Project based on 16S and 23S rRNA sequence analyses. Syst Appl Microbiol 33:291–299

# 24 The Family *Peptostreptococcaceae*

*Alexander Slobodkin*
Winogradsky Institute of Microbiology, Russian Academy of Sciences, Moscow, Russia



**Abstract**
*Peptostreptococcaceae*, a family within the order *Clostridiales*, includes the genera *Peptostreptococcus*, *Acetoanaerobium*, *Filifactor*, *Proteocatella*, *Sporacetigenium*, and *Tepidibacter*. Genera *Acetoanaerobium*, *Proteocatella*, and *Sporacetigenium* are monospecific. Representatives of the family have different cell morphology which varies among the genera from cocci to rods and filaments. Species of *Filifactor*, *Proteocatella*, *Sporacetigenium*, and *Tepidibacter* form endospores. All members of the family are anaerobes with fermentative type of metabolism. The genus *Tepidibacter* contains moderately thermophilic species. Members of *Peptostreptococcaceae* are found in different habitats including human body, manure, soil, and sediments. Species of *Peptostreptococcus* and *Filifactor* are components of the human oral microbiome. *Tepidibacter* spp. inhabit deep-sea hydrothermal vents. Strains of *Filifactor* are pathogenic.

## Taxonomy: Historical and Current

### Short Description of the Family

Pep.to.strep.to.coc.ca'ce.ae. N.L. n. *Peptostreptococcus* a bacterial genus, the type genus of the family; *-aceae* ending to denote a family; N.L. fem. pl. n. *Peptostreptococcaceae* the family of *Peptostreptococcus* (from *Bergey's Manual*). The emendation of the original description (Ezaki 2009a) is given bellow.

Phylogenetically a member of the order *Clostridiales* (Prévot 1953), phylum Firmicutes. The family contains the type genus *Peptostreptococcus*, and genera *Acetoanaerobium*, *Filifactor*, *Proteocatella*, *Sporacetigenium*, and *Tepidibacter* (see comment below). Morphology of the cells varies from cocci to rods and filaments. Motile by means of peritrichous flagella or nonmotile. Some species form round or ovoid subterminal or terminal endospores. Anaerobic. Fermentative type of metabolism. Utilize proteinaceous substrates and carbohydrates; some species are asaccharolytic. Some species grow on amino acids using Stickland reactions. Catalase-negative or occasionally weakly positive. Mesophilic, moderately thermophilic or phychrotolerant. Neutrophilic. G+C values of DNA range between 24 and 54 mol%. Isolated from human and animal clinical samples, soil, sediments, manure, anaerobic sludge, and deep-sea hydrothermal vents.

Comment: At the time of the original description (Ezaki 2009a) the family contained the type genus *Peptostreptococcus* (Kluyver and van Neil 1936; emended by Ezaki et al. 2001), *Filifactor* (Collins et al. 1994), and *Tepidibacter* (Slobodkin et al. 2003; emended by Tan et al. 2012). Since then, ribosomal RNA sequence databases, SILVA (Pruesse et al. 2007; http://www.arb-silva.de), Ribosomal Database Project II (Cole et al. 2009; http://rdp.cme.msu.edu), and EzTaxon-e (Kim et al. 2012; http://eztaxon-e.ezbiocloud.net) classified three other genera—*Acetoanaerobium* (Sleat et al. 1985), *Sporacetigenium* (Chen et al. 2006), and *Proteocatella* (Pikuta et al. 2009) as members of *Peptostreptococcaceae*. These databases also include to *Peptostreptococcaceae* several validly published *Clostridium* and *Eubacterium* species: *Clostridium bartlettii*, *C. bifermentans*, *C. difficile*, *C. ghoni*, *C. glycolicum*, *C. hiranonis*, *C. irregulare*, *C. litorale*, *C. lituseburense*, *C. mangenotii*, *C. mayombei*, *C. paradoxum*, *C. sordellii*, *C. thermoalcaliphilum* and *Eubacterium acidaminophilum*, *E. tenue*, and *E. yurii*. Inclusion of these microorganisms to *Peptostreptococcaceae* was also proposed by Ludwig et al. 2009. Abovementioned species of

E. Rosenberg et al. (eds.), *The Prokaryotes – Firmicutes and Tenericutes*, DOI 10.1007/978-3-642-30120-9_217,

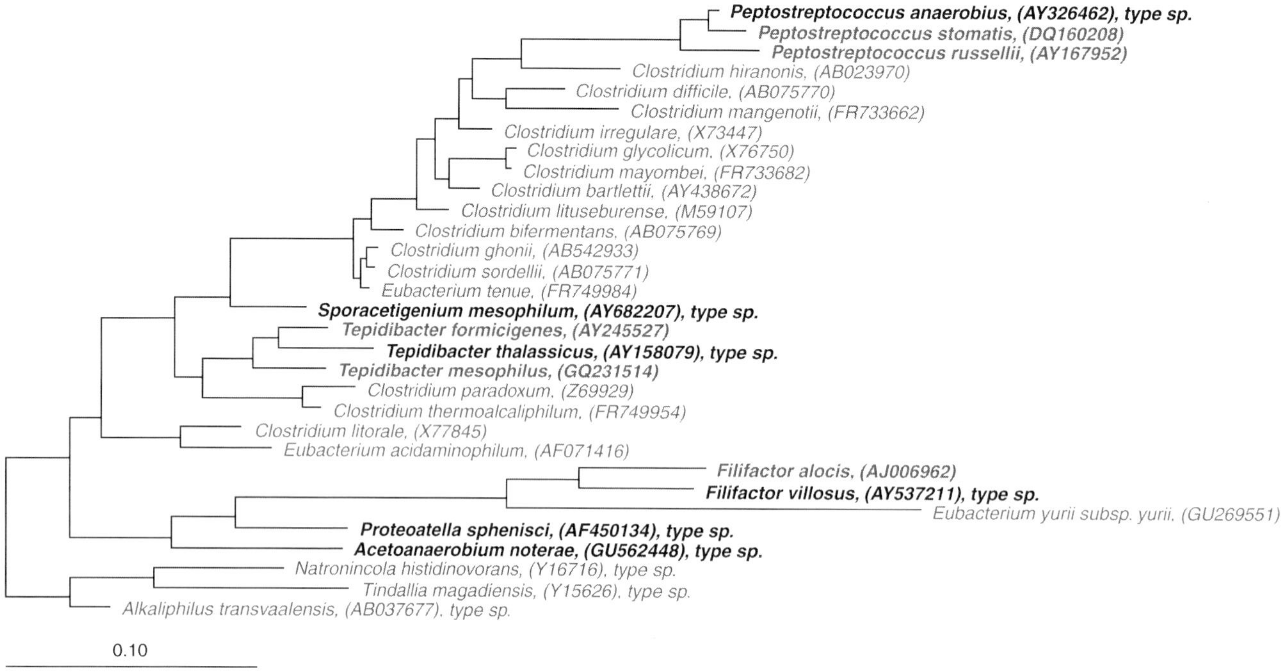


**Fig. 24.1**
**Dendrogram showing the phylogenetic structure of the family *Peptostreptococcaceae* and the closest phylogenetic neighbors**

*Clostridium* and *Eubacterium* need taxonomic revision and will not be covered in this chapter.

### Phylogenetic Structure of the Family and Its Genera

*Peptostreptococcaceae* forms separate, well-defined clade in the order *Clostridiales*. The closest phylogenetic neighbors are genera *Alkaliphilus, Natronincola, and Tindallia* belonging to "Clostridiaceae 2" (LTP nomenclature; http://www.arb-silva.de) (➲ *Fig. 24.1*). Within the family the genera *Acetoanaerobium, Filifactor*, and *Proteocatella* (and *Eubacterium yurii*) form deeply branching cluster; another cluster embrace the genera *Peptostreptococcus, Sporacetigenium*, and *Tepidibacter* (and 14 misclassified *Clostridium* species together with *Eubacterium acidaminophilum and Eubacterium tenue*). The genus *Filifactor* is most distantly related to other members of the family (83–87 % 16S rRNA gene sequence similarity).

## Molecular Analyses

### DNA–DNA Hybridization Studies

DNA–DNA hybridization studies between different species of the family were performed only for two species of the genus *Peptostreptococcus*. Hybridization between *P. anaerobius* NCTC 11460$^T$ and *P. stomatis* strains W2278$^T$ and W3855 was 8 % and 14 %, respectively (Downes and Wade 2006). *Intraspecies* DNA–DNA hybridization between 8 strains of *Filifactor alocis* isolated from cats and the type strain of *F. alocis* ATCC 35896$^T$ isolated from human oral samples was in the range of 77–100 % (Love et al. 1987). For all others members of the family DNA–DNA hybridization studies were not carried out due to low values of 16S rRNA gene sequence similarity between the species (*P. russellii* and other species of *Peptostreptococcus*—93–96 %, *Filifactor* species—93 %, *Tepidibacter* species—94–95 %). *Genera Acetoanaerobium, Proteocatella*, and *Sporacetigenium* are represented by one species each.

### Genome Comparison

The complete genomes of three species of *Peptostreptococcaceae* have been sequenced as the reference genomes for the Human Microbiome Project (Human Microbiome Jumpstart Reference Strains Consortium et al. 2010). The genome of *Peptostreptococcus anaerobius* 653-L (not a type strain of the species) (GenBank: ADJN00000000.1) has a size of 2.08 Mb, contains 1,930 genes (1,871 protein-coding genes), and its G+C content of DNA is 35.9 mol%. *Peptostreptococcus stomatis* DSM 17687$^T$ (GenBank: ADGQ00000000.1) has the genome size of 1.99 Mb with 1,659 genes, (1,600 protein-coding genes). The genome of *Filifactor alocis* ATCC 35896$^T$ (GenBank: CP002390.1) has a size of 1.93 Mb, contains 1,709 genes (1,641 protein-coding genes), and the mol% G+C of DNA is 35.4 %.

## Phenotypic Analyses

The family *Peptostreptococcaceae* is morphologically diverse and includes cocci, rods, or long filaments. Most strains have the

**Table 24.1**
**Morphological and chemotaxonomic characteristics of genera of *Peptostreptococcaceae***

| Characteristic | *Peptostreptococcus*[a–c] | *Acetoanaerobium*[d] | *Filifactor*[e–g] | *Proteocatella*[h] | *Sporacetigenium*[i] | *Tepidibacter*[j–l] |
|---|---|---|---|---|---|---|
| Morphology | Cocci in pairs, irregular masses, or chains | Straight rods | Rods with rounded ends or filaments | Straight rods | Rods | Straight to slightly curved rods |
| Gram-stain | positive | negative | variable | positive | positive | positive |
| Motility (flagellation) | – | + (peritrichous flagella) | – | +(peritrichous flagella) | + (peritrichous flagella) | + (peritrichous flagella) |
| Spore formation | – | – | –/+ | + | + | + |
| Growth temperature (optimum) (°C) | 25–45 (37) | (37) | 30–45 (37) | 2–37 (29) | 20–42 (37–39) | 10–60 (28–50) |
| pH range (optimum) | NR | 6.6–8.4 (7.6) | NR | 6.7–9.7 (8.3) | 6.0–9.5 (7.5) | 4.8–8.9 (6.0–7.3) |
| Fermentation of sugars | w | + | – | – | + | + |
| Peptidoglycan (position 3, bridge) | Lys, D-Asp | NR | Orn, D-Asp | NR | *meso*-DAP | NR |
| Major fatty acids | iso-$C_{14:0}$, iso-$C_{16:0}$, $C_{16:0}$, $C_{18:1}\ \omega 9c$ | NR | NR | $C_{14:0}$, $C_{16:0}$ | $C_{14:0}$, $C_{16:1}\ \omega 7c$, $C_{16:0}$ | iso-$C_{15:0}$ |
| G+C content | 34–36 | 37 | 34 | 39.5 | 53.9 | 24–30 |

Symbols and abbreviations: + positive, – negative, *w* weakly positive, *NR* not reported
Data from: [a]Ezaki 2009b; [b]Downes and Wade 2006; [c]Whitehead et al. 2011; [d]Sleat et al. 1985; [e]Love et al. 1979; [f]Cato et al. 1985; [g]Jalava and Eerola 1999; [h]Pikuta et al. 2009; [i]Chen et al. 2006; [j]Slobodkin et al. 2003; [k]Urios et al. 2004; [l]Tan et al. 2012

diameter of the cells in the range of 0.6–0.9 μm, irrespective of the cell shape. The family contains spore-forming as well as non-spore-forming species. Most species are Gram-stain-positive. The majority of the strains are obligately anaerobic chemo-organotrophs. The key metabolic property for all members of the family is anaerobic growth via fermentation of proteinaceous substrates and some carbohydrates. *Peptostreptococcaceae* includes mesophilic, psychrotolerant, and moderate thermophilc species. All strains grow at pH close to 7.0. The main morphological and chemotaxonomic characteristics of genera of *Peptostreptococcaceae* are listed in ➲ *Table 24.1*.

## *Peptostreptococcus* Kluyver and van Niel 1936, emend. Ezaki, Kawamura, Li, Li, Zhao and Shu 2001

Pep.to.strep.to.co'ccus. Gr. adj. *peptos* cooked, digested; N.L. masc. n. *Streptococcus* a bacterial genus name; N.L. masc. n. *Peptostreptococcus* the digesting streptococcus.

Cells of all three species of *Peptostreptococcus* are non-spore-forming Gram-stain-positive cocci, 0.8–1.0 μm in diameter. Cells may occur in pairs, irregular masses, or chains. Colonies of *P. anaerobius* are convex, 2.2–4.0 mm in diameter. Colonies of *P. russellii* are 2.0–3.0 mm in diameter, convex, opaque, smooth, and whitish in color. *P. stomatis* forms circular, high convex to pyramidal, opaque, shiny, and cream to off-white in color colonies, 0.8–1.8 mm in diameter, with a narrow, gray, peripheral outer ring. Optimal cultivation temperature for all species is 37 °C (Downes and Wade 2006; Ezaki 2009b; Whitehead et al. 2011). Members of *Peptostreptococcus* are obligately anaerobic chemo-organotrophs and metabolize peptone and amino acids to acetic, butyric, isobutyric, caproic, and isocaproic acid (Holdeman Moore et al. 1986; Ezaki et al. 2006). Carbohydrates are weakly fermented by all strains (➲ *Table 24.2*). Urea is not hydrolyzed and indole is not produced by all species. *P. russellii* produces prodigious amounts of ammonia (>40 mM) from various nitrogen sources (Tryptone and Casamino acids) (Whitehead et al. 2011). Diamino acid of peptidoglycan is lysine (Ezaki 2009b). Comparison of other selected characteristics of species of the genus *Peptostreptococcus* is given in ➲ *Table 24.2*.

## *Acetoanaerobium* Sleat, Mah and Robinson 1985

A.ce.to.an.ae. ro'bi.um. L. n. *acetum* vinegar; Gr. pref. *an* not; Gr. n. *aer* air; Gr. n. *bios* life; M. L. neut. n. *Acetoanaerobium* vinegar anaerobe

The genus *Acetoanaerobium* includes so far only one species *A. noterae*—represented by one strain NOT-3$^T$ (= ATCC 35199$^T$). Cell of *A. noterae* are straight rods, 0.8-μm wide and 1.0–5.0-μm long, motile with three or four peritrichous flagella. Cells stain gram-negative; the cell wall is atypical and is composed of two distinct layers, a darker inner layer and lighter outer layer.

**Table 24.2**
**Comparison of selected characteristics of species of the genus *Peptostreptococcus***

| Characteristic | *P. anaerobius* DSM 2949[T a, b] | *P. russellii* DSM 23041[T b] | *P. stomatis* DSM 17678[T b, c] |
|---|---|---|---|
| Temperature range (°C) | NR | 25–45 | NR |
| Carbohydrates weakly fermented | Glucose, mannose | Glucose | Glucose, fructose, maltose |
| Carbohydrates not fermented | Arabinose, lactose, mannitol, raffinose, sorbitol, sucrose | Mannose, raffinose | Arabinose, cellobiose, lactose, mannitol, mannose, melezitose, melibiose, raffinose, rhamnose, ribose, salicin, sorbitol, sucrose, trehalose |
| Nitrate reduction | NR | – | – |
| Catalase | NR | – | – |
| α-glucosidase | + | – | + |
| Proline arylamidase | + | + | – |
| Fermentation products | From PYG: Caproate, acetate, butyrate, iso-butyrate, iso-valerate | From glucose: Acetate, lactate, formate, citrate | From PYG: Acetate, Iso-caproate, iso-butyrate, iso-valerate |
| Peptidoglycan (position 3, bridge) | Lys, D-Asp | Lys, D-Asp | NR |
| Whole-cell-wall sugars | NR | Glucose, xylose, and traces of mannose | NR |
| Respiratory quinones | NR | Not detected | NR |
| Polar lipids | NR | Aminoglycolipid, diphosphatidylglycerol, glycolipids, phosphatidylglycerol and phospholipids | NR |
| Predominant cellular fatty acids (>10 %) | $C_{16:0}$, $C_{18:1}$ ω9c | iso-$C_{16:0}$ | iso-$C_{14:0}$, iso-$C_{16:0}$ |
| DNA G+C content (mol%) | 34–36 | 35.6 | 36 |

Symbols and abbreviations: + positive, – negative, *NR* not reported. Fermentation products: starting with upper-case letter = major product, starting with lower-case letter = minor product
Data from: [a]Ezaki 2009b; [b]Whitehead et al. 2011; [c]Downes and Wade 2006

Colonies are rhizoid, opaque, and granular. Young colonies are white, but older colonies are brownish and up to 2 cm in diameter after 1 month of incubation. Obligately anaerobic. Yeast extract, maltose, and glucose are used for heterotrophic growth. Vitamins are not required. Compounds not supporting growth include arabinose, rhamnose, ribose, xylose, fructose, galactose, cellobiose, lactose, mannose, sucrose, melezitose, trehalose, erythritol, adonitol, arabitol, dulcitol, inositol, mannitol, sorbitol, formate, acetate, pyruvate, lactate, malate, fumarate, succinate, citrate, glutamate, methylamine, trimethylamine, and methanol. The strain produces acetate, propionate, iso-butyrate, butyrate, and iso-valerate (and little or no $H_2$) during growth on yeast extract alone. When either glucose or maltose serves as a substrate, acetate is the only fermentation product. The main physiological feature that distinguishes *Acetoanaerobium* from other *Peptostreptococcaceae* is its capacity for lithotrophic acetogenic growth with $H_2$:$CO_2$. Yeast extract is required for growth and $H_2$ utilization. Growth on yeast extract and $H_2$:$CO_2$ is biphasic, with an initial rapid growth phase independent of the presence of $H_2$. This is followed by $H_2$-dependent acetate production during the second slower growth phase (Sleat et al. 1985). Temperature and pH ranges and optima for growth and G+C content of DNA are presented in ➋ *Table 24.1.*

## *Filifactor* Collins, Lawson, Willems, Cordoba, Fernández-Garayzábal, Garcia, Cai, Hippe and Farrow 1994

Fi.li.fac'tor. L. n. *filum* thread; L. masc. n. *factor* maker; N.L. masc. n. *Filifactor* thread-maker.

Cells of *Filifactor* are rods 0.4–0.7 μm in diameter and 1.5–7.0 μm in length with rounded to tapered ends that occur singly, in pairs, and occasionally in chains or filaments. *F. villosus* can form filaments up to 30 μm in length. Cells of *F. villosus* have variable Gram-staining properties—positive in young (24–48 h) cultures and negative in old (7 d) cultures. However, thin-section electron microscopy studies have shown that the

**Table 24.3**
**Comparison of selected characteristics of species of the genus *Filifactor***

| Characteristic | *F. villosus* NCTC 11220[T a–c] | *F. alocis* ATCC 35896[T b–d] |
|---|---|---|
| Gram-stain | variable | negative |
| Spore formation | + | – |
| Temperature range (°C) | NR | 30–45 |
| Optimum temperature (°C) | 37 | 37 |
| Utilization of pyruvate | + | – |
| Fermentation products | Acetate, iso-butyrate, butyrate, iso-valerate, formate and traces of lactate, methylmalonate, succinate, and $H_2$[e] | Butyrate, acetate, $H_2$[f] |
| DNA G+C content (mol%) | NR | 34 |

Symbols and abbreviations: + positive, – negative, *NR* not reported
Data from: [a]Love et al. 1979; [b]Cato et al. 1985, [c]1986; [d]Jalava and Eerola 2009
[e]From cooked meat-carbohydrate and peptone-yeast extract cultures supplemented with 5 % horse serum
[f]From PYG broth

structure of the cell wall and the mode of the division are consistent with the Gram-positive bacteria. Cells of *F. alocis* stain Gram-negative. Cells of both *Filifactor* species are nonmotile and do not have flagella, but some of the strains have been shown to have twitching or end-over-end type of motility. *F. villosus* forms oval subterminal endospores that caused a slight distention of the sporangium. Formation of endospores by *F. alocis* was not observed. Colonies of *Filifactor* are small (0.5–1.0 mm) and nonhemolytic. Members of *Filifactor* are obligately anaerobic chemo-organotrophs. All strains produce acetate, butyrate, and $H_2$ as fermentation products (➲ *Table 24.3*). Both *Filifactor* species do not utilize sugars; no acid is produced from esculin, fructose, glucose, maltose, mannitol, mannose, melibiose, ribose, sucrose, or xylose. Threonine and lactate are not utilized. Nitrate is not reduced. Indole, lecithinase and lipase are not produced. Esculin is not hydrolyzed. Milk and meat are not digested (Love et al. 1979; Cato et al. 1985, 1986; Jalava and Eerola 2009). Peptidoglycan (position 3, bridge) is Orn, D-Asp (Ezaki 2009a). Comparison of other selected characteristics of species of the genus *Filifactor* is given in ➲ *Table 24.3.*

## *Proteocatella* Pikuta, Hoover, Marsic, Whitman, Lupa, Tang, and Krader 2009

Pro.te.o.ca.tel'la. N.L. n. *proteinum* protein; N.L. pref. *proteo-* prefix referring to protein used in compound words; L. fem. n. *catella* small chain; N.L. fem. n. *Proteocatella* a small chain using proteins.

The genus *Proteocatella* includes so far only one species, *P. sphenisci*, represented by one strain PPP2[T] (=ATCC BAA-755[T]). Cells of *P. sphenisci* are flexible, motile rods, 0.7–0.8 × 3.0–5.0 μm, that tend to form long chains. Multiplies by fission, sometimes unequally with the formation of terminally round mini-cells. Motile by means of flagella. Forms spherical endospores in non-swollen sporangia. Cell wall has Gram-positive structure. Colonies are creamy yellow and rounded lens-shaped with a diameter of 1–2 mm. Obligate anaerobe with fermentative type of metabolism. Catalase-negative. Grows with peptone, bacto-tryptone, Casamino acids, oxalate, starch, chitin, and yeast extract. The strain grows on sodium oxalate only in medium supplemented with selenium as a trace element; cell morphology on this substrate was atypical, with a tendency for the cells to appear swollen and with a hexagonal crystalline shape. No growth is observed on formate, acetate, lactate, pyruvate, propionate, butyrate, citrate, methanol, ethanol, glycerol, acetone, D-mannitol, D-glucose, D-fructose, D-ribose, trehalose, D-arabinose, maltose, D-mannose, lactose, sucrose, cellobiose, pectin, N-acetylglucosamine, urea, trimethylamine, triethylamine, or betaine. Separate amino acids on mineral medium supplemented with yeast extract (0.1 g/l) do not support growth. The Stickland reaction is negative. End products of peptone fermentation are acetate, butyrate, ethanol, and minor amounts of hydrogen and carbon dioxide. NaCl range for growth is 0–4 % (w/v); optimal growth at 0.5 % (w/v) NaCl. Alkalitolerant, psychrotolerant mesophile (Pikuta et al. 2009). Temperature and pH ranges and optima for growth, major cellular fatty acids and G+C content of DNA are presented in ➲ *Table 24.1.*

## *Sporacetigenium* Chen, Song and Dong 2006

Spo.ra.ce.ti.ge'ni.um. Gr. n. *spora* seed; L. n. *acetum* vinegar; Gr. v. *gennao* to produce; N.L. neut. n. *Sporacetigenium* spored vinegar (acetate) producer.

The genus *Sporacetigenium* includes so far only one species, *S. mesophilum*, represented by two strains. The physiological characteristics of the type strain ZLJ115[T] (=DSM 16796[T]) are described below. Cells of *S. mesophilum* are rods 0.9–1.0 × 3.6–7.3 μm in size, occurring singly or in short chains and motile by peritrichous flagella. The cells have Gram-positive wall structure; the peptidoglycan of the cell wall contains meso-DAP. Ovoid endospores are formed in the ends of cells, resulting in swollen cells. Colonies on PYG agar are milk white, smooth, circular, entire and translucent, slightly convex, and reaches 1 mm in diameter after cultivation at 37 °C for 48 h. Obligately anaerobic and chemo-organotrophic. Oxidase and catalase are not produced. Acid is produced from D-glucose, D-fructose, L-arabinose, D-xylose, and D-maltose. D-Galactose, D-mannose, cellobiose, sucrose, rhamnose, trehalose, melibiose, melezitose, and raffinose are fermented weakly. Acid is not produced from sorbose, starch, inulin, glycogen, salicin, amygdalin, glycerol, adonitol, dulcitol, erythritol, inositol, mannitol, or sorbitol. Fermentation of D-lactose and ribose is variable. The following compounds are not utilized: methanol, ethanol, 1-propanol, citrate, fumarate, malate, succinate, malonic acid, hippurate, sodium gluconate, butanedioic acid, b-hydroxybutyric acid, phenylacetic acid, cellulose, and xylan. The major fermentation products from glucose are acetate, ethanol, hydrogen, and carbon dioxide. Peptone may serve as nitrogen source. Starch and aesculin are hydrolyzed, whereas gelatin is not. Milk is not curdled. Urease, lecithinase, and lipase are not produced. Methyl red test is positive while Voges–Proskauer test is negative. Nitrate, sulfate, and sulfur are not reduced. $H_2S$ and $NH_3$ are produced from PYG. The strain could grow in the presence of 0–4 % (w/v) NaCl (Chen et al. 2006). Temperature and pH ranges and optima for growth, major cellular fatty acids and G+C content of DNA are presented in ➲ *Table 24.1.*

## *Tepidibacter* Slobodkin, Tourova, Kostrikina, Chernyh, Bonch-Osmolovskaya, Jeanthon and Jones 2003, emend. Tan, Wu, Zhang, Wu and Zhu 2012

Te.pi.di.bac'ter. L. adj. *tepidus* warm; N.L. *bacter* masc. equivalent of Gr. neut. dim. n. *bakterion* rod; N.L. masc. n. *Tepidibacter* a warm rod.

Cells of *Tepidibacter* are straight to slightly curved rods 0.7–1.6 μm in diameter and 2.3–6.0 μm in length occurring singly, in pairs, or in short chains. Cells exhibit tumbling motility due to peritrichous flagellation. All three species of *Tepidibacter* have Gram-positive type cell wall and form round or ovoid refractile terminal or subterminal endospores. In the late-exponential phase of growth, up to 30 % of the cells of *T. thalassicus* contain spores. All *Tepidibacter* species form colonies in anaerobic agar. *T. thalassicus* and *T. formicigenes* are moderate thermophiles with temperature range for growth 30–60 °C; *T. mesophilus* is a mesophile with upper temperature limit of 40 °C. *T. thalassicus* and *T. formicigenes* grow at marine salinity; *T. mesophilus* shows the best growth with 0.5–1.0 % of NaCl but can tolerate up to 9 % of NaCl or sea salts. Members of *Tepidibacter* are anaerobic or aerotolerant (*T. mesophilus*) chemo-organotrophs. They ferment a number of proteinaceous substrates and carbohydrates and are able to perform the Stickland reaction. The best growth of *T. thalassicus* can be obtained on complex proteinaceous substrates such as tryptone, casein, and peptone or on starch. Carbohydrates, in the presence of yeast extract, slightly stimulate growth of *T. thalassicus*. Growth of *T. formicigenes* and *T. mesophilus* on sugars is more efficient. Differences in substrates utilization by *Tepidibacter* species are shown in ➲ *Table 24.4*. *T. thalassicus* and *T. mesophilus* reduce elemental sulfur to hydrogen sulfide, but sulfur reduction does not stimulate growth. All strains of *Tepidibacter* do not use nitrate, nitrite, Fe(III), sulfate, sulfite, and thiosulfate as electron acceptors. Oxidase and catalase activities are negative for all strains. All *Tepidibacter* species produce ethanol and acetate from glucose; in addition *T. thalassicus* forms moderate amounts of $H_2$ and $CO_2$, and *T. formicigenes* produces formate as a main fermentation product (Slobodkin et al. 2003; Slobodkin 2009; Urios et al. 2004; Tan et al. 2012). The major cellular fatty acid in the three species of *Tepidibacter* is iso-$C_{15:0}$, it constitutes 97.0, 77.7, and 51.8 % of the total fatty acids in *T. thalassicus*, *T. formicigenes*, and *T. mesophilus*, respectively (Tan et al. 2012). Differentiating characteristics of species of the genus *Tepidibacter* is given in ➲ *Table 24.4.*

## Isolation, Enrichment, and Maintenance Procedures

Members of the family *Peptostreptococcaceae* can be enriched and isolated at anaerobic conditions in media that are rich in proteinaceous substances. All species have complex growth requirements, which may include vitamins, cofactors, and amino acids.

*Peptostreptococcus anaerobius* and *P. russellii* grow well on anaerobic chopped meat medium (DSM medium 78; http://dsmz.de). *P. stomatis* strains can be cultivated on fastidious anaerobe agar supplemented with 5 % horse blood in the atmosphere of 80 % $N_2$, 10 % $H_2$, 10 %, $CO_2$ or on anaerobic PYG medium supplemented with glucose (4.0 g/l), cellobiose (1.0 g/l), maltose (1.0 g/l), and soluble starch (1.0 g/l) (DSM medium 104 modified) (Downes and Wade 2006; Ezaki 2009b). Strain of *P. stomatis* was recovered from subgingival plaque using single-cell long-term cultivation method to minimize the effect of fast-growing microorganisms (Sizova et al. 2012). Colonies of *P. russellii* can be obtained on agar plates incubated in an anaerobic chamber in an atmosphere of $CO_2$:$H_2$ (96:4) on the medium containing buffer, salts, yeast extract (0.3 %), Bacto-Tryptone (1 %), and Casamino Acids (1 %) (Whitehead et al. 2011). All strains of *Peptostreptococcus* also grow on brain heart infusion medium under anaerobic conditions. In most cases, the temperature for enrichment, isolation, and cultivation of *Peptostreptococcus* species was 37 °C.

*Acetoanaerobium noterae* was isolated using anaerobic reduced mineral medium with $H_2$:$CO_2$ in the gas phase supplemented with 2.0 g/l of yeast extract. In the complete absence of yeast extract, $H_2$ was not utilized; at least 0.5 g/l was required for sustainable growth. *A. noterae* can also be cultivated without molecular hydrogen, on yeast extract (2.0 g/l) alone, or on maltose or glucose in the presence of 0.5 g/l of yeast extract. Vitamins are not required for the growth of this microorganism (Sleat et al. 1985).

*Filifactor villosus* can be isolated and grown on sheep blood agar plates and brain heart infusion agar plates incubated anaerobically. Cooked meat plus peptic digest of meat broth alone or supplemented with 0.4 % glucose, 0.1 % cellobiose, 0.1 % maltose, and 0.1 % starch can be used for cultivation of pure cultures (Love et al. 1979). Cultures of this microorganism can also be maintained on anaerobic chopped meat medium or chopped meat medium with carbohydrates (DSM medium 78 and DSM medium 110; http://dsmz.de). Fastidious anaerobic agar plates with and without 7 % (w/v) bovine blood have been used for cultivation of pure cultures of *F. alocis* under anaerobic conditions (Jalava and Eerola 1999).

**Table 24.4**
**Differentiating characteristics of the species of the genus *Tepidibacter***

| Characteristic | *T. thalassicus* DSM 15285[T a, c] | *T. formicigenes* DSM 15518[T b, c] | *T. mesophilus* JCM 16806[T c] |
|---|---|---|---|
| Temperature range (°C) | 33–60 | 35–55 | 10–40 |
| Optimum temperature (°C) | 50 | 28–32 | 45 |
| pH range | 4.8–8.5 | 5.0–8.0 | 6.0–8.9 |
| Optimum pH | 6.5–6.8 | 6.0 | 7.3 |
| NaCl concentration (%, w/v) range | 1.5–6.0 | 2.0–6.0 | 0–9.0 |
| Optimum NaCl concentration (%, w/v) | 2.0 | 3.0 | 0.5–1.0 |
| $S^0$ reduction | + | – | + |
| Utilization of | | | |
| Albumin | + | – | – |
| Casein | + | – | + |
| Gelatin | – | + | – |
| Peptone | + | w | + |
| Yeast extract | + | w | + |
| D-fructose | – | – | + |
| D-galactose | – | – | + |
| D-glucose | w | + | + |
| Maltose | w | + | + |
| Mannose | – | w | – |
| D-ribose | – | – | + |
| L-rhamnose | – | – | + |
| Sucrose | – | + | – |
| Trehalose | – | – | + |
| D-xylose | – | – | + |
| L-arginine | w | – | – |
| L-valine | w | + | + |
| D-mannitol | – | + | – |
| Ethanol | – | w | – |
| Pyruvate | w | + | – |
| Fermentation products from glucose | Ethanol, acetate, $H_2$, $CO_2$ | Formate, acetate, ethanol | Acetate, ethanol |
| Predominant cellular fatty acids (>5 %) | iso-$C_{15:0}$ | iso-$C_{15:0}$, $C_{16:0}$ | iso-$C_{15:0}$, $C_{14:0}$, $C_{16:0}$, $C_{16:1}$ cis9 |
| DNA G+C content (mol%) | 24 | 29 | 29.8 |

All strains utilized beef extract, tryptone, starch, DL-alanine plus L-proline, and DL-alanine plus L-glycine
None of the strains used L-arabinose, lactose, DL-alanine, L-glycine, acetate, betaine, butyrate, formate, fumarate, glycerol, lactate, methanol, D-sorbitol, succinate, urea, chitin, filter paper, or olive oil
Symbols and abbreviations: + positive, – negative, *w* weakly positive, *NR* not reported
Data from: [a]Slobodkin et al. 2003; [b]Urios et al. 2004; [c]Tan et al. 2012

*Proteocatella sphenisci* was enriched and isolated in a pure culture at the temperature +2 °C. Anaerobic mineral medium used for isolation included peptone (3 g/l) and yeast extract (0.2 g/l). Initial salinity of enrichments was 30 g/l of NaCl; however, the optimal NaCl concentration for pure culture of *P. sphenisci* was 5 g/l (Pikuta et al. 2009).

Strains of *Sporacetigenium mesophilum* were isolated and routinely cultivated in pre-reduced peptone/yeast extract/glucose medium (Chen et al. 2006). Type strain could be maintained in DSM medium 104b (PYX-medium, http://dsmz.de).

Thermophilic *Tepidibacter* species can be enriched in the temperature range of 45–55 °C in anaerobic medium of marine salinity supplemented with proteinaceous substrates (Slobodkin et al. 2003; Urios et al. 2004). *T. thalassicus, T. formicigenes*, and *T. mesophilus* were isolated in the presence of 0.2–1.0 g/l of yeast extract with casein (10 g/l), peptone (0.5 g/l), or Casamino acids (3 g/l) as a main carbon source, respectively. *T. thalassicus* rapidly hydrolyzes casein (Hammerstein grade) that results in visual disappearance of the casein flocks and may help in the detection of growth. All *Tepidibacter* strains form colonies in 1.5 % (w/v) anaerobic agar, and Hungate roll tube or agar block techniques can be used for the isolation of a single colony (Hungate 1969). Members of the genus *Tepidibacter* may be maintained on the medium of Slobodkin et al. (2003) with peptone or casein as a substrate or on the glucose/yeast extract/peptone medium of Urios et al. (2004). All three species of *Tepidibacter* show a good growth on DSM medium 985 with 1 % peptone (Tan et al. 2012; http://dsmz.de). Reproducible growth of *T. thalassicus* and *T. mesophilus* can be obtained in liquid anaerobic medium lacking sulfide as a reducing agent (Slobodkin et al. 2003; Tan et al. 2012). Freeze-drying of the cultures results in good recovery. *T. mesophilus* may be preserved in 25 % glycerol at −80 °C. Liquid cultures of *T. thalassicus* may be stored at +4 °C for 10–12 months without loss of viability.

## Ecology

Members of the family *Peptostreptococcaceae* were isolated from various environments including clinical human and animal samples, manure, soil, marine and terrestrial sediments, and deep-sea hydrothermal vents.

*Peptostreptococcus* species have been found in body and feces of humans and vertebrates. Taking into consideration that the group of the anaerobic Gram-positive cocci was subjected to major revision (Ezaki et al. 2006), it is difficult to determine the exact taxonomic status of the strains referred as the members of the genus *Peptostreptococcus* in reports dated before 1990s because of the absence of 16S rRNA gene sequence data. Numerous studies point to the presence of *Peptostreptococcus* strains in human oral cavity—*P. anaerobius* is a component of the human oral microbiome (Chen et al. 2010; http://www.homd.org); eight strains of *P. stomatis*, including the type strain, have been isolated from oral cavity (Downes and Wade 2006). Different phylotypes of *Peptostreptococcaceae* related to *Peptostreptococcus* species have been detected in oral samples by culture-independent methods (Paster et al. 2001; Munson et al. 2002; Sakamoto et al. 2004; Dewhirst et al. 2010). Strains of *P. anaerobius* have been isolated also from various human non-oral infection and abscesses and from intestine, vagina, and skin of healthy individuals (Downes and Wade 2006; Ezaki 2009b; Human Microbiome Jumpstart Reference Strains Consortium et al. 2010). Besides the human body, species of the genus *Peptostreptococcus* have been found in other vertebrate hosts. The presence of *Peptostreptococcus* spp. in canine oral cavity has been proven by culture-dependent and culture-independent techniques (Elliott et al. 2005; Dewhirst et al. 2012). *Peptostreptococcus* strains have been isolated from the rumen of dairy cows, deer, and sheep (Russell et al. 1988; Paster et al. 1993; Attwood et al. 1998). Most probably in the rumen ecosystem, *Peptostreptococcus* spp., that produce very high concentration of ammonia, but are not able to hydrolyze intact proteins and do not use carbohydrates, occupy a niche of peptide- and amino acid-degrading microorganism and depend on proteolytic bacteria (Attwood et al. 1998). One isolate of *Peptostreptococcus* has been obtained from feces of the mallard duck (Murphy et al. 2005). Seven strains of *P. russellii* have been isolated from a swine manure storage pit located near Peoria, IL, USA. In this habitat, concentration of the cells of *P. russellii* was at least $10^8$ cells per ml and constituted approximately 0.1–1 % of the culturable bacterial population present in the swine manure samples (Whitehead et al. 2011).

The type strain of *Acetoanaerobium noterae* has been isolated from sediment of the Notera 3 oil exploration drilling site in the Hula swamp area of Galilee, Israel. The most probable number analysis of the drilling site sample yielded $1.75 \times 10^5$ $H_2$-oxidizing acetogens per gram (wet weight) (Sleat et al. 1985). Uncultured *Acetoanaerobium* clones (>98 % 16S rRNA gene sequence similarity with the type strain, sequence length >1,300 bp, retrieved from NCBI databases using BLAST, http://blast.ncbi.nlm.nih.gov) have been detected in production waters and sewage of oil reservoirs (accession numbers AY570564, DQ011249), wastewater treatment systems (FJ167476, AF234746, HE576030), and dechorinating microbial consortia (AJ488068, GQ377124).

Species of the genus *Filifactor* so far have been found only in human and animal samples. Eleven strains of *F. villosus* including the type strain have been isolated from subcutaneous abscesses of cats (Love et al. 1979). *F. alocis* is associated with oral cavity, and a large number of strains of this microorganism have been isolated from human gingival sulcus of patients with gingivitis or periodontitis, from oral cavities of cats, and from soft tissue infections of cats caused by contamination from oral cavities (Cato et al 1985; Love et al. 1987; Jalava and Eerola. 2009). *Filifactor* spp. are the components of the human and canine microbiomes where they have been detected by culture-independent methods (Dewhirst et al. 2010, 2012; Kong et al. 2012).

The type strain of *Proteocatella sphenisci* has been isolated from a sample of guano of the Magellanic penguin (*Spheniscus magellanicus*) in Chilean Patagonia. The physiological characteristics of *P. sphenisci*—tolerance to low temperature (down to 2 °C), high pH, and marine concentrations of NaCl—reflect the environmental conditions of the habitat. Magellanic penguins are endemic to the southern tip of South America, a region with a very cold climate. The ability of the strain to use exclusively products of proteolysis and oxalate (but not sugars) is probably due to the restricted diet of these penguins that feed on marine fish and crustaceans (Pikuta et al. 2009). Three different 16S rRNA gene sequences belonging to *Proteocatella* have been found in canine oral microbiome (Dewhirst et al. 2012) and human skin microbiome (GenBank accession number HM266866) (Kong et al. 2012). Other uncultured *Proteocatella* clones have been detected in various wastewater treatment systems (HM467987, FJ645707, CU925306), in river estuary of Northern Taiwan (DQ234248), and in a drinking water reservoir in Greece (GQ340217) (>98 % 16S rRNA gene sequence similarity with the type strain, sequence length >1,300 bp, retrieved from NCBI databases using BLAST, http://blast.ncbi.nlm.nih.gov).

Both strains of *Sporacetigenium mesophilum* have been isolated from the sludge of an anaerobic digester treating municipal solid waste and sewage in Zhangzhou city, Fujian province, PR China (Chen et al. 2006). Uncultured clones of *Sporacetigenium*

(>98 % 16S rRNA gene sequence similarity with the type strain, sequence length >1,300 bp, retrieved from NCBI databases using BLAST) have been detected in such different habitats such as anaerobic zones of Tinto River (JQ815621), shallow-sea hydrothermal vent Tutum Bay, Papua New Guinea (JN881597), alkaline lake Alchichica, Mexico (JN825560), tallgrass prairie soil (EU134685), and rhizosphere of reed (AB240265).

Thermophilic strains of *Tepidibacter* inhabit deep-sea hydrothermal vents. The type strain of *T. thalassicus* has been isolated from the outer wall of a actively venting hydrothermal sulfidic chimney-like deposit ("black smoker") covered with the polychetous annelid *Alvinella* spp. (13° N to the East-Pacific Rise, hydrothermal site Genesis, depth 2,650 m) (Slobodkin et al. 2003). Strain NS55-A that is closely phylogenetically related to *T. thalassicus* (16S rRNA gene sequence similarity—98.9 %) have been obtained from black exterior surface layer of hydrothermal chimney structure (North Big Chimney, Iheya North field in the Mid-Okinawa Trough, depth ca. 1,000 m) (Nakagawa et al. 2005). *T. formicigenes* has been isolated from hydrothermal fluid (the Menez-Gwen hydrothermal site, Mid-Atlantic Ridge, 37° 51′ N 31° 31′ W, depth 800–1,000 m) (Urios et al. 2004). Location of these vents, two of which are in different parts of Pacific Ocean and one is in Atlantic Ocean, suggests wide geographical distribution of *Tepidibacter* species in marine hydrothermal environments where they probably function as decomposers of organic matter produced by deep-sea biota. *Tepidibacter* spp. also inhabit terrestrial geothermal environments; *T. formicigenes* strain JB2 (99 % 16S rRNA gene sequence similarity with the type strain of *T. formicigenes*) has been obtained from Tunisian hot spring with salinity 1.8–2.0 % (Sayeh et al. 2010). Mesophilic representatives of the genus have been found in cold marine sediments and soils. Strain UXO3-5 (96.8–97.3 % 16S rRNA gene sequence similarity with *T. thalassicus* and *T. formicigenes*) has been isolated from marine sediments in unexploded ordnance disposal sites (800-m offshore Oahu Island, Hawaii, depth 10–21 m). This microorganism, obtained in pure culture under mesophilic conditions (21 °C), accounts for 4.5 % of total anaerobes in sediment (Zhao et al. 2007). *T. mesophilus* has been isolated from the completely different habitat—the soil polluted by crude oil (the Karamay Oil Field, 45° 36′ N 84° 57′ E, northwestern China). Ecological functions of *T. mesophilus* are currently unknown, but the aerotolerance of this strain suggests its adaptation to soil environments (Tan et al. 2012). Uncultured *Tepidibacter* clones (>95 % 16S rRNA gene sequence similarity with the type strains of *T. thalassicus, T. formicigenes*, and *T. mesophilus*, sequence length >1,300 bp, retrieved from NCBI databases using BLAST, http://blast.ncbi.nlm.nih.gov) have been detected in environmental samples and enrichment cultures obtained from cold and hydrothermal marine ecosystems: seaweed bed associated with marine hot springs on East Coast of Kalianda Island, Indonesia (accession number JQ670702); coastal soil of Gulf of Khambhat, India (JX240907); superficial sediments of Milazzo Harbor, Italy (AJ810557); and polluted coastal seawater in Tunisia (CU914830 to CU914837).

## Pathogenicity: Clinical Relevance

Among the members of the family *Peptostreptococcaceae* pathogenicity are definitely shown only for the species of the genus *Filifactor*. Strains of this genus have been isolated from human gingival sulcus of patients with gingivitis or periodontitis, from oral cavities of cats, and from subcutaneous wound abscesses of cats; therefore, a pathogenic role of *Filifactor* in mixed anaerobic infections was suggested (Cato et al. 1985; Love et al. 1979, 1987). Association of *F. alocis* with periodontal diseases was also confirmed by culture-independent studies (Kumar et al. 2005; Siqueira and Rocas 2003). *F. alocis* is involved in the formation of periodontal biofilms in patients suffering from generalized aggressive periodontitis and chronic periodontitis and can be considered an excellent marker organism for periodontal disease. *F. alocis* predominantly colonized apical parts of the pocket in close proximity to the soft tissues and was involved in numerous structures that constitute characteristic architectural features of subgingival periodontal biofilms (Schlafer et al. 2010). *F. alocis* has virulence attributes that can enhance its persistence under oxidative stress conditions and mediate invasion of epithelial cells by *Porphyromonas gingivalis* (Aruni et al. 2011). Recently, the pathogenic mechanisms of *F. alocis* in periodontal diseases have been investigated. When infected with *F. alocis*, primary cultures of gingival epithelial cells (GECs) stimulate the secretion of the pro-inflammatory cytokines interleukin-1β, interleukin-6, and tumor necrosis factor-α. *F. alocis* also induced apoptosis in GECs through pathways that involved caspase-3 but not caspase-9. Apoptosis was coincident with inhibition of mitogen-activated protein kinase kinase activation (Moffatt et al. 2011).

Pathogenicity of the members of the genus *Peptostreptococcus* currently is difficult to assess. In the Internet-available medical literature, there are about 2,000 references about involvement of *Peptostreptococcus* in clinical cases. However, in the majority of these studies, the data on 16S rRNA gene sequence are not provided, so it is impossible to determine if the authors are dealing with the species of the genus *Peptostreptococcus sensu stricto* or with the other Gram-positive anaerobic cocci. Information about clinical relevance and pathogenicity of Gram-positive anaerobic cocci before taxonomic revision of this group is summarized in review by Murdoch (1998). Representatives of the genus *Peptostreptococcus* are frequently isolated from clinical samples of healthy and sick individuals (see section "➋ Ecology"). Strains of *P. anaerobius* have been isolated from leg ulcer, urinary tract infection, ankle wound, buttock abscess, and vaginal infection (Downes and Wade 2006). *P. anaerobius* can cause primary sternal osteomyelitis (Chen et al. 2012). Strains of *P. stomatis* have been isolated from dento-alveolar abscesses, endodontic infections, a periodontal pocket, and from a pericoronal infection (Downes and Wade 2006). Recent culture-independent studies show that acute noma disease (gangrenous disease that leads to severe disfigurement of the face) and necrotizing gingivitis are associated with large increase in counts of members of the *Peptostreptococcus* genus (Bolivar et al. 2012). On the other hand, species of *Peptostreptococcus* are a part of the normal oral and vaginal microflora (Zhou et al. 2004; Aas et al. 2005).

**Table 24.5**
**Antibiotic sensitivity of the members of the family *Peptostreptococcaceae***

| Microorganism | Number of strains tested | Sensitive | Resistant |
|---|---|---|---|
| *Peptostreptococcus anaerobius*[a] | 9 | Amoxicillin-clavulanic acid (0.12), cefoxitin (0.25), ciprofloxacin (0.5), clindamycin (0.06), imipenem (0.03), metronidazole (0.06), penicillin G (0.03), piperacillin-tazobactam (0.5), trovafloxacin (0.06) | NR |
| *Acetoanaerobium noterae*[b] | 1, the type strain | Cephalosporin, chloramphenicol, cycloserine, erythromycin, penicillin, (all at 100) | NR |
| *Filifactor alocis*[c] | 20 strains, including the type strain | Chloramphenicol (12), clindamycin (1.6), erythromycin (3), tetracycline (6) | Penicillin (2)[d] |
| *Filifactor villosus*[e] | 1, the type strain | Amoxycillin (2.5), carbenicillin (100), chloramphenicol (12), doxycycline (6), erythromycin (3), penicillin (2) | NR |
| *Proteocatella sphenisci*[f] | 1, the type strain | Gentamicin, kanamycin, rifampicin, tetracycline, vancomycin (250), chloramphenicol (125) | Ampicillin (250) |

Concentrations of antibiotics in parentheses are given in μg/ml except for penicillin for which U/ml is used. For *Peptostreptococcus anaerobius*, minimal inhibitory concentration ($MIC_{50}$) is presented
*NR* not reported
Data from: [a]Bowker et al. 1996; [b]Sleat et al. 1985; [c]Cato et al. 1985; [d]One of 20 strains; [e]Love et al. 1979; [f]Pikuta et al. 2009

There are no reports on pathogenicity and medical relevance of the representatives of *Acetoanaerobium*, *Proteocatella*, *Sporacetigenium*, and *Tepidibacter*. No strains of these genera were found in clinical samples.

Antibiotic sensitivity of the members of genera *Peptostreptococcus*, *Acetoanaerobium*, *Filifactor*, and *Proteocatella* is shown in ➋ *Table 24.5*. Majority of the strains are susceptible to penicillin and chloramphenicol. It is worth to note that *Proteocatella sphenisci* is resistant to ampicillin, a characteristic that is rare in environmental bacterial strains.

## Application

To date, microorganisms belonging to *Peptostreptococcaceae* did not find any application in industrial or bioremediation processes; however, there are a number of reports about biotechnological potential of the members of the family.

*Peptostreptococcus russellii* may play a role in swine manure management. It produces prodigious amounts of ammonia (> 40 mM) from different nitrogen sources (Tryptone and Casamino acids) and belongs to the so-called hyper-ammonia-producing microorganisms. This group of organisms may be important in the digestion and fermentation of proteinaceous material in the manure and the production of ammonia and other compounds (Attwood et al. 1998; Whitehead and Cotta 2004). The fermentation of amino acids such as tryptophan, phenylalanine, and tyrosine may give rise to the production of indole and phenolic compounds (such as skatole), contributing to the foul odors associated with swine facilities (Whitehead et al. 2011).

*Tepidibacter thalassicus* has the potential for immobilization of radionuclides such as technetium(VII). Washed cell suspensions of *T. thalassicus* completely reduced technetium [$^{99}$Tc(VII)], supplied as soluble pertechnetate with molecular hydrogen or peptone as an electron donor, forming highly insoluble Tc(IV)-containing grayish-black precipitate. This capacity can be used during bioremediation of thermally insulated contaminated environments and in biotechnological treatment of the heated nuclear waste streams (Chernyh et al. 2007). Under mesophilic conditions, *Tepidibacter* sp. strain UXO-3-5 can metabolize octahydro-1,3,5,7-tetranitro-1,3,5,7-tetrazocine (HMX), a toxic explosive known to be resistant to biodegradation. This organism plays a significant role in HMX removing in sites of undersea deposition of unexploded ordnance in Hawaii (Zhao et al. 2007).

Production of acetate from molecular hydrogen by *Acetoanaerobium noterae* and formation of $H_2$ during fermentation of glucose by *Sporacetigenium mesophilum* also deserve attention for biotechnological applications (Sleat et al. 1985; Chen et al. 2006).

## References

Aas JA, Paster BJ, Stokes LN, Olsen I, Dewhirst FE (2005) Defining the normal bacterial flora of the oral cavity. J Clin Microbiol 43:5721–5732
Aruni AW, Roy F, Fletcher HM (2011) *Filifactor alocis* has virulence attributes that can enhance its persistence under oxidative stress conditions and mediate invasion of epithelial cells by *Porphyromonas gingivalis*. Infect Immun 79:3872–3886
Attwood GT, Klieve AV, Ouwerkerk D, Patel BK (1998) Ammonia-hyper producing bacteria from New Zealand ruminants. Appl Environ Microbiol 64:1796–1804
Bolivar I, Whiteson K, Stadelmann B, Baratti-Mayer D, Gizard Y, Mombelli A, Pittet D, Schrenze J, Geneva Study Group on Noma (GESNOMA) (2012) Bacterial diversity in oral samples of children in niger with acute noma, acute necrotizing gingivitis, and healthy controls. PLoS Negl Trop Dis 6:e1556
Bowker KE, Wootton M, Holt HA, Reeves DS, MacGowan AP (1996) The in-vitro activity of trovafloxacin and nine other antimicrobials against 413 anaerobic bacteria. J Antimicrob Chemother 38:271–281

Cato EP, Moore LVH, Moore WEC (1985) *Fusobacterium alocis* sp. nov. and *Fusobacterium sulci* sp. nov. from the human gingival sulcus. Int J Syst Bacteriol 35:475–477

Cato EP, George WL, Finegold SM (1986) Genus *Clostridium* Prazmowski 1880, 23AL. In: Sneath PHA, Mair NS, Sharpe ME, Holt JG (eds) Bergey's manual of systematic bacteriology, vol 2. The Williams & Wilkins, Baltimore, pp 1141–1200

Chen S, Song L, Dong X (2006) *Sporacetigenium mesophilum* gen. nov., sp. nov., isolated from an anaerobic digester treating municipal solid waste and sewage. Int J Syst Evol Microbiol 56:721–725

Chen T, W-Han Yu, Izard J, Baranova OV, Lakshmanan A, Dewhirst FE (2010) The human oral microbiome database: a web accessible resource for investigating oral microbe taxonomic and genomic information. Database 2010. doi: 10.1093/database/baq013. Article ID baq013

Chen YL, Tsai SH, Hsu KC, Chen CS, Hsu CW (2012) Primary sternal osteomyelitis due to *Peptostreptococcus anaerobius*. Infection 40:195–197

Chernyh NA, Gavrilov SN, Sorokin VV, German KE, Sergeant C, Simonoff M, Robb F, Slobodkin AI (2007) Characterization of technetium(VII) reduction by cell suspensions of thermophilic *Bacteria* and *Archaea*. Appl Microbiol Biotechnol 76:467–472

Cole JR, Wang Q, Cardenas E, Fish J, Chai B, Farris RJ, Kulam-Syed-Mohideen AS, McGarrell DM, Marsh T, Garrity GM, Tiedje JM (2009) The Ribosomal Database Project: improved alignments and new tools for rRNA analysis. Nucleic Acids Res 37 (Database issue):D141–145

Collins MD, Lawson PA, Willems A, Cordoba JJ, Fernández-Garayzábal J, Garcia P, Cai J, Hippe H, Farrow JAE (1994) The phylogeny of the genus *Clostridium*: proposal of five new genera and eleven new species combinations. Int J Syst Bacteriol 44:812–826

Dewhirst FE, Chen T, Izard J, Paster BJ, Tanner ACR, Yu W-H, Lakshmanan A, Wade WG (2010) The human oral microbiome. J Bacteriol 192:5002–5017

Dewhirst FE, Klein EA, Thompson EC, Blanton JM, Chen T, Milella L, Buckley CM, Davis IJ, Bennett ML, Marshall-Jones ZV (2012) The canine oral microbiome. PLoS One 7:e36067

Downes J, Wade WG (2006) *Peptostreptococcus stomatis* sp. nov., isolated from the human oral cavity. Int J Syst Evol Microbiol 56:751–754

Elliott DR, Wilson M, Buckley CM, Spratt DA (2005) Cultivable oral microbiota of domestic dogs. J Clin Microbiol 43:5470–5476

Ezaki T (2009a) *Peptostreptococcaceae* fam. nov. In: DeVos P, Garrity GM, Jones D, Krieg NR, Ludwig W, Rainey FA, Schleifer K-H, Whitman WB (eds) Bergey's manual of systematic bacteriology, vol 3, 2nd edn. Springer, New York, p 1008

Ezaki T (2009b) Genus *Peptostreptococcus*. In: DeVos P, Garrity GM, Jones D, Krieg NR, Ludwig W, Rainey FA, Schleifer K-H, Whitman WB (eds) Bergey's manual of systematic bacteriology, vol 3, 2nd edn. Springer, New York, pp 1008–1009

Ezaki T, Kawamura Y, Li N, Li ZY, Zhao L, Shu S (2001) Proposal of the genera *Anaerococcus* gen. nov., *Peptoniphilus* gen. nov. and *Gallicola* gen. nov. for members of the genus *Peptostreptococcus*. Int J Syst Evol Microbiol 51:1521–1528

Ezaki T, Li N, Kawamura Y (2006) The anaerobic Gram-positive cocci. In: Dworkin M, Falkow S, Rosenberg E, Schleifer K-H, Stackebrandt E (eds) Prokaryotes, vol 4, 3rd edn. Springer, New York, pp 785–808

Holdeman Moore LV, Johnson JL, Moore WEC (1986) Genus *Peptostreptococcus*. In: Sneath PHA, Mair NS, Holt JG (eds) Bergey's manual of systematic bacteriology, vol 2. The Williams & Wilkins, Baltimore, pp 1083–1092

Human Microbiome Jumpstart Reference Strains Consortium et al (2010) A catalog of reference genomes from the human microbiome. Science 328:994–999

Hungate RE (1969) A roll tube method for cultivation of strict anaerobes. In: Norris JR, Ribbons RW (eds) Methods in microbiology, vol 3B. Academic Press, London, pp 117–132

Jalava J, Eerola E (1999) Phylogenetic analysis of *Fusobacterium alocis* and *Fusobacterium sulci* based on 16S rRNA gene sequences: proposal of *Filifactor alocis* (Cato, Moore and Moore) comb. nov. and *Eubacterium sulci* (Cato, Moore and Moore) comb. nov. Int J Syst Bacteriol 49:1375–1379

Jalava J, Eerola E (2009) Genus *Filifactor*. In: DeVos P, Garrity GM, Jones D, Krieg NR, Ludwig W, Rainey FA, Schleifer K-H, Whitman WB (eds) Bergey's manual of systematic bacteriology, vol 3, 2nd edn. Springer, New York, pp 1009–1013

Kim OS, Cho YJ, Lee K, Yoon SH, Kim M, Na H, Park SC, Jeon YS, Lee JH, Yi H, Won S, Chun J (2012) Introducing EzTaxon-e: a prokaryotic 16S rRNA Gene sequence database with phylotypes that represent uncultured species. Int J Syst Evol Microbiol 62:716–721

Kluyver AJ, van Neil CB (1936) Prospects for a natural classification of bacteria. Zentbl Bacteriol Parasitenkd Infektionskr Hyg Abt II 94:369–403

Kong HH, Oh J, Deming C, Conlan S, Grice EA, Beatson MA, Nomicos E, Polley EC, Komarow HD, Murray PR, Turner ML, Segre JA (2012) Temporal shifts in the skin microbiome associated with disease flares and treatment in children with atopic dermatitis. Genome Res 22:850–859

Kumar PS, Griffen AL, Moeschberger ML, Leys EJ (2005) Identification of candidate periodontal pathogens and beneficial species by quantitative 16S clonal analysis. J Clin Microbiol 43:3944–3955

Love DN, Jones RF, Bailey M (1979) *Clostridium villosum* sp. nov. from subcutaneous abscesses in cats. Int J Syst Bacteriol 29:241–244

Love DN, Cato EP, Johnson JL, Jones RF, Bailey M (1987) Deoxyribonucleic acid hybridization among strains of Fusobacteria isolated from soft tissue infections of cats: comparison with human and animal type strains from oral and other sites. Int J Syst Bacteriol 37:23–26

Ludwig W, Schleifer K-H, Whitman WB (2009) Revised road map to the phylum Firmicutes. In: DeVos P, Garrity GM, Jones D, Krieg NR, Ludwig W, Rainey FA, Schleifer K-H, Whitman WB (eds) Bergey's manual of systematic bacteriology, vol 3, 2nd edn. Springer, New York, pp 1–13

Moffatt CE, Whitmore SE, Griffen AL, Leys EJ, Lamont RJ (2011) *Filifactor alocis* interactions with gingival epithelial cells. Mol Oral Microbiol 26:365–373

Munson MA, Pitt-Ford T, Chong B, Weightman AJ, Wade WG (2002) Molecular and cultural analysis of the microflora associated with endodontic infections. J Dent Res 81:761–766

Murdoch DA (1998) Gram-positive anaerobic cocci. Clin Microbiol Rev 11:81–120

Murphy J, Devane ML, Robson B, Gilpin BJ (2005) Genotypic characterization of bacteria cultured from duck faeces. J Appl Microbiol 99:301–309

Nakagawa S, Takai K, Inagaki F, Chiba H, Ishibashi J, Kataoka S, Hirayama H, Nunoura T, Horikoshi K, Sako Y (2005) Variability in microbial community and venting chemistry in a sediment-hosted backarc hydrothermal system: impacts of subseafloor phase-separation. FEMS Microbiol Ecol 54:141–155

Paster BJ, Russell JB, Yang CMJ, Chow JM, Woese CR, Tanner R (1993) Phylogeny of the ammonia-producing ruminal bacteria *Peptostreptococcus anaerobius*, *Clostridium sticklandii*, and *Clostridium aminophilum* sp. nov. Int J Syst Bacteriol 43:107–110

Paster BJ, Boches SK, Galvin JL, Ericson RE, Lau CN, Levanos VA, Sahasrabudhe A, Dewhirst FE (2001) Bacterial diversity in human subgingival plaque. J Bacteriol 183:3770–3783

Pikuta EV, Hoover RB, Marsic D, Whitman WB, Lupa B, Tang J, Krader P (2009) *Proteocatella sphenisci* gen. nov., sp. nov., a psychrotolerant, spore-forming anaerobe isolated from penguin guano. Int J Syst Evol Microbiol 59:2302–2307

Prévot AR (1953) In: Hauduroy P, Ehringer G, Guillot G, Magrou J, Prévot CAR, Rosset D, Urbain A (eds) Dictionnaire des Bactéries Pathogènes, 2nd edn. Paris, Masson

Pruesse E, Quast C, Knittel K, Fuchs B, Ludwig W, Peplies J, Glöckner FO (2007) SILVA: a comprehensive online resource for quality checked and aligned ribosomal RNA sequence data compatible with ARB. Nuc Acids Res 35:7188–7196

Russell JB, Strobel HJ, Chen G (1988) Enrichment and isolation of a ruminal bacterium with a very high specific activity of ammonia production. Appl Environ Microbiol 54:872–877

Sakamoto M, Huang Y, Ohnishi M, Umeda M, Ishikawa I, Benno Y (2004) Changes in oral microbial profiles after periodontal treatment as determined by molecular analysis of 16S rRNA genes. J Med Microbiol 53:563–571

Sayeh R, Birrien JL, Alain K, Barbier G, Hamdi M, Prieur D (2010) Microbial diversity in Tunisian geothermal springs as detected by molecular and culture-based approaches. Extremophiles 14:501–514

Schlafer S, Riep B, Griffen AL, Petrich A, Hübner J, Berning M, Friedmann A, Göbel UB, Moter A (2010) Filifactor alocis–involvement in periodontal biofilms. BMC Microbiol 10:66

Siqueira JF, Rocas IN (2003) Detection of *Filifactor alocis* in endodontic infections associated with different forms of periradicular diseases. Oral Microbiol Immunol 18:263–265

Sizova MV, Hohmann T, Hazen A, Paster BJ, Halem SR, Murphy CM, Panikov NS, Epstein SS (2012) New approaches for isolation of previously uncultivated oral bacteria. Appl Environ Microbiol 78:194–203

Sleat R, Mah RA, Robinson R (1985) *Acetoanaerobium noterae* gen. nov., sp. nov.: an anaerobic bacterium that forms acetate from $H_2$ and $CO_2$. Int J Syst Bacteriol 35:10–15

Slobodkin A (2009) Genus *Tepidibacter*. In: DeVos P, Garrity GM, Jones D, Krieg NR, Ludwig W, Rainey FA, Schleifer K-H, Whitman WB (eds) Bergey's manual of systematic bacteriology, vol 3, 2nd edn. Springer, New York, pp 1013–1015

Slobodkin AI, Tourova TP, Kostrikina NA, Chernyh NA, Bonch-Osmolovskaya EA, Jeanthon C, Jones BE (2003) *Tepidibacter thalassicus* gen. nov., sp. nov., a novel moderately thermophilic, anaerobic, fermentative bacterium from a deep-sea hydrothermal vent. Int J Syst Evol Microbiol 53:1131–1134

Tan HQ, Wu XY, Zhang XQ, Wu M, Zhu XF (2012) *Tepidibacter mesophilus* sp. nov., a mesophilic fermentative anaerobe isolated from soil polluted by crude oil, and emended description of the genus *Tepidibacter*. Int J Syst Evol Microbiol 62:66–70

Urios L, Cueff V, Pignet P, Barbier G (2004) *Tepidibacter formicigenes* sp. nov., a novel spore-forming bacterium isolated from a Mid-Atlantic Ridge hydrothermal vent. Int J Syst Evol Microbiol 54:439–443

Whitehead TR, Cotta MA (2004) Isolation and identification of hyper-ammonia producing bacteria from swine manure storage pits. Curr Microbiol 48:20–26

Whitehead TR, Cotta MA, Falsen E, Moore E, Lawson PA (2011) *Peptostreptococcus russellii* sp. nov., isolated from a swine-manure storage pit. Int J Syst Evol Microbiol 61:1875–1879

Zhao JS, Manno D, Hawari J (2007) Abundance and diversity of octahydro-1,3,5,7-tetranitro-1,3,5,7-tetrazocine (HMX)-metabolizing bacteria in UXO-contaminated marine sediments. FEMS Microbiol Ecol 59:706–717

Zhou X, Bent SJ, Schneider MG, Davis CC, Islam MR, Forney LJ (2004) Characterization of vaginal microbial communities in adult healthy women using cultivation-independent methods. Microbiology 150:2565–2573

# 25 The Family *Planococcaceae*

*S. Shivaji*[1] · *T. N. R. Srinivas*[2] · *G. S. N. Reddy*[1]
[1]CSIR-Centre for Cellular and Molecular Biology, Hyderabad, India
[2]Regional Centre, CSIR-National Institute of Oceanography, Visakhapatnam, India



**Abstract**

*Planococcaceae*, a family within the order *Bacillales*, embraces 14 genera: *Bhargavaea, Caryophanon, Chryseomicrobium, Filibacter, Jeotgalibacillus, Kurthia, Paenisporosarcina, Planococcus, Planomicrobium, Rummeliibacillus, Solibacillus, Sporosarcina, Ureibacillus*, and *Viridibacillus*. Members of the family are Gram-variable, spore forming or nonspore forming, and motile or nonmotile; morphology varies from trichomes in case of *Caryophanon*, filamentous in case of *Filibacter* to rods or rod-cocci, or spherical rods in case of other genera. Diagnostic amino acid in the peptidoglycan is L-lysine with a peptidoglycan variation of A4α type. Most dominating fatty acids of the family are iso-$C_{15:0}$ or anteiso-$C_{15:0}$ or iso-$C_{16:0}$ or $C_{16:1(\omega11c)}$ or anteiso-$C_{17:0}$ or $C_{16:1(\omega7c)}$ alcohol. Diphosphatidylglycerol, phosphatidylglycerol, phosphatidylethanolamine, phosphatidylserine and unidentified phospholipids and glycolipids except in case of *Jeotgalibacillus* are the predominant lipids. The G+C values of DNA for the family *Planococcaceae* ranges from 34 to 54 %. Phylogenetically a member of *Firmicutes* group and is closely related to *Bacillaceae*.

## Short Description of the Family

### Taxonomy, Historical, and Current

Plan.o.coc.ca'ce.ae. N.L. masc. n. *Planococcus*, type genus of the family; L. suff. *-aceae*, ending denoting family; N.L. fem. pl. n. *Planococcaceae*, the *Planococcus* family. The family was described by Krasil'nikov in 1949.

Phylogenetically the family *Planococcaceae* is a member of the order *Bacillales* (Prévot 1953) and the phylum is Firmicutes. *Planococcaceae* embraces the type genus *Planococcus* (Migula 1894; emend. Nakagawa et al. 1996; emend. Yoon et al. 2010b), and other genera are *Bhargavaea* (Manorama et al. 2009; emend Verma et al. 2012), *Caryophanon* (Peshkoff 1939), *Chryseomicrobium* (Arora et al. 2011), *Filibacter* (Maiden and Jones 1985), *Jeotgalibacillus* (Yoon et al. 2001a), *Kurthia* (Trevisan 1885), *Paenisporosarcina* (Krishnamurthi et al. 2009b), *Planomicrobium* (Yoon et al. 2001b), *Rummeliibacillus* (Vaishampayan et al. 2009), *Solibacillus* (Krishnamurthi et al. 2009a), *Sporosarcina* (De Vos et al. 2009; Kluyver and van Niel 1936; emend. Yoon et al. 2001c), *Ureibacillus*

E. Rosenberg et al. (eds.), *The Prokaryotes – Firmicutes and Tenericutes*, DOI 10.1007/978-3-642-30120-9_351,

(Fortina et al. 2001), and *Viridibacillus* (Albert et al. 2007) (http://www.bacterio.cict.fr/classifgenerafamilies.html#Planococcaceae). Most of the genera are Gram-positive except the genera *Filibacter* (Maiden and Jones 1984) and *Ureibacillus* (Fortina et al. 2001), which stained Gram-negative. Besides these, the genera *Jeotgalibacillus* (Yoon et al. 2001a) and *Planomicrobium* (Yoon et al. 2001b) are Gram-variable, where the older cells stained Gram-positive and the younger cells stained Gram-negative. Morphological forms vary from trichomes (curved to straight multicellular rods) in case of *Caryophanon* (Peshkoff 1939), filaments (composed of cylindrical straight or curved) in case of *Filibacter* (Maiden and Jones 1984), to rods or rod-cocci (coccoid when the cells are old) in case of other genera. Most of the genera are motile, except *Bhargavaea* (Manorama et al. 2009), *Chryseomicrobium* (Arora et al. 2011), and *Paenisporosarcina* (Krishnamurthi et al. 2009b), and the motility is due to a single or two or peritrichous flagella or by gliding in case of *Filibacter.* Members of the family are aerobic to facultatively anaerobic and produce acid from a variety of carbohydrates. Cross-linking of the peptidoglycan is by the A4α type; diagnostic amino acid is L-lysine and the interpeptide bridges contain either aspartic acid or glutamic acid or in some cases the linkage is direct. Iso-methyltetradecanoic acid (iso-$C_{15:0}$) or anteiso-methyltetradecanoic acid (anteiso-$C_{15:0}$) or both are the major fatty acids present in all the genera, and other prominent fatty acids are iso-$C_{16:0}$, $C_{16:1(\omega 11c)}$, anteiso-$C_{17:0}$, and $C_{16:1(\omega 7c)}$ alcohol. Menaquinone MK-7 is the predominant isoprenoid quinone, but MK-6 and MK-8 are also present in some genera. Polar lipids present were diphosphatidylglycerol, phosphatidylglycerol, phosphatidylethanolamine, phosphatidylserine, and unidentified phospholipids and glycolipids except in case of *Jeotgalibacillus* (Yoon et al. 2001a) where most of the lipids are aminophospholipids. The G+C values of DNA for the family *Planococcaceae* ranges from 34 to 54 %. Characteristics that differentiate the genera of the family *Planococcaceae* are listed in ➲ *Table 25.1.*

## Phylogenetic Structure of the Family and Its Genera

According to the phylogenetic affiliation of the type strains of *Firmicutes* in the RaxML 16S rRNA gene tree of the Living Tree Project (Yarza et al. 2008; Yarza et al. 2010; Munoz et al. 2011), the family *Planococcaceae* is closely related to *Bacillaceae* and *Bacillaceae* 3, moderately related to a broad group containing the families *Staphylococcaceae, Listeriaceae, Carnobacteriaceae* 1, *Carnobacteriaceae* 2, *Enterococcaceae, Aerococcaceae, Streptococcaceae, Lactobacillaceae,* and *Leuconostocaceae* (Yarza et al. 2010; http://www.arb-silva.de/projects/living-tree) (➲ *Fig. 25.1*). Among the *Bacillaceae,* the closely related groups are *Bacillus aquimaris, B. vietnamensis, B. marisflavi, B. coahuilensis,* and *B. seohaeanensis* (Yarza et al. 2010) (➲ *Fig. 25.1*). Families *Bacillaceae* 2, *Bacillaceae* 1, and *Sporolactobacillaceae* are deeply rooted and distantly related. Other families of the order *Bacillales* such as *Paenibacillaceae, Thermoactinomycetaceae, Alicyclobacillaceae, Tenericutes, Acidithiobacillaceae,* and *Clostridiaceae* and genera *Ammoniphilus, Oxalophagus, Caldalkalibacillus, Microaerobacter, Tumebacillus, Alicyclobacillus, Calditerricola,* and *Thermicanus* belonging to the family *Bacillaceae* formed sister clades (Yarza et al. 2010; http://www.arb-silva.de/projects/living-tree) (➲ *Fig. 25.1*).

Phylogenetic reconstruction of the family *Planococcaceae* based on 16S rRNA gene sequences using the neighbor-joining algorithm (NJ), (Yarza et al. 2010; http://www.arb-silva.de/projects/living-tree) resulted in the formation of a monophyletic branch for the family *Planococcaceae* (➲ *Fig. 25.2*). However, phylogenetic tree created by Yarza et al. (2010), using maximum likelihood (ML) (tree not shown), resulted in formation of *Jeotgalibacillus* as a separate sister clade in the family *Planococcaceae,* and the genus *Jeotgalibacillus* appeared to be closely related to *Listeriaceae.* Phylogenetic analyses of *Planococcaceae* were also performed using minimum evolution (ME) and maximum parsimony (DNAPARS), and the topology of the phylogenetic trees was similar to NJ. In the phylogenetic analyses of the family *Planococcaceae,* the genera *Bhargavaea, Caryophanon, Chryseomicrobium, Jeotgalibacillus, Kurthia, Paenisporosarcina, Rummeliibacillus, Solibacillus, Ureibacillus,* and *Viridibacillus* formed distinct clades. The remaining genera, *Planococcus, Planomicrobium, Filibacter,* and *Sporosarcina,* did not resolve into distinct clades (➲ *Fig. 25.3*). Phylogenetic analysis of the genera *Planococcus, Marinococcus,* and *Sporosarcina* and their relationships to members of the genus *Bacillus* were also attempted by Farrow et al. (1992).

Genus *Planococcus* was described by Migula (1894) based on phylogenetic and chemotaxonomic characteristics. Subsequently, Yoon et al. (2001b) transferred several species of the genus *Planococcus* to a new genus *Planomicrobium* based on cell morphology and phylogeny. Genus *Planococcus* forms two major clades with clade 1 including the species *P. antarcticus* (Reddy et al. 2002), *P. kocurii* (Hao and Komagata 1985), *P. donghaensis* (Choi et al. 2007), and *P. halocryophilus* (Mykytczuk et al. 2012), and clade 2 contains *P. rifietensis* (Romano et al. 2003), *P. citreus* (Migula 1894), *P. columbae* (Suresh et al. 2007), *P. maritimus* (Yoon et al. 2003), *P. plakortidis* (Kaur et al. 2012), and *P. maitriensis* (Alam et al. 2003) (➲ *Fig. 25.3*). Out of the 11 species described, under the genus *Planococcus, P. salinarum* (Yoon et al. 2010b) formed a separate branch and is closely related to clade 1 of *Planococcus.*

*Planococcus* is closely related to *Planomicrobium,* but it can be differentiated from the nearest phylogenetic neighbor, on the basis of 16S rRNA gene signature nucleotides. Signature nucleotides were identified based on their presence in all the species of the particular genus and were not present in more than four genera of the family *Planococcaceae* in the 16S rRNA gene. In the genus, *Planococcus* T and A are conserved at positions 183 and 190, respectively (Dai et al. 2005) whereas in *Planomicrobium* C and G are present, respectively, at identical positions. Besides these, the genus *Planococcus* contains C (170); T (192); T, T, G, C, G, and G (between 192 and 209); C (209); T (210); T (212); A (217); G (218); C, T, and G (between 218 and 220); A (220);

**Table 25.1**

Diagnostic characteristics that differentiate the genera of the family *Planococcaceae*[a]. 1. *Bhargavaea* (Manorama et al. 2009; Verma et al. 2012); 2. *Caryophanon* (Peshkoff 1939); 3. *Chryseomicrobium* (Arora et al. 2011); 4. *Filibacter* (Maiden and Jones 1984); 5. *Jeotgalibacillus* (Yoon et al. 2001a); 6. *Kurthia* (Trevisan 1885); 7. *Paenisporosarcina* (Krishnamurthi et al. 2009b); 8. *Planococcus* (Migula 1894); 9. *Planomicrobium* (Yoon et al. 2001b); 10. *Rummeliibacillus* (Vaishampayan et al. 2009); 11. *Solibacillus* (Krishnamurthi et al. 2009a); 12. *Sporosarcina* (Kluyver and van Niel 1936); 13. *Ureibacillus* (Fortina et al. 2001); 14. *Viridibacillus* (Albert et al. 2007)

| Characteristic | 1 | 2 | 3 | 4 | 5 | 6 | 7 | 8 | 9 | 10 | 11 | 12 | 13 | 14 |
|---|---|---|---|---|---|---|---|---|---|---|---|---|---|---|
| Morphology | Rod to coccoid | Trichome rods | Rods | Filamentous, composed of cylindrical cells, straight or curved | Rods | Rods | Rod/coccus | Cocci/short rods/rods | Cocci/short rods/rods | Rods | Rods | Rods or sphere | Rods | Rods |
| Spore formation | – | – | – | – | + | – | + | – | – | + | + | + | + | + |
| Endospore shape | – | – | – | – | Ro/el | – | Ro | – | – | Ro | Ro | Ro | Sp | Ro |
| Gram-stain | + | + | + | – | v | + | + | +/v | v | + | + | + | – | + |
| Motility | nm | m | nm | m, gliding | m | m/nm | nm | m/nm | m | m | m | m | m | m |
| Flagella type | – | PT | – | – | PT | PT | – | One or two flagella | Single flagellum/PT | NR | PT | Single or PT | PT | NR |
| Peptidoglycan type | A4α | A4α | A4α | A4α | A1α | A4α | A4α | A4α | A4α | A4α | A4α | A4α | A4α | A4α |
| Diagnostic peptidoglycan amino acids | L-Lys-D-Glu | L-Lys-D-Glu | L-Lys-D-Asp | L-Lys-D-Glu | L-Lys | L-Lys-D-Asp | L-Lys-D-Asp | L-Lys-D-Glu | L-Lys-D-Glu | L-Lys-D-Glu or L-Lys-D-Asp | L-Lys-D-Glu | L-Lys-L-Gly-D-Glu | L-Lys-D-Asn | L-Lys-D-Glu or L-Lys-D-Asp |
| Major polar lipids | PG, DPG | NR | PG, DPG, PE, | NR | APL-1, APL-2, APL-3, APL-4 | PG, DPG, PE | PG, DPG, PE | PG, DPG, PE | PG, DPG, PE, | PG, DPG, PE | PG, DPG, PE, PS | NR | PG, DPG, PL, GL | PG, DPG, PE |
| Major fatty acids | Iso-$C_{15:0}$, iso-$C_{16:0}$, anteiso-$C_{15:0}$ | Iso-$C_{15:0}$, $C_{16:1(\omega 11c)}$ | Iso-$C_{15:0}$, iso-$C_{16:0}$, $C_{16:1(\omega 7c)}$ OH, | Anteiso-$C_{15:0}$, anteiso-$C_{17:0}$ | Iso-$C_{15:0}$ | Iso-$C_{15:0}$, ante-iso-$C_{15:0}$ | Iso-$C_{15:0}$, anteiso-$C_{15:0}$, $C_{16:1(\omega 7c)}$ OH | Iso-$C_{15:0}$, anteiso-$C_{15:0}$, iso-$C_{16:0}$, $C_{16:1(\omega 7c)}$ OH | Iso-$C_{15:0}$, anteiso-$C_{15:0}$, iso-$C_{16:0}$, $C_{16:1(\omega 7c)}$_ OH | Iso-$C_{15:0}$, anteiso-$C_{15:0}$ | Iso-$C_{15:0}$, iso-$C_{16:1}$ | Anteiso-$C_{15:0}$ | Iso-$C_{15:0}$, iso-$C_{17:0}$ | Iso-$C_{15:0}$, anteiso-$C_{15:0}$ |
| Menaquinone | MK-6, MK-8 | MK-6 | MK-6, MK-7, MK-7($H_2$), MK-8 | MK-7 | MK-7, MK-8 | MK-7 | MK-7, MK-8 | MK-7, MK-8 | MK-6, MK-7, MK-8 | MK-7 | MK-7 | MK-7 | MK-7 | MK-7, MK-8 |
| DNA G+C content | 50.2–53.7 | 41–46 | 53.4 | 44.0 | 44 | 36–38 | 46 | 39–52 | 35–47 | 34.3 | 39.3 | 40–42 | 35–45 | 35–40.4 |

[a]Some data was also obtained from *Bergey's Manual of Systematic Bacteriology*, vol. 3

*Ro* round, *Sp* spherical, *el* ellipsoid, + positive, – negative, *v* variable, *nm* nonmotile, *m* motile, *PT* peritrichous, *NR* not reported, *l-Lys* L-lysine, *d-Glu* D-glutamic acid, *d-Asp* D-aspartic acid, *Gly* glycine, *PG* phosphatidylglycerol, *DPG* diphosphatidylglycerol, *PE* phosphatidylethanolamine, *PS* phosphatidylserine, *PL* unidentified phospholipids, *GL* unidentified glycolipids, *APL* unidentified aminophospholipids

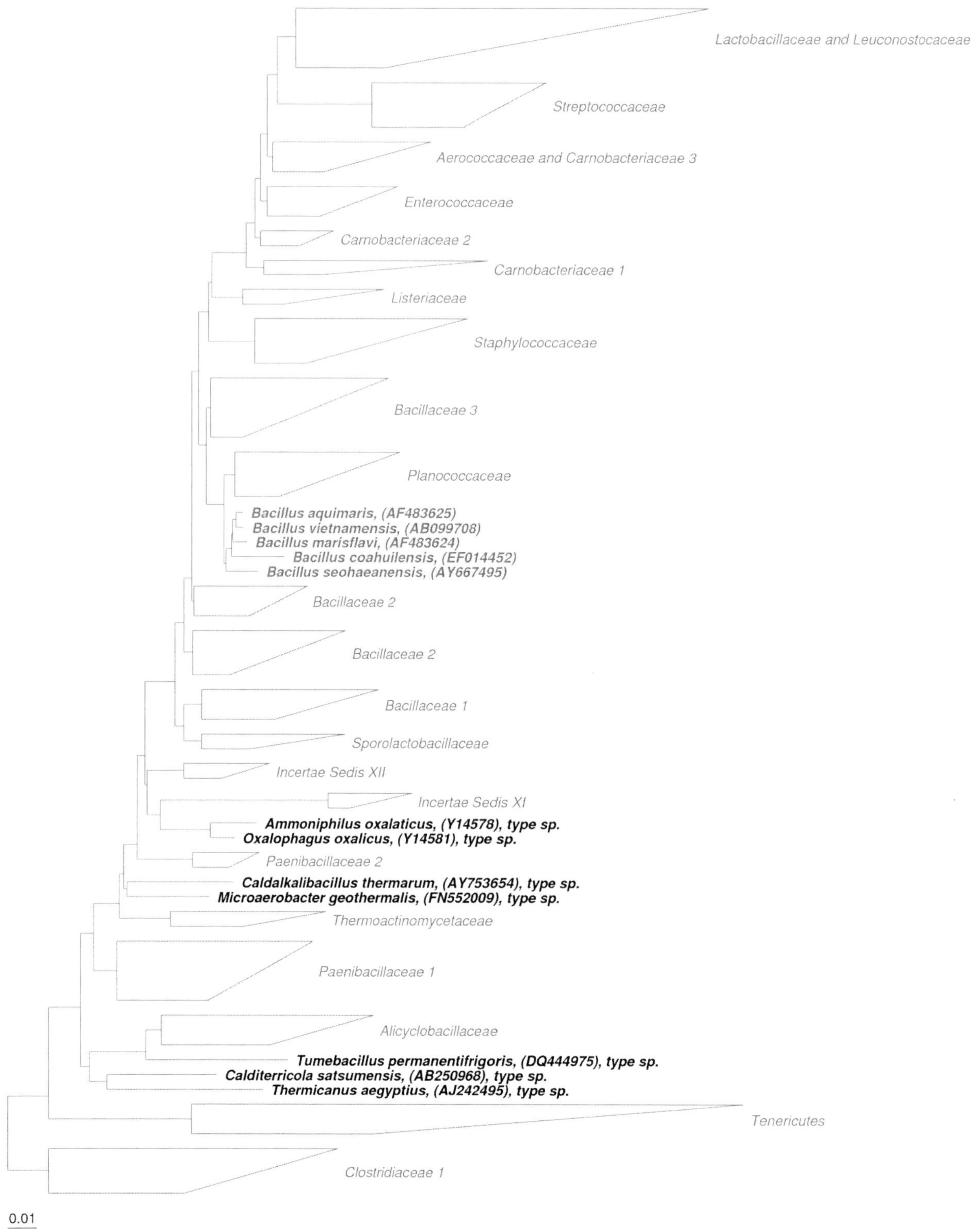


**Fig. 25.1**

**Phylogenetic reconstruction of the Order *Bacillales* based on 16S rRNA and created using the neighbor-joining algorithm with the Jukes-Cantor correction. The sequence datasets and alignments were used according to the All-Species Living Tree Project (LTP) database (Yarza et al. 2010; http://www.arb-silva.de/projects/living-tree). The tree topology was stabilized with the use of a representative set of nearly 750 high quality type strain sequences proportionally distributed among the different bacterial and archaeal phyla. In addition, a 40 % maximum frequency filter was applied in order to remove hypervariable positions and potentially misplaced bases from the alignment. Scale bar indicates estimated sequence divergence**

**Fig. 25.2**

Phylogenetic reconstruction of the family *Planococcaceae* based on 16S rRNA and created using the neighbor-joining algorithm with the Jukes-Cantor correction. The sequence datasets and alignments were used according to the All-Species Living Tree Project (LTP) database (Yarza et al. 2010; http://www.arb-silva.de/projects/living-tree). The tree topology was stabilized with the use of a representative set of nearly 750 high quality type strain sequences proportionally distributed among the different bacterial and archaeal phyla. In addition, a 40 % maximum frequency filter was applied in order to remove hypervariable positions and potentially misplaced bases from the alignment. Scale bar indicates estimated sequence divergence

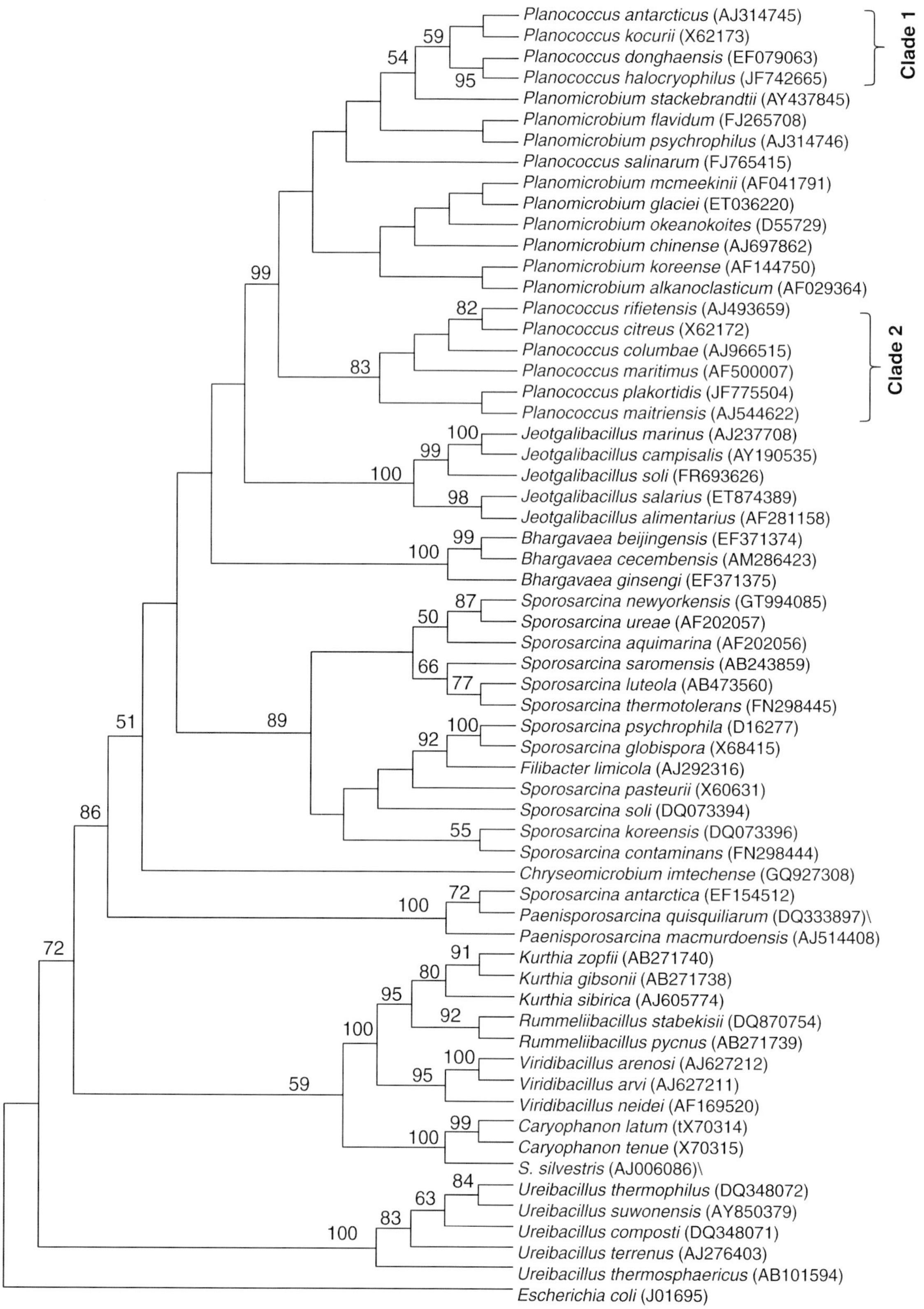


**Fig. 25.3**

**Neighbour-joining tree based on 16S rRNA gene sequences, showing the phylogenetic relationship among the genera of family *Planococcaceae*. Bootstrap values (expressed as percentages of 1,000 replications) greater than 50 % are given at nodes**

C (221); G (222); A (228); T (295); G (397); C (416); C (485); A (496); G (499); C (519); C (520); T (618); A (761); G (1273); and G (1395) as signature nucleotides. The signature nucleotides at positions 210, 217, 618, and 1395 differentiate the genus *Planococcus* from *Planomicrobium*. Besides the above signatures, nucleotides G, C, and G (between 192 and 209) and C, T, and G (between 218 and 220) are highly conserved among the members of *Planococcus* compared to *Planomicrobium*.

Genus *Planomicrobium* has nine species which on phylogenetic analyses form two clades closely related to clade 1 of *Planococcus*. *Planomicrobium mcmeekinii* (Junge et al. 1998), *P. glaciei* (Zhang et al. 2009), *P. okeanokoites* (ZoBell and Upham 1944), *P. chinense* (Dai et al. 2005), *P. koreensae* (Yoon et al. 2001b), and *P. alkanoclasticum* (Engelhardt et al. 2001) formed a tight and a distinct cluster, whereas *P. flavidum* (Jung et al. 2009) and *P. psychrophilum* (Reddy et al. 2002) formed a separate clade closely affiliated with clade 1 of *Planococcus*. Genus *Planomicrobium* contains the following 16S rRNA gene signature nucleotides: C (170), T (192), C (210), C (214), T (212), G (217), A (220), C (221), G (222), A (228), C (295), G (397), C (416), C (485), A (496), G (499), C (519), C (520), C (618), G (1273), and A (1395).

Genus *Bhargavaea* was created by Manorama et al. (2009) and emended by Verma et al. (2012). The phylogenetic analysis of the genus was based on both 16S rRNA and Gyr B genes of three species: *B. cecembensis* (Manorama et al. 2009), *B. beijingensis*, and *B. ginsengi* (Verma et al. 2012). NJ-based phylogenetic analysis indicated that the genus is closely related to *Planococcus*, *Planomicrobium*, and *Jeotgalibacillus* (➲ *Fig. 25.3*). According to Verma et al. (2012), genus *Bhargavaea* is closely related to the genus *Sporosarcina*. However, the RaxML 16S rRNA gene tree of the Living Tree Project (Yarza et al. 2010; Munoz et al. 2011) indicated that the genus *Bhargavaea* forms a deeply rooted branching from other members of the family (➲ *Fig. 25.2*) and its sister clade is the family *Listeriaceae*. Genus *Bhargavaea* contains the following signature nucleotides in its 16S rRNA gene sequence: C (116), T (162), G (163), T (164), A (165), G (206), C (444), C (445), G (451), T (488), C (505), T (1027), A (1034), C (between 1053 and 1054) and G (between 1061 and 1062), A (1068), G (1278), and C (1324). Out of these, the signature nucleotides at positions 116, 206, 444, 451, 488, and 1068 are unique to this genus.

Genus *Caryophanon* was described by Peshkoff (1939) based on its unique cell morphology. Stackebrandt et al. (1987) performed the 16S rRNA oligonucleotide analyses of spore-forming *Bacilli* and nonspore-forming genus *Caryophanon* and found that the oligomers AAUAAG (position 456), AAUCUG, CCCCCG (617), ACAUAG (1002), UCCUUAG, CACUCUCG (1132), CCAAUCCG, UACCUUUG (462), UUACCUUG (480), ACUUCAUCG, AUUUCUUCG, CUACAAACG, CCAUCCCACUG, UAACCCUUUUAG, and UAACCUACCUUAUAG (124) are uniquely found in the genus *Caryophanon*. Phylogenetic analysis performed by Farrow et al. (1994) indicated that the genus *Caryophanon* branches out with *Bacillus fusiformis* and *Bacillus sphaericus*. Present phylogenetic analyses based on 16S rRNA gene sequence indicated that the genus is closely associated with *Solibacillus* and their sister clade contains the genera *Viridibacillus*, *Rummeliibacillus*, and *Kurthia*. Clade represented by above genera form a deeply rooted branching with other members of the family (➲ *Fig. 25.3*). A comparison of 16S rRNA gene sequences of the family *Planococcaceae* resulted in identification of following signature nucleotides: T (154), T (250), A (444), T (451), G (484), T (487), T (696), G (862), C (878), T (923), A (930), T (1145), C (1146), T (1158), T (1164), G (1179), C (1280), G (1311), T (1313), and G (1317). The nucleotide G at position 1179 is present only in *Caryophanon*.

The genus *Chryseomicrobium* is represented by a single species (*Chryseomicrobium imtechense*) which phylogenetically forms a deep branching from the genera *Planococcus*, *Sporosarcina*, *Jeotgalibacillus*, *Bhargavaea*, and *Planomicrobium* (present analysis; Arora et al. 2011). However, the RaxML 16S rRNA gene tree of the Living Tree Project (Yarza et al. 2010; Munoz et al. 2011) indicated that the closest genera are *Caryophanon*, *Kurthia*, *Paenisporosarcina*, and *Sporosarcina* (➲ *Fig. 25.2*). In all the above phylogenetic methods, the rooting is not well supported as the bootstrap values were low (<50 %). Though the phylogeny is based on a single species, it did possess some signature nucleotides in its 16S rRNA, and they are as follows: C (588), A (640), T (652), T (923), A (930), C (1029), C (1050), T (1271), and T (1363) (the positions are with respect to *E. coli* 16S rRNA gene sequence accession number J01695) and C at position 1029 is uniquely present in this genus compared to the other genera of the family.

Genus *Filibacter* was originally described by Maiden and Jones (1984), and 16S rRNA analysis of the lone species *Filibacter limicola* was compared with species of *Bacillus* (Clausen et al. 1985). However, no reports are available on phylogenetic analyses of *Filibacter*. Present analyses and the RaxML 16S rRNA gene tree of the Living Tree Project (Yarza et al. 2010; Munoz et al. 2011) indicated that the genus is closely related to *Sporosarcina* (➲ *Figs. 25.2* and ➲ *25.3*). Further, its affiliation is supported by the riboprinting analysis (Stackebrandt 2009). A comparison of 16S rRNA gene sequence of *Filibacter* with other members of the family *Planococcaceae* resulted in the identification of A (505), A-T (640–652), A (956), T (1271), A (1272), G (1278), and T (1319) as diagnostic nucleotides, and A (1272) and T (1319) are present only in this genus.

Genus *Jeotgalibacillus* was described by Yoon et al. (2001a), and it is represented, presently, by five species. Phylogenetic analysis of all species of the genus was performed by Cunha et al. (2012), but the study did not compare their relationship with other members of the family, *Planococcaceae*. Phylogenetic analyses using the RaxML 16S rRNA gene tree of the Living Tree Project (Yarza et al. 2010; Munoz et al. 2011) resulted in clustering of the genus with *Caryophanon* and *Kurthia*. However, the RaxNJ 16S rRNA gene tree of the Living Tree Project (Yarza et al. 2010; Munoz et al. 2011) indicated that it forms a clade by itself and is deeply rooted from other members of the family (➲ *Fig. 25.2*). In our analyses, genus *Jeotgalibacillus* clustered with *Planococcus*, *Planomicrobium*, and *Bhargavaea* in methods NJ (➲ *Fig. 25.3*) and ME (data not shown), but the bootstrap values were below 50 %, thus implying a poor phylogenetic relationship with members of *Planococcaceae*. Our analysis identified the nucleotides T (164), A (165), C (169), G (203), T (229), T (294). G (397), C (416), A (486) T (498), C (519), C (520), T (612), C (640) G (652), A (675), G (688), T (732), C (770), A (782), G (862), C (878), C (1037), A (1290), T (1299), and T (1363) as 16S rRNA gene signatures and of these T (732), A (1290), and T (1299) were unique to this genus.

Phylogenetic analyses based on 16S rRNA gene sequence indicated that genus *Kurthia* clusters with the genus *Caryophanon* (➲ *Fig. 25.2*). However, our analyses indicated the affiliation of *Kurthia* with the genus *Rummeliibacillus* in the ME, NJ (➲ *Fig. 25.3*), and DNA parsimony trees. Genus *Kurthia* is characterized by the presence of the following diagnostic nucleotides in their 16S rRNA gene: T (250), T (266), A (312), T (342), A (363), T (438), T (486), G (490), A (493), A (498), T (586), T (617), A (675), T (696), G (764), A (922), T (931), A (1028), T (1066), A (1144), T (1164), A (1281), T (1291), T (1305), and T (1310).

Genus *Paenisporosarcina* was created by Krishnamurthi et al. (2009b) and emended by Reddy et al. (2013), and presently it contains four species. Phylogeny of the genus based on 16S rRNA gene sequence indicated that the genus forms a separate branch rooted from the clade represented by *Planococcus, Planomicrobium, Sporosarcina, Bhargavaea, Jeotgalibacillus, Chryseomicrobium*, and *Filibacter* (➲ *Fig. 25.3*). In RaxML and RaxNJ 16S rRNA gene trees of the Living Tree (Yarza et al. 2010; Munoz et al. 2011), the phylogenetic affiliation of *Paenisporosarcina* was different, and it formed a clade with *Sporosarcina* (in case of RaxML) and with *Kurthia* and *Caryophanon* (in case of RaxNJ). Genus *Paenisporosarcina* is characterized by the presence of the following diagnostic nucleotides in their 16S rRNA gene: A (159), T (162), T (165), T (211), G (228), C (229), A (252), A (283), G (487), A (684), and C (699), and nucleotides at positions 211 and 283 are unique to the genus.

Genus *Rummeliibacillus* was described by Vaishampayan et al. (2009) and is presently represented by two species, *Rummeliibacillus stabekisii* and *Rummeliibacillus pycnus*. Phylogenetic analyses based on 16S rRNA gene sequence using the methods ME, NJ, and DNA parsimony indicated that it is closely affiliated with the genus *Kurthia* (➲ *Fig. 25.3*). This clade is part of a bigger cluster represented by the genera *Caryophanon, Solibacillus*, and *Viridibacillus*. Genus *Rummeliibacillus* contains the signature nucleotides C (169), G (203), T (266), A (312), T (342), A (363), T (438), T (486), G (490), A (493), A (498), T (586), T (617), T (735), A (922), T (931), T (1066), T (1158), A (1281), T (1291), T (1305), and T (1310) in its 16S rRNA gene.

Genus *Solibacillus* was carved by Krishnamurthi et al. (2009a) from the genus *Bacillus* and is represented by a single species *Solibacillus silvestris* that was originally described by Rheims et al. (1999). Our 16S rRNA genes sequence-based analyses using the phylogenetic methods ME, NJ (➲ *Fig. 25.3*), and DNA parsimony indicated that *Solibacillus* and *Caryophanon* cluster together as part of the clade represented by *Kurthia, Rummeliibacillus*, and *Viridibacillus*. Genus contains the 16S rRNA gene signature nucleotides T, A, T, A, T, G, T, A, G, C, T, A, A, A, T, C, and T at positions 106, 113, 154, 444, 451, 484, 612, 782, 862, 878, 923, 930, 1034, 1066, 1145, 1280, and 1305, and nucleotides at positions 106 and 113 are unique to this genus.

Genus *Sporosarcina* was proposed by Kluyver and van Niel (1936) and presently contains 12 species. In RaxML and RaxNJ, 16S rRNA gene trees of the Living Tree (Yarza et al. 2010; Munoz et al. 2011) and our analyses indicated the formation of two major branches within the genus *Sporosarcina* and also contain the genus *Filibacter* within the major clade (➲ *Figs. 25.2* and ➲ *25.3*). Clade represented by *Sporosarcina* is well separated from the genera *Planococcus, Planomicrobium, Bhargavaea*, and *Jeotgalibacillus* (➲ *Figs. 25.2* and ➲ *25.3*). Genus *Sporosarcina* contains the following signature nucleotides in its 16S rRNA gene sequence: G, G, G, A, T, A, C, G, A, G, T, and T at positions 170, 486, 487, 640, 652, 684, 709, 735, 956, 1035, 1284, and 1415, respectively.

Genus *Ureibacillus* was created by Fortina et al. (2001) with *Ureibacillus thermosphaericus* as the type species and has five species. In the phylogenetic analyses based on RaxML and RaxNJ 16S rRNA gene trees of the Living Tree (Yarza et al. 2010; Munoz et al. 2011), all the species cluster together with a single branching and are related to a major clade represented by all other genera of the family (➲ *Fig. 25.2*). Our analysis, based on NJ, NE, and DNA parsimony, indicated a deeply rooted branching of all five species (➲ *Fig. 25.3*). Evaluation of 16S rRNA gene sequence with respect to other genera of the family *Planococcaceae* resulted in the identification of T, G, C, G, T/C, C, G, T, A, G, C, C, G, G, A, C, C, C, A, C, A, T, A, A, G, and C at positions 115, 181, 182, 189, 190, 208, 210, 229, 242, 404, 411, 501, 688, 695, 735, 765, 770, 1146, 1193, 1194, 1195, 1198, 1199, 1313, 1317, and 1321, respectively, as the signature nucleotides.

Genus *Viridibacillus* was carved from *Bacillus* by reclassifying three species: *Bacillus arvi, Bacillus arenosi*, and *Bacillus neidei* (Albert et al. 2007). Previous studies have not reported the affiliation of *Viridibacillus* with members of the family *Planococcaceae*. Present analysis, based on 16S rRNA gene sequence, indicated that the species are closely related to each other and clustered within the clade represented by *Caryophanon, Solibacillus, Rummeliibacillus*, and *Kurthia* (➲ *Fig. 25.3*). Genus contains the following diagnostic nucleotides in its 16S rRNA gene: A (163), T (251), T (266), A (312), T (342), A (363), T (438), T (486), A (498), T (586), T (612), A (674), A (709), T (931), A (1028), T (1066), A (1281), T (1291), T (1300), A (1301), T (1302), T (1305), and T (1310). Nucleotides at positions 163, 251, and 1302 are unique to this genus.

## Molecular Analysis

### DNA-DNA Hybridization Studies

DNA-DNA hybridization (DDH) between species of the 14 genera of *Planococcaceae* indicated that similarity values ranged from 13 to <70 %. However, DNA-DNA relatedness was high within strains of the same species (>70 %) (Adcock et al. 1976). DNA-DNA relatedness between *Caryophanon latum* and *C. tenue* was 13–30 % (Adcock et al. 1976), between *Jeotgalibacillus* species <21 % (Yoon et al. 2004, 2010a); between *Kurthia* species <41 % (Cherevach et al. 1983); between *Paenisporosarcina quisquiliarum* and *P. macmurdoensis* 18 % (Krishnamurthi et al. 2009b); between *Planococcus* species <62 % (Nakagawa et al. 1996; Suresh et al. 2007; Choi et al. 2007; Mykytczuk et al. 2012; Yoon et al. 2003, 2001b; Kaur et al. 2012); between *Planomicrobium* species <62 % (Mayilraj et al. 2005; Dai et al. 2005; Jung et al. 2009; Zhang et al. 2009; Yoon et al. 2001b); between *Rummeliibacillus stabekisii* and *R. pycnus* 13 % (Vaishampayan et al. 2009); between *Sporosarcina* species <70 % (Yu et al. 2008; Yoon et al. 2001c; Kämpfer et al. 2010; Kwon et al. 2007; Tominaga et al. 2009; Wolfgang et al. 2012; An et al. 2007); between *Ureibacillus* species <52 % (Weon et al. 2007; Fortina et al. 2001; Kim et al. 2006); and between *Viridibacillus* species <42.0 % (Heyrman et al. 2005).

DNA-DNA relatedness experiments were not carried out between a number of novel taxa with their respective nearest neighbor, as the level of 16S rRNA gene sequence similarity between the novel taxa with the nearest neighbor (Reddy et al. 2003; Manorama et al. 2009; Arora et al. 2011; Yoon et al. 2001a; Cunha et al. 2012; Rheims et al. 1999) was less than 96.9 %, which is below the cutoff value (97 %) suggested by Stackebrandt and Goebel (1994) for genomic distinction of species.

The species *Filibacter limicola* which was established by Maiden and Jones (1985) was mainly based on phenotypic characteristics and was not based on either DNA-DNA hybridization or 16S rRNA gene sequence analysis. Phylogenetically *Filibacter limicola* is closer to the species of the genus *Sporosarcina* especially *S. psychrophila* and *S. globispora* (➲ *Figs. 25.2* and ➲ *25.3*).

## Genome Comparison

Whole genome sequencing of nine strains in the family *Planococcaceae* has been reported (NCBI Database). Out of these nine strains, only five strains are completely processed by NCBI GenBank. The whole genome sequence of *Kurthia* sp. JC8E (GOLD ID: Gi13538) is 2,976,740 bp long, has 3,006 genes including one 16S rRNA gene, and has a G+C mol% of 38.22 %. The genome has 2,916 proteins including some antibiotic resistance proteins like glyoxalase/bleomycin resistance protein (providing resistance against bleomycin), several penicillin-binding proteins (providing resistance against penicillin), and lantibiotic ABC transporter ATP-binding protein (providing resistance against lantibiotics). The bacterium may also be resistant to camphor and tellurium due to the presence of camphor resistance CrcB protein and tellurium resistance protein, respectively. NTP pyrophosphohydrolase (including oxidative damage repair enzymes, which hydrolyze noncanonical NTPs) can be used for "house cleaning" work (Galperin et al. 2006). Fatty acid desaturase enzyme was also identified in *Kurthia* sp. JC8E which is responsible for desaturation of fatty acids (Ford 2010).

Another reported whole genome sequence of *Kurthia* sp. JC30 (GOLD ID: Gi13535) is 3,201,696 bp long and has 3,296 genes including one 16S rRNA gene and a G+C mol% of 39.26 %. The genome has 3,201 proteins. As in *Kurthia* sp. JC8E, the *Kurthia* sp. JC30 also has antibiotics resistance proteins like glyoxalase/bleomycin resistance protein and several penicillin-binding proteins. Apart from these two, *Kurthia* sp. JC30 also has other antibiotic resistance proteins like tetracycline resistance protein and teicoplanin resistance associated membrane protein. *Kurthia* sp. JC30 shows the presence of some metal resistance proteins like copper resistance protein and tellurium resistance protein. The annotation of the genome also reports the presence of a lipase enzyme.

The genome of *Planococcus donghaensis* MPA1U2 (GOLD ID: Gi08597) (Choi et al. 2007) is 3,303,464 bp long, has 3,331 genes including one 16S rRNA gene, and has a G+C mol% of 39.68 %. The genome codes for 3,267 proteins including some antibiotics resistance proteins like glyoxalase/bleomycin resistance protein, acriflavin resistance protein (providing resistance against acriflavin), and quinolone resistance protein (providing resistance against quinolone). The bacterium also shows the presence of protein CopC (providing resistance against copper), mercuric resistance operon regulatory protein MerR (providing resistance against mercury), and aluminum resistance protein (providing resistance against aluminum). The annotation of the genome also reports the presence of multiple resistance and pH homeostasis protein; this protein is involved in pH homeostasis and resistance against cholate and $Na^+$ (Ito et al. 1999).

The reported whole genome sequence of *Planococcus antarcticus* DSM 14505$^T$ (GOLD ID: Gi17702) (Margolles et al. 2012) is 3,772,109 bp long and has 3,750 genes, including one 16S rRNA gene and a G+C mol% of 42.14 %. The annotation of the genome reports 3,825 proteins. *P. antarcticus* DSM 14505$^T$ has some antibiotic resistance proteins like glyoxalase/bleomycin resistance protein and quinolone resistance protein that have been discussed earlier. The bacterium also has an organic hydroperoxide resistance protein that provides resistance against organic hydroperoxides (Cussiol et al. 2010). The annotation of *P. antarcticus* DSM 14505$^T$ also reports to have many other resistance proteins like tellurium resistance protein, copper resistance protein, aluminum resistance protein, arsenical resistance proteins, mercuric resistance operon regulatory protein MerR, and camphor resistance protein. Abortive infection bacteriophage resistance protein is also present in the genome of the bacterium; this protein is involved in resistance against bacteriophage infections (Fineran et al. 2009).

The whole genome sequence of *Sporosarcina newyorkensis* 2681 (GOLD ID: Gi05603) is 3,627,407 bp long, has 3,899 genes, including one 16S rRNA gene, and has a G+C mol% of 42.14 %. The genome reports the presence of 3,825 proteins. *S. newyorkensis* 2681 also has some antibiotic resistance proteins like glyoxalase/bleomycin resistance protein, quinolone resistance protein, tetracycline resistance protein, and several penicillin resistance proteins. *S. newyorkensis* 2681 has many other resistance proteins like aluminum resistance protein, copper resistance protein, mercuric resistance operon regulatory protein MerR, tellurite resistance protein, and camphor resistance protein. The genome annotation also reports the presence of bacitracin resistance protein. BCCT family osmoprotectant transporter is also found in the genome of *S. newyorkensis* 2681(Kuhlmann et al. 2011). The bacterium may also show lantibiotic resistance due to the presence of MutG family lantibiotic protection ABC superfamily in the genome.

All the five whole genome sequences show similarities like presence of antibiotics resistance and metal resistance. Some of them even show resistance against lantibiotics and bacitracin apart from peculiar features like osmotic protection and protection against bacteriophage infections.

Apart from whole genome analyses, there were several other molecular studies targeting a few housekeeping genes and some operons such as *gyr*B gene (gyrase subunit B) in *Bhargavaea beijingensis* and *Bhargavaea ginsengi* (Verma et al. 2012); *sec*Y (preprotein translocase SecY), *adk* (adenylate kinase), and *map* (methionine aminopeptidase) genes in *Jeotgalibacillus marinus* (AY690426); chitinase gene in *Kurthia gibsonii* and *Kurthia zopfii* (JQ739168; D63702); *tuf* gene (encoding elongation factor Tu) in *Planomicrobium chinense* (AB472778); *DHFR-PRL*, *fokIR*, *fokIM*, *MFokl*, and *RFokl* genes encoding dihydrofolate reductase-Prolaction fusion protein, endonuclease, methyltransferase, methylase, and endonuclease, respectively, in *Planomicrobium okeanokoites* (E02430; Looney et al. 1989; Kita et al. 1989); cspB gene (for major cold shock protein), *ctsU*, *ctsV*, *ctsW*, *ctsX*, *ctsY*, *ctsZ*, *ORF*-7 genes (encoding cyclic tetrasaccharide-synthesizing enzymes), *ctsX*, *ctsY*, *ctsZ*, *ORF*-4 genes (for hypothetical protein, 3-α-isomaltosyltransferase, 6-α-glucosyltransferase, hypothetical protein) in *Sporosarcina globispora* (Schröder et al. 1993; Aga et al. 2002, 2003); urease operon (comprising *ureA*, *ureB*, *ureC*, *ureD*, *ureE*, *ureF*, and *ureG* genes), osmotically regulated ectoine biosynthesis gene cluster (*ectA*, *ectB*, *and ectC* genes) (Kuhlmann and Bremer 2002); *cytc* gene (for cytochrome c553) in *Sporosarcina pasteurii* (X78411, AF361945, AF316874, AJ318066, respectively); *aph* (encoding amino acyl peptidase), *cdd* (encoding cytidine deaminase), *leuB* (3-isopropylmalate dehydrogenase), *pyr*P (uracil permease), *pyr*B (aspartate transcarbamoylase), and *pyr*C (dihydroorotase) genes in *Sporosarcina psychrophila* (Brunialti et al. 2011; AJ237978; AB706401; HQ625362; AY147014); *gyr*B gene (encoding DNA gyrase B) in *Sporosarcina saromensis* (AB243078); *Su-1* gene (encoding small, acid-soluble spore protein), *spo*IIIE gene (sporulation protein SpoIIIE), *ssl*A gene (encoding S-layer protein SslA), *tet*L gene (encoding tetracycline efflux pump), *tet*M gene (encoding tetracycline resistance protein), and phenylalanine dehydrogenase gene in *Sporosarcina ureae* (Magill et al. 1990; Chary et al. 2000; Ryzhkov et al. 2007; You et al. 2012; GU584212; AB001031).

## Riboprinting

Conventional methods based on polyphasic taxonomy are not always sufficient to delineate strains at the species level. Under these circumstances, additional molecular methods like riboprinting and multilocus sequence typing (MLST) can help to resolve the taxonomic status. Riboprinting involves polymerase chain reaction (PCR) amplification of small subunit ribosomal RNA genes followed by restriction digestion and analysis of the DNA fingerprint of the ribosomal RNA genes. The DNA fingerprint obtained is known as a riboprint pattern. The advantage of this method compared to other typing methods is that this is automated and the catalog of fingerprints is dynamic, as it changes and expands with the addition of each new sample that is processed. The RiboPrinter system combines automation with the power of DNA to not only identify the genus/species of contaminants but also provide deeper, strain-level typing at the same time.

Riboprint patterns have been generated for some members of the family *Planococcaceae* including species of the genus *Kurthia*, *Sporosarcina*, *Filibacter*, and *Ureibacillus*. Comparison of the riboprint patterns of *Filibacter limicola*, *Sporosarcina globispora*, *Sporosarcina pasteurii*, *Sporosarcina psychrophila*, and *Sporosarcina ureae* showed that they are all different. But the pattern of *Sporosarcina psychrophila* and *Sporosarcina globisprora* was closely related which is also supported by 16S rRNA gene sequence analysis (➋ *Fig. 25.4*). Riboprint patterns of all the three type strains of the genus *Kurthia* are different (➋ *Fig. 25.5*) (Pukall and Stackebrandt 2009), implying that the method could also be used effectively for discriminating at the species level.

Riboprint patterns of *Bacillus thermosphaericus* DSM $10633^T$, strains TH29 and TU1A, and *Ureibacillus terrenus* $TH9A^T$ indicated that the riboprint patterns for strains $TH9A^T$ and TU1A were very similar and could thus be considered characteristic of *Ureibacillus terrenus*, and riboprint patterns for strains *Bacillus thermosphaericus* DSM $10633^T$ and isolate TH29 displayed similar pattern and could thus be considered characteristic of *Bacillus thermosphaericus* (*Ureibacillus thermosphaericus*) (Fortina et al. 2001) (➋ *Fig. 25.6*).

## Multilocus Sequence Typing

Multilocus sequence typing (MLST) is an explicit method for characterizing isolates of bacterial strains using the sequences of internal fragments of up to seven housekeeping genes. Using an automated DNA sequencer, internal fragments of about seven genes (~450–500 bp) are sequenced on both strands. Distinct alleles are assigned to each housekeeping gene based on the diversity of the sequence present within a bacterial species. The alleles define the sequence type (ST), and each isolate of a species is explicitly exemplified by a series of multiple integers which match the alleles. Using seven housekeeping loci, it has been

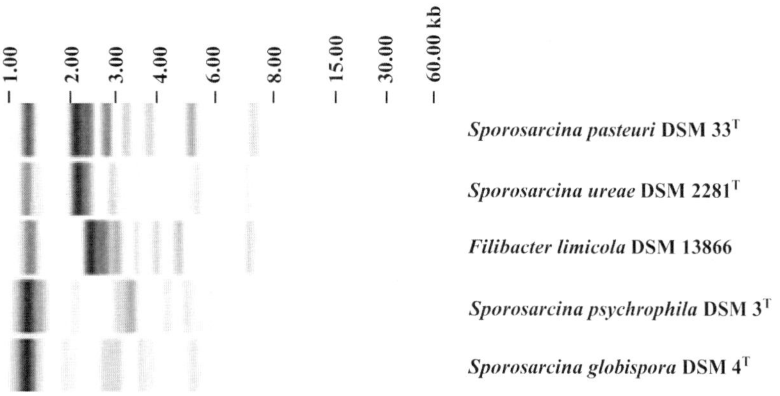


**Fig. 25.4**
Riboprint™ patterns of the DNA of *Filibacter limicola* and the type species of the genus *Sporosarcina*, *Sporosarcina urea* and three closely related species of the genus *Sporosarcina* (Data taken from Stackebrandt (2009))

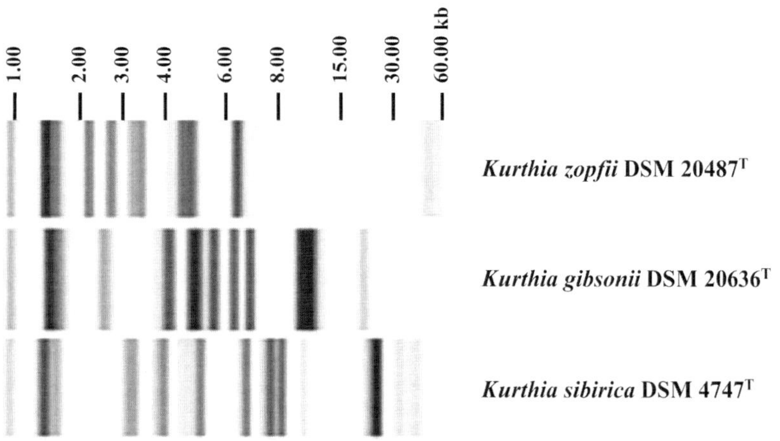


**Fig. 25.5**
Diversity of normalized *Eco*RI ribotype patterns found within the type strains of the genus *Kurthia* (Data taken from Pukall and Stackebrandt (2009))

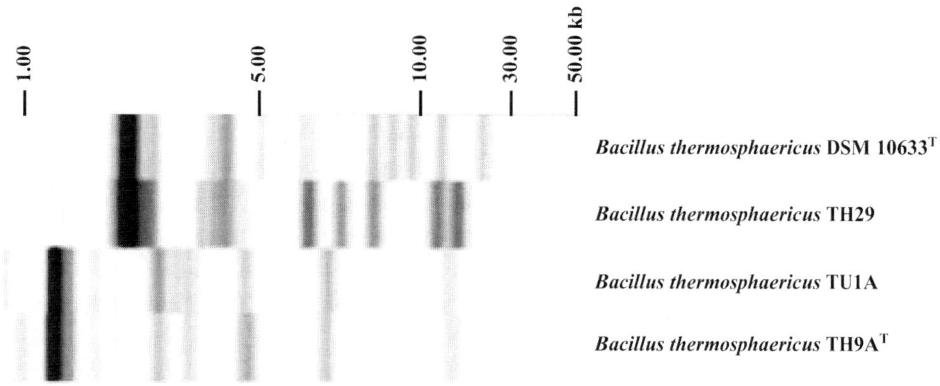


**Fig. 25.6**
Diversity of normalized ribotype patterns found within isolates related to strains of *Bacillus thermosphaericus* (Data taken from Fortina et al. (2001))

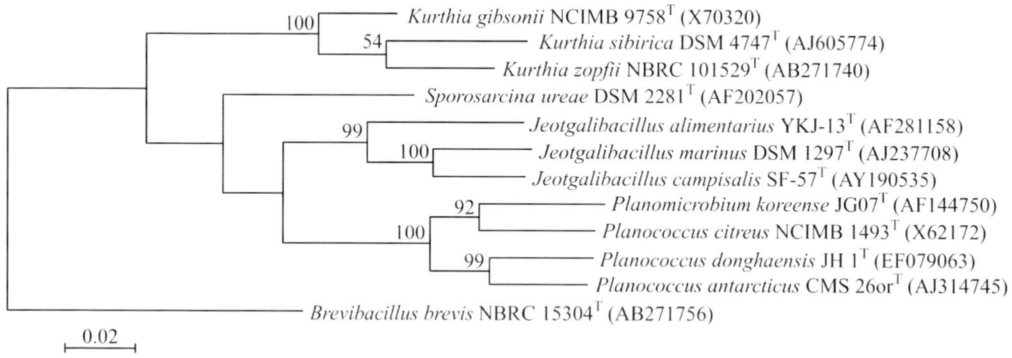


**Fig. 25.7**

**Maximum-likelihood tree generated on the basis of concatenated 16S rRNA and *gro*EL gene sequences based on the Tamura-Nei model. The tree with the highest log likelihood (8924.2140) is shown. The tree is drawn to scale, with branch lengths measured as the number of substitutions per site. The analysis involved 11 nucleotide sequences of members of the family *Planococcaceae*. All positions containing gaps and missing data were eliminated. There were a total of 1,906 positions in the final dataset**

observed that majority of the bacterial species have ample disparity with respect to the number of alleles per locus, allowing billions of discrete allelic profiles to be distinguished. MLST is based on multilocus enzyme electrophoresis but varies in that it assigns alleles at multiple housekeeping loci directly by sequence information rather than circuitously through the banding pattern of their gene products.

Vema et al. (2012) reclassified *Bacillus beijingensis* and *B. ginsengi* (Qiu et al. 2009) as *Bhargavaea beijingensis* comb. nov. and *Bhargavaea ginsengi* comb. nov. based on the phylogenetic analysis of 16S rRNA gene sequence and *GyrB* amino acid sequence. *Bhargavaea cecembensis*, *Bacillus beijingensis*, and *Bacillus ginsengi* constituted a deeply rooted cluster separated from the clades represented by the genera *Bacillus*, *Planococcus*, *Planomicrobium*, *Sporosarcina*, *Lysinibacillus*, *Viridibacillus*, *Kurthia*, and *Geobacillus*, supporting their placement in the genus *Bhargavaea* (Vema et al. 2012).

The phylogenetic tree constructed with concatenated 16S rRNA and *gro*EL gene sequences of 11 members of the family *Planococcaceae* was similar to the clustering observed when only 16S rRNA gene sequences were used for the analysis (➋ *Figs. 25.2* and ➋ *25.7*).

## Plasmids

Species of the genera *Bhargavaea* and *Sporosarcina* contain a mobilizable plasmid carrying *tet* (L) gene (You et al. 2012). Both *Bhargavaea* sp. DMV46A and *Bhargavaea* sp. DMV9 contain plasmids pBSDMV46A and pBSDMV9, respectively, with relaxase and tetracycline efflux pump gene in plasmid pBSDMV46A and rep gene on plasmid pBSDMV9. Chanda et al. (2010) confirmed that in *Kurthia gibsonii*, resistance to ampicillin, bacitracin, and cefotaxime is due to genes located in the plasmid. Curing of the plasmid from the *Planococcus* sp. strain S5 showed that plasmid pLS5 was involved in salicylate degradation (Labuzek et al. 2003).

## Phages

Bacteriophages specific to the genus *Caryophanon* include $Csl_{x13}b$ (*Myoviridae*, morphotype A1), $Csl_{x13}a$, $Ct_{kas}$ (*Siphoviridae*, morphotype B1), and øCVL-29 (*Siphoviridae*, morphotype B2) (Bacteriophage Names 2000). These phages have been shown to be either specific to *C. latum* or *C. tenue* or active against both species (Trentini 1978). These bacteriophages produced either clear or turbid center plaques on native isolates of *C. latum* and produced clear plaques on *C. tenue* but were inactive against *C. latum* isolated from cow dung (Peshkoff et al. 1966; 1967). Nauman and Wilkie (1974) isolated a plaque-forming bacteriophage (øCL-29) and a clear plaque mutant (øCLV-29) with the aid of the mutagen *N*-methyl-*N*′-nitro-*N*-nitrosoguanidine (NG) from C. *latum* (Nauman and Wilkie 1974).

Rocourt et al. (1984) isolated 6/K27 a bacteriophage of *Kurthia zopfii* belonging to the family *Podoviridae* (http://ictvdb.bio-mirror.cn/WIntkey/Images/em_6-K27-HWA.htm). There are no predicted phage-related sequences, transposons, insertion elements, extrachromosomal elements, or pseudogenes in *Planococcus donghaensis* strain MPA1U2 (Pearson and Noller 2011). But *Planococcus antarcticus* DSM 14505$^T$ contains a phage protein in the genome (Margolles et al. 2012). *Sporosarcina newyorkensis* 2681 contains *mal*R (gene encoding phage integrase family site-specific recombinase), phage protein, phage methyltransferase, and phage infection protein in the genome (EGQ26312, EGQ26741, EGQ21095, EGQ26766) (Data from the NCBI database, protein).

## Phenotypic Analyses

The main features of members of *Planococcaceae* are listed in ➲ *Table 25.1.*

### *Planococcus* (Migula 1894, Emend. Nakagawa et al. 1996; Emend. Yoon et al. 2010)

*Planococcus* (*Plan.o.coc'cus.* Gr.n. *planos* wanderer; Gr. N. *coccus* a grain, berry; N.L. masc. N. *Planococcus* motile coccus).

The genus *Planococcus* was originally described by Migula (1894) and later on emended twice by Nakagawa et al. (1996) and then by Yoon et al. (2010b). Strains belonging to the genus *Planococcus* can be identified based on the following characteristics: cells are cocci or short rods or rods, motile (possess one or more flagella) or nonmotile, Gram-positive to Gram-variable, nonspore forming, and occur singly, in pairs, in threes, or in tetrads. The color of the cell mass is yellow to orange. Strictly aerobes, halotolerant, catalase positive, and urease negative. The cell wall peptidoglycan is L-Lys-D-Glu with a peptidoglycan variation of A4α; the major isoprenoid quinones are MK-6, MK-7, and MK-8; the cellular fatty acids are mainly branched fatty acids and dominated by iso-$C_{15:0}$, anteiso-$C_{15:0}$, and iso-$C_{16:0}$; and predominant polar lipids are phosphatidylglycerol and diphosphatidylglycerol. The G+C content of the DNA is 39–52 mol%. In addition to the above characteristics, the genus *Planococcus* can be differentiated from the nearest phylogenetic neighbor, the *Planomicrobium*, on the basis of 16S rRNA gene signature nucleotides wherein at position 183 and 190, the nucleotides, respectively, T and A are conserved in case of *Planococcus* (Dai et al. 2005) whereas in case of *Planomicrobium* C and G are conserved at positions 183 and 190.

The type species of the genus is *Planococcus citreus* (Migula 1894), and the type strain is ATCC 14404$^T$ = CIP 81.74$^T$ = DSM 20549$^T$ = IFO (now NBRC) 15849$^T$ = JCM 2532$^T$ = LMG 17319 = VKM B-1307$^T$.

Presently, the genus contains 11 species: *P. antarcticus* (Reddy et al. 2002), *P. citreus* (Migula 1894), *P. columbae* (Suresh et al. 2007), *P. donghaensis* (Choi et al. 2007), *P. halocryophilus* (Mykytczuk et al. 2012), *P. kocurii* (Hao and Komagata 1985), *P. maitriensis* (Alam et al. 2003), *P. maritimus* (Yoon et al. 2003; Ivanova et al. 2006), *P. plakortidis* (Kaur et al. 2012), *P. rifietoensis* (Romano et al. 2003), and *P. salinarum* (Yoon et al. 2010b). The discriminative characteristics are listed in the ➲ *Tables 25.2* and ➲ *25.3.* All the 11 species are negative for growth at pH 5 and positive for growth at pH 7, at 15, 25, and 30 °C, and in 2–7 % NaCl. Except *P. kocurii* (Hao and Komagata 1985) and *P. halocryophilus* (Mykytczuk et al. 2012), nine species are negative for the following biochemical tests performed using VITEK2-GP plates: cyclodextrin, α-mannosidase, phosphatase, leucine arylamidase, β-glucuronidase, D-sorbitol, urease, polymyxin B resistance, D-galactose, D-ribose, L-lactate alkalinization, lactose, methyl β-D-glucopyranoside, pullulan, trehalose, dihydrolase, and resistance to optochin. Negative for acid production from xylose, trehalose, inositol, adonitol, dulcitol, inulin, sorbitol, rhamnose, lactose, melibiose, and arabinose. Positive for oxidation of D-sorbitol using Biolog GP2 MicroPlates; negative for α-cyclodextrin, β-cyclodextrin, glycogen, inulin, mannan, *N*-acetyl-D-glucosamine, *N*-acetyl-β-D-mannosamine, amygdalin, D-arabitol, L-fucose, gentiobiose, *myo*-inositol, lactose, melezitose, melibiose, methyl α-D-galactoside, methyl β-D-galactoside, methyl α-D-glucoside, methyl β-D-glucoside, methyl α-D-mannoside, L-rhamnose, sedoheptulosan, D-tagatose, xylitol, α-hydroxybutyric acid, *p*-hydroxyphenylacetic acid, α-ketoglutaric acid, lactamide, D-lactic acid methyl ester, L-lactic acid, D-malic acid, *N*-acetyl-L-glutamic acid, glycyl-L-glutamic acid, L-pyroglutamic acid, AMP, D-fructose 6-phosphate, α-D-glucose 1-phosphate, and D-glucose 6-phosphate.

Most of the characteristics for species *P. halocryophilus* (Mykytczuk et al. 2012) are included in ➲ *Tables 25.2* and ➲ *25.3* except the following: utilizes rhamnose, lactose, pectin, dextrin, acetic acid, gelatin, and gluconic acid but not melibiose, inositol, methyl pyruvate, aminobutyric acid, ketobutyric acid, and bromosuccinic acid. Cells produce acid from rhamnose and L-arabinose but not from inositol, sorbitol, and amygdalin. *P. halocryophilus* is positive for arginine decarboxylase, lysine decarboxylase, and production of $H_2S$ but negative for ornithine decarboxylase and tryptophan deaminase.

Species *P. kocurii* was poorly characterized compared to the other species of the genus. Most of the characteristics are listed in ➲ *Tables 25.2* and ➲ *25.3*, and in addition to them, it exhibits following features as characterized by Jung et al. (1998): *P. kocurii* is positive for utilization of propionate, pyruvate, benzoate, α-ketoglutarate, acetate, aspartate, Na-butyrate, glucolic acid, and methanol but not fumarate, glutamate, lactose, L-proline, L-leucine, and *N*-acetyl-glucosamine.

### *Bhargavaea* (Manorama et al. 2009 Emend. Verma et al. 2012)

*Bhargavaea* (Bhar.ga.va'e.a. N.L. fem. n. *Bhargavaea* named in honor of Pushpa Mittra Bhargava, the renowned Indian biologist).

Genus *Bhargavaea* was described by Manorama et al. (2009) and emended by Verma et al. (2012). The genus presently contains three species: *Bhargavaea cecembensis* (Manorama et al. 2009), *Bhargavaea beijingensis*, and *Bhargavaea ginsengi* (Verma et al. 2012). The latter two species were originally described as *Bacillus beijingensis* and *Bacillus ginsengi* (Qiu et al. 2009) but were transferred to the genus *Bhargavaea* based on their 16S rRNA gene sequence and *GyrB* amino acid sequence.

The emended description of the genus is as follows: cells are Gram-positive, aerobic, nonspore forming, rod to coccoid, and rod shaped that are positive for catalase, oxidase, and lipase

**Table 25.2**

Diagnostic phenotypic characteristics of species of the genus *Planococcus* (Data from Kaur et al. (2012); Mykytczuk et al. (2012); Hao and Komagata (1985); Nakagawa et al. (1996); Engelhardt et al. (2001); Yoon et al. (2003); Shivaji (2009)). Strains: 1, *P. plakortidis* AS/ASP6 (II)$^{T\ a}$; 2, *P. maritimus* TF-9$^{T\ a}$; 3, *P. rifietoensis* M8$^{T\ a}$; 4, *P. maitriensis* S1$^{T\ a}$; 5, *P. citreus* NCIMB 1493$^{T\ a}$; 6, *P. salinarum* ISL-16$^{T\ a}$; 7, *P. columbae* PgEx11$^{T\ a}$; 8, *P. donghaensis* JH1$^{T\ a}$; 9, *P. antarcticus* CMS 26or$^{T\ a}$; 10. *P. halocryophilus* Or1$^{T\ b}$; 11. *P. kocurii* DSM 20747$^{T\ c}$

| Characteristic | 1 | 2 | 3 | 4 | 5 | 6 | 7 | 8 | 9 | 10 | 11 |
|---|---|---|---|---|---|---|---|---|---|---|---|
| **Morphological characteristics** | | | | | | | | | | | |
| Colony color | Dull yellow/ orange | Yellow/ orange | Orange | Orange | Orange/yellow | Pale yellow | Orange | Orange | Orange | Bright orange | Orange/ yellow |
| Cell shape | Cocci | Cocci | Cocci | Cocci | Cocci | Cocci/short rods/rods | Cocci | Cocci | Cocci | Cocci | Cocci |
| Cell arrangement | Pairs or clusters | NR | Single, pairs, tetrads, clumps | Single, pairs, in threes, tetrads | Single, pairs, in threes, tetrads | NR | NR | Single, pairs, in threes, tetrads | Single, pairs, in threes, tetrads | Single or pairs | Single, pairs, in threes |
| Motility | + | + | + | + | + | – | + | + | + | + | + |
| **Growth characteristics** | | | | | | | | | | | |
| pH Range | 6–10 | 5–8 | 6–10.5 | 6–12 | 6–11 | 6–7.5 | 6–11 | 7–8 | 6–12 | 6–11 | NR |
| Temperature range | 15–37 | 4–41 | 5–42 | 0–30 | 4–37 | 4–38 | 8–42 | 4–37 | 0–30 | 10–37 | 4–37 |
| 37 °C | + | + | + | – | + | + | + | + | – | + | + |
| 42 °C | – | + | + | – | – | – | + | – | – | – | – |
| NaCl tolerance (%) | 7 | 17 | 15 | 12.5 | 15 | 13 | 14 | 12 | 12 | 19 | 10 |
| 10 % NaCl | – | + | + | + | + | + | + | – | + | + | + |
| 15 % NaCl | – | + | + | – | + | – | – | – | – | + | – |
| **Biochemical characteristics** | | | | | | | | | | | |
| Methyl red test | + | + | – | – | – | – | – | – | + | NR | – |
| Voges-Proskauer test | – | – | – | – | – | – | – | + | – | – | – |
| Indole test | – | – | – | – | – | – | – | – | – | + | – |
| Arginine dihydrolase | – | – | – | – | – | – | – | – | – | + | – |
| Nitrate reduction | – | – | – | + | – | – | – | + | + | – | – |
| Oxidase | + | – | + | – | – | + | + | + | + | + | – |
| **Hydrolysis of** | | | | | | | | | | | |
| Starch | – | – | – | – | – | – | – | + | – | – | – |
| Casein | – | + | – | – | – | + | – | + | – | – | NR |
| Gelatin | – | + | + | – | + | – | – | + | – | + | v |
| Aesculin | – | – | – | – | – | – | – | + | – | – | – |

| | | | | | | | | | | | |
|---|---|---|---|---|---|---|---|---|---|---|---|
| **VITEK2-GP plate tests** | | | | | | | | | | | |
| β-Galactosidase | − | + | + | − | + | − | − | − | − | + | NR |
| α-Glucosidase | − | − | − | − | + | − | − | − | − | NR | NR |
| Ala-Phe-Pro arylamidase | − | + | + | − | + | + | − | − | − | NR | NR |
| L-Aspartate arylamidase | − | + | + | + | − | − | − | − | − | NR | NR |
| β-Galactopyranosidase | − | + | + | − | + | − | − | + | − | NR | NR |
| L-Proline arylamidase | + | + | + | + | + | + | + | − | + | NR | NR |
| α-Galactosidase | − | − | + | − | + | − | − | − | − | NR | NR |
| Alanine arylamidase | − | + | − | − | − | − | − | − | − | NR | NR |
| Tyrosine arylamidase | − | + | − | − | − | − | − | + | − | NR | NR |
| L-Pyrrolidonyl arylamidase | + | − | + | − | − | − | − | + | − | NR | NR |
| **Acid production from sugars** | | | | | | | | | | | |
| Glucose | + | + | − | − | + | − | − | − | − | + | + |
| Galactose | − | + | − | − | − | − | − | + | − | NR | − |
| Fructose | − | + | − | + | − | − | − | + | − | NR | W |
| Mannitol | − | + | − | − | − | − | + | + | − | + | − |
| Cellobiose | − | + | − | − | − | − | − | − | − | NR | − |
| Salicin | − | + | − | − | + | + | − | − | + | NR | NR |
| Mannose | − | + | − | + | − | − | − | + | − | NR | − |
| Maltose | + | − | − | − | + | − | − | + | − | NR | NR |
| Raffinose | − | − | − | − | + | − | − | − | − | − | − |
| Sucrose | + | − | − | − | + | − | + | − | − | + | − |
| Fructose | − | −* | − | + | − | + | − | + | − | NR | NR |
| **Substrate oxidation (Biolog GP MicroPlates)** | | | | | | | | | | | |
| L-Arabinose | + | − | − | − | − | − | + | − | + | NR | NR |
| Arbutin | − | − | − | − | − | − | − | + | − | NR | NR |
| Cellobiose | − | + | − | − | − | − | − | − | − | + | NR |
| Dextrin | − | + | − | + | + | − | − | + | + | NR | NR |
| D-Fructose | − | + | + | + | + | − | − | + | + | + | − |
| D-Galactose | − | + | − | − | − | − | − | + | + | + | NR |
| D-Galacturonic acid | − | − | − | − | − | − | − | − | + | NR | − |
| D-Glucose | − | + | − | − | + | − | − | + | − | + | + |
| D-gluconic acid | − | − | − | − | − | − | − | − | − | + | |

**Table 25.2 (continued)**

| Characteristic | 1 | 2 | 3 | 4 | 5 | 6 | 7 | 8 | 9 | 10 | 11 |
|---|---|---|---|---|---|---|---|---|---|---|---|
| *N*-acetyl-D-glucosamine | – | – | – | – | – | – | – | – | – | + | – |
| Lactulose | – | + | + | – | – | – | – | – | – | NRA | NR |
| Maltose | + | – | – | – | + | – | + | + | – | + | NR |
| Maltotriose | – | – | – | – | – | – | + | – | – | NR | NR |
| D-Mannitol | – | – | + | + | + | – | – | + | – | + | NR |
| D-Mannose | – | – | + | – | + | – | – | + | – | + | NR |
| 3-Methyl D-glucose | – | + | – | – | – | – | – | – | – | NR | NR |
| Palatinose | – | – | – | + | – | – | – | – | + | NR | NR |
| D-Psicose | – | + | + | + | – | – | – | + | – | NR | NR |
| Raffinose | – | + | – | – | – | – | – | – | – | – | NR |
| D-Ribose | + | + | + | + | + | + | + | + | + | + | – |
| Stachyose | – | – | – | – | – | – | + | – | – | NR | NR |
| Salicin | – | + | – | – | + | + | – | – | + | NR | NR |
| Sucrose | + | – | – | – | + | – | + | – | – | + | – |
| Trehalose | – | – | – | – | – | – | – | + | – | + | NR |
| Turanose | – | – | – | – | – | – | – | + | – | NR | NR |
| Tween 40 | + | – | – | – | – | + | – | + | – | + | NR |
| Tween 80 | + | – | – | – | – | + | + | – | – | – | – |
| D-Xylose | + | – | + | – | – | – | – | + | + | NR | NR |
| Acetic acid | + | – | + | + | + | + | – | – | + | NR | NR |
| β-Hydroxybutyric acid | – | – | – | – | – | + | – | + | – | NR | NR |
| γ-Hydroxybutyric acid | – | – | – | – | – | + | – | – | – | NR | NR |
| α-Ketovaleric acid | + | – | + | + | + | + | – | + | – | NR | NR |
| L-Malic acid | + | – | – | – | – | – | – | – | – | NR | NR |
| Pyruvic acid methyl ester | + | – | + | – | + | + | – | + | – | NR | NR |
| Succinic acid monomethyl ester | – | + | + | – | – | – | + | + | – | NR | NR |
| Propionic acid | – | – | – | + | – | – | – | – | – | NR | NR |
| Pyruvic acid | + | – | + | – | + | + | + | – | – | NR | NR |
| Succinamic acid | + | – | – | – | – | – | – | + | – | NR | NR |

| | | | | | | | | | | | |
|---|---|---|---|---|---|---|---|---|---|---|---|
| Succinic acid | – | – | – | – | – | – | – | + | – | NR | NR |
| L-Alaninamide | + | – | + | – | + | + | – | – | – | NR | NR |
| D-Alanine | + | – | – | – | + | – | – | – | – | NR | NR |
| L-Alanine | + | – | – | – | + | – | – | – | – | + | NR |
| L-Alanyl – glycine | + | – | – | – | + | – | – | – | – | NR | NR |
| L-Arginine | – | – | – | – | – | – | – | – | – | + | NR |
| L-Asparagine | + | – | – | – | – | – | – | – | – | NR | NR |
| L-Glutamic acid | + | – | – | – | – | – | – | + | – | + | NR |
| L-Serine | + | – | – | – | + | – | – | + | – | + | NR |
| Putrescine | – | – | – | – | – | – | + | – | – | NR | NR |
| 2, 3-Butanediol | – | – | – | – | – | + | – | – | – | NR | NR |
| Glycerol | – | + | – | – | + | + | – | + | – | + | NR |
| Adenosine | – | + | – | – | – | + | – | – | – | NR | NR |
| Deoxyadenosine | – | + | – | – | – | + | – | – | – | NRD | NR |
| Inosine | – | + | + | – | – | + | – | – | – | NR | NR |
| Thymidine | – | + | – | + | – | + | – | + | – | NR | NR |
| Uridine | – | + | + | – | + | + | – | – | – | NR | NR |
| TMP | – | + | – | – | – | + | – | – | – | NR | NR |
| UMP | – | – | – | – | – | + | – | – | – | NR | NR |
| DL-α-Glycerol phosphate | – | + | – | – | – | – | – | – | – | NR | NR |
| **Chemotaxonomic characteristics** | | | | | | | | | | | |
| Menaquinone | MK-6, MK-7, MK-8 | MK-6, MK-7, MK-8 | MK-8 | MK-7, MK-8 | MK-6, MK-7, MK-8 | MK-7, MK-8 | MK-7, MK-7($H_2$), MK-8 | MK-7, MK-8 | MK-7, MK-8 | MK-6, MK-7, MK-8 | MK-7, MK-8 |
| Polar lipids | PG, DPG, PE | PG, DPG, PE | PG, DPG, PC | NR | PG, DPG, PE | NR | PG, DPG, PC | PG, DPG, PE | PG, DPG, PE | NR | PG, DPG, PE |
| G+C content of DNA (mol%) | 51 | 48 | 48 | 39 | 48.5 | 48.3 | 50.5 | 47 | 41.5 | 40.5 | 40-43 |
| Source | Marine sponge | Sea water | Algal mat | Cyanobacterial mat | Shrimp and sea food | Marine sediment | Pigeon feces | Deep-sea sediment | Cyanobacterial mat | Perma-frost | Fish and frozen foods |

[a] Kaur et al. (2012)
[b] Mykytczuk et al. (2012)
[c] Hao and Komagata (1985); Nakagawa et al. (1996); Engelhardt et al. (2001); Yoon et al. (2003); Shivaji (2009)
\+ positive, – negative, *v* variable, *NR* not reported, *S* sensitive, *R* resistant, *MK* menaquinone, *PG* phosphatidylglycerol, *DPG* diphosphatidylglycerol, *PE* phosphatidylethanolamine, *PC* phosphatidylcholine. For antibiotic concentration, check the original papers since for different species different concentrations have been used

**Table 25.3**

**Cellular fatty acid composition (%) of species of the genus *Planococcus* (Data from Kaur et al. (2012); Mykytczuk et al. (2012)). Strains: 1, *P. plakortidis* AS/ASP6 (II)[T a]; 2, *P. maritimus* TF-9[T a]; 3, *P. rifietoensis* M8[T a]; 4, *P. maitriensis* S1[T a]; 5, *P. citreus* NCIMB 1493[T a]; 6, *P. salinarum* ISL-16[T a]; 7, *P. columbae* PgEx11[T a]; 8, *P. donghaensis* JH1[T a]; 9, *P. antarcticus* CMS 26or[T a]; 10. *P. halocryophilus* Or1[T b]; 11. *P. kocurii* DSM 20747[T b]**

| Fatty acid | 1 | 2 | 3 | 4 | 5 | 6 | 7 | 8 | 9 | 10 | 11 |
|---|---|---|---|---|---|---|---|---|---|---|---|
| **Straight chain** | | | | | | | | | | | |
| $C_{12:0}$ | NR | NR | NR | 2.9 | NR | 9.6 | NR | 3.2 | NR | NR | NR |
| $C_{14:0}$ | 1.5 | NR | 0.6 | 2.4 | NR | 5.5 | NR | 2.3 | NR | 0.5 | NR |
| $C_{16:0}$ | 1.9 | 1.9 | 6.8 | 9 | 4 | 5.4 | NR | 6.3 | 2.3 | 7.0 | 1–2.7 |
| $C_{18:0}$ | 0.5 | NR | 5.9 | 9.2 | 2 | 7.4 | NR | 8.9 | NR | 2.9 | 0–2.2 |
| **Branched chain** | | | | | | | | | | | |
| iso-$C_{14:0}$ | 9.6 | 2.8 | 2.8 | 3.4 | NR | 5 | 11.1 | 5.4 | 1 | 1.2 | 6–16 |
| iso-$C_{15:0}$ | 13.2 | 12.1 | 6.9 | 5.5 | 1.9 | NR | 13.6 | 2.6 | 2.5 | 1.7 | 6–14 |
| anteiso-$C_{15:0}$ | 34.2 | 40.5 | 37 | 35.5 | 56 | 35.1 | 41 | 32.5 | 53.1 | 46.0 | 40–48 |
| iso-$C_{16:0}$ | 11.7 | 6.1 | 7.6 | 8.2 | 3.6 | 11.8 | 11.1 | 7.7 | 3.2 | 2.9 | 4–11 |
| iso-$C_{17:0}$ | 2.9 | 5.6 | 4.2 | 3.1 | NR | NR | 3 | 2 | 3.1 | 2.3 | 0–3.0 |
| anteiso-$C_{17:0}$ | 5.5 | 9.3 | 8.6 | 7.5 | 18.8 | 7.7 | 5.3 | 7.4 | 12.4 | 18.0 | 2–14 |
| **Unsaturated** | | | | | | | | | | | |
| $C_{16:1\ (\omega7c)}$ alcohol | 7 | 0.9 | 4 | 5 | 4.5 | 10.8 | 8.4 | 6.8 | 3.8 | 2.1 | 1–11 |
| **Summed feature 4** | NR | NR | NR | NR | NR | NR | NR | NR | NR | 5.2 | NR |
| **Summed feature 5** | 1.2 | 4.5 | 1.9 | NR | 5.4 | NR | 2.5 | 5.1 | 9.4 | NR | NR |

[a]Kaur et al. (2012)
[b]Mykytczuk et al. (2012)
– negative, *NR* not reported, summed feature 4; iso-$C_{17:1}$ I and/or anteiso-$C_{17:1}$ B; Summed feature 5: $C_{18:2(\omega6,9c)}$ and/or anteiso-$C_{18:0}$. If a given fatty acid is present above 5.0 % in at least one species, the composition in the other species is given

activities and some produce urease. The major fatty acids are iso-$C_{15:0}$ and anteiso-$C_{15:0}$. MK-8 is the predominant respiratory quinone. Cell wall peptidoglycan type is A4α and contains L-lysine as the diagnostic diamino acid. The major polar lipids are phosphatidylglycerol and diphosphatidylglycerol. The type species *Bhargavaea cecembensis*, DSE10$^T$, contains the signature nucleotides G, A, C, T, C, A, G, C, and T at positions 182, 444, 480, 492, 563, 931, 1253, 1300, and 1391, respectively, in the 16S rRNA gene sequence. The G+C content of the genomic DNA is 50–54.0 mol%.

Initially, the genus description was based on a single species, and the type strain of the genus, *Bhargavaea cecembensis*, was characterized phenotypically using the HiMedia *Enterobacteriaceae* identification kits KB003 and KB009 A, B, and C. Verma et al. (2012) characterized the three species using the VITEK-2 GP compact system, API ZYM (bioMérieux) and Biolog GP2 MicroPlates along with HiMedia kits besides using the conventional phenotyping methods (Gordon et al. 1973; Kovacs 1956; Simmons 1926).

The type species of the genus is *Bhargavaea cecembensis* (Manorama et al. 2009; emend Verma et al. 2012), and the type strain is DSE10$^T$, = LMG 24411$^T$ = JCM 14375$^T$.

All the three species of *Bhargavaea* are Gram-positive; nonmotile; nonspore forming; positive for oxidase, catalase, caseinase, gelatinase, nitrate reduction, and β-hemolysis; and negative for indole production, Simmons' citrate, methyl red and Voges-Proskauer reactions, phenylalanine deaminase, glucose fermentation, and $H_2S$ production. All the three species are positive for esterase (C4), esterase lipase (C8), leucine arylamidase, and naphthol-AS-BI-phosphohydrolase but negative for alkaline phosphatase, lipase (C14), valine arylamidase, cystine arylamidase, trypsin, α-chymotrypsin, acid phosphatase, α-galactosidase, β-galactosidase, β-glucuronidase, α-glucosidase, β-glucosidase, *N*-acetyl-β-glucosaminidase, α-mannosidase, and α-fucosidase. None of the strains produced acid from adonitol, arabinose, glucose, dulcitol, inositol, inulin, lactose, mannitol, melibiose, raffinose, sorbitol, sucrose, or xylose; however, acid was produced from trehalose and fructose. Oxidation of different carbon substrates (Biolog GP2) shows that all three species are positive for Tween 40, β-hydroxybutyric acid, α-ketoglutaric acid, α-ketovaleric acid, acetic acid, pyruvic acid methyl ester, succinic acid monomethyl ester, pyruvic acid, succinamic acid, L-alaninamide, L-alanyl glycine, L-glutamic acid, β-cyclodextrin, L-serine, and glycerol, while negative for α-cyclodextrin, dextrin, glycogen, inulin, mannan, amygdalin, Tween 80, L-arabinose, D-arabitol, arbutin, cellobiose, D-fructose, L-fucose, D-galactose, D-galacturonic acid, gentiobiose, D-gluconic acid, D-glucose, *myo*-inositol, L-asparagine, lactose, lactulose, maltose, maltotriose, D-mannitol, D-mannose, melezitose, melibiose, methyl α-D-galactoside, methyl β-D-galactoside, 3-methyl glucose, *N*-acetyl-L-glutamic acid, methyl α-D-glucoside, methyl β-D-glucoside, methyl α-D-mannoside, palatinose, D-psicose, raffinose, L-rhamnose, D-ribose, salicin, sedoheptulosan, D-sorbitol, stachyose, sucrose, D-tagatose, trehalose, turanose, xylitol, D-xylose, γ-hydroxybutyric acid, *p*-hydroxyphenylacetic acid, lactamide, D-lactic acid methyl ester, L-lactic acid, propionic acid, L-pyroglutamic acid, putrescine, 2,3-butanediol, D-fructose 6-phosphate, α-D-glucose 1-phosphate, D-glucose 6-phosphate, and D L-glycerol phosphate. Results from the GP VITEK-2 compact system indicated that all strains are positive for ala-phe-pro arylamidase, Ellman test, L-pyrrolidonyl arylamidase, L-aspartate arylamidase, leucine arylamidase, phenylalanine arylamidase, and tyrosine arylamidase and negative for β-xylosidase, L-lysine arylamidase, L-proline arylamidase, α-galactosidase, β-galactosidase, alanine arylamidase, *N*-Acetyl-β-glucosaminidase, β-mannosidase, cyclodextrin, D-galactose, glycogen, *myo*-inositol, methyl α-D-glucopyranoside acidification, methyl D-xyloside, maltotriose, glycine arylamidase, D-mannitol, melezitose, *N*-acetyl-D-glucosamine, palatinose, L-rhamnose, β-glucosidase, phosphorylcholine, pyruvate, α-glucosidase, D-tagatose, inulin, D-glucose, D-ribose, putrescine assimilation, and aesculin hydrolysis. The differentiating characteristics of species of the genus *Bhargavaea* are listed in ❯ *Tables 25.4* and ❯ *25.5.*

## *Caryophanon* (Peshkoff 1939)

*Caryophanon* (Ca.ry.o'pha.non. Gr. N. *Karyon* nut, kernel, nucleus; Gr. adj. *Phaneros* bright, conspicuous; N. L. neut. N. *Caryophanon* that which has a conspicuous nucleus).

The genus presently contains two species, *Caryophanon latum* and *Caryophanon tenue* (Peshkoff 1939; Trentini 1978; Pringsheim and Robinow 1947). The genus *Caryophanon* is a unique prokaryote because cells are extremely large, exhibit unusual structural complexity, and present in specialized ecological niches. This giant bacterium forms trichomes (multicellular rods) with rounded or tapered ends, and the size of the trichome is 1.0–3.5 μm in diameter and 10–40 μm in length (Pringsheim and Robinow 1947). Several trichomes may form short chains without any branching. Peshkoff first isolated this organism from fresh cow dung in Russia in 1939. Growing the organism was a challenging task. The organism grows poorly on routine nutrient agar but thrives on yeast extract plus meat-extract agar, especially if supplemented with small amounts of sodium acetate. A medium containing 0.5 % Bacto yeast extract, 0.5 % Bacto peptone, and 0.01 % sodium acetate, adjusted to pH 7.4–7.6, proved very favorable for isolation and maintenance (Pringsheim and Robinow 1947). Cow dung agar with lactalbumin hydrolysate (Moran and Witter 1976) or a semisynthetic medium containing (gm per liter) lactalbumin hydrolysate (10.0), sodium acetate (5.0), $MgCl_2{\cdot}6H_2O$ (20.0), $CaCl_2$ (11.0), $Cu_2SO_4$ (0.4 mg), $FeSO_4$ (0.152 mg), biotin (0.02 mg), thiamin (0.05 mg), and nitrilotriacetic acid (19.1 mg) (Smith and Trentini 1973; Trentini 1978) supported good growth. Recently Fritze and Claus (2009) could grow both the species of *Caryophanon* using the DSM medium no.34 (per liter: yeast extract [2.0 g], trypticase [2.0 g], soya peptone [2.0 g], sodium acetate [1.0 g], sodium glutamate [0.1 g], thiamine-HCl × 2 $H_2O$ [0.2 mg], biotin [0.05 mg], $K_2HPO_4$ [1.0 g], $MgSO_4.7\ H_2O$ [0.27 g], Tris–HCl buffer [10 mM] and pH 7.8) for the purpose of fatty acid analyses. In all the cases, biotin is required and thiamine is a stimulator.

Cells of *Caryophanon* can grow from 10 to 40 °C with an optimum of 25–37 °C and can tolerate 4 °C for 4 weeks. In liquid media, it does not multiply, but very small concentrations of agar are sufficient to enable it to grow. After isolating in pure form, the typical cell morphology can be preserved only in liquid media containing cow dung. However, change in morphology of both species during the extended cultivation on agar medium was demonstrated by Peshkov et al. (1978). Cells of *Caryophanon* are Gram-positive, motile rods by means of peritrichous flagella and asporogenous. However, Fritze and Claus (2009) reported that the type strains of the genus *Caryophanon, C. latum* (DSM 14151), and *C. tenue* (DSM 14152) lost motility on culture media. Phenotypic characteristics (❯ *Table 25.6*) of the genus are based on the study by Fritze and Claus (2009), Pringsheim and Robinow (1947), and Trentini (1978). Cells are strictly aerobic and chemoorganotrophic. Carbohydrates are not utilized as substrates, but acetate and other organic acids are utilized as major carbon sources. Cells are catalase positive and negative for oxidase, urease, nitrate reduction, phenylalanine deaminase, and indole production. Both the species can hydrolyze gelatin but not starch, tyrosine, and hippurate. *C. tenue* can hydrolyze Tween 20, 40, 60, and 80. *C. latum* and *C. tenue* are negative for acid production from L-arabinose, D-glucose, mannitol, and xylose, can utilize citrate and propionate, and were resistant to the antibiotics streptomycin, polymyxin B, and nalidixic acid. The other characteristics are listed in ❯ *Tables 25.6* and ❯ *25.7.* According to Adcock et al. (1976), both species of *Caryophanon* had similar antibiotic resistance pattern. The general appearance of the organism and the character of the nuclear structures suggested the existence of a direct relationship to blue-green algae (Peshkoff 1940). However, features such as motility due to the presence of typical, peritrichous flagella, regular binary fission of the composite rods, and the absence of a distinctive slime layer and the 16S rRNA gene sequence placed the genus in Eubacteria. The close phylogenetic relationship between *Caryophanon* and related genera of the family *Planococcaceae* was described by Stackebrandt et al. (1987). Placement of both *Caryophanon* species within rRNA group two (Ash et al. 1991) was confirmed by Farrow et al. (1994) and Fritze and Claus (2009).

The peptidoglycan of *C. latum* and *C. tenue* was very similar and was found to contain glucosamine, muramic acid, alanine, glutamic acid, and lysine in the molar ration of about 2:2: 1:1 (Becker et al. 1967; Trentini 1978). One of the glutamic acid residues is bound to the €-amino group of lysine and was responsible for the cross-linking of the peptide subunit, and therefore, it is of A4α type (Stackebrandt et al. 1987). Cell walls of *Caryophanon* species are sensitive to lysozyme. Menaquinones with six isoprene units (MK-6) are the major isoprenologues present in *Caryophanon*

**Table 25.4**
**Diagnostic phenotypic characteristics of species of the genus *Bhargavaea* (Data from Manorama et al. (2009) and Verma et al. (2012))**

| Characteristic | *B. cecembensis* LMG 24411[Ta] | *B. beijingensis* DSM 19037 [Ta] | *B. ginsengi* DSM 19038 [Ta] |
|---|---|---|---|
| **Morphological characteristics** | | | |
| Cell shape | Rods | Coccoid rods | Coccoid rods |
| Cell arrangement | Single, pairs, or short chains | Single or pairs | Single, pairs, or short chains |
| Cell dimensions (μm) | 1.0 × 2.0–8.0 | 1.0 × 1.2 | 1.0 × 1.2–2.0 |
| **Growth characteristics** | | | |
| Temp. range (°C) | 15–55 | 7–45 | 4–45 |
| Optimum growth (°C) | 37 | 30 | 30 |
| Growth pH range | 7–7.5 | 5.5–11 | 6–11 |
| **Biochemical characteristics** | | | |
| Urease | + | + | − |
| Sucrose fermentation | − | + | + |
| Lysozyme resistance | − | − | + |
| **Acid production from** | | | |
| Galactose | − | − | + |
| Salicin | − | + | + |
| Mannose | + | − | − |
| Rhamnose | + | − | − |
| Cellobiose | + | − | − |
| Maltose | − | − | + |
| **Utilization of substrates (Biolog)** | | | |
| *N*-Acetyl-D-glucosamine | − | − | + |
| *N*-Acetyl-β-D-mannosamine | − | − | + |
| α-Hydroxybutyric acid | + | + | − |
| D-Malic acid | − | + | − |
| L-Malic acid | + | + | − |
| Succinic acid | − | + | + |
| D-Alanine | + | + | − |
| L-Alanine | + | − | + |
| Glycyl-L-glutamic acid | − | + | − |
| Adenosine | − | + | + |
| 2′-Deoxyadenosine | − | + | + |
| Inosine | − | + | + |
| Thymidine | − | + | + |
| **Chemotaxonomic characteristics** | | | |
| Menaquinone | MK-8, MK-9 | MK-8 | MK-7, MK-8 |
| Polar lipids | PG, DPG, PC, UPL | PG, DPG, UPL | PG, DPG, PC, UPL |
| Peptidoglycan composition | L-Lys-D-Asp | L-Lys-D-Glu | L-Lys-D-Glu |
| G+C content of DNA (mol%) | 51 | 48 | 48 |
| Source | Deep-sea sediment | Tissue of ginseng roots | Tissue of ginseng roots |

[a]Data from Manorama et al. (2009) and Verma et al. (2012)
+ positive, − negative, *MK* menaquinone, *PG* phosphatidylglycerol, *DPG* diphosphatidylglycerol, *PC* phosphatidylcholine, *UPL* unidentified phospholipid, l-*Lys* L-lysine, d-*Asp* D-aspartic acid, d-*Glu* D-glutamic acid

(Collins and Jones 1981). The major fatty acids present are iso-$C_{15:0}$ and $C_{16:1(\omega7c)}$ alcohol, and composition of other fatty acids of *C. latum* and *C. tenue* is given in *Table 25.7* (Fritze and Claus 2009).

The main difference between *C. latum* and *C. tenue* is in their nature of trichome. Cells of *C. latum* in the trichome are larger in width than in length and show cross wall formation. In *C. tenue*, cells within trichomes are slightly larger in length than in

**Table 25.5**
**Cellular fatty acid composition (%) of species of the genus *Bhargavaea* [a]**

| Fatty acid | *B. cecembensis* LMG 24411[T] | *B. beijingensis* DSM 19037 [T] | *B. ginsengi* DSM 19038 [T] |
|---|---|---|---|
| **Branched chain** | | | |
| iso-$C_{14:0}$ | 8.9 | 4.2 | 4.9 |
| iso-$C_{15:0}$ | 18.5 | 37.4 | 29 |
| iso-$C_{16:0}$ | 14 | 5.9 | 8.6 |
| anteiso-$C_{15:0}$ | 33.4 | 27 | 32.6 |
| anteiso-$C_{17:0}$ | 9.5 | 11.6 | 9.9 |

[a]Data from Manorama et al. (2009) and Verma et al. (2012). If a given fatty acid is present above 5.0 % in at least one species, the composition in the other species is given

**Table 25.6**
**Diagnostic phenotypic characteristics of species of the genus *Caryophanon*[a]**

| Characteristic | *C. latum* DSM14151[T] | *C. tenue* DSM 14152[T] |
|---|---|---|
| **Morphological characteristics** | | |
| Trichome width (μm) | 2.3–3.5 | 1.0–2.0 |
| Trichome length (μm) | 6.0–20.0 | 4.0–10.0 |
| No. of cells per trichome | 4–15 | 2–3 |
| **Biochemical characteristics** | | |
| Indole production | + | – |
| **Hydrolysis of** | | |
| Tween 80 | – | + |
| Tween 60 | – | + |
| Tween 40 | – | + |
| Tween 20 | – | + |
| **Chemotaxonomic characteristics** | | |
| G+C content of DNA (mol%) | 44–45.6 | 41.2–41.6 |
| Genome length | $1100–1200 \times 10^6$ | $900–1000 \times 10^6$ |
| Source | Cow dung | Cow dung |

[a]Data from Fritze and Claus (2009), Trentini (1978), and Pringsheim and Robinow (1947)
+ positive, – negative

width and lack the multiple septation typical of *C. latum* and show mostly only one cross septum in a trichome (Peshkov and Marek 1972). Further, the size of trichomes of *C. latum* varies between $2.3 \times 6.0$ and $3.5 \times 20$ μm, and the cell number ranges from 4 in the shorter to 15 in the longer trichomes. The trichome dimensions of *C. tenue* range between $1.0 \times 4.0$ and $2.0 \times 10.0$ μm, and the cell number range from 2 to 3. In addition, *C. latum* has a preference for acetate as a carbon source followed by butyrate (Provost and Doetsch 1962), and in contrast,

**Table 25.7**
**Cellular fatty acid composition (%) of species of the genus *Caryophanon* (Data from Fritze and Claus (2009))**

| Fatty acid | *C. latum* DSM14151[Ta] = NCIMB 9533[Ta] | *C. tenue* DSM 14152[Ta] |
|---|---|---|
| **Straight chain** | | |
| $C_{16:0}$ | 4.0 | 5.5 |
| **Branched chain** | | |
| $C_{14:0\ iso}$ | 4.3 | 5.4 |
| $C_{15:0\ iso}$ | 39.4 | 28.8 |
| $C_{16:0\ iso}$ | 1.7 | 10.6 |
| **Unsaturated** | | |
| $C_{16:1(\omega7c)}$ alcohol | 8.7 | 12.5 |
| $C_{16:1(\omega11c)}$ | 23.3 | 18.5 |

[a]Fritze and Claus (2009)
Note: DSM medium no. 34 was used for growing *Caryophanon*. If a given fatty acid is present above 5.0 % in at least one species, the composition in the other species is given
–, absent

*C. tenue* has a preference for valerate and capronate to acetate (Rowenhagen 1987).

The type species of the genus is *Caryophanon latum* (Peshkoff 1939), and the type strain is ATCC 33407[T] = LMG 17312[T] = NCIMB 9533[T] = VKM B-105[T].

## *Chryseomicrobium* (Arora et al. 2011)

*Chryseomicrobium* (Chry'se.o.mi.cro'bi.um. Gr. adj. *chruseos* golden; N.L. neut. n. microbium microbe from Gr. adj. mikros small and Gr. n. bios life; N.L. neut. n. *Chryseomicrobium* yellow microbe).

The type strain of *Chryseomicrobium* was isolated from sea water of the Bay of Bengal, India (Arora et al. 2011). The description of the genus is based on a single species, *Chryseomicrobium imtechense.* Cells of *Chryseomicrobium* are Gram-positive, nonmotile, nonspore forming, and rod shaped, and the cells size is 0.3–0.7 μm wide and 1.7–2.9 μm long. Grows from 4 to 45 °C with an optimum of 30 °C, pH range is 6–9, and can tolerate 6 % NaCl. Catalase positive, casein is hydrolyzed but not gelatin, starch, ONPG, Tween 40, Tween 80, and aesculin. Negative for egg yolk reaction, oxidase, urease, $H_2S$ production, indole formation, methyl red and Voges-Proskauer tests, and nitrate is not reduced to nitrite. No growth on MacConkey or Simmons' citrate agar. Acid is produced from glucose, salicin, fructose, maltose, sucrose, inulin, melibiose, and cellobiose.

In the Biolog GP2 system, cells of *Chryseomicrobium* are positive for oxidation of α-ketovaleric acid, glycerol, γ-hydroxybutyric acid, pyruvic acid, L-serine, D-ribose, pyruvic acid methyl ester, L-alanine, 2-deoxyadenosine, and L-arabinose and negative for oxidation of α- and β-cyclodextrin, glycogen, inulin, mannan, lactose, methyl β-D-glucoside, L-alaninamide, trehalose, D-lactic acid methyl ester, D-alanine, arbutin, L-lactic acid, xylitol, D- and L-malic acid, raffinose, L-asparagine, L-fucose, L-rhamnose, adenosine 5′-monophosphate, melezitose, α-hydroxybutyric acid, glycyl-L-glutamic acid, D-galacturonic acid, melibiose, salicin, propionic acid, L-pyroglutamic acid, uridine 5′-monophosphate, gentiobiose, methyl α-D-galactoside, sedoheptulosan, D-fructose 6-phosphate, *N*-acetyl-b-glucosamine, D-gluconic acid, methyl β-D-galactoside, D-sorbitol, *p*-hydroxyphenylacetic acid, succinamic acid, putrescine, α-D-glucose 1-phosphate, amygdalin, *myo*-inositol, methyl α-D-glucoside, sucrose, *N*-acetyl L-glutamic acid, and DL-α-glycerol phosphate.

The cell wall peptidoglycan is of the A4α type with an interpeptide bridge containing L-Lys-D-Asp, major menaquinones are MK-7 and MK-8, which contribute to 80.7 % and MK-7 ($H_2$), and MK-6 are present but in lesser amounts. The cellular fatty acid profile includes $C_{16:0}$, $C_{16:1(\omega 7c)}$ alcohol, $C_{16:1(\omega 11c)}$, iso-$C_{14:0}$, iso-$C_{15:0}$, anteiso-$C_{15:0}$, iso-$C_{16:0}$, iso-$C_{17:0}$, and iso-$C_{17:1(\omega 10c)}$ of which iso-$C_{15:0}$ (68.6 %) and $C_{16:1(\omega 7c)}$ alcohol (10.9 %) are the predominant fatty acids. The polar lipids present are phosphatidylglycerol, diphosphatidylglycerol, phosphatidylethanolamine, phosphatidylcholine, an unknown phospholipid, an unknown lipid, and an unknown glycolipid. The mol% G+C content of DNA is 53.4.

Cell shape, nonspore-forming nature, presence of high amounts of iso-$C_{15:0}$ and $C_{16:1(\omega 7c)}$ alcohol, and relatively high DNA G+C content make the genus *Chryseomicrobium* unique. Phylogenetically, the genus is closely related to the clades representing *Planococcus* and *Planomicrobium*.

The type specie of the genus *Chryseomicrobium* is *Chryseomicrobium imtechense* (Arora et al. 2011), and the type strain is MW $10^T$ = MTCC $10098^T$ = JCM $16573^T$ and isolated from a seawater.

## *Filibacter* (Maiden and Jones 1984)

*Filibacter* (*Fi.li.bac'ter.* L. n. *filum* a thread; N. L. masc. N. *bacter* masculine form of Gr. neut. N. *bactron* a rod; N. L. masc. N. *Filibacter* thread rod).

The type strain of *Filibacter* was isolated from the sediment of a eutrophic lake on a dilute peptone medium by Maiden and Jones (1984). Cells of *Filibacter* are composed of cylindrical filaments that are straight or curved, filaments are neither sheathed nor branched, and cell junctions are clearly marked. Filaments are 1.1 μm in width and 8–150 μm in length, and the cell dimensions within the filaments are 1.1 μm in width and 3–30 μm in length. Filaments are flexible but do not show any active flexing. Cells of *Filibacter* spread widely over the surface of agar producing whorls of growth and spiral colonies because of the gliding nature of the cell. Cells stained Gram-negative motile by means of gliding, nonspore forming, strictly aerobic, and slimy. They grow from 4 to 26 °C with an optimum at 20 °C and pH optimum of 7–7.4 and had a requirement for vitamins. No single amino acid is required for growth, but all are required for maximum biomass production, and growth does not occur in absence of vitamins. The unique feature of this organism is its inability to utilize organic compounds in the absence of amino acids. However, histidine is inhibitory due to the possible inhibition of glutamine synthetase. Phenotypic characteristics were described by Maiden and Jones (1984) by testing on API 20B strips, and most of the tests showed negative results. But cells of *Filibacter* are positive for catalase, oxidase, β-galactosidase, lipase, chymotrypsin, phosphoamidase, and urease. Hydrolyzes gelatin but not casein or starch and reduces nitrate to nitrite. Could not utilize glucose, fructose, sucrose, lactose, galactose, mannitol, glucosamine, galacturonic acid, acetate, pyruvate, lactate, citrate, glutamate, butyrate, ethanol, and glycerol as carbon sources (Maiden and Jones 1984). However, acetate, butyrate, and glycerol (at 2 mg L) enhanced the growth in presence of amino acids. Among the amino acids, the combination of the following families supported the growth: the glutamate family (glutamate, glutamine, arginine, and proline) plus aspartate family (aspartate, asparagines, lysine, threonine isoleucine, and methionine) or serine family (serine, glycine, and cysteine) and the aspartate family plus the pyruvate family (alanine, valine, and leucine) and serine or aromatic family (tryptophan, phenylalanine, and tyrosine). The interdependence of combination of amino acids from different families could be due to the coupled oxidation-reduction reactions. No single amino acid or a mixture of amino acids from a single family supported the growth (Maiden and Jones 1984). In addition, none of the substrates of API20 NE and none of carbohydrates (sugars and acids) provided by API 50 CHE panels supported the growth. Peptidoglycan contains lysine and the murein type is L-Lys-D-Glu and peptidoglycan variation is A4α type and the major menaquinone is MK-7. Membrane fatty acids are dominated by anteiso-$C_{15:0}$ and anteiso-$C_{17:0}$ (Nichols et al. 1986), and the lone β-hydroxy fatty acid found was $C_{12:0}$ β-OH. Cytochromes present are of *c* type, while *b* type also occurs in lesser amount (Maiden and Jones 1984). The mol% G+C content of DNA of strain *Filibacter* is 44.0 %. Morphologically, the genus *Filibacter* is closely related to *Vitreoscilla* with respect to formation of Gram-negative, unpigmented, multicellular, and gliding filaments. However, the genomic riboprinting (Stackebrandt 2009) and chemotaxonomic characteristics supported the association of *Filibacter* with members of *Bacillus* rRNA group two (Ash et al. 1991). Comparison of 16S rRNA oligonucleotide of the genus *Filibacter* with gliding organisms, *Vitreoscilla, Cytophaga*, and *Flexibacter* and representative of Gram-positive organisms, further, indicated a close phylogenetic relation with rRNA group two (Clausen et al. 1985).

Fatty acid composition of strain *Filibacter limicola* $NCIB11923^T$ was determined by Nichols et al. (1986) by growing the strain in mineral medium (Maiden and Jones 1985; Jones 1983) containing trypticase soy (27 g L). The composition of the medium was (mg per liter): $K_2HPO_4$, 28; $MgCl_2 \cdot 6H_2O$, 1.27; $KNO_3$, 4; $(NH_4)_2SO_4$, 60; $MnSO_4 \cdot 4H_20$, 8; ferric citrate, 6. The pH of the medium was adjusted with 1 M $KHCO_3$, before autoclaving, and the final pH was 7.0–7.2. Composition (%) of fatty acids is $C_{12:0}$ (0.3), anteiso-$C_{17:0}$ (21.3), $C_{14:0}$ (1.8), $C_{15:0}$ (0.1), $C_{16:0}$ (9.2), $C_{17:0}$ (tr), $C_{18:0}$ (0.2), $C_{14:1}$ (0.1), $C_{14:1(\omega5c)}$(0.1), $C_{16:1(\omega5c)}$ (0.1), $C_{16:1(\omega7c)}$ (2.8), $C_{16:1(\omega7t)}$ (0.1), $C_{16:1(\omega9c)}$ (5.4), br $C_{17:1}$ (0.2), iso-$C_{17:1}$ (6.5), $C_{18:1(\omega9c)}$ (0.1), $C_{18:1(\omega7c)}$ (0.3), $C_{18:1(\omega7t)}$ (0.1), and iso-$C_{14:0}$ (tr) (Nichols et al. 1986).

The type species of the genus is *Filibacter limicola* (Maiden and Jones 1985), and the type strain is 1SS101$^T$ = ATCC 43646$^T$ = NCIMB 11923$^T$. Bacterium was isolated from lake sediment.

## *Jeotgalibacillus* (Yoon et al. 2001)

*Jeotgalibacillus* (*Je.ot.ga.li.ba.cil'lus.* Korean n. *jeotgal* jeotgal, traditional Korean food; Gr. n. *baktron* rod; N.L. masc. n. *Jeotgalibacillus* rod from jeotgal).

Genus *Jeotgalibacillus* was described by Yoon et al. (2001a), and the genus presently contains five species, of which two were transferred from *Marinibacillus*. The species present are *Jeotgalibacillus alimentarius* (Yoon et al. 2001a), *Jeotgalibacillus campisalis* (Yoon et al. 2004, 2010a), *Jeotgalibacillus marinus* (Rüger and Richter 1979; Yoon et al. 2010a), *Jeotgalibacillus salarius* (Yoon et al. 2010a), and *Jeotgalibacillus soli* (Cunha et al. 2012). According to the genus description proposed by Yoon et al. (2001a), cells are rod shaped, form round endospores, present in swollen sporangia, and catalase and oxidase positive but urease negative, and nitrate is reduced to nitrite. Majority of the species are negative for lipases, valine arylamidase, cystine arylamidase, trypsin, acid phosphatase, α-galactosidase, β-glucuronidase, α-mannosidase, α-fucosidase, and *N*-acetyl-β-glucosaminidase. Cell wall peptidoglycan contains L-lysine at position three of the peptide subunit. Predominant menaquinones are MK-7 and MK-8. Major fatty acid is iso-$C_{15:0}$. DNA GC content is 44 mol%.

Data is available for all the species with respect to their growth, morphology, enzymes, and acid production characteristics. Species of *Jeotgalibacillus* exhibit white to yellowish orange pigmentation and are strictly aerobic. *J. alimentarius* is facultatively anaerobic, Gram-variable, and spore forming, and the spores are round or ellipsoidal, located terminally or subterminally, and in case of *J. soli*, spores can also be present in the center. The growth temperature ranges from 4 °C to 55 °C, most of them are mesophilic, whereas *J. marinus* is psychrophilic. All of them are halotolerant, and *J. alimentarius* can tolerate up to 20 % of NaCl. Cells are motile by means of peritrichous or single polar or subpolar flagella. Characteristics such as catalase, oxidase, urease, and nitrate reduction are variable among the species. However, in majority of cases, the above properties matched with the genus description of Yoon et al. (2001a). Besides these, all five species shared the following common characteristics: negative for alkaline phosphatase and positive for production of acid from D-fructose, D-glucose, maltose, sucrose, and trehalose but not from L-arabinose, inositol, rhamnose, and sorbitol. The peptidoglycan variation is A1α where the linkage between the peptide chains is direct, major menaquinone is MK-7, and the major lipids present were aminophospholipids APL1 to 4. Most of the strains are characterized with respect to their acid production, enzymes, and antibiotic susceptibility. The differentiating characteristics and fatty acid profiles of all the type species are given in ➲ *Tables 25.8* and ➲ *25.9*.

Phylogenetically, *Jeotgalibacillus* is closely related to *Lysinibacillus*, *Solibacillus*, and *Bacillus*. The genus is unique compared to the members of the family *Planococaceae* in containing the A1α type of peptidoglycan variant and presence of aminophospholipids.

Type species is *Jeotgalibacillus alimentarius* (Yoon et al. 2001a), and the type strain is YKJ-13$^T$ = JCM 10872$^T$ = KCCM 80002$^T$.

## *Kurthia* (Trevisan 1885)

*Kurthia* (*Kurth'i.a.* N.L. fem. Gen. N. *Kurthia* named for H. Kurth, the German bacteriologist who described the type species).

Genus *Kurthia* was described by Trevisan (1885) to accommodate species that are characterized by bird's feather like morphology on gelatin nutrient agar. Cells are aerobic, Gram-positive rods, motile by means of peritrichous flagella, nonspore forming, containing L-Lys-D-Asp type of peptidoglycan, MK-7 as the major menaquinone, iso-$C_{15:0}$ and anteiso-$C_{15:0}$ as major fatty acids, and DNA mol % G+C range of 36–38. Presence of PG, DPG, and PE as predominant polar lipids was characterized in *K. zopfii* and *K. gibsonii*. Besides the above characteristics, 16S rRNA gene sequence analysis is another unambiguous marker in assigning the strains to this genus. The rods are approximately 0.6–1.2 μm in diameter and the length varies from 2 to 5 μm in young cultures and 5–10 μm in old cultures that form filaments. The genus contains three species: *K. zopfii* (Kurth 1883; Trevisan 1885), *K. gibsonii* (Shaw and Keddie 1983), and *K. sibirica* (Belikova et al. 1986). Optimum temperature for growth is 20–30 °C, and *K. gibsonii* can grow up to 45 °C.

Phenotypic characterization of *K. zopfii* and *K. gibsonii* was done by Shaw and Keddie (1983) and Keddie and Shaw (1986), respectively, and that of *K. sibirica* was by Belikova et al. (1986). Pukall and Stackebrandt (2009) characterized all the three species of *Kurthia* using the Biolog GP2 microtiter plates for substrate utilization, and the results are given in ➲ *Table 25.5*. The phenotypic characteristics that differentiate the three species of *Kurthia as* determined by Biolog GP2 plates (Pukall and Stackebrandt 2009) and fatty acids composition (Pukall and Stackebrandt 2009; Goodfellow et al. 1980; Collins et al. 1979) of all the three species are listed in ➲ *Tables 25.10* and ➲ *25.11*, respectively. Polar lipid analysis was done in case of *K. zopfii* and *Kurthia gibsonii*, and they are phosphatidyl glycerol, diphosphatidyl glycerol, and phosphatidylethanolamine.

**Table 25.8**

**Diagnostic phenotypic characteristics of species of the genus *Jeothgalibacillus* (Data from Yoon et al. (2001a); Yoon et al. (2004); Rüger and Richter (1979); Yoon et al. (2010a); Cunha et al. (2012))**

| Characteristic | *J. alimentarius* JCM 10872[Ta] | *J. campisalis* JCM 11810[Tb,d]; | *J. marinus* DSM 1297[Tc,d] | *J. salaries* DSM 23492[Td] | *J. soli* DSM 23228[Te] |
|---|---|---|---|---|---|
| **Morphological characteristics** | | | | | |
| Colony color | Orange-yellow | Light orange-yellow | White | Light yellow | White |
| Gram-stain | Variable | Variable | + | Variable | + |
| Spore shape | Round | Round or ellipsoidal | Round | Ellipsoidal | NR |
| Spore position | ST or T | ST or T | ST or T | ST | C or PC |
| Swollen sporangia | + | Slightly | - or slightly | Slightly | NR |
| Flagella type | Peritrichous | Single polar | Peritrichous | Peritrichous | Single polar or subpolar |
| **Growth characteristics** | | | | | |
| Anaerobic growth | + | – | – | – | – |
| Optimum pH | 7–8 | 7–8 | 7 | 7 | 8–8.5 |
| Temperature range (°C) | 10–55 | 5–39 | 5–30 | 5–40 | 15–40 |
| Optimum temperature (°C) | 30–35 | 30 | 12–23 | 30 | 30–37 |
| NaCl tolerance (%) | 20 | 15 | 7 | 18 | 9 |
| **Biochemical characteristics** | | | | | |
| Nitrate reduction | + | + | – | – | – |
| Catalase | + | – | – | + | + |
| Oxidase | + | – | + | + | + |
| **APIZYM** | | | | | |
| Esterase (C4) | NR | – | – | + | W |
| Esterase (C8) | NR | W | – | + | W |
| Leucine arylamidase | NR | – | – | – | + |
| Alpha-Chymotrypsin | NR | – | – | – | + |
| Naphthol-AS-BI-phosphohydrolase | – | W | + | – | – |
| β-Galactosidase | + | + | – | – | – |
| α -Glucosidase | + | – | – | – | – |
| β -Glucosidase | NR | – | – | – | + |
| **Hydrolysis of** | | | | | |
| Gelatin | + | v | + | – | – |
| Starch | – | + | – | – | + |
| Tween 80 | + | – | – | + | – |
| **Acid production from sugars** | | | | | |
| Cellobiose | – | + | – | W | – |
| Galactose | + | – | – | + | – |
| Lactose | – | – | + | – | – |
| Mannitol | + | + | – | + | + |
| D-Mannose | – | – | + | – | – |
| Melibiose | + | + | – | – | – |
| Melezitose | + | W | NR | + | – |
| Raffinose | + | NR | – | – | + |
| Ribose | + | – | NR | + | + |
| Xylose | – | – | + | – | – |

**Table 25.8 (continued)**

| Characteristic | *J. alimentarius* JCM 10872[Ta] | *J. campisalis* JCM 11810[Tb,d]; | *J. marinus* DSM 1297[Tc,d] | *J. salaries* DSM 23492[Td] | *J. soli* DSM 23228[Te] |
|---|---|---|---|---|---|
| **Chemotaxonomic characteristics** | | | | | |
| Menaquinone | MK-7, MK-8 | MK-7, MK-8 | MK-7 | MK-7, MK-8 | MK-7 |
| Polar lipids | NR | APL-1, APL-2, APL-3, APL-4 | APL-1, APL-2, APL-3, APL-4 | NR | APL-1, APL-2, APL-3, APL-4 |
| G+C content of DNA (mol%) | 44 | 41.8 | 37–42 | 42.9 | 39.4 |
| Source | Jeotgal, sea food | Marine solar saltern | Deep-sea water | Marine sediment | Soil |

[a]Yoon et al. (2001a)
[b]Yoon et al. (2004)
[c]Rüger and Richter (1979)
[d]Yoon et al. (2010a)
[e]Cunha et al. (2012)
+positive, – negative, *W* weakly positive, *v* variable, *NR* not reported, *ST* subterminal, *T* terminal, *C* central, *PC* paracentral, *S* sensitive, *R* resistant, *MK* menaquinone, *APL* aminophospholipid

**Table 25.9**
**Cellular fatty acid composition (%) of species of the genus *Jeotgalibacillus* (Data from Yoon et al. (2010a); Cunha et al. (2012))**

| Fatty acid | *J. alimentarius* JCM 10872[Ta] | *J. campisalis* JCM 11810[Ta] | *J. marinus* DSM 1297[Ta] | *J. salarius* DSM 23492[Ta] | *J. soli* DSM 23228[Tb] |
|---|---|---|---|---|---|
| **Branched chain** | | | | | |
| iso-$C_{14:0}$ | 4.2 | 14.1 | 5.3 | 10.8 | 1.6. |
| iso-$C_{15:0}$ | 46.3 | 2.5 | 22.2 | 9.4 | 22.0 |
| anteiso-$C_{15:0}$ | 15.8 | 49.8 | 47.7 | 35.9 | 45.5 |
| iso-$C_{16:0}$ | 6.3 | 5.2 | 5.1 | 12 | 2.5 |
| iso-$C_{17:0}$ | 7.6 | – | 2.6 | 1.7 | 5.8 |
| anteiso-$C_{17:0}$ | 5.8 | 7.4 | 10.7 | 12.2 | 11.3 |
| **Unsaturated** | | | | | |
| $C_{16:1\ (\omega7c)}$ alcohol | 6.7 | 12.3 | 2.6 | 12.6 | 1.8 |

[a]Yoon et al. (2010a)
[b]Cunha et al. (2012)
– absent. If a given fatty acid is present above 5.0 % in at least one species, the composition in the other species is given

The type species of the genus is *K. zopfii* (Trevisan 1885), and the type strain is ATCC 33403$^T$ = CCUG 38890$^T$ = CIP 103249$^T$ = DSM 20580$^T$ = JCM 6101$^T$ = LMG 17318$^T$ = NBRC 101529$^T$ = NCIMB 9878$^T$ = NCTC 10597$^T$ = VKM B-1568$^T$.

## *Paenisporosarcina* (Krishnamurthi et al. 2009)

*Paenisporosarcina* (*Pae'ni.spo'ro.sar.ci'na*. L. adv. *paeni* almost; N.L. fem. n. *Sporosarcina* a bacterial genus name; N.L. fem. n. *Paenisporosarcina* almost a *Sporosarcina*, because it is closely related to this genus but is phylogenetically distinct).

Genus *Paenisporosarcina* was described by Krishnamurthi et al. (2009b) and emended by Reddy et al. (2013). Originally the genus was carved from *Sporosarcina* based on the presence of L-Lys-D-Asp in the peptidoglycan, iso-$C_{14:0}$, iso-$C_{15:0}$, anteiso-$C_{15:0}$, and $C_{16:1(\omega7c)}$ alcohol as major fatty acids. During the description of *Paenisporosarcina indica*, Reddy et al. (2013) found that *Sporosarcina antarctica* (Yu et al. 2008) and *Paenisporosarcina indica* were phylogenetically closely related to the members of *Paenisporosarcina* but contained L-Lys-D-Glu in their peptidoglycan. In order to accommodate the above two species, Reddy et al. (2013) emended the genus *Paenisporosarcina*, described *Paenisporosarcina indica*, and transferred *Sporosarcina antarctica* (Yu et al. 2008) to the genus *Paenisporosarcina* as *Paenisporosarcina antarctica*. The emended description of the genus *Paenisporosarcina* is as follows: cells of *Paenisporosarcina* are Gram-positive rods and/or cocci, strictly aerobic, nonmotile, form round

**Table 25.10**
**Diagnostic phenotypic characteristics of species of the genus *Kurthia* (Data from Shaw and Keddie (1983); Belikova et al. (1986); Pukall and Stackebrandt (2009))**

| Characteristic | *K. zopfii* DSM 20580[Ta] | *K. gibsonii* DSM 20636[Ta] | *K. sibirica* DSM 4747[Tb] |
|---|---|---|---|
| **Morphological characteristics** | | | |
| Cell shape | Rods or chains | Rods or chains | Rods |
| **Growth characteristics** | | | |
| Temperature range (°C) | 5–35 | 5–45 | 5–37 |
| Optimum temperature (°C) | 25 | 25 | 20–25 |
| NaCl tolerance (%) | 5 | 7.5 | 6.5 |
| **Biochemical characteristics** | | | |
| Phosphatase | – | + | + |
| DNAase | – | + | NR |
| RNAase | + | – | NR |
| **Acid production** | | | |
| Glycerol | – | + | + |
| Ethanol | + | – | – |
| **Substrate utilization**[c#] | | | |
| *N*-acetyl-β-D-glucosamine | ++ | – | – |
| D-Fructose | – | – | ++ |
| D-Mannose | – | – | + |
| D-Psicose | – | – | ++ |
| Sedoheptulosan | – | – | W |
| γ-Hydroxybutyric acid | + | – | – |
| Pyruvatic acid methyl ester | + | ++ | – |
| Succinic acid monomethyl ester | ++ | ++ | – |
| L-Alanine | + | W | – |
| L-Glutamic acid | W | – | + |
| Glycerol | W | + | – |
| Inosine | + | + | – |
| Uridine | + | + | – |
| Adenosine-5′-monophosphate | W | + | – |
| Thymidine-5′-monophosphate | – | + | – |
| Uridine-5′-monophosphate | + | ++ | – |
| Mol % G+C content of DNA | 36–38 | 36–38 | 37 |
| Source | Intestinal contents | Hen and cow dung | Intestinal tract |

[a]Shaw and Keddie (1983)
[b]Belikova et al. (1986)
[c]Pukall and Stackebrandt (2009)
+ positive, ++ strongly positive, – negative, *NR* not reported, *W* weakly positive

endospores in a terminal or subterminal position, and catalase positive. The major fatty acids are iso-$C_{14:0}$, anteiso-$C_{15:0}$, and $C_{16:1\omega7c}$ alcohol. MK-7 and/or MK-8 are the major menaquinones, the cell wall peptidoglycan is of the A4α type with L-Lys-D-Asp or L-Lys-D-Glu, and the polar lipids present are diphosphatidylglycerol, phosphatidylethanolamine, and an aminophospholipid (APL1). The G+C content of the genus ranges from 38.0 to 46.0 mol%. The type species is *Paenisporosarcina quisquiliarum*.

The other characteristics of the genus are as follows: all the described species were negative for nitrate reduction, lysine decarboxylase, ornithine decarboxylase, Voges-Proskauer test, indole production, $H_2S$ production, and production of acid from L-arabinose, D-lactose, D-mannose, D-mannitol, and L-rhamnose. Positive for utilization of D-fructose, D-maltose, pyruvate, D-xylose, and D-glucose, but not trehalose, D-lactose, L-rhamnose, melezitose, melibiose, *N*-acetyl-B-glucosamine, D-sorbitol, myo-inositol, and sucrose. Three species

**Table 25.11**
**Cellular fatty acid composition (%) of species of the genus *Kurthia* (Data from Pukall and Stackebrandt (2009))**

| Fatty acid | *K. zopfii* DSM 20580[Ta] | *K. gibsonii* DSM 20636[Ta] | *K. sibirica* DSM 4747[Ta] |
|---|---|---|---|
| **Straight chain** | | | |
| $C_{16:0}$ | 4.6 | 6.0 | 2.4 |
| **Branched chain** | | | |
| iso-$C_{14:0}$ | 6.8 | 8.9 | 2.0 |
| iso-$C_{15:0}$ | 42.9 | 26.8 | 65.4 |
| anteiso-$C_{15:0}$ | 39.3 | 31.1 | 12.2 |
| iso-$C_{16:0}$ | – | 8.8 | 1.1 |
| **Unsaturated** | | | |
| iso-$C_{17:1(\omega\ 10c)}$ | – | – | 6.5 |

[a]Pukall and Stackebrandt (2009)
–, absent. If a given fatty acid is present above 5.0 % in at least one species, the composition in the other species is given

*Paenisporosarcina quisquiliarum*, *Paenisporosarcina macmurdoensis*, and *Paenisporosarcina indica* were positive for hydrolysis of starch but negative for hydrolysis of tyrosine, aesculin, Tween 80; production of acid from D-xylose; and utilization of xylitol, salicin, and methyl α-D-glucoside.

The two species of the Genus *Paenisporosarcina*, viz., *Paenisporosarcina quisquiliarum* and *Paenisporosarcina macmurdoensis*, were characterized using the Biolog GP2 system (Krishnamurthi et al. 2009b), and it was found that both species were positive for oxidation of adenosine 2′-deoxyadenosine, L-alanine, dextrin, D-fructose, D-mannose, maltotriose, D-ribose, turanose, thymidine, uridine, and D-xylose but not methyl β-D-glucoside; L-alaninamide; α-cyclodextrin; D-lactic acid methyl ester; D-alanine; β-cyclodextrin; arbutin; L-lactic acid; xylitol; D- and L-malic acid; glycogen; raffinose; L-asparagine; inulin; L-fucose; adenosine 5′-monophosphate; mannan; melezitose; α-hydroxybutyric acid; glycyl-L-glutamic acid; Tweens 20, 40, and 60; D-galacturonic acid; melibiose; salicin; propionic acid; L-pyroglutamic acid; uridine 5′-monophosphate; gentiobiose; methyl D-galactoside; sedoheptulosan; γ-hydroxybutyric acid; L-serine; D-fructose 6-phosphate; D-gluconic acid; methyl β-D-galactoside; D-sorbitol; *p*-hydroxyphenylacetic acid; succinamic acid; putrescine; α-D-glucose 1-phosphate; amygdalin; myo-inositol; methyl α-D-glucoside; sucrose; *N*-acetyl-L-glutamic acid; and DL-glycerol phosphate. *Paenisporosarcina antarctica* was characterized by API strips, and besides sharing some common characteristic with above three species, it exhibited the following characteristics: positive for naphthol-AS-BI-phosphohydrolase and negative for leucine arylamidase, valine arylamidase, cystine arylamidase, lecithinase, acid phosphatase, esterase (C4), esterase lipase (C8), lipase (C14), α-galactosidase, β-galactosidase, β-glucuronidase, α-glucosidase, β-glucosidase, *N*-acetyl-β-glucosaminidase, α-mannosidase, α-fucosidase, and α-chymotrypsin; hydrolysis of trypsin; production of acid from glycerol, maltose, melibiose, melezitose, raffinose, and D-sorbitol; and utilization of malate, succinate, capric acid, adipic acid, and phenyl acetic acid. The diagnostic characteristics of four species of the genus are listed in *Tables 25.12* and *25.13*.

## *Planomicrobium* (Yoon et. 2001)

*Planomicrobium* (*Pla.no.mi.cro'bi.um.* Gr. n. *planos* wanderer; Gr. adj. *micros* small; Gr. n. *bios* life; M.L. n. *Planomicrobium* motile microbe).

Genus *Planomicrobium* was created by Yoon et al. (2001b) to accommodate strains that are coccoid or short rods in the early growth period but later on change to rods. The genus includes strains that are Gram-positive to Gram-variable, aerobic, nonspore forming, motile by means of a single polar flagellum or peritrichous flagella, catalase positive, and urease negative and contain L-Lys-D-Asp or L-Lys-D-Glu as the peptidoglycan type, preponderance of MK-8 followed by MK-7 or by the predominance of MK-8 followed by MK-7 and MK-6, and iso-$C_{14:0}$, anteiso-$C_{15:0}$, $C_{16:1(\omega 7c)}$ alcohol, and iso-$C_{16:0}$ as the dominant cellular fatty acids. The G+C content of the genomic DNA is 35–47 mol%. At the time of description of the genus, there were only three species (Yoon et al. 2001b), but currently, the genus contain nine species, *Planomicrobium alkanoclasticum* (Engelhardt et al. 2001; Dai et al. 2005), *Planomicrobium chinense* (Dai et al. 2005), *Planomicrobium flavidum* (Jung et al. 2009), *Planomicrobium glaciei* (Zhang et al. 2009), *Planomicrobium koreense* (Yoon et al. 2001b), *Planomicrobium mcmeekinii* (Junge et al. 1998; Yoon et al. 2001b), *Planomicrobium okeanokoites* (Nakagawa et al. 1996; Yoon et al. 2001b), *Planomicrobium psychrophilum* (Reddy et al. 2002; Dai et al. 2005), and *Planomicrobium stackebrandtii* (Mayilraj et al. 2005; Jung et al. 2009). All the nine species form a monophyletic clade clearly separating out from the parent genus, the *Planococcus* (Jung et al. 2009), from which it was carved. In addition, the genus *Planomicrobium* contains the signature nucleotides C and G at positions 183 and 190, respectively, which unambiguously distinguishes it from the phylogenetically closest genus, the *Planococcus* (Dai et al. 2005). The phenotypic characteristics that were specific for the genus at the time of description, such as oxidase, nitrate reduction, and hydrolysis of aesculin, casein, gelatin, starch, and Tween 80, are variable among the nine species and thus cannot be used as characteristics of the genus (*Table 25.10*). However, the morphology, chemotaxonomic markers, 16S rRNA gene signature nucleotides can easily distinguish the genus *Planomicrobium* from other genera of the family.

The type species is *Planomicrobium koreense* (Yoon et al. 2001b). The type strain is JG07$^T$ = CIP 107134$^T$ = JCM 10704$^T$ = KCTC 3684$^T$.

The growth characteristics of species belonging to the genus *Planomicrobium* are listed in *Table 25.10.* Phenotypic characterization, such as acid production, carbon utilization, and qualitative enzymes tests, for type strains, were carried out, in majority of the cases, using noncommercial methods for *P. chinense* (Dai et al. 2005), *P. flavidum* (Jung et al. 2009),

**Table 25.12**

**Diagnostic phenotypic characteristics of species of the genus *Paenisporosarcina* (Data from Reddy et al. (2003) and Krishnamurthi et al. (2009b))**

| Characteristic | *Paenisporosarcina quisquiliarum* | *Paenisporosarcina macmurdoensis* | *Paenisporosarcina indica* | *Paenisporosarcina antarctica* |
|---|---|---|---|---|
| **Morphological characteristics** | | | | |
| Cell shape | Rods | Rods/cocci | Rods | Rods |
| Colony color | Cream | Creamish white | Red/white | Light yellow |
| Spore position | T | ST | ST | NA |
| **Growth characteristics** | | | | |
| Growth temperature range (°C) | 15–37 | 0–30 | 0–25 | 0–23 |
| Optimum temperature (°C) | 25–30 | 20 | 20 | 17–18 |
| pH Range | 6–9 | 6–9 | 6–8 | 5–10 |
| pH Optimum | 7–8 | 7 | 7 | 6–8 |
| **Biochemical characteristics** | | | | |
| Oxidase | + | – | – | + |
| Arginine dihydrolase | + | – | – | – |
| **Hydrolysis of** | | | | |
| Gelatin | – | + | + | – |
| **Acid production from** | | | | |
| D-Fructose | + | – | – | – |
| D-Galactose | + | – | – | – |
| D-Glucose | + | – | – | – |
| Sucrose | + | – | – | – |
| **Utilization of carbon compounds** | | | | |
| D-Galactose | – | + | – | – |
| D-Ribose | + | – | – | NR |
| Succinate | + | – | – | – |
| L-Alanine | + | – | – | NR |
| **Oxidation of carbon compounds (Biolog GP2)** | | | | |
| Maltose | + | – | NR | – |
| Succinic acid monomethyl ester | + | – | NR | NR |
| Pyruvic acid | + | – | NR | NR |
| Methyl D-glucose | + | – | NR | NR |
| α-Ketovaleric acid | + | – | NR | NR |
| D-Tagatose | – | + | NR | NR |
| D-Arabitol | – | + | NR | NR |
| Lactulose | – | + | NR | NR |
| Methyl α-D-mannoside | – | + | – | NR |
| Cellobiose | – | + | – | – |
| D-Mannitol | – | + | – | – |
| **Chemotaxonomic characteristics** | | | | |
| Menaquinone (major) | MK-7, MK-8 | MK-7 | MK-7 | MK-7 |
| Menaquinone (minor) | MK-9, MK-10, MK-11 | – | – | – |
| Polar lipids | DPG, PG, PE, APL, UL | DPG, PG, PE, APL, UL | DPG, PG, APL, UL1-UL4, PL | NR |
| G+C content of DNA (mol %) | 46.0 | 44 | NR | 39.2 |
| Source | Landfill | Cyanobacterial mats | Soil | Soil |

+ Positive, – negative, *NR* not reported, *MK* menaquinone, *DPG* diphosphatidylglycerol, *PG* phosphatidylglycerol, *PE* phosphatidylethanolamine, *PL* phospholipid. *APL* unknown phospholipid, *UL* unknown lipid

**Table 25.13**

**Cellular fatty acid composition (%) of species of the genus *Paenisporosarcina* (Data are from Krishnamurthi et al. (2009b); Reddy et al. (2013); Yu et al. (2008))**

| Fatty acid | *P. quisquiliarum* SK55[Ta] | *P. macmurdoensis* CMS21w[Ta] | *P. indica* PN2[Tb] | *P. antarctica* N-05[Tc] |
|---|---|---|---|---|
| **Branched chain** | | | | |
| $C_{14:0\ iso}$ | 4.7 | 11.6 | 7.4 | 7.6 |
| $C_{15:0\ iso}$ | 39.6 | 15.1 | 12.0 | 1.4 |
| $C_{15:0\ ante}$ | 19.3 | 22.2 | 50.2 | 39.8 |
| $C_{16:0\ iso}$ | 4.0 | 9.3 | 5.5 | 7.0 |
| **Unsaturated** | | | | |
| $C_{16:1(\omega7c)}$ alcohol | 9.6 | 23.8 | 5.6 | 18.9 |
| **Summed feature 4** | 5.5 | 6.5 | 2.5 | 11.9 |

[a]Krishnamurthi et al. (2009b)
[b]Reddy et al. (2013)
[c]Yu et al. (2008)
Summed feature 4: anteiso-$C_{17:1}$ B and/or iso I; If a given fatty acid is present above 5.0 % in at least one species, the composition in the other species is given

*P. koreense* (Yoon et al. 2001b), *P. psychrophilum* (Reddy et al. 2002), *P. mcmeekinii* (Junge et al. 1998), *P. okeanokoites* (Nakagawa et al. 1996), and *P. stackebrandtii* (Mayilraj et al. 2005). Biolog GP MicroPlate system was used for *P. alkanoclasticum* (Engelhardt et al. 2001) and API ZYM, API 20E and API 20NE for *P. glaciei* (Zhang et al. 2009). Since phenotypic characterization was performed by various methods, most of the species are lacking the uniform characteristics that can be used to draw the similarities or differences. However, certain tests, such as methyl red, Voges-Proskauer, indole production, phenylalanine deaminase, were either negative or not reported for majority of species. In acid production, most of the type strains were negative for L-arabinose, D-mannose, xylose, cellobiose, lactose, melibiose, raffinose, rhamnose, and sucrose. Substrate utilization and antibiotic susceptibility were done only in few strains. The diagnostic and differential characteristics are listed in *Tables 25.14* and *25.15.*

## *Rummeliibacillus* (Vaishampayan et al. 2009)

*Rummeliibacillus* (*Rum.me'li.i.ba.cil'lus.* N.L. n. *Rummelius* Rummel; L. masc. n. *bacillus* a rod, and also a bacterial genus name; N.L. masc. n. *Rummeliibacillus* a bacterium close to the genus *Bacillus* and named in honor of former NASA Planetary Protection Officer Dr. John Rummel, an astrobiologist responsible for bringing planetary protection into the public domain).

*Rummeliibacillus* is strictly aerobic, Gram-positive, and spore forming. The spores are round and terminally located in swollen sporangia. Strains are motile by means of peritrichous flagella. Besides these, the most straightforward placement of an unidentified strain into the genus *Rummeliibacillus* is a combination of 16S rRNA gene sequence analysis and presence of the peptidoglycan type, L-Lys-D-Glu or L-Lys-D-Asp (variation A4α), MK-7 as the menaquinone type and phosphatidylglycerol, diphosphatidylglycerol, and phosphatidylethanolamine as major polar lipid with moderate amounts of an unknown aminophospholipid (APL1), minor amounts of two unknown phospholipids (PL1, PL2), and an unknown aminolipid (AL). Fatty acid profiles consist largely of anteiso-$C_{15:0}$ (approximately 50 %) and iso-$C_{15:0}$ (approximately 25 %) acids. The main characteristic that differentiates *Rummeliibacillus* from *Viridibacillus* (genus that is closely related to *Rummeliibacillus*) is the absence of green pigment. The G+C content of the genomic DNA of the genus is approximately 35.0 mol%.

The genus presently contains two species, *Rummeliibacillus stabekisii* and *Rummeliibacillus pycnus* (Vaishampayan et al. 2009). *Rummeliibacillus pycnus* was originally described as *Bacillus pycnus* (Nakamura et al. 2002), but due to its 16S rRNA gene sequence-based phylogenetic closeness, the species was transferred to the genus *Rummeliibacillus*. Both the species were characterized using API 20NE and API 50 CH strips, and both the species are positive for catalase and negative for nitrate reduction and Voges-Proskauer test. *R. stabekisii* is negative for oxidase, can hydrolyze gelatin, and produces acid from adonitol, L-arabinose, cellobiose, D-galactose, D-lactose, D-mannose, D-mannitol, D-ribose, D-xylose, and citrate. *R. pycnus* produces acid from citrate and pyruvate, does not produce indole from tryptophan, and does not decompose casein; tyrosine; urea; Tweens 40 and 80; and lecithin. Oxidizes pyruvate, β-hydroxybutyrate but not deoxyadenosine, inosine, AMP, and UMP. The other prominent characteristics are listed in *Tables 25.16* and *25.17*. The genus is phylogenetically closely related to the genera *Kurthia* and *Viridibacillus*.

The type species of the genus is *Rummeliibacillus stabekisii* (Vaishampayan et al. 2009), and the type strain is KSC-SF6g$^T$ = NBRC 104870$^T$ = NRRL B-51320$^T$.

**Table 25.14**

**Diagnostic phenotypic characteristics of species of the genus *Planomicrobium* (Data from Engelhardt et al. (2001); Dai et al. (2005); Jung et al. (2009); Zhang et al. (2009); Yoon et al. (2001b); Junge et al. (1998); Reddy et al. (2002); Mayilraj et al. (2005)). 1. *P. alkanoclasticum* MAE2$^{T}$ [a,b]; 2. *P. chinense* JCM 12466$^{T}$ [b,c]; 3. *P. flavidum* KCTC 13261$^{T}$ [c]; 4. *P. glaciei* JCM 15088$^{T}$ [c,d]; 5. *P. koreense* JCM 10704$^{T}$ [c,e]; 6. *P. mcmeekinii* ATCC 700539$^{T}$ [c,e,f]; 7. *P. okeanokoites* ATCC 33414$^{T}$ [c,e]; 8. *P. psychrophilum* CMS 53or$^{T}$ [b,c,g]; 9. *P. stackebrandtii* DSM 16419$^{T}$ [c,h]**

| Characteristic | 1 | 2 | 3 | 4 | 5 | 6 | 7 | 8 | 9 |
|---|---|---|---|---|---|---|---|---|---|
| **Morphological characteristics** | | | | | | | | | |
| Colony color | Orange | Yellow-orange | Light yellow | Yellow-orange | Yellow-orange | Pale orange | Yellow-orange | Orange | Orange |
| Cell shape | Rods | Cocci/short rods | Cocci/short rods | Cocci/short rods | Cocci/short rods | Cocci/short rods | Rods | Rods | Cocci |
| Cells size (μm) | 0.4–0.8 × 1.7–2.6 | 0.8 × 1.0 | 0.4–0.8 × 0.4–1.6 | NR | NR | 0.8–1.2 × 0.8–10 | 0.4–0.8 × 1.0–20 | NR | NR |
| Gram-stain | v | + | v | + | v | + | v | + | + |
| **Growth characteristics** | | | | | | | | | |
| pH Range | NR | 5–10 | 6–8 | 5–10 | 6–9 | NR | NR | 6–12 | 5.6–11 |
| Temperature range (°C) | 15–41 | 10–45 | 4–37 | 4–28 | 4–38 | 0–37 | 20–37 | 2–30 | 15–30 |
| NaCl tolerance (%) | 15 | 10 | 13 | 11 | 7 | 7 | 6 | 12 | 7 |
| **Biochemical characteristics** | | | | | | | | | |
| Oxidase | – | NR | + | – | – | – | + | + | – |
| **Hydrolysis of** | | | | | | | | | |
| Starch | + | – | – | – | – | – | – | – | – |
| Casein | NR | – | W | + | + | + | + | + | – |
| Gelatin | + | + | – | + | + | + | + | + | + |
| Aesculin | NR | – | – | NR | + | – | – | + | NR |
| Tween 80 | – | – | + | – | – | – | – | – | NR |
| Nitrate reduction | – | + | NR | + | – | + | – | – | – |
| **Acid production from sugars: Jung et al.** (2009) | | | | | | | | | |
| Cellobiose | + | – | – | NR | + | – | – | – | NR |
| D-Fructose | + | NR | + | – | – | + | + | + | + |
| D-Glucose | – | + | – | – | + | + | – | – | – |
| Lactose | – | – | – | – | + | – | – | – | + |
| Maltose | – | NR | – | NR | + | W | – | – | NR |
| Mannitol | + | – | – | NR | – | – | – | – | – |
| Melibiose | – | – | – | NR | + | – | – | – | NR |
| Raffinose | – | – | – | NR | – | – | – | – | + |
| Ribose | – | NR | – | + | – | – | + | – | NR |
| L-Rhamnose | – | – | – | NR | – | – | – | – | + |
| Sucrose | – | – | – | NR | – | – | – | – | + |
| D-Xylose | – | – | – | NR | – | – | + | – | – |
| **Chemotaxonomic characteristics** | | | | | | | | | |
| Menaquinone | MK-7, MK-8 | MK-8 | MK-7, MK-8 | MK-7, MK-8 | MK-6, MK-7, MK-8 | MK-7, MK-8 | MK-7, MK-8 | MK-7, MK-8 | MK-7, MK-8 |
| Peptidoglycan composition | NR | L-Lys-D-Glu | NR | L-Lys-D-Glu | L-Lys-D-Glu | L-Lys-D-Asp | Lys-D-Asp | L-Lys-D-Glu | L-Lys-D-Glu |

**Table 25.14 (continued)**

| Characteristic | 1 | 2 | 3 | 4 | 5 | 6 | 7 | 8 | 9 |
|---|---|---|---|---|---|---|---|---|---|
| G+C content of DNA (mol%) | 45.3 | 34.8 | 45.9 | 49 | 47 | 35 | 46 | 44.5 | 40 |
| Source | Intertidal beach sediment | Coastal sediment | Marine solar saltern | Glacier water | Sea food | Sea ice | Ocean bed | Cyanobacterial mat sample | Cold desert of Himalayas |

[a]Engelhardt et al. (2001)
[b]Dai et al. (2005)
[c]Jung et al. (2009)
[d]Zhang et al. (2009)
[e]Yoon et al. (2001b)
[f]Junge et al. (1998)
[g]Reddy et al. (2002)
[h]Mayilraj et al. (2005)
+ positive, – negative, *v* variable, *W* weakly positive, *NR* not reported, *MK* menaquinone, l-*Lys* L-lysine, d-*Glu* D-Glutamate, d-*Asp* D-Aspartate

**Table 25.15**

**Cellular fatty acid composition (%) of species of the genus *Planomicrobium* (Data from Engelhardt et al. (2001); Zhang et al. (2009); Jung et al. (2009); Mayilraj et al. (2005)). 1. *P. alkanoclasticum* MAE2$^{T}$ [a]; 2. *P. chinense* CGMCC 1.3454$^{T}$ [b]; 3. *P. flavidum* ISL-41$^{T}$ [c]; 4. *P. glaciei* 0423$^{T}$ [b]; 5. *P. koreense* DSM 15895$^{T}$ [b]; 6. *P. mcmeekinii* DSM 13963$^{T}$ [b]; 7. *P. okeanokoites* NBRC 12536$^{T}$ [b]; 8. *P. psychrophilum* CMS 53or$^{T}$ [d]; 9. *P. stackebrandtii* DSM 16419$^{T}$ [d]**

| Fatty acid | 1 | 2 | 3 | 4 | 5 | 6 | 7 | 8 | 9 |
|---|---|---|---|---|---|---|---|---|---|
| **Straight chain** | | | | | | | | | |
| $C_{15:0}$ | NR | NR | NR | NR | NR | NR | NR | NR | 5.5 |
| **Branched chain** | | | | | | | | | |
| iso-$C_{14:0}$ | 6.0 | 22.9 | 8.0 | 15.2 | 16.2 | 19.4 | 42.9 | 5.3 | 4.7 |
| iso-$C_{15:0}$ | 7.5 | 4.9 | 1.8 | 1.6 | 6.3 | 4.9 | Tr | 5.8 | 2.9 |
| anteiso-$C_{15:0}$ | 45.5 | 40.6 | 39.0 | 37.0 | 43.6 | 36.5 | 5.1 | 53.5 | 49.8 |
| iso-$C_{16:0}$ | 17.1 | 6.8 | 11.5 | 5.8 | 11.5 | 12.9 | 27.2 | 7.1 | 5.7 |
| iso-$C_{17:0}$ | 7.4 | Tr | 2.8 | NR | Tr | 1.2 | NR | - | 2.8 |
| anteiso-$C_{17:0}$ | 10.2 | 1.8 | 11.3 | 3.0 | 2.9 | 2.7 | Tr | 4.0 | 4.6 |
| **Unsaturated** | | | | | | | | | |
| $C_{16:1(\omega7c)}$ alcohol | NR | 12.0 | 11.0 | 4.6 | 9.6 | 9.4 | 15.1 | 10.1 | 8.5 |
| $C_{16:1(\omega11c)}$ | 6.4 | 5.5 | 1.0 | 3.2 | 4.0 | 2.5 | 3.6 | NR | 1.7 |
| **Summed feature 4** | 9.3 | NR | 8.4 | NR | NR | NR | NR | 8.6 | 8.6 |

[a]Engelhardt et al. (2001)
[b]Zhang et al. (2009)
[c]Jung et al. (2009)
[d]Mayilraj et al. (2005)
- absent, *NR* not reported, *Tr* traces; summed feature 4: iso-$C_{17:1}$ I and/or anteiso-$C_{17:1}$ B. If a given fatty acid is present above 5.0 % in at least one species, the composition in the other species is given

## *Solibacillus* (Krishnamurthi et al. 2011)

*Solibacillus* (*So.li.ba.cil'lus.* L. n. solum soil; L. n. Bacillus a bacterial genus; N.L. n. *Solibacillus* a Bacillus-like organism isolated from soil).

The type strain of *Solibacillus silvestris* was isolated from the top soil layer of a beech forest soil near Braunschweig, Germany. It was originally described as *Bacillus silvestris* by Rheims et al. (1999). But *Bacillus silvestris* differed from other members of the genus *Bacillus*, in that it contained A4α type of peptidoglycan instead of A4γ type of peptidoglycan present in *Bacillus* strains (Schleifer and Kandler 1972; Kämpfer et al. 2006). In addition, *Bacillus silvestris* also differs in that it contains iso-$C_{15:0}$ and iso-$C_{16:1}$ as predominant fatty acids in contrast to the type species of *Bacillus* and other members which contains iso- and anteiso-$C_{15:0}$. Based on these differences, Krishnamurthy et al.

**Table 25.16**
**Diagnostic phenotypic characteristics of species of the genus *Rummeliibacillus* (Data from Vaishampayan et al. (2009); Nakamura et al. (2002))**

| Characteristic | *R. stabekisii* KSC-SF6g[Ta] | *R. pycnus* JCM 11075[Ta,b] |
|---|---|---|
| **Morphological characteristics** | | |
| Cell size (μm) | 1.07–1.14 × 2.64–3.32 | 1.0 × 3.0–5.0 |
| **Growth characteristics** | | |
| Temperature range (°C) | 28–55 | 10–45 |
| Optimum temperature (°C) | 28–32 | 28 |
| Tolerance to NaCl (%) | 7 | 5 |
| **Biochemical characteristics** | | |
| Indole production | – | + |
| **Hydrolysis of** | | |
| Gelatin | + | – |
| Starch | + | – |
| **Acid production** | | |
| D-Glucose | – | + |
| **Chemotaxonomic characteristics** | | |
| Peptidoglycan composition | L-Lys-D-Asp | L-Lys-D-Glu |
| G+C content of DNA (mol%) | 34.3 | 35 |
| Source | Payload Hazardous Servicing Facility | Soil |

[a]Vaishampayan et al. (2009)
[b]Nakamura et al. (2002)
+ positive, – negative, *S* sensitive, *l-Lys* L-lysine, *d-Asp* D-Aspartate, *d-Glu* D-Glutamate

**Table 25.17**
**Cellular fatty acid composition (%) of species of the genus *Rummeliibacillus* (Data from Vaishampayan et al. (2009))**

| Fatty acid | *R. stabekisii* KSC-SF6g[Ta] | *R. pycnus* JCM 11075[Ta] |
|---|---|---|
| **Branched chain** | | |
| iso-$C_{14:0}$ | 6.7 | 4.7 |
| iso-$C_{15:0}$ | 26 | 74.2 |
| anteiso-$C_{15:0}$ | 49.9 | 9.9 |
| anteiso-$C_{17:0}$ | 7.0 | 1.2 |

[a]Vaishampayan et al. (2009)
Strains were grown in TSB at 25–32 °C for 30–48 h. If a given fatty acid is present above 5.0 % in at least one species, the composition in the other species is given

(2009a) created the genus *Solibacillus* to accommodate the species *Bacillus silvestris* and reclassified it as *Solibacillus silvestris.*

Cells of *Solibacillus silvestris* are Gram-positive, strictly aerobic, rod shaped and the rods measure 0.5–0.7 μm in width by 0.9–2.0 μm in length, motile by means of peritrichous flagella, endospore forming, and the spores are round and lie terminally in a swollen sporangium. Cells can grow from 10 to 40 °C with an optimum temperature of 20–30 °C, pH range is 6–7, and can tolerate a NaCl concentration of 5 %. *Solibacillus silvestris* is catalase positive and negative for oxidase, Voges-Proskauer test, indole production, arginine dihydrolase, phenylalanine deaminase, and nitrate-to-nitrite reduction test and could not hydrolyze casein, gelatin, starch, tyrosine, aesculin, and Tween 80.

Oxidation of substrates provided in the API 50CH panel indicates that glycerol and ribose are utilized as sole carbon sources, but no acid is produced. Negative for utilization of D-glucose, erythritol, D- and L-arabinose, D- and L-xylose, adonitol, methyl β D-xyloside, galactose, mannose, D-mannitol, D-fructose, L-sorbose, rhamnose, dulcitol, inositol, sorbitol, methyl α-D-mannoside, methyl α-D-glucoside, *N*-acetylglucosamine, amygdalin, arbutin, aesculin, salicin, cellobiose, maltose, lactose, melibiose, sucrose, trehalose, inulin, melezitose, D-raffinose, starch, glycogen, xylitol, *p*-gentiobiose, D-turanose, D-lyxose, D-tagatose, D- and L-fucose, D- and L-arabitol, gluconate, 2-ketogluconate, 5-ketogluconate, citrate, and propionate. The peptidoglycan contains L-Lys-D-Glu and represents peptidoglycan type A4α. Polar lipids comprise phosphatidylglycerol, diphosphatidylglycerol, phosphatidylethanolamine, minor amounts of phosphatidylserine, and one unknown phospholipid. The predominant isoprenoid quinone is of the MK-7 type. The fatty acid composition is characterized by the presence of iso-$C_{15:0}$ and iso-$C_{16:1}$, contributing to 61.0 %, and the other fatty acids are $C_{17:0}$, $C_{16:1\ (\omega7c)}$ alcohol, iso-$C_{17:1(\omega\ 10c)}$ anteiso-$C_{17:1}$, iso-$C_{14:0}$, iso-$C_{15:0}$, anteiso-$C_{15:0}$, iso-$C_{16:0}$, iso-$C_{17:0}$, and anteiso-$C_{17:0}$. The G+C content of the DNA is 39.3 mol % and phylogenetically a member of the *Bacillus* RNA group 2 (Ash et al. 1991).

The type species of the genus is *Solibacillus silvestris* (Rheims et al. 1999; Krishnamurthi et al. 2009) and the type strain is HR3-23[T] = DSM 12223[T] = ATCC BAA-269[T] = CIP 106059[T].

## *Sporosarcina* (Kluyver and van Niel 1936 Emend. Yoon et al. 2001)

*Sporosarcina*. (*Spo.ro.sar.ci'na.* M.L. n. *spora* a spore; M.L. fem. n. *Sarcina* generic name; M.L. fem. n. *Sporosarcina* spore-forming *Sarcina*).

The genus *Sporosarcina* was described by Kluyver and van Niel in 1936 and emended by Yoon et al. (2001c). The genus presently contains 12 validly described species (http://www.bacterio.cict.fr/s/sporosarcina.html), namely, *S. aquimarina* (Yoon et al. 2001c), *S. contaminans* (Kämpfer et al. 2010), *S. globispora* (Larkin and Stokes 1967; Yoon et al. 2001c), *S. koreensis* (Kwon et al. 2007), *S. luteola* (Tominaga et al. 2009), *S. newyorkensis* (Wolfgang et al. 2012), *S. pasteurii* (Miquel 1889; Yoon et al. 2001c), *S. psychrophila* (Nakamura 1984; Yoon et al. 2001c), *S. saromensis* (An et al. 2007), *S. soli* (Kwon et al. 2007), *S. thermotolerans* (Kämpfer et al. 2010), and *S. ureae* (Beijerinck 1901). *S. macmurdoensis* (Reddy et al. 2003) and *S. antarctica*

(Yu et al. 2008) were transferred to the genus *Paenisporosarcina* as *Paenisporosarcina macmurdoensis* (Krishnamurthi et al. 2009b) and *Paenisporosarcina antarctica* (Reddy et al. 2013).

Members of the genus *Sporosarcina* are characterized by the presence of spherical or rod-shaped cells that are Gram-positive to Gram-variable, motile or nonmotile, endospore forming, spores are spherical or round, and location is central or terminal or subterminal. Facultatively anaerobic or strictly aerobic, catalase and oxidase positive (not determined for *S. pasteurii* but positive for the type strain of *S. pasteurii*). The diagnostic amino acid at position three of the peptide subunit of the peptidoglycan is L-lysine, and peptidoglycan composition in nine out of thirteen species is L-Lys-D-Glu. In other species, it is L-Lys-L-Ala-L-Asp (*S. aquimarina*) or L-Lys-Gly-D-Glu (*S. newyorkensis* and *S. ureae*) or L-Lys-D-Asp (*S. pasteurii*). However, the peptidoglycan variation in all the species is A4α. The predominant menaquinone in all the species is MK-7, and the major fatty acid is anteiso-$C_{15:0}$. The genus-specific phenotypic characteristics that were described by Yoon et al. (2001c) have deviated with the addition of new species wherein the hydrolysis of gelatin, urea, and starch are variable. The G+C content of the genomic DNA ranges from 38 to 47 mol%.

The type species is *Sporosarcina ureae* (Beijerinck 1901; Kluyver and van Niel 1936). The type strain is ATCC $6473^T$ = DSM $2281^T$ = IFO (now NBRC) $12699^T$ = JCM $2577^T$ = LMG $17366^T$ = VKM B-$595^T$.

Phylogenetically, based on 16S rRNA gene sequence, the genus is closely related to the clade represented by *Filibacter, Bacillus, Paenibacillus, Lysinbacillus,* and *Viridibacillus* (Wolfgang et al. 2012). The genus *Sporosarcina* can be differentiated from other members of the family based on its morphology, chemotaxonomic markers, and 16S rRNA genes sequence.

Phenotypic characterization of type strains were carried out in majority of the cases, and the diagnostic characteristics are listed in ➲ *Tables 25.18* and ➲ *25.19*. Species of the genus are psychrophilic (Yu et al. 2008; Nakamura 1984) to moderately thermotolerant (Kämpfer et al. 2010) and halotolerant. Majority of the species are negative (exceptions are listed in brackets), for hydrolysis of casein (*S. pasteurii*), aesculin, and Tween 80 (*S. luteola* and *S. ureae*), acetate (*S. soli*) and citrate utilization, Voges-Proskauer test, indole production, $H_2S$ production, arginine dihydrolase (*S. koreensis*), lysine decarboxylase, ornithine decarboxylase, and phenylalanine deaminase (*S. soli* and *S. ureae*). Acid is not produced from L-arabinose, lactose (*S. globispora; S. ureae*), mannitol (*S. psychrophila*), maltose (*S. luteola*) and sucrose (*S. ureae*).

## *Ureibacillus* (Fortina et al. 2001)

*Ureibacillus* (Ur.e.i.ba.cil'lus. L. n. *urea* urea; L. dim. n. *bacillus* from *Bacillus*, a genus of aerobic endospore-forming bacteria; *Ureibacillus* a ureolytic aerobic bacillus).

Genus *Ureibacillus* was created by Fortina et al. (2001) to accommodate a number of moderately thermophilic sporeformers that are phylogenetically distinct and from a separate clade in the genus *Bacillus*. Cells are Gram-negative rods, motile by means of peritrichous flagella, and bear spherical endospores which lie in terminal or subterminal positions in swollen sporangia. Members of the genus are strictly aerobic and moderately thermophilic. The cross-linkage of peptidoglycan type is of the L-Lys-D-Asp (variation A4α). The polar lipids are phosphatidylglycerol, diphosphatidylglycerol, phospholipids, and glycolipids of unknown composition. The major cellular fatty acid is iso-$C_{16:0}$ that contributes more than 60.0 % in so far described species. The G+C content of DNA ranges from 35 to 42 mol%.

The type species is *Ureibacillus thermosphaericus* (Andersson et al. 1995; Fortina et al. 2001), and the type strain is *P*-$11^T$ = CIP $104857^T$ = DSM $10633^T$ = HAMBI $1900^T$ = LMG $17959^T$.

Following strains were described till today under this genus: *U. composti* (Weon et al. 2007), *U. suwonensis* (Kim et al. 2006), *U. terrenus* (Fortina et al. 2001), *U. thermophilus* (Weon et al. 2007), and *U. thermosphaericus* (Andersson et al. 1995; Fortina et al. 2001). Among all these, *U. thermosphaericus* was originally described as *Bacillus thermosphaericus.*

Phenotypic characterization of all the isolates was done traditionally, without using any commercial kits. All the five isolates exhibited similar characteristics with respect to Gram-staining, spore formation, motility, and anaerobic growth. In addition, the following characteristics were shared by all the isolates: positive for catalase and oxidase; negative for Voges-Proskauer test; indole production; nitrate reduction; hydrolysis of casein, gelatin, and starch; production of acid from L-arabinose, D-glucose, mannitol, and xylose. Diagnostic characteristics that differentiate the species are listed in ➲ *Tables 25.20* and ➲ *25.21*. Genus *Ureibacillus* can be distinguished from the other members of the family *Planococcaceae* based on the growth temperature, Gram-staining, fatty acid methyl esters, and polar lipids.

## *Viridibacillus* (Albert et al. 2001)

*Viridibacillus* (*Vi.ri.di.ba.cil'lus.* L. adj. *viridis* green; L. masc. n. *bacillus* rod: N.L. masc. n. *Viridibacillus* the green bacillus/rod).

Genus *Viridibacillus* was established by Albert et al. (2007) to accommodate strains that belong to *Bacillus* rRNA group 2 (Ash et al. 1991) and produce green pigment on R2A medium after 48 h growth. Cells of the genus are Gram-positive, spore forming, motile rods. Endospores are round and are located terminally in swollen or slightly swollen sporangia. Growth occurs in the presence of 2 % NaCl but not in 7 % NaCl (w/w). Sporulation and release of endospores (free spores) are abundant on R2A agar after 24 and 48 h at 25 °C growth temperature. The quinone system consists of MK-8 (69–81 %) and moderate amounts of MK-7 (19–30.5 %). In the fatty acid profile, the major fatty acids are iso-$C_{15:0}$ are anteiso-$C_{15:0}$, and both of them contribute close to 70.0 %. The polar lipids present are phosphatidylglycerol, diphosphatidylglycerol, and phosphatidylethanolamine and moderate amounts of an unknown

**Table 25.18**

Diagnostic phenotypic characteristics of species of the genus *Sporosarcina* (Data from Yoon et al. (2001c); Kämpfer et al. (2010); Kwon et al. 2007; Tominaga et al. (2009); Wolfgang et al. (2012); An et al. (2007); Kwon et al. (2007); Kämpfer et al. (2010); An et al. (2007); Wolfgang et al. (2012)). 1. *S. aquimarina* JCM 10887$^{T}$ [a]; 2. *S. contaminans* DSM 22204$^{T}$ [b]; 3. *S. globispora* ATCC 23301$^{T}$ [a]; 4. *S. koreensis* JCM 16400$^{T}$ [c]; 5. *S. luteola* JCM 15791$^{T}$ [d]; 6. *S. newyorkensis* DSM 23544$^{T}$ [e]; 7. *S. pasteurii* DSM 33$^{T}$ [a]; 8. *S. psychrophila* JCM 9075$^{T}$ [a]; 9. *S. saromensis* JCM 23205$^{T}$ [f]; 10. *S. soli* DSM 16920$^{T}$ [c]; 11. *S. thermotolerans* DSM 22203$^{T}$ [b]; 12. *S. ureae* DSM 2281$^{T}$ [f]

| Characteristics | 1 | 2 | 3 | 4 | 5 | 6 | 7 | 8 | 9 | 10 | 11 | 12 |
|---|---|---|---|---|---|---|---|---|---|---|---|---|
| **Morphological characteristics** | | | | | | | | | | | | |
| Colony color | Light orange | Beige | White | Light orange | Yellow | Gray | White | White | Beige | Light orange | Beige | White |
| Cell shape | R | R | R | R | R | R | Sp | R | R | R | R | Sp |
| Gram-stain | v | + | v | + | v | + | + | | + | + | + | + |
| Spore shape | Round | Spherical | Round | Oval | NR | Round to oval | NR | Round | Spherical | Round | Spherical | NR |
| Spore position | T | T | ST | T | T | ST | T | ST | T | C | T | NR |
| Cell arrangement | NR | NR | Single or pairs | Single or short chains | NR | Single or palisades | NR | Single or pairs | NR | Single or pairs | NR | NR |
| Motility | + | + | + | + | + | + | + | + | + | – | – | + |
| **Growth characteristics** | | | | | | | | | | | | |
| Anaerobic growth | + | – | + | – | + | + | + | + | – | – | – | – |
| Optimum pH | 6.5–7 | 7 | 7 | 7 | 7 | 7.2–9.5 | 9 | 7 | 6.5 | 8 | 7 | 7 |
| Temperature range (°C) | 4–37 | 15–50 | 5–30 | 15–40 | 10–40 | 10–42 | 5–37 | 0–30 | 5–40 | 15–37 | 15–50 | –37 |
| Optimum temperature (°C) | 25 | 30–37 | 25 | 30 | 25–30 | 22–28 | 30 | 20 | 27 | 30 | 30–37 | 25 |
| NaCl tolerance (%) | 13 | 10 | 5 | 7 | 7.5 | 13 | 10 | 5 | 9 | 5 | 3 | 3 |
| **Biochemical characteristics** | | | | | | | | | | | | |
| Nitrate reduction | + | – | + | – | + | – | + | + | – | + | – | + |
| Urease | + | – | + | + | – | + | + | + | + | + | – | + |
| DNase | + | + | NR | + | – | + | NR | NR | NR | + | + | – |
| **Hydrolysis of** | | | | | | | | | | | | |
| Gelatin | + | – | + | + | + | – | + | + | + | – | + | – |
| Starch | – | – | – | + | – | – | – | – | + | – | + | – |

| | | | | | | | | | | | | |
|---|---|---|---|---|---|---|---|---|---|---|---|---|
| Tyrosine | – | + | NR | + | – | – | NR | – | NR | + | – | – |
| **Acid production from sugars** | | | | | | | | | | | | |
| D-Fructose | + | NR | + | – | + | – | | + | – | | | – |
| D-Glucose | – | – | + | – | + | – | NR | + | – | – | – | + |
| Galactose | – | NR | + | – | + | – | NR | + | – | NR | NR | – |
| Glycerol | + | NR | + | – | + | – | NR | NR | NR | NR | NR | NR |
| Lactose | – | NR | + | | | – | NR | – | – | | | + |
| Xylose | – | – | + | – | + | – | NR | + | – | – | – | – |
| **Chemotaxonomic characteristics** | | | | | | | | | | | | |
| Polar lipids | NR | PG, DPG, PE | NR | PG, DPG, PE | PG, DPG, PE | PG, DPG, PE | NR | NR | PG, DPG, PE | PG, DPG, UPL | PG, DPG, PE | PG, DPG, PE |
| Peptidoglycan composition | L-Lys-L-Ala-L-Asp | L-Lys-D-Glu | L-Lys-D-Glu | L-Lys-D-Glu | L-Lys-D-Glu | L-Lys-L-Gly-D-Glu | L-Lys-D-Asp | L-Lys-D-Glu | L-Lys-D-Glu | L-Lys-D-Glu | L-Lys-D-Glu | L-Lys-L-Gly-D-Glu |
| G+C content of DNA (mol%) | 40 | NR | 40 | 46.5 | 43.6 | 42.4 | 39 | 44 | 46 | 44.5 | NR | 40–42 |
| Source | Sea water | Industrial clean room | Soil, river water | Soil | Surface of soya sauce instrument | Patient's blood | Soil, water, sewage | Soil and river water | Sediment | Soil | Human blood | Conta-minant from clean-room floor |

[a] Yoon et al. (2001c)
[b] Kämpfer et al. (2010)
[c] Kwon et al. 2007
[d] Tominaga et al. (2007)
[e] Wolfgang et al. (2012)
[f] An et al. (2007)

+ positive, – negative, *v* variable, *NR* not reported, *R* rod, *Sp* spherical, *T* terminal, *ST* subterminal, *C* central, *UPL* unidentified phospholipid, *PG* phosphatidylglycerol, *DPG* diphosphatidylglycerol, *PE* phosphatidylethanolamine, l-*Lys* L-lysine, l-*Ala* L-Alanine, d-*Glu* D-Glutamate, d-*Asp* D-Aspartate, l-*Asp* L-Aspartate, l-*Gly* L-Glycine

**Table 25.19**

**Cellular fatty acid composition (%) of species of the genus *Sporosarcina* (Data from Wolfgang et al. (2012); Yoon et al. (2001c); An et al. (2007)). Taxa: 1. *S. newyorkensis* 6062$^{T\ a}$; 2. *S. ureae* DSM 2281$^{T\ a}$; 3. *S. aquimarina* DSM 14554$^{T\ a}$; 4. *S. soli* DSM 16920$^{T\ a}$; 5. *S. koreensis* DSM 16921$^{T\ a}$; 6. *S. thermotolerans* CCUG 53480$^{T\ a}$; 7. *S. contaminans* CCUG 53915$^{T\ a}$; 8. *S. luteola* DSM 23150$^{T\ a}$; 9. *S. globispora* DSM 4$^{T\ b}$; 10. *S. psychrophila* KCTC 3446$^{T\ b}$; 11. *S. pasteurii* KCTC 3558$^{T\ b}$; 12. *S. saromensis* HG645$^{T\ c}$. Fatty acid methyl esters for which values were less than 0.1 % have been omitted from the table**

| Fatty acid | 1 | 2 | 3 | 4 | 5 | 6 | 7 | 8 | 9 | 10 | 11 | 12 |
|---|---|---|---|---|---|---|---|---|---|---|---|---|
| **Straight chain** | | | | | | | | | | | | |
| $C_{14:0}$ | 0.6 | 0.3 | 2.4 | 2.5 | 0.5 | 9 | 0.4 | 1.2 | 0.6 | 0.9 | 1.1 | - |
| $C_{16:0}$ | 0.8 | 0.6 | 3 | 2.2 | 0.2 | 10.6 | 0.5 | 0.5 | 0.9 | 1.5 | 4.7 | 1.4 |
| **Branched chain** | | | | | | | | | | | | |
| iso-$C_{14:0}$ | 26.1 | 1.8 | 4.3 | 5.5 | 20.9 | 7.5 | 4.7 | 11 | 3.4 | 4.1 | 15.4 | 5.6 |
| iso-$C_{15:0}$ | 32.3 | 7.6 | 4.3 | 45.5 | 34.6 | 27.7 | 55.8 | 37.6 | 4 | 6.4 | 6.9 | 49.5 |
| anteiso-$C_{15:0}$ | 15.2 | 62.7 | 76.8 | 37.7 | 30.8 | 39.6 | 23.5 | 41.1 | 61.8 | 68.4 | 48.6 | 33.3 |
| iso-$C_{16:0}$ | 10 | 0.6 | 1.6 | 1.1 | 4.4 | – | 3.6 | 2.6 | 1.5 | 2.1 | 7.5 | 4.2 |
| anteiso-$C_{17:0}$ | 1.4 | 11.3 | 3.2 | 1.1 | 1 | – | 2.8 | 1.6 | 6.9 | 6.6 | 4.0 | 2.4 |
| **Unsaturated** | | | | | | | | | | | | |
| $C_{16:1(\omega7c)}$ alcohol | 6.6 | 2.7 | 1 | 0.5 | 3.9 | – | 3.6 | 2.2 | 5.5 | 2.4 | 2.8 | – |
| $C_{16:1(\omega11c)}$ | 3.4 | 3.5 | 2.4 | 1.5 | 1.2 | 5.6 | 1.4 | 1.1 | 3.1 | 1.8 | 3.2 | – |
| **Summed feature 4** | 0.8 | 6.7 | 0.7 | – | 1 | – | 1.1 | 0.4 | 10.6 | 3.9 | 0.7 | – |

[a]Wolfgang et al. (2012)
[b]Yoon et al. (2001c)
[c]An et al. (2007)
–, absent; summed feature 4 contains iso-$C_{17:1}$ I and/or anteiso-$C_{17:1}$ B. If a given fatty acid is present above 5.0 % in at least one species the composition in the other species is given

aminophospholipid (APL1), two unknown phospholipids (PL1, PL2), and three unknown polar lipids. Cell wall peptidoglycan variation is A4a, and it is either L-Lys-D-Glu or L-Lys-D-Asp as the peptidoglycan type. The G+C content of genomic DNA of species of the genus ranges from 35 to 40.4 mol%.

The type species of the genus is *Viridibacillus arvi* (Heyrman et al. 2005), and the type strain is DSM 16319$^T$ = LMG 22166$^T$.

The genus currently contains three species, *V. neidei* (Nakamura et al. 2002; Albert et al. 2007), *V. arenosi* (Heyrman et al. 2005; Albert et al. 2007), and *V. arvi* (Heyrman et al. 2005; Albert et al. 2007). Species *V. arenosi* (Heyrman et al. 2005; Albert et al. 2007) and *V. arvi* were characterized using API Biotype 100 kit, and substrate utilization were carried out using API 50CHB kits. Strain *V. neidei* was studied based on Biolog GP system. All three species are positive for catalase and gelatin hydrolysis; negative for hydrolysis of aesculin, casein, starch, and Tween 80; for tests of Voges-Proskauer, indole production, methyl red; for $H_2S$ production, arginine dihydrolase, lysine decarboxylase, ornithine decarboxylase, citrate utilization; and for nitrate reduction. Both *V. arvi* and *V. arenosi* (Heyrman et al. 2005; Albert et al. 2007) produced acid only from D-fructose in the API 50CHB strips, and no substrate was used as a sole carbon source in the API Biotype 100 kit. Species *V. neidei* (Nakamura et al. 2002; Albert et al. 2007) is negative for phenylalanine deaminase and requires biotin, thiamin, and cystine for growth. It oxidizes pyruvate and β-hydroxybutyrate but not citrate, propionate, L-alanine, glycyl L-glutamate, 2′-deoxyadenosine, inosine, AMP, and UMP in the Biolog GP system. Species *V. neidei* has been partially characterized for antibiotic susceptibility, and it is sensitive to chloramphenicol, tobramycin, streptomycin, erythromycin, and tetracycline. Three species are characterized by the presence of MK-8 as the major menaquinone contributing 70–80 % followed by MK-7 (20–39 %) and trace amounts of MK-6. Polar lipids are nearly identical, and the peptidoglycan is L-Lys-D-Asp in case of *V. arvi* and *V. arenosi*, whereas *V. neidi* contains L-Lys-D-Asp type. The characteristics that diagnose the three species are listed in *Tables 25.22* and *25.23*.

## Isolation, Enrichment, and Maintenance Procedures

Species of the genus *Planococcus* have been isolated using various kinds of media such as antarctic bacterial medium (0.5 % peptone, 0.2 % yeast extract, and 1.5 % agar, pH 6.4) with 1.5 % NaCl or in the absence of salt, nutrient agar containing 1–5 % NaCl, sea water agar (1.0 % beef extract, 1.0 % peptone, 2.0 % agar, tap water 250 mL, and sea water 750 mL; pH 7.2), Zobell marine agar 2216 (Difco Laboratories, Detroit, USA) with 6 % NaCl, 100-fold-diluted marine agar (MA; Difco) containing 1 % starch, trypticase soy agar (TSA) or broth (TSB; Difco) with or

**Table 25.20**
**Diagnostic phenotypic characteristics of species of the genus *Ureibacillus* (Data from Weon et al. (2007); Kim et al. (2006); Fortina et al. (2001); Weon et al. (2007); Andersson et al. (1995))**

| Characteristic | *U. composti* DSM 17951[Ta] | *U. suwonensis* DSM 16752[Tb] | *U. terrenus* DSM 12654[Tc] | *U. thermophilus* DSM 17952[Td] | *U. thermosphaericus* DSM 10633[Tc, e] |
|---|---|---|---|---|---|
| **Morphological characteristics** | | | | | |
| Cell shape | Rods | Rods | Rods | Rods | Cocci/short rods |
| Cell size | 0.7–0.9 × 2.5–4.0 | 0.5–0.7 × 1.5–2.0 | NR | 0.8–1.2 × 2.5–3.5 | NR |
| Spore shape | Spherical | Spherical or oval | Round or oval | Round | Round or oval |
| Spore position | ST or T | ST or T | ST or T | ST or T | T |
| Cell arrangement | NR | Single or chains | Single or chains | NR | Single or chains |
| **Growth characteristics** | | | | | |
| Temperature range (°C) | 37–60 | 35–60 | 42–65 | 30–65 | 32–64 |
| NaCl tolerance (%) | 5 | 5 | Variable | 5 | 5 |
| **Biochemical characteristics** | | | | | |
| Phenylalanine deaminase | + | W | – | + | – |
| Urease | – | – | + | – | + |
| **Hydrolysis of** | | | | | |
| Aesculin | + | – | + | + | + |
| Tyrosine | – | – | + | + | – |
| **Chemotaxonomic characteristics** | | | | | |
| Menaquinone | MK-7, MK-8, MK-9 | MK-7, MK-8, MK-9, MK-10, MK-6 | MK-7, MK-8, MK-9, MK-10, MK-11 | MK-7, MK-8, MK-9 | MK-7 |
| Mol% G+C content of DNA | 42.4 | 41.5 | 39.6–41.5 | 38.5 | 35.7–39.2 |
| Source | Compost | Compost | Soil | Compost | Air |

[a]Weon et al. (2007)
[b]Kim et al. (2006)
[c]Fortina et al. (2001)
[d]Weon et al. (2007)
[e]Andersson et al. (1995)
+ positive, – negative, *W* weakly positive, *NR* not reported, *ST* subterminal, *T* terminal, *MK* menaquinone

without salt (5 % NaCl), and R2A solid media (Difco, USA) supplemented with 0.4 g $L^{-1}$ sodium acetate.

Species of the genus *Bhargavaea* were enriched and isolated on Zobell marine agar and Luria-Bertani (LB) agar plates that had been seeded with a tissue suspension of ginseng roots using the standard dilution plating technique (Manorama et al. 2009; Verma et al. 2012).

*Caryophanon* strains have been isolated and grown on different agar media adjusted to pH 7.5–8.5. Clarified manure extract agar (Peshkov 1967), cow dung agar (Smith and Trentini 1972), cow dung agar with lactalbumin hydrolysate (Moran and Witter 1976), and peptone-yeast extract-acetate agar (Pringsheim and Robinow 1947). A semisynthetic medium (Smith and Trentini 1973; Trentini 1978) was also found to support good growth of most strains of *C. latum* and *C. tenue.* Biotin has also been found to be essential for growth, whereas thiamine appears to stimulate the growth of both species. Kele and McCoy (1971) also developed a defined liquid medium for *C. latum.*

*Chryseomicrobium imtechense* was isolated on TSA by dilution plating method (Arora et al. 2011).

*Filibacter limicola* was isolated using MYP medium [mg $L^{-1}$: yeast extract (Difco), 10; peptone (Oxoid L37), 100; $K_2HPO_4$, 28; $MgCl_2{\cdot}6H_2O$, 127; $KNO_3$, 4; $(NH_4)_2SO_4$, 60; $MnSO_4{\cdot}4H_2O$, 8; Ferric citrate, 6] with the final pH adjusted to 7.0–7.2 with 1 M $KHCO_3$ before autoclaving (Stackebrandt 2009). Strains of *Filibacter limicola* also grow well, though less abundant, in different media such as tryptic soy broth (Oxoid CM129), CASO agar (g L: casein peptone, 15.0; soy peptone, 5.0; sodium chloride, 5.0; agar, 15.0, and pH 7.3 at 20 °C) supplemented with inosine (Stackebrandt 2009), casamino acids, or complete amino acid mixture in the presence of vitamins (Maiden and

**Table 25.21**
**Fatty acid composition (%) of species of the genus *Ureibacillus* (Data from Weon et al. (2007))**

| Fatty acids | *U. composti* DSM 17951[Ta] | *U. suwonensis* DSM 16752[Ta] | *U. terrenus* DSM 12654[Ta] | *U. thermophilus* DSM 17952[Ta] | *U. thermosphaericus* DSM 10633[Ta] |
|---|---|---|---|---|---|
| **Straight chain** | | | | | |
| $C_{16:0}$ | 5.4 | 6.4 | 7.9 | 3.7 | 8.1 |
| **Branched chain** | | | | | |
| iso-$C_{15:0}$ | 2.9 | – | 1.4 | 4.0 | 10.2 |
| iso-$C_{16:0}$ | 72.2 | 77.3 | 67.5 | 60 | 68.4 |
| iso-$C_{17:0}$ | 2.3 | 1.2 | 5.5 | 5.9 | 2.5 |
| anteiso-$C_{17:0}$ | 2.7 | – | 3.4 | 6.8 | – |
| **Unsaturated** | | | | | |
| $C_{16:1(\omega7c)}$ alcohol | 7.2 | 8.5 | 4.3 | 8.4 | 3.0 |

[a]Weon et al. (2007)
For the fatty acid analyses, cells for all strains were harvested after growth on R2A agar at 50 °C for 2 days. If a given fatty acid is present above 5.0 % in at least one species, the composition in the other species is given

**Table 25.22**
**Diagnostic phenotypic characteristics of species of the genus *Viridibacillus* (ata from Heyrman et al. (2005); Albert et al. (2007); Nakamura et al. (2002))**

| Characteristic | *V. arvi* DSM 16317[Ta, b] | *V. arenosi* DSM 16319[Ta, b] | *V. neidei* DSM 15031[Tb, c] |
|---|---|---|---|
| **Morphological characteristics** | | | |
| Colony color | Cream | Cream | White |
| Gram-stain | Variable | Variable | + |
| Spore shape | Spherical | Spherical | Round |
| Swollen sporangia | Slightly | v | + |
| **Growth characteristics** | | | |
| Anaerobic growth | W | – | – |
| Temperature range (°C) | 5–40 | 5–37 | 5–45 |
| Optimum temperature (°C) | 30 | 30 | 28–33 |
| pH range | 6–9 | 6–9 | 5–10 |
| **Biochemical characteristics** | | | |
| Nitrate reduction | – | + | – |
| Urease | + | – | – |
| **Acid production from sugars** | | | |
| D-Fructose | + | – | + |

[a]Heyrman et al. (2005)
[b]Albert et al. (2007)
[c]Nakamura et al. (2002)
+ positive, – negative, *v* variable, *W* weakly positive

Jones 1984), anaerobe agar (BBL, Becton-Dickinson), and in all the above cases, no growth occurred in absence of vitamins (Stackebrandt 2009).

Strains of the genus *Jeotgalibacillus* were isolated by a dilution plating/serial dilution using marine agar (Difco) or marine agar supplemented with up to 10 % (w/v) NaCl. Strains also grow on sea water medium (De Clerck and De Vos 2009; Cunha et al. 2012).

*Kurthia* strains were enriched on nutrient gelatin plates [per liter of distilled water: meat extract (Lab Lemco powder, Oxoid), 4 g; peptone (Difco), 5 g; yeast extract (Difco), 2.5 g; NaCl, 5 g; gelatin (BDH Chemical Co., Poole, UK), 100 g; pH 7.0] (YNG; Keddie 1981) and incubated at 20 °C with the lids upright and observed daily to check for filamentous outgrowths beyond the zone of gelatin liquefaction around the streak. To get pure cultures, a small piece of nutrient gelatin containing the outgrowth is then streaked on a nutrient agar (NA) medium.

*Paenisporosarcina quisquiliarum* was isolated by dilution plating of a surface soil sample onto TSBA plates (tryptic soy broth solidified with 1.5 % agar), whereas *Paenisporosarcina macmurdoensis* was isolated using ABM agar (0.5 % peptone, 0.2 % yeast extract, and 1.5 % agar, pH 7.2; Shivaji et al. 1988; Reddy et al. 2003). *Planomicrobium* strains were isolated using media containing NaCl (varying concentrations), marine agar (Difco), or artificial sea water basal medium with 1 % peptone and 0.5 % yeast extract (Eguchi et al. 1996). *Planomicrobium alkanoclasticum* was isolated on Bushnell-Haas medium (g L: Magnesium sulfate, 0.2; calcium chloride, 0.02; monopotassium phosphate, 1.0; dipotassium phosphate, 1.0; ammonium nitrate, 1.0; ferric chloride, 0.050; agar 20.0; pH 7.0) (Engelhardt et al. 2001).

**Table 25.23**
**Fatty acid composition (%) of species of the genus *Viridibacillus* (Data from Albert et al. (2007))**

| Fatty acid | *V. arvi* DSM $16317^T$ | *V. arenosi* DSM $16319^T$ | *V. neidei* NRRL BD-$87^T$ | | | |
|---|---|---|---|---|---|---|
| **Time (h)** | **21** | **30** | **21** | **30** | **21** | **30** |
| **Branched chain** | | | | | | |
| iso-$C_{15:0}$ | 44.7 | 46.2 | 55.2 | 55.5 | 52.4 | 54.7 |
| anteiso-$C_{15:0}$ | 21.6 | 21.7 | 11.3 | 11.2 | 15.4 | 15.3 |
| anteiso-$C_{17:0}$ | 5.3 | 4.8 | 1.8 | 1.8 | 2.8 | 2.4 |
| **Unsaturated** | | | | | | |
| $C_{16:1(\omega 11c)}$ | 3.8 | 3.9 | 8.2 | 7.0 | 4.1 | 3.8 |
| iso-$C_{17:1(\omega 10c)}$ | 5.4 | 5.2 | 7.0 | 7.1 | 7.0 | 6.8 |
| **Summed feature 4** | 7.2 | 6.8 | 5.1 | 5.1 | 6.3 | 5.8 |

Albert et al. (2007). Summed feature 4: iso-$C_{17:1}$ I/anteiso-$C_{17:1}$ B
Strains were grown on TSBA at 23 °C for 21 or 30 h as indicated; If a given fatty acid is present above 5.0 % in at least one species, the composition in the other species is given

Strains of the genus *Rummeliibacillus* were isolated by streaking serially diluted samples on TSA (Vaishampayan et al. 2009; Nakamura et al. 2002).

*Solibacillus silvestris* (Rheims et al. 1999; Krishnamurthi et al. 2009a) was isolated by a dilution plating/serial dilution using TSA (Vaishampayan et al. 2009).

Strains of the genus *Sporosarcina* were isolated by streaking serially diluted samples on different media. NA plates supplemented with 30–100 g urea/L, tryptic soy yeast extract agar (Difco), which was adjusted to pH 8.5 with NaOH before sterilization and filter-sterilized urea solution (1 %), trypticase soy agar supplemented with artificial sea water (pH 7.5) (Yoon et al. 2001c), TSA (Kwon et al. 2007), ABM agar (Reddy et al. 2003), JCM57 medium (10 g of glucose, 1.0 g of asparagine, 0.5 g of $K_2HPO_4$, 2.0 g of yeast extract, 15 g of agar per liter of distilled water, pH adjusted to 7.3), 1/10 diluted MA 2216 (An et al. 2007), PYG medium (0.5 % Bacto peptone, 0.02 % yeast extract, 0.5 % glucose, 0.3 % beef extract, 0.05 % NaCl, and 0.15 % $MgSO_4$. $7H_2O$, w/v, pH adjusted to 7.0) (Zhang et al. 2007), blood agar (Kämpfer et al. 2010), brain-heart infusion (BHI) (Oxoid, Hampshire, United Kingdom) agar (Tominaga et al. 2009), and TSA plates supplemented with 5 % sheep blood (Wolfgang et al. 2012).

Strains of *Ureibacillus* were isolated by using TSA (Weon et al. 2007; Kim et al. 2006) and CESP agar (casitone, 15 g; yeast extract, 5 g; soytone, 3 g; peptone, 2 g; $MgSO_4$, 0.015 g; $FeCl_3$, 0.007 g; $MnCl_2$, 0.002 g; made up to 1 L with distilled water, pH 7.2) (Fortina et al. 2001).

Strains of *Viridibacillus* were isolated on TSA plates (Heyrman et al. 2005; Nakamura et al. 2002).

Some members of family *Planococcaceae* cannot be isolated without prior enrichment. Species of the genus *Caryophanon* have been isolated only after an enrichment step. Samples of fresh cow dung are collected, placed in closed bottles, and kept at room temperature (Pringsheim and Robinow 1947; Weeks and Kelley 1958); after 1–2 days, when *Caryophanon* had multiplied in the sample, a loopful of dung is suspended in a few drops of tap water, and samples that are microscopically rich in *C. latum* are streaked on cow dung agar or peptone-yeast extract-acetate agar (Fritze and Claus 2009). For enrichment of *Planococcus* strains, sea water agar or NA containing 5–7 % NaCl can be used (Shivaji 2009). *P. halocryophilus* was enriched in R2A medium (Difco, USA) supplemented with 0.4 g $L^{-1}$ sodium acetate incubated at 5 °C. Following 11 months of incubation, 100 μL of culture was transferred to R2A (Difco, USA) solid media at 5 °C, resulting in the isolation (Mykytczuk et al. 2012). *Planomicrobium alkanoclasticum* was enriched in Bushnell-Haas medium (Brown and Braddock 1990) containing crude oil (0.5 g), at 15 °C for 14 days (Dai et al. 2005). *S. antarctica* was enriched prior isolation in PYG medium (0.5 % Bacto peptone, 0.02 % yeast extract, 0.5 % glucose, 0.3 % beef extract, 0.05 % NaCl, and 0.15 % $MgSO_4$. $7H_2O$, w/v, pH adjusted to 7.0) (Zhang et al. 2007) supplemented with cycloheximide, nystatin, and nalidixic acid (all at 25 mg $mL^{-1}$) and shaken at 10 °C for 4 days at 150 r.p.m. The culture was further diluted (1:10) and spread onto PYG agar plates and incubated at 4 °C for 2 weeks for isolation (Yu et al. 2008).

The members of the family are preserved using routine techniques like cryopreservation. For short-term preservation, the log-phase cultures on agar plates can be preserved at 4 °C for 4–6 weeks. *Planococci* cultures can be stored at 4 °C in screw-capped tubes or on plates containing semisolid medium after inoculating and overnight growth at optimum temperature. Species of the genus *Kurthia* can be maintained on yeast extract nutrient agar (YNA) medium at 25 °C for at least 6 months at room temperature (20 °C). *Caryophanon* cultures can be maintained for 4 weeks at 4 °C on clarified manure extract agar or cow dung agar after incubation at about 25 °C for 48 h. The psychrophilic or psychrotolerant cultures, grown at 20 °C, can be maintained at 10 °C for some weeks. Vegetative cultures of *S. ureae* grown on nutrient agar are viable for up to a year when stored at 4–10 °C in the dark. Endospores survive several years in screw-capped tubes under the same conditions (Claus and Fahmy 1986). Strains of *S. ureae* form spores in nutrient agar supplemented with urea (final concentration 0.2 %) if incubated at 25 °C. Sporulation of *Sporosarcina* cultures can be enhanced by adding 50 mg of $MnSO_4$ $H_2O$ per liter to the medium of MacDonald and MacDonald (1962) (Claus and Fahmy 1986). For long-term preservation, they can be freeze-dried using cryoprotectants such as skim milk (10–20 %, w/v), serum containing 5 % *meso*-inositol, glycerol (10–25 %), and DMSO (5 %) prepared in the appropriate medium. The cultures could also be stored in liquid nitrogen at −196 °C or at −20 or −70 or −80 °C in freezers.

## Ecology

### Habitat

Species of the genus *Planococcus* have been isolated from various aquatic and terrestrial habitats such as sea water (Yoon et al. 2003; Zobell and Upham 1944), Antarctic sea (Wang et al. 2011b), Antarctic sea ice (Bowman et al. 1997), Antarctic soil (Shivaji et al. 1988), Antarctic cyanobacterial mats (Alam et al. 2003; Reddy et al. 2002), sub-Antarctic sediments of Isla de Los Estados, Argentina (Olivera et al. 2007), high Arctic permafrost (Mykytczuk et al. 2012), high Arctic hypersaline spring channels (Lay et al. 2012), northeast Siberian sea coast permafrost sample (Hinsa-Leasure et al. 2010), coastal soil (Siddikee et al. 2010), mangrove soil (Kannan et al. 2006), glacial soil (Mayilraj et al. 2005), Palk Bay sediment (Nithya et al. 2011), marine sediment (Choi et al. 2007), marine solar saltern (Yoon et al. 2010b), salt water of Lake Red (Sovata, Romania) (Borsodi et al. 2010), saline and hyperalkaline Lonar Lake (Surakasi et al. 2010), sulfur spring (Romano et al. 2003), spacecraft assembly facility (Venkateswaran et al. 2001), and fish-brining tanks (Georgala 1957). Species of *Planococcus* have also been isolated from various life-forms including marine clams and fish (Hao and Komagata 1985; Novitsky and Kushner 1976), pigeon feces (Suresh et al. 2007), a shrimp (Alvarez 1982), a marine sponge *Plakortis simplex* (Schulze)(Kaur et al. 2012), from *Agaricus bisporus* composting phase II (He et al. 2009), a cuttlefish (*Sepia pharaonis*) fillets (Jeyasekaran et al. 2012), a sea urchin *Hemicentrotus pulcherrimus* (Huang et al. 2009), a sea anemone (Xiao et al. 2009), a snail (*Nassarius semiplicatus*) (Wang et al. 2008), reed (*Phragmites australis*), periphyton communities (two Hungarian soda ponds) (Rusznyák et al. 2008), fish (Radwan et al. 2007), brown and red algae (Beleneva and Zhukova 2006), and corals in (Beleneva et al. 2005). *Planococcus* sp. are pathogenic (10 %) to *Hylesia metabus* larvae, at doses of $3\text{–}4 \times 10^7$ cells.

*Bhargavaea cecembensis* was isolated from a deep-sea sediment sample (depth of 5904 m) from the Chagos-Laccadive ridge system in the Indian Ocean (Manorama et al. 2009). *Bacillus beijingensis* and *Bacillus ginsengi* were isolated from the internal tissue of ginseng roots (Qiu et al. 2009).

*C. latum* and *C. tenue* were both isolated from fresh cattle dung. In most successive studies, *C. latum* was found only in cattle manure or cattle manure-contaminated materials like bedding straw, barn dust, or barnyard soil. *C. latum* has not been found as part of the natural flora of the intestinal tract of cattle (Trentini 1978) and seems to be a natural, specific, and temporary coprophilic resident of cattle dung. It seems to be dispersed to new droppings by contaminated air and by flying insects (Trentini 1978; Trentini and Machen 1973). Recently *Caryophanon* strains have been reported from *Agaricus bisporus* composting phase II (He et al. 2009) and gingival scrapings from dogs (Saphir and Carter 1976).

*Chryseomicrobium imtechense* was isolated from a marine sample from Bay of Bengal, India (Arora et al. 2011).

*Filibacter limicola* has been isolated from sediments of Blelham Tarn, a eutrophic freshwater lake (Maiden 1983).

*Jeotgalibacillus alimentarius* was isolated from jeotgal, traditional Korean fermented seafood (Yoon et al. 2001a). The other species of the genus have been isolated from various habitats such as from forest soil sample (*J. soli*; Chen et al. 2010b), from a marine solar saltern (*J. salaries* and *J. campisalis*; Yoon et al. 2010a), and from sediments of the Iberian deep sea, the tropical Atlantic, and the Arctic and Antarctic Oceans (*J. marinus*, Yoon et al. 2010a). *Jeotgalibacillus* strains have also been reported from non-saline soil samples collected from Xiaoxi National Natural Reserve (Chen et al. 2010a), decomposing reed rhizomes in a Hungarian soda lake (Borsodi et al. 2005), and hull of a ship in the form of biofilms (Inbakandan et al. 2010).

Strains of *Kurthia* have been isolated from the feces of patients suffering from diarrhea, but there is no evidence of pathogenicity in members of the genus (Keddie 1981). *K. zopfii* and *K. gibsonii* were isolated from meat (Gardner 1969) and meat products (Keddie 1981) and from feces of farm animals (chickens and pigs) (Keddie 1981). Authentic *Kurthia* species have also been isolated sporadically from milk, soil, and surface waters, presumably as a result of contamination with animal dung (Keddie 1981), from bottled drinking water (Jeena et al. 2006), from the nasal cavity of sea lions (Hernández-Castro et al. 2005), from the stomach and intestinal contents of the Susuman mammoth (Belikova et al. 1980), from the housefly (Wei et al. 2012), from Maari, a Baobab seed fermented product (Parkouda et al. 2010), from cigarettes (Rooney et al. 2005), from activated sludge (Kahru et al. 1998), from refrigerated meat (Pin and Baranyi 1998), in dental plaque from the beagle dog (Wunder et al. 1976), and from sloughing spoilage of California ripe olives (Patel and Vaughn 1973).

*Paenisporosarcina macmurdoensis* was isolated from cyanobacterial mat samples collected from ponds of Wright Valley, McMurdo Region, Antarctica (Matsumoto 1993), and *Paenisporosarcina quisquiliarum* from surface soil at a landfill site in Chandigarh, India (Krishnamurthi et al. 2009b).

Species of *Planomicrobium* have been isolated from diverse habitats including fresh water, marine habitats, cold regions, and mesophilic climates. *Planomicrobium koreense* was isolated from the Korean traditional fermented seafood jeotgal (Yoon et al. 2001b). A few of them have been isolated from marine habitats like *Planomicrobium alkanoclasticum*, which was isolated from intertidal beach sediment (Engelhardt et al. 2001), *Planomicrobium chinense* from coastal sediment from the Eastern China Sea (Dai et al. 2005), *Planomicrobium mcmeekinii* from Antarctic sea ice (Junge et al. 1998), *Planomicrobium okeanokoites* from marine mud (Zobell and Upham 1944), and *Planomicrobium flavidum* from a marine solar saltern of the Yellow Sea (Jung et al. 2009). A few of the species were also isolated from cold habitats like *Planomicrobium glaciei* from frozen soil collected from the China no. 1 glacier (Zhang et al. 2009), *Planomicrobium stackebrandtii* from a cold desert of the Himalayas (Mayilraj et al. 2005), and *Planomicrobium psychrophilum* from a cyanobacterial mat sample from McMurdo Dry Valleys, Antarctica (Reddy et al. 2002). *Planomicrobium* strains were also isolated from biofilms of full-scale drinking water distribution systems (Liu et al. 2012) and ancient permafrost sediments from the Kolyma lowland of Northeast Eurasia (Vishnivetskaya et al. 2006).

*Rummeliibacillus stabekisii* was isolated from the Payload hazardous servicing facility at the Kennedy Space Center, FL, USA (Vaishampayan et al. 2009), and *Rummeliibacillus pycnus* was isolated from soil (Nakamura et al. 2002).

*Solibacillus silvestris* (Krishnamurthi et al. 2009a) was Isolated from the top soil layer of a beech forest soil near Braunschweig, Lower Saxony, Germany (Rheims et al. 1999).

*Sporosarcina* species have been mostly isolated from soils and water from different habitats. *S. antarctica* was isolated from soil samples collected off King George Island, West Antarctica (Yu et al. 2008); *S. globispora* and *S. psychrophila* from soil and river water (Larkin and Stokes 1967; Nakamura 1984); *S. koreensis* and *S. soli* from upland soil in Suwon, Korea (Kwon et al. 2007); *S. pasteurii* from soil, sewage, and incrustrations on urinals (Miquel 1889); *S. saromensis* isolated from surface water in Lake Saroma (An et al. 2007); *S. aquimarina* from sea water in Korea (Yoon et al. 2001c); *S. contaminans* and *S. thermotolerans* from an industrial clean-room floor and from a human blood sample, respectively (Kämpfer et al. 2010); *S. luteola* from soy sauce (Tominaga et al. 2009); and *S. newyorkensis* from clinical specimens and raw cow's milk (Wolfgang et al. 2012). *Sporosarcina* strains were also isolated from chicken-waste-impacted farm soil (You et al. 2012), ancient algal mats from the McMurdo Dry Valleys, Antarctica (Antibus et al. 2012), ice and brine (Bakermans and Skidmore 2011), surimi seafood products (Coton et al. 2011), petroleum hydrocarbon-contaminated soil (Wan et al. 2011), arctic terrestrial and marine environments (Kim et al. 2010), sea urchin *Hemicentrotus pulcherrimus* (Huang et al. 2009), from bovine slurry waste (Murayama et al. 2010), sea anemone *Stichodactyla haddoni* (Williams et al. 2007), farm-packaged product (Huck et al. 2008), and deep-sea sediment of the South China Sea (Liu and Shao 2007).

Species of *Ureibacillus* have also been isolated from a diverse range of habitats including air, soil, compost, and human waste. Most species have been isolated from various different types of compost like *U. composti* and *U. thermophilus* from livestock-manure composts (Weon et al. 2007) and *U. suwonensis* from cotton waste composts (Kim et al. 2006). *U. terrenus* was isolated from soil (Fortina et al. 2001) and *U. thermosphaericus* from air (Fortina et al. 2001). *Ureibacillus* strains were also isolated from the Ramsar hot springs in Iran (Abbasalizadeh et al. 2012), oyster mushroom (*Pleurotus* sp.) (Vajna et al. 2012), terrestrial systems (Portillo et al. 2012), dried human feces (Hoyles et al. 2012), rice straw (Wang et al. 2011a), autothermal thermophilic aerobic digestion of sewage sludge (Liu et al. 2010), personal-use composting reactor (Watanabe et al. 2010), pig manure composting (Xie et al. 2009), anaerobic sludge (Nakasaki et al. 2009), conventional dairy farms (Coorevits et al. 2008), hot aerobic poultry and cattle manure composts (Wang et al. 2007), and aerobic digestion of swine waste (Gagné et al. 2001).

*Viridibacillus* species have been isolated from soil samples (Heyrman et al. 2005; Nakamura et al. 2002).

## Pathogenicity, Clinical Relevance

The members of the family *Planococcaceae* are reported to be nonpathogenic to humans, although a few reports indicate their presence in diseased individuals. A few species and strains of *Planococcus* have been demonstrated to be pathogenic to other animals. For instance, a strain of *Planococcus* sp. is pathogenic (10 %) to *Hylesia metabus* larvae (Osborn et al. 2002), and *Planococcus* W29 has strong inhibitory activity against *Serratia liquefaciens* which is a fish pathogen. Abdel Gabbar et al. (1995) implicated *P. halophilus* in an outbreak of necrotic hepatitis in chickens, whereas Oeding (1971) reported that *P. citreus* has no antigenic relationship to *Staphylococci* and *Micrococci*. The principle components responsible for the pathogenic activity have been identified as aromatic rings, phenolic groups, and a covalently bonded glycopeptide which is water soluble with a molecular weight of about 2800 (Austin and Billaud 1990). *Caryophanon latum* and *C. tenue* are not known to be pathogenic. The *Caryophanon* sp. 52E5 and *Planomicrobium* sp. 34D8 inhibited swarming in *Serratia marcescens* MG1 (Alagely et al. 2011). There is no evidence of pathogenicity in authentic members of the genus *Kurthia*, although a number of strains of *Kurthia* have been solated from various clinical sources (Faoagali 1974; Severi 1946; Yang et al. 1985) and most frequently from the feces of patients suffering from diarrhea (Elston 1961; Jarumilinta et al. 1976). The genus *Planomicrobium* is essentially a marine genus with no previously reported pathogenicity (Skerratt et al. 2002). Chomarat et al. (1990) reported the isolation of *S. ureae* from a bronchial biopsy in a child with cystic fibrosis, but the strain is not pathogenic.

Antibiotic susceptibility has been studied in few species of the genus *Planococcus*. *P. citreus*, *P. maritimus*, *P. maitriensis*, and *P. antarcticus* are susceptible to chloramphenicol and tetracycline.

*P. maritimus*, *P. maitriensis*, and *P. antarcticus* were sensitive to bacitracin, lincomycin, and streptomycin but varied in their susceptibility to ampicillin, erythromycin, gentamicin, penicillin-G, and rifampin (Alam et al. 2003; Jeffries 1969; Kocur et al. 1970; Novitsky and Kushner 1976; Reddy et al. 2002; Yoon et al. 2003). *P. columbae*, *P. halocryophilus*, *P. halophilus*, *P. plakortidis*, and *P. salinarum* are sensitive to a number of antibiotics (Suresh et al. 2007; Mykytczuk et al. 2012; Novitsky and Kushner 1976; Kaur et al. 2012; Yoon et al. 2010b).

*J. campisalis*, *J. marinus*, and *J. salarius* are susceptible to carbenicillin, cephalothin, chloramphenicol, gentamicin, neomycin, oleandomycin, and penicillin-G; resistant to polymyxin B; and are variable to ampicillin, kanamycin, lincomycin, novobiocin, streptomycin, and tetracycline (Yoon et al. 2010a).

*Planomicrobium flavidum* is susceptible to ampicillin, carbenicillin, cephalothin, chloramphenicol, gentamicin, kanamycin, lincomycin, neomycin, novobiocin, oleandomycin, penicillin-G, and streptomycin, but not to polymyxin B or tetracycline.

*Bacillus pycnus* (*Rummeliibacillus pycnus*; Vaishampayan et al. 2009) has been partially characterized for the antibiotic susceptibility, and it is susceptible to chloramphenicol, tobramycin, streptomycin, erythromycin, and tetracycline.

## Application

### Bioremediation

*Planococcus* sp. strain S5 isolated from activated sludge, which is able to grow on salicylate or benzoate as sole carbon source has been used for sodium salicylate degradation. This strain harbor a plasmid pLS5 which has genes coding for catechol 1, 2-dioxygenase and catechol 2,3-dioxygenase involved in degradation of salicylate and benzoate (Labuzek et al. 2003). Strain S5 is also able to utilize phenol as the sole carbon and energy source and can grow on 1 or 2 mM phenol, and this was attributed to both catechol 1,2- and catechol 2,3-dioxygenase. Catechol 2, 3-dioxygenase was optimally active at 60 °C and pH 8.0 and showed meta-cleavage activities for various catechols like catechol (100 %), 3-methylcatechol (13.67 %), 4-methylcatechol (106.33 %), and 4-chlorocatechol (203.80 %). The high reactivity of this enzyme towards 4-chlorocatechol is different from that observed for other catechol 2,3-dioxygenases (Hupert-Kocurek et al. 2012). Strain S5 can thus be used to bioremidiate sites contaminated with the above aromatic hydrocarbons and phenolic compounds. *P. citreus* has also been used for the bioremediation of Cr (VI) (Cheng et al. 2010).

*S. ginsengisoli* CR5 tolerates high concentrations (50 mM) of As (III), a highly toxic metalloid. Microbially induced calcite precipitation (MICP)-based bioremediation by *S. ginsengisoli* is a viable, environmental friendly technology for remediation of arsenic-contaminated sites (Achal et al. 2012).

Mercuric reductase plays a significant role in biogeochemical cycling and detoxification of Hg and is useful in clean up of Hg-contaminated effluents. *Sporosarcina* sp. strain G3 tolerates up to 40, 525, 210, 2900, and 370 μM of Cd, Co, Zn, Cr, and Hg, respectively, and reduces and detoxifies redox-active metals like Cr and Hg. The chromate reductase and MerA activities in the crude cell extract of the strain G3 were 1.5 and 0.044 units/mg protein, respectively (Bafana 2011).

Wang et al. (2011a) established a microbial consortium, designated WCS-6, which is capable of degrading lignocellulose. The consortium included nine bacterial isolates related to *Bacillus thermoamylovorans* BTa, *Paenibacillus barengoltzii* SAFN-016, Proteobacterium S072, *Pseudoxanthomonas taiwanensis* CB-226, *Rhizobiaceae* M100, *Bacillus* sp. E53-10, betaproteobacterium HMD444, *Petrobacter succinimandens* 4BON, and *Tepidiphilus margaritifer* N2-214 and five DNA sequences related to *U. thermosphaericus*, uncultured bacterium clone GC3, uncultured *Clostridium* sp. clone A1-3, *Clostridium thermobutyricum*, and *Clostridium thermosuccinogenes*.

Vargas-García et al. (2007) studied the effect of inoculation of microbial isolates, *Bacillus shackletonni*, *Streptomyces thermovulgaris*, and *U. thermosphaericus* in composting processes to improve lignocellulose degradation and reported that *U. thermosphaericus* decreased the final lignin content in a range between 17.2 % and 24.3 % with high efficiency. The biological detoxification of lignocellulosic hydrolysate by *U. thermosphaericus* is gentle, more precise in their action, and the efficiency is quite comparable to other physical and chemical methods (Okuda et al. 2008; Parawira and Tekere 2011). *U. thermosphaericus* was also resistant to $Hg^{2+}$ salts with minimum inhibitory concentration (MIC) value of 30 μg $mL^{-1}$. The resistance to mercury in *U. thermosphaericus* was attributed to mercuric reductase activity which removed $Hg^{2+}$ from the medium by the formation of a black precipitate, identified as HgS. According to Glendinning et al. (2005), this is a new mechanism of Hg tolerance, based on the production of nonvolatile thiol species.

An unidentified species of *Sporosarcina* sp. along with two unknown species of *Bacillus* sp. and *Pseudomonas* sp. were demonstrated to be very effective in the bioleaching of aluminum, arsenic, copper, manganese, iron, and zinc in Pb/Zn smelting at 65 °C and pH 1.5 (Cheng et al. 2009).

*P. maitriensis* produces extracellular polymeric substance (EPS) with oil spreading potential comparable to that of Triton X100 and Tween 80 and could emulsify xylene, hexane, and oils from jatropha, paraffin, and silicone and reduce the surface tension (from 72 to 46.07 mN $m^{-1}$) and interfacial tension. EPS could thus be used for bioremediation, enhanced oil recovery, and in cosmetics due to its emulsifying and tensiometric properties (Kumar et al. 2007).

### Formation of Added-Value Products

*Kurthia* produces two enzymes, namely, carbamoylase and hydantoinase, which are key enzymes for the production of optically pure amino acids from dl-5-substituted hydantoins

(SSH) (Mei et al. 2009). *Kurthia* sp. has also been used to produce L-proline from glutamic acid or aspartic acid with the aid of detergents (Kato et al. 1972a, b).

Eleven biotin biosynthetic genes have been identified in *Kurthia* sp. strain 538-KA26 (Kiyasu et al. 2001) with potential use in biotech industry.

*P. maritimus* produces a number of carotenoids such as the red-pigmented glyco-carotenoic acid ester and methyl glucosyl-3,4-dehydro-apo-8′-lycopenoate with antioxidative activity (Shindo et al. 2008). These antioxidants could be used as drugs for stroke and neurodegenerative diseases, as dietary supplements and also as preservatives in food and cosmetics and to prevent the degradation of rubber and gasoline. Recently, Krishnaveni and Jayachandran (2009) demonstrated that crude extracts of both *P. maritimus* KP8 and *Staphylococcus arlettae* KP2 downregulated the synthesis of inflammatory mediators such as tumor necrosis factor-alpha (TNF-alpha), interleukin-1beta (IL-1beta), and cyclooxygenase-2 (COX-2), besides markedly inhibiting p38 mitogen-activated protein (MAP) kinase suggesting that the crude ethyl acetate extracts from these isolates have the potential of inhibiting inflammation in mitogen-induced peripheral blood mononuclear cells (Krishnaveni and Jayachandran 2009).

Endophytic *S. aquimarina* SjAM16103 are known to have plant growth-promoting activity due to their ability to produce indole acetic acid and siderophore which help to solubilize phosphate molecules and fix atmospheric nitrogen (Janarthine and Eganathan 2012). In fact more and more agriculturists are now resorting to the use of endophytic bacteria to promote plant growth due to their abilities to fix nitrogen, to produce phytohormones, to solubilize phosphate, and to control disease (Janarthine and Eganathan 2012). Zhang et al. (2010) demonstrated that *Sporosarcina* sp. strain N52 produces intracellular glucan from L-arabinose, and additionally strain N52 could be used for waste management and bioconversion of organic materials to the valuable alpha-glucan which can be used as a food additive.

## Enzymes

Cold-active enzymes have a high catalytic efficiency at low temperature and thus are suitable for processes that need to be accomplished at low temperature. For instance, the catalytic efficiency of β-galactosidase of *Planococcus* sp. L4 at 5 and 20 °C is 14 and 47 times more than that of *Escherichia coli* β-galactosidase at 20 °C, respectively. Hence, cold-active β-galactosidase from the strain L4 can be used for removal of lactose from milk and dairy products at cold conditions or as a reporter enzyme for psychrophilic genetic systems (Hu et al. 2007). *Sporosarcina* sp. RRLJ1 produces a protease with potential for commercial use as it has high activity at pH 6.5 (Boruah and Bezbaruah 2000). *S. ureae* produces an alkaline stable (pH 7.75–12.5) urease which could be used for several applications at alkaline pH (McCoy et al. 1992).

## Other Applications

*P. rifietoensis* and several other strains can control gray moulds (which cause disease in strawberry fruits) and eventually increase the yield of fruits and can be a key substitute to synthetic fungicides. These strains can also be used under storage and greenhouse conditions (Essghaier et al. 2009). *Planococcus* strains from Antarctica have the ability to produce cold-active antimicrobial compounds with potential use in chilled food preservation. Thus, Antarctic soils could represent an untapped reservoir of novel, cold-active antimicrobial-producers (O'Brien et al. 2004).

Geologically stored carbon dioxide ($CO_2$) needs to be trapped to avoid increase in $CO_2$ in the atmosphere. Microbes also have a role to play in this important aspect. For instance, microbially induced calcium carbonate precipitation has the potential in remediation of a broad array of structural damages including sealing of concrete cracks with $CaCO_3$, a property which has been extensively studied in *S. pasteurii* (Achal et al. 2009a). Thus, microbially induced mineral precipitation (MICCP) technologies may effectively seal and strengthen fractures to alleviate $CO_2$-leakage potential (Phillips et al. 2012). Martin et al. (2012) suggested the use of an anaerobic strain instead of *Sporosarcina pasteurii* to precipitate $CaCO_3$ in the anoxic subsurface via ureolysis since *S. pasteurii* cannot grow anaerobically. Bergdale et al. (2012) used genetically engineered microorganisms with dual abilities of producing extracellular polymeric substances (EPSs) as well as inducing MICCP to remediate the structural damages. They transformed *Pseudomonas aeruginosa* strain 8821 capable of producing EPS with the entire *Sporosarcina pasteurii* urease gene cluster including *ure*A, *ure*B, *ure*C, *ure*D, *ure*E, *ure*F, and *ure*G and used the recombinant strain to induce calcite precipitation in the presence of EPS to provide a stronger matrix than MICCP alone as a biosealant.

Another strategy developed by Okwadha and Li (2011) used a biosealant which was obtained by using PCB, urea, $Ca^{2+}$, and bacteria (*S. pasteurii* strain ATCC 11859). It was envisaged that the biosealant would have reduced water permeability and high resistance to carbonation and would be thermally stable up to 840 °C for environmentally sustainable development (Okwadha and Li 2011).

Achal et al. (2011) developed a mutant of *S. pasteurii* strain Bp M-3 which could utilize corn steep liquor or lactose mother liquor (low-cost industrial waste) from starch/diary industry as a nutrient source for the growth and production of calcite (Achal et al. 2009a, b, 2011).

A novel carotenoid from *S. aquimarina* strain SF238 acetyl-4,4′-diapolycopene-4,4′-dioate has excellent antioxidative properties which can protect the cells from photosensitized peroxidation reactions like other related 4,4′-diapolycopene-4,4′-dioate derivatives (Steiger et al. 2012).

Ranganathan et al. (2006) used bacteriotherapy to alleviate uremic intoxication by ingestion of live *S. pasteurii* (nonpathogenic alkalophilic urease positive) which are able to catabolize uremic solutes in the gut. The bacterium was able to remove urea from 21 ± 4.7 mg to 228 ± 6.7 mg per hour using 10 cfu (colony forming units) in in vitro study. *S. pasteurii* does not disturb the microbial community in the human intestine, thus implying that *S. pasteurii* reduced blood urea-nitrogen levels and significantly prolonged the lifespan of uremic animals (Ranganathan et al. 2006).

## Acknowledgement

The authors wish to thank Council of Scientific and Industrial Research (CSIR), New Delhi, India, for the financial support.

## References

Abbasalizadeh S, Salehi Jouzani G, Motamedi Juibari M, Azarbaijani R, Parsa Yeganeh L, Ahmad Raji M, Mardi M, Salekdeh GH (2012) Draft genome sequence of *Ureibacillus thermosphaericus* strain thermo-BF, isolated from ramsar hot springs in Iran. J Bacteriol 194:4431

Abdel Gabbar KM, Dewani P, Junejo BM (1995) Possible involvement of *Planococcus halophilus* in an outbreak of necrotic hepatitis in chickens. Vet Rec 136:74

Achal V, Mukherjee A, Basu PC, Reddy MS (2009a) Lactose mother liquor as an alternative nutrient source for microbial concrete production by *Sporosarcina pasteurii*. J Ind Microbiol Biotechnol 36:433–438

Achal V, Mukherjee A, Basu PC, Reddy MS (2009b) Strain improvement of *Sporosarcina pasteurii* for enhanced urease and calcite production. J Ind Microbiol Biotechnol 36:981–988

Achal V, Mukherjee A, Reddy MS (2011) Effect of calcifying bacteria on permeation properties of concrete structures. J Ind Microbiol Biotechnol 38:1229–1234

Achal V, Pan X, Fu Q, Zhang D (2012) Biomineralization based remediation of As (III) contaminated soil by *Sporosarcina ginsengisoli*. J Hazard Mater 201–202:178–184

Adcock KA, Seidler RJ, Trentini WC (1976) Deoxyribonucleic acid studies in the genus *Caryophanon*. Can J Microbiol 22:1320–1327

Aga H, Maruta K, Yamamoto T, Kubota M, Fukuda S, Kurimoto M, Tsujisaka Y (2002) Cloning and sequencing of the genes encoding cyclic tetrasaccharide-synthesizing enzymes from *Bacillus globisporus* C11. Biosci Biotechnol Biochem 66:1057–1068

Aga H, Nishimoto T, Kuniyoshi M, Maruta K, Yamashita H, Higashiyama T, Nakada T, Kubota M, Fukuda S, Kurimoto M, Tsujisaka Y (2003) 6-Alpha-glucosyltransferase and 3-alpha-isomaltosyltransferase from *Bacillus globisporus* N75. J Biosci Bioeng 95:215–224

Alagely A, Krediet CJ, Ritchie KB, Teplitski M (2011) Signaling-mediated cross-talk modulates swarming and biofilm formation in a coral pathogen *Serratia marcescens*. ISME J 5:1609–1620

Alam SI, Singh L, Dube S, Reddy GSN, Shivaji S (2003) Psychrophilic *Planococcus maitriensis* sp.nov. from Antarctica. Syst Appl Microbiol 26:505–510

Albert RA, Archambault J, Lempa M, Hurst B, Richardson C, Gruenloh S, Duran M, Worliczek HL, Huber BE, Rosselló-Mora R, Schumann P, Busse HJ (2007) Proposal of *Viridibacillus* gen. nov. and reclassification of *Bacillus arvi*, *Bacillus arenosi* and *Bacillus neidei* as *Viridibacillus arvi* gen. nov., comb. nov., *Viridibacillus arenosi* comb. nov. and *Viridibacillus neidei* comb. nov. Int J Syst Evol Microbiol 57:2729–2737

Alvarez RJ (1982) Role of *Planococcus citreus* in the spoilage of *Penaeus* shrimp. Zentralbl Bakteriol Mikrobiol Hyg Abt I Orig C 3:503–512

An SY, Haga T, Kasai H, Goto K, Yokota A (2007) *Sporosarcina saromensis* sp. nov., an aerobic endospore-forming bacterium. Int J Syst Evol Microbiol 57:1868–1871

Andersson M, Laukkanen M, Nurmiaholassila EL, Rainey FA, Niemelä SI, Salkinoja-Salonen M (1995) *Bacillus thermosphaericus* sp. nov., a new thermophilic ureolytic *Bacillus* isolated from air. Syst Appl Microbiol 18:203–220

Antibus DE, Leff LG, Hall BL, Baeseman JL, Blackwood CB (2012) Cultivable bacteria from ancient algal mats from the McMurdo Dry Valleys, Antarctica. Extremophiles 16:105–114

Arora PK, Chauhan A, Pant B, Korpole S, Mayilraj S, Jain RK (2011) *Chryseomicrobium imtechense* gen. nov., sp. nov., a new member of the family *Planococcaceae*. Int J Syst Evol Microbiol 61:1859–1864

Ash C, Farrow JAE, Wallbanks S, Collins MD (1991) Phylogenetic heterogeneity of the genus *Bacillus* revealed by comparative analysis of small subunit ribosomal RNA sequences. Lett Appl Microbiol 13:202–206

Austin B, Billaud AC (1990) Inhibition of the fish pathogen, *Serratia liquefaciens*, by an antibiotic-producing isolate of *Planococcus* recovered from sea water. J Fish Dis 13:553–556

Bafana A (2011) Mercury resistance in *Sporosarcina* sp. G3. Biometals 24:301–309

Bakermans C, Skidmore ML (2011) Microbial metabolism in ice and brine at −5 °C. Environ Microbiol 13:2269–2278

Becker B, Wortzel EM, Nelson JH (1967) Chemical composition of the cell wall of *Caryophanon latum*. Nature 213:300

Beijerinck MW (1901) Anhäufungsversuche mit Ureumbakterien, Uremspaltung durch Urease und durch Katabolismus. Zentralbl Bakteriol Parasitenkd Infektionskr Hyg Abt II 7:33–61

Beleneva IA, Zhukova NV (2006) Bacterial communities of brown and red algae from Peter the Great Bay, the Sea of Japan. Mikrobiologiia 75:410–419

Beleneva IA, Dautova TI, Zhukova NV (2005) Characterization of communities of heterotrophic bacteria associated with healthy and diseased corals in Nha Trang Bay (Vietnam). Mikrobiologiia 74:667–676

Belikova VL, Cherevach NV, Baryshnikova LM, Kalakutskii LV (1980) Morphologic, physiologic and biochemical characteristics of *Kurthia zopfii*. Microbiologiya 49:51–55

Belikova VA, Cherevach NV, Kalakoutskii LV (1986) *Kurthia sibirica*, a new bacterial species of the *Kurthia* genus. Mikrobiologiya 55:831–835

Bergdale TE, Pinkelman RJ, Hughes SR, Zambelli B, Ciurli S, Bang SS (2012) Engineered biosealant strains producing inorganic and organic biopolymers. J Biotechnol 161:181–189

Borsodi AK, Micsinai A, Rusznyák A, Vladár P, Kovács G, Tóth EM, Márialigeti K (2005) Diversity of alkaliphilic and alkalitolerant bacteria cultivated from decomposing reed rhizomes in a Hungarian soda lake. Microbiol Ecol 50:9–18

Borsodi AK, Kiss RI, Cech G, Vajna B, Tóth EM, Márialigeti K (2010) Diversity and activity of cultivable aerobic planktonic bacteria of a saline Lake located in Sovata. Romania Folia Microbiol (Praha) 55:461–466

Boruah HP, Bezbaruah B (2000) Protease from *Sporosarcina* sp. RRLJ 1. Ind J Exp Biol 38:293–296

Bowman JP, McCammon SA, Brown MV, Nichols DS, McMeekin TA (1997) Diversity and association of psychrophilic bacteria in Antarctic sea ice. Appl Environ Microbiol 63:3068–3078

Brown EJ, and Braddock JF (1990) Sheen screen; a miniaturised most probable number method for the enumeration of oil-degrading microorganisms. Applied and Environmental Microbiology 52, 149–156

Brunialti EA, Gatti-Lafranconi P, Lotti M (2011) Promiscuity, stability and cold adaptation of a newly isolated acylaminoacyl peptidase. Biochimie 93:1543–1554

Chanda VB, Priya PB, Sushama RC (2010) Plasmids of encoded bacteria as vectors for transformation in plants. Int J Integr Biol 9:113–118

Chary VK, Hilbert DW, Higgins ML, Piggot PJ (2000) The putative DNA translocase SpoIIIE is required for sporulation of the symmetrically dividing coccal species *Sporosarcina ureae*. Mol Microbiol 35:612–622

Chen Q, Liu Z, Peng Q, Huang K, He J, Zhang L, Li W, Chen Y (2010a) Diversity of halophilic and halotolerant bacteria isolated from non-saline soil collected from Xiaoxi National Natural Reserve, Hunan Province. Wei Sheng Wu Xue Bao 50:1452–1459

Chen YG, Peng DJ, Chen QH, Zhang YQ, Tang SK, Zhang DC, Peng QZ, Li WJ (2010b) *Jeotgalibacillus soli* sp. nov., isolated from non-saline forest soil, and emended description of the genus *Jeotgalibacillus*. Antonie Van Leeuwenhoek 98:415–421

Cheng Y, Guo Z, Liu X, Yin H, Qiu G, Pan F, Liu H (2009) The bioleaching feasibility for Pb/Zn smelting slag and community characteristics of indigenous moderate-thermophilic bacteria. Bioresour Technol 100:72737–72740

Cheng Y, Yan F, Huang F, Chu W, Pan D, Chen Z, Zheng J, Yu M, Lin Z, Wu Z (2010) Bioremediation of Cr (VI) and immobilization as Cr (III) by *Ochrobactrum anthropi*. Environ Sci Technol 44:6357–6363

Cherevach NV, Tourova TP, Belikova VL (1983) DNA-DNA homology studies among strains of *Kurthia zopfii*. FEMS Microbiol Lett 19:243–245

Choi JH, Im WT, Liu QM, Yoo JS, Shin JH, Rhee SK, Roh DH (2007) *Planococcus donghaensis* sp. nov., a starch-degrading bacterium isolated from the East Sea, South Korea. Int J Syst Evol Microbiol 57:2645–2650

Chomarat M, de Montclos M, Flandrois JP, Breysse F (1990) Isolation of *Sporosarcina ureae* from a bronchial biopsy in a child with cystic fibrosis. Eur J Clin Microbiol Infect Dis 9:302–303

Claus D, Fahmy F (1986) Genus *Sporosarcina*. In: Kreig NR, Holt JG (eds) Bergey's manual of systematic bacteriology, vol 2. Williams & Wilkins, Baltimore, pp 1202–1206

Clausen V, Jones JG, Stackebrandt E (1985) 16S ribosomal RNA analysis of *Filibacter limicola* indicates a close relationship to the genus *Bacillus*. J Gen Microbiol 131:2659–2663

Collins MD, Jones D (1981) Distribution of isoprenoid quinone structural type in bacteria and their taxonomic implication. Microbiol Rev 45:316–354

Collins MD, Goodfellow M, Minnikin DE (1979) Isoprenoid quinones in the classification of coryneform and related bacteria. J Gen Microbiol 110:127–136

Coorevits A, De Jonghe V, Vandroemme J, Reekmans R, Heyrman J, Messens W, De Vos P, Heyndrickx M (2008) Comparative analysis of the diversity of aerobic spore-forming bacteria in raw milk from organic and conventional dairy farms. Syst Appl Microbiol 31:126–140

Coton M, Denis C, Cadot P, Coton E (2011) Biodiversity and characterization of aerobic spore-forming bacteria in surimi seafood products. Food Microbiol 28:252–260

Cunha S, Tiago I, Paiva G, Nobre F, Da Costa MS, Veríssimo A (2012) *Jeotgalibacillus soli* sp. nov., a gram-stain-positive bacterium isolated from soil. Int J Syst Evol Microbiol 62:608–612

Cussiol JR, Alegria TG, Szweda LI, Netto LE (2010) Ohr (organic hydroperoxide resistance protein) possesses a previously undescribed activity, lipoyl-dependent peroxidase. J Biol Chem 285:21943–21950

Dai X, Wang YN, Wang BJ, Liu SJ, Zhou YG (2005) *Planomicrobium chinense* sp. nov., isolated from coastal sediment, and transfer of *Planococcus psychrophilus* and *Planococcus alkanoclasticus* to *Planomicrobium* as *Planomicrobium psychrophilum* comb. nov. and *Planomicrobium alkanoclasticum* comb. nov. Int J Syst Evol Microbiol 55:699–702

De Clerck E, De Vos P (2009) Genus IV. *Jeotgalibacillus* Yoon, Weiss, Lee, Lee, Kang and Park 2001c, $2092^{VP}$. In: De Vos P, Garrity GM, Jones D, Krieg NR, Ludwig W, Rainey FA, Schleifer K-H, Whitman WB (eds) Bergey manual of systematic bacteriology, vol 3, 2nd edn. Springer, New York, p 364

De Vos P, Garrity GM, Jones D, Krieg NR, Ludwig W, Rainey FA, Schleifer KH, Whitman WB (2009) Genus VIII. *Sporosarcina* Kluyver and Van Neil, 1936, emend. Yoon, Lee, Weiss, Kho, Kang and Park 2001. In: De Vos P, Garrity GM, Jones D, Krieg NR, Ludwig W, Rainey FA, Schleifer KH, Whitman WB (eds) Bergey's manual of systematic bacteriology, vol 3, 2nd edn. Springer, New York, pp 359–362

Eguchi M, Nishikawa T, Macdonald K, Cavicchioli R, Gottschal JC, Kjelleberg S (1996) Responses to stress and nutrient availability by the marine ultramicrobacterium *Sphingomonas* sp. strain RB2256. Appl Environ Microbiol 62:1287–1294

Elston HR (1961) *Kurthia bessonii* isolated from clinical material. J Pathol Bacteriol 81:245–247

Engelhardt MA, Daly K, Swannell RPJ, Head IM (2001) Isolation and characterization of a novel hydrocarbon-degrading, gram-positive bacterium, isolated from intertidal beach sediment, and description of *Planococcus alkanoclasticus* sp. nov. J Appl Microbiol 90:237–247

Essghaier B, Fardeau ML, Cayol JL, Hajlaoui MR, Boudabous A, Jijakli H, Sadfi-Zouaoui N (2009) Biological control of grey mould in strawberry fruits by halophilic bacteria. J Appl Microbiol 106:833–846

Faoagali JL (1974) *Kurthia*, an unusual isolate. Am J Clin Pathol 62:604–606

Farrow JAE, Ash C, Wallbanks S, Collins MD (1992) Phylogenetic analysis of the genera *Planococcus*, *Marinococcus* and *Sporosarcina* and their relationships to members of the genus *Bacillus*. FEMS Microbiol Lett 72:167–172

Farrow JAE, Wallbanks S, Collins MD (1994) Phylogenetic interrelationships of round-spore-forming bacilli containing cell walls based on lysine and the non-spore-forming genera *Caryophanon*, *Exiguobacterium*, *Kurthia*, and *Planococcus*. Int J Syst Bacteriol 44:74–82

Fineran PC, Blower TR, Foulds IJ, Humphreys DP, Lilley KS, Salmond GP (2009) The phage abortive infection system, ToxIN, functions as a protein-RNA toxin-antitoxin pair. Proc Natl Acad Sci 106:894–899

Ford JH (2010) Saturated fatty acid metabolism is key link between cell division, cancer, and senescence in cellular and whole organism aging. Age (Dordr) 32:231–237

Fortina MG, Pukall R, Schumann P, Mora D, Parini C, Manachini PL, Stackebrandt E (2001) *Ureibacillus* gen. nov., a new genus to accommodate *Bacillus thermosphaericus* (Andersson et al., 1995), emendation of *Ureibacillus thermosphaericus* and description of *Ureibacillus terrenus* sp. nov. Int J Syst Evol Microbiol 51:447–455

Fritze D, Claus D (2009) Genus II. *Caryophanon* Peshkoff 1939, $244^{AL}$. In: De Vos P, Garrity GM, Jones D, Krieg NR, Ludwig W, Rainey FA, Schleifer K-H, Whitman WB (eds) Bergey manual of systematic bacteriology, vol 3, 2nd edn. Springer, New York, pp 354–359

Gagné A, Chicoine M, Morin A, Houde A (2001) Phenotypic and genotypic characterization of esterase-producing *Ureibacillus thermosphaericus* isolated from an aerobic digestor of swine waste. Can J Microbiol 47:908–915

Galperin MY, Moroz OV, Wilson KS, Murzin AG (2006) House cleaning, a part of good housekeeping. Mol Microbiol 59:5–19

Gardner GA (1969) Physiological and morphological characteristics of *Kurthia zopfii* isolated from meat products. J Appl Bacteriol 32:371–380

Georgala DL (1957) Quantitative and qualitative aspects of the skin flora of North Sea cod and the effect thereon of handling on ship and on shore. PhD thesis, University of Aberdeen

Glendinning KJ, Macaskie LE, Brown NL (2005) Mercury tolerance of thermophilic *Bacillus* sp. and *Ureibacillus* sp. Biotechnol Lett 27:1657–1662

Goodfellow M, Collins MD, Minnikin DE (1980) Fatty acid and polar lipid composition in the classification of *Kurthia*. J Appl Bacteriol 48:269–276

Gordon RE, Haynes WC, Pang CHN (1973) The genus *Bacillus*. Agriculture handbook no. 427. United States Department of Agriculture, Washington, DC

Hao MV, Komagata K (1985) A new species of *Planococcus*, *Planococcus kocurii* isolated from fish, frozen foods, and fish curing brine. J Gen Appl Microbiol 31:441–455

He L, Chen M, Pan Y (2009) Bacterial communities in the phase II of *Agaricus bisporus* compost by denaturing gradient gel electrophoresis. Wei Sheng Wu Xue Bao 49:227–232

Hernández-Castro R, Martínez-Chavarría L, Díaz-Avelar A, Romero-Osorio A, Godínez-Reyes C, Zavala-González A, Verdugo-Rodríguez A (2005) Aerobic bacterial flora of the nasal cavity in Gulf of California sea lion (*Zalophus californianus*) pups. Vet J 170:359–363

Heyrman J, Rodríguez-Díaz M, Devos J, Felske A, Logan NA, De Vos P (2005) *Bacillus arenosi* sp. nov., *Bacillus arvi* sp. nov. and *Bacillus humi* sp. nov., isolated from soil. Int J Syst Evol Microbiol 55:111–117

Hinsa-Leasure SM, Bhavaraju L, Rodrigues JL, Bakermans C, Gilichinsky DA, Tiedje JM (2010) Characterization of a bacterial community from a Northeast Siberian seacoast permafrost sample. FEMS Microbiol Ecol 74:103–113

Hoyles L, Honda H, Logan NA, Halket G, La Ragione RM, McCartney AL (2012) Recognition of greater diversity of *Bacillus* species and related bacteria in human faeces. Res Microbiol 163:3–13

Hu JM, Li H, Cao LX, Wu PC, Zhang CT, Sang SL, Zhang XY, Chen MJ, Lu JQ, Liu YH (2007) Molecular cloning and characterization of the gene encoding cold-active β-galactosidase from a psychrotrophic and halotolerant *Planococcus* sp. L4. J Agric Food Chem 55:2217–2224

Huang K, Zhang L, Liu Z, Chen Q, Peng Q, Li W, Cui X, Chen Y (2009) Diversity of culturable bacteria associated with the sea urchin *Hemicentrotus pulcherrimus* from Naozhou Island. Wei Sheng Wu Xue Bao 49:1424–1429

Huck JR, Sonnen M, Boor KJ (2008) Tracking heat-resistant, cold-thriving fluid milk spoilage bacteria from farm to packaged product. J Dairy Sci 91:1218–1228

Hupert-Kocurek K, Guzik U, Wojcieszyńska D (2012) Characterization of catechol 2,3-dioxygenase from *Planococcus* sp. strain S5 induced by high phenol concentration. Acta Biochim Pol 59:345–351

Inbakandan D, Murthy PS, Venkatesan R, Khan SA (2010) 16S rDNA sequence analysis of culturable marine biofilm forming bacteria from a ship's hull. Biofouling 26:893–899

Ito M, Guffanti AA, Oudega B, Krulwich TA (1999) mrp, a multigene, multifunctional locus in *Bacillus subtilis* with roles in resistance to cholate and to Na+ and in pH homeostasis. J Bacteriol 181:2394–2402

Ivanova EP, Wright JP, Lysenko AM, Zhukova NV, Alexeeva YV, Buljan V, Kalinovskaya NI, Nicolau DV, Christen R, Mikhailov VV (2006) Characterization of unusual alkaliphilic Gram-positive bacteria isolated from degraded brown alga thalluses. Mikrobiol Z 68:10–20

Janarthine SRS, Eganathan P (2012) Plant growth promoting of endophytic *Sporosarcina aquimarina* SjAM16103 isolated from the pneumatophores of *Avicennia marina* L. Int J Microbiol doi:10.1155/2012/532060 (in press)

Jarumilinta R, Miranda M, Villarejos VM (1976) A bacteriological study of the intestinal mucosa and luminal fluid of adults with acute diarrhoea. Ann Trop Med Parasitol 70:165–179

Jeena MI, Deepa P, Rahiman KMM, Shanthi RT, Hatha AA (2006) Risk assessment of heterotrophic bacteria from bottled drinking water sold in Indian markets. Int J Hyg Environ Health 209:191–196

Jeffries L (1969) Menaquinones in the classification of *Micrococcaceae* with observations on the application of lysozyme and novobiocin sensitivity test. Int J Syst Bacteriol 19:183–187

Jeyasekaran G, Shakila RJ, Sukumar D (2012) Microbiological quality of Cuttlefish (*Sepia pharaonis*) fillets stored in dry and wet ice. Food Sci Technol Int 18:455–464

Jones JG (1983) A note on the isolation and enumeration of bacteria which deposit and reduce ferric iron. J Appl Bacteriol 54:305–310

Jung YT, Kang SJ, Oh TK, Yoon JH, Kim BH (2009) *Planomicrobium flavidum* sp. nov., isolated from a marine solar saltern, and transfer of *Planococcus stackebrandtii* Mayilraj et al., 2005 to the genus *Planomicrobium* as *Planomicrobium stackebrandtii* comb. nov. Int J Syst Evol Microbiol 59:2929–2933

Junge K, Gosink JJ, Hoppe HG, Staley JT (1998) Arthrobacter, Brachybacterium and Planococcus isolates identified from Antarctic sea ice brine. Description of *Planococcus mcmeekinii*, sp. nov. Syst Appl Microbiol 21:306–314

Kahru A, Reiman R, Rätsep A (1998) The efficiency of different phenol-degrading bacteria and activated sludges in detoxification of phenolic leachates. Chemosphere 37:301–318

Kämpfer P, Rossello-Mora R, Falsen E, Busse HJ, Tindall BJ (2006) *Cohnella thermotolerans* gen. nov., sp. nov., and classification of *Paenibacillus hongkongensis* as *Cohnella hongkongensis* sp. nov. Int J Syst Evol Microbiol 56:781–786

Kämpfer P, Falsen E, Lodders N, Schumann P (2010) *Sporosarcina contaminans* sp. nov. and *Sporosarcina thermotolerans* sp. nov., two novel endospore-forming species. Int J Syst Evol Microbiol 60:1353–1357

Kannan P, Ignacimuthu S, Paulraj MG (2006) Facultative alkalophilic bacteria from mangrove soil with varying buffering capacity and $H^+$ conductance. Ind J Biochem Biophys 43:382–385

Kato J, Fukushima H, Kisumi M, Chibata I (1972a) Mechanism of proline production by *Kurthia catenaforma*. Appl Microbiol 23:699–703

Kato J, Kisumi M, Chibata I (1972b) Effect of L-aspartic acid and L-glutamic acid on production of L-proline. Appl Microbiol 23:758–764

Kaur I, Das AP, Acharya M, Klenk HP, Sree A, Mayilraj S (2012) *Planococcus plakortidis* sp. nov., isolated from the marine sponge *Plakortis simplex* (Schulze). Int J Syst Evol Microbiol 62:883–889

Keddie RM (1981) The genus *Kurthia*. In: Starr MP, Stolp H, Trüper HG, Balows A, Schlegel HG (eds) The prokaryotes: a handbook on habitats, isolation and identification of bacteria. Springer, Berlin, pp 1888–1893

Keddie RM, Shaw S (1986) Genus *Kurthia*. In: Mair NS, Sneath PHA, Sharpe ME, Holt JG (eds) Bergey's manual of systematic bacteriology, vol 2. Williams & Wilkins, Baltimore, pp 1255–1258

Kele RA, McCoy E (1971) Defined liquid minimal medium for *Caryophanon latum*. Appl Microbiol 22:728–729

Kim BY, Lee SY, Weon HY, Kwon SW, Go SJ, Park YK, Schumann P, Fritze D (2006) *Ureibacillus suwonensis* sp. nov., isolated from cotton waste composts. Int J Syst Evol Microbiol 56:663–666

Kim EH, Cho KH, Lee YM, Yim JH, Lee HK, Cho JC, Hong SG (2010) Diversity of cold-active protease-producing bacteria from arctic terrestrial and marine environments revealed by enrichment culture. J Microbiol 48:426–432

Kita K, Kotani H, Sugisaki H, Takanami M (1989) The *fokI* restriction-modification system. I. Organization and nucleotide sequences of the restriction and modification genes. J Biol Chem 264:5751–5756

Kiyasu T, Nagahashi Y, Hoshino T (2001) Cloning and characterization of biotin biosynthetic genes of *Kurthia* sp. Gene 265:103–113

Kluyver AJ, Van Niel CB (1936) Prospects for a natural system of classification of bacteria. Zentralbl Bakteriol Mikrobiol Hyg Abt II 94:369–403

Kocur M, Pacova Z, Hodgkiss W, Martinec T (1970) The taxonomic status of the genus *Planococcus* Migula 1894. Int J Syst Bacteriol 20:241–248

Kovacs N (1956) Identification of *Pseudomonas pyocyanea* by the oxidase reaction. Nature 178:703

Krasil'nikov NA (1949) Opredelitelv Bakterii i Actinomicetov. Akademii Nauk SSSR, Moscow, 328

Krishnamurthi S, Chakrabarti T, Stackebrandt E (2009a) Re-examination of the taxonomic position of *Bacillus silvestris* Rheims et al., 1999 and proposal to transfer it to *Solibacillus* gen. nov. as *Solibacillus silvestris* comb. nov. Int J Syst Evol Microbiol 59:1054–1058

Krishnamurthi S, Bhattacharya A, Mayilraj S, Saha P, Schumann P, Chakrabarti T (2009b) Description of *Paenisporosarcina quisquiliarum* gen. nov., sp. nov., and reclassification of *Sporosarcina macmurdoensis* Reddy et al., 2003 as *Paenisporosarcina macmurdoensis* comb. nov. Int J Syst Evol Microbiol 59:1364–1370

Krishnaveni M, Jayachandran S (2009) Inhibition of MAP kinases and down regulation of TNF-alpha, IL-beta and COX-2 genes by the crude extracts from marine bacteria. Biomed Pharmacother 63:469–476

Kuhlmann AU, Bremer E (2002) Osmotically regulated synthesis of the compatible solute ectoine in *Bacillus pasteurii* and related *Bacillus* sp. Appl Environ Microbiol 68:772–783

Kuhlmann AU, Hoffmann T, Bursy J, Jebbar M, Bremer E (2011) Ectoine and hydroxyectoine as protectants against osmotic and cold stress: uptake through the SigB-controlled betaine-choline- carnitine transporter-type carrier EctT from *Virgibacillus pantothenticus*. J Bacteriol 193:4699–4708

Kumar SA, Mody K, Jha B (2007) Evaluation of biosurfactant/bioemulsifier production by a marine bacterium. Bull Environ Contam Toxicol 79:617–621

Kurth H (1883) Über *Bacterium zopfii*, eine neue Bacterienart. Ber Deutsch Bot Ges 1:97–100

Kwon SW, Kim BY, Song J, Weon HY, Schumann P, Tindall BJ, Stackebrandt E, Fritze D (2007) *Sporosarcina koreensis* sp. nov. and *Sporosarcina soli* sp. nov., isolated from soil in Korea. Int J Syst Evol Microbiol 57:1694–1698

Labuzek S, Hupert-Kocurek KT, Skurnik M (2003) Isolation and characterisation of new *Planococcus* sp. strain able for aromatic hydrocarbons degradation. Acta Microbiol Pol 52:395–404

Larkin JM, Stokes JL (1967) Taxonomy of psychrophilic strains of *Bacillus*. J Bacteriol 94:889–895

Lay CY, Mykytczuk NC, Niederberger TD, Martineau C, Greer CW, Whyte LG (2012) Microbial diversity and activity in hypersaline high arctic spring channels. Extremophiles 16:177–191

Liu Z, Shao ZZ (2007) The diversity of alkane degrading bacteria in the enrichments with deep sea sediment of the South China Sea. Wei Sheng Wu Xue Bao 47:869–873

Liu S, Song F, Zhu N, Yuan H, Cheng J (2010) Chemical and microbial changes during autothermal thermophilic aerobic digestion (ATAD) of sewage sludge. Bioresour Technol 101:9438–9444

Liu R, Yu Z, Zhang H, Yang M, Shi B, Liu X (2012) Diversity of bacteria and mycobacteria in biofilms of two urban drinking water distribution systems. Can J Microbiol 58:261–270

Looney MC, Moran LS, Jack WE, Feehery GR, Benner JS, Slatko BE, Wilson GG (1989) Nucleotide sequence of the FokI restriction-modification system: separate strand-specificity domains in the methyltransferase. Gene 80:193–208

MacDonald RE, MacDonald SW (1962) The physiology and natural relationships of the motile, sporeforming sarcinae. Can J Microbiol 8:795–808

Magill NG, Loshon CA, Setlow P (1990) Small, acid-soluble, spore proteins and their genes from two species of *Sporosarcina*. FEMS Microbiol Lett 60:293–297

Maiden MFJ (1983) The biology of filamentous bacteria in freshwater sediments. PhD thesis. University of Wales Institute of Science and Technology

Maiden MFJ, Jones JG (1984) A new filamentous, gliding bacterium, *Filibacter limicola* gen. nov. sp. nov., from lake sediment. J Gen Microbiol 130:2943–2959

Maiden MFJ, Jones JG (1985) *In* Validation of the publication of new names and new combinations previously effectively published outside the IJSB. List no. 18. Int J Syst Bacteriol 35:375–376

Manorama R, Pindi PK, Reddy GSN, Shivaji S (2009) *Bhargavaea cecembensis* gen. nov., sp. nov., isolated from the Chagos-Laccadive ridge system in the Indian Ocean. Int J Syst Evol Microbiol 59:2618–2623

Margolles A, Gueimonde M, Sánchez B (2012) Genome sequence of the Antarctic psychrophile bacterium *Planococcus antarcticus* DSM 14505. J Bacteriol 194:4465

Martin D, Dodds K, Ngwenya BT, Butler IB, Elphick SC (2012) Inhibition of *Sporosarcina pasteurii* under anoxic conditions: implications for subsurface carbonate precipitation and remediation via ureolysis. Environ Sci Technol 46:8351–8355

Matsumoto GI (1993) Geochemical features of the McMurdo Dry valley lakes, Antarctica. Physical and biogeochemical processes in Antarctic lakes. Antarct Res Ser 49:95–118

Mayilraj S, Prasad GS, Suresh K, Saini HS, Shivaji S, Chakrabarti T (2005) *Planococcus stackebrandtii* sp. nov., isolated from a cold desert of the Himalayas. Ind Int J Syst Evol Microbiol 55:91–94

McCoy DD, Cetin A, Hausinger RP (1992) Characterization of urease from *Sporosarcina ureae*. Arch Microbiol 157:411–416

Mei Y, He B, Liu N, Ouyang P (2009) Screening and distributing features of bacteria with hydantoinase and carbamoylase. Microbiol Res 164:322–329

Migula W (1894) Über ein neues system der bakterien. Arb Bakteriol Inst Karlsr 1:235–238

Miquel P (1889) Étude sur la fermantation ammoniacale et sur les ferments de l'urée. Ann Microgr 1:506–519

Moran JW, Witter LD (1976) Effect of temperature and pH on the growth of *Caryophanon latum* colonies. Can J Microbiol 22:1401–1403

Munoz R, Yarza P, Ludwig W, Euzéby J, Amann R, Schleifer KH, Glöckner FO, Rosselló-Móra R (2011) Release LTPs104 of the all-species living tree. Syst Appl Microbiol 34:169–170

Murayama M, Kakinuma Y, Maeda Y, Rao JR, Matsuda M, Xu J, Moore PJ, Millar BC, Rooney PJ, Goldsmith CE, Loughrey A, McMahon MA, McDowell DA, Moore JE (2010) Molecular identification of airborne bacteria associated with aerial spraying of bovine slurry waste employing 16S rRNA gene PCR and gene sequencing techniques. Ecotoxicol Environ Saf 73:443–447

Mykytczuk NCS, Wilhelm RC, Whyte LG (2012) *Planococcus halocryophilus* sp. nov., an extreme sub-zero species from high Arctic permafrost. Int J Syst Evol Microbiol 62:1937–1944

Nakagawa Y, Sakane T, Yokata A (1996) Emendation of the genus *Planococcus* and transfer of *Flavobacterium okeanokoites* Zobell and Upham 1944 to the genus *Planococcus* as *Planococcus okeanokoites* comb. nov. Int J Syst Bacteriol 46:866–870

Nakamura LK (1984) *Bacillus psychrophilus* sp. nov., nom. rev. Int J Syst Bacteriol 34:121–123

Nakamura LK, Shida O, Takagi H, Komagata K (2002) *Bacillus pycnus* sp. nov. and *Bacillus neidei* sp. nov., round-spored bacteria from soil. Int J Syst Evol Microbiol 52:501–505

Nakasaki K, le Tran TH, Idemoto Y, Abe M, Rollon AP (2009) Comparison of organic matter degradation and microbial community during thermophilic composting of two different types of anaerobic sludge. Bioresour Technol 100:676–682

Nauman RK, Wilkie EF (1974) Isolation of bacteriophage for *Caryophanon latum*. J Virol 13:1151–1152

Nichols P, Stulp BK, Jones JG, White DC (1986) Comparison of fatty acid content and DNA homology of the filamentous gliding bacteria *Vitreoscilla Flexibacter, Filibacter*. Arch Microbiol 146:1–6

Nithya C, Gnanalakshmi B, Pandian SK (2011) Assessment and characterization of heavy metal resistance in Palk Bay sediment bacteria. Mar Environ Res 71:283–294

Novitsky TJ, Kushner DJ (1976) *Planococcus halophilus* sp. nov., a facultatively halophilic coccus. Int J Syst Bacteriol 26:53–57

O'Brien A, Sharp R, Russell NJ, Roller S (2004) Antarctic bacteria inhibit growth of food-borne microorganisms at low temperatures. FEMS Microbiol Ecol 48:157–167

Oeding P (1971) Serological investigation of *Planococcus* strains. Int J Syst Bacteriol 21:323–325

Okuda N, Soneura M, Ninomiya K, Katakura Y, Shioya S (2008) Biological detoxification of waste house wood hydrolysate using *Ureibacillus thermosphaericus* for bioethanol production. J Biosci Bioeng 106:128–133

Okwadha GD, Li J (2011) Biocontainment of polychlorinated biphenyls (PCBs) on flat concrete surfaces by microbial carbonate precipitation. J Environ Manage 92:2860–2864

Olivera NL, Sequeiros C, Nievas ML (2007) Diversity and enzyme properties of protease-producing bacteria isolated from sub-Antarctic sediments of Isla de Los Estados, Argentina. Extremophiles 11:517–526

Osborn F, Berlioz L, Vitelli-Flores J, Monsalve W, Dorta B, Rodríguez Lemoine V (2002) Pathogenic effects of bacteria isolated from larvae of Hylesia metabus Crammer (*Lepidoptera*: Saturniidae). J Invertebr Pathol 80:7–12

Parawira W, Tekere M (2011) Biotechnological strategies to overcome inhibitors in lignocellulose hydrolysates for ethanol production: review. Crit Rev Biotechnol 31:20–31

Parkouda C, Thorsen L, Compaoré CS, Nielsen DS, Tano-Debrah K, Jensen JS, Diawara B, Jakobsen M (2010) Microorganisms associated with Maari, a Baobab seed fermented product. Int J Food Microbiol 142:292–301

Patel IB, Vaughn RH (1973) Cellulolytic bacteria associated with sloughing spoilage of California ripe olives. Appl Microbiol 25:62–69

Pearson MD, Noller HF (2011) The draft genome of *Planococcus* donghaensis MPA1U2 reveals nonsporulation pathways controlled by a conserved Spo0A regulon. J Bacteriol 193:6106

Peshkoff MA (1939) Cytology, karyology and cycle of development of new microbes, *Caryophanon latum* and *Caryophanon tenue*. Dokl Akad Nauk SSSR 25:239–242

Peshkoff MA (1940) Phylogenesis of new microbes, *Caryophanon latum* and *Caryophanon tenue*, organisms which are intermediate between blue-green algae and the bacteria. J Gen Biol (Russ) 1:598, Detailed summary in English

Peshkoff MA, Tikhonenko AS, Marek BI (1966) A phage for the multicellular bacterium, *Caryophanon tenue* Peshkoff. Mikrobiologiva 35:684–687

Peshkoff MA, Stefanov SB, Shadrina IA, Marek BI (1967) Virulent phage for Can, ophanon sublatum X13. Mikrobiologiva 36:136–139

Peshkov MA (1967) New high productivity nutrient media for Caryophanon and a method for quantitative evaluation of bacterial yield. Microbiology (En. Trans. from Mikrobiologiya) 37:548–551

Peshkov MA, Marek BI (1972) Fine structure of *Caryophanon latum* and *Caryophanon tenue* Peshkov. Microbiology 41:941–945 (En. transl from Mikrobiologiya)

Peshkov MA, Shekhovtsov VP, Zharikova GG (1978) Morphology of the multicellular, trichome bacterium, *Caryophanon*, in the process of growth on an agarized medium. Mikrobiologiia 47:539–544

Phillips AJ, Lauchnor E, Eldring JJ, Esposito R, Mitchell AC, Gerlach R, Cunningham AB, Spangler LH (2012) Potential $CO_2$ leakage reduction through biofilm-induced calcium carbonate precipitation. Environ Sci Technol. doi:10.1021/es301294q

Pin C, Baranyi J (1998) Predictive models as means to quantify the interactions of spoilage organisms. Int J Food Microbiol 41:59–72

Portillo MC, Santana M, Gonzalez JM (2012) Presence and potential role of thermophilic bacteria in temperate terrestrial environments. Naturwissenschaften 99:43–53

Prévot AR (1953) In: Hauduroy P, Ehringer G, Guillot G, Magrou J, Prévot AR, Rosset and Urbain A. Dictionnaire des Bactéries Pathogènes, 2nd ed., Masson, Paris, pp. 1-692.

Pringsheim EG, Robinow CF (1947) Observation on two very large bacteria, *Caryophanon latum* Peshkov and *Lineola longa* (*nomen provisorum*). J Gen Microbiol 1:267–278

Provost PJ, Doetsch RN (1962) An appraisal of *Caryophanon latum*. J Gen Microbiol 28:547–557

Pukall R, Stackebrandt E (2009) Genus V. *Kurthia* Trevisan 1885, 92[AL] Nom. cons. Opin. 13 Jud. Comm. 1954, 152. In: De Vos P, Garrity GM, Jones D, Krieg NR, Ludwig W, Rainey FA, Schleifer K-H, Whitman WB (eds) Bergey manual of systematic bacteriology, vol 3, 2nd edn. Springer, New York, pp 364–370

Qiu F, Zhang X, Liu L, Sun L, Schumann P, Song W (2009) *Bacillus beijingensis* sp. nov. and *Bacillus ginsengi* sp. nov., isolated from ginseng root. Int J Syst Evol Microbiol 59:729–734

Radwan SS, Al-Hasan RH, Mahmoud HM, Eliyas M (2007) Oil-utilizing bacteria associated with fish from the Arabian Gulf. J Appl Microbiol 103:2160–2167

Ranganathan N, Patel BG, Ranganathan P, Marczely J, Dheer R, Pechenyak B, Dunn SR, Verstraete W, Decroos K, Mehta R, Friedman EA (2006) In vitro and in vivo assessment of intraintestinal bacteriotherapy in chronic kidney disease. ASAIO J 52:70–79

Reddy GSN, Prakash JSS, Vairamani M, Prabhakar S, Matsumoto GI, Shivaji S (2002) *Planococcus antarcticus* and *Planococcus psychrophilus* sp. nov. isolated from cyanobacterial mat samples collected from ponds in Antarctica. Extremophiles 6:253–261

Reddy GSN, Matsumoto GI, Shivaji S (2003) *Sporosarcina macmurdoensis* sp. nov., from a cyanobacterial mat sample from a pond in the McMurdo Dry Valleys, Antarctica. Int J Syst Evol Microbiol 53:1363–1367

Reddy GSN, Poorna Manasa B, Singh SK, Shivaji S (2013) *Paenisporosarcina indica* sp. nov., a psychrophilic bacterium from Pindari Glacier of the Himalayan mountain ranges and reclassification of *Sporosarcina antarctica* Yu et al., 2008 as *Paenisporosarcina antarctica* comb. nov. and emended description of the genus *Paenisporosarcina*. Int J Syst Evol Microbiol 63:2927–2933

Rheims H, Fruhling A, Schumann P, Rohde M, Stackebrandt E (1999) *Bacillus silvestris* sp. nov., a new member of the genus *Bacillus* that contains lysine in its cell wall. Int J Syst Bacteriol 49:795–802

Rocourt J, Ackermann HW, Brault J (1984) Isolation and morphology of bacterio phages of *Kurthia-zopfii*. Ann Virol (Paris) 134:557–568

Romano I, Giordano A, Lama L, Nicolaus B, Gambacorta A (2003) *Planococcus rifietoensis* sp. nov. isolated from algal mat collected from a sulphurous spring in Campania (Italy). Syst Appl Microbiol 26:357–366

Rooney AP, Swezey JL, Wicklow DT, McAtee MJ (2005) Bacterial species diversity in cigarettes linked to an investigation of severe pneumonitis in U.S. Military personnel deployed in operation Iraqi freedom. Curr Microbiol 51:46–52

Rowenhagen B (1987) Fettsäureverwertung durch Caryophanon latum und Caryophanon tenue sowie morphologische und cytologische Unterschiede dieser *Bakterien* species. PhD thesis, University of Braunschweig

Rüger HJ, Richter G (1979) *Bacillus globisporus* subsp. *marinus* subsp. nov. Int J Syst Bacteriol 29:196–203

Rusznyák A, Vladár P, Szabó G, Márialigeti K, Borsodi AK (2008) Phylogenetic and metabolic bacterial diversity of phragmites australis periphyton communities in two Hungarian soda ponds. Extremophiles 12:763–773

Ryzhkov PM, Ostermann K, Rödel G (2007) Isolation, gene structure, and comparative analysis of the S-layer gene sslA of *Sporosarcina ureae* ATCC 13881. Genetica 131:255–265

Saphir DA, Carter GR (1976) Gingival flora of the dog with special reference to bacteria associated with bites. J Clin Microbiol 3:344–349

Schleifer KH, Kandler O (1972) Peptidoglycan types of bacterial cell walls and their taxonomic implications. Bacteriol Rev 36:407–477

Schröder K, Zuber P, Willimsky G, Wagner B, Marahiel MA (1993) Mapping of the *Bacillus subtilis* cspB gene and cloning of its homologs in thermophilic, mesophilic and psychrotrophic bacilli. Gene 136:277–280

Severi R (1946) L'azione patogena delle *Kurthia* e la loro sistematica. Una nuova specie: "*Kurthia variabilis*". Giorn Batteriol Immunol 24:107–114

Shaw S, Keddie RM (1983) A numerical taxonomic study of the genus *Kurthia* with a revised description of *Kurthia zopfii* and a description of *Kurthia gibsonii* sp. nov. Syst Appl Microbiol 4:253–276

Shindo K, Endo M, Miyake Y, Wakasugi K, Morritt D, Bramley PM, Fraser PD, Kasai H, Misawa N (2008) Methyl glucosyl-3,4-dehydro-apo-8′-lycopenoate, a novel antioxidative glyco-C(30)-carotenoic acid produced by a marine bacterium *Planococcus maritimus*. J Antibiot (Tokyo) 61:729–735

Shivaji S (2009) Genus I. *Planococcus* Migula 1894, 236[AL]. In: De Vos P, Garrity GM, Jones D, Krieg NR, Ludwig W, Rainey FA, Schleifer K-H, Whitman WB (eds) Bergey manual of systematic bacteriology, vol 3, 2nd edn. Springer, New York, pp 348–354

Shivaji S, Shymala Rao N, Saisree L, Sheth V, Reddy GSN, Bhargava PM (1988) Isolation and identification of *Micrococcus roseus* and *Planococcus* sp. from schirmacher oasis, Antarctica. J Biosci 13:409–414

Siddikee MA, Chauhan PS, Anandham R, Han GH, Sa T (2010) Isolation, characterization, and use for plant growth promotion under salt stress, of ACC deaminase-producing halotolerant bacteria derived from coastal soil. J Microbiol Biotechnol 20:1577–1584

Simmons JS (1926) A culture medium for differentiating organisms of typhoid-colon aerogenes groups and for isolation of certain fungi. J Infect Dis 39:209–214

Skerratt JH, Bowman JP, Hallegraeff G, James S, Nichols PD (2002) Algicidal bacteria associated with blooms of a toxic dinoflagellate in a temperate Australian estuary. Mar Ecol Prog Ser 244:1–15

Smith DL, Trentini WC (1972) Enrichment and selective isolation of *Caryophanon latum*. Can J Microbiol 18:1197–1200

Smith DL, Trentini WC (1973) On the gram reaction of *Caryophanon latum*. Can J Microbiol 19:757–760

Stackebrandt E (2009) Genus III. *Filibacter* Maiden and Jones 1985, 375[VP] (Effective publication: Maiden and Jones 1984, 2957.). In: De Vos P, Garrity GM, Jones D, Krieg NR, Ludwig W, Rainey FA, Schleifer K-H, Whitman WB (eds) Bergey manual of systematic bacteriology, vol 3, 2nd edn. Springer, New York, pp 359–363

Stackebrandt E, Goebel BM (1994) Taxonomic note: a place for DNA-DNA reassociation and 16S rRNA sequence analysis in the present species definition in bacteriology. Int J Syst Bacteriol 44:846–849

Stackebrandt E, Ludwig W, Weizenegger M, Dorn S, Mcgill TJ, Fox GE, Woese CR, Schubert W, Schleifer KH (1987) Comparative 16S ribosomal RNA oligonucleotide analyses and murein types of round spore-forming bacilli and non-spore-forming relatives. J Gen Microbiol 133:2523–2529

Steiger S, Perez-Fons L, Fraser PD, Sandmann G (2012) The biosynthetic pathway to a novel derivative of 4,4′-diapolycopene-4,4′-oate in a red strain of *Sporosarcina aquimarina*. Arch Microbiol 194:779–784

Surakasi VP, Antony CP, Sharma S, Patole MS, Shouche YS (2010) Temporal bacterial diversity and detection of putative methanotrophs in surface mats of Lonar crater lake. J Basic Microbiol 50:465–474

Suresh K, Mayilraj S, Bhattacharya A, Chakrabarti T (2007) *Planococcus columbae* sp. nov., isolated from pigeon faeces. Int J Syst Evol Microbiol 57:1266–1271

Tominaga T, An SY, Oyaizu H, Yokota A (2009) *Sporosarcina luteola* sp. nov. isolated from soy sauce production equipment in Japan. J Gen Appl Microbiol 55:217–223

Trentini WC (1978) Biology of the genus *Caryophanon*. Annu Rev Microbiol 32:123–141

Trentini WC, Machen C (1973) Natural habitat of *Caryophanon latum*. Can J Microbiol 19:689–694

Trevisan V (1885) Carratteri di alcuni nuovi generi di Batteriacee. Atti Della Accad Fis-Med-Stast Milano 3:92–107. (ser 4)

Vaishampayan P, Miyashita M, Ohnishi A, Satomi M, Rooney A, La Duc MT, Venkateswaran K (2009) Description of *Rummeliibacillus stabekisii* gen. nov., sp. nov. and reclassification of *Bacillus pycnus* Nakamura et al., 2002 as *Rummeliibacillus pycnus* comb. nov. Int J Syst Evol Microbiol 59:1094–1099

Vajna B, Szili D, Nagy A, Márialigeti K (2012) An improved sequence-aided T-RFLP analysis of bacterial succession during oyster mushroom substrate preparation. Microb Ecol 64:702–713

Vargas-García MC, Suárez-Estrella F, López MJ, Moreno J (2007) Effect of inoculation in composting processes: modifications in lignocellulosic fraction. Waste Manag 2:1099–1107

Venkateswaran K, Satomi M, Chung S, Kern R, Koukol R, Basic C, White D (2001) Molecular microbial diversity of a spacecraft assembly facility. Syst Appl Microbiol 24:311–320

Verma P, Pandey PK, Gupta AK, Seong CN, Park SC, Choe HN, Baik KS, Patole MS, Shouche YS (2012) Reclassification of *Bacillus beijingensis* Qiu et al. 2009 and *Bacillus ginsengi* Qiu et al. 2009 as *Bhargavaea beijingensis* comb. nov. and *Bhargavaea ginsengi* comb. nov. and emended description of the genus Bhargavaea. Int J Syst Evol Microbiol 62:2495–2504

Vishnivetskaya TA, Petrova MA, Urbance J, Ponder M, Moyer CL, Gilichinsky DA, Tiedje JM (2006) Bacterial community in ancient Siberian permafrost as characterized by culture and culture-independent methods. Astrobiology 6:400–414

Wan C, Du M, Lee DJ, Yang X, Ma W, Zheng L (2011) Electrokinetic remediation and microbial community shift of β-cyclodextrin-dissolved petroleum hydrocarbon-contaminated soil. Appl Microbiol Biotechnol 89:2019–2025

Wang CM, Shyu CL, Ho SP, Chiou SH (2007) Species diversity and substrate utilization patterns of thermophilic bacterial communities in hot aerobic poultry and cattle manure composts. Microb Ecol 54:1–9

Wang X, Yu R, Luo X, Zhou M, Shen J, Gu Z (2008) Toxicity screening and identification of bacteria isolated from snails Nassarius semiplicatus and their habitat. Wei Sheng Wu Xue Bao 48:911–916

Wang W, Yan L, Cui Z, Gao Y, Wang Y, Jing R (2011a) Characterization of a microbial consortium capable of degrading lignocellulose. Bioresour Technol 102:9321–9324

Wang YB, Miao JL, He BJ, Liang Q, Liu FM, Zheng Z (2011b) Laser tweezers Raman spectroscopy analysis of cold-adapted aromatic hydrocarbons-degradating strains isolated from Antarctic Sea. Guang Pu Xue Yu Guang Pu Fen Xi 31:418–421

Watanabe K, Nagao N, Toda T, Kurosawa N (2010) Bacterial community in the personal-use composting reactor revealed by isolation and cultivation-independent method. J Environ Sci Health B 45:372–378

Weeks OB, Kelley LM (1958) Observations on the growth of the bacterium *Caryophanon latum.* J Bacteriol 75:326–330

Wei T, Hu J, Miyanaga K, Tanji Y (2012) Comparative analysis of bacterial community and antibiotic-resistant strains in different developmental stages of the housefly (*Musca domestica*). Appl Microbiol Biotechnol. doi:10.1007/s00253-012-4024-1

Weon HY, Lee SY, Kim BY, Noh HJ, Schumann P, Kim JS, Kwon SW (2007) *Ureibacillus composti* sp. nov. and *Ureibacillus thermophilus* sp. nov., isolated from livestock-manure composts. Int J Syst Evol Microbiol 57:2908–2911

Williams GP, Babu S, Ravikumar S, Kathiresan K, Prathap SA, Chinnapparaj S, Marian MP, Alikhan SL (2007) Antimicrobial activity of tissue and associated bacteria from benthic sea anemone *Stichodactyla haddoni* against microbial pathogens. J Environ Biol 28:789–793

Wolfgang WJ, Coorevits A, Cole JA, De Vos P, Dickinson MC, Hannett GE, Jose R, Nazarian EJ, Schumann P, Landschoot AV, Wirth SE, Musser KA (2012) *Sporosarcina newyorkensis* sp. nov. from clinical specimens and raw cow's milk. Int J Syst Evol Microbiol 62:322–329

Wunder JA, Briner WW, Calkins GP (1976) Identification of the cultivable bacteria in dental plaque from the beagle dog. J Dent Res 55:1097–1102

Xiao H, Chen Y, Liu Z, Huang K, Li W, Cui X, Zhang L, Yi L (2009) Phylogenetic diversity of cultivable bacteria associated with a sea anemone from coast of the Naozhou Island in Zhanjiang, China. Wei Sheng Wu Xue Bao 49:246–250

Xie KZ, Xu PZ, Zhang FB, Chen JS, Tang SH, Huang X, Yan C, Gu WJ (2009) Effects of microbial agent inoculation on bacterial community diversity in the process of pig manure composting. Ying Yong Sheng Tai Xue Bao 20:2012–2018

Yang M, Sun Y, Ge P, Dong Q, Ma Z (1985) A case of infant septicaemia caused by *Kurthia zopfii.* Chinese J Microbiol Immunol 5:485

Yarza P, Richter M, Peplies J, Euzeby J, Amann R, Schleifer KH, Ludwig W, Glöckner FO, Rossello-Mora R (2008) The all-species living tree project: a 16S rRNA-based phylogenetic tree of all sequenced type strains. Syst Appl Microbiol 31:241–250

Yarza P, Ludwig W, Euzéby J, Amann R, Schleifer KH, Glöckner FO, Rosselló-Móra R (2010) Update of the all species living tree project based on 16S and 23S rRNA sequence analyses. Syst Appl Microbiol 33:291–299

Yoon JH, Lee KC, Weiss N, Kho YH, Kang KH, Park YH (2001a) *Sporosarcina aquimarina* sp. nov., a bacterium isolated from seawater in Korea, and transfer of *Bacillus globisporus* (Larkin and Stokes 1967), *Bacillus psychrophilus* (Nakamura 1984) and *Bacillus pasteurii* (Chester 1898) to the genus *Sporosarcina* as *Sporosarcina globispora* comb. nov., *Sporosarcina psychrophila* comb. nov. and *Sporosarcina pasteurii* comb. nov., and emended description of the genus *Sporosarcina.* Int J Syst Evol Microbiol 51:1079–1086

Yoon JH, Kang SS, Lee KC, Lee ES, Kho YH, Kang KH, Park YH (2001b) *Planomicrobium koreense* gen. nov., sp. nov., a bacterium isolated from the Korean traditional fermented seafood jeotgal, and transfer of *Planococcus okeanokoites* (Nakagawa et al., 1996) and *Planococcus mcmeekinii* (Junge et al., 1998) to the genus *Planomicrobium.* Int J Syst Evol Microbiol 51:1511–1520

Yoon JH, Weiss N, Lee KC, Lee IS, Kang KH, Park YH (2001c) *Jeotgalibacillus alimentarius* gen. nov., sp. nov., a novel bacterium isolated from jeotgal with L-lysine in the cell wall, and reclassification of *Bacillus marinus* Rüger 1983 as *Marinibacillus marinus* gen. nov., comb. nov. Int J Syst Evol Microbiol 51:2087–2093

Yoon JH, Weiss N, Kang KH, Oh TK, Park YH (2003) *Planococcus maritimus* sp. nov., isolated from sea water of a tidal flat in Korea. Int J Syst Evol Microbiol 53:2013–2017

Yoon JH, Kim IG, Schumann P, Oh TK, Park YH (2004) *Marinibacillus campisalis* sp. nov., a moderate halophile isolated from a marine solar saltern in Korea, with emended description of the genus *Marinibacillus.* Int J Syst Evol Microbiol 54:1317–1321

Yoon JH, Kang SJ, Schumann P, Oh TK (2010a) *Jeotgalibacillus salarius* sp. nov., isolated from a marine saltern, and reclassification of *Marinibacillus marinus* and *Marinibacillus campisalis* as *Jeotgalibacillus marinus* comb. nov. and *Jeotgalibacillus campisalis* comb. nov., respectively. Int J Syst Evol Microbiol 60:15–20

Yoon JH, Kang SJ, Lee SY, Oh KH, Oh TK (2010b) *Planococcus salinarum* sp. nov., isolated from a marine solar saltern, and emended description of the genus *Planococcus.* Int J Syst Evol Microbiol 60:754–758

You Y, Hilpert M, Ward MJ (2012) Detection of a common and persistent tet (L)-carrying plasmid in chicken-waste-impacted farm soil. Appl Environ Microbiol 78:3203–3213

Yu Y, Xin YH, Liu HC, Chen B, Sheng J, Chi ZM, Zhou PJ, Zhang DC (2008) *Sporosarcina antarctica* sp. nov., a psychrophilic bacterium isolated from the Antarctic. Int J Syst Evol Microbiol 58:2114–2117

Zhang DC, Wang HX, Cui HL, Yang Y, Liu HC, Dong XZ, Zhou PJ (2007) *Cryobacterium psychrotolerans* sp. nov., a novel psychrotolerant bacterium isolated from the China No. 1 glacier. Int J Syst Evol Microbiol 57:866–869

Zhang DC, Liu HC, Xin YH, Yu Y, Zhou PJ, Zhou YG (2009) *Planomicrobium glaciei* sp. nov., a psychrotolerant bacterium isolated from a glacier. Int J Syst Evol Microbiol 59:1387–1390

Zhang Z, Srichuwong S, Kobayashi T, Arakane M, Park JY, Tokuyasu K (2010) Bioconversion of L-arabinose and other carbohydrates from plant cell walls toalpha-glucan by a soil bacterium, *Sporosarcina* sp. N52. Bioresour Technol 101:9734–9741

Zobell CE, Upham HC (1944) A list of marine bacteria including descriptions of sixty new species. Bull Scripps Inst Oceanogr Univ Calif 5:239–292

# 26 The Family *Sporolactobacillaceae*

*Young-Hyo Chang*[1] · *Erko Stackebrandt*[2]
[1]Korean Collection for Type Cultures, Biological Resource Centre, Korea Research Institute of Bioscience and Biotechnology, Daejeon, Republic of Korea
[2]Leibniz Institute DSMZ-German Collection of Microorganisms and Cell Cultures GmbH, Braunschweig, Germany



**Abstract**

With *Sporolactobacillus*, *Tuberibacillus*, and *Pullulanibacillus*, the family *Sporolactobacillaceae* (Ludwig W, Schleifer K-H, Whitman WB. Family VII *Sporolactobacillaceae*. In: De Vos P, Garrity GM, Jones D, Krieg NR, Ludwig W, Rainey FA, Schleifer K-H, Whitman WB (eds) *Bergey's Manual of Systematic Bacteriology*, The Firmicutes, 2nd edn. Springer, Dordrecht, p 386, Validation List N° 132 Int J Syst Evol Microbiol, 2010, 60:469–472, 2009) embraces three genera of Gram-positive, spore-forming rods with identical peptidoglycan and menaquinone composition and similar fatty acid composition. A fourth genus with no standing in nomenclature, *Scopulibacillus*, should be validated and added to the family. On the other hand, the genus, *Sinobaca*, considered a member of the family by some systematists, shows a separate phylogenetic position and differs in phenotypic properties from members of *Sporolactobacillaceae*. The biotechnological important genus is *Sporolactobacillus*, as members are potentially probiotic and some members produce high amount of D(−)-lactic acid in batch and continuous cultures containing inexpensive agricultural raw material. The stereocomplex of D- and L-lactic acid is of industrial importance for the production of polylactic acid, widely used in the packaging, food, cosmetic, pharmaceutical, and leather industries as well as in agriculture and medicine.

## Taxonomy, Historical and Current

The history of the genus *Sporolactobacillus* and some of its species described in the pre-molecular, pre-16S rRNA sequencing period is somewhat checkered as the definition of the genus based upon a few morphological and phenotypic properties gave misleading hints for the affiliation of new taxa. The discovery of spore-forming Gram-positive or Gram-variable rods which were metabolically defined by producing lactic acid homofermentatively, thus being intermediate to bacilli and lactobacilli, was first reported by Kitahara (1940). These so named strains "wild lactobacilli" were investigated by Nakayama (1960) who clustered them into *Bacillus coagulans*.

Kitahara and Suzuki (1963), studying bacteria from assorted chicken feed, isolated strain EU[T] and found on the basis of a taxonomic assessment that this strain cannot be classified into any of the known genera. Actually, this taxon was considered an example of a living fossil linking *Lactobacillus*, *Clostridium*, and *Bacillus* (Kitahara and Suzuki 1963, p. 2).The type strain of the newly created monospecific subgenus *Sporolactobacillus* of the genus *Lactobacillus* was a highly motile rod, possessing a small number of peritrichous flagella and formed oval endospores in the terminal position in slightly swollen sporangia in glucose-yeast extract-$CaCO_3$ medium. It produced D(−)-lactic acid homofermentatively on several carbohydrates and fermented inulin but not fructose. Metabolically, though the nutritional requirements were less complex than those of authentic *Lactobacillus* species, it resembled *Lactobacillus leichmannii*, also *Bacillus leichmannii* I according to Henneberg (1903). At the time of the description of *Sporolactobacillus inulinus*, the homofermentative lactobacilli were classified into the subgenera *Streptobacterium* and *Thermobacterium*. As indicated in the comment of Jean-Paul Euzeby (http://www.bacterio.cict.fr/), the creation of *Sporolactobacillus* inadvertently reduced *Lactobacillus* to subgeneric rank. A few years later, Kitahara and Toyota (1972) proposed informally that the subgenus *Sporolactobacillus* should be raised to genus status within the family *Bacillaceae*. The Approved Lists of Bacterial Names (Skerman et al. 1980) cite Kitahara and Suzuki (1963) as the authors of the generic name and Kitahara and Lai (1967) as authors of the type species name. The genus was included as the only genus in the family *Sporolactobacillaceae* (Ludwig et al. 2009, validated in 2010), based solely upon the phylogenetic position of its members.

E. Rosenberg et al. (eds.), *The Prokaryotes – Firmicutes and Tenericutes*, DOI 10.1007/978-3-642-30120-9_348,

Another example for following the winding road of taxonomic affiliation is given by the species *Sporolactobacillus laevolacticus*. This species was described by Nakayama and Yanoshi (1967) for mesophilic, catalase-positive, motile, and spore-forming strains (type strain M-8$^T$) from rhizosphere which lowered the pH of glucose broth 6.4 to 3.8–3.2. The cells could grow aerobically with the amount of lactic acid decreasing and that of volatile acids. These properties matched the description of the genus *Bacillus* (peritrichous flagellation, sporulation, catalase reaction, and aerobic property). Together with a second group of strains with similar properties but producing DL-lactic acid, the strains were described as *Bacillus laevolacticus* and *B. racemilacticus*, respectively. Thus the range of properties defining the genus *Bacillus* was expanded by the formation of D(−)- and DL-lactic acid. However, none of the two names were included in the Approved Lists of Bacterial Names (Skerman et al. 1980). Based upon chemotaxonomic evidence, Collins and Jones (1979) pointed out that the transfer of *Sporolactobacillus* (*S. inulinus* and the nonvalidly named *S. laevus* and *S. racemicus*) into *Bacillus* would be premature as long as the classification of *Bacillus* itself remains unsatisfying and unresolved. Andersch et al. (1994) validated the name *Bacillus laevolacticus* by providing an in-depth DNA-based and phenotype-based characterization. The name *Bacillus racemilacticus* was not validated because a single strain only was included in the study. The species *B. laevolacticus* was reclassified as *Sporolactobacillus laevolacticus* not before 2006 when a comparative 16S rRNA gene sequence-based study saw the type strain to clearly fell into the radiation of other members of the genus (>95 % sequence similarity) showing a close relatedness to *S. nakayamae* subspecies (>99 %) (Hatayama et al. 2006).

The first molecular indication that *Sporolactobacillus inulinus* ATCC 6473 is related to members of *Bacillus* originate from an early 16S rRNA cataloging study (Fox et al. 1977) in which the *S. inulinus* was used as a root to relate some members of *Bacillus* and *Sporosarcina ureae*. In a subsequent study (Fox et al. 1980), *S. inulinus* was branching next to *B. subtilis*, *B. sphaericus*, and *Thermoactinomyces vulgaris*. In yet another report, *S. inulinus* was found as an individual line of descent branching next to *Bacillus alcalophilus* (Ash et al. 1991; Farrow et al. 1994) and the same neighbor was found in a more extensive study on *Sporolactobacillus* (Suzuki and Yamasato (1994), including strains which were later described as novel *Sporolactobacillus* species (Yanagida et al. 1997). In the revised roadmap to the phylum Firmicutes (Ludwig et al. 2009), *Sporolactobacillus*, together with *Pullulanibacillus* and *Tuberibacillus* (Hatayama et al. 2006) clustered as a sister clade to the *Bacillaceae* group 2, which embraces a wide range of genera which were either described recently to reclassify former *Bacillus* species. The RaXML tree depicted in ➲ *Fig. 26.1* sees *Pullulanibacillus* and *Tuberibacillus* to form an individual line separate from *Sporolactobacillus* species, while *Sinobaca qinghaiensis* groups with *Marinococcus* and *Salsuginibacillus*. The latter situation is also seen in a NJ tree, while *Sporolactobacillus*, *Pullulanibacillus*, and *Tuberibacillus* cluster together (not shown). Though *Pullulanibacillus* and *Tuberibacillus* were indicated as members of *Sporolactobacillaceae* in the roadmap chapter (Ludwig et al. 2009), they were not covered in individual chapters in *Bergey's Manual of Systematic Bacteriology*, 2nd ed. (De Vos et al. 2009). According to Euzeby (http://www.bacterio.cict.fr/s/sporolactobacillaceae.html), the family contains not only the three genera indicated above but also *Sinobaca* (the renamed *Sinococcus* (Li et al. 2006; Li et al. 2008)) which, in its original description, clustered with members of the genus *Marinococcus*. The same position within *Bacillaceae* group 2 is shown in the consensus dendrogram of Ludwig et al. (2009). A rational and phylogenetic evidence is missing, why *Sinobaca* has been included in *Sporolactobacillaceae*. Nevertheless, for comparative reasons *Sinobaca* will be included in this chapter. Another genus, *Scopulibacillus* has been described (Lee and Lee 2009) but not yet validated. It branches adjacent to *Pullulanibacillus* and *Tuberibacillus* in the 16S rRNA gene neighbor-joining tree. Its properties are included in ➲ *Table 26.1*, but the type species *Scopulibacillus darangshiensis* is not further considered in this chapter.

➲ *Table 26.1* lists some common and differentiating properties of the four genera covered in this chapter. It is especially the morphology, absence of spore formation, and different menaquinone type that distinguish *Sinobaca* from other family members.

## Molecular Analyses

In a DNA–rRNA hybridization study (Miller et al. 1970) on members of *Lactobacillus Sporolactobacillus inulinus* was used as an outgroup and showed zero hybridization with *Lactobacillus leichmannii* among which a possible relationship had been suggested (Suzuki and Kitahara 1964). DNA–DNA reassociation (DDH) studies (Dellaglio et al. 1975) between the type strains of *Sporolactobacillus inulinus* and various species of the subgenus *Streptobacterium* Orla-Jensen (*L. casei*, *L. plantarum*, *L. coryniformis*, *L. curvatus*, *L. xylosus*, *L. alimentarius*, *L. farciminis*) showed low similarities (<30 % DDH similarity) and confirmed the species status of *S. inulinus*.

Intrageneric relatedness, obtained with different hybridization formats, showed species to be well separated from each other genomically. For example, low levels of DDH relatedness were found between the type strain of *S. vineae* and those of *S. inulinus*, *S. terrae*, and *S. kofuensis* (18.5, 18.0, and 17.0 %, respectively) (Chang et al. 2008), while those obtained between the two subspecies of *S. nakayamae* were in the range of 47–67 % (Yanagida et al. 1997). At the intraspecific level strains, extensive hybridization showed the coherency of the two subspecies of *S. nakayamae*, *S. terrae*, and *S. lactosus* (Yanagida et al. 1997) as well as strains of *S. laevolacticus* (Andersch et al. 1994).

In addition to 16S rRNA gene sequence analysis, the phylogenetic position of *S. putidus* was determined by partial *gyr* B gene sequence analysis. Both markers showed the isolated position of the type strain among other *Sporolactobacillus*-type strains though the nearest neighbors differed (Fujita et al. 2010).

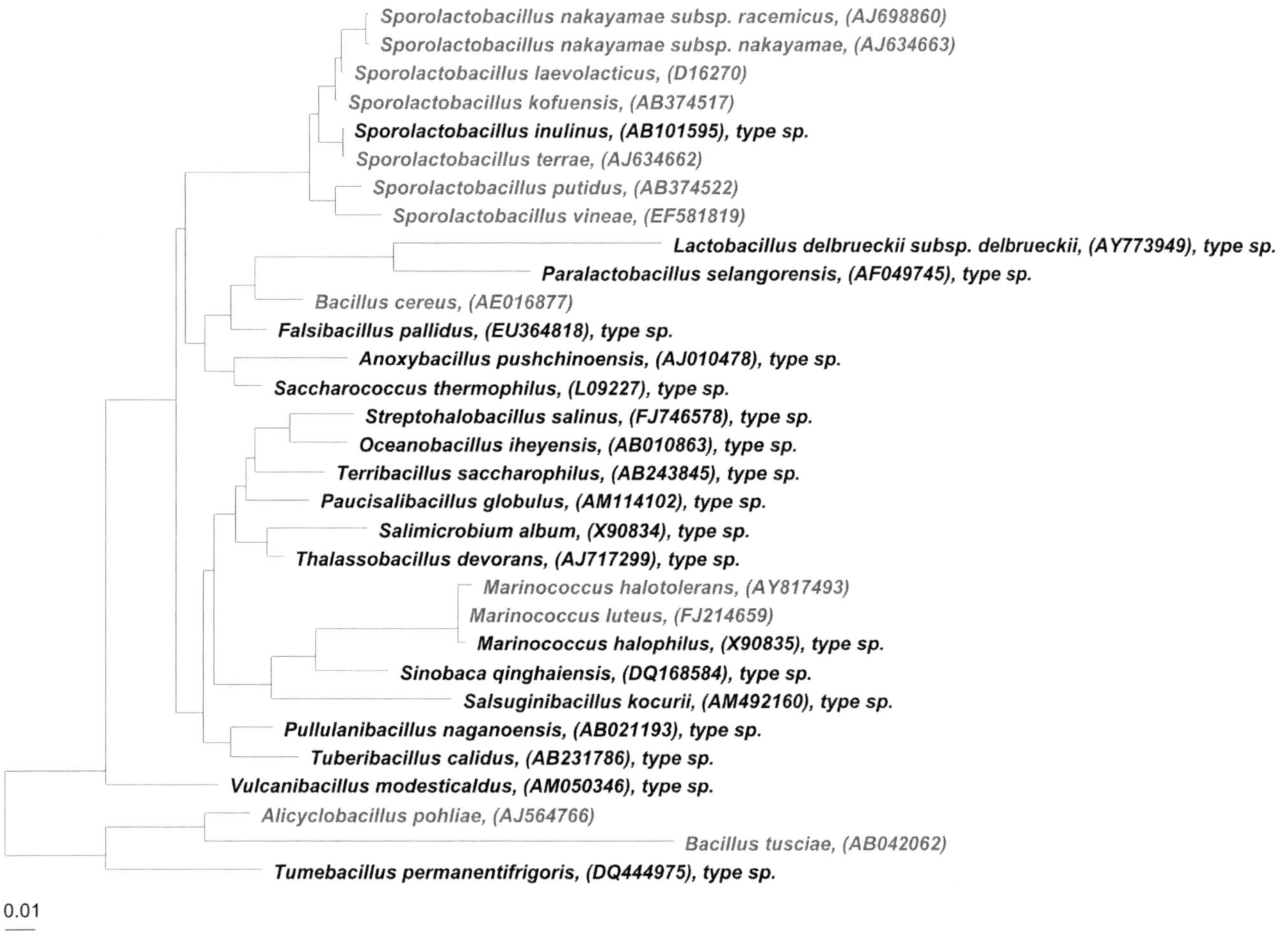


**Fig. 26.1**

**Maximum likelihood genealogy reconstruction based on the RAxML algorithm (Stamatakis 2006) of the sequences of all members of the family *Sporolactobacillaceae* and some neighboring taxa present in the LTP_106 (Yarza et al. 2010). The trees were reconstructed by using a subset of sequences representative of close relative genera to stabilize the tree topology. In addition, a 60 % conservational filter for the whole bacterial domain was used to remove hypervariable positions. The bar indicates 1 % sequence divergence. The sequences refer to the following strains: *S. inulinus* IFO13595$^T$, *S. kofuensis* JCM 3419$^T$, *S. laevolacticus* IAM 12321$^T$, *S. nakayamae* subsp. *nakayamae* DSM 11696$^T$, S . *nakayamae* subsp. *racemicus* DSM 16324$^T$, *S. terrae* DSM 1169$^T$, *S. putidus* QC81-06$^T$, *S. vineae* SL153$^T$ (KCTC 5376$^T$), *P. naganoensis* ATCC 53909$^T$, *T. calidus* 607$^T$, and *S. qinghaiensis* YIM 70212$^T$**

The draft genome sequence of *S. inulinus* CASD (CGMCC 2185) has been published by Yu et al. (2011). This strain was chosen for its high concentrations of D-lactic acid (207 g $L^{-1}$, see section on ❯ Application). The draft genome 2,930,096 bp with 3,476 predicted coding sequences and a G+C content of 44.97 mol% (accession number AFVQ00000000). Analysis of the annotated genes via KEGG pathway analysis indicated the presence of proteins for glycolysis, amino acids, nucleotides, lipids, cofactor/vitamin metabolism, and importantly D-lactic acid formation. The findings of many fructose- and mannose-metabolizing genes are in accord with the ability of all *Sporolactobacillus* species to ferment (among others) these carbohydrates. A large number of two-component system genes (allowing organisms to sense and respond to changes in different environmental conditions) were annotated from the draft genome sequence, which, according to the authors, gives a good chance to explore the mechanism of the extraordinary high concentration lactate tolerance of *S. inulinus* CASD.

The presence of an operon consisting of genes of a tyramine-producing pathway has been reported in *Sporolactobacillus* sp. P3J [tyrosyl-tRNA synthetase (tyrS), tyrosine decarboxylase (tdc), tyrosine permease (tyrP), and Na+/H+ antiporter (nhaC)] which were organized as already described in other tyramine-producing lactic acid bacteria (Coton et al. 2011). The downstream presence of a putative phage terminase and the upstream presence of a putative transposase let the authors suggest that the operon of this biogenic amine pathway was acquired, like in other lactic acid bacteria, through horizontal gene transfer.

The draft genome sequence of *S. vineae* SL153$^T$ (KCTC 5376 $^T$) has also been published by Kim et al. (2012). This strain showed a high level of cell adhesion activity as well as an

**Table 26.1**

**Some phenotypic properties of *Sporolactobacillaceae* genera (Data were compiled from *Table 26.2* for *Sporolactobacillus*, Hatayama et al. (2006) for *Tuberibacillus* and *Pullulanibacillus*, Li et al. (2006) for *Sinobaca* (published as *Sinococcus*), and Lee and Lee (2009) for *Scopulibacillus*)**

| Properties | *Sporolactobacillus* | *Tuberibacillus* | *Pullulanibacillus* | *Sinobaca* | *Scopulibacillus* |
|---|---|---|---|---|---|
| Morphology | Rods | Rods | Rods | Coccus | Rods |
| Motility | + | – | – | + | – |
| Spore formation | + | + | + | – | + |
| Facultative anaerobic growth | + | – | – | – | – |
| Catalase | – | + | + | + | + |
| Oxidase | – | + | – | – | – |
| Temperature range (°C) | 15–45 | 40–60 | 28–33 | 10–45 | 25–30 |
| pH range | 5–8[a] | 5–7 | 4–6 | 7.5–11 | 7–9 |
| NaCl range (%) | 0–10 | 0–4 | 0–2 | 1–25[b] | 0–4 |
| Production of D- or DL-lactic acid | + | – | – | nd | nd |
| Acid production from | | | | | |
| Fructose | + | nd | nd | – | – |
| Cellobiose | v to – | nd | nd | + | - |
| Mannitol | v to + | nd | + | + | – |
| Maltose | v to + | nd | nd | + | – |
| Xylose | –[a] | v | + | nd | nd |
| Peptidoglycan type | A1γ | A1γ | A1γ | A1γ | A1γ |
| Major menaquinone | MK7 | MK7 | MK7 | MK5 | MK7 |
| Major fatty acids >20 % | ai-$C_{15:0}$, ai-$C_{17:0}$[c] | ai-$C_{17:0}$ | $C_{16:0}$, i-$C_{16:0}$ | ai-$C_{15:0}$, ai-$C_{17:0}$ | ai-$C_{15:0}$, ai-$C_{17:0}$ |
| Minor fatty acids >3 <20 % | i-$C_{16:0}$, $C_{16:0}$ | $C_{16:0}$, i-$C_{15:0}$, i-$C_{16:0}$, i-$C_{17:0}$, i-$C_{17:0}$ 3-OH[d] | $C_{14:0}$, i-$C_{14:0}$, i-$C_{15:0}$, ai-$C_{15:0}$, ai-$C_{17:0}$ [e] | i-$C_{16:0}$, $C_{16:0}$ | $C_{16:0}$, i-$C_{15:0}$, i-$C_{16:0}$, i-$C_{17:0}$ |
| Whole-cell sugar | Glu, man, gal or gal, man, rham[a] | Not detected | Gal, glu, rham | Rib | Glu |
| Mol% DNA G+C | 43–50 | 46–47 | ~45 | 47 | 51 |

[a]Not all strains investigated
[b]Also in KCl and $MgCl_6 \times 6H_2O$
[c]Composition is different in the analysis of *S. terrae* (Fujita et al. 2010)
[d]Type strain only, when grown in CYC medium (see section on "Isolation, Enrichment, and Maintenance Procedures") at 55 °C
[e]When grown in nutrient broth containing 1 % starch

antagonistic activity against pathogens such as *Vibrio* sp. The percentage of G+C content in all 92 contigs was 49 mol% (accession numbers BAEY01000001 to BAEY01000092). Analysis of the annotated genes via BLAST and the RAST server indicated the genome contained 2,933 protein-coding genes, three copies of the small-subunit rRNA, 61 tRNA genes, one copy of the large-subunit rRNA, and two copies of 5S RNA. The genome contains methionine biosynthesis, a lysine biosynthesis DAP pathway, cAMP signaling in bacteria, dehydrogenase complexes, and bacterial translation termination factors. There are many metabolism and carbohydrate proteins, including those involved in central carbohydrate metabolism, acetyl-coenzyme A (CoA) fermentation to butyrate, pyruvate metabolism II, and pyruvate alanine serine interconversions.

## Phenotypic Analyses

The properties of the family *Sporolactobacillaceae*, order *Bacillales*, are as follows: Gram-positive endospore-forming, mesophilic, and thermophilic rods; motile or nonmotile; and facultatively anaerobic or aerobic. D(−)- or DL-lactic acid is

**Table 26.2**

**Properties differentiating the species and subspecies of the genus *Sporolactobacillus*. The numbers refer to the taxa which are listed according to phylogenetic position: (1) *S. inulinus* (Kitahara and Suzuki 1963); (2) *S. putidus* (Fujita et al. 2010); (3) *S. vineae* (Chang et al. 2008); (4) *S. terrae*; (5) *S. nakayamae* subsp. *nakayamae*; (6) *S. nakayamae* subsp. *racemicus*; (7) *S. kofuensis*; (8) *S. lactosus*[a] (all Yanagida et al. 1997); and (9) *S. laevolacticus* (Nakayama and Yanoshi 1967; Andersch et al. 1994; Hatayama et al. 2006) (Data are compiled from the original description and recent comparisons (Fujita et al. 2010; Chang et al. 2008))**

| Properties | 1 | 2 | 3 | 4 | 5 | 6 | 7 | 8 | 9 |
|---|---|---|---|---|---|---|---|---|---|
| Motility | + | + | + | + | + | + | –[b] | + | + |
| Catalase | – | – | – | – | – | – | – | – | +[c] |
| Temperature range (°C) | 25–40 | 30–45 | 25–40 | 15–40 | 15–40 | 15–40 | 25–40 | 15–45 | 15–40 |
| NaCl range (%) | 0–7 | 0–3 | 0–7 | 0–4 (few 7–10) | 0–5 (few 7) | 3–5 (most) | 0–4 (few 7–10) | 0–4 (few 7) | 0–5 |
| Lactic acid isomer | D(–) | D(–) | DL | D(–) | D(–) | DL | D(–) | D(–) | D(–) |
| Acid production from | | | | | | | | | |
| Galactose | –[d] | + | – | + | + | + | + | + | + |
| Sorbose | – | – | v | + | nd | nd | nd | nd | nd |
| Lactose | – | nd | nd | v | – | v | – | + | ± |
| Melibiose | – | – | – | – | – | + | – | + | nd |
| Cellobiose | – | – | – | v | – | v | – | v | nd |
| Raffinose | + | – | – | v | v | v | + | nd | nd |
| Starch | – | – | – | – | – | v | – | v | + |
| Inulin | + | – | – | + | – | + | + | + | + |
| Salicin | – | – | – | v | – | v | – | v | nd |
| Aesculin | – | – | – | – | + | – | – | nd | nd |
| α-Methyl-glucoside | + | – | + | v | v | v | v | v | – |
| Sorbitol | + | – | + | – | v | v | – | v | ± |
| Tagatose | + | + | – | + | + | + | + | nd | nd |
| DNA mol% G+C | 47–50 | 48 | 51–52 | 43–46 | 43–47 | 43–46 | 43 | 43–46 | 43–45 |

[a]The strain JCM 9690[T] is no longer available at the JCM catalog and most probably there is no alternative collection (see Euzeby, http://www.bacteriocict.fr/s/sporolactobacillus.html

[b]The type strain X1-1[T] is nonmotile, while all other 9 strains of the species were motile (Yanagida et al. 1997)

[c]According to Hatayama et al. (2006) the strain is catalase negative

[d]+ positive (>805 of strains), – negative (0–20 %), *v* variable (21–79 %), *w* weak, ± obscure according to Nakayama and Yanoshi (1967), *nd* not determined

produced by homofermentation from several carbohydrates by *Sporolactobacillus*. Other members produce undetermined acids under aerobic conditions, but not lactic acid. Peptidoglycan contains meso-diaminopimelic acid ($A_2$pm) of the direct linkage type; MK7 is the predominant quinone; major fatty acids are anteiso-$C_{15:0}$ and anteiso-$C_{17:0}$; iso-$C_{15:0}$ and iso-$C_{16:0}$ occur in smaller amounts. Type genus is *Sporolactobacillus*.

## *Sporolactobacillus* Kitahara and Suzuki 1963, 69[AL]

Spo.ro.lac.to.ba.cil'lus.Gr. n. *spora* seed. L. n. lac, *lactis* milk; L. dim. n. *bacillus* a small rod; N. L. masc. n. *Sporolactobacillus* sporing milk rodlet.

Most generic properties are shown in *Table 26.1*. Strains consist of facultatively anaerobic or microaerophilic rods (0.4–1.0 × 2.0–4.0 μm); occurring singly, in pairs, and, rarely, in short rods; and motile by peritrichous flagella or a few polarly and laterally inserted flagella. Good growth occurs in media containing glucose. For some strains, poor or no growth is reported to occur in nutrient broth. Mesophilic. Acid is produced from glucose, fructose, mannose, sucrose, trehalose, maltose, and mannitol. The type species is *Sporolactobacillus inulinus* (Kitahara and Suzuki 1963) Kitahara and Lai 1967, 197[AL]. *S. inulinus* and *S. laevolacticus* can convert L-lactic acid to D-lactic acid after the late log phase (Sawai et al. 2011).

The major differences between the species and the two subspecies of *S. nakayamae* are indicated in *Table 26.2*. Additional information can be taken from the original species descriptions (for authors, see legend of *Table 26.2*) or (except for *S. putidus* and *S. vineae*) from the compilation of Yanagida and Suzuki (2009).

The type strain of *S. lactosus* JCM 9690[T] is no longer available at the JCM catalog and most probably there is no alternative collection [Yarza pers. communication to Jean Euzeby (http://www.bacterio.cict.fr/s/sporolactobacillus.html)]. The species *Sporolactobacillus cellulosolvens*, used for lactic acid production from molasses (Kanwar et al. 1995), has not been validly named. This applies also to *Sporolactobacillus laevus*, an invalid species for which several 16S rRNA gene sequences are found in public databases.

## *Tuberibacillus* Hatayama, Shoun, Ueda and Nakamura 2006, 2549[VP]

Tu.be.ri.ba.cil'lus. L. neut. n. *tuber* swelling; L. masc. n. *bacillus* a small staff; N.L. masc. n. *Tuberibacillus* a small staff with a swelling.

Most generic properties are shown in ➋ *Table 26.1.* Cells are aerobic, thermophilic rods that are 3–7 × 0.3–0.5 μm, occurring singly or in chains. Terminally formed endospores within swollen sporangia are oval (0.7–1.0 × 0.5–0.7 μm). The type species *Tuberibacillus calidus* displays the additional following properties (Hatayama et al. 2006): Colonies are round, obscure-edged, translucent, and cream in color. Casein is hydrolyzed. Negative reactions for the deamination of phenylalanine, hydrolysis of starch and tyrosine, utilization of citrate and propionate, and the production of lactic acid. Acid is produced from glucose and arabinose, but not from lactose. Nitrate reduction and acid production from xylose are dependent on the strain. The type strain is strain 607$^T$.

## *Pullulanibacillus* Hatayama, Shoun, Ueda and Nakamura 2006, 2549[VP]

Pul.lu.la.ni.ba.cil'lus. N.L. n. pullulanum pullulan; L. masc. n. *bacillus* a small rod; N.L. masc. n. *Pullulanibacillus* a small rod hydrolyzing pullulan.

Most generic properties are shown in ➋ *Table 26.1.* Cells are aerobic rods (0.5–1.0 × 2.1–10.0 μm) with rounded or square ends, occurring singly or in chains. Endospores are oval and cause swelling of the sporangia. Mesophilic and moderately acidophilic. Nitrate and nitrite reduction negative. Menaquinone-7 is a major component. Menaquinone-5 is a minor component. The type species *Pullulanibacillus naganoensis* displays the additional following properties given to *Bacillus naganoensis* by Tomimura et al. (1990): Rod-shaped cells (0.5–1.0 × 2.1–10.0 μm), occurring singly or in chains, have rounded or square ends. Terminally formed endospores within swollen sporangia are oval. Colonies are about 2–3 mm in diameter, opaque, smooth, glistening, convex, and circular with entire margins. Produces acid (after >14 days of incubation) from L-arabinose, D-glucose, and lactose (weakly). No gas is produced from glucose. Starch hydrolysis is positive. Gelatin hydrolysis, casein hydrolysis, phenylalanine deaminase, lecithinase, indole, Voges–Proskauer reactions, citrate utilization, and propionate utilization are negative. Does neither decompose tyrosine or hippurate nor produce dihydroxyacetone from glycerol. The type strain is D39$^T$.

## *Sinobaca* Li, Zhi and Euzéby 2008

(Si.no.ba'ca. M.L. n. *Sina*, China; L. fem. n. *baca*, a grain or berry, and in bacteriology a coccus; N.L. fem. n. *Sinobaca*, coccus-shaped microbe isolated from places in China).

Most generic properties are shown in ➋ *Table 26.1.* Cells are strictly aerobic cocci (diameter 0.8–1.0 μm) and Gram-positive. Non-spore-forming, motile with multiple flagella. Mesophilic and moderately alkalophilic. Catalase positive and oxidase negative. Menaquinone-5 is a major component. The peptidoglycan type is A1$\gamma$ (*meso*-diaminopimelic acid, directly cross-linked). The major whole-cell wall sugar is ribose; galactose is present in minor amounts. Polar lipids contain diphosphatidylglycerol, phosphatidylglycerol, and some unidentified components, including one phospholipid, one glycolipid, and two aminoglycolipids. Major cellular fatty acids are ai-$C_{15:0}$ and ai-$C_{17:0}$. The G+C content is about 47 mol%. The type species *Sinobaca qinghaiensis* displays the additional following properties by Li et al. (2006): Colony is orange color, circular, and opaque (1.5–1.8 mm in diameter) after 24 h at 28 °C. The optimum concentration of KCl for growth is 10 % (w/v). Optimum growth occurs at pH 8.0–9.5 and 28 °C. Grows in 1–25 % KCl, $MgCl_2$, $6H_2O$, and NaCl. Positive for lipase, $\beta$-glucosidase, $\beta$-galactosidase, $\alpha$-glucosidase, and casein hydrolysis, but negative for arginine dihydrolase, lysine decarboxylase, ornithine decarboxylase, $\alpha$-galactosidase, $\alpha$-maltosidase, urease, *N*-acetylglucosaminidase, nitrate reduction, gelatin liquefaction, ammonia production, methyl red and Voges–Proskauer tests, milk peptonization and coagulation, growth on cellulose, $H_2S$ and melanin production, and starch hydrolysis. Maltose, mannitol, glucose, mannose, fructose, galactose, sucrose, cellobiose, and trehalose can be utilized as carbon sources; adonitol, arabinose, arabitol, rhamnose, inositol, and sorbitol cannot be utilized. Acid is produced from glucose, maltose, sucrose, cellobiose, and trehalose. The type species is *Sinobaca qinghaiensis* YIM 70212$^T$.

# Isolation, Enrichment, and Maintenance Procedures

Most of the early work has been done on *Sporolactobacillus inulinus* and other strains which were later described as new species of *Sporolactobacillus*. The basic procure includes of heat treatment (10 min at 80 °C) of samples to suppress asporogeneous cells (Yanagida and Suzuki 2009). For the enrichment, a glucose broth has been used for the cultivation of *Bacillus laevolacticus* and *B. racemilacticus* (Nakayama and Yanoshi 1967). The following ingredients were filled up to 1 L with autoclaved tap water, pH 6.4: soil extract, 100 mL; yeast extract, 100 mL; polypeptone, 10 g; glucose, 10 g; salt A, 10 mL; salt B, 10 mL; and salt C, 1 mL.

For obtaining the soil extract, 50 g of garden soil was mixed with 100 mL of water, autoclaved for 20 min at 130 °C, and filtered. Yeast extract was obtained by mixing 10 g of dried brewery yeast with 100 mL of water, heated on a boiling water bath, and centrifuged. Salt A: $KH_2PO_4$ 5 %; $K_2HPO_4$ 5 %. Salt B: $MgSO_4 \times 7H_2O$ 3 %; NaC1 0.1 %; $MnSO_4 \times 5H_2O$, 0.1 %; $CuSO_4 \times 5H_2O$ 0.01 %; $CoCl_2 \times 6H_2O$ 0.01 %. Salt C: $FeSO_4 \times 7H_2O$ 0.1 %; $Na_3$ citrate 2.7 %.

GYP medium for enrichment and further growth is given by Kitahara and Suzuki (1963). GYP is glucose 2 %; yeast extract (Difco) 0.5 %; and peptone 0.5 %, 1 % sodium acetate, and 0.5 % (v/v) salts solution (pH 6.8). The salts solution contained per liter of distilled water 4 % $MgSO_4 \times 7H_2O$ 4 %, 0.2 % $MnSO_4 \times 4H_2O$, 0.2 % $FeSO_4 \times 7H_2O$, 0.2 % NaCl plus $CaCO_3$ 1 % pH 6.8. Cultures were incubated in an anaerobic jar in a 100 % $CO_2$ atmosphere. Acid-producing bacteria were recognized by the appearance of clear zones around colonies.

Growth is also supported by MRS medium at pH 5.5 (medium used by the DSMZ) and has been used in the isolation of *S. putidus*.

A selective detection method was described by Doores and Westhoff (1983) who successfully enriched *Sporolactobacillus* spp. from a variety of foods, feed, soil and environmental samples. The authors used a modified MRS medium [(including 1.0 % (w/v) α-methyl glucoside, 0.1 % (w/v) potassium sorbate, 0.00224 % (w/v) bromocresol green indicator, adjusted to pH 5.5 with acetic acid)] and rinsed and incubated a sample at 37 °C for 7 days under 5 % $CO_2$. Volumes of 2 mL from each sample were heat shocked at 80 °C for 5 min and 0.1 mL spread onto plates of Lactobacilli MRS agar (Difco), pH 5.5, and APT agar (BBL), pH 5.5. Incubation was for 5 days.

Isolates from vineyards (Yanagida et al. 2005) were isolated on GYP medium and on medium BM (polypeptone 2 %, tryptose 0.5 %, Bacto-liver extract 2 %, yeast extract 0.5 %, Tween 80 0.001 %, glucose 1 %, fructose 0.5 %, DL-malic acid 0.1 %, $MnCl_2{\cdot}4H_2O$ 0.008 %, filtered tomato juice 25 % and distilled water 75 %, adjusted to pH 5.4) under anaerobic conditions (BBLTM GasPakTM, $H_2 + CO_2$) at 30 °C for 2–3 days. Variations of MRS medium have been used in a different study on isolated from vineyards (Bae et al. 2006) in which MRS plus ethanol (5 %), MRS broth supplemented with 15 % (v/v) tomato juice, pH 5.5 and 3.5, and autoenrichment in grape juice homogenate were applied.

Formation of endospores is rare in most media but not tomato–meat (TM) medium (Kitahara and Lai 1967). When the strain was incubated under 5 % $CO_2$ at 37 °C in TM medium devoid of ammonium sulfate and calcium carbonate, about 20 % of total cells form tadpole-like structures. TM is yeast extract 0.1 %, meat extract 0.5 %, α-methylglucoside 0.5 %, ammonium sulfate 1 %, and tomato serum 20 % (pH 5.5). The other authors reported that starch stimulated the sporulation frequency (Nakayama and Yanoshi 1967) and TM medium without tomato serum was superior (Doores and Westhoff 1981). Spores of *Sporolactobacillus* are ellipsoidal and swollen and appear in the terminal to subterminal position. Spores contain dipicolinic acid and the ultrastructure is similar to those of *Bacillus* spores (Kitahara and Lai 1967).

*Tuberibacillus calidus* was isolated on MC agar, consisting of corpses of microbes obtained during the later piling step of a hyperthermal composting process (Hatayama et al. 2006). Though isolated on MC agar, this medium did not support their maintenance nor did LB medium, nutrient medium (nutrient broth; Difco), or GYP medium. Growth was achieved in CYC medium [Czapek-Dox liquid medium, modified (Oxoid) 33.4 g; Bacto yeast extract (Difco) 2.0 g; Bacto vitamin assay Casamino acids (Difco) 6.0, 1 L of distilled water, pH 7.2] (Lacey and Cross 1989) at 60 °C.

The presence of *Pullulanibacillus naganoensis* was screened on a medium containing per liter of distilled water: yeast extract (Oxoid) 0.1 %, tryptone (Difco) 0.2 %; $(NH_4)_2SO_4$ 0.2 %, $KH_2PO_4$ 0.03 %, $MgSO_4 \times 7H_2O$ 0.02 %, $FeSO_4 \times 7H_2O$ 10 mg, $CaCl_2 \times 2H_2O$ 0.02 %, $MnCl_2 \times 4H_2O$ 1 mg, agar 2 %, soluble starch (Sigma) 1 %, blue-colored soluble starch (Rinderknecht et al. 1967) 0.3 %, and red-colored pullulan 0.75 %. The pH was adjusted with 0.2 N sulfuric acid to pH 4.0. Red-colored pullulan was prepared (Tomimura et al. 1990) by dissolving 100 g of pullulan PFIO (Hayashibara Co., Ltd., Okayama, Japan) in 2 L of distilled water. The solution temperature was increased to 50 °C, and 10 g of Mikacion Brilliant Red 5BS (Mitsubishi Kasei Co., Ltd., Tokyo, Japan) and 100 mL of 10 % $Na_3PO_4$ were added. After a 75-min incubation, the colored pullulan was precipitated by adding 1.6 L of ethanol (99.5 %) and collected by decantation. The precipitate was washed twice with 60 % ethanol, washed once with 99.5 % ethanol, and air dried. The color of the agar plate is dark violet, and the presence of pullanase active strains is visible by blue zones surrounding colonies. A secondary screening medium contained the ingredients described above except that amylopectin (1 % per liter) was substituted for the soluble starches and colored pullulan. Colonies of a pullanase active strain (hydrolyzing the α-1,6 linkages of amylopectin, releasing long amylose chains) are surrounded by a dark blue zone when exposed to iodine vapors.

All strains can be maintained long term at 4 °C as 20 % (v/v) glycerol suspensions. Long-term preservation include freezing at −80 °C in 10 % (v/v) skimmed milk, lyophilization, or in straws under $N_2$ vapor.

## Ecology

The original description of *Sporolactobacillus inulinus* strain $EU^T$ was performed on a strain isolated from chicken feed (Kitahara and Suzuki 1963), but the majority of strains of other species of the family originate from soil habitats (Tomimura et al. 1990; Yanagida et al. 2005; Chang et al. 2008), mainly the rhizosphere (Nakayama and Yanoshi 1967; Yanagida et al. 1997). The occurrence in fruit juice has also been reported (Fujita et al. 2010). Doores and Westhoff (1983), using a selective isolation medium, screened almost 700 foods, feed, soil, and environmental samples but found *Sporolactobacillus* strains in only two samples. Thus members of this genus can be considered "rare species." The original of two strains of *Tuberibacillus* from compost (Hatayama et al. 2006) does not allow conclusions about their natural habitat. BLAST similarity search reveals that highest scores are among cultured organisms of validly names species taken mainly from culture collections. Clone sequences of uncultured organisms originate from arctic streams (Larouche et al., unpublished; e.g., accession number J849516) and from lake sediment samples used for repeated batch fermentation processes (Romano et al., unpublished; e.g., JF312665).

## Application

### Polylactic Acid

Lactic acid is a monomer of polylactic acid (PLA), which is widely used in packaging food and cosmetic, pharmaceutical, and leather industries; it is naturally degradable and environmentally harmless (Avinc and Khoddami 2009). PLA is also considered to be one of the most promising biodegradable polymers that can be applied to drug delivery systems, orthopedic screws, textiles, packaging materials, and agricultural films (Shukla et al. 2004; Wang et al. 2010). It is estimated that by 2020 the PLA market will have surged to 3 million tons annually with a market value of US$ 6 billion. Production from raw materials such as sugar and tapioca or rice starch (Fukushima et al. 2004) will reduce the $CO_2$ emission as compared to the production from conventional plastic production. Usually PLA is produced from pure L-lactic acid (Inkinen et al. 2011) which has the disadvantage of having a melting point that is lower than that of petroleum-based polymers (Fukushima et al. 2007). The melting point of a stereocomplex of poly L-lactic acid and poly D-lactic acid (230 °C) is not only higher than that of poly L-lactic acid (180 °C) (Ikada et al. 1987) but also improves mechanical and thermal resistance (Fukushima et al. 2007). A yield of D-lactic acid production through fermentation is lower than that of L-lactic acid; the search for higher D-lactic acid-producing strains is high on the agenda. However, the wild-type *Sporolactobacillus* strains are not regarded to be efficient D-lactic acid producers.

A high D(−)-lactic acid-producing strain of *Sporolactobacillus*, strain CASD, has recently been reported with increased growth rate at the early stage of repeated batch fermentation cycles (Zhao et al. 2010). When grown on 40 g $L^{-1}$ of peanut meal and $CaCO_3$ (80 g $L^{-1}$) in 30-L fed-batch fermentation, a maximum D-lactate production (120 g $L^{-1}$) and optical purity of 99.3 % were achieved (Wang et al. 2010). Lactate production was improved at 42 °C and with the addition of 0.3 g $L^{-1}$ of neutral protease to the batch. The concentration of D-lactate obtained by strain CASD was almost as double as high as that found for *Lactobacillus delbrueckii* or *Corynebacterium glutamicum* (Wang et al. 2010; Gao et al. 2011; Sawai et al. 2011). Strain CASD (see ➋ Molecular Analyses) and its application in D-lactic acid production have been patented (Xu et al. 2007). D-lactate production improvement was obtained by mutating *Sporolactobacillus* sp. DX12 by an N+ ion beam (Xu et al. 2010). The mutant showed higher 6-phosphofructokinase, pyruvate kinase, and D-lactate dehydrogenase activities as compared to the wild strain, resulting in the production of 121.6 g $L^{-1}$ of D-lactic acid with the molar yields of 162.1 % to glucose (198.8 % higher yield than the wild strain). An additional attempt to improve the yield of D-lactic acid was done with *Sporolactobacillus inulinus* ATCC 15538 which was subjected to recursive protoplast fusion in a genome shuffling format (Zheng et al. 2010). As compared to the original strain, the acid-resistant mutant F3–4 increased the D-lactic acid production by 119 % (93.4 g $L^{-1}$ in a 5-L bioreactor). A further improvement of yield and optical purity of D-lactic acid was achieved by changing from batch to continuous culture procedures.

### Probiotics

Besides well-known spore-forming probiotics such as *Bacillus* species included in commercial probiotic products (e.g., *B. cereus*, *B. pumilus*, *B. subtilis*) and *Brevibacillus laterosporus* (Sanders et al. 2003) also *Sporolactobacillus* spp. have been investigated for their potential to serve as probiotics. The bile and acid resistance of strains of *S. inulinus*, *Bacillus laevolacticus* (later described as *S. laevolacticus*), and of *S. racemicus* (later described as *S. terrae* (Yanagida et al. 1997)) IAM 12395 were evaluated by Hyronimus et al. (2000) on MRS agar plates. The only strain that grew well at pH 2.5 even for 6 h was *S. laevolacticus*, while none of the *Bacillus* and other *Sporolactobacillus* strains grew at pH 2.0 for 3 h. The survival rate of *S. inulinus* strains on 0.3 % oxgall medium decreased to <26 % after 3 h and <15 % after 6 h. The minimal inhibition concentration of bile was 0.7 % for *S. laevolacticus* strains, while those for *B. cereus* strains was >1.0 %. Strains *S. terrae* IAM 12395 did not grew at all under these conditions. These data suggest that strains surviving low pH (*S. laevolacticus* DSM6475 and *S. inulinus strains*) are sensitive to oxgall, while *S. terrae* with resistance to oxgall do not survive in acidic medium. As the authors point out, 3-h incubation is longer than experienced under in vivo conditions which would be no longer than 30–60 min; also, spores were not tested for their long-time survival. These tests were performed by Huang et al. (2007) who found that at pH 2.0–4.0 conditions spores of *S. inulinus* BCRC 14647 survived better than their vegetative cells and those of bifidobacteria strains (e.g., 45 %, 80 %, and 92 % for 3 h at pH 2.0, 3.0, and 4.0, respectively). Better survival of spores was also determined for oxgall tolerance (27 %, 34 %, and 14 % for 3 h at concentrations of 0.1 %, 0.2 %, and 0.4 % gall, respectively). Concerning the adhesiveness to $CaCO_2$ cells most *Lactobacillus* and *Bifidobacteria* cells adhered better than the *Sporolactobacillus* strain, while the latter performed better than the reference lactic acid strains in the antagonistic activity of culture supernatant against a strain of *Salmonella enteritidis*. The authors conclude from the lack of invasion of caco-2 cells that *S. inulinus* show high safety properties, and they suggest that the vegetative cell form of *S. inulinus* presents of great probiotic potential.

*S. vineae* SL153$^{T}$ (KCTC 5376 $^{T}$) has also been investigated for their potential probiotics by Chang et al. (2011). The strain showed a high level of bile tolerance, cell adhesion activity, as well as an antagonistic activity against pathogens such as *Vibrio* sp. Growth of the strain was examined on GYP medium containing porcine bile extract (Sigma) concentration from 0.1 % to 5 % (w/v). The strain also showed highly growth inhibition toward *V. cholerae*, *V. alginolyticus*, *V. fluvialis*, *V. parahaemolyticus*, *Aeromonas bivalvium*, and *Listonella anguillarum*. Cell adherence to HT-29 cell (ATCC HT-38) indicated that the strain similar to the *S. inulinus*.

Effects of *Sporolactobacillus* P44 on postprandial porto-arterial concentration differences (PACD) of glucose, galactose, L-lactic acid, amino nitrogen, and urea in the growing pig were studied by Rychen and Simões Nunes (1993). The areas of PACD of glucose, galactose, and amino nitrogen of probiotics diet for the first 3 h after the meal were significantly higher after probiotics diet ingestion than those of basal diet intake. Plasma concentrations of urea and PACD of urea and L-lactic acid were not modified by the probiotic. These effects disappeared immediately after probiotics diet interruption, suggesting that added bacteria presence in the intestinal lumen was fundamental to the modifications observed in apparent absorption.

Effects of three microbial probiotics (*Sporolactobacillus* P44, *Bacillus cereus* IP5832, or a combination of *Lactobacillus acidophilus*, *L. fermentum*, and *L. brevis*) on postprandial PACD of glucose, galactose, L-lactic acid, and amino nitrogen in the young pig were also studied by the same authors Rychen and Simões Nunes (1993). Areas of PACD of glucose, galactose, and L-lactic acid were not influenced by the probiotics supplements. Areas of PACD of amino nitrogen were significantly higher after the ingestion of the *Sporolactobacillus* P44 diet than that of basal diet.

## Acknowledgements

Parts of this chapter has been prepared under the EMbaRC project [EU Seventh Framework Programme Research Infrastructures INFRA-2008-1.1.2.9: Biological Resources Centres (BRCs) for microorganisms (Grant agreement number: FP7-228310)] to support science in BRCs. Part of this work was supported by grant of the KRIBB Research Initiative Program funded by the Ministry of Education, Science and Technology, Republic of Korea.

## References

Andersch I, Pianka S, Fritze D, Claus D (1994) Description of *Bacillus laevolacticus* (ex Nakayama and Yanoshi 1967) sp. nov., nom. rev. Int J Syst Bacteriol 44:659–664

Ash C, Farrow JAE, Wallbanks S, Collins MD (1991) Phylogenetic heterogeneity of the genus *Bacillus* as revealed by comparative analysis of small subunit-ribosomal RNA sequences. Lett Appl Microbiol 13:202–206

Avinc O, Khoddami A (2009) Overview of poly(lactic acid) (PLA) fibre. Fibre Chemist 41:391–401

Bae S, Fleet GH, Heard GM (2006) Lactic acid bacteria associated with wine grapes from several Australian vineyards. J Appl Microbiol 100:712–727

Chang YH, Jung MY, Park IS, Oh HM (2008) *Sporolactobacillus vineae* sp. nov., a spore-forming lactic acid bacterium isolated from vineyard soil. Int J Syst Evol Microbiol 58:2316–2320

Chang YH, Jung MY, Park IS (2011) Probiotics spore-forming lactic acid bacteria SL153. US Patent 20110014166A1

Collins MD, Jones D (1979) Isoprenoid quinone composition as a guide to the classification of *Sporolactobacillus* and possibly related bacteria. J Appl Microbiol 47:293–297

Coton M, Fernández M, Trip H, Ladero V, Mulder NL, Lolkema JS, Alvarez MA, Coton E (2011) Characterization of the tyramine-producing pathway in *Sporolactobacillus* sp. P3J. Microbiology 157:1841–1849

De Vos P, Garrity GM, Jones D, Krieg NR, Ludwig W, Rainey FA, Schleifer K-H, Whitman WB (eds) (2009) Bergey's manual of systematic bacteriology, vol 3, 2nd edn, The Firmicutes. Springer, Dordrecht

Dellaglio F, Bottazzi V, Vescovo M (1975) Deoxyribonucleic acid homology among *Lactobacillus* species of the subgenus *Streptobacterium* Orla-Jensen. Int J Syst Bacteriol 25:160–172

Doores S, Westhoff DC (1981) Heat resistance of *Sporolactobacillus inulinus*. J Food Sci 46:810–812

Doores S, Westhoff DC (1983) Selective method for the isolation of *Sporolactobacillus* from food and environmental sources. J Appl Microbiol 54:273–280

Farrow JAE, Wallbanks S, Collins MD (1994) Phylogenetic interrelationships of round-spore-forming bacilli containing cell walls based on lysine and the non-spore-forming genera *Caryophanon*, *Exiguobacterium*, Kurthia, and *Planococcus*. Int J Syst Bacteriol 44:74–82

Fox GE, Pechman KR, Woese CR (1977) Comparative cataloging of 16S ribosomal ribonucleic acid: molecular approach to procaryotic systematics. Int J Syst Bacteriol 27:44–57

Fox GE, Stackebrandt E, Hespell RB, Gibson J, Maniloff J, Dyer TA, Wolfe RS, Balch WE, Tanner RS, Magrum LJ, Zablen LB, Blakemore R, Gupta R, Bonen L, Lewis BJ, Stahl DA, Luehrsen KR, Chen KN, Woese CR (1980) The phylogeny of prokaryotes. Science 209:457–463

Fujita R, Mochida K, Kato Y, Goto K (2010) *Sporolactobacillus putidus* sp. nov., an endospore-forming lactic acid bacterium isolated from spoiled orange juice. Int J Syst Evol Microbiol 60:1499–1503

Fukushima K, Sogo K, Miura S, Kimura Y (2004) Production of D-lactic acid by bacterial fermentation of rice starch. Macromol Biosci 4:1021–1027

Fukushima K, Chang YH, Kimura Y (2007) Enhanced stereocomplex formation of poly(L-lactic acid) and poly(D-lactic acid) in the presence of stereoblock poly(lactic acid). Macromol Biosci 7:829–835

Gao C, Ma CQ, Xu P (2011) Biotechnological routes based on lactic acid production from biomass. Biotechnol Adv 29:930–939

Hatayama K, Shoun H, Ueda Y, Nakamura A (2006) *Tuberibacillus calidus* gen. nov., sp. nov., isolated from a compost pile and reclassification of *Bacillus naganoensis* Tomimura et al. 1990 as *Pullulanibacillus naganoensis* gen. nov., comb. nov. and *Bacillus laevolacticus* Andersch et al. 1994 as *Sporolactobacillus laevolacticus* comb. nov. Int J Syst Evol Microbiol 56:2545–2551

Henneberg W (1903) Zur Kenntniss der Milchsäurebakterien der Brenneriemaische, der Milch, des Bieres, der Presshefe, der Melasse, der Sauerkohls, der sauren Gurken und des Sauerteigs, sowie einige Bemerkungen über die Milchsäurebakterien des menschlichen Magens. Zeitschr Spiritusind 26:329–332

Huang HY, Huang SY, Chen PY, King VA, Lin YP, Tsen JH (2007) Basic characteristics of *Sporolactobacillus inulinus* BCRC 14647 for potential probiotic properties. Curr Microbiol 54:396–404

Hyronimus B, Le Marrec C, Sassi AH, Deschamps A (2000) Acid and bile tolerance of spore-forming lactic acid bacteria. Int J Food Microbiol 61:193–197

Ikada Y, Jamshidi K, Tsuji H, Hyon SH (1987) Stereocomplex formation between enantiomeric poly(lactides). Macromolecular 20:904–906

Inkinen S, Hakkarainen M, Albertsson A-C, Södergård A (2011) From lactic acid to poly(lactic acid) (PLA): characterization and analysis of PLA and its precursors. Biomacromolecules 12:523–532

Kanwar SS, Tewari HK, Chadha BS, Punj V, Sharma VK (1995) Lactic acid production from molasses by *Sporolactobacillus cellulosolvens*. Acta Microbiol Immunol Hung 42:331–338

Kim D-S, Sin Y, Kim D-W, Paek J, Kim RN, Jung MY, Park I-S, Kim A, Kang A, Park H-S, Choi S-H, Chang Y-H (2012) Genome sequence of the probiotic bacterium *Sporolactobacillus vineae* SL153$^T$. J Bacteriol 194:3015–3016

Kitahara K (1940) Studies on the lactic acid bacteria isolated from mashes of various kinds of cereals. J Agric Chem Soc Jpn 16:123, article in Japanese, cited by Kitahara and Suzuki, 1963

Kitahara K, Lai CL (1967) On the spore formation of *Sporolactobacillus inulinus*. J Gen Appl Microbiol 13:97–203

Kitahara K, Suzuki J (1963) *Sporolactobacillus* nov. subgen. J Gen Appl Microbiol 9:59–71

Kitahara K, Toyota T (1972) Auto-spheroplastization and cell-permeation in *Sporolactobacillus inulinus*. J Gen Appl Microbiol 18:99–107

Lacey J, Cross T (1989) Genus *Thermoactinomyces* Tsilinsky 1899, 501$^{AL}$. In: Williams ST, Sharpe ME, Holt JG (eds) Bergey's manual of systematic bacteriology, vol 4. Williams & Wilkins, Baltimore, pp 2574–2585

Lee SD, Lee DW (2009) *Scopulibacillus darangshiensis* gen. nov., sp. nov., isolated from rock. J Microbiol 47:710–715

Li W-J, Zhang Y-Q, Schumann P, Tian X-P, Zhang Y-Q, Xu L-H, Jiang C-L (2006) *Sinococcus qinghaiensis* gen. nov., sp. nov., a novel member of the order *Bacillales* from a saline soil in China. Int J Syst Evol Microbiol 56:1189–1192

Li WJ, Zhi XY, Euzéby JP (2008) Proposal of *Yaniellaceae* fam. nov., *Yaniella* gen. nov. and *Sinobaca* gen. nov. as replacements for the illegitimate prokaryotic names *Yaniaceae* Li et al. 2005, *Yania* Li et al. 2004, emend Li et al. 2005, and *Sinococcus* Li et al. 2006, respectively. Int J Syst Evol Microbiol 58:525–527

Ludwig W, Schleifer K-H, Whitman WB (2009) Family VII *Sporolactobacillaceae*. In: De Vos P, Garrity GM, Jones D, Krieg NR, Ludwig W, Rainey FA, Schleifer K-H, Whitman WB (eds) Bergey's manual of systematic bacteriology, 2nd edn, The Firmicutes. Springer, Dordrecht, p 386, Validation List N° 132 Int J Syst Evol Microbiol, 2010, 60:469–472

Miller A III, Sandine WE, Elliker PR (1970) Deoxyribonucleic acid homology in the genus *Lactobacillus*. Can J Microbiol 17:625–634

Nakayama O (1960) Bull Fac Agr Tamagawa Univ 1:73 (cited by Kitahara and Suzuki, 1963)

Nakayama O, Yanoshi M (1967) Spore-bearing lactic acid bacteria isolated from rhizosphere. I. Taxonomic studies on *Bacillus laevolacticus* nov. sp. and *Bacillus racemilacticus* nov. sp. J Gen Appl Microbiol 13:139–153

Rinderknecht H, Wilding P, Haverback BJ (1967) A new method for the determination of α-amylase. Experientia 23:805

Rychen G, Simões Nunes C (1993) Effects of a microbial probiotic (*Sporolactobacillus* P44) on postprandial porto-arterial concentrations differences of glucose, galactose and amino-nitrogen in the growing pig. Reprod Nutr Dev 33(53):1–539

Sanders ME, Morelli L, Tompkins TA (2003) Sporeformers as human probiotics: *Bacillus*, *Sporolactobacillus* and *Brevibacillus*. Compr Rev Food Sci Food Saf 2:101–110

Sawai H, Na K, Sasaki N, Mimitsuka T, Minegishi S, Henmi M, Yamada K, Shimizu S, Yonehara T (2011) Membrane-integrated fermentation system for improving the optical purity of D-lactic acid produced during continuous fermentation. Biosci Biotechnol Biochem 75:2326–2332

Shukla VB, Zhou S, Yomano LP, Shanmugam KT, Preston JF, Ingram LO (2004) Production of D(−)-lactic acid from sucrose and molasses. Biotechnol Lett 26:689–693

Skerman VBD, McGowan V, Sneath PHA (1980) Approved lists of bacterial names. Int J Syst Bacteriol 30:225–420

Stamatakis A (2006) RAxML-VI-HPC: maximum likelihood-based phylogenetic analyses with thousands of taxa and mixed models. Bioinformatics 22:2688–2690

Suzuki J, Kitahara K (1964) Base compositions of deoxyribonucleic acid in *Sporolactobacillus inulinus* and other lactic acid bacteria. J Gen Appl Microbiol 10:305–311

Suzuki T, Yamasoto K (1994) Phylogeny of spore-forming lactic acid bacteria based on 16S rRNA gene sequences. FEMS Microbiol Lett 115:13–18

Tomimura E, Zeman NW, Frankiewicz JR, Teague WM (1990) Description of *Bacillus naganoensis* sp. nov. Int J Syst Bacteriol 40:123–125

Wang L, Zhao B, Li F, Xu K, Ma C, Tao F, Li Q, Xu P (2010) Highly efficient production of D-lactate by *Sporolactobacillus* sp. CASD with simultaneous enzymatic hydrolysis of peanut meal. Appl Microbiol Biotechnol 89:1009–1017

Xu P, Ma Y, Zhao B, Qin J, Yu B, Wang L, Ma C, Yan S, Zhou S (2007) Method for producing D-lactic acid and brood-cell Lactobacillus special for the same. China Patent 200710176056

Xu TT, Bai ZZ, Wang LJ, He BF (2010) Breeding of D(−)-lactic acid high producing strain by low-energy ion implantation and preliminary analysis of related metabolism. Appl Biochem Biotechnol 160:314–321

Yanagida F, Suzuki K-I (2009) Genus I *Sporolactobacillus* Kitahara and Suzuki 1963, 69$^{AL}$. In: De Vos P, Garrity GM, Jones D, Krieg NR, Ludwig W, Rainey FA, Schleifer K-H, Whitman WB (eds) Bergey's manual of systematic bacteriology, vol 3, 2nd edn, The Firmicutes. Springer, Dordrecht, pp 386–391

Yanagida F, Suzuki KI, Kozaki M, Komagata K (1997) Proposal of *Sporolactobacillus nakayamae subsp nakayamae* sp. nov., subsp. nov., *Sporolactobacillus nakayamae* subsp. *racemicus* subsp. nov., *Sporolactobacillus terrae* sp. nov., *Sporolactobacillus kofuensis* sp. nov., and *Sporolactobacillus lactosus* sp. nov. Int J Syst Bacteriol 47:499–504

Yanagida F, Chen Y-S, Shinohara T (2005) Isolation and characterization of lactic acid bacteria from soils in vineyards. J Gen Appl Microbiol 51:313–318

Yarza P, Ludwig W, Euzéby J, Amann R, Schleifer K-H, Glöckner FO, Rosselló-Móra R (2010) Update of the All-Species Living-Tree Project based on 16S and 23S rRNA sequence analyses. Syst Appl Microbiol 33:291–299

Yu B, Su F, Wang L, Xu K, Zhao B, Xu P (2011) Draft genome sequence of *Sporolactobacillus inulinus* Strain CASD, an efficient D-lactic acid-producing bacterium with high-concentration lactate tolerance capability. J Bacteriol 193:5864–5865

Zhao B, Wang L, Li F, Hua D, Ma C, Ma Y, Xu P (2010) Kinetics of D-lactic acid production by *Sporolactobacillus* sp. strain CASD using repeated batch fermentation. Bioresour Technol 101:6499–6505

Zheng H, Gong J, Chen T, Chen X, Zhao X (2010) Strain improvement of *Sporolactobacillus inulinus* ATCC 15538 for acid tolerance and production of D-lactic acid by genome shuffling. Appl Microbiol Biotechnol 85:1541–1549

# 27 The Family *Staphylococcaceae*

*Stephen Lory*
Department of Microbiology and Immunobiology, Harvard Medical School, Boston, MA, USA

**Abstract**
The brief discussion in this chapter will cover the genera *Staphylococcus*, *Macrococcus*, *Salinicoccus*, *Jeotgalicoccus*, and *Nosocomiicoccus*.

The *Staphylococcaceae* is a family of Gram-positive bacteria that includes the genera *Staphylococcus*, *Macrococcus*, *Salinicoccus*, *Jeotgalicoccus*, and *Nosocomiicoccus* with a closely related *Jeotgalicoccus pinnipedialis*. Their phylogenetic relationship is shown in ➲ *Fig. 27.1*.

The genus *Staphylococcus* consists of 54 species (➲ *Fig. 27.2*) with a wide distribution and, frequently, in association with a variety of human and animal hosts. Microscopically, they are Gram-positive, with an appearance of a grape-like clusters, due to the formation of perpendicular division planes during cell division of individual cocci. They are facultative anaerobes, using a variety of carbohydrates for energy and as carbon sources. They can utilize respiratory or fermentative pathways; however, most strains prefer oxygen-based energy generation. In the vast majority of *Staphylococcus* species, the end product of anaerobic fermentation is lactic acid, while aerobically, carbohydrates are oxidized to acetate and carbon dioxide. Although many *Staphylococcaceae* can also oxidize amino acids, sugars such as glucose are preferred for energy generation. Growth requirements of staphylococci include over half of the 20 amino acids and several vitamins, explaining in part their high prevalence on skin or other mucosal surfaces of mammals. Staphylococci are routinely cultured on rich media (blood or chocolate agar), and when it is necessary to suppress the growth of other bacteria found in the same isolation site, sodium chloride is added to the media. Further inclusion of specific carbohydrates allows for differentiation of individual strains based on their fermentation patterns. Another diagnostic test is based on the determination of coagulase production (an enzyme capable of converting fibrinogen to fibrin), which is made only by the more virulent species: *S. aureus*, a human pathogen and two animal pathogens *S. intermedius* and *S. hyicus*.

Although occasionally causing mastitis in cattle, the major infections associated with *Staphylococcus aureus* are in human hosts (Gordon and Lowy 2008). It is a commensal organism, with a ca. 30 % carriage rates among the healthy, primarily on the skin, intranasally or in the intestinal tract. However, *S. aureus* is also a potentially dangerous pathogen capable of causing a variety of superficial or invasive diseases, as well as toxigenic infections following ingestion of contaminated food. Virulent *S. aureus* can be transmitted from person to person either by aerosol or physical contact; the same routes may also apply to nosocomial transmission. The range of diseases caused by *S. aureus* is extremely wide and they include skin and soft tissue infections such as abscesses or cellulitis, osteomyelitis, endocarditis, toxic shock syndrome, sinusitis, pneumonia, and food poisoning.

The substantial arsenal of virulence factors produced by *S. aureus* accounts for its success as pathogen including adhesins and secreted toxins; these molecules can be either degradative enzymes or specific modulators of host immune defenses (Zecconi and Scali 2013; Otto 2014). Most strains of *S. aureus* express various adhesins, facilitating their binding to the components of extracellular matrix such as collagen, fibronectin, and elastin. A number of *S. aureus* secreted proteins prevent neutrophil migration, while the expression of a surface immunoglobulin-binding protein A interferes with antibody-mediated opsonic killing. Another staphylococcal secreted protein, SCIN, and a polysaccharide capsule surrounding the bacteria block complement-mediated killing by the infected host. Secreted toxins, particularly those capable of causing membrane damage, often preferentially target immune cells including neutrophils, and these include the α-, γ-, and β-hemolysin and several leukocidins (the Panton-Valentine leukocidin, LukED, LukGH/AB). A class of secreted protein immunomodulators, called superantigen toxins, produced by *S. aureus*, are responsible for the disease toxic shock syndrome and the symptoms of staphylococcal food poisoning (Xu and McCormick 2012). These toxins act through the activation of T cells resulting in an overproduction of cytokines and systemic inflammation. A distinct group of staphylococcal proteins, the exfoliative toxins, are proteases capable of degrading cutaneous tissues and cause a severe skin disease called staphylococcal scalded skin syndrome.

Genome sequencing of *S. aureus* isolates from human and animal infections revealed a complex population structure, dominated by certain lineages of highly virulent strains (Lindsay 2014). The virulence potential of individual strains is usually determined by the acquisition of horizontally transferred genes from related species or different genera and various combinations of these genetic elements shape specific lineages. All superantigen toxin genes are transmitted by bacteriophages. During the past decade outbreaks of serious *S. aureus* infections were caused by these virulent strains that have also acquired mobile genetic determinants for the synthesis of a variant of penicillin-binding protein 2a (pBP2a) and a modified cell wall, making them resistant to the antibiotic methicillin (the so-called MRSA strains) thus eliminating the ability of this useful drug to treat serious infections.

E. Rosenberg et al. (eds.), *The Prokaryotes – Firmicutes and Tenericutes*, DOI 10.1007/978-3-642-30120-9_350,

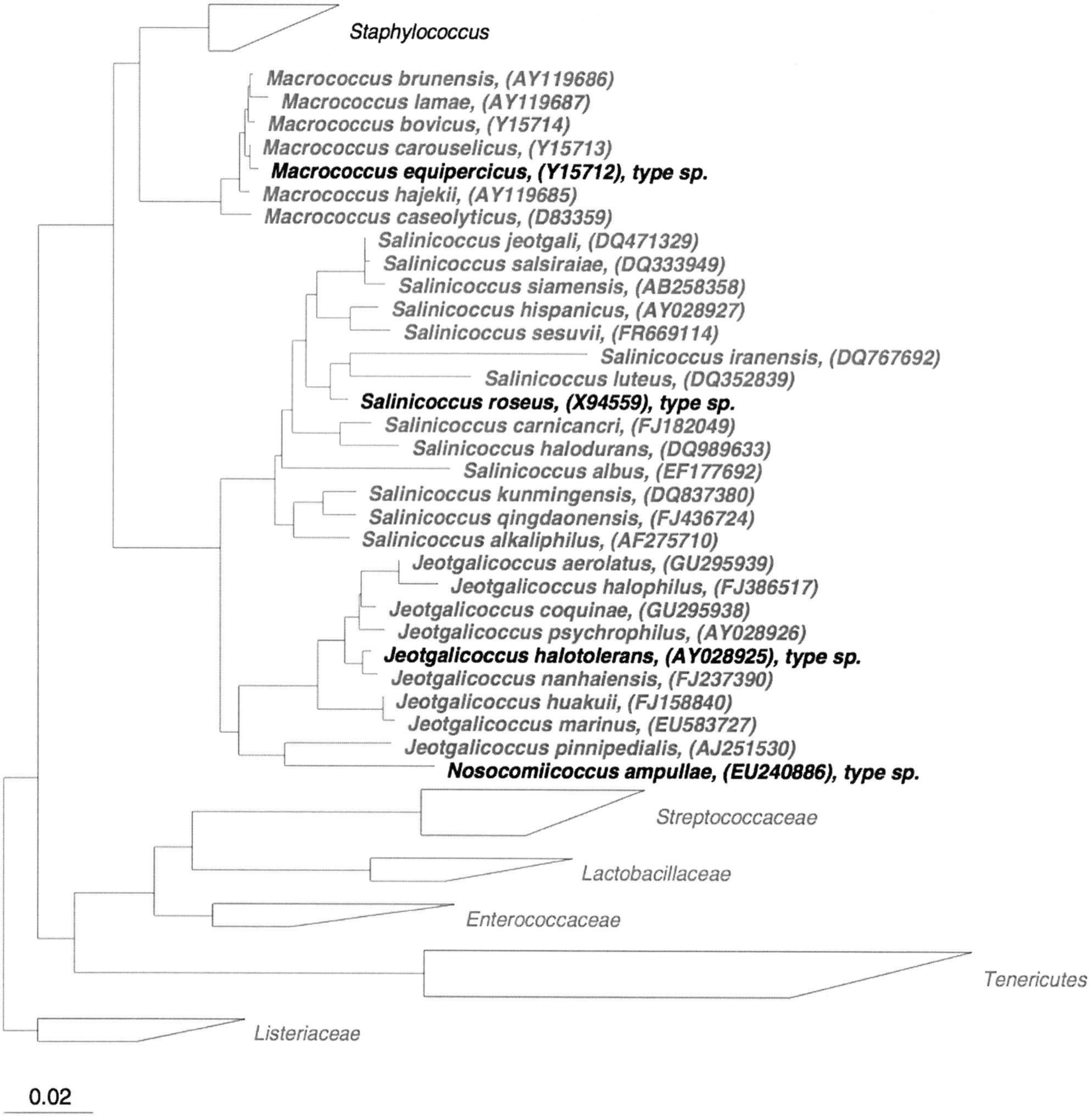


**Fig. 27.1**

**Phylogenetic reconstruction of the family *Staphylococcaceae* based on 16S rRNA and created using the neighbor-joining algorithm with the Jukes-Cantor correction. The sequence datasets and alignments were used according to the All-Species Living Tree Project (*LTP*) database (Yarza et al. 2010; http://www.arb-silva.de/projects/living-tree). The tree topology was stabilized with the use of a representative set of nearly 750 high-quality type strain sequences proportionally distributed among the different bacterial and archaeal phyla. In addition, a 40 % maximum frequency filter was applied in order to remove hypervariable positions and potentially misplaced bases from the alignment. Scale bar indicates estimated sequence divergence**

Human infections with other (coagulase-negative) members of the genus *Staphylococcus* are becoming increasingly frequent in hospital and community settings (von Eiff et al. 2002). Infections caused by *Staphylococcus epidermidis*, *Staphylococcus lugdunensis*, and *Staphylococcus haemolyticus* are responsible for a large fraction of bloodstream infections including endocarditis involving damaged heart tissues (Rogers et al. 2009). *Staphylococcus saprophyticus* is the most prevalent Gram-positive organism implicated in urinary tract infections in women. Many species can form robust biofilms on abiotic

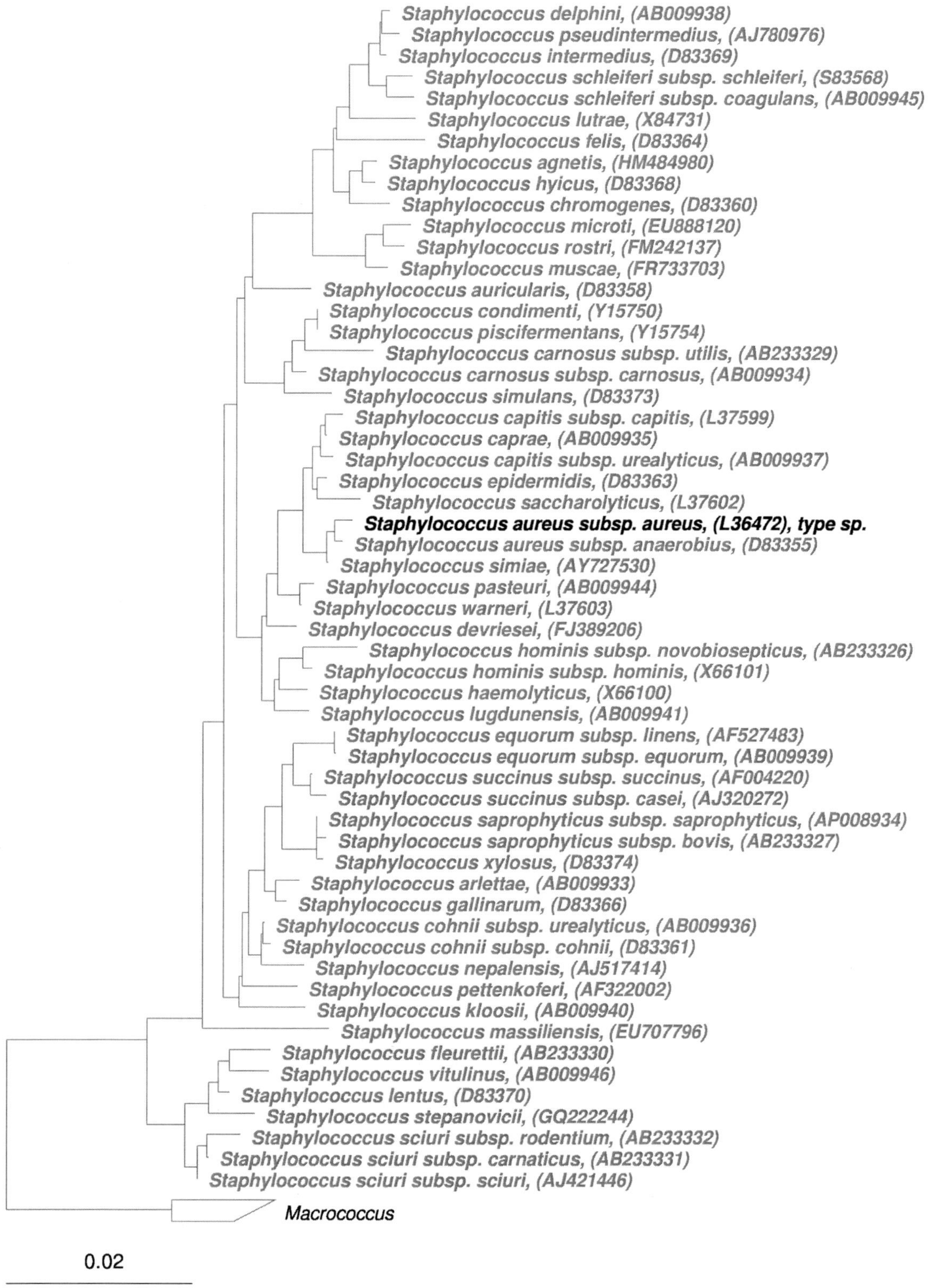


**Fig. 27.2**

Relationships of the various members of the genus *Staphylococcus*. Phylogenetic reconstruction of the genus *Staphylococcus* based on 16S rRNA and created using the neighbor-joining algorithm with the Jukes-Cantor correction. The sequence datasets and alignments were used according to the All-Species Living Tree Project (*LTP*) database (Yarza et al. 2010; http://www.arb-silva.de/projects/living-tree). The tree topology was stabilized with the use of a representative set of nearly 750 high-quality type strain sequences proportionally distributed among the different bacterial and archaeal phyla. In addition, a 40 % maximum frequency filter was applied in order to remove hypervariable positions and potentially misplaced bases from the alignment. Scale bar indicates estimated sequence divergence surfaces, and when adhering to medical devices such as intravascular catheters, they can become the source of the infecting organisms. A number of animal infections are associated with coagulase-negative staphylococci, and they can be transmitted to humans through food or by physical contact. Many strains of coagulase-negative pathogenic staphylococci, just like MRSA, also frequently carry the gene cassette encoding the variant of pBP2a providing these bacteria with a resistance mechanism to a number of ß-lactam antibiotics.

*Salinicoccus* species are halophilic organisms typically isolated from environments of moderate to high salinity, ranging from sea water and salt lakes to salt mines, salted fish, and fermented seafood. Genome sequence analysis of *Salinicoccus carnicancri* identified a large number of genes involved in osmoprotection and transport of osmolites, such as a choline-glycine betaine transporter and glycine betaine/choline-binding proteins (Hyun et al. 2013). Therefore, *S. carnicancri* and likely the other halophilic *Staphylococcaceae* have evolved a special capacity to modulate their intracellular salinity, allowing them to thrive in locales containing high concentrations of salt, usually not accessible to the majority of other environmental organisms.

All seven species of *Macrococcus* are frequently isolated from animal sources; however, there has not been a documented case of human infection caused by the members of this genus. In several species, the genes for the alternative penicillin-binding protein responsible for ß-lactam resistance in MRSA was found on a transposon, suggesting that some of the bacteria belonging to the *Macrococcus* genus were the likely source of this gene, which was acquired by pathogenic staphylococci in response to methicillin treatment of infections.

*Jeotgalicoccus* species are distributed in a wide range of environments. They are frequently found in fermented food and in association with animals. They are halotolerant or moderately halophilic facultative anaerobic organisms growing as individual cocci or in pairs. Different *Jeotgalicoccus* derive energy primarily from fermentation of different sugars. There has not been any disease or infection attributed to the organisms in this genus.

There are two species in the genus *Nosocomiicoccus: Nosocomiicoccus massiliensis* and *Nosocomiicoccus ampulae,* isolated from the feces of a patient with AIDS and from the surface of saline bottles in a hospital, respectively (Mishra et al. 2013, Alves et al. 2008). They are both aerobic cocci forming small clusters, grow on rich media with *N. ampulae* showing halophilic properties. In spite of their isolation in a clinical setting, there is no evidence that they are responsible for human infections. Within the *Staphylococcaceae*, *Nosocomiicoccus* species are most closely related to those classified as *Jeotgalicoccus*, with a very strong 16SrRNA sequence similarity to *Jeotgalicoccus pinnipedialis*.

## References

Alves M, Nogueira C, de Magalhães-Sant'ana A, Chung AP, Morais PV, da Costa MS (2008) *Nosocomiicoccus ampullae* gen. nov., sp. nov., isolated from the surface of bottles of saline solution used in wound cleansing. Int J Syst Evol Microbiol. Dec;58(Pt 12):2939–2944

Gordon RJ, Lowy FD (2008) Pathogenesis of methicillin-resistant *Staphylococcus aureus* infection. Clin Infect Dis 46(Suppl 5):S350–S359

Hyun DW, Whon TW, Cho YJ, Chun J, Kim MS, Jung MJ, Shin NR, Kim JY, Kim PS, Yun JH, Lee J, Oh SJ, Bae JW (2013) Genome sequence of the moderately halophilic bacterium *Salinicoccus carnicancri* type strain Crm(T) (=DSM 23852(T)). Stand Genomic Sci 8(2):255–263

Lindsay JA (2014) *Staphylococcus aureus* genomics and the impact of horizontal gene transfer. Int J Med Microbiol 304(2):103–109

Mishra AK, Edouard S, Dangui NP, Lagier JC, Caputo A, Blanc-Tailleur C, Ravaux I, Raoult D, Fournier PE (2013) Non-contiguous finished genome sequence and description of *Nosocomiicoccus massiliensis* sp. nov. Stand Genomic Sci 9(1):205–219

Otto M (2014) *Staphylococcus aureus* toxins. Curr Opin Microbiol 17C:32–37

Rogers KL, Fey PD, Rupp ME (2009) Coagulase-negative staphylococcal infections. Infect Dis Clin North Am 23(1):73–98

von Eiff C, Peters G, Heilmann C (2002) Pathogenesis of infections due to coagulase-negative staphylococci. Lancet Infect Dis 2(11):677–685

Xu SX, McCormick JK (2012) Staphylococcal superantigens in colonization and disease. Front Cell Infect Microbiol 2:52

Yarza P, Ludwig W, Euzeby J, Amann R, Schleifer KH, Glöckner FO, Rossello-Mora R (2010) Update of the All-Species Living Tree Project based on 16S and 23S rRNA sequence analyses. Syst Appl Microbiol 33:291–299. doi:10.1016/j.syapm.2010.08.001

Zecconi A, Scali F (2013) *Staphylococcus aureus* virulence factors in evasion from innate immune defenses in human and animal diseases. Immunol Lett 150(1–2):12–22

# 28 The Family *Streptococcaceae*

*Stephen Lory*
Department of Microbiology and Immunobiology, Harvard Medical School, Boston, MA, USA

**Abstract**
This chapter discusses briefly the family *Streptococcaceae* and its three genera *Streptococcus*, *Lactococcus*, and *Lactovum*.

The order *Lactobacillales* includes the family *Streptococcaceae* with three genera *Streptococcus*, *Lactococcus*, and *Lactovum*, in addition to closely related species currently referred to as the "*Lactococcus piscium* group" (➋ *Fig. 28.1*). Their ecological distribution is diverse; in addition to thriving in many natural habitats, it includes human and animal hosts. They are gram-positive cocci, arranged in linear chains ranging from two to dozens.

The genus *Streptococcus* encompasses 78 recognized species (➋ *Fig. 28.2*), and a number of them are frequent human and animal commensal colonizers. However, the same species are also capable of causing serious infections in their hosts. The various *Streptococcus* species are sometimes referred to by their serological characteristics (the Lancefield grouping) based on a major polysaccharide surface antigen. *Streptococcus pyogenes* is a group A Streptococcus, *Streptococcus agalactiae* is group B, and *Streptococcus bovis*, group D; these are common names still used in scientific and medical literature, while group O (*Streptococcus pneumoniae*) is rarely used.

Streptococci are facultative anaerobes and require rich, extensively supplemented media for growth. Depending on the availability, most species express transporters for a relatively large variety of sugars, which are the source of energy and carbon. Energy is generated by fermentation of carbohydrates to pyruvate via glycolysis, with lactic acid as the primary and organic acids and ethanol as the secondary end products. Some streptococci (i.e., *Streptococcus agalactiae*) possess cytochromes and can also generate energy by oxidative phosphorylation. Streptococcal species lack biosynthetic enzymes for many essential amino acids and most vitamin and cofactors; these need to be also acquired from the environment to support growth. In the laboratory, they are usually grown on solid media containing sheep blood in an atmosphere enriched for carbon dioxide.

The most significant species relevant to a disease are the human host-restricted *S. pyogenes* and *S. pneumoniae*. The carriage rate of both of these organisms in healthy individuals, without any symptoms of disease, is relatively high. Other streptococci can cause infections in both animals and humans. For example, *S. agalactiae* is a commensal colonizer of the intestinal track of animals and is occasionally responsible for bovine mastitis, but it can also be the causative agent of human neonatal pneumonia, septicemia, and meningitis.

Infections by *S. pyogenes* can lead to a wide range of diseases depending on a specific strain, route of entry, existing host factors, and effectiveness of rapid antibiotic therapy (Cunningham 2008). *S. pyogenes* can cause acute primary infections and a number of postinfection sequelae. Primary infections include superficial skin and soft tissue infections (potentially leading to a more serious necrotizing fasciitis), upper and lower respiratory infections such as pharyngitis and pneumonia, as well as invasive diseases including sepsis, toxic shock syndrome, and endocarditis. Complications following *S. pyogenes* infections, particularly if the antibiotic intervention was not prompt or effective, include glomerulonephritis, acute rheumatic fever, and scarlet fever.

Diagnosis of *S. pyogenes* infections is based on the recognition of likely symptoms in conjunction with laboratory tests. These include culturing of the organisms from an infected site (i.e., a throat swab) on media containing sheep blood agar, with the appearance of clearing around colonies (ß-hemolysis). While other streptococci can be ß-hemolytic, a bacitracin sensitivity test (inclusion of a bacitracin impregnated disk on the blood agar plate) is used for their differentiation, since *S. pyogenes* is sensitive to the antibiotic while other ß-hemolytic bacteria are not. Gram staining of *S. pyogenes* grown on plates demonstrates the appearance of very long chains.

Although a number of different environmental factors contribute to the incidence and severity of infections, recent molecular analyses of *S. pyogenes* genomes from different outbreaks and their comparison to commensal isolates point towards the evolution of related hypervirulent, highly invasive strains. These are characterized by the presence of particular sequence variants of a surface protein (the M-protein, encoded by the *emm* gene). While the M-protein sequence is only an epidemiological marker, whole genome sequence analyses of these *S. pyogenes* clones identified genetic determinants most likely responsible for the hypervirulent, invasive phenotypes, namely, lysogenization by bacteriophage carrying specific toxin genes and mutations in transcriptional regulators of virulence gene expression (Cole et al. 2011).

*S. pneumoniae* (also referred to as the pneumococcus) is very likely the most important human bacterial pathogen based on the estimation of worldwide morbidity and mortality from infections caused by this organism (Henriques-Normark and Tuomanen 2013). Infections by *S. pneumoniae* can lead to otitis media, pneumonia, sinusitis, bacteremia, and meningitis. The bacteria can be visualized directly by Gram staining of respiratory secretions, where they have a characteristic lancet-shaped diplococcus appearance and are near or within an unusually large number of neutrophils. Subsequent culture on blood agar in a carbon dioxide atmosphere and inclusion of a disk containing the chemical optochin are used to differentiate pneumococci from different α-hemolytic streptococci

E. Rosenberg et al. (eds.), *The Prokaryotes – Firmicutes and Tenericutes*, DOI 10.1007/978-3-642-30120-9_345,

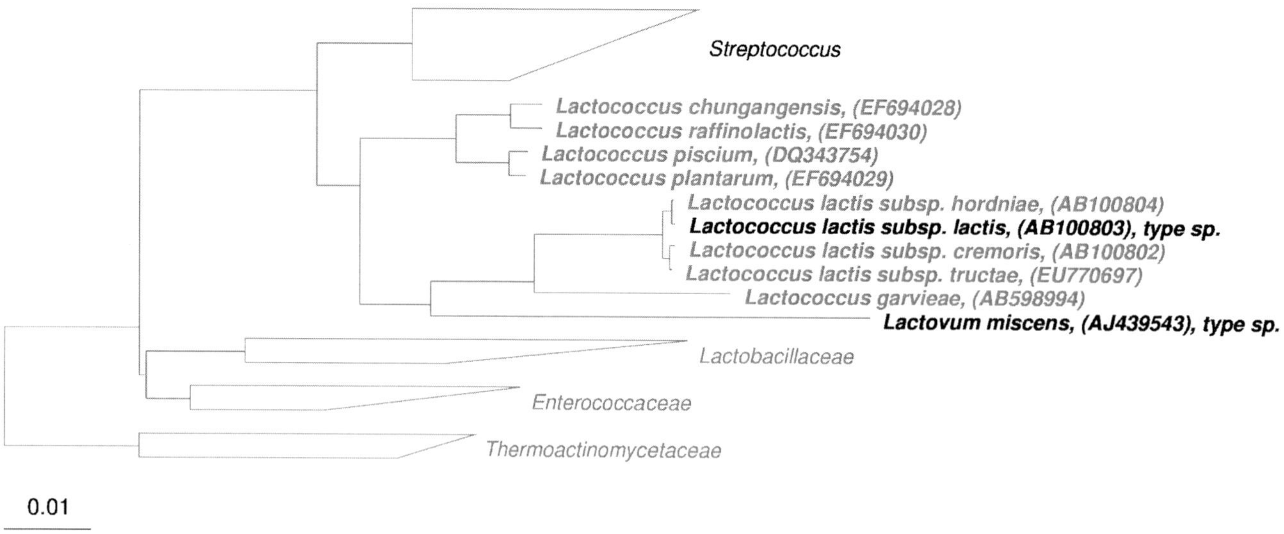


**Fig. 28.1**
**Neighbor-joining tree of the *Streptococcaceae* genera. Phylogenetic reconstruction of the family *Streptococcaceae* based on 16S rRNA and created using the neighbor-joining algorithm with the Jukes-Cantor correction. The sequence datasets and alignments were used according to the All-Species Living Tree Project (LTP) database (Yarza et al. 2010; http://www.arb-silva.de/projects/living-tree). The tree topology was stabilized with the use of a representative set of nearly 750 high-quality type strain sequences proportionally distributed among the different bacterial and archaeal phyla. In addition, a 40 % maximum frequency filter was applied in order to remove hypervariable positions and potentially misplaced bases from the alignment. Scale bar indicates estimated sequence divergence**

(an area of incomplete clearing around the colonies), since *S. pneumoniae* are resistant, while the others are not.

Finally, Gram staining is used to confirm the diagnosis of pneumococcal infection. Specialized staining of bacteria can further reveal the presence of a characteristic capsule, and type-specific anti-capsular antibodies can be used during immunostaining for the identification of specific serotypes.

Pneumococci are naturally competent for DNA uptake, and genetic transformation plays an important role in their adaptability to the host immune responses and to antibiotic therapy. Whole genome sequencing of multiple strains showed extensive sequence polymorphisms, indicative of recombination between the bacterial chromosome and exogenous genetic material, presumably DNA acquired by transformation. DNA uptake and recombination are important for the generation of a large number of capsular polysaccharide types allowing the bacteria to avoid the host immune system which may include protective antibodies generated through vaccination (Brueggemann et al. 2007). Recombination between DNA segments from resistant and susceptible strains (including closely related commensal streptococci) also provides the means of escaping antibiotic killing. One common mechanism is based on the introduction of specific resistance-determining mutations into the genes of susceptible strains encoding penicillin-binding proteins, the targets of antibiotic penicillin, reducing or eliminating its ability to kill the bacteria (Hakenbeck et al. 2012).

The oral cavity of healthy individuals has an extensive bacterial community including over 25 species of streptococci. There is strong evidence that their presence can be protective against dental diseases, and at least one species (*Streptococcus salivarius*) has been tested as a probiotic (Burton et al. 2011). However, perturbation of the oral microbiome can lead to tooth decay involving *Streptococcus mutans*, and a number of other *Streptococcus* species have been implicated in endodontic infections (Anderson et al. 2013). Following dental procedures, oral streptococci can cause infective endocarditis, an infection and inflammation of the inner surface of the heart. Species associated with this serious infection include *Streptococcus gordonii*, as well as *Streptococcus mutans* and *Streptococcus mitis*.

Other pathogenic members of the *Streptococcus* genus show a more restricted animal host with a disease-causing potential, ranging from various house pets (*Streptococcus canis*), pigs (*Streptococcus suis*) to cows (*Streptococcus bovis*), dogs and horses (various subspecies of *Streptococcus equi*). It is not uncommon that contact with infected animals leads to human infections.

Streptococci play an important role in facilitating fermentation of milk products during the production of cheeses, yogurt, sour cream, and buttermilk. By a complex mixed culture fermentation process, *Streptococcus salivarius* subsp. *thermophilus*, together with *Lactobacillus delbrueckii subps. bulgaricus or Lactobacillus acidophilus*, converts milk to yogurt and Swiss or mozzarella cheese, by coagulating casein and breakdown of lactose and galactose, respectively (Sieuwerts et al. 2008).

The *Lactococcus piscium* group includes *Lactococcus piscium* a pathogen of salmonoid fish. Interestingly, *L. piscium* growth in cooked seafood has been shown to have protective properties against bacterial food spoilage and contamination by pathogenic food-borne organism such as *Listeria monocytogenes* and *Vibrio*

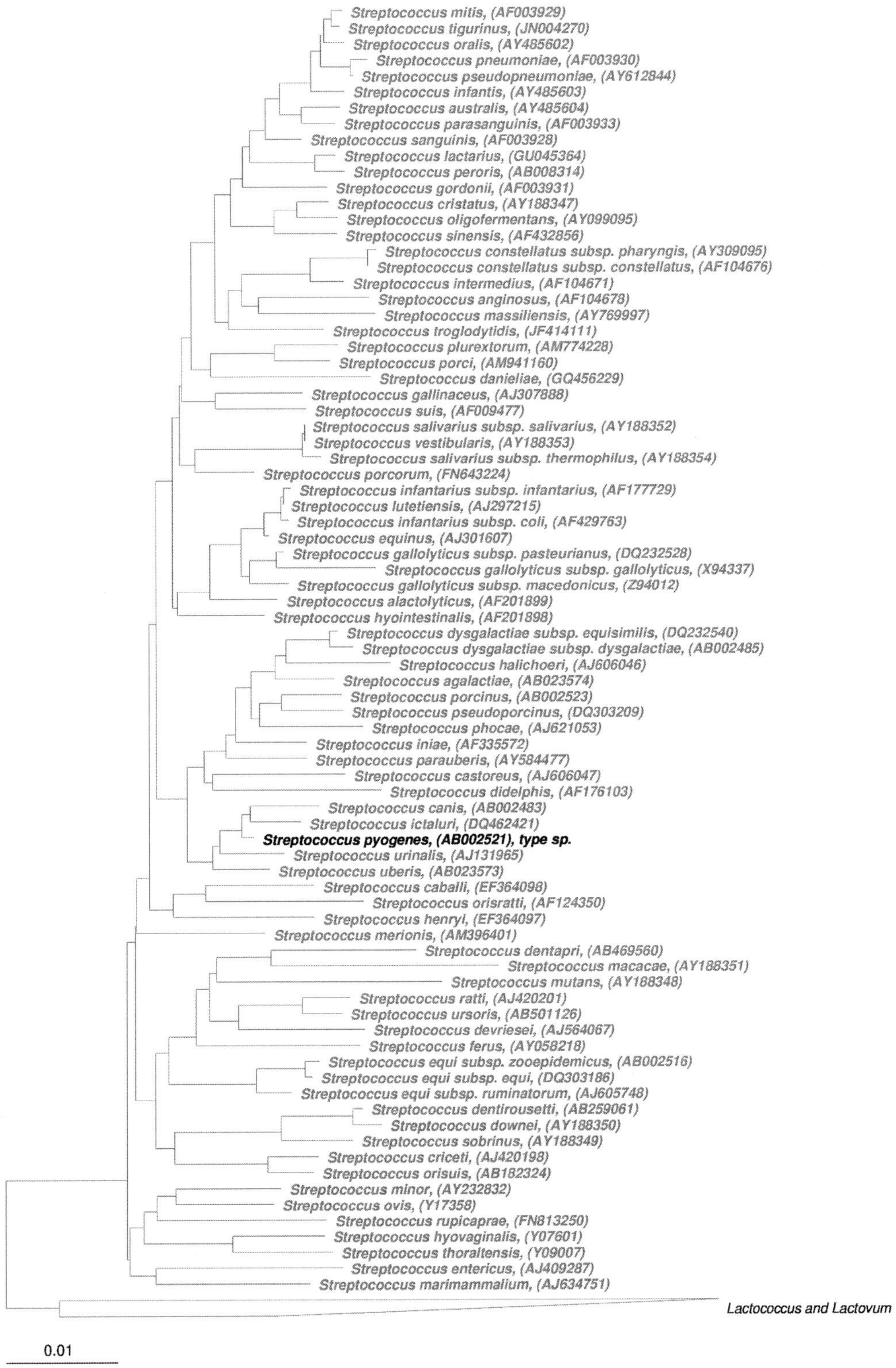


**Fig. 28.2**

**Phylogeny of *Streptococcus* species.** Phylogenetic reconstruction of the genus *Streptococcus* based on 16S rRNA and created using the neighbor-joining algorithm with the Jukes-Cantor correction. The sequence datasets and alignments were used according to the All-Species Living Tree Project (LTP) database (Yarza et al. 2010; http://www.arb-silva.de/projects/living-tree). The tree topology was stabilized with the use of a representative set of nearly 750 high-quality type strain sequences proportionally distributed among the different bacterial and archaeal phyla. In addition, a 40 % maximum frequency filter was applied in order to remove hypervariable positions and potentially misplaced bases from the alignment. Scale bar indicates estimated sequence divergence *cholerae* (Matamoros et al. 2009). Although found occasionally in the human intestinal microbiota, there have been no infections, other than those of fish, reported to date. Other members of this group include *Lactococcus plantarum*, *Lactococcus raffinolactis*, and *Lactococcus chungangensis*. Their precise ecological distribution is unknown, and they are frequently recovered from the intestinal tracks of domesticated animals, from unpasteurized milk products or spoiled food.

The remaining members of the *Streptococcaceae* are either free living in the environment (mainly on plant surfaces) or are associated with the food supply. The largest number of species is found within the genus *Lactococcus*. Various subspecies of *L. lactis* have been extensively exploited by the dairy industry for the production of buttermilk, sour cream, and a variety of cheeses. *L. lactis* subsp. *lactis* is used in the manufacture of hard cheeses, while *L. lactis* subsp. *cremoris* is the preferred organism for the production of soft cheeses. In addition to generation of lactic acid from sugars, specific strains of *L. lactis* hydrolyze casein into specific peptides and amino acids, which provide individual cheeses their unique flavor (Smit et al. 2005).

Because of their lack of virulence and tolerance of low pH, *L. lactis* are now considered as vehicles for mucosal delivery of recombinant human therapeutics and vaccines via an oral route (Bahey-El-Din et al. 2010). Moreover, certain strains of *L. lactis* secrete a potent lantibiotic, a bacteriocin nisin with a broad antimicrobial activity; this property adds to their value as a food preservative, and they have been included as key microbial components of probiotic formulations (de Juarez et al. 2009).

Zoonotic infections by a number of *Lactococcus* species can occur sporadically. For example, *Lactococcus garvieae* is responsible for the disease lactococcosis-afflicting fish and shellfish and has been associated with bovine mastitis. It can be readily isolated from rivers, sewage waters, vegetables, and meat and dairy products. It is also an opportunistic human pathogen capable of causing superficial skin infections as well septicemia and infectious endocarditis either in immunocompromised patients or healthy individuals who had contact with infected fish. Additionally, consumption of contaminated raw seafood can lead to *L. garviea* gastrointestinal infections (Wilbring et al. 2011). Another aquatic disease streptococcosis is caused by *Streptococcus iniae*, resulting in a high mortality (over 50 %) in infected fish, and can be occasionally transmitted to humans handling infected fish through skin abrasions where the infection may lead to an invasive disease (Weinstein et al. 1997).

*Lactovum miscens* is the only species in the genus *Lactovum* (Matthies et al. 2004). It is a classical lactic acid producer, converting a number of different hexose to lactate, formate, acetate, and ethanol. They are aerotolerant and grow well at reduced temperature and pH.

## References


Anderson AC, Al-Ahmad A, Elamin F, Jonas D, Mirghani Y, Schilhabel M, Karygianni L, Hellwig E, Rehman A (2013) Comparison of the bacterial composition and structure in symptomatic and asymptomatic endodontic infections associated with root-filled teeth using pyrosequencing. PLoS One 8(12):e84960

Bahey-El-Din M, Gahan CG, Griffin BT (2010) *Lactococcus lactis* as a cell factory for delivery of therapeutic proteins. Curr Gene Ther 10(1):34–45

Brueggemann AB, Pai R, Crook DW, Beall B (2007) Vaccine escape recombinants emerge after pneumococcal vaccination in the United States. PLoS Pathog 3(11):e168

Burton JP, Wescombe PA, Cadieux PA, Tagg JR (2011) Beneficial microbes for the oral cavity: time to harness the oral streptococci? Benef Microbes 2(2):93–101

Cole JN, Barnett TC, Nizet V, Walker MJ (2011) Molecular insight into invasive group A streptococcal disease. Nat Rev Microbiol 9(10):724–736

Cunningham MW (2008) Pathogenesis of group A streptococcal infections and their sequelae. Adv Exp Med Biol 609:29–42

de Juarez LJ, Jozala AF, Mazzola PG, Penna TCV (2009) Nisin biotechnological production and application: a review. Trends Food Sci Technol 20:146–154

Hakenbeck R, Brückner R, Denapaite D, Maurer P (2012) Molecular mechanisms of β-lactam resistance in *Streptococcus pneumoniae*. Future Microbiol 7(3):395–410

Henriques-Normark B, Tuomanen EI (2013) The pneumococcus: epidemiology, microbiology, and pathogenesis. Cold Spring Harb Perspect Med 3(7): a010215

Matamoros S, Leroi F, Cardinal M, Gigout F, Kasbi Chadli F, Cornet J, Prévost H, Pilett MF (2009) Psychrotrophic lactic acid bacteria used to improve the safety and quality of vacuum-packaged cooked and peeled tropical shrimp and cold-smoked salmon. J Food Prot 72(2):365–374

Matthies C, Gössner A, Acker G, Schramm A, Drake HL (2004) *Lactovum miscens* gen. nov., sp. nov., an aerotolerant, psychrotolerant, mixed-fermentative anaerobe from acidic forest soil. Res Microbiol 155(10):847–854

Sieuwerts S, de Bok FAM, Hugenholtz J, van Hylckama Vlieg JET (2008) Unraveling microbial interactions in food fermentations: from classical to genomics approaches. Appl Environ Microbiol 74:4997–5007

Smit G, Smit BA, Engels WJ (2005) Flavour formation by lactic acid bacteria and biochemical flavour profiling of cheese products. FEMS Microbiol Rev 29(3):591–610

Weinstein MR, Litt M, Kertesz DA, Wyper P, Rose D, Coulter M, McGeer A, Facklam R, Ostach C, Willey BM, Borczyk A, Low DE (1997) Invasive infections due to a fish pathogen, *Streptococcus iniae*. S. iniae Study Group. N Engl J Med 337(9):589–594

Wilbring M, Alexiou K, Reichenspurner H, Matschke K, Tugtekin SM (2011) *Lactococcus garvieae* causing zoonotic prosthetic valve endocarditis. Clin Res Cardiol 100:545–546

Yarza P, Ludwig W, Euzeby J, Amann R, Schleifer KH, Glöckner FO, Rossello-Mora R (2010) Update of the All-Species Living Tree Project based on 16S and 23S rRNA sequence analyses. Syst Appl Microbiol 33:291–299. doi:10.1016/j.syapm.2010.08.001

# 29 The Family *Syntrophomonadaceae*

*Bernhard Schink*[1] · *Raúl Muñoz*[2]
[1]Department of Biology, University of Konstanz, Konstanz, Germany
[2]Marine Microbiology Group, Department of Ecology and Marine Resources, Institut Mediterráni d'Estudis Avancats (CSIC-UIB), Esporles, Illes Balears, Spain



**Abstract**
The family *Syntrophomonadaceae* comprises the genera *Syntrophomonas, Pelospora, Syntrophothermus*, and *Thermosyntropha*. All these bacteria are strictly anaerobic and depend on reducing conditions for growth. They are Gram-positive with low DNA content, but in most cases the murein layer is thin and an outer membrane appears, resembling the cell wall architecture of Gram-negative bacteria. Also in Gram-staining, these bacteria mostly behave Gram-negative. Except for *Pelospora*, all members of this family degrade fatty acids of four carbon atoms or more by beta oxidation, in close association with hydrogen- or formate-utilizing partner organisms, and depend on this association for thermodynamic reasons. Most representatives of this family can be grown in pure culture with crotonate which is dismutated to acetate and butyrate. *Pelospora* sp. grows by decarboxylation of glutarate or succinate.

## Taxonomy, Historical and Current

Syn.tro.pho.mo.na.da.ce'ae. Gr. prefix syn together, Gr. v. trepho, to feed; Gr. n. monas, a unit; N.L. ending –aceae, ending to denote a family, N.L. fem. pl. n. *Syntrophomonadaceae* the *Syntrophomonas* family.

The family *Syntrophomonadaceae* includes organisms that oxidize monocarboxylic acids with 4–18 carbons syntrophically, and are unable to use alternate electron acceptors such as sulfate, thiosulfate, nitrate, Fe(III), fumarate, or dimethylsulfoxide. The *Syntrophomonadaceae* are phylogenetically part of the phylum of Gram-positive bacteria with low DNA G+C – content. Although many of the members of this family have cell walls typical of Gram-positive bacteria, many representatives have cell walls typical of Gram-negative bacteria, with the outer membrane of *Syntrophomonas* confirmed by electron microscopy. The division between Gram-negative and Gram-positive microbes has historically been considered an indicator of the deepest taxonomic separations (Buchanan and Gibbons 1974; Gibbons and Murray 1978), so it is unusual to find members of the same family containing both of these cell wall structures. Another example of this can be found in the family *Acidaminococcaceae*, also a family of low-G+C Gram-positive bacteria that contains some genera which have ultrastructures typical of Gram-negative bacteria.

Sequence analysis of the 16S rRNA genes (➲ *Fig. 29.1*) indicates that the family *Syntrophomonadaceae* is monophyletic and clearly separated from other families such as the *Ruminococcaceae*. The type genus *Syntrophomonas* with the type species *S. wolfei* is clearly separated from the genera *Thermosyntropha* and *Syntrophothermus*. Only the genus *Pelospora* is closely related to *Syntrophomonas* species, e. g. *S. bryantii, S. palmitatica, S. sapovorans*, or *S. curvata*. The only species of *Pelospora* described so far, *P. glutarica*, differs substantially from all other members of the *Syntrophomonadaceae*, especially with respect to its metabolism (see below) but appears to be closely related on the basis of 16S rRNA sequence comparison; the sequence similarity is closest to *S. curvata* (91.77 %). The genus *Syntrophospora* was created in 1990 (Zhao et al. 1990) in order to separate the spore-forming syntrophic fatty acid oxidizer *S. bryantii* (formerly *Clostridium bryantii*; Stieb and Schink 1985) from the (so far) non-sporeforming *Syntrophomonas* representatives. However, after spore formation was discovered also in several representatives of the *Syntrophomonas* genus (see below) and the 16S rRNA sequence similarity between *Syntrophospora bryantii* and other *Syntrophomonas* species turned out to be very high it appeared no longer justified to maintain *Syntrophospora* as a genus separate from *Syntrophomonas* (Wu et al. 2006b).

E. Rosenberg et al. (eds.), *The Prokaryotes – Firmicutes and Tenericutes*, DOI 10.1007/978-3-642-30120-9_365,

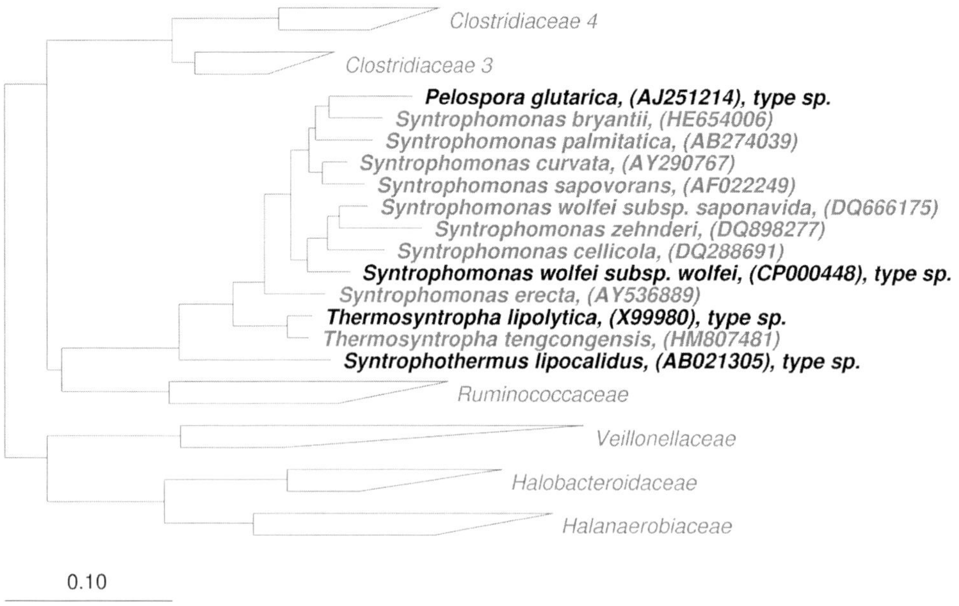


**Fig. 29.1**

**Phylogenetic reconstruction of the family *Syntrophomonadaceae* based on 16S rRNA and created using the maximum likelihood algorithm RAxML (Stamatakis 2006). The sequence datasets and alignments were used according to the All-Species Living Tree Project (LTP) database (Yarza et al. 2010; http://www.arb-silva.de/projects/living-tree). Representative sequences from closely related taxa were used as outgroups. In addition, a 40 % maximum frequency filter was applied in order to remove hypervariable positions and potentially misplaced bases from the alignment. Scale bar indicates estimated sequence divergence**

## Genera

The descriptions below include some of the most important and distinctive characteristics of the genera, and more detailed descriptions may be found in the original citations which are included in the footnotes to ➋ *Table 29.1*. Whenever possible, the descriptions of the genera summarize the properties of all described species as well as the type species. The description of many of the species of this family is based on a single strain. Additionally, caution must be taken in the evaluation of results from different laboratories because many of the growth descriptions depend greatly upon experimental conditions. In particular, variation in growth optima may occur depending on whether growth is measured by turbidity, cell counts, methane formation, growth rate, or growth yield.

### *Syntrophomonas*

Members of the genus *Syntrophomonas* contain Gram-negatively staining weakly motile, rod-shaped cells. The species and subspecies of *Syntrophomonas* are differentiated from each other on the basis of their substrate utilization patterns (➋ *Table 29.1*). *Syntrophomonas wolfei* comprizes two subspecies, one of which (*S. wolfei* subsp. *wolfei*) syntrophically degrades normal saturated fatty acids with 4–8 carbons and isoheptanoate; whereas *S. wolfei* subsp. *saponavida* syntrophically degrades normal saturated fatty acids with 4–18 carbons and the iso-fatty acids isoheptanoate and longer. *S. wolfei* cells possess 2–8 flagella that are laterally inserted in a linear fashion on the concave side of the cell. Among the more recently isolated species, *S. zehnderi* (Sousa et al. 2007) and *S. palmitatica* (Hatamoto et al. 2007) use fatty acids of 4–18 carbon atoms, whereas *S. erecta* (Zhang et al. 2005), *S. curvata* (Zhang et al. 2004), *S. cellicola* (Wu et al. 2006b) and *S. bryantii* use only short-chain fatty acids (4–9 carbon atoms). Even-numbered fatty acids are fermented to acetate only, odd-numbered fatty acids to acetate plus propionate (➋ *Table 29.2*). Spore formation was observed with *S. erecta, S. cellicola, S. zehnderi*, and *S. bryantii*; whether the other species are definitively unable to form spores remains still to be examined. Further subspecies have been described (*S. wolfei* subsp. *methylbutyratica, S. erecta* subsp. *sporosyntropha*) which differ only marginally from the described species (Wu et al. 2006a, 2007). All species and subspecies of *Syntrophomonas* except for *S. zehnderi* grow in pure culture with crotonate and are stimulated by the addition of B vitamins, amino acids, or rumen fluid. Most rapid growth is observed at 35–37 °C and at near neutral pH. The G+C content of the DNA is in the range of 45–49 %.

### *Pelospora*

Only one species, the type species *Pelospora glutarica*, has been described. The ability to grow on dicarboxylic acids such as

**Table 29.1**
**Phenotypic characteristics of genera in the family *Syntrophomonadaceae***

| Characteristic | *Pelospora* | *Syntrophomonas* | *Syntrophothemus* | *Thermosyntropha* |
|---|---|---|---|---|
| **Cytology** | | | | |
| Cell shape | Rod | Rod | Rod | Rod |
| Cell size (μm) | 0.8 × 4.5–6.5 | 0.4–0.7 × 2.0–3.7 | 0.4–0.5 × 2.0–4.0 | 0.3–0.4 × 3.0–3.5 |
| Gram stain | – | – | – | – |
| Flagella | + | + | + | – |
| Spore formation | + | +/– | – | – |
| **Substrates utilized** | | | | |
| **In pure culture** | | | | |
| Pyruvate | – | – | – | +[a] |
| Yeast extract | – | – | – | + |
| Tryptone | – | – | – | + |
| Casamino acids | – | – | – | + |
| Crotonate | – | +/–[b] | + | + |
| Succinate | + | – | – | – |
| Glutarate | + | – | – | – |
| **In coculture w/H2-using bacterium:** | | | | |
| Triacylglycerides | – | – | – | + |
| Propionate | – | – | – | – |
| Isobutyrate | – | – | + | – |
| C4–C10 | – | + | + | + |
| C11–C18 | – | +/–[c] | – | + |
| Elaidate | – | + | ND | ND |
| Isovalerate | – | – | – | – |
| Isoheptanoate | – | + | – | ND |
| Oleate | – | +/–[d] | – | + |
| Linoleate | – | +/–[e] | – | + |
| Olive oil | – | ND | ND | + |
| **Growth requirements** | | | | |
| Organic growth factors | Rumen fluid | PABA + B-vitamins[f] | None | Yeast extract |
| **Conditions supporting most rapid growth:** | | | | |
| Temperature (°C) | 37 | 37–40 | 55 | 60–66 |
| pH | 7.1–8.2 | 5.0–7.0 | 6.5–7.0 | 8.1–8.9 |

Data compiled from: Beaty and McInerney (1990); Lorowitz et al. (1989); Matthies et al. (2000); McInerney et al. (1979, 1981); Roy et al. (1986); Sekiguchi et al. (2000); Stieb and Schink (1985); Svetlitshnyi et al. (1996), and Zhao et al. (1990, 1993)
Abbreviations: + present in all species, – absent in all species, *ND* not determined, *PABA* p-aminobenzoic acid
[a]This genus weakly degrades pyruvate in pure culture
[b]Most species within this genus can grow in pure culture on crotonate, however, *Syntrophomonas sapovorans* cannot
[c]Some species in this genus can only degrade fatty acids up to C12
[d]Some species in this genus cannot degrade oleate in syntrophic co-cultures
[e]Some species in this genus cannot degrade linoleate in syntrophic co-cultures
[f]The B-vitamins include biotin, thiamine, cyanocobalamin, and lipoic acid

glutarate, methylsuccinate, and succinate, without syntrophic interactions is the most distinctive phenotypic feature of *Pelospora*. No other genus in the family metabolizes such substrates either alone or in syntrophic associations. The product of glutarate fermentation is a mix of butyrate and isobutyrate; succinate fermentation leads to propionate as sole product. Isomerization of butyrate and isobutyrate is a feature that this bacterium shares with *Syntrophothermus lipocalidus* (see below). Cells are rod-shaped and stain Gram-negative, are motile by one subpolar flagellum, and form

**Table 29.2**
**The major reactions involved in the anaerobic degradation of mono- and dicarboxylic acids by members of the family *Syntrophomonadaceae***

| Reactions | Free energy yield $\Delta G_0'$ (kJ/mol) |
|---|---|
| **Some reactions of $H_2$/formate-using bacteria** | |
| Methanogens | |
| $4H_2 + HCO_3^- + H^+ \rightarrow CH_4 + 3H_2O$ | −135.6 |
| $4HCO_2^- + H_2O + H^+ \rightarrow CH_4 + 3HCO_3^-$ | −130.4 |
| Sulfate-reducing bacteria | |
| $4H_2 + SO_4^{2-} + H^+ \rightarrow HS^- + 4H_2O$ | −151.0 |
| **Some reactions of members in the family *Syntrophomonadaceae*:** | |
| Without $H_2$/formate-using bacterium | |
| $CH_3CH_2CH_2COO^- + 2H_2O \rightarrow 2CH_3COO^- + H^+ + 2H_2$ | +48.1 |
| $CH_3CH_2CH_2CH_2COO^- + 2H_2O \rightarrow CH_3COO^- + CH_3CH_2COO^- + H^+ + 2H_2$ | +48.1 |
| $2CH_3HC{=}CH{-}COO^- + 2H_2O \rightarrow 2CH_3COO^- + CH_3CH_2CH_2COO^- + H^+$ | −101.9[a] |
| $^-OOCCH_2CH_2COO^- + H_2O \rightarrow CH_3CH_2COO^- + HCO_3^-$ | −20.5[b] |
| With $H_2$/formate-using bacterium | −39.4 |
| $2CH_3CH_2CH_2COO^- + HCO_3^- + H_2O \rightarrow 4CH_3COO^- + CH_4 + H^+$ | |
| $2CH_3CH_2CH_2CH_2COO^- + HCO_3^- + H_2O \rightarrow 2CH_3COO^- + 2CH_3CH_2COO^- + CH_4 + H^+$ | −39.4 |
| $2CH_3CH_2CH_2COO^- + SO_4^{2-} \rightarrow 4CH_3COO^- + HS^- + H^+$ | −54.8 |
| $2CH_3CH_2CH_2CH_2COO^- + SO_4^{2-} \rightarrow 2CH_3COO^- + 2CH_3CH_2COO^- + HS^- + H^+$ | −54.8 |

[a]Most members of the family *Syntrophomonadaceae* grow on crotonate in pure culture
[b]*Pelospora glutarica* is the only member of the *Syntrophomonadaceae* family that is capable of growing on succinate. Data compiled from Schink (1997) and Thauer et al. (1977), or calculated from data therein

terminal endospores. This mesophilic bacterium grows most rapidly at pH values of 7.1–8.2 and salt concentrations lower than 100 mM.

### *Syntrophothermus*

The rod-shaped cells of *Syntrophothermus* stain Gram-negative, are weakly motile, and non-sporeforming. They can grow in pure culture only on crotonate. In syntrophic cooperation with H2-utilizing microorganisms, cells of *Syntrophothermus* can metabolize saturated fatty acids with 4–10 carbon atoms by β-oxidation. The ability to degrade isobutyrate by isomerization to butyrate in syntrophic cooperation with an $H_2$-utilizing organism is the most distinguishing phenotypic feature of *Syntrophothermus*. Most rapid growth is observed at 55 °C and at near neutral pH. *Syntrophothermus lipocalidus* is the type species of this genus.

### *Thermosyntropha*

*Thermosyntropha lipolytica* is the type species of this genus. The ability to hydrolyze triglycerides and utilize the liberated short- and long-chain fatty acids in syntrophic cooperation with an $H_2$-utilizing methanogen is the most distinctive and perhaps the most important phenotypic feature of *Thermosyntropha*. *Th. lipolytica* can be cultured on a number of substrates (refer to *Table 29.1*), whereas most species in the family grow in pure culture only with crotonate. Oleate, linoleate, and saturated fatty acids (butyrate up to stearate) are metabolized in coculture with $H_2$-utilizing microorganisms. The rod-shaped cells of *Thermosyntropha* stain Gram-negative, are nonmotile, and non-sporeforming. It grows most rapidly at pH 8.1–8.9 and at temperatures between 60 °C and 66 °C. More recently, a new species, *Th. tengcongensis* was described which was isolated from a Chinese hot spring (Zhang et al. 2012). With respect to their physiological properties *Th. lipolytica* and *Th. tengcongensis* are nearly identical, but comparison of the 16SrRNA sequences (*Fig. 29.1*) justifies separation of two different species (96.7 % similarity).

## Enrichment and Isolation

All species of *Syntrophomonadaceae* are extremely sensitive to oxygen. Stringent anoxic techniques as described by Hungate (1969) or modifications of those techniques (Sowers and Noll 1995) are used to prepare anoxic media and solutions for cultivation. These methods usually involve the replacement of air with oxygen-free gases and the addition of strong chemical reducing agents such as cysteine, mercaptoethanesulfonate, sodium sulfide, or sodium dithionite (alone or in combinations) to maintain low oxidation-reduction potentials.

*Syntrophomonadaceae* often grow in the presence of large numbers of heterotrophic bacteria. Therefore, it is difficult to isolate them directly without a preliminary enrichment step. Fatty acid-degrading syntrophs can be selectively enriched in a reduced medium that contains a fatty acid as electron donor and carbon dioxide as electron acceptor. Oxygen, ferric iron, sulfate, sulfur, nitrate, and nitrite should be avoided because such ions would support growth of nonsyntrophic bacteria. The culture must contain a suitable $H_2$-utilizing methanogen to maintain low concentrations of hydrogen and make the degradation of fatty acids thermodynamically favorable. Such methanogens may be present in starting sample material, or a suitable methanogen may be added.

Enrichment media like the media for cultivation of pure cultures of syntrophs are usually buffered by carbon dioxide plus bicarbonate to a pH similar to that of the environment, and contain inorganic ions. A small amount of organic compounds that may include possible growth factors is often added. Rumen fluid (2–20 % of volume) or yeast extract and peptone (0.05–0.5 g per liter) are sometimes added for this purpose. However, additions of organic matter other than the syntrophic substrate may support the growth of non-syntrophic bacteria, thus, these should be excluded from enrichment cultures or be minimized. Additionally, the use of organic reducing agents should be avoided. The use of cysteine may enrich for cysteine degraders. The catabolic substrate is added at a concentration of about 1–2 g per liter. Certain species of the *Syntrophomonadaceae* family can be selected for by addition of a specific substrate. For instance, *Syntrophothermus lipocalidus* uses isobutyrate which no other species in the family can metabolize.

To obtain numerically important syntrophs from an environment, serial dilutions of the original sample should be prepared in enrichment media in the presence of a background lawn of hydrogen-scavenging partner organisms. After a suitable time of incubation, the highest dilution (smallest inoculum) that produced a successful enrichment culture should contain the syntrophic organism that was most numerous in the original sample. This enrichment culture may be used as inoculum to isolate syntrophic cocultures.

## Growth Media and Solutions

A large number of anoxic media have been formulated for the growth of fatty-acid-degrading syntrophic bacteria. Many are found in the original species description papers cited throughout this chapter. We describe only a generalized medium here. Optimal growth of these microorganisms may require modification of the concentrations of some of the components or special additions. The following basal medium can be used to cultivate the various species of the family *Syntrophomonadaceae* in cocultures by changing the catabolic substrate added to the medium (McInerney et al. 1979):

| Basal medium | |
|---|---|
| Mineral solution (see below) | 50 ml |
| Trace metal solution (see below) | 1 ml |
| Vitamin solution (see below) | 5 ml |
| Bicarbonate solution | 30 ml |
| Resazurin | 1 mg |
| Sulfide solution (see below) | 2 ml |
| Rumen fluid | 20 ml |
| $H_2O$ dist. | ad 1.0 l |

The bicarbonate solution is prepared by dissolving $NaHCO_3$ (8.4 g per 100 ml) in oxygen-free distilled water and equilibrating the preparation with 100 % carbon dioxide. The sulfide solution (3 g $Na_2S \cdot 9\,H_2O$ dissolved in 25 ml of oxygen-free boiled $H_2O$) is autoclaved separately under an oxygen-free $N_2/CO_2$ (4:1) gas mixture in a tightly closed bottle in a separate safety container (develops high pressure in the autoclave!). All the constituents above (except the sulfide solution) are added at the indicated final concentrations in percent (v/v), and the pH of the medium is adjusted to 7.2–7.4 under a stream of oxygen-free $N_2/CO_2$ (4:1) gas mixture. The medium is then dispensed (under a continuous stream of the above anoxic gas mixture) into 27-ml serum tubes fitted with butyl rubber stoppers, sealed, and autoclaved (121 °C, 20 min). The medium is cooled, and before use, the sulfide solution is added individually to each tube. For solid media, purified agar (18 g per l liter) is added to the medium and is maintained in suspension by utilizing a magnetic stirrer as it is dispensed into serum tubes.

| Mineral solution | |
|---|---|
| $KH_2PO_4$ | 10.0 g |
| $MgCl_2 \cdot 6\,H_2O$ | 6.6 g |
| NaCl | 8.0 g |
| $NH_4Cl$ | 8.0 g |
| $CaCl_2 \cdot 2\,H_2O$ | 1.0 g |
| Distilled water | 1 l |

| Trace metal solution | |
|---|---|
| Nitrilotriacetic acid | 2.0 g |
| $MnSO_4 \cdot H_2O$ | 1.0 g |
| $Fe(NH_4)_2(SO_4)_2 \cdot 6\,H_2O$ | 0.8 g |
| $CoCl_2 \cdot 6\,H_2O$ | 0.2 g |
| $ZnSO_4 \cdot 7\,H_2O$ | 0.2 g |
| $CuCl_2 \cdot 2\,H_2O$ | 0.02 g |
| $NiCl_2 \cdot 6\,H_2O$ | 0.02 g |
| $Na_2MoO_4 \cdot 2\,H_2O$ | 0.02 g |
| $Na_2SeO_4$ | 0.02 g |
| $Na_2WO_4$ | 0.02 g |
| Distilled water | 1 l |

To prepare the trace metal solution, dissolve the nitrilotriacetic acid in 800 ml of distilled water and adjust the pH to 6.0 with KOH. Then dissolve the minerals and bring the volume to 1 l.

| Vitamin solution | |
|---|---|
| Nicotinic acid | 20 mg |
| Cyanocobalamin | 20 mg |
| Thiamin • 2 HCl | 10 mg |
| p-Aminobenzoic acid | 10 mg |
| Pyridoxamine • 2 HCl | 50 mg |
| Calcium D (+) pantothenate | 5 mg |
| Distilled water | 1 l |

The substrate is added at a concentration which will not inhibit growth. For short-chain fatty acids, a concentration of 20 mM is used, while lower concentrations (<5 mM) are used for long-chain fatty acids which can easily become toxic. Higher concentrations of long-chain fatty acids require the supplementation of equimolar calcium chloride into the medium.

## Culturing $H_2$-Utilizing Microorganisms

Several mesophilic anaerobes that syntrophically oxidize fatty acids have been isolated in cocultures with $H_2$-utilizing strains such as *Desulfovibrio vulgaris* and *Methanospirillum hungatei*. *Syntrophomonas wolfei* subsp. *wolfei* was the first documented mesophilic strain isolated in coculture with a methanogen or a sulfate-reducer. Thermophilic fatty acid-oxidizing bacteria such as *Thermosyntropha* or *Syntrophothermus spp.* have been obtained in cocultures with strains of *Methanothermobacter thermautotrophicus* or related strains. Thus, choosing the appropriate $H_2$-utilizing microbe is essential to obtain cocultures of syntrophic bacteria. Fatty acid β-oxidation is thermodynamically more favorable if it is coupled to $H_2$ utilization by a sulfate reducer rather than by a methanogen because the reduction of sulfate to sulfide by $H_2$ is thermodynamically more favorable than the reduction of $CO_2$ to $CH_4$ (➋ *Table 29.2*). Thus, sulfate-reducing syntrophic cocultures grow faster and to higher yields than the methanogenic cocultures, but it may be easier later to obtain axenic cultures when methanogens are used as partners.

The basal medium above may be sufficient for the growth of the $H_2$-utilizing microorganisms. A gas mixture of $H_2$ and $CO_2$ (4:1) is added after dispensing and autoclaving the medium. The culture tubes are pressurized to 100 kPa above atmospheric pressure. For the growth of $H_2$-utilizing sulfate-reducing bacteria, the basal medium is supplemented with 3 g of $Na_2SO_4$. Since most $H_2$-utilizing methanogens or sulfate reducers are not real autotrophs but depend on acetate as a co-substrate for cell matter synthesis a low amount (about 5 mM) of acetate should be provded in these cultures.

During growth on $H_2$ and $CO_2$, methanogens quickly consume gas to produce $CH_4$ (5 mol of gas go to 1 mol of $CH_4$); therefore a negative pressure develops in the culture tubes. Additionally, as the partial pressure of $CO_2$ decreases, the medium becomes alkaline, which may inhibit growth and cause cell lysis. To minimize these problems, the volume of the headspace should be pressurized periodically throughout the growth period with $N_2/CO_2$ (4:1) gas mixture.

## Isolation of Defined Co-cultures and Pure Cultures

Serial ten-fold dilutions of the enrichment culture are prepared while avoiding exposure of the cultures to air, and a culture of the $H_2$-using partner organism is added to each tube. These tubes are mixed and inoculated into molten anaerobic roll-tube media (45 °C) with and without the catabolic substrate. The roll tubes are rolled to coat the agar on the inside of the tube and incubated at an appropriate temperature. Colonies may take several months to develop, and result in syntrophic cocultures. Colonies that appear within the first few weeks of incubation are probably growing too fast to be syntrophic cocultures, and these may be marked so that late-forming colonies can be easily distinguished. Syntrophic colonies contain a mixture of the syntroph and its partner, so when a methanogen is used as the syntrophic partner, the syntrophic colonies may be distinguished by the presence of the methanogens' epifluorescence. Colonies are selected, picked, and inoculated into sterile enrichment medium. These cultures are immediately diluted and re-inoculated into roll-tube media together with the syntrophic partner (as described above). This process is repeated until a single colony type remains. At this point, the culture contains a single strain of syntrophic bacteria plus the syntrophic partner that was selected.

To obtain axenic cultures from these cocultures, it is necessary to find a suitable substrate that supports the growth of the syntroph without its partner. Crotonate has been successfully used for butyrate-degrading bacteria. The coculture is grown in medium with crotonate as the sole catabolic substrate; this enriches for the butyrate-degrading syntroph and limits the growth of the methanogen. A pure culture of the syntroph may be obtained by serial dilution of the culture (with higher dilutions having the syntroph present but no methanogens) or by preparing roll tube media with crotonate to obtain pure colonies of the syntroph.

## Maintenance Procedures

Pure cultures or cocultures may be stored for several weeks as liquid suspensions, or suspensions can be frozen in the presence of a cryoprotectant (5 % glycerol) by cooling at 1 °C/min and stored at liquid nitrogen temperatures (Boone 1995), or cultures can be maintained by regular subculturing (Hippe 1984).

# Ecology and Habitat

Members of the family *Syntrophomonadaceae* are predominantly found in methanogenic environments in syntrophic associations with methanogens. In such environments, various electron acceptors are not readily available and methanogenesis is the dominant metabolism (McInerney 1986). In such environments, primary fermentative bacteria degrade organic matter to extracellular products including $H_2$, formate, and acetate. These are consumed directly by methanogens in the production of methane and carbon dioxide. Other products of the primary fermentative bacteria, such as monocarboxylic acids, cannot be used directly by methanogens. Rather, these compounds are degraded by syntrophic interactions between methanogens and syntrophs such as the *Syntrophomonadaceae*. The important syntrophs that degrade fatty acids of 4–18 carbon atoms are in this family, whereas those that degrade propionate (3 carbons) belong to the genera *Syntrophobacter* (Boone and Bryant 1980) or *Smithella* (Liu et al. 1999).

Thus, syntrophic bacteria represent an essential trophic guild that converts the fatty acids produced by fermentative bacteria from complex organic matter to methanogenic substrates such as acetate, $H_2$, and formate. Examples of such environments include sewage digestors, waterlogged soils, aquifers and sediments. Syntrophic bacteria are generally found where organic matter is degraded and inorganic electron acceptors are absent. In shallow marine sediments, sulfate reduction is the dominant metabolism, and many fatty acids may be degraded directly by sulfate-reducing bacteria. However, syntrophy does still occur in such environments (Stieb and Schink 1985; Tschech and Schink 1985a, b). *Syntrophomonadaceae* can also grow in nonmethanogenic anoxic environments: *Pelospora glutarica* grows on compounds such as succinate in the absence of a syntrophic partner (Matthies et al. 2000).

Whereas most of the members of *Syntrophomonadaceae* grow by syntrophically oxidizing monocarboxylic fally acids, *Pelospora* sp. do not require syntrophy (they ferment succinate or glutarate). All other members of this family obtain their energy for growth from the degradation of a variety of fatty acids ranging from 4–18 carbons in length. Oxidation of these compounds is thermodynamically unfavorable unless the products ($H_2$ and/or formate, or acetate) are maintained at low concentrations by the action of the syntrophic partner, such as a methanogen (➋ *Table 29.1*). When the methanogenic partner is absent, the concentrations of $H_2$ and formate rapidly increase to values that thermodynamically inhibit fatty acid oxidation. Therefore, syntrophic bacteria in their natural environment are obligately dependent on the activity of the methanogens. There are no alternate fermentative pathways for energy conservation from compounds such as butyrate. When these organisms first were isolated, no mechanism other than syntrophy was known that would allow their growth, and they were called "obligately proton-reducing acetigens" (McInerney et al. 1979). Since that time, butyrate-degrading syntrophs were found capable of growth on crotonate (Beaty and McInerney 1987). This eliminates the need for a $H_2$/formate-using bacterium, allowing their growth as pure cultures.

## Biochemical and Physiological Properties

Fermentation of butyrate to acetate plus $H_2$ is an endergonic reaction (see ➋ *Table 29.1*) under standard conditions. Degradation of butyrate is feasible only at a low $H_2$ partial pressure (below $10^{-4}$–$10^{-5}$ atm; McInerney et al. 1981; Schink 1997), which can be maintained by methanogenic partners. The pathway of butyrate oxidation in syntrophic butyrate-oxidizing bacteria has been tentatively elucidated with *Syntrophomonas wolfei*. It proceeds through classical fatty acid β-oxidation which involves seven steps (Wofford et al. 1986). After activation of butyrate with acetyl-CoA by a CoA-transferase, butyryl-CoA is oxidized to crotonyl-CoA and further via L(+)-3-hydroxybutyryl-CoA to acetoacetyl-CoA. Thiolytic cleavage yields two acetyl-CoA molecules. One acetyl-CoA activates the butyrate molecule at the beginning, the other one is converted to acetyl phosphate by phosphotransacetylase. The terminal reaction, catalyzed by acetate kinase, yields an acetate molecule and one ATP by substrate-level phosphorylation.

Since the electrons obtained in the butyryl-CoA dehydrogenase reaction arise at a comparably high redox potential, i.e., at −125 mV, release of these electrons as molecular hydrogen ($E^{\circ\prime} = -414$ mV) or formate ($E^{\circ\prime} = -430$ mV) requires that part of the metabolic energy has to be reinvested to fuel a reversed electron transport system to make these redox reactions thermodynamically feasible (Thauer and Morris 1984). Experimental evidence of a proton motive force-driven reversed electron transport in butyrate degradation was obtained with intact cell suspensions of *S. wolfei* (Wallrabenstein and Schink 1994). After the genome of *S. wolfei* was completely sequenced (Sieber et al. 2010) biochemical and proteomic experiments became possible which indicated that this reversed electron transport is accomplished by an externally oriented formate dehydrogenase system which shifts electrons from the flavin level to that of $CO_2$ reduction by simultaneous consumption of the inside-oriented proton gradient (Müller et al. 2009, 2010; Schmidt et al. 2013). This is consistent with the calculated overall amount of energy available in this reaction which is in the range of −20 kJ per mol, corresponding to about one-third of an ATP equivalent (Thauer et al. 1977; Schink 1997). These findings indicate that actually formate rather than hydrogen may be the primary electron carrier between the fatty acid oxidizer and its methanogenic partner. Nonetheless, *S. wolfei* contains a cytoplasmic enzyme able to convert formate to $CO_2$ plus $H_2$ and vice versa (Müller et al. 2009), thus, actually both channels may be used for interspecies electron transfer, either alternatively or simultaneously.

Syntrophic degradation of long-chain fatty acids probably involves several rounds of β-oxidation with the concomitant release of electrons as $H_2$ via a reversed electron transport, probably analogous to the pathway described above for syntrophic butyrate degradation. Also the genome of *Syntrophothermus lipocalidus* has been sequenced in the meantime (Djao et al. 2010) but biochemical studies on this bacterium have not been reported so far.

### Biochemistry of Crotonate Metabolism

Most members of the family *Syntrophomonadaceae* are capable of growing in pure culture on crotonate, without a syntrophic partner (Beaty and McInerney 1987; McInerney and Wofford 1992). Crotonate utilization bypasses an unfavorable step, the oxidation of butyryl-CoA to crotonyl-CoA in the butyrate degradation pathway. Therefore, the bacterium generates ATP without the dependence upon interspecies electron transfer because crotonate is metabolized by a disproportionation reaction in which part of the substrate is oxidized to acetate and the remainder reduced to butyrate (Beaty and McInerney 1987; McInerney and Wofford 1992). It was reported that *Syntrophomonas wolfei* contains a *c*-type cytochrome

(McInerney and Wofford 1992) which was assumed to be involved an energy-conserving electron transport chain. Genomic evidence speaks for the presence of a *b*-type cytochrome Sieber et al. 2010), and the path of a possible electron-transport-linked energy conservation in this step is still unclear. Biochemical evidence indicates that also hydrogenases are involved in this metabolic activity (Schmidt et al. 2013).

## Acknowledgment

This manuscript is largely based on its predecessor that was written by Martin Sobieraj and David R. Boone and published in the last edition of the Prokaryotes. The author wants to dedicate this manuscript to the late David Boone who made substantial contributions to microbial taxonomy in general and to our understanding of syntrophic fatty acid oxidation in particular. Unfortunately, David passed away in 2005 far too early, at 53 years of age.

## References


Beaty PS, McInerney MJ (1987) Growth of *Syntrophomonas wolfei* in pure culture on crotonate. Arch Microbiol 147:389–393

Beaty PS, McInerney MJ (1990) Nutritional features of *Syntrophomonas wolfei* Appl. Environ Microbiol 50:3223–3224

Boone DR (1995) Short-and long-term maintenance of methanogen stock cultures. In: Sowers KR, Schreier HJ (eds) Archaea: Methanogens: a laboratory manual. Cold Spring Harbor Laboratory Press, New York, pp 79–83

Boone DR, Bryant MP (1980) Propionate-degrading bacterium, *Syntrophobacter wolinii* sp. nov. gen. nov., from methanogenic ecosystems. Appl Environ Microbiol 40:626–632

Buchanan RE, Gibbons NE (eds) (1974) Bergey's manual of determinative bacteriology. Williams and Wilkins, Baltimore

Djao ODN et al (2010) Complete genome sequence of *Syntrophothermus lipocalidus* type strain (TGB-C1(T)). Stand Genomic Sci 3:267–275

Gibbons NE, Murray RGE (1978) Proposals concerning the higher taxa of bacteria. Int J Syst Bacteriol 28:1–6

Hatamoto M, Imachi H, Fukayo S, Ohashi A, Harada H (2007) *Syntrophomonas palmitatica* sp nov., an anaerobic, syntrophic, long-chain fatty-acid-oxidizing bacterium isolated from methanogenic sludge. Int J Syst Evol Microbiol 57:2137–2142

Hippe H (1984) Maintenance of methanogenic bacteria. In: Kirsop BE, Snell JJS (eds) Maintenance of microorganisms. Academic Press, London, pp 69–81

Hungate RE (1969) A roll tube method for cultivation of strict anaerobes. In: Norris R, Ribbons DW (eds) Methods in microbiology. Academic Press, New York, pp 117–132

Liu Y, Balkwill DL, Aldrich HC, Drake GR, Boone DR (1999) Characterization of the anaerobic propionate-degrading syntrophs *Smithella propionica* gen. nov., sp. nov. and *Syntrophobacter wolinii*. Int J Syst Bacteriol 49:545–556

Lorowitz WH, Zhao H, Bryant MP (1989) *Syntrophomonas wolfei* subsp. *saponavida* subsp. nov., a long-chain fatty-acid-degrading, anaerobic, syntrophic bacterium; *Syntrophomonas wolfei* subsp. *wolfei* subsp. nov.; and emended descriptions of the genus and species. Int J Syst Bacteriol 39:122–126

Matthies C, Springer N, Ludwig W, Schink B (2000) *Pelospora glutarica* gen. nov., sp. nov., a glutarate-fermenting, strictly anaerobic, spore-forming bacterium. Int J Syst Evol Microbiol 50:645–648

McInerney MJ (1986) Transient and persistent associations among prokaryotes. In: Poindexter JS, Leadbetter ER (eds) Bacteria in nature. Plenum Press, New York, pp 293–338

McInerney MJ, Wofford NQ (1992) Enzymes involved in crotonate metabolism in *Syntrophomonas wolfei*. Arch Microbiol 158:344–349

McInerney MJ, Bryant MP, Pfennig N (1979) Anaerobic bacterium that degrades fatty acids in syntrophic association with methanogens. Arch Microbiol 122:129–135

McInerney MJ, Bryant MP, Hespell RB, Costerton JW (1981) *Syntrophomonas wolfei* gen. nov. sp. nov., an anaerobic, syntrophic, fatty acid-oxidizing bacterium. Appl Environ Microbiol 41:1029–1039

Müller N, Schleheck D, Schink B (2009) Involvement of NADH: acceptor oxidoreductase and butyryl-CoA dehydrogenase in reversed electron transport during syntrophic butyrate oxidation by *Syntrophomonas wolfei*. J Bacteriol 191:6167–6177

Müller N, Worm P, Schink B, Stams AJM, Plugge CM (2010) Syntrophic butyrate and propionate oxidation: from genomes to reaction mechanisms. Environ Microbiol Rep 2:489–499

Roy F, Samain E, Dubourguier HC, Albagnac G (1986) *Syntrophomonas sapovorans* sp. nov., a new obligately proton reducing anaerobe oxidizing saturated and unsaturated long chain fatty acids. Arch Microbiol 145:142–147

Schink B (1997) Energetics of syntrophic cooperation in methanogenic degradation. Microbiol Mol Biol Rev 61:262–280

Schmidt A, Müller N, Schink B, Schleheck D (2013) A proteomic view at the biochemistry of syntrophic butyrate oxidation in *Syntrophomonas wolfei*. Plos One http://www.plosone.org/article/info%3Adoi%2F10.1371%2Fjournal.pone.0056905

Sekiguchi Y, Kamagata Y, Nakamura K, Ohashi A, Harada H (2000) *Syntrophothermus lipocalidus* gen. nov., sp. nov., a novel thermophilic, syntrophic, fatty-acid-oxidizing anaerobe which utilizes isobutyrate. Int J Syst Evol Microbiol 50:771–779

Sieber JR et al (2010) The genome of *Syntrophomonas wolfei*: new insights into syntrophic metabolism and biohydrogen production. Environ Microbiol 12:2289–2301

Sousa DZ, Smidt H, Alves MM, Stams AJM (2007) *Syntrophomonas zehnderi* sp nov., an anaerobe that degrades long-chain fatty acids in co-culture with *Methanobacterium formicicum*. Int J Syst Evol Microbiol 57:609–615

Sowers KR, Noll KM (1995) Techniques for anaerobic growth. In: Robb FT, Sowers KR, Schreier HJ, DasSarma S, Fleischmann EM (eds) Archaea: a laboratory manual. Cold Spring Harbor Laboratory Press, New York, pp 15–47

Stamatakis A (2006) RAxML-VI-HPC: Maximum likelihood-based phylogenetic anlyses with thousands of taxa and mixed models. Bioinformatics 22 (21):2688–2690

Stieb M, Schink B (1985) Anaerobic oxidation of fatty acids by *Clostridium bryantii* sp. nov., a sporeforming, obligately syntrophic bacterium. Arch Microbiol 140:387–390

Svetlitshnyi V, Rainey F, Wiegel J (1996) *Thermosyntropha lipolytica* gen. nov., sp. nov., a lipolytic, anaerobic, alkalitolerant, thermophilic bacterium utilizing short-and long-chain fatty acids in syntrophic coculture with a methanogenic archaeum. Int J Syst Bacteriol 46:1131–1137

Thauer RK, Morris JG (1984) Metabolism of chemotrophic anaerobes: old views and new aspects. Symp Soc Gen Microbiol 36:123–374

Thauer RK, Jungermann K, Decker K (1977) Energy conservation in chemotrophic anaerobic bacteria. Bacteriol Rev 41:100–180

Tschech A, Schink B (1985a) Fermentative degradation of resorcinol and resorcylic acids. Arch Microbiol 143:52–59

Tschech A, Schink B (1985b) Fermentative metabolism of monohydroxybenzoates by defined syntrophic cocultures. Arch Microbiol 145:396–402

Wallrabenstein C, Schink B (1994) Evidence of reversed electron transport in syntrophic butyrate or benzoate oxidation by *Syntrophomonas wolfei* and *Syntrophus buswellii*. Arch Microbiol 162:136–142

Wofford NQ, Beaty PS, McInerney MJ (1986) Preparation of cell-free extracts and the enzymes involved in fatty acid metabolism in Syntrophomonas wolfei. J Bacteriol 167:179–185

Wu CG, Liu XL, Dong XZ (2006a) *Syntrophomonas erecta* subsp sporosyntropha subsp nov., a spore-forming bacterium that degrades short chain fatty acids in co-culture with methanogens. Syst Appl Microbiol 29:457–462

Wu CG, Liu XL, Dong XZ (2006b) *Syntrophomonas cellicola* sp nov., a spore-forming syntrophic bacterium isolated from a distilled-spirit-fermenting cellar, and assignment of *Syntrophospora bryantii* to *Syntrophomonas bryantii* comb. nov. Int J Syst Evol Microbiol 56:2331–2335

Wu CG, Dong XZ, Liu XL (2007) *Syntrophomonas wolfei* subsp *methylbutyratica* subsp nov., and assignment of *Syntrophomonas wolfei* subsp *saponavida* to *Syntrophomonas saponavida* sp nov comb. nov. Syst Appl Microbiol 30:376–380

Yarza P, Ludwig W, Euzeby J, Amann R, Schleifer KH, Glockner FO, Rossello-Mora R (2010) Update of the All-Species Living Tree Project based on 16S and 23S rRNA sequence analyses. Syst Appl Microbiol 33: 291–299

Zhang CY, Liu XL, Dong XZ (2004) *Syntrophomonas curvata* sp nov., an anaerobe that degrades fatty acids in coculture with methanogens. Int J Syst Evol Microbiol 54:969–973

Zhang CY, Liu XL, Dong XX (2005) *Syntrophomonas erecta* sp nov., a novel anaerobe that syntrophically degrades short-chain fatty acids. Int J Syst Evol Microbiol 55:799–803

Zhang F, Liu XL, Dong XZ (2012) *Thermosyntropha tengcongensis* sp nov., a thermophilic bacterium that degrades long-chain fatty acids syntrophically. Int J Syst Evol Microbiol 62:759–763

Zhao H, Yeng D, Woese CR, Bryant MP (1990) Assignment of *Clostridium bryantii* to *Syntrophospora bryantii* gen. nov. comb., based on 16S rRNA sequence analysis of its crotonate-grown pure culture. Int J Syst Bacteriol 40:40–44

Zhao H, Yang D, Woese CR, Bryant MP (1993) Assignment of fatty acid-b-oxidizing syntrophic bacteria to *Syntrophomonadaceae* fam. nov. on the basis of 16S rRNA sequence analyses. Int J Syst Bacteriol 43:278–286

# 30 The Family *Thermodesulfobacteriaceae*

*Koji Mori*
Biological Resource Center, National Institute of Technology and Evaluation (NBRC), Kisarazu, Chiba, Japan



**Abstract**
*Thermodesulfobacteriaceae* is the only family belonging to the order *Thermodesulfobacteriales*, class *Thermodesulfobacteria*, and phylum Thermodesulfobacteria and consists of the 4 valid genera, *Thermodesulfobacterium*, *Thermodesulfatator*, *Caldimicrobium*, and *Thermosulfurimonas* and the invalid genus *Geothermobacterium*. They are Gram-negative, rod-shaped, non-sporulating, anaerobic thermophiles that grow with sulfur compounds or Fe(III) as an energy source. Phylogenetic analysis based on the 16S rRNA gene demonstrates that this family is one of the deeply branching lineages within the domain *Bacteria*. They inhabit thermal environments such as terrestrial hot springs, oil reservoirs, and hydrothermal fields.

## Taxonomy, Historical and Current

### Short Description of the Family

Ther.mo.de.sul.fo.bac.te.ri.a'ce.ae N.L. neut. n. *Thermodesulfobacterium* type genus of the family; -aceae ending to denote a family; N.L. fem. pl. n. *Thermodesulfobacteriaceae* the family of *Thermodesulfobacterium*.

The family *Thermodesulfobacteriaceae* belongs to the order *Thermodesulfobacteriales*, class *Thermodesulfobacteria*, and phylum Thermodesulfobacteria and was created by Hatchikian et al. for the type genus *Thermodesulfobacterium* (Hatchikian et al. 2001). The genus *Thermodesulfobacterium* was proposed by Zeikus et al. in 1983 and validated in 1995 as a non-sporulating rod-shaped thermophilic dissimilatory sulfate-reducing bacterium, *Thermodesulfobacterium commune* (Zeikus et al. 1983). However, *Thermodesulfobacterium thermophilum* was initially isolated as *Desulfovibrio thermophilus* in 1974 (Rozanova and Khudyakova 1974), and its name was validated by its inclusion in the Approved Lists of Bacterial Names (Skerman et al. 1980). Although *D. thermophilus* was tentatively renamed as *Thermodesulfobacterium mobile*, it is an illegitimate latter synonym of *T. thermophilum* in accordance with the Bacteriological Code (Rozanova and Pivovarova 1988; Tao et al. 1996; Trüper 2003). Only the type genus *Thermodesulfobacterium* has been validated in this family, and, to date, 3 genera, *Thermodesulfatator* (Moussard et al. 2004), *Caldimicrobium* (Miroshnichenko et al. 2009), and *Thermosulfurimonas* (Slobodkin et al. 2012), have been added. In addition, the genus *Geothermobacterium* was proposed by Kashefi et al. (2002) but has not been validated yet.

Morphologically, the cells are ovals or short rods. The Gram reaction is negative and spore formation is not reported. Most cells are motile with single or some polar flagella except for *T. commune* and *Thermodesulfobacterium hveragerdense* (Hatchikian et al. 2001; Sonne-Hansen and Ahring 1999). The growth properties of the genera are summarized in ➋ *Table 30.1*. They are thermophilic, neutrophilic, and strictly anaerobic. Except for some species belonging to the genus *Thermodesulfobacterium*, chemolithoautotrophic growth occurs by the use of sulfur compounds or Fe(III) as an energy source and $CO_2$ as a carbon source. In addition, some species of the genera *Thermodesulfobacterium* and *Thermodesulfatator* can use some organic compounds for growth.

### Phylogenetic Structure of the Family and Its Genera

The phylogenetic tree of the family *Thermodesulfobacteriaceae*, based on the 16S rRNA gene sequence, is shown in ➋ *Fig. 30.1*. Although *Thermodesulfobacteriaceae* is the only family in the phylum Thermodesulfobacteria, the lineage is completely separated from all known phyla and represents one of the deeply branching groups in the domain *Bacteria*. The family consists of 5 genera including 1 invalid genus, and the sequence divergence of the 16S rRNA gene for the family is 11.6 %.

E. Rosenberg et al. (eds.), *The Prokaryotes – Firmicutes and Tenericutes*, DOI 10.1007/978-3-642-30120-9_366,

**Table 30.1**
**Differential characteristics among the genera *Thermodesulfobacteriaceae***

| Genus | *Thermodesulfobacterium* | *Thermodesulfatator* | *Caldimicrobium* | *Thermosulfurimonas* | *Geothermobacterium* |
|---|---|---|---|---|---|
| Type species | *T. commune* | *T. indicus* | *C. rimae* | *T. dismutans* | *G. ferrireducens* |
| Growth | | | | | |
| Autotrophy | ± | + | + | + | + |
| Organotrophy | ± | – | + | – | – |
| Optimum temp. (°C) | 65–75 | 65–70 | 75 | 74 | 85–90 |
| Optimum pH | 6.5–7.0 | 6.3–7.0 | 7.0–7.2 | 7.0 | 6.8–7.0 |
| Electron acceptor | | | | | |
| $SO_4^{2-}$ | + | + | – | – | – |
| $S_2O_3^{2-}$ | ± | – | + | +[a] | – |
| $SO_3^{2-}$ | ± | – | – | (+)[a] | – |
| $S^0$ | – | – | + | (+)[a] | – |
| $Fe^{3+}$ | – | ND | – | – | + |
| G+C content (mol%) | 28–40 | 45–46 | 35.2 | 52 | ND |
| Habitat | Terrestrial/marine | Marine | Terrestrial | Marine | Terrestrial |

Data from Sonne-Hamsen and Ahring (1999), Hatchikian et al. (2001), Jeanthon et al. (2002), Kashefi et al. (2002), Moussard et al. (2004), Miroshnichenko et al. (2009), Alain et al. (2010), and Slobodkin et al. (2012)
*ND* not determined
[a]*T. dismutans* grows by disproportionation of elemental sulfur, thiosulfate, and sulfite to sulfide and sulfate or by the use of $H_2$ and thiosulfate

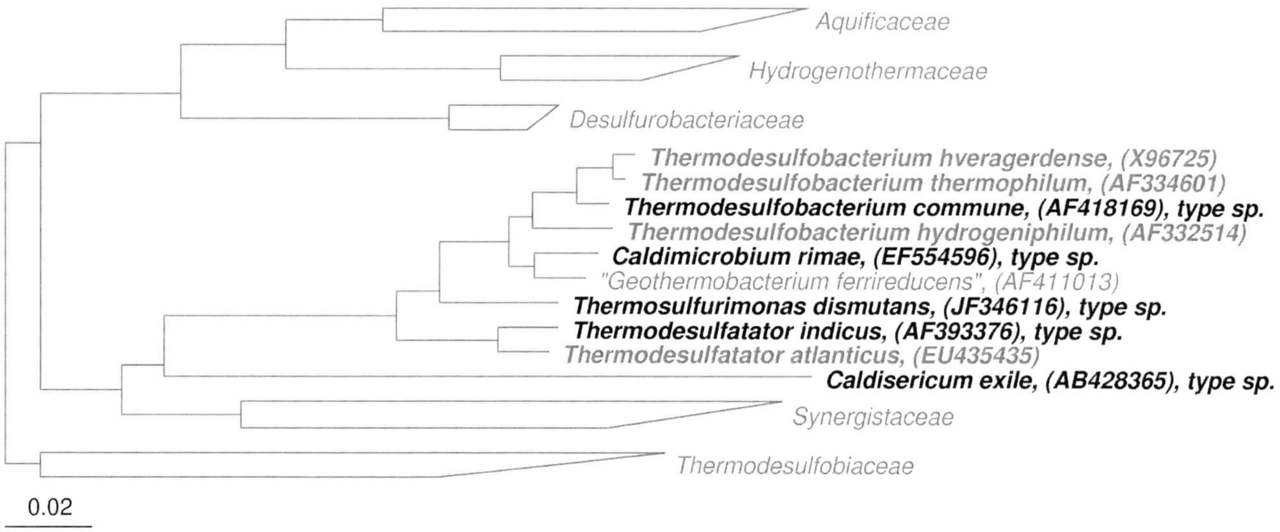


**Fig. 30.1**
**Phylogenetic reconstruction of the family *Thermodesulfobacteriaceae* based on the 16S rRNA gene and created using the neighbor-joining algorithm with the Jukes-Cantor correction. The sequence datasets and alignments were used according to the All-Species Living Tree Project (LTP) database (Yarza et al. 2010) (http://www.arb-silva.de/projects/living-tree). Tree topology was stabilized with the use of a representative set of nearly 750 high-quality type strain sequences that were distributed proportionally among the different bacterial and archaeal phyla. In addition, a 30 % maximum frequency filter was applied in order to remove hypervariable positions and potentially misplaced bases from the alignment. The scale bar indicates estimated sequence divergence**

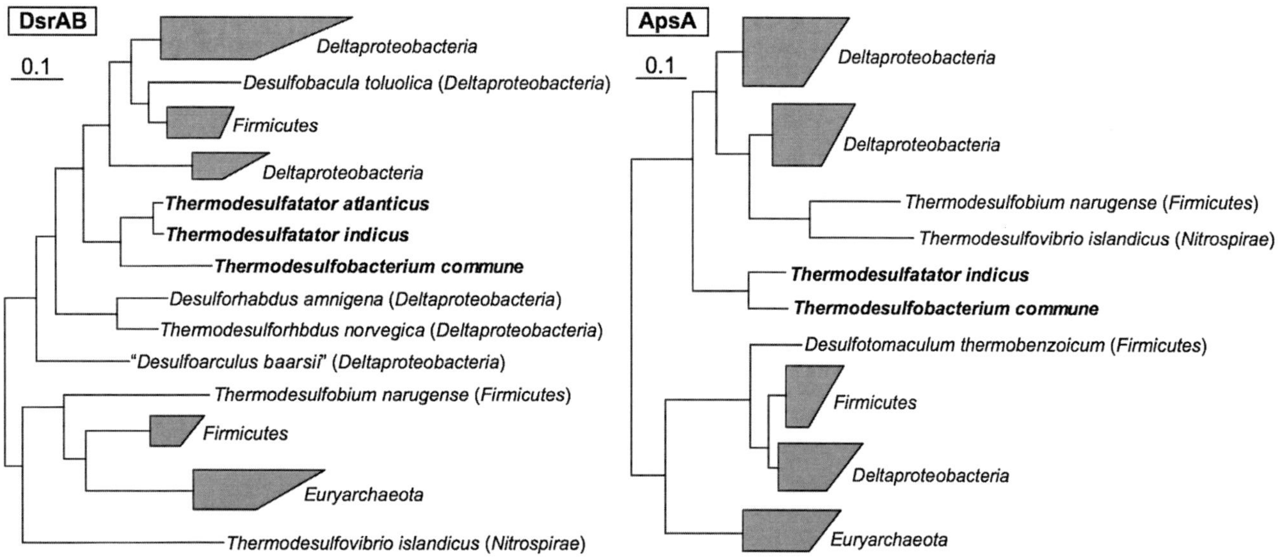


**Fig. 30.2**

**Phylogenetic relationships based on the deduced amino acid sequences of DsrAB and ApsA of sulfate-reducing prokaryotes inferred from the maximum-likelihood method after alignment with Clustal W (Mori et al. 2003). The scale bar indicates estimated sequence divergence. The sequence accession numbers used in the analyses are as follows (*dsrAB*, *apsA*, respectively): *T. commune* (AF334596, AF418114), *T. indicus* (CP002683), *T. atlanticus* (FN186055, none), *Archaeoglobus fulgidus* VC-16$^T$ (M95624, AE000988), *Archaeoglobus profundus* (AF071499, AF418134), *Archaeoglobus veneficus* (AB274311, AB418132), *Archaeoglobus infectus* (AB274309, AB274310), *Thermodesulfovibrio islandicus* (AF334599, AF418113), *Thermodesulfobium narugense* (AB077818, AB080361), *Desulfotomaculum acetoxidans* (AF271768, AF418153), *Desulfotomaculum geothermicum* (AF273029, AF418115), *Desulfotomaculum putei* (AF273032, AF418147), *Desulfotomaculum ruminis* (U58118, AF418164), *Desulfotomaculum thermobenzoicum* (AF273030, AF418161), *Desulfoarculus baarsii* (AF334600, AF418149), *Desulfobacter postgatei* (AF418198, AF418157), *Desulfobacula toluolica* (AF271773, AF418128), *Desulfobulbus rhabdoformis* (AJ250473, AF418110), *Desulfococcus multivorans* (U58126, AF418136), *Desulfofaba gelida* (AF334593, AF418118), *Desulfomonile tiedjei* (AF334595, AF418162), *Desulfonatronovibrio hydrogenovorans* (AF418197, AF418111), *Desulfonatronum lacustre* (AF418189, AF418137), *Desulforhabdus amnigena* (AF337901, AF418139), *Desulforhopalus singaporensis* (AF418196, AF418163), *Desulfovibrio africanus* (AF271772, AF418140), *Desulfovibrio desulfuricans* (AJ249777, AF226708) and *Thermodesulforhabdus norvegica* (AF334597, AF418159)**

## Molecular Analyses

### Genome Comparison

The complete genome sequence of *T. commune* DSM 2178$^T$ is currently being determined, but has not been released yet.

*T. indicus* CIR29812$^T$ is the only released complete genome sequence of the family *Thermodesulfobacteriaceae* (CP002683; Anderson et al. 2012). The 2,322,244 base pair circular genome comprises 42.4 mol% G+C content, 2 copies of the rRNA operon, and 2,233 predicted protein-coding sequences. The majority of the protein-coding genes (73.2 %) were assigned a putative function, while the remaining ones were annotated as hypothetical proteins.

### *dsrAB* and *apsA* Genes

Anaerobic sulfate-reducing respiration is phylogenetically spread in 5 lineages in the domain *Bacteria*, i.e., family *Thermodesulfobacteriaceae*, genus *Thermodesulfovibrio*, family *Thermodesulfobiaceae*, class *Deltaproteobacteria*, and family *Peptococcaceae* and in 1 lineage in the domain *Archaea*, i.e., genus *Archaeoglobus*. As for the evolutionary history of anaerobic sulfate (sulfite) respiration, 2 genes have been investigated in detail: (i) the *dsrAB* genes, which encode the alpha and beta subunits of dissimilatory sulfite reductase and catalyze the reduction of sulfite to sulfide, and (ii) the *apsA* gene, which encodes the alpha subunit of adenosine-5′-phosphosulfate reductase. Lateral gene transfer of the *dsrAB* genes has occurred frequently between the major lineages of *Bacteria* and *Archaea*, and the genus *Thermodesulfobacterium* supposedly possesses non-orthologous *dsrAB* genes because of the inconsistency between the *dsrAB* and 16S rRNA trees (Klein et al. 2001; Zverlov et al. 2005). The *dsrAB* genes of the genus *Thermodesulfatator* were also determined (CP002683 and FN186055) and placed phylogenetically in the same lineage of the genus *Thermodesulfobacterium* (➋ *Fig. 30.2*). As for the *apsA* gene, the lateral gene transfer has also occurred in *Bacteria*. The *apsA* gene of the genus *Thermodesulfobacterium* has a monophyletic lineage, but is not as distant as the 16S rRNA gene (Friedrich 2002). The *apsA* gene of *T. indicus* (CP002683) is close to those of *Thermodesulfobacterium* species (➋ *Fig. 30.2*).

**Table 30.2**
**Differential characteristics among the genus *Thermodesulfobacterium***

| Species | *T. commune* | *T. thermophilum* | *T. hveragerdense* | *T. hydrogeniphilum* |
|---|---|---|---|---|
| Type strain | YSRA-1$^T$ | 7$^T$ | JSP$^T$ | SL6$^T$ |
| Cell size (μm) | 0.3 × 0.9 | 0.6 × 2.0 | 0.5 × 2.8 | 0.4–0.5 × 0.5–0.8 |
| G+C content (mol%) | 34 | 38 | 40 | 28 |
| Motility | – | + | – | + |
| Growth | | | | |
| Optimum temp. (°C) | 70 | 65 | 70–74 | 75 |
| Optimum pH | 7.0 | ND | (7.0)[a] | 6.5 |
| Optimum salinity (g/L) | 0 | 1 | 0 | 30 |
| Chemolithoautotrophy | – | – | – | + |
| Chemoheterotrophy | + | + | + | – |
| Electron acceptors: | | | | |
| $SO_4^{2-}$ | + | + | + | + |
| $S_2O_3^{2-}$ | + | + | + | – |
| $SO_3^{2-}$ | – | + | + | – |
| Electron donors: | | | | |
| $H_2$ | + | + | + | + |
| Formate | ND | + | – | – |
| Lactate | + | + | + | – |
| Pyruvate | + | + | + | – |
| Pyruvate fermentation | + | + | + | – |
| Habitat | Hot springs | Stratal water from oil deposits | Alkaline hot springs | Deep-sea hydrothermal vents |

Data from Zeikus et al. (1983), Sonne-Hamsen and Ahring (1999), Hatchikian et al. (2001), and Jeanthon et al. (2002)
*ND* not determined
[a]Indicated as pH of cultivation

## Phenotypic Analyses

### *Thermodesulfobacterium* Zeikus et al. 1995, 197[VP]

Ther.mo.de.sul.fo.bac.te′ri.um. Gr. adj. *thermos* hot; L. pref. *de* from; L. neut. n. *sulfur* sulfur; L. neut. n. *bacterium* a small rod; N.L. neut. n. *Thermodesulfobacterium* a thermophilic rod reducing sulfate.

The genus *Thermodesulfobacterium* consists of the type species *T. commune* (Hatchikian et al. 2001; Zeikus et al. 1983) and *T. thermophilum* (Hatchikian et al. 2001), *T. hveragerdense* (Sonne-Hansen and Ahring 1999), and *T. hydrogeniphilum* (Jeanthon et al. 2002). The characteristics of each species are summarized in ➋ *Table 30.2*. Straight rod-shaped cells, which possess an outer wall membrane layer, occur singly, in pairs, or in chains. The Gram reaction is negative and spore formation is not reported. The cells form extrusions or blebs next to the outer membrane layer. The cells of *T. commune* and *T. hveragerdense* are nonmotile, while those of *T. thermophilum* and *T. hydrogeniphilum* are motile with a single polar flagellum. They grow under thermophilic and neutrophilic conditions. Most species prefer the absence of NaCl, except *T. hydrogeniphilum*. They utilize strictly anaerobic and dissimilatory sulfate-reducing metabolism. They grow chemolithoheterotrophically, except for *T. hydrogeniphilum*. Three species, i.e., *T. commune*, *T. thermophilum*, and *T. hveragerdense*, grow by the oxidation of molecular hydrogen using sulfate as an electron acceptor and require organic components such as acetate and yeast extract as carbon sources. *T. hydrogeniphilum* grows chemolithoautotrophically using sulfate under an $H_2/CO_2$ atmosphere, and its growth is stimulated by acetate, fumarate, 3-methylbutyrate, glutamate, yeast extract, peptone, and tryptone. As a substitution for sulfate, some species use thiosulfate and sulfite as electron acceptors. Except for *T. hydrogeniphilum*, pyruvate is fermented. *T. commune* and *T. thermophilum* utilize menaquinone-7 as their major respiratory quinone (Collins and Weddel 1986). The following cellular components of *T. commune* have been reported: the major fatty acid is anteiso-$C_{17:0}$ (Langworthy et al. 1983); the polar lipids are predominantly phosphatidylinositol and phosphatidylethanolamine (Moussard et al. 2004); and the main polyamine is $N^4$-bis(aminopropyl) spermidine (Hamana et al. 1996; Hamana et al. 1999).

**Table 30.3**
**Differential characteristics among the genus *Thermodesulfatator***

| Species | *T. indicus* | *T. atlanticus* |
|---|---|---|
| Type strain | CIR29812$^T$ | AT1325$^T$ |
| Cell size (μm) | 0.4–0.5 × 0.8–1.0 | 0.3–0.8 × 1.0–6.1 |
| G+C content (mol%) | 46.0 | 45.6 |
| Major fatty acids | $C_{18:0}$, $C_{18:1}\omega7c$, $C_{16:0}$, $C_{17:0}$ | $C_{18:0}$, $C_{18:1}\omega7c$, $C_{16:0}$ |
| Growth | | |
| Optimum temp. (°C) | 70 | 65–70 |
| Optimum pH | 6.25 | 6.5–7.5 |
| Optimum salinity (g/L) | 25 | 25 |
| Carbon source | $CO_2$ | $CO_2$, organic compounds |
| Nitrogen source | $NH_4^+$, peptone, nitrate, tryptone | $NH_4^+$, glutamate, gelatin, yeast extract, tryptone, urea |
| Habitat | Active hydrothermal chimneys | Active hydrothermal sulfide chimneys |

Data from Moussard et al. (2004) and Alain et al. (2010)

Brock reported that it contains lipids combining bacterial and archaeal properties: glycerol is ether-linked to a unique C17 hydrocarbon side chain along with some fatty acids instead of phytanyl side chains (Brock 2000). They contain cytochrome $c_3$ (Hatchikian et al. 1984) and $\alpha_2\beta_2$ type of desulfofuscidin (sulfite reductase) but no desulfoviridin (Hatchikian and Zeikus 1983). Some proteins from *T. commune* and *T. thermophilum* were purified and characterized: APS reductase (Fauque et al. 1986), desulfofuscidin (Fauque et al. 1990; Hatchikian 1994; Hatchikian and Zeikus 1983), cytochrome $c_3$ (Fauque et al. 1991; Hatchikian et al. 1984; LeGall and Fauque 1988), rubredoxin (LeGall and Fauque 1988; Shimizu et al. 1989), and [NiFe] hydrogenase (Fauque et al. 1992).

## *Thermodesulfatator* Moussard et al. 2004, 231$^{VP}$

Ther.mo.de.sul.fa.ta'tor. Gr. adj. *thermos* hot; N.L. masc. n. *desulfatator* sulfate-reducer; N.L. masc. n. *Thermodesulfatator* thermophilic sulfate-reducer.

The genus *Thermodesulfatator* consists of the type species *T. indicus* (Moussard et al. 2004) and *T. atlanticus* (Alain et al. 2010). The characteristics of each species are summarized in *Table 30.3*. Rod-shaped cells occur singly, in pairs, or in 3 chains. The Gram reaction is negative, and spore formation is not reported. The cells are motile with a single polar flagellum, and they grow under thermophilic, neutrophilic, and halophilic conditions. Anaerobic chemolithoautotrophic growth occurs with $H_2$ as an electron donor, sulfate as an electron acceptor, and $CO_2$ as a carbon source. Organic compounds are not used as electron donors, and elemental sulfur, thiosulfate, sulfite, cysteine, and nitrate are not used as electron acceptors. In addition to $CO_2$, *T. atlanticus* can use formate, acetate, propionate, methanol, pyruvate, glucose, monomethylamine, peptone, and yeast extract as carbon sources with sulfate under an $H_2$ atmosphere; however, *T. indicus* is unable to. Conversely, some organic components can stimulate their growth. Ammonium and tryptone are used as nitrogen sources. Cells are resistant to streptomycin and kanamycin, but sensitive to ampicillin. The G+C content is approximately 46 mol%. The major fatty acids consist of $C_{18:0}$, $C_{18:1}\omega7c$, and $C_{16:0}$. The following cellular components of *T. indicus* have been reported: menaquinone-7 is the major respiratory quinone; the polar lipids are predominantly phosphatidylinositol and phosphatidylethanolamine; and the main polyamine is $N^4$-bis(aminopropyl) spermidine (Hosoya et al. 2004).

## *Caldimicrobium* Miroshnichenko et al. 2009, 1042$^{VP}$

Cal.di.mi.cro'bi.um. L. adj. *caldus* hot; N.L. neut n. *microbium* microbe; N.L. neut. n. *Caldimicrobium* microbe living in hot places.

The type species *C. rimae* is only the member of the genus *Caldimicrobium* (Miroshnichenko et al. 2009). The cells are ovals or rods (0.5 μm in width and 1.0–1.2 μm in length) with 2 polar flagella. The Gram reaction is negative and spore formation is not reported. Optimal growth occurs at 75 °C and pH 7.0–7.2. It grows as a facultative chemolithoautotroph using $H_2$ as an electron donor, thiosulfate and elemental sulfur as electron acceptors, and $CO_2$ as a carbon source. Sulfate, sulfite, nitrate, nitrite, amorphous Fe(III) oxide, and Fe(III) citrate are not used as electron acceptors. In addition to $H_2$, organic compounds such as ethanol, fumarate, succinate, and malate are used as electron donors for chemoheterotrophic growth. Fermented growth by pyruvate, lactate, or fumarate is not observed.

Yeast extract is not required for growth and does not stimulate growth. The G+C content is 35 mol%. The type strain DS$^T$ of *C. rimae* was isolated from a terrestrial neutral hot spring.

### *Thermosulfurimonas* Slobodkin et al. 2012, 2569$^{VP}$

Ther.mo.sul.fu.ri.monas. Gr. adj. *thermos* hot; L. neut. n. *sulfur* sulfur; Gr. fem. n. *monas* a unit, monad; N.L. fem. n. *Thermosulfurimonas* thermophilic sulfur monad.

The type species *T. dismutans* is the only member of the genus *Thermosulfurimonas* (Slobodkin et al. 2012). The cells are ovals or rods (0.5–0.6 μm in width and 1.0–1.5 μm in length) with a single polar flagellum. The Gram reaction is negative and endospores are not formed. Optimal growth occurs at 74 °C and pH 7.0. Growth occurs at an NaCl concentration ranging from 15 to 35 g/L. It grows strictly anaerobically and chemolithoautotrophically using elemental sulfur, thiosulfate, and sulfite as energy sources and $CO_2$ as a carbon source. Elemental sulfur, thiosulfate, and sulfite are disproportionated to sulfide and sulfate (Bak and Cypionka 1987; Thamdrup et al. 1993). Their growth is enhanced by using poorly crystalline Fe(III) oxide (ferrihydrite) as a sulfide-scavenging agent. In the presence of elemental sulfur and ferrihydrite, malate and maleinate stimulate growth. It is able to grow with thiosulfate and $H_2/CO_2$ in the absence of ferrihydrite. It does not reduce sulfate, nitrate, ferrihydrite, Fe(III) citrate, 9,10-anthraquinone, fumarate, or oxygen with acetate, lactate, ethanol, pyruvate, malate, peptone, and $H_2$ as electron donors. The G+C content is approximately 52 mol%. The major fatty acids consist of $C_{18:0}$, anteiso-$C_{17:0}$, and $C_{17:0}$. Type strain S95$^T$ of *T. dismutans* was isolated from a deep-sea hydrothermal vent.

### *Geothermobacterium*

Ge.o.ther.mo.bac.te'ri.um. Gr. n. *ge* the earth; Gr. adj. *thermos* hot; L. neut. n. *bacterium* a small rod; N.L. neut. n. *Geothermobacterium* rod from hot earth.

*G. ferrireducens* is the only species belonging to this genus, but has not been validated yet (Kashefi et al. 2002). Rod-shaped cells (0.5 μm in width and 1.0–1.2 μm in length) occur singly or in pairs. Spore formation is not observed. Electron microscopic observation indicates that the cell wall structure is typical of Gram-negative cells. The cells are highly motile with a single polar flagellum. Optimal growth occurs at 85–90 °C, pH 6.8–7.0, and 0–0.5 g/L NaCl. It grows strictly chemolithoautotrophically using $H_2$ as an electron donor and poorly crystalline Fe(III) oxide (ferrihydrite) as an electron acceptor. Sulfate, thiosulfate, sulfite, elemental sulfur, nitrate, fumarate, Fe(III) citrate, Fe(III) pyrophosphate, and structural Fe(III) within phyllosilicate minerals (Kashefi et al. 2008) are not used as electron acceptors. Its growth is inhibited by chloramphenicol, puromycin, rifampin, erythromycin, kanamycin, phosphomycin, vancomycin, and trimethoprim. It inhabits the sediment of terrestrial hot springs.

## Isolation, Enrichment, and Maintenance Procedures

Members of the family *Thermodesulfobacteriaceae* can be obtained on different anaerobic media. Cultivation can be performed in the presence of suitable energy sources (sulfur compounds and Fe(III)) and carbon sources ($CO_2$ and organic compounds) for each genus. Incubation temperatures ranging from 65 °C to 90 °C are recommended. Enrichment and several transfer procedures may be necessary before isolation. Isolation is possible by picking a colony from agar or gellan gum by the roll-tube method and by repeating serial dilutions in liquid medium. Immunomagnetic cultivation using specific antisera as capture agents was reported as a successful enrichment procedure (Christensen et al. 1992).

Liquid cultures retain their viability after storage at room temperature for several months. Long-term preservation in liquid nitrogen in the presence of 10 % (v/v) glycerol or 5 % (v/v) DMSO under anaerobic conditions is possible.

## Ecology

### Habitat

Members of the family *Thermodesulfobacteriaceae* are found in thermal environments such as terrestrial hot springs, oil reservoirs, and hydrothermal fields. Culture-dependent studies have been limited, and most isolates belonging to this family have been described as new species. They were isolated from terrestrial hot springs at Ink Pot Spring and Obsidian Pool in Yellowstone, Icelandic alkaline hot springs, and the Treshchinnyi Spring in Kamchatka (Kashefi et al. 2002; Miroshnichenko et al. 2009; Sonne-Hansen and Ahring 1999; Zeikus et al. 1983), from stratal waters of the oil deposit in the Caspian Sea (Hatchikian et al. 2001), and from deep-sea hydrothermal fields at Guaymas Basin, Central Indian Ridge, Mid-Atlantic Ridge and Eastern Lau Spreading Center (Alain et al. 2010; Jeanthon et al. 2002; Moussard et al. 2004; Slobodkin et al. 2012). In addition, *Thermodesulfobacterium* species were retrieved from thermal anaerobic oil water originating from porous rock formations located at 2–4 km below the sea floor in the Norwegian sector of the North Sea (Christensen et al. 1992) and an oil reservoir (L'Haridon et al. 1995). In culture-independent studies, 16S rRNA gene sequences belonging to the family have been identified in terrestrial hot springs (Everroad et al. 2012; Lau et al. 2009; Meyer-Dombard et al. 2011; Otaki et al. 2012; Skirnisdottir et al. 2000; Spear et al. 2005), hydrothermal fields (Nakagawa et al. 2005; Nunoura and Takai 2009; Postec et al. 2007), oil reservoirs (Kobayashi et al. 2012; Pham et al. 2009), and biomat that developed in a hot-water-springing underground mine (Nakagawa et al. 2002). In addition to the

16S rRNA gene, *dsrAB* genes, which were inferred to be close to the family, have been detected from an active deep-sea hydrothermal chimney (Nakagawa et al. 2004). Culture-dependent and culture-independent studies have indicated that the family inhabits determinate thermal environments. Conversely, in the last decade, its inhabitable area gradually expanded to hydrothermal fields (Alain et al. 2010; Jeanthon et al. 2002; Moussard et al. 2004; Slobodkin et al. 2012) and higher temperature sites (Kashefi et al. 2002). Further study may provide a comprehensive understanding of the contributors to sulfur and iron cycling in thermal environments.

## References


Alain K, Postec A, Grinsard E, Lesongeur F, Prieur D, Godfroy A (2010) *Thermodesulfatator atlanticus* sp. nov., a thermophilic, chemolithoautotrophic, sulfate-reducing bacterium isolated from a Mid-Atlantic Ridge hydrothermal vent. Int J Syst Evol Microbiol 60:33–38

Anderson I, Saunders E, Lapidus A, Nolan M, Lucas S, Tice H, Del Rio TG, Cheng JF, Han C, Tapia R, Goodwin LA, Pitluck S, Liolios K, Mavromatis K, Pagani I, Ivanova N, Mikhailova N, Pati A, Chen A, Palaniappan K, Land M, Hauser L, Jeffries CD, Chang YJ, Brambilla EM, Rohde M, Spring S, Goker M, Detter JC, Woyke T, Bristow J, Eisen JA, Markowitz V, Hugenholtz P, Kyrpides NC, Klenk HP (2012) Complete genome sequence of the thermophilic sulfate-reducing ocean bacterium *Thermodesulfatator indicus* type strain (CIR29812$^T$). Stand Genomic Sci 6:155–164

Bak F, Cypionka H (1987) A novel type of energy metabolism involving fermentation of inorganic sulphur compounds. Nature 326:891–892

Brock TD (2000) Biology of microorganisms, 9th edn. Prentice-Hall International, London

Christensen B, Torsvik T, Lien T (1992) Immunomagnetically captured thermophilic sulfate-reducing bacteria from north sea oil field waters. Appl Environ Microbiol 58:1244–1248

Collins MD, Weddel F (1986) Respiratory quinones of sulphate-reducing and sulphur-reducing bacteria: a systematic investigation. Syst Appl Microbiol 8:8–18

Everroad RC, Otaki H, Matsuura K, Haruta S (2012) Diversification of bacterial community composition along a temperature gradient at a thermal spring. Microbes Environ 27:374–381

Fauque G, Czechowski M, Kang-Lissolo L, DerVartanian DV, Moura JJG, Moura I, Lampreia J, Xavier AV, LeGall J (1986) Purification of adenylyl sulfate (APS) reductase and desulfofuscidin from a thermophilic sulfate reducer: *Desulfovibrio thermophilus*. In: Annual meeting of the Society for Industrial Microbiology, San Francisco, p 92

Fauque G, Lino AR, Czechowski M, Kang L, DerVartanian DV, Moura JJ, LeGall J, Moura I (1990) Purification and characterization of bisulfite reductase (desulfofuscidin) from *Desulfovibrio thermophilus* and its complexes with exogenous ligands. Biochim Biophys Acta 1040:112–118

Fauque G, LeGall J, Barton LL (1991) Sulfate-reducing and sulfur-reducing bacteria. In: Shively JM, Barton LL (eds) Variations in autotrophic life. Academic, London, pp 271–337

Fauque G, Czechowski M, Berlier YM, Lespinat PA, LeGall J, Moura JJG (1992) Partial purification and characterization of the first hydrogenase isolated from a thermophilic sulfate-reducing bacterium. Biochem Biophys Res Commun 184:1256–1260

Friedrich MW (2002) Phylogenetic analysis reveals multiple lateral transfers of adenosine-5′-phosphosulfate reductase genes among sulfate-reducing microorganisms. J Bacteriol 184:278–289

Hamana K, Hamana H, Niitsu M, Samejima K, Itoh T (1996) Distribution of long linear and branched polyamines in thermophilic eubacteria and hyperthermophilic archaebacteria. Microbios 85:19–33

Hamana K, Hamana H, Shinozawa T, Niitsu M, Samejima K, Itoh T (1999) Polyamines of the thermophilic eubacteria belonging to the genera *Aquifex*, *Thermodesulfobacterium*, *Thermus* and *Meiothermus*, and the thermophilic archaebacteria belonging to the genera *Sulfurisphaera*, *Sulfophobococcus*, *Stetteria*, *Thermocladium*, *Pyrococcus*, *Thermococcus*, *Methanopyrus* and *Methanothermus*. Microbios 97:117–130

Hatchikian EC (1994) Desulfofuscidin: dissimilatory, high-spin sulfite reductase of thermophilic, sulfate-reducing bacteria. In: Peck HD Jr, LeGall J (eds) Inorganic microbial sulfur metabolism. Academic, San Diego, pp 276–295

Hatchikian EC, Zeikus JG (1983) Characterization of a new type of dissimilatory sulfite reductase present in *Thermodesulfobacterium commune*. J Bacteriol 153:1211–1220

Hatchikian EC, Papavassiliou P, Bianco P, Haladjian J (1984) Characterization of cytochrome $c_3$ from the thermophilic sulfate reducer *Thermodesulfobacterium commun*. J Bacteriol 159:1040–1046

Hatchikian EC, Ollivier B, Garcia J-L (2001) Family I. *Thermodesulfobacteriaceae* fam. nov. In: Boon DR, Castenholz RW (eds) Bergey's manual of systematic bacteriology, 2nd edn, The *Archaea* and the deeply branching and phototrophic *Bacteria*. Springer, New York, pp 390–393

Hosoya R, Hamana K, Niitsu M, Itoh T (2004) Polyamine analysis for chemotaxonomy of thermophilic eubacteria: polyamine distribution profiles within the orders *Aquificales*, *Thermotogales*, *Thermodesulfobacteriales*, *Thermales*, *Thermoanaerobacteriales*, *Clostridiales* and *Bacillales*. J Gen Appl Microbiol 50:271–287

Jeanthon C, L'Haridon S, Cueff V, Banta A, Reysenbach AL, Prieur D (2002) *Thermodesulfobacterium hydrogeniphilum* sp. nov., a thermophilic, chemolithoautotrophic, sulfate-reducing bacterium isolated from a deep-sea hydrothermal vent at Guaymas Basin, and emendation of the genus *Thermodesulfobacterium*. Int J Syst Evol Microbiol 52:765–772

Kashefi K, Holmes DE, Reysenbach AL, Lovley DR (2002) Use of Fe(III) as an electron acceptor to recover previously uncultured hyperthermophiles: isolation and characterization of *Geothermobacterium ferrireducens* gen. nov., sp. nov. Appl Environ Microbiol 68:1735–1742

Kashefi K, Shelobolina ES, Elliott WC, Lovley DR (2008) Growth of thermophilic and hyperthermophilic Fe(III)-reducing microorganisms on a ferruginous smectite as the sole electron acceptor. Appl Environ Microbiol 74:251–258

Klein M, Friedrich M, Roger AJ, Hugenholtz P, Fishbain S, Abicht H, Blackall LL, Stahl DA, Wagner M (2001) Multiple lateral transfers of dissimilatory sulfite reductase genes between major lineages of sulfate-reducing prokaryotes. J Bacteriol 183:6028–6035

Kobayashi H, Endo K, Sakata S, Mayumi D, Kawaguchi H, Ikarashi M, Miyagawa Y, Maeda H, Sato K (2012) Phylogenetic diversity of microbial communities associated with the crude-oil, large-insoluble-particle and formation-water components of the reservoir fluid from a non-flooded high-temperature petroleum reservoir. J Biosci Bioeng 113:204–210

L'Haridon S, Reysenbacht AL, Glenat P, Prieur D, Jeanthon C (1995) Hot subterranean biosphere in a continental oil reservoir. Nature 377:223–224

Langworthy TA, Holzer G, Zeikus JG, Tornabene TG (1983) Iso- and anteiso-branched glycerol diethers of the thermophilic anaerobe *Thermodesulfotobacterium commune*. Syst Appl Microbiol 4:1–17

Lau MC, Aitchison JC, Pointing SB (2009) Bacterial community composition in thermophilic microbial mats from five hot springs in central Tibet. Extremophiles 13:139–149

LeGall J, Fauque G (1988) Dissimilatory reduction of sulfur compounds. In: Zehnder AJB (ed) Biology of anaerobic microorganisms. Wiley, New York, pp 587–639

Meyer-Dombard DR, Swingley W, Raymond J, Havig J, Shock EL, Summons RE (2011) Hydrothermal ecotones and streamer biofilm communities in the Lower Geyser Basin, Yellowstone National Park. Environ Microbiol 13:2216–2231

Miroshnichenko ML, Lebedinsky AV, Chernyh NA, Tourova TP, Kolganova TV, Spring S, Bonch-Osmolovskaya EA (2009) *Caldimicrobium rimae* gen. nov., sp. nov., an extremely thermophilic, facultatively lithoautotrophic, anaerobic bacterium from the Uzon Caldera, Kamchatka. Int J Syst Evol Microbiol 59:1040–1044

Mori K, Kim H, Kakegawa T, Hanada S (2003) A novel lineage of sulfate-reducing microorganisms: *Thermodesulfobiaceae* fam. nov., *Thermodesulfobium narugense*, gen. nov., sp. nov., a new thermophilic isolate from a hot spring. Extremophiles 7:283–290

Moussard H, L'Haridon S, Tindall BJ, Banta A, Schumann P, Stackebrandt E, Reysenbach AL, Jeanthon C (2004) *Thermodesulfatator indicus* gen. nov., sp. nov., a novel thermophilic chemolithoautotrophic sulfate-reducing bacterium isolated from the Central Indian Ridge. Int J Syst Evol Microbiol 54:227–233

Nakagawa T, Hanada S, Maruyama A, Marumo K, Urabe T, Fukui M (2002) Distribution and diversity of thermophilic sulfate-reducing bacteria within a Cu-Pb-Zn mine (Toyoha, Japan). FEMS Microbiol Ecol 41:199–209

Nakagawa T, Nakagawa S, Inagaki F, Takai K, Horikoshi K (2004) Phylogenetic diversity of sulfate-reducing prokaryotes in active deep-sea hydrothermal vent chimney structures. FEMS Microbiol Lett 232:145–152

Nakagawa S, Takai K, Inagaki F, Chiba H, Ishibashi J, Kataoka S, Hirayama H, Nunoura T, Horikoshi K, Sako Y (2005) Variability in microbial community and venting chemistry in a sediment-hosted backarc hydrothermal system: impacts of subseafloor phase-separation. FEMS Microbiol Ecol 54:141–155

Nunoura T, Takai K (2009) Comparison of microbial communities associated with phase-separation-induced hydrothermal fluids at the Yonaguni Knoll IV hydrothermal field, the Southern Okinawa Trough. FEMS Microbiol Ecol 67:351–370

Otaki H, Everroad RC, Matsuura K, Haruta S (2012) Production and consumption of hydrogen in hot spring microbial mats dominated by a filamentous anoxygenic photosynthetic bacterium. Microbes Environ 27:293–299

Pham VD, Hnatow LL, Zhang S, Fallon RD, Jackson SC, Tomb JF, DeLong EF, Keeler SJ (2009) Characterizing microbial diversity in production water from an Alaskan mesothermic petroleum reservoir with two independent molecular methods. Environ Microbiol 11:176–187

Postec A, Lesongeur F, Pignet P, Ollivier B, Querellou J, Godfroy A (2007) Continuous enrichment cultures: insights into prokaryotic diversity and metabolic interactions in deep-sea vent chimneys. Extremophiles 11:747–757

Rozanova EP, Khudyakova AI (1974) Participation of microorganisms in sulphur tunover in Pomaretzkoye Lake. Mikrobiologiya 43:908–912

Rozanova EP, Pivovarova TA (1988) Reclassification of *Desulfovibrio thermophilus* (Rozanova and Khudykova, 1974). Mikrobiologiya 57:102–106

Shimizu F, Ogata M, Yagi T, Wakabayashi S, Matsubara H (1989) Amino acid sequence and function of rubredoxin from *Desulfovibrio vulgaris* Miyazaki. Biochimie 71:1171–1177

Skerman VDB, McGowan C, Sneath PHA (1980) Approved lists of bacterial names. Int J Syst Bacteriol 30:225–420

Skirnisdottir S, Hreggvidsson GO, Hjorleifsdottir S, Marteinsson VT, Petursdottir SK, Holst O, Kristjansson JK (2000) Influence of sulfide and temperature on species composition and community structure of hot spring microbial mats. Appl Environ Microbiol 66:2835–2841

Slobodkin AI, Reysenbach AL, Slobodkina GB, Baslerov RV, Kostrikina NA, Wagner ID, Bonch-Osmolovskaya EA (2012) *Thermosulfurimonas dismutans* gen. nov., sp. nov., an extremely thermophilic sulfur-disproportionating bacterium from a deep-sea hydrothermal vent. Int J Syst Evol Microbiol 62:2565–2571

Sonne-Hansen J, Ahring BK (1999) *Thermodesulfobacterium hveragerdense* sp. nov., and *Thermodesulfovibrio islandicus* sp. nov., two thermophilic sulfate reducing bacteria isolated from a Icelandic hot spring. Syst Appl Microbiol 22:559–564

Spear JR, Walker JJ, McCollom TM, Pace NR (2005) Hydrogen and bioenergetics in the Yellowstone geothermal ecosystem. Proc Natl Acad Sci USA 102:2555–2560

Tao T-S, Yue Y-Y, Fang C-X (1996) Irregularities in the validation of the genus *Thermodesulfobacterium* and its species: request for an opinion. Int J Syst Bacteriol 46:622

Thamdrup B, Finster K, Hansen JW, Bak F (1993) Bacterial disproportionation of elemental sulfur coupled to chemical reduction of iron or manganese. Appl Environ Microbiol 59:101–108

Trüper HG (2003) Valid publication of the genus name *Thermodesulfobacterium* and the species names *Thermodesulfobacterium commune* (Zeikus, et al. 1983) and *Thermodesulfobacterium thermophilum* (ex *Desulfovibrio thermophilus* Rozanova and Khudyakova 1974). Opinion 71. Int J Syst Evol Microbiol 53:927

Yarza P, Ludwig W, Euzeby J, Amann R, Schleifer KH, Glockner FO, Rossello-Mora R (2010) Update of the All-Species Living Tree Project based on 16S and 23S rRNA sequence analyses. Syst Appl Microbiol 33:291–299

Zeikus JG, Dawson MA, Thompson TE, Ingvorsen K, Hatchikian EC (1983) Microbial ecology of volcanic sulphido-genesis: isolation and characterization of *Thermodesulfobacterium commune* gen. nov. and sp. nov. J Gen Microbiol 129:1159–1169

Zverlov V, Klein M, Lucker S, Friedrich MW, Kellermann J, Stahl DA, Loy A, Wagner M (2005) Lateral gene transfer of dissimilatory (bi)sulfite reductase revisited. J Bacteriol 187:2203–2208

# 31 The Family *Thermoactinomycetaceae*

*Leonor Carrillo · Marcelo Rafael Benítez-Ahrendts*
National University of Jujuy, Jujuy, Argentina



**Abstract**

The family *Thermoactinomycetaceae* is a member of the order *Bacillales*, Gram-positive bacteria that form endospores and mycelia, are non-acid-fast, and do not contain mycolic acids in their cell wall. At the time of writing, it encompasses 13 genera and only 20 species.

The genera are *Thermoactinomyces, Laceyella, Seinonella, Thermoflavimicrobium, Planifilum, Mechercharimyces, Shimazuella, Desmospora, Kroppenstedtia, Marininema, Melghirimyces, Lihuaxuella*, and *Polycladomyces.*

Strains of this family have been isolated from various environmental samples, such as soil, marine sediments, sugar cane, compost, sputa, and other sources.

E. Rosenberg et al. (eds.), *The Prokaryotes – Firmicutes and Tenericutes*, DOI 10.1007/978-3-642-30120-9_355,

## Short Description

### *Thermoactinomycetaceae* Matsuo et al., 2006, emend. Yassin et al., 2009, von Jan et al., 2011, and Li et al. 2012

Aerial mycelium may be produced and form substrate mycelia. Aerial mycelia are white or yellow. Form well-developed, branched, and septate substrate mycelia. Form sessile spores, singly on aerial and substrate hyphae, or on simple or branched sporophores, with the structure and properties of bacterial endospores. Gram-positive, chemo-organotrophic, and aerobic. The genera form a coherent phylogenetic unit on the basis of partial 16S rRNA gene sequences and comprise the family. The pattern of 16S rRNA gene signatures consists of (C–G) at positions 415:428, (C–G) at 441:493, (C–G) at 681:709, (G–C) at 682:708, and (G) at 694. Cell walls contain *meso*-diaminopimelic acid (*meso*-DAP) or LL-diaminopimelic acid (LL-DAP). Major menaquinones are unsaturated with seven or nine isoprene units. The G+C content of the DNA ranges from 40 mol% to 60.3 mol%. The type genus is *Thermoactinomyces.*

Because of the morphological characteristics of the genus *Thermoactinomyces* (Tsilinsky 1899) aerobic, Gram-positive, and showing filamentous growth, members of this genus were earlier considered as actinomycetes. Meanwhile, due to dipicolinic-acid-containing endospore (Cross et al. 1968), DNA G+C content (Lacey and Cross 1989), menaquinone profiles (Collins et al. 1982), phylogenetic studies of the 5S rRNA (Park et al. 1993), 16S rRNA (Stackebrandt and Woese 1981), and 16S rDNA sequences (Yoon and Park 2000), it was subsequently placed within the family *Bacillaceae.* Later, based on phylogenetic analysis and chemotaxonomic characteristics, the genus *Thermoactinomyces* was assigned to a new family, *Thermoactinomycetaceae* (Matsuo et al. 2006), containing other five genera: *Laceyella, Thermoflavimicrobium, Seinonella* (Yoon et al. 2005), *Planifilum* (Hatayama et al. 2005), and *Mechercharimyces* (Matsuo et al. 2006). Yassin et al. (2009), Von Jan et al. (2011) and Li et al. (2012) have recently emended the description of the family.

At the time of writing, the family *Thermoactinomycetaceae* accommodates 13 genera including: *Thermoactinomyces* (Tsilinsky 1899), *Laceyella, Seinonella, Thermoflavimicrobium* (Yoon et al. 2005), *Planifilum* (Hatayama et al. 2005), *Mechercharimyces* (Matsuo et al. 2006), *Shimazuella* (Park et al. 2007), *Desmospora* (Yassin et al. 2009), *Kroppenstedtia* (Von Jan et al. 2011), *Marininema* (Li et al. 2012), *Melghirimyces* (Addou et al. 2012), *Lihuaxuella* (Yu et al. 2012), *Polycladomyces* (Tsubouchi et al. 2013).

The main features of 13 members of *Thermoactinomycetaceae* are listed in ➲ *Table 31.1.*

## Phylogenetic Tree of Type Strains of Species

List of type strains used for dendrogram construction (➲ *Fig. 31.1*): *Thermoactinomyces vulgaris* KCTC 9076$^T$, *Thermoactinomyces intermedius* KCTC 9646$^T$, *Laceyella sacchari* DSM 43356$^T$, *Laceyella putida* KCTC 3666$^T$, *Laceyella sediminis* RHA1$^T$, *Laceyella tengchongensis* YIM 10002$^T$, *Thermoflavimicrobium dichotomica* KCTC 3667$^T$, *Seinonella peptonophila* KCTC 9740$^T$, *Planifilum yunnanense* LA5$^T$, *Planifilum fimeticola* H0165$^T$, *Planifilum fulgidum* 500275$^T$, *Mechercharimyces mesophilus* YM3-251$^T$, *Mechercharimyces asporophorigenens* YM11-542$^T$, *Kroppenstedtia eburnea* DSM 45196$^T$, *Desmospora activa* IMMIB L-1269$^T$, *Shimazuella kribbensis* A 9500$^T$, *Marininema mesophilum* SCSIO 10219$^T$, *Melghirimyces algeriensis* NariEX$^T$.

## DNA–DNA Hybridization

DNA–DNA relatedness levels among the type species of *Thermoactinomyces vulgaris* KCTC 9076$^T$-*Thermoactinomyces intermedius* KCTC 9646$^T$ and *Laceyella sacchari* KCTC 9790$^T$-*Laceyella putida* KCTC 3666$^T$ were, respectively, 47.4–48.7 % and 7.7–11.9 %. The strains of *Thermoflavimicrobium dichotomica* KCTC 3667$^T$ and *Seinonella peptonophila* KCTC 9740$^T$ exhibited levels of relatedness ranging from 2.5 % to 5.6 % with *Thermoactinomyces* and *Laceyella* species (Yoon et al. 2000). The DNA–DNA relatedness levels between *Planifilum yunnanense* LA5$^T$ and *P.fimeticola* H0165$^T$ and *P. fulgidum* 500275$^T$ were among 37.0 and 64.7% (Hatayama et al. 2005; Zhang et al. 2007). *Mechercharimyces mesophilus* YM3-251$^T$ and *Mechercharimyces asporophorigenens* YM11-542$^T$ had a DNA–DNA relatedness of 49 % (Matsuo et al. 2006). *Melghirimyces algeriensis* NariEX$^T$ hybridization was done with *Kroppenstedtia eburnea* DSM 45196$^T$ and the two strains shared only 9.6 % DNA–DNA relatedness (Addou et al. 2012). All these values are below the 70 % threshold for the delineation of genomic species (Stackebrandt and Goebel 1994). In the DNA–DNA hybridization assay between *Polycladomyces abyssicola* JIR-001$^T$ and its phylogenetic neighbors, *Melghirimyces algeriensis* NariEXT, *Planifilum fimeticola* H0165T, *P. fulgidum* 500275 T, and *P. yunnanense* LA5T, the mean hybridization levels were 5.3–7.5 %, 2.3–4.7 %, 2.1–4.8 %, and 2.5–4.9 %, respectively (Tsubouchi et al. 2013). Descriptions of genera *Desmospora, Shimazuella, Marininema* and *Lihuaxuella* did not include results of DNA–DNA hybridization.

## Genome Comparison

16S rRNA oligonucleotide sequencing revealed that the genus *Thermoactinomyces* was more closely related to *Bacillus* species than to actinomycetes (Stackebrandt and Woese 1981). The type strain of *Thermoactinomyces vulgaris* was phylogenetically related to the genus *Bacillus* based on 5S rRNA sequences (Park et al. 1993). Also, each genus of the family *Thermoactinomycetaceae* possesses its distinctive signature nucleotides at positions 154:167, 203:214, and 693 (➲ *Table 31.1*) (Yu et al. 2012).

*Thermoactinomyces vulgaris* KCTC 9076$^T$ and *Thermoactinomyces intermedius* KCTC 9646$^T$ exhibited relatively high

**Table 31.1**
Differential characteristics of 13 genera in the family *Thermoactinomycetaceae*

| Characteristic \ Taxa | | *Thermoactinomyces* | *Thermoflavimicrobium* | *Laceyella* | *Seinonella* | *Mechercharimyces* | *Shimazuella* | *Desmospora* | *Planifilum* | *Lihuaxuella* | *Kroppenstedtia* | *Melghirimyces* | *Marininema* | *Polycladomyces* |
|---|---|---|---|---|---|---|---|---|---|---|---|---|---|---|
| Color of aerial mycelia | | white | yellow | white | white | white | white | yellow | – | – | white | yellow | – | white |
| Dichotomously branched sporophores | | – | + | – | – | – | – | – | – | – | – | – | – | – |
| Growth on novobiocin (25 µg/ml) | | + | + | + | – | + | + | + | + | nd | + | nd | nd | + |
| Degradation of: | Gelatin | + | + | – | – | + | – | + | + | + | + | + | – | + |
| | Starch | – | – | – | – | – | + | + | + | – | – | – | – | – |
| Optimal temperature for growth °C | | 50-55 | 55 | 48-55 | 35 | 30 | 32 | 37-50 | 55-70 | 50 | 45 | 40-55 | 30 | 60 |
| Diaminopimelic acid | | *meso*-DAP | *meso*-DAP | *meso*-DAP | *meso*-DAP | *meso*-DAP | *meso*-DAP | *meso*-DAP | *meso*-DAP | *meso*-DAP | *LL*-DAP | *LL*-DAP | *LL*-DAP | *meso*-DAP |
| Predominant menaquinone | | MK-7 | MK-7 | MK-9 | MK-7 | MK-9 | MK-9 | MK-7 | MK-7 | MK-7 | MK-7 | MK-7 | MK-7 | MK-7 |
| Other menaquinones | | MK-8 or MK-9 | nr | MK-8 or MK-10 | MK-8 or MK-9 or MK-10 | MK-8 | MK-10 | nr | nr | nr | nr | MK-6, MK-8 | nr | MK-8 |
| Major fatty acids | | iso-$C_{15:0}$, anteiso-$C_{15:0}$ | iso-$C_{15:0}$, anteiso-$C_{15:0}$, iso-$C_{16:0}$ | iso-$C_{15:0}$, anteiso-$C_{15:0}$ | iso-$C_{14:0}$, ante iso-$C_{15:0}$, iso-$C_{16:0}$ | iso-$C_{15:0}$, iso-$C_{17:1}$ ω11c | anteiso-$C_{15:0}$, iso-$C_{16:0}$, iso-$C_{15:0}$, anteiso-$C_{17:0}$ | iso-$C_{15:0}$, $C_{16:0}$, iso-$C_{17:0}$ | iso-$C_{17:0}$, anteiso-$C_{17:0}$, iso-$C_{15:0}$ or $C_{16:0}$ | iso-$C_{15:0}$, anteiso-$C_{15:0}$, anteiso-$C_{17:0}$ | iso-$C_{15:0}$, anteiso-$C_{15:0}$ | iso-$C_{15:0}$, anteiso-$C_{15:0}$ | anteiso-$C_{15:0}$, iso-$C_{15:0}$ | iso-$C_{15:0}$, iso-$C_{17:0}$, iso-$C_{16:0}$, $C_{16:0}$ |
| Polar lipids components | | nd | nd | DPG, PE, PG, PI, PIM | nd | nd | PE | DPG, PG, PE, PME | nd | DPG, PG, PME, PE, APL | DPG, PG, PE | PG,DPG, PE, PME, PL | DPG, PME, PE, PG, PL | DPG, PE, PME, PG, PS |
| DNA G+C content (mol%) | | 48.0 | 43.0 | 47.9-49 | 40 | 45.0 | 39.4 | 49,3 | 56.8-60.3 | 55.6 | 54.6 | 47.3 | 46.5 | 55.1 |
| 16rRNA gene signature nucleotides at positions | 154:167 | U-G | U-G | U-U | U-A | C-A | U-G | U-G | U-A | A-U | U-G | C-A | U-A | nd |
| | 203:214 | A-G | A-C | -C | A-U | A-C | A-U | A-A | U-C | G-G | A-C | A-C | A-G | nd |
| | 693 | G | G | G | G | G | G | G | G | U | G | G | G | nd |

Data from: Addou et al. 2012; Chen et al. 2012; Hatayama et al. 2005; Li et al. 2012; Matsuo et al. 2006; Park et al. 2007; Tsubouchi et al. 2013; von Jan et al. 2011; Yassin et al. 2009; Yoon et al. 2005; Yu et al. 2012; Zhang et al. 2007; Zhang et al. 2010

*nr* not recorded, *nd* no data, + positive reaction, – negative reaction, *v* variable, *meso-DAP* meso-diaminopimelic acid, *LL*-diaminopimelic acid, *APL* aminophospholipid, *DPG* diphosphatidylglycerol, *PE* phosphatidylethanolamine, *PG* phosphatidylglycerol, *PI* phosphatidylinositol, *PIM* phosphatidylinositolmannosides, *PL* unknown phospholipid, *PME* phosphatidyl-monomethylethanolamine

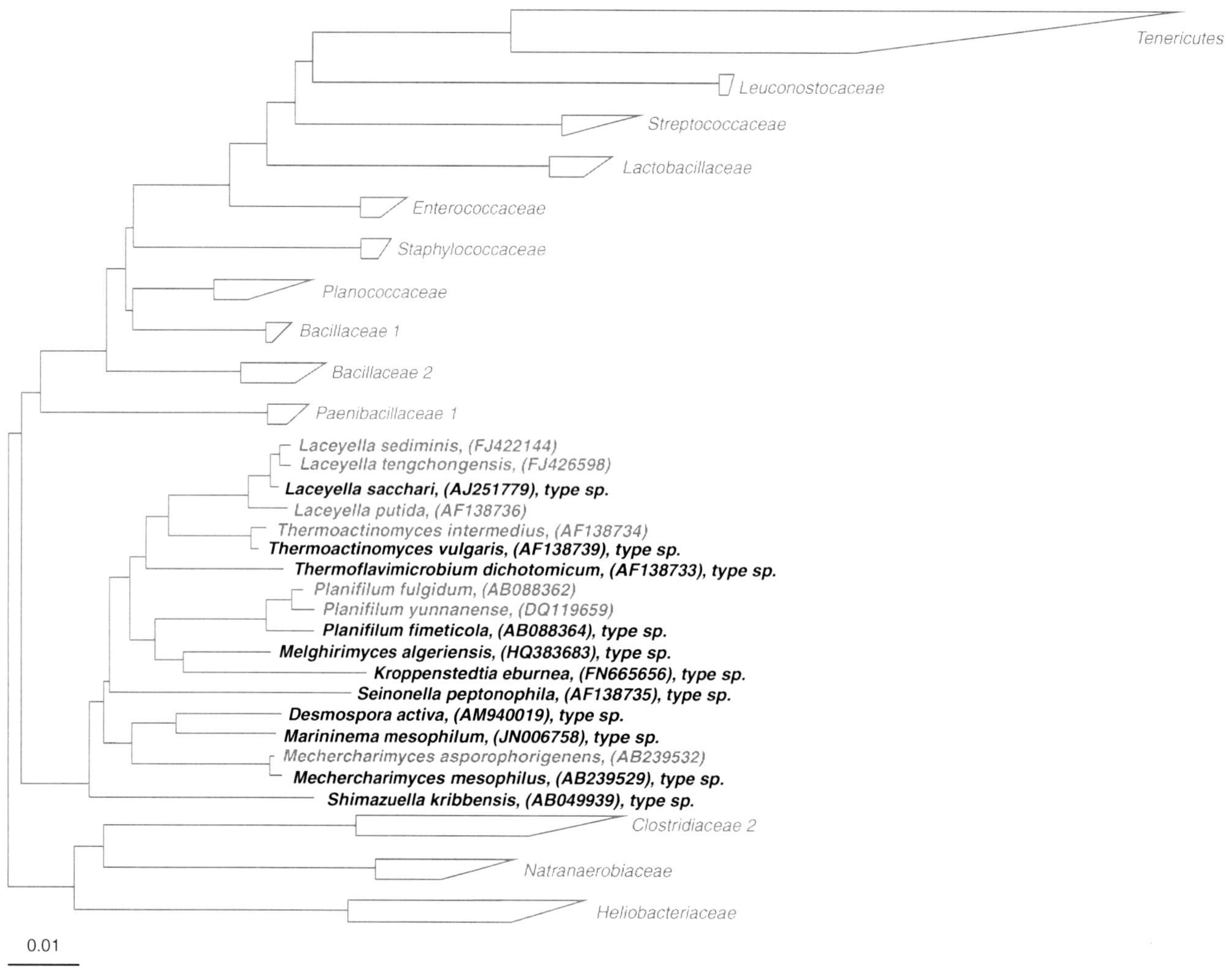


**Fig. 31.1**

**Neighbor joining phylogenetic tree based on 16S rNA gene sequence (before descriptions of *Lihuaxuella thermophila* YIM 77831$^T$ and *Polycladomyces abyssicola* JIR 001$^T$). *Bar*, 0.01 nucleotide substitutions per site (Rosselló-Mora et al. 2012)**

16S rDNA similarity value. *Thermoflavimicrobium dichotomicum* KCTC 3667$^T$ and *Seinonella peptonophila* KCTC 9740$^T$ exhibited relatively low 16S rRNA gene similarity levels, 90.1–94.2 % and 90.1–91.6 %, respectively, to the type strains of *Thermoactinomyces* and *Laceyella* species (Yoon et al. 2005).

The almost complete 16S rRNA gene sequences of *Laceyella sediminis* RHA1$^T$ and *Laceyella tengchongensis* YIM 10002$^T$ had 1,467 bp and 1,398 bp, respectively. *L. sediminis* RHA1$^T$ shared 99.8, 99.6, and 97.8 % 16S rRNA gene sequence similarities, respectively, with *L. sacchari* DSM 43356$^T$, *L. tengchongensis* YIM 10002$^T$, and *L. putida* DSM 44608$^T$. The *L. tengchongensis* YIM 10002$^T$ showed <95.2 % sequence similarity with other members of the family *Thermoactinomycetaceae* (Zhang et al. 2010; Chen et al. 2012). *L. sacchari* KCTC 9790$^T$ and *L. sacchari* KCTC 9789 (previously *T. thalpophilus*) had the same 16S rDNA sequences, except a single position corresponding to one ambiguous nucleotide (Yoon et al. 2005).

The results of alignment with 16S rRNA gene sequences obtained from GenBank showed that *Planifilum yunnanense* strain LA5$^T$ had similarities of 99 % and 97 % with *Planifilum fulgidum* 500275$^T$ and *Planifilum fimeticola* H0165$^T$ (Zhang et al. 2007). Hatayama et al. (2005) found that the closest related species to the thermophilic strains H0165$^T$ and 500275$^T$ were those of the genus *Thermoactinomyces sensu stricto* (91.2–91.6 % 16S rRNA gene sequence similarity), *Laceyella* (88.8–89.4 %), *Thermoflavimicrobium* (89.3–89.7 %), and *Seinonella* (87.8–88.1 %).

The 16S rRNA gene sequences of the strains *Mechercharimyces mesophilus* YM3-251$^T$ and *Mechercharimyces asporophorigenens* YM11-542$^T$ had similarity values in the range 99.4–99.9 %, values that correspond to 2–9 nucleotide differences, respectively, at 1,428 locations. The gene encoding the DNA gyrase B subunit was used to highlight differences between the strains. The sequences had similarity values within the range 96.5–99.8 % corresponding, respectively, to 2–40 nucleotide differences at 1,167 locations. Phylogenetic analyses based on 16S rRNA and gyrB gene sequences showed that the strains constituted an independent clade within the family *Thermoactinomycetaceae* (Matsuo et al. 2006).

Based on 1,459-bp long 16S rRNA gene sequences, the highest 16S rRNA gene similarity values found between *Shimazuella kribbensis* A 9500$^T$ and other genera of the family were with *T. vulgaris* KCTC 9557 (90.38 %), *T. dichotomicum* KCTC 3667$^T$ (90.22 %), *L. sacchari* KCTC 9790$^T$ (90.04 %), *T. intermedius* KCTC 9646$^T$ (89.98 %), *L. putida* KCTC 3666$^T$ (89.67 %), *M. asporophorigenens* DSM 44955$^T$ (89.50 %), *M. mesophilus* DSM 44894$^T$ (89.41 %), *S. peptonophila* KCTC 9740$^T$ (89.41 %), *P. fimeticola* JCM 12507$^T$ (88.43 %), and *P. fulgidum* JCM 12508$^T$ (88.35 %) (Park et al. 2007).

The comparison of the 16S rRNA gene sequence of *Kroppenstedtia eburnea* JFMB-ATE$^T$ with those of the type species within the family *Thermoactinomycetaceae* revealed the highest degree of sequence similarity with *P. fimeticola* (93.1 %), *D. activa* (93.0 %), *P. yunnanense* (93.0 %), *P. fulgidum* (92.9 %), *M. mesophilus* (92.1 %), *L. putida* (91.9 %), *M. asporophorigenens* (91.8 %), *L. sacchari* (91.7 %), *T. intermedius* (91.3 %), *T. dichotomicum* (91.0 %), and *T. vulgaris* (91.0 %). The strain clustered nearest to the genus *Planifilum*; however, the high degree of 16S rRNA gene sequence dissimilarity (6.9 %) indicated only a distant relationship (von Jan et al. 2011).

The almost complete gene sequence of *Desmospora activa* IMMIB L-1269$^T$ had 1497 nt. The phylogenetic analysis showed that strain IMMIB L-1269$^T$ formed a distinct sub-line within of the family *Thermoactinomycetaceae* and also displayed a highest sequence similarity with respect to the type strains of *Thermoflavimicrobium dichotomicum* (92.3 %), *M. asporophorigenens* (92.3 %), *M. mesophilus* (92.1 %), *Laceyella sacchari* (92.2 %), *Laceyella putida* (91.9 %), *Thermoactinomyces vulgaris* (91.9 %), *Thermoactinomyces intermedius* (91.7 %), *Planifilum* species (91.9 % or less), *Shimazuella kribbensis* (90.4 %), and *Seinonella peptonophila* (89.0 %). Strain IMMIB L-1269$^T$ clustered with members of the genus *Planifilum*, but divergence values were relatively high (>7.7 %) (Yassin et al. 2009).

The 16S rRNA gene sequence similarity values between *Marininema mesophilum* SCSIO 10219$^T$ and other genera of the family *Thermoactinomycetaceae* were among 88.9 and 94.9%. In the phylogenetic tree, strain SCSIO 10219$^T$ formed a stable clade with *D. activa* IMMIB L-1269$^T$ but the relatively high sequence divergence values (>5.1 %) showed that the isolate is distantly related to it (Li et al. 2012).

Based on a 16S rRNA gene consensus sequence of 1,344 nt, *Melghirimyces algeriensis* NariEX$^T$ belongs to the family *Thermoactinomycetaceae* with the type strains of *Kroppenstedtia eburnea* and *Desmospora activa* as its closest phylogenetic neighbors (95.38 % and 95.28 % similarities, respectively) (Addou et al. 2012).

The highest 16S rRNA gene sequence similarity values between *Lihuaxuella thermophila* YIM 77831$^T$ and other type strains of the family *Thermoactinomycetaceae* were among 91.2 and 95.5%. However, the phylogenetic analysis revealed relatively high sequence divergence values (>4.5 %), which clearly distinguished the strain YIM 77831$^T$ from other members of the family. The almost complete 16S rRNA gene sequence had 1,516 bp (Yu et al. 2012).

Similarity grades between the 16S rRNA gene sequence of *Polycladomyces abyssicola* JIR-001$^T$ and those of the type genus of *Thermoactinomycetaceae* species were as follows: *M. algeriensis*, 93.5 %; *P. fimeticola*, 92.9 %; *P. fulgidum*, 92.5 %; *D. activa*, 92.4 %; *P. yunnanense*, 92.3 %; *T. intermedius*, 92.3 %; *T. vulgaris*, 92.0 %; *T. dichotomicum*, 90.8 %; *K. eburnea*, 90.6 %; *M. asporophorigenens*, 90.5 %; *L. putida*, 89.8 %; *L. sacchari*, 89.8 %; *L. tengchongensis*, 89.5 %; *L. sediminis*, 89.7 %; *M. mesophilus*, 88.0 %; *S. peptonophila*, 87.5 %; and *S. kribbensis*, 85.5 %. *M. algeriensis* and *P. fimeticola* are the closest neighbors (Tsubouchi et al. 2013).

## MALDI-TOF

Only a number of clinical isolates *Kroppenstedtia eburnea* were identified by proteomic profiles obtained by matrix-assisted laser desorption/ionization time of flight mass spectrometry (Barker et al. 2012).

## Phages

Phages are commonly associated with genera *Thermoactinomyces* and *Laceyella* and they are species specific (Treuhaft 1977). Some *T. vulgaris* phages show cross-infectivity with members of the genus *Bacillus* but not with cell wall chemotype III actinomycetes (Kurtboke and Sivasithamparam 1993). The phage genome was incorporated into spores early in their formation in a heat-stable state and only multiplied on germination (Kretschmer 1980).

## Phenotypic Analyses

### *Thermoactinomyces* Tsilinsky, 1899 emend. Yoon et al. 2005

*Thermoactinomyces* species are aerobic, Gram-positive, non-acid-fast, chemo-organotrophic, thermophilic, and grow at 55 °C but not at 30 °C (optimum 50–55 °C). The aerial mycelium is abundant and white. The substrate mycelium consists of stable, branched, septate hyphae, from which aerial hyphae arise, forming a loose network of almost straight hyphae over the substrate. Spores are formed singly on both substrate and aerial mycelium and may be either sessile or on sporophores. The spores are spheroidal, 0.5–1.5 μm in diameter, with a ridged surface that gives an angular appearance, refractile and phase-bright, staining only with endospores stains. Growth occurs in the presence of novobiocin (25 μg/ml), and mycelia degrade casein, gelatin, arbutin, and esculin but not starch, hypoxanthine, or xanthine. A pH greater than 7.0 is essential to spore germination. The wall peptidoglycan contains *meso*-diaminopimelic acid (6.5–7 %) but not diagnostic sugars in the cell wall, indicating that the wall chemotype is type III. This genus has unsaturated menaquinones with seven, or eight and nine isoprene units and the predominant one is MK-7. The cellular fatty acid profile contains major amounts of iso-C(15:0), and significant amounts of iso-C(17:0) and anteiso-C(15:0) (Lacey and Cross 1989; Yoon et al. 2005). The GC content of DNA is 48 mol %. The pattern of 16S rRNA gene signatures consists of (U–G) at positions 154:167, (A–G) at 203:214 and (G) at 693 (Yu et al. 2012).

Two species are assigned to the genus: *Thermoactinomyces vulgaris* (type species) and *Thermoactinomyces intermedius* (➲ *Table 31.2*).

### *Thermoactinomyces vulgaris* Tsilinsky 1899

Basonym: *Thermoactinomyces albus* Orlowska 1969; *Thermoactinomyces candidus* Kurup et al. 1975

The type strain KCTC 9076$^T$ (= KCC A-0162$^T$ = ATCC 43649$^T$ = DSM 43016$^T$ = NBRC 13606$^T$ = JCM 3162$^T$ = VKM Ac-1195$^T$) was isolated from compost. GenBank accession number (16 rRNA gene): AF 138739.

Colonies are fast-growing, flat on nutrient and CYC agars at 55 °C, with a moderate covering of white mycelium and, often, a feathery margin on CYC agar. Endospores are produced on short, unbranched sporophores. The colony reverse is white or cream, never pink or brown. No soluble pigments are produced. pH of usual media 7.2–7.4. Grows on and produces hemolysis in blood agar. Elastin; DNA; RNA; and Tweens 20, 40, 60, and 80 are degraded, but not adenine, cellulose, chitin, guanine, keratin, testosterone, L-tyrosine, or xylan. D-glucose, D-mannose, D-ribose, and glycerol are used as a sole carbon source. Nitrate is not reduced to nitrite. Growth occurs in media containing lysozyme (0.005 % w/v) or 3 % (w/w) NaCl. Produces $C_4$ esterase, $C_8$ lipase, phosphoamidase, alkaline phosphatase, leucine aminopeptidase, and acid phosphatase. Whole-cell hydrolysates contain glucose and mannose. The GC content of DNA is 48 mol %. (Kurup et al. 1975; Lacey and Cross 1989). The cellular fatty acid profile is iso-$C_{15:0}$ (54.3 %), iso-$C_{17:0}$ (18.6 %), anteiso-$C_{15:0}$ (10.2 %), and anteiso-$C_{17:0}$ (5.4 %) (Yoon et al. 2005) (➲ *Table 31.2*).

### *Thermoactinomyces intermedius* Kurup et al. 1980

The type strain KCTC 9646$^T$ (= ATCC 33205$^T$ = T-323$^T$ = DSM 43846$^T$ = JCM 3312$^T$ = NBRC 14230$^T$ = NRRL B-16979$^T$ = VKM Ac-1427$^T$) was isolated from air conditioner filter. GenBank accession number (16 rRNA gene): AJ251775.

Colonies have white aerial mycelium and yellowish to yellowish-brown substrate mycelium. Brown, water-soluble melanin pigments produced on media with 0.5 % (w/v) L-tyrosine. Endospores are sessile or produced on short sporophores. The width of hyphae varied from 0.6 to 1.0 μm. Growth is good at 50–55 °C but poor at 37 °C. pH of usual media is 7.2–7.4. The strains degrade L-tyrosine, DNA, elastin, but not chitin or cellulose, and show resistance to lysozyme. Nitrate is not reduced to nitrite. The spores are resistant to heating at 100 °C for 2 h. The cellular fatty acid profile is iso-$C_{15:0}$ (55.1 %), iso-$C_{17:0}$ (16.3 %), and anteiso-$C_{15:0}$ (11.4 %). The GC content of DNA is 48 mol % (Lacey and Cross 1989; Yoon et al. 2005) (➲ *Table 31.2*).

## *Laceyella* Yoon et al. 2005

Members of the genus *Laceyella* are aerobic, chemo-organotrophic, Gram-positive, and thermophilic filamentous bacteria (optimal temperature 48–55 °C) and grow on media supplemented with novobiocin (25 μg/ml). Substrate and aerial mycelia are formed, and the aerial mycelium is white. Sessile endospores may be produced on sporophores. Grayish-yellow or yellow-brown diffusible pigment may be produced. The predominant menaquinone is MK-9. The cell-wall peptidoglycan contains meso-diaminopimelic acid (Yoon et al. 2005; Chen et al. 2012). The pattern of 16S rRNA gene signature nucleotides consists of (U–U) at positions 154:167, (-C) at 203:214 and (G) at 693 (Yu et al. 2012).

The genus comprises four species: *Laceyella sacchari* (type species), *Laceyella putida, Laceyella tengchongensis,* and *Laceyella sediminis* (➲ *Table 31.2*).

### *Laceyella sacchari* (Waksman and Cork 1953) Yoon et al. 2005

Basonym: *Thermoactinomyces sacchari* Lacey 1971; *Thermoactinomyces thalpophilus* Waksman and Cork 1953; *Thermoactinomyces thalpophilus* Lacey and Cross 1989.

The type strain DSM 43356$^T$ (= ATCC 27375$^T$ = KCTC 9790$^T$ = NCIMB 10486$^T$ = CCUG 7967$^T$ = JCM 3137$^T$ = JCM 3214$^T$ = NBRC 13920$^T$ = NCTC 10721$^T$ = NRRL B-16981$^T$ = VKM Ac-1360$^T$) was isolated from sugar cane bagasse. GenBank accession number (16S rRNA gene): AJ251779.

A sparse, transient, tufted aerial mycelium, rapidly autolyzing and depositing endospores in a thick layer, is produced on the surface of yeast-malt or nutrient agar supplemented with 1 % (w/v) glucose. Growth on nutrient agar is poor, restricted, and thin, with no aerial mycelium and few spores. Endospores are produced on sporophores up to 3-mm long. Yellow-brown soluble pigment may be produced. Growth occurs at 35–65 °C (optimum 55–60 °C). Water-soluble melanin may be produced on CYC agar with 0.5 % (w/v) L-tyrosine. pH of usual media is 7.2–7.4. Starch; elastin; DNA; RNA; and Tweens 20, 40, 60, and 80 are degraded, but not adenine, cellulose, guanine, or keratin. Degradation of gelatin, aesculin, arbutin, chitin, and tyrosine is variable. D-fructose, maltose, and D-mannitol are utilized as carbon sources. Cellulose, meso-inositol, D-raffinose, L-rhamnose, and D-xylose are not utilized. Utilization of L-arabinose, D-mannose, and sucrose is variable. Growth in the presence of 1 % (w/v) NaCl is variable. *L. sacchari* strains produce alkaline phosphatase, $C_4$ esterase, and $C_8$ lipase but not

**Table 31.2**
**Differentiating characteristics of species of *Thermoactinomyces* and *Laceyella***

| Characteristics \ Species | | | *Thermoactinomycesvulgaris* [a, d] | *Thermoactinomyces intermedius* [a, c,d] | *Laceyella sacchari* [a, d] | *Laceyella putida* [a, d] | *Laceyella tengchongensis* [e] | *Laceyella sediminis* [f] |
|---|---|---|---|---|---|---|---|---|
| Abundant white aerial mycelium | | | + | + | − | + | + | + |
| Colour of substrate mycelium | | | white or cream | yellowish -brown | olive-buff | yellowish -brown | yellow-white | yellow-white |
| Soluble pigment | | | − | − | +/− (yellow -brown) | greyish yellow | − | − |
| Melanin production | | | − | + | +/− | + | − | − |
| Degradation | Starch | | − | − | + | + | − | + |
| | Gelatin | | + | + | +/− | + | + | + |
| | Hypoxanthine | | − | − | − | − | + | + |
| | Xanthine | | − | − | − | + | − | + |
| | L-Tyrosine | | − | + | +/− | + | + | nd |
| | Adenine | | − | nd | − | − | − | + |
| Growth conditions | Temperature °C | range | 35–60 | 35–60 | 35–65 | 36–58 | 28–70 | 28–65 |
| | | opt. | 55 | 50–55 | 55–60 | 48 | 55 | 55 |
| | pH | range | nd | nd | nd | nd | 6.0–8.0 | 5.0–9.0 |
| | | opt. | 7.2–7.4* | 7.2–7.4* | 7.2–7.4* | 7.2–7.4* | nd | 7.0 |
| | NaCl (%) w/v | 5 | + | nd | − | − | nd | − |
| | | 1 | + | nd | +/− | − | nd | + |
| Utilization | Lactose | | − | nd | + | − | + | + |
| | Maltose | | + | nd | + | + | − | + |
| | Trehalose | | + | nd | + | − | + | + |
| | Raffinose | | nd | nd | + | − | − | − |
| | D-Mannitol | | + | nd | + | − | + | − |
| | D-Fructose | | + | nd | + | − | − | nd |
| | D-Mannose | | + | nd | +/− | − | − | − |
| | L-Rhamnose | | − | nd | − | − | + | − |
| | D-Ribose | | + | nd | + | − | − | − |
| | D-Xylose | | − | nd | − | nd | − | − |
| | Glycine | | nd | nd | + | + | − | + |
| | L-Cysteine | | nd | nd | + | − | + | − |
| | L-Lysine | | nd | nd | − | − | + | − |
| | L-Proline | | nd | nd | − | + | − | + |
| | L-Serine | | nd | nd | + | + | − | − |
| | L-Threonine | | nd | nd | − | + | + | − |
| | L-Valine | | nd | nd | − | + | − | + |
| DNA G+C content (mol%) | | | 48 | 48 | 48.0 | 49.0 | 48.6 | 47.9 |
| Cellular fatty acids | | | iso-$C_{15:0}$, iso-$C_{17:0}$, anteiso-$C_{15:0}$ | iso-$C_{15:0}$, iso-$C_{17:0}$, anteiso-$C_{15:0}$ | iso-$C_{15:0}$, anteiso-$C_{15:0}$, iso-$C_{16:0}$ | iso-$C_{15:0}$, anteiso-$C_{15:0}$, iso-$C_{14:0}$, iso-$C_{16:0}$ | iso-$C_{15:0}$, anteiso-$C_{15:0}$, iso-$C_{16:0}$, iso-$C_{14:0}$ | iso-$C_{15:0}$, anteiso-$C_{15:0}$ |
| Whole-cell sugars | | | Glu, Man[b] | nd | Xyl, Ara, Glu | Xyl, Ara, Glu | Rib, Xyl, Glu | Rib, Glu |
| Phospholipids | | | nd | nd | DPG, PE, PME, PI, PIM, PL | DPG, PE, PG, PI, PIM, PL | DPG, PE, PG, PI, PIM, PL | DPG, PE, PG, PI, PIM, PL |
| Menaquinones | | | MK-7 | MK-7 | MK-9, MK-8, MK-10 | MK-9, MK-8 | MK-9, MK-8 | MK-9, MK-8 |

*pH of recommended media, *nd* no data, + positive reaction, − negative reaction, +/− variable reaction. *Ara* arabinose, *Glu* glucose, *Man* mannose, *Rib* ribose, *Xyl* xylose. *DPG* diphosphatidylglycerol, *PE* phosphatidylethanolamine, *PI* phosphatidylinositol, *PIM* phosphatidylinositolmannosides, *PME* phosphatidylmethylethanolamine, *PG* phosphatidylglycerol, *PL* unknown phospholipids

Data from: [a]Lacey and Cross 1989
[b]McCarthy and Cross 1984
[c]Kurup et al. 1980
[d]Yoon et al. 2005
[e]Zhang et al. 2010
[f]Chen et al. 2012

α- or β-glucosidase or β-glucuronidase. Whole-cell sugars are xylose, arabinose, and glucose. Menaquinones (peak area ratio) are MK-9 (75 %), MK-8, and MK-10. The major cellular fatty acids are iso-$C_{15:0}$ (66.61 %) and anteiso-$C_{15:0}$ (12.02 %). The polar lipids are diphosphatidylglycerol, phosphatidylethanolamine, phosphatidylmethylethanolamine phosphatidylinositol, phosphatidylinositolmannosides, and other phospholipids. DNA G+C content of the type strain is 48 mol% (Lacey and Cross 1989; Yoon et al. 2005) (➲ *Table 31.2*).

### *Laceyella putida* (Lacey and Cross 1989) Yoon et al. 2005

Basonym: *Thermoactinomyces putidus* Lacey and Cross 1989.

The type strain is KCTC 3666$^T$ (= NCIMB 12324$^T$ = ATCC 49853$^T$ = DSM 44608$^T$ = JCM 8091$^T$). GenBank accession number (16S rRNA gene): AF138736.

Colonies are often very wrinkled and puckered with endospores formed on short and unbranched sporophores. Aerial mycelium white, but may appear cream, pale yellow, or yellowish-brown due to yellowish-brown substrate mycelium. Sporing hyphae lyze quickly, leaving spores on the surface of agar. Grayish-yellow soluble pigment may be produced. Brown, water-soluble melanin pigment is produced on CYC agar supplemented with 0.5 % (w/v) L-tyrosine. Growth between 36 °C and 58 °C, optimally at 48 °C. Sensitive to 0.5–1 % (w/v) NaCl. pH of usual media is 7.2–7.4. Degradation of aesculin, arbutin, and chitin is variable. L-Tyrosine, gelatin, and starch are degraded, but not hypoxanthine and DNA. Sucrose and maltose are utilized as carbon sources. D-Fructose, glycerol, D-mannitol, D-mannose, D-ribose, and D-trehalose are not utilized. *L. putidus* strains produce acid phosphatase, chymotrypsin, and leucine aminopeptidase. Whole-cell hydrolysates contain glucose and ribose. Whole-cell sugars are xylose, arabinose, and glucose. The menaquinones are MK-9 and MK-8. The cellular fatty acids are iso-$C_{15:0}$ (50.66 %), anteiso-$C_{15:0}$ (18.34 %), iso-$C_{14:0}$ (9.89 %), iso-$C_{16:0}$ (9.68 %), and others. The polar lipids are diphosphatidylglycerol, phosphatidylethanolamine, phosphatidylglycerol, phosphatidylinositol, phosphatidylinositolmannosides, and other phospholipids. The DNA G+C content of the type strain is 49 mol% (Lacey and Cross 1989; Yoon et al. 2005) (➲ *Table 31.2*).

### *Laceyella tengchongensis* Zhang et al. 2010

The type strain YIM 10002$^T$ (= DSM 45262$^T$ = CCTCC AA 208050$^T$) was isolated from a soil sample collected from a big empty volcano. GenBank/EMBL/DDBJ accession number (16S rRNA gene): FJ426598.

Gram-positive, aerobic, chemo-organotrophic, thermophilic, and filamentous bacteria. Substrate and aerial mycelia are well developed and form endospores (0.7–0.8 μm). Aerial and substrate mycelia are white to yellow-white. No soluble pigment is produced. Strain grows well at pH 6.0–8.0 and 55 °C, but not below 28 °C or above 70 °C. Type strains grow well on oatmeal agar, moderately well on Czapek's, yeast-malt agar and inorganic salts starch agar, weakly on glycerol-asparagine agar, and not at all on nutrient agar. Positive for gelatin liquefaction and milk peptonization and coagulation, and negative for nitrate reduction and $H_2S$ and melanin production. Casein, hypoxanthine, gelatin, and L-tyrosine are degraded, but not adenine, xanthine, starch, and urea. L-Fucose, lactose, mannitol, and L-rhamnose are utilized as carbon sources, but not D-arabinose, D-fructose, D-galactose, maltose, D-mannose, raffinose, D-ribose, and D-xylose. L-Arginine, L-asparagine, L-cysteine, L-lysine, and L-threonine are used as nitrogen sources, but not adenine, glycine, L-hydroxyproline, L-proline, L-serine, L-valine, and xanthine. The whole-cell sugars are ribose, xylose, and glucose. The menaquinones (peak area ratio) are MK-9 (87 %) and MK-8 (13 %). The polar lipids are diphosphatidylglycerol, phosphatidylethanolamine, phosphatidylglycerol, phosphatidylinositol, phosphatidylinositolmannosides, and other phospholipids. The cellular fatty acids profile contains major amounts of branched fatty acids and minor amounts of straight-chain and unsaturated fatty acids. The fatty acids are iso-$C_{15:0}$ (57.63 %), anteiso-$C_{15:0}$ (13.79 %), iso-$C_{16:0}$ (8.88 %), iso-$C_{14:0}$ (7.11 %), and others. The DNA G+C content of the type strain is 48.6 mol% (Zhang et al. 2010) (➲ *Table 31.2*).

### *Laceyella sediminis* Chen et al. 2012

The type strain RHA1$^T$ (= DSM 45263$^T$ = CCTCC AA 208058$^T$) was isolated from a sediment sample of a hot spring. GenBank/EMBL/DDBJ accession number (16S rRNA gene): FJ422144.

Gram-positive, aerobic, thermophilic, and filamentous bacteria. White aerial and yellow-white substrate mycelia are produced, bearing single endospores on short sporophores. No soluble pigments are produced. Growth occurs at 28–65 °C (optimum 55 °C), at pH 5.0–9.0 (optimum pH 7.0), and in the presence of 0–1 % (w/v) NaCl (optimum 0 %). Positive for gelatin liquefaction, degradation of starch, and milk peptonization and coagulation, but negative for nitrate reduction, and $H_2S$ and melanin production. Utilizes lactose, trehalose, maltose, and gelatin as carbon sources, but not D-mannitol, L-rhamnose, raffinose, D-mannose, D-xylose, or D-ribose. Degrades L-proline, xanthine, L-valine, adenine, glycine, L-asparagine, and L-arginine, but not L-lysine, L-cysteine, L-threonine, L-serine, or L-hydroxyproline. Whole-cell hydrolysates contain ribose and glucose. The phospholipids comprised diphosphatidylglycerol, phosphatidylethanolamine, phosphatidylglycerol, phosphatidylinositol, phosphatidylinositolmannosides, and other phospholipids. The predominant menaquinone MK-9, MK-8 was detected as a minor component. The cellular fatty acids include branched, straight-chain, and unsaturated components. The major fatty acids are iso-$C_{15:0}$ (62.39 %) and anteiso-$C_{15:0}$ (17.55 %). The DNA G+C content of the type strain is 47.9 mol% (Chen et al. 2012).

## *Planifilum* Hatayama et al. 2005

Gram-positive, aerobic, and thermophilic. Substrate mycelia are formed on Luria-Bertani agar, CYC agar, SY agar, and Bacto nutrient agar, but aerial mycelia are not formed. Endospores are produced singly along mycelia. The cell-wall peptidoglycan contains meso-diaminopimelic acid, alanine, and glutamic acid but no diagnostic sugars (Hatayama et al. 2005). Grows with novobiocin (25 μg/ml) (von Jan et al. 2011). The pattern of 16S rRNA gene signatures consists of (U–A) at positions 154:167, (U–C) at 203:214 and (G) at 693 (Yu et al. 2012).

The genus comprises three species: *Planifilum fimeticola* (type species), *Planifilum fulgidum*, and *Planifilum yunnanense* (➋ *Table 31.3*).

### *Planifilum fimeticola* Hatayama et al. 2005

The type strain H0165[T] (=ATCC BAA-969[T] = JCM 12507[T]) was isolated from a hyperthermal composting process. GenBank/EMBL/DDBJ accession number (16S rRNA gene): AB088364.

Colonies are lustrous, cream-yellow with radial wrinkles. Growth occurs at 50–65 °C, optimally at 55–63 °C. Casein, starch, and L-tyrosine are degraded. Lactose, D-raffinose, trehalose, D-arabinose, D-fucose, D-galactose, D-mannose, D-sorbitol, xylitol, L-threonine, and inosine are utilized as carbon sources. The menaquinone is MK-7 (98.8 %). The cellular fatty acids are iso-$C_{16:0}$ (45.1 %), iso-$C_{17:0}$ (13.9 %), anteiso-C(17:0) 13.2 %, iso-C(18:0) 6.1 %, and others. DNA G+C content of the type strain is 60.3 mol% (Hatayama et al. 2005; Zhang et al. 2007) (➋ *Table 31.3*).

### *Planifilum fulgidum* Hatayama et al. 2005

The type strain 500275[T] (=ATCC BAA-970[T] = JCM 12508[T]) was isolated from a hyperthermal composting process. GenBank/EMBL/DDBJ accession number (16S rRNA gene): AB088362.

Colonies are lustrous, cream-yellow with radial wrinkles. Growth occurs at 50–67 °C, optimally at 60–65 °C. Casein and starch are degraded but not L-tyrosine. Sucrose, D-arabinose, D-xylose, i-erythritol, xylitol, methyl-β-D-glucoside, α-hydroxybutyric acid, itaconic acid, quinic acid, L-aspartic acid, L-ornithine, and L-proline are utilized as carbon sources. Major menaquinone (peak area ratio) is MK-7 (98.1 %). MK-8 is at trace level. The cellular fatty acids are iso-$C_{17:0}$ (34.8 %), anteiso-$C_{17:0}$ (17.6 %), iso-$C_{15:0}$ (11.4 %), iso-$C_{16:0}$ (9.6 %), $C_{16:0}$ (8.8 %), and others. The DNA G+C content is 60.0 mol% (Hatayama et al. 2005; Zhang et al. 2007) (➋ *Table 31.3*).

### *Planifilum yunnanense* Zhang et al. 2007

The type strain LA5[T] (=CCTCC AA206002[T] = KCTC 13052[T]) was isolated from a hot spring. GenBank/EMBL/DDBJ accession number (16S rRNA gene): DQ119659.

Colonies are rough and cream-yellow with radial wrinkles. Growth occurs at 50–75 °C (optimum growth at 60–70 °C) and pH 6.0–10.0 (optimum growth at pH 8.5) on Luria-Bertani agar, potato dextrose agar, Czapek and starch-casein but only weak growth was observed on glycerol-asparagine and Gause's synthetic medium. Casein, gelatin, and starch are degraded. Lactose, D-raffinose, gentibiose, sucrose, D-xylose, D-fucose, D-galactose, D-allulose, *myo*-inositol, i-erythritol, D-sorbitol, *N*-acetyl-D-glucosamine, methyl-β-D-glucoside, succinic acid monomethyl ester, D-galacturonic acid, *p*-hydroxyphenylacetic acid, itaconic acid, succinic acid, L-aspartic acid, L-glutamic acid, L-leucine, L-ornithine, L-proline, L-threonine, inosine, and uridine are utilized as carbon sources. The predominant menaquinone is MK-7. MK-8 is detected at trace level. The fatty acids are iso-$C_{17:0}$ (27.7 %), $C_{16:0}$ (22.4 %), iso-$C_{16:0}$ (8.8 %), anteiso-$C_{17:0}$ (8.6 %), and others. The DNA G+C content is 56.8 mol% (Zhang et al. 2007) (➋ *Table 31.3*).

## *Mechercharimyces* Matsuo et al. 2006

Cells are aerobic, Gram-positive, and mesophilic. Form aerial mycelia and substrate mycelia. Aerial mycelia are abundant and white. Form well-developed, branched, and septate substrate mycelia on marine agar 2216. Do not produce soluble pigment. Cell-wall peptidoglycan contains meso-DAP, glutamic acid, and alanine, but no characteristic sugars (Matsuo et al. 2006). The pattern of 16S rRNA gene signatures consists of (C-A) at positions 154:167, (A–C) at 203:214 and (G) at 693 (Yu et al. 2012).

The genus comprises two species: *Mechercharimyces mesophilus* (type) and *Mechercharimyces asporophorigenens* (➋ *Table 31.3*).

### *Mechercharimyces mesophilus* Matsuo et al. 2006

The type strain YM3-251[T] (= MBIC06230[T] = DSM 44894[T]) was isolated from a sediment from a marine lake. GenBank/EMBL/DDBJ accession number (16S rRNA gene): AB239529.

Exhibits the following properties in addition to those given in the genus description. Colonies are fast-growing, lightly ridged, with a moderate covering of white mycelia and a feathery margin on marine agar 2216 at 27 °C. Growth occurs at 15–37 °C, with optimum growth at 30 °C and pH 7.6–8.0. Forms endospores singly on short, unbranched sporophores. Casein and gelatin are degraded, but not starch, hypoxanthine, xanthine, or L-tyrosine. Produces dark-brown pigment on L-tyrosine-containing marine agar. Growth occurs in the presence of novobiocin (25 mg/ml). Presents alkaline

**Table 31.3**
**Differentiating characteristics of species of *Planifilum* and *Mechercharimyces***

| Characteristics / Species | | | *Planifilum fimeticola* [b] | *Planifilum fulgidum* [b] | *Planifilum yunnanense* [c] | *Mechercharimyces mesophilus* [a] | *Mechercharimyces asporophorigenens* [a] |
|---|---|---|---|---|---|---|---|
| Aerial mycelium | | | – | – | – | white | white |
| Substrate mycelium | | | Lustrous, cream-yellow with radial wrinkles | | Rough and cream-yellow | Inconspicuous | |
| Endosporos | | | + | + | + | + | – |
| Dark-brown pigment on L-tyrosine media | | | – | – | – | + | + |
| Growth conditions | Temperature °C | range | 50–65 | 50–67 | 50–75 | 15–37 | 20–37 |
| | | opt. | 55–63 | 60–65 | 60–70 | 30 | 30 |
| | pH | range | nd | nd | 6.0–10.0 | nd | nd |
| | | opt. | 7.5* | 7.5* | 8.5 | 7.6–8.0* | 7.6–8.0* |
| | Novobiocin 25 µg/ml | | nd | nd | nd | + | + |
| Degradation | Casein | | + | + | + | + | + |
| | Starch | | + | + | + | – | – |
| | Gelatin | | nd | nd | + | + | + |
| | Hypoxanthine | | – | – | nd | – | – |
| | Xanthine | | – | – | nd | – | – |
| | Aesculin | | + | – | nd | – | – |
| | L-Tyrosine | | + | – | – | – | – |
| Utilization | D-Raffinose | | + | – | + | nd | nd |
| | Gentiobiose | | – | – | + | nd | nd |
| | Trehalose | | + | – | – | nd | nd |
| | Lactose | | + | – | + | nd | nd |
| | Sucrose | | – | + | + | nd | nd |
| | D-Arabinose | | + | + | – | nd | nd |
| | D-Xylose | | – | + | + | nd | nd |
| | D-Galactose | | + | – | + | nd | nd |
| | D-Mannose | | + | – | – | nd | nd |
| | D-Allulose | | – | + | + | nd | nd |
| | *myo*-Inositol | | – | – | + | nd | nd |
| | *i*-Erythritol | | – | + | + | nd | nd |
| | D-Sorbitol | | + | – | + | nd | nd |
| | Xylitol | | + | + | – | nd | nd |
| | N-Acetyl-D-glucosamine | | – | – | + | nd | nd |
| | Methyl β-D-glucoside | | – | + | + | nd | nd |
| | Succinic acid monomethyl ester | | – | – | + | nd | nd |
| | D-Galacturonic acid | | – | – | + | nd | nd |
| | α-Hydroxybutyric acid | | – | + | – | nd | nd |
| | *p*-Hydroxyphenylacetic acid | | – | – | + | nd | nd |
| | Itaconic acid | | – | + | + | nd | nd |
| | Quinic acid | | – | + | – | nd | nd |
| | Succinic acid | | – | – | + | nd | nd |
| | L-Aspartic acid | | – | + | + | nd | nd |
| | L-Glutamic acid | | – | – | + | nd | nd |
| | L-Leucine | | – | – | + | nd | nd |
| | L.Ornithine | | – | + | + | nd | nd |
| | L-Proline | | – | + | + | nd | nd |
| | L-Threonine | | + | – | + | nd | nd |
| | Inosine | | + | – | + | nd | nd |
| | Uridine | | – | – | + | nd | nd |
| DNA G+C content (mol%) | | | 60.3 | 60.0 | 58.6 | 45.1 | 45.2 |
| Major celular fatty acids | | | iso-$C_{16:0}$, iso-$C_{17:0}$, anteiso-$C_{17:0}$, iso-$C_{18:0}$ | iso-$C_{17:0}$, anteiso-$C_{17:0}$, iso-$C_{15:0}$ | iso-$C_{17:0}$, $C_{16:0}$, iso-$C_{16:0}$, anteiso-$C_{17:0}$ | iso-$C_{15:0}$, iso-$C_{(17:1)}$ω11c, iso-$C_{16:0}$, iso-$C_{17:0}$ | iso-$C_{15:0}$, iso-$C_{17:1}$ ω11c, iso-$C_{17:0}$, iso-$C_{16:0}$ |
| Menaquinones | | | MK-7 | MK-7 | MK-7 | MK-9, MK-8 | MK-9, MK-8 |

Datafrom: [a]Matsuo et al. 2006
[b]Hatayama et al. 2005
[c]Zhang et al. 2007
*nd* no data, + positive reaction, – negativereaction, +/– variable reaction, *pH of usual media

phosphatase and trypsin activities. The predominant menaquinones (peak area ratio) are MK-9 (73 %) and MK-8 (21.7 %). Major cellular fatty acids are iso-$C_{15:0}$ (57.3 %), iso-$C_{17:1}$ ω11c (18.0 %), iso-$C_{16:0}$ (8.9 %), and iso-$C_{17:0}$ (8.1 %). The DNA G+C content of the type strain is 45.1 mol % (Matsuo et al. 2006) (➲ *Table 31.3*).

### *Mechercharimyces asporophorigenens* Matsuo et al. 2006

The type strain YM11-542$^T$(= MBIC06487$^T$ = DSM 44955$^T$) was isolated from a sediment from a marine lake. GenBank/EMBL/DDBJ accession number (16S rRNA gene): AB239532.

Exhibits the following properties in addition to those given in the genus description. Colonies are fast-growing, lightly ridged, with a moderate covering of white mycelia and a feathery margin on marine agar 2216 at 27 °C. Growth occurs at 20–37 °C, with optimum growth at 30 °C and pH 7.6–8.0. Forms oval-shaped endospores in substrate mycelia or aerial mycelia. Does not form sessile endospores or sporophores. Casein and gelatin are degraded, but not starch, hypoxanthine, xanthine, or L-tyrosine. Produces dark-brown pigment on L-tyrosine-containing marine agar. Growth occurs in the presence of novobiocin (25 μg/ml). Presents alkaline phosphatase activity. The menaquinones (peak area ratio) are MK-9 (76.1 %) and MK-8 (23.9 %). Major cellular fatty acids are iso-$C_{15:0}$ (50.6 %), iso-$C_{17:1}$ ω11c (16.5 %), iso-$C_{17:0}$ (12.4 %), and iso-$C_{16:0}$ (10.2 %). The DNA G+C content of the type strain is 45.2 mol% (Matsuo et al. 2006).

## *Thermoflavimicrobium* Yoon et al. 2005

Aerobic, Gram-positive, non-acid-fast, chemo-organotrophic, and thermophilic. Members of the genus can be distinguished from those classified in the family *Thermoactinomycetaceae* by their ability to produce single spores on dichotomously branched sporophores and yellow pigmented colonies. Sessile endospores are round with surface ridges giving angular appearance. Growth occurs at 35–65 °C. Aerial mycelium is abundant at 55 °C. The cell-wall peptidoglycan contains meso-DAP but no characteristic sugars (Yoon et al. 2005). The pattern of 16S rRNA gene signatures consists of (U–G) at positions 154:167, (A–G) at 203:214 and (G) at 693 (Yu et al. 2012).

The type species is *Thermoflavimicrobium dichotomicum.*

### *Thermoflavimicrobium dichotomicum* (Krasil'nikov and Agre 1964) Yoon et al. 2005

Basonym: *Thermoactinomyces dichotomicus* corrig. (Krasil'nikov and Agre 1964) Cross and Goodfellow 1973, *Actinobifida dichotomica* Krasil'nikov and Agre 1964, *Thermomonospora citrina* Manachini et al. 1966

The type strain KCTC 3667$^T$ (= ATCC 49854$^T$ = JCM 9688$^T$ = DSM 44778$^T$ = NCIMB 10211$^T$ = NRRL B-16978$^T$) was isolated from soil. GenBank accession number (16S rRNA gene): AF138733.

Colonies are yellow to orange, distinctively fast-growing with dichotomously branched mycelium and sporophores on nutrient agar and CYC agar at 55 °C (optimal temperature). Margins of colonies are entire on CYC agar. Exosporium surrounding the spores is present. Casein, gelatin, hypoxanthine, and xanthine are degraded, but not starch. Elastin; DNA; guanine; RNA; and Tweens 20, 40, 60, and 80 are degraded, but not arbutin, aesculin, tyrosine, adenine, cellulose, hippurate, or keratin. Growth occurs in the presence of 0.5 % (w/v) NaCl, but not at 1.0 % (w/v) NaCl. Nitrate is not reduced to nitrite. L-Arabinose, D-galactose, D-glucose, glycerol, lactose, maltose, mannitol, meso-inositol, D-raffinose, L-rhamnose, D-sorbitol, starch, sucrose, and D-xylose are utilized as carbon sources. Produces alkaline phosphatase, $C_4$ esterase, and $C_8$ lipase but not chymotrypsin, α- or β-glucosidase, β-glucuronidase, or leucine aminopeptidase. Grows in the presence of novobiocin (25 μg/ml) and lysozyme (0.005 % w/v). pH of usual media 7.2–7.4. MK-7 (85 %) is the predominant menaquinone. The fatty acids are iso-$C_{15:0}$ (46.7 %), anteiso-$C_{15:0}$ (12.7 %), iso-$C_{16:0}$ (10.3 %), iso-$C_{17:0}$ (7.0 %), $C_{16:0}$ (6.7 %), iso-$C_{14:0}$ (5.8 %), and others. The DNA G+C content is 43 mol % (Lacey and Cross 1989; Yoon et al. 2005).

## *Desmospora* Yassin et al. 2009

Gram-positive, non-acid-fast, aerobic, catalase-positive, and chemo-organotrophic. Non-fragmentary vegetative mycelium forms leathery colonies that are covered with aerial mycelium. Aerial mycelia are long, moderately branched, straight, or flexuous. On agar media, aerial mycelium is yellow in color. Aerial mycelium bears both long chains of arthrospores and sessile endospores that are formed singly on simple unbranched sporophores. Motile elements are not produced. Thermotolerant. The peptidoglycan contains *meso*-diaminopimelic acid; no characteristic sugars are detected in whole-cell hydrolysates. The muramic acid residues of the peptidoglycan are N-glycolated. The predominant menaquinone is MK-7. The phospholipid pattern consists predominantly of diphosphatidylglycerol, phosphatidylglycerol, phosphatidylethanolamine, phosphatidyl-monomethylethanolamine, and two additional unknown phospholipids (Yassin et al. 2009). Grows with novobiocin (25 μg/ml) (von Jan et al. 2011). The pattern of 16S rRNA gene signatures consists of (U–G) at positions 154:167, (A–A) at 203:214, and (G) at 693 (Yu et al. 2012).

The type species is *Desmospora activa.*

### *Desmospora activa* Yassin et al. 2009

The type strain, IMMIB L-1269$^T$ (= DSM 45169$^T$ = CCUG 55916$^T$), was isolated from sputa from a patient. GenBank/EMBL/DDBJ accession number (16S rRNA gene): AM940019.

Colonies are yellow with radial wrinkles. No diffusible pigments are produced on yeast-malt agar, oatmeal agar, or inorganic salts-starch agar. Melanoid pigments are not formed in peptone-iron agar or tyrosine agar. Growth at 37–50 °C. Casein, elastin, aesculin, gelatin, and urea are hydrolyzed, but not adenine, guanine, hypoxanthine, keratin, testosterone, tyrosine, and xanthine. Assimilates acetate, isoamyl alcohol, 2,3-butanediol, citrate, D-galactose, D-glucose, D-gluconate, myo-inositol, L-lactate, D-lactose, 1,2-propanediol, D-sorbitol, sucrose, trehalose, and D-xylose as carbon sources, but not adipate, adonitol, L-arabinose, cellobiose, meso-erythritol, *m*-hydroxybenzoate, *p*-hydroxybenzoate, maltose, D-mannitol, melezitose, raffinose, or L-rhamnose. Acetamide, L-alanine, arginine, gelatin, L-proline, and L-serine are utilized as simultaneous sources of carbon and nitrogen, but ornithine is not. The major fatty acids are iso-$C_{15:0}$ (41.35 %), $C_{16:0}$ (14.48 %), iso-$C_{17:0}$ (12.48 %), $C_{16:1}$ ω7*c* (8.51 %), and $C_{14:0}$ (7.17 %). The DNA G+C content is 49.3 mol % (Yassin et al. 2009).

## *Melghirimyces* Addou et al. 2012

Aerobic, Gram-positive, forms extensively branched yellow aerial and substrate mycelia after 72 h at 55 °C on tryptone yeast broth and yeast-malt agar supplemented with 10 % (w/v) NaCl. The major menaquinone is MK-7. Cell-wall peptidoglycan contains LL-DAP. The phospholipid pattern consists of phosphatidylglycerol, diphosphatidylglycerol, phosphatidylethanolamine, phosphatidyl-monomethylethanolamine, unknown phospholipids, and an unknown lipid (Addou et al. 2012). The pattern of 16S rRNA gene signatures consists of (C-A) at positions 154:167, (A–C) at 203:214 and (G) at 693 (Yu et al. 2012).

The type species is *Melghirimyces algeriensis.*

### *Melghirimyces algeriensis* Addou et al. 2012

The type strain, NariEX$^T$ (= DSM 45474$^T$ = CCUG 59620$^T$), was isolated from a soil collected from a salt lake. GenBank/EMBL/DDBJ accession number (16S rRNA gene): HO383683.

On yeast-malt agar containing 10 % (w/v) NaCl, colonies are yellow, flat, dull with regular margins and radial wrinkles formed on the surface. It tolerates 0–21 % (w/v) NaCl, with optimal growth occurring at 7–12 % (w/v) NaCl. Growth occurs at 37–60 °C and pH 5.0–9.5, optimal at 40–55 °C and pH 6–8 with 10 % NaCl. Glycerol, maltose, cellobiose, rhamnose, sucrose, *myo*-inositol, fructose, ribose, mannose, glucose, raffinose, alanine, threonine, proline, glycine, asparagine, glutamic acid, aspartic acid, acetate, oxalate, succinate, and malate are used as sole carbon sources for growth. The following compounds are not utilized: erythritol, lactose, arabinose, xylose, ornithine, serine, cysteine, isoleucine, methionine, lysine, histidine, valine, glutamine, citrate, formate, benzoate, fumarate, and propionate. Utilization of arginine as a sole carbon source is doubtful. Acids are not produced from organic compounds. Decomposes casein and gelatin. Cellulose, tyrosine, xanthine, hypoxanthine, adenine, urea, and starch are not hydrolyzed. Nitrate reduction is positive. Indole production and Voges–Proskauer reaction are negative. Major fatty acids are iso-$C_{15:0}$ (59.13 %), anteiso-$C_{15:0}$ (18.18 %), iso-$C_{17:0}$ (6.66 %), and $C_{17:0}$ (4.62 %). The peptidoglycan hydrolysates contain LL-DAP, alanine, glycine, and glutamic acid in a molar ratio of 1.1:1.4:0.5:1.0. Cell wall sugars are xylose, mannose, galactose, and traces of glucose. The DNA G+C content of the type strain is 47.3 mol% (Addou et al. 2012).

## *Seinonella* Yoon et al. 2005

Aerobic, Gram-positive, non-acid-fast, chemo-organotrophic. Mesophilic. Aerial mycelium is white. The substrate mycelium is white to yellowish-brown. Sessile endospores are produced on flexuous branches of the aerial mycelium and on the substrate mycelium. Growth at 25–45 °C, and optimally at 35 °C. Optimal pH for growth is 7.6–8.0; no growth at pH 5.0. The cell-wall peptidoglycan contains meso-DAP (Yoon et al. 2005). The pattern of 16S rRNA gene signatures consists of (U–A) at positions 154:167, (A–U) at 203:214 and (G) at 693 (Yu et al. 2012).

The type species is *Seinonella peptonophila.*

### *Seinonella peptonophila* (Nonomura and Ohara 1971) Yoon et al. 2005

Basonym: *Thermoactinomyces peptonophilus* Nonomura and Ohara 1971.

The type strain KCTC 9740$^T$ (= ATCC 27302$^T$ = JCM 10113$^T$ = DSM 44666$^T$) was isolated from soil. GenBank accession number (16 rRNA gene): AF138735.

Mesophilic species, with endospores less heat resistant ($D_{90°C}$ = 45 min). Endospores sessile on flexuous branches of the aerial mycelium and on the substrate mycelium. Aerial mycelium white and substrate mycelium white to yellowish-brown. High concentrations of peptone or yeast extract (3 % w/v) in addition to B vitamins are essential for good growth. The strains are not resistant to novobiocin (25 μg/ml). A low concentration of glycerol or glucose (0.2 % w/v) is favorable for aerial mycelium production. No distinct soluble pigments are produced. Casein, gelatin, and L-tyrosine are not hydrolyzed. Nitrate reduction is negative. Predominant menaquinone is MK-7 (59 %), and significant amounts of MK-9 (17 %), MK-10 (14 %) and MK-8 (10 %) are present. The fatty acids are iso-$C_{14:0}$ (26.9 %), anteiso-$C_{15:0}$ (26.5 %), iso-$C_{16:0}$ (15.7 %), $C_{16:0}$ (9.2 %) and $C_{16:1}$ω11*c* (7.0 %), and others. The DNA G+C content is 40 mol% (Lacey and Cross 1989; Yoon et al. 2005).

## *Shimazuella* Park et al. 2007

Cells are Gram-positive, aerobic, and mesophilic. Aerial mycelium is abundant and white and is not fragmented. Forms well-developed, branched, and septate substrate mycelia on Bennett's agar and yeast extract-malt extract (pH 7.3). Soluble pigments are not produced. Spiny endospores grow singly (1.0–1.4 × 0.7–0.9 mm) on aerial mycelium and are nonmotile. The cell-wall peptidoglycan contains *meso*-diaminopimelic acid, glutamic acid, and alanine, but no characteristic sugars. The diagnostic phospholipid is phosphatidylethanolamine (Park et al. 2007). The pattern of 16S rRNA gene signatures consists of (U–G) at positions 154:167, (A–U) at 203:214 and (G) at 693 (Yu et al. 2012).

The type species is *Shimazuella kribbensis.*

### *Shimazuella kribbensis* Park et al. 2007

The type strain, A 9500$^T$ (= KCTC 9933$^T$ = DSM 45090$^T$), was isolated from soil collected in Sobaek Mountain, South Korea. GenBank/EMBL/DDBJ accession number (16S rRNA gene): AB049939.

Colonies are fast-growing, ridged with white mycelia and a feathery margin on Bennett's agar at 28 °C. Growth occurs between 20 °C and 50 °C, with optimum growth at 32 °C. Forms endospores singly on unbranched sporophores. No pigments are observed. Casein and starch are degraded, but not gelatin, hypoxanthine, xanthine, or L-tyrosine. Growth occurs in the presence of novobiocin (25 μg/ml). The major menaquinone is MK-9 and is found at a ratio of 7:3 with MK-10. Major cellular fatty acids are anteiso-$C_{15:0}$ (43.34 %), iso-$C_{16:0}$ (14.23 %), $C_{16:0}$ (7.90 %), iso-$C_{15:0}$ (7.40 %), and anteiso-$C_{17:0}$ (7.17 %). The DNA G+C content is 39.4 mol% (Park et al. 2007).

## *Kroppenstedtia* von Jan et al. 2011

Aerobic, Gram-positive, and chemo-organotrophic. Thermotolerant. Substrate and aerial mycelia are formed, both producing chains of arthrospores and heat-resistant endospores, the latter are formed singly on unbranched sporophores. The predominant menaquinone is MK-7. Diagnostic sugars are not detected in whole-cell hydrolysates. The phospholipid pattern predominantly consists of diphosphatidylglycerol, phosphatidylethanolamine, phosphatidylglycerol, and unknown phospholipids (von Jan et al. 2011). The pattern of 16S rRNA gene signatures consists of (U–G) at positions 154:167, (A–C) at 203:214 and (G) at 693 (Yu et al. 2012).

The only species is *Kroppenstedtia eburnea.*

### *Kroppenstedtia eburnea* von Jan et al. 2011

The type strain is JFMB-ATE$^T$ (= DSM 45196$^T$ = NRRL B-24804$^T$ = CCUG 59226$^T$) was isolated from a plastic surface. GenBank/EMBL/DDBJ accession number (16S rRNA gene): FN665656.

Colonies are ivory colored, cloudy, flat and of irregular shape with undulate margin and a dull surface, forming radial wrinkles and sparse white aerial mycelia. Growth occurs between 25 °C and 50 °C with an optimum at 45 °C, and at pH 5.0–8.5. No diffusible pigments are produced on brain-heart infusion and glucose yeast-malt media and no melanoid pigments on peptone-iron agar. No growth occurs on yeast-malt agar, oatmeal agar, inorganic salts-starch agar, glycerol-asparagine agar, GPHF medium, and trypticase soy broth agar. Casein, gelatin, and aesculin are hydrolyzed but not starch, xanthine, hypoxanthine, chitin, and L-tyrosine. Urease reaction is positive. Alcohol dehydrogenase, indole production, nitrate reduction/denitrification, and Voges–Proskauer reaction are negative. Major fatty acids are iso-$C_{15:0}$ (73.3 %), anteiso-$C_{15:0}$ (13.1 %), iso-$C_{16:0}$ (4.5 %), and iso-$C_{14:0}$ (3.9 %). The cell-wall peptidoglycan contains LL-DAP, alanine, glycine, and glutamic acid in a molar ratio of 1.2:0.9:0.4:1.0. Diagnostic sugars are not detected in whole-cell hydrolysates. The DNA G+C content is 54.6 mol% (von Jan et al. 2011).

## *Marininema* Li et al. 2012

Cells are aerobic, Gram-positive, oxidase-negative, and catalase-positive. Growth occurs at 25–35 °C and pH 5.0–8.0, with optimum growth at 30 °C and pH 6.0–7.0. Aerial mycelium is not produced. Endospores are formed on the substrate mycelium. The cell wall contains LL-DAP as the diamino acid. Whole-cell hydrolysates contain mannose, ribose, rhamnose, and glucose. The phospholipids are diphosphatidylglycerol, phosphatidylmethylethanolamine, phosphatidylethanolamine, phosphatidylglycerol, and five unknown phospholipids. The predominant menaquinone is MK-7. Major fatty acids are anteiso-$C_{15:0}$ and iso-$C_{15:0}$ (Li et al. 2012). The pattern of 16S rRNA gene signatures consists of (U–A) at positions 154:167, (A–G) at 203:214, and (G) at 693 (Yu et al. 2012).

The type species is *Marininema mesophilum.*

### *Marininema mesophilum* Li et al. 2012

The type strain SCSIO 10219$^T$ (=CCTCC AA 2011006$^T$ = DSM 45610$^T$) was isolated from a sediment sample collected in the South China Sea. GenBank/EMBL/DDBJ accession number (16S rRNA gene): JN006758.

Aerial mycelium is not produced. Grows well on nutrient agar, TSA, and potato-glucose agar media, forming yellow-white colonies with radial wrinkles. No growth occurs on yeast-malt agar, oatmeal agar, inorganic salts-starch agar, glycerol-asparagine agar or Czapek's agar. Soluble pigment is not produced on any of the tested media. Growth occurs at 25–35 °C and pH 5.0–8.0. Tolerates up to 7 % (w/v) NaCl. The optimal temperature and pH value for growth are 30 °C and pH 6.0–7.0, respectively. Positive for catalase and Tween 20 hydrolysis. Negative for oxidase, urease, gelatin liquefaction, milk coagulation, milk peptonization, $H_2S$ production, hydrolysis of cellulose, starch, Tween 40 and Tween 80, and nitrate reduction. Hypoxanthine and adenine are not hydrolyzed. Utilizes D-galactose,

lactose, sodium pyruvate, and sucrose as sole carbon sources. D-Arabinose, cellobiose, D-fructose, glycerol, inositol, maltose, D-mannitol, D-mannose, raffinose, L-rhamnose, D-ribose, sodium acetate, sorbitol, xylitol, and D-xylose are not utilized. L-Alanine, L-arginine, L-glutamic acid, glycine, L-histidine, L-lysine, L-proline, L-serine, L-threonine, and L-valine can be used as sole nitrogen sources, but not L-asparagine or L-cysteine. The strain contains LL-diaminopimelic acid, glutamic acid, alanine, lysine, and glycine in the cell wall. Major fatty acids are anteiso-$C_{15:0}$ (43.04 %), iso-$C_{15:0}$ (29.48 %), anteiso-$C_{17:0}$ (6.86 %), iso-$C_{16:0}$ (6.20 %), and iso-$C_{17:0}$ (5.15 %). The DNA G+C content is 46.5 mol% (Li et al. 2012).

## *Lihuaxuella* Yu et al. 2012

Cells are Gram-positive, aerobic, and thermotolerant. Growth occurs at 28–65 °C and pH 6.0–8.0, with optimum growth at 50 °C and pH 7.0. Endospores are produced on the well-developed branched substrate mycelium. Hyphae are not fragmented. Aerial mycelium is not formed. The diagnostic acid of the peptidoglycan is *meso*-diaminopimelic acid. Whole-cell hydrolysates contain glucose, galactose, mannose, ribose, and rhamnose. The polar lipid pattern consists of phosphatidylmethylethanolamine, diphosphatidylglycerol, phosphatidylethanolamine, phosphatidylglycerol, an unidentified aminophospholipid, and four unknown phospholipids. The pattern of 16S rRNA gene signatures consists of (A–U) at positions 154:167, (G–G) at 203:214 and (U) at 693 (Yu et al. 2012).

The type species is *Lihuaxuella thermophila.*

### *Lihuaxuella thermophila* Yu et al. 2012

The type strain YIM $77831^T$ (=CCTCC AA $2011024^T$ = JCM $18059^T$) was isolated from a geothermal soil collected at Tengchong, Yunnan, China. GenBank accession number (16S rRNA gene): JX045707.

Grows well and forms brown substrate mycelia on yeast-malt agar and nutrient agar. Grows weakly and forms milk white substrate mycelium on oatmeal agar, inorganic salts-starch agar, glycerol-asparagine agar, and Czapek's agar media, but not on tryptic soy agar and potato-glucose agar media. Aerial mycelium and soluble pigment are not produced on any of the test media. Growth occurs from 28 °C to 65 °C, at pH 6.0–8.0 and 0–1 % NaCl (w/v). The optimal temperature and pH value for growth are 50 °C and pH 7.0, respectively. Positive for catalase, gelatin liquefaction, milk coagulation, milk peptonization, reduction of nitrate, hydrolysis of Tween 40, Tween 60, and Tween 80. Negative for urease, oxidase, and hydrolysis of cellulose, starch, and Tween 20. Lactose, mannose, xylose, inositol, maltose, rhamnose, raffinose, galactose, glucose, and cellobiose can be utilized as sole carbon sources, but ribose, xylitol, fructose, mannitol, and L-arabinose are not utilized. L-histidine, L-asparagine, L-ornithine, L-phenylalanine, L-cysteine, L-tryptophan, L-arginine, L-lysine, L-threonine, L-tyrosine, and L-hydroxyproline are utilized as sole nitrogen sources, but not L-glycine, L-serine, L-methionine, L-proline, L-alanine, L-valine, and L-cystine. The cell wall contains meso-DAP, and whole-cell hydrolysates contain glucose, galactose, mannose, ribose, and rhamnose. The only menaquinone is MK-7. Major fatty acids are iso-$C_{15:0}$ (47.51 %), anteiso-$C_{15:0}$ (22.26 %), and anteiso-$C_{17:0}$ (12.43 %). The G+C content is 55.6 mol% (Yu et al. 2012).

## *Polycladomyces* Tsubouchi et al. 2013

Aerobic, Gram-positive, thermophilic, forms branched white aerial mycelium with single endospores at 60 °C. The predominant menaquinone is MK-7, but MK-8 is also present. Cell-wall peptidoglycan contains *meso*-diaminopimelic acid, alanine, and glutamic acid in addition to glucosamine and muramic acid but no characteristic sugars. Major cellular fatty acids are iso-$C_{15:0}$, iso-$C_{17:0}$, and iso-$C_{16:0}$. The polar lipids profile consists of diphosphatidylglycerol, phosphatidylethanolamine, phosphatidyl-monomethylethanolamine, phosphatidylglycerol, glucolipid, phosphatidylserine, an unidentified aminophospholipid, an unknown phospholipid, and two unknown lipids.

The type species is *Polycladomyces abyssicola.*

### *Polycladomyces abyssicola* Tsubouchi et al. 2013

The type strain, JIR-$001^T$ (= JCM $18147^T$ = CECT $8074^T$), was isolated from a deep seafloor sediment collected from the Shimokita Peninsula of Japan. DDBJ accession number (16S rRNA gene): AB688114.

White aerial mycelium is formed on yeast-malt agar. No diffusible pigment is detected. Growth occurs at temperatures ranging from 55 °C to 73 °C, at pH ranging from 6.5 to 8.5, and at NaCl concentrations between 1 % and 2 % (w/v). Ribose, glycerol, xylose mannose, alanine, glycine, asparagine, arginine, and fumarate are used as sole carbon sources for growth. The following compounds are not utilized: maltose, cellobiose, erythritol, sucrose, fructose, arabinose, raffinose, lactose, cysteine, isoleucine, threonine, methionine, proline, serine, glutamic acid, valine, glutamine, benzoate, oxalate, malate, and propionate. Acid productions from organic compounds are not observed. Casein, gelatin, and esculin are degraded, whereas starch, xanthine, hypoxanthine, and tyrosine are not hydrolyzed. Major cellular fatty acids are iso-$C_{15:0}$ (35.5 %), iso-$C_{17:0}$ (28.11 %), iso-$C_{16:0}$ (22.54 %), and $C_{16:0}$ (4.74 %). The DNA G+C content of the type strain is 55.1 mol% (Tsubouchi et al. 2013).

## Isolation, Enrichment, and Maintenance Procedures

The main problem in isolation of members of the family *Thermoactinomycetaceae* is the exclusion of organisms that cover large areas of the isolation plates, e.g., swarming bacilli and streptomycetes. The method used to collect samples may influence the growth of other bacteria. The isolation of most species may be achieved on agar media containing 25 μg novobiocin/ml and 50 μg cycloheximide/ml incubated, but *Seinonella* is sensitive (von Jan et al. 2011) and there are no data about *Marininema*, *Melghirimyces*, *Laceyella sediminis*, and *Lihuaxuella*.

Compost, bagasse, hay, soil, or mud samples may be suspended in a common aqueous diluent containing peptone (0.1 %) and Tween 80 (0.05 %) or a 50 mM potassium phosphate buffer (pH 8) and mixed thoroughly. The liquid phase is diluted to 1:100 or 1:1,000 in triplicate for each sample and aliquots (100 μl) are spread onto the surface of suitable agar medium in prepoured plates. Usually, the media are supplied by commercial laboratories. The plates are incubated at the suitable temperature for 3–7 days. The colonies are identified by using a microscope fitted with an ×40 long-working-distance objective. The optimum isolation of some *Thermoactinomycetaceae* species is attained using a low nutrient organic media because the overgrowth of media by other bacteria is dramatically reduced. For isolation of thermophilic species, plates are enclosed in polyethylene bags or in sealed containers with some water (Amner et al. 1988).

Compost, bagasse, and hay samples may be dried at 37 °C and agitated in a small wind tunnel or sedimentation chamber to produce a spore cloud which, after sedimentation for 90 min, is passed through an Andersen sampler loaded with plates of isolation media for 10 s at a rate of 25 l/min (Lacey 1997).

Sediments or soils may be suspended in a liquid medium supplement with NaCl for halophilic species (Addou et al. 2012). As for the enrichment, samples can be put into a suitable liquid media and incubated with shaking at 100–140 rpm before the isolation (Chen et al. 2012).

Usually, animal infected material is crushed and washed in sterile saline solution and samples are placed on plates of a suitable medium, or stabbed into tubes of medium. Among the media most commonly chosen are brain-heart infusion agar and blood agar (Williams and Cross 1971).

*Thermoactinomycetaceae* species can be cultured on the same media as used for isolation. After a purification step, the strains are maintained on agar slopes in screw-capped tubes at room temperature or 4 °C. For long-term preservation, the isolates are maintained as suspensions of spores and mycelium fragments in glycerol (20 %, v/v) at –80 °C, or by lyophilization with spores suspended in double strength skim milk (Lacey and Cross 1989) (➋ *Table 31.4*).

## Media

Bennett's agar (g/l): yeast extract, 1; beef extract, 1; enzymatic digest of casein, 2; agar, 15; glucose, 10; pH 7.3 (Williams and Cross 1971).

Brain-heart infusion agar (g/l): infusion from 200 g calf brains, 7.7; infusion from 250 g beef heart, 9.8; proteose peptone, 10; sodium chloride, 5; disodium phosphate 2.5; agar 12, pH 7.4 (DSMZ n°215).

BSW-1: 0.1 × artificial seawater; 0.5 % (w/v) marine broth 2216; 0.5 % (w/v) NaCl (Tsubouchi et al. 2013).

Columbia agar + 5 % blood (g/l): pancreatic digest of casein, 10; heart pancreatic digest, 3; meat peptic digest, 5; NaCl, 5; maize starch, 1; yeast extract, 1; agar, 13.5; pH 7.3. Cool to 50 °C and aseptically add 5 % citrated sheep blood (DSMZ n°715).

Czapek yeast casaminoacids agar (g/l): sucrose, 30; $K_2HPO_4$, 1; $MgSO_4.7H_2O$, 0.5; KCl, 0.5; $FeSO_4.7H_2O$, 0.01; yeast extract, 2; vitamin-free casaminoacids, 6; agar, 15; pH 7.2 (Lacey and Cross 1989).

DSM 88 agar: $(NH_4)_2SO_4$, 1.3 g; $KH_2PO_4$, 0.28 g; $MgSO_4.7H_2O$, 0.25 g; $CaCl_2.2H_2O$, 0.07 g; $FeCl_3.6H_2O$, 0.02 g; $MnCl_2.4H_2O$, 1.8 mg; $Na_2B_4O_7.10H_2O$, 4.5 mg; $ZnSO_4.7H_2O$, 2.2 mg; $CuCl_2.2H_2O$, 0.05; $Na_2MoO_4.2H_2O$, 0.03; $VOSO_4.2H_2O$, 0.03; $CoSO_4$, 0.01; yeast extract, 1; demineralized water, 1 l (DSMZ n°88).

Half-strength nutrient agar (g/l): beef extract, 2.5 g; peptone, 2.5 g; NaCl, 2.5 g; agar, 20 g (Williams and Cross 1971).

Luria Bertani (g/l): casein enzymatic hydrolysate, 10; yeast extract, 5; NaCl, 10; pH 7.5. For solid medium, add agar or gelrite, 15 (DSMZ n°381).

Marine agar 2216 (g/l): peptone, 5; yeast extract, 1; ferric citrate, 0.1; sodium chloride, 19.45; magnesium chloride, 8.8; sodium sulfate, 3.24; calcium chloride, 1.8; potassium chloride, 0.55; sodium bicarbonate, 0.16; potassium bromide, 0.08; strontium chloride, 0.034; boric acid, 0.022; sodium silicate, 0.004; sodium fluoride, 0.0024; ammonium nitrate, 0.0016; disodium phosphate, 0.008; agar; 15; pH 7.6 (DSMZ n°604).

1/10 MYGS-AF medium: malt extract, 1 g; yeast extract, 500 mg; glucose, 500 mg; seawater, 1 l; agar 20 g; cycloheximide 100 mg; nystatin, 50 mg; griseofulvin, 20 mg; pH 7.8–8.0 (Matsuo et al. 2006).

Nutrient agar (g/l): beef extract, 5; peptone, 5; NaCl, 5; agar 20 (Williams and Cross 1971).

Oatmeal agar ISP-3: oatmeal, 20 g; distilled water, 1 l; cook for 20 min, filter through a cloth and restore volume of filtrate; trace salts solution, 1 ml; agar, 18 g; pH 7.2. Trace salts solution (g/l): $FeSO_4.7H_2O$, 0.1; $MnCl_2.4H_2O$, 0.1; $ZnSO_4.7H_2O$, 0.1 (Shirling and Gottlieb 1966).

PY agar (g/l): peptone, 20; yeast extract, 20; glycerol, 2; $MgSO_4.7H_2O$, 0.3; agar, 20; pH 7.6 (Lacey and Cross 1989).

1/10 PYGS-AF medium: metal mix X, 250 ml; distilled water, 750 ml; agar, 20 g; peptone, 1 g; yeast extract, 0.5 g; C solution, 5 ml; cycloheximide, 50 mg; griseofulvin 25 mg; nalidixic acid, 20 mg; aztreonam, 40 mg.

Metal mix X: NaCl, 500 g; $MgSO_4.7H_2O$, 180 g; $CaCl_2.2H_2O$, 2.8 g; KCl, 14 g; $Na_2HPO_4.12H_2O$, 5 g; $FeSO_4.7H_2O$, 200 g; PII metals, 600 ml; S2 metals, 100 ml; distilled water, 4.3 l; pH 7·6.

C solution: sodium pyruvate, 25 g; mannitol, 50 g; glucose, 50 g; distilled water, 500 ml; pH 7 · 5; sterilized by filtration.

**Table 31.4**
**Media used for isolation and maintenance of *Thermoactinomycetaceae* species**

| Species | Isolation | Maintenance | References |
|---|---|---|---|
| *Desmospora activa* | Columbia agar supplemented with 5 % sheep blood at 37 °C | Yeast-malt agar ISP-2 | Yassin et al. 2009 |
| *Laceyella putida*[a] | Czapek yeast casaminoacids (CYC) agar at 50 °C | CYC agar | Lacey and Cross 1989 |
| *Laceyella sacchari* | Yeast-malt agar ISP-2 at 55 °C | Yeast-malt agar ISP 2 | Lacey and Cross 1989 |
| *Laceyella sediminis* | DSM 88 agar with soluble starch (1 %, w/v) at 55 °C | DSM 88 agar Oatmeal agar ISP-3 | Chen et al. 2012 |
| *Laceyella tengchongensis* | Oatmeal agar ISP-3 at 55 °C | Oatmeal agar ISP-3 | Zhang et al. 2010 |
| *Lihuaxuella thermophila* | R2A medium at 50 °C | Yeast-malt agar ISP-2 | Yu et al. 2012 |
| *Kroppenstedtia eburnea* | Trypticase soy agar at 37 °C | Brain-heart infusion agar | von Jan et al. 2011 |
| *Marininema mesophilum* | R2A medium with 20 μg nalidixic acid/ml + 20 μg cycloheximide/ml at 30 °C | Nutrient agar | Li et al. 2012 |
| *Mechercharimyces mesophilus*, *M. asporophorigenens* | 1/10 MYGS-AF, 1/10 PYGS-AF or skimmed milk media at 30 °C | Marine agar 2216 | Matsuo et al. 2006 |
| *Melghirimyces algeriensis* | Yeast-malt agar ISP-2 + 10 % NaCl at 55 °C | Yeast-malt agar ISP-2+ 10 % NaCl | Addou et al. 2012 |
| *Planifilum fimeticola*, *Planifilum fulgidum* | Luria-Bertani agar at 60 °C | Luria-Bertani agar | Hatayama et al. 2005 |
| *Planifilum yunnanense* | Luria Bertani gelrite at 60 °C | Luria Bertani gelrite | Zhang et al. 2007 |
| *Polycladomyces abyssicola* | BSW-1 agar at 60 °C | Yeast-malt agar ISP-2 + 1–2 % NaCl | Tsubouchi et al. 2013 |
| *Seinonella peptonophila*[b] | PY agar at 35 °C | PY agar | Lacey and Cross 1989 |
| *Shimazuella kribbensis* | Bennett's agar at 30 °C | Bennett's agar or Yeast-malt agar ISP-2 | Park et al. 2007 |
| *Thermoactinomyces vulgaris*, *T. intermedius* | Half-strength nutrient or Czapek yeast casaminoacids (CYC) at 50–55 °C | Nutrient agar, CYC agar, or SY medium | Lacey and Cross 1989; Yoon et al. 2005 |
| *Thermoflavimicrobium dichotomicum* | Half-strength nutrient agar or Czapek yeast casaminoacids (CYC) agar at 55 °C | SY medium, CYC agar, or nutrient agar | Lacey and Cross 1989; Yoon et al. 2005 |

[a]Sensitive to 0.5 % (w/v) NaCl
[b]High concentration of vitamins B and peptone or yeast extract are essential
*ISP* International *Streptomyces* Project (Shirling and Gottlieb 1966)

PII metals: $Na_2$-EDTA, 1 g, $H_3BO_3$, 1.13 g; Fe solution 1 ml [$FeCl_3.6H_2O$, 2 · 42 g/50 ml]; Mn solution 1 ml [$MnCl_2.4H_2O$, 7.2 g/50 ml]; Zn solution 1 ml [$ZnCl_2$, 0.52 g/50 ml (+ HCl)]; Co solution, 1 ml [$CoCl_2.6H_2O$, 0.2 g/50 ml]; distilled water 996 ml; pH 7.5.

S2 metals: NaBr, 1.28 g; Mo solution, 10 ml [$Na_2MoO_4.2H_2O$, 0.63 g/50 ml]; Sr solution, 10 ml [$SrCl_2.6H_2O$, 3.04 g/50 ml]; Rb solution, 10 ml [RbCl, 141 · 5 mg/50 ml]; Li solution, 10 ml [LiCl, 0.61 g/50 ml]; I solution, 10 ml [KI, 6 · 55 mg/50 ml]; V solution, 10 ml [$V_2O_5$, 1.785 mg/50 ml (+ NaOH)]; distilled water 940 ml; pH 7.5 (Matsuo et al. 2006).

R2A medium (g/l): yeast extract, 0.5; proteose peptone, 0.5; casaminoacids, 0.5; glucose, 0.5; soluble starch, 0.5; Na-pyruvate, 0.3; $K_2HPO_4$, 0.3; $MgSO_4.7H_2O$, 0.05; agar, 15; pH 7.2 (DSMZ n°830).

Skimmed milk medium: skimmed milk, 5 g; distilled water, 200 ml; yeast extract, 500 mg; seawater, 800 ml; agar, 20 g; pH 7.8–8.0 (Matsuo et al. 2006).

SY medium (g/l): starch, 15; yeast extract, 10; $MgSO_4$, 0.5; tap water (Yoon et al. 2005).

Trypticase soy agar (g/l): peptone from casein, 15; peptone from soymeal, 5; NaCl, 5; agar, 15; pH 7.3 (DSMZ n°535).

Yeast-malt agar ISP-2 (g/l): yeast extract, 4; malt extract, 10; dextrose, 4; pH 7.3 (Shirling and Gottlieb 1966).

## Habitat

*Thermoactinomyces* "*sensu lato*" is abundant in moldy fodders and other vegetable matter including straw, cereal grains, cotton, composts, hay, and manure. They are favored by spontaneous heating to temperatures up to 65 °C, often resulting in production of more than $10^6$ spores/g dry weight if the aeration is not restricted. The spores easily become airborne when the substrate is disturbed (Williams et al. 1984). *Laceyella sacchari* is isolated from sugar cane, filter press muds, sugar mills, and soil (Carrillo et al. 2009) but it is abundant in self-heated sugar cane bagasse (Lacey 1974). *Thermoactinomyces vulgaris* is most abundant in moldy hay. *Thermoflavimicrobium dichotomicum* is also common in composts. All three species have also been isolated from soil, mud, and peat, although usually in small numbers seldom exceeding $10^4$/g dry weight of soil (Lacey and Cross 1989). *Planifilum fimeticola* and *Planifilum fulgidum* were isolated from a composting process in Japan and identified using a polyphasic approach (Hatayama et al. 2005).

Kurup et al. (1983) isolated *T. vulgaris* from marijuana cigarettes and Huuskonen et al. (1984) observed the presence of *T. vulgaris* spores on raw tobacco in a cigarette factory. The spores do not appear to be inactivated during the combustion (Cunnington et al. 2000). Airborne *Thermoactinomyces* species were detected using fluorescently labeled oligonucleotide probes as sensors for whole cells (Neef et al. 2003).

The mesophilic *Seinonella peptonophila* was isolated from soil, but *Thermoactinomyces intermedius* was found in air conditioners, humidifiers, house dust, and grass compost where it occurs with other species (Kurup et al. 1980). *Laceyella putida* was isolated from a narrow range of substrates, including soil and deep mud cores (Lacey and Cross 1989). *Shimazuella kribbensis* was obtained from soil of Sobaek Mountain, South Korea, and its taxonomy position was investigated by using a polyphasic approach (Park et al. 2007).

Erosion of soil may result in the accumulation of spores in lake muds and marine sediments, giving counts of $10^4$ to $10^6$ spores/g dry weight. The occurrence in deep mud cores and in archaeological excavations suggests that *L. sacchari* and *T. vulgaris* spores may remain viable for thousands of years (Nilsson and Renberg 1990).

*L. sacchari* strains were isolated from geothermal spring sediments and soil samples from West Anatolia in Turkey (Uzel et al. 2011). A *Thermoactinomyces* strain was isolated from a warm geothermal spring in Armenia containing >20 % hydrocarbonate + sulfate and >20 % sodium + magnesium (28.4–32.0 °C, pH 6.2) (Panosyan 2010). The identification of isolates was made by cultural-physiological characteristics and 16S rDNA or 16S sRNA gene sequences similarity.

Phylogenetic analysis based on 16S rRNA gene sequence data indicated that 34.4 % of strains obtained during a culture-dependent approach applied to tropical marine sediments were members of the class *Bacilli* and some strains closely related to the genera *Laceyella* were found (Gontang et al. 2007).

*Mechercharimyces mesophilus* and *Mechercharimyces asporophorigenens* were isolated from sediment samples collected from a marine lake in Mecherchar Island, Pacific Ocean (Matsuo et al. 2006), and *Marininema mesophilum* from a sediment sample at a depth of 2,105 m in the South China Sea (Li et al. 2012). *Polycladomyces abyssicola* was isolated from a sediment sample at a depth of 48 m below the seafloor off the Shimokita Peninsula of Japan in the northwestern Pacific Ocean (water depth 1,180 m) (Tsubouchi et al. 2013). The halotolerant and thermotolerant *Melghirimyces algeriensis* was isolated from soil of a salt lake in Algeria (Addou et al. 2012). The taxonomy position of these species was investigated by using a polyphasic approach.

A wide physicochemical diversity of springs (97 °C; pH from <1.8 to >9.3) in Rehai Geothermal Field (Yunnan, China) provides a multitude of niches for microorganisms. Cultivation-independent studies using 16S rRNA gene sequences, shotgun metagenomics, or "functional gene" sequences have revealed a much broader diversity of microorganisms than represented in culture (Hedlung et al. 2011). Four thermophilic species of *Thermoactinomycetaceae* family were isolated from hot springs and soils in Yunnan province: *Planifilum yunnanense* (Zhang et al. 2007), *Laceyella tengchongensis* (Zhang et al. 2010), *Laceyella sediminis* (Chen et al. 2012), and *Lihuaxuella thermophila* (Yu et al. 2012).

*Desmospora activa*, isolated from sputum of a patient, was characterized using phenotypic and molecular taxonomy methods (Yassin et al. 2009). *T. vulgaris* and *L. sacchari* strains associated with mushroom worker's lung were identified by 16S rDNA sequence typing (Xu et al. 2002). *Kroppenstedtia eburnea* was first obtained by surface sampling (von Jan et al. 2011), after Barker et al. (2012) reported clinical isolates identified by partial 16S rRNA gene sequences, MALDI-TOF, and antimicrobial susceptibility.

## Pathogenicity

Members of the *Thermoactinomycetaceae* cause human hypersensitivity pneumonitis (farmer's lung, mushroom worker's lung, bagassosis, air conditioner-associated lung, or humidification-system-induced disease), but they are rarely encountered in the clinical microbiology laboratory, because they produce allergic reactions rather than productive infections (Lacey 1971; Kurup et al. 1980). The immunopathological reactions in the lungs involve a specific hypersensitivity response to the spore antigens and a nonspecific response to spore-associated biologically active components, including proteases (Pauwels et al. 1978).

Environments in which disease occurs could bear more than $10^6$ spores per $m^3$ in the atmosphere. *T. vulgaris* and *L. sacchari* have been associated to disease, but the importance of *T. vulgaris* as a causative agent has been underestimated because while testing patient sera, *L. sacchari* isolates have been commonly used (McNeil and Brown 1994; Carrillo et al. 1987). Immunodiffusion tests using antigens according to Edwards (1972) have shown cross-reactivity between these species (Kurup et al. 1976). In addition, *L. putida* has been isolated from a lung biopsy of a patient with farmer's lung (Lacey and Cross 1989).

Few studies have investigated IgG levels in workers exposed to grain dust (Swan et al. 2007). Boiron et al. (1985) and Huuskonen et al. (1984) used an ELISA for the detection of antibodies against the species associated with bagassosis and hypersensitivity pneumonitis in the tobacco industry, respectively.

The use of 16S rRNA gene sequencing permitted the identification of *T. vulgaris* and *L. sacchari* isolates associated with mushroom worker's lung (Xu et al. 2002). *Desmospora activa* was isolated from sputa of a patient with suspected pulmonary tuberculosis and characterized using phenotypic and molecular taxonomic methods (Yassin et al. 2009). Barker et al. (2012) obtained clinical isolates of *Kroppenstedtia eburnea* from blood, cerebrospinal fluid, peritoneal fluid, and skin, identified by 16S rRNA gene sequencing, MALDI-TOF mass spectrometry, and antimicrobial susceptibility profiles.

### Antimicrobial Susceptibility Profiles

*Desmospora activa.* Susceptible to amikacin, ciprofloxacin, erythromycin, fusidic acid, gentamicin, imipenem, levofloxacin, linezolid, meropenem, teicoplanin, and vancomycin (Yassin et al. 2009).

*Kroppenstedtia eburnea.* Susceptible to trimethoprim/sulfamethoxazole, ceftriaxone, ciprofloxacin, linezolid, minocycline, moxifloxacin, imipenem, tobramycin, amikacin, cefepime, tigecycline, and amoxicillin/clavulanic acid (Barker et al. 2012).

*L. sacchari* and *T. vulgaris.* Susceptible to ampicillin, cephaloridine, chloramphenicol, colistin sulfate, demethylchlortetracycline, erythromycin, gentamicin, kanamycin, neomycin, nitrofurantoin, oleandomycin, penicillin, streptomycin, sulfafurazole, and vancomycin (Lacey and Cross 1989).

## Application

### Biofertilizer

Thermophilic species of the family *Thermoactinomycetaceae* appear mainly during the maturation phase and cooling of composting (Tuomela et al. 2000). "*Thermoactinomyces sensu lato*" shows β-glucosidase activity on solid cultures with microcrystalline cellulose at 55 °C, pH 7.4 (Hagerdal et al. 1978). Strains with cellulase activity were isolated from *Brassica* waste showing high efficiency and bioactivity during composting (Chang et al. 2009).

*Laceyella* and some actinomycetes showed vigorous growth on nonsterilized fresh swine and poultry feces without any additives in order to obtain a biofertilizer. The feces were deodorized very rapidly at pH 8.0–8.5. Organic nitrogen was gradually mineralized, so there was no inhibitory effect on plant growth (Hayashida et al. 1988).

### Enzymes

Thermitase (GenBank accession n° SUMYTV) is a thermostable extracellular serine protease secreted by *T. vulgaris.* The enzyme (EC 3.4.21.66) shows maximal stability between pH 6.0 and 7.5 and maximal activity between pH 7.5 and 9.5. The temperature optimum was 60 °C. It is capable of efficient degradation of the insoluble proteins elastin and collagen (Kleine 1982). It contains three $Ca^{++}$-binding sites. The crystal structure of thermitase was determined (Teplyakov et al. 1990).

*T. vulgaris* subtilase (GenBank accession n° EF108326) belongs to the subtilisin family. The enzyme exhibited optimal proteolytic activity at 50–55 °C and pH 10.5–11.0 and was stable under high pH conditions (pH 11.0–12.0), and NaCl could stabilize the enzyme at lower pH values. The enzyme was not dependent on calcium for either maturation or stability. The protease is located on the surface of the spore coat (Cheng et al. 2009).

A *L. sacchari* strain identified by 16S rRNA sequencing produced cyclomaltodextrinase (EC 3.2.1.54) capable of degrading cyclomaltodextrins, which are cyclic, nonreducing oligosaccharides, built up from six, seven, or eight glucopyranose units (Turner et al. 2005). Other enzymes of glycoside hydrolase family were "*T. vulgaris*" R-47 neopullulanase (EC 3.2.1.135) and cyclodextrinase, that hydrolyze specific ($\alpha 1 \rightarrow 4$)-glucosidic linkages of pullulan to produce panose and hydrolyze cyclodextrins or catalyze transglycosylation to the C-4 and C-6 positions of acceptors like glucose (Tonozuka et al. 2002). A malto-oligosaccharide-metabolizing enzyme from "*T. vulgaris*" R-47 homologous to glucoamylase (1,4-α-D-glucan glucohydrolase; EC 3.2.1.3) degraded malto-oligosaccharides more efficiently than starch, releasing β-D-glucose from the nonreducing ends (Ichikawa et al. 2004).

Xylanase-producing *L. sacchari* B42 was isolated from bagasse and identified using 16S rDNA sequence data. The molecular mass of the purified xylanase was 30.0 kDa. The optimal temperature was 70 °C. The enzyme retained 72 % of its activity at 70 °C and 48 % activity at 80 °C after 6 h of incubation. The optimal pH was 10.0 and enzyme appeared to be stable over a broad range (pH 11.0–12.0). Approximately 68 % and 64 % of the original activity was retained after 5 h of incubation at pH 10.0 and 11.0, respectively. The enzymatic biobleaching of kraft pulp reduced 26 % kappa number, decreased 1.68 % lignin content, and released by 24-fold reducing sugars (Singh et al. 2012).

Several strains of *Laceyella sacchari* and *Planifilum fimeticola*, with identity to the type strain of 99.7–99.9 % and 93 %, respectively, have the ability to degrade polylactic acid (PLA) plastic. The specific activity of the enzyme *Laceyella sacchari* LP175 was 328 U/mg with purity 15.3-folds increased and 48.1 % yield obtained. SDS-PAGE analysis indicated that the molecular weight of purified PLA-degrading enzyme was approximately 28 kDa (Kitpreechavanich 2011).

Malate dehydrogenase (L-malate:$NAD^+$ oxidoreductase EC 1.1.1.37) from *L. sacchari* was tetrameric (MW 130,000) and exhibited a high degree of structural homology to *Bacillus caldotenax* malate dehydrogenase as judged by immunological cross-reactivity (Smith et al. 1984).

Leucine dehydrogenase was purified from *T. intermedius.* The enzyme retained about 90 % of activity on incubation at 70 °C for at least 40 min in the presence of 3 M NaCl.

The enzyme showed pro-S stereospecificity for hydrogen transfer of NADH in the reductive amination. The complete DNA sequence of enzyme gene was determined. The amino acid sequence of the enzyme showed 80.7 % similarity with that of the *Bacillus stearothermophilus* enzyme (Ohshima et al. 1994).

Phenylalanine dehydrogenase (L-phenylalanine:NAD oxidoreductase, deaminating; EC 1.4.1.-) was found in *T. intermedius* IFO 14230. The enzyme consists of six subunits identical in molecular weight (41,000) and is highly thermostable: It is not inactivated by incubation at pH 7.2 and 70 °C for at least 60 min or in the range of pH 5–10.8 at 50 °C for 10 min. The enzyme preferably acts on L-phenylalanine and phenylpyruvate, in the presence of NAD and NADH, respectively (Ohshima et al. 1991). This enzyme is required for the synthesis of a diabetes medicine (Hanson et al. 2007).

*L. sacchari*, isolated from soil and identified by 16S rDNA sequence analysis, showed a tyrosinase extracellular enzymatic activity. The purified enzyme was a molecular mass of 30 910 Da. Maximal activities of the purified enzyme were found to occur at pH 6.8 (Dolashki et al. 2012).

*Thermoactinomyces vulgaris* carboxypeptidase T (CpT) is capable of hydrolyzing both hydrophobic and positively charged substrates. Although there is a considerable structural similarity between CpT and pancreatic carboxypeptidases, the mechanisms underlying their substrate specificities are different (Akparov et al. 2007).

Elwan et al. (1978) found strong lipolytic activity in *T. vulgaris*. Lipase is produced optimally at 55 °C and pH 6.8 in a medium containing corn oil. Activity of the enzyme is greatest at 55 °C and pH 8.0. Inactivation occurs in 45 min at 80 °C. The lipase of the isolate *Thermoactinomyces* HRK-1 (Al-Khudary et al. 2004) showed maximum activity at 60 °C and pH 8.0. The enzyme was highly thermostable since it retained 100 % of its activity after boiling for two hours. The lipase was slightly inhibited by $Mg^{++}$ and $Co^{++}$. Calcium, ferrous, and ferric ions enhanced its activity.

## Drugs

The cytotoxic substance mechercharmycin A was isolated from the marine-derived *Mechercharimyces mesophilus* YM3-251. The structure of mechercharmycin A was determined by an X-ray crystallographic analysis to be cyclic peptide-like. It exhibited relatively strong antitumor activity on human lung cancer and leukemia cells. The cyclic structure must have been essential for its strong antitumor activity (Kanoh et al. 2005).

Urukthapelstatin A, a cyclic peptide, was isolated from the cultured mycelia of marine-derived *Mechercharimyces asporophorigenens* YM11-542. It inhibited the growth of human lung cancer A549 cells with an IC(50) value of 12 nM (Matsuo et al. 2007).

## Latest Descriptions (➲ *Table 31.5*)

### *Thermoactinomyces daqus* Yao et al. 2014

The type strain H-18$^T$ (=DSM 45914$^T$=CICC 10681$^T$) was isolated from a fermentation starter used in liquors. GenBank/EMBL/DDBJ accession number for the 16S rRNA gene sequence is KF590624.

The thermophilic bacterium formed white aerial mycelium and greyish-yellow substrate mycelium, bearing single endospores on aerial and substrate hyphae or on unbranched short sporophores. The cell-wall peptidoglycan contained *meso*-diaminopimelic acid. The major fatty acids were iso-$C_{15:0}$ and iso-$C_{17:0}$. The predominant menaquinone was MK-7. The G+C content of the genomic DNA was 49.1 mol%. 16S rRNA gene sequence comparisons indicated that the strain was related to *Thermoactinomyces vulgaris* KCTC 9076$^T$ (96.42 % similarity), *Thermoactinomyces intermedius* KCTC 9646$^T$ (96.06 %), *Laceyella putida* KCTC 3666$^T$ (96.32 %) and *Laceyella sacchari* KCTC 9790$^T$ (95.55 %). Strain H-18$^T$ showed low DNA–DNA relatedness (40.8 %, 33.4 %, 20.0 % and 14.4 %) with the above strains (Yao et al. 2014).

### *Planifilum composti* Han et al. 2013

The type strain P8$^T$ (=KACC 16581$^T$=NBRC 108858$^T$) was isolated from compost. GenBank/EMBL/DDBJ accession number for the 16S rRNA gene sequence is JN793954

The isolates grew aerobically from 50 °C to 75 °C (optimum at 55 °C) pH 4.0–9.0 (optimum pH 6.5). Aerial mycelia were not observed. Single spores were produced along the substrate hypha. The predominant menaquinone was MK-7. Major fatty acids were iso-$C_{17:0}$, iso-$C_{15:0}$ and iso-$C_{16:0}$. The cell wall contained meso-diaminopimelic acid and the polar lipids were phosphatidylethanolamine, aminophospholipid and sphingoglycolipid. The G+C contents were 55.9–56.5 mol%. The strains belonged to the genus *Planifilum* with 16S rRNA gene sequence identities of 96.1–97.2 %. Levels of DNA–DNA relatedness between strain P8$^T$and the type strains of recognized species of the genus *Planifilum* ranged from 28.9 % to 38.2 % (Han et al. 2013).

### *Melghirimyces thermohalophilus* Addou et al. 2013

The type strain Naril1A$^T$ (DSM 45514$^T$ =CCUG 60050$^T$) was isolated from soil of a salt lake. GenBank/EMBL/DDBJ accession number for the 16S rRNA gene sequence is JX861508.

The strain was an aerobic, halophilic, thermotolerant, Gram-positive bacterium, growing at NaCl concentrations between 5 % and 20 % w/v and temperature and pH ranges between 43–60 °C and 5.0–10.0, respectively. The major fatty acids were iso-$C_{15:0}$, anteiso-$C_{15:0}$ and iso-$C_{17:0}$. The G+C value was 53.4 mol %. LL-diaminopimelic acid was the diamino acid of

**Table 31.5**
**New species published in 2013–2014**

| Characteristic | *Thermoactinomyces daqus* | *Planifilum composti* | *Melghirimyces thermohalophilus* | *Melghirimyceso profundicolus* | *Kroppenstedtia guangzhouensis* | *Marininema halotolerans* |
|---|---|---|---|---|---|---|
| Aerial mycelia | White | Absent | Ivory | Absent | Ivory | absent |
| Temperature °C | optimum 50 – 55 | 50 – 75 optimum 55 | 43 – 60 | optimum 50 – 55 | 30 – 60 | optimum 28 |
| pH | nd | 4.0 – 9.0 | 5.0 – 10.0 | optimum 7.0 | 5.5 – 9.5 | optimum 7.0 |
| NaCl (%) w/v | nr | nr | 5 – 20 | optimum 3.0 | 0 – 3.0 | 0 – 5 |
| Diaminopimelic acid | *meso*-DAP | *meso*-DAP | LL-DAP | nd | LL-DAP | nd |
| Predominant menaquinone | MK-7 | MK-7 | MK-7 | MK-7 | MK-7 | MK-7 |
| Major fatty acids | iso-$C_{15:0}$<br>iso-$C_{17:0}$ | iso-$C_{17:0}$<br>iso-$C_{15:0}$<br>iso-$C_{16:0}$ | iso-$C_{15:0}$<br>anteiso-$C_{15:0}$<br>iso-$C_{17:0}$ | iso-$C_{15:0}$<br>anteiso-$C_{15:0}$<br>iso-$C_{17:0}$ | iso-$C_{15:0}$<br>iso-$C_{16:0}$<br>iso-$C_{17:0}$<br>anteiso-$C_{15:0}$ | anteiso-$C_{15:0}$<br>iso-$C_{15:0}$<br>anteiso-$C_{17:0}$<br>iso-$C_{16:0}$ |
| Polar lipids components | nd | PE, APL, SG | PG, DPG, PE, PIM, PME, PL | DPG, PME, PE, PG | DPG, PE, PME, PG | DPG, PG, PE |
| DNA G+C content (mol%) | 49.1 | 55.9–56.5 | 53.4 | 52.6 | 56.3 | nd |
| Reference | Yao et al. 2014 IJSEM[a] 64: 206 | Han et al. 2013 IJSEM 63: 4557 | Addou et al. 2013 IJSEM 63: 1717 | Li et al. 2013 IJSEM 63: 4552 | Yang et al. 2013 IJSEM 63: 4077 | Zhang et al. 2013 IJSEM 63: 4562 |

*nr* not recorded, *nd* no data, *APL* aminophospholip, *DPG* diphosphatidylglycerol, *PE* phosphatidylethanolamine, *PG* phosphatidylglycerol, *PL* unknow phospholipid, *PME* phosphatidylmonomethylethanolamine, *SG* sphingoglycolipid
[a] *International Journal of Systematic and Evolutionary Microbiology*

the peptidoglycan. The major menaquinone was MK-7, but MK-6 and MK-8 were also present in trace amounts. The polar lipids profile consisted of phosphatidylglycerol, diphosphatidylglycerol, phosphatidylethanolamine, phosphatidylmonomethylethanolamine and three unidentified phospholipids. The strain showed a 16S rRNA gene sequence similarity of 96.7% with the type strain of *Melghirimyces algeriensis* (Addou et al. 2013).

## *Melghirimyces profundicolus* Li et al. 2013

The type strain is SCSIO 11153$^T$ (=DSM 45787$^T$= CCTCC AA 2012007$^T$ = NBRC 109068$^T$) was isolated from a marine sediment. GenBank/EMBL/DDBJ accession number for the 16S rRNA gene sequence is JX555981.

Good growth of the filamentous bacterium was observed at 50–55 °C and pH 7.0 with 3 % NaCl. It formed ivory-white colonies with radial wrinkles but aerial mycelium was absent. It exhibited 96.4 % and 96.2 % 16S rRNA gene sequence similarities to the type strains of *Melghirimyces algeriensis* and *Melghirimyces thermohalophilus*, respectively. The menaquinone type was MK-7. Major cellular fatty acids were iso-$C_{15:0}$, anteiso-$C_{15:0}$ and iso-$C_{17:0}$. The polar lipids were diphosphatidylglycerol, phosphatidylmethylethanolamine, phosphatidylethanolamine and phosphatidylglycerol. The DNA G+C content of strain SCSIO 11153$^T$ was 52.6 mol% (Li et al. 2013).

## *Kroppenstedtia guangzhouensis* Yang et al. 2013

The type strain GD02$^T$ (=CGMCC 1.12404$^T$=KCTC 29149$^T$) was isolated from soil. GenBank/EMBL/DDBJ accession number for the 16S rRNA gene sequence is KC311557.

The Gram-stain-positive, spore-forming, aerobic and filamentous thermoactinomycete could grow in the presence of 0–3.0 % NaCl (w/v), at temperatures of 30–60 °C and at pH 5.5–9.5, forming ivory-coloured colonies. The 16S rRNA gene sequence similarity with *Kroppenstedtia eburnea* DSM 45196$^T$ was 96.1 %. The G+C content of the genomic DNA was 56.3 mol%. The cell-wall peptidoglycan contained LL-diaminopimelic acid, the main polar lipids were diphosphatidylglycerol, phosphatidylethanolamine, phosphatidylmethylethanolamine and phosphatidylglycerol, and the major menaquinone was MK-7. The major cellular fatty acids (>5 %) were iso-$C_{15:0}$, iso-$C_{16:0}$, iso-$C_{17:0}$ and anteiso-$C_{15:0}$ (Yang et al. 2013).

## *Marininema halotolerans* Zhang et al. 2013

The type strain is YIM M11385$^T$ (=CCTCC AB 2012052$^T$=DSM 45789$^T$) was isolated from a marine sediment. GenBank/EMBL/DDBJ accession number for the 16S rRNA gene sequence is KC684888.

The strain grew optimally at 28 °C, pH 7.0 and in the presence of 0–5 % (w/v) NaCl. It exhibited a quinone system with only MK-7, the polar lipid profile included diphosphatidylglycerol, phosphatidylglycerol and phosphatidylethanolamine as major components, and the major fatty acids were anteiso-$C_{15:0}$, iso-$C_{15:0}$, anteiso-$C_{17:0}$ and iso-$C_{16:0}$. The 16S rRNA gene sequence similarity with *Marininema mesophilum* SCSIO 10219$^T$ was 98.3 %. The level of DNA–DNA relatedness between strain YIM M11385$^T$ and *M. mesophilum* SCSIO 10219$^T$ was 59.36 % (Zhang et al. 2013).

## References


Addou AN, Schumann P, Spröer C, Bouanane-Darenfed A, Amarouche-Yala S, Hacene H, Cayol JL, Fardeau ML (2013) *Melghirimyces thermohalophilus* sp. nov., a novel thermoactinomycete isolated from an Algerian salt lake. Int J Syst Evol Microbiol 63:1717–1722

Addou AN, Schumann P, Spröer C, Hacene H, Cayol JL, Fardeau ML (2012) *Melghirimyces algeriensis* gen. nov., sp. nov., a member of the family *Thermoactinomycetaceae*, isolated from a salt lake. Int J Syst Evol Microbiol 62:1491–1498

Al-Khudary R, Hashwa F, Mroueh M (2004) A novel olive oil degrading *Thermoactinomyces* species with a high extremely thermostable lipase activity. Eng Life Sci 4:78–82

Akparov VK, Grishin AM, Yusupova MP, Ivanova NM, Chestukhina GG (2007) Structural principles of the wide substrate specificity of *Thermoactinomyces vulgaris* carboxypeptidase T. Reconstruction of the carboxypeptidase B primary specificity pocket. Biochemistry (Mosc) 72:416–423

Amner W, McCarthy AJ, Edwards C (1988) Quantitative assessment of factors affecting the recovery of indigenous and released thermophilic bacteria from compost. Appl Environ Microbiol 54:3107–3112

Barker AP, Simmon KE, Cohen S, Slechta ES, Fisher MA, Schlaberg R (2012) Isolation and identification of *Kroppenstedtia eburnea* from multiple patient samples. J Clin Microbiol. doi:10.1128/JCM.01186-12

Boiron P, Delga JM, Puel B, Drouhet E (1985) Étude sérologique de la bagassose par ELISA. Comparaison avec l'immuno-electro-diffusion. Bull Soc Fr Mycol Méd 14:309–314

Carrillo L, Romano F, Alderete EC (1987) Determinacion de la inmunidad a *Thermoactinomyces thalpophilus* en Jujuy, Argentina. Acta Bioq Clin Lat Am 21:321–327

Carrillo L, Maldonado MJ, Benitez Ahrendts MR (2009) Alkalithermophilic actinomycetes of subtropical area of Jujuy, Argentina. Rev Arg Microbiol 41:112–116

Chang C-C, Ng C-C, Wang C-Y, Shyu T-T (2009) Activity of cellulase from *Thermoactinomycetes* and *Bacillus* spp. isolated from *Brassica* waste compost. Sci Agric (Piracicaba, Braz) 66:304–308

Chen JJ, Lin LB, Zhang LL, Zhang J, Tang SK, Wei YL, Li WJ (2012) *Laceyella sediminis* sp. nov., a thermophilic bacterium isolated from a hot spring. Int J Syst Evol Microbiol 62:38–42

Cheng G, Zhao P, Tang X-F, Tang B (2009) Identification and characterization of a novel spore-associated subtilase from *Thermoactinomyces* sp. DCF. Microbiology 155:3661–3672

Collins MD, Mackillop GC, Cross T (1982) Menaquinone composition of members of the genus *Thermoactinomyces*. FEMS Microbiol Lett 13:151–153

Cross T, Walker PD, Gould GW (1968) Thermophilic actinomycetes producing resistant endospores. Nature (London) 220:352–354

Cunnington D, Teichtahl H, Hunt JM, Dow C, Valentine R (2000) Necrotizing pulmonary granulomata in a marijuana smoker. Chest 117:1511–1515

Dolashki A, Voelter W, Gushterova A, Van Beeumen J, Devreese B, Tchorbanov B (2012) Isolation and characterization of novel tyrosinase from *Laceyella sacchari*. Protein Pept Lett 19:538–543

DSMZ (2012) List of recommended media for microorganisms. http://www.dsmz.de/catalogues/catalogue-microorganisms/culture-technology/list-of-media-for-microorganisms.html

Elwan SH, Mostafa SA, Khodair AA, Ali O (1978) Lipase productivity of a lipolytic strain of *Thermoactinomyces vulgaris*. Zentralbl Bakteriol Naturwiss 133:706–712

Edwards JM (1972) The double dialysis method of producing farmer's lung antigens. J Lab Clin Med 79:683–688

Gontang EA, Fenical W, Jensen PR (2007) Phylogenetic diversity of gram-positive bacteria cultured from marine sediments. Appl Environ Microbiol 73:3272–3282

Hagerdal BGR, Ferchak JD, Pye EK (1978) Cellulolytic enzyme system of *Thermoactinomyces* sp. grown on microcrystalline cellulose. Appl Environ Microbiol 36:606–612

Han SI, Lee JC, Lee HJ, Whang KS (2013) *Planifilum composti* sp. nov., a thermophile isolated from compost. Int J Syst Evol Microbiol 63:4557–4561

Hanson RL, Goldberg SL, Brzozowski DB, Tully TP, Cazzulino D, Parker WL, Lyngberg OK, Vu TC, Wong MK, Patel RN (2007) Preparation of an amino acid intermediate for the dipeptidyl peptidase iv inhibitor, saxagliptin, using a modified phenylalanine dehydrogenase. Advan Syn Catal 349:1369–1378

Hatayama K, Shoun H, Ueda Y, Nakamura A (2005) *Planifilum fimeticola* gen. nov., sp. nov. and *Planifilum fulgidum* sp. nov., novel members of the family '*Thermoactinomycetaceae*' isolated from compost. Int J Syst Evol Microbiol 55:2101–2104

Hayashida S, Nanri N, Teramoto Y, Nishimoto T, Ohta K, Miyaguchi M (1988) Identification and characteristics of actinomycetes useful for semicontinuous treatment of domestic animal feces. Appl Environ Microbiol 54:2058–2063

Hedlund BP, Cole JK, Williams AJ, Hou W, Zhou E, Li W, Dong H (2012) A review of the microbiology of the Rehai geothermal field in Tengchong, Yunnan Province, China. Geoscience Frontiers 3:273–288

Huuskonen MS, Husman K, Jarvisalo J, Korhonen O, Kotimaa M, Kuusela T, Nordman H, Zitting A, Mantyjarvi R (1984) Extrinsic allergic alveolitis in the tobacco industry. Br J Ind Med 41:77–83

Ichikawa K, Tonozuka T, Mizuno M, Tanabe Y, Kamitori S, Nishikawaa A, Sakano Y (2005) Crystallization and preliminary X-ray analysis of *Thermoactinomyces vulgaris* R-47 maltooligosaccharide-metabolizing enzyme homologous to glucoamylase. Acta Crystallographica F61:302–304

Kalakoutskii LV, Agre N (1973) Endospores of Actinomycetes: dormancy and germination. In: Sykes G, Skinner FA (eds) *Actinomycetales*. Characteristics and practical importance. Academic Press, London, pp 179–195

Kanoh K, Matsuo Y, Adachi K, Imagawa H, Nishizawa M, Shizuri Y (2005) Mechercharmycins A and B, cytotoxic substances from marine-derived *Thermoactinomyces* sp. YM3-251. J Antibiot 58:289–292

Kitpreechavanich V (2011) Phylogenetic of PLA-degrading thermophilic bacteria and characterization of PLA- degrading enzyme. PS093. BISMiS 2011, 19–23 May 2011, Beijing, China

Kleine R (1982) Properties of thermitase, a thermostable serine protease from *Thermoactinomyces vulgaris*. Acta Biol Med Ger 41:89–102

Kretschmer S (1980) Transinfection in *Thermoactinomyces vulgaris*. Z Allg Mickrobiol 20:73–75

Kurtboke DI, Sivasithamparam K (1993) Taxonomic implications of the reactions of representative *Bacillus* strains to *Thermoactinomyces*-phage. Actinomycetes 4:1–7

Kurup VP, Barboriak JJ, Fink JN, Lechevalier MP (1975) *Thermoactinomyces candidus* a new species of thermophilic actinomycetes. Int J Syst Bacteriol 25:150–154

Kurup VP, Barboriak JJ, Fink JN, Scribner G (1976) Immunologic cross-reactions among thermophilic actinomycetes associated with hypersensitivity pneumonitis. J Allergy Clin Immunol 57:417–421

Kurup VP, Hollick GE, Pagan EF (1980) *Thermoactinomyces intermedius*, a new species of amylase negative thermophilic actinomycetes. Sci Ciencia 7:104–108

Kurup VP, Resnick A, Kagen SL, Cohen SH, Fink JN (1983) Allergenic fungi and actinomycetes in smoking materials and their health implications. Mycopathologia 82:61–64

Lacey J (1971) *Thermoactinomyces sacchari* sp. nov., a thermophilic actinomycete causing bagassosis. J Gen Microbiol 66:327–338

Lacey J (1974) Moulding of sugar-cane bagasse and its prevention. Ann Appl Biol 76:61–76

Lacey J (1997) Actinomycetes in composts. Ann Agric Environ Med 4:113–121

Lacey J, Cross T (1989) Genus *Thermoactinomyces* Tsiklinsky 1899. In: Williams ST, Sharpe ME, Holt JG (eds) Bergey's manual of systematic bacteriology, vol 4. Williams and Wilkins, Baltimore, pp 2574–2585

Li J, Qin S, You ZQ, Long LJ, Tian XP, Wang FZ, Zhang S (2013) *Melghirimyces profundicolus* sp. nov., isolated from a deep sea sediment. Int J Syst Evol Microbiol 63:4552–4556

Li J, Zhang G-T, Yang J, Tian X-P, Wang F-Z, Zhang CS, Zhang S, Li W-J (2012) *Marininema mesophilum* gen. nov., sp. nov., a thermoactinomycete isolated from deep sea sediment, and emended description of the family *Thermoactinomycetaceae*. Int J Syst Evol Microbiol 62:1383–1388

Matsuo J, Katsuta A, Matsuda S, Shizuri Y, Yokota A, Kasai H (2006) *Mechercharimyces mesophilus* gen. nov., sp. nov. and *Mechercharimyces asporophorigenens* sp. nov., antitumour substance-producing marine bacteria, and description of *Thermoactinomycetaceae* fam. nov. Int J Syst Evol Microbiol 56:2837–2842

Matsuo Y, Kanoh K, Yamori T, Kasai H, Katsuta A, Adachi K, Shin-Ya K, Shizuri Y (2007) Urukthapelstatin A, a novel cytotoxic substance from marine-derived *Mechercharimyces asporophorigenens* YM11-542. I. Fermentation, isolation and biological activities. J Antibiotics 60:251–255

McCarthy AJ, Cross T (1984) A taxonomic study of *Thermomonospora* and other monosporic actinomycetes. J Gen Microbiol 130:5–25

McNeil MM, Brown JM (1994) The medically important aerobic actinomycetes: epidemiology and microbiology. Clin Microbiol Rev 7:357–417

Neef A, Schäfer R, Beimfohr C, Kämpfer P (2003) Fluorescence based rRNA sensor systems for detection of whole cells of *Saccharomonospora* spp. and *Thermoactinomyces* spp. Biosens Bioelectron 18:565–569

Nilsson M, Renberg I (1990) Viable endospores of *Thermoactinomyces vulgaris* in lake sediments as indicators of agricultural history. Appl Environ Microbiol 56:2025–2028

Ohshima T, Takada H, Yoshimura T, EsakiI N, Soda K (1991) Distribution, purification, and characterization of thermostable phenylalanine dehydrogenase from thermophilic Actinomycetes. J Bacteriol 173:3943–3948

Ohshima T, Nishida N, Bakthavatsalam S, Kataoka K, Takada H, Yoshimura T, Esaki N, Soda K (1994) The purification, characterization, cloning and sequencing of the gene for a halostable and thermostable leucine dehydrogenase from *Thermoactinomyces intermedius*. Eur J Biochem 222:305–312

Panosyan HH (2010) Phylogenetic diversity based on 16S rRNA gene sequence analysis of aerobic thermophilic endospore-forming bacteria isolated from geothermal springs in Armenia. Biolog J Armenia 4:73–90

Park Y-H, Kim E, Yim D-G, Kho Y-H, Mheen T-I, Goodfellow M (1993) Supragenic classification of *Thermoactinomyces vulgaris* by nucleotide sequencing of 5S ribosomal RNA. Zentbl Bakteriol 278:469–478

Park DJ, Dastager SG, Lee JC, Yeo SH, Yoon JH, Kim CJ (2007) *Shimazuella kribbensis* gen. nov., sp. nov., a mesophilic representative of the family *Thermoactinomycetaceae*. Int J Syst Evol Microbiol 57:2660–2664

Pauwels R, Devos M, Callens L, van der Straeten M (1978) Respiratory hazards from proteolytic enzymes. Lancet 1:669

Rosselló-Mora R, Yarza P, Muñoz R (2012) The PK4 tree editing team. pk4@imedea.uib-csic.es

Shirling EB, Gottlieb D (1966) Methods for characterization of *Streptomyces* species. Int J Syst Bacteriol 16:313–340

Singh V, Chandra Pandey V, Pathak DC, Agrawal S (2012) Purification and characterization of *Laceyella sacchari* strain B42 xylanase and its potential for pulp biobleaching. Afr J Microbiol Res 6:1397–1410

Smith K, Sundaram TK, Kernick M (1984) Malate dehydrogenases from Actinomycetes: structural comparison of *Thermoactinomyces* enzyme with other Actinomycete and *Bacillus* Enzymes. J Bacteriol 157:684–687

Stackebrandt E, Woese CR (1981) Towards a phylogeny of the actinomycetes and related organisms. Curr Microbiol 5:197–202

Stackebrandt E, Goebel BM (1994) Taxonomic note: a place for DNA–DNA reassociation and 16S rRNA sequence analysis in the present species definition in bacteriology. Int J Syst Bacteriol 44:846–849

Swan JRM, Blainey D, Crook B (2007) The HSE Grain Dust Study workers' exposure to grain dust contaminants, immunological and clinical response. RR540. Health and Safety Laboratory. Buxton, Derbyshire

Teplyakov AV, Kuranova IP, Harutyunyan EH, Vainshtein BK, Frommel C, Hohne WE, Wilson KS (1990) Crystal structure of thermitase at 1.4 Å resolution. J Mol Biol 214:261–279

Tonozuka T, Yokota T, Ichikawa K, Mizuno M, Kondo S, Nishikawa A, Kamitori S, Sakano Y (2002) Crystal structures and substrate specificities of two α-amylases hydrolyzing cyclodextrins and pullulan from *Thermoactinomyces vulgaris* R-47. Biologia Bratislava 57(Suppl 11):71–76

Treuhaft MW (1977) Isolation of bacteriophage from *Thermoactinomyces*. J Clin Microbiol 6:420–424

Tsilinsky P (1899) Sur les mucedinéés thermophiles. Ann Inst Pasteur 13:500–505

Tsubouchi T, Shimane Y, Mori K, Usui K, Hiraki T, Tame A, Uematsu K, Maruyama T, Hatada Y (2013) *Polycladomyces abyssicola* gen. nov., sp. nov., a thermophilic filamentous bacterium isolated from hemipelagic sediment in Japan. Int J Syst Evol Microbiol 63:1972–1981

Tuomela M, Vikman M, Hatakka A, Itävaara M (2000) Biodegradation of lignin in a compost environment: a review. Bioresour Technol 72:169–183

Turner P, Nilsson C, Svensson D, Holst O, Gorton L, Nordberg Karlsson E (2005) Monomeric and dimeric cyclomaltodextrinases reveal different modes of substrate degradation. Biologia, Bratislava, 60. Suppl 16:79–87

Uzel A, Hameş Kocabaş EE, Bedir E (2011) Prevalence of *Thermoactinomyces thalpophilus* and *T. sacchari* strains with biotechnological potential at hot springs and soils from West Anatolia in Turkey. Turk J Biol 35:195–202

von Jan M, Riegger N, Pötter G, Schumann P, Verbarg S, Spröer C, Rohde M, Lauer B, Labeda DP, Klenk H-P (2011) *Kroppenstedtia eburnea* gen. nov., sp. nov., a thermoactinomycete isolated by environmental screening, and emended description of the family *Thermoactinomycetaceae* Matsuo et al. 2006 emend. Yassin et al. 2009. Int J Syst Evol Microbiol 61:2304–2310, International Journal of Systematic and Evolutionary Microbiologyijs.sgmjournals.org

Williams ST, Cross T (1971) Actinomycetes. In: Booth C (ed) Methods in microbiology, vol 4. Academic, London, pp 295–334

Williams ST, Lanning S, Wellington EMH (1984) Ecology of actinomycetes. In: Goodfellow M, Mordarski M, Williams ST (eds) The biology of Actinomycetes. Academic, London, pp 481–528

Xu J, Rao JR, Millar BC, Elborn JS, Evans J, Barr JG, Moore JE (2002) Improved molecular identification of *Thermoactinomyces* spp. associated with mushroom worker's lung by 16S rDNA sequence typing. J Med Microbiol 51:1117–1127

Yang G, Qin D, Wu C, Yuan Y, Zhou S, Cai Y (2013) *Kroppenstedtia guangzhouensis* sp. nov., a thermoactinomycete isolated from soil. Int J Syst Evol Microbiol 63:4077–4080

Yao S, Liu Y, Zhang M, Zhang X, Li H, Zhao T, Xin C, Xu L, Zhang B, Cheng C (2014) *Thermoactinomyces daqus* sp. nov., a thermophilic bacterium isolated from high-temperature Daqu. Int J Syst Evol Microbiol 64:206–210

Yassin AF, Hupfer H, Klenk H-P, Siering C (2009) *Desmospora activa* gen. nov., sp. nov., a thermoactinomycete isolated from sputum of a patient with suspected pulmonary tuberculosis, and emended description of the family *Thermoactinomycetaceae* Matsuo et al. 2006. Int J Syst Evol Microbiol 59:454–459

Yoon JH, Park YH (2000) Phylogenetic analysis of the genus *Thermoactinomyces* based on 16S rDNA sequences. Int J Syst Evol Microbiol 50:1081–1086

Yoon JH, Shin YK, Park TH (2000) DNA–DNA relatedness among *Thermoactinomyces* species: *Thermoactinomyces candidus* as a synonym of

*Thermoactinomyces vulgaris* and *Thermoactinomyces thalpophilus* as a synonym of *Thermoactinomyces sacchari*. Int J Syst Evol Microbiol 50:1905–1908

Yoon JH, Kim IG, Shin YK, Park YH (2005) Proposal of the genus *Thermoactinomyces sensu stricto* and three new genera, *Laceyella, Thermoflavimicrobium* and *Seinonella*, on the basis of phenotypic, phylogenetic and chemotaxonomic analyses. Int J Syst Evol Microbiol 55:395–400

Yu TT, Zhang BH, Yao JC, Tang SK, Zhou EM, Yin YR, Wei DQ, Ming H, Li WJ (2012) *Lihuaxuella thermophila* gen. nov., sp. nov., isolated from a geothermal soil sample in Tengchong, Yunnan, south-west China. Antonie Van Leeuwenhoek (doi: 10.1007/s10482-012-9771-6)

Zhang XM, He J, Zhang DF, Chen W, Jiang Z, Sahu MK, Sivakumar K, Li WJ (2013) *Marininema halotolerans* sp. nov., a novel thermoactinomycete isolated from a sediment sample, and emended description of the genus Marininema Liet al. 2012. Int J Syst Evol Microbiol 63:4562–4567

Zhang Y-X, Dong C, Biao S (2007) *Planifilum yunnanense* sp. nov., a thermophilic thermoactinomycete isolated from a hot spring. Int J Syst Evol Microbiol 57:1851–1854

Zhang J, Tang S-K, Zhang Y-Q, Yu L-Y, Klenk H-P, Li W-J (2010) *Laceyella tengchongensis* sp. nov., a thermophile isolated from soil of a volcano. Int J Syst Evol Microbiol 60:2226–2230

# 32 The Family *Thermoanaerobacteraceae*

*Erko Stackebrandt*
Leibniz Institute DSMZ-German Collection of Microorganisms and Cell Cultures GmbH, Braunschweig, Germany



**Abstract**

The family *Thermoanaerobacteraceae* is a family of the order *Thermoanaerobacterales*, phylum Firmicutes, comprising several genera of strictly anaerobic, rod-shaped, spore-forming bacteria which were mostly isolated from hot springs. The metabolism is variable, ranging from carbohydrate fermentation to chemolithoautotrophy. Reduction of thiosulfate varies. This brief overview concentrates on genera and species described since 2006, and which are not covered in the chapter *Thermoanaerobacteraceae* in *Bergey's Manual of Systematic Bacteriology*, 2nd edition.

## Taxonomy

The family *Thermoanaerobacteraceae* (Wiegel 2009) encompasses the genera *Ammonifex, Brockia, Caldanaerobacter, Caldanaerobius, Caloribacterium, Carboxydothermus, Desulfovirgula, Gelria, Moorella, Tepidanaerobacter, Thermacetogenium, Thermanaeromonas,* and the type genus *Thermoanaerobacter.* The genus list of Euzeby (http://www.bacterio.net/) lists a few more genera though the type species were transferred to other genera; as the taxa were validly published, they did keep their taxonomic status: *Acetogenium kivui* was reclassified as *Thermoanaerobacter kivui* (Collins et al. 1994), *Carboxydibrachium pacificum* as *Caldanaerobacter subterraneus* subsp. *pacificus* (Fardeau et al. 2004), *Thermoanaerobium acetigenum* as *Caldicellulosiruptor acetigenus* (Onyenwoke et al. 2006), *Thermoanaerobium brockii* as *Thermoanaerobacter brockii* subsp. *brockii* (Lee et al. 1993), *and species of Thermobacteroides* to *Thermoanaerobacter* (Rainey and Stackebrandt 1993), *Clostridium* (Fardeau et al. 2001), and *Coprothermobacter* (Rainey and Stackebrandt 1993).

The last comprehensive coverage of the family *Thermoanaerobacteraceae* has been presented by 2009 in *Bergey's Manual of Systematic Bacteriology*, 2nd ed (Wiegel 2009), covering descriptions since 2006. Since then, new genera and new species of genera described until 2006 were proposed. This communication concentrates on recently described genera and species, and the reader should consult the chapter of 2009 for obtaining a more comprehensive overview of the biology of family members. ➲ *Table 32.1* is a list of species belonging to genera described until 2006, together with some of the salient feature of these taxa. ➲ *Table 32.2* compiles recently described genera of *Thermoanaerobacteraceae.*

The family contains anaerobic, mainly heterotrophic, but also chemolithoautotrophic members. Cells are mainly Gram positive, rod shaped, and spore forming, though same species stain Gram negatively, while no spores have been reported for other species. The family is phylogenetically heterogeneous, forming individual clades which are related to other families according to the ML tree (➲ *Fig. 32.1*). The heterogeneity has been observed by Sekiguchi et al. (2006), while other species description exclusively included members of the family as reference strains in depicting phylogenetic dendrograms. *Thermoanaerobacter* and *Caldanaerobacter* are closely related and group adjacent to a clade consisting of *Coprothermobacter, Dictyoglomus,* and other *Thermoanaerobacteraceae* members such as *Ammonifex, Tepidanaerobacter,* and *Brockia. Desulfovirgula* and *Thermanaeromonas* form a third clade. A forth clade is phylogenetically heterogeneous in itself as it embraces authentic clostridia, symbiotic and synthrophic taxa, *Heliobacteriaceae* and *Peptococcaceae,* as well as the *Thermoanaerobacteraceae* members *Caldanaerobius, Caloribacterium, Symbiobacterium, Gelria,* and *Moorella. Syntrophaceticus schinkii and Symbiobacterium thermophilum,* not reported to be members of *Thermoanaerobacteraceae,* cluster closely with *Thermacetogenium schinkii and Gelria glutamica,* respectively. A fifth clade consists of *Carboxydothermus.* This family is certainly in need of a taxonomic revision.

## Genome Sequences

Several representatives of *Thermoanaerobacteraceae,* especially from the type genus, were subjected to the analysis of genome sequences. Only a few examples of published or deposited sequences of strains of the various genera are given in ➲ *Table 32.3.* More information is available in the GOLD database (genomes.org/cgi-bib/Gold/Search.cgi).

E. Rosenberg et al. (eds.), *The Prokaryotes – Firmicutes and Tenericutes,* DOI 10.1007/978-3-642-30120-9_367,

**Table 32.1**
**Species published since 2006, for genera described before 2006**

| Genus | *Moorella* | *Carboxydothermus* | | |
|---|---|---|---|---|
| **Species** | *humiferrea* | *pertinax* | *islandicus* | *siderophilus* |
| Gram-stain | Positive | n.r. | Positive | Positive |
| Motility | Peritrichous flagella | Peritrichous flagella | Peritrichous flagella | Lateral flagella |
| Morphology | Straight rods, singly or short chains | Rods | Short, slightly curved rods | Short, straight rods |
| Spore formation | + in terminal swollen sporangia | n.r. | n.r. | – |
| Growth range °C optimum | 46.0–70.0 (65.0) | 50.0–70.0 (65.0) | 50.0–70.0 (65.0) | 40.0–78.0 (70.0–72.0) |
| pH range optimum | 5.5–8.5 (7.0) | 4.6–8.6 (6.0–6.5) | 5–8 (5.5–6.0) | 6.6–8.0 (7.0) |
| Metabolism | Lactate, malate succinate, glycerol, YE as electron donors, 9,10-anthraquinon-2,6-disulfonate as electron shuttle to Fe(III) oxide. Various hexose, pyruvate, and peptone are fermented | YE, peptone, pyruvate, glucose, or chemolithoautotrophic on CO as electron donor, and ferric citrate, amorphous Fe (III) oxide, and 9,10-anthraquinon-2, 6-disulfonate as electron acceptors. $CO + H_2O \rightarrow CO_2 + H_2$ | YE, lactate, pyruvate, or chemolithoautotrophic on CO as electron donor, 9,10-anthraquinon-2, 6-disulfonate as electron acceptor. $CO + H_2O \rightarrow CO_2 + H_2$ | Chemoheterotrophic with glucose, xylose, lactate, or YE; chemolithotrophic with CO, in the presence of Fe(III) or 9,10-anthraquinon-2,6-disulfonate hydrogen, $CO_2$ and Fe(II), or $AQDSH_2$, respectively, are produced |
| Endproducts of carbohydrate fermentation | Fructose: acetate | Check | Pyruvate: acetate, $CO_2$, $H_2$ | None |
| Sulfur metabolism | Thiosulfate reduced to sulfide | Thiosulfate and elemental sulfur not reduced | Thiosulfate and elemental sulfur not reduced | Thiosulfate and elemental sulfur not reduced |
| Major fatty acids >10 % | n.r. | iso-$C_{15:0}$,$C_{15:0}$, $C_{15:0}$DMA and/or $C_{14:0}$3-OH | $C_{14:0}$, iso-$C_{15:0}$, $C_{16:0}$ | n.r. |
| Mol% G+C | 51 | 42.2 | 37.7 | 41.5 |
| Type strain | 64-FGQ$^T$ | Ug1$^T$ | SET IS-9$^T$ | 1315$^T$ |
| Habitat | Terrestrial hypothermic spring, Kamchatka, Russia | Hot spring, Kyushu Island, Japan | Hot spring, Iceland | Hot spring, Kamchatka, Russia |
| Publication | Nepomnyashchaya et al. 2012 | Yoneda et al. 2012 | Novikov et al. 2011 | Slepova et al. 2009 |

| Genus | *Thermoanaerobacter* | | | *Ammonifex* |
|---|---|---|---|---|
| **Species** | *pseudethanolicus*[a] | *thiophilus* | *uzonensis* | *pentosaceus* |
| Gram stain | Positive | Positive | Type positive, stain negative | Stain negative |
| Motility | + | Peritrichous flagella | Peritrichous flagella | – |
| Morphology | Rods, older cells form chains of coccoid cells to long filamentous cells | Rods, singly or pairs | Straight to slightly curved rods | Rods, single, pairs, short chains |
| Spore formation | Round, terminal | Round, terminal | Oval, subterminal | Terminal |

**Table 32.1 (continued)**

| Genus | *Thermoanaerobacter* | | | *Ammonifex* |
|---|---|---|---|---|
| Species | *pseudethanolicus*[a] | *thiophilus* | *uzonensis* | *pentosaceus* |
| Growth range °C optimum | 37.0–76.0 (65.0–70.0) | 60.0–82.0 (75.0) | 32.5–69.0 (61.0) | 50.0–80.0 (70.0) |
| pH range optimum | 5.8–8.5 | 6.0–7.5 (6.8–7) | 4.2–8.9 (7.1) | 5.4–8.9 (7.0) |
| Metabolism | Chemoorganotroph. Fermentation of glucose, fructose, mannose, galactose, ribose, xylose, lactose, sucrose, maltose, cellobiose, starch, and pyruvate in the presence of YE | Autotrophic growth with $CO_2$ as carbon source and $H_2$ or formate as electron donors. Facultative chemolithoautotrophic | Chemoorganotroph. YE, various hexoses, cellobiose, inulin, mannitol, pyruvate, crotonate | Chemoorganotroph. YE, various hexoses, pectin, starch, xylose |
| Endproducts of carbohydrate fermentation | Hexoses, starch: ethanol and $CO_2$; lactate and acetate in minor amounts | Not fermentative | Glucose: acetate, ethanol, $CO_2$, $H_2$ | Xylose: ethanol, $H_2$, lactate, acetate. Presence of thiosulfate shifts the endproducts toward acetate, sulfite enhances formation of ethanol |
| Sulfur metabolism | Thiosulfate and sulfite reduced to $H_2S$ | Weak growth with sulfate, sulfur, thiosulfate reduced | Thiosulfate reduced to sulfide. In the presence of YE, glucose, and thiosulfate, $H_2S$ and elemental sulfur are formed | Sulfite and thiosulfate reduced |
| Major fatty acids >10 % | n.r. | n.r. | iso-$C_{15:0}$, $C_{15:0}$ | iso-$C_{15:0}$, iso-$C_{14:0}$ 3-OH, iso-$C_{170}$ |
| Mol% G+C | 32–34.4 | 56.2 | 33.6 | 34.2 |
| Type strain | $39E^T$ | $SR^T$ | $JW/IWO10^T$ | $DTU01^T$ |
| Habitat | Octopus spring, Yellowstone, USA | Hot spring, Kamchatka, Russia | Hot spring, Kamchatka, Russia | Household waste reactor |
| Publication | Onyenwoke et al. 2007 | Miroshnichenko et al. 2008 | Wagner et al. 2008 | Tomas et al. 2013 |

*n.r.* not recorded, *n.a.* not applicable, *YE* yeast extract

[a]According to Hollaus and Sleytr (1972) based upon *Clostridium thermohydrosulfuricum* ATCC 7956$^T$ and on *Thermoanaerobacter ethanolicus* (Wiegel and Ljungdahl 1981; Lee et al. (1993))

**Table 32.2**
**New genera of *Thermoanaerobacteraceae* and their species, described since 2006**

| Genus | *Caldanaerobius* | *Brockia* | *Caloribacterium* | *Tepidanaerobacter* |
|---|---|---|---|---|
| Species | *fijiensis* | *lithotrophica* | *cisternae* | *syntrophicus* |
| Gram type/ stain | Type positive, stain negative | Type positive | Type positive | Type positive, stain negative |
| Motility | Peritrichous flagella | Peritrichous flagella | Nonpolar flagella | – |
| Morphology | Straight to slightly curved | Thin regular rods, singly or pairs | Straight rods, singly or pairs | Irregular rods |
| Spore formation | Spherical, terminal | | – | – |
| Growth range °C (optimum) | 40.0–67.0 (60.0–63.0) | 46.0–78.0 (60.0–65.0) | 28.0–65.0 (50.0) | 25.0–60.0 (45.0–50.0) |
| pH range (optimum) | 4.5–8.4 (6.8) | 5.5–8.5 (6.5) | 5.5–8.0 (7–7.5) | 5.5–8.5 (6–7) |
| Substrates used | YE, various pentoses, hexoses, cellobiose | H2, formate as electron donors, elemental sulfur, thiosulfate, polysulfide as electron acceptors | YE, peptone, various hexoses, pyruvate, citrate | YE, various riboses, hexoses, starch, pectin, crotonate. In coculture with *Methanothermobacter thermoautotrophicus*, ethanol, glycerol, and lactate are utilized |
| Endproducts of glucose fermentation | Ethanol, acetate formate | Not applicable | Acetate, hydrogen, $CO_2$ | (Plus YE), acetate, $H_2$ |
| Sulfur metabolism | Thiosulfate reduced to elemental sulfur | $H_2$-sulfur metabolism | Thiosulfate reduced to sulfide | Thiosulfate reduced |
| Major fatty acids >10 % | n.r. | $C_{16:0}$, iso-$C_{16:0}$, $C_{18:0}$, iso-$C_{17:0}$ | iso-$C_{15:0}$, $C_{16:0}$, iso-$C_{17:1}\omega8$, $C_{18:0}$ | iso-$C_{15:0}$, $C_{16:1}\omega9c$, $C_{15:1}$ |
| Mol% G+C | 37.6 | 63 | 43.1 | 37–38 |
| Habitat | Sediment hot spring, Fiji | Sediment hot spring, Kamchatka, Russia | Underground gas storage reservoir, Siberia, Russia | Thermophilic sludge digester, Japan |
| Type strain | JW/YJL-F3$^T$ | Kam1851$^T$ | SGL43$^T$ | JL$^T$ |
| Publication | Lee et al. 2008 | Perevalova et al. 2013 | Slobodkina et al. 2012 | Sekiguchi et al. 2006 |

For abbreviations see *Table 32.1*

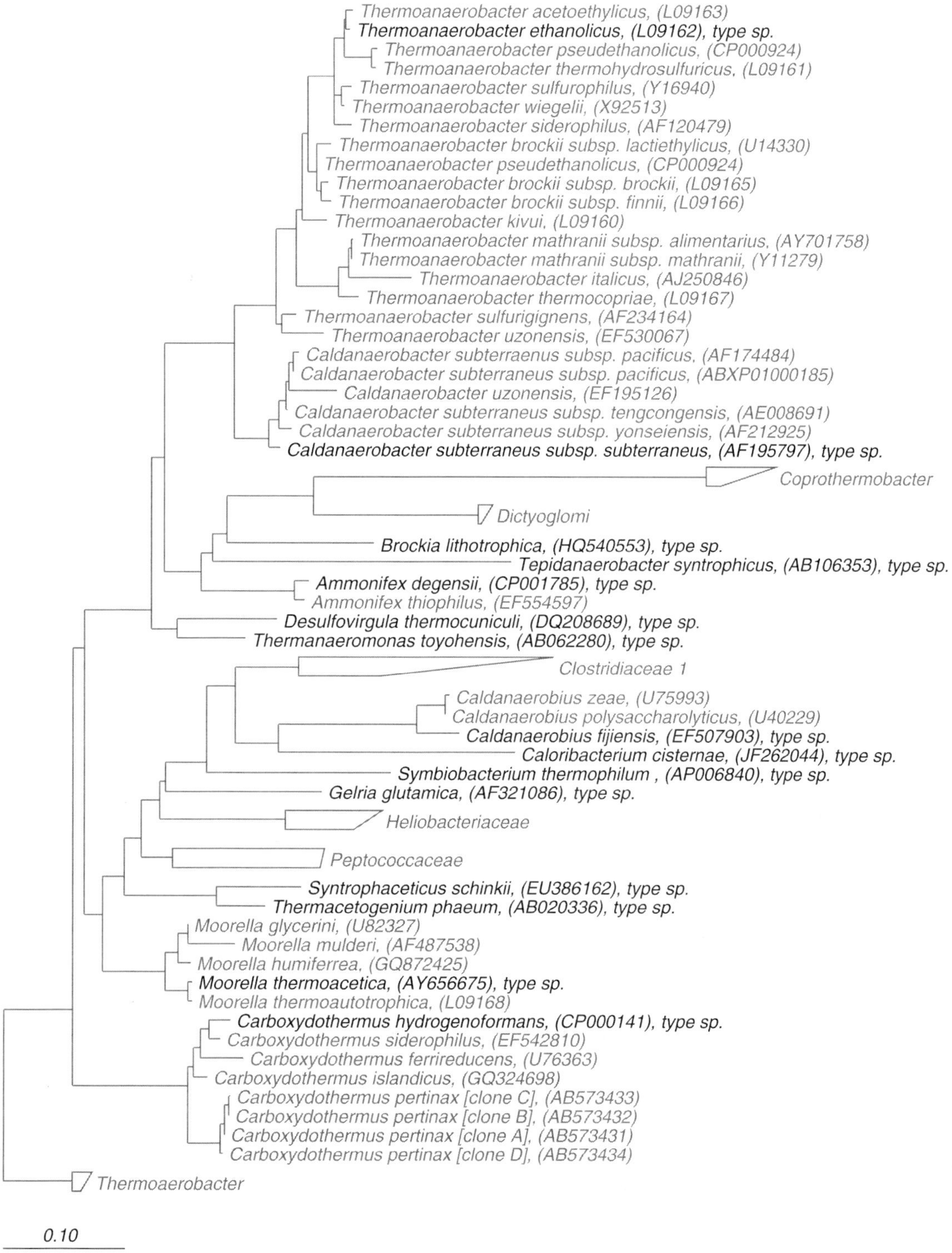


**Fig. 32.1**

**Neighbor-joining genealogy reconstruction based on the RAxML algorithm (Stamatakis 2006) of the sequences of members of the family *Thermoanaerobacteraceae* and some neighboring taxa present in the LTP_106 (Yarza et al. 2010). The tree was reconstructed by using a subset of sequences. Representative sequences from closely relative genera were used to stabilize the tree topology. In addition, a 40 % maximum frequency filter was applied to remove hypervariable positions from the alignment. Scale bar indicates estimated sequence divergence**

**Table 32.3**
**Examples of complete, incomplete, and draft genome sequences of *Thermoanaerobacteraceae* members as listed in the GOLD genome database**

| Taxon | Strain number | GOLD identification | Status |
|---|---|---|---|
| *Ammonifex degensii* | KC4 | Gc02209 | Complete, unpublished |
| *Caldanaerobacter subterraneus*<br>Sequenced as *Thermoanaerobacter tengcongensis* | MB4$^T$ | Gc00086 | Bao et al. 2002 |
| *Caldanaerobius polysaccharolyticum* | DSM 13641$^T$ | Gi02942 | Incomplete, unpublished |
| *Carboxydothermus ferrireducens* | DSM 11255$^T$ | Gr00123 | Permanent draft, unpublished |
| *Carboxydothermus hydrogenoformans* | DSM 6008$^T$ | Gc00307 | Wu et al. 2005 |
| *Desulfovirgula thermocuniculi* | DSM 16036$^T$ | Gi11444 | Permanent draft, unpublished |
| *Moorella thermoacetica* | ATCC 39073 | Gc00397 | Pierce et al. 2008 |
| *Tepidanaerobacter acetatoxydans* | Re1 | Gc02475 | Manzoor et al. 2013 |
| *Thermacetogenium phaeum* | PB | Gc0028002 | Oehler et al. 2012 |
| *Thermanaeromonas toyohensis* | DSM 14490 | Gi04363 | Incomplete, unpublished |
| *Thermoanaerobacter mathranii* | DSM 11426$^T$ | Gc01347 | Complete, unpublished |
| *Thermoanaerobacter thermocopriae* | ATCC 51646 | Gi22384 | Complete, unpublished |
| *Thermoanaerobacter pseudethanolicus* | ATCC 33223$^T$ | Gc00718 | Hemme et al. 2011 |

## References


Bao Q, Tian Y, Li W, Xu Z, Xuan Z, Hu S, Dong W, Yang J, Chen Y, Xue Y, Xu Y, Lai X, Huang L, Dong X, Ma Y, Ling L, Tan H, Chen R, Wang J, Yu J, Yang H (2002) A complete sequence of the *T. tengcongensis* genome. Genome Res 12:689–700

Collins MD, Lawson PA, Willems A, Cordoba JJ, Fernandez-Garayzabal J, Garcia P, Cai J, Hippe H, Farrow JAE (1994) The phylogeny of the genus *Clostridium*: proposal of five new genera and eleven new species combinations. Int J Syst Bacteriol 44:812–826

Fardeau ML, Ollivier B, Garcia JL, Patel BKC (2001) Transfer of *Thermobacteroides leptospartum* and *Clostridium thermolacticum* as *Clostridium stercorarium* subsp. *leptospartum* subsp. nov., comb. nov. and *C. stercorarium* subsp. *thermolacticum* subsp. nov., comb. nov. Int J Syst Bacteriol 51:1127–1131

Fardeau ML, Bonilla Salinas M, L'Haridon S, Jeanthon C, Verhé F, Cayol JL, Patel BKC, Garcia JL, Ollivier B (2004) Isolation from oil reservoirs of novel thermophilic anaerobes phylogenetically related to *Thermoanaerobacter subterraneus*: reassignment of *T. subterraneus*, *Thermoanaerobacter yonseiensis*, *Thermoanaerobacter tengcongensis* and *Carboxydibrachium pacificum* to *Caldanaerobacter subterraneus* gen. nov., sp. nov., comb. nov. as four novel subspecies. Int J Syst Evol Microbiol 54:467–474

Hemme CL, Fields MW, He Q, Deng Y, Lin L, Tu Q et al (2011) Correlation of genomic and physiological traits of *Thermoanaerobacter* species with biofuel yields. Appl Environ Microbiol 77:7998–8008

Hollaus F, Sleytr U (1972) On the taxonomy and fine structure of some hyperthermophilic saccharolytic clostridia. Arch Microbiol 86:129–146

Lee YE, Jain MK, Lee C, Lowe SE, Zeikus JG (1993) Taxonomic distinction of saccharolytic thermophilic anaerobes: description of *Thermoanaerobacterium xylanolyticum* gen. nov., sp. nov., and *Thermoanaerobacterium saccharolyticum* gen. nov., sp. nov.; reclassification of *Thermoanaerobium brockii*, *Clostridium thermosulfurogenes*, and *Clostridium thermohydrosulfuricum* E100-69 as *Thermoanaerobacter brockii* comb. nov., *Thermoanaerobacterium thermosulfurigenes* comb. nov., and *Thermoanaerobacter thermohydrosulfuricus* comb. nov., respectively; and transfer of *Clostridium thermohydrosulfuricum* 39E to *Thermoanaerobacter ethanolicus*. Int J Syst Bacteriol 43:41–51

Lee YJ, Mackie RI, Cann IKO, Wiegel J (2008) Description of *Caldanaerobius fijiensis* gen. nov., sp. nov., an inulin-degrading, ethanol-producing, thermophilic bacterium from a Fijian hot spring sediment, and reclassification of *Thermoanaerobacterium polysaccharolyticum* and *Thermoanaerobacterium zeae* as *Caldanaerobius polysaccharolyticus* comb. nov. and *Caldanaerobius zeae* comb. nov. Int J Syst Evol Microbiol 58:666–670

Manzoor S, Bongcam-Rudloff E, Schnürer A, Müller B (2013) First genome sequence of a syntrophic acetate-oxidizing bacterium, *Tepidanaerobacter acetatoxydans* strain Re1. Genome Announc. doi:10.1128/genomeA.00213-12, 1pii: e00213-12

Miroshnichenko ML, Tourova TP, Kolganova TV, Kostrikina NA, Chernych N, Bonch-Osmolovskaya EA (2008) *Ammonifex thiophilus* sp. nov., a hyperthermophilic anaerobic bacterium from a Kamchatka hot spring. Int J Syst Evol Microbiol 58:2935–2938

Nepomnyashchaya YN, Slobodkina GB, Baslerov RV, Chernyh NA, Bonch-Osmolovskaya EA, Netrusov AI, Slobodkin AI (2012) *Moorella humiferrea* sp. nov., a thermophilic, anaerobic bacterium capable of growth via electron shuttling between humic acid and Fe(III). Int J Syst Evol Microbiol 62:613–617

Novikov AA, Sokolova TG, Lebedinsky AV, Kolganova TV, Bonch-Osmolovskaya EA (2011) *Carboxydothermus islandicus* sp. nov., a thermophilic, hydrogenogenic, carboxydotrophic bacterium isolated from a hot spring. Int J Syst Evol Microbiol 61:2532–2537

Oehler D, Poehlein A, Leimbach A, Müller N, Daniel R, Gottschalk G, Schink B (2012) Genome-guided analysis of physiological and morphological traits of the fermentative acetate oxidizer *Thermacetogenium phaeum*. BMC Genomics 13:723

Onyenwoke RU, Lee Y-J, Dabrowski S, Ahring BK, Wiegel J (2006) Reclassification of *Thermoanaerobium acetigenum* as *Caldicellulosiruptor acetigenus* comb. nov. and emendation of the genus description. Int J Syst Evol Microbiol 56:1391–1395

Onyenwoke RU, KevbrinV V, Lysenko AM, Wiegel J (2007) *Thermoanaerobacter pseudethanolicus* sp. nov., a thermophilic heterotrophic anaerobe from Yellowstone National Park. Int J Syst Evol Microbiol 57:2191–2193

Perevalova AA, Kublanov IV, Baslerov RV, Zhang G, Bonch-Osmolovskaya EA (2013) *Brockia lithotrophica* gen. nov., sp. nov., an anaerobic thermophilic bacterium from a terrestrial hot spring. Int J Syst Evol Microbiol 63:479–483

Pierce E, Xie G, Barabote RD, Saunders E, Han CS, Detter JC, Richardson P, Brettin TS, Das A, Ljungdahl LG, Ragsdale SW (2008) The complete genome sequence of *Moorella thermoacetica* (f. *Clostridium thermoaceticum*). Environ Microbiol 10:2550–2573

Rainey FA, Stackebrandt E (1993) Transfer of the type species of the genus *Thermobacteroides* to the genus *Thermoanaerobacter* as *Thermoanaerobacter acetoethylicus* (Ben-Bassat and Zeikus 1981) comb. nov., description of *Coprothermobacter* gen. nov., and reclassification of *Thermobacteroides proteolyticus* as *Coprothermobacter proteolyticus* (Ollivier et al. 1985) comb. Nov. Int J Syst Bacteriol 43:857–859

Sekiguchi Y, Imachi H, Susilorukmi A, Muramatsu M, Ohashi A, Harada H, Hanada S, Kamagata Y (2006) *Tepidanaerobacter syntrophicus* gen. nov., sp. nov., an anaerobic, moderately thermophilic, syntrophic alcohol- and lactate-degrading bacterium isolated from thermophilic digested sludges. Int J Syst Evol Microbiol 56:1621–1629

Slepova TV, Sokolova TG, Kolganova TV, Tourova TP, Bonch-Osmolovskaya EA (2009) *Carboxydothermus siderophilus* sp. nov., a thermophilic, hydrogenogenic, carboxydotrophic, dissimilatory FeIII-reducing bacterium from a Kamchatka hot spring. Int J Syst Evol Microbiol 59:213–217

Slobodkina GB, Kolganova TV, Kostrikina NA, Bonch-Osmolovskaya EA, Slobodkin AI (2012) *Caloribacterium cisternae* gen. nov., sp. nov., an anaerobic thermophilic bacterium from an underground gas storage reservoir. Int J Syst Evol Microbiol 62:1543–1547

Stamatakis A (2006) RAxML-VI-HPC maximum likelihood-based phylogenetic analyses with thousands of taxa and mixed models. Bioinformatics 22:2688–2690

Tomas AF, Karakashev D, Angelidaki I (2013) *Thermoanaerobacter pentosaceus* sp. nov., an anaerobic, extremely thermophilic, high ethanol-yielding bacterium isolated from household waste. Int J Syst Evol Microbiol 63:2396–2404

Wagner ID, Zhao W, Zhang CL, Romanek CS, Rohde M, Wiegel J (2008) *Thermoanaerobacter uzonensis* sp. nov., an anaerobic thermophilic bacterium isolated from a hot spring within the Uzon Caldera, Kamchatka, Far East Russia. Int J Syst Evol Microbiol 58:2565–2573

Wiegel J (2009) Family I *Thermoanaerobacteraceae* fam. nov. In: De Vos P, Garrity GM, Jones D, Krieg NR, Ludwig W, Rainey FA, Schleifer KH, Whitman WB (eds) Bergey's manual of systematic bacteriology, 2nd edn. Springer, New York, p 1225, Validation List no. 132 (2010) Int J Syst Evol Microbiol 60, 469–472

Wiegel J, Ljungdahl LG (1981) *Thermoanaerobacter ethanolicus* gen. nov., spec. nov., a new, extreme thermophilic, anaerobic bacterium. Arch Microbiol 128:343–348

Wu M, Ren Q, Durkin AS, Daugherty SC, Brinkac LM, Dodson RJ et al (2005) Life in hot carbon monoxide: the complete genome sequence of *Carboxydothermus hydrogenoformans* Z-2901. PLoS Genet 1:e65

Yarza P, Ludwig W, Euzéby J, Amann R, Schleifer K-H, Glöckner FO, Rosselló-Móra R (2010) Update of the All-Species Living-Tree Project based on 16S and 23S rRNA sequence analyses. Syst Appl Microbiol 33:291–299

Yoneda Y, Yoshida T, Kawaichi S, Daifuku T, Takabe K, Sako Y (2012) *Carboxydothermus pertinax* sp. nov., a thermophilic, hydrogenogenic, FeIII-reducing, sulfur-reducing carboxydotrophic bacterium from an acidic hot spring. Int J Syst Evol Microbiol 62:1692–1697

# 33 The Family *Thermodesulfobiaceae*

*Wajdi Ben Hania*[1] · *Bernard Ollivier*[2] · *Marie-Laure Fardeau*[2]
[1]CNRS/INSU, IRD, Aix-Marseille Université, Université du Sud Toulon-Var, Marseille, France
[2]Aix-Marseille Université, Université du Sud, Toulon-Var, CNRS/INSU, IRD, MIO, UM 110, Marseille, France



**Abstract**
The family *Thermodesulfobiaceae* is a polyphyletic family within the order *Thermoanaerobacterales*, class *Clostridia*; it embraces the type genus *Thermodesulfobium* which contains one species, *T. narugense* and the genus *Coprothermobacter* which includes two species: *C. platensis* and *C. proteolyticus*. Members of the family are defined by a wide range of morphological and chemotaxonomic properties, such as fatty acids, quinones, etc. They are all strictly anaerobic. Members of the family are found in anaerobic digesters, but they have been isolated from aquatic environment as well.

## Taxonomy: Historical and Current

### Short Description of the Family

Ther.mo.de.sul.fo.bi.a'ce.ae. M. L. fem.n. *Thermodesulfobium* type genus of the family; -aceae ending to denote a family; M. L. fern. pl.n. *Thermodesulfobiaceae*, the family of *Thermodesulfobium* (Modified from *Bergey's Manual*). The description is an emended version of the description given in *Bergey's Manual*, 2nd edition (Mori and Hanada 2009).

The family *Caldicoprobacteraceae* is a member of the order *Thermoanaerobacterales*, phylum Firmicutes. It contains the type genus *Thermodesulfobium* (Mori et al. 2003) and *Coprothermobacter* (Rainey and Stackebrandt 1993a). Gram-negative. Cells are rods of varying lengths. Nonmotile. Strictly anaerobic. Thermophilic or moderately thermophilic.

$C_{16:0}$ is the prominent fatty acid; $C_{15:0}$, iso- $C_{14:0}$ 3 OH, iso-$C_{17:0}$ may also occur. Menaquinone MK-7 (H2) and MK-7 are the predominant quinones (when mentioned). Polar lipids were not analyzed. G+C values of DNA range between 35 and 45 mol%. Isolated from microbial mats in a hot spring, and from anaerobic mesothermic or hot digesters.

### Phylogenetic Structure of the Family and Its Genera

According to the phylogenetic branching of *Firmicutes* type strains in the RaxML 16S rRNA gene tree of the Living Tree Project (Yarza et al. 2008), the family is moderately related to the families *Synergistetes* and *Aquificae* (➋ *Fig. 33.1*).

The family *Thermodesulfobiaceae* contains two genera *Thermodesulfobium* and *Coprothermobacter.* The genus *Coprothermobacter* contains two species: *C. proteolyticus* and *C. platensis. C. proteolyticus* was first described by Ollivier et al. (1985) as *Thermobacteroides proteolyticus*, but, later, phylogenetic studies (Rainey and Stackebrandt 1993a; Rainey et al. 1993) of anaerobic thermophilic bacteria demonstrated that some of them had to be reclassified. It was the case for members of the genus *Thermobacteroides* (Ben-Bassat and Zeikus 1981; Ollivier et al. 1985) which belonged to phylogenetically very diverse taxa (Rainey and Stackebrandt 1993b). Within this genus, *Thermobacteroides proteolyticus* represented a deep root adjacent to members of the order *Thermotogales*, showing only 81.9 % sequence similarity with *Thermobacteroides acetoethylicus* over the stretch of about 1,200 analyzed nucleotides (Rainey and Stackebrandt 1993a). On the basis of these phylogenetic findings, supported by phenotypic characteristics, the reclassification of the species investigated was evident. In consequence, the genus *Thermobacteroides* was invalidated, and the description of the genus *Coprothermobacter* and the assignment of *Thermobacteroides proteolyticus* as the type species *Coprothermobacter proteolyticus* was proposed (Rainey and Stackebrandt 1993a).

Another strain of *C. proteolyticus* was also isolated by Kersters et al. (1994), showing the same metabolic properties of the type strain of the species isolated by Ollivier et al. (1985).

E. Rosenberg et al. (eds.), *The Prokaryotes – Firmicutes and Tenericutes*, DOI 10.1007/978-3-642-30120-9_221,

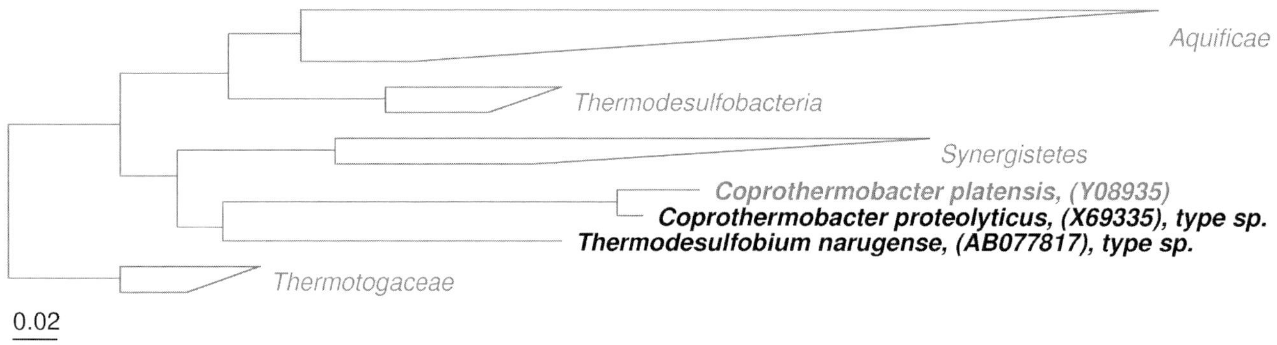


**Fig. 33.1**
**Phylogenetic reconstruction of the family *Thermodesulfobiaceae* based on 16S rRNA and created using the maximum likelihood algorithm RAxML (Stamatakis 2006). The sequence datasets and alignments were used according to the All-Species Living Tree Project (LTP) database (Yarza et al. 2008; http://www.arb-silva.de/projects/living-tree). Representative sequences from closely related taxa were used as outgroups. In addition, a 40 % maximum frequency filter was applied in order to remove hypervariable positions and potentially misplaced bases from the alignment. Scale bar indicates estimated sequence divergence**

## Molecular Analyses

### DNA-DNA Hybridization Studies

Results of DNA-DNA hybridization studies is only reported for one species of *Coprothermobacter* and shows less than 12 % similarity between the chromosomal DNAs of the two species of *Coprothermobacter*, *C. proteolyticus* and *C. platensis* (Etchebehere et al. 1998).

### Genome Comparison

The genome of one species has been released. The genome of the type strain *Thermodesulfobium narugense* Na82, DSM 14796 T is 1,898,865 bp long, contains 1,807 protein genes, including 56 RNA genes, and the mol% G+C of DNA is 33.8 %. The latter value falls in the range of 96.3 % determined for the species (➋ *Table 33.1*) by HPLC. http://www.genome.jp/kegg-bin/show_organism?org, http://genome.jgi-psf.org/thenr/thenr.info.html.

## Phenotypic Analyses

The main features of members of *Thermodesulfobiaceae* are listed in ➋ *Tables 33.1* and ➋ *33.2.*

### *Thermodesulfobium* Mori et al. 2003, 288

Ther.mo.de.sul.fo'bi.um. Gr. adj. thermos hot; L. pref. de from; L. n. sulfur sulfur; Gr. n. bios life; L. neut. n. *Thermodesulfobium* a thermophilic organism that reduces a sulfur compound.

**Table 33.1**
**Morphological and chemotaxonomic characteristics of genera of *Thermodesulfobiaceae***

| | *Thermodesulfobium* | *Coprothermobacter* |
|---|---|---|
| Morphology | Rods | Rods |
| Gram stain | Negative | Negative |
| Metabolism | Anaerobic growth | Anaerobic growth |
| Motility | – | – |
| Major fatty acids | C16 :0 | iso-C15 : 0 and C16 :0 |
| Menaquinone | MK-7 (H2) and MK-7 | nd |
| G+C content | 35.1 | 43–45 |

As the genus *Thermodesulfobium* contains only one species, the description of the genus is also the description of the type species, Na82$^T$ *Thermodesulfobium narugense*$^T$.

Cells of *Thermodesulfobium narugense*$^T$ are rod-shaped (0.5 · 2–4 μm diameter-length). This bacterium shows no motility under the microscope. Spore formation is not observed. Gram staining is negative with cell wall having an outer membrane. Neither storage compounds nor extensive internal membranes are observed. The polymyxin B-LPS test (Wiegel and Quandt 1982) also suggests that it possesses the typical Gram-negative cell wall, since the fibrous structure and blebs of lipopolysaccharides are observed around the surface of polymyxin-B-treated cells.

Growth occurs only in strictly anaerobic conditions under an $H_2/CO_2$ atmosphere and cannot occur under aerobic conditions. Growth occurs between 37 °C and 65 °C, with an optimum at 50–55 °C. Anaerobic growth is always coupled to sulfate reduction. In the presence of sulfate, the bacterium also uses formate as electron donor, but growth on formate is clearly

**Table 33.2**
**Comparison of selected characteristics of members of the genus *Coprothermobacter***

| Characteristics | *C. proteolyticus*[a] | *C. proteolyticus*[b] | *C. platensis*[c] |
|---|---|---|---|
| Dimensions (μm) | 0.5 × 1–6 | 0.5 × 1–6 | 0.5 × 1.5–2.0 |
| Optimum temperature (°C) | 63 | 63–70 | 55 |
| *Growth on* | | | |
| Fructose | + | + | + |
| Sucrose | + | + | + |
| Melibiose | – | – | ND[d] |
| Xylose | + | + | – |
| *Resistance to* | | | |
| Penicillin G (20 U/ml) | – | – | + |
| Polymixin (20 mg/l) | + | + | – |
| G+C % | 45 | 43–44 | 43 |
| Major fatty acid | Iso-$C_{15:0}$ and $C_{16:0}$ | Iso-$C_{15:0}$ and $C_{16:0}$ | ND[d] |

[a]Data from Ollivier et al. (1985)
[b]Data from Kersters et al. (1994)
[c]Data from Etchebbehere et al. (1998)
[d]*ND* Not determined

lower than that on $H_2$. When $H_2$ and $CO_2$ are provided as energy and carbon sources, thiosulfate, nitrate, or nitrite also serves as an electron acceptor. The isolate, however, does not use sulfite, elemental sulfur, Fe (III), citrate, fumarate, dimethyl sulfoxide, or $O_2$ as electron acceptor. No growth occurs in the presence of glucose, acetate, lactate, pyruvate, malate, propionate, butyrate, fumarate, succinate, citrate, ethanol, propanol, or methanol (Mori et al. 2003). Menaquinone (MK)-7(H2) and MK-7 (53.6 % and 35.8 %, respectively, of total quinones) are the major quinones. MK-7(H4) (5.1 %) and MK-8 (5.5 %) are detected as minor fractions. Hexadecanoic (C16:0) acid is the dominant component of fatty acid pattern (45.7 % of the total fatty acids). The following fatty acids are also detected: cyclo-$C_{19:0}$(9,10)cis (15.2 %), $C_{18:0}$ (14.8 %), $C_{18:1}$(M9)cis (13.9 %), $C_{20:0}$ (3.8 %), $C_{14:0}$ (3.2 %), $C_{12:0}$-3OH (2.6 %), and $C_{12:0}$ (0.7 %). No data on the analysis of polar lipids are available.

## *Coprothermobacter* Rainey and Stackebrandt. 1993a, 857

Co pro. ther mo-bac'ter. Gr. fem. n. kopros manure; Gr. adj. thermos warm; Gr. hyp. mas. n. bakter rod; N.L. mas. n. Coprothermobacter, because it is a thermophilic rod-shaped bacterium isolated from cattle manure.

Cells of *Coprothermobacter* are rod-shaped ranging from 1 to 6 μm in length, occurring singly or in pairs in young cultures. Colonies are white, circular (diameter 1–2 mm), convex, smooth with entire edges. Gram staining is negative. Electron micrographs of thin sections of the two species of the genus reveal a thin inner wall layer and a heavy outer wall (Rainey and Stackebrandt 1993a). Thermophilic temperature range for growth was between 35 °C and 75 °C, with optimum at 55–70 °C, 55 °C for *C. platensis* and 65–70 °C for *C. proteolyticus.*

Cells are strictly anaerobic. They are proteolytic using gelatin and peptones. Sugars are used poorly unless yeast extract or rumen fluid is added. *C. proteolyticus* ferments peptone, gelatin, casein, and Trypticase peptone in the presence of 0.1 % yeast extract. It grows on the following sugars when yeast extract and either rumen fluid or Trypticase is added: glucose, fructose, maltose, sucrose, and mannose. The fermentation products from gelatin or glucose in the presence of yeast extract are acetate, $H_2$, and $CO_2$ along with smaller quantities of isobutyrate, isovalerate, and propionate (Ollivier et al. 1985, Rainey and Stackebrandt 1993b). *Coprothermobacter platensis* ferments peptone, gelatin, casein, bovine albumin, and yeast extract. The addition of 0.02 % of yeast extract in the culture medium is necessary to stimulate growth on sugars. Glucose, fructose, sucrose, maltose, and starch are poorly fermented. Fermentation products from glucose are acetate, $H_2$, and $CO_2$. The major fermentation products from gelatin are acetate, propionate, $H_2$, and $CO_2$ (Etchebehere et al. 1998).

Both species reduce thiosulfate, but not sulfate, to sulfide with a concomitant increase in growth and glucose utilization (Etchebehere et al. 1998).

An extracellular protease activity is observed for cells grown on gelatin (Etchebehere et al. 1998; Kersters et al. 1994).

The major polyamines synthesized by *Coprothermobacter proteolyticus* are putrescine, spermidine, and spermine (Hamana et al. 1996).

The main features of members of *Thermodesulfobiaceae* are listed in ➲ *Tables 33.1* and ➲ *33.2*.

Growth of *C. proteolyticus* is inhibited by neomycin (0.15 g/l) and penicillin G (20 U/ml); vancomycin, polymyxin B, sodium azide, and kanamycin are not effective inhibitors (Kersters et al. 1994). Vancomycin (2.5 mg/l), neomycin (0.15 g/l), and polymyxin (20 mg/l) inhibit growth of *Coprothermobacter platensis* (Etchebehere et al. 1998).

## Biochemical Characteristics

### Protease Assays

Protease activity is assayed using the azocasein method under anaerobic conditions as described by Brock et al. (1982), (Etchebehere et al. 1998) or by the method of Twining (1984), (Kersters et al. 1994).

*Coprothermobacter proteolyticus* possesses a thermostable protease with optimal temperature of 85 °C and optimal pH of 9.5. The protease retains about 90 % of its activity at pH 10.0 and appears quite specific as compared to enzymes from other thermophilic or hyperthermophilic proteolytic microorganisms (Klingeberg et al. 1991).

### Isolation, Enrichment, and Maintenance Procedures

Members of the family *Thermodesulfobiaceae* are mainly isolated on two types of culture medium under anaerobic conditions.

For enrichment of *Thermodesulfobium narugense*, the sulfate is used as terminal electron acceptor as described by Mori et al. (2003), with $H_2$ used as the energy source. The enrichment culture is transferred several times to new culture medium of the same composition. Single colonies are formed after 2 weeks of incubation at 55 °C on the culture medium solidified with 2 % agar in vials. After a second purification step on agar, a uniformly shaped axenic culture is obtained.

*T. narugense* is maintained in the enrichment medium under a $H_2/CO_2$ (4:1, v/v) atmosphere at 55 °C. The culture should be transferred every 2 weeks. After growth, the culture can also be stored at room temperature for several weeks. For long-term storage, it can be preserved in liquid nitrogen (−196 °C) under strictly anaerobic conditions with 5 % dimethylsulfoxide or 10 % glycerol. Liquid drying is also successful with a protective medium composed of 0.1 M potassium buffer (pH 7.0), 3 % sodium glutamate, 1.5 % ribitol, and 0.05 % cysteine hydrochloride monohydrate.

*Coprothermobacter* species can be enriched using culture media and procedures similar to those described by Ollivier et al. (1985) with gelatin as the energy source and $Na_2S$ and cysteine as the reductive agents. In parallel, peptone-yeast medium (Holdeman et al. 1977) may be used for the enrichment of *Coprothermobacter proteolyticus* (Kersters et al. 1994). At least, three subcultures in the same growth conditions, at temperature from 55 °C up to 70 °C, are needed before isolation.

After several transfers, the enrichment cultures are serially diluted using the method of Hungate (1969), with roll tubes containing the basal medium, gelatin as the energy source, and purified agar at a concentration of 2 %. For isolation, agar medium can also be poured into plates within an anaerobic chamber. In order to detect selectively proteolytic colonies during the first step of isolation, casein can be used as substrate. Colonies, surrounded with large clearing zones, are picked and re-streaked on gelatin agar plates or roll tubes. At least two colonies are picked and the process of serial dilution in roll tubes is repeated in order to purify the cultures.

Stock cultures can be maintained on the medium described by Ollivier et al. (1985) and transferred at least monthly. Liquid cultures retain viability after several weeks of storage at room temperature. Cultures have also to be refrigerated. Because of lysis of cells, it is recommended to stock cultures before the end of the exponential phase.

## Ecology

### Habitat

The members of *Thermodesulfobiaceae* family were mainly isolated from digesters or hot spring.

*Thermodesulfobium narugense* was isolated from a vent of Narugo hot spring located in the prefecture of Miyagi in Japan (Mori et al. 2003). This vent harbored white microbial mats, mainly formed by sulfur-oxidizing bacteria. A mat sample was taken at a site presenting a temperature and a pH of 58 °C and 6.9, respectively (Mori et al. 2003). Several clones close to *Thermodesulfobium* were also detected across the planet. For example, clones were retrieved from anaerobic sediments at Rio Tinto, located in the core of the Iberian Pyritic Belt having a constant acidic pH (Sanchez-Andrea et al. 2011). Other clones were detected from hot springs in Yellowstone National Park (USA) (unpublished), from thermal pools in the uzon caldera, Kamchatka, Russia (Burgess et al. 2012), and mud and water from Los Azufres geothermic belt, Michoacan, Mexico (unpublished) (➲ *Table 33.3*).

Strains of *Coprothermobacter proteolyticus* were isolated from anaerobic hot digesters. The type strain BT (ATTCC 35245 = OCM 4 = DSMZ 5265 = LMG 11567) was isolated from a digester fermenting tannery wastes and cattle manure. Strain 18 was isolated from a biokitchen waste digester. Strain 18 has been deposited in the LMG culture Collection (LGM 14268).

*Coprothermobacter platensis* was isolated from a methanogenic mesothermic reactor treating a protein-rich wastewater (Kersters et al. 1994). Clones close to *Coprothermobacter* were also retrieved mainly in anaerobic digesters (Riviere et al. 2009; Tandishabo et al. 2012). Some clones were indigenous to petroleum reservoirs (Kobayashi et al. 2012).

**Table 33.3**
**Clones close to *Thermodesulfobium* or *Coprothermobacter***

| GenBank accession # | Sequences origin | Reference | Closest genus |
|---|---|---|---|
| | DNA from environmental samples | | |
| **HQ730662.1** | Anaerobic zones of Tinto River, Iberian Pyritic Belt | Sanchez-Andrea et al. (2011) | *Thermodesulfobium* |
| **JQ815731.1** | | | |
| **JQ420033.1** | | | |
| **DQ834002.1** | Hot Springs in Yellowstone National Park | Unpublished | |
| **GQ328297.1** | Thermal pools in the uzon caldera, kamchatka, Russia | Burgess et al. (2012) | |
| **GQ328292.1** | | | |
| **GQ328235.1** | | | |
| **GQ328409.1** | | | |
| **GQ328239.1** | | | |
| **GQ328359.1** | | | |
| **GQ328317.1** | | | |
| **GQ328359.1** | | | |
| **HF677523.1** | Mud and water from Los Azufres geothermic belt, Michoacan, Mexico | Unpublished | *Coprothermobacter* |
| **JF808034.1** | Depleted oil reservoir | Kobayashi, H., | |
| **GU363592.1** | Anaerobic digesters | Tandishabo et al. (2012) | |
| **CU924340.1** | Anaerobic digesters | Riviere et al. (2009) | |

Other clones were found in microbial mats, mud, and water from Los Azufres geothermic belt, Michoacan, Mexico (unpublished) (➋ *Table 33.3*).

## References


Ben-Bassat A, Zeikus JG (1981) *Thermobacteroides acetoethylicus* gen. nov. sp. nov., a new chemoorganotrophic, anaerobic, thermophilic bacterium. Arch Microbiol 128:365–370

Brock FM, Forsberg CW, Buchanan-Smith JG (1982) *Coprothermobacter platensis* sp. nov. Proteolytic activity of rumen microorganisms and effects of proteinase Inhibitors. Appl Environ Microbiol 44:561–569

Burgess EA, Unrine JM, Mills GL, Romanek CS, Wiegel J (2012) Comparative geochemical and microbiological characterization of two thermal pools in the uzon caldera, kamchatka, Russia. Microb Ecol 63:471–489

Etchebehere C, Pavan ME, Zorzopulos J, Soubes M, Muxi L (1998) *Coprothermobacter platensis* sp. nov., a new anaerobic proteolytic thermophilic bacterium isolated from an anaerobic mesophilic sludge. Int J Syst Bacteriol 48:1297–1304

Hamana H, Niitsu M, Samejima K (1996) Polyamines of thermophilic Gram-positive anaerobes belonging the genera *Caldicellulosiruptor, Caloramator, Clostridium, Coprothermobacter, Moorella, Thermoanaerobacter* and *Thermoanaerobacterium.* Microbios 85:213–222

Holdeman LV, Cato EP, Moore WEC (1977) Culture methods; use of pre-reduced media. In: Holdeman LV, Cato EP, Moore WEC (eds) Anaerobic laboratory manual, 4th edn. Anaerobe Laboratory, Virginia Polytechnic Institute and State University, ∞, pp 117–149

Hungate RE (1969) A roll tube method for the cultivation of strict anaerobes. In: Norris JR, Ribbons DW (eds) Methods in microbiology, vol 3B. Academic Press, New York, pp 117–132

Kersters I, Maestrojuan GM, Torck U, Vancanneyt M, Kersters K, Verstraete W (1994) Isolation of *Coprothermobacter proteolyticus* from an anaerobic digester and further characterization of the species. Syst Appl Microbiol 17:289–295

Klingeberg M, Hashwa F, Antranikian G (1991) Properties of extremely thermostable proteases from anaerobic hyperthermophilic bacteria. Appl Microbiol Biotechnol 34:715–719

Kobayashi H, Kawaguchi H, Endo K, Mayumi D, Sakata S, Ikarashi M, Miyagawa Y, Maeda H, Sato K (2012) Analysis of methane production by microorganisms indigenous to a depleted oil reservoir for application in microbial enhanced oil recovery. J Biosci Bioeng 113:84–87

Mori K, Hanada S (2009) Thermodesulfobium. In: Whitman WB, de Vos P, Garrity GM, Jones D, Krieg NR, Ludwig W, Rainey FA, Schleifer K-H, Whitman WB (eds.) Bergeys manual of systematic bacteriology, 2nd edn, vol 3. Springer, New York, p 1268

Mori K, Kim H, Kakegawa T, Hanada S (2003) A novel lineage of sulfate-reducing microorganisms: Thermodesulfobiaceae fam. nov., *Thermodesulfobium narugense*, gen. nov., sp. nov., a new thermophilic isolate from a hot spring. Extremophiles 7:283–290

Ollivier BM, Mah RA, Ferguson TJ, Boone DR, Garcia JL, Robinson R (1985) Emendation of the genus *Thermobacteroides. Thermobacteroides proteolyticus* sp. nov., a proteolytic acetogen from a methanogenic enrichment. Int J Syst Bacteriol 35:425–428

Rainey FA, Stackebrandt E (1993a) A transfer of the type species of the genus *Thermobacteroides* to the genus *Themoanaerobacter* as *Themoanaerobacter acetoethylicus* (Ben-Bassat and Zeikus 1981) comb. nov., description of *Coprothermobacter* gen. nov., and reclassification of *Thermobacteroides proteolyticus* as *Coprothermobacter proteolyticus* (Ollivier et al. 1985) comb. nov. Int J Syst Bacteriol 43:857–859

Rainey FA, Stackebrandt E (1993b) Phylogenetic analysis of the bacterial genus *Thermobacteroides* indicates an ancient origin of *Thermobacteroides proteolyticus.* Lett Appl Microbiol 16:282–286

Rainey FA, Ward NL, Morgan HW, Toalster R, Stackebrandt E (1993) Phylogenetic analysis of anaerobic thermophilic bacteria: aid for their reclassification. J Bacteriol 175:4772–4779

Riviere D, Desvignes V, Pelletier E, Chaussonneri S, Guermaz S, Weissenbach J, Li T, Camacho P, Sghir A (2009) Towards the definition of a core of microorganisms involved in anaerobic digestion of sludge. ISME J 3:700–714

Sanchez-Andrea I, Rodriguez N, Amils R, Sanz JL (2011) Microbial diversity in anaerobic sediments at rio tinto, a naturally acidic environment with a high heavy metal content. Appl Environ Microbiol 77:6085–6093

Stamatakis A (2006) RAxML-VI-HPC: maximum likelihood-based phylogenetic analyses with thousands of taxa and mixed models. Bioinformatics 22:2688–2690

Tandishabo K, Nakamura K, Umetsu K, Takamizawa K (2012) Distribution and role of *Coprothermobacter* spp. in anaerobic digesters. J Biosci Bioeng 114:518–520

Twining SS (1984) Fluorescein isothiocyanate-labeled casein assay for proteolytic enzymes. Anal Biochem 143:30–34

Wiegel J, Quandt L (1982) Determination of Gram type using the reaction between polymyxin B and lipopolysaccharides of the outer cell wall of whole bacteria. J Gen Microbiol 128:2261–2270

Yarza P, Richter M, Peplies J, Euzéby J, Amann R, Schleifer KH, Ludwig W, Glöckner FO, Rossello-Mora R (2008) The all species living-tree project: a 16S rRNA-based phylogenetic tree of all sequenced type strains. Syst Appl Microbiol 31:241–250. http://www.arb-silva.de/projects/living-tree

# 34 The Family *Thermolithobacteriaceae*

*Tatyana Sokolova*[1] · *Juergen Wiegel*[2]
[1]Winogradsky Institute of Microbiology, Russian Academy of Sciences, Moscow, Russia
[2]Department of Microbiology, The University of Georgia, Athens, GA, USA



**Abstract**
*Thermolithobacteraceae* is a family within an order *Thermolithobacterales.* It is a monogeneric family and contains two species *Thermolithobacter ferrireducens* and *T. carboxydivorans.* Both species are strict anaerobes and extreme thermophiles isolated from terrestrial hot springs. *Thermolithobacter* species are very close phylogenetically to each other possessing 16S rRNA genes with 99 % of similarity. *T. ferrireducens* and *T. carboxydivorans* cells are short rods 1.8–2.0 by 0.5 μm. They are extreme thermophiles and neutrophiles. Both species can grow chemolithoautotrophically. While *T. ferrireducens* grows at the expense of $H_2$ oxidation with Fe(III) as e-acceptor, *T. carboxydivorans* grows on CO. The nearest phylogenetically neighboring family is Incertae Sedis XVI which consists of *Carboxydocella* species, also chemolithoautotrophic CO utilizers.

## Taxonomy: Historical and Current

*Thermolithobacteraceae* is a family within the order *Thermolithobacterales.* It is a monogeneric family and consists of two species *Thermolithobacter ferrireducens* and *T. carboxydivorans.* Both species are strict anaerobes and extreme thermophiles isolated from terrestrial hot springs. These two species are combined in one genus on the basis of 16SrRNA gene sequence similarity and their preferred chemolithotrophic growth. Originally *T. ferrireducens* has been described as "*Ferribacter thermoautotrophicus*" gen. nov., sp. nov., on the ground of its physiological features and 16S rRNA gene sequence (Wiegel et al. 2003) and *T. carboxydivorans*—as "*Carboxydothermus restrictus*" (Svetlichny et al. 1994). The results from 16S rRNA oligonucleotide analysis and DNA-DNA hybridization revealed that these species are closely related and constitute one genus. According to maximum likelihood (➲ *Fig. 34.1*) and neighbor joining analysis (not shown), the nearest neighboring family is Incertae Sedis XVI which consists of *Carboxydocella* species (➲ *Fig. 34.1*). Remarkably, similar to *Thermolithobacter* species, all *Carboxydocella* species are able to grow chemolithoautotrophically at the expense of either CO (Sokolova et al. 2002; Slepova et al. 2006) or $H_2$ (Slobodkina et al. 2012) oxidation.

## Molecular Analyses

Phylogenetic analysis revealed high similarity between two species of the genus *Thermolithobacter*, the only member of *Thermolithobacteraceae.* The 16S rRNA gene sequences of *T. carboxydivorans*R1$^T$ and *T. ferrireducens* JW/KA-2$^T$, JW/KA-1, and JW/JH-Fiji-2 possessed 99 % sequence similarity. Notably, two *T. ferrireducens* strains, JW/KA-1 and JW/KA-2 T, possess two 16S rRNA genes, each with 98.9 % of similarity. DNA-DNA hybridization studies revealed 35 % of similarity between the two type strains, and thus, *T. carboxydivorans* and *T. ferrireducens* were only distantly related and constitute two separate species. However, the characterization of future isolates has to elucidate whether there is a coherent and clear separation of the species based on DNA-DNA hybridization and physiological and biochemical properties or the differences present a continuum of the variety of those properties.

## Phenotypic Analyses

*Thermolithobacteriaceae* is presently represented by the only genus *Thermolithobacter* (Sokolova et al. 2007; Validation list N 116. 2007).

*Thermolithobacter* (*Ther.mo.li.tho.bac'ter.* Gr. adj. *thermos*—hot; Gr. masc. n. *lithos*—stone; N. L. masc. n.*bacter*—equivalent of Gr. neut. n. *baktron*, rod staff; N. L. masc. n. *Thermolithobacter*—thermophilic lithotrophically growing rods).

The genus contains two species—*T. ferrireducens* and *T. carboxydivorans.* The type species *T. ferrireducens* was isolated from terrestrial hot springs at Yellowstone National Park (type strain) and at Fiji. Cultures of *Thermolithobacter* contain short rods approximately 1.8–2 by 0.5 μm. *T. ferrireducens* cells

E. Rosenberg et al. (eds.), *The Prokaryotes – Firmicutes and Tenericutes*, DOI 10.1007/978-3-642-30120-9_220,

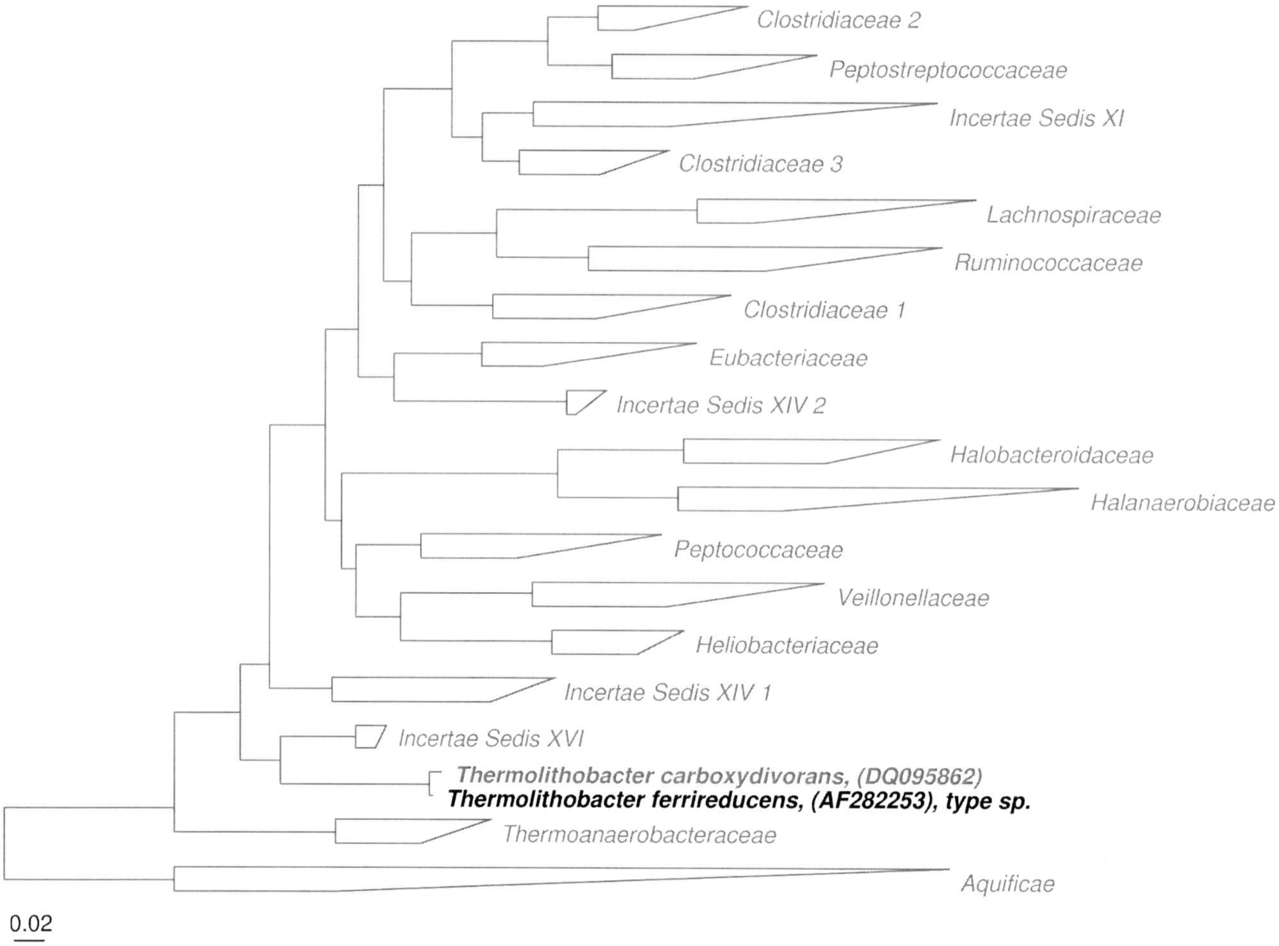


**Fig. 34.1**
**Phylogenetic reconstruction of the family *Thermolithobacteraceae* basedon16S rRNA and created using the maximum likelihood algorithm RAxML (Stamatakis 2006). The sequence datasets and alignments were used according to the All-Species Living Tree Project (LTP) database (Yarza et al. 2010; http://www.arb-silva.de/projects/living-tree). Representative sequences from closely related taxa were used as out groups. In addition, a 40 % maximum frequency filter was applied in order to remove hypervariable positions and potentially misplaced bases from the alignment. Scale bar indicates estimated sequence divergence**

occur as single, in V-shaped pairs, or in chains. *T. carboxydivorans* cells occur as single or in short chains. Cells are motile due to 2–3 peritrichously located flagella. The cell wall is of Gram-positive type. *Thermolithobacter* species are obligate anaerobes. They are extreme thermophiles and neutrophiles. Both species can grow chemolithoautotrophically at the expense of $H_2$ or CO oxidation, respectively.

The temperature range for growth of *T. ferrireducens* is 50–75 °C, optimum is 73 °C, and the $pH^{60\ °C}$ range is 6.5–8.5, with an optimum at 7.1–7.3. After 24 h incubation in roll tubes with solidified 9,10-anthraquinone-2,6-disulfonic acid (AQDS)-containing medium, *T. ferrireducens* produces flat, round, white, 1–2 mm in diameter colonies.

*T. ferrireducens* grows chemolithoautotrophically on $H_2/CO_2$ (4/1 v/v) mixture with ferric iron hydromorphic oxide. The oxidation of hydrogen was coupled to the reduction of ferric iron hydromorphic oxide to magnetite and siderite. An increase in cell number was observed with a concomitant increase in magnetite formation (*Fig. 34.2*).

*T. ferrireducens* uses $H_2$ or formate as electron donors and ferric iron hydromorphic oxide, AQDS, thiosulfate, or fumarate as electron acceptors. Formate may be used as the only source of energy and carbon. *T. ferrireducens* strains do not grow on CO or $CO/H_2$ mixture. No growth occurs on $H_2/CO_2$ with sulfur (precipitated or sublimated), nitrate, sulfate, ferric iron citrate, crystalline ferric hydroxide, Mn(IV), U(VI), Se(VI), or As(V). In the presence of ferric iron hydromorphic oxide as an electron acceptor and $H_2$ as an electron donor, *T. ferrireducens* JW/KA-$2^T$ utilized $CO_2$, fumarate (20 mM), yeast extract (1 % w/v), casamino acids (1 % w/v), or crotonate (20 mM) as carbon sources. Tryptone (1.0 %), glucose (20 mM), galactose (20 mM), xylose (20 mM), sucrose (20 mM), propionate

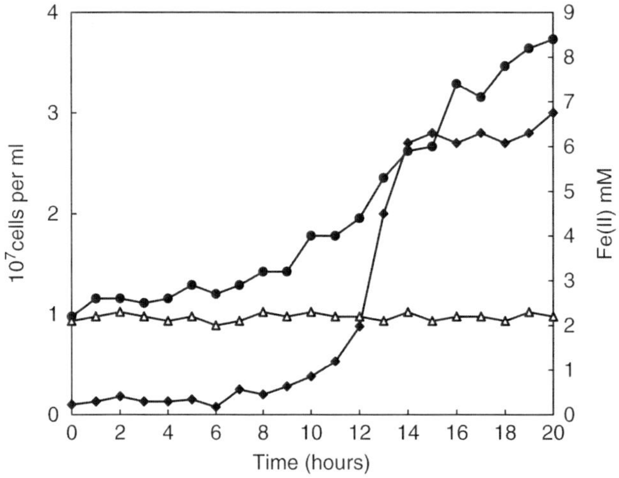


**Fig. 34.2**
**Cell growth (*diamond*) and Fe (II) formation (*circle*) by *T. ferrireducens* during the growth at 60 °C at pH$^{25}$ 7.2 under $H_2/CO_2$ (4/1 v/v)**

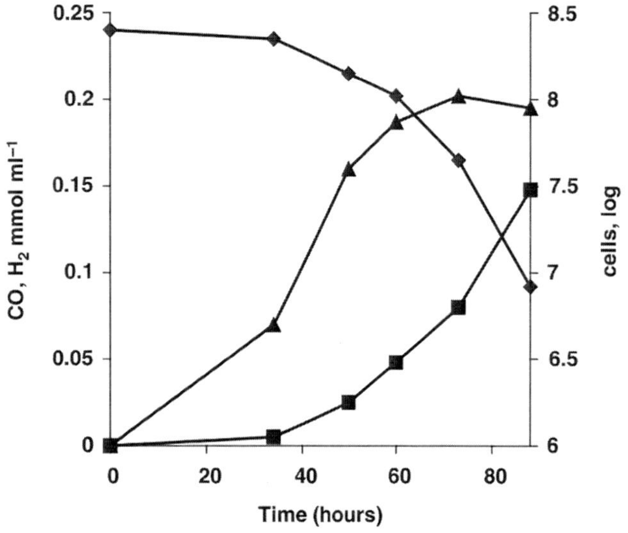


**Fig. 34.3**
**Cell growth of *T. carboxydivorans* R1$^T$ (*triangle*), CO consumption (*diamond*), and $H_2$ production (*square*) during the growth at pH$^{25}$ 7.2 at 70 °C**

(20 mM), starch (5 g/l), sodium acetate (30 mM), succinate (20 mM), lactic acid (20 mM), glycerol (130 mM), cellobiose (20 mM), ethanol (20 mM), 1-butanol (20 mM), 2-propanol (20 mM), acetone (20 mM), phenol (10 mM), ethylene glycol (20 mM), 1,3 propanediol (20 mM), catechol (20 mM), or olive oil (10 ml/l) was not utilized as carbon or energy sources neither in the presence nor in the absence of ferric iron hydromorphic oxide.

The phospholipid fatty acid (PLFA) profile of *T. ferrireducens* strain JW/JH-Fiji-2, which showed 99 % 16S rRNA gene sequence similarity to JW/KA-1 and JW/KA-2 T, contained the following fatty acids (mol%): i15:0 (32.2 ± 6.4), a15:0 (4.1 ± 0.9), i16:0 (0.8 ± 0.1), 16:0 (15.9 ± 1.0), i17:0 (35.0 ± 4.5), a17:0 (11.2 ± 1.5), and 18:0 (0.8 ± 0.2). The neutral lipid fraction of strain JW/JH-Fiji-2 contained the following respiratory quinones (mol%): demethylmenaquinone-9 (82.7) as the major compound and menaquinone (MK)-4 (2.2), MK-5 (0.7), and MK-6 (2.2) and ubiquinone (UQ)-6 (9.0), UQ-7 (0.6), UQ-9 (1.4), and UQ-10 (1.2) as minor compounds. The minor quinones, except UQ-9, were also found in the control (culture media) and could derive from passive accumulation in the membrane (R. Geyer, unpublished results).

The G+C content of the DNA of the *T. ferrireducens* type strain is 52.7 ± 0.3 mol%. Type strain is JW/KA-2$^T$ (= ATCC 700985 T = DSM 13639 T), isolated from a mixed sample of geothermally heated black sediment and water collected from a hot spring outflow at Calcite Springs (close to the river) in the Yellowstone National Park (Wyoming, USA).

*T. carboxydivorans* grows within the temperature range from 40 °C to 78 °C with an optimum at 70 °C. The pH$^{70}$ for growth ranges from 6.6 to 7.6 with an optimum at 6.8–7.0.

*T. carboxydivorans* R1$^T$ grows chemolithoautotrophically on CO (100 % CO in the gas phase) producing hydrogen as the sole reduced product (➲ *Fig. 34.3*). However, it does not grow on CO in the presence of Fe(III) citrate, ferric iron hydromorphic oxide, AQDS, $SO_4^{2-}$, $S_2O_3^{2-}$, fumarate, or $NO_3^-$. In contrast to *T. ferrireducens*, this species also does not grow on a $H_2/CO_2$ mixture neither in the presence nor in the absence of ferric iron hydromorphic oxide, AQDS, $NO_3^-$, $SO_3^{2-}$, or fumarate. Neither does this species reduce $SO_4^{2-}$ and $S_2O_3^{2-}$ nor elemental sulfur in the presence of yeast extract, formate, acetate, pyruvate, citrate, succinate, and lactate in the growth medium or $H_2/CO_2$ as head gas.

During the growth on solid medium under an atmosphere of 100 % CO, *T. carboxydivorans* produces round, white, semitranslucent colonies, 1 mm in diameter.

The G+C content of the DNA of *T. carboxydivorans* type strain is 52 ± 1 mol%. The type strain is strain R1$^T$ (= DSM 7242$^T$, = VKM 2359$^T$), isolated from a terrestrial hot spring at Raoul Island, Archipelago Kermadec.

## Isolation, Enrichment, and Maintenance Procedures

Members of family *Thermolithobacteriaceae* grow in a rather narrow range of media; they do not grow on any common organic substrate. The growth conditions for the two *Thermolithobacter* species differ from one another. *T. ferrireducens* can be enriched and isolated on ferric iron hydromorphic oxide-containing mineral medium.

For this species $H_2$ is the best substrate for enrichment and isolation which avoids the growth of unwanted organotrophic prokaryotes. *T. carboxydivorans* can be enriched and isolated in the mineral medium under 100 % of CO in the gas phase. The medium for *Thermolithobacter* species enrichment and isolation is prepared anaerobically by the Hungate technique (Ljungdahl and Wiegel 1986) under 4/1 mixture of hydrogen and carbon dioxide gases or 100 % CO and contained per liter of deionized water: 0.33 g of $KH_2PO_4$, 0.33 g of $NH_4Cl$, 0.33 g of KCl, 0.33 g of $MgCl_2 \cdot 2H_2O$, 0.33 g of $CaCl_2 \cdot 2H_2O$, 2.0 g of $NaHCO_3$, 1 ml of vitamin solution (Wolin et al. 1963), and 1.2 ml of trace element solution (Slobodkin and Wiegel 1997) with a $pH^{25\,°C}$ adjusted to 7.0 with 10 % (w/v) NaOH. For *T. ferrireducens* enrichment and cultivation, 90 mM of ferric iron hydromorphic oxide is required. Ferric iron hydromorphic oxide is prepared as it was previously described (Slobodkin and Wiegel 1997). *T. ferrireducens* is typically grown in Hungate or Balch tubes at 60 °C. Since the reducing agents sodium sulfide, dithionite, cysteine, Ti(III) citrate (Gaspard et al. 1998), and HCl-cysteine reduce Fe(III), no reducing agents are added to the medium. A similar medium reduced with sodium sulfide (500 mg $l^{-1}$) with 100 % CO in the gas phase but without ferric iron hydromorphic oxide is suitable to cultivate *T. carboxydivorans*. Pure cultures can be isolated by serial dilutions and sequential single colony isolations.

## Ecology

*T. ferrireducens* strains were isolated from terrestrial hot springs at Yellowstone National Park and at Fiji. A sample from Yellowstone which was a combined sample of water, organic filamentous material, and sediment was collected from a runoff of a hot spring close to the Yellowstone river at the Calcite Spring area (44°54.291' N, 110°24.242' W). The sample contained white and black bacterial filaments. The thermal spring sampling point had a temperature gradient from 60 °C to 85 °C with a pH of 7.6. The sample from Fiji was collected from the main spring of the Nakama Springs in Vanua Levu. The sample contained water and sediment. Estimated original underground temperature was around 170 °C with a pH 7.5. Strain with properties similar to *T. ferrireducens* could have been involved in the formation of the low temperature banded iron formations (Slobodkin and Wiegel 1997; Onyenwoke and Wiegel 2011)

*T. carboxydivorans* strain $R1^T$ was isolated from a neutral terrestrial hot spring at the Raoul Island, Archipelago Kermadec (New Zealand). The temperature at the sampling point was 70 °C. A not-further-characterized strain with 99 % similarity in 16S rRNA gene sequence was detected in the enrichment culture growing in ferric iron oxide-containing medium with $H_2$ in the gas phase, obtained from a hot spring at Geyser Valley, Kamchatka (Slobodkin A.I., personal communication). At the sampling point the pH was 6.5 and the temperature was 73 °C. The BLAST search in NCBI database gives only eight environmental 16S rRNA gene clones which could be considered to be the members of *Thermolithobacteriaceae* since they have more than 90 % of similarity in nucleotide sequence with those in *T. ferrireducens* and *T. carboxydivorans*.

## Pathogenicity: Clinical Relevance

There is no data on *Thermolithobacteriaceae* pathogenicity. All currently known representatives of the family are extreme thermophiles not growing at the temperature below 50 °C. Thus, they should not be pathogenic.

*T. ferrireducens* grows on $H_2/CO_2$ at 60 °C in mineral ferric iron-containing medium supplied with 100 μg/ml of tetracycline or ampicillin or with 10 μg/ml but not with 100 μg/ml of rifampicin, erythromycin, streptomycin, and chloramphenicol. Growth and CO consumption by *T. carboxydivorans* are inhibited by 100 μg/ml of penicillin, erythromycin, or chloramphenicol but not by 100 μg/ml of streptomycin, rifampicin, or tetracycline.

## Acknowledgments

This work was supported by Molecular Cell Biology Program Russian Academy of Sciences.

## References


Gaspard S, Vazquez F, Holliger C (1998) Localization and solubilization of the iron(III) reductase of *Geobacter sulfurreducans*. Appl Environ Microbiol 64:3188–3194

Ljungdahl LG, Wiegel J (1986) Working with anaerobic bacteria. In: Demain AL, Solomon NA (eds) Manual of industrial microbiology and biotechnology. American Society for Microbiology, Washington, DC, pp 84–94

Onyenwoke R, Wiegel J (2011) Chapter 8. In: Angrove DM (ed) Magnetite: structure, properties and applications. Nova, New York, pp 297–316

Slepova TV, Sokolova TG, Lysenko AM, Tourova TP, Kolganova TV, Kamzolkina OV, Karpov GA, Bonch-Osmolovskaya EA (2006) *Carboxydocella sporoproducens* sp. nov., a novel anaerobic CO-utilizing/H2-producing thermophile bacterium from Kamchatka hot spring. Int J Syst Evol Microbiol 56:797–800

Slobodkin AI, Wiegel J (1997) Fe(III) as an electron acceptor for $H_2$ oxidation in thermophilic anaerobic enrichment cultures from geothermal areas. Extremophiles 1:106–109

Slobodkina GB, Panteleeva AN, Sokolova TG, Bonch-Osmolovskaya EA, Slobodkin AI (2012) *Carboxydocellamanganica* sp. nov., a thermophilic, dissimilatory Mn(IV)- and Fe(III)-reducing bacterium from a Kamchatka hot spring. Int J Syst Evol Microbiol 62:890–894

Sokolova TG, Kostrikina NA, Chernyh NA, Tourova TP, Kolganova TV, Bonch-Osmolovskaya EA (2002) *Carboxydocella thermoautotrophica* gen. nov., sp. nov., a novel anaerobic CO-utilizing thermophile from a Kamchatkan hot spring. Int J Syst Evol Microbiol 52:1961–1967

Sokolova T, Hanel J, Onyenwoke RU, Reysenbach AL, Banta A, Geyer R, González JM, Whitman WB, Wiegel J (2007) Novel chemolithotrophic, thermophilic, anaerobic bacteria *Thermolithobacter ferrireducens* gen. nov., sp. nov. and *Thermolithobacter carboxydivorans* sp. nov. Extremophiles 11:145–157

Stamatakis A (2006) RAxML-VI-HPC: maximum likelihood-based phylogenetic analyses with thousands of taxa and mixed models. Bioinformatics 22:2688–2690

Svetlichny VA, Sokolova TG, Kostrikina NA, Lysenko AM (1994) A new thermophilic anaerobic carboxydotrophic bacterium *Carboxydothermus restrictus* sp. nov. Microbiology 63:294–297 (English translation of Mikrobiologiya)

Validation list N 116 (2007) Int J Syst Evol Microbiol 57:1371–1373

Wiegel J, Hanel J, Aygen K (2003) Chemolithoautotrophic thermophilic iron(III)-reducer. In: Ljungdahl LG, Adams MW, Barton LL, Ferry JG, Johnson MK (eds) Biochemistry and physiology of anaerobic bacteria. Springer, New York

Wolin EA, Wolin MJ, Wolfe RS (1963) Formation of methane by bacterial extracts. J Biol Chem 238:2882–2886

Yarza P, Ludwig W, Euzéby J, Amann R, Schleifer K-H, Glöckner FO, Rosselló-Móra R (2010) Update of the all-species living-tree project based on 16S and 23S rRNA sequence analyses. System Appl Microbiol 33:291–299

# 35 The Family *Veillonellaceae*

*Hélène Marchandin*[1,2] · *Estelle Jumas-Bilak*[1,3]
[1]Equipe Pathogènes et Environnements, UMR5119 ECOSYM, Université Montpellier 1, Montpellier, France
[2]Laboratoire de Bactériologie, Centre Hospitalier Régional Universitaire de Montpellier, Montpellier, France
[3]Laboratoire d'Hygiène hospitalière, Centre Hospitalier Régional Universitaire de Montpellier, Montpellier, France



**Abstract**

The family *Veillonellaceae* belongs to the phylum Firmicutes, the class *Negativicutes*, and the order *Selenomonadales*. Delineation of the family was established in 2010 on the basis of 16S rRNA gene phylogenetic analyses, and to date, the family includes 6 genera of Gram-negative, anaerobic, or microaerophilic cocci and coccobacilli and 25 species, i.e., the genera *Veillonella* (12 species), *Megasphaera* and *Dialister* (5 species each), *Allisonella*, *Anaeroglobus*, and *Negativicoccus* (1 species each). The most striking particularity of this family, and more generally of the class *Negativicutes*, is to group bacteria with Gram-negative cell wall structure within a phylum of Gram-positive bacteria. Genera can be distinguished based on their phenotypic, genetic, genomic, and phylogenetic characteristics, while molecular-based methods may be required for species affiliation, particularly in the genus *Veillonella*. The isolates displayed various resistance patterns to antimicrobial agents. The family includes three beer-spoilage species belonging to the genus *Megasphaera*, other species being representatives of several human and other animal microbiotae, and some of them can act as opportunistic pathogens for animals including humans being usually responsible for polymicrobial infections and more rarely for monomicrobial severe infections like osteoarticular infections or endocarditis.

## Taxonomy, Historical, and Current

### *Veillonellaceae* Rogosa 1971b, emend. Marchandin et al. 2010

Veil.lo.nel.la′ce.ae. M.L. fem. n. *Veillonella* type genus of the family;-*aceae* ending to denote a family; M.L. fem. pl. n. *Veillonellaceae*, the *Veillonella* family.

The family *Veillonellaceae* was proposed by Rogosa in 1971 to group anaerobic Gram-negative cocci belonging to three genera, i.e., *Veillonella*, *Acidaminococcus*, and *Megasphaera*, *Veillonella* being the type genus of the family (Rogosa 1971b). Since this date, several novel species were characterized within

E. Rosenberg et al. (eds.), *The Prokaryotes – Firmicutes and Tenericutes*, DOI 10.1007/978-3-642-30120-9_361,

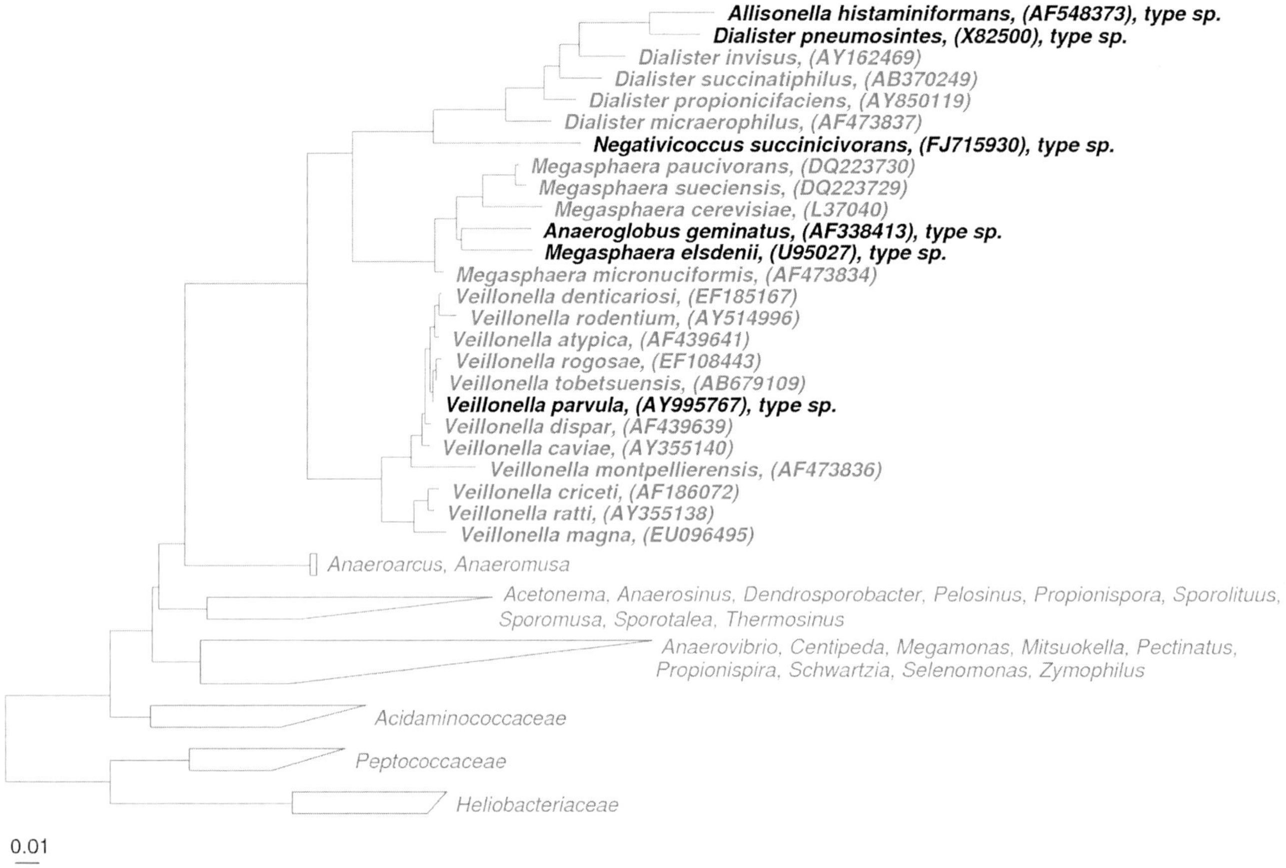


**Fig. 35.1**

**Phylogenetic reconstruction of the family *Veillonellaceae* based on the maximum likelihood algorithm RAxML (Stamatakis 2006). Sequence dataset and alignments according to the All-Species Living Tree Project, release LTPs108 (Yarza et al. 2008). Representative sequences from close relative genera were used to stabilize the tree topology. In addition, a 40 % maximum frequency filter was applied to remove hypervariable positions from the alignment. Scale bar indicates estimated sequence divergence. Sequence of the *Selenomonadales incertae sedis Quinella ovalis* was not included in the tree because of 27 undetermined positions. The position of the branch supporting the genera *Anaeroarcus* and *Anaeromusa* varied according the method used to generate the tree (neighbor joining or maximum likelihood)**

these three genera and novel genera belonging to this family were described, so in the last edition of the *Bergey's Manual of Systematic Bacteriology*, 28 genera have been considered as members of the family *Veillonellaceae* and an expanded description of the family members' characteristics has been edited (Rainey 2009a). Of note, during this period, both terms family "*Acidaminococcaceae*" and *Sporomusa* subbranch of the *Bacillus/Clostridium* group or of the low G+C Gram-positive bacteria could be found as designing the same taxon. Until 2010, the family was classified in the phylum Firmicutes (low G+C Gram-positive bacteria), in the class *Clostridia*, and in the order *Clostridiales*. However, reorganization was awaited for this taxon (Ludwig et al. 2009), and in 2010, on the basis of 16S rRNA gene-based phylogeny, Marchandin et al. proposed to elevate the family *Veillonellaceae* to class rank in the phylum Firmicutes. The name *Negativicutes* has been proposed for the novel class with reference to the typical Gram-negative cell wall structure with an outer membrane as observed in electron microscopy of members of the taxon (Marchandin et al. 2010). In this novel class, the order *Selenomonadales* grouped all the known genera previously classified in the family *Veillonellaceae* and two clades were observed in the order. The first one grouped the four genera *Acidaminococcus*, *Phascolarctobacterium*, *Succinispira*, and *Succiniclasticum* and was referred to as family *Acidaminococcaceae*; the second one grouped the six genera *Allisonella*, *Anaeroglobus*, *Dialister*, *Megasphaera*, *Negativicoccus*, and *Veillonella* and represents the emended family *Veillonellaceae*. Phylogenetic position of other genera previously classified in the family *Veillonellaceae* (*Acetonema*, *Anaeroarcus*, *Anaeromusa*, *Anaerosinus*, *Anaerovibrio*, *Centipeda*, *Dendrosporobacter*, *Megamonas*, *Mitsuokella*, *Pectinatus*, *Pelosinus*, *Propionispira*, *Propionispora*, *Quinella*, *Schwartzia*, *Selenomonas*, *Sporolituus*, *Sporomusa*, *Sporotalea*, *Thermosinus*, and *Zymophilus*) could not be determined and should currently be considered as *Selenomonadales incertae sedis* (Marchandin et al. 2010; ➲ *Fig. 35.1*). Summarized presentation of species and genera belonging to the family *Veillonellaceae* in its current definition is shown in ➲ *Table 35.1*. The family includes

**Table 35.1**
**Alphabetical ordered presentation of taxa currently composing the family *Veillonellaceae***

| Genus[a] | Species | Reference |
|---|---|---|
| *Allisonella* | *Allisonella histaminiformans* (type species of the genus) | Garner et al. (2002) |
| *Anaeroglobus* | *Anaeroglobus geminatus* (type species of the genus) | Carlier et al. (2002) |
| *Dialister* | *Dialister invisus* | Downes et al. (2003) |
| | *Dialister pneumosintes* (type species of the genus) | Moore and Moore (1994) |
| | *Dialister micraerophilus* | Jumas-Bilak et al. (2005) |
| | *Dialister propionicifaciens* | Jumas-Bilak et al. (2005) |
| | *Dialister succinatiphilus* | Morotomi et al. (2008) |
| *Megasphaera* | *Megasphaera cerevisiae* | Engelmann and Weiss (1985) |
| | *Megasphaera elsdenii* (type species of the genus) | Rogosa (1971a) |
| | *Megasphaera micronuciformis* | Marchandin et al. (2003a) |
| | *Megasphaera paucivorans* | Juvonen and Suihko (2006) |
| | *Megasphaera sueciensis* | Juvonen and Suihko (2006) |
| *Negativicoccus* | *Negativicoccus succinicivorans* (type species of the genus) | Marchandin et al. (2010) |
| *Veillonella* (type genus of the family) | *Veillonella atypica* | Rogosa (1985) |
| | *Veillonella caviae* | Mays et al. (1982) |
| | *Veillonella criceti* | Rogosa (1985) |
| | *Veillonella denticariosi* | Byun et al. (2007) |
| | *Veillonella dispar* | Rogosa (1985) |
| | *Veillonella magna* | Kraatz and Taras (2008) |
| | *Veillonella montpellierensis* | Jumas-Bilak et al. (2004) |
| | *Veillonella parvula* (type species of the genus) | Prevot (1933) |
| | *Veillonella ratti* | Rogosa (1965) |
| | *Veillonella rodentium* | Rogosa (1965) |
| | *Veillonella rogosae* | Arif et al. (2008a) |
| | *Veillonella tobetsuensis* | Mashima et al. (2013) |

[a]Reference for the genus description is as for the type species of the genus

25 species belonging to 6 genera, i.e., the genera *Veillonella* (12 species), *Megasphaera* and *Dialister* (5 species each), *Allisonella*, *Anaeroglobus*, and *Negativicoccus* (1 species each).

Current definition of the family *Veillonellaceae* is as described by Rogosa in 1971 and emended by Marchandin et al. in 2010 (Rogosa 1971b; Marchandin et al. 2010).

Gram-negative bacteria with a typical Gram-negative cell wall structure with an outer membrane observed in electron microscopy. Cocci or coccobacilli. Anaerobic or microaerophilic. No endospores. Nonmotile. Cytochrome oxidase negative. Catalase negative, but some strains decompose peroxide by means of a nonheme-containing catalase, so-called pseudocatalase. Chemoorganotrophic. Possess complex nutritional requirements. Gas may or may not be produced. Carbohydrates may or may not be fermented. Lactic acid may not be produced and if present is not a major product; lactate is fermented by some genera with the production of $CO_2$, $H_2$, and various lower volatile fatty acids containing 2–6 C atoms. Found in homothermic animals such as man, ruminants, rodents, and pigs samples, particularly from the alimentary tract.

The type genus is *Veillonella* Prévot 1933.

Characteristics further detailed in this chapter for the species included in the table came from the corresponding reference as listed in the *Table 35.1*.

## Molecular Analyses

### Phylogeny of Type Strains of Species

Type strains of most species belonging to the family *Veillonellaceae* are clearly differentiated, each being supported by a robust branch in 16S rRNA gene-based phylogeny. However, no clear delineation could be observed for some closely related species sharing 99 % or more of their *rrs* gene nucleotide positions. This was observed for *Megasphaera paucivorans* and *Megasphaera sueciensis* sharing nearly identical 16S rRNA gene sequences (99.3 %) (Juvonen and Suihko 2006), for *Veillonella denticariosi* and *Veillonella rodentium*, for *Veillonella ratti* and *Veillonella criceti*, and for *Veillonella dispar* and *Veillonella parvula*, each pair of species sharing 99 % of their 16S rDNA nucleotide positions (Marchandin et al. 2005; Byun et al. 2007; Michon et al. 2010; ➲ *Fig. 35.1*). Furthermore, none of the variable positions between *V. dispar* and *V. parvula* could be retained as species-specific because they are subject to intraspecies variability, mostly due intrachromosomal heterogeneity existing between the four 16S rRNA gene copies found in the genus *Veillonella*. Indeed, in this genus, it has been shown that intragenomic and intraspecific variability in *rrs* may surpass interspecific variability. Intragenomic *rrs* V3 region heterogeneity, as well as recombination events among strains or isolates of different species, clearly impacted both phylogeny and taxonomy and impaired the 16S rRNA-based species identification (Marchandin et al. 2003b; Michon et al. 2010). Molecular studies are required for species identification in the genus *Veillonella* due to the lack of discriminating routine phenotypic tests and should therefore be based on the analysis of genes other than 16S rRNA for some *Veillonella* species.

## DNA–DNA Hybridization Studies

Multispecies genera were subjected to DNA–DNA hybridization (DDH) studies except for the genus *Dialister*, probably because fastidious growth did not allow to get enough biomass to perform DDH. In these cases, alternative genomic approaches were usually proposed in addition to phenotypic, genotypic, and phylogenetic data included in taxonomic studies (Downes et al. 2003; Jumas-Bilak et al. 2005). DDH clearly delineated species only for some genera. For instance, DDH values between *Megasphaera* species ranged from 3.1 % to 41 % (Engelmann and Weiss 1985; Juvonen and Suihko 2006), whereas DDH did not clearly separate species from each other in the genus *Veillonella* (Mays et al. 1982). In the latter study, DDH was performed on 111 isolates affiliated to *Veillonella* species or subspecies by serotyping. DDH values were usually higher in a definite serogroup than observed among isolates of different serogroups, but DDH data were hardly interpretable for some groups of isolates. As an example, in the *V. parvula* serogroups, the 44 isolates showed DDH values ranging from 53 % to 100 % while showing up to 44 % of DDH with some members of the *V. dispar* serogroup. Overlapping the threshold of DDH value of 70 % was also observed for the 8 isolates belonging to the *V. dispar* serogroups and for the 31 isolates of the *Veillonella atypica* serogroups (Mays et al. 1982). Consequently, these DDH experiments seem to be insufficient to support any taxonomic conclusions at least regarding these three species. More recently, DDH values supported the separation of *V. magna* from *V. ratti*, *V. criceti*, and *V. parvula* (DDH values of 40.0 %, 39.0 %, and 27.0 %, respectively) (Kraatz and Taras 2008) and of *V. denticariosi* from *V. rodentium* (highest DDH value to *V. rodentium* of 48–49 %) and all the species described at the time of publication (Byun et al. 2007). In the latter study, all the species described at the time of publication. Again, all the species appeared clearly separated on the basis of DDH values below 50 % except for *V. dispar* and *V. parvula* showing a DDH value of 64 %, while two *V. parvula* strains showed a DDH value of 74 % (Byun et al. 2007). Finally, DDH in the genus *Veillonella* confirmed the existence of very related species or species complexes, such as *V. dispar* and *V. parvula*, as also suggested by 16S rRNA gene analysis.

## Multilocus Sequence Analysis (MLSA)

As a rule in bacterial taxonomy, the most studied molecular marker in the family *Veillonellaceae* is 16S rRNA gene. However, insufficient discrimination between some species in the genus *Veillonella*, as well as intragenomic heterogeneity of 16S rRNA gene copies, led to the use of other housekeeping genes for establishing phylogenetic relationships between closely related species and or characterizing novel species within the genus. The *dnaK* and *rpoB* markers were used for taxonomic purpose (Jumas-Bilak et al. 2005; Arif et al. 2008a; Marchandin et al. 2010; Michon et al. 2010; Mashima et al. 2013) but no multilocus study was formally published for members of the family. However, multilocus genetics of *Veillonellaceae* coming from whole genome sequencing data should become available in the near future.

## Fingerprinting Methods

Fingerprinting methods have been scarcely applied to members of the family *Veillonellaceae*.

Pulsed-field gel electrophoresis (PFGE) after DNA macrorestriction by the endonuclease *Sma*I has been successfully applied to *Veillonella* typing to explore the source of infection in a case of prosthetic joint infection (Marchandin et al. 2010). PCR-based typing methods like REP-PCR (Arif et al. 2008b) or enterobacterial repetitive intergenic consensus sequences-PCR (ERIC-PCR) (Versalovic et al. 1991) (unpublished personal data) could also be performed on members of this genus. Besides epidemiological aim, other fingerprinting methods were applied to the members of the family members for taxonomic and identification purposes. In 1997, Sato et al. proposed an identification method for *Veillonella* spp. based on 16S rDNA amplification followed by restriction fragment length polymorphism (RFLP) analysis. Using this approach, the authors were able to identify the seven *Veillonella* species yet described (Sato et al. 1997a, b). However, this approach was subsequently ineffective to discriminate *Veillonella* species because of numerous atypical restriction patterns further related to the high proportion of *Veillonella* isolates displaying intragenomic *rrs* heterogeneity (Marchandin et al. 2003b).

In the genus *Megasphaera*, discrimination between strains or species varied according to the fingerprint method used and the species studied. Automated ribotyping allowed subgrouping of *Megasphaera cerevisiae* isolates (Suihko and Haikara 2001), and ribotyping fingerprints also supported the characterization of the two species *M. sueciensis* and *M. paucivorans*, each *Megasphaera* species displaying distinct patterns (Juvonen and Suihko 2006). Juvonen et al. a few years later proposed an

**Table 35.2**
**Genome size, G + C%, and gene and protein number from whole genome sequencing data available at the NCBI Microbial genomes entry for members of the family *Veillonellaceae***

| Genus (number of genomes/number of taxa or putative taxa) | Genome size range (in Mb) | GC content range (%) | Putative gene number | Putative protein number |
|---|---|---|---|---|
| *Anaeroglobus* (1/1) | 1.79 | – | 2,199 | 2,148 |
| *Dialister* (3/3) | 1.26–2.46 | 45.5 | 1,296–2,219 | 1,243–2,155 |
| *Megasphaera* (7/6) | 1.64–2.47 | 46.1–52.8 | 1,572–2,304 | 1,513–2,219 |
| *Veillonella* (15/11) | 1.75–2.18 | 38.5–39 | 1,647–2,060 | 1,58–2,000 |

approach for routine study of occurrence and diversity of strictly anaerobic beer spoilers in the brewing process; this included a group-specific amplification for detecting the different beer-spoilage genera followed by RFPL for discriminating involved species (Juvonen et al. 2008). For *Megasphaera elsdenii*, PCR fingerprint techniques targeting the ribosomal RNA operon (amplified ribosomal DNA restriction analysis, ribosomal RNA intergenic spacer analysis) or the whole genome (ERIC-PCR, random amplification of polymorphic DNA) suggested that genetic variability could be low in *M. elsdenii* because of a close genetic relatedness between seven isolates originating from vastly different habitats worldwide (Piknová et al. 2006).

Finally, protein profile analysis was part of the polyphasic differentiation of *Dialister invisus* from *Dialister pneumosintes* (Downes et al. 2003).

## MALDI-TOF

Members of the family *Veillonellaceae* were scarcely included in MALDI-TOF studies. More usually, a few strains of *Veillonella*, which represents the genus the most frequently isolated from human clinical samples, were studied with variable success (correct identification – only observed for *V. parvula* for five out of the eight isolates in La Scola et al. 2011, identification to the genus level only, no identification). Whatever the system used, databases do not include all members of the family *Veillonellaceae* or all species of the genus *Veillonella*. The incompleteness of the databases did not allow any accurate identification of the family members, particularly clinical isolates of several *Veillonella* species, *D. pneumosintes* and *Acidaminococcus intestini* could not be identified by mass spectrometry-based methods even in recent published studies (Justesen et al. 2011; La Scola et al. 2011; Veloo et al. 2011; Fedorko et al. 2012).

## Genomes Analyses

Increasing number of whole sequenced genomes is available mainly due to the interest either in species with biotechnological interest or in sequencing the human microbiome. Two complete genome sequences have been published for members of the family *Veillonellaceae*, i.e., *V. parvula* and *M. elsdenii* (Gronow et al. 2010; Marx et al. 2011). The single chromosome of *V. parvula* type strain includes 2,132,142 bases with a G+C content of 38.63 %. DNA coding region represented 88.46 % of the total genome. Other genome characteristics are a total of 1,920 genes of which 1,859 (96.82 %) were protein-coding genes, 61 RNA genes, 15 pseudogenes, 4 rRNA operons, and 5 CRISPR repeats (Gronow et al. 2010). The genome of the type strain of *M. elsdenii* includes 2,474,718 bases with a G+C content of 53 %. The number of putative genes was 2,220. Seven *rrn* operons and 64 predicted tRNAs of which one is a pseudogene were found in the genome (Marx et al. 2011). For 77.06 % and 88 % of the genes of *V. parvula* and *M. elsdenii*, respectively, a significant protein family was found when searching against Pfam. The sequences annotated by COG fell into 21 and 18 out of the 25 functional COG classes, for these 2 genomes, respectively (Gronow et al. 2010; Marx et al. 2011). Both genomes did not include genes associated with RNA processing and modification, nuclear structure, cytoskeleton, and extracellular structures (COG functional categories A, Y, Z, and W, respectively). In addition, the *M. elsdenii* genomes did not include genes associated with chromatin structure and dynamics, and intracellular trafficking and secretion (COG classes B and U, respectively), while these two categories represented 0.1 % and 1.9 % of the genes associated to a COG class in the *V. parvula* genome (Gronow et al. 2010; Marx et al. 2011). *V. parvula* and *M. elsdenii* genomes display 28 and 9 outer membrane protein-encoding genes, respectively, confirming their atypical cell wall structure in the Firmicutes phylum.

Besides, several genomes are currently being sequenced for additional strains belonging to *V. parvula* and for other species in the family. According to the NCBI Microbial genomes entry (available at http://www.ncbi.nlm.nih.gov/genomes/MICROBES/microbial_taxtree.html), 25 additional genomes are currently sequenced (status September 2012) including representative isolates of known species (*Anaeroglobus geminatus*, *Dialister micraerophilus*, *D. invisus*, *Dialister succinatiphilus*, *Megasphaera micronuciformis*, *V. atypica*, *V. dispar*, and *V. ratti*) and uncharacterized isolates thought to represent novel taxa in both genera *Megasphaera* and *Veillonella*. The majority of them are considered as reference genomes in the NIH Human Microbiome Project (Human Microbiome Project Consortium 2012). Current available data are summarized in the *Table 35.2*.

Comparative analysis of these complete genome sequences when available with those currently available for members of other classes in the phylum Firmicutes as well as for members of other phyla will probably contribute to increased knowledge on the atypical phylogenetic position of the family within the low G+C Gram-positive phylum and on evolutionary history originating in this position. Similarly, from multigene and multiprotein sequence-based comparative analyses for members of the class *Negativicutes*, the position and classification of genera currently standing as *Selenomonadales incertae sedis* could certainly be reevaluated. Further analyses of the complete genome sequences are also needed to precise the genome content in putative virulence genes and in resistance to antimicrobial agent-encoding genes.

### Genome Structure

From whole genome sequence, *V. parvula* was shown to harbor a unique replicon with no associated extrachromosomal elements and four rRNA operons (Gronow et al. 2010). Besides data from genomes with full sequence genome, genome structure was mainly investigated using PFGE-based approaches including migration of undigested DNA in agarose plugs and mapping experiments with the intron-encoded restriction enzyme I-*Ceu*I to measure the bacterial chromosomes and to determine the *rrn* skeletons. Large-scale genome structure of members of the family *Veillonellaceae* (i) revealed a unique and circular chromosome for all members of the family with no large-sized plasmids visible on PFGE gels; (ii) allowed first estimations of genome size (➲ *Table 35.3*), and *rrn* copy number and distribution across the chromosome, all confirmed by subsequent full genome sequencing where available; and (iii) showed taxonomic value with low variability observed at the intraspecific level and specific patterns observed for most species within each genus (Marchandin et al. 2003a; Jumas-Bilak et al. 2005; Marchandin et al. 2010). Among species that were investigated using this approach, only *V. atypica*, *V. dispar*, and *V. parvula* could not be distinguished based on their *rrn* pattern after PFGE of I-*Ceu*I-restricted DNA, while the other five species studied (V. *montpellierensis*, *V. ratti*, *V. rodentium*, *V. criceti*, and *V. caviae*) displayed a specific distribution of *rrn* operons across their chromosome (unpublished personal data), as observed for *M. micronuciformis* compared to *M. elsdenii* (Marchandin et al. 2003a) and for the four species *D. invisus*, *D. micraerophilus*, *D. pneumosintes*, and *Dialister propionicifaciens* (Jumas-Bilak et al. 2005). Again, these results underlined the absence of clear delineation between these three *Veillonella* species as also suggested by DDH studies and 16S rRNA gene sequencing.

Genomic data either from whole genome sequencing or from other approaches are summarized in the ➲ *Table 35.3*. Examples of genomic size estimation using PFGE of undigested DNA and large-scale genome structure determination by I-*Ceu*I genome restriction are shown in ➲ *Fig. 35.2*. When several isolates were studied for a species, mean values together with the number of strains studied were given in ➲ *Table 35.3*. In addition, we investigated a collection of 26 clinical isolates of *Veillonella* sp. from human origin unidentified to the species level and showed that all the strains possessed 4 *rrn* operon copies on a unique circular chromosome with mean size 2.17 MB (range 1.75–2.55 Mb) (Marchandin et al. 2001a).

Plasmids have been found in a high proportion of *Veillonella* spp. isolates from oral cavity (Arai et al. 1984). More recently, a unique endogenous plasmid sizing 4,813 pb was isolated from a strain of *V. parvula* out of the 12 *Veillonella* spp. isolates from saliva samples studied (Liu et al. 2012). These plasmids may encode metabolic functions such as fructose fermentation in *V. criceti* (Mays et al. 1982) or support antibiotic resistance determinants (Marchandin et al. 2004).

### Phages

Members of the family *Veillonellaceae* are all part of animal microbiotae, which are known as ecosystems containing a lot of bacteriophages. However, interactions between phages and *Veillonellaceae* remained rarely studied and the few available data were for *Veillonella* isolated from the oral cavity. Several phages were shown to infect *V. rodentium*, and further investigation allowed the identification of the receptor of the veillonellophage $N_2$ as cell wall polysaccharides of the type strain of the species (Totsuka and Ono 1989).

In *M. elsdenii*, Piknova et al. found a restriction-modification system, thought to represent the main defense tool against phage infection. This system was shown different from that identified in another species of the class *Negativicutes*, *Selenomonas ruminantium* suggesting that original strategies are developed for bacteriophage protection in *M. elsdenii* (Piknová et al. 2004).

## Phenotypic Analyses

The family includes Gram-negative bacteria with various cell morphologies, i.e., coccoid-shaped, ovoid-shaped, and coccobacillary-shaped bacteria. Common characteristics of members of the family are to be nonmotile, non-endospore-forming, and Gram-negative. The main features of members of *Veillonellaceae* are listed in ➲ *Table 35.4*. All species that were observed in electron microscopy displayed a typical Gram-negative cell wall structure with an outer membrane (➲ *Fig. 35.3*). Of note, all the genera including anaerobic Gram-negative cocci are included in the family *Veillonellaceae*, except for the genus *Syntrophococcus* belonging to the family *Lachnospiraceae* (Krumholz and Bryant 1986).

### *Allisonella* Garner et al. 2003, 373[VP] (Effective Publication: Garner et al. 2002, 504)

Al.li.son.ella. N.L. fem. dim. n. *Allisonella*, named after the American microbiologist Milton J. Allison, a prominent rumen microbiologist who isolated *Oxalobacter formigenes*, a ruminal bacterium that decarboxylates oxalate.

**Table 35.3**
**Summarized genomic structures for species in the family *Veillonellaceae***

| Species[a] | Genome size (in Mb) | DNA G + C content (mol%) | *rrn* operon number |
|---|---|---|---|
| *A. histaminiformans* | NA | 45–48 (HPLC) | NA |
| *A. geminatus* | 1.79 (WG) | 51.8 ($T_m$) | 4 (GM) |
| | 1.84 (PFGE) (mean for 2 strains) | | |
| *D. invisus* | 1.9 (WG) | 45.5 (WG) | 4 (GM) |
| | 1.97 (PFGE) | 45–46 | |
| *D. pneumosintes* | 1.34 (PFGE) | 35 | 4 (GM) |
| *D. micraerophilus* | 1.26–1.4 (WG) | 36.3 | 4 (GM) |
| | 1.35 (PFGE) (mean for 7 strains) | | |
| *D. propionicifaciens* | 1.7 (PFGE) (mean for 4 strains) | ND | 4 (GM) |
| *D. succinatiphilus* | 2.46 (WG) | 51.9 | NA |
| *M. cerevisiae* | NA | 42.4–44.8 (Tm) | NA |
| *M. elsdenii* | 2.47 (WG) | 52.8 (WG) | 7 (WG) |
| | 2.58 (PFGE) | 53.1–54.1 (Bd) | 7 (GM) |
| *M. micronuciformis* | 1.73 (genomo sp. 1) (WG) | 46.1 (WG) (genomo sp.1) | 4 |
| | 1.77 (WG) | 46.4 (Tm) | |
| | 1.81 (PFGE) (mean for 2 strains) | | |
| *M. paucivorans* | NA | 40.5 (Tm) | NA |
| *M. sueciensis* | NA | 43.1 (Tm) | NA |
| *N. succinicivorans* | 1.62 (PFGE) (mean for 3 strains) | NA | 4 |
| *V. atypica* | 2.05–2.15 (WG) | 39 (WG) | 4[b] |
| | 2.05 (PFGE) | 36–40, mean 39 (Tm) | |
| *V. caviae* | NA | 37–39, mean 39 (Tm) | 4 |
| *V. criceti* | NA | 38–40, mean 39 (Tm) | 4 |
| *V. denticariosi* | NA | NA | 4 |
| *V. dispar* | 2.12 (WG) | 38.8 (WG) | 4 |
| | 2.13 (PFGE) | 38–40, mean 39 (Tm) | |
| *V. magna* | NA | NA | NA |
| *V. montpellierensis* | NA | NA | 4 |
| *V. parvula* | 2.13–2.16 (WG) | 38.5–38.6 (WG) | 4 (WG) |
| | 2.13 (PFGE) | 37–40, mean 38 (Tm) | 4 (GM) |
| *V. ratti* | NA | 41–43, mean 42 (Tm) | 4 |
| *V. rodentium* | NA | 42–43, mean 43 (Tm) | 4 |
| *V. rogosae* | NA | NA | 4 |
| *V. tobetsuensis* | NA | NA | NA |

*PFGE* pulsed-field gel electrophoresis of undigested DNA,; *GM* genome mapping with intron-encoded endonuclease I-*CeuI*, *NA* not available
[a]Reference for data listed in this table were either from the species characterization publication as referenced in *Table 35.1*, from personal data (Marchandin 2001a), or from whole genome (WG) sequence data available on line
[b]Except for *V. atypica* strain ATCC 17744 (Marchandin et al. 2003b)

*Allisonella histaminiformans* is the only species in the genus. Cells are facultative anaerobic, ovoid bacteria arranged in pairs or chains. *A. histaminiformans* was cultivated from histidine enrichments of feces of cattle fed grain and of cecum of a horse on carbonate-based medium containing 50 mM histidine without glucose incubated in anaerobic conditions. Despite lysine can be used, this amino acid alone did not allow growth of *A. histaminiformans* that requires histidine and butyrate. The species do not utilize carbohydrate or organic acid and histidine decarboxylation represents the sole source of energy. Histamine and $CO_2$ are produced as end products from histidine. Resistance to the ionophore monensin at 25 μM is observed. Catalase, oxidase, and indole production are negative (Rainey 2009b).

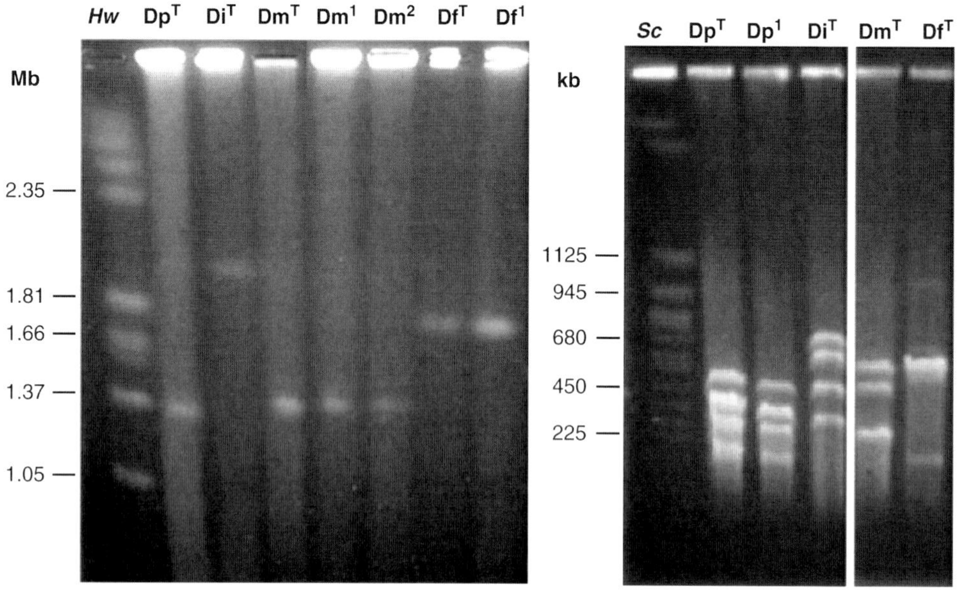


**Fig. 35.2**
**Pulsed-field gel electrophoresis patterns of undigested DNA (*left*) and large-scale genome structure determination by I-*Ceu*I genome restriction (*right*) for four *Dialister* species (including type strain (*superscript T*) and one or several clinical isolates (*superscript 1, 2*) identified by 16S rRNA gene sequencing: *Dp D. pneumosintes*, *Di D. invisus*, *Dm D. micraerophilus*, *Df D. propionicifaciens*. *Hw* (*Hansenula wingei* DNA) and *Sc* (*Saccharomyces cerevisiae* chromosomes) as molecular weight markers. Sizes are indicated in kilobases (kb) or megabases (Mb)**

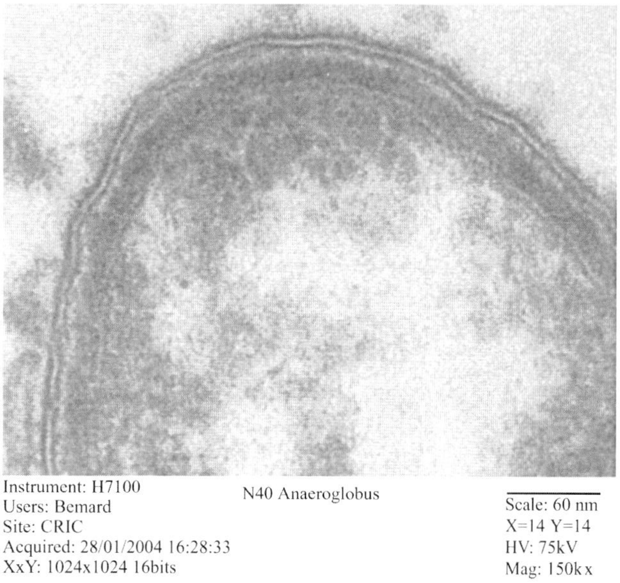


**Fig. 35.3**
**Ultrastructure of type strain of *Anaeroglobus geminatus* CIP 106856$^{T}$ showing Gram-negative-type cell wall structure with outer membrane, observed for members of the family *Veillonellaceae*, after electron microscopy of ultrathin sections**

## *Anaeroglobus* Carlier et al. 2002, 986$^{VP}$

An.ae.ro.glo′bus. Gr. pref. *an*, not; Gr. n. *aer aeros*, air; L. masc. n. *globus*, globe, sphere; N.L. masc. n. *Anaeroglobus*, a sphere not living in air.

To date, the genus includes a single species (➋ *Table 35.1*), *A. geminatus*, and characteristics of the genus are those of the species (Carlier et al. 2002; Carlier 2009b). The cells are Gram-negative cocci of 0.5–1.1 μm in diameter, usually arranged in pairs. Nonpigmented and nonhemolytic colonies are observed in blood agar after 48 h of incubation at 37 °C. These colonies are tiny, circular, convex, and translucent with a smooth surface. By presumptive identification tests, the strain was resistant to 5 μg vancomycin disk and susceptible to 1 mg kanamycin, 10 μg colistin, 4 μg metronidazole, and bile disks. Mainly unreactive in most conventional phenotypic tests (nitrate reduction, gas production, catalase production), *A. geminatus* is weakly saccharolytic, being able to ferment galactose and mannose only among the carbohydrates tested. These characteristics are unique for a member of the family *Veillonellaceae*. The metabolic end products are shown in ➋ *Table 35.4*. A positive reaction was also observed for the type strains in the β-glucosidase test (rapid API ID 32 A, bioMérieux). The type strain was isolated from a postoperative fluid collection diagnosed after gastrectomy and esophago-jejunal anastomosis.

**Table 35.4**
**Characteristics differentiating the six genera of the family *Veillonellaceae* (adapted from Marchandin 2007)**

| Characteristic | *Allisonella* | *Anaeroglobus* | *Dialister* | *Megasphaera* | *Negativicoccus* | *Veillonella* |
|---|---|---|---|---|---|---|
| Cell morphology | Ovoid | Cocci | Coccobacilli | Cocci | Cocci | Cocci |
| Cell size (mm) | 1–8 length<br>0.4–0.8 diameter | 0.5–1.1 | 0.2–0.9 × 0.3–2.0 | 0.4–2.6 | 0.4 | 0.3–0.85 |
| Growth atmosphere conditions | Facultative anaerobic | Anaerobic | Anaerobic or anaerobic and microaerophilic | Anaerobic | Anaerobic and microaerophilic | Anaerobic |
| Catalase production | – | – | – | – | – | +/–[a] |
| Nitrate reduction | – | – | – | – | – | + |
| Ability to ferment carbohydrates | – | + (Galactose, mannose) | – | +/– | – | +/– |
| Lactate fermentation | – | – | – | +/– | – | + |
| Succinate decarboxylation | – | – | +/– | – | + | + |
| Amino acids as main source of energy | + (Histidine) | – | – | – | – | – |
| Gas production | + | – | – | +/– | – | +/– |
| Metabolic end products | –[b] | A, P, iB, B, iV | (A), (P), (L), (S) | (A), (P), (iB), B, iV, (V), (C) | A, P, (L) | A, P |
| Major fatty acids | $C_{16:1}$, $C_{18:1}$, $C_{14:0}$ | ND | $C_{16:0}$, $C_{18:1cis9}$ | $C_{12:0}$, $C_{16:0}$, $C_{16:1}$, $C_{18:1}$ $C_{17\ cyclo}$, $C_{19\ cyclo}$, $C_{12:0\ 3OH}$, $C_{14:0\ 3OH}$ [c] | ND | $C_{13:0}$, $C_{17:1\omega8}$ |

+/– Characteristic variable among species (see following table for each multispecies genera) or among strains; parentheses indicate an inconstant production of the metabolic end product
[a]Some species produce a pseudocatalase consisting of an atypical catalase lacking porphyrin
[b]Volatile acids are not produced
[c]Investigated for the species *M. elsdenii* and *M. cerevisiae* only and showing almost identical results

## *Dialister* (ex Bergey et al. 1923) Moore and Moore 1994, 191[VP] emend. Downes et al. 2003, 1939, emend. Jumas-Bilak et al. 2005, 2478 emend. Morotomi et al. 2008, 2716

Di.a.lis′ter. Etymology unknown.

The genus *Dialister* includes Gram-negative coccobacilli (0.2–0.9 μm × 0.3–2.0 μm) that can grow in anaerobic and/or microaerophilic conditions (Wade 2009). Convoluted cell surface is observed after negative staining. Growth in broth media produced no visible turbidity or slightly turbidity at best. Circular, convex, translucent, and tiny colonies (less than 0.5 mm in diameter) are formed on Columbia sheep-blood agar plates. *D. succinatiphilus* was recovered on Gifu anaerobic medium (Nissui Pharmaceutical) (Morotomi et al. 2008). Except for *D. succinatiphilus* for which no data are available, presumptive identification tests showed *Dialister* strains to be resistant to 5 μg vancomycin disk and susceptible to 1 mg kanamycin and bile disks. All strains tested except one *D. micraerophilus* were susceptible to a 4 μg metronidazole disk (Jumas-Bilak et al. 2005). Indole and catalase are not produced. Esculin and urea are not hydrolyzed.

The phenotypic differences between the five currently recognized *Dialister* species are summarized in the ➲ *Table 35.5*.

## *Megasphaera* Rogosa 1971a, 187[AL] emend. Engelmann and Weiss 1985, emend. Marchandin et al. 2003a, 552

Me.ga.phae′ra. Gr. adj. *megâs*, big; L. fem. n. *sphaera*, a sphere; N.L. fem. n. *Megasphaera*, big sphere.

The genus includes strictly anaerobic, Gram-negative cocci with cell size ranging from 0.4 to 2.6 μm in diameter.

Main characteristics of the genus are listed in ➲ *Table 35.4*. Glucose, fructose, and lactate may or may not be fermented. Gas is produced and gluconate is fermented by all species but *M. micronuciformis*. Weak production of acetoin is observed for *M. elsdenii*, *M. cerevisiae*, and *M. sueciensis*, and desulfoviridin is produced by the type strain of *M. micronuciformis*.

The genus includes three beer-spoilage species (*M. cerevisiae*, *M. paucivorans*, and *M. sueciensis*), the ruminal species *M. elsdenii* further isolated from human clinical samples, and

◻ Table 35.5
Characteristics differentiating the five species of the genus Dialister (adapted from Wade 2009)

| Characteristic | D. pneumosintes | D. invisus | D. micraerophilus | D. propionicifaciens | D. succinatiphilus |
|---|---|---|---|---|---|
| Growth in microaerophilic conditions | V[a] | − | + | V | − |
| Colistin disk (10 µg) | R | S | R[b] | R[b] | ND |
| Enhancement of growth by sodium succinate | − | + | − | + | + |
| Metabolic end products | A, P, L, S (trace amounts) | A, P (trace amounts) | Not detected | A, P, L | L, P (trace amounts) |
| Rapid ID 32 A Profile | 0000012401 | 0000000000 | 2000013305 | 0000000000 | 0000400000 |
| Detected activities | Arginine, leucine, glycine and histidine arylamidase activities | None | Arginine, leucine, phenylalanine, tyrosine, alanine and serine arylamidase activities, arginine dihydrolase activity | None | Alkaline phosphatase |
| Major CFA[c] | $C_{12:0}$ | ND | $C_{16:0}$ | ND | ND |
|  | $C_{16:0}$ |  | $C_{18:1 cis9}$ |  |  |
|  | $C_{18:1 cis9}$ |  | $C_{18:0}$ |  |  |

V variable, S susceptible, R resistant, ND not determined, CFA cellular fatty acids
[a]depending on the strain growth may occur either in anaerobic conditions or in both anaerobic and microaerophilic conditions (personal data)
[b]one strain of each species displayed susceptibility to the colistin disk (Jumas-Bilak et al. 2005)
[c]representing more than 10 % of total fatty acids (Jumas-Bilak et al. 2005)

*M. micronuciformis* recovered from human specimens. Several growth media may be used for *Megasphaera* cultivation, colony description is given below for the most commonly used media depending on the species. Growth of *M. elsdenii* in peptone-yeast extract (PY) supplemented with lactate (PYL) yielded round, smooth, slightly raised, with a glistening, mucoid appearance and were 0.2–1 mm in diameter after 48 h of incubation and 3–4 mm in diameter after prolonged incubation. Growth occurs from 25 °C to 40 °C (optimal temperature, 37–40 °C) but not at 45 °C. *M. cerevisiae* colonies on PYL or PY supplemented with fructose (PYF) are whitish, smooth, opaque, flat, shiny, and 0.5–2.0 mm in diameter after 4 days at incubation. The optimum temperature for growth is around 30 °C (28–37 °C). Growth of *M. paucivorans* and *M. sueciensis* is obtained in PYF and PY supplemented with glucose (PYG), MRS, and SMMP media after 3–4 days of incubation at 30 °C, which is the optimal growth temperature (range 15–37 °C). After 7 days, colonies are yellowish or slightly yellowish, glossy, circular, convex, and opaque with entire margins, and diameter is about 1–1.5 mm for *M. paucivorans* and 0.5–0.8 mm for *M. sueciensis*. Growth of *M. micronuciformis* is obtained on blood agar media after 2–3 days of incubation at 37 °C. The colonies are circular, convex, shiny, and translucent with a smooth surface and approximately 0.5–1.0 mm in diameter, nonpigmented, and nonhemolytic.

The phenotypic differences between the five currently recognized *Megasphaera* species are summarized in the ❷ *Table 35.6*.

## *Negativicoccus* Marchandin et al. 2010, 1271[VP]

Ne.ga.ti.vi.coc′cus. L. adj. *negativus*, negative; N.L. masc. n. *coccus* (from Gr. masc. n. *kokkos*), grain or berry; N.L. masc. n. *Negativicoccus*, coccus with a typical Gram-negative cell wall structure with an outer membrane observed by electron microscopy.

*N. succinicivorans* is the only species currently characterized in the genus. It includes Gram-negative coccoid-shaped bacteria of 0.4 mm in diameter. After 48 h of incubation at 37 °C in anaerobic conditions, colonies on blood agar are very tiny (less than 0.5 µm in diameter), circular, convex, and translucent. Susceptibility to a bile disk (1 mg) and to kanamycin (500 µg) and metronidazole (50 µg) disks and resistance to vancomycin (5 µg) and to colistin (10 µg) disks are observed. The species is nonreactive towards conventional biochemical tests (nitrate reduction, nitrite reduction, gas production, catalase, urease, and indole production) and is asaccharolytic. Enzymic profile showed arginine arylamidase activity and inconstant alkaline phosphatase activity. Major metabolic end products are acetic and propionic acids. Trace amounts of 2-hydroxyvaleric acid are

**Table 35.6**
**Characteristics differentiating the five species of the genus *Megasphaera***

| Characteristic | *M. elsdenii* | *M. cerevisiae* | *M. micronuciformis* | *M. paucivorans* | *M. sueciensis* |
|---|---|---|---|---|---|
| Cell size (μm) | 1.6–2.6 | 1.3–2.1 | 0.4–0.6 | 1.2–1.9 × 1.0–1.4 | 1.0–1.4 × 0.8–1.2 |
| Growth at 45 °C | + | – | ND | – | – |
| Ability to ferment | | | | | |
| Arabinose | – | +/– | – | – | – |
| Fructose | + | + | – | – | – |
| Glucose | + | – | – | – | – |
| Maltose | + | – | – | – | – |
| Mannitol | + | – | – | – | – |
| Sucrose | + | – | – | ND | ND |
| Lactate | + | + | – | – | – |
| Gluconate | + | + | – | + | + |
| Vancomycin disk (5 μg) | R | R | S | R | R |
| Colistin disk (10 μg) | S | S | S | R | R |
| Gas production | + | + | – | + | + |
| $H_2S$ production | + | + | – | + | + |
| Volatile fatty acids | A, (P), (iB), B, iV, C, C | A, P, (iB), B, iV, V, C | A, P, (iB), B, iV, (V), PhA | A, (P), iB, B, iV, V, (iC), C | (A), P, iB, B, iV, V, C |

+/– variable, *R* resistant, *S* susceptible, *ND* not determined

also produced. Inconstant production of lactic acid is observed. Growth is enhanced by sodium succinate. Strains supporting the species characterization were from skin and soft tissue human clinical samples.

## *Veillonella* Prévot 1933, 118[AL] emend. Mays et al. 1982, 34

Veil.lo.nel′la. N.L. fem. dim. n. *Veillonella*, named after Adrien Veillon, the French microbiologist who isolated the type species.

*Veillonella* are anaerobic, Gram-negative cocci usually arranged in pairs, masses, or short strains with convoluted cell surface after negative staining. Main characteristics of the genus are as listed in ➋ *Table 35.4*. Cell diameter ranged from 0.3 to 0.85 μm, *V. magna* being the species with the largest cell size (0.65–0.85 μm) compared with other species (cells with 0.3–0.5 μm in diameter). Colonies are observed after 48 h of incubation at 37 °C on blood-containing agar media incubated anaerobically. Colony morphology depends on the species but their common characteristics are to be nonhemolytic, smooth, opaque, and grayish-white with size ranging from 1 to 3 mm in diameter. A pigment responsible for red fluorescence of colonies under ultraviolet light (360 m) may be produced. Catalase activity corresponding to a pseudocatalase lacking porphyrin is observed in about one third of the strains. Resistance to vancomycin disk (5 μg) and susceptibility to bile (1 mg), kanamycin (500 μg), and colistin (10 μg) disks are the most commonly observed profile for *Veillonella* spp. However, a few strains are resistant to the kanamycin disk, and *V. montpellierensis* and *V. ratti* and related strains are resistant to the colistin disk (Marchandin et al. 2005). Gas is produced by most strains except *V. denticariosi* and *V. tobetsuensis*. Lactate is fermented and succinate decarboxylated. Carbohydrates are not fermented except for strains of the *V. ratti–V. criceti* group showing the ability to ferment fructose. Major metabolic end products in TGY broth are acetic and propionic acid. Major cellular fatty acids (CFA) in all species are $C_{13:0}$ and $C_{17:1}{}^{\omega 8}$; other less encountered CFA are variable among species (Carlier 2009a; Mashima et al. 2013). No species-specific pattern is observed except for *V. tobetsuensis* harboring $C_{15:1}{}^{\omega 8c}$, $C_{15:1}{}^{\omega 6c}$, and $C_{15:0}{}^{3OH}$ previously not identified in other *Veillonella* species (Mashima et al. 2013). Reduction of nitrate allowed the differentiation of the genus *Veillonella* from other genera of the family. However, phenotypic tests did not clearly distinguish among species. Therefore, molecular-based species identification is required, and despite some species could be differentiated by 16S rRNA gene sequencing, *dnaK* and *rpoB* genes were demonstrated as more discriminatory between *Veillonella* species, with *rpoB* being the most discriminatory one.

# Isolation, Enrichment, and Maintenance Procedures

Isolation of species recovered from human clinical samples can be done on Columbia, *Brucella*, or Wilkins-Chalgren blood agar media, incubated at 37 °C either anaerobically or microaerobically depending on the species for 2–3 days (see above). Cultivation in brain–heart infusion (BHI) or trypticase–glucose–yeast extract (TGY) broth could be considered for most species. Considering metabolic properties listed above, cultivation in lactate-based broth media can be done for *Veillonella* and *Megasphaera elsdenii*. For *Dialister* spp., poor growth is observed in liquid media but may be enhanced by addition of sodium succinate for some species. Subcultivation should be done weekly for maintenance in these conditions.

*A. histaminiformans* was recovered from MRS medium inoculated with rumen further subcultured onto carbonate agar or broth medium plus histidine (Garner et al. 2002).

Media containing fructose or lactate instead of glucose are required for *M. cerevisiae* growth, while pyruvate or gluconate is required for good growth of *M. paucivorans* and *M. sueciensis* (Juvonen and Suihko 2006; Marchandin et al. 2009). Enrichment methods are part of beer quality control procedures and a selective medium (SMMP) for enrichment of both *Megasphaera* and *Pectinatus* in beer is used (Juvonen and Suihko 2006; Haikara and Juvonen 2009).

Long-term storage can be done either after lyophilization, freezing at −70 °C in broth containing 10 % glycerol or in liquid nitrogen.

## Ecology

Except for the three beer-spoilage *Megasphaera* species (*M. cerevisiae*, *M. paucivorans*, and *M. sueciensis*), the common feature of the members of the family *Veillonellaceae* is to belong to complex ecosystems represented by human and/or other animal microbiotae. Depending on the species, *Veillonellaceae* were primarily found in the rumen (*A. histaminiformans*, *M. elsdenii*) and in buccal, vaginal, and/or intestinal tract of human and animals and may belong to human skin microbiota.

*Allisonella* was the only of the six genus of the family not yet cultured from samples of human origin; however, two sequences are deposited in the GenBank database as *Allisonella* sp. uncultured clones corresponding to clones from human vaginal microbiome (clone 0B040A1_00322) and human forearm superficial skin bacterial biota (clone BL34) (accession numbers JF475534 and DQ130020, respectively), suggesting that *Allisonella* sp. could be a minor or infrequent colonizer of human skin and vaginal microbiotae. Otherwise, *A. histaminiformans* is a ruminal bacterium.

*A. geminatus* primarily isolated from human intestinal tract was also found in human feces and gastrointestinal specimens (Li et al. 2012), in vulval and vaginal samples (Brown et al. 2007), in the skin microbiome (Grice et al. 2009), and in the canine oral microbiome (Dewhirst et al. 2012) by cultivation-independent methods, but the majority of available 16S rRNA gene sequences were from human oral microbiome studies (Paster et al. 2001; Bik et al. 2010; Dewhirst et al. 2010).

*N. succinicivorans* was demonstrated in cultivation-dependent and cultivation-independent studies on the human vaginal epithelium (Hyman et al. 2005) and in cultivation-independent studies in the human skin microbiota (Grice et al. 2009).

*Dialister* species are mainly found in the human oral cavity (*D. pneumosintes*, *D. invisus* and *D. micraerophilus*). While uncultured *D. micraerophilus* clones were at present found in the human oral microbiome only, *D. pneumosintes* and *D. invisus* and Dialister sp. uncultured clones were also found in the canine oral cavity (Elliott et al. 2005; Dewhirst et al. 2010, 2012). Uncultured *Dialister* sp. clones were found in the human gastrointestinal tract (fecal samples, gastric mucosa), in the vaginal microbiota of human and rhesus macaque (Zhou et al. 2007; Spear et al. 2010), and to a lesser extent in the skin microbiome (Grice et al. 2009).

For *D. propionicifaciens* and *D. succinatiphilus*, the two species with less cultured and uncultured representatives, retrieving 16S rRNA gene sequences (>99 % similarity) deposited in the GenBank database, allowed to precise the habitat of the two species. Clones that could be affiliated to the species *D. propionicifaciens* originated from gastrointestinal specimens, vaginal and skin microbiome, and indoor dust, suggesting that the species could not be present in the oral cavity. The type and currently unique strain of *D. succinatiphilus* was isolated from fecal samples of healthy subjects (Morotomi et al. 2008). Uncultured clones that were related to *D. succinatiphilus* supported the presence of the species in the mammal gastrointestinal tract because they originated from fecal samples, swine intestine, pig ordure-based biogas digester, and anaerobic sludge from a methanogenic bioreactor. A single sequence is currently deposited for an uncultured clone from skin origin (clone nbu192h03c1, GenBank accession number GQ018968), and no sequence was available for uncultured *Dialister* sp. oral clones suggesting that this species may have more restricted habitat than other *Dialister* species. In addition, several sequences were deposited for clones found during environmental studies of municipal waste compost, of a riverbank and in bacterial communities in penicillin G production wastewater.

Other sequences deposited for clones of environmental origin related to the genus *Dialister* are rare (hydrogen-producing anaerobic reactor, raw liquid sewage, soil, grease hat within grease trap).

In the genus *Megasphaera*, *M. elsdenii* is a normal inhabitant of the mammal intestines. The species is considered as one of the most important microorganisms in the rumen (Marounek et al. 1989). The species has been isolated from intestinal contents and feces of cattle, sheep, and pigs (Elsden et al. 1956; Gutierrez et al. 1959; Giesecke et al. 1970) and from normal human feces (Sugihara et al. 1974; Werner 1973; Sato et al. 2008). Cultivation-independent studies confirmed the finding of *M. elsdenii* in the intestinal tract and extended the range of mammal species harboring *M. elsdenii* in their gut (Leser et al. 2002; Gill et al. 2006; Ley et al. 2008). For example, *M. elsdenii* clones were identified in the Indian rhinoceros and in the banteng feces (Leser et al. 2002; Gill et al. 2006; Ley et al. 2008). In addition, it has been demonstrated in the normal microflora of the human conjunctiva (Thiel and Schumacher 1994).

From cultivation-independent studies, *M. micronuciformis* primary habitat is believed to be the oral cavity. Indeed, numerous sequences sharing more than 99 % of their nucleotides with the 16S rRNA gene sequence of the species type strain are deposited in the GenBank database for oral clones (Bik et al. 2010; Dewhirst et al. 2010) during studies on the human oral microbiome or on biofilm on health denture weavers. More rare deposits are for clones originating from the human vaginal epithelium (Hyman et al. 2005) and from the human skin microbiome (Grice et al. 2009).

For the three beer-spoilage *Megasphaera* species, a unique sequence deposited in databases matched the 16S rRNA gene sequences of *M. paucivorans* and *M. sueciensis.* The sequence was from the uncultured bacterium clone FL23 (GenBank accession number HM481311) recovered during a study of bacterial and archaeal populations through different phases of remediation at Ft. Lewis, WA, a trichloroethene (TCE)-contaminated groundwater site by using high-density phylogenetic microarray (PhyloChip) (Lee et al. 2012).

The genus *Veillonella* currently groups 12 species belonging to the oral, genitourinary, respiratory, and/or intestinal microbiota of humans and other animals. Depending on the species, isolates were recovered from human samples (*V. denticariosi, V. dispar, V. montpellierensis, V. rogosae,* and *V. tobetsuensis*) and from nonhuman animal samples (*V. caviae, V. criceti, V. magna, V. ratti,* and *V. rodentium*) or were from both human and other animal origin (*V. atypica* and *V. parvula*) (Rogosa 1984; Jumas-Bilak et al. 2004; Byun et al. 2007; Arif et al. 2008a; Kraatz and Taras 2008; Mashima et al. 2013). Despite each species may have a preferential host, it has been suggested that host restriction may not exist in the genus *Veillonella* after the isolation of a human isolate belonging to the *V. ratti–V. criceti* group in mixed aerobic–anaerobic flora from a semen sample (Marchandin et al. 2005).

The majority of species were isolated and also frequently detected during culture-independent studies of the diversity of oral cavity and digestive microbiota in healthy individuals (Paster et al. 2001; Wang et al. 2005; Aas et al. 2005; Quintanilha et al. 2007; Zilberstein et al. 2007; Preza et al. 2009). Relative importance of the different *Veillonella* species has been studied from the tongue of healthy adults targeting the *rpoB* gene (Beighton et al. 2008; Mashima et al. 2011). Whether *Veillonella* were cultivated from the dorsum surface of the tongue and identified by using *rpoB* sequencing (Beighton et al. 2008) or were detected directly from tongue biofilm using species-specific primers designed from a highly variable region in the *rpoB* gene (Mashima et al. 2011), the predominant species were *V. atypica, V. dispar,* and *V. rogosae. V. parvula* was isolated from only one subject and *V. denticariosi* and *V. montpellierensis* were not isolated (Beighton et al. 2008). *V. montpellierensis* may have a different and restricted habitat represented by the genital tract. Indeed, the species was to date only isolated from a gastric fluid sample in a neonate and from amniotic fluid samples (Jumas-Bilak et al. 2004) and found in the human vagina during cultivation-independent studies (Zhou et al. 2007).

Species of the genus *Veillonella* are considered as early colonizers of the human mouth and gut microbiota colonizing the human mouth within the three first months of life and appearing in gastric aspirates and neonates' feces on the first days of life (Jones et al. 2010). After first finding, *Veillonella* spp. persistently colonize the oral cavity and the intestine (Sato et al. 1993; Könönen et al. 1999; Favier et al. 2002; Jacquot et al. 2011).

Despite the count of *Veillonella* spp. in oral and gut microbiota is variable according to diet, oral hygiene, pathology, or antimicrobial treatment (Al-Ahmad et al. 2010; Albert et al. 1978; Malinen et al. 2005; Casarin et al. 2012), *Veillonella* are usually highly represented members of these microbiotae. For example, they were predominant in the upper digestive tract (Zilberstein et al. 2007) and represented up to 36.3 % of the total anaerobic colony count from healthy adult tongue (Beighton et al. 2008) and more than 50 % of the Gram-negative bacteria found during cultivation-independent study of the digestive microbiota of very premature infants (Jacquot et al. 2011).

*Veillonella* is believed to be a critical genus that guides the development of multispecies communities (Kolenbrander 2011; Palmer et al. 2006), and central role of the early colonizer *Veillonella* sp. in forming multispecies biofilm communities with initial, middle, and late colonizers of enamel has been established (Periasamy and Kolenbrander 2010). In oral flora, *Veillonella* spp. establish a nutrition chain with other bacteria (Egland et al. 2004) and were statistically associated with periodontal health (Kumar et al. 2006). *Veillonella* obtain energy from lactic acid produced by the streptococci as an end product of carbohydrate fermentation (Diaz et al. 2006). Similarly, *Veillonella* spp. may play a role in gut microbiota development at an early age (Harmsen et al. 2000; Jacquot et al. 2011).

## Pathogenicity, Clinical Relevance

*A. geminatus* is uncommonly isolated from human clinical samples. The type species was isolated as part of a polymicrobial culture from a postoperative fluid collection probably due to anastomosis suture slacking in a 70-year-old woman (Carlier et al. 2002). Uncultured *Anaeroglobus* sp. clone CAP32 (HQ914729) was found in sputum from patients with community-acquired pneumonia. Due to the few strains and sequences currently available for this species, pathogenicity and antimicrobial susceptibility pattern remain largely unknown.

The particular properties of the ruminal bacterium *A. histaminiformans,* i.e., histamine production and histidine utilization, could favor the development of bovine and equine laminitis (Garner et al. 2002). No data on antimicrobial susceptibility are available.

*Dialister* species were cultivated from a large variety of human clinical samples (Morio et al. 2007; Morotomi et al. 2008) from diverse origin, i.e., cutaneous and soft tissues, intra-abdominal collections, respiratory and gynecological tracts, blood, bone, semen, and gastric fluid. The species *D. pneumosintes* was the main recovered species before

*D. micraerophilus*, while *D. propionicifaciens* and *D. invisus* represented minor species in these specimens (Morio et al. 2007).

Except for *D. succinatiphilus* for which uncultured clones were found in intestinal mucosal biopsy from ulcerative colitis patient and in feces samples in patients with irritable bowel disease but remained of unknown signification, the involvement of *Dialister* spp. in human infections, mainly in polymicrobial infections, is now clearly established. Both *D. pneumosintes* and *D. invisus* have been mainly implicated in oral diseases such as advanced caries (Chhour et al. 2005), periodontitis, acute necrotizing ulcerative gingivitis, endodontic infections, and peri-implant diseased pocket by cultivation or cultivation-independent studies. *D. pneumosintes* was detected in high prevalence and/or counts in necrotizing periodontal lesions in both HIV-negative and HIV-positive patients (Brito et al. 2012; Ramos et al. 2012). Positive associations of *D. pneumosintes* with other pathogens such as *Treponema denticola* were demonstrated (Siqueira and Rôças 2003). *D. invisus* was the most frequent or one of the most frequent taxa in asymptomatic and symptomatic apical periodontitis and in chronic apical abscesses. The species was also thought to interact with other bacterial and/or viral species in pathological community (Rôças et al. 2011).

In respiratory tract infections, *D. pneumosintes*, *D. invisus*, or uncultivated clones affiliated to the genus *Dialister* were detected in samples from patients with community-acquired or ventilator-associated pneumonia (Bahrani-Mougeot et al. 2007; Bousbia et al. 2012), in airway specimens from children with cystic fibrosis (Harris et al. 2007), and in explanted cystic fibrosis lungs (Rudkjøbing et al. 2011).

In urogenital tract infections, *Dialister* species and new phylotypes of *Dialister* sp. were found in patients with bacterial vaginosis while absent from samples taken in control subjects and after successful treatment of patients with bacterial vaginosis (Fredricks et al. 2005). *Dialister* sp. was also recovered from amniotic fluid and placenta samples (Evaldson et al. 1980) and from Fallopian tube of a confirmed acute salpingitis patient (Hebb et al. 2004). *D. invisus* has also been identified in urinary tract specimens from renal transplant recipients (Domann et al. 2003). Systemic disease may originate from the urogenital tract as illustrated by a case of bacteremia of vaginal origin reported by Lepargneur et al. (2006).

More rarely, *D. pneumosintes* has been isolated from bite wound infections (Goldstein et al. 1984) and in other severe diseases like human brain abscess (Rousée et al. 2002).

*D. micraerophilus* strains have been characterized from various human clinical specimens, including bone and blood cultures, whereas the four *D. propionicifaciens* isolates currently reported were from cutaneous infections and semen (Morio et al. 2007).

Antimicrobial susceptibility of 55 human clinical isolates of *Dialister* spp. was described against 14 antimicrobial agents. Global susceptibility was reported in this study; however, depending on the breakpoints used, some strains could be considered as either susceptible or with decreased susceptibility to piperacillin, metronidazole, macrolides, fluoroquinolones, and rifampin. The clinical impact of these decreased susceptibilities remains unknown but antimicrobial susceptibility testing has been recommended for clinically important *Dialister* sp. isolates (Morio et al. 2007).

In the genus *Megasphaera*, two species were isolated from human clinical samples, *M. elsdenii* and *M. micronuciformis*. However, their isolation remains a rare finding. In addition, *M. elsdenii* was recovered in foot rot lesions in goats (Duran et al. 1990).

Pathogenicity of *M. micronuciformis* remains unknown, only few representatives of the species being currently reported. Isolates were recovered from a liver abscess, pus samples (Marchandin et al. 2003a), the airways of patients with cystic fibrosis (CF) (Sibley et al. 2011), and sinus and throat samples (personal unpublished data). Cultivation-independent studies revealed clones with sequences matching that of the type strain of *M. micronuciformis* in sputum samples from adult patients with CF (van der Gast et al. 2011), in biofilm of extubated endotracheal tube of ICU patient (Perkins et al. 2010), in skin microbiome of children with atopic dermatitis (Kong et al. 2012), in gastrointestinal specimens of which intestinal biopsies from patients with diverse inflammatory bowel diseases including Crohn's disease and ulcerative colitis (Gophna et al. 2006; Li et al. 2012), and from the human mouth in microflora associated with dental caries (Munson et al. 2004).

Haralambie isolated *M. elsdenii* from fecal samples of 12 % of adults and 7.4 % of children suffering from gastrointestinal disorders in the range of greater than or equal to $10^7$ UFC/g feces. Since, *M. elsdenii* has not been isolated from fecal samples of healthy people, it was believed that *M. elsdenii* in high load does not belong to the resident anaerobe microbiota of the human intestine (Haralambie 1983). *M. elsdenii* was also cultured from gastric and amniotic fluid samples (Marchandin and Jumas-Bilak 2006; personal unpublished data), from rectal drainage, and from a frontal lobe tumor (Brancaccio and Legendre 1979), but the corresponding clinical cases were not documented. Isolation of *M. elsdenii* from a putrid lung abscess as part of mixed flora first suggested a possible role of this species in human pathological processes (Sugihara et al. 1974). A few years later, primary human pathogenicity was further confirmed by the report of a case of human endocarditis due to *M. elsdenii* in a 49-year-old apparently immunocompetent patient (Brancaccio and Legendre 1979).

Finally, in urogenital tract infections, new phylotypes of *Megasphaera* sp. were found in patients with bacterial vaginosis while absent from samples taken in control subjects and in patients successfully treated suggesting that *Megasphaera* sp. could play a role in this dysbiotic polymicrobial disease (Fredricks et al. 2005).

Antimicrobial susceptibility data are available for *M. elsdenii* strains of animal origin only (Marounek et al. 1989; Piriz et al. 1992). Piriz et al. tested the susceptibility of 20 isolates of *M. elsdenii* from ovine foot rot against 28 antimicrobial agents that showed a great variability of susceptibility according to the strain owing to the wide range of minimal inhibitory concentrations (MICs) observed for most of the agents tested

(Piriz et al. 1992). Full susceptibility was noted against azlocillin, piperacillin, mezlocillin, cefoperazone, and cefotaxime. Variable rates of strains showing resistance or decreased susceptibility were observed against the other drugs ranging from 5 % for tetracycline, cefoxitin, and ampicillin to 95 % for trimethoprim. Of note, 80 % of the strains were susceptible to metronidazole (Piriz et al. 1992). Examining four *M. elsdenii* ruminal strains, Marounek et al. showed that they were rather insensitive to many antimicrobial compounds, especially to ionophores and other antimicrobial feed additives (Marounek et al. 1989). However, the four strains were susceptible to tetracycline. By contrast Stanton et al. found that *M. elsdenii* strains are among the most numerous tetracycline-resistant populations in swine intestinal tracts (Stanton and Humphrey 2003). *M. elsdenii* strains resistant to tetracycline were detected at high population levels in cecal samples from healthy swine (Stanton and Humphrey 2003). Characterizing the genetic support for this resistance revealed known *tet* genes as well as novel mosaic gene combinations of tet(O) and tet(W). To date, seven tetracycline resistance genotypes of *M. elsdenii* have been detected (Stanton and Humphrey 2003; Stanton et al. 2004). Strains carrying the recombinant mosaic gene consistently exhibited the highest tetracycline MICs suggesting that these strains may have a selective advantage in the swine gut. Moreover, Stanton et al. suggested that these resistance-encoding genes may be communicable among intestinal anaerobes. The role for mutualist bacteria is highlighted not only in the preservation and dissemination of antibiotic resistance in the intestinal tract but also in the evolution of resistance genes (Stanton and Humphrey 2003).

*N. succinicivorans* was isolated from skin and soft tissue human clinical samples in polymicrobial cultures (Marchandin et al. 2010). *N. succinicivorans* uncultured clones were also found in skin microbiome of children with atopic dermatitis (Kong et al. 2012) and in gastrointestinal specimens from patients with inflammatory bowel diseases (Li et al. 2012). In all these cases, clinical relevance of the presence of *N. succinicivorans* could not be established. However, a recent case of bacteremia due to *N. succinicivorans* has been reported in a 57-year-old woman with hemochromatosis and pancreatitis, N. *succinicivorans* being isolated from two anaerobic blood culture vials (Church et al. 2011). This case first documented human pathogenicity of the species. The only available data on antimicrobial susceptibility pattern are for this invasive isolate, which was found susceptible to penicillin but resistant to clindamycin and metronidazole (Church et al. 2011).

Among the *Veillonellaceae* family, *Veillonella* spp. are the most frequently isolated from human clinical samples and were described in a large variety of human infections. Members of the genus *Veillonella* have been more frequently isolated from human clinical specimens in aerobic–anaerobic polymicrobial cultures and were involved in periodontal disease, head and neck and respiratory tract infections, and in skin and soft tissue infections from animal and human bites (Brook 2006; Goldstein et al. 1984). They were also recovered from lower respiratory airways in lung cancer and CF patients but their pathogenic role remained unknown (Rybojad et al. 2011; Tunney et al. 2008). However, their pathogenic potential in pure culture has been proved in a mouse model developed to study the abscessogenic potential of pure and mixed cultures of oral anaerobes associated with infections of endodontic origin (Baumgartner et al. 1992). Several case reports also documented the isolation of *Veillonella* sp. in pure culture during severe opportunistic human infections including bacteremia, endocarditis, bone and joint infections like osteomyelitis and prosthetic joint infections, myositis, and meningitis (Barnhart et al. 1983; Beumont et al. 1995; Bhatti and Frank 2000; Boo et al. 2005; Brook 1996; Brook and Frazier 1992; Fisher and Denison 1996; Houston et al. 1997; Isner-Horobeti et al. 2006; Liu et al. 1998; Marchandin et al. 2001b; Marriott et al. 2007; Pouchot et al. 1992; Rovery et al. 2005; Singh and Yu 1992; Strach et al. 2006; Zaninetti-Schaerer et al. 2004). It is generally believed that the infections were the result of hematogenous spreading from oral, respiratory, genitourinary, or digestive source. Among them, bone and joint infections were the more frequently reported, case reports for other listed infection types remaining relatively rare. As an example, eight endocarditis cases are currently reported in the literature to our knowledge (Rovery et al. 2005; Oh et al. 2005). Because of the lack of conventional phenotypic and biochemical tests that allow species discrimination and to the low level of 16S rRNA gene sequence variation among several *Veillonella* species, species identification should not be considered as reliable in many published cases and whether some *Veillonella* species might be specifically associated with particular diseases remains unelucidated. Among the five *V. montpellierensis* strains currently described in the literature, one has been responsible for a case of endocarditis (Rovery et al. 2005). Most recent studies using molecular methods based on the *rpoB* gene showed that *V. denticariosi* was only cultured from caries lesions in children (Arif et al. 2008b). More generally, these approaches revealed that *Veillonella* spp. were more diverse from caries-free sites than those from caries lesions in children (Arif et al. 2008b).

Cultivation-independent studies also revealed that some *Veillonella* phylotypes could also be part of dysbiotic microbiota associated to polymicrobial diseases, as supported by statistically significant difference in recovery of some of these phylotypes in severe caries and caries-free children (Kanasi et al. 2010) and in patients with refractory periodontitis (Colombo et al. 2009).

Pathogenicity of these infections remained largely unknown, notably because of a lack of knowledge on *Veillonella* virulence factors. The majority of studies were conducted on the *Veillonella* lipopolysaccharide (LPS) and revealed that it is highly toxic and that endotoxinic activity may be as high as that observed for *Fusobacterium* spp. and enterobacteria (Botta et al. 1994; Delwiche et al. 1985; Hofstad and Kristoffersen 1970; Matera et al. 1991). In fetal rat bone model, lipopolysaccharide from *Veillonella* stimulated the osteoclasts to bone resorption (Sveen and Skaug 1980). More recent investigations on molecular mechanisms responsible for innate immune response against *Veillonella* revealed that *Veillonella* LPS (i) is able to induce cytokine (tumor necrosis factor alpha and interleukin (IL)-1 beta, IL-6, and IL-10) production in vitro but in a 10- to 100-fold less effective manner than does

*Enterobacteriaceae* LPS, (ii) is able to activate PBMC p38 mitogen-activated protein kinase, and (iii) has Toll-like receptor 4 dependant action (Matera et al. 2009). On another hand, it has been hypothesized that during pathogenesis of spinal osteomyelitis, poor tissue perfusion may be the primary cause of lactic acid production permitting the growth of *Veillonella* (Bongaerts et al. 2004). The ability of veillonellae to form polymicrobial biofilm, for example, demonstrated on the surface of biliary stents (Scheithauer et al. 2009), represents a potentially important attachment and survival strategy for further pathogenicity.

*Veillonella* were for a long time considered as largely susceptible to antimicrobial agents and penicillin G was considered as the treatment of choice against infections due to these bacteria (Barnhart et al. 1983). Since then, several studies revealed acquired resistance to penicillin in *Veillonella*. Two mechanisms were shown to account for this resistance: β-lactamase production noted in up to 12.5 % of the isolates (Alou et al. 2009; Valdés et al. 1982) or penicillin-binding proteins with low β-lactam affinity (Reig et al. 1997). Resistance to other β-lactams included resistance to ampicillin, amoxicillin (Nyfors et al. 2003; Ready et al. 2004), amoxicillin plus clavulanic acid (Reig et al. 1997), piperacillin, piperacillin plus tazobactam (Theron et al. 2003; Tunney et al. 2008), ticarcillin plus clavulanic acid (Singer et al. 2008), and cefoxitin (Reig et al. 1997). Most of these resistances appeared to be increasing. A recent study showed that about 64 % and 40 % of the 158 oral *Veillonella* spp. strains isolated from healthy children dental plaque samples displayed resistance to penicillin and ampicillin, respectively (Ready et al. 2012). Depending on the study, penicillin-resistant isolates accounted for up to 85 % of the *Veillonella* strains (Baquero and Reig 1992; Nyfors et al. 2003; Ready et al. 2004; Reig et al. 1997; Roberts et al. 2006) and all isolates may be resistant to amoxicillin (Ready et al. 2004). These studies were all conducted in children and resistant isolates were demonstrated in children as young as 2 years (Nyfors et al. 2003). Resistance towards other antimicrobial agents was detected in *Veillonella* spp. including resistance to erythromycin, trimethoprim/sulfamethoxazole, metronidazole, and tetracycline (Baquero and Reig 1992; Finegold et al. 2004; Ready et al. 2004). In the study by Lancaster et al. 10 % of the *Veillonella* were resistant to tetracycline representing the second most commonly identified tetracycline-resistant bacteria oral bacteria in healthy 4- and 6-year-old children who had not received antibiotics during the 3 months prior to sampling (Lancaster et al. 2003). Ready et al. showed that 12.5 % of *Veillonella* spp. isolated from the dental plaque of 52 healthy subjects who had not received antibiotics in the previous 3 months harbored tetracycline resistance (*tet*) genes. Resistance determinants were diverse because five *tet* genes were detected, the most commonly found being *tet*(M) and four strains (4.2 %) harboring more than one *tet* gene (Ready et al. 2006). The Tn916 transposon carrying the *tet*(M) gene was successfully transferred from a *tet*(M)-positive *Veillonella* strain to four *Streptococcus* spp. by conjugation (Ready et al. 2006), and transfer of the conjugative transposon-encoded tetracycline resistance was further shown to occur by transformation in biofilms (Ready et al. 2006; Hannan et al. 2010). Altogether, these observations suggested the potential role of *Veillonella* spp. as a reservoir of transferable tetracycline resistance in the multispecies oral biofilm. This role of reservoir for antimicrobial resistance genes may also be illustrated by the recovery of a metronidazole-susceptible *Veillonella* clinical isolate harboring the nitroimidazole resistance gene *nimE* (Marchandin et al. 2004). *Veillonella* may also modify the survival of other bacterial species to various antimicrobial agents in biofilms independently of resistance determinant transfer. Luppens et al. showed that *Streptococcus mutans* showed an increase in survival after exposure to various antimicrobials in the presence of *Veillonella* and suggested that growing in a biofilm together Veillonella changes the physiology of *S. mutans* and gives this bacterium an advantage in surviving antimicrobial treatment (Luppens et al. 2008). This study underlines the limit of reporting individual data on susceptibility or resistance to antimicrobial agents for *Veillonella* spp. owing to the unique physiology of members of this genus and their frequent recovery in polymicrobial diseases, such as caries and periodontitis.

## Application

### Probiotic Effect in Animal Husbandry

Members of the family *Veillonellaceae* are of particular interest for probiotic activities of *M. elsdenii* and of a mixed polymicrobial population including a *Veillonella* sp. strain from fowl origin. *M. elsdenii* is thought to play a major role in preventing or reducing acidosis due to lactic acid accumulation in cattle introduced to a high grain diet. Indeed, based on its capacity to ferment 74–97 % of ruminal lactate, *M. elsdenii* is considered as a probiotic microorganism providing benefits for energy balance and animal productivity (Counotte et al. 1981; Stewart and Bryant 1988; Ouwerkerk et al. 2002). More recently, another application has been suggested for this species as it may delay the colonization of swine by antibiotic-resistant strains (Stanton and Humphrey 2011). A *Veillonella* isolated from the cecal contents of adult chickens showed inhibitory activity on the growth of *Salmonella enterica* subsp. *enterica* serotypes Typhimurium and Enteritidis, *Escherichia coli* 0157:H7, and *Pseudomonas aeruginosa* when cocultured with *Bacteroides fragilis* (Hinton and Hume 1995). This isolate is part of a defined probiotic or composition of anaerobic bacteria effective for controlling or inhibiting *Salmonella* colonization of fowl. The probiotic includes populations or cultures of 29 substantially biologically pure bacteria patented by Nisbet et al. in 1997 (United States Patent 5,604,127).

### Food and Beverage Industry

*Megasphaera* has emerged in breweries along with *Pectinatus* and is responsible for 3–7 % of bacterial beer spoilage, which makes this bacterium one of the most feared organisms for brewers. It is the most anaerobic species known to exist in the brewing environment and its role in beer spoilage has increased because the improved technology in modern breweries has resulted in significant reduction of oxygen content in the final products (Sakamoto and Konings 2003). *M. cerevisiae* grows with an optimum at 28 °C at pH values above 4 and its growth is still possible at ethanol concentrations up to 5.5 (w/v) (Sakamoto and Konings 2003). Beer spoilage caused by this organism results in extreme turbidity, the production of considerable amount of butyric acid and hydrogen sulfide that causes a fecal odor in beer.

## Biotechnological Interest and Bioremediation

*M. elsdenii* is of biotechnological interest because of its production of various volatile fatty acids and of its enzymatic content. It may be of interest as a possible biocatalyst in chemical industry, particularly for its ability to metabolize acrylate (Prabhu et al. 2012). Propionyl-CoA transferase from *M. elsdenii* is considered for the production of a variety of biopolymers (Matsumoto et al. 2011) and its inositol polyphosphatases are of biotechnological interest for their ability to reduce the metabolically unavailable organic phosphate content of feedstuffs and to produce lower inositol polyphosphates for research and pharmaceutical applications (Puhl et al. 2009). Considering bioremediation, *M. elsdenii* was able to degrade trinitrotoluene as observed for several ruminal bacteria (De Lorme and Craig 2009).

## Acknowledgments

Two collaborators have been particularly precious in our work on the family *Veillonellaceae*, particularly in the characterization of 2 of the 6 genera and of 6 of the 25 species of the family. So, we would like to sincerely thank Bernard Gay, recently retired, for performing electron microscopy studies on all the novel taxa we characterized within this family. We would also like to posthumously dedicate this chapter to Jean-Philippe Carlier working at the National Reference Center of the Pasteur Institute in Paris for his constant help and long, faithful, and fruitful collaboration with our team. And of course, no adventure in the unknown world of *Veillonellaceae* would have been possible without first isolation and recognition of these unknown bacteria and the authors are very grateful to Dr Hélène Jean-Pierre and to the technical teams from the anaerobe Laboratory of the Montpellier Teaching Hospital and from the bacteriology laboratory of the Faculty of Pharmacy of Montpellier.

## References

Aas JA, Paster BJ, Stokes LN, Olsen I, Dewhirst FE (2005) Defining the normal bacterial flora of the oral cavity. J Clin Microbiol 43:5721–5732

Al-Ahmad A, Roth D, Wolkewitz M, Wiedmann-Al-Ahmad M, Follo M, Ratka-Krüger P, Deimling D, Hellwig E, Hannig C (2010) Change in diet and oral hygiene over an 8-week period: effects on oral health and oral biofilm. Clin Oral Investig 14:391–396

Albert MJ, Bhat P, Rajan D, Maiya PP, Pereira SM, Baker SJ (1978) Faecal flora of South Indian infants and young children in health and with acute gastroenteritis. J Med Microbiol 11:137–143

Alou L, Giménez MJ, Manso F, Sevillano D, Torrico M, González N, Granizo JJ, Bascones A, Prieto J, Maestre JR, Aguilar L (2009) Tinidazole inhibitory and cidal activity against anaerobic periodontal pathogens. Int J Antimicrob Agents 33:449–452

Arai T, Kusakabe A, Komatsu S, Kitasato S (1984) A survey of plasmids in *Veillonella* strains isolated from human oral cavity. Arch Exp Med 57:233–237

Arif N, Do T, Byun R, Sheehy E, Clark D, Gilbert SC, Beighton D (2008a) *Veillonella rogosae* sp. nov., an anaerobic, Gram-negative coccus isolated from dental plaque. Int J Syst Evol Microbiol 58:581–584

Arif N, Sheehy EC, Do T, Beighton D (2008b) Diversity of *Veillonella* spp. from sound and carious sites in children. J Dent Res 87:278–282

Bahrani-Mougeot FK, Paster BJ, Coleman S, Barbuto S, Brennan MT, Noll J, Kennedy T, Fox PC, Lockhart PB (2007) Molecular analysis of oral and respiratory bacterial species associated with ventilator-associated pneumonia. J Clin Microbiol 45:1588–1593

Baquero F, Reig M (1992) Resistance of anaerobic bacteria to antimicrobial agents in Spain. Eur J Clin Microbiol Infect Dis 11:1016–1020

Barnhart RA, Weitekamp MR, Aber RC (1983) Osteomyelitis caused by *Veillonella*. Am J Med 74:902–904

Baumgartner JC, Falkler WA Jr, Beckerman T (1992) Experimentally induced infection by oral anaerobic microorganisms in a mouse model. Oral Microbiol Immunol 7:253–256

Beighton D, Clark D, Hanakuka B, Gilbert S, Do T (2008) The predominant cultivable *Veillonella* spp. of the tongue of healthy adults identified using rpoB sequencing. Oral Microbiol Immunol 23:344–347

Beumont MG, Duncan J, Mitchell SD, Esterhai JL Jr, Edelstein PH (1995) *Veillonella* myositis in an immunocompromised patient. Clin Infect Dis 21:678–679

Bhatti MA, Frank MO (2000) *Veillonella parvula* meningitis: case report and review of *Veillonella* infections. Clin Infect Dis 31:839–840

Bik EM, Long CD, Armitage GC, Loomer P, Emerson J, Mongodin EF, Nelson KE, Gill SR, Fraser-Liggett CM, Relman DA (2010) Bacterial diversity in the oral cavity of 10 healthy individuals. ISME J 4:962–974

Bongaerts GP, Schreurs BW, Lunel FV, Lemmens JA, Pruszczynski M, Merkx MA (2004) Was isolation of *Veillonella* from spinal osteomyelitis possible due to poor tissue perfusion? Med Hypotheses 63:659–661

Boo TW, Cryan B, O'Donnell A, Fahy G (2005) Prosthetic valve endocarditis caused by *Veillonella parvula*. J Infect 50:81–83

Botta GA, Arzese A, Minisini R, Trani G (1994) Role of structural and extracellular virulence factors in Gram-negative anaerobic bacteria. Clin Infect Dis 18(Suppl 4):S260–S264

Bousbia S, Papazian L, Saux P, Forel JM, Auffray JP, Martin C, Raoult D, La Scola B (2012) Repertoire of intensive care unit pneumonia microbiota. PLoS One 7:e32486

Brancaccio M, Legendre GG (1979) *Megasphaera elsdenii* endocarditis. J Clin Microbiol 10:72–74

Brito L, Sobrinho AR, Teles R, Socransky S, Haffajee A, Vieira L, Teles F (2012) Microbiologic profile of endodontic infections from HIV− and HIV+ patients using multiple-displacement amplification and checkerboard DNA-DNA hybridization. Oral Dis 18:558–567

Brook I (1996) *Veillonella* infections in children. J Clin Microbiol 34:1283–1285

Brook I (2006) Microbiology of intracranial abscesses associated with sinusitis of odontogenic origin. Ann Otol Rhinol Laryngol 115:917–920

Brook I, Frazier EH (1992) Infections caused by *Veillonella* species. Infect Dis Clin Pract (Baltim Md) 1:377–381

Brown CJ, Wong M, Davis CC, Kanti A, Zhou X, Forney LJ (2007) Preliminary characterization of the normal microbiota of the human vulva using cultivation-independent methods. J Med Microbiol 56:271–276

Byun R, Carlier J-P, Jacques NA, Marchandin H, Hunter N (2007) *Veillonella denticariosi* sp. nov., isolated from human carious dentine. Int J Syst Evol Microbiol 57:2844–2848

Carlier J-P (2009a) Genus I. *Veillonella*. In: De Vos P, Garrity G, Jones D, Krieg NR, Ludwig W, Rainey FA, Schleifer K-H, Whitman WB (eds) Bergey's manual of systematic bacteriology, vol 3, 2nd edn, The *Firmicutes*. Springer, New York, pp 1039–1045

Carlier J-P (2009b) Genus VI. *Anaeroglobus*. In: De Vos P, Garrity G, Jones D, Krieg NR, Ludwig W, Rainey FA, Schleifer K-H, Whitman WB (eds) Bergey's manual of systematic bacteriology, vol 3, 2nd edn, The *Firmicutes*. Springe, New York, pp 1051–1052

Carlier J, Marchandin H, Jumas-Bilak E, Lorin V, Henry C, Carriere C, Jean-Pierre H (2002) *Anaeroglobus geminatus* gen. nov., sp. nov., a novel member of the family *Veillonellaceae*. Int J Syst Evol Microbiol 52:983–986

Casarin RC, Barbagallo A, Meulman T, Santos VR, Sallum EA, Nociti FH, Duarte PM, Casati MZ, Gonçalves RB (2012) Subgingival biodiversity in subjects with uncontrolled type-2 diabetes and chronic periodontitis. J Periodontal Res. doi:10.1111/j.1600-0765.2012.01498.x

Chhour KL, Nadkarni MA, Byun R, Martin FE, Jacques NA, Hunter N (2005) Molecular analysis of microbial diversity in advanced caries. J Clin Microbiol 43:843–849

Church DL, Simmon KE, Sporina J, Lloyd T, Gregson DB (2011) Identification by 16S rRNA gene sequencing of *Negativicoccus succinicivorans* recovered from the blood of a patient with hemochromatosis and pancreatitis. J Clin Microbiol 49:3082–3084

Colombo AP, Boches SK, Cotton SL, Goodson JM, Kent R, Haffajee AD, Socransky SS, Hasturk H, Van Dyke TE, Dewhirst F, Paster BJ (2009) Comparisons of subgingival microbial profiles of refractory periodontitis, severe periodontitis, and periodontal health using the human oral microbe identification microarray. J Periodontol 80:1421–1432

Counotte GHM, Prins RA, Janssen RHAM, DeBie MJA (1981) Role of *Megasphaera elsdenii* in the fermentation of DL-[2-13C]lactate in the rumen of dairy cattle. Appl Environ Microbiol 42:649–655

De Lorme M, Craig M (2009) Biotransformation of 2,4,6-trinitrotoluene by pure culture ruminal bacteria. Curr Microbiol 58:81–86

Delwiche EA, Pestka JJ, Tortorello ML (1985) The *Veillonellae*: gram-negative cocci with a unique physiology. Annu Rev Microbiol 39:175–193

Dewhirst FE, Chen T, Izard J, Paster BJ, Tanner AC, Yu WH, Lakshmanan A, Wade WA (2010) The human oral microbiome. J Bacteriol 192:5002–5017

Dewhirst FE, Klein EA, Thompson EC, Blanton JM, Chen T, Milella L, Buckley CM, Davis IJ, Bennett ML, Marshall-Jones ZV (2012) The canine oral microbiome. PLoS One 7:e36067

Diaz PI, Chalmers NI, Rickard AH, Kong C, Milburn CL, Palmer RJ Jr, Kolenbrander PE (2006) Molecular characterization of subject-specific oral microflora during initial colonization of enamel. Appl Environ Microbiol 72:2837–2848

Domann E, Hong G, Imirzalioglu C, Turschner S, Kühle J, Watzel C, Hain T, Hossain H, Chakraborty T (2003) Culture-independent identification of pathogenic bacteria and polymicrobial infections in the genitourinary tract of renal transplant recipients. J Clin Microbiol 41:5500–5510

Downes J, Munson M, Wade WG (2003) *Dialister invisus* sp. nov., isolated from the human oral cavity. Int J Syst Evol Microbiol 53:1937–1940

Duran SP, Manzano JV, Valera RC, Machota SV (1990) Obligately anaerobic bacterial species isolated from foot-rot lesions in goats. Br Vet J 146:551–558

Egland PG, Palmer RJ Jr, Kolenbrander PE (2004) Interspecies communication in *Streptococcus gordonii-Veillonella atypica* biofilms: signaling in flow conditions requires juxtaposition. Proc Natl Acad Sci USA 101:16917–16922

Elliott DR, Wilson M, Buckley CM, Spratt DA (2005) Cultivable oral microbiota of domestic dogs. J Clin Microbiol 43:5470–5476

Elsden SR, Volcani BE, Gilchrist FMC, Lewis D (1956) Properties of a fatty acid forming organism isolated from the rumen of sheep. J Bacteriol 72:681–689

Engelmann U, Weiss N (1985) *Megasphaera cerevisiae* sp. nov.: a new gram-negative obligately anaerobic coccus isolated from spoiled beer. Syst Appl Microbiol 6:287–290

Evaldson G, Carlström G, Lagrelius A, Malmborg AS, Nord CE (1980) Microbiological findings in pregnant women with premature rupture of the membranes. Med Microbiol Immunol 168:283–297

Favier CF, Vaughan EE, de Vos WM, Akkermans DL (2002) Molecular monitoring of succession of bacterial communities in human neonates. Appl Environ Microbiol 68:219–226

Fedorko DP, Drake SK, Stock F, Murray PR (2012) Identification of clinical isolates of anaerobic bacteria using matrix-assisted laser desorption ionization-time of flight mass spectrometry. Eur J Clin Microbiol Infect Dis 31:2257–2262

Finegold SM, John SS, Vu AW, Li CM, Molitoris D, Song Y, Liu C, Wexler HM (2004) In vitro activity of ramoplanin and comparator drugs against anaerobic intestinal bacteria from the perspectiveof potential utility in pathology involving bowel flora. Anaerobe 10:205–211

Fisher RG, Denison MR (1996) *Veillonella parvula* bacteremia without an underlying source. J Clin Microbiol 34:3235–3236

Fredricks DN, Fiedler TL, Marrazzo JM (2005) Molecular identification of bacteria associated with bacterial vaginosis. N Engl J Med 353:1899–1911

Garner MR, Flint JF, Russell JB (2002) *Allisonella histaminiformans* gen. nov., sp. nov. a novel bacterium that produces histamine, utilizes histidine as its sole energy source, and could play a role in bovine and equine laminitis. Syst Appl Microbiol 25:498–506

Giesecke D, Wiesmayr S, Ledinek M (1970) *Peptostreptococcus elsdenii* from the caecum of pigs. J Gen Microbiol 64:123–126

Gill SR, Pop M, Deboy RT, Eckburg PB, Turnbaugh PJ, Samuel BS, Gordon JI, Relman DA, Fraser-Liggett CM, Nelson KE (2006) Metagenomic analysis of the human distal gut microbiome. Science 312:1355–1359

Goldstein EJ, Citron DM, Finegold SM (1984) Role of anaerobic bacteria in bite-wound infections. Rev Infect Dis 6(suppl 1):S177–S183

Gophna U, Sommerfeld K, Gophna S, Doolittle WF, Veldhuyzen van Zanten SJ (2006) Differences between tissue-associated intestinal microfloras of patients with Crohn's disease and ulcerative colitis. J Clin Microbiol 44:4136–4141

Grice EA, Kong HH, Conlan S, Deming CB, Davis J, Young AC, NISC Comparative Sequencing Program, Bouffard GG, Blakesley RW, Murray PR, Green ED, Turner ML, Segre JA (2009) Topographical and temporal diversity of the human skin microbiome Science 324:1190–1192

Gronow S, Welnitz S, Lapidus A, Nolan M, Ivanova N, Glavina Del Rio T, Copeland A, Chen F, Tice H, Pitluck S, Cheng JF, Saunders E, Brettin T, Han C, Detter JC, Bruce D, Goodwin L, Land M, Hauser L, Chang YJ, Jeffries CD, Pati A, Mavromatis K, Mikhailova N, Chen A, Palaniappan K, Chain P, Rohde M, Göker M, Bristow J, Eisen JA, Markowitz V, Hugenholtz P, Kyrpides NC, Klenk HP, Lucas S (2010) Complete genome sequence of *Veillonella parvula* type strain (Te3). Stand Genomic Sci 2:57–65

Gutierrez J, Davis RE, Lindahl IH, Warwick EJ (1959) Bacterial changes in the rumen during the onset of feed-lot bloat of cattle and characteristics of *Peptostreptococcus elsdenii* n. sp. Appl Microbiol 7:16–22

Haikara A, Juvonen R (2009) Genus XV. *Pectinatus*. In: De Vos P, Garrity G, Jones D, Krieg NR, Ludwig W, Rainey FA, Schleifer K-H, Whitman WB (eds) Bergey's manual of systematic bacteriology, vol 3, 2nd edn, The *Firmicutes*. Springer, New York, pp 1074–1079

Hannan S, Ready D, Jasni AS, Rogers M, Pratten J, Roberts AP (2010) Transfer of antibiotic resistance by transformation with eDNA within oral biofilms. FEMS Immunol Med Microbiol 59:345–349

Haralambie E (1983) *Megasphaera elsdenii*, occurrence in 2,255 fecal samples from men, chimpanzees and mice. Zentbl Bakteriol Mikrobiol Hyg Ser A 253:489–494

Harmsen HJ, Wildeboer-Veloo AC, Raangs GC, Wagendorp AA, Klijn N, Bindels JG, Welling GW (2000) Analysis of intestinal flora development in breast-fed and formula-fed infants by using molecular identification and detection methods. J Pediatr Gastroenterol Nutr 30:61–67

Harris JK, De Groote MA, Sagel SD, Zemanick ET, Kapsner R, Penvari C, Kaess H, Deterding RR, Accurso FJ, Pace NR (2007) Molecular identification of bacteria in bronchoalveolar lavage fluid from children with cystic fibrosis. Proc Natl Acad Sci USA 104:20529–20533

Hebb JK, Cohen CR, Astete SG, Bukusi EA, Totten PA (2004) Detection of novel organisms associated with salpingitis, by use of 16S rDNA polymerase chain reaction. J Infect Dis 190:2109–2120

Hinton A Jr, Hume ME (1995) Antibacterial activity of the metabolic by-products of a *Veillonella* species and *Bacteroides fragilis*. Anaerobe 1:121–127

Hofstad T, Kristoffersen T (1970) Chemical composition of endotoxin from oral *Veillonella*. Acta Path Microbiol Scand 78:760–764

Houston S, Taylor D, Rennie R (1997) Prosthetic valve endocarditis due to *Veillonella dispar*: successful medical treatment following penicillin desensitization. Clin Infect Dis 24:1013–1014

Human Microbiome Project Consortium (2012) A framework for human microbiome research. Nature 486:215–221

Hyman RW, Fukushima M, Diamond L, Kumm J, Giudice LC, Davis RW (2005) Microbes on the human vaginal epithelium. Proc Natl Acad Sci USA 102:7952–7957

Isner-Horobeti ME, Lecocq J, Dupeyron A, De Martino SJ, Froehlig P, Vautravers P (2006) *Veillonella* discitis. A case report. Joint Bone Spine 73:113–115

Jacquot A, Neveu D, Aujoulat F, Mercier G, Marchandin H, Jumas-Bilak E, Picaud J-C (2011) Dynamics and clinical evolution of bacterial gut microflora in extremely premature patients. J Pediatr 158:390–396

Jones V, Wilks M, Johnson G, Warwick S, Hennessey E, Kempley S, Millar M (2010) The use of molecular techniques for bacterial detection in the analysis of gastric aspirates collected from infants on the first day of life. Early Hum Dev 86:167–170

Jumas-Bilak E, Carlier J-P, Jean-Pierre H, Teyssier C, Gay B, Campos J, Marchandin H (2004) *Veillonella montpellierensis* sp. nov., a novel, anaerobic, Gram-negative coccus isolated from human clinical samples. Int J Syst Evol Microbiol 54:1311–1316

Jumas-Bilak E, Jean-Pierre H, Carlier J-P, Teyssier C, Bernard K, Gay B, Campos J, Morio F, Marchandin H (2005) *Dialister micraerophilus* sp. nov. and *Dialister propionifaciens* sp. nov., isolated from human clinical samples. Int J Syst Evol Microbiol 55:2471–2478

Justesen US, Holm A, Knudsen E, Andersen LB, Jensen TG, Kemp M, Skov MN, Gahrn-Hansen B, Møller JK (2011) Species identification of clinical isolates of anaerobic bacteria: a comparison of two matrix-assisted laser desorption ionization-time of flight mass spectrometry systems. J Clin Microbiol 49:4314–4318

Juvonen RJ, Suihko M-L (2006) *Megasphaera paucivorans* sp. nov., *Megasphaera sueciensis* sp. nov. and *Pectinatus haikarae* sp. nov., isolated from brewery samples, and emended description of the genus *Pectinatus*. Int J Syst Evol Microbiol 56:695–702

Juvonen R, Koivula T, Haikara A (2008) Group-specific PCR-RFLP and real-time PCR methods for detection and tentative discrimination of strictly anaerobic beer-spoilage bacteria of the class *Clostridia*. Int J Food Microbiol 125:162–169

Kanasi E, Dewhirst FE, Chalmers NI, Kent R Jr, Moore A, Hughes CV, Pradhan N, Loo CY, Tanner AC (2010) Clonal analysis of the microbiota of severe early childhood caries. Caries Res 44:485–497

Kolenbrander PE (2011) Multispecies communities: interspecies interactions influence growth on saliva as sole nutritional source. Int J Oral Sci 3:49–54

Kong HH, Oh J, Deming C, Conlan S, Grice EA, Beatson MA, Nomicos E, Polley EC, Komarow HD, NISC Comparative Sequence Program, Murray PR, Turner ML, Segre JA (2012) Temporal shifts in the skin microbiome associated with disease flares and treatment in children with atopic dermatitis. Genome Res 22:850–859

Könönen E, Kanervo A, Takala A, Asikainen S, Jousimies-Somer H (1999) Establishment of oral anaerobes during the first year of life. J Dent Res 78:1634–1639

Kraatz M, Taras D (2008) *Veillonella magna*, isolated from the jejunal mucosa of a healthy pig, and emended description of *Veillonella ratti*. Int J Syst Evol Microbiol 58:2755–2761

Krumholz LR, Bryant MP (1986) *Syntrophococcus sucromutans* sp. nov. gen. nov. uses carbohydrates as electron donors and formate, methoxymonobenzoids or *Methanobrevibacter* as electron acceptor systems. Arch Microbiol 143:313–318

Kumar PS, Leys EJ, Bryk JM, Martinez FJ, Moeschberger ML, Griffen AL (2006) Changes in periodontal health status are associated with bacterial community shifts as assessed by quantitative 16S cloning and sequencing. J Clin Microbiol 44:3665–3673

La Scola B, Fournier PE, Raoult D (2011) Burden of emerging anaerobes in the MALDI-TOF and 16S rRNA gene sequencing era. Anaerobe 17:106–112

Lancaster H, Ready D, Mullany P, Spratt D, Bedi R, Wilson M (2003) Prevalence and identification of tetracycline-resistant oral bacteria in children not receiving antibiotic therapy. FEMS Microbiol Lett 228:99–104

Lee PK, Warnecke F, Brodie EL, Macbeth TW, Conrad ME, Andersen GL, Alvarez-Cohen L (2012) Phylogenetic microarray analysis of a microbial community performing reductive dechlorination at a TCE-contaminated site. Environ Sci Technol 46:1044–1054

Leser TD, Amenuvor JZ, Jensen TK, Lindecrona RH, Boye M, Møller K (2002) Culture-independent analysis of gut bacteria: the pig gastrointestinal tract microbiota revisited. Appl Environ Microbiol 68:673–690

Ley RE, Hamady M, Lozupone C, Turnbaugh PJ, Ramey RR, Bircher JS, Schlegel ML, Tucker TA, Schrenzel MD, Knight R, Gordon JI (2008) Evolution of mammals and their gut microbes. Science 320:1647–1651

Li E, Hamm CM, Gulati AS, Sartor RB, Chen H, Wu X, Zhang T, Rohlf FJ, Zhu W, Gu C, Robertson CE, Pace NR, Boedeker EC, Harpaz N, Yuan J, Weinstock GM, Sodergren E, Frank DN (2012) Inflammatory bowel diseases phenotype, *C. difficile* and NOD2 genotype are associated with shifts in human ileum associated microbial composition. PLoS One 7:e26284

Liu JW, Wu J-J, Wang LR, Teng LJ, Huang TC (1998) Two fatal cases of *Veillonella* bacteremia. Eur J Clin Microbiol Infect Dis 17:62–64

Liu J, Xie Z, Merritt J, Qi F (2012) Establishment of a tractable genetic transformation system in *Veillonella* spp. Appl Environ Microbiol 78:3488–3491

Ludwig W, Schleifer K-H, Whitman WB (2009) Revised road map to the phylum Firmicutes. In: De Vos P, Garrity G, Jones D, Krieg NR, Ludwig W, Rainey FA, Schleifer K-H, Whitman WB (eds) Bergey's manual of systematic bacteriology, vol 3, 2nd edn, The *Firmicutes*. Springer, New York, pp 1–13

Luppens SB, Kara D, Bandounas L, Jonker MJ, Wittink FR, Bruning O, Breit TM, Ten Cate JM, Crielaard W (2008) Effect of *Veillonella parvula* on the antimicrobial resistance and gene expression of *Streptococcus mutans* grown in a dual-species biofilm. Oral Microbiol Immunol 23:183–189

Malinen E, Rinttilä T, Kajander K, Mättö J, Kassinen A, Krogius L, Saarela M, Korpela R, Palva A (2005) Analysis of the fecal microbiota of irritable bowel syndrome patients and healthy controls with real-time PCR. Am J Gastroenterol 100:373–382

Marchandin H (2001a) Organisation génomique, phylogéenie et taxonomie polyphasique des bactéries du genre *Veillonella* et des genres apparentés du sous-groupe *Sporomusa*. PhD Thesis, Université Montpelier I, Montpelier

Marchandin H, Jean-Pierre H, Carrière C, Canovas F, Darbas H, Jumas-Bilak E (2001b) Prosthetic joint infection due to *Veillonella dispar*. Eur J Clin Microbiol Infect Dis 20:340–342

Marchandin H (2007) Les cocci à Gram négatif anaérobies. In: Freney J, Renaud F, Leclercq R, Riegel P (eds) Précis de bactériologie clinique, 2nd edn. ESKA, Paris, pp 1733–1738

Marchandin H, Jumas-Bilak E (2006) 16S rRNA gene sequencing: interest and limits for identification and characterization of novel taxa within the family *Acidaminococcaceae*. In: McNamara PA (ed) Trends in RNA research. Nova Publishers, Hauppauge, pp 225–251

Marchandin H, Jumas-Bilak E, Gay B, Teyssier C, Jean-Pierre H, Simeon de Buochberg M, Carriere C, Carlier J-P (2003a) Phylogenetic analysis of some *Sporomusa* sub-branch members isolated from human clinical specimens: description of *Megasphaera micronuciformis* sp. nov. Int J Syst Evol Microbiol 53:547–553

Marchandin H, Teyssier C, Siméon de Buochberg M, Jean-Pierre H, Carriere C, Jumas-Bilak E (2003b) Intra-chromosomal heterogeneity between the four 16S rRNA gene copies in the genus *Veillonella*: implications for phylogeny and taxonomy. Microbiology 149:1493–1501

Marchandin H, Jean-Pierre H, Campos J, Dubreuil L, Teyssier C, Jumas-Bilak E (2004) *nimE* gene in a metronidazole-susceptible *Veillonella* sp. strain. Antimicrob Agents Chemother 48:3207–3208

Marchandin H, Teyssier C, Jumas-Bilak E, Robert M, Artigues A-C, Jean-Pierre H (2005) Molecular identification of the first human isolate belonging to the *Veillonella ratti*-*Veillonella criceti* group based on 16S rDNA and *dnaK* gene sequencing. Res Microbiol 156:603–607

Marchandin H, Juvonen R, Haikara A (2009) Genus XIII *Megasphaera*. In: De Vos P, Garrity G, Jones D, Krieg NR, Ludwig W, Rainey FA, Schleifer K-H, Whitman WB (eds) Bergey's manual of systematic bacteriology, vol 3, 2nd edn, The *Firmicutes*. Springer, New York, pp 1082–1089

Marchandin H, Teyssier C, Campos J, Jean-Pierre H, Roger F, Gay B, Carlier J-P, Jumas-Bilak E (2010) *Negativicoccus succinicivorans* gen. nov., sp. nov., isolated from human clinical samples, emended description of the family *Veillonellaceae* and description of *Negativicutes* classis nov., *Selenomonadales* ord. nov. and *Acidaminococcaceae* fam. nov. in the bacterial phylum Firmicutes. Int J Syst Evol Microbiol 60:1271–1279

Marounek M, Fliegrova K, Bartos S (1989) Metabolism and some characteristics of ruminal strains of *Megasphaera elsdenii*. Appl Environ Microbiol 55:1570–1573

Marriott D, Stark D, Harkness J (2007) *Veillonella parvula* discitis and secondary bacteremia: a rare infection complicating endoscopy and colonoscopy? J Clin Microbiol 45:672–674

Marx H, Graf AB, Tatto NE, Thallinger GG, Mattanovich D, Sauer M (2011) Genome sequence of the ruminal bacterium *Megasphaera elsdenii*. J Bacteriol 193:5578–5579

Mashima I, Kamaguchi A, Nakazawa F (2011) The distribution and frequency of oral *Veillonella* spp. in the tongue biofilm of healthy young adults. Curr Microbiol 63:403–407

Mashima I, Kamaguchi A, Miyakawa H, Nakazawa F (2013) *Veillonella tobetsuensis* sp. nov., a novel, anaerobic, Gram-negative coccus isolated from human tongue biofilm. Int J Syst Evol Microbiol 63:1443–1449

Matera G, Liberto MC, Berlinghieri MC, Focà A (1991) Biological effects of *Veillonella parvula* and *Bacteroides intermedius* lipopolysaccharides. Microbiologica 14:315–323

Matera G, Muto V, Vinci M, Zicca E, Abdollahi-Roodsaz S, van de Veerdonk FL, Kullberg BJ, Liberto MC, van der Meer JW, Focà A, Netea MG, Joosten LA (2009) Receptor recognition of and immune intracellular pathways for *Veillonella parvula* lipopolysaccharide. Clin Vaccine Immunol 16:1804–1809

Matsumoto K, Ishiyama A, Sakai K, Shiba T, Taguchi S (2011) Biosynthesis of glycolate-based polyesters containing medium-chain-length 3-hydroxyalkanoates in recombinant *Escherichia coli* expressing engineered polyhydroxyalkanoate synthase. J Biotechnol 156:214–217

Mays TD, Holdeman LV, Moore WEC, Rogosa M, Johnson JL (1982) Taxonomy of the genus *Veillonella* Prevot. Int J Syst Bacteriol 32:28–36

Michon A-L, Aujoulat F, Roudière L, Soulier O, Zorgniotti I, Jumas-Bilak E, Marchandin H (2010) Intragenomic and intraspecific heterogeneity in *rrs* may surpass interspecific variability in a natural population of *Veillonella*. Microbiology 156:2080–2091

Moore LV, Moore WE (1994) *Oribaculum catoniae* gen. nov., sp. nov.; *Catonella morbi* gen. nov., sp. nov.; *Hallella seregens* gen. nov., sp. nov.; *Johnsonella ignava* gen. nov., sp. nov.; and *Dialister pneumosintes* gen. nov., comb. nov., nom. rev., anaerobic gram-negative bacilli from the human gingival crevice. Int J Syst Bacteriol 44:187–192

Morio F, Jean-Pierre H, Dubreuil L, Jumas-Bilak E, Calvet L, Mercier G, Devine R, Marchandin H (2007) Antimicrobial susceptibilities and clinical sources of *Dialister* species. Antimicrob Agents Chemother 51:4498–4501

Morotomi M, Nagai F, Sakon H, Tanaka R (2008) *Dialister succinatiphilus* sp. nov. and *Barnesiella intestinihominis* sp. nov., isolated from human faeces. Int J Syst Evol Microbiol 58:2716–2720

Munson MA, Banerjee A, Watson TF, Wade WG (2004) Molecular analysis of the microflora associated with dental caries. J Clin Microbiol 42:3023–3029

Nyfors S, Kononen E, Bryk A, Syrjanen R, Jousimies-Somer H (2003) Age-related frequency of penicillin resistance of oral *Veillonella*. Diagn Microbiol Infect Dis 46:279–283

Oh S, Havlen PR, Hussain N (2005) A case of polymicrobial endocarditis caused by anaerobic organisms in an injection drug user. J Gen Intern Med 20:C1–C2

Ouwerkerk D, Klieve AV, Forster RJ (2002) Enumeration of *Megasphaera elsdenii* in rumen contents by real-time *Taq* nuclease assay. J Appl Microbiol 92:753–758

Palmer RJ Jr, Diaz PI, Kolenbrander PE (2006) Rapid succession within the *Veillonella* population of a developing human oral biofilm in situ. J Bacteriol 188:4117–4124

Paster BJ, Boches SK, Galvin JL, Ericson RE, Lau CN, Levanos VA, Sahasrabudhe A, Dewhirst FE (2001) Bacterial diversity in human subgingival plaque. J Bacteriol 183:3770–3783

Periasamy S, Kolenbrander PE (2010) Central role of the early colonizer *Veillonella* sp. in establishing multispecies biofilm communities with initial, middle, and late colonizers of enamel. J Bacteriol 192:2965–2972

Perkins SD, Woeltje KF, Angenent LT (2010) Endotracheal tube biofilm inoculation of oral flora and subsequent colonization of opportunistic pathogens. Int J Med Microbiol 300:503–511

Pierre Lepargneur J, Dubreuil L, Levy J (2006) Isolation of *Dialister pneumosintes* isolated from a bacteremia of vaginal origin. Anaerobe 12:274–275

Piknová M, Filova M, Javorský P, Pristas P (2004) Different restriction and modification phenotypes in ruminal lactate-utilizing bacteria. FEMS Microbiol Lett 236:91–95

Piknová M, Bíres O, Javorský P, Pristas P (2006) Limited genetic variability in *Megasphaera elsdenii* strains. Folia Microbiol (Praha) 51:299–302

Piriz S, Cuenca R, Valle J, Vadillo S (1992) Susceptibilities of anaerobic bacteria isolated from animals with ovine foot rot to 28 antimicrobial agents. Antimicrob Agents Chemother 36:198–201

Pouchot J, Vinceneux P, Michon C, Mathieu A, Boussougant Y (1992) Pyogenic sacroiliitis due to *Veillonella parvula*. Clin Infect Dis 15:175

Prabhu R, Altman E, Eiteman MA (2012) Lactate and acrylate metabolism by *Megasphaera elsdenii* under batch and steady state conditions. Appl Environ Microbiol 78:8564–8570

Prevot AR (1933) Etude de systématique bactérienne. I. Lois générales. II. Cocci anaérobius. Ann Sci Nat Zool Biolo Anim 15:23–260

Preza D, Olsen I, Willumsen T, Grinde B, Paster BJ (2009) Diversity and site-specificity of the oral microflora in the elderly. Eur J Clin Microbiol Infect Dis 28:1033–1040

Puhl AA, Greiner R, Selinger LB (2009) Stereospecificity of myo-inositol hexakisphosphate hydrolysis by a protein tyrosine phosphatase-like inositol polyphosphatase from *Megasphaera elsdenii*. Appl Microbiol Biotechnol 82:95–103

Quintanilha AG, Zilberstein B, Santos MA, Pajecki D, Moura EG, Alves PR, Maluf-Filho F, Cecconello I (2007) A novel sampling method for the investigation of gut mirobiota. World J Gastroenterol 13:3990–3995

Rainey FA (2009a) Family X. *Veillonellaceae*. In: De Vos P, Garrity G, Jones D, Krieg NR, Ludwig W, Rainey FA, Schleifer K-H, Whitman WB (eds) Bergey's manual of systematic bacteriology, vol 3, 2nd edn, The *Firmicutes*. Springer, New York, pp 1039–1110

Rainey FA (2009b) Genus IV. *Allisonella*. In: De Vos P, Garrity G, Jones D, Krieg NR, Ludwig W, Rainey FA, Schleifer K-H, Whitman WB (eds) Bergey's manual of systematic bacteriology, vol 3, 2nd edn, The *Firmicutes*. Springer, New York, pp 1048–1049

Ramos MP, Ferreira SM, Silva-Boghossian CM, Souto R, Colombo AP, Noce CW, de Gonçalves LS (2012) Necrotizing periodontal diseases in HIV-infected Brazilian patients: a clinical and microbiologic descriptive study. Quintessence Int 43:71–82

Ready D, Lancaster H, Qureshi F, Bedi R, Mullany P, Wilson M (2004) Effect of amoxicillin use on oral microbiota in young children. Antimicrob Agents Chemother 48:2883–2887

Ready D, Pratten J, Roberts AP, Bedi R, Mullany P, Wilson M (2006) Potential role of *Veillonella* spp. as a reservoir of transferable tetracycline resistance in the oral cavity. Antimicrob Agents Chemother 50:2866–2868

Ready D, Bedi R, Mullany P, Wilson M (2012) Penicillin and amoxicillin resistance in oral *Veillonella* spp. Int J Antimicrob Agents 40:188–189

Reig M, Mir N, Baquero F (1997) Penicillin resistance in *Veillonella*. Antimicrob Agents Chemother 41:1210

Roberts SA, Shore KP, Paviour SD, Holland D, Morris AJ (2006) Antimicrobial susceptibility of anaerobic bacteria in New Zealand: 1999–2003. J Antimicrob Chemother 57:992–998

Rôças IN, Siqueira JF Jr, Debelian GJ (2011) Analysis of symptomatic and asymptomatic primary root canal infections in adult Norwegian patients. J Endod 37:1206–1212

Rogosa M (1965) The genus *Veillonella*: IV. Serological groupings, and genus and species emendations. J Bacteriol 90:704–709

Rogosa M (1971a) Transfer of *Peptostreptococcus elsdenii* Gutierrez et al. to a new genus, *Megasphaera* (*M. elsdenii* (Gutierrez et al.) comb. nov.). Int J Syst Bacteriol 21:187–189

Rogosa M (1971b) Transfer of *Veillonella* Prevot and *Acidaminococcus* Rogosa from *Neisseriaceae* to *Veillonellaceae* fam. nov. and the inclusion of *Megasphaera* Rogosa in *Veillonellaceae*. Int J Syst Bacteriol 21:231–233

Rogosa, M (1984) Anaerobic Gram-negative cocci. In: Krieg NR, Holt JG (eds) Bergey's manual of systematic bacteriology, vol 1, Williams & Wilkins, Baltimore, pp 680–685

Rousée JM, Bermond D, Piémont Y, Tournoud C, Heller R, Kehrli P, Harlay ML, Monteil H, Jaulhac B (2002) *Dialister pneumosintes* associated with human brain abscesses. J Clin Microbiol 40:3871–3873

Rovery C, Etienne A, Foucault C, Berger P, Brouqui P (2005) *Veillonella montpellierensis* endocarditis. Emerg Infect Dis 11:1112–1114

Rudkjøbing VB, Thomsen TR, Alhede M, Kragh KN, Nielsen PH, Johansen UR, Givskov M, Høiby N, Bjarnsholt T (2011) True microbiota involved in chronic lung infection of cystic fibrosis patients found by culturing and 16S rRNA gene analysis. J Clin Microbiol 49:4352–4355

Rybojad P, Los R, Sawicki M, Tabarkiewicz J, Malm A (2011) Anaerobic bacteria colonizing the lower airways in lung cancer patients. Folia Histochem Cytobiol 49:263–266

Sakamoto K, Konings WN (2003) Beer spoilage bacteria and hop resistance. Int J Food Microbiol 89:105–124

Sato M, Hoshino E, Nomura S, Ishioka K (1993) Salivary microflora of geriatric edentulous persons wearing dentures. Microb Ecol Health Dis 6:293–299

Sato T, Matsuyama J, Sato M, Hoshino E (1997a) Differentiation of *Veillonella atypica*, *Veillonella dispar* and *Veillonella parvula* using restricted fragment-length polymorphism analysis of 16S rDNA amplified by polymerase chain reaction. Oral Microbiol Immunol 12:350–353

Sato T, Sato M, Matsuyama J, Hoshino E (1997b) PCR-restriction fragment length polymorphism analysis of genes coding for 16S rRNA in *Veillonella* spp. Int J Syst Bacteriol 47:1268–1270

Sato T, Matsumoto K, Okumura T, Yokoi W, Naito E, Yoshida Y, Nomoto K, Ito M, Sawada H (2008) Isolation of lactate-utilizing butyrate-producing bacteria from human feces and in vivo administration of *Anaerostipes caccae* strain L2 and galacto-oligosaccharides in a rat model. FEMS Microbiol Ecol 66:528–536

Scheithauer BK, Wos-Oxley ML, Ferslev B, Jablonowski H, Pieper DH (2009) Characterization of the complex bacterial communities colonizing biliary stents reveals a host-dependent diversity. ISME J 3:797–807

Sibley CD, Grinwis ME, Field TR, Eshaghurshan CS, Faria MM, Dowd SE, Parkins MD, Rabin HR, Surette MG (2011) Culture enriched molecular profiling of the cystic fibrosis airway microbiome. PLoS One 6:e22702

Singer E, Calvet L, Mory F, Muller C, Chomarat M, Bézian MC, Bland S, Juvenin ME, Drugeon H, Fosse T, Goldstein F, Jaulhac B, Monteil H, Marchandin H, Jean-Pierre H, Dubreuil L (2008) Monitoring of antibiotic resistance of gram negative anaerobes. Med Mal Infect 38:256–263

Singh N, Yu VL (1992) Osteomyelitis due to *Veillonella parvula*: case report and review. Clin Infect Dis 14:361–363

Siqueira JF, Rôças IN (2003) Positive and negative bacterial associations involving *Dialister pneumosintes* in primary endodontic infections. J Endod 29:438–441

Spear GT, Gilbert D, Sikaroodi M, Doyle L, Green L, Gillevet PM, Landay AL, Veazey RS (2010) Identification of rhesus macaque genital microbiota by 16S pyrosequencing shows similarities to human bacterial vaginosis: implications for use as an animal model for HIV vaginal infection. AIDS Res Hum Retroviruses 26:193–200

Stamatakis A (2006) RAxML-VI-HPC: maximum likelihood-based phylogenetic analyses with thousands of taxa and mixed models. Bioinformatics 22:2688–2690

Stanton TB, Humphrey SB (2003) Isolation of tetracycline-resistant *Megasphaera elsdenii* strains with novel mosaic gene combinations of tet(O) and tet(W) from swine. Appl Environ Microbiol 69:3874–3882

Stanton TB, Humphrey SB (2011) Persistence of antibiotic resistance: evaluation of a probiotic approach using antibiotic-sensitive *Megasphaera elsdenii* strains to prevent colonization of swine by antibiotic-resistant strains. Appl Environ Microbiol 77:7158–7166

Stanton TB, McDowall J-S, Rasmussen MA (2004) Diverse tetracycline resistance genotypes of *Megasphaera elsdenii* strains selectively cultured from swine feces. Appl Environ Microbiol 70:3754–3757

Stewart CS, Bryant MP (1988) The rumen bacteria. In: Hobson PN (ed) The rumen microbial ecosystem. Elsevier Applied Science, London, pp 21–71

Strach M, Siedlar M, Kowalczyk D, Zembala M, Grodzicki T (2006) Sepsis caused by *Veillonella parvula* infection in a 17-year-old patient with X-linked agammaglobulinemia (Bruton's disease). J Clin Microbiol 44:2655–2656

Sugihara PT, Sutter VL, Attebery HR, Bricknell KS, Finegold SM (1974) Isolation of *Acidaminococcus fermentans* and *Megasphaera elsdenii* from normal human feces. Appl Microbiol 27:274–275

Suihko M-L, Haikara A (2001) Characterization of *Pectinatus* and *Megasphaera* strains by automated ribotyping. J Inst Brew 107:175–184

Sveen K, Skaug N (1980) Bone resorption stimulated by lipopolysaccharides from *Bacteroides*, *Fusobacterium* and *Veillonella*, and by the lipid A and the polysaccharide part of *Fusobacterium* lipopolysaccharide. Scand J Dent Res 88:535–542

Theron MM, Janse van Rensburg MN, Chalkley LJ (2003) Penicillin-binding proteins involved in high-level piperacillin resistance in *Veillonella* spp. J Antimicrob Chemother 52:120–122

Thiel HJ, Schumacher U (1994) Normal flora of the human conjunctiva: a study of 135 persons of various ages. Klin Monatsbl Augenheilkd 205:348–357

Totsuka M, Ono T (1989) Purification and characterization of bacteriophage receptor on *Veillonella rodentium* cells. Phage-receptor on *Veillonella*. Antonie Van Leeuwenhoek 56:263–271

Tunney MM, Field TR, Moriarty TF, Patrick S, Doering G, Muhlebach MS, Wolfgang MC, Boucher R, Gilpin DF, McDowell A, Elborn JS (2008) Detection of anaerobic bacteria in high numbers in sputum from patients with cystic fibrosis. Am J Respir Crit Care Med 177:995–1001

Valdés MV, Lobbins PM, Slots J (1982) β-lactamase producing bacteria in the human oral cavity. J Oral Pathol 11:58–63

van der Gast CJ, Walker AW, Stressmann FA, Rogers GB, Scott P, Daniels TW, Carroll MP, Parkhill J, Bruce KD (2011) Partitioning core and satellite taxa from within cystic fibrosis lung bacterial communities. ISME J 5:780–791

Veloo AC, Knoester M, Degener JE, Kuijper EJ (2011) Comparison of two matrix-assisted laser desorption ionisation-time of flight mass spectrometry methods for the identification of clinically relevant anaerobic bacteria. Clin Microbiol Infect 17:1501–1506

Versalovic J, Koeuth T, McCabe ER, Lupski JR (1991) Use of the polymerase chain reaction for physical mapping of *Escherichia coli* genes. J Bacteriol 173:5253–5255

Wade WG (2009) Genus XII. *Dialister*. In: De Vos P, Garrity G, Jones D, Krieg NR, Ludwig W, Rainey FA, Schleifer K-H, Whitman WB (eds) Bergey's manual of systematic bacteriology, vol 3, 2nd edn, The *Firmicutes*. Springer, New York, pp 1060–1062

Wang M, Ahrné S, Jeppsson B, Molin G (2005) Comparison of bacterial diversity along the human intestinal tract by direct cloning and sequencing of 16S rRNA genes. FEMS Microbiol Ecol 54:219–231

Werner H (1973) *Megasphaera elsdenii* – a normal inhabitant of the human intestines. Zentralbl Bakteriol Parasitenkd Infektionskr Hyg Abt I Orig A 223:343–347

Yarza P, Richter M, Peplies J, Euzeby J, Amann R, Schleifer KH, Ludwig W, Glöckner FO, Rossello-Mora R (2008) The All-Species Living Tree project: a 16S rRNA-based phylogenetic tree of all sequenced type strains. Syst Appl Microbiol 31:241–250

Zaninetti-Schaerer A, Van Delden C, Genevay S, Gabay C (2004) Total hip prosthetic joint infection due to *Veillonella* species. Joint Bone Spine 71:161–163

Zhou X, Brown CJ, Abdo Z, Davis CC, Hansmann MA, Joyce P, Foster JA, Forney LJ (2007) Differences in the composition of vaginal microbial communities found in healthy Caucasian and black women. ISME J 1:121–133

Zilberstein B, Quintanilha AG, Santos MA, Pajecki D, Moura EG, Alves PR, Maluf Filho F, de Souza JA, Gama-Rodrigues J (2007) Digestive tract microbiota in healthy volunteers. Clinics 62:47–54

# 36 The Genus *Virgibacillus*

*Cristina Sánchez-Porro · Rafael R. de la Haba · Antonio Ventosa*
Department of Microbiology and Parasitology, University of Sevilla, Sevilla, Spain



**Abstract**
The genus *Virgibacillus*, belongs to the family *Bacillaceae*, within the phylum Firmicutes. The members of this genus are widely found in many habitats and the currently described species have been mostly isolated from saline environments. As of January 2014, this genus includes a total of 27 species with validly published names. In this chapter, the historical and current taxonomy have been reviewed along with molecular and phenotypic analyses. Also the preservation procedures and maintenance are described. Finally applications of members of the genus *Virgibacillus* are also addressed.

## Introduction

In 1998, Heyndrickx et al. characterized a set of 12 strains of the species *Bacillus pantothenticus* (Proom and Knight 1950) genotypically using amplified rDNA restriction analysis (ARDRA), phenotypically using biochemical tests and morphological observations, and chemotaxonomically using fatty acid methyl esters analysis and SDS-PAGE. The polyphasic data indicated that *B. pantothenticus* lies at the periphery of rRNA group 1 of Ash et al. (1991), supporting its recognition as a separate genus within the family *Bacillaceae*, for which the name *Virgibacillus* was proposed (Heyndrickx et al. 1998).

## Taxonomy, Historical and Current

### Short Description of the Genus

Vir.gi.ba.cil'lus. L. n. *virga* a green twig, transf., a branch in a family tree; L. dim. n. *bacillus* from *Bacillus*, a genus of aerobic endospore-forming bacteria; *Virgibacillus* a branch of the genus *Bacillus*. The genus was originally described by Heyndrickx et al. (1998) and lately emended by Wainø et al. (1999) and Heyrman et al. (2003).

This genus belongs to the family *Bacillaceae*, within the phylum Firmicutes (Heyrman et al. 2009). Currently, it includes a total of 27 species with validly published names (Parte 2014) and two other species, designated '*Virgibacillus zhanjiangensis*' (Peng et al. 2009) and '*Virgibacillus natachei*' (Amziane et al. 2013) that have been proposed, but these names have not yet been validly published (➋ *Table 36.1*). The type species is *Virgibacillus pantothenticus* (Proom and Knight 1950; Heyndrickx et al. 1998). Phylogenetically related genera are *Oceanobacillus*, *Ornithinibacillus*, *Lentibacillus*, *Paucisalibacillus* and *Cerasibacillus*.

Gram-staining-positive motile rods ($0.3 - 0.8 \times 2.0 - 8.0$ μm) that occur singly, in pairs or, especially in older cultures, forming chains and/or filaments. Oval to ellipsoidal endospores in terminal (sometimes subterminal or paracentral) positions in swollen sporangia are present. Colonies are small, 0.5–5 mm in diameter after 2 days on marine agar or trypticase soy agar, circular and slightly irregular, smooth, glossy or sometimes matt, flat or low-convex, butyrous, slightly transparent to opaque, and usually creamy to yellowish white or unpigmented, but two species possess pink pigmentation (Heyrman et al. 2003; Niederberger et al. 2009). Aerobic or facultatively anaerobic. Catalase positive, except for one species (Kim et al. 2011). In the API 20E system and in conventional tests the Voges-Proskauer reaction is negative, indole is not produced, nitrate reduction to nitrite is variable. Aesculin, casein, and gelatin hydrolysis positive in most species, but urease and hydrogen sulphide are usually not produced (except for a few strains which give weak positive reactions for the latter in the API 20E system). Some species are positive for arginine dihydrolase, citrate utilization and *o*-nitrophenyl-β-D-galactoside activity in the API 20E system. Growth is optimal at 4–10 % NaCl and at pH around 7–10. Growth may occurs between 5 °C and 50 °C (optimum at 28 °C or 37 °C) (➋ *Table 36.2*). Raffinose can be used as sole carbon and energy source, but no growth was observed on D-arabinose,

E. Rosenberg et al. (eds.), *The Prokaryotes – Firmicutes and Tenericutes*, DOI 10.1007/978-3-642-30120-9_353,

**Table 36.1**
**Types strains, GenBank/EMBL/DDBJ accession numbers of the 16S rRNA gene sequence and isolation source of *Virgibacillus* species**

| Name of *Virgibacillus* species | Type strain | GenBank/EMBL/DDBJ accession number | Isolation source | Reference |
|---|---|---|---|---|
| *V. pantothenticus* | NCTC 8162 | D16275 | Soil samples (Southern England) | Heyndrickx et al. (1998, 1999) |
| *V. albus* | YIM 93624 | JQ680032 | Soil sample of Lop Nur salt lake (China) | Zhang et al. (2012) |
| *V. alimentarius* | J18 | GU202420 | Salt-fermented seafood made from gizzard shad (Korea) | Kim et al. (2011) |
| *V. arcticus* | Hal 1 | EF675742 | Permafrost core of 9 m depth (Canadian high Arctic) | Niederberger et al. (2009) |
| *V. byunsanensis* | ISL-24 | FJ357159 | Marine solar saltern of the Yellow Sea (Korea) | Yoon et al. (2010) |
| *V. campisalis* | IDS-20 | GU586225 | Marine solar saltern (West coast of Korea) | Lee et al. (2012b) |
| *V. carmonensis* | LMG 20964 | AJ316302 | Mural paintings of necropolis at Carmona (Spain) | Heyrman et al. (2003) |
| *V. chiguensis* | NTU-101 | EF101168 | Chigu, a disused salt field (Southern Taiwan) | Wang et al. (2008) |
| *V. dokdonensis* | DSW-10 | AY822043 | Dokdo Island (East Sea, Korea) | Yoon et al. (2005) |
| *V. halodenitrificans* | ATCC 49067 | AY543169 | Saltern in France | Yoon et al. (2004) |
| *V. halophilus* | 5B73C | AB243851 | Field soil in Kakegawa (Japan) | An et al. (2007) |
| *V. halotolerans* | WS 4627 | HE577174 | Dairy product sample in Babaria (Germany) | Seiler and Wenning (2013) |
| *V. kekensis* | YIM kkny16 | AY121439 | Saline mud sample from the Keke salt lake (North-West China) | Chen et al. (2008) |
| *V. koreensis* | BH30097 | AY616012 | Salt field near Taean-Gun on the Yellow Sea (Korea) | Lee et al. (2006) |
| *V. litoralis* | JSM 089168 | FJ425909 | Saline soil sample from the shore of Naozhou Island (South China Sea) | Chen et al. (2009b) |
| *V. marismortui* | 123 | AJ009793 | Dead Sea water | Heyrman et al. (2003) |
| '*V. natechei*' | FarD | JX435821 | Saline lake in Algeria | Amziane et al. (2013) |
| *V. necropolis* | LMG 19488 | AJ315056 | Mural paintings of necropolis at Carmona (Spain) | Heyrman et al. (2003) |
| *V. olivae* | $E_{30}8$ | DQ139839 | Wastewater of Spanish green olive processing | Quesada et al. (2007) |
| *V. proomi* | LMG 12370 | AJ012667 | Soil samples (Southern England) | Heyndrickx et al. (1999) |
| *V. salarius* | SA-Vb1 | AB197851 | Salt crust from Chott el Gharsa (Tunisia) | Hua et al. (2008) |
| *V. salexigens* | C-20Mo | Y11603 | Ponds of solar salterns or hypersaline soils (Spain) | Heyrman et al. (2003) |
| *V. salinus* | XH-22 | FM205010 | Sediment of a salt lake near Xilinhot (Inner Mongolia, China) | Carrasco et al. (2009) |
| *V. sediminis* | YIM kkny3 | AY121430 | Saline sediment sample from the Keke salt lake (North-West China) | Chen et al. (2009a) |
| *V. siamensis* | MS3-4 | AB365482 | Fermented fish (*pla-ra*) in Thailand | Tanasupawat et al. (2010) |
| *V. soli* | CC-YMP-6 | EU213011 | Soil samples from Yang-Ming Mountain (Taiwan) | Kämpfer et al. (2011) |
| *V. subterraneus* | H57B72 | FJ746573 | Subsurface saline soil sample of the Qaidam basin (China) | Wang et al. (2010) |
| *V. xinjiangensis* | SL6-1 | DQ664543 | Salt lake in Xin-jiang province (China) | Jeon et al. (2009) |
| '*V. zhanjiangensis*' | JSM 079157 | FJ425904 | Sea water from a tidal flat of Naozhou Island (South China Sea) | Peng et al. (2009) |

D-fructose, or D-xylose. The predominant cellular fatty acids are iso-$C_{15:0}$, anteiso-$C_{15:0}$ and anteiso-$C_{17:0}$ (➋ *Table 36.3*). The major polar lipids are diphosphatidylglycerol and phosphatidylglycerol. Other five phospholipids and one polar lipid of unknown structure are present in all species examined. The presence of phosphatidylethanolamine and other minor lipids can vary among the species (Heyrman et al. 2003) (➋ *Table 36.3*). The major respiratory menaquinone is MK-7, with minor to trace amounts of MK-6 and MK-8 (➋ *Table 36.3*). The cell wall contains *meso*-diaminopimelic acid as the diagnostic diamino acid in the peptidoglycan, with the exception of *Virgibacillus arcticus*, which possess A1α peptidoglycan type (Niederberger et al. 2009) (➋ *Table 36.3*). The DNA G+C content ranges between 36.0 and 43.0 mol% (➋ *Table 36.3*).

The Subcommittee on the Taxonomy of *Bacillus* and related organisms, a member of the International Committee on Systematics of Prokaryotes, published in 2009 the minimal standards for describing new taxa of aerobic, endospore-forming bacteria (including the genus *Virgibacillus*) in order to assist authors working in the field (Logan et al. 2009).

## Phylogenetic Structure of the Genus

The first phylogenetic study concerning the genus *Virgibacillus* was conducted in 1991 by Ash et al. Sequencing and comparative analysis of the 16S rRNA from 51 species of *Bacillus* was performed revealing five phylogenetically distinct clusters. Group 1 contained a total of 28 species, including *B. subtilis*, the type species of the genus, and *B. pantothenticus*, which would be further transferred to the new genus *Virgibacillus*. This study demonstrated the heterogeneity of the genus *Bacillus* and provided a firm basis for the division of the genus into several phylogenetically distinct genera. Later, representative strains of *B. pantothenticus* and other species from rRNA groups 1 and 2 and related genera (*Aneurinibacillus*, *Brevibacillus* and *Paenibacillus*) were characterized by amplified rDNA restriction analysis (ARDRA) (Heyndrickx et al. 1996), concluding that *B. pantothenticus* formed a phylogenetic group sufficiently different from other *Bacillus* species to warrant their status as a separate genus. After a polyphasic study carried out to confirm this hypothesis, the genus *Virgibacillus* was proposed to accommodate *B. pantothenticus* and related strains (LMG 12370 and LMG 17369) which appear to belong to an as-yet-undescribed new *Virgibacillus* species (Heyndrickx et al. 1998). Additional chemotaxonomic data provided by Wainø et al. (1999) supported the transfer of *B. pantothenticus* to the genus *Virgibacillus*, as *V. pantothenticus*.

Later, a polyphasic study of strains LMG 12370 and LMG 17369, originally assigned as *Bacillus* (now *Virgibacillus*) *pantothenticus*, along with strains representing species belonging to *Bacillus*, *Halobacillus* and *Paenibacillus*, was undertaken using amplified rDNA restriction analysis (ARDRA), fatty acid methyl esters (FAME) analysis, SDS-PAGE of whole-cell proteins and biochemical test, and revealed the presence within *Virgibacillus* of an as-yet-undescribed new species, for which the name *Virgibacillus proomii* was proposed (Heyndrickx et al. 1999) and also emended the description of *V. pantothenticus*. In 2003 Heyrman and coworkers studied a group of 13 strains isolated from samples of biofilms on the mural painting of the Servilia tomb (necropolis in Carmona, Spain) and the Saint-Catherine chapel (castle at Herberstein, Austria). These strains were subjected to a polyphasic taxonomic study, including $(GTG)_5$-PCR, 16S rRNA sequence analysis, DNA-DNA hybridizations, DNA base ratio determination, analysis of fatty acids, polar lipids and menaquinones and morphological and biochemical characterization. In a clustering based on the 16S rRNA gene sequence data, these species were placed in an intermediate position between the genera *Virgibacillus* and *Salibacillus*. The genus *Salibacillus* (Wainø et al. 1999) was proposed for a single species previously assigned to the genus *Bacillus*, *Bacillus salexigens* (Garabito et al. 1997), and *Bacillus marismortui* (Arahal et al. 1999) was later transferred to the genus *Salibacillus* (Arahal et al. 2000). Additionally, the study of Heyrman et al. (2003) demonstrated that those 13 strains showed intermediate DNA G+C contents and shared phenotypic properties with both genera. Therefore, a proposal to combine *Virgibacillus* and *Salibacillus* in a single genus was carried out and according to the rules of the Bacteriological Code, the two *Salibacillus* species were transferred to the genus *Virgibacillus*. In addition, those authors proposed three novel *Virgibacillus* species, *V. carmonensis*, *V. necropolis* and *V. picturae*, but the later was subsequently transferred to the genus *Oceanobacillus*, as *Oceanobacillus picturae* (Lee et al. 2006).

In 2004, Yoon et al. isolated a moderately halophilic bacterial strain, SF-121, that was found to have the closest phylogenetic affiliation to *Bacillus halodenitrificans* according to the results of 16S rRNA gene sequence analysis. However, strain SF-121 and *B. halodenitrificans* were found to be phylogenetically more closely related to the genus *Virgibacillus* than to the genus *Bacillus* and, a combination of phenotypic properties, phylogenenic analysis based on 16S rRNA gene sequence and genotypic relatedness indicated that strain SF-121 was a member of the species *Bacillus halodenitrificans* and, moreover, it was transferred to the genus *Virgibacillus* as *Virgibacillus halodenitrificans* (Yoon et al. 2004).

Currently, the genus *Virgibacillus* comprises 27 species with validly publised names (Parte 2014). All of them have been included in this genus on the basis of exhaustive polyphasic approaches.

➋ *Table 36.1* shows the type strains of species of *Virgibacillus* together with the GenBank/EMBL/DDBJ accession numbers of the 16S rRNA gene sequence of these type species.

With regard to the current phylogenetic structure and according to the 16S rRNA gene sequence analysis, the genus *Virgibacillus* is polyphyletic and does not form a well-defined group (➋ *Fig. 36.1*). The species within this genus cluster together to species of the genera *Lentibacillus*, *Oceanobacillus*, *Ornithinibacillus*, *Paucisalibacillus*, *Cerasibacillus*, and *Sediminibacillus*. Even the group containing the type species of the genus, *V. pantothenticus*, is not monophyletic, forming a branch with the species *V. chiguensis*,

## Table 36.2

**Differential phenotypic characteristics of species of the genus *Virgibacillus***

| | *V. pantothenticus* | *V. albus* | *V. alimentarius* | *V. arcticus* | *V. byunsanensis* | *V. campisalis* | *V. carmonensis* | *V. chiguensis* | *V. dokdonensis* | *V. halodenitrificans* | *V. halophilus* | *V. halotolerans* | *V. kekensis* | *V. koreensis* |
|---|---|---|---|---|---|---|---|---|---|---|---|---|---|---|
| Cell size (μm) | 0.5 – 0.7 × 2.0 – 8.0 | 0.3 – 0.5 × 2.0 – 6.0 | 0.5 × 1.2 | 0.5 – 0.8 × 2.0 – 5.0 | 0.2 – 0.4 × 1.0 – 10.0 | 0.5 – 1.0 × 1.0 – 4.0 | 0.5 – 0.7 × 2.0 – 7.0 | 0.7 – 0.9 × 2.5 – 5.0 | 0.6 – 0.8 × 2.5 – 5.0 | 0.6 – 0.8 × 2.5 – 4.0 | 0.5 × 1.75 | 0.7 – 0.8 × 3.0 – 5.0 | 2.0 – 3.0 × 0.3 – 0.5 | 0.5 – 0.7 × 2.0 – 7.0 |
| Endospore shape | E, S | E, S | E | E | S, E | S | E, (S) | S, E | S, E | E | E | E | E | E |
| Endospore position | T, (S) | T | T, S | C, S | T | T | S | T, S | T, S | T, S | S | C, S | T | T |
| Colony size (mm) | 0.5–2 | ND | 1–2 | 2 | 0.7–1 | 0.5–1.5 | 0.5–1 | 5–7 | 3–5 | 2–3 | ND | 2–4 | 2–3 | ND |
| Colony pigmentation | ND | ND | ND | ND | ND | ND | ND | Milky white | Milky white | Cream | Pale yellow | Cream | Creamy grey | Cream |
| Motility | ND | ND | + | + | + | + | ND | + | + | + | + | w | + | + |
| Temperature range (optimum) (°C) | 10–50 (37) | 15–45 (25–30) | 4–40 (37) | 0–30 (ND) | 4–40 (ND) | 15–40 (37) | 10–40 (25–30) | 15–50 (40) | 15–50 (37) | 10–45 (35–40) | 5–45 (ND) | 8–35 (27–30) | 10–50 (37) | 10–45 (25) |
| pH range (optimum) | ND (7.0) | 4.0–9.0 (7.0) | 7.0–11.0 (10.0) | 4.4–9.1 (7.0) | 6.0–ND (7.0–8.0) | 6.0–9.0 (7.5–8.0) | ND | 5.0–9.0 (7.5) | 7.0–8.0 (5.5) | 5.8–9.6 (ND) | 5.0–10.0 (ND) | 6.5–8.5 (7.0–8.0) | 6.0–10.0 (7.0) | 5.5–9.0 (7.0) |
| NaCl range (optimum) (% w/v) | ND (4–10) | 1–17 (5–10) | 0–30 (9–10) | 0–20 (5) | 0–20 (8) | 0.5–20 (4–5) | 2–25 (5–10)[a] | 0–30 (5–10) | 0–23 (4–5) | 2–23 (3–7) | 0–18 (ND) | 0.5–16.5 (3–5) | 0–25 (10) | 0–20 (5–10) |
| Anaerobic growth | + | – | – | + | – | + | – | – | + | + | – | – | – | + |
| Catalase | + | + | – | + | | + | + | ND | ND | ND | + | + | + | + |
| Oxidase | ND | + | + | – | + | + | ND | ND | + | + | + | + | + | + |
| Voges-Proskauer | – | + | ND | – | ND | ND | – | – | – | – | ND | – | – | ND |
| Indole production | – | – | – | – | ND | ND | – | – | – | – | – | – | – | – |
| Nitrate reduction | v | + | ND | + | ND | ND | + | + | ND | + | + | – | + | – |
| Aesculin hydrolysis | + | ND | ND | ND | ND | – | w | + | ND | ND | ND | + | – | + |
| Casein hydrolysis | + | – | – | + | ND | – | + | + | ND | ND | ND | + | – | ND |
| Gelatin hydrolysis | +[b] | – | – | + | ND | – | – | + | ND | ND | + | + | – | – |
| Urease | – | – | – | – | ND | – | – | ND | – | – | + | + | – | – |
| $H_2S$ production | –[c] | – | – | – | ND | ND | – | – | ND | – | – | – | – | ND |
| Arginine dihydrolase | –[c] | – | – | – | ND | – | – | – | – | – | – | ND | ND | – |
| Utilization as a sole carbon and energy sources: | | | | | | | | | | | | | | |
| D-Cellobiose | w | + | ND | – | ND | – | w | + | + | ND | ND | ND | –[e] | ND |
| D-Fructose | – | + | ND | | ND | – | – | + | ND | ND | ND | ND | ND | ND |
| D-Glucose | w | | ND | + | ND | + | – | + | ND | ND | ND | ND | + | ND |
| Sucrose | – | + | ND | + | ND | – | + | + | ND | ND | ND | ND | ND | ND |
| D-Trehalose | – | + | ND | – | ND | – | + | – | – | ND | ND | ND | + | ND |
| D-Xylose | – | – | ND | ND | ND | – | – | ND | ND | ND | ND | ND | + | ND |
| Acid production from: | | | | | | | | | | | | | | |
| D-Galactose | – | ND | – | ND | ND | – | – | + | ND | + | w | + | – | – |
| D-Glucose | – | – | + | w | ND | – | – | + | ND | + | + | + | + | ND |
| D-Fructose | – | + | – | w | ND | – | – | + | ND | + | + | + | – | + |
| Glycerol | + | + | + | ND | ND | ND | – | ND | ND | w[f] | w | + | – | – |
| Myo-inositol | – | ND | – | – | + | – | – | ND | + | – | – | ND | – | – |
| D-Mannose | – | ND | – | w | ND | – | – | + | ND | + | + | + | + | – |
| L-Rhamnose | + | ND | – | – | ND | – | – | ND | ND | – | – | ND | – | – |
| D-Trehalose | + | + | – | + | ND | – | – | – | ND | + | + | + | – | ND |

Except when indicated, data were taken from the original articles describing the taxa

Endospore shape: *E* ellipsoidal, *S* spherical. Endospore position: *T* terminal, *S* subterminal, *C* central. Shapes or positions in frequently observed are shown in parentheses. *ND* not determined, *w* weakly positive, *v* variable

[a]Data from Chen et al. (2009b)

[b]Positive for most of the strains

[c]Negative, except for a few strains

[d]Conflicting results between table and description in Arahal et al. (1999)

[e]Data from Jeon et al. (2009)

[f]Data from Lee et al. (2006)

| *V. litoralis* | *V. marismortui* | *V. natechi* | *V. necropolis* | *V. olivae* | *V. proomii* | *V. salarius* | *V. salexigens* | *V. salinus* | *V. sediminis* | *V. siamensis* | *V. soli* | *V. subterraneus* | *V. xinjiangensis* | *V. zhanjiangensis* |
|---|---|---|---|---|---|---|---|---|---|---|---|---|---|---|
| 0.5 – 0.7 × 2.0 – 5.0 | 0.5 – 0.7 × 2.0 – 3.6 | 0.5 – 0.7 × 2.0 – 5.0 | 0.5 – 0.7 × 2.0 – 5.0 | 1.8 × 0.26 | 0.5 – 0.7 × 2.0 – 8.0 | 0.6 – 0.9 × 1.8 – 3.5 | 0.3 – 0.6 × 1.5 – 3.5 | 0.9 × 1.5 – 6.0 | 0.4 – 0.7 × 2.5 – 4.0 | 0.5 – 0.7 × 2.0 – 5.0 | 0.5 × 5.0 | 0.1 – 0.3 × 1.0 – 3.0 | 0.8 – 1.2 × 1.4 – 2.4 | 0.4 – 0.7 × 2.5 – 5.5 |
| S, E | E | S | E | S, E | E, S | S, E | E | S, E | E | E | S, E | S, E | E, S | E, S |
| T | T, S | S | C, S, T | T, S | T, (S) | T, S | C, S, T | T, S | S | T, S | T | T | T, S | T |
| 0.5–1 | ND | 2–3 | 0.2–0.5 | 0.5–2 | 1–4 | 2–2.5 | ND | 1 | 2–3 | 2–5 | ND | 2–3 | ND | 0.5–1 |
| Cream | Cream | ND | ND | Yellow cream | ND | White | ND | Cream | White to pale yellow | Red | Orange to brownish | Cream to slightly ligh-yellow | Cream | Cream |
| + | + | + | ND | + | ND | + | ND | + | + | + | ND | + | – | + |
| 10–45 (30) | 15–50 (37) | 10–40 (35) | 10–40 (25–35) | 20–45 (30) | 15–50 (37) | 10–50 (30–35) | 15–45 (37) | 15–40 (37) | 10–55 (35–40) | 15–40 (37) | 15–40 (25–30) | 10–50 (30) | 8–52 (32–35) | 10–45 (30) |
| 6.0–10.0 (8.0) | 6.0–9.0 (7.5) | 6.0–12.0 (7.0) | ND | 4.0–8.8 (7.0) | ND | 5.5–10.0 (7.5) | 6.0–10.0 (7.5) | 6.0–10.0 (7.5) | 6.0–10.5 (7.5–8.0) | 7.0–8.0 (7.0) | 7.0–10.0 (7.5–8.5) | 6.0–9.0 (7.5) | 6.5–9.5 (7.5–8.0) | 6.0–10–0 (6.5) |
| 2–25 (5–10) | 5–25 (10) | 1–20 (10) | 2–25 (5–10)[a] | 0–10 (5) | 0.5–10 (ND) | 0.5–25 (7–10) | 7–20 (10) | 3–20 (10) | 1–20 (5–10) | 1–20 (5) | 0–6 (ND) | 0–25 (9) | 0–20 (5–7) | 1–15 (4–7) |
| – | – | + | – | – | + | – | – | – | – | + | + | – | – | – |
| + | + | + | + | + | + | + | + | + | + | + | + | + | + | + |
| + | + | + | ND | + | ND | + | ND | – | + | + | + | + | – | + |
| – | – | ND | – | – | – | ND | – | – | – | ND | ND | – | ND | – |
| – | – | – | – | – | – | – | – | – | – | ND | – | – | ND | – |
| + | + | ND | + | + | – | – | – | + | + | – | + | – | + | + |
| – | + | ND | – | + | + | + | + | + | + | – | ND | ND | – | – |
| – | + | – | + | + | + | + | + | – | – | w | ND | ND | + | – |
| + | + | – | w | + | +[b] | + | + | – | + | + | + | – | – | + |
| – | + | + | – | ND | ND | – | ND | ND | – | ND | – | – | – | – |
| – | ND[d] | – | – | ND | – | – | + | – | – | – | – | – | ND | – |
| ND | – | + | – | ND | –[c] | – | – | – | ND | – | – | ND | ND | – |
| | | | | | | | | | | | | | | |
| + | – | ND | + | + | w | ND | + | – | + | ND | ND | + | + | – |
| + | – | ND | – | + | – | + | – | – | – | ND | ND | + | + | – |
| + | – | ND | + | + | w | ND | w | – | + | ND | + | + | + | + |
| – | + | ND | + | – | + | ND | + | + | – | ND | + | + | – | – |
| + | + | ND | + | – | – | ND | + | – | + | ND | ND | – | + | – |
| + | – | ND | – | ND | – | ND | – | + | – | ND | ND | + | ND | + |
| | | | | | | | | | | | | | | |
| – | – | – | – | – | + | – | w | + | ND | – | – | + | – | – |
| + | + | + | w | – | + | + | w | + | + | + | w | + | + | + |
| – | + | – | w | + | + | + | w | + | + | ND | – | + | + | – |
| – | w | – | w | –[e] | – | + | – | – | ND | – | ND | + | – | – |
| – | – | – | – | – | + | – | – | ND | ND | – | ND | + | – | – |
| + | + | + | w | – | + | + | w | ND | + | – | – | + | – | – |
| – | – | – | – | – | V | – | – | ND | ND | ND | ND | + | ND | – |
| – | – | – | w | ND | + | – | – | + | ND | – | – | – | ND | – |

**Table 36.3**
**G+C content of DNA and chemotaxonomic features of *Virgibacillus* species. Except when indicated, data were taken from the original articles describing the taxa**

| Name of *Virgibacillus* species | G+C content of DNA (mol%) | Major polar lipids | Major fatty acids | Major menaquinones | Peptidoglycan type |
|---|---|---|---|---|---|
| *V. pantothenticus* | 38.3 | ND | iso-$C_{15:0}$ and anteiso-$C_{15:0}$ | MK-7 | A1γ with *m*-DAP |
| *V. albus* | 37.9 | DPG, PG, PI, a GL and two unidentified PLs | anteiso-$C_{15:0}$ and $C_{16:0}$ | MK-7 | A1γ with *m*-DAP |
| *V. alimentarius* | 37.0 | DPG, PG, PE and two unknown lipids | anteiso-$C_{15:0}$ and anteiso-$C_{17:0}$ | MK-7 | A1γ with *m*-DAP |
| *V. arcticus* | 38.2 | DPG, PG and two unidentified PLs | iso-$C_{14:0}$, iso-$C_{15:0}$, anteiso-$C_{15:0}$, iso-$C_{16:0}$, $C_{16:0}$ and anteiso-$C_{17:0}$ | MK-7 | A1α |
| *V. byunsanensis* | 37.6 | DPG, PG and two unidentified PLs | anteiso-$C_{15:0}$ | MK-7 | A1γ with *m*-DAP |
| *V. campisalis* | 39.5 | DPG, PG and two unidentified PLs | anteiso-$C_{15:0}$ and anteiso-$C_{17:0}$ | MK-7 | A1γ with *m*-DAP |
| *V. carmonensis* | 38.9 | DPG and PG | anteiso-$C_{15:0}$ and anteiso-$C_{17:0}$ | MK-7 | ND |
| *V. chiguensis* | 37.3 | DPG, PE and PG | anteiso-$C_{15:0}$, anteiso-$C_{17:0}$ and iso-$C_{15:0}$ | MK-7 | A1γ with *m*-DAP |
| *V. dokdonensis* | 36.7 | DPG, PG, PE and unidentified PLs | anteiso $C_{15:0}$, iso-$C_{15:0}$, anteiso-$C_{17:0}$ and iso-$C_{16:0}$ | MK-7 | A1γ with *m*-DAP |
| *V. halodenitrificans* | 38.0-39.0 | PG, DPG and two unidentified PLs | ND | MK-7 | A1γ with *m*-DAP |
| *V. halophilus* | 42.6 | ND | anteiso-$C_{15:0}$, iso-$C_{15:0}$, anteiso-$C_{17:0}$ and iso-$C_{16:0}$ | MK-7 | A1γ with *m*-DAP |
| *V. halotolerans* | 39.1 | DPG and PG | anteiso-$C_{15:0}$, anteiso-$C_{17:0}$ | MK-7 | A1γ with *m*-DAP |
| *V. kekensis* | 41.8 | DPG, PG and two unknown PLs | anteiso-$C_{15:0}$, iso-$C_{14:0}$, $C_{16:1}\omega 7c$ alcohol, anteiso-$C_{17:0}$ and iso-$C_{16:0}$ | MK-7 | A1γ with *m*-DAP |
| *V. koreensis* | 41.0 | PG, DPG and two unidentified PLs [a] | anteiso-$C_{15:0}$ and iso-$C_{16:0}$ | MK-7 | A1γ with *m*-DAP |
| *V. litoralis* | 40.2 | DPG and PG | anteiso-$C_{15:0}$, iso-C $_{15:0}$ and anteiso-$C_{17:0}$ | MK-7 | A1γ with *m*-DAP |
| *V. marismortui* | 40.7 | DPG, PG, PE, five PLs, one APL and one polar lipids of unknown structure | ND | MK-7 | A1γ with *m*-DAP |
| '*V. natechi*' | 42.1 | DPG, PG and three PLs | anteiso-$C_{15:0}$, anteiso-$C_{17:0}$, $C_{20:0}$ and anteiso-$C_{19:0}$ | MK-7 | ND |
| *V. necropolis* | 37.3 | DPG and PG | anteiso-$C_{15:0}$ and anteiso-$C_{17:0}$ | MK-7 | ND |
| *V. olivae* | 33.4 | ND | iso-$C_{15:0}$, anteiso-$C_{15:0}$, iso-$C_{17:0}$ and anteiso-$C_{17:0}$ | ND | ND |
| *V. proomi* | 37.0 | ND | iso-$C_{15:0}$ and anteiso-$C_{15:0}$ | ND | ND |
| *V. salarius* | 37.3 | PG, DPG, PE, two unknown PLs and cellular polar lipids | iso-$C_{15:0}$ and anteiso-$C_{15:0}$ | Mk-7 | A1γ with *m*-DAP |
| *V. salexigens* | 39.5 | PG, DPG and two PLs of unknown structure | iso-$C_{15:0}$ and anteiso-$C_{15:0}$ | MK-7 | A1γ with *m*-DAP |
| *V. salinus* | 38.8 | DPG, PG, GL and two unidentified PLs | anteiso-$C_{15:0}$ and iso-$C_{14:0}$ | MK-7 | A1γ with *m*-DAP |
| *V. sediminis* | 40.9 | DPG, PG, PE and unknown PLs | anteiso-$C_{15:0}$ and anteiso-$C_{17:0}$ | MK-7 | A1γ with *m*-DAP |
| *V. siamensis* | 38.8 | PG, DPG and an unidentified GL | anteiso-$C_{15:0}$ and anteiso-$C_{17:0}$ | MK-7 | A1γ with *m*-DAP |

**Table 36.3 (continued)**

| Name of *Virgibacillus* species | G+C content of DNA (mol%) | Major polar lipids | Major fatty acids | Major menaquinones | Peptidoglycan type |
|---|---|---|---|---|---|
| *V. soli* | ND | DPG, PG, PE and one unidentified PL | iso-$C_{15:0}$ and anteiso-$C_{15:0}$ | MK-7 | ND |
| *V. subterraneus* | 37.1 | DPG, PG and GL | anteiso-$C_{15:0}$ and anteiso-$C_{17:0}$ | MK-7 | A1γ with *m*-DAP |
| *V. xinjiangensis* | 44.5 | PG and DPG | anteiso-$C_{15:0}$, anteiso-$C_{17:0}$ and $C_{16:0}$ | MK-7 | A1γ with *m*-DAP |
| '*V. zhanjiangensis*' | 39.5 | DPG and PG | anteiso-$C_{15:0}$ and anteiso-$C_{17:0}$ | MK-7 | A1γ with *m*-DAP |

[a]Data from Zhang et al. (2012)

*MK* menaquinone, *m*-DAP meso-diaminopimelic acid, *DPG* diphosphatidylglycerol, *PG* phosphatidylglycerol, *PI* phosphatidylinositol, *GL* glycolipid, *PE* phosphatidylethanolamine, *PL* phospholipid, *APL* aminophospholipid, *ND* not determined

*V. dokdonensis*, *V. proomii*, *Ornithinibacillus bavariensis*, and *Paucisalibacillus globulus* (➲ *Fig. 36.1*). Therefore, a deep revision of the phylogeny of the genus *Virgibacillus* is required. According to ➲ *Fig. 36.1*, it seems that the monophyletic branch containing 25 species of *Virgibacillus* and the following seven species: *Lenitibacillus lacisalsi*, *Lentibacillus salicampi*, *Oceanobacillus iheyensis*, *Oceanobacillus picturae*, *Ornithinobacillus bavariensis*, *Paucisalibacillus globulus* and *Cerasibacillus quisquiliarum* may constitute a stable and coherent group, which might be regarded as the true genus *Virgibacillus*. On the other hand, the species *V. albus* and *V. koreensis*, which are quite distant from the other species of the genus *Virgibacillus* and closely related to the genus *Sediminibacillus* (➲ *Fig. 36.1*), do not belong apparently to the genus *Virgibacillus*. Nevertheless, further in-depth studies are required in order to confirm the taxonomic status and the coherence of the genus *Virgibacillus* and the proposals for transferring those current species which are not phylogenetically coherent.

## Molecular Analyses

### Genome Analysis

*Virgibacillus halodenitrificans* 1806 is the only strain of the genus for which the full genome sequence has been published (ALEF00000000) (Lee et al. 2012a). However, this strain is not the type strain of the species; *Virgibacillus halodenitrificans* 1806 is an endospore-forming halophilic bacterium isolated from salterns in Korea (Lee et al. 2012a). The single replicon genome, determined by a whole-genome shotgun strategy using an Illumina HiSeq 2,000 instrument, is 3,920,549 bp long with a 37.4 % G+C content. This value is only slightly lower than those determined by the Tm method (38.0–39.0 mol%, ➲ *Table 36.3*). Of the predicted 3,949 protein-coding genes, 45 % were assigned to subsystem categories. Strain 1806 contains genes related to osmolarity for the uptake of compatible solutes, genes involved in fructose metabolism and also genes involved in alternative respiration pathways that supported its ability to grow anaerobically (Lee et al. 2012a).

## Phenotypic Analyses

As stated on previous sections the species of the genus *Virgibacillus* have been extensively studied on the basis of polyphasic approaches including their phenotypic characterization. ➲ *Table 36.2* includes the differential features of the species of *Virgibacillus*. Other approaches have been suggested for the characterization or identification of members of *Virgibacillus*. The VITEK2 *Bacillus* identification card (BCL) can be used for the identification of *Bacillus* species and members of related genera as *Virgibacillus* (Halket et al. 2010).

## Isolation, Enrichment and Maintenance Procedures

### Isolation

The members of the genus *Virgibacillus* are widely found in many habitats and the currently described species have been mostly isolated from saline environments. The places of isolation of the species of the genus *Virgibacillus* are shown in ➲ *Table 36.1*.

*Virgibacillus* strains have been isolated from salterns in southern Taiwan (Wang et al. 2007). Huang et al. (2009) isolated representatives of the genus *Virgibacillus* in the sea urchin *Hemicentrotus pulcherrimus*, collected from Naozhou Island in Zhanjiang, China, and also Xiao et al. (2009) isolated members of *Virgibacillus* associated with a sea anemone from the coast of the same island.

It has been shown that endospores of *Virgibacillus* species are present in non-saline environments such as ordinary garden soils, yards, fields and roadways in an area surrounding Tokyo, Japan. A possible source of the endospores originating from Asian dust storm has been reported (Echigo et al. 2005).

### Maintenance

Suitable culture media for the growth of the *Virgibacillus* strains as well as the range of pH and temperature and the optimal values are shown in ➲ *Table 36.2*.

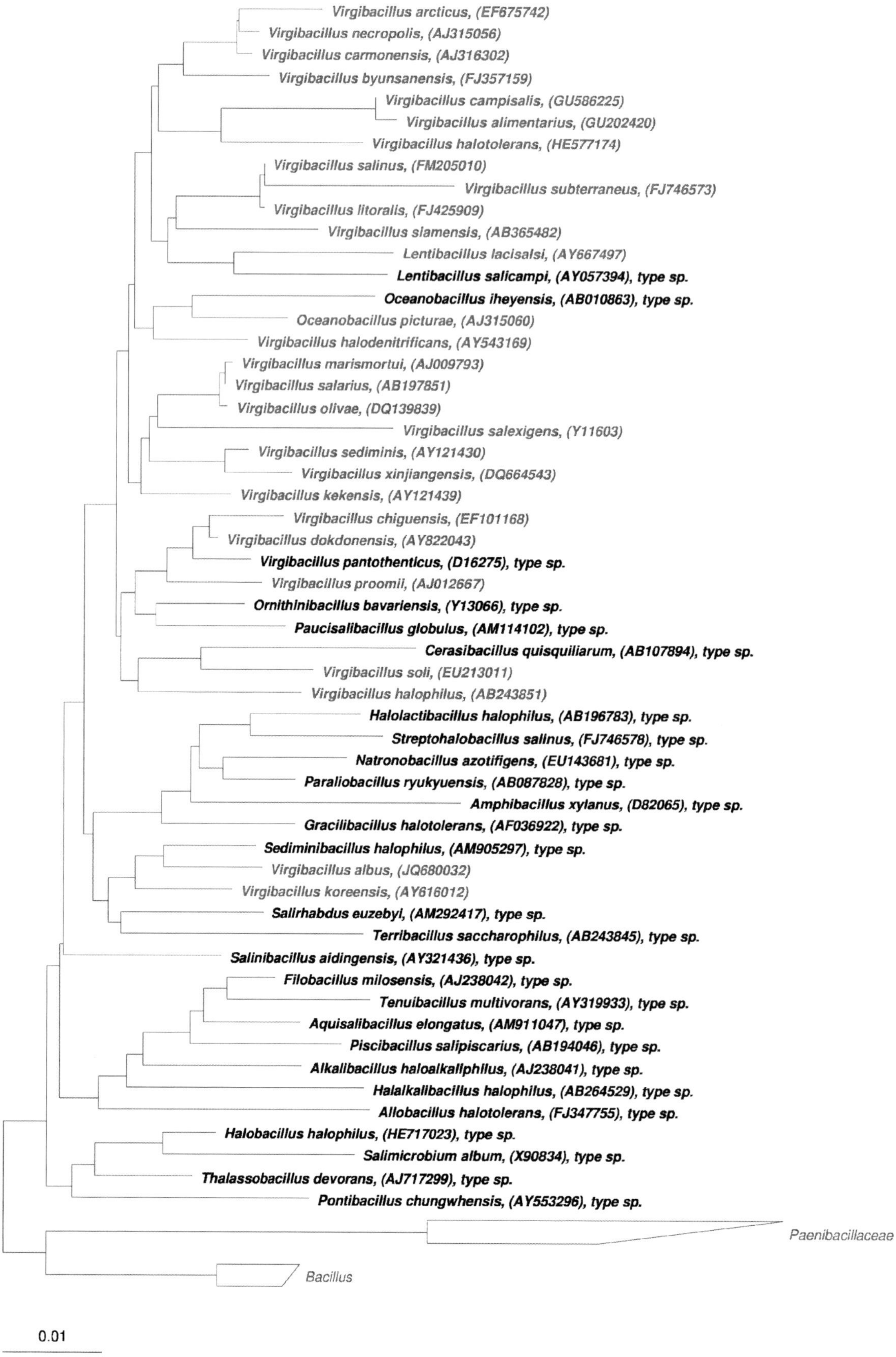
Virgibacillus arcticus, (EF675742)
Virgibacillus necropolis, (AJ315056)
Virgibacillus carmonensis, (AJ316302)
Virgibacillus byunsanensis, (FJ357159)
Virgibacillus campisalis, (GU586225)
Virgibacillus alimentarius, (GU202420)
Virgibacillus halotolerans, (HE577174)
Virgibacillus salinus, (FM205010)
Virgibacillus subterraneus, (FJ746573)
Virgibacillus litoralis, (FJ425909)
Virgibacillus siamensis, (AB365482)
Lentibacillus lacisalsi, (AY667497)
Lentibacillus salicampi, (AY057394), type sp.
Oceanobacillus iheyensis, (AB010863), type sp.
Oceanobacillus picturae, (AJ315060)
Virgibacillus halodenitrificans, (AY543169)
Virgibacillus marismortui, (AJ009793)
Virgibacillus salarius, (AB197851)
Virgibacillus olivae, (DQ139839)
Virgibacillus salexigens, (Y11603)
Virgibacillus sediminis, (AY121430)
Virgibacillus xinjiangensis, (DQ664543)
Virgibacillus kekensis, (AY121439)
Virgibacillus chiguensis, (EF101168)
Virgibacillus dokdonensis, (AY822043)
Virgibacillus pantothenticus, (D16275), type sp.
Virgibacillus proomii, (AJ012667)
Ornithinibacillus bavariensis, (Y13066), type sp.
Paucisalibacillus globulus, (AM114102), type sp.
Cerasibacillus quisquiliarum, (AB107894), type sp.
Virgibacillus soli, (EU213011)
Virgibacillus halophilus, (AB243851)
Halolactibacillus halophilus, (AB196783), type sp.
Streptohalobacillus salinus, (FJ746578), type sp.
Natronobacillus azotifigens, (EU143681), type sp.
Paraliobacillus ryukyuensis, (AB087828), type sp.
Amphibacillus xylanus, (D82065), type sp.
Gracilibacillus halotolerans, (AF036922), type sp.
Sediminibacillus halophilus, (AM905297), type sp.
Virgibacillus albus, (JQ680032)
Virgibacillus koreensis, (AY616012)
Salirhabdus euzebyi, (AM292417), type sp.
Terribacillus saccharophilus, (AB243845), type sp.
Salinibacillus aidingensis, (AY321436), type sp.
Filobacillus milosensis, (AJ238042), type sp.
Tenuibacillus multivorans, (AY319933), type sp.
Aquisalibacillus elongatus, (AM911047), type sp.
Piscibacillus salipiscarius, (AB194046), type sp.
Alkalibacillus haloalkaliphilus, (AJ238041), type sp.
Halalkalibacillus halophilus, (AB264529), type sp.
Allobacillus halotolerans, (FJ347755), type sp.
Halobacillus halophilus, (HE717023), type sp.
Salimicrobium album, (X90834), type sp.
Thalassobacillus devorans, (AJ717299), type sp.
Pontibacillus chungwhensis, (AY553296), type sp.
Paenibacillaceae
Bacillus
0.01

■ Fig. 36.1 (continued)

*Virgibacillus* species do not require special procedures for maintenance.

Medium-term maintenance can be performed in 20 % (v/v) glycerol suspensions at −20 °C or at −80 °C. For long-term preservation liquid nitrogen or lyophilization is recommended.

## Applications

Members of the genus *Virgibacillus* are involved in different aspects in food processing. Jeotgal or jeot is a traditional Korean salted and fermented food made by adding 20–30 % (w/v) salts to various types of seafoods. It has been shown that members of the genus *Bacillus* and its relatives (especially *Virgibacillus*) may be the major group of organisms involved in jeotgal fermentation (Guan et al. 2011). *Virgibacillus kekensis* was isolated in Dongcai, a traditional pickled mustard product in Sichuan Province, China (Dong et al. 2012). Fish sauce production relies on a natural fermentation process requiring 12–18 months for process completion. Two isolates identified as *Virgibacillus* sp. SK33 and *Virgibacillus* sp. SK37 have been shown to be potential strains for accelerating the fish sauce production and could be promising strains for starter culture development used in fish sauce fermentation (Sinsuwan et al. 2007, 2012; Yongsawatdigul et al. 2007). Recently, the effect of *Virgibacillus* sp. SK37, together with reduced salt content, on fish sauce quality, particularly free amino acids and odor-active compounds was investigated. The results suggested that the inoculation of fish sauce samples with this strain under reduced salt contents of 15–20 % likely contributed to stronger malty or dark chocolate notes (Lapsongphon et al. 2013). On the other hand, Essghaier et al. (2009) showed that *Virgibacillus marismortui* may be useful in biological control against grey mould caused by *Botrytis cinerea*, an economically important disease of strawberries.

*Virgibacillus*, together with isolates belonging to the genus *Bacillus* and other related bacteria were isolated in a study focused on the culturable aerobic bacteria associated with the human gastrointestinal tract and it was suggested that these bacilli with potential as probiotics could be isolated from the human gut (Hoyles et al. 2012).

Several screenings have been carried out in order to isolate bacteria able to produce hydrolytic enzymes such as amylases, proteases, DNAses, phosphatases or lipases. Members of the genus *Virgibacillus* able to produce extracellular enzymes have been obtained from different areas of Howz Soltan playa, a hypersaline lake in the central desert zone of Iran (Rohban et al. 2009), in sediment underlying the oxygen minimum zone of the eastern Arabian Sea (Divya et al. 2010) and from various saline habitats of India (Kumar et al. 2012).

*Virgibacillus halodenitrificans* produces a salt-inducible peptide with putative kinase activity (Rafiee et al. 2007). The above mentioned strains *Virgibacillus* sp. SK33 and *Virgibacillus* sp. SK37 were isolated from Thai fish sauce and are able to produce different enzymes with potential biotechnological applications (Sinsuwan et al. 2010, 2011; Phrommao et al. 2011a, b). A NaCl-activated and organic solvent-stable heterotrimer proteinase from *Virgibacillus* sp. SK33 has been purified. This proteinase could have a potential application in high ionic strength environments and aqueous-organic solvent systems (Sinsuwan et al. 2010). Besides, cell-associated proteinases from *Virgibacillus* sp. SK33 have been extracted and characterized. The extracted enzymes could be used to hydrolyze food proteins at a NaCl content as high as 25 % (Sinsuwan et al. 2011). The three major proteinases produced by *Virgibacillus* sp. SK37 are member of bacillopeptidase F-like enzymes exhibiting thermophilic and halotolerant characteristics with high stability at 30 % NaCl. They showed potential to be a processing-aid for food and biotechnological applications, particulary at high salt conditions (Phrommao et al. 2011a). Besides, *Virgibacillus* sp. SK37 produces a proteinase with a potential processing-aid for the production of a mungbean meal hydrolyzate with antioxidant properties (Lapsongphon and Yongsawatdigul 2013). Finally, a gene encoding a novel member of the subtilase superfamily was isolated from *Virgibacillus* sp. SK37. The stability toward $H_2O_2$ and moderately halo- and thermo-tolerant properties of this enzyme make it very attractive for various biotechnological applications (Phrommao et al. 2011b).

*Virgibacillus* sp. Rob has been described as the producer of a bioflocculant. Chemical analysis of the bioflocculant revealed it to be a polysaccharide (Cosa et al. 2011).

Strains of the genus *Virgibacillus*, as well as from other genera, were obtained in a study carried out to isolate and characterize bacteria with antimicrobial activities from Brazilian sponges. These findings suggest that the identified strains may contribute to the search for new sources of antimicrobial substances (Santos et al. 2010).

*Virgibacillus* sp. H4 was isolated from a mangrove soil of Bhitarkanila, India and has a high Cr (VI) reducing ability under saline conditions. This feature suggests that *Virgibacillus* sp. H4 could be a new and efficient strain capable of remediating highly saline Cr (VI) polluted industrial effluents (Mishra et al. 2012). Moreover, other strain of *Virgibacillus* was

**Fig. 36.1**

**Phylogenetic reconstruction of the genus *Virgibacillus* based on 16S rRNA gene sequence comparison obtained using the neighbour-joining algorithm with the Jukes-Cantor correction. The sequence datasets and alignments were used according to the All-Species Living Tree Project (LTP) database (Yarza et al. 2010; http://www.arb-silva.de/projects/living-tree). The tree topology was stabilized with the use of a representative set of nearly 750 high quality type strain sequences proportionally distributed among the different bacterial and archaeal phyla. In addition, a 40 % maximum frequency filter was applied in order to remove hypervariable positions and potentially misplaced bases from the alignment. Scale bar indicates estimated sequence divergence**

isolated under anoxic conditions in highly copper-contaminated Chilean marine sediments (Besaury et al. 2013) and its use in biorremediation processes has been suggested. Finally, *Virgibacillus* sp. J2 was isolated from activated sludge in an epoxy wastewater treatment system and the best degradation conditions of the organic epoxy wastewater were determined (Wang et al. 2013).

Like many halophilic and halotolerant bacteria, members of this genus synthesize various types of compatible solutes, including ectoine. *Virgibacillus salexigens*, *Virgibacillus marismortuii* and *Virgibacillus pantothenticus* have been described as producers of ectoine in response to high salinity or low growth temperature (He et al. 2005; Kuhlmann et al. 2008, 2011; Reuter et al. 2010). Besides, the crystal structure of the ectoine hydroxylase EctD, enzyme responsible for ectoine hydroxylation, from *Virgibacillus salexigens* has been reported (Reuter et al. 2010; Widderich et al. 2013).

## Acknowledgments

The work of authors was supported by grants from the Spanish Ministry of Science and Innovation (CGL2010-19303, CGL2013-46941-P) and the Junta de Andalucía (P10-CVI-6226). FEDER funds also supported authors' research.

## References

Amziane M, Metiaz F, Darenfed-Bouanane A, Djenane Z, Selama O, Abderrahmani A, Cayol JL, Fardeau ML (2013) *Virgibacillus natechei* sp. nov., a moderately halophilic bacterium isolated from sediment of a saline lake in southwest of Algeria. Curr Microbiol 66:462–466

An SY, Asahara M, Goto K, Kasai H, Yokota A (2007) *Virgibacillus halophilus* sp. nov., spore-forming bacteria isolated from soil in Japan. Int J Syst Evol Microbiol 57:1607–1611

Arahal DR, Márquez MC, Volcani BE, Schleifer KH, Ventosa A (1999) *Bacillus marismortui* sp. nov., a new moderately halophilic species from the Dead Sea. Int J Syst Bacteriol 49:521–530

Arahal DR, Márquez MC, Volcani BE, Schleifer KH, Ventosa A (2000) Reclassification of *Bacillus marismortui* as *Salibacillus marismortui* comb. nov. Int J Syst Evol Microbiol 50:1501–1503

Ash C, Farrow JAE, Wallbanks S, Collins MD (1991) Phylogenetic heterogeneity of the genus *Bacillus* revealed by comparative analysis of small-subunit-ribosomal RNA sequences. Lett Appl Microbiol 13:202–206

Besaury L, Marty F, Buquet S, Mesnage V, Muyzer G, Quillet L (2013) Culture-dependent and independent studies of microbial diversity in highly copper-contaminated Chilean marine sediments. Microb Ecol 65:311–324

Carrasco IJ, Márquez MC, Ventosa A (2009) *Virgibacillus salinus* sp. nov., a moderately halophilic bacterium from sediment of a saline lake. Int J Syst Evol Microbiol 59:3068–3073

Chen YG, Cui XL, Fritze D, Chai LH, Schumann P, Wen ML, Wang YX, Xu LH, Jiang CL (2008) *Virgibacillus kekensis* sp. nov., a moderately halophilic bacterium isolated from a salt lake in China. Int J Syst Evol Microbiol 58:647–653

Chen YG, Cui XL, Wang YX, Zhang YQ, Tang SK, Li WJ, Liu ZX, Wen ML, Peng Q (2009a) *Virgibacillus sediminis* sp. nov., a moderately halophilic bacterium isolated from a salt lake in China. Int J Syst Evol Microbiol 59:2058–2063

Chen YG, Liu ZX, Peng DJ, Zhang YQ, Wang YX, Tang SK, Li WJ, Cui XL, Liu YQ (2009b) *Virgibacillus litoralis* sp. nov., a moderately halophilic bacterium isolated from saline soil. Antonie van Leeuwenhoek 96:323–329

Cosa S, Mabinya LV, Olaniran AO, Okoh OO, Bernard K, Deyzel S, Okoh AI (2011) Bioflocculant production by *Virgibacillus* sp. Rob isolated from the bottom sediment of Algoa Bay in the Eastern Cape, South Africa. Molecules 16:2431–2442

Divya B, Soumya KV, Nair S (2010) 16S rRNA and enzymatic diversity of culturable bacteria from the sediments of oxygen minimum zone in the Arabian Sea. Antonie van Leeuwenhoek 98:9–18

Dong L, Pu B, Ao X, Zhang X, Zheng Y, Li X (2012) Bacterial biodiversity in Dongcai, a traditional pickled mustard product in Sichuan Province, China. Wei Sheng Wu Xue Bao 52:519–525

Echigo A, Hino M, Fukushima T, Mizuki T, Kamekura M, Usami R (2005) Endospores of halophilic bacteria of the family *Bacillaceae* isolated from non-saline Japanese soil may be transported by Kosa event (Asian dust storm). Saline Systems 1:8

Essghaier B, Fardeau ML, Cayol JL, Hajlaoui MR, Boudabous A, Jijakli H, Sadfi-Zouaoui N (2009) Biological control of grey mould in strawberry fruits by halophilic bacteria. J Appl Microbiol 106:833–846

Garabito MJ, Arahal DR, Mellado E, Márquez MC, Ventosa A (1997) *Bacillus salexigens* sp. nov., a new moderately halophilic *Bacillus* species. Int J Syst Bacteriol 47:735–741

Guan L, Cho KH, Lee JH (2011) Analysis of the cultivable bacterial community in jeotgal, a Korean salted and fermented seafood, and identification of its dominant bacteria. Food Microbiol 28:101–113

Halket G, Dinsdale AE, Logan NA (2010) Evaluation of the VITEK2 BCL card for identification of *Bacillus* species and other aerobic endosporeformers. Lett Appl Microbiol 50:120–126

He J, Wang T, Sun JQ, Gu LF, Li SP (2005) Isolation and characteristic of a moderately halophilic bacterium accumulated ectoine as main compatible solute. Wei Sheng Wu Xue Bao 45:900–904

Heyndrickx M, Lebbe L, Kersters K, De Vos P, Forsyth G, Logan NA (1998) *Virgibacillus*: a new genus to accommodate *Bacillus pantothenticus* (Proom and Knight 1950). Emended description of *Virgibacillus pantothenticus*. Int J Syst Bacteriol 48:99–106

Heyndrickx M, Vauterin L, Vandamme P, Kersters K, De Vos P (1996) Applicability of combined amplified 16S rDNA restriction analysis (ARDRA) patterns in bacterial phylogeny and taxonomy. J Microbiol Methods 26:247–259

Heyndrickx M, Lebbe L, Kersters K, Hoste B, De Wachter R, De Vos P, Forsyth G, Logan NA (1999) Proposal of *Virgibacillus proomii* sp. nov. and emended description of *Virgibacillus pantothenticus* (Proom and Knight 1950) Heyndrickx et al. 1998. Int J Syst Bacteriol 49:1083–1090

Heyrman J, Logan NA, Busse HJ, Balcaen A, Lebbe L, Rodriguez-Diaz M, Swings J, De Vos P (2003) *Virgibacillus carmonensis* sp. nov., *Virgibacillus necropolis* sp. nov. and *Virgibacillus picturae* sp. nov., three novel species isolated from deteriorated mural paintings, transfer of the species of the genus *Salibacillus* to *Virgibacillus*, as *Virgibacillus marismortui* comb. nov. and *Virgibacillus salexigens* comb. nov., and emended description of the genus *Virgibacillus*. Int J Syst Evol Microbiol 53:501–511

Heyrman J, De Vos P, Logan N (2009) Genus XIX. *Virgibacillus*. In: De Vos P, Garrity GM, Jones D, Krieg NR, Ludwig W, Rainey FA, Schleifer K-H, Whitman WB (eds) Bergey's manual of systematic bacteriology, vol 3, 2nd edn. Springer, New York, pp 193–228

Hoyles L, Honda H, Logan NA, Halket G, La Ragione RM, McCartney AL (2012) Recognition of greater diversity of *Bacillus* species and related bacteria in human faeces. Res Microbiol 163:3–13

Hua NP, Hamza-Chaffai A, Vreeland RH, Isoda H, Naganuma T (2008) *Virgibacillus salarius* sp. nov., a halophilic bacterium isolated from a Saharan salt lake. Int J Syst Evol Microbiol 58:2409–2414

Huang K, Zhang L, Liu Z, Chen Q, Peng Q, Li W, Cui X, Chen Y (2009) Diversity of culturable bacteria associated with the sea urchin *Hemicentrotus pulcherrimus* from Naozhou Island. Wei Sheng Wu Xue Bao 49:1424–1429

Jeon CO, Kim JM, Park DJ, Xu LH, Jiang CL, Kim CJ (2009) *Virgibacillus xinjiangensis* sp. nov., isolated from a Salt Lake of Xin-jiang Province in China. J Microbiol 47:705–709

Kämpfer P, Arun AB, Busse HJ, Langer S, Young CC, Chen WM, Syed AA, Rekha PD (2011) *Virgibacillus* soli sp. nov., isolated from mountain soil. Int J Syst Evol Microbiol 61:275–280

Kim J, Jung M-J, Roh SW, Nam Y-D, Shin K-S, Bae J-W (2011) *Virgibacillus alimentarius* sp. nov., isolated from a traditional Korean food. Int J Syst Evol Microbiol 61:2851–2855

Kuhlmann AU, Bursy J, Gimpel S, Hoffmann T, Bremer E (2008) Synthesis of the compatible solute ectoine in *Virgibacillus pantothenticus* is triggered by high salinity and low growth temperature. Appl Environ Microbiol 74:4560–4563

Kuhlmann AU, Hoffmann T, Bursy J, Jebbar M, Bremer E (2011) Ectoine and hydroxyectoine as protectants against osmotic and cold stress: uptake through the SigB-controlled betaine-choline- carnitine transporter-type carrier EctT from *Virgibacillus pantothenticus*. J Bacteriol 193:4699–4708

Kumar S, Karan R, Kapoor S, Singh SP, Khare SK (2012) Screening and isolation of halophilic bacteria producing industrially important enzymes. Braz J Microbiol 43:1595–1603

Lapsongphon N, Yongsawatdigul J (2013) Production and purification of antioxidant peptides from a mungbean meal hydrolysate by *Virgibacillus* sp. SK37 proteinase. Food Chem 141:992–999

Lapsongphon N, Cadwallader KR, Rodtong S, Yongsawatdigul J (2013) Characterization of protein hydrolysis and odor-active compounds of fish sauce inoculated with *Virgibacillus* sp. SK37 under reduced salt content. J Agric Food Chem 61:6604–6613

Lee JS, Lim JM, Lee KC, Lee JC, Park YH, Kim CJ (2006) *Virgibacillus koreensis* sp. nov., a novel bacterium from a salt field, and transfer of *Virgibacillus picturae* to the genus *Oceanobacillus* as *Oceanobacillus picturae* comb. nov. with emended descriptions. Int J Syst Evol Microbiol 56:251–257

Lee SJ, Lee YJ, Jeong H, Lee SJ, Lee HS, Pan JG, Kim BC, Lee DW (2012a) Draft genome sequence of *Virgibacillus halodenitrificans* 1806. J Bacteriol 194:6332–6333

Lee SY, Kang CH, Oh TK, Yoon JH (2012b) *Virgibacillus campisalis* sp. nov., from a marine solar saltern. Int J Syst Evol Microbiol 62:347–351

Logan NA, Berge O, Bishop AH, Busse H-J, De Vos P, Fritze D, Heyndrickx M, Kämpfer P, Rabinovitch L, Salkinoja-Salonen MS, Seldin L, Ventosa A (2009) Proposed minimal standards for describing new taxa of aerobic, endospore-forming bacteria. Int J Syst Evol Microbiol 59:2114–2121

Mishra RR, Dhal B, Dutta SK, Dangar TK, Das NN, Thatoi HN (2012) Optimization and characterization of chromium (VI) reduction in saline condition by moderately halophilic *Virgibacillus* sp. isolated from mangrove soil of Bhitarkanika, India. J Hazard Mater 227–228:219–226

Niederberger TD, Steven B, Charvet S, Barbier B, Whyte LG (2009) *Virgibacillus arcticus* sp. nov., a moderately halophilic, endospore-forming bacterium from permafrost in the Canadian high Arctic. Int J Syst Evol Microbiol 59:2219–2225

Parte AC (2014) List of prokaryotic names with standing in nomenclature. http://www.bacterio.net

Peng QZ, Chen J, Zhang YQ, Chen QH, Peng DJ, Cui XL, Li WJ, Chen YG (2009) *Virgibacillus zhanjiangensis* sp. nov., a marine bacterium isolated from sea water. Antonie Van Leeuwenhoek 96:645–652

Phrommao E, Rodtong S, Yongsawatdigul J (2011a) Identification of novel halotolerant bacillopeptidase F-like proteinases from a moderately halophilic bacterium, *Virgibacillus* sp. SK37. J Appl Microbiol 110:191–201

Phrommao E, Yongsawatdigul J, Rodtong S, Yamabhai M (2011b) A novel subtilase with NaCl-activated and oxidant-stable activity from *Virgibacillus* sp. SK37. BMC Biotechnol 11:65

Proom H, Knight BCJG (1950) *Bacillus pantothenticus* (n.sp.). J Gen Microbiol 4:539–541

Quesada T, Aguilera M, Morillo JA, Ramos-Cormenzana A, Monteoliva-Sánchez M (2007) *Virgibacillus olivae* sp. nov., isolated from waste wash-water from processing of Spanish-style green olives. Int J Syst Evol Microbiol 57:906–910

Rafiee MR, Sokhansanj A, Yoosefi M, Naghizadeh MA (2007) Identification of salt-inducible peptide with putative kinase activity in halophilic bacterium *Virgibacillus halodenitrificans*. J Biosci Bioeng 104:178–181

Reuter K, Pittelkow M, Bursy J, Heine A, Craan T, Bremer E (2010) Synthesis of 5-hydroxyectoine from ectoine: crystal structure of the non-heme iron (II) and 2-oxoglutarate-dependent dioxygenase EctD. PLoS One 5:e10647

Rohban R, Amoozegar MA, Ventosa A (2009) Screening and isolation of halophilic bacteria producing extracellular hydrolyses from Howz Soltan Lake, Iran. J Ind Microbiol Biotechnol 36:333–340

Santos OC, Pontes PV, Santos JF, Muricy G, Giambiagi-deMarval M, Laport MS (2010) Isolation, characterization and phylogeny of sponge-associated bacteria with antimicrobial activities from Brazil. Res Microbiol 161:604–612

Seiler H, Wenning M (2013) *Virgibacillus halotolerans* sp. nov., isolated from a dairy product. Int J Syst Evol Microbiol 63:3358–3363

Sinsuwan S, Rodtong S, Yongsawatdigul J (2007) NaCl-activated extracellular proteinase from *Virgibacillus* sp. SK37 isolated from fish sauce fermentation. J Food Sci 72:C264–C269

Sinsuwan S, Rodtong S, Yongsawatdigul J (2010) Purification and characterization of a salt-activated and organic solvent-stable heterotrimer proteinase from *Virgibacillus* sp. SK33 isolated from Thai fish sauce. J Agric Food Chem 58:248–256

Sinsuwan S, Rodtong S, Yongsawatdigul J (2011) Evidence of cell-associated proteinases from *Virgibacillus* sp. SK33 isolated from fish sauce fermentation. J Food Sci 76:C413–C419

Sinsuwan S, Rodtong S, Yongsawatdigul J (2012) Hydrolytic activity of *Virgibacillus* sp. SK37, a starter culture of fish sauce fermentation, and its cell-bound proteinases. World J Microbiol Biotechnol 28:2651–2659

Tanasupawat S, Chamroensaksri N, Kudo T, Itoh T (2010) Identification of moderately halophilic bacteria from Thai fermented fish (pla-ra) and proposal of *Virgibacillus siamensis* sp. nov. J Gen Appl Microbiol 56:369–379

Wainø M, Tindall BJ, Schumann P, Ingvorsen K (1999) *Gracilibacillus* gen. nov., with description of *Gracilibacillus halotolerans* gen. nov., sp. nov.; transfer of *Bacillus dipsosauri* to *Gracilibacillus dipsosauri* comb. nov., and *Bacillus salexigens* to the genus *Salibacillus* gen. nov., as *Salibacillus salexigens* comb. nov. Int J Syst Bacteriol 49:821–831

Wang CY, Ng CC, Chen TW, Wu SJ, Shyu YT (2007) Microbial diversity analysis of former salterns in southern Taiwan by 16S rRNA-based methods. J Basic Microbiol 47:525–533

Wang CY, Chang CC, Ng CC, Chen TW, Shyu YT (2008) *Virgibacillus chiguensis* sp. nov., a novel halophilic bacterium isolated from Chigu, a previously commercial saltern located in southern Taiwan. Int J Syst Evol Microbiol 58:341–345

Wang X, Xue Y, Ma Y (2010) *Virgibacillus subterraneus* sp. nov., a moderately halophilic Gram-positive bacterium isolated from subsurface saline soil. Int J Syst Evol Microbiol 60:2763–2767

Wang J, Xu Z, Peng SC, Xia MS, Yue ZB, Chen TH (2013) Screening of epoxy-degrading halophiles and their application in high-salt wastewater treatment. Huan Jing Ke Xue 34:1510–1516

Widderich N, Pittelkow M, Höppner A, Mulnaes D, Buckel W, Gohlke H, Smits SH, Bremer E (2013) Molecular dynamics simulations and structure-guided mutagenesis provide insight into the architecture of the catalytic core of the ectoine hydroxylase. J Mol Biol 426:586–600

Xiao H, Chen Y, Liu Z, Huang K, Li W, Cui X, Zhang L, Yi L (2009) Phylogenetic diversity of cultivable bacteria associated with a sea anemone from coast of the Naozhou island in Zhanjiang, China. Wei Sheng Wu Xue Bao 49:246–250

Yarza P, Ludwig W, Euzeby J, Amann R, Schleifer KH, Glöckner FO, Rossello-Mora R (2010) Update of the all-species living tree project based on 16S and 23S rRNA sequence analyses. Syst Appl Microbiol 33:291–299

Yongsawatdigul J, Rodtong S, Raksakulthai N (2007) Acceleration of Thai fish sauce fermentation using proteinases and bacterial starter cultures. J Food Sci 72:M382–M390

Yoon JH, Oh TK, Park YH (2004) Transfer of *Bacillus halodenitrificans* Denariaz et al. 1989 to the genus *Virgibacillus* as *Virgibacillus halodenitrificans* comb. nov. Int J Syst Evol Microbiol 54:2163–2167

Yoon JH, Kang SJ, Lee SY, Lee MH, Oh TK (2005) *Virgibacillus dokdonensis* sp. nov., isolated from a Korean island, Dokdo, located at the edge of the East Sea in Korea. Int J Syst Evol Microbiol 55:1833–1837

Yoon JH, Kang SJ, Jung YT, Lee KC, Oh HW, Oh TK (2010) *Virgibacillus byunsanensis* sp. nov., isolated from a marine solar saltern. Int J Syst Evol Microbiol 60:291–295

Zhang YJ, Zhou Y, Ja M, Shi R, Chun-Yu WX, Yang LL, Tang SK, Li WJ (2012) *Virgibacillus albus* sp. nov., a novel moderately halophilic bacterium isolated from Lop Nur salt lake in Xinjiang province, China. Antonie van Leeuwenhoek 102:553–560

# Section II

# Tenericutes

# 37 The Family *Acholeplasmataceae* (Including Phytoplasmas)

*Marta Martini*[1] · *Carmine Marcone*[2] · *Ing-Ming Lee*[3] · *Giuseppe Firrao*[1]
[1]Dipartimento di Scienze Agrarie ed Ambientali, Università degli Studi di Udine, Udine, Italy
[2]Dipartimento di Farmacia, Università degli Studi di Salerno, Fisciano, Salerno, Italy
[3]Molecular Plant Pathology Laboratory, USDA, ARS, Beltsville, MD, USA

***Taxonomy, Historical and Current*** ........................ *469*

***Phenotypic and Molecular Analyses*** ........................ *472*
Genus *Acholeplasma* ........................................ 472
Genus "*Candidatus* Phytoplasma" ........................ 476

***Ecology and Pathogenicity*** ................................ *492*

***Application*** ................................................ *495*

**Abstract**
The family *Acholeplasmataceae* was originally established to accommodate the genus *Acholeplasma*, comprising the mollicutes that could be cultivated without the supplement of cholesterol and that use UGA as a stop codon instead of coding for tryptophan. It was later shown that the phytoplasmas, a large group of uncultivable, wall-less, non-helical mollicutes that are associated with plants and insects, shared taxonomically relevant properties with members of the genus *Acholeplasma*. Being not cultivable in vitro in axenic culture, the phytoplasmas could not be classified using the standards used for other mollicutes and are named using the category of *Candidatus*, as "*Ca.* Phytoplasma."

Although phytoplasmas are associated with habitats and ecology different from acholeplasmas, the *two* genera *Acholeplasma* and "*Candidatus* Phytoplasma" are phylogenetically related and form a distinct clade within the Mollicutes. The persisting inability to grow the phytoplasmas in vitro hinders the identification of their distinctive phenotypic traits, important criteria for mollicute classification. Until supplemental phenotypic traits become available, the genus "*Candidatus* Phytoplasma" is designated, on the basis of phylogeny, as a tentative member in the family *Acholeplasmataceae*. Phylogenetic analysis *based on* gene sequences, in particular, ribosomal sequences, has provided the major supporting evidence for the composition and taxonomic subdivision of this group of organisms with diverse habitats and ecology and has become the mainstream for the *Acholeplasmataceae* systematics. However, without the ability to determine phenotypic properties, the circumscription of related species among the non culturable members of the family remains a major issue.

The genus *Acholeplasma* comprises 14 species predominantly associated with animals and isolated from mammalian fluids but regarded as not normally pathogenic. Conversely, the genus "*Ca.* Phytoplasma" includes plant pathogens of major economic relevance worldwide. To date, 36 "*Ca.* Phytoplasma species" have been described.

## Taxonomy, Historical and Current

The family *Acholeplasmataceae*, the sole family of the class *Acholeplasmatales*, encompasses a phylogenetically well-distinct group of organisms within the class *Mollicutes*. It includes, however, a complex and diverse array of cultivated and not cultivated organisms. Historically, the establishment of this family was supported by differences in the isolation origin and in nutritional requirements. Sabin (1941) proposed the genus *Sapromyces* and the family Sapromycetaceae to distinguish organisms isolated from sewage (Laidlaw and Elford 1936) from the parasitic species of the pleuropneumonia group. Although the species *S. laidlawi* was later transferred in the genus *Mycoplasma* by Edward and Freundt (1956), the same authors (Edward and Freundt 1969) proposed the reestablishment of genus and family for those strains as they differ from other members of the Mycoplasmatales in not requiring sterol for growth, in the belief that dependence on sterol was a fundamental property. The name *Sapromyces*, however, was illegitimate, as it was the valid name of a genus of fungi of the order *Leptomitales*, and therefore, the genus was named *Acholeplasma* and the family *Acholeplasmataceae* (Edward and Freundt 1970). Since the reestablishment of genus and family, several experimental data suggested that there were profound taxonomically relevant properties that were connected with the nutritional requirement for sterol. *Acholeplasma* species are not capable of synthesizing cholesterol; however, they can grow without it because, unlike other Mollicutes, cholesterol is not an essential membrane component of *Acholeplasma* membrane (Razin et al. 1982). Thus, the difference in sterol requirement relates to other properties in lipid metabolism, and to the incorporation and location of lipids in the cell membrane, and leads to consequent major differences in the structure and composition of their cell membranes. For example, a difference between *Acholeplasma* and *Mycoplasma* is the positional distribution of fatty acids in their membrane phosphatidylglycerol molecules. In two *Acholeplasma* species, the saturated fatty acids are preferentially located at position 1 of glycerol, and the unsaturated

E. Rosenberg et al. (eds.), *The Prokaryotes – Firmicutes and Tenericutes*, DOI 10.1007/978-3-642-30120-9_387,

fatty acids are located at position 2, whereas in *Mycoplasma* the positional distribution of the fatty acids is the reverse (Rottem and Markowitz 1979). *Acholeplasma* species have also been found to synthesize saturated fatty acids and polyterpenes from acetate, to possess a nicotinamide adenine dinucleotide-dependent lactate dehydrogenase that is located in the cell membrane and is specifically activated by fructose 1,6-diphosphate, and to contain superoxide dismutase, as well as glucose-6-phosphate and 6-phosphogluconate dehydrogenases. Moreover, acholeplasmas do not share with mycoplasmas the alternative genetic code that uses the codon UGA for the amino acid tryptophan instead of the usual opal stop codon. These differences justified the establishment of a second order within the class *Mollicutes*, the order *Acholeplasmatales* (Freundt et al. 1984), with the purpose of accommodating the family *Acholeplasmataceae*. The subdivision of the Mollicutes into two major branches has later been shown to be congruent with phylogenetic analysis. The rRNA sequence analysis (Rogers et al. 1985; Johansson and Pettersson 2002; Maniloff 2002), as well as the analysis of entire genomes (Sirand-Pugnet et al. 2007), supported the separation at the root of the mollicute phylogeny of the acholeplasmas and related organisms (AAP branch: Acholeplasmas, Anaeroplasmas and Phytoplasmas), from other mollicutes (SEM branch: Spiroplasmas, Mycoplasmas, and Entomoplasmas), even though the finding of SEM mollicutes growing without sterols showed that growth requirement for sterols was not as a profound character as it was initially believed (Rose et al. 1993).

Since the late 1980s, the branch that is now called AAP was subdivided into two orders, to separate strictly anaerobic anaeroplasmas from the facultatively aerobic acholeplasmas (Robinson and Freundt 1987). It was further shown that the phytoplasmas, a large group of uncultivable, wall-less, non-helical mollicutes associated with plants and insects, shared taxonomically relevant properties with members of the genus *Acholeplasma* (Lim and Sears 1989, 1992; Lim et al. 1992). Global phylogenetic analysis of 16S rRNA gene sequences from representative gram-positive walled prokaryotes, wall-less mycoplasmas, and plant pathogenic phytoplasmas revealed that phytoplasmas together with acholeplasmas and anaeroplasmas formed a discrete subclade (Anaeroplasma subclade) paraphyletic with the Mycoplasma subclade within the Mollicute clade, which separated from other representative walled prokaryotes. Phytoplasmas formed a large discrete monophyletic group diverging from acholeplasmas (Gundersen et al. 1994; Wei et al. 2007).

Being not cultivable in vitro in axenic culture, the phytoplasmas could not be classified using the standards used for other mollicutes (ICSB-STM 1995) and are named using the category of *Candidatus*. This fact makes it difficult to list distinctive characters of the order *Acholeplasmatales* based on experimental evidence, and phylogeny based on gene sequences provides an inevitable alternative for determining the composition of this order.

Within the order *Acholeplasmatales*, cultivable organisms have been assigned to the genus *Acholeplasma*, whereas the non-cultivable to the genus "*Candidatus* Phytoplasma." Since the ability to grow in vitro depends on a wide number of biochemical, physiological, and genetic features that are absent in the non-cultivable strains, this rough split actually corresponds to substantial and effective differences in the genome composition of the members of the two genera. However, the phylogenetic relationship between some members of the two genera (such as *A. palmae* and "*Ca.* P. americanum") estimated from well-accepted molecular markers such as 16S rRNA suggests the retention of both genera in a single family. Thus, this chapter will deal with the family *Acholeplasmataceae* as comprising the two genera, *Acholeplasma* and "*Ca.* Phytoplasma." Other classification schemes, relying on the assumption that the status of *Candidatus* is not compatible with formal systematics, label the genus "*Ca.* Phytoplasma" as *incertae sedis* (Harrison et al. 2010).

Indeed, the classification of phytoplasmas within the genus "*Ca.* Phytoplasma" has a peculiar history and significance. Due to the inability to cultivate these bacteria and to provide a stable way for conservation of type strains in culture collections, the phytoplasmas cannot be named according to the rules of the bacteriological code (Lapage et al. 1992). Although rule 16a of the code states that a nonliving record may be used as a type, it would not be able to stand the continuous advances in the taxonomic methods that require that the "name-bearing"-type strain be accessed for reexamination.

The restrictions in the application of a proper formal nomenclature made it difficult to unambiguously refer to the phytoplasmas. Being plant pathogens, they were initially named by the disease they caused, but when molecular methods clarified that the same symptoms may be associated to distinct phytoplasmas, and that the same phytoplasma may affect different plants, the traditional naming approach proved to be clearly inadequate. In the attempt to address the general issue concerning several groups of uncultivable bacteria, Murray and Stackebrandt (1995) proposed the category of *Candidatus* as a way to refer to partially characterized uncultivable organisms. This introduction prompted the activity of a working team of the International Research Programme on Comparative Mycoplasmology (IRPCM), created within the frame of the International Organization for Mycoplasmology, to provide a way to refer unambiguously to phytoplasmas (ICSB-STM 1995).

While adapting this category to the phytoplasmas, the IRPCM working team realized the risk of an uncontrolled proliferation of names of phytoplasmas and defined a guidance to restrict the number of species being proposed (IRPCM 2004). Although the *Candidatus* category has been introduced with the scope of unambiguous reference to organisms bearing a specific sequence, and could therefore be applied to all different samples that display a unique sequence, the IRPCM working team agreed to introduce a threshold of 97.5 % identity in the 16S rRNA sequence for the description of novel *candidatae* species. This was clearly an artifact due only to practical reasons, as naming as many *candidatae* species as the number of 16S rDNA sequences available would have had the consequence of inundating the literature with hundreds of useless names.

Although the establishment of the 97.5 % threshold had the merit to restrict the number of species to a manageable number of species, there are several limitations of the approach that were soon realized by the phytoplasma scientific community.

First, 16S rRNA is not a good marker to establish differences at the species level. There has been general agreement in the last 40 years that the gold standard for the prokaryotic species establishment and circumscription has been the DNA-DNA hybridization (DDH) (Brown et al. 2007; Richter and Rosselló-Móra 2009). It has been the only taxonomic method that offered a numerical and relatively stable species boundary, and its use granted the absence of subjective judgment in species delimitation. As a consequence, DDH has dominated the way how the current classification has been constructed. Although 16S rRNA determination is much easier and workable than DDH, it was shown that the correlation plot of the two phylogenetic parameters DNA similarity and 16S rRNA homology is not linear (Stackebrandt and Goebel 1994). More recently, the limited resolution at the species level of the 16S rDNA was confirmed by genome sequencing, which showed extensive genotypic differences among many organisms with highly similar 16S rRNA gene sequences (Konstantinidis and Tiedje 2007).

Sequence analysis of 16S rRNA is superior from the level of domains (55 % similarity) to moderately related species, i.e., below 97.5 % similarity. Above this value, DNA reassociation values can either be low or as high as 100 % (Stackebrandt and Goebel 1994). Several groups of organisms have been identified which share almost identical 16S rRNA sequences but in which DNA hybridization is significantly lower than 70 %, thus indicating that they represent individual species. However, at sequence identity values below 97.5 %, it is unlikely that two organisms have more than 60–70 % DNA similarity and hence that they are related at the species level (Stackebrandt and Goebel 1994). Thus, a 97.5 % similarity threshold in the 16S rDNA, as it has been used for "*Ca.* P. species," may be a cautionary level, but nothing that could be used to establish species boundaries: two strains that share less than 97.5 % are certainly belonging to two species, but whether or not strains that share more than 97.5 % should be placed in the same species cannot be assessed by their similarity in the 16S rRNA sequence. Indeed, since the beginning of the implementation of the *Candidatus* category in phytoplasmas, it was necessary to introduce exceptions to accommodate the classification of phytoplasmas that are clearly poorly related to each other on an ecological and epidemiological ground, despite their high sequence similarity in the 16S rDNA. In the absence of sound molecular markers for general use, the IRPCM working team outlined some additional guidance lines for the establishment of different species for strains sharing more than 97.5 % 16S rDNA sequence identity based on epidemiological, environmental, or molecular evidences (IRPCM 2004). This agreement has been made to be preliminary, i.e., lasting the time that sufficient information would be available to the finer molecular characterization of the taxa.

A second relevant problem posed by the definition of *Candidatus* species on the sole basis of 16S rRNA is that the association with a single, one-dimensional characteristic such as the sequence of a single gene does not allow the definition of a taxonomic space. Thus, the placement of a strain in one rather than in another species is not supported by multiple elements that contribute to approximate the complexity of the similarity relationships between two organisms. By choosing a single gene, it was implicitly assumed that the entire diversity of two organisms could be summarized in the handful of nucleotide positions that distinguish the two 16S rRNA sequences compared. Not only this is obviously imprecise, but it is also prone to conflicts. For instance, if two species share less than 97.5 % similarity between each other, and a strain is discovered which has 98 % similarity to both, then the two species cannot be longer recognized as distinct because they do not share less than 97.5 % similarity with any other species.

The consequence of both problems listed above was the substantial inability to circumscribe species in the genus "*Candidatus* Phytoplasma." As the promoters of this naming initiative were aware of this difficulty, the "*Candidatus* Phytoplasma species" was defined as a point, i.e., only strains with the exact identical 16S rDNA sequence are to be retained within the same species. A taxonomic space where the species could be circumscribed was deliberately not defined, as this would not be possible using the sole 16S rDNA without extreme subjectivity. Strains with a 16S rRNA sequence similar but not identical to a given "*Candidatus* Phytoplasma species" were not named and were since then referred as "related to" (IRPCM 2004). This was a workable yet largely unsatisfactory solution, as several strains were confined to an unnamed status and only indirectly referred to.

Subsequent work aimed at the search of additional molecular markers to be used for phytoplasma classification provided the basis for the alleviation of the inconveniences introduced in this first collaborative approach to the phytoplasma taxonomy. The convergence toward a DNA sequence-based approach of the research in phytoplasmology on one side and of the addresses of general bacterial systematics on the other has greatly shortened the distance between the formal taxonomy and the tentative phytoplasma classification. Over the last 30 years, the former reliance upon morphological, physiological, and biochemical characterization in bacterial systematics has been replaced by an approach that takes into account the complexity of the organisms. At first with the introduction of a polyphasic approach (Vandamme et al. 1996) that integrates different kinds of data and information, and then with the present major focus on multilocus sequence analysis (MLSA), the bacterial taxonomy evolved a modern concept that searches the consensus of multigene analyses. MLSA is based on multilocus sequence typing (MLST), introduced by Maiden and coworkers (Maiden et al. 1998), which has been developed for microbial typing based on sequences of multiple genes. Different housekeeping genes are preferably used in this approach because they are expected to evolve at a slow and constant rate (Kämpfer and Glaeser 2012). MLSA is more powerful than 16S rRNA gene to resolve the taxa at the species level and was proposed to replace DDH studies (Gevers et al. 2005; Konstantinidis et al. 2006; Kuhnert and Korczak 2006).

On a similar track, the phytoplasmologists pursued the identification of additional sequence markers for the improvement of the taxonomic approach based on the sole 16S rDNA analysis (Lee et al. 2009). Sequences of genes evolving more rapidly than 16S rDNA may be appropriate to resolve the differences among strains that may belong to related species. Even more important, multiple gene analysis allows the construction of a taxonomic space where strains can be collocated, cluster of similar strains identified, and the boundaries that delimitated those cluster defined. Briefly, with the use of multiple molecular markers, the circumscription of the species becomes possible even for "*Ca.* Phytoplasma" species.

The several markers that have been identified in the last few years that proved useful in the analysis of the taxonomic relationships among phytoplasmas are discussed in some detail below in this chapter. Despite the large effort, they are still limited in number, and more critically, a comprehensive evaluation of their taxonomic significance was not yet carried out. Indeed, the phylogenies build on different genes may be discordant and choosing few genes without preliminary assessment of their congruity with the organism evolutionary pattern may lead to a subjective estimation of the relationships among taxa (Rokas et al. 2003). For the phytoplasmas, the number of available genes is still limited and their congruity with the organism evolution difficult to estimate due to the significant impact of horizontal gene transfer (Saccardo et al. 2012). Therefore, species circumscriptions such as those attempted in this chapter should still be seen as tentative until sufficient data in support could be gained from genome analysis. When the selection of genes for use in addition to 16S rRNA will be corroborated by the results of comparative genomics, then the phytoplasma taxonomic space will be at all similar to that of cultivable organisms and the species within the genus "*Candidatus* Phytoplasma" will be circumscribed with the same accuracy that is presently applied to formally recognized species. By starting of the process of defining the taxonomic space for the genus "*Ca.* Phytoplasma," we are aware that the *Candidatus* category is assuming the meaning of a provisional taxon, and its use as a simple way to record the properties of a poorly characterized organism has been definitely left behind.

## Phenotypic and Molecular Analyses

A comprehensive view of the characteristics of the family *Acholeplasmataceae* has been limited by the different approaches that could be adopted for the two genera that compose the family. Members of the family *Acholeplasmataceae* typically appear by electron microscopy observation as pleomorphic, with coccoid, coccobacillary, or filamentous forms and bounded by a 7–8 nm membrane, and hence cannot be distinguished from other mollicutes by their morphology. The genus "*Candidatus* Phytoplasma," being made of non-cultivable organisms, could not be characterized using the biochemical and cultural assays that have been used extensively for member of the genus *Acholeplasma*. Standing the inability to ascertain common properties that distinguish them from other mollicutes, the substantiation of this family deeply relies on DNA sequence data and analyses. According to such analyses, members of the family use UGA as a stop codon, and not as tryptophan. Extensive phylogenetic analysis of DNA sequence information has been carried out for the phytoplasmas, particularly for those associated with economically important plant diseases and, to a minor extent, to the acholeplasmas. As shown by the analysis of 16SrRNA, ITS, rpoB, and gyrB carried out comprehensively here (➋ *Fig. 37.1*) and separately for acholeplasmas and phytoplasmas by others (Lee et al. 2000; Volokhov et al. 2006, 2007), phylogeny strongly and congruently supports the position of this family within the mollicutes.

### Genus *Acholeplasma*

Although originally isolated from sewage and soil and thus regarded as saprophytes, most *Acholeplasma* strains have been isolated from the upper respiratory tract and urogenital tract of vertebrate animals so that, according to Tully (1979), they are the most common mollicutes in the vertebrates. Accordingly, such *Acholeplasma* species have an optimum growth temperature of 37 °C and grow much slower at 27 °C. A few species are of plant (*Acholeplasma brassicae* and *Acholeplasma palmae*) and insect (*Acholeplasma pleciae*) origin and have an optimum growth temperature of 30 °C. Acholeplasmas can be grown on liquid and solid media, where they form colonies with mollicute-typical fried-egg appearance, and when grown in absence of sterols, colonies may be much smaller and they may show only central zones of growth, without the less dense peripheral areas. When observed by transmission electron microscope, all species show pleomorphism, although in several cases a predominant form can be detected.

According to the most recent revision of the minimum standard document for the definition of mollicute species (Brown et al. 2007), the 14 currently recognized *Acholeplasma* species are taxonomically distinguished primarily by serology that has been documented to be congruent with DNA-DNA hybridization data and with 16S rRNA gene placements (Aulakh et al. 1983; Stephens et al. 1983a, b; Brown et al. 2007). Antisera raised against whole-cell antigens (Tully 1979) are used in growth inhibition (Clyde 1983), plate immunofluorescence (Tully 1973), and metabolism inhibition (Taylor-Robinson 1983) assays. In addition, circumscription of the species has been attempted by evaluating some biochemical tests. Most acholeplasmas are fermenters and can produce acid from different sugars, and such feature has been traditionally used in species differentiation.

Characterization and differentiation of acholeplasmas using DNA analysis were primarily performed using the 16S rDNA gene. *Acholeplasma* species are all well distinct from each other, and in most cases, they share among each other less than 97 % identity in their 16S rRNA genes. Indeed, only *A. laidlawii*, *A. pleciae*, and *A. granularum* share among themselves 97–98.7 % identity. There are, therefore, several oligonucleotide

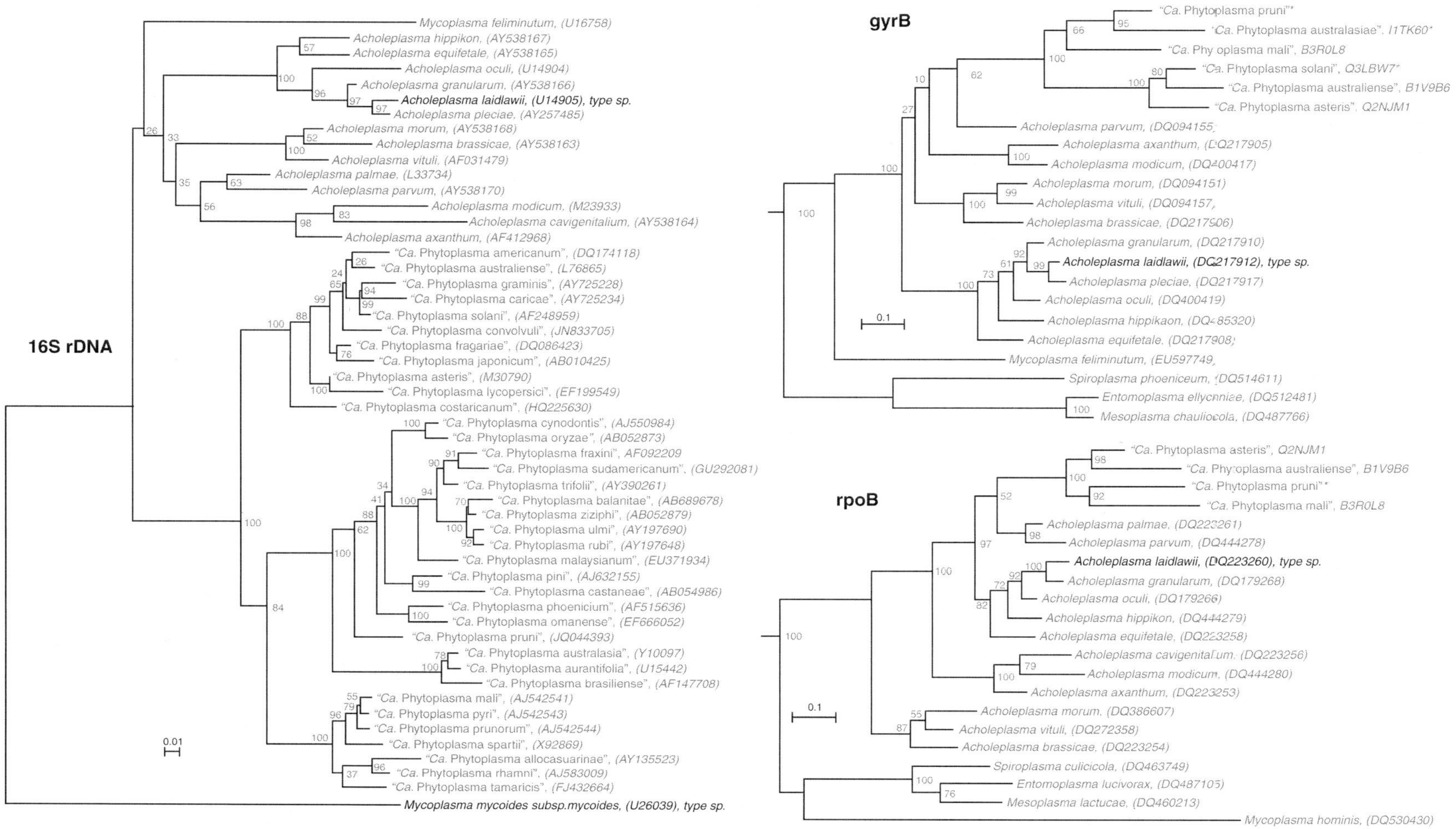


**Fig. 37.1**
**Phylogenetic trees of the *Acholeplasmataceae* (and other mollicutes introduced as out-groups) based on maximum likelihood analysis of alignments of the 16SrDNA (*left*), *gyrB* (*top right*), and *rpoB* (*bottom right*) gene sequences. Sequences labelled with an asterisk are not from the type or reference strain. Numbers at nodes are bootstrap values calculated with 100 replicates**

signatures in the 16S rDNA sequences that could be used for species definition. Species-specific signatures presently represent the only unifying criteria for the description of all species within the family *Acholeplasmataceae* and hence have been identified and provided below.

According to the analyses of 16S rDNA sequence and *rpoB* and *gyrB* genes, the genus *Acholeplasma* could be subdivided into four subclades: (1) *A. laidlawii, A. pleciae, A. granularum, A. hippikon, A. oculi,* and *A. equifetale;* (2) *A. brassicae, A. vituli,* and *A. morum;* (3) *A. palmae* and *A. parvum;* and (4) *A. modicum, A. cavigenitalium,* and *A. axanthum.* There are some additional species that have been assigned to the genus *Acholeplasma* (*A. multilocale, A. florum, A. entomophilum, A. seiffertii*), but it has been successively shown that they do not belong to this genus (Brown et al. 2010). Conversely, the species named *Mycoplasma feliminutum* is related to the genus *Acholeplasma* and is possibly a member of this or of a close, still to be named, other genus of the family (Brown et al. 2010).

***Subclade I*** is made of a group of three closely related species (*A. laidlawii, A. pleciae, A. granularum*) and three less related species (*A. hippikon, A. oculi, and A. equifetale)* that share at least 91.2 % identity between their 16S rRNA gene sequences and 90.9 % identity or less with 16S rRNA gene sequences of *Acholeplasma* species belonging to other subclades.

***Acholeplasma laidlawii*** (Edward and Freundt 1970), named to honor the microbiologist P. Laidlaw, is the type species of the genus *Acholeplasma*. It is a cosmopolite species, as strains have been isolated from many animal hosts and also from soil and plants. When cultured on agar plates containing serum, it shows large colonies with well-developed central zones and peripheral growth. It can also be cultured in serum-free media, where colonies are smaller. Electron microscopy observation reveals pleomorphism, with prevalence of filaments, branched or not branched, or coccoid forms depending on the growth medium. The growth temperature range is 20 °C–41 °C with optimum at 37 °C, also for plant and soil isolates. The species is serologically distinct from other species of the clade, but partial cross-reactions may occur with *A. granularum* strains. DNA-DNA hybridization between strains of this species ranges from 40 % to >80 %. The type strain of *A. granularum*, strain BTS-39 T, showed 20 % hybridization with *A. laidlawii* strain PG8T. *A. laidlawii* is positive to fermentation of glucose, esculin hydrolysis, production of carotenoid pigments, film and spots reaction, and benzyl viologen reduction and negative to fermentation of mannose.

Type strain: ATCC 23206, PG8, NCTC 10116, CIP 75.27, and NBRC 14400. Oligonucleotide sequences of unique regions of the 16S rRNA gene (accession U14905) are 5′-AAATAA-GTCCGGAGGCTTACAGA-3′ (pos. 977–1000), 5′-GTTG-GGCAAAAG-3′ (pos. 795–807), 5′-AGTGGTGAAGG-3′ (pos. 684–695), 5′-GAAAAATTGAAAATTGACGGTACCAT-3′ (pos. 430–456), 5′-AGCAGTAAGGGAAT-3′ (pos. 332–346), and 5′-TTGGTGGAGTAAAAGCCTACCAAGACG-3′ (pos. 231–258).

***Acholeplasma equifetale*** (Kirchhoff 1974) was isolated from the lung and liver of aborted horse fetuses, from the respiratory tract of healthy horses, and from the respiratory tract and cloacae of broiler chickens (Bradbury 1978). Cells are predominantly coccoid. It grows and forms typical fried-egg colonies on agar plates with and without horse serum. Growth temperature is 22–37 °C. *A. equifetale* is positive to fermentation of glucose, fermentation of mannose, film and spots reaction, and benzyl viologen reduction.

Type strain: C112, ATCC 29724, and NCTC 10171. Oligonucleotide sequences of unique regions of the 16S rRNA gene (accession AY538165) are 5′-GAAGGGGTGACCT-CAAGCAA-3′ (pos. 1186–1206), 5′-GTAAAGTCCTTTTA-3′ (pos. 379–393), and 5′-AGGTAAAAGCTTA-3′ (pos. 210–223).

***Acholeplasma granularum*** (Edward and Freundt 1970) was isolated frequently from the nasal cavity of swine and also from swine feces, conjunctiva and nasopharynx of horses, and the genital tract of guinea pigs and as a contaminant of eukaryotic cell cultures. Colonies on solid medium are large with clearly marked center, although they are smaller and may lack the peripheral zone on media without serum. Cells are short filaments and coccoid. Optimum growth temperature is 37 °C. *A. granularum* is positive to fermentation of glucose and benzyl viologen reduction and negative to fermentation of mannose, esculin hydrolysis, production of carotenoid pigments, and film and spots reaction.

Type strain: BTS-39, ATCC 19168, and NCTC 10128. Oligonucleotide sequences of unique regions of the 16S rRNA gene (accession no. AY538166) are 5′-TGGTTAATTAAGTTTGT-3′ (pos. 521–538), 5′-TCGGTCTAGGAGGGGT-3′ (pos. 170–186), and 5′-TAGGATATAGAGATAATTTCT-3′ (pos. 132–153).

***Acholeplasma hippikon*** (Kirchhoff 1978) was isolated from the lung of aborted horse fetuses. Cells are predominantly coccoid. Optimum growth temperature is 37 °C. *A. hippikon* is positive to fermentation of glucose, fermentation of mannose, film and spots reaction, and benzyl viologen reduction.

Type strain: C1, ATCC 29725, and NCTC 10172. Oligonucleotide sequences of unique regions of the 16S rRNA gene (accession no. AY538167) are 5′-GAGAAGCAAGAGGGT-3′ (pos. 1180–1195), 5′-ATGGCAAATACAAAGA-3′ (pos. 1166–1182), 5′-CATGTTCTTTAATTCGTCGATA-3′ (pos. 879–901), 5′-TAGAGTAAGACAGA-3′ (pos. 587–601), and 5′-GTAGGTTGTTAAT-3′ (pos. 517–530).

***Acholeplasma oculi*** corrig. (al-Aubaidi et al. 1973) was isolated from the conjunctiva of goats with keratoconjunctivitis; porcine nasal secretions; equine nasopharynx, lung, spinal fluid, joint, and semen; urogenital tract of cattle; and external genitalia of guinea pigs, ducks, and turkeys. It has also been found in amniotic fluid of pregnant women (Waites et al. 1987), palm trees, and other plants (Eden-Green and Tully 1979; Somerson et al. 1982). Cells appear with spherical, ring-shaped, and coccobacillary forms. Growth temperature range is 25–37 °C. *A. oculi* is positive to fermentation of glucose, esculin hydrolysis, production of carotenoid pigments, and benzyl viologen reduction and negative to fermentation of mannose and film and spots reaction.

Type strain: 19-L, ATCC 27350, and NCTC 10150. Oligonucleotide sequences of unique regions of the 16S rRNA gene (accession no. U14904) are 5′-AGCGATGGGTT-GACCCGGAGC-3′(pos.1231–1252),5′-ACCTGGCCTCCAAAC-3′

(pos. 1188–1203), 5′-GATGACCGTCAATCAATCATCCCCCT-3′ (pos. 1158–1184), 5′-AAACTGTTTAGCTAGAGTGAGAC-3′ (pos. 616–639), 5′-TGAATAACCCCCG-3′ (pos. 469–483), and 5′-ATGTAAGGTTCT-3′ (pos. 419–431).

***Acholeplasma pleciae*** (Knight 2004) was originally isolated from the hemolymph of a larva of the corn-root maggot (*Plecia* sp.; Tully et al. 1994a). Cells are predominantly coccoid. Temperature range for growth is 18–32 °C, with optimal growth at 30 °C. It is positive to fermentation of glucose.

Type strain: PS-1 and ATCC 49582. Oligonucleotide sequences of unique regions of the 16S rRNA gene (accession no. AY257485) are 5′-AAGCGAGAGGGT-3′ (pos. 1184–1196), 5′-GTTATTCAAGTCTGTGGT-3′ (pos. 526–544), and 5′-GGGTAAGAGCCTACCAAGACA-3′ (pos. 213–234).

***Subclade II*** is made of 3 related species (*A. brassicae, A. vituli, and A. morum*) that share at least 95.7 % identity between their 16S rRNA gene sequences and 92.8 % identity or less with 16S rRNA gene sequences of *Acholeplasma* species belonging to other subclades.

***Acholeplasma brassicae*** (Tully et al. 1994b) was isolated from the surface of broccoli plant. Cells are predominantly coccoid. Optimum growth is at 30 °C. *A. brassicae* is positive to fermentation of glucose and benzyl viologen reduction and negative to fermentation of mannose and esculin hydrolysis.

Type strain: 0502 and ATCC 49388. Oligonucleotide sequences of unique regions of the 16S rRNA gene (accession no. AY538163) are 5′-CATATTAGTACT-3′ (pos. 818–830), 5′-GTGTTGCGATAACGC-3′ (pos. 789–804), 5′-TGCAGGGCTCAACCC-3′ (pos. 571–586), and 5′-AAGAATGTCGTTGGTAGGAAA-3′ (pos. 408–429).

***Acholeplasma vituli*** (Angulo et al. 2000) was isolated from fetal bovine serum or contaminated eukaryotic cell cultures containing serum. Cells are predominantly coccoid in shape. Temperature range for growth is 25–37 °C. It is positive to fermentation of glucose and mannose and negative to esculin hydrolysis, production of carotenoid pigments, and film and spots reaction.

Type strain: FC 097–2, ATCC 700667, and CIP 107001. Oligonucleotide sequences of unique regions of the 16S rRNA gene (accession no. AF031479) are 5′-AACCGCAAGGAG-3′ (pos. 1418–1430), 5′-AGTTCAGATTGTAGTCTG-3′ (pos. 1271–1289), and 5′-AAACGTCATACA-3′ (pos. 1199–1211).

***Acholeplasma morum*** (Rose et al. 1980) was isolated from commercial fetal bovine serum and from calf kidney tissue cultures on media containing fetal bovine serum. Colonies on solid medium are smaller than those of other acholeplasmas and only with addition of fatty acids on serum-free media. Cells are predominantly coccoid and coccobacillary with some beaded filaments. Optimum growth temperature is 35–37 °C. *A. morum* is positive to fermentation of glucose, esculin hydrolysis, production of carotenoid pigments, and benzyl viologen reduction and negative to fermentation of mannose and film and spots reaction.

Type strain: 72–043, ATCC 33211, and NCTC 10188. Oligonucleotide sequences of unique regions of the 16S rRNA gene (accession no. AY538168) are 5′-GTTCGAATTGGAGT-3′ (pos. 1245–1259), 5′-GTGCAAGGCTCAA-3′ (pos. 568–581), and 5′-ATGATCTGTGAGTGAC-3′ (pos. 426–442).

***Subclade III*** is made of 2 species (*A. palmae and A. parvum*) that share 94.3 % identity between their 16S rRNA gene sequences and 93.3 % identity or less with 16S rRNA gene sequences of *Acholeplasma* species belonging to other subclades. It is the subclade most closely related to the phytoplasma, as its members share up to 91.0 % identity in the 16S rDNA sequence with members of subclade I of "*Ca.* Phytoplasma."

***Acholeplasma palmae*** (Tully et al. 1994b) was isolated from the crown tissues of a palm tree (*Cocos nucifera*) with lethal yellowing disease, although it has not been shown to be associated with the disease. Cells are predominantly coccoid. The temperature range for growth is 18–37 °C, with optimal growth occurring at 30 °C. No evidence for pathogenicity. *A. palmae* is positive to fermentation of glucose and benzyl viologen reduction and negative to fermentation of mannose and esculin hydrolysis.

Type strain: J233 and ATCC 49389. Oligonucleotide sequences of unique regions of the 16S rRNA gene (accession no. L33734) are 5′-CGAAACCCATAA-3′ (pos. 1201–1213), 5′-CGAAAGAGTGATCTGGAGCGAAACC-3′ (pos. 1183–1208), 5′-GACGCTCATGCAC-3′ (pos. 688–701), and 5′-CGGTCTATAATGT-3′ (pos. 524–537).

***Acholeplasma parvum*** (Atobe et al. 1983) was isolated from the oral cavities and vagina of healthy horses. Cells are coccobacillary. It needs special growth factor of 1 % Phytone or Soytone peptone supplements for growth and it grows better with the addition of 15 % fetal bovine serum. *A. parvum* is positive to benzyl viologen reduction and negative to fermentation of glucose and production of carotenoid pigments.

Type strain: H23M, ATCC 29892, and NCTC 10198. Oligonucleotide sequences of unique regions of the 16S rRNA gene (accession no. AY538170) are 5′-GGAAGCTAAGCAG-3′ (pos. 1198–1211), 5′-GATGCTCTGGAAACTGGATGGCTAGAGT-3′ (pos. 585–613), 5′-GCGGCCATTTAAGTCTGGGGTGTA-3′ (pos. 541–565), and 5′-GAATGGCTATTGTAGG-3′ (pos. 406–422).

***Subclade IV*** is made of 3 species (*A. modicum, A. cavigenitalium,* and *A. axanthum*) that share at least 92.2 % identity between their 16S rRNA gene sequences and 90.2 % identity or less with 16S rRNA gene sequences of *Acholeplasma* species belonging to other subclades.

***Acholeplasma modicum*** (Leach 1973) was primarily isolated from cattle (blood, bronchial lymph nodes, thoracic fluids, lungs, and semen) but also from nasal secretions of pigs, and occasionally from chickens, turkeys, and ducks.

Colonies on solid medium are smaller than those of other acholeplasmas, particularly on serum-free media. Cells have spherical, ring-shaped, and coccobacillary forms. *A. modicum* is positive to fermentation of glucose and benzyl viologen reduction and negative to fermentation of mannose, esculin hydrolysis, production of carotenoid pigments, and film and spots reaction.

Type strain: PG49, ATCC 29102, and NCTC 10134. Oligonucleotide sequences of unique regions of the 16S rRNA gene (accession no. M23933) are 5′-GAAGGAGCGATCTGGAGCA-3′ (pos. 1242–1261), 5′-GCTGTGGCGCTTCAAAAACTGA-3′ (pos. 619–641), and 5′-GATCCAATCAAGGATCGTTCTAG-3′ (pos. 198–221).

***Acholeplasma axanthum*** (Tully and Razin 1970) has been isolated in prevalence from bovine tissues but also from other animals and occasionally from plants. Cells are predominantly bacillary and coccoid, with short mycelioid elements. On agar plates with serum, colonies are large with clearly marked centers, while those on serum-free agar are smaller and usually lack the peripheral growth around. Optimum growth is at 37 °C. *A. axanthum* is positive to fermentation of glucose, esculin hydrolysis, production of carotenoid pigments, and benzyl viologen reduction and negative to fermentation of mannose and film and spots reaction.

Type strain: S-743, ATCC 25176, and NCTC 10138. Oligonucleotide sequences of unique regions of the 16S rRNA gene (accession no. AF412968) are 5′-ACTCTAACAAGACTGC-3′ (pos. 1089–1105), 5′-AGCTATGGAGACATAGT-3′ (pos. 955–972), and 5′-AACTGGGTAGCTAGAGTT-3′ (pos. 596–614).

***Acholeplasma cavigenitalium*** (Hill 1992) was isolated from the vagina of guinea pigs. Cells are predominantly coccoid. It forms typical fried-egg colonies on agar plates. Optimum growth temperature is 35–37 °C. This species grows well on medium with additions of 10–15 % fetal bovine serum, but not horse serum. *A. cavigenitalium* is positive to fermentation of glucose and benzyl viologen reduction and negative to fermentation of mannose, esculin hydrolysis, and film and spots reaction.

Type strain: GP3, NCTC 11727, and ATCC 49901. Oligonucleotide sequences of unique regions of the 16S C gene (accession no. AY538164) are 5′-TAAAAATATTCTCA-3′ (pos. 1238–1252), 5′-AATGGATAGAACAAAGGGAGGCGAAA-3′ (pos. 1190–1216), 5′-CGGTTAGTAAAGTTTAGGGTATA-3′ (pos. 550–573), and 5′-AGAATTGTTGGGATAG-3′ (pos. 413–429).

## Genus "*Candidatus* Phytoplasma"

Due to the inability to obtain host cell-free pure culture of any phytoplasma strain, phytoplasma classification and taxonomy inevitably depended largely on molecular criteria. Phylogenetic analysis by including diverse phytoplasma strains indicated that phytoplasmas formed distinct and coherent clusters that are most closely related to *Acholeplasma* spp. than to other mollicutes. A separate taxonomic rank at least at genus level was proposed to distinguish phytoplasmas from their closest relatives (Gundersen et al. 1994). Consequently, a provisional new genus "*Candidatus* Phytoplasma" was adopted and was provisionally placed under the family *Acholeplasmataceae* (ICSB-STM 1995). However, despite their relatively close relatedness, acholeplasmas and phytoplasmas have marked different ecologies and habitats, as it is discussed in the next section. The unique biological and ecological characteristics as well as genomic evidence yet to be characterized may warrant assignment of a separate taxonomic rank at family level. Thus, the ultimate taxonomic rank and position of a taxon encompassing all phytoplasmas, whether within the family *Acholeplasmataceae* or a new family in the order *Acholeplasmatales*, remain to be determined, as discussed in the previous section.

Currently, a provisional system is adopted with some modifications from that proposed by Murray and Stackebrandt (1995) and used for naming "*Candidatus* Phytoplasma" species (IRPCM 2004). A guideline using the 16S rRNA gene sequence identity, with the threshold of <97.5 %, as the primary criteria was proposed for naming a new "*Candidatus* Phytoplasma species." Thus far, a total of 36 "*Ca.* Phytoplasma" species have been officially published and 5 have been proposed. Of which, 30 were solely based on sequence similarity criteria and 6 that did not fulfill the sequence identity criteria were designated primarily on specific biological and ecological properties and other genomic criteria. Comprehensive phylogenetic analysis using 16S rRNA gene sequences from GenBank collection, which consists of sequences of extensive phytoplasma strains collected from a wide array of plant sources and from various geographical locations, revealed that the phytoplasmas clade formed three distinct phylogenetic subclades and at least 14 phylogenetic distinct clusters were identified within the main clade (Wei et al. 2007; Zhao et al. 2009a). Members in subclade I are most closely related to acholeplasmas, followed by members in subclade III, while members in subclade II are most distantly related to acholeplasmas.

"*Ca.* Phytoplasma sp." description includes, among others, the signature sequences or unique oligonucleotide regions of 16S rRNA gene. As discussed above, it will never be precise to define a taxon at species level that is based on phylogenetic criteria alone, especially based on a highly conserved genetic marker 16S rRNA gene, to draw a boundary that distinguishes two species. Often the phenotypic characters or ecological properties have to be included for validity. But, ideally, phenotypic or ecology criteria should be congruent with the phylogenetic evidence. To improve the efficiency and accuracy of phylogeny-oriented species nomenclature system, multiple genetic markers should be considered to be included for analysis. Genetic markers that are related to phenotypic or biological properties may be selected. Recent enormous advances in genome sequencing have made it feasible to conduct comparative genomics with entire genome. Multilocus analysis using multiple conserved gene markers for assessing genetic distances among prokaryotes has been proposed to represent a highly useful and practical method to replace the traditional DNA-DNA hybridization procedure (Kuhnert and Korczak 2006; Razin 1992; Richter and Rosselló-Móra 2009). Several useful gene markers for use in phytoplasma classification have been attempted in the last decade and rather comprehensive databases with rp and *sec* Y gene markers, yet limited, have been established for differentiation and classification of sufficient diverse phytoplasma strains in several important 16Sr groups (Lee et al. 2004a, b, 2006b, 2009, 2010; Martini et al. 2002, 2007). Phylogenetic analyses based on rp and *sec* Y gene sequences are congruent with that basing on 16S rRNA gene. These gene markers will provide additional sets of signature sequences that will facilitate in defining "*Candidatus* Phytoplasma spp." and related strains within a given species boundary.

Molecular tools for rapid classification of a wide array of phytoplasmas have greatly advanced in last two decades. Actual and computer-simulated RFLP analyses of 16S rRNA gene sequences using 17 restriction enzymes were developed and employed for preliminary classification of phytoplasma strains. This procedure has delineated, thus far, 31 distinct 16S rRNA RFLP (16Sr) groups and more than 200 16Sr subgroups (Lee et al. 1998; Wei et al. 2007; Zhao et al. 2009b). The collective RFLP patterns specific to each group and subgroup generated provide the basis for constructing a comprehensive scheme for phytoplasma classification. This computer-simulated program also enables identification of potential new "*Candidatus* Phytoplasma sp." that fulfills the criteria proposed in the guideline.

Based on the proposed guideline (IRPCM 2004), thus far, 36 "*Candidatus* Phytoplasma" species have been proposed and described. In the following paragraphs, descriptions of "*Candidatus* Phytoplasma" species encompassed in each of three subdivisions will be presented. According to the guideline, each of "*Candidatus* Phytoplasma" species refers only to the reference strain designated and does not encompass the strains closely related to the reference strain, although using the term "reference strain" indicates that there are more strains that are encompassed in a given "*Candidatus* Phytoplasma" species. By doing so, the taxonomic affiliation of the numerous related strains will remain uncertain. The species concepts of "*Candidatus* Phytoplasma" species envisioned by earlier authors in their report are not always consistent from one another. So the readers may encounter such a situation. Identification and classification of phytoplasma strains are mainly based on RFLP analysis of 16S rRNA gene sequences (Lee et al. 1998; Wei et al. 2007). Other genetic markers are used for differentiation of related strains within a given "*Candidatus* Phytoplasma" species.

***Subclade I*** is made of 11 species that share at least 92.7 % identity between their 16S rRNA gene sequences. Although it is phylogenetically distinct from subclade III, members of the two subclades may share up to 92.9 % identity of their 16S rRNA gene sequences.

**"*Candidatus* Phytoplasma asteris,"** the causative agent of the widespread yellows of aster and several other plants, was classified in 16SrI group (or Aster yellows group) on the basis of RFLP and sequence analysis of 16Sr RNA gene. Originally it was proposed that the "*Ca.* P. asteris," designated on the basis of 16S rRNA gene sequence similarity with threshold boundary of 97.5 %, encompasses all known subgroups within aster yellows group 16SrI (Lee et al. 2004a). Currently, 14 16SrI subgroups have been designated (Zhao et al. 2009b). Members of 14 subgroups analyzed have sequence similarities ranging from 98.3 % to 99.8 %, which fall within the threshold boundary (97.5 % and above) of "*Ca.* P. asteris." According to the guidelines for naming "*Ca.* Phytoplasma" species (IRPCM 2004), the reference strain is the only legitimate member that can be designated as "*Ca.* P. asteris." Therefore, other strains designated as members of all 14 16SrI subgroups will only be defined as "*Ca.* P. asteris"-related strains. The genetic variability among strains between 16SrI subgroups or within each subgroup is more evident when more variable gene markers, such as ribosomal protein (Lee et al. 2004a), *secY* (Lee et al. 2006b), *secA* (Hodgetts et al. 2008), and *nusA* (Shao et al. 2006) genes, are employed for analysis. Phylogenetic analysis using more variable markers clearly identifies unique lineages among various strains, which seems to coincide with unique ecological niches these strains occupy. The reference strain *Oenothera* virescence phytoplasma OAY belongs to 16SrI-B. GenBank accession number for rDNA sequences of this strain is M30790. Oligonucleotide sequences of unique regions of 16S rRNA gene are 5′-GGGAGGA-3′ (positions 226–232), 5′-CTGACGGTACC-3′ (476–485), and 5′-CACAGTGGAGGTTATCAGTTG-3′ (1008–1028).

Aster yellows and related diseases are responsible for substantial economic loss in many crops including vegetable, ornamental flower, small fruit, grains, and hay. Diseases (e.g., aster yellows, clover phyllody, strawberry green petal, lettuce yellows, and blueberry stunt) are named according to their host plants. Symptoms can vary, depending on the phytoplasma strain. Typical symptoms of AY (caused by members of subgroup 16SrI-A and 16SrI-B) include virescence (greening of flower petals) and phyllody (development of floral parts into leaflike structures) (Kunkel 1926, 1931; Lee et al. 2004a; Lee and Davis 2000; McCoy et al. 1989), flower streaking and malformation, yellowing and upright posture of leaves, elongation and etiolation of internodes, excessive branching of axillary shoots, witches' broom, and general stunting of plants. However, some infected plants may exhibit only some of these symptoms. Symptoms induced by subgroup 16SrI-C phytoplasmas generally include virescence and phyllody, without excessive shoot proliferation. Symptoms induced by members of the other 16SrI subgroups include general stunting (little leaves, small flowers, shortening of internodes), leaf curl or rolling, small and faintly colored flowers, and some symptoms that are typical of the AY syndrome. Plants infected with mild strains may show no obvious symptoms. Subgroup 16SrI-B represents the largest and most diverse strain cluster in the 16SrI group. Subgroups 16SrI-L and 16SrI-M appear to be more common in the European continent (Marcone et al. 2000).

The vast majority of strains in the AY phytoplasma group infect herbaceous dicot plant hosts (Kunkel 1926; Marwitz 1990). However, a number of strains that belong to subgroups 16SrI-A, 16SrI-B, and 16SrI-C are capable of infecting monocot plants (e.g., corn, onion, gladiolus, oat, wheat, and grass) (Lee et al. 2004a; Jomantiene et al. 2002; Wu et al. 2010). Some strains in subgroups 16SrI-A, 16SrI-B, 16SrI-D, 16SrI-E, 16SrI-F, and 16SrI-Q can cause diseases in woody plants (e.g., gray dogwood, sandalwood, blueberry, mulberry, peach, cherry, olive, and paulownia) (Lee et al. 2004a).

The aster yellows group of phytoplasmas causes diseases in countless plant species belonging to 42 families (McCoy et al. 1989). Among them, families Asteraceae, Brassicaceae, and

Apiaceae most commonly harbor strains in this group. Strains that belong to subgroups 16SrI-A, 16SrI-B, and 16SrI-C are associated with more than 80 plant species worldwide, while strains in subgroups 16SrI-D (paulownia witches'-broom phytoplasma) and 16SrI-E (blueberry stunt phytoplasma) are associated with a single species or closely related species.

The majority of group 16SrI phytoplasmas are transmitted by broad range of leafhopper species in family *Cicadellidae* (Tsai 1979). For instance, subgroups 16SrI-A, 16Sr-B, and 16SrI-C may be transmitted by leafhoppers of up to 21 different species. Vector specificity, however, varies with subgroups, although some vectors are commonly shared by members of subgroups 16SrI-A, 16Sr-B, and 16SrI-C. *Macrosteles* spp., *Euscelis* spp., *Scaphytopius* spp., and *Aphrodes* spp. are the major vectors of AY phytoplasma strains (Brcak 1979; Tsai 1979). However, subgroups 16SrI-E and 16SrI-F and specific strain, e.g., maize bushy stunt phytoplasma, a member of 16SrI-B, have a narrow range of vectors (Maramorosch 1955; Moya-Raygoza and Nault 1998).

Aster yellows phytoplasmas are distributed worldwide in all five continents, the vast majority of strains being present in the temperate regions. Several subgroups are distributed in wide geographical areas: Subgroup 16SrI-B phytoplasmas are distributed worldwide, and subgroups 16SrI-A and 16SrI-C are present in several continents. Some phytoplasma strains may be constrained to a particular geographical region: Maize bushy stunt phytoplasma (16SrI-B) and blueberry stunt phytoplasma (16SrI-E) seem to be only present in America; Paulownia witches'-broom phytoplasma (16SrI-D) was only found in Southeastern Asia.

**"*Candidatus* Phytoplasma solani"** (Quaglino et al. 2013) strains were classified in 16SrXII group, 16SrXII-A subgroup, based on RFLP analysis of 16Sr RNA gene sequence. *Tuf, rplV-rpsC, and secY* gene sequences were also employed as additional markers for separation of "*Ca.* P. solani" from other "*Ca.* Phytoplasma" species and for identification of distinct ecological types. Strain STOL11 was selected as the reference strain for "*Ca.* P. solani" species (GenBank accession number AF248959). The unique oligonucleotide sequence regions of 16S RNA gene are 15′- ATTTTTAAAAGACCTAGCAATAGGTATGCTTAG-3′, 5′-ATGGTGGAAAAACCATTATGACGGTACCT-3′, 5′-GCA-ACGCTCAACGTTGTGATGCTATA-3′, 5′- TATGGCCTCAAT-GAGTAATAATT-3′, 5′- TTTCAAAATGTTAATCCCAAT-3′, and 5′-AACTAATATTGACATCGTTAA-3′.

Phytoplasma strains classified in "*Ca.* P. solani" or 16SrXII-A subgroup infect a wide range of wild and cultivated plants, mainly in *Solanaceae* family, and cause characteristic symptoms that vary with associated plant species. Stolbur or purple-top (associated with potato), which is referred to diseases associated with potato, tomato, eggplant, *Capsicum* spp., and other plants in family *Solanaceae*, exhibits symptoms including leaf yellowing and purpling (purple-top), leaf rolling or curling, and small leaves. Bois noir, which is referred to diseases associated with grapevine, shows symptoms including downward rolling of leaves, leaf yellowing, poor fruit ripening, and stunting of plants. Leaf yellowing or purpling and general stunting are commonly exhibited in other crops or weeds infected with this phytoplasma. However, similar symptoms in the same hosts may also be caused by other phytoplasmas in 16SrI-A, 16SrI-B, 16SrII-A, or 16SrVI-A. For examples, tomato big bud can be caused by all four phytoplasmas. Symptoms cannot be used as the sole criterion for identification of associated phytoplasmas. In earlier reports prior to 1993, the identification of stolbur phytoplasma associated with "stolbur" diseases was primarily based on symptoms alone and therefore was not reliable.

Stolbur phytoplasmas (16SrXII-A) have a wide range of host plants, including some important crops, e.g., potato, tomato, eggplant, pepper, carrot, grapevine, and corn, belonging to at least 15 families (McCoy et al. 1989; Berger et al. 2009). In nature, infection is largely found in *Solanaceae, Apiaceae, Asteraceae, Convolvulaceae, Fabaceae*, and *Urticaceae*. Some weeds in *Asteraceae, Convolvulaceae* (*Convolvulus arvensis*), and *Fabaceae* (*Trifolium* spp.) may serve as important reservoirs for stolbur phytoplasmas. Stolbur phytoplasmas were usually associated with herbaceous plants including recently discovered hosts such as maize and strawberry (Duduk and Bertaccini 2004; Terlizzi et al. 2006; Jović et al. 2007), but later the phytoplasmas were also found to infect woody plants, e.g., grapevines, causing a disease, called Bois noir, which is now widespread in Europe (Belli et al. 2010; Palermo et al. 2004; Radonjić et al. 2006).

Several plant hopper species, *Hyalesthes obsoletus, Reptalus panzeri*, and *R. quinquecostatus* in family *Cixiidae* and *Dictyophara europaea* in family *Dictyopharidae*, and leafhopper species, *Macrosteles quadripunctulatus, Macrosteles* spp., and *Anaceratagallia ribauti* in family *Cicadellidae*, are known to transmit stolbur phytoplasmas in potato, tomato, and grapevine and numerous other vegetable crops and weed hosts (Batlle et al. 2009; Gatineau et al. 2001; Trivellone et al. 2005). *Lygus* spp. (*L. pratensis* and *L. rugulipennis*) may also be potential vectors for stolbur phytoplasmas (Březíková and Linhartová 2007).

Stolbur phytoplasmas are widespread in Europe and, to a lesser extent, in Asia (Březíková and Linhartová 2007; Eroglu et al. 2011; Palermo et al. 2004; Radonjić et al. 2006; Trivellone et al. 2005). There is one report in Argentina (Quaglino et al. 2013).

**"*Candidatus* Phytoplasma australiense"** (Davis RE et al. 1997) reference strain and closely related strains from other sources are classified in 16SrXII group, 16SrXII-B subgroup. These strains are most closely related to "*Ca.* P. solani." There is genetic variability among intraspecies strains (Constable and Symons 2004). Sequences of *tuf, secY*, and *rplV-rpsC* genes were employed to identify genetic variability among intraspecies strains (Quaglino et al. 2013). Strain AUSGY was selected as reference strain for "*Ca.* P. australiense" (GenBank accession number L76865). Oligonucleotide sequences of unique regions of the 16S rRNA gene are 5′-CGGTAGAAATATCGT-3′and 5‘-TTTATCTTTAAAAGACCTCGCAAGA-3′. A pair of primer AUSGYF1, 5′-ATCTTTAAAAGACCTCGCAAG-3′, and primer AUSGYR2, 5′-AGTTTTACCCAATGTTTAGTACTC, is used for priming specific amplification of DNA from this phytoplasma.

"*Ca.* P. australiense" strains are responsible for several economical diseases including Australian grapevine yellows, strawberry lethal yellowing, *Phormium* yellows, sudden decline of cabbage tree, and diseases in some other hosts (Magaray and Wachtel 1986; Constable et al. 2003, 2004; Andersen et al 1998, 2001; Beever et al. 2004; Getachew et al. 2007; Liefting et al. 1998, 2007, 2009; Padovan et al. 2000; Saqib et al. 2006; Streten et al. 2005). Symptoms associated with Australian grapevine yellows phytoplasma infection of grape varieties include irregular chlorosis, yellowing, and reddening of leaves; dieback of affected shoots; and shrived berries or unripen green and rubbery fruit. Infected potato exhibits uprolling and purpling of leaves (Liefting et al. 2009). Infected red clover, several other pasture legumes, and paddy melon (*Cucumis myriocarpus*) exhibit little leaves, leaf deformation, shoot proliferation, and stunting (Saqib et al. 2006). In other hosts, this phytoplasma causes green petal and lethal yellows (e.g., strawberry), little leaves and lethal yellows (e.g., *Phormium* sp.), yellow leaf curl (e.g., pumpkin), sudden decline (cabbage tree), dieback (papaya), and other various symptoms (Padovan et al. 2000; Andersen et al. 1998, 2001).

The leafhopper *Orosius argentatus* (Cicadellidae, Hemiptera) (Pilkington et al. 2004) and the plant hopper *Oliarus atkinsoni* (Cixiidae, Homoptera) (Liefting et al. 1997) have been confirmed to be vectors of Australian lucerne yellows phytoplasma and *Phormium* yellow leaf phytoplasma, respectively. No confirmed reports indicate that these vectors may transmit Australian grapevine yellows in grapevines.

The phytoplasma is widespread in Australia and New Zealand.

**"*Candidatus* Phytoplasma americanum"** (Lee et al. 2006a) is classified in 16SrXVIII group (American potato purple-top wilt phytoplasma group). There is genetic variability among infra-species strains. Strain APPTW12-NE was designated as the reference strain (GenBank accession number DQ-174122). Oligonucleotide sequences of unique regions of the 16S rRNA gene are 5′-GTTTCTTCGGAAA-3′, 5′-GTTAGAAATGACT-3′, and 5′-GCTGGTGGCTT-3′. "*Ca.* P. americanum" is most closely related to "*Ca.* P. australiense" species in the subdivision I.

Strains in "*Ca.* P. americanum" (16SrXVIII-A and 16SrXVIII-B subgroups) infect potatoes in northern Texas and Nebraska, causing purple-top disease syndrome similar to potato purple-top diseases found in Oregon and Washington states. The symptoms include stunting, chlorosis, slight purple discoloration of new growth, leaf curl, swollen nodes, broken axillary buds, and the formation of aerial tubers (Secor et al. 2006). Storage tubers from affected plants either did not sprout or produced spindle or hair sprouts. The extent of this disease thus far is limited to this region in the USA. Thus far, potato is the only natural host that has been reported.

**"*Candidatus* Phytoplasma costaricanum"** (Lee et al. 2011) is classified in 16SrXXXI group that includes, according to RFLP analysis of 16S rRNA gene sequence, strains SoyST (from soybean), SwPPV (from sweet pepper), and PasFBP (from passion fruit vine). There is genetic variability among these intraspecies strains, sharing with 99.2–99.9 % of 16S rRNA gene sequence similarity. Strain SoySTc1 (GenBank accession number HQ225630) was designated as the reference strain. Oligonucleotide sequences of unique regions of the 16S rRNA gene are 5′-TTAAGGAAGAAAAATTGGTGGAAA-3′, 5′-TTAGGTAAGTTTATGGTGTAA-3′, 5′-GTTCAACGCTTAACGTTGTGATG-3′, 5′-CTACAACGCAAGTTGATG-3′, and 5′- GGGGGCCTAACTCGCAAGA-3′.

Strains of "*Ca.* P. costaricanum" infect three different crops, soybean (*Glycine max*), sweet pepper (*Capsicum annuum*), and passion fruit vines (*Passiflora edulis*), in the same region of Costa Rica and cause various symptoms in each crop (Villalobos et al. 2009). The infected soybean plants exhibited symptoms that included general stunting, small leaves, excessive bud breaking, aborted seedpods, and remaining green stems at harvest time; infected sweet pepper plants exhibited purple vein syndrome (SwPPV) characterized by dark green and rugose leaves, a zigzag pattern to the midvein, and purple vein discoloration; and the infected passion fruit vine exhibited bud proliferation (PasFBP) and chlorosis. The diseases occur sporadically and are limited to the region in Costa Rica.

**"*Candidatus* Phytoplasma japonicum"** (Sawayanagi et al. 1999) was originally classified, based on RFLP analysis of 16S rRNA gene sequence, in 16SrI group and was later reclassified in 16SrXII group, 16SrXII-D subgroup, and is most closely related to "*Ca.* Phytoplasma fragariae" (16SrXII-E) (Zhao et al. 2009b). Strain JHP was selected as reference strain for "*Ca.* P. japonicum" (GenBank no. AB010425). Oligonucleotide sequences of unique regions of the 16S rRNA gene are 5′-GTGTAGCCGGGCTGAGAGGTCA-3′ and 5′-TCCAACTCTAGCTAAACACTTTTCTG-3′. Antigenic membrane protein can also be used to specifically identify this species (Kakizawa et al. 2008).

The Japanese *Hydrangea* phyllody (JHP) disease is a serious disease and spreads wherever *Hydrangea* spp. are grown in Japan. JHP phytoplasma-infected Japanese hydrangea plants show several disease symptoms involved in floral malformations, such as virescence, phyllody, and proliferation (Kanehira et al. 1996; Sawayanagi et al. 1999). Thus far, this species has only been detected in Japan and has not been reported elsewhere.

**"*Candidatus* Phytoplasma fragariae"** (Valiunas et al. 2006) was reclassified, based on RFLP analysis of 16S rRNA gene sequence, in 16SrXII group, 16SrXII-E subgroup, and is most closely related to "*Ca.* P. japonicum" (16SrXII-D). Strain StrawY is the reference strain (GenBank accession number DQ086423). Oligonucleotide sequences of unique regions of the 16S rRNA gene are 5′-GTGCAATGCTCAACGTTGTGAT-3′, 5′-AATTGCA-3′, and 5′-TGAGTAATCAAGAGGGAG-3′.

"*Ca.* P. fragariae" causes diseases in strawberry in Lithuania and in potato in China. Disease induced in infected strawberry plants, termed strawberry yellows, exhibits general stunting and yellowing of leaves (Valiunas et al. 2006). In China, infected potato plants display symptoms of curled, yellowish, and purplish leaves, shortened internodes, and aerial tuber formation, which are very similar to those induced by infection with "*Ca.* P. solani" (subgroup 16SrXII-A). The disease has become prevalent in seed and commercial fields in Yunnan Province and Inner Mongolia (Cheng et al. 2012).

**"*Candidatus* Phytoplasma convolvuli"** (Martini et al. 2012) represents a candidate for a new 16Sr group. Phylogenetic analysis of 16S rRNA gene sequence indicates strains of "*Ca.* P. convolvuli" form a monophyletic group within the subdivision I. The species is most closely related to "*Ca.* P. fragariae" (16SrXII-E) and "*Ca.* P. solani" (16SrXII-A). Reference strain is BY-S57/11 (GenBank accession number JN833705). Unique oligonucleotide sequences of the 16S rRNA gene are 5′-GCCTTTGGGC-3′, 5′-GATAGGA-AGGCATCTTCTTG-3′, 5′-ACGTTGGTTAAAACCA-3′, and 5′-GCTTTTGCAAAGCTT-3′.

"*Ca.* P. convolvuli" is associated with bindweed (*Convolvulus arvensis*) in several European countries. The infected bindweeds exhibit symptoms of undersized leaves, shoot proliferation, and yellowing. The disease is termed bindweed yellows (BY). The bindweed is also the primary host of "*Ca.* P. solani".

**"*Candidatus* Phytoplasma lycopersici"** (Arocha et al. 2007) was originally affiliated with 16SrI group, and later RFLP analysis of updated 16S rRNA sequence (GenBank accession number EF199549.2) with a complete set of 17 restriction enzymes indicated that this species represents a candidate for a new 16Sr group (Zhao et al. 2009b). Phylogenetic analysis indicates strains of "*Ca.* P. lycopersici" form a monophyletic group within the subdivision I. Strain THP was designated as reference strain (GenBank accession no. AY787136 and an updated sequence EF199549.2). Oligonucleotide sequences of unique regions of the 16S rRNA gene are 5′-CTTA-3′ (175–178), 5′-AATGGT-3′ (198–203), 5′-ATA-3′ (229–231), 5′-TTGGAGGAA-3′ (234–242), 5′-CACG-3′ (302–305), 5′-TCT-3′ (315–317), 5′-GCT-3′ (334–336), 5′-TAT-3′ (336–338), 5′-TAC-3′ (413–415), and 5′-AGC-3′ (434–436).

"*Ca.* P. lycopersici" was found to be associated with morrenia little leaf (MVLL) disease in native weed *Morrenia variegate* and with "hoja de perejil" (THP) disease in tomato in Bolivia. Tomato plants affected with "hoja de perejil" disease are characterized by proliferation of axillary buds and elongation of lateral shoots showing small and fernlike leaves, becoming large bushy plants as the season progresses (Arocha et al. 2007; Jones et al. 2005). The species has not been found outside American continent.

**"*Candidatus* Phytoplasma graminis"** (Arocha et al. 2005a) reference strain from Bermuda grass (*Cynodon dactylon;* GenBank accession no. AY742327) and its closely related strains from delphacid plant hoppers *Saccharosydne saccharivora* (98.3 % sequence similarity) and weeds (sequence similarity ranging from 99.5 % to 99.9 %) Canadian horseweed (*Conyza canadensis*; GenBank accession no.AY742328), phasey bean (*Macroptilium lathyroides;* GenBank accession no. AY742329), and Johnson grass (*Sorghum halepense;* GenBank accession no. AY742330) are classified in 16SrXVI group, according to RFLP analysis of 16S rRNA gene sequence. The reference strain is SCYLP from Cuban sugarcane (GenBank accession no. AY725228). Oligonucleotide sequences of unique regions of 16S rRNA gene are 5′-TTTG-3′, 5′-TTG-3′, 5′-GGG-3′, 5′-TAA-3′, and 5′-ATTTACGTTTCTG-3′.

Sugarcane yellow leaf disease (SCYLP), characterized by a yellowing of the midrib and lamina, is widely spread in most sugarcane-growing countries, causing severe economic loss. Up to 14 % of infected sugarcane, however, can be symptomless. The cause of this disease varies with different geographical regions. Phytoplasmas have been consistently associated with yellow leaf syndrome (Cronjé et al. 1998; Arocha et al. 1999; Aljanabi et al. 2001). In Cuba, SCYLP is associated with "*Ca.* P. graminis." *Macroptilium lathyroides* and *Sorghum halepense* are symptomless. Whether the graminaceous weed hosts will serve as reservoirs of phytoplasma that contribute to the infection of sugarcane is unknown. "*Ca.* P. graminis" species and its closely related strains, thus far, have been reported only in Cuba.

**"*Candidatus* Phytoplasma caricae"** (Arocha et al. 2005b) strain PAY and its closely related strain *Emp3* (GenBank accession no. AY725235), obtained from the putative vector *Empoasca papaya*, are classified in 16SrXVII group based on RFLP analysis of 16S rRNA gene sequence. Phylogenetic analysis indicates that 16SrXVII group is a monophyletic group within subclade I. Based on sequence similarity of 16S rRNA gene sequence, the most close relative to "*Ca.* P. caricae" is "*Ca.* P. solani". PAY (papaya phytoplasma) (GenBank no. AY725234) is the reference strain for "*Ca.* P. caricae." Oligonucleotide sequences of unique regions of 16S rRNA gene are 5′-AAA-3′ (161–163), 5′-ATT-3′ (558–560), 5′-AGGCGCC-3′ (1039–1045), and 5′-GCGGATTTAGTCACTTTTCAGGC-3′ (1324–1346).

Although a disease with yellow leaf syndrome and bunchy-top symptom similar to PBT (papaya bunchy-top) diseases was reported elsewhere to be associated with bacteria-like organism (Davis et al. 1998), the PBT-like disease that is spreading in papaya-growing areas in Cuba is associated with "*Ca.* P. caricae" (Arocha et al. 2003). The disease has an economic impact to the papaya industry in Cuba. Individuals of *Empoasca papaya* were found to be carriers of "*Ca.* P. caricae," so this leafhopper may be a potential vector that transmits PBT in Cuba. The species apparently is limited in Cuba.

***Subclade II*** is an ample and diverse group that is the less closely related to the acholeplasmas. It is made of 18 species that share more than 90.7 % identity between their 16S rRNA gene sequences, but there are members that share up to 92.8–92.9 % identity with members of subclades I or III.

**"*Candidatus* Phytoplasma brasiliense"** is associated with witches'-broom disease of hibiscus (*Hibiscus rosa-sinensis*) (HibWB) in Brazil (Montano et al. 2001a). Symptoms are characterized by excessive axillary branching, abnormally small leaves, and deformed flowers. In Brazil, alternate plant hosts of "*Ca.* P. brasiliense" (suggesting the involvement of a polyphagous insect vector) are naturally infected *Catharanthus roseus* (periwinkle) exhibiting yellowing and witches'-broom symptoms (Montano et al. 2001b); *Sida* sp. exhibiting symptoms characterized by stunting, chlorosis, small leaves, and witches' broom (Eckstein et al. 2011); and *Brassica oleracea* (cauliflower) exhibiting general stunting, malformation of inflorescence, reddening leaves, and vessel necrosis (Canale and Bedendo 2013). Recently, the identification of a *Prunus persica* (peach) phytoplasma strain from Azerbaijan as an isolate of "*Ca.* P. brasiliense" has been reported in a geographical region different from South America (Balakishiyeva et al. 2011).

Phylogenetic analysis of 16S rDNA sequences identified the hibiscus witches'-broom phytoplasma as a member of a distinct cluster of the class *Mollicutes* and indicated that the hibiscus witches'-broom phytoplasma is phylogenetically closely related to phytoplasma strains of group 16SrII (peanut witches'-broom group).

On the basis of the PCR/RFLP analyses of 16S rRNA gene, the hibiscus witches'-broom phytoplasma was classified as a member of 16S rDNA RFLP group 16SrXV (hibiscus witches'-broom group), subgroup 16SrXV-A. RFLP analysis of R16F2n/R2 fragment with restriction enzymes *Mse* I, *Hae* III, and *Hpa* II can differentiate "*Ca.* P. brasiliense" from closely related phytoplasma strains belonging to 16SrXV-B subgroup and 16SrII group (Montano et al. 2001a; Villalobos et al. 2011).

The reference strain is HibWB26, member of subgroup 16SrXV-A (GenBank accession number AF147708). Oligonucleotide sequences of unique regions of the 16S rRNA gene are 5′-GAAAAAGAAAG-3 (at positions 162–172 of the HibWB DNA, AF147708), 5′-TCTTTCTTT-3′ (176–184), 5′-CAG-3′ (575–577), 5′-ACTTTG-3′ (630–635), and 5′-GTCAAAAC-3′ (822–829).

The analysis of non-ribosomal *dna* K gene indicated the existence of two "*Ca.* P. brasiliense" groups of strains that are genetically different (19 mutations): strains infecting *Hibiscus rosa-sinensis* (hibiscus) in Brazil and strains infecting *Catharanthus roseus* (periwinkle), *Ocimum basilicum* (basil), and *Prunus persica* (peach) from Suriname, Lebanon, and Azerbaijan, respectively (Balakishiyeva et al. 2011).

**"*Candidatus* Phytoplasma aurantifolia"** is associated with witches'-broom disease of small-fruited acid lime (*Citrus aurantifolia*) (WBDL), which is a severe disease widespread in the Sultanate of Oman, the UAE, and Southern Iran, where it appeared in the late 1970s, in 1989 (Zreik et al. 1995), and in 1997 (Salehi et al. 1997; Bové et al. 2000), respectively.

Affected trees are characterized by the presence of witches' brooms, easily detected by their compactness and the small to very small leaves, which are often pale green to yellow. In the early stages of the disease, the tree shows only one, then a few, branches with witches' brooms. These early witches' brooms are soon followed by many others in various parts of the tree. In the final stage of the disease, the trees have many dead twigs, shoots, and branches with only a few witches' broom left. At this stage, the trees are almost dead. The progress of the disease is very fast; once the first symptoms (witches' broom) appear, the trees seem to decline very rapidly, and within a few years, they are totally unproductive and nearly dead. No flowers or fruits are produced on branches with witches' brooms (Bové et al. 1988).

Small-fruited acid lime (*Citrus aurantifolia*), Palestine sweet lime (*C. limetta*), sweet limetta (*C. limettioides*), and bakraee (natural *C. reticulata* hybrid) are naturally infected and show symptoms. The pathogen has been experimentally transmitted to several species of citrus by graft inoculation, *C. aurantifolia*, *C. excelsa*, *C. hystrix*, *C. ichangensis*, *C. karna*, *C. macrophylla*, Etrog citron (*C. medica*), Meyer lemon (*C. limon*), Rangpur lime (*C. limonia*), rough lemon (*C. jambhiri*), Troyer citrange (*Poncirus trifoliata* x *C. sinensis*), pear-shaped lemon (probably a lemon hybrid), and bakraee (*C. reticulata* hybrid) (Bové et al. 1996; Salehi et al. 2005), and to periwinkle (Garnier et al. 1991; Salehi et al. 2002) and a number of solanaceous plants such as tomato, eggplant, and tobacco by dodder or graft inoculation (Salehi et al. 2000, 2002). Seed transmission of "*Ca.* P. aurantifolia" and of phytoplasmas in general is a matter of controversy; however very recently it has been demonstrated that witches'-broom disease of lime affects seed germination and seedling growth but is not seed transmissible (Faghihi et al. 2011).

In Iran, "*Ca.* P. aurantifolia" is naturally spread by the leafhopper *Hishimonus phycitis* (Salehi et al. 2007), which is a common phloem-feeding insect associated with lime trees.

Among its closest relatives are FBP (faba bean phyllody, X83432, 16SrII-C) periwinkle-maintained phytoplasma strain from Sudan and CoP (cotton phyllody, EF186827, 16SrII-F) phytoplasma strain from Burkina Faso sharing between 96.7 and 98.1 % sequence homology in the 16S rRNA gene (Martini et al. 2007; Martini 2004).

On the basis of putative-restriction-site analyses of 16S rDNAs, WBDL phytoplasma strain was placed in peanut witches'-broom group 16SrII, subgroup -B (Lee et al. 1998). The restriction enzyme that permits to differentiate, on the basis of R16F2n/R2 fragment, WBDL phytoplasma from other representatives of group 16SrII is *Tsp*509I (Martini 2004). GenBank accession number is U15442. Oligonucleotide sequence complementary to unique region of 16S rRNA is 5′-GCAAGTGGTGAACCATTTGTTT-3′.

The genome size of "*Ca.* P. aurantifolia" is 720 kbp (Zreik et al. 1995). Besides the sequences of its 16S ribosomal DNA (U15442, EF186815) and the 16S-23S ribosomal DNA spacer region (U15442) (Zreik et al. 1995), sequences of other genes have been deposited in GenBank and used especially for phylogenetic studies of "*Ca.* Phytoplasma." These genes are 23S ribosomal DNA (EU168731; Hodgetts et al. 2008), ribosomal protein *rplV-rpsC* genes (EF186828; Martini et al. 2007), *tuf* gene (JQ824276; Makarova et al. 2012), *secA* gene (EU168731; Hodgetts et al. 2008), and immunodominant membrane protein (GU339497, JQ745272; Siampour et al. 2013).

Sequence analyses of *rplV-rpsC* showed that "*Ca.* P. aurantifolia" shared between 96.4 % and 97.9 % sequence homology with its closest relatives FBP and CoP phytoplasma strains (Martini 2004).

**"*Candidatus* Phytoplasma australasiae"** is associated with papaya yellow crinkle (PpYC), papaya mosaic (PpM), and tomato big bud (TBB) diseases from Australia (White et al. 1998; revised nomenclature by Firrao et al. 2005).

Transmission studies have demonstrated that the etiologic agent of yellow crinkle was the same as that causing tomato big bud in Australia (Greber 1966). Viruses were first thought to be the pathogens causing tomato big bud and papaya yellow crinkle. However, it was later shown by transmission electron microscopy that phytoplasmas are associated with both diseases (Bowyer et al. 1969; Gowanlock et al. 1976).

Recent surveys in Australia suggested that TBB is a highly successful phytoplasma associated with a remarkable variety of different plant species (Davis RE et al. 1997; Schneider et al. 1999). Most of the plant host species are members of the families *Fabaceae*, *Solanaceae*, and *Asteraceae* (Davis RI et al. 1997). More recently, in Australia, new host plants have been described: capsicum (*Capsicum annuum*), celery (*Apium graveolens*), chicory (*Cichorium intybus*) (Tran-Nguyen et al. 2003), pale purple coneflower (*Echinacea pallida*) (Pearce et al. 2011), and Arabian pea (*Bituminaria bituminosa*) (Aryamanesh et al. 2011). The wide host range of the TBB phytoplasma possibly reflects the feeding habits of its insect vector, the common brown leafhopper, *Orosius argentatus*, which is widely distributed throughout Australia (Hill 1943).

Phytoplasma strains closely related to "*Ca.* P. australasiae" (16SrII-D) have been described also outside Australia. In Oman, phytoplasma strains closely related to "*Ca.* P. australasiae" have been described associated with alfalfa (*Medicago sativa*) witches' broom (Khan et al. 2002; Al-Zadjali et al. 2007), sesame (*Sesamum indicum*) witches' broom (Al-Sakeiti et al. 2005), Arabian jasmine (*Jasminum sambac*) witches' broom (Al-Zadjali et al. 2007), and chickpea (*Cicer arietinum*) phyllody and little leaf (Al-Saady et al. 2006), eggplant (*Solanum melongena*) phyllody (Al-Subhi et al. 2011), and beach naupaka (*Scaevola taccada*) witches'-broom diseases (Al-Zadjali et al. 2012). In Egypt and Sudan, phytoplasma strains belonging to subgroup 16SrII-D have been described to infect tomato (*Lycopersicon esculentum*), eggplant (*Solanum melongena*) and squash (*Cucurbita pepo*) plants (Ayman and Foissac 2012), and chickpea (*Cicer arietinum*) and faba bean (*V icia faba*) (Alfaro-Fernández et al. 2012), respectively. In Pakistan, phytoplasma strains of 16SrII-D subgroup have been found associated with chickpea and sesame phyllody diseases, transmitted to healthy plants by grafting and through the leafhopper *Orosius orientalis* and *O. albicinctus*, respectively (Akhtar et al. 2009a, b). In India, phytoplasma strains exhibiting 100 % 16S rDNA sequence similarity with "*Ca.* P. australasiae"-reference strain have been reported to infect tomato (*Solanum lycopersicum*) plants showing leaf yellowing and curling, little leaf, severe stunting, and phyllody (Singh et al. 2012). In Iran, a work based on molecular and biological characterizations demonstrated that the 16S rRNA gene sequence of a phytoplasma associated with garden beet (*Beta vulgaris* L. ssp. *esculenta*) witches' broom was nearly identical with the one of PpYC (Y10097) and that *Orosius albicinctus* is a vector of the disease (Mirzaie et al. 2007).

The PpYC and PpM phytoplasma 16S rDNA sequences were identical to each other. In the phylogenetic analysis, PpYC was most closely related to tomato big bud strain TBB from Australia, within the peanut witches'-broom cluster described by Gundersen et al. (1994).

PpYC, together with TBB, peanut witches'-broom strain PnWB from Taiwan, sweet potato witches'-broom strain SPWB from Taiwan, sunn hemp witches'-broom strain SUNHP from Thailand, and sweet potato little leaf strain SPLL from Australia, form a clade distinct from "*Ca.* P. aurantifolia" from Oman and faba bean phyllody strain FBP from Sudan. Direct pairwise comparisons of sequences showed that the PpYC 16S rDNA sequence was most similar to those of TBB (99.7 %), PnWB (99.7 %), SUNHP (99.4 %), SPWB (99.4 %), and SPLL (99.1 %). The PpYC sequence was 98.8 % similar to the "*Ca.* P. aurantifolia" sequence and 98.6 % similar to the FBP sequence.

Based on RFLP analysis of 16S rRNA gene sequences, TBB phytoplasma strain was classified in 16S rDNA RFLP group 16SrII, subgroup -D (Lee et al. 1998; Khan et al. 2002).

Phytoplasma strains belonging to 16SrII-D subgroup can be differentiated from other closely related phytoplasma strains of 16SrII group by collective RFLP profiles obtained with *Mse* I, *Taq* I, and *Tsp* 509I restriction enzymes on R16F2n/R2 PCR products (Martini 2004).

The reference strain is PpYC, GenBank accession number Y10097. The following two unique sequences that distinguish PpYC, PpM, and TBB from other phytoplasma strains were found in 16S rDNA: 5′-TAAAAGGCATCTTTTATC-3′ (178–195; numbering corresponding to 16S rRNA gene sequence of OAY (Lim and Sears 1989)) and 5′-CAAGGAAGAAAAG-CAAATGGCGAACCATTTGTTT-3′ (444–477).

The 16S-23S spacer region DNA sequences of the PpYC and PpM phytoplasmas were identical to each other. PpYC and PpM are most similar to TBB (99.6 %) and SPLL (99.6 %) while showing 98.9 % and 98.4 % identity with "*Ca.* P. aurantifolia" and FBP, respectively.

In "*Ca.* P. australasiae" strain TBB, the nucleotide sequences of two extrachromosomal elements (3319 and 4092 bp) were also determined (Tran-Nguyen and Gibb 2006).

Besides the sequences of its 16S ribosomal DNA (EF193359), sequences of other genes of "*Ca.* P. australasiae" strain TBB have been deposited in GenBank and used especially to improve knowledge on the phylogenetic relationships within the genus "*Ca.* Phytoplasma": *rplV-rpsC* genes (EF193373; Martini et al. 2007), 23SrDNA (EU168763) and *secA* genes (EU168729) (Hodgetts et al. 2008), *secY* gene (GU004347; Lee et al. 2010), and *tuf* gene (JQ824250; Makarova et al. 2012).

**"*Candidatus* Phytoplasma pruni"** is associated with X-disease, one of the most serious diseases known in peach (*Prunus persica*) (Davis et al. 2013) (Stoddard et al. 1951). The disease was first reported in 1933 in the state of Connecticut and called the "X-disease of peach" because of its unknown nature. It was for many years believed to be of viral origin (Stoddard 1934, 1938; Stoddard et al. 1951).

Symptoms of X-disease on peach include tattered, shot-holed appearance of leaves; loss of severely affected leaves, leaving a cluster of leaves at the ends of individual branches; dieback of branches; and death of trees (Douglas 1986; Stoddard 1938; Stoddard et al. 1951).

In following years, X-disease was reported in numerous states in northeastern USA, in California and other western states, and in Canada (Stoddard et al. 1951). An important natural plant host of the pathogen was found to be wild chokecherry (*Prunus virginiana*) growing in the forest (Douglas 1986). Other *Prunus* spp. that have been described susceptible to infection by the X-disease pathogen include cherry (*Prunus avium and P. cerasus*), Japanese plum (*P. salicina*), almond (*P. dulcis*), apricot (*P. armeniaca*), nectarine (*P. persica* var. *nectarina*), Chinese bush cherry (*P. japonica*), Bessey cherry (*P. besseyi*), wild American plum (*P. americana*), wild-goose plum (*P. munsoniana*), and European plum (*P. domestica*) (Douglas 1986; Stoddard et al. 1951). The pathogen can be transmitted by several leafhopper species, including *Colladonas clitellarius*, *C. montanus*, *C. geminatus*, *Euscelidius variegatus*, *Fieberiella florii*, *Graphocephala confluens*, *Gyponana lamina*, *Keonella confluens*, *Norvellina seminuda*, *Osbornellus borealis*, *Paraphlepsius irroratus*, and *Scaphytopius delongi* (*S. acutus*) (Kirkpatrick et al. 1990; Larsen and Whalen 1988; McClure 1980; Rice and Jones 1972).

On the basis of 16S rDNA sequence, "*Ca.* P. pruni" is a homogeneous pathogen, revealing 16S rDNA sequences identical (100 %) or nearly identical among strains. RFLP analyses, cloning, and sequencing of 16S rDNA gene of six Connecticut X-disease phytoplasma strains from naturally diseased peach trees revealed the presence of two sequence-heterogeneous rRNA operons. By contrast, the same type of analyses did not indicate the presence of sequence-heterogeneous rDNA in reference strains CX-95 or WX95 maintained in periwinkle (Davis et al. 2013).

Based on RFLP analysis of 16S rRNA gene sequences, peach X-disease phytoplasma strains were classified in 16S rDNA RFLP group 16SrIII, subgroup -A (Davis et al. 2013). Phytoplasma strains belonging to 16SrIII-A subgroup can be differentiated from other closely related phytoplasma strains of 16SrIII group by composite RFLP profiles obtained with *Mse* I and *Bst* UI restriction enzymes on R16F2n/R2 PCR products (Zhao et al. 2009b). The reference strain is PX11CT1, GenBank accession number JQ044393, rrnA, and JQ044392, rrnB. Oligonucleotide sequences of unique regions in the 16S rRNA gene are 5'-CACATTAGTTAGTTGGTAGGGTAAAGGCCTACC-3′ (226–258), 5′-GTACCTCGGTATG-3′ (402–414), 5′-TTATTA-AGGAAGAAAAAAGAGTGGAAAAACTCCCT-3′ (425–459), 5′-ACGGTACTTAA-3′ (462–472), 5′-TAATAAGTCTATAGTT-TAATTTCAGTGCTTAACGCTGTTGTGCTATAG-3′ (571–618), 5′-GTTTTACTAGAGTGAG-3′ (624–639), 5′-TAAAACTGGTAC-3′ (817–828), 5′-TTTCTTGCGAAGTTA-3′ (970–984), 5′-ATGGAGGTCATCAGGAAAACAGGTGGTGC-3′ (999–1027), 5′-CTTGTCGTTAGTTGCCAGCATGTAAT-3′ (1083–1108), 5′-GATGGGGACTTTAACGA-3′ (1109–1125), 5′-GGTTG-ATACAAAG-3′ (1211–1223), and 5′-TCTCAAAAAATCAATC-3′ (1252–1267).

Davis et al. (2013) proposed that the term "*Ca.* P. pruni" be applied to phytoplasma strains whose 16S rRNA gene sequences contain the oligonucleotide sequences of unique regions, including X-disease phytoplasmas and – within the tolerance of a single base difference in one unique sequence – peach rosette (AF236121), little peach (AF236122), and peach red suture (AF236123) phytoplasmas.

Nucleotide and phylogenetic analyses of *secY* and *rps3-rpl22* gene sequences provided additional molecular markers of the "*Ca.* P. pruni" lineage within 16SrIII group. The amplified rp genomic regions of X-disease phytoplasma strains were identical in nucleotide sequence, except for a single base difference located in the L22 gene; this base difference accounted for an amino acid substitution which distinguished CX-95 and WX95 from the X-disease phytoplasma strains from Connecticut. In the *secY* genomic locus, a 9-base insertion/deletion (indel) distinguished two group 16SrIII phytoplasma strain clusters, one containing X-disease phytoplasma strains and walnut witches'-broom (WWB), *poinsettia* branch-inducing (PoiBI), and *spirea* stunt (SP1) strains (*secY* gene length of 1263 bases) and the other containing clover yellow edge (CYE), milkweed yellows (MW1), goldenrod yellows (GR1), pecan bunchy-top (PBT), potato purple-top (PPTAKpot6, PPTAKpot7, PPTM117), and *Vaccinium* witches'-broom (VAC) strains (*secY* gene length of 1272 bases). The *secY-map* intergenic region from all of the X-disease phytoplasma strains and strain PoiBI contained a 4-base insertion, compared to the *secY-map* intergenic regions from strains affiliated with diverse subgroups of group 16SrIII (Davis et al. 2013).

**"*Candidatus* Phytoplasma pini"** is associated with pine trees (*Pinus* spp.) (Schneider et al. 2005) in Germany, Spain, Poland, Czech Republic, Lithuania, and Croatia showing abnormal shoot branching, dwarfed needles, and other symptoms (Schneider et al. 2005; Śliwa et al. 2008; Valiunas et al. 2010; Ježić et al. 2012).

Schneider et al. (2005) observed, in southwestern Germany, a Scots pine (*P. sylvestris*) showing conspicuous shoot proliferation symptoms in combination with dwarfed needles on one major branch, giving the branch a dense, ball-like appearance, whereas in northeastern Spain, they observed several Aleppo pines (*P. halepensis*) showing abnormal shoot proliferation and short, yellowish, and sometimes twisted needles. However, these aberrations did not result in the ball-like structures of affected branches described above for *P. sylvestris*.

In the following years, "*Ca.* P. pini" infection has been reported in other *Pinus* spp., *P. mugo* in Croatia (Ježić et al. 2012) and *P. banksiana*, *P. mugo*, *P. nigra*, *P. tabuliformis*, as well as *Abies procera* and *Tsuga canadensis* in Poland and Czech Republic (Kamińska et al. 2011). Moreover, in China, "*Ca.* P. pini" has been associated with a disease of *Taxodium distichum* var. *imbricarium*, with major symptoms consisting of little necrotic leaves, abnormal proliferation of twigs, and overall necrotic appearance of the whole tree (Huang et al. 2011).

On the basis of 16S rDNA, "*Ca.* P. pini" is a homogeneous pathogen; nucleotide sequence comparisons revealed that 16S rDNA sequences of several strains are identical or nearly identical showing similarity between 99.8 % and 100 % (Schneider et al. 2005; Śliwa et al. 2008; Valiunas et al. 2010; Kamińska et al. 2011; Ježić et al. 2012). The pine phytoplasma is only distantly related to other phytoplasmas. The closest relatives are members of the palm lethal yellowing and rice yellow dwarf groups and "*Ca.* P. castaneae," which share between 94.5 and 96.6 % 16S rRNA gene sequence similarity with "*Ca.* P. pini."

"*Ca.* P. pini" (AJ632155) has been classified in the new 16SrXXI (pine shoot proliferation) group, subgroup -A, through the use of computer-simulated RFLP analysis of F2nR2 fragment (Wei et al. 2007). *Mse* I, *Rsa* I, and *Hinf* I are the key restriction enzymes that distinguish this phytoplasma group from the others (Wei et al. 2007). The reference strain is Pin127S, GenBank accession number AJ632155. Oligonucleotide sequences of unique regions in the 16S rRNA gene are 5′-GGAAATCTTTCGGGATTTTAGT-3′ (67–88) and 5′-TCTCAGTGCTTAACGCTGTTCT-3′ (603–624).

**"*Candidatus* Phytoplasma castaneae"** is associated in Korea with Japanese chestnut trees (*Castanea crenata*) showing symptoms of witches' broom, including abnormally small leaves and yellowing of young leaves (Jung et al. 2002). The phylogenetic analysis of 16S rDNA sequences placed the Japanese chestnut witches'-broom (CnWB) phytoplasma within a distinct subgroup in the phytoplasma clade of the class *Mollicutes* and indicated that the CnWB phytoplasma is most closely related to coconut phytoplasmas (16SrIV group) suggesting that they share a common ancestor.

All of the 16S rDNA sequences of the "*Ca.* P. castaneae" strains isolated from several independent areas in Kyongnam and Chonbuk provinces in Korea were identical. Sequence comparisons revealed that similarity between the "*Ca.* P. castaneae" and other phytoplasmas ranged from 86.8 % to 94.9 % and that between the CnWB phytoplasma and other mollicutes ranged from 70.1 % to 86.4 %.

"*Ca.* P. castaneae" has been classified in the 16SrXIX (Japanese chestnut witches'-broom) group, subgroup 16SrXIX-A, through the use of computer-simulated RFLP analysis of F2nR2 fragment (Wei et al. 2007). The reference strain is CnWB, GenBank accession number AB054986. Oligonucleotide sequences of unique regions of the 16S rDNA are 5′-CTAGTTTAAAAACAATGCTC-3′ (587–606) and 5′-CTCATCTTCCTCCAATTC-3′ (1145–1162).

**"*Candidatus* Phytoplasma malaysianum"** is associated with virescence and phyllody symptoms in naturally diseased Madagascar periwinkle (*Catharanthus roseus*) plants in western Malaysia (Nejat et al. 2013). Full-length 16S rRNA gene pairwise sequence similarities revealed that the Malaysian periwinkle virescence (MaPV) phytoplasma 16S rDNA shared less than 97.5 % sequence similarity with that of previously described "*Ca.* Phytoplasma" species. The "*Ca.* Phytoplasma" species most closely related to the MaPV phytoplasma was "*Ca.* P. trifolii" (AY390261) sharing 96.5 % 16S rDNA sequence similarity. Nucleotide sequence alignments revealed that the 16S rDNA sequence of MaPV phytoplasma shared 99.1 % and 99.2 % sequence identity with Malayan yellow dwarf (MYD, EU498727) and Malayan oil palm (MOP, EU498728) phytoplasmas, respectively. The MYD phytoplasma was discovered in diseased coconut palm (*Cocos nucifera*) trees showing yellowing symptoms in the Banting area of Selangor State, and the MOP phytoplasma was identified in oil palm (*Elaeis guineensis*) plants grown in the same area but exhibiting yellowing and necrosis symptoms. Furthermore, the 16S rRNA genes of MYD and MOP phytoplasmas possessed all the signature sequences that are unique to "*Ca.* P. malaysianum"; therefore, MYD and MOP phytoplasmas have been termed "*Ca.* P. malaysianum"-related strains (Nejat et al. 2013).

On the bases of in silico RFLP analyses of the 16S rDNA F2nR2 fragment, the new 16SrXXXII group, the Malaysian periwinkle virescence phytoplasma group, subgroup 16SrXXXII-A, was designed with MaPV phytoplasma as the representative strain. Restriction analysis with *AluI* alone was sufficient to distinguish MaPV phytoplasma from strains in all other 16Sr groups. The two "*Ca.* P. malaysianum"-related strains, MYD and MOP, also exhibited new and mutually distinct 16S rDNA F2nR2 RFLP patterns indicating that both MYD and MOP phytoplasmas are members of group 16SrXXXII but were assigned to two different subgroups 16SrXXXII-B (with MYD phytoplasma as the representative strain) and 16SrXXXII-C (with MOP phytoplasma as the representative strain), respectively (Nejat et al. in press). Restriction enzymes useful to distinguish the phytoplasma strains of the three subgroups in group 16SrXXXII are *Bst*UI or *Hha*I, *Sau* 3AI, and *Bfa*I. The reference strain is MaPVR, GenBank accession number EU371934. Oligonucleotide sequences of unique regions of the 16S rRNA gene are 5′-GAAATAGAAGGA-TAACCTTTTATTTTT-3′ (164–190), 5′- CGAAGAAG-TATTTAGGTAT-3′ (407–425), 5′-CGCTGTTCTGTT-3′ (608–619), and 5′-155 GTCTAGCTAGAGTGAG-3′ (629–644).

**"*Candidatus* Phytoplasma fraxini"** is associated with ash yellows (AshY) and lilac witches'-broom (LWB) diseases (Griffiths et al. 1999) in *Fraxinus* spp. and *Syringa* spp. (Oleaceae) respectively, in North America (Sinclair and Griffiths 1994; Sinclair et al. 1996).

Both diseases cause slow apical and radial growth, diminished apical dominance or deliquescent branching, suppressed root development, precocious flowering and/or shoot growth, and witches' brooms. Subnormal greenness and foliar malformations are common, and chlorosis occurs in occasional plants. Highly susceptible plants commonly sustain dieback of branches and roots, produce brooms and stunted deliquescent branches, and die prematurely (Sinclair et al. 1996).

The known host range of "*Ca.* P. fraxini" in nature includes 12 ash species and 19 lilac species (Sinclair et al. 1996). Experimental hosts include *Cuscuta* spp. (dodder), *Daucus carota* (carrot), *Trifolium pratense* (red clover), and *Catharanthus roseus* (periwinkle) (Hibben and Wolanski 1971). Phytoplasmas associated with Ashy and LWB are graft transmissible between *Fraxinus* and *Syringa* (Hibben et al. 1991).

Recently, in Chile, phytoplasma strains closely related to "*Ca.* P. fraxini" (16SrVII-A) have been reported to be associated with grapevine (*Vitis vinifera*) yellows (Gajardo et al. 2009) and with symptomatic murta (*Ugni molinae*, family *Myrtaceae*) and peony (*Paeonia lactiflora*, family *Paeoniaceae*) (Arismendi et al. 2011).

On the bases of RFLP analyses of P1/P7 products (16S rRNA gene and the 16S-23S spacer) obtained from 19 ash or lilac phytoplasma strains, a total of four RFLP profile types were obtained with *Alu* I , *Hha* I, or *Taq* I restriction enzymes. RFLP analyses on a portion of a ribosomal protein operon, amplified with primer pair rpF1/R1 from each of the four strains, resulted in the detection of two RFLP profiles with *Mse* I. Southern analysis, utilizing two nonspecific probes from other phytoplasma groups, revealed three RFLP profile types in anonymous chromosomal DNA of strains representing the four 16S rDNA genotypes (Griffiths et al. 1999).

Sequencing of the amplimers from strains AshY1, AshY3, AshY5, and LWB3 (which represent the four 16S rDNA RFLP profile types) revealed only three positions in the 16S rRNA gene and one position in the 16S-23S spacer at which single nucleotide substitutions occurred. Sequence similarity between any two strains was >99.8 %. In contrast, 16S rDNA sequence similarity between strain AshY1 and the most closely related phytoplasma in a different group (brinjal little leaf, BLL; 16SrVI-D) was 96.5 %.

Phylogenetic analysis based on 16S rDNA sequence from the four representative strains AshY1, AshY3, AshY5, and LWB3, together with sequences from 14 other mollicutes retrieved from GenBank, produced a tree on which the Ashy and LWB strains clustered as a discrete group; thus, the Ashy phytoplasma group is coherent but heterogeneous. The name "*Ca.* P. fraxini" was proposed for this group.

According to the classification scheme proposed by Lee et al. (1998), "Ca. P. fraxini" is a member of the subgroup 16SrVII-A, and restriction enzymes useful to distinguish "*Ca.* P. fraxini" from closely related phytoplasma strains in groups 16SrV, 16SrVI, and VII are *Mse* I, *Alu* I, *Hha* I, *Taq* I, *Hae* III, and *Hinf* I (Lee et al. 1998; Barros et al. 2002; Conci et al. 2005). The reference strain is AshY1, GenBank accession number AF092209. Oligonucleotide sequences of unique regions of the 16S rRNA gene are 5′-CGGAAACCCCTCAAAAGGTTT-3′ (66–86) and 5′-AGGAAAGTC-3′ (588–596).

The genome size of strain Ashy3 was estimated to be 645 kbp by PFGE (Griffiths et al. 1999). Besides the sequences of its 16S ribosomal DNA and the 16S-23S ribosomal DNA spacer region (AF092209), sequences of other genes of "*Ca.* P. fraxini" strain AshY1 have been deposited in GenBank and used especially to improve knowledge on the phylogenetic relationships within the genus "*Ca.* Phytoplasma": *rplV-rpsC* genes (EF183492; Martini et al. 2007), 23SrDNA (EU168779) and *secA* genes (EU168745) (Hodgetts et al. 2008), and *secY* gene (GU004329; Lee et al. 2010).

**"*Candidatus* Phytoplasma sudamericanum"** is associated with abnormal proliferation of axillary shoots resulting in formation of witches'-broom growths of passion fruit (*Passiflora edulis* f. *flavicarpa*) in Brazil (Davis et al. 2012). Passion fruit witches'-broom (PassWB) disease was first reported in the states of Rio and Pernambuco, Brazil, by Kitajima et al. (1981).

Nucleotide sequence alignments revealed that strain PassWB-Br3 shared less than 97.5 % 16S rRNA gene sequence similarity with previously described "*Ca.* Phytoplasma" species. Phylogenetic analyses of 16S rRNA gene sequences indicated that strain PassWB-Br3 is distinct from previously described "*Ca.* Phytoplasma" species forming a well-supported branch. Phylogenetic analysis indicated also that "*Ca.* P. sudamericanum," "*Ca.* P. fraxini," and "*Ca.* P. trifolii" shared a common ancestor. The unique properties of its DNA, in addition to natural host and geographical occurrence, supported the recognition of strain PassWB-Br3 as a representative of a distinct taxon, "*Ca.* P. sudamericanum" (Davis et al. 2012).

Results from iPhyClassifier analysis of virtual RFLP patterns of the 16S rRNA gene (GU292081) indicated that strain PassWB-Br3 represents a previously undescribed subgroup in group 16SrVI, 16SrVI-I. Enzymes that distinguished the PassWB-Br3 F2n/R2 fragment from that of other group 16SrVI subgroups included *Hae* III and *Taq* I (Davis et al. 2012). The reference strain is PassWB-Br3, GenBank accession number GU292081. The oligonucleotide sequences of unique regions in the 16S rRNA gene are 5′-CGAGGACAACAACTG-3′ (127–141), 5′-AGGTAAGTCTATAATTTAATTTAATTTCAGTGCTTAACGCTGTCGTGTT-3′ (580–623), 5′-AGAGACACAGGT-3′ (1018–1029), and 5′-TTGTCGTTAATTGCCAGCACAT-3′ (1095–1116).

**"*Candidatus* Phytoplasma trifolii"** is associated with virescence and proliferation of shoots of alsike clover (*Trifolium hybridum*) in Canada (Hiruki and Wang 2004).

Clover proliferation (CP) was first reported as a yellows-type virus disease of alsike clover (*Trifolium hybridum*) in Alberta, Canada, in the early 1960s (Chiykowski 1965); subsequently, CP was demonstrated to be associated with a phytoplasma. "*Ca.* P. trifolii" was transmitted by dodder from diseased alsike clover to periwinkle (*Catharanthus roseus*), tomato (Lycopersicon esculentum cv. Earliana), and potato (*Solanum tuberosum* cv. Russet Burbank). "*Ca.* P. trifolii" was transmitted *by M. fascifrons* from alsike clover (*T. hybridum*) to China aster (*Callistephus chinensis*), periwinkle (*Catharanthus roseus*), carrot (*Daucus carota*), and tobacco (*Nicotiana rustica*) (Chiykowski 1965).

According to the classification scheme by Lee et al. (1998), "*Ca.* P. trifolii" belongs to subgroup 16SrVI-A. Other phytoplasma strains belonging to the same subgroup are alfalfa witches' broom (AWB), Canada; beet leafhopper-transmitted virescence (BLTVA, e.g., strain VR), California, USA; potato witches' broom (PWB), Canada; potato yellows (PY), North Dakota, USA; and tomato big bud (TBB), California, USA (Lee et al. 1998; Hiruki and Wang 2004). Additional strains belonging to 16SrVI-A are listed in Wei et al. (2008) showing that the geographical distribution of phytoplasmas belonging to this subgroup is not limited to Canada and North America but comprise South Korea, Lebanon, Iran, Austria, and France. Moreover, other reports of strains closely related to "*Ca.* P. trifolii" (16SrVI-A) came from Turkey (Sertkaya et al. 2007), South Bohemia (Přibylová et al. 2009), China (Zhang et al. 2012), and Malaysia (Taylor et al. 2011).

On the basis of 16S rRNA gene sequences, the BLTVA phytoplasma and "*Ca.* P. trifolii" strain CP share about 99.2 % similarity (Martini et al. 2007). Phylogenetic analysis indicated clearly that the "*Ca.* P. trifolii" and its close relatives, brinjal little leaf (BLL), *Fragaria multicipita* phytoplasma (FM), and Illinois elm yellows (ILEY) phytoplasmas, formed a subcluster and were different from all other phytoplasmas, with sequence divergences of ≥2.5 % (Hiruki and Wang 2004).

Phytoplasma strains of 16SrVI-A subgroup produced, on the basis of F2nR2 fragment (1.25 kb), unique RFLP profiles with *Alu* I and *Mse* I. In clover proliferation group, at least eight different subgroups (16SrVI-A, 16SrVI-B, 16SrVI-C, 16SrVI-D, 16SrVI-E, 16SrVI-F, 16SrVI-G, 16SrVI-H) have been proposed, and the key restriction enzymes useful to distinguish "*Ca.* P. trifolii" (16SrVI-A) from closely related phytoplasma strains in group 16SrVI are *Alu* I, *Hae* III, and *Hha* I (Wei et al. 2008). On the basis of 16S rRNA gene amplified with P1/16S-SR primer pair (1.55 kb), "*Ca.* P. trifolii" strain CP (16SrVI-A) can be differentiated from BLTVA phytoplasma strain (16SrVI-A) by *Tsp* 509I restriction enzyme (Sertkaya et al. 2007); therefore, subgroup 16SrVI-A includes genetically heterogeneous phytoplasma strains. RFLP and sequence analysis of rp gene sequences confirmed the variability between strains BLTVA and CP that were classified into two rp subgroups, consistent with their differing ecological niches and biological properties (Martini et al. 2007). The reference strain is CP, GenBank accession number AY390261. Oligonucleotide sequences of unique regions of the 16S rRNA gene are 5′-TTCTTACGA-3′ (201–209) and 5′- TAGAGTAAAAGCC-3′ (252–264).

Besides the sequences of its 16S ribosomal DNA and the 16S-23S ribosomal DNA spacer region (AY390261; Hiruki and Wang 2004), sequences of other genes of "*Ca.* P. trifolii" strain CP have been deposited in GenBank and used especially to improve knowledge on the phylogenetic relationships within the genus "*Ca.* Phytoplasma." These genes are ribosomal protein *rplV-rpsC* genes (EF183486; Martini et al. 2007), *secY* gene (GU004315; Lee et al. 2010), and *tuf* gene (JQ824231, Makarova et al. 2012).

**"*Candidatus* Phytoplasma ulmi"** is responsible for yellows disease in *Ulmus* spp. in North America and Europe and induces symptoms including epinasty, yellowing, dwarfing, and premature casting of leaves, witches' brooms at the tips of twigs and branches, and precocious opening of vegetative buds (Lee et al. 2004b). Recently, in Italy, "*Ca.* P. ulmi" has been reported to infect *Zelkova serrata* showing symptoms of chlorosis which involve the whole plant or some of the branches, foliar reddening on one or more branches, attenuation of apical dominance and proliferation of lateral shoots, witches' broom, reduced growth, and stunting of the plant (Romanazzi and Murolo 2008).

The elm yellows (EY) phytoplasma is transmitted in North America by *Scaphoideus luteolus* (Baker 1949) and in Europe by *Macropsis mendax* (Carraro et al. 2004b).

This phytoplasma is phylogenetically closely related to other pathogens such as "*Ca.* P. rubi" and "*Ca.* P. ziziphi," the causative agents of rubus stunt (RuS) and jujube witches' broom (JWB), respectively (Malembic-Maher et al. 2011; Jung et al. 2003a); flavescence dorée (FD) and Palatinate grapevine yellows (PGY) phytoplasmas; cherry lethal yellows (CLY) phytoplasma; peach yellows (PY-In) phytoplasma; alder yellows (ALY) phytoplasmas; spartium witches'-broom (SpaWB)-EY phytoplasma; hemp dogbane (HD)-associated phytoplasma (Lee et al. 2004b); and *Clematis vitalba*-associated phytoplasma (Angelini et al. 2004). These pathogens form, together with a few other phytoplasmas, a distinct major cluster within the phytoplasma clade, the EY phytoplasma (16SrV) group. "*Ca.* P. ulmi" strain EY1 is the reference phytoplasma strain for this group and is assigned to subgroup 16SrV-A (Lee et al. 1998, 2004b).

Four EY phytoplasma strains (EY1, EY125, EY626, and EY627) in subgroup 16SrV-A, which are associated with EY-infected elms in North America and Europe, shared 99.9 % sequence similarity in the 16S rRNA gene, 99.7 % in the ribosomal protein genes, and 99.5 % (based on two strains) in *sec* Y. "*Ca.* P. ulmi" phytoplasmas shared <97.5 % sequence similarity with all known phytoplasmas belonging to other phytoplasma groups and showed 99.2 % 16S rRNA gene sequence similarity, 96 % similarity in the ribosomal protein genes, and 87.9 % similarity in the *sec* Y gene with respect to "*Ca.* P. ziziphi" (JWB phytoplasma strain).

Phylogenetic analyses based on the three genes clearly indicated that the 16SrV-A strain cluster (consisting of strains EY1T, EY125, EY626, and EY627) represents a distinct lineage divergent from the 16SrV-B strain cluster (consisting of CLY5, PY-In, and JWB) and the 16SrV-C, 16SrV-D, and 16SrV-E cluster (consisting of flavescence dorée, alder yellows, spartium witches'-broom, hemp dogbane, rubus stunt phytoplasma strains) in the EY group (Lee et al. 2004b).

RFLP analyses of F2nR2 fragment of "*Ca.* P. ulmi" and all other EY group strains analyzed by Lee et al. (2004a) showed identical RFLP patterns with *Mse* I and *Alu* I restriction enzymes and the patterns were unique to this group (Lee et al. 1998). On the other hand, "*Ca.* P. ulmi" (16SrV-A) was distinguished from strains of other subgroups (16SrV-B, 16SrV-C, 16SrV-D, and 16SrV-E) by collective profiles obtained from digests of F2nR2 fragment with *Bfa* I and *Rsa* I or *Hpa* II (Lee et al. 2004b). On the basis of RFLP analyses of ribosomal protein operon, "*Ca.* P. ulmi" showed unique patterns with *Tsp* 509I and *Mse* I restriction enzymes (Lee et al. 2004b). Subgroup 16SrV-A strains were classified into two secYV subgroups, secYV-A (EY1) and secYV-M (EY626). The Italian strain EY626 contains additional *Mse* I and *Tsp* 509I sites that distinguish it from strain EY1 from the USA (Daire et al. 1997; Angelini et al. 2001; Lee et al. 2004b).

Recently, phytoplasma strains closely related to "*Ca.* P. ulmi" from Serbia were characterized by means of RFLP analysis and DNA sequencing of four genomic loci: 16S rRNA, ribosomal protein *rpl22-rps3*, *secY*, and *map*. In total, five different genotypes were identified based on collective sequencing of all four genes showing a high degree of genetic variability. In particular, four of these genotypes presented significant nucleotide changes compared with the "*Ca.* P. ulmi" reference strain (Jović et al. 2011b).

The reference strain for "*Ca.* P. ulmi" is EY1, GenBank accession number AY197655. Oligonucleotide sequences of unique regions of 16S rRNA are 5′-GGAAA-3′ (827–835) and 5′-CGTTAGTTGCC-3′ (1098–1108); *rpl22–rps3* are 5′-TTACG-CTTGCC-3′ (284–294), 5′-CATTTAATAAAATTGCTATT-3′ (739–758), and 5′-AAATTCTATTTCTATGGGAAT-3′ (910–932); and *sec* Y are 5′-TTTGATCCAATGTTAA-3′ (350–365), 5′-GTCTTTCGGTCATGGATTGA-3′ (595–614), 5′-ATTTAGTCTAAT-3′ (616–627), and 5′-CAAATAGAACAA-3′ (1053–1064).

"*Ca.* P. ulmi" has been proposed as a distinct species from "*Ca.* P. ziziphi" on the basis of unique DNA and because the two phytoplasmas occupy different ecological niches (no plant hosts or vectors in common) and exhibit strikingly different geographical distributions (Jung et al. 2003a; Lee et al. 2004b).

The partial sequence of *tuf* gene of "*Ca.* P. ulmi" strain EY1 was also deposited in GenBank under accession number JQ824225 (Makarova et al. 2012).

**"*Candidatus* Phytoplasma rubi"** is associated with rubus stunt (RuS) in wild and cultivated red raspberry (*Rubus idaeus*), in wild and cultivated blackberry (*Rubus fruticosus*, *R. laciniatus*, *R. caesius*, and *Rubus* hybrids), and in loganberry (*Rubus loganobaccus*) throughout Europe and Turkey (Lee et al. 1995; Mäurer and Seemüller 1995; Sertkaya et al. 2004; Malembic-Maher et al. 2011). The presence of rubus stunt phytoplasma in great mallow (*Malva sylvestris*) and dog rose (*Rosa canina*) was also reported (Jarausch et al. 2001).

Infected *Rubus* spp. plants may show a variety of symptoms such as stunting, shoot proliferation, small leaves, short internodes, enlarged sepals, phyllody, flower proliferation, and fruit malformations (van der Meer 1987; Mäurer and Seemüller 1995).

Rubus stunt phytoplasma was transmitted by the insect vector *Macropsis fuscula* (de Fluiter and van der Meer 1953) and by grafting from loganberry to loganberry, parsley-leaved blackberry, (*R. laciniatus*) and raspberry and from wild blackberry and raspberry to loganberry (Prentice 1950). The causative agent of rubus stunt was also transmitted from naturally infected plants to the experimental host *Catharanthus roseus* (periwinkle) via dodder (*Cuscuta* spp.) bridges (Marcone et al. 1999b).

The 16S rRNA gene sequence similarity between Rubus stunt phytoplasma strain RuS and "*Ca.* P. ulmi" strain EY1 was 98.9 % and between RuS and phytoplasma strains members of subgroups 16SrV-C and 16SrV-D was 99.4 % (Malembic-Maher et al. 2011).

Sequence analysis performed on five genetic loci, the *tuf* gene (encoding the translation elongation factor EFTu), the *rplV-rpsC* locus (encoding ribosomal proteins L22 and S3), the *rplF-rplR* locus (encoding ribosomal proteins L6 and L18), the *map* gene (encoding the methionine aminopeptidase), and the *uvrB-degV* gene (encoding excinuclease B and DegV protein), demonstrated that all Rubus stunt and dog rose strains were genetically very homogeneous sharing at least 99.9 % gene sequence similarity. From comparative analysis with other members of 16SrV group, it was possible to define 24 "*Ca.* P. rubi"-specific oligonucleotides on the five genetic loci (Malembic-Maher et al. 2011).

Phylogenetic analysis of the concatenated gene sequences clearly distinguished three separate clusters supported by a bootstrap value of 100 %: The first cluster corresponded to all FD strains, AldY, PGY, Spartium, and Clematis phytoplasmas; the second cluster grouped all Rubus stunt and dog rose strains representing a distinct lineage genetically very homogeneous; and the third cluster corresponded to strains of "*Ca.* P. ulmi."

In the classification scheme proposed by Lee et al. (1998), "*Ca.* P. rubi" strains are classified in subgroup 16SrV-E (Davis and Dally 2001; Lee et al. 2004b). On the basis of 16S rRNA gene (F2nR2 fragment), "*Ca.* P. rubi" strains (16SrV-E) can be differentiated from other EY strains by *Tsp* 509I restriction enzyme, whereas on the basis of ribosomal protein genes, they can be differentiated by *Hha* I, *Mse* I, and *Tsp* 509I (Lee et al. 2004b).

The reference strain is RuS, GenBank accession number AY197648. Oligonucleotide sequence complementary to unique region of the 16S rRNA gene is 5′-AGTCAAGATAGTTTCTATAAC-3′.

Rubus stunt phytoplasma has a specific ecology when compared with "*Ca.* P. ulmi" and "*Ca.* P. ziziphi" previously described in EY group, with which it shares >97.5 % 16S rRNA gene sequence similarity. It is characterized by a different plant host range (*Rubus* spp.) and a different insect vector, the leafhopper *Macropsis fuscula*. Therefore, due to its distinct biological niche and its genomic differentiation, it was proposed that the Rubus stunt phytoplasma represents a distinct taxon: "*Ca.* P. rubi" (Malembic-Maher et al. 2011).

The partial sequence of *tuf* gene of "*Ca.* P. rubi" strain RuS was also deposited in GenBank under accession number JQ824210 (Makarova et al. 2012).

**"*Candidatus* Phytoplasma ziziphi"** is associated with jujube (*Ziziphus jujuba*) witches'-broom (JWB) disease, which is prevalent in China and Korea where it causes serious problems for the industry. The disease has also been reported in Japan (Jung et al. 2003a). JWB disease was first described, tentatively, as a graft-transmissible viral disease of jujube trees in Korea (Kim 1965). However, transmission electron microscopy showed that a phytoplasma was associated with the disease (Yi and La 1973). JWB is characterized by the excessive production of axillary and terminal buds on the branches and by the clustering of root sprouts, which produce chlorotic and spindly leaves. Symptoms always develop first on the lower branches of the main stem and then spread through the crown. Phyllody develops on flowers resulting in very low fruit production. Tip dieback occurs on infected branches and infected tree dye in few years (Wang et al. 1981).

JWB is transmitted both by grafting and by insect vector. The transmission of JWB by grafting was reported in the early 1960s (Wang et al. 1964). Insect vectors that are known to transmit JWB include *Hishimonus sellatus* and *Hishimonoides chinensis*, which are polyphagous in nature and have a wide distribution (La and Woo 1980; Wang et al. 1984; Weintraub and Beanland 2006).

Recently, phytoplasma strains closely related to "*Ca.* P. ziziphi" have been reported in India to infect *Ziziphus jujuba* and *Z. nummularia* (Khan et al. 2008) and other plant species in China such as *Spiraea salicifolia* showing yellows and small and deformed leaves (Li et al. 2010), *Sophora japonica* (Chinese scholar tree) with witches'-broom symptoms (Yu et al. 2012), *Amaranthus retroflexus* (amaranth) without symptoms (Yang et al. 2011), *Senna surattensis* (sunshine tree) with symptoms including enlargement and flattening of stems and excessive proliferation of shoots (Wu et al. 2012), *Cannabis* spp. (hemp fiber) with witches'-broom symptoms (Zhao et al. 2007), and *Broussonetia papyrifera* (paper mulberry) with witches'-broom symptoms (Liu et al. 2004).

All 16S rDNA sequences of JWB phytoplasma, isolated from four different regions in Japan and Korea, were virtually identical to each other and to the sequences of two isolates, JWB-Kor2 and JWB-Ch, which were deposited in GenBank. The sequence similarity among these strains was higher than 99.5 %.

The JWB phytoplasma 16S rDNA sequences were most closely related to that of the elm yellows (EY) phytoplasma strains (16SrV-A) in EY group. Phylogenetic analysis of the 16S rDNA sequences from the JWB phytoplasma strains, together with sequences from most of the phytoplasmas retrieved from GenBank, produced a tree in which the JWB isolates clustered as a discrete subgroup (Jung et al. 2003a). Lee et al. (2004b) demonstrated that "*Ca.* P. ziziphi" strain JWB and closely related strains (cherry lethal yellows strain CLY5 and peach yellows strain PY-In) formed a distinct lineage distantly related to all other members of the EY group. "*Ca.* P. ziziphi" strain JWB and closely related strains (CLY5 and PY-In) shared 99.9 % sequence similarity in the 16S rRNA gene, 99.1–99.7 % in the ribosomal protein genes, and 98.5–99.6 % in *secY* (Lee et al. 2004b).

RFLP analyses using 17 restriction enzymes showed identical patterns with all JWB isolates, proving that the JWB phytoplasmas were relatively homogeneous. The JWB phytoplasmas could also be distinguished from other phytoplasmas by RFLP analysis of 16S rDNA. Although JWB and other phytoplasmas of EY group had most of the restriction sites in common, either the *Hpa* II or *Rsa* I restriction sites clearly distinguished JWB phytoplasmas from all other members of the EY subgroup, supporting the hypothesis that the JWB phytoplasmas represent a distinct subgroup (Jung et al. 2003a; Lee et al. 2004b). In the classification system proposed by Lee et al. (1998), JWB has been classified in 16SrV-B subgroup together with cherry lethal yellows and peach yellows phytoplasma from China and India, respectively (Zhu et al. 1997; Lee et al. 2004b). On the basis of RFLP analyses of *rpl22-rps3* and *secY* genes, subgroup 16SrV-B strains were classified into three rpV and three secYV subgroups, rpV-C/secYV-C (JWB), rpV-M/secYV-N (PY-In), and rpV-B/secYV-B (CLY5) (Lee et al. 2004b).

The reference strain is JWB-G1, GenBank accession number AB052876. Oligonucleotide sequences of unique regions of the 16S rRNA gene are 5′-TAAAAAGGCATCTTTTTGTT-3′ and 5′-AATCCGGACTAAGACTGT-3′.

The uniqueness of the JWB phytoplasma appears to be correlated with a specific insect vector (*Hishimonus sellatus*) and the host plant (*Ziziphus jujuba*) or with a specific geographical distribution. The unique properties of the JWB phytoplasma sequences clearly indicated that it represents a distinct taxon, "*Ca.* P. ziziphi".

**"*Candidatus* Phytoplasma balanitae"** is associated with naturally infected wild *Balanites triflora* plants exhibiting typical witches'-broom symptoms (*Balanites* witches' broom, BltWB) with yellow and reduced size leaves in Myanmar (Win et al. 2013). The 16S rRNA gene sequence (1,529 bp) revealed that BltWB phytoplasma had the highest similarity with that of "*Ca.* P. ziziphi" (98.2 %) and it is also closely related to that of "*Ca.* P. ulmi" (98.0 %) and "Ca. P. rubi" (98.0 %). Phylogenetic analysis of the 16S rRNA gene sequences indicated that BltWB phytoplasma clustered with elm yellows-related phytoplasmas. In addition, sequences of *rp* and *secY* genes of BltWB phytoplasma also shared <95.3 % and <90 % similarity with previously described phytoplasmas and the sequences were deposited in GenBank (AB689679 and AB689680). Phylogenetic analysis on these latter genes of BltWB showed that this phytoplasma was clearly distinguished from those of other "*Ca.* Phytoplasma" species. RFLP analysis of the 16S rRNA gene including 16S-23S spacer region differentiated the BltWB phytoplasmas from "*Ca.* P. ziziphi," "*Ca.* P. ulmi," and "*Ca.* P. trifolii." Virtual RFLP pattern produced by *iPhyClassifier* from the BltWB 16S rDNA F2nR2 fragment is different from the reference patterns of all previously established 16Sr groups/subgroups. The most similar is the reference pattern of the 16Sr group V, subgroup V-B (GenBank accession AB052876), with a similarity coefficient of 0.91, indicating that this strain may represent a new subgroup within the 16Sr group V.

The reference strain for "*Ca.* P. balanitae" is BltWB, GenBank accession number AB689678. Oligonucleotide sequences of unique regions of the 16S rRNA gene are 5′-TTGGAAACGG-3′, 5′-ACTAACGAA-3′, 5′-CGGCC-3′, and 5′-ATCCGGACTGAGACCGTN-3′.

The BltWB phytoplasma was proposed to represent a distinct taxon, "*Ca.* P. balanitae," taking into consideration, besides the 16S rRNA similarity, the unique plant host and the restricted geographical occurrence.

**"*Candidatus* Phytoplasma cynodontis"** is the causal agent of Bermuda grass white leaf (BGWL), a destructive disease of Bermuda grass (*Cynodon dactylon*). This disease is known to occur in several Asian countries, Sudan, Kenya, Ethiopia, Italy, Cuba, and Australia (Marcone and Rao 2008a; Nejat et al. 2009a; Salehi et al. 2009; Arocha Rosete and Jones 2010; Obura et al. 2010; Bekele et al. 2011). "*Ca.* P. cynodontis" is a member of the BGWL phytoplasma group or 16SrXIV group. Other members of this group are phytoplasmas infecting mainly gramineous plants, such as brachiaria grass (*Brachiaria distachya*) white leaf (BraWL), annual bluegrass (*Poa annua*) white leaf (ABGWL), dactyloctenium (*Dactyloctenium aegyptium*) white leaf (DacWL), and carpet grass (*Axonopus compressus*) white leaf (CGWL) agents (Seemüller et al. 1998b; Marcone et al. 2004b; Marcone and Rao 2008a). Also, phytoplasmas associated with slow decline and white tip dieback diseases of date palm (*Phoenix dactylifera*) in North Africa (Cronjé et al. 2000a, b) and coconut yellow decline of coconut (*Cocos nucifera*) in Malaysia (Nejat et al. 2009a, b) are very closely related to the BGWL phytoplasma. Although this pathogen preferentially infects Bermuda grass, "*Ca.* P. cynodontis"-related strains have been detected in white leaf-diseased plants of *Dichanthium annulatum*, *Oplismenus burmannii*, and *Digitaria sanguinalis* (Rao et al. 2009, 2010). The leafhopper *Exitianus capicola* has been reported to transmit the pathogen in Iran (Salehi et al. 2009). The BGWL phytoplasma is a relatively homogeneous pathogen. Phylogenetic studies revealed that BGWL phytoplasma strains from several countries were identical or nearly identical at the 16S rDNA sequence level (Marcone et al. 2004b; Rao et al. 2007, 2009, 2010). BGWL-C1 is the reference strain.

The GenBank accession number for rDNA sequences of this strain is AJ550984. Oligonucleotide sequence complementary to unique region of the 16S RNA gene of "*Ca.* P. cynodontis" is 5'-AATTAGAAGGCATCTTTTAAT-3'. Phytoplasmas associated with BraWL, CGWL, and slow decline and white tip dieback diseases of date palm and coconut yellow decline of coconut showed 16S rDNA and/or 16S-23S rDNA spacer sequences that were identical or nearly identical to that of the BGWL phytoplasma (Cronjé et al. 2000a, b; Jung et al. 2003b; Marcone et al. 2004b; Nejat et al. 2009a, b). At the 16S rDNA sequence level, the BGWL phytoplasma differs from sugarcane white leaf (SCWL), sugarcane grassy shoot (SCGS), and rice yellow dwarf (RYD) phytoplasmas in 1.5–2.3 % of the nucleotide positions. However, from sequence analyses of *SecA* gene and 16S-23S rDNA spacer region, serological comparisons, vector transmission specificity, and plant host preferences, there is supporting evidence that the BGWL agent is sufficiently different from these phytoplasmas (Marcone et al. 2004b; Nejat et al. 2009a; Bekele et al. 2011). PFGE analysis revealed a chromosome size of 530 kb for seven BGWL phytoplasma isolates collected at different locations in southern Italy (Marcone et al. 1999a, 2004b). The estimated genome size represents not only the smallest mollicute chromosome reported to date but also the smallest genome known for any free-living, self-replicating organism.

**"*Candidatus* Phytoplasma oryzae"** is associated with RYD, a major disease of rice (*Oryza sativa*), which occurs in most rice-growing areas of Asian countries. The causative agent, the RYD phytoplasma, is naturally transmitted by the leafhoppers *Nephotettix cincticeps*, *N. virescens*, and *N. nigropictus*. This taxon is a member the RYD phytoplasma group or 16SrXI group, also named SCWL group. Other members of this group are SCWL, SCGS, sorghum (*Sorghum stipoideum*) grassy shoot (SGS), cirsium (*Cirsium arvensis*) phyllody (Cirp), galactia (*Galactia tenuiflora*) little leaf (Gall), and napier grass (*Pennisetum purpureum*) stunt phytoplasmas as well as the strain BVK obtained from the leafhopper *Psammotettix cephalotes* in Germany (Seemüller et al. 1998b; Lee et al. 2000; Jung et al. 2003b; Jones et al. 2004; Marcone et al. 2004b; Marcone and Rao 2008b). GenBank accession numbers for rDNA sequences of "*Ca.* P. oryzae" are D12581 and AB052873. Oligonucleotide sequences complementary to unique regions of the 16S rRNA gene of "*Ca.* P. oryzae" are 5′-AACTGGATAGGAAATTAAAAGGT-3′ and 5′-ATGAGACTGCCAATA-3′ (Jung et al. 2003b). At the 16S rDNA sequence level, the RYD phytoplasma differs from most other members of the RYD group, including the SCWL, SCGS, and SGS agents, as well as from BGWL group phytoplasmas, in less than 2.5 % of the nucleotide positions (Jung et al. 2003b; Marcone et al. 2004b; Rao et al. 2008). However, due to its unique properties, such as plant host and insect vector specificity and geographical distribution, RYD phytoplasma is regarded as a distinct taxonomic entity (Jung et al. 2003b). Recently, on the basis of 16S rDNA and *SecA* sequence analyses, a "*Ca.* P. oryzae"-related strain has been reported to be associated with yellow leaf disease of areca palm (*Areca catechu*) in India (Manimekalai et al. 2013).

**"*Candidatus* Phytoplasma phoenicium"** is associated with almond witches' broom (AlmWB), a destructive disease of almond, which is present in Lebanon and Iran (Choueiri et al. 2001; Abou-Jawdah et al. 2002; Verdin et al. 2003). The AlmWB agent is also known to occur in nature on peach, nectarine, and rootstock GF 677 (*P. dulcis* x *P. persica*) (Abou-Jawdah et al. 2002, 2009; Molino Lova et al. 2011; Salehi et al. 2011). This taxon belongs to pigeon pea witches'-broom phytoplasma group or 16SrIX group, subgroup 16SrIX-B (Verdin et al. 2003; Lee et al. 2012). It is most closely related to Picris echioides yellows (PEY) and Knautia arvensis phyllody (KAP) phytoplasmas, sharing a 16S rDNA sequence similarity of 98.7 and 99.0 %, respectively (Abou-Jawdah et al. 2002; Verdin et al. 2003). PEY and KAP phytoplasmas are members of subgroup 16SrIX-C (Lee et al. 2012). The isolate identified in AlmWB-affected almond trees in Lebanon is the reference strain (Verdin et al. 2003). The GenBank accession number for rDNA sequences of this strain is AF515636. Oligonucleotide sequence complementary to unique region of the 16S RNA gene of "*Ca.* P. phoenicium" is 5′-CCTTTTTCGGAAGGTA-3′ (Verdin et al. 2003). "*Ca.* P. phoenicium"-related strains were identified in declining almond, peach, and nectarine trees in Lebanon as well as in western juniper (*Juniperus occidentalis*) trees affected by the juniper witches'-broom (JunWB) disease in Oregon, USA. Based on computer-simulated RFLP analysis, the Lebanese "*Ca.* P. phoenicium"-related strains were assigned to subgroups 16SrIX-G, 16SrIX-F, and 16SrIX-D (Molino Lova et al. 2011), whereas those associated with the JunWB disease, to subgroup 16SrIX-F (Davis et al. 2010).

**"*Candidatus* Phytoplasma omanense"** (omanense, epithet pertaining to Oman) is the causative agent of cassia witches' broom (CWB), a disease affecting *Cassia italica*, which is present in Oman (Khan et al. 2007; Al-Saady et al. 2008). The CWB phytoplasma, which has been reassigned by Zhao et al. (2009a) to the 16SrXXIX group, subgroup 16SrXXIX-A, is most closely related to members of 16SrIX group sharing a 16S rDNA sequence similarity that varies from 95 to 97 % (Al-Saady et al. 2008). IM-1 is the reference strain. GenBank accession number for rDNA sequences of this strain is EF666051. Oligonucleotide sequences complementary to unique regions of the 16S rRNA gene of "*Ca.* P. omanense" are 5′-AAAAAACAGT-3′, 5′-TTGC-3′, 5′-GTTAAAG-3′, 5′-TAATT-3′, and 5′-AAATT-3′ (Al-Saady et al. 2008).

***Subclade III*** is a relatively tight clade, made of 7 species that share at least 94.7 % identity between their 16S rRNA gene sequences and at least 92.8 % identity with 16S rRNA gene sequences of phytoplasmas belonging to other subclades.

**"*Candidatus* Phytoplasma mali"** (mali, epithet referring to the plant host) is the causal agent of apple proliferation (AP), one of the most economically important phytoplasmal diseases, which is known to occur in several major apple-growing areas of western and central Europe. This phytoplasma is phylogenetically closely related to other temperate fruit tree pathogens such as "*Ca.* P. pyri" and "*Ca.* P. prunorum," the causative agents of pear decline (PD) and European stone fruit yellows (ESFY), respectively, the phytoplasma identified in decline-affected

Japanese pear (*Pyrus pyricola*) in Taiwan (PDTW), and the peach yellow leaf roll (PYLR) agent (Seemüller and Schneider 2004; Liu et al. 2007). In accordance with the widely accepted RFLP-based classification system, "*Ca.* P. mali" is assigned to subgroup 16SrX-A (Lee et al. 2007; Wei et al. 2007, 2008). In nature, "*Ca.* P. mali" is associated with cultivars and rootstocks of *Malus* x *domestica* (domestic apple) and also with a number of wild and ornamental *Malus* spp. and hybrids (Seemüller et al. 2011a). Using several suitable restriction enzymes, a phytoplasma showing the same rDNA RFLP profiles as "*Ca.* P. mali" was occasionally identified in naturally infected plants of *Corylus avellana* (European hazel), *Pyrus communis* (French pear), *P. pyrifolia* (Nashi pear), *Prunus salicina* (Japanese plum), *P. avium* (sweet cherry), *P. persica* var. *nectarina* (nectarine), *P. domestica* (European plum), *P. armeniaca* (apricot), *Crataegus monogyna* (hawthorn), *Quercus robur* and *Q. rubra* (oak), *Carpinus betulus* (hornbeam), and *Convolvulus arvensis* (wild bindweed) (Seemüller and Schneider 2004; Mehle et al. 2007; Seemüller et al. 2011a; Cieślińska and Morgaś 2011). The pathogen has also been transmitted from diseased apple tree to *Catharanthus roseus* (periwinkle), *Nicotiana occidentalis*, and *Apium graveolens* (celery) via dodder (*Cuscuta subinclusa*, *C. campestris*, *C. europea*, and *C. reflexa*) bridges. By grafting, it was also transmitted to other *Nicotiana* species and tomato (*Lycopersicon esculentum*) (Lauer and Seemüller 2000). "*Ca.* P. mali" is mainly spread in nature by the psyllids *Cacopsylla picta* (syn. *C. costalis*) and *C. melanoneura*. The leafhopper *Fieberiella florii* is also reported as a vector of the pathogen (Frisinghelli et al. 2000; Tedeschi et al. 2003; Tedeschi and Alma 2006).

On the basis of 16S rDNA and 16S/23S rDNA spacer region sequences, "*Ca.* P. mali" is a homogeneous pathogen. Nucleotide sequence comparisons revealed that the 16S rDNA sequences of several "*Ca.* P. mali" strains from various European countries are identical or nearly identical, showing similarity values between 99.9 and 100 % (Seemüller and Schneider 2004; Seemüller et al. 2011a). AP15, which is probably the most common type of AP phytoplasma, is the reference strain. The GenBank accession number for rDNA sequences of strain AP15 is AJ542541. Oligonucleotide sequence complementary to unique region of the 16S RNA gene of "*Ca.* P. mali" is 5′-AATACTCGAAACCAGTA-3′ (Seemüller and Schneider 2004). In interspecies comparisons of the AP/PD, AP/PDTW, AP/ESFY, and AP/PYLR agents, differences in 16S rDNA sequences were 1.0–1.1, 1.1–1.3, 1.3–1.5, and 1.4–1.6 %, respectively (Seemüller and Schneider 2004; Liu et al. 2007; Seemüller et al. 2011a). These differences are below the recommended threshold of 2.5 % for defining a novel species under the provisional status "*Candidatus*" (IRPCM Phytoplasma/Spiroplasma Working Team – Phytoplasma Taxonomy Group 2004; Firrao et al. 2005). However, supporting data for separation of AP, PD, and ESFY agents at putative species level were obtained by examining other molecular markers and considering insect vector and plant host specificity (Seemüller and Schneider 2004). More distantly related to "*Ca.* P. mali" are four other phytoplasmas that cluster in the same subclade as the AP group members: "*Ca.* P. spartii," "*Ca.* P. rhamni," "*Ca.* P. allocasuarinae," and "*Ca.* P. tamaricis." These phytoplasmas share between 94 and 97.2 % 16S rDNA sequence similarity with "*Ca.* P. mali" (Marcone et al. 2004a; Zhao et al. 2009a). At 16S/23S rDNA spacer region sequence level, the sequence identity values between "*Ca.* P. mali" and the other AP group fruit tree phytoplasmas range from 96.0 to 98.5 %, whereas dissimilarities with the other phytoplasmas clustering in the AP group are greater than 10 % (Marcone et al. 2004a; Seemüller and Schneider 2004; Liu et al. 2007). Analysis of ribosomal protein genes and non-ribosomal loci, including the *imp*, *aceF*, *pnp*, *secY*, and *hflB* genes and a putative nitroreductase gene, revealed a considerable genomic variability in "*Ca.* P. mali" strains (Seemüller and Schneider 2004; Danet et al. 2007, 2011; Martini et al. 2008; Schneider and Seemüller 2009; Seemüller et al. 2010, 2011b; Casati et al. 2011). The highest sequence variability occurred in the *imp* gene with similarity values ranging from 83.2 to 90.1 % (Seemüller et al. 2011b). Also, the highest dissimilarities observed between "*Ca.* P. mali" and PD and ESFY agents were 49.2 and 67.9 % on the basis of *imp* gene, 14.1 and 15.4 % in *hflB*, 10 and 11 % in *aceF*, 10 and 8 % in *secY*, and 5 and 7 % in *pnp* genes, respectively (Danet et al. 2007, 2011; Schneider and Seemüller 2009; Casati et al. 2011). Like "*Ca.* P. pyri" and "*Ca.* P. prunorum," "*Ca.* P. mali" has a linear chromosome (Kube et al. 2008). The chromosome size of "*Ca.* P. mali" strains, including AP15, varies, ranging between 600 and 690 kb (Marcone et al. 1999a; Seemüller and Schneider 2007; Kube et al. 2008). Due to the close relationships of "*Ca.* P. mali" with "*Ca.* P. pyri" and "*Ca.* P. prunorum," most of the primers located in the 16S rRNA gene and in the 16S/23S rDNA spacer region, which were designed for specific detection of the AP agent, showed cross-reactivity with the DNA of the other AP group fruit tree phytoplasmas (for review, see Seemüller et al. 1998a). However, "*Ca.* P. mali" can clearly be distinguished from the other AP group fruit tree phytoplasmas using RFLP analysis of PCR-amplified 16S rDNA sequences employing *Ssp* I and *Sfe* I restriction endonucleases (Lorenz et al. 1995; Seemüller et al. 1998a).

**"*Candidatus* Phytoplasma pyri"** is the cause of PD disease, one of the most important disorders occurring in the cultivated European or French pear *Pyrus communis*. This disease is widespread in all pear-growing areas of North America and Europe. "*Ca.* P. pyri" is a member of the subgroup 16SrX-C (Lee et al. 2007; Wei et al. 2007, 2008). This pathogen has been identified in naturally infected rootstocks and scion cultivars of *P. communis* and *P. pyricola* and in rootstocks or own-rooted trees of *P. ussuriensis*, *P. calleryana*, *P. elaeagrifolia*, and *Cydonia oblonga* (quince). By grafting, the PD phytoplasma has been transmitted to progenies of a large number of *Pyrus* spp. Therefore, it seems that most or all *Pyrus* spp. are hosts of "*Ca.* P. pyri" (Seemüller et al. 2011c). By psylla feeding, "*Ca.* P. pyri" has been transmitted to periwinkle (Kaloostian et al. 1971) and via dodder (*C. odorata*) bridges to periwinkle and tobacco (*N. occidentalis* and *N. tabacum*) (Marcone et al. 1999b). The pathogen is transmitted in nature by the psyllids *C. pyricola* (pear psylla) and *C. pyri* (Jensen et al. 1964; Carraro et al. 1998a).

"*Ca.* P. pyri" is a homogeneous pathogen at 16S rDNA sequence level. Sequences of several "*Ca.* P. pyri" strains from

Europe are identical or nearly identical, with similarity values between 99.9 and 100 % (Seemüller and Schneider 2004; Seemüller et al. 2011c). PD1 is the reference strain. The GenBank accession number for rDNA sequences of strain PD1 is AJ542543. Oligonucleotide sequence complementary to unique region of the 16S RNA gene of "*Ca.* P. pyri" is 5′-TTAATAAGTC-TATGGTCT-3′. This oligonucleotide sequence is also shared by the PYLR agent (Seemüller and Schneider 2004). In interspecies comparisons between the PD/PYLR, PD/PDTW, PD/ESFY, and PDTW/ESFY agents, differences in 16S rDNA sequences were 0.4, 0.9–1.5, 1.2–1.3, and 1.2–1.4 %, respectively (Seemüller and Schneider 2004; Liu et al. 2007; Seemüller et al. 2011c). As mentioned above, differences ranging from 1.0 to 1.6 % were observed in the comparisons of PD, PDTW, and PYLR agents with "*Ca.* P. mali". These findings indicate that differences between "*Ca.* P. pyri" and PDTW agent are of the same magnitude as those occurring between "*Ca.* P. pyri" and the other AP group fruit tree phytoplasmas. Also, at 16S/23S rDNA spacer region sequence level, PDTW phytoplasma is more closely related to ESFY agent than to "*Ca.* P. pyri," showing similarity values of 98.8 and 97.7–98.4 %, respectively (Liu et al. 2007). Therefore, the PDTW phytoplasma, which is known to occur only in Taiwan and seems to be transmitted in nature by the psyllids *C. qianli* and *C. chinensis* (Liu et al. 2007), should be regarded as a distinct taxonomic entity. The 16S/23S rDNA spacer region sequence identity values between "*Ca.* P. pyri" and the other AP group fruit tree phytoplasmas range from 95.2 to 98.8 % whereas dissimilarities of "*Ca.* P. pyri" with the other phytoplasmas clustering in the AP group are more than 10 % (Marcone et al. 2004a; Seemüller and Schneider 2004; Liu et al. 2007). Also, the highest dissimilarities observed between "*Ca.* P. pyri" and AP and ESFY agents were 49.2 and 67.5 % on the basis of *imp* gene, 14.1 and 10.4 % in *hflB*, 10 and 12 % in *aceF*, 10 and 7 % in *secY*, and 5 and 6 % in *pnp* genes, respectively (Danet et al. 2007, 2011; Schneider and Seemüller 2009). The chromosome of "*Ca.* P. pyri" strain PD1 is 660 kb in size (Marcone et al. 1999a). "*Ca.* P. pyri" can clearly be differentiated from AP and ESFY phytoplasmas using RFLP analysis of PCR-amplified 16S rDNA sequences employing *Ssp* I and *Bsa* AI restriction endonucleases (Lorenz et al. 1995; Seemüller et al. 1998a). PYLR agent, which is the cause of a major disease of peach in California, proved to be indistinguishable from "*Ca.* P. pyri" in most studies in which ribosomal and non-ribosomal DNA sequences were employed (for review, see Seemüller et al. 1998a). However, significant differences between "*Ca.* P. pyri" and PYLR agent were observed in the *imp* gene (Morton et al. 2003). This finding supports geographical and pathological evidence that PD and PYLR are caused by different organisms.

**"*Candidatus* Phytoplasma prunorum"** is an important prokaryotic pathogen that infects stone fruits in Europe. It is known to cause several disorders of *Prunus* spp. which are collectively referred to as ESFY. This phytoplasma, which is assigned to subgroup 16SrX-B (Lee et al. 2007; Wei et al. 2007, 2008), preferentially infects plants in the genus *Prunus*. It occurs in nature mainly on apricot, Japanese plum, and peach (*P. persica*). However, the pathogen is also common on almond (*P. dulcis*) and flowering cherry (*P. serrulata*) (Marcone et al. 2010a). On the basis of primer specificity and RFLP analysis of PCR-amplified DNA, "*Ca.* P. prunorum" infections have also been detected in naturally infected plants of *P. domestica*, *P. avium*, *P. cerasus* (sour cherry), *P. mahaleb*, *P. cerasifera*, *P. bokhariensis*, *P. brigantina*, *P. cocomilia*, *P. hollywood*, *P. orthosepal*, *P. simonii*, *P. spinosa*, *P. subcordata*, *P. cerasifera* x *P. munsoniana* (*P.* "Marianna" GF 8/1), and *P. besseyi* x *P. hortulana* (for review, see Marcone et al. 2010a). By grafting and insect vector, "*Ca.* P. prunorum" was experimentally transmitted to several *Prunus* taxa listed above including *P. insititia*, *P. tomentosa*, *P. padus*, *P. laurocerasus*, *P. cerasus* x *P. canescens*, *P. fruticosa* x *P. avium*, *P. fruticosa* x *P. cerasus* (Kison and Seemüller 2001; Carraro et al. 2004a). It has also been transmitted from diseased stone fruit trees to periwinkle and from periwinkle to *N. tabacum* via dodder (*C. campestris* and *C. reflexa*) bridges (Loi et al. 1995; Marcone et al. 1999b; Marcone and Seemüller 2001). Moreover, "*Ca.* P. prunorum" was transmitted by grafting from *N. tabacum* to several other *Nicotiana* species and other solanaceous plants, including tomato (Marcone and Seemüller 2001). By PCR assays using specific primers and RFLP analysis, "*Ca.* P. prunorum" was detected in naturally infected plants of *Fraxinus excelsior* (ash), *Rosa canina* (dog rose), *Celtis australis* (hackberry), *C. avellana*, and *Vitis vinifera* (grapevine) (Marcone et al. 2010a). The psyllid *Cacopsylla pruni* has been identified as a natural vector of "*Ca.* P. prunorum" (Carraro et al. 1998b).

Like the other AP group fruit tree phytoplasmas, "*Ca.* P. prunorum" is a homogeneous taxon at level of ribosomal DNA sequences. Sequence alignment revealed that the 16S rDNA sequences of five "*Ca.* P. prunorum" strains from various locations in Europe are identical or nearly identical, showing similarity values between 99.8 and 100 %. ESFY-G1 (= GSFY1) is the reference strain. The GenBank accession number for rDNA sequences of this strain is AJ542544. Oligonucleotide sequences complementary to unique regions of the 16S RNA gene of "*Ca.* P. prunorum" are 5′-AATACCCGAAACCGTA-3′ and 5′-TGAAGTTTTGAGGCATCTCGAA-3′ (Seemüller and Schneider 2004). In interspecies comparisons of the ESFY/AP, ESFY/PD, ESFY/PYLR, and ESFY/PDTW agents, differences in 16S rDNA sequences were 1.3–1.5, 1.2–1.3, 1.4–1.6, and 1.2–1.4 %, respectively (Seemüller and Schneider 2004; Liu et al. 2007). Other phytoplasmas that cluster in the same subclade as the AP group members share between 94 and 97.1 % 16S rDNA sequence similarity with "*Ca.* P. prunorum" (Marcone et al. 2004a; Zhao et al. 2009a). At 16S/23S rDNA spacer region sequence level, "*Ca.* P. prunorum" differs from the other AP group fruit tree phytoplasmas in 1.2–3.0 % of nucleotide positions and from the other phytoplasmas clustering in the AP subclade in more than 11 % of positions (Marcone et al. 2004a; Seemüller and Schneider 2004; Liu et al. 2007). Sequence alignment of several less-conserved, non-ribosomal genes has shown a considerable diversity among "*Ca.* P. prunorum" strains in the *imp*, *aceF*, *secY*, and *pnp* genes (Danet et al. 2007, 2011; Marcone et al. 2010a, b). The greatest dissimilarity values identified between "*Ca.* P. prunorum" and AP and PD agents were 67.9 and 67.5 % on the basis of *imp* gene, 15.4 and 10.4 % in *hflB*, 11 and 12 % in *aceF*, 8 and 7 % in *secY*, 7 and 6 % in *pnp*, and 5.4 and 5.9 % for ribosomal protein (*rpsV* and *rpsC*) genes, respectively (Morton et al. 2003; Danet et al. 2007, 2011; Lee et al. 2007; Martini et al. 2007; Schneider and Seemüller 2009; Marcone et al. 2010a, b). Pulsed-field gel electrophoresis (PFGE) analysis revealed a uniform chromosome size of 630 kb for three strains of "*Ca.* P. prunorum" including the reference strain GSFY1 (Marcone et al. 1999a). "*Ca.* P. prunorum" can clearly be distinguished from the other AP group fruit tree phytoplasmas using RFLP analysis of PCR-amplified 16S rDNA sequences employing *Ssp* I, *Bsa* AI, and *Rsa* I restriction endonucleases (Seemüller et al. 1998a).

**"*Candidatus* Phytoplasma spartii"** is associated with spartium witches' broom (SpaWB), a lethal disease of *Spartium junceum* (Spanish broom) that occurs in Italy and Spain (Marcone et al. 2004a). "*Ca.* P. spartii" is a member of the subgroup 16SrX-D (Lee et al. 2007; Wei et al. 2007, 2008). This taxon shares 97.1–97.2 % 16S rDNA sequence similarity with AP group fruit tree phytoplasmas. In the 16S/23S rDNA spacer region, sequence dissimilarities between "Ca. P. spartii" and AP group fruit tree phytoplasmas are greater than 12 % (Marcone et al. 2004a). SpaWB is the reference strain. The GenBank accession number for rDNA sequences of this strain is X92869. Oligonucleotide sequence complementary to unique region of the 16S RNA gene of "*Ca.* P. spartii" is 5′-TTATCCGCGTTAC-3′. A "*Ca.* P. spartii"-related strain has been identified in witches'-broom-affected plants of *Sarothamnus scoparius*. Distinction of "*Ca.* P. spartii" and the *Sarothamnus scoparius*-infecting agent is possible by RFLP analysis of rDNA sequences using *Hha* I restriction endonuclease (Marcone et al. 2004a).

**"*Candidatus* Phytoplasma rhamni"** is associated with buckthorn witches' broom (BWB), a lethal witches'-broom disease of *Rhamnus catharticus* (buckthorn). This disease has been reported in southwestern Germany and northern Italy (Mäurer and Seemüller 1996; Poggi Pollini et al. 2005). However, "*Ca.* P. rhamni" infections have also been detected in non-symptomatic plants of buckthorn in several European countries (Jović et al. 2011a). "*Ca.* P. rhamni" shares 96 % 16S rDNA sequence similarity with AP group fruit tree phytoplasmas and 95 % with "Ca. P. spartii" (Marcone et al. 2004a). Greater differences occur in the sequences of the 16S/23S rDNA spacer region, where "*Ca.* P. rhamni" differs from the AP group fruit tree phytoplasmas in 14–17 % of nucleotide positions and from "*Ca.* P. spartii" in 16 % of positions (Marcone et al. 2004a). BWB is the reference strain. GenBank accession numbers for rDNA sequences of this strain are X76431 and AJ583009. Oligonucleotide sequence complementary to unique region of the 16S RNA gene of "*Ca.* P. rhamni" is 5′-CGAAGTATTTCGATAC-3′ (Marcone et al. 2004a).

**"*Candidatus* Phytoplasma allocasuarinae"** is associated with allocasuarina yellows (AlloY), a disease that affects *Allocasuarina muelleriana* (slaty she-oak) in Australia (Marcone et al. 2004a). "*Ca.* P. allocasuarinae" is most closely related to "*Ca.* P. rhamni," sharing 96 % 16S rDNA sequence identity, whereas the 16S rDNA similarity with each the SpaWB, AP, PD, ESFY, and PYLR phytoplasmas is 94 %. At 16S/23S rDNA spacer region level, the "*Ca.* P. allocasuarinae" differs from AP fruit tree phytoplasmas in 11–15 % of the nucleotide positions and from the SpaWB and BWB phytoplasmas in 17 % and 18 % of the nucleotide positions, respectively (Marcone et al. 2004a). AlloY is the reference strain. GenBank accession number for rDNA sequences of this strain is AY135523. Oligonucleotide sequence complementary to unique region of the 16S RNA gene of "*Ca.* P. allocasuarinae" is 5′-TTTATTCGAGAGGGCG-3′ (Marcone et al. 2004a).

**"*Candidatus* Phytoplasma tamaricis"** is the causative agent of salt cedar witches' broom (SCWB), a disease affecting *Tamarix chinensis*, which occurs in China (Zhao et al. 2009a). This taxon is most closely related to AP fruit tree phytoplasmas, sharing 96.6% 16S rDNA sequence similarity with "*Ca.* P. prunorum" (Zhao et al. 2009a). Based on computer-simulated RFLP analysis, "*Ca.* P. tamaricis" was assigned to a new 16Sr group, the 16Sr XXX group (Zhao et al. 2009a). SCWB1 is the reference strain. GenBank accession number for rDNA sequences of this strain is FJ432664. Oligonucleotide sequences complementary to unique regions of the 16S RNA gene of "*Ca.* P. tamaricis" are 5′-ATTAGGCATCTAGTAACTTTG-3′, 5′-TGCTCAACATTGTTGC-3′, 5′-AGCTTTGCAAAGTTG-3′, and 5′-TAACAGAGGTTATCAGAGTT-3′ (Zhao et al. 2009a).

## Ecology and Pathogenicity

The two genera of the family *Acholeplasmataceae* have distinct ecology and habitat. Acholeplasmas are believed to be commonly present in the fluids of vertebrate animals, particularly from the upper respiratory tract and urogenital tract, and are frequently isolated from eukaryotic cell culture due to their occurrence in animal serum used in tissue culture media.

*Acholeplasma axanthum*, *A. brassicae*, and *A. palmae* as well as strains of *A. laidlawii* and *A. oculi* were isolated from plants, although they may represent contamination from other sources. In all cases, the acholeplasmas were isolated from the plant surface and were never reported as associated with phloem. Although *A. pleciae* is the sole acholeplasma isolated from insects (in addition to an unpublished report of *A. morum*, cited in Brown et al. 2010), inoculation into leafhoppers, including those known to be vectors of plant mycoplasma diseases, showed their multiplication and prolonged persistence in insect tissues (Whitcomb et al. 1973; Eden-Green and Markham 1987). Nevertheless, there is no evidence of association of the acholeplasmas isolated from plant surfaces with plant or insect disease. In general, the evidence for a pathogenic role of acholeplasmas in natural diseases is not strong: acholeplasmas were found in both healthy and diseased animal tissues and most animals share antibodies against acholeplasmas in sera. It was shown once that *A. axanthum* was pathogenic for goslings and young goose embryos (Kisary et al. 1976), but any additional evidence of *Acholeplasma* spp. as pathogens is missing.

Conversely, phytoplasmas are plant pathogens that are associated with diseases, collectively referred to as yellows diseases, in more than a thousand plant species worldwide. In diseased plants, phytoplasmas reside almost exclusively in the phloem sieve tube elements and are transmitted from plant to plant by phloem-feeding homopteran insects, mainly leafhoppers (Cicadellidae) and plant hoppers (Fulgoromorpha) and less frequently psyllids (Psyllidae) (Weintraub and Beanland 2006). A few species of heteropteran insects of the family *Pentatomidae* (stinkbugs) are also reported as phytoplasma vectors (Hiruki 1999; Weintraub and Beanland 2006). Once phytoplasmas have entered the phloem sieve tube elements, they spread systemically throughout the plant by passing through phloem sieve plate pores. Occasionally, a few phloem parenchyma cells adjacent to sieve tubes are also invaded. In their natural insect vectors, phytoplasmas must pass through a complex biological cycle in order to be transmitted to a plant. After being ingested with phloem sap from an infected plant, phytoplasmas must traverse the insect midgut lining; reach the hemolymph, where they circulate and multiply; and invade various other insect organs and tissues, including the salivary glands, where phytoplasmas multiply further. Then, phytoplasmas are introduced, along with saliva, into sieve tube elements of a new host plant during insect feeding (Hogenhout et al. 2008; Gasparich 2010). Although phytoplasma DNA has been detected in embryos of lethal yellowing diseased coconut palms and seeds from phytoplasma-infected plants of lime, alfalfa, tomato, oilseed rape, maize, and apricot, there is no clear-cut evidence that phytoplasmas are seed-borne pathogens (for reviews, see Faghihi et al. 2011; Dickinson et al. 2013). Also, phytoplasmas cannot be transmitted mechanically. However, they can be spread by the use of infected vegetative propagating material (Lee et al. 2000; Dickinson et al. 2013). Many phytoplasmas have been experimentally transmitted from naturally infected plants to periwinkle via dodder (*Cuscuta* spp.) bridges. Periwinkle is the most commonly used experimental host in which phytoplasmas are routinely maintained by periodic grafting.

Most of the phytoplasma host plants are angiosperms in which a wide range of specific and nonspecific symptoms are induced. Symptoms of affected plants may vary with the phytoplasma strain, host plant, stage of the disease, age of the plant at the time of infection, phytoplasma concentration in infected tissues, strain interactions, and environmental conditions (for reviews, see McCoy et al. 1989; Lee et al. 2000; Seemüller et al. 2002; Marcone 2010). Specific symptoms include virescence, phyllody, big bud, flower proliferation, and other flower abnormalities, all resulting in sterility, witches' brooms, rosetting, internode elongation and etiolation, shortened internodes, enlarged stipules, off-season growth, and brown discoloration of phloem tissue. Less specific and nonspecific symptoms, which are most often common in woody plants, include foliar yellowing and reddening, small leaves, leaf roll, leaf curl, vein clearing, vein enlargement, vein necrosis, premature autumn coloration, premature defoliation, undersized fruits, poor terminal growth, sparse foliage, dieback, stunting of overall plant growth, and decline. In rare instances, phytoplasma-infected plants are fully non-symptomatic over their life span whereas a temporary or permanent remission of symptoms may also occur. Fewer phytoplasmas have been detected in gymnosperms of which most hosts are from *Pinaceae* and *Cupressaceae* families. Infections usually result in yellowing symptoms, stunted growth, dwarfed needles, and proliferation of shoots (Schneider et al. 2005; Davis et al. 2010; Kamińska et al. 2011). Since phytoplasmas live and multiply in functional phloem sieve tube elements, the main effect of phytoplasmal infections apparently is the impairment of the sieve tube function. Several studies have shown that inhibition of phloem transport occurs in phytoplasma-infected plants, which, in turn, leads to an accumulation of abnormal amounts of carbohydrates in source leaves, i.e., mature leaves, and a marked reduction of these essential energy-storage compounds in sink organs, i.e., young leaves, flowers, and roots (Catlin et al. 1975; Braun and Sinclair 1976, 1978; Kartte and Seemüller 1991; Lepka et al. 1999; Maust et al. 2003). Changes in photosynthate translocation along with other impaired physiological functions, including reduced photosynthesis, stomatal conductance and root respiration, altered secondary metabolism, and disturbed plant hormone balance, could account for symptoms exhibited by infected plants (Lepka et al. 1999; Lo Gullo et al. 2000; Tan and Whitlow 2001; Bertamini et al. 2003; Maust et al. 2003; Choi et al. 2004; Ding et al. 2013). However, the exact mechanisms by which phytoplasmas induce disease in plants and the reason for different reactions of the host plants to phytoplasmal infections are still poorly understood. Recent studies have shown that symptoms of flower abnormalities occurring in phytoplasma-infected plants are associated with deregulations of key floral development genes (Pracros et al. 2006; Cettul and Firrao 2011; Himeno et al. 2011; Su et al. 2011). Also, several other plant host genes, which are differentially expressed upon phytoplasmal infections, have been identified. These include genes involved in phytohormone activity, photosynthesis, carbohydrate and lipid metabolism, amino acid transport, phenylpropanoid biosynthesis, and plant stress and/or defense response (Jagoueix-Eveillard et al. 2001; Carginale et al. 2004; Nicolaisen and Horvath 2008; Albertazzi et al. 2009; Hren et al. 2009a, b; Chen and Lin 2011; De Luca et al. 2011; Ding et al. 2013). Furthermore, the availability of complete phytoplasma genome sequences has made it possible to identify a considerable number of genes that are likely to play major roles in phytoplasma-host interactions. Among these, there are genes encoding surface membrane proteins and effector (virulence) proteins (Bai et al. 2009; Hoshi et al. 2009; MacLean et al. 2011; Sugio et al. 2011a, b; Kube et al. 2012).

Insect vectors of phytoplasmas are differently affected by the phytoplasmas they transmit. *Colladonus montanus* leafhoppers infected with the X-disease phytoplasma lived approximately half as long as uninfected leafhoppers. In the infected leafhoppers, pathological lesions of several organs, including salivary glands, were reported to occur. Also, X-disease phytoplasma-infected *C. montanus* leafhoppers produced fewer offsprings than did healthy leafhoppers, whereas increases in mortality were reported for six leafhopper species which transmit the maize bushy stunt (MBS) phytoplasma (for review, see Kirkpatrick 1991). Work by Bressan et al. (2005a, b) showed that the flavescence dorée (FD) phytoplasma greatly reduced longevity and fecundity of its natural and experimental vectors, the leafhoppers *Scaphoideus titanus* and *Euscelidius variegatus*, respectively. A beneficial effect was observed when the aster leafhopper, *Macrosteles quadrilineatus*, fed on aster yellows (AY) phytoplasma-infected plants of aster, lettuce, carrot, and periwinkle. The exposed leafhoppers lived longer and produced more offsprings than nonexposed leafhoppers (for review, see Hogenhout et al. 2008). Recent studies revealed that the reproduction of *M. quadrilineatus* increased considerably when this leafhopper was reared on either AY phytoplasma-infected *Arabidopsis thaliana* plants or transgenic *A. thaliana* plants expressing the gene *SAP11* (Sugio et al. 2011a, b). It has been shown that SAP11, which is an AY phytoplasma effector protein, interferes with plant TCP (*TEOSINTE BRANCHED1, CYCLOIDEA, PROLIFERATING CELL FACTORS 1* and *2*) transcription factor family, which is known to play roles in various aspects of plant development. In particular, SAP11 destabilizes class II TCPs (= *CINCINNATA* [CIN]-TCPs), leading to a decreased synthesis of jasmonic acid (JA), a phytohormone that is involved in the plant defense response against insect herbivores, including the AY phytoplasma vector *M. quadrilineatus* (Sugio et al. 2011a, b). Therefore, an increase in *M. quadrilineatus* population would also result in an increase in AY phytoplasma spread in nature. Adults of *Dalbulus maidis*, a maize leafhopper, confined on aster, lettuce, and *A. thaliana* plants do not attempt to lay eggs and die within a few days. However, when these plants are infected with AY phytoplasma, adults live longer and lay eggs from which nymphs hatch approximately 15 days later (Purcell 1988; Sugio et al. 2011a). Thus, phytoplasma infections can manipulate plants to convert them from being nonhosts into hosts or better hosts for a given insect vector (Hogenhout et al. 2008). There is evidence that highly specific phytoplasma-insect interactions are involved in the transmission process. In particular, specific attachment reactions between phytoplasmas and insect receptors are required for penetration of the gut and salivary gland barriers of the vector. Work by Suzuki et al. (2006) revealed that an abundant surface membrane protein of the onion yellows (OY) phytoplasma, designed as antigenic membrane protein (Amp), formed a complex with insect microfilaments, including actin, myosin heavy chain, and myosin light chain proteins, of the visceral smooth muscle surrounding the intestinal tract, in all OY phytoplasma-transmitting leafhopper species but not in those of non-OY phytoplasma-vector species. Therefore, interaction between Amp and insect microfilaments determines vector specificity of phytoplasmas. Some phytoplasmas have a low insect vector specificity, being transmitted by several vector species, e.g., subgroup 16SrI-B phytoplasmas, whereas others show a very high vector specificity, being transmitted by only one or a few vector species, e.g., temperate fruit phytoplasmas of the AP group (Seemüller et al. 1998b, 2002; Lee et al. 2000). Also, many insect vectors can transmit more than one phytoplasma. The number of insect vectors and their feeding behavior play major roles in determining the plant host range of a given phytoplasma. For example, phytoplasmas of subgroups 16SrI-A, 16SrI-B, and 16SrI-C, which are transmitted by numerous polyphagous leafhoppers including *Macrosteles* spp., *Euscelis* spp., *Scaphytopius* spp., and *Aphrodes* spp., are causing diseases in a wide range of plant species, whereas the FD phytoplasma, a member of the 16SrV group, which is transmitted by the monophagous vector *S. titanus*, is known to infect in nature only grapevine (for review, see Lee et al. 2000). Although phytoplasmas were not believed to be transmitted vertically to the progeny of vector insects for many years, PCR-based and electron microscopical studies, conducted over the last years, provided indications for transovarial transmission of AY phytoplasmas by the leafhoppers *S. titanus* and *Hishimonoides sellatiformis* (Alma et al. 1997; Kawakita et al. 2000) and SCWL and ESFY phytoplasmas by *Matsumuratettix hiroglyphicus* and *C. pruni*, respectively (Hanboonsong et al. 2002; Tedeschi et al. 2006).

Phytoplasmas occur worldwide, but there are differences in the distribution of the various taxonomic groups and subgroups. For example, 16SrI-B phytoplasmas are distributed worldwide, whereas phytoplasmas of subgroups 16SrI-L and 16SrI-M appear to be restricted to Europe. Fruit tree phytoplasmas of the AP group are known to occur in Europe with the exception of the PYLR and PDTW agents, whereas RYD phytoplasmas are only known from Asian countries. The geographical distribution of phytoplasmas appears to be correlated with that of their plant hosts and insect vectors (for reviews, see Seemüller et al. 1998b; Lee et al. 2000, 2004a). Phytoplasmas may differ considerably in their plant host specificity. As mentioned above, phytoplasmas of the subgroups 16SrI-A, 16SrI-B, and 16SrI-C have a wide plant host range which is composed of more than 80 plant species. In contrast, fruit tree phytoplasmas of the AP group preferentially infect only one host (for reviews, see Seemüller et al. 1998b, 2002; Lee et al. 2000). Plant host specificity is still poorly understood. Because most or all phytoplasmas grow in periwinkle and induce specific symptoms in this host, it seems that there is no strict plant host specificity in phytoplasmas. However, in nature, the plant host range of a given phytoplasma largely depends on the three-way interaction between pathogen, plant host, and insect vector. Over the last two decades, several studies have shown that a single plant can be doubly or multiply infected with different phytoplasmas. This phenomenon is common in perennial plants, whose long life spans provide vast opportunities to be visited and inoculated by vectors carrying various phytoplasmas. Furthermore, distinctly different phytoplasmas may induce similar symptoms in a given plant host. A well-known example of distinct phytoplasmas inducing similar symptoms in the same plant is grapevine affected by grapevine yellows disorder, which can be caused either by the 16SrV phytoplasmas including the FD agent or by phytoplasmas from the 16SrI, 16SrII, 16SrIII, and 16SrXII groups (Belli et al. 2010). There are also indications that several phytoplasmas, including AP, ESFY, ash yellows, alder yellows, and AY agents, exist as strains which greatly differ in aggressiveness, ranging from being avirulent (or nearly avirulent) to highly virulent. Interactions between distinct strains of the same taxon have been described for a number of phytoplasma-plant host combinations (for review, see Marcone 2010). Recent work has shown that multiple infections by distinctly different strains of AP phytoplasma are widespread in AP-affected apple trees (Seemüller et al. 2010, 2011b). These studies also revealed that multiple infections are of pathological relevance due to antagonistic strain interactions leading to shifts in the phytoplasma composition that drastically alter virulence.

Many of the phytoplasma diseases, especially those of woody plants, are of great economic importance. Among these, there are apple proliferation, pear decline, European stone fruit yellows, X-disease of stone fruits, grapevine yellows, and lethal yellowing of coconut and other palms.

## Application

Considering the several major pathogens of cultivated plants that are members of the *Acholeplasmataceae* and their destructive impact on human economy, it might appear inappropriate to mention applications. However, phytoplasma infection is beneficial for commercial production of free-branching poinsettia (*Euphorbia pulcherrima*). Named after Robert Poinsett who introduced poinsettia to the USA from Mexico in 1825, poinsettia has become a major ornamental potted plant in North America (Ecke et al. 1990). Two morphotypes of poinsettia cultivars are grown commercially: One is restricted branching characterized by strong apical dominance, producing few axillary shoots and "flowers" (modified leaves called bracts), and the other is free-branching characterized by weak apical dominance, producing many axillary shoots and "flowers."

Free-branching poinsettia cultivars that produce numerous axillary shoots are essential for propagating desirable multi-flowered potted poinsettias and comprise the majority of commercial cultivars propagated today. Many free-branching cultivars (>100) have been developed and propagated commercially in the last decade. In the USA, poinsettias are one of the most economically important floricultural crops. The branching factor has been a mystery to horticulturists for decades. Recent evidence has indicated that the poinsettia branching factor is a graft-transmissible biological agent. In 1997, Lee et al. (1997) using PCR and DNA fingerprinting (RFLP analysis) diagnostic procedures, provided evidence indicating that the self-branching ability of the majority of commercial free-branching cultivars of today is not due to genetic traits selected through breeding but by the grafting of new seedlings (phytoplasma-free) to a free-branching rootstock that contains phytoplasma. The presence of phytoplasma causes the induction of free-branching in these infected poinsettias. This is the first reported example of a pathogenic phytoplasma as the causal agent of a desirable and economically important trait. The finding has benefited growers and the floral industry by applying proper cultural management to improve the quality of poinsettias. Commercial poinsettia pot plants are produced by cutting from mother stock that is infected with phytoplasma.

This phytoplasma associated with poinsettia plants belongs to 16SrIII group, subgroup 16SrIII-H. Other related strains in 16SrIII group may also be able to promote induction of free-branching of poinsettia (Abad et al. 1997). Recently, Nicolaisen and Christensen (2007) reported that phytoplasma infection induced changes in gene expression in poinsettia, which may account for the induction of free-branch.

## References


Abad JA, Randall C, Moyer JW (1997) Genomic diversity and molecular characterization of *poinsettia* phytoplasmas. Phytopathology 87:S1

Abou-Jawdah Y, Karakashian A, Sobh H, Martini M, Lee I-M (2002) An epidemic of almond witches'-broom in Lebanon: classification and phylogenetic relationships of the associated phytoplasma. Plant Dis 86:477–484

Abou-Jawdah Y, Sobh H, Akkary M (2009) First report of almond witches' broom phytoplasma ('*Candidatus* Phytoplasma phoenicium') causing a severe disease on nectarine and peach in Lebanon. Bull OEPP/EPPO Bull 39:94–98

Akhtar KP, Shah TM, Atta BM, Dickinson M, Hodgetts J, Khan RA, Haq MA, Hameed S (2009a) Symptomatology, etiology and transmission of chickpea phyllody disease in Pakistan. J Plant Pathol 91:649–653

Akhtar KP, Sarwar G, Dickinson M, Ahmad M, Haq MA, Hameed S, Iqbal MJ (2009b) Sesame phyllody disease: its symptomatology, etiology, and transmission in Pakistan. Turk J Agric For 33:477–486

Al-Aubaidi JM, Dardiri AH, Muscoplatt CC, McCauley EH (1973) Identification and characterization of *Acholeplasma oculusi* spec. nov. from the eyes of goats with keratoconjunctivitis. Cornell Vet 63:117–129

Albertazzi G, Milc J, Caffagni A, Francia E, Roncaglia E, Ferrari F, Tagliafico E, Stefani E, Pecchioni N (2009) Gene expression in grapevine cultivars in response to Bois Noir phytoplasma infection. Plant Sci 176:792–804

Alfaro-Fernández A, Ali MA, Abdelraheem FM, Saeed EAE, Font San Ambrosio MI (2012) Molecular identification of 16SrII-D subgroup phytoplasmas associated with chickpea and faba bean in Sudan. Eur J Plant Pathol 133:791–795

Aljanabi S, Parmessur Y, Moutia Y, Saumtally S, Dookun A (2001) Further evidence of the association of a phytoplasma and a virus with yellow leaf syndrome in sugarcane. Plant Pathol 50:628–636

Alma A, Bosco D, Danielli A, Bertaccini A, Vibio M, Arzone A (1997) Identification of phytoplasmas in eggs, nymphs and adults of *Scaphoideus titanus* ball reared on healthy plants. Insect Mol Biol 6:115–121

Al-Saady NA, Al-Subhi AM, Al-Nabhani A, Khan AJ (2006) First report of a group 16SrII phytoplasma infecting chickpea in Oman. Plant Dis 90:973

Al-Saady NA, Khan AJ, Calari A, Al-Subhi AM, Bertaccini A (2008) '*Candidatus* Phytoplasma omanense', associated with witches'-broom of *Cassia italica* (Mill.) Spreng. in Oman. Int J Syst Evol Microbiol 58:461–466

Al-Sakeiti MA, Al-Subhi AM, Al-Saady NA, Deadman ML (2005) First report of witches'-broom disease of sesame (*Sesamum indicum*) in Oman. Plant Dis 89:530

Al-Subhi M, Al-Saady NA, Khan AJ, Deadman ML (2011) First report of a group 16SrII phytoplasma associated with witches'-broom of eggplant in Oman. Plant Dis 95:360

Al-Zadjali AD, Natsuaki T, Okuda S (2007) Detection, identification and molecular characterization of a phytoplasma associated with Arabian jasmine (*Jasminum sambac* L.) witches' broom in Oman. J Phytopathol 155:211–219

Al-Zadjali AD, Al-Sadi AM, Deadman ML, Okuda S, Natsuki T, Al-Zadjali TS (2012) Detection, identification and molecular characterization of a phytoplasma associated with beach naupaka witches'-broom. J Plant Pathol 94:379–385

Andersen MT, Longmore J, Liefting LW, Wood GA, Sutherland PW, Beck DL, Forster RLS (1998) Phormium yellow leaf phytoplasma is associated with strawberry lethal yellows disease in New Zealand. Plant Dis 82:606–609

Andersen MT, Beever RE, Sutherland PW, Forster RLS (2001) Association of '*Candidatus* Phytoplasma australiense' with sudden decline of cabbage tree in New Zealand. Plant Dis 85:462–469

Angelini E, Clair D, Borgo M, Bertaccini A, Boudon-Padieu E (2001) Flavescence dorée in France and Italy – occurrence of closely related phytoplasma isolates and their near relationships to Palatinate grapevine yellows and an alder yellows phytoplasma. Vitis 40:79–86

Angelini E, Squizzato F, Lucchetta G, Borgo M (2004) Detection of a phytoplasma associated with grapevine Flavescence dorée in *Clematis vitalba*. Eur J Plant Pathol 110:193–201

Angulo AF, Reijgers R, Brugman J, Kroesen I, Hekkens FEN, Carle P, Bové JM, Tully JG, Hill AC, Schouls LM, Schot CS, Roholl PJM, Polak-Vogelzang AA (2000) *Acholeplasma vituli* sp. nov., from bovine serum and cell cultures. Int J Syst Evol Microbiol 50:1125–1131

Arismendi N, Gonzàlez F, Zamorano A, Andrade N, Pino AM, Fiore N (2011) Molecular identification of '*Candidatus* Phytoplasma fraxini' in murta and peony in Chile. Bull Insect 64(Suppl):S95–S96

Arocha Rosete Y, Jones P (2010) Phytoplasma diseases of the gramineae. In: Weintraub PG, Jones P (eds) Phytoplasmas: genomes, plant hosts and vectors. CAB International, Wallingford/Oxfordshire, UK, pp 170–187

Arocha Y, González L, Peralta E, Jones P (1999) First report of virus and phytoplasma pathogens associated with yellow leaf syndrome of sugarcane in Cuba. Plant Dis 83:1171

Arocha Y, Horta D, Peralta E (2003) First report on molecular detection of phytoplasmas in papaya in Cuba. Plant Dis 87:1148

Arocha Y, Piñol B, Fernández M, Picornell S, Almeida R, Palenzuela I, Wilson MR, Jones P (2005a) '*Candidatus* Phytoplasma graminis' and '*Candidatus* Phytoplasma caricae', two novel phytoplasmas associated with diseases of sugarcane, weeds and papaya in Cuba. Int J Syst Evol Microbiol 55:2451–2463

Arocha Y, López M et al (2005b) Transmission of sugarcane yellow leaf phytoplasma by the delphacid leafhopper *Saccharosydne saccharivora*, a new vector of sugarcane yellow leaf disease. Plant Pathol 54:634–642

Arocha Y, Antesana O, Montellano E, Franco P, Plata G, Jones P (2007) '*Candidatus* Phytoplasma lycopersici', a phytoplasma associated with 'hoja de perejil' disease in Bolivia. Int J Syst Evol Microbiol 57:1704–1710

Aryamanesh N, Al-Subhi AM, Snowball R, Yan G, Siddique KHM (2011) First report of *Bituminaria* witches'-broom in Australia caused by a 16SrII phytoplasma. Plant Dis 95

Atobe H, Watabe J, Ogata M (1983) *Acholeplasma parvum*, a new species from horses. Int J Syst Bacteriol 33:344–349

Aulakh GS, Stephens EB, Rose DL, Tully JG, Barile MF (1983) Nucleic acid relationships among *Acholeplasma* species. J Bacteriol 153:1338–1341

Ayman FO, Foissac X (2012) Occurrence and incidence of phytoplasmas of the 16SrII-D subgroup on solanaceous and cucurbit crops in Egypt. Eur J Plant Pathol 133:353–360

Bai X, Correa VR, Toruño TY, Ammar E-D, Kamoun S, Hogenhout SA (2009) AY-WB phytoplasma secretes a protein that targets plant cell nuclei. Mol Plant Microbe Interact 22:18–30

Baker WL (1949) Notes on the transmission of the virus causing phloem necrosis of American elm, with notes on the biology of its insect vector. J Econ Entomol 42:729–732

Balakishiyeva G, Qurbanov M, Mammadov A, Bayramov S, Aliyev J, Foissac X (2011) Detection of '*Candidatus* Phytoplasma brasiliense' in a new geographic region and existence of two genetically distinct populations. Eur J Plant Pathol 130:457–462

Barros TSL, Davis RE, Resende RO, Dally EL (2002) *Erigeron* witches'-broom phytoplasma in Brazil represents new subgroup VII-B in 16S rRNA gene group VII, the ash yellows phytoplasma group. Plant Dis 86:1142–1148

Batlle A, Altabella N, Sabaté J, Laviña A (2009) Study of the transmission of stolbur phytoplasma to different crop species, by *Macrosteles quadripunctulatus*. Ann Appl Biol 152:235–242

Beever RE, Wood GA, Andersen MT, Pennycook SR, Sutherland PW, Forster RLS (2004) '*Candidatus* phytoplasma australiense' in *Coprosma robusta* in New Zealand. New Zeal J Bot 42:663–675

Bekele B, Abeysinghe S, Hoat TX, Hodgetts J, Dickinson M (2011) Development of specific secA-based diagnostics for 16SrXI and 16SrXIV phytoplasmas of the Gramineae. Bull Insect 64(suppl):15–16

Belli G, Bianco PA, Conti M (2010) Grapevine yellows: past, present, and future. J Plant Pathol 92:303–326

Berger J, Schweigkofler W, Kerschbamer C, Roschatt C, Dalla Via J, Baric S (2009) Occurrence of Stolbur phytoplasma in the vector *Hyalesthes obsoletus*, herbaceous host plants and grapevine in South Tyrol (Northern Italy). Vitis 48:185–192

Bertamini M, Nedunchezhian N, Tomasi F, Grando MS (2003) Phytoplasma [Stolbur-subgroup (Bois Noir-BN)] infection inhibits photosynthetic pigments, ribulose-1,5-bisphosphate carboxylase and photosynthetic activities in field grown grapevine (*Vitis vinifera* L. cv. Chardonnay) leaves. Physiol Mol Plant P 61:357–366

Bové JM, Dannet JL, Hassanzadeh N, Bananej K, Salehi M, Taghizadeh M, Garnier M (2000) Witches'-Broom disease of lime (WBDL) in Iran. In: Proceedings of the 14th International Organization of Citrus Virologists (IOCV), Riverside, pp 207–215

Bové JM, Garnier M, Mjeni AM, Khayrallah A (1988) Witches' broom disease of small fruited acid lime trees in Oman: first MLO disease of citrus. In: Proceedings of the 10th International Organization of Citrus Virologists (IOCV), Riverside, pp 307–309

Bové JM, Navarro L, Bonnet P, Zreik L, Garnier M (1996) Reaction of citrus cultivars to graft-inoculation of phytoplasma aurantifolia-infected lime shoots. In: Proceedings of the 13th International Organization of Citrus Virologists (IOCV), Riverside, pp 249–251

Bowyer W, Atherton G, Teakle D, Ahern GA (1969) *Mycoplasma-like* bodies in plants affected by legume little leaf, tomato big bud, and lucerne witches' broom diseases. Aust J Biol Sci 22:271–274

Bradbury JM (1978) *Acholeplasma equifetale* in broiler chickens. Vet Rec 102:516

Braun EJ, Sinclair WA (1976) Histopathology of phloem necrosis in *Ulmus americana*. Phytopathology 66:598–607

Braun EJ, Sinclair WA (1978) Translocation in phloem necrosis-diseased American elm seedlings. Phytopathology 68:1733–1737

Brcak J (1979) Leafhopper, planthopper vectors of plant disease agents in central and southern Europe. In: Maramorosch K, Harris KF (eds) Leafhopper vectors and plant disease agents. Academic, London, pp 97–146

Bressan A, Clair D, Sémété O, Boudon-Padieu É (2005a) Effect of two strains of flavescence dorée phytoplasma on the survival and fecundity of the experimental leafhopper vector *Euscelidius variegatus* kirschbaum. J Invertebr Pathol 89:144–149

Bressan A, Girolami V, Boudon-Padieu E (2005b) Reduced fitness of the leafhopper vector *Scaphoideus titanus* exposed to flavescence dorée phytoplasma. Entomol Exp Appl 115:283–290

Březíková M, Linhartová S (2007) First report of potato stolbur phytoplasma in hemipterans in southern Moravia. Plant Prot Sci 43:73–76

Brown DR, Whitcomb RF, Bradbury JM (2007) Revised minimal standards for description of new species of the class *Mollicutes* (division Tenericutes). Int J Syst Evol Microbiol 57:2703–2719

Brown DR, Bradbury JM, Johansson K-E (2010) Genus I. *Acholeplasma*. In: Garrity GM et al (eds) Bergey's manual of systematic bacteriology: volume 4: the bacteroidetes, spirochaetes, tenericutes (mollicutes), acidobacteria, fibrobacteres, fusobacteria, dictyoglomi, gemmatimonadetes, lentisphaerae, verrucomicrobia, chlamydiae, and planctomycetes. Springer, New York, pp 688–696

Canale MC, Bedendo IP (2013) '*Candidatus* Phytoplasma brasiliense' (16SrXV-A subgroup) associated with cauliflower displaying stunt symptoms in Brazil. Plant Dis 97(3):419

Carginale V, Maria G, Capasso C, Ionata E, La Cara F, Pastore M, Bertaccini A, Capasso A (2004) Identification of genes expressed in response to phytoplasma infection in leaves of *Prunus armeniaca* by messenger RNA differential display. Gene 33(2):29–34

Carraro L, Loi N, Ermacora P, Gregoris A, Osler R (1998a) Transmission of pear decline by using naturally infected *Cacopsylla pyri* L. Acta Hortic 472:665–668

Carraro L, Osler R, Loi N, Ermacora P, Refatti E (1998b) Transmission of European stone fruit yellows phytoplasma by *Cacopsylla pruni*. J Plant Pathol 80:233–239

Carraro L, Ferrini F, Ermacora P, Loi N (2004a) Transmission of stone fruit yellows phytoplasma to *Prunus* species by using vector and graft transmission. Acta Hortic 657:449–453

Carraro L, Ferrini F, Ermacora P, Loi N, Martini M, Osler R (2004b) *Macropsis mendax* as a vector of elm yellows phytoplasma of *Ulmus* species. Plant Pathol 53:90–95

Casati P, Quaglino F, Stern AR, Tedeschi R, Alma A, Bianco PA (2011) Multiple gene analyses reveal extensive genetic diversity among '*Candidatus* Phytoplasma mali' populations. Ann Appl Biol 158:257–266

Catlin PB, Olsson EA, Beutel JA (1975) Reduced translocation of carbon and nitrogen from leaves with symptoms of pear curl. J Am Soc Hortic Sci 100:184–187

Cettul E, Firrao G (2011) Development of phytoplasma-induced flower symptoms in *Arabidopsis thaliana*. Physiol Mol Plant P 76:204–211

Chen W-Y, Lin C-P (2011) Characterization of *Catharanthus roseus* genes regulated differentially by peanut witches' broom phytoplasma infection. J Phytopathol 159:505–510

Cheng M, Dong J, Zhang Z, McBeath JH (2012) Molecular characterization of stolbur group subgroup E (16SrXII-E) phytoplasma associated with potato in China. Plant Dis 96:1372

Chiykowski LN (1965) A yellows-type virus of alsike clover in Alberta. Can J Bot 43:527–536

Choi YH, Tapias EC, Kim HK, Lefeber AWM, Erkelens C, Verhoeven JTJ, Brzin J, Zel J, Verpoorte R (2004) Metabolic discrimination of *Catharanthus roseus* leaves infected by phytoplasma using H-NMR spectroscopy and multivariate data analysis. Plant Physiol 135:2398–2410

Choueiri E, Jreijiri F, Issa S, Verdin E, Bové J, Garnier M (2001) First report of a phytoplasma disease of almond (*Prunus amygdalus*) in Lebanon. Plant Dis 85:802

Cieślińska M, Morgaś H (2011) Detection and identification of '*Candidatus* Phytoplasma *prunorum*, '*Candidatus* Phytoplasma mali' and '*Candidatus* Phytoplasma pyri' in stone fruit trees in Poland. J Phytopathol 159:217–222

Clyde WAJ (1983) Growth inhibition tests. In: Tully J, Razin S (eds) Methods in mycoplasmology, vol 1. Academic, New York, pp 405–410

Conci L, Meneguzzi N, Galdeano E, Torres L, Nome C, Nome S (2005) Detection and molecular characterization of an alfalfa phytoplasma in Argentina that represents a new subgroup in the 16S rDNA ash yellows group ('*Candidatus* Phytoplasma fraxini'). Eur J Plant Pathol 113:255–265

Constable FE, Symons RH (2004) Genetic variability amongst isolates of Australian grapevine phytoplasmas. Australas Plant Path 33:115–119

Constable FE, Gibb KS, Symons RH (2003) Seasonal distribution of phytoplasmas in Australian grapevines. Plant Pathol 52:267–276

Constable FE, Jones J, Gibb KS, Chalmers YM, Symons RH (2004) The incidence, distribution and expression of Australian grapevine yellows, restricted growth and late season curl diseases in selected Australian vineyards. Ann Appl Biol 144:205–218

Cronjé P, Tymon A, Bailey R, Jones P (1998) Association of a phytoplasma with a yellow leaf syndrome of sugarcane in Africa. Ann Appl Biol 133:177–186

Cronjé P, Dabek AJ, Jones P, Tymon AM (2000a) First report of a phytoplasma associated with a disease of date palms in North Africa. Plant Pathol 49:801

Cronjé P, Dabek AJ, Jones P, Tymon AM (2000b) Slow decline: a new disease of mature date palms in North Africa associated with a phytoplasma. Plant Pathol 49:804

Daire X, Clair D, Reinert W, Boudon-Padieu E (1997) Detection and differentiation of grapevine yellows phytoplasmas belonging to elm yellows group and to the stolbur subgroup by PCR amplification of non-ribosomal DNA. Eur J Plant Pathol 103:507–514

Danet JL, Bonnet P, Jarausch W, Carraro L, koric D, Foissac X (2007) Imp and secY, two new markers for MLST (multilocus sequence typing) in the 16SrX phytoplasma taxonomic group. Bull Insect 60:339–340

Danet JL, Balakishiyeva G, Cimerman A, Sauvion N, Marie-Jeanne V, Labonne G, Laviña A, Battle A, Križanac I, Škorić D, Ermacora P, Ulubaş Serçe Ç, Çağlayan K, Jarausch W, Foissac X (2011) Multilocus sequence analysis reveals the genetic diversity of European fruit tree phytoplasmas and supports the existence of inter-species recombination. Microbiology 157:438–450

Davis MJ, Ying Z, Brunner BR, Pantoja A, Ferwerda FH (1998) Rickettsial relative associated with Papaya Bunchy Top disease. Curr Microbiol 36:80–84

Davis RE, Dally EL (2001) Revised subgroup classification of group 16SrV phytoplasmas and placement of flavescence dorée-associated phytoplasmas in two distinct subgroups. Plant Dis 85:790–797

Davis RE, Dally EL, Gundersen DE, Lee I-M, Habili N (1997) '*Candidatus* Phytoplasma australiense' a new phytoplasma taxon associated with Australian grapevine yellows. Int J Syst Bacteriol 47:262–269

Davis RE, Dally EL, Zhao Y, Lee I-M, Jomantiene R, Detweiler AJ, Putnam ML (2010) First report of a new subgroup 16SrIX-E ('*Candidatus* Phytoplasma phoenicium'-related) phytoplasma associated with juniper witches' broom disease in Oregon, USA. Plant Pathol 59:1161

Davis RE, Zhao Y, Dally EL, Jomantiene R, Lee I-M, Wei W, Kitajima EW (2012) '*Candidatus* Phytoplasma sudamericanum', a novel taxon, and strain PassWB-Br4, a new subgroup 16SrIII-V phytoplasma, from diseased passion fruit (*Passiflora edulis* f. *flavicarpa* Deg.). Int J Syst Evol Microbiol 62:984–989

Davis RE, Zhao Y, Dally EL, Lee I-M, Jomantiene R, Douglas SM (2013) '*Candidatus* Phytoplasma pruni', a novel taxon associated with X-disease of stone fruits, *Prunus* spp.: multilocus characterization based on 16S rRNA, secY, and ribosomal protein genes. Int J Syst Evol Microbiol 63:766–776

Davis RI, Schneider B, Gibb KS (1997) Detection and differentiation of phytoplasmas in Australia. Aust J Agr Res 48:535–544

De Fluiter HJ, van der Meer FA (1953) Rubus stunt, a leafhopper borne virus disease. Tijdschr Plantenziekte 59:195–197

De Luca V, Capasso C, Capasso A, Pastore M, Carginale V (2011) Gene expression profiling of phytoplasma-infected Madagascar periwinkle leaves using differential display. Mol Biol Rep 38:2993–3000

Dickinson M, Tuffen M, Hodgetts J (2013) The phytoplasmas: an introduction. In: Dickinson M, Hodgetts J (eds) Phytoplasma: methods and protocols, methods in molecular biology, vol 938. Springer, New York, pp 1–14

Ding Y, Wu W, Wei W, Davis RE, Lee IM, Hammond RW, Sheng JP, Shen L, Jiang Y, Zhao Y (2013) Potato purple top phytoplasma-induced disruption of gibberellin homeostasis in tomato plants. Ann Appl Biol 162:131–139

Douglas SM (1986) Detection of mycoplasma-like organisms in peach and chokecherry with X-disease by fluorescence microscopy. Phytopathology 76:784–787

Duduk B, Bertaccini A (2004) Corn with symptoms of reddening: new host of stolbur phytoplasma. Plant Dis 90:1313–1319

Ecke P Jr, Matkin OA, Hartley DE (1990) The poinsettia manual, 3rd edn. Paul Ecke Poinsettias, Encinitas

Eckstein B, Barbosa JC, Rezende JAM, Bedendo IP (2011) A *Sida* sp. is a new host for "*Candidatus* Phytoplasma brasiliense" in Brazil. Plant Dis 95:363

Eden-Green SJ, Markham PG (1987) Multiplication and persistence of *Acholeplasma* spp. in leafhoppers. J Invertebr Pathol 49:235–241

Eden-Green SJ, Tully JG (1979) Isolation of *Acholeplasma* spp. from coconut palms affected by lethal yellowing disease in Jamaica. Curr Microbiol 2:311–316

Edward DG, Freundt EA (1956) The classification and nomenclature of organisms of the pleuropneumonia group. J Gen Microbiol 14:197–207

Edward DG, Freundt EA (1969) Proposal for classifying organisms related to *Mycoplasma laidlawii* in a family *Sapromycetaceae*, genus *Sapromyces*, within the Mycoplasmatales. J Gen Microbiol 57:391–395

Edward DG, Freundt EA (1970) Amended nomenclature for strains related to *Mycoplasma laidlawii*. J Gen Microbiol 62:1–2

Eroglu S, Ozbek H, Sahin F (2011) First report of group 16SrXII phytoplasma causing stolbur disease in potato plants in the eastern and southern Anatolia regions of Turkey. Plant Dis 94:1374

Faghihi MM, Bagheri AN, Bahrami HR, Hasanzadeh H, Rezazadeh R, Siampour M, Samavi S, Salehi M, Izadpanah K (2011) Witches'-broom disease of lime affects seed germination and seedling growth but is not seed transmissible. Plant Dis 95:419–422

Firrao G, Gibb KS, Streten C (2005) Short taxonomic guide to the genus '*Candidatus* Phytoplasma'. J Plant Pathol 87:249–263

Freundt EA, Whitcomb RF, Barile MF, Razin S, Tully JG (1984) Proposal for elevation of the family *Acholeplasmataceae* to ordinal rank: Acholeplasmatales. Int J Syst Bacteriol 34:346–349

Frisinghelli C, Delaiti L, Grando MS, Forti D, Vindimian ME (2000) *Cacopsylla costalis* (Flor 1861), as a vector of apple proliferation in Trentino. J Phytopathol 148:425–431

Gajardo A, Fiore N, Prodan S, Paltrinieri S, Botti S, Pino AM, Zamorano A, Montealegre J, Bertaccini A (2009) Phytoplasmas associated with grapevine yellows disease in Chile. Plant Dis 93:789–796

Garnier M, Zreik L, Bové JM (1991) Witches' broom, a lethal mycoplasmal disease of lime trees in the Sultanate of Oman and the United Arab Emirates. Plant Dis 75546–551

Gasparich GE (2010) Spiroplasmas and phytoplasmas: microbes associated with plant hosts. Biologicals 38:193–203

Gatineau F, Larrue J, Clair D, Lorton F, Richard-Molard M, Boudon-Padieu E (2001) A new natural planthopper vector of stolbur phytoplasma in the genus *Pentastiridius* (Hemiptera: Cixiidae). Eur J Plant Pathol 10:263–271

Getachew MA, Mitchell A, Gurr GM, Fletcher MJ, Pilkington LJ, Nikandrow A (2007) First report of a '*Candidatus* Phytoplasma australiense'—related strain in lucerne in Australia. Plant Dis 91:111

Gevers D, Cohan FM, Lawrence JG, Spratt BG, Coenye T, Feil EJ, Stackebrandt E, Van de Peer Y, Vandamme P, Thompson FL, Swings J (2005) Opinion: re-evaluating prokaryotic species. Nat Rev Microbiol 3:733–739

Gowanlock DH, Greber RS, Behncken GM, Finlay J (1976) Electron microscopy of mycoplasma-like bodies in several Queensland crop species. Aust Plant Pathol Soc Newslett. 5: abstract 223

Greber RS (1966) Identification of the virus causing papaya yellow crinkle with tomato big bud virus by transmission tests. Qld J Agric Anim Sci 23:147–153

Griffiths HM, Sinclair WA, Davis CD, Smart RE (1999) The phytoplasma associated with ash yellows and lilac witches'-broom: '*Candidatus* Phytoplasma fraxini'. Int J Syst Bacteriol 49:605–1614

Gundersen DE, Lee I-M, Rehner SA, Davis RE, Kingsbury DT (1994) Phylogeny of mycoplasma like organisms (phytoplasmas): a basis for their classification. J Bacteriol 176:5244–5254

Hanboonsong Y, Choosai C, Panyim S, Damak S (2002) Transovarial transmission of sugarcane white leaf phytoplasma in the insect vector *Matsumuratettix hiroglyphicus* (Matsumura). Insect Mol Biol 11:97–103

Harrison NA, Gundersen-Rindal D, Davis RE (2010) Genus I. "*Candidatus* Phytoplasma". In: Garrity GM et al (eds) Bergey's manual of systematic bacteriology: volume 4: the bacteroidetes, spirochaetes, tenericutes (mollicutes), acidobacteria, fibrobacteres, fusobacteria, dictyoglomi, gemmatimonadetes, lentisphaerae, verrucomicrobia, chlamydiae, and planctomycetes. Springer, New York, pp 696–719

Hibben CR, Wolanski B (1971) Dodder transmission of a mycoplasma from ash witches'-broom. Phytopathology 61:151–156

Hibben CR, Sinclair WA, Davis RE, Alexander JHIII (1991) Relatedness of mycoplasmalike organisms associated with ash yellows and lilac witches'-broom. Plant Dis 75:1227–1230

Hill AV (1943) Insect transmission and host plants of big bud of tomato. J Counc Sci Ind Res Aust 16:85–90

Hill AC (1992) *Acholeplasma cavigenitalium* sp. nov., isolated from the vagina of guinea pigs. Int J Syst Bacteriol 42:589–592

Himeno M, Neriya Y, Minato N, Miura C, Sugawara K, Ishii Y, Yamaji Y, Kakizawa S, Oshima K, Namba S (2011) Unique morphological changes in plant pathogenic phytoplasma-infected petunia flowers are related to transcriptional regulation of floral homeotic genes in an organ-specific manner. Plant J 67:971–979

Hiruki C (1999) Paulownia witches'-broom disease important in East Asia. Acta Horticulturae 496:63–68

Hiruki C, Wang KR (2004) Clover proliferation phytoplasma: '*Candidatus* Phytoplasma trifolii'. Int J Syst Evol Microbiol 54:1349–1353

Hodgetts J, Boonham N, Mumford R, Harrison N, Dikinson M (2008) Phytoplasma phylogenetics based on analysis of secA and 23S rRNA gene sequences for improved resolution of candidates species of '*Candidatus* Phytoplasma'. Int J Syst Evol Microbiol 58:1826–1837

Hogenhout SA, Oshima K, Ammar E-D, Kakizawa S, Kingdom HN, Namba S (2008) Phytoplasmas: bacteria that manipulate plants and insects. Mol Plant Pathol 9:403–423

Hoshi A, Oshima K, Kakizawa S, Ishii Y, Ozeki J, Hashimoto M, Komatsu K, Kagiwada S, Yamaji Y, Namba S (2009) A unique virulence factor for proliferation and dwarfism in plants identified from a phytopathogenic bacterium. Proc Natl Acad Sci USA 106:6416–6421

Hren M, Ravnikar M, Brzin J, Ermacora P, Carraro L, Bianco PA, Casati P, Borgo M, Angelini E, Rotter A, Gruden K (2009a) Induced expression of sucrose synthase and alcohol dehydrogenase I genes in phytoplasma-infected grapevine plants grown in the field. Plant Pathol 58:170–180

Hren M, Nikolić P, Rotter A, Blejec A, Terrier N, Ravnikar M, Dermastia M, Gruden K (2009b) 'Bois noir' phytoplasma induces significant reprogramming of the leaf transcriptome in the field grown grapevine. BMC Genomics 10:460

Huang S, Tiwari AK, Rao GP (2011) '*Candidatus* Phytoplasma pini' affecting *Taxodium distichum* var . *imbricarium* in China. Phytopathogen Mollicutes 1:91–94

ICSB-STM (International Committee on Systematic Bacteriology—Subcommittee on the Taxonomy of Mollicutes) (1995) Minutes of the Interim Meeting, 17 and 26 July 1994, Bordeaux, France. Int J Syst Evol Microbiol 45:415–417

IRPCM (The IRPCM Phytoplasma/Spiroplasma Working Team–Phytoplasma taxonomy group) (2004) "*Candidatus* Phytoplasma", a taxon for the wall-less, non-helical prokaryotes that colonize plant phloem and insects. Int J Syst Evol Micr 54:1243–1255

Jagoueix-Eveillard S, Tarendeau F, Guolter K, Danet J-L, Bové J, Garnier M (2001) *Catharanthus roseus* genes regulated differentially by mollicute infections. Mol Plant Pathol 14:225–233

Jarausch W, Jarausch-Wehrheim B, Danet JL, Broquaire JM, Dosba F, Saillard C, Garnier M (2001) Detection and identification of European stone fruit yellows and other phytoplasmas in wild plants in the surroundings of apricot chlorotic leaf roll-affected orchards in southern France. Eur J Plant Pathol 107:209–217

Jensen DD, Griggs WH, Gonzales CQ, Schneider H (1964) Pear decline virus transmission by pear psylla. Phytopathology 54:1346–1351

Ježić M, Poljak I, Idžojtić M, Ćurković-Perica M (2012) First report of '*Candidatus* Phytoplasma pini' in Croatia. In: Book of abstracts of the 19th congress of the International Organisation for Mycoplasmology, Toulouse, pp 158–159

Johansson K-E, Pettersson B (2002) Taxonomy of mollicutes. In: Razin S, Hermann R (eds) Molecular biology and pathogenicity of mycoplasmas. Kluwer, New York, pp 31–44

Jomantiene R, Davis RE, Alminaite A, Valiunas D, Jasinskaite R (2002) First report of oat as host of a phytoplasma belonging to group 16SrI, subgroup A. Plant Dis 86:443

Jones P, Devonshire BJ, Holman TJ, Ajanga S (2004) Napier grass stunt: a new disease associated with a 16SrXI group phytoplasma in Kenya. Plant Pathol 53:519

Jones P, Arocha Y, Antesana O, Montellano E, Franco P (2005) 'Hoja de perejil' (parsley leaf) of tomato, morrenia little leaf, two new diseases associated with a phytoplasma in Bolivia. Plant Pathol 54:235

Jović J, Cvrković T, Mitrović M, Krnjanjić S, Redingbaugh MG, Pratt RC, Gingery RE, Hogenhout SA, Toševski I (2007) Roles of stolbur phytoplasma and *Reptalus panzeri* (Cixiinae, Auchenorrhyncha) in the epidemiology of Maize redness in Serbia. Eur J Plant Pathol 118:85–89

Jović J, Krstić O, Toševski I, Gassmann A (2011a) The occurrence of '*Candidatus* Phytoplasma rhamni' in *Rhamnus cathartica* L. without symptoms. Bull Insect 64(suppl):227–228

Jović J, Cvrković T, Mitrović M, Petrović A, Krstić O, Krnjajić S, Toševski I (2011b) Multigene sequence data and genetic diversity among '*Candidatus* Phytoplasma ulmi' strains infecting *Ulmus* spp. in Serbia. Plant Pathol 60:356–368

Jung H-Y, Sawayanagi T, Kakizawa S, Nishigawa H, Miyata S, Oshima K, Ugaki M, Lee J-T, Hibi T, Namba S (2002) '*Candidatus* Phytoplasma castaneae', a novel phytoplasma taxon associated with chestnut witches' broom disease. Int J Syst Evol Microbiol 52:1543–1549

Jung H-Y, Sawayanagi T, Wongkaew P, Kakizawa S, Nishigawa H, Wei W, Oshima K, Miyata SI, Ugaki M, Hibi T, Namba S (2003a) '*Candidatus* Phytoplasma oryzae', a novel phytoplasma taxon associated with rice yellow dwarf disease. Int J Syst Evol Microbiol 53:1925–1929

Jung HY, Sawayanagi T, Kakizawa S, Nishigawa H, Wei W, Oshima K, Miyata S, Ugaki M, Hibi T, Namba S (2003b) '*Candidatus* Phytoplasma ziziphi', a novel phytoplasma taxon associated with jujube witches'-broom disease. Int J Syst Evol Microbiol 53:1037–1041

Kakizawa S, Ishii Y, Hoshi A, Jung HY, Kagiwada S, Yamaji Y, Oshima K, Namba S (2008) Cloning and characterization of the antigenic membrane protein (Amp) gene and in situ detection of Amp from malformed flowers infected with Japanese hydrangea phyllody phytoplasma. Phytopathology 98:769–775

Kaloostian GH, Hibino H, Schneider H (1971) Mycoplasmalike bodies in periwinkle: their cytology and transmission by pear psylla from pear trees affected with pear decline. Phytopathology 61:1177–1179

Kamińska M, Berniak H, Obdrzalek J (2011) New natural host plants of '*Candidatus* Phytoplasma pini' in Poland and the Czech Republic. Plant Pathol 60:1023–1029

Kämpfer P, Glaeser SP (2012) Prokaryotic taxonomy in the sequencing era—the polyphasic approach revisited. Environ Microbiol 14:291–317

Kanehira T, Horikoshi N, Yamakita Y, Shinohara M (1996) Occurrence of hydrangea phyllody in Japan and detection of the causal phytoplasma. Ann Phytopathol Soc Japan 62:537–540

Kartte S, Seemüller E (1991) Histopathology of apple proliferation in *Malus* taxa and hybrids of different susceptibility. J Phytopathol 131:149–160

Kawakita H, Saiki T, Wei W, Mitsuhashi W, Watanabe K, Sato M (2000) Identification of mulberry dwarf phytoplasmas in the genital organs and eggs of leafhopper *Hishimonoides sellatiformis*. Phytopathology 90:909–914

Khan AJ, Botti S, Al-Subhi AM, Gundersen-Rindal DE, Bertaccini A (2002) Molecular identification of a new phytoplasma associated with alfalfa witches'-broom in Oman. Phytopathology 92:1038–1047

Khan AJ, Al-Subhi AM, Calari A, Al-Saady NA, Bertaccini A (2007) A new phytoplasma associated with witches' broom of *Cassia italica* in Oman. Bull Insect 60:269–270

Khan MS, Raj SK, Snehi SK (2008) Natural occurrence of '*Candidatus* Phytoplasma ziziphi' isolates in two species of jujube trees (*Ziziphus* spp.) in India. Plant Pathol 57:1173

Kim C-J (1965) Witches' broom of jujube tree (*Zizyphus jujube* Mill. var. inermis Rehd.). Transmission by grafting. Korean J Microbiol 3:1–6

Kirchhoff H (1974) New species of the families *Acholeplasmataceae* and *Mycoplasmataceae* in horses. Zentralblatt fur Veterinarmedizin—Reihe B 21:207–210

Kirchhoff H (1978) *Acholeplasma equifetale* and *Acholeplasma hippikon*, two new species from aborted horse fetuses. Int J Syst Bacteriol 28:76–81

Kirkpatrick BC (1991) Mycoplasma-like organisms—plant and invertebrate pathogens. In: Balows A, Trüper HG, Dworkin M, Harder W, Schleifer KS (eds) The Prokaryotes, vol 3. Springer, New York, pp 4050–4067

Kirkpatrick BC, Fisher GA, Fraser JD, Purcell AH (1990) Epidemiological and phylogenetic studies on western X-disease mycoplasma-like organisms. In: Stanek G, Cassell, GH, Tully JG, Whitcomb RF (eds) Recent advances in mycoplasmology. Gustav Fischer Verlag, Stuttgart, Germany, pp 288–297

Kisary J, El-Ebeedy AA, Stipkovits L (1976) Mycoplasma infection of geese. II. Studies on pathogenicity of mycoplasmas in goslings and goose and chicken embryos. Avian Pathol 5:15–20

Kison H, Seemüller E (2001) Differences in strain virulence of the European stone fruit yellows phytoplasma and susceptibility of stone fruit trees on various rootstocks to this pathogen. J Phytopathol 149:533–541

Kitajima EW, Robbs CF, Kimura O, Wanderley LJG (1981) O irizado do chuchuzeiro e o superbrotamento do maracuja'—duas enfermidades associadas a microrganismos do tipo mycoplasma constatadas nos estados do Rio de Janeiro e Pernambuco. Fitopatologia Brasileira 6:115–122

Knight TF (2004) Reclassification of *Mesoplasma pleciae* as *Acholeplasma pleciae* comb. nov. on the basis of 16S rRNA and gyrB gene sequence data. Int J Syst Evol Microbiol 54:1951–1952

Konstantinidis KT, Ramette A, Tiedje JM (2006) Toward a more robust assessment of intraspecies diversity, using fewer genetic markers. Appl Environ Microbiol 72:7286–7293

Konstantinidis KT, Tiedje JM (2007) Prokaryotic taxonomy and phylogeny in the genomic era: advancements and challenges ahead. Curr Opin Microbiol 10:504–509

Kube M, Schneider B, Kuhl H, Dandekar T, Heitmann K, Migdoll AM, Reinhardt R, Seemüller E (2008) The linear chromosome of the plant-pathogenic mycoplasma '*Candidatus* Phytoplasma mali'. BMC Genomics 9:306

Kube M, Mitrovic J, Duduk B, Rabus R, Seemüller E (2012) Current view on phytoplasma genomes and encoded metabolism. Scientific World J., Volume 2012, article ID 185942, 25 pages.

Kuhnert P, Korczak BM (2006) Prediction of whole-genome DNA-DNA similarity, determination of G+C content and phylogenetic analysis within the family *Pasteurellaceae* by multilocus sequence analysis (MLSA). Microbiology 152:2537–2548

Kunkel LO (1926) Studies on aster yellows. Am J Bot 23:646–705

Kunkel LO (1932) Celery yellows of California not identical with the aster yellows of New York. Contrib Boyce Thompson Inst 4:405–414

La Y-J, Woo K-S (1980) Transmission of jujube witches' broom mycoplasma by the leaf hopper *Hishimonus sellatus* Uhler. J Korean For Soc 48:29–39

Laidlaw PP, Elford WJ (1936) A new group of filterable organisms. Proc R Soc B 20:292–303

Lapage SP, Sneath PHA, Lessel EF, Skerman VBD, Seeliger HPR, Clark WA (1992) International code of nomenclature of bacteria. Bacteriological code, 1990 edn. ASM Press, Washington, DC

Larsen KJ, Whalen ME (1988) Dispersal of *Paraphlopsius irroratus* (Say) (Homoptera: Cicadellidae) in peach and cherry orchards. Environ Entomol 17:842–851

Lauer U, Seemüller E (2000) Physical map of the chromosome of the apple proliferation phytoplasma. J Bacteriol 182:1415–1418

Leach RH (1973) Further studies on classification of bovine strains of mycoplasmatales, with proposals for new species, *Acholeplasma modicum* and *Mycoplasma alkalescens*. J Gen Microbiol 75:135–153

Lee I-M, Bertaccini A, Vibio M, Gundersen DE, Davis RE, Mittempergher L, Conti M, Gennari F (1995) Detection and characterization of phytoplasmas associated with disease in *Ulmus* and *Rubus* in northern and central Italy. Phytopathol Mediterr 34:174–183

Lee I-M, Klopmeyer M, Bartoszyk IM, Gundersen-Rindal DE, Chou T-S, Thomson KL, Eisenreich R (1997) Phytoplasma induced free-branching in commercial poinsettia cultivars. Nat Biotechnol 15:178–182

Lee I-M, Gundersen-Rindal DE, Davis RE, Bartoszyk I-M (1998) Revised classification scheme of phytoplasmas based on RFLP analysis of 16S rRNA, ribosomal protein gene sequences. Int J Syst Bacteriol 48:1153–1169

Lee I-M, Davis RE, Gundersen-Rindal DE (2000) Phytoplasma: phytopathogenic mollicutes. Annu Rev Microbiol 54:221–255

Lee I-M, Gundersen-Rindal DE, Davis RE, Bottner KD, Marcone C, Seemüller E (2004a) '*Candidatus* Phytoplasma asteris', a novel phytoplasma taxon associated with aster yellows and related diseases. Int J Syst Evol Microbiol 54:1037–1048

Lee I-M, Martini M, Marcone C, Zhu SF (2004b) Classification of phytoplasma strains in the elm yellows group (16SrV) and proposition of '*Candidatus* Phytoplasma ulmi' for the phytoplasma associated with elm yellows. Int J Syst Evol Microbiol 54:337–347

Lee I-M, Davis RE (2000) Aster yellows. In: Maloy OC, Murray TD (eds) Encyclopedia of plant pathology. Wiley, New York, pp 60–63

Lee I-M, Bottner KD, Secor G, Rivera-Varas V (2006a) '*Candidatus* Phytoplasma americanum', a phytoplasma associated with a potato purple top wilt disease complex. Int J Syst Evol Microbiol 56:1593–1597

Lee IM, Zhao Y, Bottner KD (2006b) SecY gene sequence analysis for finer differentiation of diverse strains in the aster yellows phytoplasma group. Mol Cell Probes 20:87–91

Lee I-M, Zhao Y, Davis RE, Wei W, Martini M (2007) Prospects of DNA-based systems for differentiation and classification of phytoplasmas. Bull Insect 60:239–244

Lee IM, Zhao Y, Davis RE (2009) Prospects of multiple gene-based systems for differentiation and classification of phytoplasmas. In: Weintraub PG, Jones P (eds) Phytoplasmas: genomes, plant hosts and vectors. CAB International, Wallingford/Oxfordshire, UK, pp 51–63

Lee I-M, Bottner-Parker KD, Zhao Y, Davis RE, Harrison N (2010) Phylogenetic analysis and delineation of phytoplasmas based on *secY* gene sequences. Int J Syst Evol Microbiol 60:2887–2897

Lee I-M, Bottner-Parker KD, Zhao Y, Villalobos W, Moreira L (2011) '*Candidatus* Phytoplasma costaricanum' a new phytoplasma associated with a newly emerging disease in soybean in Costa Rica. Int J Syst Evol Microbiol 61:2822–2826

Lee I-M, Bottner-Parker KD, Zhao Y, Bertaccini A, Davis RE (2012) Differentiation and classification of phytoplasmas in the pigeon pea witches'-broom group (16SrIX): an update based on multiple gene sequence analysis. Int J Syst Evol Microbiol 62:2279–2285

Lepka P, Stitt M, Moll E, Seemüller E (1999) Effect of phytoplasmal infection on concentration and translocation of carbohydrates and amino acids in periwinkle and tobacco. Physiol Mol Plant Pathol 55:59–68

Li Z, Wu Z, Liu H, Hao X, Zhang C, Wu Y (2010) *Spiraea salicifolia*: a new plant host of "*Candidatus* Phytoplasma ziziphi"-related phytoplasma. J Gen Plant Pathol 76:299–301

Liefting LW, Beever RE, Winks CJ, Pearson MN, Forster RLS (1997) Planthopper transmission of *Phormium* yellow leaf phytoplasma. Aust Plant Pathol 26:148–154

Liefting LW, Padovan AC, Gibb KS, Beever RE, Andersen MT, Newcomb RD, Beck DL, Foster RLS (1998) '*Candidatus*' Phytoplasma australiense' is the phytoplasma associated with Australian Grapevine yellows, papaya dieback and *Phormium* yellow leaf disease. Eur J Plant Pathol 104:619–623

Liefting LW, Beever RE, Andersen MT, Clover GRG (2007) Phytoplasma diseases in New Zealand. Bull Insect 60:165–166

Liefting LW, Veerakone S, Ward LI, Clover GRG (2009) First report of '*Candidatus* Phytoplasma australiense' in potato. Plant Dis 93:969

Lim P-O, Sears BB (1989) 16S rRNA sequence indicates that plant-pathogenic mycoplasmalike organisms are evolutionarily distinct from animal mycoplasmas. J Bacteriol 171:5901–5906

Lim P-O, Sears BB (1992) Evolutionary relationships of a plant-pathogenic mycoplasma-like organism and *Acholeplasma laidlawii* deduced from two ribosomal protein gene sequences. J Bacteriol 174:2606–2611

Lim P-O, Sears BB, Klomparens KL (1992) Membrane properties of a plant-pathogenic mycoplasma-like organism. J Bacteriol 174:682–686

Liu Q, Wu T, Davis RE, Zhao Y (2004) First report of witches'-broom disease of *Broussonetia papyrifera* and its association with a phytoplasma of elm yellows group (16SrV). Plant Dis 88:770

Liu H-L, Chen C-C, Lin C-P (2007) Detection and identification of the phytoplasma associated with pear decline in Taiwan. Eur J Plant Pathol 117:281–291

Lo Gullo MA, Trifilò P, Raimondo F (2000) Hydraulic architecture and water relations of *Spartium junceum* branches affected by a mycoplasm disease. Plant Cell Environ 23:1079–1088

Loi N, Carraro L, Musetti R, Pertot I, Osler R (1995) Dodder transmission of two different MLOs from plum trees affected by "Leptonecrosis". Acta Horticulturae 386:465–470

Lorenz KH, Schneider B, Ahrens U, Seemüller E (1995) Detection of the apple proliferation and pear decline phytoplasmas by PCR amplification of ribosomal and nonribosomal DNA. Phytopathology 85:771–776

MacLean AM, Sugio A, Makarova OV, Findlay KC, Grieve VM, Tóth R, Nicolaisen M, Hogenhout SA (2011) Phytoplasma effector SAP54 induces indeterminate leaf-like flower development in Arabidopsis plants. Plant Physiol 157:831–841

Magaray PA, Wachtel MF (1986) Grapevine yellows, a widespread apparently new disease in Australia. Plant Dis 70:694

Maiden MC, Bygraves JA, Feil E, Morelli G, Russell JE, Urwin R, Zhang Q, Zhou J, Zurth K, Caugant DA, Feavers IM, Achtman M, Spratt BG (1998) Multilocus sequence typing: a portable approach to the identification of clones within populations of pathogenic microorganisms. Proc Natl Acad Sci USA 95:3140–3145

Makarova O, Contaldo N, Paltrinieri S, Kawube G, Bertaccini A (2012) DNA Barcoding for identification of '*Candidatus* Phytoplasmas' using a Fragment of the elongation factor tu gene. PLoS ONE 7:e52092

Malembic-Maher S, Salar P, Filippin L, Carle P, Angelini E, Foissac X (2011) Genetic diversity of European phytoplasmas of the 16SrV taxonomic group and proposal of '*Candidatus* Phytoplasma rubi'. Int J Syst Evol Microbiol 61:2129–2134

Maniloff J (2002) Phylogeny and evolution. In: Razin S, Hermann R (eds) Molecular biology and pathogenicity of mycoplasmas. Kluwer, New York, pp 31–44

Manimekalai R, Nair S, Soumya VP, Thomas GV (2013) Phylogenetic analysis identifies '*Candidatus* Phytoplasma oryzae'-related strain associated with yellow leaf disease of areca palm (*Areca catechu* L.) in India. Int J Syst Evol Microbiol 63:1376–1382

Maramorosch K (1955) Transmission of blueberry stunt virus by *Scaphytopius magdalensis*. J Econ Entomol 48:106

Marcone C (2010) Movement of phytoplasmas and the development of disease in the plant. In: Weintraub PG, Jones P (eds) Phytoplasmas: genomes, plant hosts and vectors. CAB International, Wallingford/Oxfordshire, UK, pp 114–131

Marcone C, Rao GP (2008a) '*Candidatus* Phytoplasma cynodontis': the causal agent of Bermuda grass white leaf disease. In: Harrison NA, Rao GP, Marcone C (eds) Characterization, diagnosis and management of phytoplasmas. Studium Press, Houston, pp 353–364

Marcone C, Rao GP (2008b) White leaf and grassy shoot diseases of sugarcane. In: Harrison NA, Rao GP, Marcone C (eds) Characterization, diagnosis and management of phytoplasmas. Studium Press, Houston, pp 293–305

Marcone C, Seemüller E (2001) A chromosome map of the European stone fruit yellows phytoplasma. Microbiology 147:1213–1221

Marcone C, Hergenhahn F, Ragozzino A, Seemüller E (1999a) Dodder transmission of pear decline, European stone fruit yellows, rubus stunt, picris echioides yellows and cotton phyllody phytoplasmas to periwinkle. J Phytopathol 147:187–192

Marcone C, Neimark H, Ragozzino A, Lauer U, Seemüller E (1999b) Chromosome sizes of phytoplasmas composing major phylogenetic groups and subgroups. Phytopathology 89:805–810

Marcone C, Lee I-M, Davis RE, Ragozzino A, Seemüller E (2000) Classification of aster yellows-group phytoplasmas based on combined analyses of rRNA and *tuf* gene sequences. Int J Syst Evol Microbiol 50:1703–1713

Marcone C, Schneider B, Seemüller E (2004a) '*Candidatus* Phytoplasma cynodontis', the phytoplasma associated with Bermuda grass white leaf disease. Int J Syst Evol Microbiol 54:1077–1082

Marcone C, Gibb KS, Streten C, Schneider B (2004b) '*Candidatus* Phytoplasma spartii', '*Candidatus* Phytoplasma rhamni' and *Candidatus* Phytoplasma allocasuarinae', respectively associated with spartium witches'-broom, buckthorn witches'-broom and allocasuarina yellows diseases. Int J Syst Evol Microbiol 54:1025–1029

Marcone C, Jarausch B, Jarausch W (2010a) '*Candidatus* Phytoplasma prunorum', the causal agent of European stone fruit yellows: an overview. J Plant Pathol 92:19–34

Marcone C, Schneider B, Seemüller E (2010b) Comparison of European stone fruit yellows phytoplasma strains differing in virulence by multi-gene sequence analyses. Julius-Kühn-Archiv 427:193–196

Martini M (2004) Ribosomal protein gene-based phylogeny: a basis for phytoplasma classification. PhD thesis, University of Udine, Italy

Martini M, Botti S, Marcone C, Marzachì C, Casati P, Bianco PA, Benedetti R, Bertaccini A (2002) Genetic variability among Flavescence dorée phytoplasmas from different origins in Italy and France. Mol Cell Probes 16:197–208

Martini M, Lee I-M, Bottner KD, Zhao Y, Botti S, Bertaccini A, Harrison NA, Carraro L, Marcone C, Khan AJ, Osler R (2007) Ribosomal protein gene-based phylogeny for finer differentiation and classification of phytoplasmas. Int J Syst Evol Microbiol 57:2037–2051

Martini M, Ermacora P, Falginella L, Loi N, Carraro L (2008) Molecular differentiation of '*Candidatus* Phytoplasma mali' and its spreading in Friuli Venezia Giulia region (North-East Italy). Acta Hortic 781:395–402

Martini M, Marcone C, Maixner M, Mitrović J, Myrta A, Delić D, Bertaccini A, Ermacora P, Duduk B (2012) '*Candidatus* Phytoplasma convolvuli', a new phytoplasma taxon associated with bindweed yellows in four European countries. Int J Syst Evol Microbiol 62:2010–2015

Marwitz R (1990) Diversity of yellows disease agents in plant infections. Zbl Bakt: Suppl 20:431–434

Mäurer R, Seemüller E (1995) Nature and genetic relatedness of the mycoplasma-like organism causing rubus stunt in Europe. Plant Pathol 44:244–249

Mäurer R, Seemüller E (1996) Witches' broom of *Rhamnus catharticus*: a new phytoplasma disease. J Phytopathol 144:221–223

Maust BE, Espadas F, Talavera C, Aguilar M, Santamaría JM, Oropeza C (2003) Changes in carbohydrate metabolism in coconut palms infected with the lethal yellowing phytoplasma. Phytopathology 93:976–981

McClure MS (1980) Spatial and seasonal distribution of leafhopper vectors of peach X-disease in Connecticut. Environ Entomol 9:668–672

McCoy RE, Caudwell A, Chang CJ, Chen TA, Chiykowski LN, Cousin MT, Dale JL, de Leeuw GTN, Golino DA, Hackett KJ, Kirkpatrick BC, Marwitz R, Petzold H, Sinha RC, Sugiura M, Whitcomb RF, Yang IL, Zhu BM, Seemüller E (1989) Plant diseases associated with mycoplasma-like organisms. In: Whitcomb RF, Tully JG (eds) The Mycoplasmas, vol 5. Academic, San Diego, pp 545–640

Mehle N, Brzin J, Boben J, Hren M, Frank J, Petrović N, Gruden K, Dreo T, Žežlina I, Seljak G, Ravnicar M (2007) First report of '*Candidatus* Phytoplasma mali' in *Prunus avium*, *P. armeniaca* and *P. domestica*. Plant Pathol 56:721

Mirzaie A, Esmailzadeh-Hosseini SA, Jafari-Nodoshan A, Rahimian H (2007) Molecular characterization and potential insect vector of a phytoplasma associated with garden beet witches'- broom in Yazd, Iran. J Phytopathol 155:198–203

Molino Lova M, Quaglino F, Abou-Jawdah Y, Choueiri E, Sobh H, Alma A, Tedeschi R, Casati P, Bianco PA (2011) '*Candidatus* Phytoplasma phoenicium'-related strains infecting almond, peach and nectarine in Lebanon. Bull Insect 64(suppl):267–268

Montano HG, Dally EL, Davis RE, Pimentel JP, Brioso PST (2001a) First report of natural infection by '*Candidatus* Phytoplasma brasiliense' in *Catharanthus roseus*. Plant Dis 85:1209

Montano HG, Davis RE, Dally EL, Hogenhout S, Pimentel JP, Brioso PST (2001b) '*Candidatus* Phytoplasma brasiliense', a new phytoplasma taxon associated with hibiscus witches' broom disease. Int J Syst Evol Microbiol 51:1109–1118

Morton A, Davies DL, Blomquist CL, Barbara DJ (2003) Characterization of homologues of the apple proliferation immunodominant membrane protein gene from three related phytoplasmas. Mol Plant Pathol 4:109–114

Moya-Raygoza G, Nault R (1998) Transmission biology of maize bushy stunt phytoplasma by the corn leafhopper (Homoptera: Cicadellidae). Ann Entomol Soc Am 91:668–676

Murray RG, Stackebrandt E (1995) Taxonomic note: implementation of the provisional status *Candidatus* for incompletely described procaryotes. Int J Syst Bacteriol 45:186–187

Nejat N, Sijam K, Abdullah SNA, Vadamalai G, Dickinson M (2009a) Phytoplasmas associated with disease of coconut in Malaysia: phylogenetic groups and host plant species. Plant Pathol 58:1152–1160

Nejat N, Sijam K, Abdullah SNA, Vadamalai G, Dickinson M (2009b) First report of a 16SrXIV, '*Candidatus* cynodontis' group phytoplasma associated with coconut yellow decline in Malaysia. Plant Pathol 58:389

Nejat N, Vadamalai G, Davis RE, Harrison NA, Sijam K, Dickinson M, Abdullah SNA, Zhao Y (2013) '*Candidatus* Phytoplasma malaysianum', a novel taxon associated with virescence and phyllody of Madagascar periwinkle (*Catharanthus roseus*). Int J Syst Evol Microbiol 63:540–548

Nicolaisen M, Christensen NM (2007) Phytoplasma induced changes in gene expression in poinsettia. Bull Insect 60:215–216

Nicolaisen M, Horvath DP (2008) A branch-inducing phytoplasma in *Euphorbia pulcherrima* is associated with changes in expression of host genes. J Phytopathol 156:403–407

Obura E, Masiga D, Midega CAO, Wachira F, Pickett JA, Deng AL, Khan ZR (2010) First report of a phytoplasma associated with Bermuda grass white leaf disease in Kenya. New Dis Rep 21:23

Padovan A, Gibb K, Persley D (2000) Association of '*Candidatus* Phytoplasma australiense' with green petal and lethal yellows diseases in strawberry. Plant Pathol 49:362–369

Palermo S, Elekes M, Botti S, Ember I, Oroz A, Bertaccini A, Kölber M (2004) Presence of stolbur phytoplasma in Cixiidae in Hungarian vineyards. Vitis 43:201–203

Pearce TL, Scott JB, Pethybridge SJ (2011) First report of a 16SrII-D subgroup phytoplasma associated with pale purple coneflower witches'-broom disease in Australia. Plant Dis 95:773

Pilkington LJ, Gurr GM, Fletcher MJ, Nikandrow A, Elliott E (2004) Vector status of three leafhopper species for Australian lucerne yellows phytoplasma. Aust J Entomol 42:366–373

Poggi Pollini C, Giunchedi L, Gobber M, Miorelli P, Pignatta D, Terlizzi F (2005) Indagini sulla presenza del giallume europeo delle drupacee (ESFY) e di altri fitoplasmi in piante spontanee in provincia di Trento. Petria 15:201–204

Pracros P, Renaudin J, Eveillard S, Mouras A, Hernould M (2006) Tomato flower abnormalities induced by stolbur phytoplasma infection are associated with changes of expression of floral development genes. Mol Plant Microbe Interact 19:62–68

Prentice IW (1950) Rubus stunt: a virus disease. J Hortic Sci 26:35–42

Přibylová J, Petrzik K, Špak J (2009) The first detection of '*Candidatus* Phytoplasma trifolii' in *Rhododendron* hybridum. Eur J Plant Pathol 124:181–185

Purcell AH (1988) Increased survival of *Dalbulus maidis*, a specialist on maize, on non-host plants infected with mollicute plant pathogens. Entomol Exp Appl 46:187–196

Quaglino F, Zhao Y, Casati P, Bulgari D, Bianco PA, Wei W, Davis RE (2013) '*Candidatus* Phytoplasma solani', a novel taxon associated with stolbur and bois noir related diseases of plants. Int J Syst Evol Microbiol 63:2879–2894

Radonjić S, Hrncic S, Jovic J, Cvrkovic T, Krstic O, Krnjajic S, Tosevski I (2006) Occurrence and distribution of grapevine yellows caused by Stolbur phytoplasma in Montenegro. J Phytopathol 157:682–685

Rao GP, Raj SK, Snehi SK, Mall S, Singh M, Marcone C (2007) Molecular evidence for the presence of '*Candidatus* Phytoplasma cynodontis', the Bermuda grass white leaf agent, in India. Bull Insect 60:145–146

Rao GP, Srivastava S, Gupta PS, Sharma SR, Singh A, Singh S, Singh M, Marcone C (2008) Detection of sugarcane grassy shoot phytoplasma infecting sugarcane in India and its phylogenetic relationships to closely related phytoplasmas. Sugar Tech 10:74–80

Rao GP, Mall S, Singh M, Marcone C (2009) First report of a '*Candidatus* Phytoplasma cynodontis'-related strain (group 16SrXIV) associated with white leaf disease of *Dichanthium annulatum* in India. Aust Plant Dis Note 4:56–58

Rao GP, Mall S, Marcone C (2010) '*Candidatus* Phytoplasma cynodontis' (16SrXIV group) affecting *Oplismenus burmannii* (Retz.) *P. Beauv.* and *Digitaria sanguinalis* (L.) Scop. in India. Aust Plant Dis Note 5:93–95

Razin S (1992) Mycoplasma taxonomy and ecology. In: Maniloff J, McElhansey RN, Finch LR, Baseman JB (eds) Mycoplasmas: molecular biology and pathogenesis. American Society for Microbiology, Washington, DC, pp 1–22

Razin S, Efrati H, Kutner S, Rottem S (1982) Cholesterol and phospholipid uptake by mycoplasmas. Rev Infect Dis 4(Suppl):S85–S92

Rice RE, Jones RA (1972) Leafhopper vectors of the western X-disease pathogen: collections in central California. Environ Entomol 1:726–730

Richter M, Rosselló-Móra R (2009) Shifting the genomic gold standard for the prokaryotic species definition. Proc Natl Acad Sci USA 106:19126–19131

Robinson IM, Freundt EA (1987) Proposal for an amended classification of anaerobic mollicutes. Int J Syst Bacteriol 37:78–81

Rogers MJ, Simmons J, Walker RT, Weisburg WG, Woese CR, Tanner RS, Robinson IM, Stahl DA, Olsen G, Leach RH (1985) Construction of the mycoplasma evolutionary tree from 5S rRNA sequence data. Proc Natl Acad Sci USA 82:1160–1164

Rokas A, Williams BL, King N, Carroll SB (2003) Genome-scale approaches to resolving incongruence in molecular phylogenies. Nature 425:798–804

Romanazzi G, Murolo S (2008) *Candidatus* Phytoplasma ulmi' causing yellows in *Zelkova serrata* newly reported in Italy. Plant Pathol 57:1174

Rose DL, Tully JG, Del Giudice RA (1980) *Acholeplasma morum*, a new non-sterol-requiring species. Int J Syst Bacteriol 30:647–654

Rose DL, Tully JG, Bové JM, Whitcomb RF (1993) A test for measuring growth responses of mollicutes to serum and polyoxyethylene sorbitan. Int J Syst Bacteriol 43:527–532

Rottem S, Markowitz O (1979) Unusual positional distribution of fatty acids in phosphatidylglycerol of sterol-requiring mycoplasmas. FEBS Lett 107:379–382

Sabin A (1941) The filterable micro-organisms of the pleuropneumonia group. Bacteriol Rev 5:1–66

Saccardo F, Martini M, Palmano S, Ermacora P, Scortichini M, Loi N, Firrao G (2012) Genome drafts of four phytoplasma strains of the ribosomal group 16SrIII. Microbiology 158:2805–2814

Salehi M, Izadpanah K, Rahimian H (1997) Witches'-broom disease of lime in Sistan, Baluchistan. Iran J Plant Pathol 33:76

Salehi M, Izadpanah K, Taghizadeh M (2000) A study on host range and possible vector of lime witches' broom in Iran. In: Abstract book of the 14th Iranian Plant Protection Congress, September 2000, Isfahan

Salehi M, Izadpanah K, Taghizadeh M (2002) Witches'-broom disease of lime in Iran: new distribution areas, experimental herbaceous hosts and transmission trials. In: Proceedings of the 15th International Organization of Citrus Virologists (IOCV), Riverside, pp 293–296

Salehi M, Nejat N, Tvakoli AR, Izadpanah K (2005) Reaction of citrus cultivars to '*Candidatus* Phytoplasma aurantifolia' in Iran. Iran J Plant Pathol 41:147–149

Salehi M, Izadpanah K, Siampour M, Bagheri AN, Faghihi M (2007) Transmission of '*Candidatus* Phytoplasma aurantifolia' to Bakraee (*Citrus reticulata* hybrid) by feral *Hishimonus phycitis* leafhoppers in Iran. Plant Dis 91:466

Salehi M, Izadpanah K, Siampour M, Taghizadeh M (2009) Molecular characterization and transmission of Bermuda grass white leaf phytoplasma in Iran. J Plant Pathol 91:655–661

Salehi M, Haghshenas F, Khanchezar A, Esmailzadeh-Hosseini SA (2011) Association of '*Candidatus* Phytoplasma phoenicium' with GF-677 witches' broom in Iran. Bull Insect 64(suppl):113–114

Saqib M, Jones MGK, Jones RAC (2006) '*Candidatus* Phytoplasma australiense' is associated with diseases of red clover and paddy melon in south-west Australia. Aust Plant Pathol 35:283–285

Sawayanagi T, Horikoshi N, Kanehira T, Shinohara M, Bertaccini A, Cousin MT, Hiruki C, Namba S (1999) '*Candidatus* Phytoplasma japonicum', a new phytoplasma taxon associated with Japanese *Hydrangea phyllody*. Int J Syst Bacteriol 49:1275–1285

Schneider B, Seemüller E (2009) Strain differentiation of '*Candidatus* Phytoplasma mali' by SSCP and sequence analyses of the hflB gene. J Plant Pathol 91:103–112

Schneider B, Padovan A, De la Rue S, Eichner R, Davis R, Bernuetz A, Gibb K (1999) Detection and differentiation of phytoplasmas in Australia: an update. Aust J Agr Res 50:333–342

Schneider B, Torres E, Martín MP, Schröder M, Behnke HD, Seemüller E (2005) '*Candidatus* Phytoplasma pini', a novel taxon from *Pinus silvestris* and *Pinus halepensis*. Int J Syst Evol Microbiol 55:303–307

Secor GA, Lee I-M, Bottner KD, Rivera-Varas V, Gudmestad C (2006) First report of a defect of processing potatoes in Texas and Nebraska associated with a new phytoplasma. Plant Dis 90:377

Seemüller E, Schneider B (2004) '*Candidatus* Phytoplasma mali', '*Candidatus* Phytoplasma pyri' and '*Candidatus* Phytoplasma prunorum', the causal agents of apple proliferation, pear decline and European stone fruit yellows, respectively. Int J Syst Evol Microbiol 54:1217–1226

Seemüller E, Schneider B (2007) Differences in virulence and genomic features of strains of '*Candidatus* Phytoplasma mali', the apple proliferation agent. Phytopathology 97:964–970

Seemüller E, Marcone C, Lauer U, Ragozzino A, Göschl M (1998a) Current status of molecular classification of the phytoplasmas. J Plant Pathol 80:3–26

Seemüller E, Kison H, Lorenz K-H, Schneider B, Marcone C, Smart CD, Kirkpatrick BC (1998b) Detection and identification of fruit tree phytoplasmas by PCR amplification of ribosomal and nonribosomal DNA. In: Manceau C, Spak J (eds) COST 823 -New technologies to improve phytodiagnosis. Advances in the detection of plant pathogens by polymerase chain reaction. Office for Official Publications of the European Communities, Luxembourg, pp 56–66

Seemüller E, Garnier M, Schneider B (2002) Mycoplasmas of plants and insects. In: Razin S, Hermann R (eds) Molecular biology and pathogenicity of mycoplasmas. Kluwer, New York, pp 91–115

Seemüller E, Kiss E, Sule S, Schneider B (2010) Multiple infection of apple trees by distinct strains of '*Candidatus* Phytoplasma mali' and its pathological relevance. Phytopathology 100:863–870

Seemüller E, Schneider B, Jarausch B (2011a) Pear decline phytoplasma. In: Hadidi A, Barba M, Candresse T, Jelkmann W (eds) Virus and virus-like diseases of pome and stone fruits. APS Press, St. Paul, pp 77–84

Seemüller E, Carraro L, Jarausch W, Schneider B (2011b) Apple proliferation phytoplasma. In: Hadidi A, Barba M, Candresse T, Jelkmann W (eds) Virus and virus-like diseases of pome and stone fruits. APS Press, St. Paul, pp 67–73

Seemüller E, Kampmann M, Kiss E, Schneider B (2011c) HflB gene-based phytopathogenic classification of '*Candidatus* Phytoplasma mali' strains and evidence that strain composition determines virulence in multiply infected apple trees. Mol Plant Microbe Interact 24:1258–1266

Sertkaya G, Osler R, Musetti R, Ermacora P, Martini M (2004) Detection of phytoplasmas in *Rubus* spp. by microscopy and molecular techniques in Turkey. Acta Hortic 656:181–186

Sertkaya G, Martini M, Musetti R, Osler R (2007) Detection and molecular characterization of phytoplasmas infecting sesame and solanaceous crops in Turkey. Bull Insect 60:141–142

Shao J, Jomantiene R, Dally EL, Zhao Y, Lee I-M, Nuss DL, Davis RE (2006) Phylogeny and characterization of phytoplasmal nusA and use of the nusA gene in detection of group 16SrI strains. J Plant Pathol 88:193–201

Siampour M, Izadpanah K, Galetto L, Salehi M, Marzachì C (2013) Molecular characterization, phylogenetic comparison and serological relationship of the Imp protein of several '*Candidatus* Phytoplasma aurantifolia' strains. Plant Pathol 62:452–459

Sinclair WA, Griffiths HM (1994) Ash yellows and its relationship to dieback and decline of ash. Annu Rev Phytopathol 32:49–60

Sinclair WA, Griffiths HM, Davis RE (1996) Ash yellows and lilac witches'-broom: phytoplasmal diseases of concern in forestry and horticulture. Plant Dis 80:468–475

Singh J, Rani A, Kumar P, Baranwal VK, Saroj PL, Sirohi A (2012) First report of a 16SrII-D phytoplasma '*Candidatus* Phytoplasma australasia' associated with a tomato disease in India. New Dis Rep 26:14

Sirand-Pugnet P, Lartigue C, Marenda M, Jacob D, Barré A, Barbe V, Schenowitz C, Mangenot S, Couloux A, Segurens B, de Daruvar A, Blanchard A, Citti C (2007) Being pathogenic, plastic, and sexual while living with a nearly minimal bacterial genome. PLoS Genet 3:e75

Śliwa H, Kaminska M, Korszun S, Adler P (2008) Detection of '*Candidatus* Phytoplasma pini' in *Pinus sylvestris* trees in Poland. J Phytopathol 156:88–92

Somerson NL, Kocka JP, Rose D, Del Giudice RA (1982) Isolation of acholeplasmas and a mycoplasma from vegetables. Appl Environ Microbiol 43:412–417

Stackebrandt E, Goebel BM (1994) Taxonomic note: a place for DNA-DNA reassociation and 16S rRNA sequence analysis in the present species definition in bacteriology. Int J Syst Bacteriol 44:846–849

Stephens EB, Aulakh GS, Rose DL (1983a) Interspecies and intraspecies DNA homology among established species of *Acholeplasma*: a review. Yale J Biol Med 56:729–735

Stephens EB, Aulakh GS, Rose DL (1983b) Intraspecies genetic relatedness among strains of *Acholeplasma laidlawii* and of *Acholeplasma axanthum* by nucleic acid hybridization. J Gen Microbiol 129:1929–1934

Stoddard EM (1934) Progress report on the investigation of a new peach disease. Conn Pomol Soc Proc 44:31–36 (1933)

Stoddard EM (1938) The "X disease" of peach. Connecticut Agricultural Experiment Station, New Haven. June 1938 Circular 122, pp 54–60

Stoddard EM, Hildebrand EM, Palmiter DH, Parker KG (1951) X-disease. In: Virus diseases and other disorders with viruslike symptoms of stone fruits in North America, vol 10, Agriculture handbook. United States Department of Agriculture, United States Government Printing Office, Washington, DC, pp 37–42

Streten C, Conde B, Herrington M, Moulden J, Gibb KS (2005) '*Candidatus* Phytoplasma australiense' is associated with pumpkin yellow leaf curl disease in Queensland, Western Australia and the Northern Territory. Australas Plant Pathol 34:103–105

Su Y-T, Chen J-C, Lin C-P (2011) Phytoplasma-induced floral abnormalities in *Catharanthus roseus* are associated with phytoplasma accumulation and transcript repression of floral organ identity genes. Mol Plant Microbe Interact 24:1502–1512

Sugio A, MacLean AM, Kingdom HN, Grieve VM, Manimekalai R, Hogenhout SA (2011a) Diverse targets of phytoplasma effectors: from plant development to defense against insects. Annu Rev Phytopathol 49:175–195

Sugio A, Kingdom HN, MacLean AM, Grieve VM, Hogenhout SA (2011b) Phytoplasma protein effector SAP11 enhances insect vector reproduction by manipulating plant development and defense hormone biosynthesis. Proc Natl Acad Sci USA 108:4252–4257

Suzuki S, Oshima K, Kakizawa S, Arashida R, Jung H-Y, Yamaji Y, Nishigawa H, Ugaki M, Namba S (2006) Interaction between the membrane protein of a pathogen and insect microfilament complex determines insect-vector specificity. Proc Natl Acad Sci USA 103:4252–4257

Tan PY, Whitlow T (2001) Physiological responses of *Catharanthus roseus* (periwinkle) to ash yellows phytoplasmal infection. New Phytol 150:757–769

Taylor P, Arocha-Rosete Y, Scott J (2011) First report of '*Candidatus* Phytoplasma trifolii (group 16SrVI) infecting S *auropus androgynus*. New Dis Rep 24:23

Taylor-Robinson D (1983) Metabolism inhibition tests. In: Razin S, Tully J (eds) Methods in mycoplasmology, vol 1. Academic, New York, pp 411–421

Tedeschi R, Alma A (2006) *Fieberiella florii* (Homoptera: Auchenorrhyncha) as a vector of "*Candidatus* Phytoplasma mali". Plant Dis 90:284–290

Tedeschi R, Visentin C, Alma A, Bosco D (2003) Epidemiology of apple proliferation (AP) in northwestern Italy: evaluation of the frequency of AP-positive psyllids in naturally infected populations of *Cacopsylla melanoneura* (Homoptera: Psyllidae). Ann Appl Biol 142:285–290

Tedeschi R, Ferrato V, Rossi J, Alma A (2006) Possible phytoplasma transovarial transmission in the psyllids *Cacopsylla melanoneura* and *Cacopsylla pruni*. Plant Pathol 55:18–24

Terlizzi F, Babini AR, Credi R (2006) First report of stolbur phytoplasma (16SrXII-A) on strawberry in northern Italy. Plant Dis 90:831

Tran-Nguyen LT, Gibb KS (2006) Extrachromosomal DNA isolated from tomato big bud and '*Candidatus* Phytoplasma australiense' phytoplasma strains. Plasmid 56:153–166

Tran-Nguyen LT, Persley DM, Gibb KS (2003) First report of phytoplasma disease in capsicum, celery and chicory in Queensland, Australia. Australas Plant Pathol 32:559–560

Trivellone V, Pinzauti F, Bagnoli B (2005) *Reptalus quinquecostatus* (Dufour) (Auchenorrhyncha Cixiidae) as a possible vector of stolbur-phytoplasma in a vineyard in Tuscany. Redia 88:103–108

Tsai JH (1979) Vector transmission of mycoplasmal agents of plant diseases. In: Whitcomb RF, Tully JG (eds) The mycoplasmas, vol III. Academic, San Diego, pp 266–307

Tully JG (1973) Biological and serological characteristics of the acholeplasmas. Ann N Y Acad Sci 225:74–93

Tully JG (1979) Special features of the acholeplasmas. In: Whitcomb RF, Tully JG (eds) The mycoplasmas, vol I. Academic, San Diego, pp 431–449

Tully JG, Razin S (1970) *Acholeplasma axanthum*, sp. nov.: a new sterol-nonrequiring member of the Mycoplasmatales. J Bacteriol 103:751–754

Tully JG, Whitcomb RF, Rose DL, Bove JM, Carle P, Somerson NL, Williamson DL, Eden-Green S (1994a) *Acholeplasma brassicae* sp. nov. and *Acholeplasma palmae* sp. nov., two non- sterol-requiring mollicutes from plant surfaces. Int J Syst Bacteriol 44:680–684

Tully JG, Whitcomb RF, Hackett KJ, Rose DL, Henegar RB, Bove JM, Carle P, Williamson DL, Clark TB (1994b) Taxonomic descriptions of eight new non-sterol-requiring mollicutes assigned to the genus *Mesoplasma*. Int J Syst Bacteriol 44:685–693

Valiunas D, Staniulis J, Davis RE (2006) '*Candidatus* Phytoplasma fragariae', a novel phytoplasma taxon discovered in yellows diseased strawberry, *Fragaria* x *ananassa*. Int J Syst Evol Microbiol 56:277–281

Valiunas D, Jomantiene R, Ivanauskas A, Sneideris D, Staniulis J, Davis RE (2010) A possible threat to the timber industry: '*Candidatus* Phytoplasma pini' in Scots pine (*Pinus sylvestris* L.) in Lithuania. In: Bertaccini A, Laviña A, Torres E (eds) Current status and perspectives of phytoplasma disease research and management (Abstract book of the COST Action FA0807 Meeting). February 1-2, Sitges, Spain, p 38

Van der Meer FA (1987) Virus and viruslike diseases of *Rubus* (raspberry and blackberry). Leafhopper-borne diseases. Rubus stunt. In: Converse RH (ed) Virus diseases of small fruits, vol 631, Agriculture handbook. United States Department of Agriculture, Washington, DC, pp 197–203

Vandamme P, Pot B, Gillis M, de Vos P, Kersters K, Swings J (1996) Polyphasic taxonomy, a consensus approach to bacterial systematics. Microbiol Rev 60:407–438

Verdin E, Salar P, Danet J-L, Choueiri E, Jreijiri F, El Zammar S, Gélie B, Bové J, Garnier M (2003) '*Candidatus* Phytoplasma phoenicium' sp. nov., a novel phytoplasma associated with an emerging lethal disease of almond trees in Lebanon and Iran. Int J Syst Evol Microbiol 53:833–838

Villalobos W, Moreira L, Rivera C, Lee I-M (2009) First report of new phytoplasma diseases associated with soybean, sweet pepper, and passion fruit in Costa Rica. Plant Dis 93:201

Villalobos W, Martini M, Garita L, Muñoz M, Osler R, Moreira L (2011) *Guazuma ulmifolia* (Sterculiaceae), a new natural host of 16SrXV phytoplasma in Costa Rica. Trop Plant Path 36:110–115

Volokhov DV, George J, Liu SX, Ikonomi P, Anderson C, Chizhikov V (2006) Sequencing of the intergenic 16S-23S rRNA spacer (ITS) region of *Mollicutes* species and their identification using microarray-based assay and DNA sequencing. Appl Microbiol Biotechnol 71:680–698

Volokhov DV, Neverov AA, George J, Kong H, Liu SX, Anderson C, Davidson MK, Chizhikov V (2007) Genetic analysis of housekeeping genes of members of the genus *Acholeplasma*: phylogeny and complementary molecular markers to the 16S rRNA gene. Mol Phylogenet Evol 44:699–710

Waites KB, Tully JG, Rose DL (1987) Isolation of *Acholeplasma oculi* from human amniotic fluid in early pregnancy. Curr Microbiol 15:325–327

Wang TH, Zhu HC, Zhao ZR, Tong QQ (1964) Investigation of the pathogens of jujube witches'-broom. Acta Phytopathol Sin 3:195–198

Wang QK, Xu SH, Chen ZW, Zhang JQ (1981) On the witches'-broom disease of *Zizyphus jujuba* Mill. Acta Phytopathol Sin 11:15–18

Wang Z, Zhou P, Yu B, Jiang X, Chang C (1984) On the bionomics and control of *Hishimonoides chinensis* Anufriev, the vector of jujube witches' broom disease. Acta Phytopathol Sin 11:247–252

Wei W, Davis RE, Lee I-M, Zhao Y (2007) Computer-simulated RFLP analysis of 16S rRNA genes: identification of ten new phytoplasma groups. Int J Syst Evol Microbiol 57:1855–1867

Wei W, Lee I-M, Davis RE, Suo X, Zhao Y (2008) Automated RFLP pattern comparison and similarity coefficient calculation for rapid delineation of new and distinct phytoplasma 16S subgroup lineages. Int J Syst Evol Microbiol 58:2368–2377

Weintraub PG, Beanland L (2006) Insect vectors of phytoplasmas. Annu Rev Entomol 51:91–111

Whitcomb RF, Tully JG, Bove JM, Saglio P (1973) Spiroplasmas and acholeplasmas: multiplication in insects. Science 182:1251–1253

White DT, Blackall LL, Scott PT, Walsh KB (1998) Phylogenetic positions of phytoplasmas associated with dieback, yellow crinkle and mosaic diseases of papaya, and their proposed inclusion in '*Candidatus* Phytoplasma australiense' and a new taxon, '*Candidatus* Phytoplasma australasia'. Int J Syst Bacteriol 48:941–951

Win NKK, Lee S-Y, Bertaccini A, Namba S, Jung H-Y (2013) '*Candidatus* Phytoplasma balanitae' associated with witches' broom disease of *Balanites triflora*. Int J Syst Evol Microbiol 63:636–640

Wu Y, Hao X, Li Z, Gu P, An F, Xiang J, Wang H, Luo Z, Liu J, Xiang Y (2010) Identification of the phytoplasma associated with wheat blue dwarf disease in China. Plant Dis 94:977–985

Wu W, Cai H, Wei W, Davis RE, Lee I-M, Chen H, Zhao Y (2012) Identification of two new phylogenetically distant phytoplasmas from *Senna surattensis* plants exhibiting stem fasciation and shoot proliferation symptoms. Annals of Applied Biology 160:25–34

Yang Y, Zhao W, Li Z, Zhu S (2011) Molecular identification of a 'Candidatus Phytoplasma ziziphi'-related strain infecting amaranth (*Amaranthus retroflexus* L.) in China. J Phytopathol 159:635–637

Yi C-K, La Y-J (1973) Mycoplasma-like bodies found in the phloem elements of jujube infected with witches' broom disease. J Korean For Soc 20:111–114

Yu Z-C, Cao Y, Zhang Q, Deng D-F, Liu Z-Y (2012) '*Candidatus* Phytoplasma ziziphi' associated with *Sophora japonica* witches'-broom disease in China. J Gen Plant Pathol 78:298–300

Zhang L, Li Z, Du C, Fu Z, Wu Y (2012) Detection and identification of group 16SrVI phytoplasma in willows in China. J Phytopathol 160:755–757

Zhao Y, Sun Q, Davis RE, Lee IM (2007) First report of witches'-broom disease in a *Cannabis* spp. in China and its association with a phytoplasma of elm yellows group (16SrV). Plant Dis 91:227

Zhao Y, Sun Q, Wei W, Davis RE, Wu W, Liu Q (2009a) '*Candidatus* Phytoplasma tamaricis', a novel taxon discovered in witches'-broom-diseased salt cedar (*Tamarix chinensis* Lour.). Int J Syst Evol Microbiol 59:2496–2504

Zhao Y, Wei W, Lee I-M, Shao J, Suo X, Davis RE (2009b) Construction of an interactive online phytoplasma classification tool, iPhyClassifier, and its application in analysis of the peach X-disease phytoplasma group (16SrIII). Int J Syst Evol Microbiol 59:2582–2593

Zhu SF, Hadidi A, Gundersen DE, Lee I-M, Zhang CL (1997) Characterization of the phytoplasmas associated with cherry lethal yellows and jujube witches'-broom diseases in China. Acta Hortic 472:701–714

Zreik L, Carle P, Bove JM, Garnier M (1995) Characterization of the mycoplasmalike organism associated with witches'-broom disease of lime and proposition of a *Candidatus* taxon for the organism, '*Candidatus* Phytoplasma aurantifolia'. Int J Syst Bacteriol 45:449–453

# 38 The Family *Entomoplasmataceae*

*Gail E. Gasparich*
Fisher College of Science and Mathematics, Towson University, Towson, MD, USA

*Taxonomy: Historical and Current* ........................ 505
Family I. *Entomoplasmataceae* Tully, Bové, Laigret and Whitcomb 1993, 28VP ........................ 506

*Molecular Analyses* ........................ 507

*Phenotypic Analyses* ........................ 507
Genus I. *Entomoplasma* Tully, Bové, Laigret and Whitcomb 1993, 379$^{VP}$ ........................ 507
Genus II. *Mesoplasma* Tully, Bové, Laigret and Whitcomb 1993, 379$^{VP}$ ........................ 508

*Isolation, Enrichment, and Maintenance Procedures* ..... 510
Isolation from Arthropod Hosts (e.g., Gut and Hemolymph) and Plant Surfaces ........................ 510
Initial Cultivation and Filtration Cloning ........................ 510
Growth and Maintenance ........................ 510
Growth on Solid Media ........................ 510

*Ecology* ........................ 510

*Pathogenicity: Clinical Relevance* ........................ 510

*Application* ........................ 512
Waste Management Application ........................ 512

**Abstract**

The *Entomoplasmataceae* is a family within the class *Mollicutes* and the order *Entomoplasmatales* with two genera, *Entomoplasma* and *Mesoplasma*. Originally, many of the strains now within the *Entomoplasmataceae* were designated as belonging to the genus *Mycoplasma* or the genus *Acholeplasma* based on morphological, biological, and metabolic characteristics. In 1993, Tully and colleagues proposed a major revision to the taxonomic classification in which the Order *Entomoplasmatales* was divided into two families based on cell shape: *Entomoplasmatacea* for nonhelical bacteria and the *Spiroplasmataceae* for helical bacteria. The *Entomoplasmatacea* family was then divided into two genera based on sterol requirement: *Entomoplasma* for those that required sterol and *Mesoplasma* for those that did not require sterol, but were able to grow in serum-free medium supplemented with polyoxyethylene sorbitan (PES – normally 0.04 % Tween 80) (Tully et al. Int J Syst Bacteriol 43:378–385, 1993). Subsequent phylogenetic analyses based on 16S rRNA gene sequence consistently showed that the *Entomoplasmataceae* is a sister clade to, and appears to be derived from, the *Spiroplasmatacea* lineage and that the genera are distinct phylogenetically. Phylogenetic analyses also clearly show that the two genera do not form distinct clades but are intermixed. For this reason, it is clear that the requirement for sterol is not a characteristic that can be used to distinguish the two genera and thus, it has been proposed that the two genera be combined under the *Entomoplasma* genus designation (Johansson K-E, Pettersson B (2002) Taxonomy of *Mollicutes*. In: Razin S, Herrman R (eds) Molecular biology and pathogenicity of mycoplasmas. Kluwer, London, pp 1–29; Gasparich et al. Int J Syst Evol Microbiol 54:893–891, 2004). Currently, there are six *Entomoplasma* species and eleven *Mesoplasma* species formally described. They have been isolated from arthropod hosts or plant surfaces (most likely deposited by arthropod hosts) and have not been found to be pathogenic to either host. Species from both genera appear as nonhelical, nonmotile, pleomorphic coccoid cells of various sizes under dark-field microscopic examination, were able to be filtered through 220-nm filters, lacked a cell wall (and thus are resistant to penicillin), and all were chemo-organotropic with the ability to ferment glucose using a PEP-dependent carbohydrate phosphotransferase system. There was variable ability to hydrolyze arginine and none were able to hydrolyze urea. The genome size ranged from 613 to 1,030 kbp, the G+C content ranged from 26.4 to 34.1 mol%, and the growth temperature range was from 10 °C to 37 °C with the common optimal growth temperature being 30 °C. The *Entomoplasmataceae* family as a whole is understudied with little information available for most species beyond the original description.

## Taxonomy: Historical and Current

Prior to the use of DNA sequence analyses to separate members of the class *Mollicutes*, a variety of characteristics, such as morphology, growth requirements, and host organism, were used for taxonomic classification. Historically, in 1967, Doi and colleagues (Doi et al. 1967) were the first to discover members of the class *Mollicutes* in the microbial flora of insects. They characterized a "mycoplasma-like" organism vectored by a leafhopper as the causative agent for mulberry dwarf disease. These organisms were subsequently placed into the genus "*Candidatus* Phytoplasma" due to their inability to be cultivated (IRPCM 2004). Helical, wall-less members of the *Mollicutes* which were also transmitted by leafhoppers causing corn stunt disease and citrus stubborn disease were cultivated in the early 1970s (Davis et al. 1972; Saglio et al. 1973). In 1979, nonhelical mollicutes designated as belonging to the genus *Acholeplasma* were isolated from the surface of a coconut palm suffering from lethal

E. Rosenberg et al. (eds.), *The Prokaryotes – Firmicutes and Tenericutes*, DOI 10.1007/978-3-642-30120-9_390,

yellowing disease (Eden-Green and Tully 1979). *Acholeplasma florum* was characterized in 1984 and found to only be associated with plant hosts (McCoy et al. 1984). Clark and colleagues subsequently made a clear connection between Acholeplasmas and non-sterol-requiring mollicutes isolated from five different insect species (Clark et al. 1986). At this point in time, new species designations were characterized by sterol requirement; with nonhelical mollicutes that required sterol for growth placed in the *Mycoplasma* genus and those that did not require sterol for growth placed in the *Acholeplasma* genus.

In 1993, Tully and colleagues proposed a revision to the *Mollicute* taxonomy (Tully et al. 1993). The class *Mollicutes* would contain four Orders as indicated below:

Order I: *Mycoplasmatales* (as described by Razin and Freundt 1984)
- Family I: *Mycoplasmatacea*
  - Genus I: *Mycoplasma*
  - Genus II: *Ureaplasma*

Order II: *Entomoplasmatales*
- Family I: *Entomoplasmataceae*
  - Genus I: *Entomoplasma*
  - Genus II: *Mesoplasma*
- Family II: *Spiroplasmataceae*
  - Genus I: *Spiroplasma*

Order III: *Acholeplasmatales* (as described by Freundt et al. 1984)
- Family I: *Acholeplasmataceae*
  - Genus I: *Acholeplasma*

Order IV: *Anaeroplasmatales* (as described by Robinson and Freundt 1987)
- Family I: *Anaeroplasmatales*
  - Genus I: *Anaeroplasma*
  - Genus II: *Asteroleplasma*

In this revised classification, the *Mycoplasmataceae* family contained the two genera of *Mollicutes* that were sterol-requiring and associated with vertebrates. The two families in the Order *Entomoplasmatales* were divided into the nonhelical *Entomoplasmataceae* and the helical *Spiroplasmataceae* (transferred from the *Mycoplasmatacea*). The proposed *Entomoplasmataceae* Family had two new genera: *Entomoplasma* for nonhelical, sterol-requiring mollicutes primarily associated with arthropods and *Mesoplasma* for nonhelical, sterol non-requiring mollicutes primarily associated with arthropods. The orders *Acholeplasmatales* (Freundt et al. 1984) and *Anaeroplasmatales* (Robinson and Freundt 1987) remained unchanged.

## Family I. *Entomoplasmataceae* Tully, Bové, Laigret and Whitcomb 1993, 28VP

En.to.mo.plas.ma.ta.ce'ae. N.L. neut. N. *Entomoplasma*, atos type genus of the family; -aceae ending to denote a family; N.L. fem. Pl. n. *Entomoplasmataceae* the *Entomoplasma* family (Brown et al. 2011a).

As a result of the reclassification by Tully and colleagues (Tully et al. 1993), several *Mycoplasma* and *Acholeplasma* species were renamed (Tully et al. 1993). The nonhelical, sterol-requiring insect and plant mollicutes were changed as follows:

*Mycoplasma ellycniae* (Tully et al. 1989) to *Entomoplasma ellycniae*
*Mycoplasma melaleucae* (Tully et al. 1990) to *Entomoplasma melaleucae*
*Mycoplasma somnilux* (Williamson et al. 1990) to *Entomoplasma somnilux*
*Mycoplasma luminosum* (Williamson et al. 1990) to *Entomoplasma luminosum*
*Mycoplasma lucivorax* (Williamson et al. 1990) to *Entomplasma lucivorax*

The nonhelical, non-sterol-requiring insect and plant mollicutes were changed as follows:

*Acholeplasma florum* (McCoy et al. 1984) to *Mesoplasma florum*
*Acholeplasma entomophilum* (Tully et al. 1988) to *Mesoplasma entomophilum*
*Acholeplasma seiffertii* (Bonnet et al. 1991) to *Mesoplasma seiffertii*
*Mycoplasma lactucae* (Rose et al. 1990) to *Mesoplasma lactucae*

The movement of *Acholeplasma florum* to *Mesoplasma florum* was reinforced by a phylogenetic analysis of the *rps*3 ribosomal protein gene (Toth et al. 1994). The trees derived from the deduced amino acid sequence were consistent with those produced using 5S and 16S rRNA sequence comparisons. Clearly *Acholeplasma florum* should be moved to the *Entomoplasmatacea* branch as opposed to that of the *Acholeplasmataceae*. Additionally, it was shown that the UGA triplet encoded tryptophan, rather than a stop codon, in the *rps*3 gene in the now designated *Mesoplasma florum*, as it does in the mycoplasmas and spiroplasmas. Similarly, the movement of *Acholeplasma seiffertii* to *Mesoplasma seiffertii* was supported by 16S rRNA sequence analysis, which showed that *A. seiffertii* was more closely related to *A. entomophilum* and not to *A. laidlawii* (Navas-Castillo et al. 1993). The common metabolic profile and common habitat (floral surfaces) also supported the movement to the *Mesoplasma* genus.

In 1994, Tully and colleagues published the characterization of eight new *Mesoplasma* species identified from 28 strains isolated from different insect hosts (Tully et al. 1994). All were able to grow in serum-free medium and were not serologically related to any of the previously characterized *Mesoplasma*, *Entomoplasma*, *Acholeplasma*, or *Mycoplasma* species. These eight species included *M. pleciae*, *M. photuris*, *M. corruscae*, *M. grammopterae*, *M. syrphidae*, *M. chauliocola*, *M. coleopterae*, and *M. tabanidae*. A more detailed study of *M. pleciae* was conducted which included 16S rRNA and *gyr*B gene sequence analyses, and it was determined that this species was much more closely related to *Acholeplasma laidlawii* and *A. oculi*, than to any member of the *Entomoplasmatales*, and so was subsequently changed to *Acholeplasma pleciae* (Knight 2004).

In 1998, *E. freundtii* was added to the genus *Entomoplasma* (Tully et al. 1998) and named in recognition of Dr. Eyvind Freundt, a Danish microbiologist who was involved in the initial taxonomy and classification of mollicutes. Phylogenetic analysis using 16S rRNA sequence analysis indicated that this strain definitely belonged in the Entomoplasma-Mesoplasma-Mycoplasma clade derived from the Spiroplasmas, but did not clearly separate the *Mesoplasma* and *Entomoplasma* genera into two distinct clusters.

With the advent of 16S rDNA sequence-based phylogenetic analyses, some additional taxonomic questions have arisen concerning the phylogenetic position of the type strain for the genus *Mycoplasma* (*Mycoplasma mycoides* subsp. *mycoides*) and the splitting of the *Entomoplasmatales* into two genera. Phylogenetic analyses clearly showed that the *M. mycoides* cluster arose from *Spiroplasma* through the *Entomoplasmataceae* (the nonhelical descendents of spiroplasmas) (Woese et al. 1980; Weisburg et al. 1989; Gasparich et al. 2004). The order *Mycoplasmatales* is polyphyletic, in that *Mycoplasma* species are split into two phylogenetically separate sections that do not share a common ancestor. However, the two orders, *Mycoplasmatales* and *Entomoplasmatales*, were shown to have derived from a single common ancestor (Gasparich et al. 2004).

Shortly after the 1993 taxonomic revision (Tully et al. 1993) was published, a more extensive survey of sterol requirement and non-requirement within the *Mollicutes* was performed (Rose et al. 1993). Sterol requirements were found to be polyphyletic and could vary within a genus as was determined for strains within the *Spiroplasma* genus. Given this information, along with additional 16S rDNA sequences available for use in phylogenetic analyses for the study of the *Entomoplasmataceae* family, it has been suggested that the two genera be combined into a single genus, *Entomoplasma* (Johansson and Pettersson 2002).

A phylogenetic relationship among species in the orders *Entomoplasmatales* and *Acholeplasmatales* was conducted recently using the complete 16S-23S rRNA intergenic transcribed spacer region sequence and partial nucleotide sequences of the *rpo*B and *gyr*B genes (Volokhov et al. 2007). One interesting finding was that *Acholeplasma multilocale* (ATCC = 49900-type strain PN525) was positioned with *M. seiffertii*, *M. syrphidae*, and *M. photuris* in the *Entomoplasmatacea* cluster. *A. multilocale* was first described in 1992 and assigned to the genus *Acholeplasma* based upon a G+C content of 31 mol% and lack of a requirement for sterol for growth (Hill et al. 1992). At the time there was no antigenic cross-reaction with any *Acholeplasma* species nor was the 16S rRNA sequence available for phylogenetic analyses. *A. multilocale* was found to use UGA as a tryptophan codon in its *gyr*B and *gyr*A sequences which is also found in the *Entomplasmatales* and *Mycoplasmatales*, but not the *Acholeplasmatales*. In addition, *A. multilocale* had only one band for the rrn operon, as opposed to the two observed in all other *Acholeplasma* species tested, and did not have any tRNA inserts in the ITS region as did all other *Acholeplasma* species tested. Volokhov and colleagues thus suggested the reclassification of *A. multilocale* as a member of the family *Entomoplasmatacea*, although no formal submission for a genus change has been made at this time (Volokhov et al. 2007). This study also supported the placement of all the *Mesoplasma* and *Entomoplasma* species into a single genus as suggested by others (Johansson and Pettersson 2002; Gasparich et al. 2004).

## Molecular Analyses

➋ *Figure 38.1* shows the *Entomplasmatacea* phylogeny based on a maximum likelihood analysis of 16S rRNA sequences. The *Entomoplasmataceae* clade must have arisen from an ancestor in the *Spiroplasmataceae*. Phylogenetic studies that involve non-16S rRNA genes have resulted in similar phylogenetic constructs. A recent study compared the phylogenetic relationship among species in the orders *Entomoplasmatales* and *Acholeplasmatales* using the complete 16S-23S rRNA intergenic transcribed spacer region sequence and partial nucleotide sequences of *rpo*B and *gyr*B genes (Volokhov et al. 2007). The results clearly indicated the separation into two distinct groups with the exception of *Acholeplasma multilocale* (ATCC = 49900-type strain PN525) that was positioned with *M. seiffertii*, *M. syrphidae*, and *M. photuris* in the *Entomoplasmatacea* cluster (bootstrap value of 88 %) in the same phylogenetic group including *E. luminosum*, *E. lucivorax*, *E. somnilux*, and *E. freundtii*. This same clustering of *Mesoplasma* and *Entomoplasma* species supports the phylogeny observed in the analysis using 16S rRNA sequences in ➋ *Fig. 38.1*.

Additional genetic characteristics of the *Entomoplasmataceae* include a genome composed of one circular chromosome, only one rrn operon, no tRNA genes inserted into the ITS region, and the use of UGA as a tryptophan codon (Volokhov et al. 2007).

## Phenotypic Analyses

### Genus I. *Entomoplasma* Tully, Bové, Laigret and Whitcomb 1993, 379[VP]

En.to.mo.plas'ma. Gr. N. *entomon* insect; Gr. Neut.n. *plasma* something formed or molded, a form; N.L. neut. N. *Entomoplasma* name intended to show association with insects (Brown et al. 2011b)

The genus *Entomoplasma* currently has six described species with *E. ellychniae* being designated as the type species. All appear microscopically as nonhelical, nonmotile, pleomorphic coccoid cells of various sizes after growth in liquid culture. All were able to be filtered through 220-nm filters. As with other members of the class *Mollicutes*, all lack a cell wall and thus are resistant to penicillin. All required serum or cholesterol for growth. All have the ability to catabolize glucose with variable ability to hydrolyze arginine, none have the ability to hydrolyze urea. Serologically, each species was shown to be unrelated to the type strains

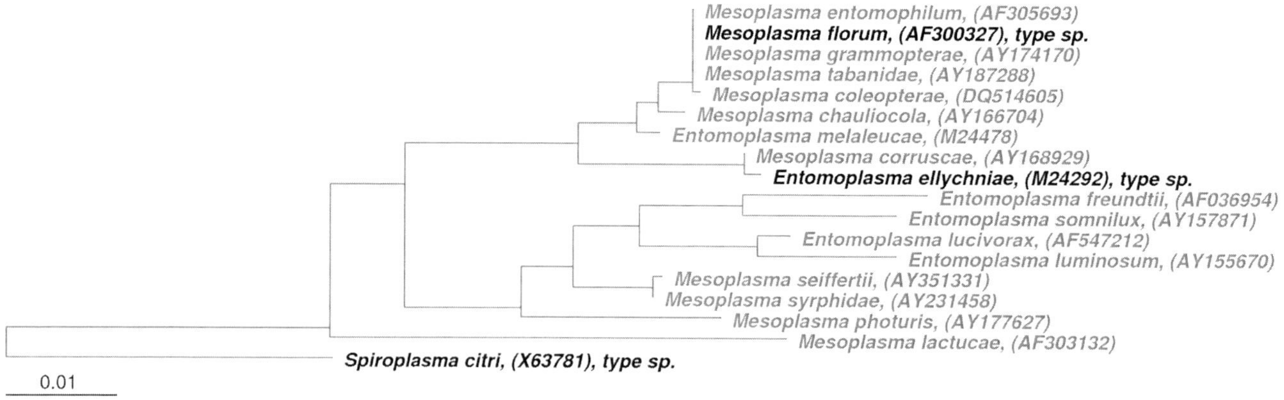


**Fig. 38.1**

**Maximum likelihood phylogenetic analysis using 16S rRNA sequence of all type strains of the *Entomoplasmatacea*. *Spiroplasma citri* was used as an outgroup. Each branch contains the species name and sequence accession number used in the analysis. A 40 % conservational filter was used to remove hypervariable positions**

**Table 38.1**

**Biological and genomic characteristics for the *Entomoplasma* species type strains**

| Genus species | Type strain | ATCC number | Optimal temp (temp range) | Ability to hydrolyze arginine | Sterol requirement | G + C content | Genome size |
|---|---|---|---|---|---|---|---|
| *E. ellychniae* | ELCN-1 | 43707 | 30 °C (18–32 °C) | No | Yes | 27.5 mol% | 680 kbp |
| *E. freundtii* | BARC 318 | 51999 | 30 °C (15–32 °C) | Yes | Yes | 34.1 mol% | 870 kbp |
| *E. luminosum* | PIMN-1 | 49195 | 30 °C /32 °C (10–32 °C) | No | Yes | 28.8 mol% | 663 kbp |
| *E. lucivorax* | PIPN-2 | 49196 | 30 °C (18–32 °C) | No | Yes | 27.4 mol% | 886 kbp |
| *E. melaleucae* | M1 | 49191 | 23 °C (10–30 °C) | No | Yes | 27.0 mol% | 652 kbp |
| *E. somnilux* | PYAN-1 | 49194 | 30 °C (10–32 °C) | No | Yes | 27.4 mol% | 613 kbp |

of all previously characterized *Mycoplasma*, *Acholeplasma*, *Mesoplasma*, *Entomoplasma*, or to any other unclassified sterol-requiring isolates from various animal, plant or insect sources. ➲ *Table 38.1* below shows some of the distinguishing biological and genomic characteristics for the type strains for each of the *Entomoplasma* species.

## Genus II. *Mesoplasma* Tully, Bové, Laigret and Whitcomb 1993, 379$^{VP}$

Me.so.plas'ma. Gr. adj. *mesos* middle; Gr. neut.n. *plasma* something formed or molded, a form; N.L. neut. N. *Mesoplasma* name intended to denote a middle position with respect to sterol or cholesterol requirement (Brown et al. 2011c)

The genus *Mesoplasma* currently has 11 described species with *M. florum* being designated as the type species. All appear microscopically as nonhelical, nonmotile, pleomorphic coccoid cells of various sizes after growth in liquid culture. All were able to be filtered through 220-nm filters. As with other members of the class *Mollicutes*, all lack a cell wall and thus are resistant to penicillin. All were able to grow without the addition of serum or cholesterol, provided there was the addition of PES (0.04 % Tween 80). All have the ability to catabolize glucose with variable ability to hydrolyze arginine, none have the ability to hydrolyze urea. Serologically, each species was shown to be unrelated to the type strains of all previously characterized *Mycoplasma, Acholeplasma, Mesoplasma, or Entomoplasma* species as well as to unclassified non-sterol-requiring isolates from animals, plants, or insect hosts. ➲ *Table 38.2* below shows some of the distinguishing biological and genomic characteristics for the type strains for each of the *Mesoplasma* species.

Metabolic studies on members of the *Entomoplasmataceae* family have been conducted as part of a comparison among members of the class *Mollicutes* and have not included all type strains for all species. Pollack and colleagues (Pollack et al. 1996) compared *E. ellychniae* ELCN-1$^{T}$, *E. melaleucae* M-1$^{T}$, *M. seiffertii* F7$^{T}$, *M. entomophilum* TAC$^{T}$, and *M. florum* L1$^{T}$ with *Mycoplasma fermentans* PG18$^{T}$ and *Acholeplasma multilocale* PN525$^{T}$ and found them to all be similar in several

◘ Table 38.2

Biological and genomic characteristics for the *Mesoplasma* species type strains

| Genus species | Type strain | ATCC number | Optimal temp (temp range) | Ability to hydrolyze arginine | Sterol requirement | G + C content | Genome size |
|---|---|---|---|---|---|---|---|
| M. chauliocola | CHPA-2 | 49578 | 32 °C (10–37 °C) | No | No | 28.3 mol% | 930 kbp |
| M. coleopterae | BARC 779 | 4953 | 30–37 °C (10–37 °C) | No | No | 27.7 mol% | 870 kbp |
| M. corruscae | ELCA-2 | 49579 | 30 °C (10–37 °C) | No | No | 26.4 mol% | 920 kbp |
| M. entomophilum | TAC | 43706 | 30 °C (23–32 °C) | No | No | 30.0 mol% | ND |
| M. florum | L1 | 33453 | 28–30 °C (18–37 °C) | No | No | 27–28 mol% | 790 kbp |
| M. grammopterae | GRUA-1 | 49580 | 30 °C (10–37 °C) | No | No | 29.1 mol% | 885 kbp |
| M. lactucae | 31-C4 | 49193 | 30 °C (18–37 °C) | No | No | 30.0 mol% | 662 kbp |
| M. photuris | PUPA-2 | 49581 | 30 °C (10–32 °C) | Yes | No | 28.8 mol% | 825 kbp |
| M. seiffertii | F7 | 49495 | 28 °C (20–35 °C) | No | No | 30.0 mol% | 1,030 kbp |
| M. syrphidae | YJS | 43706 | 23 °C (10–32 °C) | No | No | 27.6 mol% | 905 kbp |
| M. tabanidae | BARC 857 | 49584 | 37 °C (10–37 °C) | No | No | 28.3 mol% | 930 kbp |

metabolic characteristics. For example, the NADH oxidase activity was localized within the cytoplasm. Additionally, all of these strains had ATP-dependent phosphofructokinase (PFK), malate dehydrogenase (MDH), lactate dehydrogenase (LDH), and ATP- and PPi-dependent deoxyguanosine kinase, but no dUTPase or glucose-6-phosphate dehydrogenase (G6PD) activity. The lack of dUTPase activity was unexpected as it is found in the spiroplasmas and in *Mycoplasmas mycoides* which are both phylogenetically very closely related to the *Entomoplasmatacea* cluster. All possessed hypoxanthine-guanine phosphoribosyl transferase and phosphoenolpyruvate carboxylase and uracil-N-Glycosylase (UNG) activities (with the exception of *M. entomophilum* TAC$^T$). The fact that *M. entomophilum* TAC$^T$ had no UTPase or UNG activities indicates limitations on its ability to synthesize dTTP which means that it might have to take up phosphorylated pyrimidines and purines from the external environment, much like *Mycoplasma mycoides*. The presence of ATP-dependent PFK, LDH, and MDH and the absence of G6PD activities in the *Entomoplasmatacea* suggests the use of glycolysis and fermentation for energy production.

Some additional molecular studies have specifically focused on some interesting molecular and metabolic characters of *M. florum*. One such study examined the HinT proteins which belong to the HIT protein superfamily. Although their function in prokaryotes is still unclear, it has been determined that in eukaryotes, these proteins function as intracellular receptors and hydrolyze purine mononucleotides. In an attempt to better understand the prokaryotic function of HinT homologues, Hopfe and colleagues (Hopfe et al. 2005) determined that a polycistronic cluster of hitABL genes in *Mycoplasma hominis* was homologous to the genes identified in *M. florum*, and that the mollicute HinT proteins were found to be linked to membrane proteins. This is interesting, because in the *Chlamydiaceae*, the HinT proteins are all associated with cytoplasm proteins, indicating a possible function in regulation. Future studies plan to determine the function of bacterial intra- and extracellular HinT proteins.

In the transfer-messenger RNA (tmRNA) system in bacteria, nascent polypeptides on a stalled ribosome are normally tagged with an ssrA tag (short C-terminal sequence encoded by the tmRNA) that then allows for normal termination and release of ribosomal subunits. In most bacteria, this tmRNA-mediated termination leads to degradation of the ssrA-tagged proteins by the AAA protease ClpXP. However, members of the *Entomoplasmataceae* family lack the ClpXP gene. A recent study reported that ssrA-tagged proteins in *Mesoplasma florum* are degraded by the AAA Lon protease (*mf*-Lon) (Gur and Sauer 2008). Experiments showed that the ssrA tag sequence of *M. florum* was specifically and efficiently recognized by the *mf*-Lon. When an *E. coli* ssrA tag was used to label proteins, they were not efficiently degraded by *mf*-Lon. On the other hand, *E. coli* Lon proteases did not efficiently degrade proteins bearing the *M. florum* ssrA tag. This suggests that in order for members of the *Entomoplasmatacea* to retain the ability to use the ssrA-tag-mediated protein degradation, coevolution occurred in both the ssrA tag and the Lon protease to be able to function optimally together.

Recently, riboswitch activity has been described in *M. florum*. A riboswitch is an integral part of an mRNA molecule that specifically binds a small molecule, leading to self-regulation of that gene's expression. A recent study identified mRNA aptamers in *M. florum* that were similar to previously characterized guanine and adenine riboswitches (Kim et al. 2007). A subset of the aptamers selectively bound guanine or adenine. One aptamer variant (designated *mfl*-riboswitch), found in the 5′ untranslated region of an operon containing ribonucleotide reductase genes, bound selectively to 2-deoxyguansine. A subsequent study on the *mfl*-riboswitch found that binding to 2-deoxyguanosine terminated transcription of the operon (Wacker et al. 2011). Spectroscopic comparison of the *mfl*-riboswitch to purine sensing riboswitches revealed that the *mfl* aptamer can form a more flexible binding pocket than normally found in the purine sensing riboswitches. Additionally, several differences in the *mfl* aptamer consensus sequence cause the truncation of a hairpin loop normally found in purine riboswitches. The study of *M. florum* riboswitch variants has greatly expanded knowledge concerning the diversity of bacterial riboswitches.

## Isolation, Enrichment, and Maintenance Procedures

### Isolation from Arthropod Hosts (e.g., Gut and Hemolymph) and Plant Surfaces

The techniques used for the primary isolation of *Entomoplasmatacea* strains from their arthropod hosts and plant surfaces have been described in detail. The strategies used for primary isolation from insects are similar to those used for the isolation of spiroplasmas (Markham et al. 1983) and insect-derived acholeplasmas (Tully et al. 1987). Procedures for the primary isolation of mollicutes from plant or floral surfaces have been described by Bové and colleagues (Bové et al. 1983).

### Initial Cultivation and Filtration Cloning

Primary cultures were grown statically in traditional mollicute media, such as M1D (Whitcomb 1983), SM-1 (Whitcomb 1983), and SP-4 (Whitcomb 1983), at a temperature range from 26 °C to 30 °C. Cultures were passed three to five times prior to storage either using lyophilization or direct freezing in −70 °C. After initial cultivation, all strains were purified by established filtration cloning techniques (Tully 1983).

### Growth and Maintenance

Following cloning, cultures were grown on a variety of media to determine growth requirements. For example, in several cases, strains were grown in three different media: the Edward formulation of conventional 20 % horse serum mycoplasma broth (Edward 1947); serum fraction broth containing 1 % bovine serum fraction (Tully 1984); and serum-free media with or without supplementation with fatty acid mixtures such as 0.01–0.04 % (vol/vol) Tween 80 (Tully et al. 1988). Liquid cultures were typically maintained by serial tenfold dilutions with a phenol red indicator with growth measured by color change. The sterol requirement test was described by Rose and colleagues (Rose et al. 1993) in which strains were screened for their ability to maintain growth in media containing 15–20 % fetal bovine serum or in serum-free media with or without 0.04 % TWEEN 80. In that study, all *Entomoplasma* strains did not grow in either serum-free medium alone or when Tween 80 was added and all *Mesoplasma* strains grew in serum-free media only when Tween 80 was added. Long-term storage is achieved through lyophilization or direct freezing of liquid cultures in −70 °C.

### Growth on Solid Media

A solid medium was prepared by adding 0.8 % Noble agar or 1.0 % agarose to the broth base prior to autoclaving (all autoclave sensitive components are filter-sterilized and added after autoclaving). Agar cultures usually were incubated at 30 °C under aerobic (with 5 % carbon dioxide in a GasPak system) and anaerobic (hydrogen GasPak system) environments. On solid media, the colony size ranged from 200 to 300 pm in diameter and exhibited a typical fried-egg morphology, with the exception of *E. freundtii* and *E. somnilux* which had a more granular appearance with many small colonies clustered together (❯ *Table 38.3*).

## Ecology

All members of the *Entomoplasmataceae* are associated with insect or plant hosts. ❯ *Table 38.4* shows the host organisms for each of the *Entomoplasma* and *Mesoplasma* type strains and, when available, information about non-type strain hosts. Several of the isolates were found associated with plant surfaces. Although there is no direct evidence, it is thought that these isolates were deposited on the plant surfaces by insect hosts. All insect host organisms were infected primarily in the gut, although there was one example of an isolate from the hemolymph of a firefly beetle (Tully et al. 1989). There has been no direct experimental study of transmission, but given the similarity in hosts, the mechanism for transmission may be similar to that proposed for the closely related spiroplasmas. Clark proposed that spiroplasmas may be deposited on plant surfaces as their insect host eats, leaving microbes behind as it regurgitates and defecates on the plant surface (Clark 1982). This would explain how the microbes are spread from insect to insect and also explain isolation from plant surfaces. The isolation *of M. coleoptera* reflects this possible mode of transmission as the isolate was obtained from the gut of an adult soldier beetles (*Chauliognatus* sp.) feeding on Canada horseweed (*Conyza canadensis*) and the flowers of the Vara Dulce tree (*Eysenhardtia texana*) in Texas (Tully et al. 1994). However, more isolations of *Mesoplasma* and *Entomoplasma* strains would be required to clarify the ecology and biological cycle of these organisms in their hosts.

## Pathogenicity: Clinical Relevance

Due to their lack of a cell wall, members of the *Entomoplasmataceae* are resistant to any antibiotic that targets the cell wall, such as all ß-lactams, glycopeptides, and polymyxins. None of the species designated to the *Entomoplasmatacea* family have been shown to have any pathogenicity in their insect or plant host organisms.

◘ Table 38.3
Characteristics of growth in liquid and on solid media for the type strains in the family *Entomoplasmataceae*

| Genus species designation | Liquid culture | Growth on solid media (aerobic vs. anaerobic) |
|---|---|---|
| *E. ellychniae* | Grew only in SP-4 or M1D with fetal bovine serum (not in conventional mycoplasma media with horse serum (Edward formulation) or bovine serum fraction supplements) | Grew on SP-4 solid media aerobically or in 5 % carbon dioxide, did not grow anaerobically |
| *E. freundtii* | Grew in SP-4, M1D, or Edward formulation of conventional mycoplasma medium with 15–20 % horse serum or with 1 % bovine serum fraction | Exhibited cluster of many small colonies (not typical fried-egg morphology) in anaerobic environment |
| *E. luminosum* | Grew in SP-4 as well as conventional mycoplasma media with horse serum and mycoplasma broth base with bovine serum fraction | Grew on solid SP-4 media under aerobic or 5 % carbon dioxide conditions, but not anaerobic conditions |
| *E. lucivorax* | Grew in SP-4 and conventional mycoplasma media with horse serum and mycoplasma broth base with bovine serum fraction | Grew on solid SP-4 media aerobically and anaerobically |
| *E. melaleucae* | Grew in SP-4 or modified Edward media with 20 % fetal bovine serum, but not in mycoplasma media supplemented with horse serum | Colonies only grew on SP-4 or mycoplasma broth base with fetal bovine serum fraction, with best growth under anaerobic conditions – exhibited normal fried-egg morphology, but when grown aerobically or in 5 % carbon dioxide grew into very small and centerless colonies |
| *E. somnilux* | Grew in SP-4 and in conventional mycoplasma media containing horse serum (Edward formulation), but not in mycoplasma broth base supplemented with bovine serum fraction | Grew on solid SP-4 media under aerobic or anaerobic conditions, but exhibited a cluster of small colonies as opposed to the typical fried-egg morphology |
| *M. chauliocola* | Grew in mycoplasma media with fetal bovine serum, horse serum or bovine serum fraction; also grew in serum-free mycoplasma broth when supplemented with 0.04 % Tween 80 fatty acid mixture | Colonies grew on SP-4 or horse serum agar plates and exhibited fried-egg morphology under anaerobic conditions |
| *M. coleopterae* | Grew in mycoplasma media with fetal bovine serum, horse serum, or bovine serum fraction; also grew in serum-free mycoplasma broth when supplemented with 0.04 % Tween 80 fatty acid mixture | Colonies grew on SP-4 or horse serum agar plates; exhibited fried-egg morphology under anaerobic conditions |
| *M. corruscae* | Grew in mycoplasma media with fetal bovine serum, horse serum, or bovine serum fraction; also grew in serum-free mycoplasma broth when supplemented with 0.04 % Tween 80 fatty acid mixture | Colonies grew on SP-4 or horse serum agar plates and exhibited fried-egg morphology under anaerobic conditions |
| *M. entomophilum* | Grew rapidly in M1D serum-containing media, but grew less well with bovine serum fraction supplement or if supplemented with 0.04 % Tween 80 fatty acid mixture; did not require cholesterol | Grew on M1D agar and exhibited fried-egg morphology under anaerobic conditions |
| *M. florum* | Grew well in MC broth, serum fraction broth, or serum-free Tween 80 broth; did not require serum or cholesterol for growth, but did require supplementary fatty acids (Tween 80) for growth | Exhibited typical fried-egg morphology under aerobic conditions |
| *M. grammopterae* | Grew in mycoplasma medium with fetal bovine serum, horse serum, or bovine serum fraction; also in serum-free mycoplasma broth when supplemented with 0.04 % Tween 80 fatty acid mixture | Colonies on SP-4 or horse serum agar plates exhibited fried-egg morphology under anaerobic conditions |
| *M. lactucae* | Grew well in all mycoplasma broth media, including those containing fetal bovine serum (SP-4), horse serum, or bovine serum fraction supplements; also grew well in serum-free mycoplasma broth when supplemented with 0.04 % Tween 80 fatty acid mixture | Colonies with fried-egg morphology visible on solid media from all media used in liquid growth under both aerobic and anaerobic conditions |
| *M. photuris* | Grew in mycoplasma medium with fetal bovine serum, horse serum, or bovine serum fraction; also in serum-free mycoplasma broth when supplemented with 0.04 % Tween 80 fatty acid mixture | Grew on solid SP-4 or horse serum agar plates and exhibited fried-egg morphology under anaerobic conditions |
| *M. seiffertii* | Grew in mycoplasma medium formulations containing serum (fetal bovine or horse serum) or bovine serum fraction; growth in serum-free mycoplasma broth only if supplemented with 0.04 % Tween 80 fatty acid mixture | Grew on all media used in liquid culture but very poorly in serum-free base medium with Tween 80 supplement; grew in either anaerobic or aerobic conditions, but best growth in anaerobic conditions |

**Table 38.3 (continued)**

| Genus species designation | Liquid culture | Growth on solid media (aerobic vs. anaerobic) |
|---|---|---|
| *M. syrphidae* | Grew in mycoplasma medium with fetal bovine serum, horse serum, or bovine serum fraction; also in serum-free mycoplasma broth when supplemented with 0.04 % Tween 80 fatty acid mixture | Colonies grew on SP-4 or horse serum agar plates and exhibited fried-egg morphology under anaerobic conditions |
| *M. tabanidae* | Grew in mycoplasma medium with fetal bovine serum, horse serum, or bovine serum fraction; also in serum-free mycoplasma broth when supplemented with 0.04 % Tween 80-fatty acid mixture | Colonies grew on SP-4 or horse serum agar plates and exhibited fried-egg morphology under anaerobic conditions |

**Table 38.4**
**Host organisms for members of the Family *Entomoplasmataceae***

| Genus species designation | Type strain host | Non-type strain hosts |
|---|---|---|
| *E. ellychniae* | Hemolymph of the firefly beetle *Ellychniae corrusca* | |
| *E. freundtii* | Gut green tiger beetle (Coleoptera: Cicindelidae) | |
| *E. luminosum* | Gut of firefly beetle (*Photinus marginata*) | |
| *E. lucivorax* | Gut of firefly beetle (*Photinus pyralis*) | Meadowsweet flower surface (*Spirea ulmaria*) |
| *E. melaleucae* | Punktree flower surface (*Melaleuca quinquenervia*) | White Feather Honey Myrtle tree flower surface (*Melaleuca decora*), the silk oak tree flower surface (*Grevillea robusta*), and from an anthoporine bee (*Xylocopa micans*) |
| *E. somnilux* | Gut of pupal stage firefly beetle (*Pyractomena angulata*) | |
| *M. chauliocola* | Gut of adult goldenrod soldier beetle (*Chauliognathus pennsylvanicus*) | |
| *M. coleopterae* | Gut of adult soldier beetle (*Chauliognathus* sp.) | |
| *M. corruscae* | Gut of an adult firefly (*Ellycnia corrusca*) | |
| *M. entomophilum* | Gut of a tabanid fly (*Tabanus catenatus*) | Gut of nine other insect genera and from the *Bidens* sp. flower surface |
| *M. florum* | Lemon tree flower surface (*Citrus limon*) | Grapefruit flower surface (*Citrus paradisi*), powderpuff tree flower surface (*Albizia julibrissin*), and from the gut of a variety of insects |
| *M. grammopterae* | Gut of an adult long-horned beetle (*Grammoptera* sp.) | Gut of an adult soldier beetle (*Cantharidae* sp.) and the gut of an adult mining bee (*Andrena* sp.) |
| *M. lactucae* | Lettuce surface (*Lactuca sativa*) | |
| *M. photuris* | Gut of larval and adult firefly beetles (*Photuris lucicrescens* and other *Photuris* sp.) | Gut of horse fly (*Tabanus americanus*) |
| *M. seiffertii* | Sweet orange tree flower surface (*Citrus sinensis*) and wild angelica flower surface (*Angelica sylvestris*) | Mosquitoes (*Aedes detritus* and *Aedes caspius*) and tabanid fly (*Chrysops pictus*) (Gros et al. 1996)[a] |
| *M. syrphidae* | Gut of an adult syrphid fly (Diptera: Syrphidae) | Gut of bumblebee (*Bombus* sp.) and gut of a skipper (Lepidoptera: Hesperiidae) |
| *M. tabanidae* | Gut of an adult horse fly (*Tabanus abactor*) | |

[a]Citation provided if information obtained from source other than the original species description

## Application

### Waste Management Application

Vermicomposting is a form of composting that utilizes worms and insect larvae to decompose organic waste (e.g., food waste or animal bedding material) into nutrient-rich organic fertilizer primarily made up of insect castings (frass). Recent studies have explored the microflora of vermicomposting organisms to determine which microbes are dominant. Hong and colleagues (Hong et al. 2011) determined that *E. somnilux* (along with *Bacillus licheniformis*) was a dominant member of the microbial community in vermicomposting earthworms (*Eisenia fetida*). PCR-DGGE was used to determine which of the 57 bacterial 16S rDNA clones were dominant, and subsequent sequence analysis on one of the two dominant bands showed 96–98 % similarity to *E. somnilux*. It was not determined what impact *E. somnilux* had on the vermicomposting as the study went on to inoculate with *Photobacterium motobuensis* (WN9) and *Aeromonas hydrophila* (WA40) and were able to show an increase in growth of the earthworms and a subsequent increase in cast production.

In another study by Zhang and colleagues (Zhang et al. 2012), a vermireactor was developed using housefly larvae (*Musca domestica*) as the vermicomposting organism to determine effectiveness for swine manure reduction. In this instance, the researchers explored the microbial diversity of the vermicompost itself. Analysis of DGGE bands generated from extractions of the vermicompost showed that the dominant microbes were *E. somnilux*, *Proteobacterium*, and *Clostridiaceae* bacteria. The actual contribution of the *E. somnilux* in the vermicomposting process remains to be elucidated.

## References


Bonnet F, Saillard C, Vignault JC, Garnier M, Carle P, Bové JM, Rose DL, Tully JG, Whitcomb RF (1991) *Acholeplasma seiffertii* ap. nov., a mollicute from plant surfaces. Int J Syst Bacteriol 41:45–49

Bové JM, Whitcomb RF, McCoy RE (1983) Culture techniques for spiroplasmas from plants. Methods Mycoplasmol 2:225–234

Brown DR, Bradbury JM, Whitcomb RF (2011a) Family I. *Entomoplasmataceae* Tully, Bové, Laigret and Whitcomb 1993, 380[VP]. In: Krieg NR, Staley JT, Brown DR, Hedlund BP, Pastor BJ, Ward NL, Ludwig W, Whitman WB (eds) Bergey's manual of systematic bacteriology, 2nd edn. Springer, New York, p 645

Brown DR, Bradbury JM, Whitcomb RF (2011b) Genus I. *Entomoplasma* Tully, Bové, Laigret and Whitcomb 1993, 379[VP]. In: Krieg NR, Staley JT, Brown DR, Hedlund BP, Pastor BJ, Ward NL, Ludwig W, Whitman WB (eds) Bergey's manual of systematic bacteriology, 2nd edn. Springer, New York, pp 646–649

Brown DR, Bradbury JM, Whitcomb RF (2011c) Genus II. *Mesoplasma* Tully, Bové, Laigret and Whitcomb 1993, 380[VP]. In: Krieg NR, Staley JT, Brown DR, Hedlund BP, Pastor BJ, Ward NL, Ludwig W, Whitman WB (eds) Bergey's manual of systematic bacteriology, 2nd edn. Springer, New York, pp 649–653

Clark TB (1982) Spiroplasmas: diversity of arthropod reservoirs and host-parasite relationships. Science 217:57–59

Clark TB, Tully JG, Rose DL, Henegar R, Whitcomb RF (1986) Acholeplasmas and similar nonsterol-requiring mollicutes from insects: missing link in microbial ecology. Curr Microbiol 13:11–16

Davis RE, Worley JF, Whitcomb RF, Ishijima T, Steere RL (1972) Helical filaments produced by a mycoplasma-like organism associated with corn stunt disease. Science 176:521–523

Doi Y, Tersanaka M, Yora K, Assuyama H (1967) Mycoplasma-or PLT group-like microorganisms found in the phloem elements of plants infected with mulberry dwarf, potato witches' broom, aster yellows or Paulownia witches' broom. Ann Phytopathol Soc Jpn 33:259–266

Eden-Green S, Tully JG (1979) Isolation of *Acholeplasma* spp. from coconut palms affected by lethal yellowing disease in Jamaica. Curr Microbiol 2:311–316

Edward DG (1947) A selective medium for pleuropneumonia-like organisms. J Gen Microbiol 1:238–243

Freundt EA, Whitcomb RF, Barile MF, Razin S, Tully JG (1984) Proposal for elevation of the family *Acholeplasmataceae* to ordinal rank: *Acholeplasmatales*. Int J Syst Bacteriol 34:346–349

Gasparich GE, Whitcomb RF, Dodge D, French FE, Glass J, Williamson DL (2004) The genus *Spiroplasma* and its non-helical descendents: phylogenetic classification, correlation with phenotype and roots of the *Mycoplasma mycoides* clade. Int J Syst Evol Microbiol 54:893–918

Gros O, Saillard C, Helias C, Le Goff F, Marjolet M, Bové JM, Chastel C (1996) Serological and molecular characterization of *Mesoplasma seiffertii* strains isolated from hematophagous dipterans in France. Int J Syst Bacteriol 46:112–115

Gur E, Sauer RT (2008) Evolution of the ssrA degradation tag in *Mycoplasma*. Specificity switch to a different protease. Proc Natl Acad Sci USA 106:16113–16118

Hill AC, Polak-Vogelzang AA, Angulo AF (1992) *Acholeplasma multilocale* sp. nov., isolated from a horse and a rabbit. Int J Syst Bacteriol 42:513–517

Hong SW, Lee JS, Chung KS (2011) Effect of enzyme producing microorganisms on the biomass of epigeic earthworms (*Eisenia fetida*) in vermicompost. Bioresour Technol 102:6344–6347

Hopfe M, Hegemann JH, Henrich B (2005) HinT proteins and their putative interaction partners *Mollicutes* and *Chlamydiaceae*. BMC Microbiol 5:27

IRPCM Phytoplasma/Spiroplasma Working Team-Phytoplasma Taxonomy Group (2004) 'Candidatus Phytoplasma', a taxon for the wall-less, non-helical prokaryotes that colonize plant phloem and insects. Int J Syst Evol Microbiol 54:1243–1255

Johansson K-E, Pettersson B (2002) Taxonomy of *Mollicutes*. In: Razin S, Herrman R (eds) Molecular biology and pathogenicity of mycoplasmas. Kluwer, London, pp 1–29

Kim JN, Roth A, Breaker RR (2007) Guanine riboswitch variants from *Mesoplasma florum* selectively recognize 2′-deoxyguanosine. Proc Natl Acad Sci USA 104:16092–16097

Knight TF Jr (2004) Reclassification of *Mesoplasma pleciae* as *Acholeplasma pleciae* comb. nov. on the basis of 16S rRNA and *gyrB* gene sequence data. Int J Syst Evol Microbiol 54:1951–1952

Markham PG, Clark TB, Whitcomb RF (1983) Culture techniques for spiroplasmas from arthropods. Methods Mycoplasmol 2:217–223

McCoy RE, Caudwell A, Chang CJ, Tully JG, Rose DL, Carle P, Bové JM (1984) *Acholeplasma florum*: a new species isolated from plants. Int J Syst Bacteriol 34:11–15

Navas-Castillo J, Laigret F, Bové JM (1993) 16SrDNA sequence analysis of *Acholeplasma seifertii*, a mollicute from plant surfaces, and its transfer *Mesoplasma*, a new genus in the spiroplasma phylogenetic group. Nucl Acids Res 21:2249

Pollack JD, Williams MV, Banzon J, Jones MA, Harvey L, Tully JG (1996) Comparative metabolism of *Mesoplasma*, *Entomoplasma*, *Mycoplasma* and *Acholeplasma*. Int J Syst Bacteriol 46:885–890

Razin S, Freundt EA (1984) Class I. *Mollicutes* Edward and Freundt 1967, 267[AL] Order I. *Mycoplasmatales* Freundt 1955, 71[AL]. In: Krieg NR, Holt JG (eds) Bergey's manual of systematic bacteriology, vol 1. Williams & Wilkins, Baltimore, pp 740–742

Robinson IM, Freundt EA (1987) Proposal for an amended classification of anaerobic *Mollicutes*. Int J Syst Bacteriol 37:78–81

Rose DL, Kocka JP, Somerson NL, Tully JG, Whitcomb RF, Carle P, Bové JM, Colflesh DE, Williamson DL (1990) *Mycoplasma lactucae* sp. nov., a sterol-requiring mollicute from plant surface. Int J Syst Bacteriol 40:138–142

Rose DL, Tully JG, Bové JM, Whitcomb RF (1993) A test for measuring growth responses of *Mollicutes* to serum and polyoxyethylene sorbitan. Int J Syst Bacteriol 43:527–532

Saglio P, L'hospital M, Lafleche D, Dupont G, Bové JM, Tully JG, Freundt EA (1973) *Spiroplasma citri* gen. and sp. n.: a mycoplasmalike organisms associated with "stubborn" disease of citrus. Int J Syst Bacteriol 23:191–204

Toth KF, Harrison N, Sears BB (1994) Phylogenetic relationships among members of the class *Mollicutes* deduced from *rps3* gene sequences. Int J Syst Bacteriol 44:119–124

Tully JG (1983) Cloning and filtration techniques for mycoplasmas. Methods Mycoplasmol 1:173–177

Tully JG (1984) The family *Acholeplasmataceae*, genus *Acholeplasma*. In: Krieg NR, Holt JM (eds) Bergey's manual of systematic bacteriology, vol 1. Williams & Wilkins, Baltimore, pp 781–787

Tully JG, Rose DL, Whitcomb RF, Hackett KJ, Clark TB, Henegar RB, Clark E, Carle P, Bové JM (1987) Characterization of some new insect-derived acholeplasmas. Isr J Med Sci 23:699–703

Tully JG, Rose DL, Carle P, Bové JM, Hackett KJ, Whitcomb RF (1988) *Acholeplasma entomophilum* sp. nov., from gut contents of a wide range of host insects. Int J Syst Bacteriol 38:164–167

Tully JG, Rose DL, Hackett KJ, Whitcomb RF, Carle P, Bové JM, Colflesh DE, Williamson DL (1989) *Mycoplasma ellychniae* sp. nov., a sterol-requiring Mollicute from the firefly beetle *Ellychnia corrusca*. Int J Syst Bacteriol 39:284–289

Tully JG, Rose DL, McCoy RE, Carle P, Bové JM, Whitcomb RF, Weisburg WG (1990) *Mycoplasma melaleucae* sp. nov., a sterol-requiring mollicute from flowers of several tropical plants. Int J Syst Bacteriol 40:143–147

Tully JG, Bové JM, Laigret F, Whitcomb RF (1993) Revised taxonomy of the Class *Mollicutes*: proposed elevation of a monophyletic cluster of arthropod-associated Mollicutes to ordinal rank (*Entomoplasmatales* ord. nov.) with provision for familial rank to separate species with nonhelical morphology (*Entomoplasmatacea* fam. nov.) from helical species (*Spiroplasmatacea*), and emended description of the order *Mycoplasmatales*, Family *Mycoplasmatacea*. Int J Syst Bacteriol 43:378–385

Tully JG, Whitcomb RF, Hackett KJ, Rose DL, Henegar RB, Bové JM, Carle P, Williamson DL, Clark TB (1994) Taxonomic descriptions of eight new non-sterol-requiring mollicutes assigned to the Genus *Mesoplasma*. Int J Syst Bacteriol 44:685–693

Tully JG, Whitcomb RF, Hackett KJ, Williamson DL, Laigret F, Carle P, Bové JM, Henegar RB, Ellis NM, Dodge DE, Adams J (1998) *Entomoplasma freundtii* sp. nov., a new species from a green tiger beetle (Coleoptera: Cicindelidae). Int J Syst Bacteriol 48:1197–1204

Volokhov DV, Neverov AA, George J, Kong H, Liu SX, Anderson C, Davidson MK, Chizhikov V (2007) Genetic analysis of housekeeping genes of members of the genus *Acholeplasma*: phylogeny and complementary molecular markers to the 16S rRNA gene. Mol Phylogenet Evol 44:699–710

Wacker A, Buck J, Mathieu D, Richter C, Wohnert J, Schwalbe H (2011) Structure and dynamics of the deoxyguanosine-sensing riboswitch studies by NMR-spectroscopy. Nucl Acids Res 39:6802–6812

Weisburg WG, Tully JG, Rose DL, Petzel JP, Oyaizu H, Yang D, Mandelco L, Sechrest J, Lawrence TG, Van Etten J (1989) A phylogenetic analysis of the mycoplasmas: basis for their classification. J Bacteriol 171:6455–6467

Whitcomb RF (1983) Culture media for spiroplasmas. Methods Mycoplasmol 1:147–158

Williamson DL, Tully JG, Rose DL, Hackett KJ, Henegar R, Carle P, Bové JM, Colflesh DE, Whitcomb RF (1990) *Mycoplasma somnilux* sp. nov., *Mycoplasma luminosum*, and *Mycoplasma lucivorax* sp. nov., new sterol-requiring Mollicutes from firefly beetles (Coleoptera: Lampyridae). Int J Syst Bacteriol 40:160–164

Woese CR, Maniloff J, Zablen LB (1980) Phologenetic analysis of the mycoplasmas. Proc Natl Acad Sci USA 77:494–498

Zhang Z, Wang H, Zhu J, Suneethi S, Zheng J (2012) Swine manure vermicomposting via housefly larvae (*Musca domestica*): the dynamics of biochemical and microbial features. Bioresour Technol 118:563–571

# 39 The Order *Mycoplasmatales*

*Meghan May*[1] · *Mitchell F. Balish*[2] · *Alain Blanchard*[3]
[1]Department of Biological Sciences, Towson University, Towson, MD, USA
[2]Department of Microbiology, Miami University, Oxford, OH, USA
[3]INRA, UMR 1332 Biologie du Fruit et Pathologie, University of Bordeaux, Villenave d'Ornon, France

*Introduction* .... *516*

*Taxonomy and Phylogeny* .... *516*
Taxonomy .... 516
Phylogeny .... 516

*Isolation, Identification, and Typing* .... *518*
Culture Media .... 520
Freezing and Other Methods to Preserve Cultures .... 522
Detection and Identification of Mycoplasmas by PCR Assays .... 522
Typing Methods for Differentiating Mycoplasma Species or Strains .... 522

*Genetics, Cell Biology, and Physiology* .... *523*
Minimal Genomes .... 523
Horizontal Gene Transfer and Mobile Genetic Elements .... 523
Morphology .... 524
Cytoskeletal Elements .... 526
Motility .... 527
Cell Membrane .... 528

*Habitats and Ecology of the* Mycoplasmataceae .... *528*
Human and Animal Hosts .... 528
Host and Tissue Specificity .... 529
Mycoplasmas Infecting Cell Cultures .... 529
Mycoplasma Associations with Protozoa .... 529

*Epidemiology and Control* .... *530*
Transmission .... 530
Antibiotic Sensitivity .... 530
Mechanisms of Antibiotic Resistance .... 530
Vaccines .... 531

*Pathogenicity* .... *532*
Disease Manifestations .... 532
Human Diseases .... 532
Veterinary Diseases: Agricultural Animals .... 533
Veterinary Diseases: Companion and Research Animals .... 533
Veterinary Diseases: Wildlife Diseases .... 533
Mechanisms of Pathogenicity of the Mycoplasmatales .... 535
Secreted Toxins and Toxic Metabolites .... 535
Cytoadherence Mechanisms .... 537
Immune Evasion .... 537
Polysaccharides (Capsules and Biofilms) .... 538
Glycosidases .... 539
Uncharacterized Virulence-Associated Genes .... 539

*Societal Impacts of the* Mycoplasmatales .... *539*
Agricultural Losses .... 539
Genomics and Synthetic Cells .... 541

**Abstract**

Within the class *Mollicutes*, the order *Mycoplasmatales* contains more than 160 distinct *Mycoplasma* species and 8 *Ureaplasma* species. All these species are characterized by a small genome, the result of regressive evolution from a common ancestor with *Firmicutes*. This limited genetic information is associated with numerous growth requirements and for most of them, only undefined media are available. As some of the mycoplasmas illustrate among bacteria the best concept of a minimal cell, they were selected among the first bacteria for which the genome was completely sequenced. More recently, their cell biology was further explored by a combination of "omics" approaches, and they were used as platforms for the development of new methods of synthetic biology such as genome cloning in yeast, genome transplantation, and engineering. One of the goals of these studies is the de novo assembly of a minimal cell. The availability of genome sequences for several species, including in some cases for several strains from the same species, resulted in new methods for typing isolates, allowing improved epidemiological studies. The lack of a cell wall and a minimal genome lead to innovative solutions for cellular organization. In some species, there is a cell polarity with a tip involved in adhesion to host cells and in motility. The molecular structure of this tip is complex, and its assembly is coordinated with DNA replication and cell division. A number of *Mycoplasma* and *Ureaplasma* species are associated with significant pathologies of humans and animals. The infections caused by these organisms are chronic for most of them, which means that the bacteria have developed means to evade from the immune system of their hosts. Immune evasion by mycoplasmas takes several forms. A major one for which abundant evidence exists is variation of surface antigens. A second one, established more

E. Rosenberg et al. (eds.), *The Prokaryotes – Firmicutes and Tenericutes*, DOI 10.1007/978-3-642-30120-9_289,

indirectly, is molecular mimicry by surface proteins. Finally, invasion of host cells is employed by many *Mycoplasma* species and may have a role in immune evasion. A number of virulence factors have been identified for different species; they include a small number of secreted toxins, surface polysaccharides, and several enzymes that interfere with their host's metabolism and produce toxic products. Finally, there is a need for improved methods to control mycoplasmoses as their societal impact, in particular, in agriculture is important and in some cases increasing.

## Introduction

More than a century ago, in 1898 Edmond Nocard and Emile Roux, both colleagues of Louis Pasteur, succeeded to cultivate the causal agent of contagious bovine pleuropneumonia (CBPP) (Nocard and Roux 1898), now known as *Mycoplasma mycoides* subsp. *mycoides* (Manso-Silvan et al. 2009). This success was due not only to a collaboration involving Amédée Borrel, Taurelli Salimbeni, and Edouard Dujardin-Beaumetz but also to technical innovations (Bove 1999). The first one of these innovations, introduced by Metchnikoff, Roux, and Taurelli Salimbeni, was the use of a sterile collodion pouch in which broth was inoculated by pulmonary serous fluid from an infected animal. The pouch then was introduced into the peritoneal cavity of a rabbit and was removed after several days. The broth was now opalescent and microscopic examination revealed indication of bacterial multiplication; animal experimental infections confirmed the identity of the CBPP agent. The second innovation came from Dujardin-Beaumetz who modified the solid medium by adding serum, which was thought to be necessary for the growth of this fastidious organism (Borrel et al. 1910). This supplementation allowed the first cultivation of a mycoplasma on agar plates and the observation of typical fried egg-shaped colonies. These investigators from the Institute Pasteur also found that the CBPP agent belonged to the category of filterable agents, as it went through Chamberland filters. This early work had already clearly established a number of specific characteristics that are shared by all mycoplasmas: a very small size, the requirement for serum in the culture medium, passage through bacterial filters, and fried egg-shaped colonies. Other agents similar to the CBPP agent, named *Mycoplasma peripneumoniae* by Nowak (1929), were discovered in the following decades and named collectively pleuropneumonia-like organisms (PPLOs). This term was replaced by mollicutes (mollis for soft, cutis for skin) in 1967 (Edward et al. 1967), to designate this class of bacteria.

## Taxonomy and Phylogeny

### Taxonomy

The *Mycoplasmatales* fall within class *Mollicutes*, phylum Tenericutes, and domain *Bacteria*. The order *Mycoplasmatales* legitimately contains two families: *Mycoplasmataceae* and "Incertae sedis." The latter accommodates genus *Eperythrozoon* and genus *Haemobartonella*, which are now recognized as part of the order *Mycoplasmatales* (see section on "➋ Phylogeny"). Their formal nomenclature has yet to be legitimately resolved by the International Committee for the Systematics of Prokaryotes, and thus, they remain in the temporary family "Incertae sedis." The family *Mycoplasmataceae* contains more than 160 distinct *Mycoplasma* species and 8 distinct *Ureaplasma* species. Because of this large number of species, further classification of *Mycoplasma* spp. into informal taxa (groups and clusters) was undertaken (see section on "➋ Phylogeny").

Taxonomic assignment to the *Mollicutes* was originally based on the unifying characteristic of the absence of a cell wall. Despite this phenotype, phylogenetic analysis based on the 16S rRNA gene sequence clearly indicates that the *Mollicutes* belong to the clade of Gram-positive bacteria. Affiliation of species with one of the four orders in the class (i.e., *Mycoplasmatales*, *Entomoplasmatales*, *Acholeplasmatales*, and *Anaeroplasmatales*) was based on a combination of aerobic versus anaerobic growth, sterol requirement, and host range. Further taxonomic assignments at the family and genus levels were similarly made (➋ *Fig. 39.1*). It is important to note that advances in understanding from phylogenetics and phylogenomics have since challenged some of these strict definitions, particularly as it applies to morphology, host range, and metabolic characterization (see section on "➋ Phylogeny").

### Phylogeny

Elucidation of the phylogeny of the mollicutes greatly benefited from Carl Woese's pioneering work. Indeed, in a 1980 collaboration with the mycoplasmologist Jack Maniloff, the comparative analysis of 16S rRNA oligonucleotide catalogs clearly established that these peculiar organisms arose by degenerative evolution, as a deep branch of the subline of clostridial ancestry that led to *Bacillus* and *Lactobacillus* (Woese et al. 1980). The progress in DNA sequencing quickly enabled a more precise analysis, initially from the 5S rRNA (Rogers et al. 1985) and then from 16S rRNA (Weisburg et al. 1989). The latter publication is a seminal publication in mycoplasmology. The genes encoding 16S rRNA from 26 *Mycoplasma* species were sequenced, aligned, and compared phylogenetically to related bacteria from the same class and from the *Firmicutes*. The *Mollicutes* phylogenetic tree could be subdivided into five main branches: spiroplasma, pneumoniae, hominis, anaeroplasma, and asteroleplasma (Weisburg et al. 1989). The taxon *Asteroleplasma*, which includes the single species *Asteroleplasma anaerobium* is marginal, intertwined with phyla of other *Firmicutes*, which leads to the question of whether it belongs to the class *Mollicutes* (Johansson and Pettersson 2002). Therefore, with the exception of asteroleplasmas, the mollicutes represent a monophyletic group of bacteria that is thought to have diverged from ancestors shared with Gram-positive relatives at about 605 Myr (million years) (Maniloff 2002). In this tree, *Mycoplasma* species are

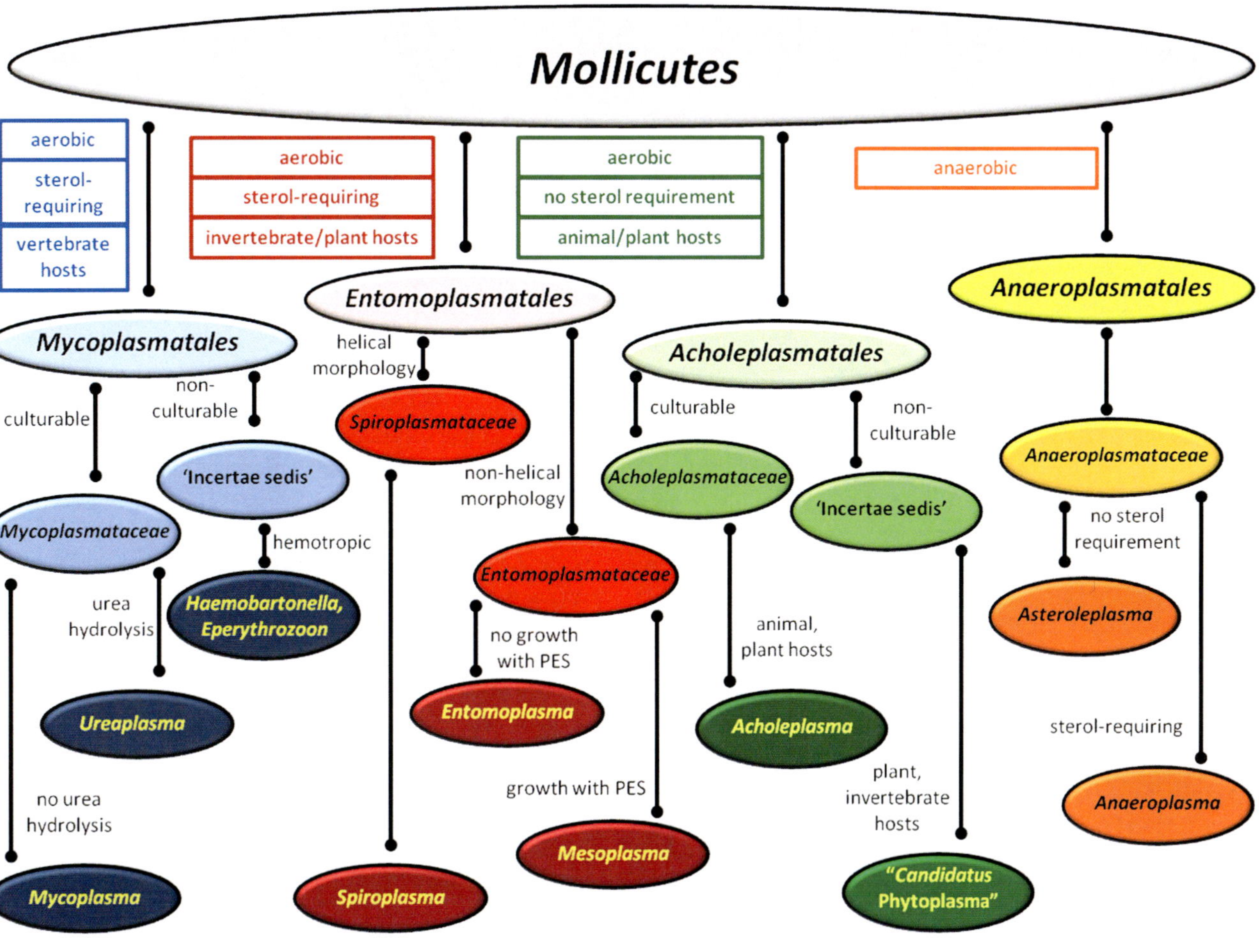


**Fig. 39.1**
***Mollicutes* taxonomy by classical definitions.** The four orders in the class *Mollicutes* are separated by defining features into their respective families and genera. It should be noted that phylogenetic analysis has challenged some of the classical definitions. Most notably, members of the family *Spiroplasmataceae* by definition parasitize invertebrate or plant hosts and have a helical morphology; however, members of the *M. mycoides* cluster (vertebrate hosts, pleomorphic shape) are appropriately affiliated with this family. Additionally, molecular detection of apparent *Mycoplasmatales* members has been reported in invertebrate hosts. Finally, "*Candidatus* Phytoplasma" species cannot be grown in axenic culture; therefore, their requirement for sterols or lack thereof cannot be assessed at this time

distributed among three groups, spiroplasma, pneumoniae, and hominis. The *Mycoplasma* genus is not monophyletic, and its type species *Mycoplasma mycoides* subsp. *mycoides* belongs to the spiroplasma group along with *Spiroplasma*, *Entomoplasma*, and *Mesoplasma* spp., which obviously is of some concern for taxonomists. The topology reported initially from the 16S rRNA sequences with a cluster of two groups, spiroplasma and pneumoniae, branching with the hominis group, has been challenged by using improved methods of phylogeny (Johansson and Pettersson 2002) (➋ *Fig. 39.2*) or by using sequences other than 16S rRNA such as concatenated genes from genome sequences (Sirand-Pugnet et al. 2007a). In these topologies, which are better supported by statistical analyses, the two groups pneumoniae and hominis cluster as a sister clade of the spiroplasma group. The hominis group only includes *Mycoplasma* species, whereas the two other groups include other genera such as *Spiroplasma* and *Mesoplasma* for the spiroplasma group and *Ureaplasma* for the pneumoniae group. Subsequent analyses of other bacteria revealed that the uncultured genera *Haemobartonella* and *Eperythrozoon*, previously classified as rickettsiae (order *Rickettsiales*) because of their small size, obligate parasitism and association with red blood cells, are actually members of the pneumoniae group, and are being renamed accordingly (Neimark et al. 2001). The new taxonomic position of these uncultured bacteria, colloquially referred to as "hemoplasmas" or "hemotropic mycoplasmas," has since been confirmed by comparative genomics (Barker et al. 2011; Messick et al. 2011).

Although the sequencing of 16S rRNA genes provided a great leap in understanding the evolution of the *Mollicutes*, and it is still used for identification of *Mycoplasma* species, it shows some limits in particular within specific groups of species and also in building trees with statistically supported branching at the group level (see above). For these reasons, a number of other conserved genes have been used for phylogenetic studies. As the five species included in the *M. mycoides* cluster show very little variation within their 16S rRNA sequences, other genes such as housekeeping genes (*fusA*, *glpQ*, *gyrB*, *lepA*, and *rpoB*) have been used to infer precise taxonomic and phylogenetic relationships within this cluster (Manso-Silvan et al. 2007). The potential difficulty of selecting conserved genes in phylogenetic analyses comes from the recognition that lateral gene transfer (LGT) was a major force in bacterial evolution and that very few genes would have been spared by these transfers. However, there is now a general consensus to recognize that LGT does not hamper phylogenetic reconstructions (Ochman et al. 2005). The use of carefully selected DNA markers and the development of phylogenetic methods have allowed the reconstruction of a tree of life that includes most of the important phyla (Ciccarelli et al. 2006). In this tree of life, the long branches of the *Mycoplasma* species confirmed Woese's foresight (Woese et al. 1984) that these organisms evolved at great speed from their last ancestors shared with the *Firmicutes*; among mycoplasmas, the hemoplasmas are the organisms for which the tree branches are the longest, which suggests that they have evolved at an even faster rate than other mycoplasmas. An increased number of *Mycoplasma* species with sequenced genomes allowed construction of trees with increased statistical strength (Sirand-Pugnet et al. 2007a). Genomic sequences now also provide the means of reconstructing the natural history of divergence within a single species such as *M. mycoides* subsp. *mycoides* SC (Dupuy et al. 2012) or following the genetic changes associated with an emerging epidemics as was performed for *Mycoplasma gallisepticum* (Delaney et al. 2012).

## Isolation, Identification, and Typing

The reputation of mycoplasmas being fastidious organisms comes mainly from the real difficulties in primary isolation from clinical samples. In these conditions, there are several factors, including the presence of inhibitors in the sample, the low number of mycoplasma cells, and the suitability of the culture medium that can hamper the success of isolation (Tully 1995). In addition, culturing certain species, including *Mycoplasma genitalium*, requires considerable effort and expertise, leaving only a few laboratories in the world that still try to recover isolates from clinical samples (Hamasuna et al. 2007). Most of these difficulties are species specific. A textbook case is that of *Mycoplasma hyorhinis* isolation from contaminated cell cultures. *M. hyorhinis* cultivar alpha strains that are adapted to growth in cell cultures were found to be non-cultivable on standard mycoplasma media because they are sensitive to high levels of inhibition activity

■ Fig. 39.2 (continued)

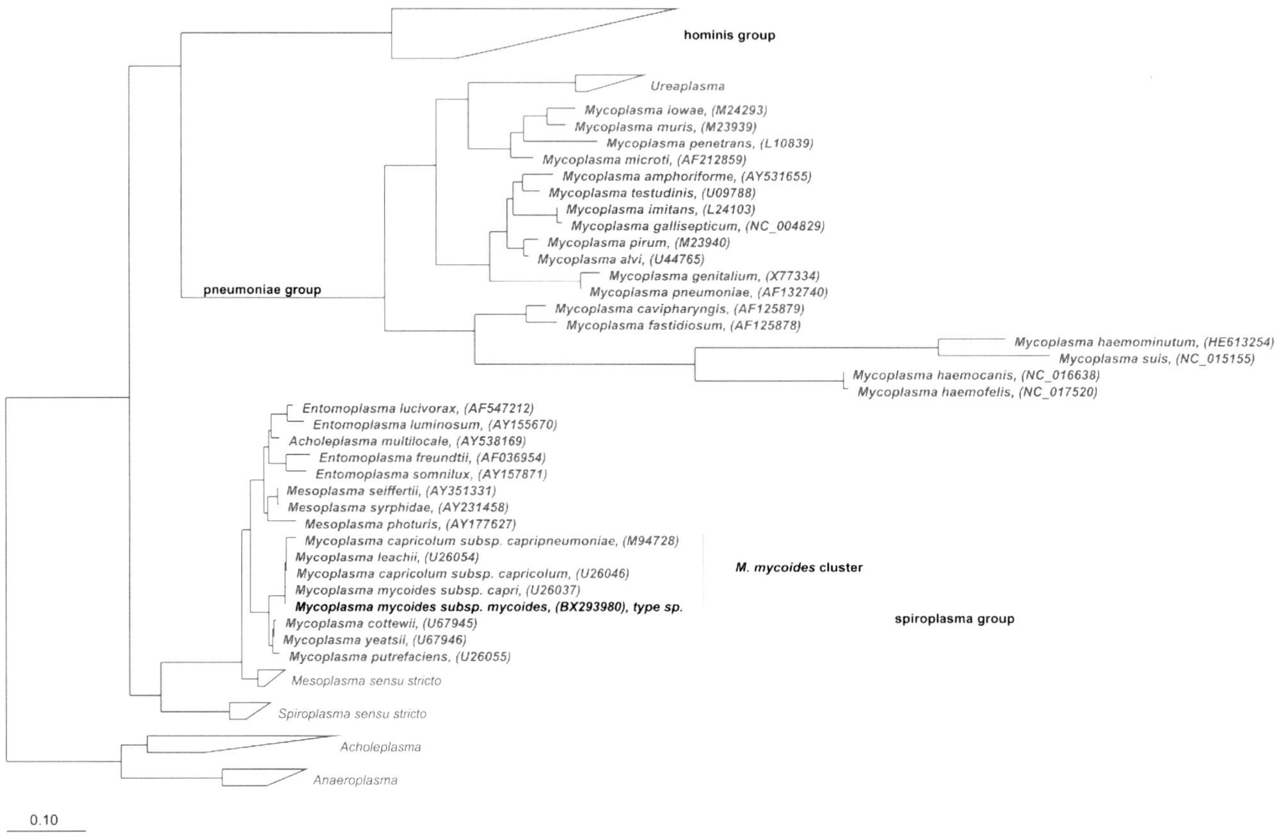

**◘ Fig. 39.2**

(a) **Phylogenetic reconstruction of the family *Mycoplasmataceae* based on 16S rRNA and created using the maximum likelihood algorithm RAxML** (Stamatakis 2006). The sequence datasets and alignments were used according to the All-Species Living Tree Project (LTP) database (Yarza et al. 2008; http://www.arb-silva.de/projects/living-tree). Representative sequences from closely related taxa were used as outgroups. In addition, a 40 % maximum frequency filter was applied in order to remove hypervariable positions and potentially misplaced bases from the alignment. Scale bar indicates estimated sequence divergence. The three phylogenetic groups (hominis, pneumoniae, and spiroplasma) and the *M. mycoides* cluster are indicated. The phylogenetic position of species belonging to the hominis group is indicated on the next figure that was obtained using the same experimental conditions, see section on "➋ Phylogeny". (**b**) Phylogenetic reconstruction of the hominis group belonging to the family *Mycoplasmataceae* based on 16S rRNA and created using the maximum likelihood algorithm RAxML (Stamatakis 2006)

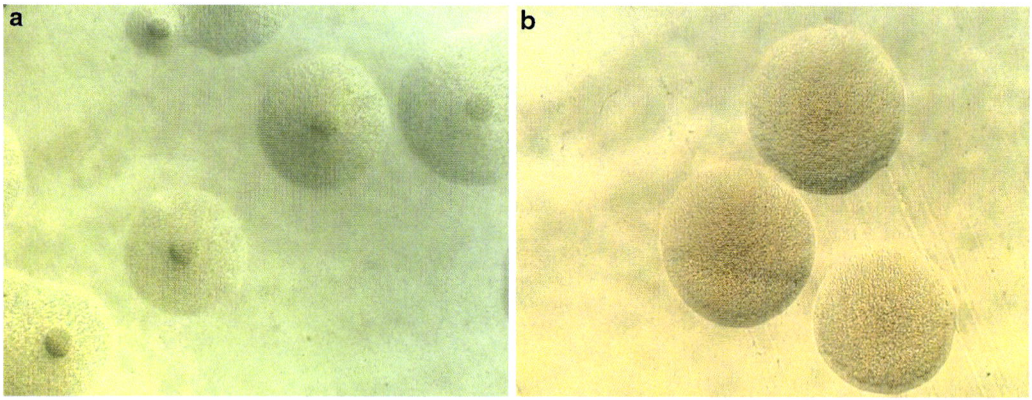

◘ Fig. 39.3
**Typical *Mycoplasma* colony morphology.** Images of single colonies from two *Mycoplasma* strains are shown. Colonies in *panel A* (*Mycoplasma mycoides* subspecies *capri* strain GM12) display the classic "fried egg" morphology, and colonies in *panel B* (*Mycoplasma mycoides* subspecies *capri* strain GM684) display umbonate morphology. Colonies were approximately 1.5 mm in diameter and were imaged using a Sargent-Welch light microscope (4× objective lens). The distinct differences in morphology between strains demonstrate the plasticity of this phenotype

by medium components (Gardella and Del Giudice 1995). Therefore, instead of requiring more nutrients than those provided in the standard mycoplasma medium, the use of a less rich medium such as modified CMRL-1066 liquid medium resulted in the growth of these specific strains.

With the difficulties taken into account, most *Mycoplasma* strains grow quite well once adapted to laboratory conditions. Instead of turbidity for most bacteria, the usual indicator of mycoplasma growth is a change of pH, most often recorded as a change of the color of the phenol red as pH indicator. The use of microscopy to check cultures is usually not performed because of the minute size of mycoplasma cells; microscopic observation is used to exclude overgrowth of mycoplasmas by other bacteria. On agar plates, the small size of the mycoplasma colonies requires the use of a stereo- or inverted microscope to observe fried egg-like colonies (*Fig. 39.3*).

## Culture Media

Mycoplasmas have numerous growth requirements, and for most of them, only undefined media are available. The quality of various peptones, sera, and yeast extracts that are included in the media must be carefully controlled, and each new batch must be validated by performing a number of controls (Tully 1995).

The composition of standard media used for *Mycoplasma* species is indicated below; a wiki site from the International Research Programme on Comparative Mycoplasmology (http://openwetware.org/wiki/IRPCM:Mycoplasma_Methods) provides a platform to exchange methods commonly used in mycoplasmology, including the composition of some of the growth media.

**Modified Hayflick medium** is used for many *Mycoplasma* species:

| | |
|---|---|
| Heart infusion broth | 28.5 g |
| Deionized water | 900 mL |
| Fresh yeast extract (25 %, w/v) | 100 mL |
| Horse serum | 200 mL |
| Calf thymus DNA (0.2 %, w/v) | 12 mL |
| Penicillin G (20,000 U/mL) | 2.5 mL |
| Phenol red solution 1 % (w/v) | 4 mL |

Dissolve the heart infusion broth in deionized water and autoclave (121 °C for 20 min). Aseptically add the remaining sterile solutions. Adjust final pH to 7.8 with sterile NaOH solution. The original medium composition included thallium acetate, which is now generally omitted due to its high toxicity for humans and the environment. For specific mycoplasmas, the heart infusion broth is replaced by Difco PPLO broth (20 g $L^{-1}$). In order to improve the yield of fastidious mycoplasmas, the medium is often supplemented with glucose (50 %, w/v; add 10 mL $L^{-1}$), a mixture of vitamins and cofactors such as Isovitalex™ Enrichment (BD) (2 mL $L^{-1}$), and/or pyruvate (4 g $L^{-1}$). For some species such as *M. mycoides* subsp. *mycoides*, the addition of a buffer system based on *N*-[2-hydroxyethyl] piperazine-*N′*-[2-ethanesulfonic acid] (HEPES) has been reported to increase growth and survival (Waite and March 2001). The avian species *Mycoplasma synoviae* has an absolute requirement for 0.005 % (w/v) L-cysteine and nicotinamide adenine dinucleotide (Waite and March 2001). For solid medium, 0.8–1.0 % (w/v) of noble agar is added to the base broth component before autoclaving, and phenol red may be omitted.

**SP-4 Medium** (Tully et al. 1977)

Difco PPLO broth
Tryptone
Peptone
Deionized water
Phenol red solution 1 % (w/v)

This medium is used to grow the most fastidious mycoplasmas. Dissolve the ingredients and adjust pH to 7.8 with NaOH solution. Add ingredients below as sterile supplements.

CMRL 1066 tissue culture medium with glutamine (10×) 20 mL
Glucose 4 mL
Fresh yeast extract (25 %, w/v) 14 mL
Fetal bovine serum 68 mL
Yeastolate (4 % aqueous solution, Difco) 20 mL
Penicillin G (20,000 U/mL) 2 mL

Adjust final pH to 7.6–7.8 and to final volume of 400 mL. For solid medium, 0.8–1.0 % (w/v) of noble agar is added to the base broth component before autoclaving, and phenol red may be omitted.

Although a large number of *Mycoplasma* species are glucose fermenters, other species are non-fermentative and use arginine hydrolysis as a source of cellular energy; a particularly noteworthy case is *M. hominis* (Pereyre et al. 2009). Therefore, arginine must be added to the growth medium of these mycoplasmas and the pH adjusted to 6.0–6.5 as the hydrolysis of this amino acid will result in alkalinization of the medium. Most mycoplasmas are facultative anaerobes and usually favor an anaerobic or microaerophilic atmosphere for primary isolation.

The *Ureaplasma* species have a unique energy metabolism that relies on urea hydrolysis. Indeed, the ATP-generating system is coupled to urea hydrolysis by the cytosolic urease via an ammonia chemical potential (Smith et al. 1993). Therefore, urea must be provided in the culture medium.

**U9C Urea Medium** (Shepard and Lunceford 1976)

| Trypticase soy broth | 15.0 g |
|---|---|
| Magnesium chloride (MgCl$_2$·6H$_2$O) | 0.2 g |
| Yeast extract | 1.0 g |
| Deionized water | 900 mL |

Adjust pH to 5.5 and autoclave at 121 °C for 15 min. Add aseptically the following sterile solutions:

| Urea | 3 mL |
|---|---|
| L-cysteine HCl (2 %, w/v) | 5 mL |
| GHL tripeptide (20 mg mL$^{-1}$) | 1 mL |
| Horse serum | 100 mL |
| Penicillin G (100,000 U mL$^{-1}$) | 10 mL |
| Phenol red (1 % w/v) | 1 mL |

Adjust the final pH to about 6.0

This medium is recommended for detection of ureaplasmas in clinical samples. Another medium, bromothymol blue broth, has been proposed for the general cultivation of ureaplasmas of human and animal origin (Robertson 1978). The addition of a divalent cation (manganous sulfate or calcium chloride) in the agar medium allows the identification of ureaplasma colonies. The cations act as an indicator of the urease activity, and the ureaplasma colonies develop a deep brown or brown-black color.

The optimum temperature for mycoplasmas growth ranges from 25 °C to 40 °C. Most mycoplasmas colonizing homeothermic hosts grow at 37 °C. For mycoplasmas infecting cold-blood animals, the optimum temperature can be lower: 25 °C for *M. mobile* (fish), 30 °C for *Mycoplasma testudinis* (tortoise), and 30–34 °C for *Mycoplasma alligatoris* (alligator).

Partly defined medium have been reported by *Mycoplasma mycoides* subsp. *capri* and *M. capricolum* subsp. *capricolum* (Rodwell 1983). These media were developed on an empirical basis. Recently, large-scale analyses of the metabolism of *M. pneumoniae*, combining both in silico modeling and experimental evaluation of the produced metabolites, provided new means to define growth requirements for mycoplasmas (Yus et al. 2009). The composition of this medium also confirms the multiple requirements of mycoplasmas for growth; all the amino acids, some vitamins, precursors for nucleic acids, and lipids must be provided.

The composition of the defined medium for *M. pneumoniae* is indicated below:

| **Minerals** | |
|---|---|
| Na$_2$HPO$_4$ | 2 mM |
| NaCl | 100 mM |
| KCl | 5 mM |
| MgSO$_4$ | 0.5 mM |
| CaCl$_2$ | 0.2 mM |
| **Carbon sources** | |
| Glucose | 10 g L$^{-1}$ |
| Glycerol | 0.5 g L$^{-1}$ |
| **Vitamins** | |
| Spermine | 0.1 mM |
| Nicotinic acid | 1 mg L$^{-1}$ |
| Thiamin | 1 mg L$^{-1}$ |
| Pyridoxal | 1 mg L$^{-1}$ |
| Thioctic acid | 0.2 mg L$^{-1}$ |
| Riboflavin | 1 mg L$^{-1}$ |
| Choline | 1 mg L$^{-1}$ |
| Folic acid | 1 mg L$^{-1}$ |
| Coenzyme A/pantothenate | 1 mg L$^{-1}$ |
| **Bases** | |
| Guanine | 20 mg L$^{-1}$ |
| Uracil | 20 mg L$^{-1}$ |
| Thymine | 10 mg L$^{-1}$ |
| Cytidine | 20 mg L$^{-1}$ |
| Adenine | 20 mg L$^{-1}$ |

| **Lipids** | |
|---|---|
| Cholesterol | 20 mg $L^{-1}$ |
| Palmitic acid | 10 mg $L^{-1}$ |
| Oleic acid | 12 mg $L^{-1}$ |
| Linoleic acid | 10 mg $L^{-1}$ |
| BSA (fatty acid free; carrier) | 2 g $L^{-1}$ |
| **Amino acids** | |
| Alanine | 4 mM |
| Arginine | 4 mM |
| Asparagine | 4 mM |
| Cysteine | 4 mM |
| Glutamine | 4 mM |
| Glycine | 4 mM |
| Histidine | 4 mM |
| Isoleucine | 4 mM |
| Leucine | 4 mM |
| Lysine | 4 mM |
| Methionine | 8 mM |
| Phenylalanine | 1 mM |
| Proline | 4 mM |
| Serine | 1 mM |
| Threonine | 4 mM |
| Tryptophan | 0.5 mM |
| Tyrosine | 0.5 mM |
| Valine | 8 mM |
| **Peptides** | |
| Peptone | 2.5 g $L^{-1}$ |
| **Others** | |
| HEPES | 50 mM |
| Penicillin | 1,000 U $mL^{-1}$ |
| Phenol red | 2.5 mg $L^{-1}$ |

(0.4 % final EtOH with lipids)
PH was adjusted to pH 7.8 with NaOH

## Freezing and Other Methods to Preserve Cultures

Mycoplasma cultures usually survive very poorly unless stored under special conditions. Broth cultures and agar plates can be kept for a few days or weeks at 4 °C; the addition of HEPES in the growth medium has been shown to increase the survival of mycoplasma cultures (Waite and March 2001). For longer preservation, it is necessary either to freeze the mycoplasma cultures or to lyophilize them. For prolonged periods of time, temperatures lower than −70 °C are preferable than the range −20 °C to −30 °C. The percentage of surviving cells is increased by the addition of dimethyl sulfoxide or glycerol as cryoprotective agents (Raccach et al. 1975). For freeze-drying, there is no evidence that a particular method will provide better results. In the specific cases of the preservation of vaccine strains, there are some data indicating that an excipient containing trehalose could protect the *M. mycoides* subsp. *mycoides* viability during the freeze-drying process (Litamoi et al. 2005).

## Detection and Identification of Mycoplasmas by PCR Assays

As is the case for most bacteria that are fastidious to grow, PCR-based assays were developed quickly for the detection and the identification of mycoplasmas and ureaplasmas. The first one, for the detection of *M. pneumoniae* in 1989 (Bernet et al. 1989), was followed by several other ones (Markham and Noormohammadi 2005; Waites et al. 2012). Different genes were targeted by PCR assays, including the 16S rDNA, which allowed development of assays with different levels of specificity (van Kuppeveld et al. 1992). PCR offers a great sensitivity for detection that is usually much greater than culture for most mycoplasmas in the clinical setting. PCR assays have also been developed for some mycoplasmas that present specific problems of identification; this is the case of *M. mycoides* subsp. *mycoides*, which belongs to a cluster of closely related species or subspecies (Dedieu et al. 1994) and for the human *Ureaplasma* spp. (Vancutsem et al. 2011).

A new generation of real-time PCR assays (qPCR) that alleviate most of the contamination problems is now widely used. These assays are essential for the most fastidious of the mycoplasmas, the uncultivated hemoplasmas (Barker et al. 2010). In the clinical setting, some of the commercial qPCR developments allow the detection of clinically relevant mycoplasmas such as *M. pneumoniae* in laboratories without experience in mycoplasmology (Dumke and Jacobs 2009; Touati et al. 2009).

The genetic PCR-based amplification allows not only detecting the presence of a specific mycoplasma, but it can also target antibiotic resistance genes. Therefore, PCR assays have been developed to detect the genetic determinants associated with antibiotic resistance in these bacteria. This includes the detection of *tetM* associated with the tetracycline resistance (Blanchard et al. 1992) and of mutations in the rRNA genes associated with macrolide resistance (Jensen 2012; Peuchant et al. 2009).

## Typing Methods for Differentiating *Mycoplasma* Species or Strains

Typing bacteria is of utmost importance for diagnosis, treatment, and epidemiological surveillance (Li et al. 2009). As phenotypic markers that distinguish *Mycoplasma* strains or even related species are extremely limited, molecular DNA methods were developed as early as 1982 using probes from the ribosomal RNA operon (ribotyping) (Amikam et al. 1982). Several DNA banding pattern-based methods, which classify bacteria according to the size of fragments generated by amplification and/or enzymatic digestion of genomic DNA, have been developed. Some of these methods evaluate the genetic diversity at the genome scale such as by analysis of genomic restriction

fragments separated by pulse-field gel electrophoresis (PFGE) (De la Fe et al. 2012) or by amplified-fragment length polymorphism (AFLP) (Kokotovic et al. 1999). Included among the DNA banding pattern methods is multilocus variable number tandem-repeat analysis, which uses the information provided by whole-genome sequencing of bacterial species to enable selection of regions encompassing short sequence repeat (SSR) motifs that are known to undergo frequent variation in the number of repeated units. Variation in the length of these DNA repeats is the basis for differentiating strains of the same species (Lindstedt 2005). MLVAs have been developed for several mycoplasmas including *M. genitalium* (Cazanave et al. 2012; Ma et al. 2008), *M. pneumoniae* (Dumke and Jacobs 2011), and *Mycoplasma bovis* (Pinho et al. 2012).

Other typing methods are based on the sequence polymorphism of specific DNA sequences. For several significant bacterial pathogens, multilocus sequence typing (MLST; Enright and Spratt 1999) which is based on polymorphisms detected at the level of conserved genes (so-called housekeeping genes) is considered the method of choice. It has several advantages, including the possibility of comparing results with those stored by other laboratories in an online database (http://www.mlst.net/) and of using dedicated methods and software for analysis (Spratt et al. 2004). However, mycoplasma typing by MLST has been limited to a few species, including *M. hyopneumoniae* (Mayor et al. 2008), because most of the housekeeping genes in mycoplasmas show very limited or no variability. An MLST approach has been also used to evaluate the phylogenetic relationships among the members of the *M. mycoides* cluster. Using this method to type more than 100 strains of this cluster, it was found that the establishment and spread of the cluster occurred about 10,000 years ago, which coincides with the origin of livestock domestication (Fischer et al. 2012). In order to overcome the limitations of MLST with mycoplasmas showing little genetic variation in housekeeping genes, multilocus sequence analysis (MLSA) was developed. In MLSA, other genes showing a higher degree of genetic variation are included in the analysis. The target genes include those encoding surface antigens or adhesins. MLSA was developed for several mycoplasmas including *M. mycoides* subsp. *mycoides* (Lorenzon et al. 2003), *M. capricolum* subsp. *capripneumoniae* (Manso-Silvan et al. 2011), and *M. agalactiae* (McAuliffe et al. 2011).

## Genetics, Cell Biology, and Physiology

### Minimal Genomes

In line with the minute size of their cells, the mycoplasma genomes are among the smallest among living organisms, with a range from 580 kbp for *M. genitalium* to 1,359 kbp for *M. penetrans* (● Table 39.1). Overall, comparative genomics strongly suggests that most of the gene losses from their common ancestor with *Firmicutes* occurred at an early stage of evolution. Indeed, all mycoplasmas share common features such as the lack of a cell wall and the inability to synthesize amino acids. However, ongoing gene loss can be observed by comparing gene sets within each of the phylogenetic branches (Sirand-Pugnet et al. 2007a). The *M. genitalium* genome is considered to be the best illustration of the concept of a minimal cell (Koonin 2000), which has generated a lot of interest. As this mycoplasma encodes proteins that play a role in the interaction with its host, its genome could be further downsized. With this aim in mind, the essential genes of this organism have been identified by inactivating individual genes using random transposon mutagenesis (for review, see Juhas et al. 2011); about 100 genes out of the 482 genes encoding proteins can be inactivated, which suggests that a minimal genome would contain ~350–400 genes. There are two principal ways to build such a genome, either by iterative removal of nonessential genes or by building a cassette-based artificial chromosome. This latter approach was followed by investigators at the J. Craig Venter Institute and resulted in the complete chemical synthesis, assembly, and cloning of the *M. genitalium* genome (Gibson et al. 2008). The next step was to transplant a version of this genome that had been modified using yeast genetic tools back into a recipient cell using the method of genome transplantation developed by the same group for mycoplasmas of the *M. mycoides* cluster (Lartigue et al. 2007). However, genome transplantation could not be achieved with *M. genitalium* and related species, which led to the choice of *M. mycoides* subsp. *capri* as the species from which the trimming of nonessential genes could be started (Gibson et al. 2010). The methods developed in these studies constitute part of the nascent field of synthetic biology (Montague et al. 2012) and offer a great opportunity for mycoplasmology. Indeed, the cloning of mycoplasma genomes in yeast offers the possibility to modify the chromosome at a level that was previously unimagined given the limited genetic tools for mycoplasmas (Halbedel and Stulke 2007). One bolt that remains to be unlocked is extension of the genome transplantation methods to other *Mycoplasma* species.

### Horizontal Gene Transfer and Mobile Genetic Elements

As the main characteristic of the mycoplasma evolution was genome downsizing, it overshadowed for a while the possibility that gene gain by horizontal gene transfer (HGT) could also be a significant component in their evolution (for review, see Sirand-Pugnet et al. 2007a). HGT events between several *Mycoplasma* species that share a common host have now been documented. The highest level of HGT events was found between *Mycoplasma* species that colonize ruminants. Indeed, it was predicted that 18 % of the *M. agalactiae* genome had undergone HGT with mycoplasmas of the *M. mycoides* cluster (Sirand-Pugnet et al. 2007b). The recognition of these genetic transfers brings new questions about the nature of the mobile genetic elements (MGE) that could mediate HGT. Compared to other bacterial species, there are far fewer MGEs in mycoplasmas. The main MGE categories that have been found

### Table 39.1
Main characteristics of representative mycoplasma genomes[a]

| Phylogenetic group | Mycoplasma species | Strain | Natural host | Genome size (Mbp) | % G+C | #CDS | # proteins | Cultivability | Gen bank acc. numb. |
|---|---|---|---|---|---|---|---|---|---|
| Spiroplasma | M. leachii | PG50 | Ruminant | 1.009 | 23.7 | 940 | 905 | Yes | CP002108.1 |
| | M. mycoides subsp. mycoides | PG1 | Ruminant | 1.212 | 24.0 | 1,053 | 1,017 | Yes | BX293980.2 |
| | M. mycoides subsp. capri | 95010 | Ruminant | 1.154 | 23.8 | 962 | 922 | Yes | FQ377874.1 |
| | M. capricolum subsp. capricolum | ATCC27343 | Ruminant | 1.010 | 23.8 | 867 | 812 | Yes | CP000123.1 |
| Hominis | M. arthritidis | 158 L3-1 | Rodent | 0.820 | 30.7 | 671 | 631 | Yes | CP001047.1 |
| | M. hominis | PG21 | Human | 0.665 | 27.1 | 577 | 523 | Yes | FP236530.1 |
| | M. mobile | 163 K | Fish | 0.777 | 25.0 | 667 | 633 | Yes | AE017308.1 |
| | M. hyorhinis | HUB-1 | Pig | 0.840 | 25.9 | 711 | 658 | Yes | CP002170.1 |
| | M. hyopneumoniae | 232 | Pig | 0.893 | 28.6 | 727 | 691 | Yes | AE017332.1 |
| | M. pulmonis | UAB CTIP | Rodent | 0.964 | 26.6 | 813 | 782 | Yes | AL445566.1 |
| | M. synoviae | 53 | Poultry | 0.799 | 28.5 | 715 | 659 | Yes | AE017245.1 |
| | M. crocodyli | MP145 | Crocodilian | 0.934 | 26.9 | 763 | 689 | Yes | CP001991.1 |
| | M. fermentans | JER | Human | 0.977 | 26.9 | 866 | 797 | Yes | CP001995.1 |
| | M. bovis | PG45 | Ruminant | 1.003 | 29.3 | 868 | 765 | Yes | CP002188.1 |
| | M. agalactiae | PG2 | Ruminant | 0.877 | 29.7 | 792 | 742 | Yes | CU179680.1 |
| Pneumoniae | M. penetrans | HF2 | Human | 1.359 | 25.7 | 1,069 | 1,037 | Yes | BA000026.2 |
| | M. pneumoniae | M129 | Human | 0.816 | 40.0 | 732 | 688 | Yes | U00089.2 |
| | M. genitalium | G37 | Human | 0.580 | 32.0 | 524 | 482 | Yes | L43967.2 |
| | M. gallisepticum | R (Low) | Poultry | 0.996 | 31.4 | 817 | 763 | Yes | AE015450.2 |
| | M. haemofelis | Langford | Cat | 1.147 | 38.8 | 1,58 | 1,545 | No | FR773153.2 |
| | M. haemocanis | Illinois | Dog | 0.920 | 35.3 | 1,191 | 1,156 | No | CP003199.1 |
| | M. suis | KI3806 | Pig | 0.709 | 31.1 | 844 | 794 | No | FQ790233.1 |
| | Ureaplasma parvum | ser 3 ATCC 700970 | Human | 0.75 | 25.5 | 653 | 614 | Yes | AF222894.1 |
| | Ureaplasma urealycum | ser 10 ATCC 33699 | Human | 0.87 | 25.8 | 695 | 646 | Yes | CP001184.1 |
| Acholeplasma | Acholeplasma laidlawii | PG-8A | Ubiquitous | 1.497 | 31.9 | 1,433 | 1,38 | Yes | CP000896.1 |

[a]Additional information and tools for comparative genomics are found on MolliGen (http://molligen.org), the genomics database dedicated to *Mollicutes*

in mycoplasmas include insertion elements (IS), integrative conjugal elements (ICE), bacteriophage, and plasmids. Though they are ubiquitous, mycoplasma MGEs are heterogeneously distributed within genera and even within species. The IS elements are extremely abundant in some species such as *M. mycoides* subsp. *mycoides* strain PG1, in which they account for 13 % of the genome (Westberg et al. 2004). The distribution and number of these IS elements also account for the genome plasticity found within a species (Bischof et al. 2006). Until recently, only a few plasmids were described in the *Mycoplasma* genus, which includes over 100 species. A new study has shown that several species of ruminant mycoplasmas carry plasmids that are members of a large family of cryptic replicons and replicate via a rolling circle mechanism (Breton et al. 2012). ICEs have been proposed to account for lateral gene flow in the prokaryotes (Wozniak and Waldor 2010). In mycoplasmas, the first ICEs were described in *M. fermentans* (Calcutt et al. 2002) and in *Mycoplasma agalactiae* (Marenda et al. 2006). It was shown that circular forms could be detected, and the heterogeneity of their distribution among strains of the same species suggested a role in genomic variation. The ability of these elements to contribute to genetic exchanges that shape the mycoplasma evolution remains to be determined.

## Morphology

The absence of a cell wall potentially leads to the expectation that mycoplasma cells tend toward having no regular shape, despite the fact that this is not the case for animal cells. Indeed, mycoplasmas are frequently described as "pleomorphic," a term which indicates a certain degree of flexibility in shape but can also be interpreted to suggest the absence of a regular shape. Moreover, published images of mycoplasma cells often include a wide range of shapes, which could potentially be attributed to real differences in shape from cell to cell, or equally to artifacts introduced during processing of these delicate organisms. Amid all this uncertainty, there are some matters that are clear: first, the ability of at least mycoplasma cells within a population to pass through 0.2-μm filters belies a certain degree of physical flexibility and second, different microscopy techniques often reveal different information about mycoplasma cell morphology, raising questions of which are the most appropriate techniques for describing cell shape.

Although some interesting cellular characteristics, including aspects of morphology, can be uncovered through light microscopy of unfixed mycoplasma cells, electron microscopy offers substantially higher resolution and is therefore a more commonly employed source of morphological information. Of the two principal forms of this technique, transmission electron microscopy (TEM) has the potential to offer somewhat higher magnification and can be used for both whole cells and sections. On the other hand, scanning electron microscopy (SEM) uniquely results in images with depth that provides a three-dimensional appearance. However, as an illustration of the weakness of TEM, a widely used tool to describe the morphological features of newly described mycoplasmas, *M. testudinis* was originally described as pseudococcoidal based on TEM images (Hill 1985), whereas SEM later revealed this organism to have a distinct polarized shape (Hatchel and Balish 2008). Nonetheless, both of these standard electron microscopy techniques provide valuable and often complementary information (➋ *Fig. 39.4*). Atmospheric SEM is variation of SEM that has been used to obtain images of mycoplasmas in an aqueous environment and allows visualization of the interior of the cell as well as permitting visualization of immunocytochemical reagents (Sato et al. 2012). However, this technique is not widely available. Both standard TEM and SEM also require dehydration of the cells, which, in addition to the cross-linking associated with fixation, can introduce significant artifacts. An interesting compromise is cryo-electron tomography (CET), a TEM technique that avoids dehydration and fixation, presumably preserving the native structure of the sample to a much greater degree than the other techniques, and captures a tilt series through a whole cell which can be used to reconstruct a three-dimensional image of the cell (Milne and Subramaniam 2009). Limitations of CET include the inherently incomplete reconstruction of the cell that accompanies a finite range of tilt, and the expense and comparative scarcity of CET equipment and expertise.

A wide range of shapes is represented among *Mycoplasma* species. Many species appear coccoidal or pseudococcoidal, although whether the irregularities in shape suggested by the latter category result from processing artifacts is unclear. Few of these have been imaged by SEM, but unpublished images of several *Mycoplasma* species of the hominis group suggest that this morphology is widespread across this clade, and *M. mycoides* subsp. *capri* JCVI-syn1.0, the product of a synthetically created genome based on that of *M. mycoides* subsp. *capri*, also appears to have this shape (Gibson et al. 2010). Within the pneumoniae group, hemoplasmas are suggested to be coccoidal or pseudococcoidal as well, sometimes flattened,

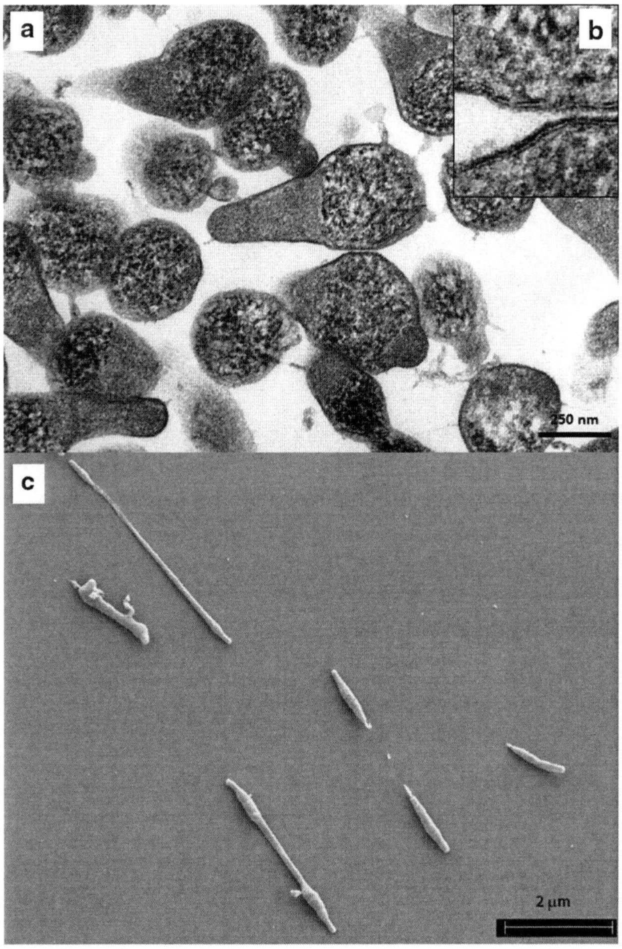


**Fig. 39.4**
**Comparison of electron microscopic techniques for visualization of *Mycoplasma penetrans* (a, b) Transmission electron micrograph of thin section of *Mycoplasma penetrans* cells. After glutaraldehyde fixation, the mycoplasma pellets were washed with a buffer containing ruthenium red before post-fixation in osmium tetroxide (Neyrolles et al. 1998). The ruthenium red allowed visualization of the thin capsule (~10 nm) that surrounds the mycoplasma cells. (b) Higher-magnification view of the mycoplasma envelope showing the trilamellar structure of the plasma membrane and the fuzzy thin capsular material. (c) Scanning electron micrograph of *Mycoplasma penetrans* cells grown on a glass cover slip. After fixation in glutaraldehyde and formaldehyde, cells were ethanol dehydrated, critical point-dried, and coated with 15 nm of gold. Some cells are connected by filaments associated with the cell division process. The cell poles are terminal organelles associated with both adherence and motility**

based on images of these organisms in association with red blood cell surfaces (Groebel et al. 2009; Willi et al. 2011). The morphology of ureaplasmas has been poorly described but might also fall into this category. On the other hand, TEM and/or SEM suggests a bacillar, perhaps twisted morphology for the three species of the *M. fastidiosum* cluster within the pneumoniae group (Bolca Topal et al. 2007; Hill 1984; Lemcke and Pland 1980). Finally, several groups of *Mycoplasma* species within both the pneumoniae and hominis groups have polarized cell structures with visually, compositionally, and functionally distinct cell poles. Within the pneumoniae group, these include all the species of the *M. pneumoniae* cluster (Hatchel and Balish 2008) and at least some species of the *Mycoplasma muris* cluster (Jurkovic et al. 2012). Species of the hominis group with this polarized morphology include members of the *Mycoplasma sualvi* and *M. pulmonis* clusters (Kirchhoff et al. 1984). In most of the species with this type of morphology, the distinctive cell pole, called the terminal organelle, attachment organelle, or headlike structure, is associated with adherence and gliding motility (Balish 2006); one pole is also associated with adherence and motility in the rod-shaped *Mycoplasma insons*, even though how this pole is distinct from the opposite pole has not yet been determined (Relich et al. 2009). The appearance of some species tends toward a filamentous network, at least sometimes due to the presence of division intermediates, such as in *M. pneumoniae* (Hasselbring et al. 2006) and *M. penetrans* (Jurkovic et al. 2012), and no doubt in some cases due to processing-associated artifacts.

## Cytoskeletal Elements

The best-characterized components of the cytoskeletons of bacterial cells, namely, the actin-related MreB and tubulin-related FtsZ, are associated functionally with modeling of cell shape via interactions with peptidoglycan synthesis enzyme complexes (Typas et al. 2011). The absence of a cell wall in the *Mycoplasmatales* is associated with the absence of the cell elongation-associated cytoskeletal proteins MreB and MreC (Balish and Krause 2006). Interestingly, other than in ureaplasmas, hemoplasmas, and *M. mobile*, mycoplasmas retain the cytoskeletal protein FtsZ (Balish and Krause 2006). In other bacteria, FtsZ, which forms transient polymers at the cell division site, has two distinct roles in cell division. One, which has no place in mycoplasmas, is as a scaffold for the assembly of peptidoglycan synthesis proteins at the appropriate place and time for cell division (Typas et al. 2011). The second is the GTP hydrolysis-dependent pinching in of the cell membrane at the same location, which results in cell division because of the spatially coordinated synthesis of peptidoglycan on the external side of the membrane, keeping the membrane from snapping back (Mingorance et al. 2010). In the absence of the accompanying peptidoglycan synthesis ratchet, it is unclear how FtsZ can promote cell division in most *Mycoplasma* species; however, in *M. genitalium*, FtsZ and adherence and/or gliding motility, which requires adherence, operate in the same pathway (Lluch-Senar et al. 2010 and see below). FtsZ-mediated membrane constriction might cooperate with cell motility to divide cells. Interestingly, in motile *Mycoplasma* species, FtsZ tends to be either highly divergent in sequence or, in the case of *M. mobile*, altogether absent, suggesting that motility partly or completely substitutes for FtsZ in cell division. In all but the *Mycoplasma neurolyticum* cluster, a highly divergent gene apparently derived from *ftsA*, which encodes a protein required for interaction of FtsZ with the cell membrane, is present immediately upstream of *ftsZ* (unpublished observations). FtsZ from *M. pulmonis*, one of the few non-divergent such proteins from a motile organism, is capable of substituting for *E. coli* FtsZ, but the divergent *M. pneumoniae* FtsZ is not (Osawa and Erickson 2006).

Bacteria also contain cytoskeletal filaments consisting of members of the ParA/MinD family, which are ATPases that, either through oscillation mediated by depolymerization and repolymerization or by interaction with other proteins, promotes the appropriate positioning of molecules or structures with which they interact (Lutkenhaus 2012). MinD, an important regulator of FtsZ positioning in many bacteria, is absent from the mycoplasmas. ParA/Soj is involved in positioning plasmids and segregating both chromosomes and organelles; its interaction with centromeres on DNA molecules is mediated by ParB/Spo0J (Mierzejewska and Jagura-Burdzy 2012). Whereas a *parA*/*soj* gene is present near the origin of replication in all mycoplasma genomes analyzed, suggesting involvement in chromosome segregation, no *parB*/*spo0J* homolog has been identified. However, the *parA*/*soj* gene is always located upstream of a hypothetical gene, suggesting that this gene may fulfill the same cellular role as *parB*/*spo0J*. No analysis of mycoplasma ParA/Soj has been reported.

The best-studied mycoplasma cytoskeletal elements are those that are mycoplasma specific and associated with terminal structures, particularly that of *M. pneumoniae*. The electron-dense core of the *M. pneumoniae* attachment organelle has been analyzed by various TEM-related methods, including CET (Henderson and Jensen 2006; Seybert et al. 2006), and its composition has been inferred from both proteomic analysis (Catrein and Herrmann 2011) and study of mutants that fail to build either structurally normal cores or any cores at all (Malandain 2004). It consists of a set of about ten proteins that are idiosyncratic to the *M. pneumoniae* cluster, with only one having any meaningful homology to known proteins, namely, TopJ, which includes a J domain whose putative co-chaperone activity is important for normal disposition of the rest of the core as well as for normal folding of the adhesin P1 (Cloward and Krause 2010, 2011). The rest of the cytoskeletal proteins of the *M. pneumoniae* core are proteins, mostly fairly large, each with some combination of alpha-helical coiled coil regions and acidic proline-rich (APR) domains whose characteristic amino acid composition but not sequence or size are conserved. Several of those with the latter also include EAGR (enriched in aromatic and glycine residues) boxes, uniquely folded domains that might be involved in protein-protein interactions required for normal gliding motility (Calisto et al. 2012). A putative order of assembly based on the stabilization and localization of each of these proteins in a series of well-characterized mutants (Cloward and Krause 2009; Hasselbring and Krause 2007; Krause et al. 1982; Willby and Krause 2002) has been proposed (Malandain 2004), but mapping of them onto this elongated structure, which is based principally on immunofluorescence and green fluorescent protein fluorescence microscopy, is fairly general, with only proximal and distal regions of the core having been assigned to each (Seto and Miyata 2003), and two different models proposed for the orientation of protein HMW2 (Balish et al. 2003; Bose et al. 2009). There are also presumably interactions between these proteins and the transmembrane adhesins that localize to the attachment organelle. For example, localization of the P1 adhesin is associated specifically with cytoskeletal proteins HMW1 and HMW2 (Balish et al. 2003), and mutual stabilization of the P30 adhesin and the cytoskeletal proteins HMW3 and P65 further suggests physical interactions (Hasselbring et al. 2012; Willby et al. 2004). Structures consistent with duplication of the cytoskeletal structure using a preexisting structure as a template have been observed in *M. gallisepticum* (Nakane and Miyata 2009).

Although similar structures and proteins are found in other members of the *M. pneumoniae* phylogenetic cluster, these proteins are absent in other *Mycoplasma* species with terminal organelles, and the cytoskeletal elements that occupy these structures are structurally dissimilar to those of the *M. pneumoniae* cluster. Among these, the cytoskeletons within the terminal structures of *M. mobile*, *M. penetrans*, and *M. iowae* have been visualized (Jurkovic et al. 2012, 2013; Nakane and Miyata 2007). Some proteins of the *M. mobile* structure have been identified; although two are related to components of the membrane ATPase, these proteins are not widespread among the mycoplasmas (Nakane and Miyata 2007). Rod-shaped *M. insons* has cytoskeletal filaments that run the length of the cells, but no evidence of a relationship with polarity has been demonstrated (Relich et al. 2009). It stands to reason that the terminal organelle and its accompanying cytoskeleton are homoplastic, having evolved independently in different lineages.

## Motility

Ten species of the pneumoniae group and two of the hominis group have been described as exhibiting gliding motility, defined as relatively smooth movement along a surface; unpublished data attest to the movement of one and two more from each respective group. In vitro gliding is observed by phase-contrast microscopy using cells attached to glass or plastic in an environmental chamber that allows for control of temperature. All the species that are motile exhibit polarized cell morphology, and all but *M. insons* have terminal organelles which are at the leading end during motility (Hatchel and Balish 2008; Relich et al. 2009). Likewise, only two species, *M. alvi* and *M. sualvi*, are established to exhibit polarized morphology but to be immotile (Hatchel and Balish (2008) and unpublished data). These data suggest that polarization is associated with motility and that motility may be secondarily lost.

Except for unidirectionality, which is universal among mycoplasmas, gliding characteristics differ considerably among mycoplasmas. At optimal temperature, which is species specific, average gliding speeds range from $\sim$30 to over 3,000 nm $s^{-1}$. Average speeds across strains within a species are also subject to considerable variation. Some species glide very persistently, and others quite discontinuously; whether this difference reflects physiologically significant differences in gliding mechanism or different degrees of species-specific optimization is unclear. Chemotaxis does not appear to play a role in the choice of direction; rather, mycoplasmas generally appear to glide in paths of ever-changing curves. Two species with distinctly curved terminal organelles, *M. genitalium* and *M. testudinis*, glide in broadly circular paths (Hatchel and Balish 2008).

Mechanisms for gliding differ from those observed in other bacteria, but they are unlikely to be the same across *Mycoplasma* species. A mechanism for carrying out gliding motility has been proposed for *M. mobile* (Miyata 2010), and the proteins involved are also present in *M. pulmonis* (Seto et al. 2005b). The proposed mechanism involves cyclic binding and release of an adhesin whose ligand is identical or closely related to a particular sialic acid modification on host cell surface proteins. Between binding and release, the adhesin undergoes a conformational change, one step of which is powered directly by ATP hydrolysis, which is carried out by a separate protein. As in *M. mobile*, adherence to a sialic acid protein modification is also required for motility of *M. pneumoniae*, although both the form of sialic acid (Kasai et al. 2013; Nagai and Miyata 2006) and the adhesins (Relich and Balish 2011; Seto et al. 2005a) involved appear to be entirely unrelated. Although it is unclear whether all mycoplasma gliding motility is associated with sialic acid binding, the ubiquity of sialic acid modifications on animal cell surfaces is likely to have provided a target for evolution of motility-associated adhesins at least two and perhaps multiple times during the history of mycoplasmas. Gliding of both *M. mobile* and *M. penetrans* is responsive to temperature, with faster movement at higher temperatures up to the point at which the cells are no longer capable of adhering to the surface (Jurkovic et al. 2013; Miyata et al. 2002); however, speeds of *M. pneumoniae* and *M. insons* are relatively temperature insensitive (Radestock and Bredt 1977; Relich et al. 2009). Although depletion of ATP results in somewhat reduced gliding speed for *M. penetrans*, it does not bring cells to a halt as for *M. mobile*, suggesting that although it contributes to some aspect of *M. penetrans* gliding motility, ATP does not directly power a conformational change in a motor protein as it does in *M. mobile* (Jurkovic et al. 2013). Thus, gliding, like terminal organelle structure, appears to have evolved independently in different mycoplasmas.

The motile behavior of mycoplasmas in association with host cells in vivo is not characterized, making it difficult to be certain about the role of motility. The relatively slow speed and absence of chemotaxis associated with mycoplasma motility make roles in evading immune cells or seeking nutrients unlikely. However, observation of both *M. pneumoniae* and *M. penetrans* supports a role for gliding motility in cell division, which is supported by the inability to generate non-adherent and therefore nonmotile mutants in *M. genitalium* cells engineered not to express the cell division protein FtsZ (Lluch-Senar et al. 2010). The observation of cell motility in vitro might reflect a force normally used to effectuate cell division, but unchecked by interactions with surfaces achieved only in vivo.

### Cell Membrane

The absence of a cell wall imposes certain demands on the mycoplasma cell membrane, particularly with regard to osmotic resistance. Presumably, the sheltered, relatively constant environment in which mycoplasmas are found contributes to the stability of these cells. Cells of the genera *Mycoplasma* and *Ureaplasma*, along with some of their wall-less relatives, additionally require incorporation of host- (or media-) derived sterols into their membranes in significant quantities, adding further osmotic resistance (Rottem 2002). Although the mechanisms by which mycoplasmas acquire cholesterol and cholesteryl esters are uncertain, the molar ratio of sterols to phospholipids can approach or exceed 1:1, and the ratio of cholesterol to cholesteryl esters is highly variable across species (Kornspan and Rottem 2012; Razin et al. 1982). Phospholipids themselves are also incorporated from the host (Rottem et al. 1986), making mycoplasma membranes compositionally much more similar to animal cell membranes than to membranes of other bacteria. Mycoplasma membranes are rich in sphingomyelin, phosphatidylcholine, or both, with the former being unmodified and the latter often being modified by either removal or addition of fatty acids (Rottem et al. 1986). Certain species selectively incorporate sphingomyelin (Hirai et al. 1992; Salman and Rottem 1995). Phosphatidylglycerol and/or cardiolipin are also present in mycoplasma membranes, the latter synthesized from the former, which is itself synthesized de novo (Rottem 1980). Some mycoplasmas contain carotenoids, also synthesized de novo (Maquelin et al. 2009). Glycolipids are abundant in mycoplasma membranes, with different species exhibiting different profiles (Kornspan and Rottem 2012). In *M. pneumoniae*, a glycosyltransferase involved in synthesis of the carbohydrate moieties of cell surface-associated carbohydrates has quite broad specificity, perhaps allowing synthesis of different kinds of carbohydrate chains under different conditions (Klement et al. 2007). The data concerning lipid diversity among mycoplasmas suggests that mycoplasmas use a relatively small number of enzymes to create, acquire, and/or modify a diverse array of lipids and lipid-containing molecules.

The protein milieu of mycoplasma membranes consists of both peripherally associated proteins and integral membrane proteins, including both transmembrane proteins and lipoproteins. More than half of the mycoplasma membrane by weight is protein, with at least a quarter of the genes of *Mycoplasma agalactiae* encoding membrane-associated proteins (Cacciotto et al. 2010). Lipoproteins constitute a significant fraction of mycoplasma membrane proteins. In common with the Firmicutes from which mycoplasmas evolved, several ligand-binding components of mycoplasma ABC transporters are lipoproteins, a necessity of the absence of a periplasmic space (Schmidt et al. 2007; Theiss and Wise 1997). Lipoproteins are also significant factors in immune evasion, with various mechanisms for phase variation present in different *Mycoplasma* species (see below). Although mycoplasma genomes also encode a large number of proteins with membrane insertion signal sequences and alpha-helical transmembrane domains, as well as the sec pathway for membrane insertion of these proteins, they lack *lepB* or *sipS* genes encoding signal peptidase I, normally involved in cleaving the signal sequence (Catrein et al. 2005), although genes similar to *sipS*, with unestablished functions, have been identified in a handful of *Mycoplasma* species (Moitinho-Silva et al. 2012). This leaves open the possibility that they rely on other proteins for this activity, possibly including signal peptidase II, which is normally used for removing signal peptides from lipoproteins.

## Habitats and Ecology of the *Mycoplasmataceae*

### Human and Animal Hosts

Mycoplasmas and ureaplasmas are obligate parasites and, by species definition, are found in vertebrate host animals. To date, there has been infection (natural or experimental) demonstrated by isolation of pure cultures in 96 different host species (Brown et al. 2011). An additional three hosts have been reported by molecular methods or microscopy of non-cultivatable hemotropic mycoplasmas (Messick et al. 2002; Stoffregen et al. 2006). The currently known host range consists of 49 mammalian species including humans, 39 avian species, 10 reptilian species, 1 piscine species, and no amphibian species. The overrepresentation of mammals and birds in the known host range likely reflects the intensity of study of these species rather than a natural predilection for homoeotherms, although this possibility cannot be fully excluded.

Recent metagenomic studies have indicated that the ubiquity of *Mollicutes* species across the spectrum of potential hosts may be much more widespread than is currently understood. The only known member of the *Mycoplasmataceae* with a piscine host habitat is *Mycoplasma mobile*; however, recent molecular examinations of the Atlantic mackerel (Svanevik and Lunestad 2011) and the Eastern oyster (King et al. 2012) found sequences of unnamed *Mollicutes* species in high relative abundance to other members of the microbiota. Additionally, a metagenomic study of ambrosia beetles identified three unnamed *Mycoplasma* species (Hulcr et al. 2012). Molecular detection of mycoplasmas in the Eastern oyster and the ambrosia beetle is particularly notable because members of the genus *Mycoplasma* are defined as having vertebrate hosts (Brown et al. 2011), with the exception of transient vector association (see section on "➋ Transmission"). As the microbiotas of additional animal species are characterized, the known habitat of the *Mycoplasmataceae* will undoubtedly expand.

## Host and Tissue Specificity

As highly adapted parasites, mycoplasmas tend to exhibit host specificity and tissue tropism. Many notable exceptions have been described, however, and caution should be taken not to exclude mycoplasmas as potential causal agents for what would seem to be an atypical clinical presentation. While many species have a demonstrated ability to infect multiple hosts, the hosts tend to be relatively close evolutionary relatives to one another. The degree of relatedness across the host range varies between *Mycoplasmataceae* species, with some infecting additional hosts within the same family (e.g., *M. mycoides* subsp. *capri* infecting multiple members of the family *Bovidae*, but no other members of the order *Ruminantia*), and others infecting multiple hosts within the same class (e.g., *M. gallisepticum* infecting multiple members of the class *Aves*, but no other members of the phylum Chordata). The widest known host range is that of *Mycoplasma gallinarum*, which is the only species described as sharing multiple hosts at the phylum level (i.e., between class *Mammalia* and *Aves*). Though the potential exists for many *Mycoplasmataceae* species to infect multiple hosts, most of these species appear to have a preferred host habitat. A small number of species including *Mycoplasma arginini*, *Mycoplasma canis*, *Mycoplasma felis*, and *Mycoplasma gateae* seem to be regularly isolated from multiple hosts, indicating that the ideal habitat of these species is less specific.

Multiple epizootic events have resulted in the permanent establishment of a *Mycoplasma* species in a novel host. Notable events in recent years include the introduction and establishment of *M. gallisepticum* into the American house finch population from poultry (Ley et al. 1996), the introduction and establishment of *M. agalactiae* into the French ibex population from small ruminants (Tardy et al. 2012), and the introduction and establishment of *M. ovipneumoniae* into the American bighorn sheep population from small ruminants (Besser et al. 2013). In each case, the epizootic strains had demonstrable changes in their accessory genomes and new clinical presentations in the affected animals (Besser et al. 2013; Tardy et al. 2012; Tulman et al. 2012).

The primary sites of most mycoplasmal infections are mucosal surfaces, where the bacteria attach to the epithelium and exist as surface pathogens. Multiple species have invasive capabilities, and intracellular phases may be part of the infectious process (Andreev et al. 1995; Dallo and Baseman 2000; Tajima et al. 1982; Tarshis et al. 1994; Taylor-Robinson et al. 1991; Winner et al. 2000; Yavlovich et al. 2004). At least one *Mycoplasmataceae* species has been isolated in culture from all major body sites except the stomach in their respective hosts, including the brain, kidney, joints, bone, mammary glands, skin, liver, pancreas, intestinal tract, eye, sinus, all tissues of the urogenital tract, all tissues of the respiratory tract, heart, spleen, oral cavity, ear canals, and lymph nodes. In addition, pure cultures of at least one species have been isolated from all major body fluids including blood, urine, semen, cerebrospinal fluid, synovial fluid, milk, and saliva (Brown et al. 2011). Colonization of the stomach with *M. hyorhinis* has been detected both by immunohistochemistry and PCR in gastritis patients (Huang et al. 2001; Kwon et al. 2004), but the organism has yet to be recovered by axenic culture from gastric tissue.

The ability to spread to multiple tissues is common in some species (e.g., *M. alligatoris*), but the majority of species have a tropism for select anatomical sites. Rarely certain species can be found in atypical tissues following either primary infection at that site or invasion from a more classical site (e.g., *M. pneumoniae*) infection of the central nervous system rather than (or following) the respiratory tract (reviewed by Waites et al. 2005). Some instances of more diverse tissue tropism appear to be due to strain differences (Lockaby et al. 1999; Walter et al. 2008); however, this is likely to result from a complex interaction between both the mycoplasma cells themselves, additional organisms present at the affected site, and the host tissue and immune response.

## Mycoplasmas Infecting Cell Cultures

Due to the presence of serum in most cell culture media, the intrinsic resistance of mycoplasmas to many antibiotics that inhibit eubacteria (see section on "➋ Antibiotic Sensitivity"), and the subtlety of the effect of infections, contamination of cell cultures with certain *Mycoplasma* species is widespread. Though mycoplasmas at large are notorious among researchers performing cell culture, only a small number of species are associated with the majority of culture contamination. The most common species isolated are *M. arginini*, *M. fermentans*, *Mycoplasma orale*, *Mycoplasma salivarium*, and *M. hyorhinis*. Other mycoplasmas are often too pathogenic for the culture to persist, and incidental contamination thus do not often go beyond a single passage. Numerous methods for the detection of mycoplasmas have been published and are commercially available. Most are based on detection of ATP depletion, bacterial DNA by fluorescence staining, or *Mollicutes* DNA by molecular methods. Eradication of mycoplasmas from irreplaceable cell cultures can be accomplished by treatment with aminoglycosides and fluoroquinolones (Uphoff and Drexler 2002). Because of their requirement for sterols, mycoplasmas do not contaminate standard bacterial cultures.

## Mycoplasma Associations with Protozoa

The intracellular location and presumed concordance of transmission between *M. hominis* and the human urogenital tract protozoal parasite *Trichomonas vaginalis* was first reported in 1975 (Nielsen and Nielsen 1975). Numerous studies have confirmed these observations, and prevalence rates of *M. hominis*-infected *T. vaginalis* strains found in humans have been reported as high as 90 % (Rappelli et al. 1998), though most studies indicate the infection rate is much lower. A single report indicates that *T. vaginalis* cells infected with *M. hominis* have a greater cytopathic effect on epithelial cells than uninfected *T. vaginalis* cells (Vancini et al. 2008). While the in vivo relevance of this finding is unclear, the role of the *T. vaginalis* habitat in both *M. hominis* transmission and protection from host immunity and antibiotic therapy is apparent (Dessi et al. 2005). Additionally, some reports have indicated an association between *M. hominis* infection and *T. vaginalis* metronidazole resistance (Wang and Xie 2012; Xiao et al. 2006), while others have indicated no such relationship between the two organisms (Butler et al. 2010; Fraga et al. 2012). A recent report by Fraga et al. indicated a significant correlation between *T. vaginalis* phylotype and *M. hominis* infection, indicating that certain clades are more permissive to *M. hominis* than others (Fraga et al. 2012).

## Epidemiology and Control

### Transmission

Mycoplasmas are most commonly transmitted to new hosts by direct contact (e.g., nose-to-nose contact, sexual contact, or milk feeding), although several other modes of transmission have been documented in certain species. Aerosol transmission of respiratory mycoplasmosis between humans and animals housed in high density has been widely observed. Such instances include dormitory settings, military barracks, or livestock and poultry rearing (Atkinson et al. 2008; Feberwee et al. 2006; Tanskanen 1987). Fomite transmission has also been documented particularly for veterinary mycoplasmas and is mediated by contaminated bedding, feeding equipment, milking equipment, insemination tools, shearing tools, ear-tagging equipment, and wild bird feeders (Dhondt et al. 2007; Ma et al. 2010; Mason and Statham 1991b; Wilson et al. 2011; Woeste and Grosse Beilage 2007).

Vertical and perinatal transmission of mycoplasmas has been reported in commercial layers, small ruminants, cattle, and humans (Filioussis et al. 2011; Fujihara et al. 2011; Romero and Garite 2008; Stipkovits and Kempf 1996; Waites et al. 2005). Transovarian transmission of *M. gallisepticum*, *Mycoplasma synoviae*, *Mycoplasma lipofaciens*, *Mycoplasma iowae*, and *Mycoplasma meleagridis* has also been described in galliform birds, geese, and ducks (reviewed by Stipkovits and Kempf 1996; Stipkovits and Szathmary 2012), but has not been documented in any non-avian host. Perinatal transmission during birth or following premature rupture of the membranes likely results in congenital mycoplasmosis seen in mammals, with the exception of transplacental transmission of *Mycoplasma wenyonii* in cattle (Hornok et al. 2011). It is noteworthy that in many cases the clinical signs of vertically transmitted mycoplasmosis are fundamentally distinct from those displayed by the infected mother. Congenital infection with *M. hominis* and *Ureaplasma* species often generates pneumonia, bronchopulmonary dysplasia, meningitis, or encephalitis in (frequently premature) newborns born to mothers with either clinical or subclinical urogenital tract infections (Viscardi et al. 2002, reviewed by Waites et al. 2005). Infected poultry eggs exhibit reduced hatchability (indicating embryo lethality) and can display an emerging pathology known as eggshell apex abnormality ("glass-top eggs"), which is a distinctive thinning of the sharp end of the egg. It should also be noted that vertical transmission of avian mycoplasmosis presents the practical concern of establishing infection in mycoplasma-free flocks by the introduction of contaminated poultry-hatching eggs (Cobb 2011).

Arthropod vectors appear to be the primary mode of transmission for hemotropic mycoplasmas, although fomite and vertical/perinatal transmission have also been reported (Fujihara et al. 2011; Mason and Statham 1991a). Ticks, fleas, lice, biting flies, and mosquitoes have all been reported to mediate transmission of hemoplasmas (Brown et al. 2011; Daddow 1980; Hornok et al. 2011; Nikol'skii and Slipchenki 1969; Prullage et al. 1993; Woods et al. 2005). Additionally, there is incidental evidence that ear mites may be capable of transmitting *M. agalactiae*, *M. capricolum* subsp. *capricolum*, *M. mycoides* subsp. *capri*, *Mycoplasma putrefaciens*, *Mycoplasma cottewii*, and *M. yeatsii* between the ear canals of infected goats on their surface (Cottew and Yeats 1982; Jimena et al. 2009). ➋ *Table 39.2* describes specific arthropod vectors and the *Mycoplasma* species they are capable of transmitting.

### Antibiotic Sensitivity

Mycoplasmas are intrinsically resistant to many antibiotics, making the treatment of mycoplasmosis inherently challenging. Mycoplasmal infections can often require long-term antimicrobial therapy, as it is not unusual for disease to recur following the cessation of short-term treatment. This is speculated to be due to location of a small proportion of bacteria in a privileged site, such as inside a host cell (Jensen 2004). Treatment of mycoplasmosis most frequently employs antibiotics inhibiting DNA replication or protein synthesis including tetracyclines, macrolides, fluoroquinolones, aminoglycosides, and certain ketolides. Tetracyclines (most often doxycycline) are used most commonly worldwide to treat human and animal mycoplasmoses due to their low cost and high efficacy (Bébéar and Kempf 2005). Aminoglycosides, pleuromutilins, and phenicols are frequently used in veterinary medicine, but are not currently used to treat human mycoplasmosis under normal circumstances (Bébéar and Kempf 2005). Antimicrobials are also used against mycoplasmas in their eradication from cell culture. Aminoglycosides and fluoroquinolones have been used successfully for this purpose (Uphoff and Drexler 2002).

### Mechanisms of Antibiotic Resistance

As discussed above, *Mycoplasmataceae* species are intrinsically resistant to entire classes of antibiotics due to their atypical cell biology. Lacking a cell wall, mycoplasmas are resistant to β-lactams, vancomycin and fosfomycin. Mycoplasmas and ureaplasmas are also resistant to sulfonamides and trimethoprim because they do not synthesize their own nucleotides, rifampicin because they express an aberrant form of RNA polymerase unrecognized by the antibiotic, and polymyxins

## Table 39.2
Arthropod vectors associated with *Mycoplasma* transmission

| Vector | Common name | *Mycoplasma* species | Host |
|---|---|---|---|
| *Aedes aegypti* | Yellow fever mosquito | *M. suis* | Pigs |
| *Aedes camptorhynchus* | Southern salt marsh mosquito | *M. ovis* | Sheep |
| *Ctenocephalides felis* | Cat flea | *M. haemofelis*, "Candidatus Mycoplasma haemominutum" | House cat |
| *Culex annulirostris* | Freshwater mosquito | *M. ovis* | Sheep |
| *Haematobia irritans* | Horn fly | *M. wenyonii*, "Candidatus Mycoplasma haemobos" | Cattle |
| *Polyplax serrata* | Mouse louse | *M. coccoides* | Mice |
| *Polyplax spinulosa* | Rat louse | *M. coccoides, M. haemomuris* | Mice |
| *Psoroptes cuniculi* | Goat ear mite | *M. agalactiae, M. capricolum* subsp. *capricolum, M. mycoides* subsp. *capri, M. putrefaciens, M. cottewii,* and *M. yeatsii* | Goats |
| *Rhipicephalus bursa* | Brown dog tick | *M. ovis, M. haemocanis* | Sheep, dogs |
| *Stomoxys calcitrans* | Stable fly | *M. wenyonii*, "Candidatus Mycoplasma haemobos" | Cattle |
| *Tabanus bovinus* | Horse fly | *M. wenyonii*, "Candidatus Mycoplasma haemobos" | Cattle |

(Bébéar and Kempf 2005). Additional resistance profiles are exhibited by individual species based on unique sequences within the target of a given antibiotic. Examples include the resistance to macrolides such as erythromycin and azithromycin exhibited by strains of *M. hominis*, *Mycoplasma fermentans*, *Mycoplasma hyopneumoniae*, *Mycoplasma flocculare*, *Mycoplasma pulmonis*, *M. hyorhinis*, *Mycoplasma hyosynoviae*, *Mycoplasma meleagridis*, and *Mycoplasma bovis*, which arises from changes in the 23S rRNA sequence that prevent proper association of the antibiotic molecule with the ribosome (reviewed in Bébéar and Kempf 2005; Garcia-Castillo et al. 2008; Pereyre et al. 2002). Additionally, fluoroquinolone resistance exhibited by strains of *M. genitalium*, *M. gallisepticum*, *Mycoplasma bovirhinis*, *M. hyopneumoniae*, *M. hominis*, *M. pneumoniae*, and *M. synoviae* is driven by point mutations in the quinolone-resistance determining region (QRDR) of *gyrA*, *gyrB*, and/or *parC* (Bébéar et al. 1997; Govorun et al. 1998; Gruson et al. 2005; Hirose et al. 2004; Le Carrou et al. 2006; Lysnyansky et al. 2008; Reinhardt et al. 2002; Shimada et al. 2010b; Vicca et al. 2007). This adds complexity to the already limited antibiotic selections available for the treatment of mycoplasmosis and makes a standard treatment recommendation difficult. Further, there is an increasing trend toward prevalence of antibiotic-resistant strains of human and agricultural mycoplasmas. Macrolide resistance in human mycoplasmosis is growing at an alarming rate, accounting for 80–90 % of clinical isolates in China and Japan, resulting in an increase of fluoroquinolone usage (Bébéar 2012). Accordingly, reports of fluoroquinolone resistance are starting to increase (Shimada et al. 2010a). Agricultural mycoplasmas have been isolated that display macrolide and tetracycline resistance (Uemura et al. 2010).

Antibiotic resistance genes have been reported in clinical isolates of prominent pathogens. The tetracycline resistance protein gene *tetM*, originally discovered in *Staphylococcus aureus*, has been detected in some clinical isolates of human urogenital pathogens including *M. hominis*, *Ureaplasma urealyticum*, and *Ureaplasma parvum* (Dégrange et al. 2008; Mardassi et al. 2012; Sanchez-Pescador et al. 1988). The putative efflux pumps Md1 and Md2 of *M. hominis* are associated with resistance to ciprofloxacin (Raherison et al. 2002, 2005). Antibiotic resistance genes have also been artificially introduced into numerous *Mycoplasma* species as selectable markers, indicating that most species are capable of expressing mobile resistance genes and synthesizing functional proteins. The most commonly utilized resistance genes include *tetM*, the chloramphenicol acetyltransferase gene *cat*, the aminoglycoside acetyltransferase gene *aacA*, and the puromycin N-acetyltransferase gene *pac* (Algire et al. 2009; Hahn et al. 1999; Mahairas and Minion 1989; Pour-El et al. 2002).

## Vaccines

Vaccination is a widely used control strategy for multiple veterinary mycoplasmoses. Several vaccines are commercially available for the protection of livestock and poultry against mycoplasmosis. The efficacy of these vaccines remains questionable in many cases, however, emphasizing the continuing issue of mycoplasmosis in veterinary medicine. Specific vaccines (commercial and experimental) targeting veterinary mycoplasmas are described in ● *Table 39.3*. Despite numerous experimental vaccines targeting *M. pneumoniae*, vaccination against human mycoplasmosis is not currently available.

### Table 39.3
Vaccines against mycoplasmosis

| Vaccine strain | Vaccine trade name/source | Target species | Host | Efficacy? (Y/N)[a] | Origin |
|---|---|---|---|---|---|
| T1/44 | PANVAC | M. mycoides subsp. mycoides | Cattle | Y | Serial passage, in ovo |
| F38 | Caprivax, KEVEVAPI | M. capricolum subsp. capripneumoniae | Goats | Y | Bacterin |
| AIK 40 | Pendik Veterinary Control | M. agalactiae | Sheep, goats | Y | Serial passage, in vitro |
| MAC | N/A | M. pneumoniae | Hamsters, humans | Y (hamsters), N (humans) | Serial passage, in vitro |
| proprietary | Pulmo-Guard MpB; BI Vetmedica | M. bovis | Cattle | Y/N | Bacterin |
| MS-H | Vaxsafe MS; Bioproperties | M. synoviae | Chickens | Y | Chemical mutagenesis |
| 86B/96 | Pending | M. bovis | Cattle | Y | Naturally occurring |
| F | F-Vax MG; Intervet Schering Plough Animal Health | M. gallisepticum | Chickens | Y | Naturally occurring |
| Ts-11 | Vaxsafe MG; Bioproperties | M. gallisepticum | Chickens | Y | Chemical mutagenesis |
| 6/85 | Mycovac-L; Intervet Schering Plough Animal Health | M. gallisepticum | Chickens | Y | Serial passage |
| GT-5 | N/A | M. gallisepticum | Chickens | Y | Serial passage ("$R_{high}$") + gapA gene |
| K5054 | N/A | M. gallisepticum | House finches | Y | Naturally occurring |
| P1 | N/A | M. pneumoniae | Hamsters | N | Subunit |
| P-5722-3 | Stellamune; Pfizer Animal Health | M. hyopneumoniae | Pigs | Y/N | Bacterin |
| B-3745 | IngelVac MycoFLEX; BI Vetmedica | M. hyopneumoniae | Pigs | Y | Bacterin |
| Proprietary | M + PAC; Merck Animal Health | M. hyopneumoniae | Pigs | Y | Bacterin |
| P97 | N/A | M. hyopneumoniae | Mice, pigs | Y | Subunit |
| CPS | N/A | M. mycoides subsp. mycoides | Mice, pigs | Y | Subunit |

[a]Yes, efficacy in both/all animals

## Pathogenicity

Mycoplasmosis includes a spectrum of clinical manifestations ranging from commensalism to fulminant inflammatory diseases with high mortality rates. The classical manifestation of mycoplasmosis is a chronic inflammatory illness that is not typically fatal, though this is the normal state for the highly virulent species *M. alligatoris* and *M. mycoides* subsp. *mycoides*. Lesions developed during mycoplasmosis can be superinfected by secondary pathogens, resulting in more severe disease with higher rates of morbidity and mortality.

## Disease Manifestations

### Human Diseases

Human mycoplasmosis is associated with *M. pneumoniae*, *M. genitalium*, *M. hominis*, *U. urealyticum*, *U. parvum*, and potentially *Mycoplasma amphoriforme*. Causal roles for *M. fermentans* and *M. penetrans* in human disease remain unclear. The overwhelming majority of human mycoplasmosis involves the respiratory tract or the urogenital tract; however, numerous atypical manifestations particularly of *M. pneumoniae* infection have been reported and affect the central nervous system, skin, eyes, ears, heart, circulatory system, liver, pancreas, kidneys, and joints (reviewed by Atkinson et al. 2008; Waites et al. 2005).

Upper respiratory tract infections by *M. pneumoniae* frequently present as pharyngitis and/or laryngitis, and lower respiratory tract include tracheitis and interstitial pneumonia. Interstitial pneumonia due to *M. pneumoniae* is clinically referred to as primary atypical pneumonia and commonly referred to as "community-acquired pneumonia" or "walking pneumonia." Symptoms of primary atypical pneumonia include low-grade fever, nonproductive cough, and chest pain. Community-acquired pneumonia outbreaks due to *M. pneumoniae* are often associated with close quarters and stress, and notoriously occur in military barracks or student dormitories. Postinfectious complications following community-acquired pneumonia disease can include the development or exacerbation of reactive airway disease/asthma and secondary ear infections. Rarely, *M. pneumoniae* infection can lead to Guillain-Barré syndrome, Stevens-Johnson syndrome, Bell's palsy, optic neuritis, or demyelinating disorders (reviewed by Waites et al. 2005). The initial isolation of *M. amphoriforme* from the human respiratory tract occurred in 2005, and a subsequent report has associated this species with respiratory tract infections (Pereyre et al. 2010; Pitcher et al. 2005). At this time there is not an established, widely available diagnostic test to differentiate between *M. amphoriforme* and *M. pneumoniae*, and so it is plausible that *M. amphoriforme* is a more prominent human pathogen than currently realized.

Numerous studies have explored the correlation of patient infection with several urogenital lesions in patients infected with *M. genitalium*, *M. hominis*, *U. urealyticum*, and *U. parvum*. All four organisms can be detected in patients lacking any overt symptoms, but the association of mycoplasmas with urogenital tract disease is strengthening. Clinical findings associated with *Ureaplasma* spp. and *M. hominis* appear to be strain or patient specific and can include nongonococcal urethritis (*Ureaplasma* spp.), bacterial vaginosis (*M. hominis*), spontaneous abortion, or preterm labor (both organisms) (Waites et al. 2005). In contrast, *M. genitalium* infection is strongly associated with nongonococcal urethritis, pelvic inflammatory disease, spontaneous abortion, and infertility, and it is consequently considered an emerging urogenital pathogen (McGowin and Anderson-Smits 2011). Postinfectious arthritis ("sexually acquired reactive arthritis") has been reported in association with *M. hominis* and *Ureaplasma* species (Taylor-Robinson and Furr 1997; Taylor-Robinson et al. 1983).

As described above (see section on "❯ Transmission"), urogenital mycoplasmosis of pregnant woman has been associated with preterm labor and premature rupture of membranes. Premature interactions between *Ureaplasma* spp. present in the vagina and a perinatal infant can result in neonatal respiratory tract or central nervous system infections. Chronic lung conditions can persist in these infants following the resolution of infection. Neonatal meningitis associated with *M. hominis* has also been reported (Hata et al. 2008).

Finally, several reports indicate that there may be synergy between infection with certain *Mycoplasma* species and human immunodeficiency virus (HIV). Increased vaginal shedding of HIV has been associated with *M. genitalium* infection, indicating that this organism may have an enormous public health consequence that extends beyond acute mycoplasmosis (Manhart et al. 2008; Perez et al. 1998). It has long been postulated that *M. penetrans* may act as a cofactor in the progression from subclinical HIV infection to the development of acquired immune deficiency syndrome (AIDS) by its ability to mediate T cell proliferation and thus HIV replication (Sasaki et al. 1995; reviewed by Blanchard 1997). Initial in vitro results further supported this by demonstrating that the cytocidal effect of HIV is enhanced in the presence of contaminating mycoplasmas (Lemaitre et al. 1992; Lo et al. 1991b). Serological evidence indicating that *M. penetrans* was far more prevalent in AIDS patients as opposed to HIV-positive, non-AIDS patients, or HIV-negative patients (Grau et al. 1998; Wang et al. 1992) is consistent with this hypothesis, but does not definitively establish *M. penetrans* as a driver of AIDS progression. Prospective studies are necessary to further elucidate the synergy between HIV and *M. penetrans*.

## Veterinary Diseases: Agricultural Animals

The impact of mycoplasmosis on the health of agricultural animals is substantial and has a secondary impact on human health via malnutrition from both lack of animal production and culling of infected herds or flocks (Boonstra et al. 2001). *Mycoplasma* species cause major diseases of cattle, swine, poultry, goats, and sheep, necessitating several of these organisms to be intensively regulated at the Federal and international levels. Mycoplasmoses of farmed horses, fish, and alligators have also been described but are considered minor contributors to disease. Disease states of agricultural animals associated with mycoplasmas are described in ❯ *Table 39.4*.

## Veterinary Diseases: Companion and Research Animals

Mycoplasmosis of mice, rats, rabbits, hamsters, dogs, cats, and iguanas has been reported. Many of these disease states have not yet been definitively attributed to the mycoplasmas present, but some are major health concerns (Brown et al. 2011). Rodents infected with *M. pulmonis* develop severe respiratory and arthritic disease, and those infected with *M. arthritidis* can develop arthritic and systemic disease. These two pathogens are serious concerns for captive mouse and rat colonies, as well as for pet mice and rats (Keystone et al. 1982; Simecka et al. 1987). Failed canine matings (classified as infertilities and stillbirths) putatively associated with mycoplasmas are a concern for dog breeders, and respiratory mycoplasmosis contributes to sometimes fatal infections in dogs and house cats. Infectious anemia due to hemotropic mycoplasmas can cause chronic illness in dogs and cats that impacts their quality of life (Brown et al. 2011). Additional disease states associated with mycoplasmas are described in ❯ *Table 39.5*.

## Veterinary Diseases: Wildlife Diseases

A large number of mycoplasmoses have been described in wildlife or zoo animals, but many are single cases or single outbreaks. Some epidemics of wildlife mycoplasmosis have emerged as excellent models for studying principles of disease emergence and the evolution of virulence. The recent emergence of *M. gallisepticum* in house finches represents a real-time experimental system in which to study epizoonosis and

**Table 39.4**
**Agricultural mycoplasmosis**

| Species | Host | Disease | Signs | Vaccine |
|---|---|---|---|---|
| *M. alkalescens* | Cattle | URTD | Nasal discharge, lameness, decreased milk production | No |
| *M. arginini* | Cattle, sheep, goats | LRTD, URTD, RTI | Nasal discharge, coughing, ocular discharge, infertility | No |
| *M. bovigenitalium* | Cattle | LRTD, URTD, RTI | Nasal discharge, coughing, infertility | No |
| *M. leachii* | Cattle | Mastitis, arthritis, RTI | Lameness, decreased milk production, pneumonia, APO | No |
| *M. bovis* | Cattle | URTD, LRTD | Coughing, head tilt, purulent discharge, lameness, decreased milk production | Yes |
| *M. bovoculi* | Cattle | Conjunctivitis | Ocular discharge | No |
| *M. californicum* | Cattle | Mastitis | Decreased milk production | No |
| *M. canadense* | Cattle | Mastitis | Decreased milk production | No |
| *M. dispar* | Cattle | LRTD | Coughing | No |
| *M. mycoides mycoides* | Cattle | CBPP | Coughing, fever, torticollis | Yes |
| *M. verecundum* | Cattle | Conjunctivitis | Ocular discharge | No |
| *M. wenyonii* | Cattle | Hemolytic anemia | Malaise, jaundice | No |
| *M. gallisepticum* | Galliform | CRD, IS, neurological disease, arthritis | Rales, coughing, fever, sinus swelling, ataxia, ocular discharge, lameness | Yes |
| *M. synoviae* | Galliform | Arthritis, LRTD, RTI | Rales, coughing, sinus swelling, PEQ, reduced hatchability, lameness | Yes |
| *M. anatis* | Duck | URTD | Nasal discharge | No |
| *M. adleri* | Goat | Arthritis | Lameness | No |
| *M. agalactiae* | Goat, sheep | LRTD, RTI, contagious agalactia | Coughing, ocular discharge, decreased milk production, lameness, APO | Yes |
| *M. capricolum capricolum* | Goat, sheep | Contagious agalactia, LRTD, encephalitis | Lameness, decreased milk production, coughing, sudden death | No |
| *M. capricolum capripneumoniae* | Goat, sheep | CCPP | Coughing, fever, torticollis | Yes |
| *M. mycoides capri* | Goat, sheep | Contagious agalactia, LRTD | Lameness, decreased milk production, ocular discharge, coughing | No |
| *M. putrefaciens* | Goat | Septicemia, contagious agalactia | Fever, malaise, cataracts, blindness, lameness, decreased milk production | No |
| *M. subdolum* | Horse | RTI | APO, infertility | No |
| *M. haemolamae* | Llama | Infectious anemia | Malaise | No |
| *M. haemosuis* | Pig | Hemolytic anemia | Malaise, jaundice | No |
| *M. hyopneumoniae* | Pig | PEP | Coughing, malaise, arrhythmia | Yes |
| *M. hyorhinis* | Pig | URTD, LRTD | Nasal discharge, coughing, arrhythmia, Lameness | No |
| *M. hyosynoviae* | Pig | arthritis | Lameness | No |
| *M. conjunctivae* | Goats, sheep | Conjunctivitis | Cataracts | No |
| *M. ovipneumoniae* | Sheep | URTD, LRTD | Nasal discharge, coughing | No |
| *M. ovis* | Sheep | Hemolytic anemia | Malaise, jaundice | No |
| *M. meleagridis* | Turkey | LRTD, RTI | PEQ, reduced hatchability, rickets, skeletal deformity | No |
| *M. iowae* | Turkey | Arthritis, osteitis, RTI | PEQ, reduced hatchability, ocular discharge, lameness, skeletal deformity | No |
| *M. pullorum* | Turkey | RTI | Reduced hatchability | No |

*Abbreviations*: *URTD* upper respiratory tract disease, *LRTD* lower respiratory tract disease, *RTI* reproductive tract infection, *CBPP* contagious bovine pleuropneumonia, *CCPP* contagious caprine pleuropneumonia, *CRD* chronic respiratory disease, *IS* infectious sinusitis, *PEP* porcine enzootic pneumonia

**Table 39.5**
**Mycoplasmosis of companion and laboratory animals**

| Species | Host | Commercial interest | Disease | Signs |
|---|---|---|---|---|
| *M. oxoniensis* | Chinese hamster | None | Conjunctivitis | Ocular discharge |
| *M. cricetuli* | Chinese hamster | None | Conjunctivitis | Ocular discharge |
| *M. canis* | Dog | Dog breeding | RTI, UTI, URTD | Frequency, coughing |
| *M. cynos* | Dog | Dog breeding | Pneumonia, APO | Coughing, malaise |
| *M. edwardii* | Dog | None | Septicemia | Fever, lameness |
| *M. haemocanis* | Dog | None | Infectious anemia | Malaise |
| *M. maculosum* | Dog | None | LRTD, UTI | Coughing, frequency |
| *M. spumans* | Dog | None | Arthritis | Lameness |
| *M. haemominutum* | Domestic cat | None | Infectious anemia | Malaise |
| *M. felis* | Cat, Horse | None | URTD (Fe), LRTD (h) | Nasal discharge, labored breathing, ocular discharge |
| *M. gateae* | Cat | None | URTD | Anorexia, lethargy, ocular discharge, lameness |
| *M. haemofelis* | Cat | None | Infectious anemia | Malaise |
| *M. equigenitalium* | Horse | Horse breeding | RTI | APO, infertility |
| *M. equirhinis* | Horse | None | Inflammatory airway disease | Nasal discharge, coughing, rales |
| *M. coccoides* | Mouse | None | Infectious anemia | Malaise |
| *M. haemomuris* | Mouse | None | Infectious anemia | Malaise |
| *M. neurolyticum* | Mouse | None | Rolling disease | Uncoordinated movement |
| *M. pulmonis* | Mouse, rat | Laboratory animals | Murine respiratory mycoplasmosis | Nasal discharge, coughing, anorexia, lameness |
| *M. ravipulmonis* | Mouse | None | Grey lung disease | Coughing |
| *M. arthriditis* | Rat | Laboratory animals | Septicemia, arthritis | Lameness |

*Abbreviations*: *URTD* upper respiratory tract disease, *LRTD* lower respiratory tract disease, *RTI* reproductive tract infection, *UTI* urinary tract infection

pathogen behavior and transmission in a new host species. Similarly, a highly lethal inflammatory disease of American alligators has been attributed to *M. alligatoris*, while the closely related *Mycoplasma crocodyli* causes a more classical presentation of mycoplasmosis in Nile crocodiles (Brown et al. 2004). These two recently diverged species with vastly different clinical outcomes present a unique opportunity for long-term comparative genomics, as opposed to the rapid genomic changes occurring in house finch strains of *M. gallisepticum*. In both cases, genome plasticity is readily apparent against the backdrop of a conserved core genome (Delaney et al. 2012; Tulman et al. 2012). Additional wildlife disease states associated with mycoplasmas are described in *Table 39.6*.

## Mechanisms of Pathogenicity of the *Mycoplasmatales*

### Secreted Toxins and Toxic Metabolites

The potential for toxic metabolites to contribute to disease is well known. Many *Mycoplasma* species have been demonstrated to produce hydrogen peroxide ($H_2O_2$) during glycerol metabolism, and many of the same species that encode a catabolic pathway for glycerol utilization have mechanisms for $H_2O_2$ resistance such as peroxidases that protect them from adverse effects (Jenkins et al. 2008; Machado et al. 2009). Mycoplasma-generated $H_2O_2$ has been shown to have a direct toxic effect on cultured cells (Hames et al. 2009; Pilo et al. 2005; Schmidl et al. 2011).

A small number of species have been shown to produce secreted virulence-mediating proteins that are completely unique within the genus. The superantigen MAM (*Mycoplasma arthritidis* mitogen) is produced by the rodent pathogen *M. arthritidis*, and the evolutionary origin of the encoding gene is unknown. MAM is an immunomodulatory protein, but its effect as defined by clinical outcome is highly variable based on the genetic background of the host animal (Kirchner et al. 1986; Mu et al. 2000). Levels of MAM correlate with the development of lethal toxicity; however, it does not appear to directly mediate the development of arthritis during *M. arthritidis* infection (Luo et al. 2008). The community-acquired respiratory distress syndrome (CARDS) toxin of *M. pneumoniae* has also been reported only in this organism.

**Table 39.6**
**Wildlife mycoplasmosis**

| Species | Host | Epizootic potential | Disease | Signs |
|---|---|---|---|---|
| *M. alligatoris* | Alligator | High[a] | LRTD, nephritis, ME, fulminant inflammatory disease | Rapid death, lethargy, lameness |
| *M. felifaucium* | Cheetah, puma | High[b] | Gastroenteritis | Emesis, wasting |
| *M. agassizii* | Desert tortoise | Low | URTD | Nasal exudates |
| *M. buteonis* | Falcon | Low | LRTD, Unclear | Uncoordinated movement, lameness, skeletal deformity |
| *M. gallisepticum* | Passerines (primarily finches) | High[b] | Conjunctivitis | Ocular discharge |
| *M. pneumoniae* | Hominids (chimpanzee, rhesus monkeys) | High[b] | PAP, URTD | Coughing, malaise |
| *M. buccale* | Orangutan | High[b] | Airsacculitis, pneumonia | Coughing, malaise |
| *M. iguanae* | Iguana | Low | Osteitis | Lameness |
| *M. sphenisci* | Jackass penguin | Low | URTD | Halitosis, choanal discharge |
| *M. crocodyli* | Nile crocodile | Low | LRTD | Lameness |
| *M. haemodidelphidis* | Opossum | Low | Infectious anemia | Malaise |
| *M. orale* | Orangutan | Low | URTD | Rales, nasal discharge |
| *M. columborale* | Pigeon | Low | LRTD | Rales, coughing, lethargy |
| *M. phocicerebrale* | Seal, human | High[b] | Eye infection, seal finger (hu) | Cataracts, blindness |
| *M. phocirhinis* | Seal | Low | URTD | Rhinitis |
| *M. zalophi* | Sea lion | High[a] | Necrotizing LRTD | Lameness, coughing |
| *M. conjunctivae* | Sheep, chamois, ibex | High[b] | Conjunctivitis | Ocular discharge, cataracts |
| *M. kahanei* | Squirrel monkey | | Hemolytic anemia | Malaise, jaundice, arrhythmia |
| *M. sturni* | Starling | High[b] | Conjunctivitis | Ocular discharge |
| *M. mobile* | Tench | Low | Red gill disease | Ulceration, labored breathing |
| *M. testudineum* | Tortoise | Low | URTD | Nasal discharge |
| *M. corogypsi* | Vulture | Low | Abscess | Chronic exudate |
| *M. vulturii* | Vulture | Low | LRTD | Rales, coughing |

*Abbreviations*: *URTD* upper respiratory tract disease, *LRTD* lower respiratory tract disease, *ME* meningoencephalitis, *PAP* primary atypical pneumonia
[a]Presumed epizootic from an unknown host
[b]Established epizootic from a known host

CARDS is part of a superfamily of toxins (ADP-ribosylating toxins; EC 2.4.2.-) that includes pertussis toxin, diphtheria toxin, and cholera toxin. Infection of mice with purified CARDS toxin generates severe lesions in the absence of *M. pneumoniae* cells. CARDS has also been implicated in hypoxemia and refractory asthma due to *M. pneumoniae* disease (Hardy et al. 2009; Muir et al. 2011; Peters et al. 2011). Unlike MAM, the amount of detectable CARDS toxin correlates directly with the degree of inflammation in the lungs during infection rather than the final clinical outcome alone (Kannan et al. 2011). A second ADP-ribosylating toxin (MYPE9110) has been reported in *M. penetrans*. Although ADP-ribosylation activity was demonstrated in vitro, a role for this toxin in virulence remains speculative (Johnson et al. 2009).

Hemolysis has been described for some *Mycoplasmataceae* species, but a common mechanism does not appear to exist throughout the genus. The activity has been attributed to a specific protein only in *M. pulmonis* (HlyA) (Minion and Goguen 1985). Hemolysis occurred in the presence of membrane fractions of numerous *Mycoplasmatales* species, indicating that the activity is cell surface associated (Minion and Jarvill-Taylor 1994). Given the absence of specific associated proteins in most species, it should be noted that bacterial hemolysis can be due exclusively to $H_2O_2$ production.

Degradation of human IgA1 by *U. urealyticum* has been reported by multiple investigators (Kilian and Freundt 1984; Robertson et al. 1984). The activity is due to an apparent serine protease that cleaves IgA1 at the hinge region, generating intact $F_c$ $F_{ab}$ fragments (Spooner et al. 1992). Isolates of *Ureaplasma* species cleaved IgA1 of their host species, but not IgA1 of other species (Kapatais-Zoumbos et al. 1985). The complete genomes of *U. urealyticum* and *U. parvum* do not encode a gene recognizable as an IgA1 protease (Glass et al. 2000; Paralanov et al. 2012), and thus, the source of this activity remains cryptic.

## Cytoadherence Mechanisms

Although some mycoplasmas may not adhere to host cell surfaces, those that do so generally require this step to gain access to the cell's nutrients and protective environment. Known receptors for mycoplasmas consist of both carbohydrate components of membrane molecules and extracellular matrix-related proteins and carbohydrates deposited onto epithelial cell surfaces from body fluids.

Different varieties of sialic acid, a common modification to cell surface glycoproteins, are recognized by transmembrane proteins of *M. pneumoniae*, *M. gallisepticum*, and *M. mobile* that act as adhesins. The *M. mobile* sialic acid-binding adhesin, Gli349, is restricted to the base of the headlike structure (Uenoyama et al. 2004). Although the *M. pneumoniae* adhesins responsible for this activity have not directly been identified, an adhesin associated with the terminal organelle, P1, has been implicated (Kornspan et al. 2011). P1 orthologs are present in *M. gallisepticum* and other members of the *M. pneumoniae* cluster (Goh et al. 1998), supporting a general role for this family of proteins in terminal organelle-mediated binding to a common ligand, possibly sialic acid. However, sialic acid binding is not limited to mycoplasmas with terminal organelles, being found additionally in *M. bovis*, where it is mediated by a poorly described adhesin, P26 (Sachse et al. 1996). Sialic acid binding in these lineages is likely the result of convergent evolution, as the adhesins appear to be unrelated.

Binding to extracellular matrix-related proteins and carbohydrates on host cell surfaces has been documented for a variety of *Mycoplasma* species. Fibronectin, plasminogen, mucin, surfactant protein A, heparin, and fibrinogen have all been described as ligands for mycoplasmas (Alvarez et al. 2003; Dumke et al. 2011; Giron et al. 1996; Jenkins et al. 2006; Piboonpocanun et al. 2005; Yavlovich et al. 2001). In many, though not all, cases in which the identity of the mycoplasma receptors for these molecules functioning as adhesins have been identified, they are proteins with already-established roles in other cellular processes including central metabolic enzymes, translation factors, structural proteins of terminal organelles, and even an ADP-ribosylating toxin. In most cases these proteins have dual cytoplasmic and membrane localization, and the means by which they achieve surface localization are unknown (Dallo et al. 2002; Thomas et al. 2013). These mechanisms may coexist with sialic acid-binding mechanisms, where the latter is restricted to the terminal organelle, but the receptors for extracellular matrix-related molecules are located over the entire surface of the mycoplasma cell. In other cases, including the P97/P102 family of *Mycoplasma hyopneumoniae*, the adhesins are a family of paralogous proteins that are endoproteolytically processed to a vast, complex array of protein fragments with the ability to bind multiple extracellular matrix-related molecules (Adams et al. 2005; Djordjevic et al. 2004; Seymour et al. 2012).

For still other mycoplasma adhesins, the substrates to which they bind are unknown. *Mycoplasma hominis* employs a wide variety of adhesins with unknown ligands, including the Vaa adhesin, which is highly variable in sequence across isolates (Chernov et al. 2005; Zhang and Wise 1996); P100, an ABC transporter component with extracellular ATPase activity whose adherence is ATPase-dependent (Henrich et al. 1999; Hopfe et al. 2011); and P60 and P80, which interact with a histidine triad nucleotide-binding protein in the *M. hominis* cytosol (Kitzerow and Henrich 2001). Another *M. hominis* adhesin is P50, whose adherence can be inhibited by dextran sulfate, suggesting binding to a related cell surface molecule (Kitzerow et al. 1999). The substrate for *M. agalactiae* P40 is not known (Fleury et al. 2002). VlhA proteins from *M. synoviae*, *M. gallisepticum*, and *M. imitans* are hemagglutinins (Markham et al. 1999; Noormohammadi et al. 1998). The substrate of at least some VlhA variants from *M. synoviae* is sialic acid, whereas other variants attach to alternative carbohydrates that have not yet been identified (May and Brown 2011). The ligands of homologous VlhA proteins from *M. gallisepticum* and *M. imitans* have not been investigated.

## Immune Evasion

Immune evasion by mycoplasmas takes several forms. A major one for which abundant evidence exists is variation of surface antigens. A second one, established more indirectly, is molecular mimicry by surface proteins. Finally, invasion of host cells is employed by many *Mycoplasma* species and may have a role in immune evasion.

Antigenic variation of surface proteins, including lipoproteins, which are often immunodominant, is widespread and perhaps ubiquitous among *Mycoplasma* species, although both the specific sets of variable proteins and the mechanisms used for generation of variation are diverse, defying a generalized description of this feature. Random variation among related gene products is generated in some species by alteration of the orientation of a promoter relative to a coding gene, in others by altering the number of nucleotides within the 5′-untranslated region of the gene, and in still others by recombination among related sequences within a coding gene, through either gene conversion or reciprocal recombination. In many, but not all, cases, the antigens that are subject to variation are associated with Cytoadherence. Some illustrative examples are described below.

The P35 antigenic lipoprotein family of *M. penetrans* is an example of promoter switching. At least 38 similar but distinct P35-encoding genes are present in *M. penetrans* strain HF-2, many of them with promoters located between inverted repeats capable of switching orientation (Horino et al. 2003). Thus, the promoter favors transcription of the gene in one orientation but not the other. Inversion is mediated by a sequence-specific recombinase (Horino et al. 2009). Similar systems are present for unrelated antigens in *M. bovis*, *M. pulmonis*, and

*M. agalactiae*, but in these cases, the recombinase is encoded adjacent to each variable locus, which often contains multiple genes for surface antigens (Ron et al. 2002; Sitaraman et al. 2002). Inversion also plays a role in generating diversity in *U. urealyticum* and *U. parvum* multiple-banded antigen, although in these organisms, it is more complicated and involves displacement of coding portions of the genes, with the XerC recombinase suggested to play a significant role (Zimmerman et al. 2009, 2013). In cases like this, the antigen in question exhibits marked size variation. A phase-locked mutant of *M. agalactiae*, in which the gene encoding Xer1 recombinase is disrupted and therefore cannot effectuate phase variation of the surface lipoproteins of the Vpma family, is capable of causing infection but exhibits reduced immune response, invasiveness, lymphopenia, and neutropenia, supporting a role for antigenic variation in dissemination and persistence (Chopra-Dewasthaly et al. 2008, 2012).

Different strains of *M. gallisepticum* contain 30–70 genes for the VlhA family of transmembrane hemagglutinins (Baseggio et al. 1996), but a limited number are expressed in a given cell. Only those with exactly 12 GAA repeats upstream of the promoter are expressed (Liu et al. 2000). Variation in *vlhA* expression arises randomly during growth, and antibodies against a particular VlhA isotype results in proliferation of cells not expressing that isotype (Glew et al. 2000; Markham et al. 1998), consistent with a role for VlhA antigenic variation in immune evasion. Although DNA strand slippage during replication is most likely what drives this variation, it is unclear whether a dedicated mechanism exists. Related antigens are found in *M. imitans* (Markham et al. 1999) and *M. synoviae* (Allen et al. 2005; Noormohammadi et al. 1998), apparently having been acquired by the latter through horizontal gene transfer. Gene conversion through recombination with pseudogenes has also been implicated in *M. synoviae* VlhA variation (Noormohammadi et al. 2000).

The *M. genitalium* adhesin MgPa and the related P1 adhesin of *M. pneumoniae*, as well as the respective genes downstream of them encoding Cytoadherence accessory proteins, experience variation through recombination of segments of their coding genes with any of several repeated DNA segments located throughout the genomes (Iverson-Cabral et al. 2007; Ma et al. 2007; Spuesens et al. 2009). The repeat regions are restricted to certain portions of the proteins, resulting in the adhesins having variable regions and constant regions (Himmelreich et al. 1997). The mechanism of recombination does not appear to involve a specific target sequence but does employ RecA (Burgos et al. 2012). Trinucleotide repeats that vary in size among isolates are also present within the coding sequence of MgPa, where they might also contribute to antigenic variation (Ma et al. 2012). Evidence for MgPa variation not only in vitro but also in vivo has been described (Iverson-Cabral et al. 2006; Ma et al. 2010), and the differences between P1 sequences are major factors in the identities of different clinical *M. pneumoniae* isolates (Kenri et al. 1999; Schwartz et al. 2009).

Although not widespread, mycoplasmas have been implicated in engaging in molecular mimicry. Some sera from *M. pneumoniae*-infected patients contain P1-reactive antigens that cross-react with cellular proteins (Jacobs et al. 1995). Infection with *M. pneumoniae* has also been reported to precede episodes of Guillain-Barre syndrome (Hughes et al. 1999) and acute motor axonal neuropathy (Susuki et al. 2004), due to the presence of antibodies against host cell gangliosides. *M. hyorhinis* has also been reported to induce antibodies against host proteins (Fernsten et al. 1987; Wise and Watson 1985). These situations likely represent extraordinary situations in which antibodies were generated against epitopes that normally resemble host epitopes too closely to generate an immune response, suggesting the possibility that the mycoplasmas evolved these regions to evade the host immune system.

Invasion of host cells is employed by at least some *Mycoplasma* species, although in no case has the mechanism of invasion been elucidated. Although it is unclear whether the ability to reside in the host cell cytoplasm has actually resulted from evolutionary pressure on the mycoplasma to evade the immune system, it seems likely that immune evasion is a consequence of this activity, as are potentially greater access to nutrients and decreased competition with other bacteria. *M. fermentans*, *M. genitalium*, *M. gallisepticum*, and *M. suis* have been identified on the interior of cells obtained directly from infected subjects, making these organisms the clearest-cut cases for mycoplasmas that are invasive in an in vivo context (Groebel et al. 2009; Lo et al. 1989; Ueno et al. 2008; Vogl et al. 2008). *M. penetrans* was observed to be capable of invading a normal human endothelial cell in vitro, making a strong case for that species also (Lo et al. 1991a); likewise, *U. diversum* invaded bovine sperm cells in vitro (Buzinhani et al. 2011). Numerous other species are able to invade cancerous tissue culture cells in vitro, but it is unclear whether this activity is initiated by the mycoplasma or by the host cell as a defense mechanism. *M. pneumoniae* can invade A549 human lung carcinoma cells, but neither HeLa cells nor normal human bronchial epithelial cells (Jordan et al. 2007; Yavlovich et al. 2004), raising the possibility that in vitro invasion is not always reflective of activity that is physiologically significant for the mycoplasma during an infection.

## Polysaccharides (Capsules and Biofilms)

Polysaccharide capsules are well-known virulence factors of many bacterial pathogens. While certain mycoplasmas generate capsules, their role in virulence is unclear. Capsular material has been associated with host cell attachment for *M. ovipneumoniae* and *M. gallisepticum*, but the relevance of this association remains unknown (Niang et al. 1998; Tajima et al. 1982). Adaptive immune responses against capsular polysaccharides in the case of *Mycoplasma dispar* appear to mediate pathological immune responses (Almeida et al. 1992). Despite this, capsular polysaccharides have been investigated as subunit vaccine antigens to protect against *M. mycoides* subsp. *mycoides* infection (Waite and March 2002) and can also be used diagnostically (March et al. 2000, 2002).

Biofilm formation in vitro has been observed for several *Mycoplasma* and *Ureaplasma* species, and increased resistance to heat, desiccation, complement-mediated lysis, phagocytosis, and antibiotics has been reported (Eterpi et al. 2011; Garcia-Castillo et al. 2008; McAuliffe et al. 2006; Shaw et al. 2012; Simmons and Dybvig 2007). Extended survival on surfaces such as agricultural equipment, bedding, dental implants, and stents attributable to biofilm formation has been reported (Henrich et al. 2010; Justice-Allen et al. 2010; Kumar et al. 2012). Ex vivo biofilm formation by *Mycoplasma pulmonis* was observed in mouse tracheal explants, indicating that biofilm formation during infection is possible (Simmons and Dybvig 2009). Although biofilm-deficient strains and species are not necessarily attenuated, the protection given to the cells demonstrated in vitro indicates the benefit that biofilm formation would have on virulence.

The mechanism of biofilm formation has been extensively explored for *M. pulmonis* and is closely associated with a family of variable surface antigens, the Vsa proteins (Bhugra et al. 1995). Surface expression of short Vsa variants results in biofilm formation, and surface expression of long Vsa variants does not. Biofilm formation in *M. pulmonis* is therefore phase variable (Simmons and Dybvig 2007). The mechanism(s) driving biofilm formation by other *Mycoplasmatales* species has not been elucidated, but is likely to be fundamentally different because the *vsa* gene family is unique to *M. pulmonis*. Biofilm formation therefore represents another excellent example, alongside the terminal organelle and gliding motility (see above), of a virulence-associated phenotype arising from independent, parallel evolutionary events.

### Glycosidases

Carbohydrate manipulation is associated with pathogenicity in many bacteria but as an element of virulence was historically thought to be absent or very rare among the *Mycoplasmatales*. The sequencing of numerous *Mycoplasma* genomes has allowed for the recognition of several previously underreported glycosidases in mycoplasmas that can be phenotypically validated. These enzymes highlight an area with potential for novel strategies for the treatment and/or prevention of mycoplasmosis, as evidenced by the successful treatment of influenza patients with the sialidase inhibitors oseltamivir and zanamivir. Most bacteria use glycoside hydrolases to cleave off nutritional substrates, but there is evidence that this is not necessarily true of mycoplasmas. The presence of a sialidase gene in *M. gallisepticum* in the absence of any other genes essential for sialic acid catabolism suggests that the organism has maintained the presence of this gene for an alternative function (Papazisi et al. 2002; Szczepanek et al. 2010). The homologous gene in most *M. synoviae* strains has been retained in association with the accessory genes in the canonical sialic acid degradation pathway; however, at least two strains with large deletions have been described, ultimately resulting in dysfunction of the sialic acid catabolism pathway (May and Brown 2008). The role of glycosidases in nutritional fitness in vivo is unknown.

The enzymatic activities sialidase (neuraminidase), hyaluronidase, β-galactosidase, *N*-acetyl-β-hexosaminidase, α-mannosidase, and β-glucosidase are associated with virulence in other pathogens. Numerous alleles of each enzyme have been described in several *Mycoplasma* species (➋ *Table 39.7*). The role for these enzymes in virulence is the subject of ongoing studies, but sialidase and β-glucosidase have been implicated in pathogenicity in prospective, quantitative studies. The disruption of the sialidase gene in *M. gallisepticum* resulted in its attenuation in the Leghorn chicken model of infection, and sialidase in *M. synoviae* significantly correlates with strain virulence (as measured by disease severity) (May et al. 2007; May et al. 2012). Sialidase activity in *M. synoviae* also correlates with host cell binding, suggesting a functional balance between the two activities (May and Brown 2011). An in vitro infection system of bovine lung cells implicated β-glucosidase in *Mycoplasma mycoides* subsp. *mycoides* virulence, most likely by increasing the nutritional fitness of the bacteria (Vilei et al. 2007).

### Uncharacterized Virulence-Associated Genes

A virulence factor can be any phenotype that confers a selective advantage to a pathogen, allowing it to colonize the host, persist, propagate, and cause disease. Certain elements have been associated with increased virulence of some *Mycoplasma* species; however, their roles in pathogenesis may be incidental or undefined. The disruption of genes by random transposon mutagenesis of multiple *M. gallisepticum* open reading frames identified the uncharacterized lipoprotein MslA (Szczepanek et al. 2010) and dihydrolipoamide dehydrogenase (Lpd) (Hudson et al. 2006) as critical contributors to in vivo fitness or virulence. Similarly, the predicted nitrogen-fixation proteins NifS and NifU are essential for interaction between *M. agalactiae* and HeLa cells (Baranowski et al. 2010). Lpd, NifS, and NifU likely represent factors of metabolic attenuation, that is, the inability to survive and persist at the nutrient-poor host cell surface and thus an inability to cause disease. Metabolic attenuation has been described for multiple pathogens (Himpsl et al. 2008; Polissi et al. 1998; Seleem et al. 2008).

## Societal Impacts of the *Mycoplasmatales*

### Agricultural Losses

Worldwide economic losses stemming from mycoplasmosis of livestock and poultry are substantial. The high mortality of contagious bovine pleuropneumonia (CBPP) notwithstanding, the morbidity associated with mycoplasmosis of poultry, swine, cattle, and small ruminants results in decreased production. Several *Mycoplasma* species cause diseases that are listed by the World Organization for Animal Health (OIE) because

## Table 39.7
### Glycosidases of mycoplasmas

| Species | Sialidase | β-Glucosidase | β-Galactosidase | β-Hexosaminidase | α-Mannosidase | Hyaluronidase |
|---|---|---|---|---|---|---|
| M. alligatoris | + | + | + | + | – | + |
| M. canis[a] | + | + | – | + | – | + |
| M. mycoides mycoides | – | + | – | – | + | – |
| M. mycoides capri[b] | – | + | – | + | + | – |
| M. leachii | – | + | – | – | + | – |
| M. crocodyli | – | – | – | + | – | + |
| M. gallisepticum | + | + | – | – | – | – |
| M. synoviae | + | – | – | – | – | – |
| M. cloacale | + | N/R | – | – | – | N/R |
| M. corogypsi | + | N/R | – | – | – | N/R |
| M. cynos | + | N/R | – | – | – | N/R |
| M. molare | + | N/R | – | – | – | N/R |
| M. meleagridis[b] | + | N/R | – | – | – | N/R |
| M. iowae[b] | + | N/R | – | – | – | N/R |
| M. anseris[b] | + | N/R | – | – | – | N/R |
| M. pullorum[b] | + | N/R | – | – | – | N/R |
| M. anatis | – | N/R | – | + | – | N/R |
| M. testudinis | – | N/R | – | – | – | + |
| M. hominis | – | + | – | – | – | – |
| M. capricolum | – | + | – | – | – | – |
| M. fermentans | – | + | – | – | – | – |
| M. florum | – | + | – | – | – | – |

N/R not reported

[a] A sialidase-negative and two β-hexosaminidase-negative strains have been described, but the majority are positive

[b] A single β-hexosaminidase-positive (M. mycoides capri) or sialidase-positive (M. meleagridis, M. iowae, M. pullorum, M. anseris) strain has been described, but the majority are negative

of the potentially devastating impact on local economies, nutrition, and food availability (❯ Table 39.8).

Porcine enzootic pneumonia (PEP) due to M. hyopneumoniae infection results in reduced feed conversion, slow growth, and downgrading of carcasses of fattening pigs. Similarly, avian mycoplasmosis (the collective term for infection of poultry with M. gallisepticum, M. synoviae, M. meleagridis, M. iowae, or complex infections with one or more of these organisms; AM) results in decreased feed conversion, downgrading of carcasses, decreased egg production, and downgrading of eggs. Losses associated with AM are estimated at $700 million USD per year in the United States (Lancaster and Fabricant 1988; Mohammed et al. 1987), and losses are unquestionably higher worldwide. Strict management policies are in place to reduce transmission of mycoplasmas in flocks and herds, because AM and PEP affect the two most consumed meat animals in the world (see ❯ Table 39.8). Although it is not uniformly effective, vaccination against both diseases is used (see section on "❯ Vaccines"). "All in/all out" husbandry, wherein all animals in a single flock or herd arrive and leave a facility as a cohort and the facility is disinfected between cohorts, is recommended, but is not always feasible because of the expense of moving large groups of animals. As a result of both limitations, prevalence and morbidity for AM and PEP remain high. Contagious agalactia and enzootic pneumonia of cattle, goats, and sheep are also major sources of economic loss due to reduction or loss of milk production and poor feed conversion.

Potentially catastrophic economic losses can occur due to outbreaks of CBPP and contagious caprine pleuropneumonia (CCPP), and their impact on the livelihood of sub-Saharan Africans in particular is widespread and multifaceted. Cattle with CBPP have a high mortality rate, and survivors produce 90 % less milk, provide substantially less meat, and are far less able to perform production work (Mariner and Catley 2003; Tambi et al. 2006; Thiaucourt 2008). The common practice of quarantine is further devastating. Quarantine of infected herds restricts their movement, preventing animals from accessing grazing grounds and water sources, and often prevents the animals from being brought to markets. In this manner, quarantine impacts both the health of uninfected animals in the herd and the ability of farmers to access the value of their cattle or goats (Barrett et al. 2004). Treatment costs, quarantine impacts,

**Table 39.8**
**Mycoplasmosis in meat-producing animals**

| Mycoplasmosis[a] | Agent(s)[a] | Meat type | Consumption rank (global)[b] | OIE list |
|---|---|---|---|---|
| EP | MH | Pork | 1 | NO |
| AM | MG, MS, MI, MM | Poultry | 2 | YES |
| CBPP, M/A, EP | Mmm, MB, ML | Beef/veal | 3 | YES |
| None known[c] | N/A | Fish | 4 | N/A |
| CCPP, CA | Mcc, MC, Mmc, MA | Goat | 5 | YES |

[a]Abbreviations used in this table: *EP* enzoonotic pneumonia, *AM* avian mycoplasmosis, *CBPP* contagious bovine pleuropneumonia, *M/A* mastitis/arthritis, *CA* contagious agalactia, *CCPP* contagious caprine pleuropneumonia, *MH Mycoplasma hyopneumoniae*, *MG Mycoplasma gallisepticum*, *MS Mycoplasma synoviae*, *MI Mycoplasma iowae*, *MM Mycoplasma meleagridis*, *Mmm Mycoplasma mycoides* subsp. *mycoides*, *MB Mycoplasma bovis*, *ML Mycoplasma leachii*, formerly "*Mycoplasma* sp. bovine group 7", *Mcc Mycoplasma capricolum* subsp. *capripneumoniae*, *MC Mycoplasma capricolum*, *Mmc Mycoplasma mycoides* subsp. *capri*, *MA Mycoplasma agalactiae*

[b]Ranks of meat consumption according to the Food and Agriculture Organization (FAOSTAT) and the World Health Organization (WHO). Egg and milk consumption are not included in ranking

[c]*Mycoplasma mobile* was isolated from a tench with "red gill disease," but a causal relationship has yet to be demonstrated

the loss of production from sick animals, and the loss of cattle themselves result in a collective annual cost of over 44.8 million euros on the African continent due to CBPP alone (Tambi et al. 2006). The loss of a valuable food source has a community-level negative impact in addition to the monetary burden on farmers (Thiaucourt 2008). The eradication of CBPP and CCPP from North America, Europe, and Australia has slowed research progress in developed nations because of the resulting restrictions on laboratory access to *M. mycoides* subsp. *mycoides and M. capricolum* subsp. *capripneumoniae.* Existing diagnostic and vaccination strategies employed in endemic countries are largely impractical or ineffective, resulting in the continued prevalence of two diseases that substantially impact the nutritional, security, and economic standing of pastoralists and their communities.

## Genomics and Synthetic Cells

The groundbreaking developments of shotgun sequencing, the minimal genome concept, and synthetic genomics all took advantage of mycoplasmas as study organisms because of the small size of their genomes. *Mycoplasma genitalium* was the second complete genome ever published (Fraser et al. 1995) and the first organism to have its minimal genome (i.e., all of the genes indispensable for viability) described (Glass et al. 2006; Hutchison et al. 1999). Shotgun sequencing inarguably changed the landscape of genetics, initiating a new era in biological science. The minimal genome concept attempted to define what is necessary for cellular life, and this notion has since been expanded to include other reductively evolved bacteria such as the symbiont *Buchnera aphidicola* (Delaye and Moya 2010; Henry et al. 2010; van Ham et al. 2003).

The first organism ever to have its complete genetic complement synthetically generated and transplanted into a membrane producing a viable, self-replicating cell was *Mycoplasma mycoides* subspecies *capri* (Gibson et al. 2010; Lartigue et al. 2009). This proof-of-concept achievement opens the possibility of designing a genome with specific features that does not naturally exist, synthesizing it, and generating a living cell. Potential benefits could include the directed generation of pathogen vaccine strains that lack all of their native virulence factors and the generation of cells specifically designed to recover hydrocarbons, generate biofuels, and generate more efficient feedstock (Synthetic Genomics, Inc). Following the announcement of this discovery by the J. Craig Venter Institute, discussions of societal risks of this new technology occurred immediately. Proposed potential risks included the release of synthetic organisms into the environment resulting in ecological disruptions or the generation of more virulent pathogens resulting in widespread outbreaks. At this point the technical barriers to successfully generating a synthetic cell are substantial (Lauran 2010; Wang 2010). The generation of the first synthetic organism was a landmark achievement with great potential benefits to society as a whole and to the understanding of biological science.

## Acknowledgments

We would like to thank our colleagues from our current and past laboratories for sharing ideas and information. Figure 39.3 was contributed by the M.M. laboratory.

## References

Adams C, Pitzer J, Minion FC (2005) In vivo expression analysis of the P97 and P102 paralog families of *Mycoplasma hyopneumoniae*. Infect Immun 73:7784–7787

Algire MA, Lartigue C, Thomas DW, Assad-Garcia N, Glass JI, Merryman C (2009) New selectable marker for manipulating the simple genomes of *Mycoplasma* species. Antimicrob Agents Chemother 53:4429–4432

Allen JL, Noormohammadi AH, Browning GF (2005) The vlhA loci of *Mycoplasma synoviae* are confined to a restricted region of the genome. Microbiology 151:935–940

Almeida RA, Wannemuehler MJ, Rosenbusch RF (1992) Interaction of *Mycoplasma dispar* with bovine alveolar macrophages. Infect Immun 60:2914–2919

Alvarez RA, Blaylock MW, Baseman JB (2003) Surface localized glyceraldehyde-3-phosphate dehydrogenase of *Mycoplasma genitalium* binds mucin. Mol Microbiol 48:1417–1425

Amikam D, Razin S, Glaser G (1982) Ribosomal RNA genes in *Mycoplasma*. Nucleic Acids Res 10:4215–4222

Andreev J, Borovsky Z, Rosenshine I, Rottem S (1995) Invasion of HeLa cells by *Mycoplasma penetrans* and the induction of tyrosine phosphorylation of a 145-kDa host cell protein. FEMS Microbiol Lett 132:189–194

Atkinson TP, Balish MF, Waites KB (2008) Epidemiology, clinical manifestations, pathogenesis and laboratory detection of *Mycoplasma pneumoniae* infections. FEMS Microbiol Rev 32:956–973

Balish MF (2006) Subcellular structures of mycoplasmas. Front Biosci 11:2017–2027

Balish MF, Krause DC (2006) Mycoplasmas: a distinct cytoskeleton for wall-less bacteria. J Mol Microbiol Biotechnol 11:244–255

Balish MF, Ross SM, Fisseha M, Krause DC (2003) Deletion analysis identifies key functional domains of the cytadherence-associated protein HMW2 of *Mycoplasma pneumoniae*. Mol Microbiol 50:1507–1516

Baranowski E, Guiral S, Sagné E, Skapski A, Citti C (2010) Critical role of dispensable genes in *Mycoplasma agalactiae* interaction with mammalian cells. Infect Immun 78:1542–1551

Barker EN, Helps CR, Peters IR, Darby AC, Radford AD, Tasker S (2011) Complete genome sequence of *Mycoplasma haemofelis*, a hemotropic mycoplasma. J Bacteriol 193:2060–2061

Barker EN, Tasker S, Day MJ, Warman SM, Woolley K, Birtles R, Georges KC et al (2010) Development and use of real-time PCR to detect and quantify *Mycoplasma haemocanis* and "*Candidatus* Mycoplasma haematoparvum" in dogs. Vet Microbiol 140:167–170

Barrett C, Osterloh S, Little PD, McPeak JG (2004) Constraints limiting marketed off-take rates among pastoralists. GL-CRSP (Global Livestock Collaborative Research Support Program)/University of California, Davis

Baseggio N, Glew MD, Markham PF, Whithear KG, Browning GF (1996) Size and genomic location of the pMGA multigene family of *Mycoplasma gallisepticum*. Microbiology 142(Pt 6):1429–1435

Bébéar C (2012) Editorial commentary: infections due to macrolide-resistant *Mycoplasma pneumoniae*: now what? Clin Infect Dis 55:1650–1651

Bébéar CM, Bové JM, Bébéar C, Renaudin J (1997) Characterization of *Mycoplasma hominis* mutations involved in resistance to fluoroquinolones. Antimicrob Agents Chemother 41:269–273

Bébéar CM, Kempf I (2005) Antimicrobial therapy and antimicrobial resistance. In: Blanchard A, Browning G (eds) Mycoplasmas. Molecular biology, pathogenicity and strategies for control. Horizon Bioscience, Norfolk, pp 535–568

Bernet C, Garret M, de Barbeyrac B, Bébéar C, Bonnet J (1989) Detection of *Mycoplasma pneumoniae* by using the polymerase chain reaction. J Clin Microbiol 27:2492–2496

Besser TE, Frances Cassirer E, Highland MA, Wolff P, Justice-Allen A, Mansfield K, Davis MA, Foreyt W (2013) Bighorn sheep pneumonia: sorting out the cause of a polymicrobial disease. Prev Vet Med 108:85–93

Bhugra B, Voelker LL, Zou N, Yu H, Dybvig K (1995) Mechanism of antigenic variation in *Mycoplasma pulmonis*: interwoven, site-specific DNA inversions. Mol Microbiol 18:703–714

Bischof DF, Vilei EM, Frey J (2006) Genomic differences between type strain PG1 and field strains of *Mycoplasma mycoides* subsp. *mycoides* small-colony type. Genomics 88:633–641

Blanchard A (1997) Mycoplasmas and HIV infection, a possible interaction through immune activation. Wien Klin Wochenschr 109:590–593

Blanchard A, Crabb DM, Dybvig K, Duffy LB, Cassell GH (1992) Rapid detection of tetM in *Mycoplasma hominis* and *Ureaplasma urealyticum* by PCR: tetM confers resistance to tetracycline but not necessarily to doxycycline. FEMS Microbiol Lett 74:277–281

Bolca Topal N, Topal U, Gokalp G, Saraydaroglu O (2007) Eosinophilic mastitis. JBR-BTR 90:170–171

Boonstra E, Lindbaek M, Fidzani B, Bruusgaard D (2001) Cattle eradication and malnutrition in under five's: a natural experiment in Botswana. Public Health Nutr 4:877–882

Borrel A, Dujardin-Beaumetz E, Jeantet P, Jouan C (1910) Le microbe de la péripneumonie. Ann Inst Pasteur 24:168–179

Bose SR, Balish MF, Krause DC (2009) *Mycoplasma pneumoniae* cytoskeletal protein HMW2 and the architecture of the terminal organelle. J Bacteriol 191:6741–6748

Bove JM (1999) The one-hundredth anniversary of the first culture of a mollicute, the contagious bovine peripneumonia microbe, by Nocard and Roux, with the collaboration of Borrel, Salimbeni, and Dujardin-Beaumetz. Res Microbiol 150:239–245

Breton M, Tardy F, Dordet-Frisoni E, Sagne E, Mick V, Renaudin J, Sirand-Pugnet P, Citti C, Blanchard A (2012) Distribution and diversity of mycoplasma plasmids: lessons from cryptic genetic elements. BMC Microbiol 12:257

Brown DR, May M, Bradbury JM, Balish MF, Calcutt MJ, Glass JI, Tasker S et al (2011) Genus I. *Mycoplasma* Nowak 1929, 1349 nom. cons. Jud. Comm. Opin. 22, 1958, 166AL, p. 575–613. In: Krieg NR, Staley JT, Brown DR, Hedlund BP, Paster BJ, Ward NL, Ludwig W, Whitman WB (eds) Bergey's manual of systematic bacteriology, vol 4, 2nd edn. Springer, New York, NY

Brown DR, Zacher LA, Farmerie WG (2004) Spreading factors of *Mycoplasma alligatoris*, a flesh-eating mycoplasma. J Bacteriol 186:3922–3927

Burgos R, Wood GE, Young L, Glass JI, Totten PA (2012) RecA mediates MgpB and MgpC phase and antigenic variation in *Mycoplasma genitalium*, but plays a minor role in DNA repair. Mol Microbiol 85:669–683

Butler SE, Augostini P, Secor WE (2010) *Mycoplasma hominis* infection of *Trichomonas vaginalis* is not associated with metronidazole-resistant trichomoniasis in clinical isolates from the United States. Parasitol Res 107:1023–1027

Buzinhani M, Yamaguti M, Oliveira RC, Cortez BA, Marques LM, Machado-Santelli GM, Assumpcao ME, Timenetsky J (2011) Invasion of *Ureaplasma diversum* in bovine spermatozoids. BMC Res Notes 4:455

Cacciotto C, Addis MF, Pagnozzi D, Chessa B, Coradduzza E, Carcangiu L, Uzzau S, Alberti A, Pittau M (2010) The liposoluble proteome of *Mycoplasma agalactiae*: an insight into the minimal protein complement of a bacterial membrane. BMC Microbiol 10:225

Calcutt MJ, Lewis MS, Wise KS (2002) Molecular genetic analysis of ICEF, an integrative conjugal element that is present as a repetitive sequence in the chromosome of *Mycoplasma fermentans* PG18. J Bacteriol 184:6929–6941

Calisto BM, Broto A, Martinelli L, Querol E, Pinol J, Fita I (2012) The EAGR box structure: a motif involved in mycoplasma motility. Mol Microbiol 86:382–393

Catrein I, Herrmann R (2011) The proteome of *Mycoplasma pneumoniae*, a supposedly "simple" cell. Proteomics 11:3614–3632

Catrein I, Herrmann R, Bosserhoff A, Ruppert T (2005) Experimental proof for a signal peptidase I like activity in *Mycoplasma pneumoniae*, but absence of a gene encoding a conserved bacterial type I SPase. FEBS J 272:2892–2900

Cazanave C, Charron A, Renaudin H, Bébéar C (2012) Method comparison for molecular typing of French and Tunisian *Mycoplasma genitalium*-positive specimens. J Med Microbiol 61:500–506

Chernov VM, Gorshkov OV, Chernova OA, Baranova NB, Akopian TA, Trushin MV (2005) Variability of the Vaa cytoadhesin genes in clinical isolates of *Mycoplasma hominis*. New Microbiol 28:373–376

Chopra-Dewasthaly R, Baumgartner M, Gamper E, Innerebner C, Zimmermann M, Schilcher F, Tichy A et al (2012) Role of Vpma phase variation in *Mycoplasma agalactiae* pathogenesis. FEMS Immunol Med Microbiol 66:307–322

Chopra-Dewasthaly R, Citti C, Glew MD, Zimmermann M, Rosengarten R, Jechlinger W (2008) Phase-locked mutants of *Mycoplasma agalactiae*: defining the molecular switch of high-frequency Vpma antigenic variation. Mol Microbiol 67:1196–1210

Ciccarelli FD, Doerks T, von Mering C, Creevey CJ, Snel B, Bork P (2006) Toward automatic reconstruction of a highly resolved tree of life. Science 311:1283–1287

Cloward JM, Krause DC (2009) *Mycoplasma pneumoniae* J-domain protein required for terminal organelle function. Mol Microbiol 71:1296–1307

Cloward JM, Krause DC (2010) Functional domain analysis of the *Mycoplasma pneumoniae* co-chaperone TopJ. Mol Microbiol 77:158–169

Cloward JM, Krause DC (2011) Loss of co-chaperone TopJ impacts adhesin P1 presentation and terminal organelle maturation in *Mycoplasma pneumoniae*. Mol Microbiol 81:528–539

Cobb SP (2011) The spread of pathogens through trade in poultry hatching eggs: overview and recent developments. Rev Sci Tech 30:165–175

Cottew GS, Yeats FR (1982) Mycoplasmas and mites in the ears of clinically normal goats. Aust Vet J 59:77–81

Daddow KN (1980) Culex annulirostris as a vector of *Eperythrozoon ovis* infection in sheep. Vet Parasitol 7:313–317

Dallo SF, Baseman JB (2000) Intracellular DNA replication and long-term survival of pathogenic mycoplasmas. Microb Pathog 29:301–309

Dallo SF, Kannan TR, Blaylock MW, Baseman JB (2002) Elongation factor Tu and E1 beta subunit of pyruvate dehydrogenase complex act as fibronectin binding proteins in *Mycoplasma pneumoniae*. Mol Microbiol 46:1041–1051

De la Fe C, Amores J, Tardy F, Sagne E, Nouvel LX, Citti C (2012) Unexpected genetic diversity of *Mycoplasma agalactiae* caprine isolates from an endemic geographically restricted area of Spain. BMC Vet Res 8:146

Dedieu L, Mady V, Lefevre PC (1994) Development of a selective polymerase chain reaction assay for the detection of *Mycoplasma mycoides* subsp. *mycoides* S.C. (contagious bovine pleuropneumonia agent). Vet Microbiol 42:327–339

Dégrange S, Renaudin H, Charron A, Bébéar C, Bébéar CM (2008) Tetracycline resistance in *Ureaplasma* spp. and *Mycoplasma hominis*: prevalence in Bordeaux, France, from 1999 to 2002 and description of two tet(M)-positive isolates of *M. hominis* susceptible to tetracyclines. Antimicrob Agents Chemother 52:742–744

Delaney NF, Balenger S, Bonneaud C, Marx CJ, Hill GE, Ferguson-Noel N, Tsai P, Rodrigo A, Edwards SV (2012) Ultrafast evolution and loss of CRISPRs following a host shift in a novel wildlife pathogen, *Mycoplasma gallisepticum*. PLoS Genet 8:e1002511

Delaye L, Moya A (2010) Evolution of reduced prokaryotic genomes and the minimal cell concept: variations on a theme. Bioessays 32:281–287

Dessi D, Delogu G, Emonte E, Catania MR, Fiori PL, Rappelli P (2005) Long-term survival and intracellular replication of *Mycoplasma hominis* in *Trichomonas vaginalis* cells: potential role of the protozoon in transmitting bacterial infection. Infect Immun 73:1180–1186

Dhondt AA, Dhondt KV, Hawley DM, Jennelle CS (2007) Experimental evidence for transmission of *Mycoplasma gallisepticum* in house finches by fomites. Avian Pathol 36:205–208

Djordjevic SP, Cordwell SJ, Djordjevic MA, Wilton J, Minion FC (2004) Proteolytic processing of the *Mycoplasma hyopneumoniae* cilium adhesin. Infect Immun 72:2791–2802

Dumke R, Hausner M, Jacobs E (2011) Role of *Mycoplasma pneumoniae* glyceraldehyde-3-phosphate dehydrogenase (GAPDH) in mediating interactions with the human extracellular matrix. Microbiology 157:2328–2338

Dumke R, Jacobs E (2009) Comparison of commercial and in-house real-time PCR assays used for detection of *Mycoplasma pneumoniae*. J Clin Microbiol 47:441–444

Dumke R, Jacobs E (2011) Culture-independent multi-locus variable-number tandem-repeat analysis (MLVA) of *Mycoplasma pneumoniae*. J Microbiol Methods 86:393–396

Dupuy V, Manso-Silvan L, Barbe V, Thebault P, Dordet-Frisoni E, Citti C, Poumarat F et al (2012) Evolutionary history of contagious bovine pleuropneumonia using next generation sequencing of *Mycoplasma mycoides* subsp. *mycoides* "Small Colony". PLoS One 7:e46821

Edward DG, Freundt EA, Chanock RM, Fabricant J, Hayflick L, Lemcke RM, Rezin S, Somerson NL, Wittler RG (1967) Recommendations on nomenclature of the order *Mycoplasmatales*. Science 155:1694–1696

Enright MC, Spratt BG (1999) Multilocus sequence typing. Trends Microbiol 7:482–487

Eterpi M, McDonnell G, Thomas V (2011) Decontamination efficacy against *Mycoplasma*. Lett Appl Microbiol 52:150–155

Feberwee A, Landman WJ, von Banniseht-Wysmuller T, Klinkenberg D, Vernooij JC, Gielkens AL, Stegeman JA (2006) The effect of a live vaccine on the horizontal transmission of *Mycoplasma gallisepticum*. Avian Pathol 35:359–366

Fernsten PD, Pekny KW, Harper JR, Walker LE (1987) Antigenic mimicry of a human cellular polypeptide by *Mycoplasma hyorhinis*. Infect Immun 55:1680–1685

Filioussis G, Giadinis ND, Petridou EJ, Karavanis E, Papageorgiou K, Karatzias H (2011) Congenital polyarthritis in goat kids attributed to *Mycoplasma agalactiae*. Vet Rec 169:364

Fischer A, Shapiro B, Muriuki C, Heller M, Schnee C, Bongcam-Rudloff E, Vilei EM, Frey J, Jores J (2012) The origin of the '*Mycoplasma mycoides* cluster' coincides with domestication of ruminants. PLoS One 7:e36150

Fleury B, Bergonier D, Berthelot X, Peterhans E, Frey J, Vilei EM (2002) Characterization of P40, a cytadhesin of *Mycoplasma agalactiae*. Infect Immun 70:5612–5621

Fraga J, Rodriguez N, Fernandez C, Mondeja B, Sariego I, Fernandez-Calienes A, Rojas L (2012) *Mycoplasma hominis* in Cuban *Trichomonas vaginalis* isolates: association with parasite genetic polymorphism. Exp Parasitol 131:393–398

Fraser CM, Gocayne JD, White O, Adams MD, Clayton RA, Fleischmann RD, Bult CJ et al (1995) The minimal gene complement of *Mycoplasma genitalium*. Science 270:397–403

Fujihara Y, Sasaoka F, Suzuki J, Watanabe Y, Fujihara M, Ooshita K, Ano H, Harasawa R (2011) Prevalence of hemoplasma infection among cattle in the western part of Japan. J Vet Med Sci 73:1653–1655

Garcia-Castillo M, Morosini MI, Galvez M, Baquero F, del Campo R, Meseguer MA (2008) Differences in biofilm development and antibiotic susceptibility among clinical *Ureaplasma urealyticum* and *Ureaplasma parvum* isolates. J Antimicrob Chemother 62:1027–1030

Gardella RS, Del Giudice RA (1995) Growth of *Mycoplasma hyorhinis* cultivar alpha on semisynthetic medium. Appl Environ Microbiol 61:1976–1979

Gibson DG, Benders GA, Andrews-Pfannkoch C, Denisova EA, Baden-Tillson H, Zaveri J, Stockwell TB et al (2008) Complete chemical synthesis, assembly, and cloning of a *Mycoplasma genitalium* genome. Science 319:1215–1220

Gibson DG, Glass JI, Lartigue C, Noskov VN, Chuang RY, Algire MA, Benders GA et al (2010) Creation of a bacterial cell controlled by a chemically synthesized genome. Science 329:52–56

Giron JA, Lange M, Baseman JB (1996) Adherence, fibronectin binding, and induction of cytoskeleton reorganization in cultured human cells by *Mycoplasma penetrans*. Infect Immun 64:197–208

Glass JI, Assad-Garcia N, Alperovich N, Yooseph S, Lewis MR, Maruf M, Hutchison CA 3rd, Smith HO, Venter JC (2006) Essential genes of a minimal bacterium. Proc Natl Acad Sci USA 103:425–430

Glass JI, Lefkowitz EJ, Glass JS, Heiner CR, Chen EY, Cassell GH (2000) The complete sequence of the mucosal pathogen *Ureaplasma urealyticum*. Nature 407:757–762

Glew MD, Browning GF, Markham PF, Walker ID (2000) pMGA phenotypic variation in *Mycoplasma gallisepticum* occurs in vivo and is mediated by trinucleotide repeat length variation. Infect Immun 68:6027–6033

Goh MS, Gorton TS, Forsyth MH, Troy KE, Geary SJ (1998) Molecular and biochemical analysis of a 105 kDa *Mycoplasma gallisepticum* cytadhesin (GapA). Microbiology 144(Pt 11):2971–2978

Govorun VM, Gushchin AE, Ladygina VG, Abramycheva Nlu, Topol' Iulu (1998) Formation of *M. hominis* and *A. laidlawii* resistance to fluoroquinolones. Mol Gen Mikrobiol Virusol 3:16–19

Grau O, Tuppin P, Slizewicz B, Launay V, Goujard C, Bahraoui E, Delfraissy JF, Montagnier L (1998) A longitudinal study of seroreactivity against *Mycoplasma penetrans* in HIV-infected homosexual men: association with disease progression. AIDS Res Hum Retroviruses 14:661–667

Groebel K, Hoelzle K, Wittenbrink MM, Ziegler U, Hoelzle LE (2009) *Mycoplasma suis* invades porcine erythrocytes. Infect Immun 77:576–584

Gruson D, Pereyre S, Renaudin H, Charron A, Bébéar C, Bébéar CM (2005) In vitro development of resistance to six and four fluoroquinolones in *Mycoplasma pneumoniae* and *Mycoplasma hominis*, respectively. Antimicrob Agents Chemother 49:1190–1193

Hahn TW, Mothershed EA, Waldo RH, Krause DC (1999) Construction and analysis of a modified Tn4001 conferring chloramphenicol resistance in *Mycoplasma pneumoniae*. Plasmid 41:120–124

Halbedel S, Stulke J (2007) Tools for the genetic analysis of *Mycoplasma*. Int J Med Microbiol 297:37–44

Hamasuna R, Osada Y, Jensen JS (2007) Isolation of *Mycoplasma genitalium* from first-void urine specimens by coculture with Vero cells. J Clin Microbiol 45:847–850

Hames C, Halbedel S, Hoppert M, Frey J, Stülke J (2009) Glycerol metabolism is important for cytotoxicity of *Mycoplasma pneumoniae*. J Bacteriol 191:747–753

Hardy RD, Coalson JJ, Peters J, Chaparro A, Techasaensiri C, Cantwell AM, Kannan TR, Baseman JB, Dube PH (2009) Analysis of pulmonary inflammation and function in the mouse and baboon after exposure to *Mycoplasma pneumoniae* CARDS toxin. PLoS One 4:e7562

Hasselbring BM, Jordan JL, Krause RW, Krause DC (2006) Terminal organelle development in the cell wall-less bacterium *Mycoplasma pneumoniae*. Proc Natl Acad Sci USA 103:16478–16483

Hasselbring BM, Krause DC (2007) Proteins P24 and P41 function in the regulation of terminal-organelle development and gliding motility in *Mycoplasma pneumoniae*. J Bacteriol 189:7442–7449

Hasselbring BM, Sheppard ES, Krause DC (2012) P65 truncation impacts P30 dynamics during *Mycoplasma pneumoniae* gliding. J Bacteriol 194:3000–3007

Hata A, Honda Y, Asada K, Sasaki Y, Kenri T, Hata D (2008) *Mycoplasma hominis* meningitis in a neonate: case report and review. J Infect 57:338–343

Hatchel JM, Balish MF (2008) Attachment organelle ultrastructure correlates with phylogeny, not gliding motility properties, in *Mycoplasma pneumoniae* relatives. Microbiology 154:286–295

Henderson GP, Jensen GJ (2006) Three-dimensional structure of *Mycoplasma pneumoniae*'s attachment organelle and a model for its role in gliding motility. Mol Microbiol 60:376–385

Henrich B, Hopfe M, Kitzerow A, Hadding U (1999) The adherence-associated lipoprotein P100, encoded by an opp operon structure, functions as the oligopeptide-binding domain OppA of a putative oligopeptide transport system in *Mycoplasma hominis*. J Bacteriol 181:4873–4878

Henrich B, Schmitt M, Bergmann N, Zanger K, Kubitz R, Häussinger D, Pfeffer K (2010) *Mycoplasma salivarium* detected in a microbial community with *Candida glabrata* in the biofilm of an occluded biliary stent. J Med Microbiol 59:239–241

Henry C, Overbeek R, Stevens RL (2010) Building the blueprint of life. Biotechnol J 5:695–704

Hill AC (1984) *Mycoplasma cavipharyngis*, a new species isolated from the nasopharynx of guinea-pigs. J Gen Microbiol 130:3183–3188

Hill AC (1985) *Mycoplasma testudinis*, a new species isolated from a tortoise. Int J Syst Bacteriol 35:489–492

Himmelreich R, Plagens H, Hilbert H, Reiner B, Herrmann R (1997) Comparative analysis of the genomes of the bacteria *Mycoplasma pneumoniae* and *Mycoplasma genitalium*. Nucleic Acids Res 25:701–712

Himpsl SD, Lockatell CV, Hebel JR, Johnson DE, Mobley HL (2008) Identification of virulence determinants in uropathogenic *Proteus mirabilis* using signature-tagged mutagenesis. J Med Microbiol 57:1068–1078

Hirai Y, Kukida S, Matsushita O, Nagamachi E, Tomochika K, Kanemasa Y (1992) Membrane lipids of *Mycoplasma orale*: lipid composition and synthesis of phospholipids. Physiol Chem Phys Med NMR 24:21–27

Hirose K, Kawasaki Y, Kotani K, Abiko K, Sato H (2004) Characterization of a point mutation in the parC gene of *Mycoplasma bovirhinis* associated with fluoroquinolone resistance. J Vet Med B Infect Dis Vet Public Health 51:169–175

Hopfe M, Dahlmanns T, Henrich B (2011) In *Mycoplasma hominis* the OppA-mediated cytoadhesion depends on its ATPase activity. BMC Microbiol 11:185

Horino A, Kenri T, Sasaki Y, Okamura N, Sasaki T (2009) Identification of a site-specific tyrosine recombinase that mediates promoter inversions of phase-variable mpl lipoprotein genes in *Mycoplasma penetrans*. Microbiology 155:1241–1249

Horino A, Sasaki Y, Sasaki T, Kenri T (2003) Multiple promoter inversions generate surface antigenic variation in *Mycoplasma penetrans*. J Bacteriol 185:231–242

Hornok S, Micsutka A, Meli ML, Lutz H, Hofmann-Lehmann R (2011) Molecular investigation of transplacental and vector-borne transmission of bovine haemoplasmas. Vet Microbiol 152:411–414

Huang S, Li JY, Wu J, Meng L, Shou CC (2001) *Mycoplasma* infections and different human carcinomas. World J Gastroenterol 7:266–269

Hudson P, Gorton TS, Papazisi L, Cecchini K, Frasca S, Geary SJ (2006) Identification of a virulence-associated determinant, dihydrolipoamide dehydrogenase (lpd), in *Mycoplasma gallisepticum* through in vivo screening of transposon mutants. Infect Immun 74:931–939

Hughes RA, Hadden RD, Gregson NA, Smith KJ (1999) Pathogenesis of Guillain-Barre syndrome. J Neuroimmunol 100:74–97

Hulcr J, Rountree NR, Diamond SE, Stelinski LL, Fierer N, Dunn RR (2012) Mycangia of ambrosia beetles host communities of bacteria. Microb Ecol 64:784–793

Hutchison CA, Peterson SN, Gill SR, Cline RT, White O, Fraser CM, Smith HO, Venter JC (1999) Global transposon mutagenesis and a minimal *Mycoplasma* genome. Science 286:2165–2169

Iverson-Cabral SL, Astete SG, Cohen CR, Rocha EP, Totten PA (2006) Intrastrain heterogeneity of the mgpB gene in *Mycoplasma genitalium* is extensive in vitro and in vivo and suggests that variation is generated via recombination with repetitive chromosomal sequences. Infect Immun 74:3715–3726

Iverson-Cabral SL, Astete SG, Cohen CR, Totten PA (2007) mgpB and mgpC sequence diversity in *Mycoplasma genitalium* is generated by segmental reciprocal recombination with repetitive chromosomal sequences. Mol Microbiol 66:55–73

Jacobs E, Bartl A, Oberle K, Schiltz E (1995) Molecular mimicry by *Mycoplasma pneumoniae* to evade the induction of adherence inhibiting antibodies. J Med Microbiol 43:422–429

Jenkins C, Samudrala R, Geary SJ, Djordjevic SP (2008) Structural and functional characterization of an organic hydroperoxide resistance protein from *Mycoplasma gallisepticum*. J Bacteriol 190:2206–2216

Jenkins C, Wilton JL, Minion FC, Falconer L, Walker MJ, Djordjevic SP (2006) Two domains within the *Mycoplasma hyopneumoniae* cilium adhesin bind heparin. Infect Immun 74:481–487

Jensen JS (2004) *Mycoplasma genitalium*: the aetiological agent of urethritis and other sexually transmitted diseases. J Eur Acad Dermatol Venereol 18:1–11

Jensen JS (2012) Protocol for the detection of *Mycoplasma genitalium* by PCR from clinical specimens and subsequent detection of macrolide resistance-mediating mutations in region V of the 23S rRNA gene. Methods Mol Biol 903:129–139

Jimena ON, Laura JM, Elena MM, Alonso NH, Teresa QM (2009) Association of *Raillietia caprae* with the presence of mycoplasmas in the external ear canal of goats. Prev Vet Med 92:150–153

Johansson KE, Pettersson B (2002) Taxonomy of *Mollicutes*. In: Razin S, Herrmann R (eds) Molecular biology and pathogenicity of mycoplasmas. Kluwer/Academic/Plenum, New York, pp 1–29

Johnson C, Kannan TR, Baseman JB (2009) Characterization of a unique ADP-ribosyltransferase of *Mycoplasma penetrans*. Infect Immun 77:4362–4370

Jordan JL, Chang HY, Balish MF, Holt LS, Bose SR, Hasselbring BM, Waldo RH 3rd, Krunkosky TM, Krause DC (2007) Protein P200 is dispensable for *Mycoplasma pneumoniae* hemadsorption but not gliding motility or colonization of differentiated bronchial epithelium. Infect Immun 75:518–522

Juhas M, Eberl L, Glass JI (2011) Essence of life: essential genes of minimal genomes. Trends Cell Biol 21:562–568

Jurkovic DA, Hughes MR, Balish MF (2013) Analysis of energy sources for *Mycoplasma penetrans* gliding motility. FEMS Microbiol Lett 338:39–45

Jurkovic DA, Newman JT, Balish MF (2012) Conserved terminal organelle morphology and function in *Mycoplasma penetrans* and *Mycoplasma iowae*. J Bacteriol 194:2877–2883

Justice-Allen A, Trujillo J, Corbett R, Harding R, Goodell G, Wilson D (2010) Survival and replication of *Mycoplasma* species in recycled bedding sand and association with mastitis on dairy farms in Utah. J Dairy Sci 93:192–202

Kannan TR, Coalson JJ, Cagle M, Musatovova O, Hardy RD, Baseman JB (2011) Synthesis and distribution of CARDS toxin during *Mycoplasma pneumoniae* infection in a murine model. J Infect Dis 204:1596–1604

Kapatais-Zoumbos K, Chandler DK, Barile MF (1985) Survey of immunoglobulin A protease activity among selected species of *Ureaplasma* and *Mycoplasma*: specificity for host immunoglobulin A. Infect Immun 47:704–709

Kasai T, Nakane D, Ishida H, Ando H, Kiso M, Miyata M (2013) Role of binding in *Mycoplasma mobile* and *Mycoplasma pneumoniae* gliding analyzed through inhibition by synthesized sialylated compounds. J Bacteriol 195:429–435

Kenri T, Taniguchi R, Sasaki Y, Okazaki N, Narita M, Izumikawa K, Umetsu M, Sasaki T (1999) Identification of a new variable sequence in the P1 cytadhesin gene of *Mycoplasma pneumoniae*: evidence for the generation of antigenic variation by DNA recombination between repetitive sequences. Infect Immun 67:4557–4562

Keystone EC, Cunningham AJ, Metcalfe A, Kennedy M, Quinn PA (1982) Role of antibody in the protection of mice from arthritis induced by *Mycoplasma pulmonis*. Clin Exp Immunol 47:253–259

Kilian M, Freundt EA (1984) Exclusive occurrence of an extracellular protease capable of cleaving the hinge region of human immunoglobulin A1 in strains of *Ureaplasma urealyticum*. Isr J Med Sci 20:938–941

King GM, Judd C, Kuske CR, Smith C (2012) Analysis of stomach and gut microbiomes of the Eastern Oyster (*Crassostrea virginica*) from Coastal Louisiana, USA. PLoS One 7:e51475

Kirchhoff H, Rosengarten R, Lotz W, Fischer M, Lopatta D (1984) Flask-shaped mycoplasmas: properties and pathogenicity for man and animals. Isr J Med Sci 20:848–853

Kirchner H, Bauer A, Moritz T, Herbst F (1986) Lymphocyte activation and induction of interferon gamma in human leucocyte cultures by the mitogen in *Mycoplasma arthritidis* supernatant (MAS). Scand J Immunol 24:609–613

Kitzerow A, Hadding U, Henrich B (1999) Cyto-adherence studies of the adhesin P50 of *Mycoplasma hominis*. J Med Microbiol 48:485–493

Kitzerow A, Henrich B (2001) The cytosolic HinT protein of *Mycoplasma hominis* interacts with two membrane proteins. Mol Microbiol 41:279–287

Klement ML, Ojemyr L, Tagscherer KE, Widmalm G, Wieslander A (2007) A processive lipid glycosyltransferase in the small human pathogen *Mycoplasma pneumoniae*: involvement in host immune response. Mol Microbiol 65:1444–1457

Kokotovic B, Friis NF, Jensen JS, Ahrens P (1999) Amplified-fragment length polymorphism fingerprinting of *Mycoplasma* species. J Clin Microbiol 37:3300–3307

Koonin EV (2000) How many genes can make a cell: the minimal-gene-set concept. Annu Rev Genomics Hum Genet 1:99–116

Kornspan JD, Rottem S (2012) The phospholipid profile of mycoplasmas. J Lipids 2012:640762

Kornspan JD, Tarshis M, Rottem S (2011) Adhesion and biofilm formation of *Mycoplasma pneumoniae* on an abiotic surface. Arch Microbiol 193:833–836

Krause DC, Leith DK, Wilson RM, Baseman JB (1982) Identification of *Mycoplasma pneumoniae* proteins associated with hemadsorption and virulence. Infect Immun 35:809–817

Kumar PS, Mason MR, Brooker MR, O'Brien K (2012) Pyrosequencing reveals unique microbial signatures associated with healthy and failing dental implants. J Clin Periodontol 39:425–433

Kwon HJ, Kang JO, Cho SH, Kang HB, Kang KA, Kim JK, Kang YS et al (2004) Presence of human mycoplasma DNA in gastric tissue samples from Korean chronic gastritis patients. Cancer Sci 95:311–315

Lancaster JE, Fabricant J (1988) The history of avian medicine in the United States. IX. Events in the history of avian mycoplasmosis 1905–70. Avian Dis 32:607–623

Lartigue C, Glass JI, Alperovich N, Pieper R, Parmar PP, Hutchison CA 3rd, Smith HO, Venter JC (2007) Genome transplantation in bacteria: changing one species to another. Science 317:632–638

Lartigue C, Vashee S, Algire MA, Chuang RY, Benders GA, Ma L, Noskov VN et al (2009) Creating bacterial strains from genomes that have been cloned and engineered in yeast. Science 325:1693–1696

Le Carrou J, Laurentie M, Kobisch M, Gautier-Bouchardon AV (2006) Persistence of *Mycoplasma hyopneumoniae* in experimentally infected pigs after marbofloxacin treatment and detection of mutations in the *parC* gene. Antimicrob Agents Chemother 50:1959–1966

Lemaitre M, Henin Y, Destouesse F, Ferrieux C, Montagnier L, Blanchard A (1992) Role of mycoplasma infection in the cytopathic effect induced by human immunodeficiency virus type 1 in infected cell lines. Infect Immun 60:742–748

Lemcke RM, Pland J (1980) *Mycoplasma fastidiosum*: a new species from horses. Int J Syst Evol Microbiol 30:151–162

Ley DH, Berkhoff JE, McLaren JM (1996) *Mycoplasma gallisepticum* isolated from house finches (*Carpodacus mexicanus*) with conjunctivitis. Avian Dis 40:480–483

Li W, Raoult D, Fournier PE (2009) Bacterial strain typing in the genomic era. FEMS Microbiol Rev 33:892–916

Lindstedt BA (2005) Multiple-locus variable number tandem repeats analysis for genetic fingerprinting of pathogenic bacteria. Electrophoresis 26:2567–2582

Litamoi JK, Ayelet G, Rweyemamu MM (2005) Evaluation of the xerovac process for the preparation of heat tolerant contagious bovine pleuropneumonia (CBPP) vaccine. Vaccine 23:2573–2579

Liu L, Dybvig K, Panangala VS, van Santen VL, French CT (2000) GAA trinucleotide repeat region regulates M9/pMGA gene expression in *Mycoplasma gallisepticum*. Infect Immun 68:871–876

Lluch-Senar M, Querol E, Pinol J (2010) Cell division in a minimal bacterium in the absence of ftsZ. Mol Microbiol 78:278–289

Lo SC, Dawson MS, Wong DM, Newton PB 3rd, Sonoda MA, Engler WF, Wang RY et al (1989) Identification of *Mycoplasma incognitus* infection in patients with AIDS: an immunohistochemical, in situ hybridization and ultrastructural study. Am J Trop Med Hyg 41:601–616

Lo SC, Hayes MM, Wang RY, Pierce PF, Kotani H, Shih JW (1991a) Newly discovered mycoplasma isolated from patients infected with HIV. Lancet 338:1415–1418

Lo SC, Tsai S, Benish JR, Shih JW, Wear DJ, Wong DM (1991b) Enhancement of HIV-1 cytocidal effects in CD4$^+$ lymphocytes by the AIDS-associated mycoplasma. Science 251:1074–1076

Lockaby SB, Hoerr FJ, Lauerman LH, Smith BF, Samoylov AM, Toivio-Kinnucan MA, Kleven SH (1999) Factors associated with virulence of *Mycoplasma synoviae*. Avian Dis 43:251–261

Lorenzon S, Arzul I, Peyraud A, Hendrikx P, Thiaucourt F (2003) Molecular epidemiology of contagious bovine pleuropneumonia by multilocus sequence analysis of *Mycoplasma mycoides* subspecies *mycoides* biotype SC strains. Vet Microbiol 93:319–333

Luo W, Yu H, Cao Z, Schoeb TR, Marron M, Dybvig K (2008) Association of *Mycoplasma arthritidis* mitogen with lethal toxicity but not with arthritis in mice. Infect Immun 76:4989–4998

Lutkenhaus J (2012) The ParA/MinD family puts things in their place. Trends Microbiol 20:411–418

Lysnyansky I, Gerchman I, Perk S, Levisohn S (2008) Molecular characterization and typing of enrofloxacin-resistant clinical isolates of *Mycoplasma gallisepticum*. Avian Dis 52:685–689

Ma L, Jensen JS, Mancuso M, Hamasuna R, Jia Q, McGowin CL, Martin DH (2010) Genetic variation in the complete MgPa operon and its repetitive chromosomal elements in clinical strains of *Mycoplasma genitalium*. PLoS One 5:e15660

Ma L, Jensen JS, Mancuso M, Hamasuna R, Jia Q, McGowin CL, Martin DH (2012) Variability of trinucleotide tandem repeats in the MgPa operon and its repetitive chromosomal elements in *Mycoplasma genitalium*. J Med Microbiol 61:191–197

Ma L, Jensen JS, Myers L, Burnett J, Welch M, Jia Q, Martin DH (2007) *Mycoplasma genitalium*: an efficient strategy to generate genetic variation from a minimal genome. Mol Microbiol 66:220–236

Ma L, Taylor S, Jensen JS, Myers L, Lillis R, Martin DH (2008) Short tandem repeat sequences in the *Mycoplasma genitalium* genome and their use in a multilocus genotyping system. BMC Microbiol 8:130

Machado CX, Pinto PM, Zaha A, Ferreira HB (2009) A peroxiredoxin from *Mycoplasma hyopneumoniae* with a possible role in $H_2O_2$ detoxification. Microbiology 155:3411–3419

Mahairas GG, Minion FC (1989) Random insertion of the gentamicin resistance transposon Tn4001 in *Mycoplasma pulmonis*. Plasmid 21:43–47

Malandain H (2004) IgE antibody in the serum – the main problem is cross-reactivity. Allergy 59:229; author reply 230

Manhart LE, Mostad SB, Baeten JM, Astete SG, Mandaliya K, Totten PA (2008) High *Mycoplasma genitalium* organism burden is associated with shedding of HIV-1 DNA from the cervix. J Infect Dis 197:733–736

Maniloff J (2002) Phylogeny and evolution. In: Razin S, Herrmann R (eds) Molecular biology and pathogenicity of mycoplasmas. Kluwer/Academic/Plenum, New York, pp 31–43

Manso-Silvan L, Dupuy V, Chu Y, Thiaucourt F (2011) Multi-locus sequence analysis of *Mycoplasma capricolum* subsp. *capripneumoniae* for the molecular epidemiology of contagious caprine pleuropneumonia. Vet Res 42:86

Manso-Silvan L, Perrier X, Thiaucourt F (2007) Phylogeny of the *Mycoplasma mycoides* cluster based on analysis of five conserved protein-coding sequences and possible implications for the taxonomy of the group. Int J Syst Evol Microbiol 57:2247–2258

Manso-Silvan L, Vilei EM, Sachse K, Djordjevic SP, Thiaucourt F, Frey J (2009) *Mycoplasma leachii* sp. nov. as a new species designation for *Mycoplasma* sp. bovine group 7 of Leach, and reclassification of *Mycoplasma mycoides* subsp. *mycoides* LC as a serovar of *Mycoplasma mycoides* subsp. *capri*. Int J Syst Evol Microbiol 59:1353–1358

Maquelin K, Hoogenboezem T, Jachtenberg JW, Dumke R, Jacobs E, Puppels GJ, Hartwig NG, Vink C (2009) Raman spectroscopic typing reveals the presence of carotenoids in *Mycoplasma pneumoniae*. Microbiology 155:2068–2077

March JB, Gammack C, Nicholas R (2000) Rapid detection of contagious caprine pleuropneumonia using a *Mycoplasma capricolum* subsp. *capripneumoniae* capsular polysaccharide-specific antigen detection latex agglutination test. J Clin Microbiol 38:4152–4159

March JB, Waite ER, Litamoi JK (2002) Re-suspension of T(1)44 vaccine cultures of *Mycoplasma mycoides* subsp *mycoides* SC in 1 molar MgSO(4) causes a drop in pH and a rapid reduction in titre. FEMS Immunol Med Microbiol 34:97–103

Mardassi BB, Aissani N, Moalla I, Dhahri D, Dridi A, Mlik B (2012) Evidence for the predominance of a single tet(M) gene sequence type in tetracycline-resistant *Ureaplasma parvum* and *Mycoplasma hominis* isolates from Tunisian patients. J Med Microbiol 61:1254–1261

Marenda M, Barbe V, Gourgues G, Mangenot S, Sagne E, Citti C (2006) A new integrative conjugative element occurs in *Mycoplasma agalactiae* as chromosomal and free circular forms. J Bacteriol 188:4137–4141

Mariner JC, Catley A (2003) Towards sustainable CBPP control programmes for Africa, The dynamics of CBPP endemism and development of effective control strategies. Food and Agriculture Organization of the United Nations Corporate Document Repository, Rome

Markham F, Noormohammadi AH (2005) Diagnosis of mycoplasmosis in animals. In: Blanchard A, Browning G (eds) Mycoplasmas. Molecular biology, pathogenicity and strategies for control. Horizon Bioscience, Norfolk, pp 355–382

Markham PF, Duffy MF, Glew MD, Browning GF (1999) A gene family in *Mycoplasma imitans* closely related to the pMGA family of *Mycoplasma gallisepticum*. Microbiology 145(Pt 8):2095–2103

Markham PF, Glew MD, Browning GF, Whithear KG, Walker ID (1998) Expression of two members of the pMGA gene family of *Mycoplasma gallisepticum* oscillates and is influenced by pMGA-specific antibodies. Infect Immun 66:2845–2853

Mason RW, Statham P (1991a) The determination of the level of *Eperythrozoon ovis* parasitaemia in chronically infected sheep and its significance to the spread of infection. Aust Vet J 68:115–116

Mason RW, Statham P (1991b) Susceptibility of sheep and goats to *Eperythrozoon ovis* infection. Aust Vet J 68:116–117

May M, Brown DR (2008) Genetic variation in sialidase and linkage to *N*-acetylneuraminate catabolism in *Mycoplasma synoviae*. Microb Pathog 45:38–44

May M, Brown DR (2011) Diversity of expressed vlhA adhesin sequences and intermediate hemagglutination phenotypes in *Mycoplasma synoviae*. J Bacteriol 193:2116–2121

May M, Kleven SH, Brown DR (2007) Sialidase activity in *Mycoplasma synoviae*. Avian Dis 51:829–833

May M, Szczepanek SM, Frasca S, Gates AE, Demcovitz DL, Moneypenny CG, Brown DR, Geary SJ (2012) Effects of sialidase knockout and complementation on virulence of *Mycoplasma gallisepticum*. Vet Microbiol 157:91–95

Mayor D, Jores J, Korczak BM, Kuhnert P (2008) Multilocus sequence typing (MLST) of *Mycoplasma hyopneumoniae*: a diverse pathogen with limited clonality. Vet Microbiol 127:63–72

McAuliffe L, Ellis RJ, Miles K, Ayling RD, Nicholas RA (2006) Biofilm formation by *Mycoplasma* species and its role in environmental persistence and survival. Microbiology 152:913–922

McAuliffe L, Gosney F, Hlusek M, de Garnica ML, Spergser J, Kargl M, Rosengarten R et al (2011) Multilocus sequence typing of *Mycoplasma agalactiae*. J Med Microbiol 60:803–811

McGowin CL, Anderson-Smits C (2011) *Mycoplasma genitalium*: an emerging cause of sexually transmitted disease in women. PLoS Pathog 7:e1001324

Messick JB, Santos AP, Guimaraes AM (2011) Complete genome sequences of two hemotropic mycoplasmas, *Mycoplasma haemofelis* strain Ohio2 and *Mycoplasma suis* Strain Illinois. J Bacteriol 193:2068–2069

Messick JB, Walker PG, Raphael W, Berent L, Shi X (2002) '*Candidatus* mycoplasma haemodidelphidis' sp. nov., '*Candidatus* Mycoplasma haemolamae' sp. nov. and *Mycoplasma haemocanis* comb. nov., haemotrophic parasites from a naturally infected opossum (*Didelphis virginiana*), alpaca (*Lama pacos*) and dog (*Canis familiaris*): phylogenetic and secondary structural relatedness of their 16S rRNA genes to other mycoplasmas. Int J Syst Evol Microbiol 52:693–698

Mierzejewska J, Jagura-Burdzy G (2012) Prokaryotic ParA-ParB-parS system links bacterial chromosome segregation with the cell cycle. Plasmid 67:1–14

Lauran N (2010) A step to artificial life: manmade DNA powers cell. Associated Press, May 20, 2010 (Also published in *The Seattle Times*, May 20, 2010. http://seattletimes.com/html/nationworld/2011916800_cell21.html)

Milne JL, Subramaniam S (2009) Cryo-electron tomography of bacteria: progress, challenges and future prospects. Nat Rev Microbiol 7:666–675

Mingorance J, Rivas G, Velez M, Gomez-Puertas P, Vicente M (2010) Strong FtsZ is with the force: mechanisms to constrict bacteria. Trends Microbiol 18:348–356

Minion FC, Goguen JD (1985) Trypsin-sensitive, bovine serum albumin-dependent hemolysis activity in *Mycoplasma pulmonis*. Infect Immun 49:440–442

Minion FC, Jarvill-Taylor K (1994) Membrane-associated hemolysin activities in mycoplasmas. FEMS Microbiol Lett 116:101–106

Miyata M (2010) Unique centipede mechanism of mycoplasma gliding. Annu Rev Microbiol 64:519–537

Miyata M, Ryu WS, Berg HC (2002) Force and velocity of *Mycoplasma mobile* gliding. J Bacteriol 184:1827–1831

Mohammed HO, Carpenter TE, Yamamoto R (1987) Economic impact of *Mycoplasma gallisepticum* and *M. synoviae* in commercial layer flocks. Avian Dis 31:477–482

Moitinho-Silva L, Heineck BL, Reolon LA, Paes JA, Klein CS, Rebelatto R, Schrank IS, Zaha A, Ferreira HB (2012) *Mycoplasma hyopneumoniae* type I signal peptidase: expression and evaluation of its diagnostic potential. Vet Microbiol 154:282–291

Montague MG, Lartigue C, Vashee S (2012) Synthetic genomics: potential and limitations. Curr Opin Biotechnol 23:659–665

Mu HH, Sawitzke AD, Cole BC (2000) Modulation of cytokine profiles by the *Mycoplasma* superantigen *Mycoplasma arthritidis* mitogen parallels susceptibility to arthritis induced by *M. arthritidis*. Infect Immun 68:1142–1149

Muir MT, Cohn SM, Louden C, Kannan TR, Baseman JB (2011) Novel toxin assays implicate *Mycoplasma pneumoniae* in prolonged ventilator course and hypoxemia. Chest 139:305–310

Nagai R, Miyata M (2006) Gliding motility of *Mycoplasma mobile* can occur by repeated binding to *N*-acetylneuraminyllactose (sialyllactose) fixed on solid surfaces. J Bacteriol 188:6469–6475

Nakane D, Miyata M (2007) Cytoskeletal "jellyfish" structure of *Mycoplasma mobile*. Proc Natl Acad Sci USA 104:19518–19523

Nakane D, Miyata M (2009) Cytoskeletal asymmetrical dumbbell structure of a gliding mycoplasma, *Mycoplasma gallisepticum*, revealed by negative-staining electron microscopy. J Bacteriol 191:3256–3264

Neimark H, Johansson KE, Rikihisa Y, Tully JG (2001) Proposal to transfer some members of the genera *Haemobartonella* and *Eperythrozoon* to the genus *Mycoplasma* with descriptions of '*Candidatus* Mycoplasma haemofelis', '*Candidatus* Mycoplasma haemomuris', '*Candidatus* Mycoplasma haemosuis' and '*Candidatus* Mycoplasma wenyonii'. Int J Syst Evol Microbiol 51:891–899

Neyrolles O, Brenner C, Prevost MC, Fontaine T, Montagnier L, Blanchard A (1998) Identification of two glycosylated components of *Mycoplasma penetrans*: a surface-exposed capsular polysaccharide and a glycolipid fraction. Microbiology 144(Pt 5):1247–1255

Niang M, Rosenbusch RF, DeBey MC, Niyo Y, Andrews JJ, Kaeberle ML (1998) Field isolates of *Mycoplasma ovipneumoniae* exhibit distinct cytopathic effects in ovine tracheal organ cultures. Zentralbl Veterinarmed A 45:29–40

Nielsen MH, Nielsen R (1975) Electron microscopy of *Trichomonas vaginalis* Donné interaction with vaginal epithelium in human trichomoniasis. Acta Pathol Microbiol Scand 83:381–389

Nikol'skii SN, Slipchenki SN (1969) Experiments in the transmission of *Eperythrozoon ovis* by the ticks *H. plumbeum* and *Rh. bursa*. Veterinariia 5:46, Russia

Nocard E, Roux E (1898) Le microbe de la péripneumonie. Ann Inst Pasteur 12:240–262

Noormohammadi AH, Markham PF, Duffy MF, Whithear KG, Browning GF (1998) Multigene families encoding the major hemagglutinins in phylogenetically distinct mycoplasmas. Infect Immun 66:3470–3475

Noormohammadi AH, Markham PF, Kanci A, Whithear KG, Browning GF (2000) A novel mechanism for control of antigenic variation in the haemagglutinin gene family of *Mycoplasma synoviae*. Mol Microbiol 35:911–923

Nowak J (1929) Morphologie, nature et cycle évolutif du microbe de la péripneumonie. Ann Inst Pasteur 43:1330–1352

Ochman H, Lerat E, Daubin V (2005) Examining bacterial species under the specter of gene transfer and exchange. Proc Natl Acad Sci USA 102(Suppl 1):6595–6599

Osawa M, Erickson HP (2006) FtsZ from divergent foreign bacteria can function for cell division in *Escherichia coli*. J Bacteriol 188:7132–7140

Papazisi L, Frasca S, Gladd M, Liao X, Yogev D, Geary SJ (2002) GapA and CrmA coexpression is essential for *Mycoplasma gallisepticum* cytadherence and virulence. Infect Immun 70:6839–6845

Paralanov V, Lu J, Duffy LB, Crabb DM, Shrivastava S, Methé BA, Inman J et al (2012) Comparative genome analysis of 19 *Ureaplasma urealyticum* and *Ureaplasma parvum* strains. BMC Microbiol 12:88

Pereyre S, Gonzalez P, De Barbeyrac B, Darnige A, Renaudin H, Charron A, Raherison S, BÉbÉar C, BÉbÉar CM (2002) Mutations in 23S rRNA account for intrinsic resistance to macrolides in *Mycoplasma hominis* and *Mycoplasma fermentans* and for acquired resistance to macrolides in *M. hominis*. Antimicrob Agents Chemother 46:3142–3150

Pereyre S, Renaudin H, Touati A, Charron A, Peuchant O, Hassen AB, Bébéar C, Bébéar CM (2010) Detection and susceptibility testing of *Mycoplasma amphoriforme* isolates from patients with respiratory tract infections. Clin Microbiol Infect 16:1007–1009

Pereyre S, Sirand-Pugnet P, Beven L, Charron A, Renaudin H, Barre A, Avenaud P et al (2009) Life on arginine for *Mycoplasma hominis*: clues from its minimal genome and comparison with other human urogenital mycoplasmas. PLoS Genet 5:e1000677

Perez G, Skurnick JH, Denny TN, Stephens R, Kennedy CA, Regivick N, Nahmias A et al (1998) Herpes simplex type II and *Mycoplasma genitalium* as risk factors for heterosexual HIV transmission: report from the heterosexual HIV transmission study. Int J Infect Dis 3:5–11

Peters J, Singh H, Brooks EG, Diaz J, Kannan TR, Coalson JJ, Baseman JG, Cagle M, Baseman JB (2011) Persistence of community-acquired respiratory distress syndrome toxin-producing *Mycoplasma pneumoniae* in refractory asthma. Chest 140:401–407

Peuchant O, Menard A, Renaudin H, Morozumi M, Ubukata K, Bébéar CM, Pereyre S (2009) Increased macrolide resistance of *Mycoplasma pneumoniae* in France directly detected in clinical specimens by real-time PCR and melting curve analysis. J Antimicrob Chemother 64:52–58

Piboonpocanun S, Chiba H, Mitsuzawa H, Martin W, Murphy RC, Harbeck RJ, Voelker DR (2005) Surfactant protein A binds *Mycoplasma pneumoniae* with high affinity and attenuates its growth by recognition of disaturated phosphatidylglycerols. J Biol Chem 280:9–17

Pilo P, Vilei EM, Peterhans E, Bonvin-Klotz L, Stoffel MH, Dobbelaere D, Frey J (2005) A metabolic enzyme as a primary virulence factor of *Mycoplasma mycoides* subsp. *mycoides* small colony. J Bacteriol 187:6824–6831

Pinho L, Thompson G, Rosenbusch R, Carvalheira J (2012) Genotyping of *Mycoplasma bovis* isolates using multiple-locus variable-number tandem-repeat analysis. J Microbiol Methods 88:377–385

Pitcher DG, Windsor D, Windsor H, Bradbury JM, Yavari C, Jensen JS, Ling C, Webster D (2005) *Mycoplasma amphoriforme* sp. nov., isolated from a patient with chronic bronchopneumonia. Int J Syst Evol Microbiol 55:2589–2594

Polissi A, Pontiggia A, Feger G, Altieri M, Mottl H, Ferrari L, Simon DC (1998) Large-scale identification of virulence genes from *Streptococcus pneumoniae*. Infect Immun 66:5620–5629

Pour-El I, Adams C, Minion FC (2002) Construction of mini-Tn4001tet and its use in *Mycoplasma gallisepticum*. Plasmid 47:129–137

Prullage JB, Williams RE, Gaafar SM (1993) On the transmissibility of *Eperythrozoon suis* by *Stomoxys calcitrans* and *Aedes aegypti*. Vet Parasitol 50:125–135

Raccach M, Rottem S, Razin S (1975) Survival of frozen mycoplasmas. Appl Microbiol 30:167–171

Radestock U, Bredt W (1977) Motility of *Mycoplasma pneumoniae*. J Bacteriol 129:1495–1501

Raherison S, Gonzalez P, Renaudin H, Charron A, Bébéar C, Bébéar CM (2002) Evidence of active efflux in resistance to ciprofloxacin and to ethidium bromide by *Mycoplasma hominis*. Antimicrob Agents Chemother 46:672–679

Raherison S, Gonzalez P, Renaudin H, Charron A, Bébéar C, Bébéar CM (2005) Increased expression of two multidrug transporter-like genes is associated with ethidium bromide and ciprofloxacin resistance in *Mycoplasma hominis*. Antimicrob Agents Chemother 49:421–424

Rappelli P, Addis MF, Carta F, Fiori PL (1998) *Mycoplasma hominis* parasitism of *Trichomonas vaginalis*. Lancet 352:1286

Razin S, Efrati H, Kutner S, Rottem S (1982) Cholesterol and phospholipid uptake by mycoplasmas. Rev Infect Dis 4(Suppl):S85–S92

Reinhardt AK, Kempf I, Kobisch M, Gautier-Bouchardon AV (2002) Fluoroquinolone resistance in *Mycoplasma gallisepticum*: DNA gyrase as primary target of enrofloxacin and impact of mutations in topoisomerases on resistance level. J Antimicrob Chemother 50:589–592

Relich RF, Balish MF (2011) Insights into the function of *Mycoplasma pneumoniae* protein P30 from orthologous gene replacement. Microbiology 157:2862–2870

Relich RF, Friedberg AJ, Balish MF (2009) Novel cellular organization in a gliding mycoplasma, *Mycoplasma insons*. J Bacteriol 191:5312–5314

Robertson JA (1978) Bromothymol blue broth: improved medium for detection of *Ureaplasma urealyticum* (T-strain mycoplasma). J Clin Microbiol 7:127–132

Robertson JA, Stemler ME, Stemke GW (1984) Immunoglobulin A protease activity of *Ureaplasma urealyticum*. J Clin Microbiol 19:255–258

Rodwell AW (1983) Defined and partly defined media. In: Razin S, Tully JG (eds) Methods in mycoplasmology. Molecular characterization, vol I. Academic, New York, pp 163–172

Rogers MJ, Simmons J, Walker RT, Weisburg WG, Woese CR, Tanner RS, Robinson IM et al (1985) Construction of the mycoplasma evolutionary tree from 5S rRNA sequence data. Proc Natl Acad Sci USA 82:1160–1164

Romero R, Garite TJ (2008) Twenty percent of very preterm neonates (23–32 weeks of gestation) are born with bacteremia caused by genital mycoplasmas. Am J Obstet Gynecol 198:1–3

Ron Y, Flitman-Tene R, Dybvig K, Yogev D (2002) Identification and characterization of a site-specific tyrosine recombinase within the variable loci of *Mycoplasma bovis*, *Mycoplasma pulmonis* and *Mycoplasma agalactiae*. Gene 292:205–211

Rottem S (1980) Membrane lipids of mycoplasmas. Biochim Biophys Acta 604:65–90

Rottem S (2002) Choline-containing lipids in mycoplasmas. Microbes Infect 4:963–968

Rottem S, Adar L, Gross Z, Ne'eman Z, Davis PJ (1986) Incorporation and modification of exogenous phosphatidylcholines by mycoplasmas. J Bacteriol 167:299–304

Sachse K, Grajetzki C, Rosengarten R, Hanel I, Heller M, Pfutzner H (1996) Mechanisms and factors involved in *Mycoplasma bovis* adhesion to host cells. Zentralbl Bakteriol 284:80–92

Salman M, Rottem S (1995) The cell membrane of *Mycoplasma penetrans*: lipid composition and phospholipase A1 activity. Biochim Biophys Acta 1235:369–377

Sanchez-Pescador R, Brown JT, Roberts M, Urdea MS (1988) The nucleotide sequence of the tetracycline resistance determinant tetM from *Ureaplasma urealyticum*. Nucleic Acids Res 16:1216–1217

Sasaki Y, Blanchard A, Watson HL, Garcia S, Dulioust A, Montagnier L, Gougeon ML (1995) In vitro influence of *Mycoplasma penetrans* on activation of peripheral T lymphocytes from healthy donors or human immunodeficiency virus-infected individuals. Infect Immun 63:4277–4283'

Sato C, Manaka S, Nakane D, Nishiyama H, Suga M, Nishizaka T, Miyata M, Maruyama Y (2012) Rapid imaging of mycoplasma in solution using atmospheric scanning electron microscopy (ASEM). Biochem Biophys Res Commun 417:1213–1218

Schmidl SR, Otto A, Lluch-Senar M, Pinol J, Busse J, Becher D, Stülke J (2011) A trigger enzyme in *Mycoplasma pneumoniae*: impact of the glycerophosphodiesterase GlpQ on virulence and gene expression. PLoS Pathog 7: e1002263

Schmidt JA, Browning GF, Markham PF (2007) *Mycoplasma hyopneumoniae* mhp379 is a $Ca^{2+}$-dependent, sugar-nonspecific exonuclease exposed on the cell surface. J Bacteriol 189:3414–3424

Schwartz SB, Thurman KA, Mitchell SL, Wolff BJ, Winchell JM (2009) Genotyping of *Mycoplasma pneumoniae* isolates using real-time PCR and high-resolution melt analysis. Clin Microbiol Infect 15:756–762

Seleem MN, Boyle SM, Sriranganathan N (2008) *Brucella*: a pathogen without classic virulence genes. Vet Microbiol 129:1–14

Seto S, Kenri T, Tomiyama T, Miyata M (2005a) Involvement of P1 adhesin in gliding motility of *Mycoplasma pneumoniae* as revealed by the inhibitory effects of antibody under optimized gliding conditions. J Bacteriol 187:1875–1877

Seto S, Uenoyama A, Miyata M (2005b) Identification of a 521-kilodalton protein (Gli521) involved in force generation or force transmission for *Mycoplasma mobile* gliding. J Bacteriol 187:3502–3510

Seto S, Miyata M (2003) Attachment organelle formation represented by localization of cytadherence proteins and formation of the electron-dense core in wild-type and mutant strains of *Mycoplasma pneumoniae*. J Bacteriol 185:1082–1091

Seybert A, Herrmann R, Frangakis AS (2006) Structural analysis of *Mycoplasma pneumoniae* by cryo-electron tomography. J Struct Biol 156:342–354

Seymour LM, Jenkins C, Deutscher AT, Raymond BB, Padula MP, Tacchi JL, Bogema DR et al (2012) Mhp182 (P102) binds fibronectin and contributes to the recruitment of plasmin(ogen) to the *Mycoplasma hyopneumoniae* cell surface. Cell Microbiol 14:81–94

Shaw BM, Simmons WL, Dybvig K (2012) The Vsa shield of *Mycoplasma pulmonis* is antiphagocytic. Infect Immun 80:704–709

Shepard MC, Lunceford CD (1976) Differential agar medium (A7) for identification of *Ureaplasma urealyticum* (human T mycoplasmas) in primary cultures of clinical material. J Clin Microbiol 3:613–625

Shimada Y, Deguchi T, Nakane K, Masue T, Yasuda M, Yokoi S, Ito S, Nakano M, Ishiko H (2010a) Emergence of clinical strains of *Mycoplasma genitalium* harbouring alterations in ParC associated with fluoroquinolone resistance. Int J Antimicrob Agents 36:255–258

Shimada Y, Deguchi T, Yamaguchi Y, Yasuda M, Nakane K, Yokoi S, Ito S, Nakano M, Ishiko H (2010b) gyrB and parE mutations in urinary *Mycoplasma genitalium* DNA from men with non-gonococcal urethritis. Int J Antimicrob Agents 36:477–478

Simecka JW, Davis JK, Cassell GH (1987) Specific vs. nonspecific immune responses in murine respiratory mycoplasmosis. Isr J Med Sci 23:485–489

Simmons WL, Dybvig K (2007) Biofilms protect *Mycoplasma pulmonis* cells from lytic effects of complement and gramicidin. Infect Immun 75:3696–3699

Simmons WL, Dybvig K (2009) Mycoplasma biofilms ex vivo and in vivo. FEMS Microbiol Lett 295:77–81

Sirand-Pugnet P, Citti C, Barre A, Blanchard A (2007a) Evolution of mollicutes: down a bumpy road with twists and turns. Res Microbiol 158:754–766

Sirand-Pugnet P, Lartigue C, Marenda M, Jacob D, Barre A, Barbe V, Schenowitz C et al (2007b) Being pathogenic, plastic, and sexual while living with a nearly minimal bacterial genome. PLoS Genet 3:e75

Sitaraman R, Denison AM, Dybvig K (2002) A unique, bifunctional site-specific DNA recombinase from *Mycoplasma pulmonis*. Mol Microbiol 46:1033–1040

Smith DG, Russell WC, Ingledew WJ, Thirkell D (1993) Hydrolysis of urea by *Ureaplasma urealyticum* generates a transmembrane potential with resultant ATP synthesis. J Bacteriol 175:3253–3258

Spooner RK, Russell WC, Thirkell DCP (1992) Characterization of the immunoglobulin A protease of *Ureaplasma urealyticum*. Infect Immun 60:2544–2546

Spratt BG, Hanage WP, Li B, Aanensen DM, Feil EJ (2004) Displaying the relatedness among isolates of bacterial species – the eBURST approach. FEMS Microbiol Lett 241:129–134

Spuesens EB, Oduber M, Hoogenboezem T, Sluijter M, Hartwig NG, van Rossum AM, Vink C (2009) Sequence variations in RepMP2/3 and RepMP4 elements reveal intragenomic homologous DNA recombination events in *Mycoplasma pneumoniae*. Microbiology 155:2182–2196

Stamatakis A (2006) RAxML-VI-HPC: maximum likelihood-based phylogenetic analyses with thousands of taxa and mixed models. Bioinformatics 22:2688–2690

Stipkovits L, Kempf I (1996) Mycoplasmoses in poultry. Rev Sci Tech 15:1495–1525

Stipkovits L, Szathmary S (2012) Mycoplasma infection of ducks and geese. Poult Sci 91:2812–2819

Stoffregen WC, Alt DP, Palmer MV, Olsen SC, Waters WR, Stasko JA (2006) Identification of a haemomycoplasma species in anemic reindeer (*Rangifer tarandus*). J Wildl Dis 42:249–258

Susuki K, Odaka M, Mori M, Hirata K, Yuki N (2004) Acute motor axonal neuropathy after *Mycoplasma* infection: evidence of molecular mimicry. Neurology 62:949–956

Svanevik CS, Lunestad BT (2011) Characterisation of the microbiota of Atlantic mackerel (*Scomber scombrus*). Int J Food Microbiol 151:164–170

Szczepanek SM, Frasca S, Schumacher VL, Liao X, Padula M, Djordjevic SP, Geary SJ (2010) Identification of lipoprotein MslA as a neoteric virulence factor of *Mycoplasma gallisepticum*. Infect Immun 78:3475–3483

Tajima M, Yagihashi T, Miki Y (1982) Capsular material of *Mycoplasma gallisepticum* and its possible relevance to the pathogenic process. Infect Immun 36:830–833

Tambi NE, Maina WO, Ndi C (2006) An estimation of the economic impact of contagious bovine pleuropneumonia in Africa. Rev Sci Tech 25:999–1011

Tanskanen R (1987) Transmission of *Mycoplasma dispar* among a succession of newborn calves on a dairy farm. Acta Vet Scand 28:349–360

Tardy F, Baranowski E, Nouvel LX, Mick V, Manso-Silvan L, Thiaucourt F, Thebault P et al (2012) Emergence of atypical *Mycoplasma agalactiae* strains harboring a new prophage and associated with an alpine wild ungulate mortality episode. Appl Environ Microbiol 78:4659–4668

Tarshis M, Katzenel A, Rottem S (1994) Use of Merocyanine 540 and Hoechst 33258 for the selective killing of contaminating mycoplasmas in cell cultures. J Immunol Methods 168:245–252

Taylor-Robinson D, Davies HA, Sarathchandra P, Furr PM (1991) Intracellular location of mycoplasmas in cultured cells demonstrated by immunocytochemistry and electron microscopy. Int J Exp Pathol 72:705–714

Taylor-Robinson D, Furr PM (1997) Genital mycoplasma infections. Wien Klin Wochenschr 109:578–583

Taylor-Robinson D, Furr PM, Tully JG (1983) Serological cross-reactions between *Mycoplasma genitalium* and *M. pneumoniae*. Lancet 1:527

Theiss P, Wise KS (1997) Localized frameshift mutation generates selective, high-frequency phase variation of a surface lipoprotein encoded by a mycoplasma ABC transporter operon. J Bacteriol 179:4013–4022

Thiaucourt F (2008) Contagious bovine pleuropneumonia. In: World Organisation for Animal Health (OIE) (ed) Terrestrial manual. World Organisation for Animal Health (OIE), pp 13–724 (Chap 2.4.9) http://www.oie.int/en/international-standard-setting/terrestrial-code/access-online/

Thomas C, Jacobs E, Dumke R (2013) Characterization of pyruvate dehydrogenase subunit B and enolase as plasminogen binding proteins in *Mycoplasma pneumoniae*. Microbiology 159:352–365

Touati A, Benard A, Hassen AB, Bébéar CM, Pereyre S (2009) Evaluation of five commercial real-time PCR assays for detection of *Mycoplasma pneumoniae* in respiratory tract specimens. J Clin Microbiol 47:2269–2271

Tully JG (1995) Culture medium formulation for primary isolation and maintenance of mollicutes. In: Razin S, Tully TG (eds) Molecular and diagnostic procedures in mycoplasmology, vol I. Academic, San Diego, pp 33–39

Tully JG, Whitcomb RF, Clark HF, Williamson DL (1977) Pathogenic mycoplasmas: cultivation and vertebrate pathogenicity of a new spiroplasma. Science 195:892–894

Tulman ER, Liao X, Szczepanek SM, Ley DH, Kutish GF, Geary SJ (2012) Extensive variation in surface lipoprotein gene content and genomic changes associated with virulence during evolution of a novel North American house finch epizootic strain of *Mycoplasma gallisepticum*. Microbiology 158:2073–2088

Typas A, Banzhaf M, Gross CA, Vollmer W (2011) From the regulation of peptidoglycan synthesis to bacterial growth and morphology. Nat Rev Microbiol 10:123–136

Uemura R, Sueyoshi M, Nagatomo H (2010) Antimicrobial susceptibilities of four species of *Mycoplasma* isolated in 2008 and 2009 from cattle in Japan. J Vet Med Sci 72:1661–1663

Ueno PM, Timenetsky J, Centonze VE, Wewer JJ, Cagle M, Stein MA, Krishnan M, Baseman JB (2008) Interaction of *Mycoplasma genitalium* with host cells: evidence for nuclear localization. Microbiology 154:3033–3041

Uenoyama A, Kusumoto A, Miyata M (2004) Identification of a 349-kilodalton protein (Gli349) responsible for cytadherence and glass binding during gliding of *Mycoplasma mobile*. J Bacteriol 186:1537–1545

Uphoff CC, Drexler HG (2002) Comparative antibiotic eradication of mycoplasma infections from continuous cell lines. In Vitro Cell Dev Biol Anim 38:86–89

van Ham RC, Kamerbeek J, Palacios C, Rausell C, Abascal F, Bastolla U, Fernandez JM et al (2003) Reductive genome evolution in *Buchnera aphidicola*. Proc Natl Acad Sci USA 100:581–586

van Kuppeveld FJ, van der Logt JT, Angulo AF, van Zoest MJ, Quint WG, Niesters HG, Galama JM, Melchers WJ (1992) Genus- and species-specific identification of mycoplasmas by 16S rRNA amplification. Appl Environ Microbiol 58:2606–2615

Vancini RG, Pereira-Neves A, Borojevic R, Benchimol M (2008) *Trichomonas vaginalis* harboring *Mycoplasma hominis* increases cytopathogenicity in vitro. Eur J Clin Microbiol Infect Dis 27:259–267

Vancutsem E, Soetens O, Breugelmans M, Foulon W, Naessens A (2011) Modified real-time PCR for detecting, differentiating, and quantifying *Ureaplasma urealyticum* and *Ureaplasma parvum*. J Mol Diagn 13:206–212

Vicca J, Maes D, Stakenborg T, Butaye P, Minion F, Peeters J, de Kruif A, Decostere A, Haesebrouck F (2007) Resistance mechanism against fluoroquinolones in *Mycoplasma hyopneumoniae* field isolates. Microb Drug Resist 13:166–170

Vilei EM, Correia I, Ferronha MH, Bischof DF, Frey J (2007) Beta-D-glucoside utilization by *Mycoplasma mycoides* subsp. *mycoides* SC: possible involvement in the control of cytotoxicity towards bovine lung cells. BMC Microbiol 7:31

Viscardi RM, Manimtim WM, Sun CC, Duffy L, Cassell GH (2002) Lung pathology in premature infants with *Ureaplasma urealyticum* infection. Pediatr Dev Pathol 5:141–150

Vogl G, Plaickner A, Szathmary S, Stipkovits L, Rosengarten R, Szostak MP (2008) *Mycoplasma gallisepticum* invades chicken erythrocytes during infection. Infect Immun 76:71–77

Waite ER, March JB (2001) Effect of HEPES buffer systems upon the pH, growth and survival of *Mycoplasma mycoides* subsp. *mycoides* small colony (MmmSC) vaccine cultures. FEMS Microbiol Lett 201:291–294

Waite ER, March JB (2002) Capsular polysaccharide conjugate vaccines against contagious bovine pleuropneumonia: immune responses and protection in mice. J Comp Pathol 126:171–182

Waites KB, Katz B, Schelonka RL (2005) Mycoplasmas and ureaplasmas as neonatal pathogens. Clin Microbiol Rev 18:757–789

Waites KB, Xiao L, Paralanov V, Viscardi RM, Glass JI (2012) Molecular methods for the detection of mycoplasma and ureaplasma infections in humans: a paper from the 2011 William Beaumont Hospital Symposium on molecular pathology. J Mol Diagn 14:437–450

Walter ND, Grant GB, Bandy U, Alexander NE, Winchell JM, Jordan HT, Sejvar JJ et al (2008) Community outbreak of *Mycoplasma pneumoniae* infection: school-based cluster of neurologic disease associated with household transmission of respiratory illness. J Infect Dis 198:1365–1374

Wang HH (2010) Synthetic genomes for synthetic biology. J Mol Cell Biol 2:178–179

Wang PJ, Xie CB (2012) *Mycoplasma hominis* symbiosis and *Trichomonas vaginalis* metronidazole resistance. Zhongguo Ji Sheng Chong Xue Yu Ji Sheng Chong Bing Za Zhi 30:210–213

Wang RY, Shih JW, Grandinetti T, Pierce PF, Hayes MM, Wear DJ, Alter HJ, Lo SC (1992) High frequency of antibodies to *Mycoplasma penetrans* in HIV-infected patients. Lancet 340:1312–1316

Weisburg WG, Tully JG, Rose DL, Petzel JP, Oyaizu H, Yang D, Mandelco L et al (1989) A phylogenetic analysis of the mycoplasmas: basis for their classification. J Bacteriol 171:6455–6467

Westberg J, Persson A, Holmberg A, Goesmann A, Lundeberg J, Johansson KE, Pettersson B, Uhlen M (2004) The genome sequence of *Mycoplasma mycoides* subsp. *mycoides* SC type strain PG1T, the causative agent of contagious bovine pleuropneumonia (CBPP). Genome Res 14:221–227

Willby MJ, Balish MF, Ross SM, Lee KK, Jordan JL, Krause DC (2004) HMW1 is required for stability and localization of HMW2 to the attachment organelle of *Mycoplasma pneumoniae*. J Bacteriol 186:8221–8228

Willby MJ, Krause DC (2002) Characterization of a *Mycoplasma pneumoniae* hmw3 mutant: implications for attachment organelle assembly. J Bacteriol 184:3061–3068

Willi B, Museux K, Novacco M, Schraner EM, Wild P, Groebel K, Ziegler U et al (2011) First morphological characterization of 'Candidatus Mycoplasma turicensis' using electron microscopy. Vet Microbiol 149:367–373

Wilson DJ, Justice-Allen A, Goodell G, Baldwin TJ, Skirpstunas RT, Cavender KB (2011) Risk of *Mycoplasma bovis* transmission from contaminated sand bedding to naive dairy calves. J Dairy Sci 94:1318–1324

Winner F, Rosengarten R, Citti C (2000) In vitro cell invasion of *Mycoplasma gallisepticum*. Infect Immun 68:4238–4244

Wise KS, Watson RK (1985) Antigenic mimicry of mammalian intermediate filaments by mycoplasmas. Infect Immun 48:587–591

Woese CR, Maniloff J, Zablen LB (1980) Phylogenetic analysis of the mycoplasmas. Proc Natl Acad Sci USA 77:494–498

Woese CR, Stackebrandt E, Ludwig W (1984) What are mycoplasmas: the relationship of tempo and mode in bacterial evolution. J Mol Evol 21:305–316

Woeste K, Grosse Beilage E (2007) Transmission of agents of the porcine respiratory disease complex (PRDC) between swine herds: a review. Part 1: diagnosis, transmission by animal contact. Dtsch Tierarztl Wochenschr 114:324–326, 328–337

Woods JE, Brewer MM, Hawley JR, Wisnewski N, Lappin MR (2005) Evaluation of experimental transmission of Candidatus Mycoplasma haemominutum and *Mycoplasma haemofelis* by *Ctenocephalides felis* to cats. Am J Vet Res 66:1008–1012

Wozniak RA, Waldor MK (2010) Integrative and conjugative elements: mosaic mobile genetic elements enabling dynamic lateral gene flow. Nat Rev Microbiol 8:552–563

Xiao JC, Xie LF, Fang SL, Gao MY, Zhu Y, Song LY, Zhong HM, Lun ZR (2006) Symbiosis of *Mycoplasma hominis* in *Trichomonas vaginalis* may link metronidazole resistance in vitro. Parasitol Res 100:123–130

Yarza P, Richter M, Peplies J, Euzeby J, Amann R, Schleifer KH, Ludwig W, Glockner FO, Rossello-Mora R (2008) The All-Species Living Tree project:

a 16S rRNA-based phylogenetic tree of all sequenced type strains. Syst Appl Microbiol 31:241–250

Yavlovich A, Higazi AA, Rottem S (2001) Plasminogen binding and activation by *Mycoplasma fermentans*. Infect Immun 69:1977–1982

Yavlovich A, Tarshis M, Rottem S (2004) Internalization and intracellular survival of *Mycoplasma pneumoniae* by non-phagocytic cells. FEMS Microbiol Lett 233:241–246

Yus E, Maier T, Michalodimitrakis K, van Noort V, Yamada T, Chen WH, Wodke JA et al (2009) Impact of genome reduction on bacterial metabolism and its regulation. Science 326:1263–1268

Zhang Q, Wise KS (1996) Molecular basis of size and antigenic variation of a *Mycoplasma hominis* adhesin encoded by divergent vaa genes. Infect Immun 64:2737–2744

Zimmerman CU, Rosengarten R, Spergser J (2013) Interaction of the putative tyrosine recombinases RipX (UU145), XerC (UU222) and CodV (UU529) of *Ureaplasma parvum* serovar 3 with specific DNA. FEMS Microbiol Lett 340:55–64

Zimmerman CU, Stiedl T, Rosengarten R, Spergser J (2009) Alternate phase variation in expression of two major surface membrane proteins (MBA and UU376) of *Ureaplasma parvum* serovar 3. FEMS Microbiol Lett 292:187–193

# 40 The Family *Spiroplasmataceae*

*Laura B. Regassa*
Department of Biology, Georgia Southern University, Statesboro, GA, USA



**Abstract**
The family *Spiroplasmataceae* is one of two in the order *Entomoplasmatales*. The family contains a single genus, *Spiroplasma*, whose members are regularly associated with arthropod or plant hosts. *Spiroplasma* species can be traced to a common ancestor; however, this lineage also includes the nonhelical *Entomoplasma*, *Mesoplasma*, and mycoides group (*Mycoplasma*) descendants. Spiroplasma cells are characterized by their helical shape, which is most common during exponential growth, and by their lack of a cell wall. They are motile due to a unique linear motor that allows for rotatory, flexional, and translational motility. Genome sizes range from 780 to 2,220 kbp in these AT-rich organisms (24–31 mol% G+C) that commonly harbor viral sequences in large areas of repetitive sequence. Spiroplasmas are chemo-organotrophic, generally fermenting glucose through the phosphoenolpyruvate-dependent sugar transferase system. Most strains require rich media for initial isolation and/or maintenance, and all spiroplasmas are resistant to penicillin. Temperature ranges (5–41 °C), growth optima, and doubling times are species specific. Due to motility, colonies are diffuse and range in size from 0.1 to 4.0 mm. Historically, *Spiroplasma* classification relied on surface serology as a surrogate for DNA-DNA hybridization assays, resulting in 49 reported serogroups and 15 subgroups. There are a total of 38 described *Spiroplasma* species, as not all serogroup/subgroup type strains have been fully characterized. Most host relationships are commensal, but cases of mutualism and pathogenicity have been reported. For example, spiroplasma infections cause citrus stubborn disease, corn stunt disease, sex ratio disorders, and honey bee mortality; spiroplasmas are pathogenic for suckling rodents and/or chicken embryos under experimental conditions.

This chapter is a modified and updated version of previous family descriptions (Williamson DL, Gasparich GE, Regassa LB, Saillard C, Renaudin J, Bové JM, Whitcomb RF (2010) Family II. *Spiroplasmataceae*. In: Krieg NR, Ludwig W, Whitman WB, Hedlund BP, Paster BJ, Staley JT, Ward N, Brown D, Parte A (eds) *Bergey's Manual of Systematic Bacteriology*, vol 4. Springer, New York, pp 654–686; Brown DR, Bradbury JM, Whitcomb RF (2010) Order II. *Entomoplasmatales*. In: Krieg NR, Ludwig W, Whitman WB, Hedlund BP, Paster BJ, Staley JT, Ward N, Brown D, Parte A (eds) *Bergey's Manual of Systematic Bacteriology*, vol 4. Springer, New York, pp 644–645)

## Taxonomy, Historical and Current

### Short Description of the Family

Spi.ro.plas.ma.ta'ce.ae. M.L. neut. n. *Spiroplasma* type genus of the family; -aceae ending to denote a family; M.L. fem. pl. n. *Spiroplasmataceae* the *Spiroplasma* family. The description is an emended version of the description given in *Bergey's Manual* (Williamson et al. 2010).

This family is one of two in the order *Entomoplasmatales*, class *Mollicutes*, phylum Tenericutes. The two families are designated *Entomoplasmataceae* for nonhelical mollicutes and *Spiroplasmataceae* for helical ones. The family *Spiroplasmataceae* contains a single genus, *Spiroplasma*, whose members are regularly associated with arthropod or plant hosts. Spiroplasmas can be clearly differentiated by their unique properties of helicity (➲ *Fig. 40.1*) and motility, combined with the absence of a cell wall (➲ *Fig. 40.2*). Spiroplasmas may be nonhelical under some growth conditions or when cultures are in the stationary phase of growth. All cells are chemo-organotrophic, usually fermenting glucose through the phosphoenolpyruvate-dependent sugar transferase system. Arginine may be hydrolyzed, but urea is not. Cells may require sterol for growth. Optimal growth usually occurs at 30–32 °C, with a few species

E. Rosenberg et al. (eds.), *The Prokaryotes – Firmicutes and Tenericutes*, DOI 10.1007/978-3-642-30120-9_389,

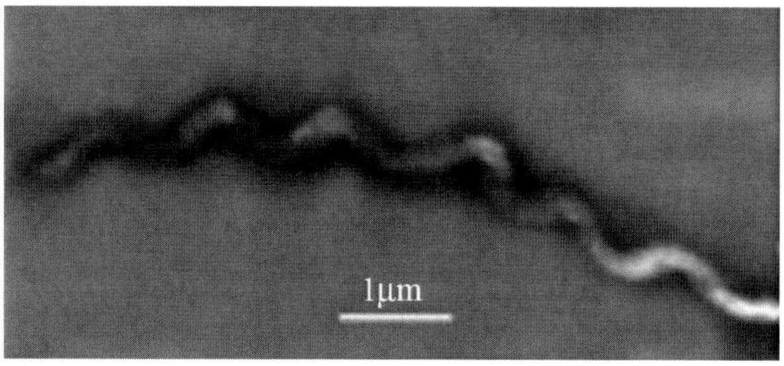


**Fig. 40.1**
**Electron micrograph of spiroplasma isolated from a Chinese mitten crab with tremor disease (Reprinted with permission from Wang 2011. Journal Invertebrate Pathology 106, 18–26.)**

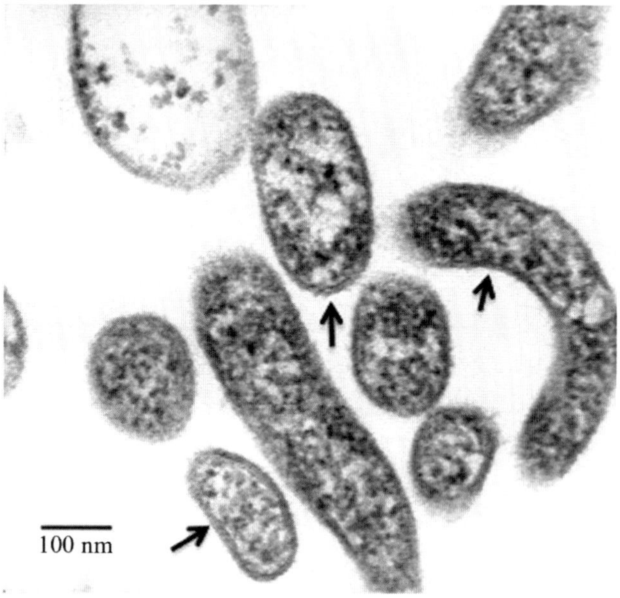


**Fig. 40.2**
**Transmission electron micrograph of spiroplasma isolated from a tabanid fly. Sections were stained with 2 % aqueous uranyl acetate and Reynolds' lead citrate. *Arrows* indicate cell membrane**

able to grow at 37 °C. Genome sizes range from 780 to 2,220 kbp and DNA G+C contents from 24 to 31 mol%. All organisms in this family are thought to utilize the UGA codon to encode tryptophan.

## Phylogenetic Structure of the Family

The term "spiroplasma" was first used as a trivial term to describe the helical organisms associated with corn stunt disease (Davis et al. 1972a, b). After similar organisms were found in association with citrus stubborn disease (Saglio et al. 1973), the trivial term was adopted as the generic name and the citrus stubborn organism was named *Spiroplasma citri. S.citri* was the first cultured spiroplasma and the first cultured mollicute of plant origin. In 1974, the genus *Spiroplasma* was elevated to the status of a family (Skripal 1974) and later added to the Approved Lists of Bacterial Names (Skripal 1983). The species concept for spiroplasmas, as for all bacteria, was based on DNA-DNA reassociation (ICSB Subcommittee on the Taxonomy of Mollicutes 1995; Johnson 1994; Rossello-Mora and Amann 2001; Stackebrandt et al. 2002; Wayne et al. 1987). In practice, DNA-DNA reassociation results with spiroplasmas proved difficult to standardize. For example, estimates of reassociation between *S. citri* (subgroup I-1) and *S. kunkelii* (subgroup I-3) varied between 30 % and 70 %, depending on the method employed and the degree of stringency (Bové and Saillard 1979; Christiansen et al. 1979; Lee and Davis 1980; Liao and Chen 1981a; Rahimian and Gumpf 1980). Given these challenges, an alternative method was identified. Surface serology of spiroplasmas was used as a surrogate for DNA-DNA hybridization assays and was a major taxonomic determinant for over 30 years (see ➲ Molecular Analyses). Although the requirement for serological testing of new *Spiroplasma* species was recently removed by the International Committee on Systematics of Prokaryotes (ICSP 2013) Subcommittee on the Taxonomy of *Mollicutes*, this chapter will reference serogroup designations to maintain continuity with the cited literature.

Woese et al. (1980) presented a 16S rDNA-based phylogenetic tree for the class *Mollicutes*, including *Spiroplasma*, which indicated that these wall-less bacteria were related to gram-positive bacteria such as *Lactobacillus* spp. and *Clostridium innocuum*. The tree suggested that *Mollicutes* might be monophyletic, but a later study by Weisburg et al. (1989) with 40 additional species of *Mollicutes* (including 10 spiroplasmas) failed to confirm the monophyly of *Mollicutes* at the deepest branching orders. Within the *Mollicutes*, the acholeplasma-anaeroplasma (*Acholeplasmatales-Anaeroplasmatales*) and spiroplasma-mycoplasma (*Mycoplasmatales-Entomoplasmatales*) lineages are monophyletic, but are separated by an ancient divergence. The Woese (1980) model also suggested that the genus *Mycoplasma* might not be monophyletic, in that the type species, *M. mycoides*, and two related species, *M. capricolum* and *M. putrefaciens*, appeared to be more closely related to the Apis clade of *Spiroplasma* than to the other *Mycoplasma* species. This conclusion was subsequently supported by more extensive analysis of mollicute 16S rRNA (Gasparich et al. 2004) and 5S rRNA (Rogers et al. 1985) gene sequences.

All characterized *Spiroplasma* species can be traced to a common ancestor; however, this lineage also includes *Entomoplasma*, *Mesoplasma*, and *Mycoplasma* species. In-depth analysis of characterized spiroplasmas and their nonhelical descendants indicates the existence of four major clades within the lineage (➲ *Fig. 40.3*; Gasparich et al. 2004). One of the four clades consists of the nonhelical species of the mycoides group (as defined by Johansson 2002) as well as the *Entomoplasma* and *Mesoplasma* species (the *Entomoplasmataceae*); this assemblage was designated the Mycoides-*Entomoplasmataceae* clade. The remaining three clades represent *Spiroplasma* species. One of these clades, the Apis clade, is a sister to the Mycoides-*Entomoplasmataceae* clade. The Apis clade contains a large number of species (serogroups III, IV, VII, IX-XIV, XVI-XXVII,

## Fig. 40.3

Phylogenetic reconstruction of the family *Spiroplasmataceae* based on 16S rRNA and created using the maximum likelihood algorithm RAxML (Stamatakis 2006). Alignments and most sequence datasets were used according to the All-Species Living Tree Project (LTP) database (Yarza et al. 2010; http://www.arb-silva.de/projects/living-tree). Sequences AB681165, AB681166, and DQ917756 are not present in the LTP database; they are replacing poor quality sequences originally assigned to their type strain. Representative sequences from closely related taxa were used as outgroups. In addition, a 40 % maximum frequency filter was applied in order to remove hypervariable positions and potentially misplaced bases from the alignment. *Scale bar* indicates estimated sequence divergence

XXIX-XLII, and XLIV-XLIX) from diverse insect hosts, many of which appear to be in insect gut-plant surface cycles. Interestingly, one of the strains (TIUS-1) that falls into the basal *S. alleghenense* and *S. sabaudiense* cluster has very poor helicity and a small genome size of 840 kbp (Williamson et al. 1998). This species diverged from the spiroplasma lineage close to the node of entomoplasmal divergence and thus may represent a "missing link" in the evolutionary development of the Mycoides-*Entomoplasmataceae* clade (Gasparich et al. 2004).

The other two *Spiroplasma* clades are the Ixodetis clade (serogroups VI and XXVIII) with *S. ixodetis* and *S. platyhelix* and the Citri-Chrysopicola-Mirum clade with representatives from serogroups I, II, V, VIII, and XLIII. The Citri-Chrysopicola-Mirum clade contains *S. mirum*, *S. poulsonii*, *S. eriocheiris*, and the serological subgroups I-1 to I-9 (citri cluster) and VIII-1 to VIII-3 (chrysopicola cluster). The two species in the Ixodetis clade are from a tick and a dragonfly, although there have been suggestions that the dragonfly may

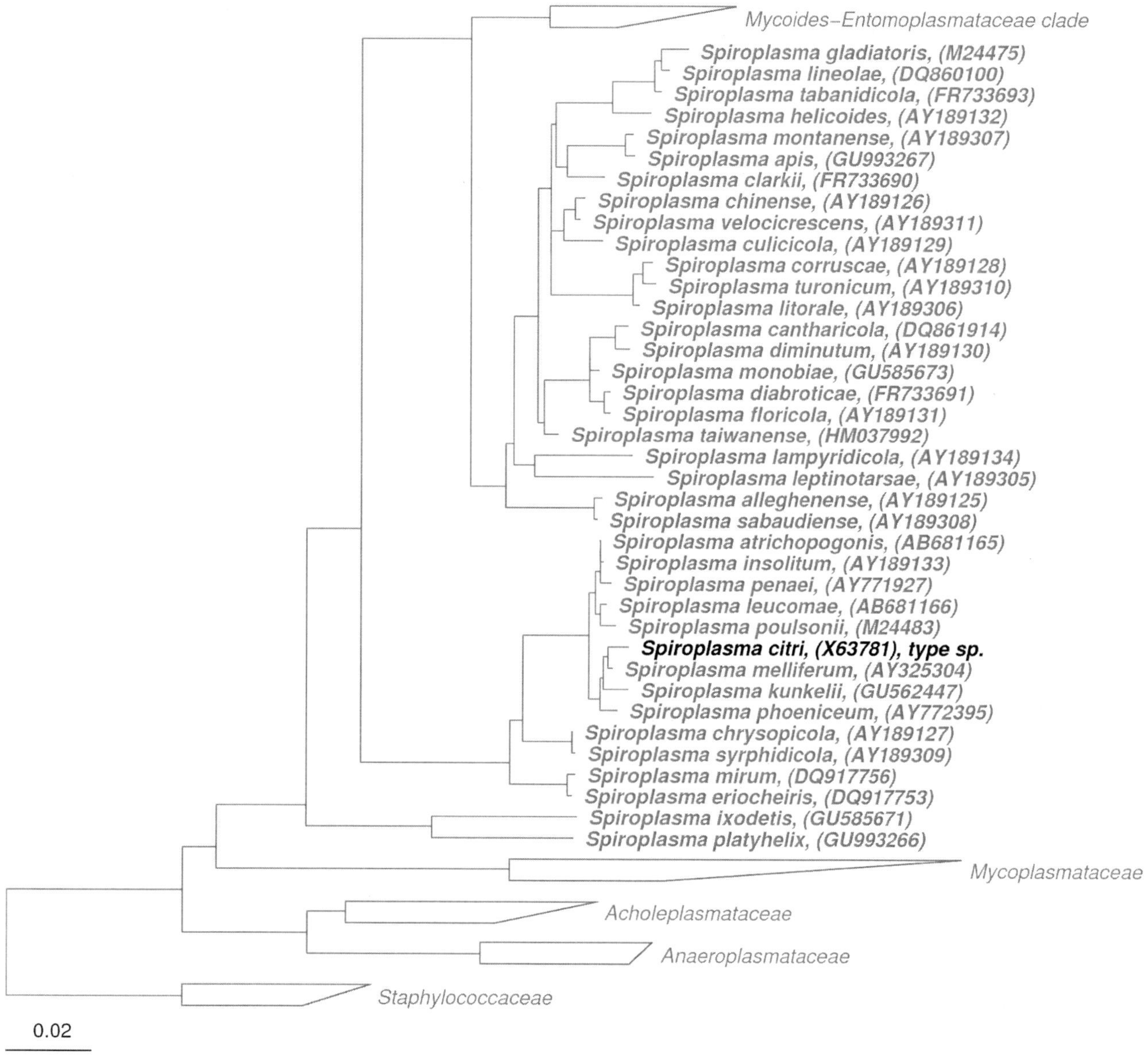

have acquired the spiroplasma via predation as it would represent the only described insect from a primitive order (*Odonata*) to harbor spiroplasmas (Hackett et al. 1990). The host range of the Citri-Chrysopicola-Mirum clade is quite diverse and includes insects, ticks, crustaceans, and plants.

Nongenetic character mapping has been completed in conjunction with *Spiroplasma* phylogenetic analyses (Gasparich et al. 2004). Genome size and G+C content were moderately conserved among closely related strains. Apparent conservation of slower growth rates in some clades was most likely attributable to host affiliation; spiroplasmas of all groups that were closely tied to a host had slower growth rates. Sterol requirements were polyphyletic, as was the ability to grow in the presence of polyoxyethylene sorbitan, but not serum. Gasparich et al. (2004) reported that serological group and subgroup classifications were generally supported by the trees. However, analysis of 27 strains within the Apis clade revealed that, although some strain clusters were maintained, trees generated from the two data sets (serological matrix and concatenated 16S-23S-ITS-*rpo*B sequences) were not congruent (Regassa, unpublished data). The apparent discrepancy may be due to the discriminatory power of serology and DNA characteristics to elucidate relationships between closely and more distantly related species.

## Molecular Analyses

### Antigenic Structure and Group Classification

A classification system that relied on the surface serology of *Spiroplasma* strains was first proposed by Junca et al. (1980) and periodically revised (Tully et al. 1987; Williamson et al. 1998) and used (Brown et al. 2007) until the requirement of serological analyses for the description of new *Spiroplasma* species was recently removed by the ICSP Subcommittee on the Taxonomy of *Mollicutes* (ICSP 2013). The classification scheme was based on serological cross-reactivity of organisms in growth inhibition (Whitcomb et al. 1982), deformation (Williamson et al. 1978), and/or metabolism inhibition (Williamson et al. 1979; Williamson 1983) tests and combined with genetic and phenotypic characteristics as part of a polyphasic approach. It should be noted that antigenic variability, which has been described for some *Mycoplasma* species (Yogev et al. 1991; Rosengarten and Wise 1990), has not been reported in spiroplasmas.

Utilization of the serological classification scheme resulted in the delineation of *Spiroplasma* groups and subgroups (❯ *Table 40.1*). Serogroups were initially defined as clusters of similar organisms, all of which possess negligible DNA/DNA homology with representatives of other groups, but moderate to high levels (20–100 %) with each other. This level of genomic differentiation correlated well with substantial differences in serology, so serology became a surrogate for DNA-DNA hybridization assays and serogroups were, therefore, putative species. Thirty-four serogroups were presented in a revised classification of *Spiroplasma* in 1998 (Williamson et al. 1998). Fifteen additional serogroups (XXXV-XLIX) were later reported based on deformation tests (Whitcomb et al. 2007; Regassa et al. 2009, 2011; Wang et al. 2011). Subgroups were defined as clusters of spiroplasma strains showing intermediate levels of intragroup DNA/DNA homology (10–70 %) and possessing corollary serological relationships (ICSB 1984). Three serogroups have been divided into a total of 15 subgroups: group I (Junca et al. 1980; Nunan et al. 2005; Saillard et al. 1987), group VIII (Gasparich et al. 1993b), and group XVI (Abalain-Colloc et al. 1993). However, with the discovery of a large number of strains for some groups, the subgroup picture has become confused. This is evident in serogroup VIII that appears to be a large strain complex (Gasparich et al. 1993b; Regassa et al. 2004). Not all of the 61 serological groups or subgroups have representative species; a total of 38 *Spiroplasma* species have been fully described in the literature.

### Genome Structure

The G+C base composition of the DNA for most *Spiroplasma* serogroups and subgroups has been determined (Carle et al. 1990, 1995; Williamson et al. 1998). A range of 26–27 mol% G+C represents the mode and mean for *Spiroplasma*, but some outlying strains suggest that base composition of spiroplasmal DNA may shift over relatively short evolutionary periods. For example, within the Citri-Chrysopicola-Mirum clade, most group I spiroplasmas and *S. poulsonii* (group II) have a base content of 26 mol% G+C. Within the same clade, the base composition of subgroups I-6 (*S. insolitum*) and I-9 (*S. penaei*) is significantly higher at 28–29 mol% G+C and the composition of group V (*S. mirum*) and group VIII strains is even higher at 29–31 mol% G+C. Codon usage reflects the A+T richness of spiroplasmal DNA (Navas-Castillo et al. 1992; Citti et al. 1992; Bové 1993). In spiroplasmas, UGA is not a stop codon but encodes tryptophan; *S. citri* uses UGA eight times more frequently than the universal tryptophan codon UGG. Overall, synonymous codons with U or A at the 5′ or 3′ ends are preferentially used over those with a C or G in that position. It should be noted that while some spiroplasmas, such as *S. citri*, have only one rRNA operon, others, such as *S. apis*, have two (Amikam et al. 1982, 1984; Razin 1985; Grau et al. 1988; Bové 1993). The three rRNA genes are linked in the classical order found in bacteria: 5′-16S-23S-5S-3′.

Genome sizes for spiroplasmas appear to be continuous (Pyle and Finch 1988) from 780 kbp for *S. platyhelix* to 2,220 kbp for *S. ixodetis* (Carle et al. 1990, 1995) based on pulsed-field gel electrophoresis. There is a general trend toward genomic simplification in *Spiroplasma* lineages. This trend culminated in loss of helicity and motility in the *Entomoplasmataceae* and eventually to the host transfer events that formed the ruminant-restricted mycoides group of mycoplasmas (Gasparich et al. 2004). Genome sizes can vary considerably for a single species; strains of *S. citri* had genomes ranging from 1,650 to 1,910 kbp (Ye et al. 1995). Genome size can fluctuate rapidly in a relatively short number of in vitro passages (Ye et al. 1996; Melcher and Fletcher 1999), and it has been hypothesized that this may be

**Table 40.1**
**Biological properties of spiroplasmas**

| Serogroup | Spiroplasma | Strain | Collection number | Morphology[a] | Genome[b] | G + C[c] | Arg[d] | Dt[e] | OptT[f] | Host |
|---|---|---|---|---|---|---|---|---|---|---|
| I-1 | *S. citri* | R8-A2[T] | ATCC 27556 | Long helix | 1,820 | 26 | + | 4.1 | 32 | Phloem/ leafhopper |
| I-2 | *S. melliferum* | BC-3 [T] | ATCC 33219 | Long helix | 1,460 | 26 | + | 1.5 | 37 | Honeybee |
| I-3 | *S. kunkelii* | E275 [T] | ATCC 29320 | Long helix | 1,610 | 26 | + | 27.3 | 30 | Phloem/ leafhopper |
| I-4 | S. sp. | 277 F | ATCC 29761 | Long helix | 1,620 | 26 | + | 2.3 | 32 | Rabbit tick |
| I-5 | S. sp. | LB-12 | ATCC 33649 | Long helix | 1,020 | 26 | - | 26.3 | 30 | Plant bug |
| I-6 | *S. insolitum* | M55 [T] | ATCC 33502 | Long helix | 1,810 | 28 | - | 7.2 | 30 | Flower surface |
| I-7 | S. sp. | N525 | ATCC 33287 | Long helix | 1,780 | 26 | + | 4.7 | 32 | Green June beetle |
| I-8 | *S. phoeniceum* | P40 [T] | ATCC 43115 | Long helix | 1,860 | 26 | + | 16.8 | 30 | Phloem/vector |
| I-9 | *S. penaei* | SHRIMP [T] | ATCC BAA-1082 (CAIM 1252) | Helix | ND[g] | 29 | + | ND | 28 | Pacific white shrimp |
| II | *S. poulsonii* | DW-1 [T] | ATCC 43153 | Long helix | 1,040 | 26 | ND | 15.8 | 30 | *Drosophila* hemolymph |
| III | *S. floricola* | OBMG [T] | ATCC 29989 | Helix | 1,270 | 26 | - | 0.9 | 37 | Plant surface |
| IV | *S. apis* | B31[T] | ATCC 33834 | Helix | 1,300 | 30 | + | 1.1 | 34.5 | Honeybee |
| V | *S. mirum* | SMCA[T] | ATCC 29335 | Helix | 1,300 | 30 | + | 7.8 | 37 | Rabbit tick |
| VI | *S. ixodetis* | Y32[T] | ATCC 33835 | Tight coil | 2,220 | 25 | - | 9.2 | 30 | Ixodid tick |
| VII | *S. monobiae* | MQ-1[T] | ATCC 33825 | Helix | 940 | 28 | - | 1.9 | 32 | *Monobia* wasp |
| VIII-1 | *S. syrphidicola* | EA-1[T] | ATCC 33826 | Minute helix | 1,230 | 30 | + | 1.0 | 32 | Syrphid fly |
| VIII-2 | *S. chrysopicola* | DF-1[T] | ATCC 43209 | Minute helix | 1,270 | 29 | + | 6.4 | 30 | Deer fly |
| VIII-3 | S. sp. | TAAS-1 | ATCC 51123 | Minute helix | 1,170 | 31 | + | 1.4 | 37 | Horsefly |
| IX | *S. clarkii* | CN-5[T] | ATCC 33827 | Helix | 1,720 | 29 | + | 4.3 | 30 | Green June beetle |
| X | *S. culicicola* | AES-1[T] | ATCC 35112 | Short helix | 1,350 | 26 | - | 1.0 | 37 | Mosquito |
| XI | *S. velocicrescens* | MQ-4[T] | ATCC 35262 | Short helix | 1,480 | 26 | - | 0.6 | 37 | *Monobia* wasp |
| XII | *S. diabroticae* | DU-1[T] | ATCC 43210 | Helix | 1,350 | 25 | + | 0.9 | 32 | Beetle |
| XIII | *S. sabaudiense* | Ar-1343[T] | ATCC 43303 | Helix | 1,175 | 29 | + | 4.1 | 30 | Mosquito |
| XIV | *S. corruscae* | EC-1[T] | ATCC 43212 | Helix | ND | 26 | - | 1.5 | 32 | Horsefly/ beetle |
| XV | S. sp. | I-25 | ATCC 43262 | Wave-coil | 1,380 | 26 | - | 3.4 | 30 | Leafhopper |
| XVI-1 | *S. cantharicola* | CC-1[T] | ATCC 43207 | Helix | ND | 26 | - | 2.6 | 32 | Cantharid beetle |
| XVI-2 | S. sp. | CB-1 | ATCC 43208 | Helix | 1,320 | 26 | - | 2.6 | 32 | Cantharid beetle |
| XVI-3 | S. sp. | Ar-1357 | ATCC 51126 | Helix | ND | 26 | - | 3.4 | 30 | Mosquito |
| XVII | *S. turonicum* | Tab4c[T] | ATCC 700271 | Helix | 1,305 | 25 | - | ND | 30 | Horsefly |
| XVIII | *S. litorale* | TN-1[T] | ATCC 43211 | Helix | 1,370 | 25 | - | 1.7 | 32 | Horsefly |
| XIX | *S. lampyridicola* | PUP-1[T] | ATCC 43206 | Unstable helix | 1,375 | 25 | - | 9.8 | 30 | Firefly |
| XX | *S. leptinotarsae* | LD-1[T] | ATCC 43213 | Motile funnel | 1,085 | 25 | + | 7.2 | 30 | Colorado potato beetle |
| XXI | S. sp. | W115 | ATCC 43260 | Helix | 980 | 24 | - | 4.0 | 30 | Flower surface |
| XXII | *S. taiwanense* | CT-1[T] | ATCC 43302 | Helix | 1,195 | 26 | - | 4.8 | 30 | Mosquito |
| XXIII | *S. gladiatoris* | TG-1[T] | ATCC 43525 | Helix | ND | 26 | - | 4.1 | 31 | Horsefly |
| XXIV | *S. chinense* | CCH[T] | ATCC 43960 | Helix | 1,530 | 29 | - | 0.8 | 37 | Flower surface |
| XXV | *S. diminutum* | CUAS-1[T] | ATCC 49235 | Short helix | 1,080 | 26 | - | 1.0 | 32 | Mosquito |

**Table 40.1 (continued)**

| Serogroup | Spiroplasma | Strain | Collection number | Morphology[a] | Genome[b] | G + C[c] | Arg[d] | Dt[e] | OptT[f] | Host |
|---|---|---|---|---|---|---|---|---|---|---|
| XXVI | *S. alleghenense* | PLHS-1[T] | ATCC 51752 | Helix | 1,465 | 31 | + | 6.4 | 30 | Scorpion fly |
| XXVII | *S. lineolae* | TALS-2[T] | ATCC 51749 | Helix | 1,390 | 25 | - | 5.6 | 30 | Horsefly |
| XXVIII | *S. platyhelix* | PALS-1[T] | ATCC 51748 | Wave-coil | 780 | 29 | + | 6.4 | 30 | Dragonfly |
| XXIX | S. sp. | TIUS-1 | ATCC 51751 | Rare helices | 840 | 28 | - | 3.6 | 30 | Tiphiid wasp |
| XXX | S. sp. | BIUS-1 | ATCC 51750 | Late helices | ND | 28 | - | 0.9 | 37 | Flower surface |
| XXXI | *S. montanense* | HYOS-1[T] | ATCC 51745 | Helix | 1,225 | 28 | + | 0.7 | 32 | Horsefly |
| XXXII | *S. helicoides* | TABS-2[T] | ATCC 51746 | Helix | ND | 27 | - | 3.0 | 32 | Horsefly |
| XXXIII | *S. tabanidicola* | TAUS-1[T] | ATCC 51747 | Helix | 1,375 | 26 | - | 3.7 | 30 | Horsefly |
| XXXIV | S. sp. | BARC 1901 | ATCC 700283 | Helix | 1,295 | 25 | - | ND | ND | Horsefly |
| XXXV | S. sp. | BARC 4886 | ATCC BAA-1183 | Helix | ND | ND | - | 0.6 | 32 | Horsefly |
| XXXVI | S. sp. | BARC 4900 | ATCC BAA-1184 | Helix | ND | ND | - | 1.0 | 30 | Horsefly |
| XXXVII | S. sp. | BARC 4908 | ATCC BAA-1187 | Helix | ND | ND | - | 1.2 | 32 | Horsefly |
| XXXVIII | S. sp. | GSU5450 | ATCC BAA-1188 | Helix | ND | ND | - | 1.5 | 32 | Horsefly |
| XXXIX | S. sp. | GSU5478 | DSM 22434 | Helix | ND | ND | - | 0.8 | 31 | Horsefly |
| XL | S. sp. | GSU5490 | DSM 22439 | Helix | ND | ND | - | 3.6 | 30 | Horsefly |
| XLI | S. sp. | GSU5508 | DSM 22438 | Helix | ND | ND | + | 0.6 | 31 | Horsefly |
| XLII | S. sp. | GSU5603 | DSM 22437 | Helix | ND | ND | + | 0.4 | 31 | Horsefly |
| XLIII | *S. eriocheiris* | TDA-040725-5[T] | DSM 21848 | Helix | ~1,500 | 30 | + | 24.0 | 30 | Chinese mitten crab |
| XLIV | S. sp. | GSU5360 | DSM 22471 | Helix | ND | ND | - | 1.1 | 31 | Horsefly |
| XLV | S. sp. | GSU5366 | DSM 22472 | Helix | ND | ND | - | 1.2 | 31 | Horsefly |
| XLVI | S. sp. | GSU5373 | DSM 22492 | Helix | ND | ND | - | 1.1 | 34.5 | Horsefly |
| XLVII | S. sp. | GSU5382 | DSM 22470 | Helix | ND | ND | - | 1.0 | 31 | Horsefly |
| XLVIII | S. sp. | GSU5405 | DSM 22494 | Helix | ND | ND | - | 0.9 | 33.5 | Horsefly |
| XLIX | S. sp. | GSU5446H | DSM 22493 | Helix | ND | ND | - | 1.7 | 34.5 | Horsefly |
| ND | *S. atrichopogonis* | GNAT3597[T] | ATCC BAA-520 (NBRC 100390) | Helix | ND | 28 | + | ND | 30 | Biting midge |
| ND | *S. leucomae* | SMA[T] | ATCC BAA-521 (NBRC 100392) | Helix | ND | 24 | + | ND | 30 | Satin moth |

[a]For detailed morphotypes, see Gasparich et al. 2004; Williamson et al. 2010
[b]Genome size (kbp)
[c]Guanine plus cytosine content of DNA (mol%)
[d]+ = catabolizes arginine
[e]Doubling time in hr
[f]Optimal growth temperature (°C); average presented for ranges
[g]*ND* not determined

partially responsible for the apparently high level of genome plasticity in *S. citri* (Carpane et al. 2012). Examination of low-passage *S. kunkelii* strains isolated over 24 years from different geographical locations revealed highly conserved genome composition, suggesting that the natural reservoir of genome variability may be lower than originally indicated (Carpane et al. 2012). Comparison of the two closely related species, *S. melliferum* and *S. citri*, has shed light on the mechanisms of genome plasticity (Ye et al. 1996; Melcher and Fletcher 1999). The genome of *S. melliferum* is 360 kbp shorter than that of *S. citri* strain R8-A2, but DNA hybridization has shown that the two *Spiroplasma* species share extensive homology (65 %). Comparison of their genomic maps revealed that the genome region containing the *S. melliferum* deletion corresponds to a variable region in the genomes of *S. citri* strains and that a large region of the *S. melliferum* genome is inverted in comparison with *S. citri*. Therefore, chromosomal rearrangements and deletions were probably major events during evolution of these genomes from the putative common ancestor. In addition, the presence of a large amount of noncoding DNA as repeat sequences (Nur et al. 1986, 1987; McIntosh et al. 1992) and integrated viral DNA (Bébéar et al. 1996) may account for differences in genome size of closely related species. In particular, viral DNA is an important source of genomic variability, as these sequences can integrate intact or fragmented and in varying copy number (Ye et al. 1996; Melcher and Fletcher 1999; Sha et al. 2000; Carle et al. 2010; Alexeev et al. 2012).

Whole genome sequencing projects for spiroplasmas have progressed slowly due in large part to areas of repetitive sequence. Three publicly available sequences for *S. citri*, *S. kunkelii*, and *S. melliferum* are described below. For *S. citri* GII3-3× (Saillard et al. 2008; Carle et al. 2010), assembly of 20,000 sequencing reads obtained from shotgun and chromosome specific libraries yielded: (i) 77 chromosomal contigs totaling 1,674 kbp of the 1,820 kbp *S. citri* GII3-3× chromosome and (ii) eight circular contigs representing seven plasmids (pSciA [7.8 kbp], pSci1-pSci6 [12.9 to 35.3 kbp]) and one replicative form SpV1-like DNA virus (SVTS2). Thirty-eight contigs were annotated and contained 1,908 putative genes or coding sequences (CDS). Twenty-nine percent of the CDS-encoded proteins are involved in cellular processes, cell metabolism, or cell structure. CDS for viral proteins and mobile elements represented 24 % of the total, while 47 % of the CDS were for hypothetical proteins with no known function. Twenty-one percent of the total CDS appeared truncated as compared to their bacterial orthologs; gene disruptions were also reported within the *S. kunkelii* genome (Zhao et al. 2003) and in *S. melliferum* but at a much lower rate (Alexeev et al. 2012). Families of paralogs were mainly clustered in a large region of the chromosome opposite the origin of replication. Eighty-four CDS were assigned to transport functions including phosphoenolpyruvate phosphotransferase systems (PTS), ATP binding cassette (ABC) transporters, and ferritin. In addition to the general PTS enzymes EI (enzyme I) and HPr (histidine protein), glucose-, fructose-, and trehalose-specific PTS permeases, and glycolytic and ATP synthesis pathways, *S. citri* possesses a Sec-dependent protein export system and a nearly complete pathway for terpenoid biosynthesis. The partial sequence of the *S. kunkelii* CR2-3X genome (1.55 Mb) is also available (http://www.genome.ou.edu/spiro.html). The physical and genetic maps have been published (Dally et al. 2006), and several studies have focused on gene content and genomic organization (Zhao et al. 2003, 2004a, b). Results show that, in addition to virus SpV1 DNA insertions, the *S. kunkelii* genome harbors more purine and amino acid biosynthesis, transcriptional regulation, cell envelope, and DNA transport/binding genes than *Mycoplasmataceae* (e.g., *M. genitalium* and *M. pneumoniae*) genomes (Bai and Hogenhout 2002). Limited comparative analysis between the annotated region of the *S. kunkellii* CR2-3X chromosome (85 Kbp; Zhao et al. 2003) and the homologous region in the *S. citri* GII3-3X chromosome showed overall synteny, but also highlighted species-specific coding regions (Carle et al. 2010). The *S. melliferum* KC3 genome has been assembled into four contigs covering 1,260 kbp and four extrachromosomal elements (4.7–14.4 kbp) that exhibit homology with the *S. citri* pSci plasmid family (Alexeev et al. 2012). Large areas of repetitive sequence with homology to SpV1 and SpV1-like (SVTS2) viruses are present in the *S. melliferum* genome, and again represent the main obstacle to closing the genome sequence. The coding density in the *S. melliferum* KC3 genome is higher than that seen for *S. citri* GII3-3× (81 % vs. 74 %), with 1,142 CDS. Sixty-three percent of the CDS have predicted functions, 29 % represent conserved hypothetical proteins, and 8 % are hypothetical. Proteogenomic profiling identified 521 proteins, including some that corresponded to CDS that were absent in the initial annotation, resulting in a total of 1,172 annotated CDS. The *S. melliferum* proteome and genome were compared to the proteomic core identified in three mollicutes (Fisunov et al. 2011). The *S. melliferum* genome contains all but 20 of the proteomic core genes and the proteome (under rich media conditions) contains all but 26 of the core proteins (Alexeev et al. 2012).

Four phage types from *Spiroplasma* have been characterized, SpV1-SpV4, some of which are capable of integrating into the chromosome as discussed above. SpV1 phages are filamentous/rod-shaped viruses (SVC1 = SpV1) that are associated with nonlytic infections (Ranhand et al. 1980; Bové et al. 1989; Renaudin and Bové 1994). They have circular, single-stranded DNA genomes (7.5–8.5 kbp), some of which have been sequenced (Renaudin and Bové 1994). SpV1 sequences also occur as prophages in the genome of the majority of *S. citri* strains studied (Renaudin and Bové 1994). These insertions take place at numerous sites in the chromosomes of *S. citri* (Ye et al. 1992; Carle et al. 2010), *S. melliferum* (Ye et al. 1994; Alexeev et al. 2012), and *S. kunkelii* (Zhao et al. 2003). Resistance of spiroplasmas to viral infection may be associated with integration of viral DNA sequences in the chromosome or extrachromosomal elements (Sha et al. 1995). The evolutionary history of these viruses is unclear, but there is some evidence for viral and plasmid coevolution in the group I *Spiroplasma* species (Gasparich et al. 1993a) and indications of potentially widespread horizontal transmission (Vaughn and de Vos 1995). SpV2 (SCV2 = SpV2), a polyhedron with a long, noncontractile tail, occurs in a small number of *S. citri* strains (Cole et al. 1973; Carle et al. 2010). It may be associated with lytic infection. SpV3, whose virions are polyhedrons with short tails, has been found in many strains of *S. citri* (Cole et al. 1974, 1977; Cole 1979) and in *Drosophila*-associated spiroplasmas (Oishi et al. 1984). The SpV3 genome is a linear double-stranded DNA molecule of 16 kbp, which can circularize to form a covalently closed molecule. There is significant diversity among SpV3 viruses, extending even to major differences in genome sizes. Dickinson and Townsend (1984) isolated an SpV3 virus that had a plaque morphology typical of temperate phages when plated on *S. citri*. In spiroplasma cells that have been lysogenized, complete viral genomes may be integrated into the spiroplasma chromosome. These cells are then immune to superinfection by the lysogenizing virus, but susceptible to other SpV3 viruses. SpV4 is a lytic phage (Chipman et al. 1998) with a naked, icosahedral nucleocapsid (Ricard et al. 1982) and a circular, single-stranded DNA (Renaudin et al. 1984a, b; Renaudin and Bové 1994). Host range studies (Renaudin et al. 1984a, b) have shown that only *S. melliferum* is susceptible to SpV4. The *S. melliferum* type strain, BC-3, and B63 are not susceptible but they can be infected by transfection, suggesting that resistance occurs at the level of phage adsorption or penetration (Renaudin et al. 1984b; Renaudin and Bové 1994).

Several spiroplasma plasmids have been described (Ranhand et al. 1980; Archer et al. 1981; Mouches et al. 1984;

Gasparich et al. 1993a; Gasparich and Hackett 1994). They are especially common in spiroplasmas of group I. Eight extrachromosomal elements, including seven plasmids, were discovered during the *S. citri* GII3-3× genome sequencing project. The six largest plasmids, pSci1 to pSci6, range from 12.9 to 35.3 kb (Saillard et al. 2008). The plasmids share extensive regions of homology, including the origin of replication; yet, they are mutually compatible in the spiroplasma cell (Breton et al. 2010a). Genes encoding proteins of the TraD-TraG, TrsE-TraE, and Soj-ParA protein families were predicted in most of the pSci sequences (Saillard et al. 2008). The presence of such genes, usually involved in chromosome integration, cell-to-cell DNA transfer, or DNA element partitioning, suggests that these molecules could be vertically as well as horizontally inherited. The largest plasmid (pSci6) encodes P32 (Killiny et al. 2006), a membrane-associated protein interestingly absent in all insect non-transmissible strains tested so far, and a protein of unknown function (pSci6_06 CDS) that is essential for *S. citri* transmission (via injection) to the leafhopper host (Breton et al. 2010a). The five remaining plasmids (pSci1 to pSci5) encode eight different *S. citri* adhesion-related proteins (ScARPs). The complete sequences of plasmids pSKU146 from *S. kunkelii* CR2-3X and pBJS-O from *S. citri* BR3 have been reported (Davis et al. 2005; Joshi et al. 2005). These large plasmids, like the above pSci plasmids, encode an adhesion and components of a type IV translocation-related conjugation system. Characterizing the replication and stability regions of *S. citri* plasmids resulted in the identification of a novel replication protein, suggesting that *S. citri* plasmids belong to a new plasmid family and that the *soj* gene is involved in segregational stability of these plasmids (Breton et al. 2008). Similar replicons were detected in various spiroplasmas of group I, such as *S.melliferum*, *S. kunkelii*, *Spiroplasma* sp. 277 F, and *S. phoeniceum*, showing that they are not restricted to plant-pathogenic spiroplasmas.

## Phenotypic Analyses

*SPIROPLASMA* Saglio, L'hospital, Laflèche, Dupont, Bové, Tully and Freundt 1973, 191[AL].

Spi.ro.plas'ma. Gr. n. *spira* a coil, spiral; Gr. n. *plasma* something formed or molded, a form; M.L. neut. n. *Spiroplasma* spiral form.

The morphology of spiroplasmas is most easily observed in suspension using dark-field microscopy (Williamson and Poulson 1979). Cells are pleomorphic, varying in size and shape from helical and branched, nonhelical filaments to spherical or ovoid. The characteristic helical form, usually 100–200 nm in diameter and 3–5 μm in length, generally occurs during the logarithmic phase of growth and in some species persists during stationary phase. Fixed and negatively stained cells usually show a blunt end and a tapered end (Williamson 1969; Williamson and Whitcomb 1974) that result from the constriction process preceding division (Garnier et al. 1981, 1984). However, the tapered end in some species has been adapted as an attachment site (Ammar et al. 2004). Spherical cells (~300 nm in diameter) and nonhelical filaments are frequently seen in the stationary phase and in all growth phases in suboptimal growth media; these cells may or may not be viable.

Helical cells are motile, with flexional and twitching movements, and often show an apparent rotatory motility (Cole et al. 1973; Davis and Worley 1973). Fimbriae and pili observed on the cell surface of some insect- and plant-pathogenic spiroplasmas appear to be involved in host-cell attachment and conjugation (Özbek et al. 2003; Ammar et al. 2004), but not in locomotion. Instead, motility is due to a contractile cytoskeleton with membrane fibrils that can be described as a "linear motor" in contrast to the near-universal, bacterial "rotary motor" (Trachtenberg 2006; Cohen-Krausz et al. 2011). Spiroplasmas exhibit temperature-dependent chemotactic movement toward higher concentrations of nutrients, such as carbohydrates and amino acids (Daniels et al. 1980; Daniels and Longland 1984), but motility is random in the absence of attractants (Daniels and Longland 1984). Both natural (Townsend et al. 1977; Townsend et al. 1980) and engineered (Cohen et al. 1989; Jacob et al. 1997; Duret et al. 1999) motility mutants have been described. These mutants form perfectly umbonate colonies on solid medium. In contrast, wild-type colonies on solid media are frequently diffuse, with irregular shapes and borders due to the motility of the cells during active growth. Overall, colony type is strongly dependent on the agar concentration and colony sizes can vary from 0.1 to 4.0 mm.

Growth rates and temperature ranges vary among *Spiroplasma* species. Enumerated microscopically, spiroplasmas reach titers of $10^8$–$10^{11}$ cells/ml in rich media (Rodwell and Whitcomb 1983). Doubling times based on media acidification can vary dramatically; Konai et al. (1996) reported doubling times ranging from 0.7 to 36.7 h. In general, spiroplasmas adapted to complex cycles or single hosts had slower growth rates than spiroplasmas known or suspected to be transmitted on plant surfaces. Temperature ranges and optima also vary for species. For example, the temperature growth range for *S. apis* was very wide (5–41 °C), but some serogroup I strains from leafhoppers and plants grew only at 25 °C and 30 °C (Konai et al. 1996). No spiroplasmas grew at 43 °C.

Biochemical tests have been used to describe *Spiroplasma* species. Spiroplasmas ferment glucose with concomitant acid production, although the utilization rates may differ. Some spiroplasmas are able to hydrolyze arginine (Hackett et al. 1996) but not urea (Razin 1983). Sterol requirements are variable (Rose et al. 1993).

The intermediary metabolism of *Mollicutes* has been reviewed (Miles 1992; Pollack et al. 1997; Pollack 2002a, b). Like all mollicutes, *Spiroplasma* species apparently lack both cytochrome pigments and, except for malate dehydrogenase, the enzymes of the tricarboxylic acid cycle. They do not have an electron-transport system, and their respiration is characterized as being flavin terminated. McElwain et al. (1988) studied *S. citri* and Pollack et al. (1989) screened 10 *Spiroplasma* species for 67 enzyme activities. All spiroplasmas were fermentative; their 6-phosphofructokinases (6-PFKs) required ATP for substrate phosphorylation during glycolysis. Additionally, nearly all

*Spiroplasma* species had dUTPase and deoxyguanosine kinase activity. More recent protein and genomic profiles in *S. melliferum* found that metabolic data were generally in good agreement with the proteogenomic results (Alexeev et al. 2012).

## Isolation, Enrichment, and Maintenance Procedure

Spiroplasmas have been isolated from a variety of arthropod and plant hosts. To isolate spiroplasmas from arthropods, an initial extract from a small arthropod (e.g., insect) in growth medium or crustacean hemolymph is passed through a 0.45 μm or 0.22 μm filter. Growth is most often monitored based on media acidification, but light turbidity may be produced in liquid cultures. An alternative to filtration involves the use of antibiotics or other inhibitors (Whitcomb et al. 1973; Markham et al. 1983; Grulet et al. 1993); spiroplasmas are resistant to 10,000 U/ml penicillin. Spiroplasma isolations from infected plants are best obtained from sap expressed from vascular bundles of hosts showing early disease symptoms. Plant sap often contains spiroplasmal substances (Liao et al. 1979) whose presence in primary cultures may necessitate blind passage or serial dilution. Many spiroplasmas envisioned by dark-field microscopy have proved to be nonculturable (Hackett and Clark 1989).

Media containing mycoplasma broth base, serum, and other supplements are required for primary growth. Success in the isolation of fastidious spiroplasmas is influenced strongly by the titer of the inoculum. M1D medium (Whitcomb 1983) has been used for primary isolations of the large majority of *Spiroplasma* species. SP-4 medium, a rich formulation derived from M1D, is necessary for isolation of *S. mirum* from embryonated eggs (Tully et al. 1982) and *S. ixodetis* from ticks (Tully et al. 1981). Isolation of some very fastidious spiroplasmas has been accomplished by cocultivation with insect cells or slow-growing yeast (Hackett et al. 1986; Hackett and Lynn 1985; Cohen and Williamson 1988).

Most spiroplasmas can be adapted to a wide variety of medium formulations, but they commonly grow more slowly upon transfer to new media. Continuous careful passaging may result in growth rate recovery to levels similar to that in the initial medium. For such adaptations, starting with a 1:1 ratio of old and new media, and gradually withdrawing the old formulation is a good strategy. For example, *S. clarkii*, after continuous passage for hundreds of generations, finally adapted to extremely simple media (Hackett et al. 1994). It should be noted that genome size can fluctuate with repeated in vitro passage (see ➲ Genome Structure). *S. citri* can be cultivated in a relatively simple medium that utilizes sorbitol to maintain osmolality (Saglio et al. 1971). A modification of this medium (BSR) has been used extensively for *S. citri* (Bové and Saillard 1979), in which the horse serum content was lowered to 10 % and the fresh yeast extract was omitted. Other simple media, such as C-3G (Liao and Chen 1977), are suitable for maintenance or large-batch cultivation of fast-growing spiroplasmas. This medium was also adequate for primary isolation of *S. kunkelii* (Alivizatos 1988). However, cultivation of more fastidious spiroplasmas is best achieved in M1D medium (Whitcomb 1983) if they derive from plant or arthropod habitats. SM-1 medium (Clark 1982) has also been successfully employed for many insect spiroplasmas. SP-4 medium (Tully et al. 1977) is suitable for spiroplasmas from tick habitats. *S. floricola* and some strains of *S. apis* have been cultivated in chemically defined media (Chang and Chen 1982; Chang 1989).

Spiroplasmas are routinely preserved by lyophilization. Most spiroplasmas can be maintained indefinitely at −70 °C; preservation success at −20 °C is irregular.

## Ecology

Almost all spiroplasmas have been found to be associated with arthropods, or an arthropod connection is strongly suspected. Numerous arthropod hosts have been identified, including insects, ticks, and crustaceans. In addition to their arthropod hosts, many of the insect-associated spiroplasmas rely upon plants for insect-to-insect dissemination on the plant surface or for growth in the plant phloem as part of their life cycle. The level of host specificity varies among *Spiroplasma* species. Most host relationships are commensal, but cases of mutualism and pathogenicity have been reported. The arthropod gut is the most common microenvironment, but some *Spiroplasma* species are able to escape the gut to infect the hemolymph or other tissues. Colonization beyond the gut is often key for pathogenicity (see ➲ Pathogenicity and Clinical Relevance).

The majority of spiroplasmas appear to be maintained in an insect gut-plant surface cycle, with tabanid-associated spiroplasmas being the most extensively surveyed. Spiroplasmas have been isolated from guts of tabanids (Diptera: Tabanidae) worldwide (French et al. 1990, 1996, 1997; Le Goff et al. 1991, 1993; Vazeille-Falcoz et al. 1997; Whitcomb et al. 1997, 2007; Regassa et al. 2009, 2011). Reported carriage rates for tabanid flies can vary dramatically, but a large-scale study that examined the carriage rates of culturable spiroplasmas for *Tabanus* species common to Australia and the southeastern United States found rates ranging from 42 % to 47 % (Regassa et al. 2009). The cumulative evidence from biodiversity studies points strongly to multiple cycles of horizontal transmission, and plant surfaces represent a major site where spiroplasmas and other microbes can be transmitted from insect to insect (Clark 1978; Davis 1978; McCoy et al. 1979). Some tabanids utilize honeydew (excreta of sucking insects) deposited on leaf surfaces, suggesting one possible mechanism for plant surface transmission. A recent study demonstrated the presence of honeydew brochosomes in the esophagus lumen of leafhoppers, consistent with the ingestion of honeydew (Ammar et al. 2011). Although several *Spiroplasma* serogroups have been isolated only from flowers or from both flowers and insects, it is not known whether any spiroplasmas can truly colonize plant surfaces. Isolations of spiroplasmas from a variety of insects suggest that it is likely that many or most of the flower isolates are deposited passively by visiting insects.

A large insect survey isolated spiroplasmas from six main orders: *Hymenoptera, Coleoptera, Diptera, Lepidoptera, Homoptera*, and *Hemiptera* (Hackett et al. 1990), although not all are involved in the insect gut-plant surface cycle.

*S. citri, S. kunkelii*, and *S. phoeniceum* have a life cycle that involves infection of plant phloem and homopterous insects (Garnier et al. 2001; Bové et al. 2003; Gasparich 2010). While no pathological effects are associated with spiroplasma multiplication in the insect host, infected plants can display a range of symptoms (see ➲ Pathogenicity and Clinical Relevance). During passage through the plant-sucking insect, spiroplasmas pass through the gut epithelial cells, multiply in the hemolymph, and then move to the salivary glands. While spiroplasmas may multiply in a number of sucking insect species that have been exposed to diseased plants, transmission of the spiroplasmas from plant to plant can often only be achieved by select vector species (Calavan and Bové 1989; Whitcomb 1989; Kersting and Sengonca 1992). The natural plant host range may be limited in part by the host specificity of the plant-sucking insects. *S. citri* is able to naturally or experimentally infect most citrus species and cultivars and many other plant species, including periwinkle and horseradish (Gasparich 2010). The natural host range of *S. kunkelii* is restricted to plants in the genus *Zea*, including maize that is susceptible to corn stunt disease (Nault 1980). *S. phoeniceum* was reported from periwinkle and was also able to infect aster plants via insect feeding (Saillard et al. 1987).

Spiroplasmas have been isolated from over 16 *Drosophila* species, where some are associated with male lethality (Haselkorn 2010; Jaenike et al. 2010a). Poulson and Sakaguchi (1961) first demonstrated that *S. poulsonii* (Williamson et al. 1999) was responsible for the sex-ratio trait in *Drosophila*. Sex-ratio spiroplasmas are vertically transmitted through female hosts, with spiroplasmas present during oogenesis (Anbutsu and Fukatsu 2003). However, horizontal transmission also occurs. A haplotype survey of 65 *Spiroplasma*-infected individuals from nine *Drosophila* species indicated at least five separate introductions of four phylogenetically distinct *Spiroplasma* haplotypes (Haselkorn et al. 2009). Ectoparasitic mites may play a role in interspecific lateral transmission of spiroplasmas (Jaenike et al. 2007). Early surveys of *Drosophila* species revealed natural infection rates for male-killing spiroplasmas of 0.1 % to 3 % for *D. hydei* in Japan and about 2.3 % for *D. melanogaster* in Brazil (Montenegro et al. 2005; Kageyama et al. 2006). Ventura et al. (2012) recently reported much higher infection rates in *D. melanogaster* in Brazil, ranging from about 0 % up to 17.7 %. While most research has focused on the sex-ratio trait in *Drosophila*, spiroplasma infections can also help protect *Drosophila* from parasites such as wasps and nematodes (Jaenike et al. 2010b; Xie et al. 2010).

Three spiroplasma serogroups have been isolated from ticks. *S. mirum* and S. sp. 277 F were isolated from the rabbit tick *Haemaphysalis leporispalustris* (Tully et al. 1982; Williamson et al. 1989), and *S. ixodetis* was isolated from *Ixodes pacificus* (Tully et al. 1995). Spiroplasmas with 16S rRNA genes highly similar to *S. ixodetis* were also found in unfed *Ixodesovatus* ticks from Japan and from an *Ixodes* tick-derived culture that was growing in a Buffalo Green Monkey mammalian cell culture line (Henning et al. 2006; Taroura et al. 2005). The ability of tick spiroplasmas, including *S. ixodetis*, to multiply at 37 °C likely reflects the role of vertebrates as tick hosts, but there is no evidence that any of these spiroplasmas are transmitted to the vertebrate host.

Within the past decade, spiroplasmas have been isolated from both freshwater and saltwater crustaceans. All cultured strains have been associated with aquaculture infections (see ➲ Pathogenicity and Clinical Relevance), including the new species *S. penaei* and *S. eriocheiris* (Nunan et al. 2005; Wang et al. 2011). *S. penaei* is lethal for Pacific white shrimp (*Penaeus vannamei*) and *S. eriocheiris* causes tremor disease in the Chinese mitten crab (*Eriocheir sinensis*). Spiroplasma-infected insects may have introduced the spiroplasmas into the aquaculture environments (Heres and Lightner 2010; Altamiranda et al. 2011). These recent discoveries have expanded the known host range for spiroplasmas and highlighted the need for further investigation of aquatic environments, as well as the interface between the terrestrial and the aquatic.

Spiroplasmas have been identified from a broad range of hosts in Africa, Asia, Australia, Europe, South America, and North America, with the greatest biodiversity seen in warm climates (Whitcomb et al. 2007). Because spiroplasmas are host associated, it seems reasonable that *Spiroplasma* species distribution would be limited by host biogeography. Early studies indicated that some tabanid-associated spiroplasmas had very discrete geographic distributions (Whitcomb et al. 1990), but the geographic scale of most ranges continues to increase as the diversity of sampling sites expands (Regassa and Gasparich 2006; Regassa et al. 2009, 2011).

## Pathogenicity and Clinical Relevance

Most spiroplasmas are commensal, but some infections adversely affect the host. Spiroplasmas have been reported as disease agents in insects, plants, and crustaceans; and, under experimental conditions, some species are pathogenic for suckling rodents or chicken embryos. Spiroplasmas have also been implicated in human disease, but contradictory reports highlight the need for more definitive evidence.

Some insect-associated spiroplasmas are entomopathogens. A critical difference between the incidental commensals and the pathogens appears to be the ability of the pathogenic spiroplasma to move from the initial site of attachment at the gut epithelial cells into the hemolymph. For example, both *S. melliferum* and *S. apis* are honeybee pathogens (Clark 1977; Mouches et al. 1982, 1983). They cross the insect gut barrier and reach the hemolymph, where they multiply abundantly and kill the bee. A recent proteogenomic study identified potential virulence factors in *S. melliferum*, including chitinase utilization enzymes and unique protein clusters with transcriptional regulators and toxins (Alexeev et al. 2012). Different modes of infection and transmission are used by spiroplasmas that cause sex ratio disorders. These sex ratio organisms are transmitted transovarially and kill the male progeny of an infected female fly. The most thoroughly studied example to date is *S. poulsonii*, which was isolated from the neotropical species *Drosophila willistoni* (Poulson and Sakaguchi 1961; Williamson and Poulson 1979; Williamson et al. 1999). In its most extreme form, a *S. poulsonii* infection is able to eliminate all male progeny in infected females. Long-term maintenance in this host is possible because spiroplasmas either do not activate or actively suppress the immune response in *Drosophila* and are not susceptible to either cellular or humoral immunity (Hurst et al. 2003; Anbutsu and Fukatsu 2010; Herren and Lemaitre 2011). In fact, activation of Toll and Imd immune pathways actually increases spiroplasma density. Spiroplasmas that cause sex ratio distortions in other hosts have also been identified, including beetles, butterflies, and moths (Hurst et al. 1999; Hurst and Jiggins 2000; Jiggins et al. 2000; Tabata et al. 2011).

The plant pathogens *S. citri*, *S. kunkelii*, and *S. phoeniceum* are maintained in a life cycle that includes both sucking insect and plant hosts (Garnier et al. 2001; Bové et al. 2003; Gasparich 2010). The spiroplasmas are maintained in the insect host and then transmitted to the plant phloem in saliva during feeding; an uninfected insect host can acquire spiroplasma during feeding when it sucks phloem sap from the sieve tubes. After ingestion, spiroplasmas in the insect host must cross the midgut barrier to infect other organs. For example, *S. kunkelii* has been observed in the leafhopper midgut, filter chamber, Malpighian tubules, hindgut, fat tissues, hemocytes, muscle, trachea, and salivary glands, but not in the brain nerve cells or nerve ganglia (Ammar and Hogenhout 2005; Ammar et al. 2011). These spiroplasmas are generally nonpathogenic for their usual insect vector. In contrast, the phytopathogenic spiroplasmas can cause a range of symptoms in the infected plant host including stunting, leaf yellowing, sterility, fruit size reduction and deformation, flower malformations, and short internodes. For commercial crops, these symptoms can result in major financial losses. In citrus stubborn disease, the severity of the disease symptoms is associated with spiroplasma density in the fruit rather than with the genotype of the infecting strain (Mello et al. 2010). *S. citri* is the causative agent of citrus stubborn disease, brittle root disease in horseradish, disease in periwinkle, and carrot purple leaf disease (Markham et al. 1974; Fletcher et al. 1981; Granett et al. 1976; Lee et al. 2006). *S. kunkelii* is the causative agent of corn stunt (Chen and Liao 1975; Williamson and Whitcomb 1975), and *S. phoeniceum* infections result in aster yellow (or Periwinkle) disease (Saillard et al. 1987). Studies to elucidate the molecular mechanisms of pathogenicity and host maintenance for the sucking insect/plant spiroplasmas have been undertaken. A full description of these studies is beyond the scope of this chapter (see Bové et al. 2003; Gasparich 2010), but some of the plasmid-encoded virulence factors were discussed above (see ➋ Genome Structure).

Spiroplasmas have been shown to cause disease in crustaceans reared in aquaculture ponds. *S. penaei* was responsible for mortality of infected Pacific white shrimp (*Penaeus vannamei*) in Columbia, South America (Nunan et al. 2004, 2005). In China, *S. eriocheiris* has been identified as the causative agent of disease in crabs, crayfish, and shrimp (Bi et al. 2008). Chinese mitten crabs (*Eriocheir sinensis*) developed tremor disease (Wang 2011; Wang et al. 2004a, b); and red swamp crayfish (*Procambarus clarkii*) that were co-reared with the Chinese mitten crabs were also infected (Bi et al. 2008; Wang et al. 2005), as were *P. vannamei* shrimp (Bi et al. 2008). In a separate incident in China, the spiroplasma strain MR-1008 was identified as the causative agent of a lethal disease in the freshwater prawn *Macrobrachium rosenbergii* (Liang et al. 2011). In all cases, spiroplasmas were isolated from the hemolymph of the infected crustaceans. Subsequent studies have shown that spiroplasmas have cytopathic effects on cultured hemocyte cells (Du et al. 2012). The spiroplasma infection induces immune gene expression and microRNA modulation in the hemocytes (Liang et al. 2012; Ou et al. 2012).

Certain species are pathogenic, under experimental conditions, for suckling rodents (rats, mice, hamsters, and rabbits) and/or chicken embryos. *S. mirum* is experimentally pathogenic for a variety of suckling animals, causing cataract and other ocular symptoms, neural pathology, and malignant transformation in cultured cells (Clark and Rorke 1979; Kotani et al. 1990). *S. melliferum* also persists and causes pathology in suckling mice (Chastel et al. 1990; Chastel and Humphery-Smith 1991), and *S. eriocheiris* is able to infect the brain tissue in embryonated chickens (Wang et al. 2003).

Spiroplasmas are among the mollicutes implicated in human disease (Baseman and Tully 1997), but two recent studies (Alexeeva et al. 2006; Hamir et al. 2011) failed to substantiate the role of spiroplasmas in animal transmissible spongiform encephalopathies (TSEs), namely, scrapie and transmissible mink encephalopathy (TME). Bastian first proposed in 1979 that spiroplasmas were associated with Creutzfeldt-Jakob disease (CJD), an extremely rare scrapie-like disease of humans (Bastian 1979). Bastian and colleagues reported finding spiroplasmas in brain samples from CJD-infected humans, scrapie-infected sheep, and chronic wasting disease-infected cervids based on 16S rDNA evidence; the sequence of the PCR amplified 16S rDNA was 96 % identical to that of *S. mirum* (Bastian and Foster 2001; Bastian et al. 2004). However, these results could not be replicated in an independent blind study of uninfected and scrapie-infected hamster brains (Alexeeva et al. 2006). In an attempt to fulfill Koch's postulate, Bastian et al. (2007) reported the transfer of spiroplasma from TSE brains and *S. mirum* to induce spongiform encephalopathy in ruminants. More recently, Hamir et al. (2011) evaluated the ability of *S. mirum* and/or TME agent to cause spongiform encephalopathy in raccoon kits. Kits intracerebrally inoculated with *S. mirum* alone did not show clinical neurological signs, their brains did not have lesions of spongiform encephalopathy, and their tissues were negative for *S. mirum* and prion protein. Spongiform encephalopathy developed in raccoon kits inoculated with TME agent. In addition, the same authors were unable to detect *S. mirum* 16S rRNA in the brains of several hundred animals with experimental or naturally occurring TSE (Hamir et al. 2011). Given the current level of understanding, it is not clear if there is a way to reconcile these seemingly contradictory reports.

Spiroplasma antimicrobial susceptibility profiles vary, but all spiroplasmas are resistant to penicillin due to the lack of a cell wall. Spiroplasmas are also resistant to rifampicin due to an asparagine residue rather than histidine at position 526 in RpoB; DNA-dependent RNA polymerases from *S. melliferum* and *S. apis* were at least 1,000 times less sensitive to rifampin than the corresponding *E. coli* enzyme (Gadeau et al. 1986; Gaurivaud et al. 1996). In early studies, spiroplasmas proved to be especially sensitive in vitro to tetracycline, erythromycin, tylosin, tobramycin, and lincomycin (Bowyer and Calavan 1974; Liao and Chen 1981b). However, strains have been isolated that are resistant to several tetracycline antibiotics, erythromycin, kanamycin, neomycin, and gentamicin (Liao and Chen 1981b; Breton et al. 2010b). Natural amphipathic peptides, such as Gramicidin S, alter the membrane potential of spiroplasma cells and induce the loss of cell motility and helicity (Bévén and Wróblewski 1997). The toxicity of the lipopeptide antibiotic globomycin was correlated with an inhibition of spiralin processing (Bévén et al. 1996). Natural18-residue peptaibols (trichorzins PA) are bacteriocidal to spiroplasmas (Bévén et al. 1998). The mode of action appears to be permeabilization of the host cell membrane.

## References


Abalain-Colloc ML, Williamson DL, Carle P, Abalain JH, Bonnet F, Tully JG, Konai M, Whitcomb RF, Bové JM, Chastel C (1993) Division of group XVI spiroplasmas into subgroups. Int J Syst Bacteriol 43:342–346

Alexeev D, Kostrjukova E, Aliper A, Popenko A, Bazaleev N, Tyakht A, Selezneva O, Akopian T, Prichodko E, Kondratov I, Chukin M, Demina I, Galyamina M, Kamashev D, Vanyushkina A, Ladygina V, Levitskii S, Lazarev V, Govorun V (2012) Application of *Spiroplasma melliferum* proteogenomic profiling for the discovery of virulence factors and pathogenicity mechanisms in host-associated spiroplasmas. J Proteome Res 11:224–236

Alexeeva I, Elliott EJ, Rollins S, Gasparich GE, Lazar J, Rohwer RG (2006) Absence of *Spiroplasma* or other bacterial 16s rRNA genes in brain tissue of hamsters with scrapie. J Clin Microbiol 44:91–97

Alivizatos AS (1988) Isolation and culture of corn stunt spiroplasma in serum-free medium. J Phytopathol 122:68–75

Altamiranda JM, Salazar MV, Briñez BR (2011) Presencia de *Spiroplasma penaei* en plancton, bentos y fauna acompañante en fincas camaroneras de Colombia. Rev MVZ Córdoba 16:2576–2583

Amikam D, Razin S, Glaser G (1982) Ribosomal RNA genes in Mycoplasma. Nucleic Acids Res 10:4215–4222

Amikam D, Glaser G, Razin S (1984) Mycoplasmas (*Mollicutes*) have a low number of rRNA genes. J Bacteriol 158:376–378

Ammar E, Hogenhout SA (2005) Use of immunofluorescence confocal laser scanning microscopy to study distribution of the bacterium corn stunt spiroplasma in vector leafhoppers (Hemiptera : Cicadellidae) and in host plants. Ann Entomol Soc Am 98:820–826

Ammar E-D, Gasparich GE, Hall DG, Hogenhout SA (2011) Spiroplasma-like organisms closely associated with the gut in five leafhopper species (Hemiptera: *Cicadellidae*). Arch Microbiol 193:35–44

Anbutsu H, Fukatsu T (2003) Population dynamics of male-killing and non-male-killing spiroplasmas in *Drosophila melanogaster*. Appl Environ Microbiol 69:1428–1434

Anbutsu H, Fukatsu T (2010) Evasion, suppression and tolerance of *Drosophila* innate immunity by a male-killing *Spiroplasma* endosymbiont. Insect Mol Biol 19:481–488

Archer DB, Best J, Barber C (1981) Isolation and restriction mapping of a spiroplasma plasmid. J Gen Microbiol 126:511–514

Bai X, Hogenhout SA (2002) A genome sequence survey of the mollicute corn stunt spiroplasma *Spiroplasma kunkelii*. FEMS Microbiol Lett 210:7–17

Baseman JB, Tully JG (1997) Mycoplasmas: sophisticated, reemerging, and burdened by their notoriety. Emerg Infect Dis 3:21–32

Bastian FO (1979) Spiroplasma-like inclusions in Creutzfeldt-Jakob Disease. Arch Pathol Lab Med 103:665–669

Bastian FO, Foster JW (2001) *Spiroplasma* sp. 16S rDNA in Creutzfeldt-Jakob disease and scrapie as shown by PCR and DNA sequence analysis. J Neuropathol Exp Neurol 60:613–620

Bastian FO, Dash S, Garry RF (2004) Linking chronic wasting disease to scrapie by comparison of *Spiroplasma mirum* ribosomal DNA sequences. Exp Mol Pathol 77:49–56

Bastian FO, Sanders DE, Forbes WA, Hagius SD, Walker JV, Henk WG, Enright FM, Elzer PH (2007) *Spiroplasma* sp. from transmissible spongiform encephalopathy brains or ticks induce spongiform encephalopathy in ruminants. J Med Microbiol 56:1235–1242

Bébéar CM, Aullo P, Bové JM, Renaudin J (1996) *Spiroplasma citri* virus SpV1: characterization of viral sequences present in the spiroplasma host chromosome. Curr Microbiol 32:134–140

Bévén L, Wróblewski H (1997) Effect of natural amphipathic peptides on viability, membrane potential, cell shape and motility of mollicutes. Res Microbiol 148:163–175

Bévén L, Le Henaff M, Fontenelle C, Wróblewski H (1996) Inhibition of spiralin processing by the lipopeptide antibiotic globomycin. Curr Microbiol 33:317–322

Bévén L, Duval D, Rebuffat S, Riddell FG, Bodo B, Wróblewski H (1998) Membrane permeabilisation and antimycoplasmic activity of the 18-residue peptaibols, trichorzins PA. Biochim Biophys Acta 1372:78–90

Bi K, Huang H, Gu W, Wang J, Wang W (2008) Phylogenetic analysis of Spiroplasmas from three freshwater crustaceans (*Eriocheir sinensis, Procambarus clarkia* and *Penaeus vannamei*) in China. J Invertebr Pathol 99:57–65

Bové JM (1993) Molecular features of mollicutes. Clin Infect Dis 17(suppl 1): S10–S31

Bové JM, Saillard C (1979) Cell biology of spiroplasmas. In: Whitcomb RF, Tully JG (eds) The mycoplasmas, vol 3. Academic, New York, pp 83–153

Bové JM, Carle P, Garnier M, Laigret F, Renaudin J, Saillard C (1989) Molecular and cellular biology of spiroplasmas. In: Whitcomb RF, Tully JG (eds) The mycoplasmas, vol 5. Academic, New York, pp 243–364

Bové JM, Renaudin J, Saillard C, Foissac X, Garnier M (2003) *Spiroplasma citri*, a plant pathogenic molligute: relationships with its two hosts, the plant and the leafhopper vector. Annu Rev Phytopathol 41:483–500

Bowyer JW, Calavan EC (1974) Antibiotic sensitivity in vitro of the mycoplasmalike organism associated with citrus stubborn disease. Phytopathology 64:346–349

Breton M, Duret S, Arricau-Bouvery N, Bévén L, Renaudin J (2008) Characterizing the replication and stability regions of *Spiroplasma citri* identifies a novel replication protein and expands the genetic toolbox for plant-pathogenic spiroplasmas. Microbiology 154:3232–3244

Breton M, Duret S, Danet LJ-L, Dubrana M-P, Renaudin J (2010a) Sequences essential for transmission of *Spiroplasma citri* by its leafhopper vector, *Circulifer haematoceps*, revealed by plasmid curing and replacement based on incompatibility. Appl Environ Microbiol 76:3198–3205

Breton M, Sagné E, Duret S, Béven L, Citti C, Renaudin J (2010b) First report of a tetracycline-inducible gene expression system for mollicutes. Microbiology 156:198–205

Brown DR, Whitcomb RF, Bradbury JM (2007) Revised minimal standards for description of new species of the class *Mollicutes* (division *Tenericutes*). Int J Syst Evol Microbiol 57:2703–2719

Calavan EC, Bové JM (1989) Ecology of *Spiroplasma citri*. In: Whitcomb RF, Tully JG (eds) The mycoplasmas. Academic, San Diego

Carle P, Tully JG, Whitcomb RF, Bové JM (1990) Size of the spiroplasmal genome and guanosine plus cytosine content of spiroplasmal DNA. Zentbl Bakteriol Suppl. 20:926–931

Carle P, Laigret F, Tully JG, Bove JM (1995) Heterogeneity of genome sizes within the genus *Spiroplasma*. Int J Syst Bacteriol 45:178–181

Carle P, Saillard C, Carrère N, Carrère S, Duret S, Eveillard S, Gaurivaud P, Gourgues G, Gouzy J, Salar P, Verdin E, Breton M, Blanchard A, Laigret F, Bové J-M, Renaudin J, Foissac X (2010) Partial chromosome sequence of *Spiroplasma citri* reveals extensive viral invasion and important gene decay. Appl Environ Microbiol 76:3420–3426

Carpane P, Melcher U, Wayadande A, Giminez MP, Laguna G, Dolezal W, Fletcher J (2012) An analysis of the genomic variability of the phytopathogenic mollicute *Spiroplasma kunkelii*. Phytopathology Sep 16 [Epub ahead of print]

Chang CJ (1989) Nutrition and cultivation of *spiroplasmas*. In: Whitcomb RF, Tully JG (eds) The mycoplasmas, vol 5. Academic, New York, pp 201–241

Chang CJ, Chen TA (1982) *Spiroplasma*: cultivation in chemically defined medium. Science 215:1121–1122

Chastel C, Humphery-Smith I (1991) Mosquito *spiroplasmas*. Adv Dis Vector Res 7:149–205

Chastel C, Gilot B, Goff FL, Divau B, Kerdraon G, Humphery-Smith I, Gruffaz R, Flohic AMS-L (1990) New developments in the ecology of mosquito *spiroplasmas*. Zentbl Bakteriol Suppl. 20:455–460

Chen TA, Liao CH (1975) Corn stunt *spiroplasma*: isolation, cultivation, and proof of pathogenicity. Science 188:1015–1017

Chipman PR, Agbandje-McKenna M, Renaudin J, Baker TS, McKenna R (1998) Structural analysis of the *Spiroplasma* virus, SpV4: implications for evolutionary variation to obtain host diversity among the *Microviridae*. Structure 6:135–145

Christiansen C, Askaa G, Freundt EA, Whitcomb RF (1979) Nucleic-acid hybridization experiments with *Spiroplasma citri* and the corn stunt and suckling mouse cataract spiroplasmas. Curr Microbiol 2:323–326

Citti C, Marechal-Drouard L, Saillard C, Weil JH, Bové JM (1992) *Spiroplasma citri* UGG and UGA tryptophan codons: sequence of the two tryptophanyl-tRNAs and organization of the corresponding genes. J Bacteriol 174:6471–6478

Clark TB (1977) *Spiroplasma* sp., a new pathogen in honey bees. J Invertebr Pathol 29:112–113

Clark TB (1978) Honey bee spiroplasmosis, a new problem for beekeepers. Am Bee J 118:18–23

Clark TB (1982) Spiroplasmas: diversity of arthropod reservoirs and host-parasite relationships. Science 217:57–59

Clark H, Rorke LB (1979) Spiroplasmas of tick origin and their pathogenicity. In: Whitcomb RF, Tully JG (eds) The mycoplasmas, vol 3. Academic, New York, pp 155–174

Cohen AJ, Williamson DL (1988) Yeast supported growth of *Drosophila* species spiroplasmas. In: Abstracts of the 7th international organization for mycoplasmology, Vienna

Cohen AJ, Williamson DL, Brink PR (1989) A motility mutant of *Spiroplasma melliferum* induced with nitrous-acid. Curr Microbiol 18:219–222

Cohen-Krausz S, Cabahug PC, Trachtenberg S (2011) The monomeric, tetrameric, and fibrillar organization of Fib: the dynamic building block of the bacterial linear motor of *Spiroplasma melleferum* BC3. J Mol Biol 410:194–213

Cole RM (1979) Mycoplasma and spiroplasma viruses: ultrastructure. In: Barile MF, Razin S (eds) The mycoplasmas, vol 1. Academic, New York, pp 385–410

Cole RM, Tully JG, Popkin TJ, Bové JM (1973) Morphology, ultrastructure, and bacteriophage infection of the helical mycoplasma-like organism (*Spiroplasma citri* gen. nov., sp. nov.) cultured from "stubborn" disease of citrus. J Bacteriol 115:367–384

Cole RM, Tully JG, Popkin TJ (1974) Virus-like particles in *Spiroplasma citri*. Colloq Inst Natl Santé Rech Med 33:125–132

Cole RM, Mitchell WO, Garon CF (1977) *Spiroplasma citri* 3: propagation, purification, proteins, and nucleic acid. Science 198:1262–1263

Dally EL, Barros TSL, Zhao Y, Lin S, Roe BA, Davis RE (2006) Physical and genetic map of the *Spiroplasma kunkelii* CR2-3x chromosome. Can J Microbiol 52:857–867

Daniels MJ, Longland JM (1984) Chemotactic behavior of spiroplasmas. Curr Microbiol 10:191–193

Daniels MJ, Longland JM, Gilbart J (1980) Aspects of motility and chemotaxis in spiroplasmas. J Gen Microbiol 118:429–436

Davis RE (1978) Spiroplasma associated with flowers of the tulip tree (*Liriodendron tulipifera* L.). Can J Microbiol 24:954–959

Davis RE, Worley JF (1973) Spiroplasma: motile helical microorganism associated with corn stunt disease. Phytopathology 63:403–408

Davis RE, Worley JF, Whitcomb RF, Ishijima T, Steere RL (1972a) Helical filaments produced by a mycoplasma-like organism associated with corn stunt disease. Science 176:521–523

Davis RE, Whitcomb RF, Chen TA, Granados RR (1972b) Current status of the aetiology of corn stunt disease. In: Elliott K, Birch J (eds) Pathogenic mycoplasmas. Elsevier Excerpta Medica, Amsterdam, pp 205–214

Davis RE, Dally EL, Jomantiene R, Zhao Y, Roe B, Lin S, Shao J (2005) Cryptic plasmid pSKU146 from the wall-less plant pathogen *Spiroplasma kunkelii* encodes an adhesin and components of a type IV translocation-related conjugation system. Plasmid 53:179–190

Dickinson MJ, Townsend R (1984) Characterization of the genome of a rod-shaped virus infecting *Spiroplasma citri*. J Gen Virol 65: 1607–1610

Du J, Ou J, Li W, Ding Z, Wu T, Meng Q, Gu W, Wang W (2012) Primary hemocyte culture of the freshwater prawn *Macrobrachium rosenbergii* and its susceptibility to the novel pathogen spiroplasma strain MR-1008. Aquaculture 330–333:21–28

Duret S, Danet JL, Garnier M, Renaudin J (1999) Gene disruption through homologous recombination in *Spiroplasma citri*: an *scm*1-disrupted motility mutant is pathogenic. J Bacteriol 181:7449–7456

El-D A, Fulton D, Bai X, Meulia T, Hogenhout SA (2004) An attachment tip and pili-like structures in insect- and plant-pathogenic spiroplasmas of the class *Mollicutes*. Arch Microbiol 18:97–105

Fisunov GY, Alexeev DG, Bazaleev NA, Ladygina VG, Galyamina MA, Kondratov IG, Zhukova NA, Serebryakova MV, Demina IA, Govorun VM (2011) Core proteome of the minimal cell: comparative proteomics of three mollicute species. PLoS One 6:e21964

Fletcher J, Schultz GA, Davis RE, Eastman EC, Goodman RM (1981) Brittle root disease of horseradish: evidence for an etiological role of *Spiroplasma citri*. Phytopathology 71:1073–1080

French FE, Whitcomb RF, Tully JG, Hackett KJ, Clark EA, Henegar RB, Wagner AG, Rose DL (1990) Tabanid spiroplasmas of the southeast USA: new groups and correlation with host life history strategy. Zbl Bakteriol (Suppl) 20:919–922

French FE, Whitcomb RF, Tully JG, Williamson DL, Henegar RB (1996) Spiroplasmas of *Tabanus lineola*. IOM Lett 4:211–212

French FE, Whitcomb RF, Tully JG, Carle P, Bové JM, Henegar RB, Adams JR, Gasparich GE, Williamson DL (1997) *Spiroplasma lineolae* sp. nov., from the horsefly *Tabanus lineola* (Diptera: Tabanidae). Int J Syst Bacteriol 47:1078–1081

Gadeau AP, Mouches C, Bové JM (1986) Probable insensitivity of mollicutes to rifampin and characterization of spiroplasmal DNA-dependent RNA polymerase. J Bacteriol 166:824–828

Garnier M, Clerc M, Bové JM (1981) Growth and division of spiroplasmas: morphology of *Spiroplasma citri* during growth in liquid medium. J Bacteriol 147:642–652

Garnier M, Clerc M, Bové JM (1984) Growth and division of *Spiroplasma citri*: elongation of elementary helices. J Bacteriol 158:23–28

Garnier M, Foissac X, Gaurivaud P, Laigret F, Renaudin J, Saillard C, Bové JM (2001) Mycoplasmas, plants, insect vectors: a matrimonial triangle. Comp Rend Acad Sci Paris Ser III 324:923–928

Gasparich GE (2010) Spiroplasmas and phytoplasmas: microbes associated with plant hosts. Biologicals 38:193–203

Gasparich GE, Hackett KJ (1994) Characterization of a cryptic extrachromosomal element isolated from the mollicute *Spiroplasma taiwanense*. Plasmid 32:342–343

Gasparich GE, Saillard C, Clark EA, Konai M, French FE, Tully JG, Hackett KJ, Whitcomb RF (1993a) Serologic and genomic relatedness of group VIII and group XVII spiroplasmas and subdivision of spiroplasma group VIII into subgroups. Int J Syst Bacteriol 43:338–341

Gasparich GE, Hackett KJ, Clark EA, Renaudin J, Whitcomb RF (1993b) Occurrence of extrachromosomal deoxyribonucleic acids in spiroplasmas associated with plants, insects, and ticks. Plasmid 29:81–93

Gasparich GE, Whitcomb RF, Dodge D, French FE, Glass J, Williamson DL (2004) The genus *Spiroplasma* and its non-helical descendants: phylogenetic classification, correlation with phenotype and roots of the *Mycoplasma mycoides* clade. Int J Syst Evol Microbiol 54:893–918

Gaurivaud P, Laigret F, Bové JM (1996) Insusceptibility of members of the class *Mollicutes* to rifampin: studies of the *spiroplasma citri* RNA polymerase beta-subunit gene. Antimicrob Agents Chemother 40:858–862

Granett AL, Blue RL, Harjung MK, Calavan EC, Gumpf DG (1976) Occurrence of *Spiroplasma citri* in periwinkle in California. Calif Agric 30:18–19

Grau O, Laigret F, Bové JM (1988) Analysis of ribosomal RNA genes in two spiroplasmas, one acholeplasma and one unclassified mollicute. Zentbl Bakteriol Suppl. 20:895–897

Grulet O, Humpherysmith I, Sunyach C, Legoff F, Chastel C (1993) "Spiromed": a rapid and inexpensive spiroplasma isolation technique. J Microbiol Method 17:123–128

Hackett KJ, Clark TB (1989) Spiroplasma ecology. In: Whitcomb RF, Tully JG (eds) The mycoplasmas, vol 5. Academic, San Diego

Hackett KJ, Lynn DE (1985) Cell-assisted growth of a fastidious spiroplasma. Science 230:825–827

Hackett KJ, Lynn DE, Williamson DL, Ginsberg AS, Whitcomb RF (1986) Cultivation of the *Drosophila* sex-ratio spiroplasma. Science 232:1253–1255

Hackett KJ, Whitcomb RF, Henegar RB, Wagner AG, Clark EA, Tully JG, Green F, McKay WH, Santini P, Rose DL, Anderson JJ, Lynn DE (1990) Mollicute diversity in arthropod hosts. Zentbl Bakteriol Suppl. 20:441–454

Hackett KJ, Hackett RH, Clark EA, Gasparich GE, Pollack JD, Whitcomb RF (1994) Development of the first completely defined medium for a spiroplasma, *Spiroplasma clarkii* strain CN-5. IOM Lett 3:446–447

Hackett KJ, Clark EA, Whitcomb RF, Camp M, Tully JG (1996) Amended data on arginine utilization by *Spiroplasma* species. Int J Syst Bacteriol 46:912–915

Hamir AN, Greenlee JJ, Stanton TB, Smith JD, Doucette S, Kunkle RA, Stasko JA, Richt JA, Kehrli ME Jr (2011) Experimental inoculation of raccoons (*Procyon lotor*) with *Spiroplasma mirum* and transmissible mink encephalopathy (TME). Can J Vet Res 75:18–24

Haselkorn TS (2010) The spiroplasma heritable bacterial endosymbiont of Drosophila. Fly 4:80–87

Haselkorn TS, Markow TA, Moran NA (2009) Multiple introductions of the spiroplasma bacterial endosymbiont into Drosophila. Mol Ecol 18:1294–1305

Henning K, Greiner-Fischer S, Hotzel H, Ebsen M, Theegarten D (2006) Isolation of *Spiroplasma* sp from an *Ixodes* tick. Int J Med Microbiol 296:157–161

Heres A, Lightner DV (2010) Phylogenetic analysis of the pathogenic bacteria *Spiroplasma penaei* based on multilocus sequence analysis. J Invertebr Pathol 103:30–35

Herren JK, Lemaitre B (2011) *Spiroplasma* and host immunity: activation of humoral immue responses increases endosymbiont load and susceptibility to certain Gram-negative bacterial pathogens in *Drosophila melanogaster*. Cell Microbiol 13:1385–1396

Hurst GD, Jiggins FM (2000) Male-killing bacteria in insects: mechanisms, incidence, and implications. Emerg Infect Dis 6:329–336

Hurst GD, Graf von der Schulenburg JH, Majerus TM, Bertrand D, Zakharov IA, Baungaard J, Volkl W, Stouthamer R, Majerus ME (1999) Invasion of one insect species, *Adalia bipunctata*, by two different male-killing bacteria. Insect Mol Biol 8:133–139

Hurst GD, Anbutsu H, Kutsukake M, Fukatsu T (2003) Hidden from the host: spiroplasma bacteria infecting Drosophila do not cause an immune response, but are suppressed by ectopic immune activation. Insect Mol Biol 12:93–97

ICSB (1984) Minutes of the interim meeting. 30 Aug and 6 Sept 1982. Tokyo. Int J Syst Bacteriol 34: 361–365

ICSB Subcommittee on the Taxonomy of Mollicutes (1995) Revised minimal standards for descriptions of new species of the class *Mollicutes* (Division *Tenericutes*). Int J Syst Bacteriol 45:605–612

ICSP Subcommittee on the Taxonomy of Mollicutes (2013) Minutes of the meetings July 15th and 19th 2012, Toulouse, France. Int J Syst Evol Microbiol 63:2361–2364

Jacob C, Nouzieres F, Duret S, Bové JM, Renaudin J (1997) Isolation, characterization, and complementation of a motility mutant of *Spiroplasma citri*. J Bacteriol 179:4802–4810

Jaenike J, Polak M, Fiskin A, Helou M, Minhas M (2007) Interspecific transmission of endosymbiotic spiroplasma by mites. Biol Lett 3:23–25

Jaenike J, Stahlhut JK, Boelia LM, Unckless RL (2010a) Association between *Wolbachia* and *Spiroplasma* within *Drosophila neotestacea*: an emerging symbiotic mutalism? Mol Ecol 19:414–425

Jaenike J, Unckless R, Cockburn SN, Boelio LM, Perlman SJ (2010b) Adaptation via symbiosis: recent spread of a Drosophila defensive symbiont. Science 329:212–215

Jiggins FM, Hurst GD, Jiggins CD, v d Schulenburg JH, Majerus ME (2000) The butterfly *Danaus chrysippus* is infected by a male-killing *Spiroplasma* bacterium. Parasitology 120:439–446

Johansson KE (2002) Taxonomy of *Mollicutes*. In: Razin S, Herrmann R (eds) Molecular biology and pathogenicity of mycoplasmas. Kluwer Academic/Plenum, New York, pp 1–29

Johnson JL (1994) Similarity analysis of DNAs. In: Gerhardt P, Murray RGE, Wood WA, Krieg NR (eds) Methods in general and molecular bacteriology. American Society for Microbiology, Washington, DC, pp 656–682

Joshi BD, Berg M, Rogers J, Fletcher J, Melcher U (2005) Sequence comparisons of plasmids pBJS-O of *Spiroplasma citri* and pSKU146 of *S. kunkelii*: implications for plasmid evolution. BMC Genomics 6:175

Junca P, Saillard C, Tully J, Garcia-Jurado O, Degorce-Dumas JR, Mouches C, Vignault JC, Vogel R, McCoy R, Whitcomb R, Williamson D, Latrille J, Bové JM (1980) Caractérization de spiroplasmes isolaté d' insectes et fleurs de France continentale, de Corse et du Maroc. Propossition pour une classification des spiroplasmes. Compt Rend Acad Sci Paris Ser 290:1209–1212

Kageyama D, Anbutsu H, Watada M, Hosokawa T, Shimada M, Fukatsu T (2006) Prevalence of a non-male-killing spiroplasma in natural populations of *Drosophila hydei*. Appl Environ Microbiol 72:6667–6673

Kersting U, Sengonca C (1992) Detection of insect vectors of the citrus stubborn disease pathogen, *Spiroplasma citri* Saglio et al., in the citrus growing area of south Turkey. J Appl Entomol 113:356–364

Killiny N, Batailler B, Foissac X, Saillard C (2006) Identification of a *Spiroplasma citri* hydrophilic protein associated with insect transmissibility. Microbiology 152:1221–1230

Konai M, Clark EA, Camp M, Koeh AL, Whitcomb RF (1996) Temperature ranges, growth optima, and growth rates of *Spiroplasma* (*Spiroplasmataceae*, class *Mollicutes*) species. Curr Microbiol 32:314–319

Kotani H, Butler GH, McGarrity GJ (1990) Malignant transformation by *Spiroplasma mirum*. Zbl Bakteriol Suppl. 20:145–152

Le Goff F, Humphery-Smith I, Leclerq M, Chastel C (1991) Spiroplasmas from European Tabanidae. Med Vet Entomol 5:143–144

Le Goff F, Marjolet M, Humphery-Smith I, Leclercq M, Helias C, Suplisson F, Chastel C (1993) Tabanid spiroplasmas from France: characterization, ecology and experimental study. Ann Parasitol Hum Comp 68:150–153

Lee IM, Davis RE (1980) DNA homology among diverse spiroplasma strains representing several serological groups. Can J Microbiol 26:1356–1363

Lee IM, Bottner KD, Munyaneza JE, Davis RE, Crosslin JM, du Toit LJ, Crosby T (2006) Carrot purple leaf: a new spiroplasmal disease associated with carrots in Washington state. Plant Dis 90:989–993

Liang T, Li X, Du J, Yao W, Sun G, Dong X, Liu Z, Ou J, Meng Q, Gu W, Wang W (2011) Identification and isolation of a spiroplasma pathogen from diseased freshwater prawns, *Macrobrachium rosenbergii*, in China: a new freshwater crustacean host. Aquaculture 318:1–6

Liang T, Ji H, Du J, Ou J, Li W, Wu T, Meng Q, Gu W, Wang W (2012) Primary culture of hemocytes from *Eriocheir sinensis* and their immune effects to the novel crustacean pathogen *Spiroplasma eriocheiris*. Mol Biol Rep 39:9747–9754

Liao CH, Chen TA (1977) Culture of corn stunt spiroplasma in a simple medium. Phytopathology 67:802–807

Liao CH, Chen TA (1981a) In vitro susceptibility and resistance of two spiroplasmas to antibiotics. Phytopathology 71:442–445

Liao CH, Chen TA (1981b) Deoxyribonucleic acid hybridization between *Spiroplasma citri* and the corn stunt spiroplasma. Curr Microbiol 5:83–86

Liao CH, Chang CJ, Chen TA (1979) Spiroplasmastatic action of plant tissue extracts. In: Proceedings of the R.O.C. U.S. Coop. Sci. Semin. Mycoplasma Dis. Plants, NSC Symp. Ser. I, Natl Sci Council, Taipei, pp 99–103

Markham PG, Townsend R, Bar-Joseph M, Daniels MJ, Plaskitt A, Meddins BM (1974) Spiroplasmas are the causal agents of citrus little-leaf disease. Ann Appl Biol 78:49–57

Markham PG, Clark TB, Whitcomb RF (1983) Culture techniques for spiroplasmas from arthropods. In: Tully JG, Razin S (eds) Methods in mycoplasmology, vol 2. Academic, New York, pp 217–223

McCoy RE, Williams DS, Thomas DL (1979) Isolation of mycoplasmas from flowers. In: Proceedings of the US-ROC plant mycoplasma seminar. National Science Council, Taipei pp 75–81

McElwain MC, Chandler DKF, Barile MF, Young TF, Tryon VV, Davis JW, Petzel JP, Chang CJ, Williams MV, Pollack JD (1988) Purine and pyrimidine metabolism in *Mollicutes* species. Int J Syst Bacteriol 38:417–423

McIntosh MA, Deng JZG, Ferrell RV (1992) Repetitive DNA sequences. In: Maniloff J (ed) Mycoplasmas, molecular biology and pathogenesis. American Society for Microbiology, Wahington, DC, pp 363–376

Melcher U, Fletcher J (1999) Genetic variation in *Spiroplasma citri*. Eur Plant Pathol 105:519–533

Mello AFS, Yokomi RK, Melcher U, Chen JC, Fletcher J (2010) Citrus stubborn severity is associated with *Spiroplasma citri* titer but not with bacterial genotype. Plant Dis 94:75–82

Miles RJ (1992) Catabolism in *Mollicutes*. J Gen Microbiol 138:1773–1783

Montenegro H, Solferini VN, Klaczko LB, Hurst GDD (2005) Male-killing spiroplasma naturally infecting *Drosophila melanogaster*. Insect Mol Biol 14:281–287

Mouches C, Bové JM, Albisetti J, Clark TB, Tully JG (1982) A spiroplasma of serogroup IV causes a May-disease-like disorder of honeybees in Southwestern France. Microbial Ecol 8:387–399

Mouches C, Bové JM, Tully JG, Rose DL, McCoy RE, Carle-Junca P, Garnier M, Saillard C (1983) *Spiroplasma apis*, a new species from the honey-bee *Apis mellifera*. Ann Microbiol (Inst Pasteur) 134A:383–397

Mouches C, Barroso G, Gadeau A, Bové JM (1984) Characterization of two cryptic plasmids from *Spiroplasma citri* and occurrence of their DNA sequences among various spiroplasmas. Ann Microbiol (Inst Pasteur) 135A:17–24

Nault LR (1980) Maize bushy stunt and corn stunt: a comparison of disease symptons, pathogen host ranges, and vectors. Phytopathology 70:659–662

Navas-Castillo J, Laigret F, Tully JG, Bové JM (1992) The mollicute *Acholeplasma florum* possesses a gene of phosphoenolpyruvate sugar phosphotransferase system and uses UGA as tryptophan codon. Compt Rend Acad Sci Paris Ser III 315:43–48

Nunan LM, Pantoja CR, Salazar M, Aranguren F, Lightner DV (2004) Characterization and molecular methods for detection of a novel spiroplasma pathogenic to *Penaeus vannamei*. Dis Aquat Organ 62:255–264

Nunan LM, Lightner DV, Oduori MA, Gasparich GE (2005) *Spiroplasma penaei* sp. nov., associated with mortalities in *Penaeus vannamei*, Pacific white shrimp. Int J Syst Evol Microbiol 55:2317–2322

Nur I, Glaser G, Razin S (1986) Free and integrated plasmid DNA in spiroplasmas. Curr Microbiol 14:169–176

Nur I, LeBlanc DJ, Tully JG (1987) Short, interspersed, and repetitive DNA sequences in *Spiroplasma* species. Plasmid 17:110–116

Oishi K, Poulson DF, Williamson DL (1984) Virus-mediated change in clumping properties of *Drosophila* SR spiroplasmas. Curr Microbiol 10:153–158

Ou J, Meng Q, Li Y, Xiu Y, Du J, Gu W, Wu T, Li W, Ding Z, Wang W (2012) Identification and comparative analysis of the *Eriocheir sinensis* microRNA transcriptome response to *Spiroplasma eriocheiris* infection using a deep sequencing approach. Fish Shellfish Immunol 32:345–352

Özbek E, Miller SA, Meulia T, Hogenhout SA (2003) Infection and replication sites of *Spiroplasma kunkelii* (Class: *Mollicutes*) in midgut and Malpighian tubules of the leafhopper *Dalbulus maidis*. J Invertebr Pathol 82:167–175

Pollack JD (2002a) Central carbohydrate pathways: metabolic flexibility and the extra role of some "housekeeping enzymes". In: Razin S, Herrmann R (eds) Molecular biology and pathogenicity of mycoplasmas. Kluwer Academic/ Plenum, New York, pp 163–199

Pollack JD (2002b) The necessity of combining genomic and enzymatic data to infer metabolic function and pathways in the smallest bacteria: amino acid, purine and pyrimidine metabolism in Mollicutes. Front Biosci 7: d1762–d1781 (Internet Journal)

Pollack JD, Mcelwain MC, Desantis D, Manolukas JT, Tully JG, Chang CJ, Whitcomb RF, Hackett KJ, Williams MV (1989) Metabolism of members of the *Spiroplasmataceae*. Int J Syst Bacteriol 39:406–412

Pollack JD, Williams MV, McElhaney RN (1997) The comparative metabolism of the mollicutes (Mycoplasmas): the utility for taxonomic classification and the relationship of putative gene annotation and phylogeny to enzymatic function in the smallest free-living cells. Crit Rev Microbiol 23:269–354

Poulson DF, Sakaguchi B (1961) Nature of "sex-ratio" agent in *Drosophila*. Science 133:1489–1490

Pyle LE, Finch LR (1988) A physical map of the genome of *Mycoplasma mycoides* subspecies *mycoides* Y with some functional loci. Nucleic Acids Res 16:6027–6039

Rahimian H, Gumpf DJ (1980) Deoxyribonucleic acid relationship between *Spiroplasma citri* and the corn stunt spiroplasma. Int J Syst Bacteriol 30:605–608

Ranhand JM, Mitchell WO, Popkin TJ, Cole RM (1980) Covalently closed circular deoxyribonucleic acids in spiroplasmas. J Bacteriol 143:1194–1199

Razin S (1983) Urea hydrolysis. In: Razin S, Tully JG (eds) Methods in mycoplasmology, vol 1. Academic, New York

Razin S (1985) Molecular biology and genetics of mycoplasmas (*Mollicutes*). Microbiol Rev 49:419–455

Regassa LB, Gasparich GE (2006) Spiroplasmas: evolutionary relationships and biodiversity. Front Biosci 11:2983–3002 (Internet Journal)

Regassa LB, Stewart KM, Murphy AC, French FE, Lin T, Whitcomb RF (2004) Differentiation of group VIII Spiroplasma strains with sequences of the 16S-23S rDNA intergenic spacer region. Can J Microbiol 50:1061–1067

Regassa LB, Murphy AC, Zarzuela AB, Jandhyam HL, Bostick DS, Bates CR, Gasparich GE, Whitcomb RF, French FE (2009) An Australian environmental survey reveals moderate *Spiroplasma* biodiversity: characterization of four new serogroups and a continental variant. Can J Microbiol 55:1347–1354

Regassa LB, French FE, Stewart KM, Murphy AC, Jandhyam HL, Beati L (2011) A Costa Rican bacterial spiroplasma biodiversity survey in tabanid flies reveals new serogroups and extends United States ranges. Int J Biodivers Conserv 3:338–344

Renaudin J, Bové JM (1994) SpV1 and SpV4, spiroplasma viruses with circular, single-stranded DNA genomes, and their contribution to the molecular biology of spiroplasmas. Adv Virus Res 44:429–463

Renaudin J, Pascarel MC, Garnier M, Carle P, Bové JM (1984a) Characterization of spiroplasma virus group 4 (SV4). Isr J Med Sci 20:797–799

Renaudin J, Pascarel MC, Garnier M, Carle-Junca P, Bové JM (1984b) SpV4, a new spiroplasma virus with circular, single-stranded DNA. Ann Virol 135E:343–361

Ricard B, Garnier M, Bové JM (1982) Characterization of spiroplasmal virus-3 from spiroplasmas and discovery of a new spiroplasmal virus (Spv4). Rev Infect Dis 4:S275

Rodwell A, Whitcomb RF (1983) Methods for direct and indirect measurement of mycoplasma growth. In: Razin S, Tully JG (eds) Methods in mycoplasmology, vol 1. Academic, New York, pp 185–196

Rogers MJ, Simmons J, Walker RT, Weisburg WG, Woese CR, Tanner RS, Robinson IM, Stahl DA, Olsen G, Leach RH (1985) Construction of the mycoplasma evolutionary tree from 5S rRNA sequence data. Proc Natl Acad Sci USA 82:1160–1164

Rose DL, Tully JG, Bové JM, Whitcomb RF (1993) A test for measuring growth responses of mollicutes to serum and polyoxyethylene sorbitan. Int J Syst Bacteriol 43:527–532

Rosengarten R, Wise KS (1990) Phenotypic switching in mycoplasmas: phase variation of diverse surface lipoproteins. Science 247:315–318

Rossello-Mora R, Amann R (2001) The species concept for prokaryotes. FEMS Microbiol Rev 25:39–67

Saglio P, Laflèche D, Bonissol C, Bové JM (1971) Isolement, culture et observation au microscope électronique des structures de type mycoplasme associées à la maladie du Stubborn des agrumes et leur comparaison avec les structures observées dans le cas de la maladie du Greening des agrumes. Physiol Vég 9:569–582

Saglio P, Lhospital M, Laflèche D, Dupont G, Bové JM, Tully JG, Freundt EA (1973) *Spiroplasma citri* gen. and sp. n.: a mycoplasma-like organism associated with "stubborn" disease of citrus. Int J Syst Bacteriol 23:191–204

Saillard C, Vignault JC, Bové JM, Raie A, Tully JG, Williamson DL, Fos A, Garnier M, Gadeau A, Carle P, Whitcomb RF (1987) *Spiroplasma phoeniceum* sp. nov., a new plant-pathogenic species from Syria. Int J Syst Bacteriol 37:106–115

Saillard C, Carle P, Duret-Nurbel S, Henri R, Killiny N, Carrére S, Gouzy J, Bové JM, Renaudin J, Foissac X (2008) The abundant extrachromosomal DNA content of the *Spiroplasma citri* GII3-3X genome. BMC Genomics 9:195

Sha Y, Melcher U, Davis RE, Fletcher J (1995) Resistance of *Spiroplasma citri* lines to the virus SVTS2 is associated with integration of viral DNA sequences into host chromosomal and extrachromosomal DNA. Appl Environ Microbiol 61:3950–3959

Sha Y, Melcher U, Davis RE, Fletcher J (2000) Common elements of spiroplasma plectroviruses revealed by nucleotide sequence of SVTS2. Virus Genes 20:47–56

Skripal IG (1974) On improvement in the systematics of the class *Mollicutes* and the establishment in the order *Mycoplasmatales* of a new family *Spiroplasmataceae* fam. nova. Mikrobiologii Zhurnal (Kiev) 36:462–467

Skripal IG (1983) Revival of the name *Spiroplasmataceae* fam. nova., nom. Rev., omitted from the 1980 Approved Lists of Bacterial Names. Int J Syst Bacteriol 33:408

Stackebrandt E, Frederiksen W, Garrity GM, Grimont PA, Kampfer P, Maiden MC, Nesme X, Rosselló-Mora R, Swings J, Trüper HG, Vauterin L, Ward AC, Whitman WB (2002) Report of the ad hoc committee for the re-evaluation of the species definition in bacteriology. Int J Syst Evol Microbiol 52:1043–1047

Stamatakis A (2006) RAxML-VI-HPC: maximum likelihood-based phylogenetic analyses with thousands of taxa and mixed models. Bioinformatics 22:2688–2690

Tabata J, Hattori Y, Sakamoto H, Yukuhiro F, Fujii T, Kugimiya S, Mochizuki A, Ishikawa Y, Kageyama D (2011) Male killing and incomplete inheritance of a novel *Spiroplasma* in the moth *Ostrinia zaguliaevi*. Microb Ecol 61:254–263

Taroura S, Shimada Y, Sakata Y, Miyama T, Hiraoka H, Watanabe M, Itamoto K, Okuda M, Inokuma H (2005) Detection of DNA of 'Candidatus *Mycoplasma haemominutum*' and *Spiroplasma* sp. in unfed ticks collected from vegetation in Japan. J Vet Med Sci 67:1277–1279

Townsend R, Markham PG, Plaskitt KA, Daniels MJ (1977) Isolation and Characterization of a non-helical strain of *Spiroplasma citri*. J Gen Microbiol 100:15–21

Townsend R, Burgess J, Plaskitt KA (1980) Morphology and ultrastructure of helical and nonhelical strains of *Spiroplasma citri*. J Bacteriol 142:973–981

Trachtenberg S (2006) The cytoskeleton of spiroplasma: a complex linear motor. J Mol Microbiol Biotechnol 11:265–283

Tully JG, Whitcomb RF, Clark HF, Williamson DL (1977) Pathogenic mycoplasmas: cultivation and vertebrate pathogenicity of a new spiroplasma. Science 195:892–894

Tully JG, Rose DL, Yunker CE, Cory J, Whitcomb RF, Williamson DL (1981) Helical mycoplasmas (spiroplasmas) from *Ixodes* ticks. Science 212:1043–1045

Tully JG, Whitcomb RF, Rose DL, Bové JM (1982) *Spiroplasma mirum*, a new species from the rabbit tick (*Haemaphysalis leporispalustris*). Int J Syst Bacteriol 32:92–100

Tully JG, Rose DL, Clark E, Carle P, Bové JM, Henegar RB, Whitcomb RF, Colflesh DE, Williamson DL (1987) Revised group classification of the genus *Spiroplasma* (class *Mollicutes*), with proposed new groups XII to XXIII. Int J Syst Bacteriol 37:357–364

Tully JG, Rose DL, Yunker CE, Carle P, Bové JM, Williamson DL, Whitcomb RF (1995) *Spiroplasma ixodetis* sp. nov., a new species from *Ixodes pacificus* ticks collected in Oregon. Int J Syst Bacteriol 45:23–28

Vaughn EE, de Vos WM (1995) Identification and characterization of the insertion element IS 1070 from *Leuconostoc lactis* NZ6009. Gene 155:95–100

Vazeille-Falcoz M, Hélias C, Goff FL, Rodhain F, Chastel C (1997) Three spiroplasmas isolated from *Haematopota* sp. (Diptera:Tabanidae) in France. J Med Entomol 34:238–241

Ventura IM, Martins AB, Lyra ML, Andrade CAC, Carvalho KA, Klaczko LB (2012) *Spiroplasma* in *Drosophila melanogaster* populations: prevalence, male-killing, molecular identification, and no association with *Wolbachia*. Microb Ecol 64:794–801

Wang W (2011) Bacterial diseases of crabs: a review. J Invertebr Pathol 106:18–26

Wang W, Rong L, Gu W, Du K, Chen J (2003) Study on experimental infections of *Spiroplasma* from the Chinese mitten crab in crayfish, mice and embryonated chickens. Res Microbiol 154:677–680

Wang W, Wen B, Gasparich GE, Zhu N, Rong L, Chen J, Xu Z (2004a) A spiroplasma associated with tremor disease in the Chinese mitten crab (*Eriocheir sinensis*). Microbiology 150:3035–3040

Wang W, Chen J, Du K, Xu Z (2004b) Morphology of spiroplasmas in the Chinese mitten crab *Eriocheir sinensis* associated with tremor disease. Res Microbiol 155:630–635

Wang W, Gu W, Ding Z, Ren Y, Chen J, Hou Y (2005) A novel spiroplasma pathogen causing systemic infection in the crayfish *Procambarus clarkii* (*Crustacea*: Decapod), in China. FEMS Microbiol Lett 249:131–137

Wang W, Gu W, Gasparich GE, Bi K, Ou J, Meng Q, Liang T, Feng Q, Zhang J, Zhang Y (2011) *Spiroplasma eriocheiris* sp. nov., associated with mortality in the Chinese mitten crab, *Eriocheir sinensis*. Int J Syst Evol Microbiol 61:703–708

Wayne LG, Brenner DJ, Colwell RR, Grimont PAD, Kandler O, Krichevsky MI, Moore LH, Moore WEC, Murray RGE, Stackebrandt E, Starr MP, Trüper HG (1987) Report of the ad hoc committee on reconciliation of approaches to bacterial systematics. Int J Syst Bacteriol 37:463–464

Weisburg WG, Tully JG, Rose DL, Petzel JP, Oyaizu H, Yang D, Mandelco L, Sechrest J, Lawrence TG, Van Etten J (1989) A phylogenetic analysis of the mycoplasmas: basis for their classification. J Bacteriol 171:6455–6467

Whitcomb RF (1983) Culture media for spiroplasmas. In: Razin S, Tully JG (eds) Methods in mycoplasmology, vol 1. Academic, New York, pp 147–158

Whitcomb, R. F. 1989.The biology of *Spiroplasma kunkelii*. *In*: Whitcomb and Tully (Ed.) The Mycoplasmas, vol. 5. Academic Press, New York, NY 487–544.

Whitcomb RF, Tully JG, Bové JM, Saglio P (1973) Spiroplasmas and acholeplasmas: multiplication in insects. Science 182:1251–1253

Whitcomb RF, Tully JG, McCawley P, Rose DL (1982) Application of the growth inhibition test to *Spiroplasma* taxonomy. Int J Syst Bacteriol 32:387–394

Whitcomb RF, Hackett KJ, Tully JG, Clark EA, French FE, Henegar RB, Rose DL, Wagner AC (1990) Tabanid spiroplasmas as a model for mollicute biogeography. Zbl Bakt Suppl. 20:931–933

Whitcomb RF, French FE, Tully JG, Carle P, Henegar R, Hackett KJ, Gasparich GE, Williamson DL (1997) *Spiroplasma* species, groups, and subgroups from north American Tabanidae. Curr Microbiol 35:287–293

Whitcomb RF, Tully JG, Gasparich GE, Regassa LB, Williamson DL, French FE (2007) *Spiroplasma* species in the Costa Rican highlands: implications for biogeography and biodiversity. Biodivers Conserv 16:3877–3894

Williamson DL (1969) The sex ratio spirochete in *Drosophila robusta*. Jpn J Genet 44:36–41

Williamson DL (1983) The combined deformation metabolism inhibition test. Method Mycoplasmol 1:477–483

Williamson DL, Poulson DF (1979) Sex ratio organisms (spiroplasmas) of *Drosophila*. In: Whitcomb RF, Tully JG (eds) The mycoplasmas, vol 3. Academic, New York, pp 175–208

Williamson DL, Whitcomb RF (1974) Helical wall-free prokaryotes in *Drosophila*, leafhoppers and plants. Colloq Inst Natl Santé Rech Med 33:283–290

Williamson DL, Whitcomb RF (1975) Plant Mycoplasmas: a cultivable spiroplasma causes corn stunt disease. Science 188:1018–1020

Williamson DL, Whitcomb RF, Tully JG (1978) The spiroplasma deformation test, a new serological method. Curr Microbiol 1:203–207

Williamson DL, Tully JG, Whitcomb RF (1979) Serological relationships of spiroplasmas as shown by combined deformation and metabolism inhibition tests. Int J Syst Bacteriol 29:345–351

Williamson DL, Tully JG, Whitcomb RF (1989) The genus *Spiroplasma*. In: Whitcomb RF, Tully JG (eds) The mycoplasmas, vol 5. Academic, San Diego, pp 71–111

Williamson DL, Whitcomb RF, Tully JG, Gasparich GE, Rose DL, Carle P, Bové JM, Hackett KJ, Adams JR, Henegar RB, Konai M, Chastel C, French FE (1998) Revised group classification of the genus *Spiroplasma*. Int J Syst Bacteriol 48:1–12

Williamson DL, Sakaguchi B, Hackett KJ, Whitcomb RF, Tuly JG, Carle P, Bové JM, Adams JR, Konai M, Henegar RB (1999) *Spiroplasma poulsonii* sp. nov., a new species associated with male-lethality in *Drosophila willistoni*, a neotropical species of fruit fly. Int J Syst Bacteriol 49:611–618

Williamson DL, Gasparich GE, Regassa LB, Saillard C, Renaudin J, Bové JM, Whitcomb RF (2010) Family II. *Spiroplasmataceae*. In: Krieg NR, Ludwig W, Whitman WB, Hedlund BP, Paster BJ, Staley JT, Ward N, Brown D, Parte A (eds) Bergey's manual of systematic bacteriology, vol 4. Springer, New York, pp 654–686

Woese CR, Maniloff J, Zablen LB (1980) Phylogenetic analysis of the mycoplasmas. Proc Natl Acad Sci USA 77:494–498

Xie J, Vilchez I, Mateos M (2010) *Spiroplasma* bacteria enhance survival of *Drosophila hydei* attacked by the parasitic wasp *Leptopilina heterotoma*. PLoS One 5:e12149

Yarza P, Ludwig W, Euzeby J, Amann R, Schleifer KH, Glöckner FO, Rossello-Mora R (2010) Update of the All-Species Living Tree Project based on 16S and 23S rRNA sequence analyses. Syst Appl Microbiol 33:291–299

Ye F, Laigret F, Whitley JC, Citti C, Finch LR, Carle P, Renaudin J, Bové JM (1992) A physical and genetic map of the *Spiroplasma citri* genome. Nucleic Acids Res 20:1559–1565

Ye FC, Laigret F, Bové JM (1994) A physical and genomic map of the prokaryote *Spiroplasma melliferum* and its comparison with the *Spiroplasma citri* map. Compt Rend Acad Sci Ser III 317:392–398

Ye FC, Laigret F, Carle P, Bové JM (1995) Chromosomal heterogeneity among various strains of *Spiroplasma citri*. Int J Syst Bacteriol 45:729–734

Ye F, Melcher U, Rascoe JE, Fletcher J (1996) Extensive chromosome aberrations in *Spiroplasma citri* Strain BR3. Biochem Genet 34:269–286

Yogev D, Rosengarten R, Watson-McKown R, Wise KS (1991) Molecular basis of mycoplasma surface antigenic variation: a novel set of divergent genes undergo spontaneous mutation of periodic coding regions and 5′ regulatory sequences. EMBO J 10:4069–4079

Zhao Y, Hammond RW, Jomantiene R, Dally EL, Lee IM, Jia H, Wu H, Lin S, Zhang P, Kenton S, Najar FZ, Hua A, Roe BA, Fletcher J, Davis RE (2003) Gene content and organization of an 85-kb DNA segment from the genome of the phytopathogenic mollicute *Spiroplasma kunkelii*. Mol Genet Genomics 269:592–602

Zhao Y, Wang H, Hammond RW, Jomantiene R, Liu Q, Lin S, Roe BA, Davis RE (2004a) Predicted ATP-binding cassette systems in the phytopathogenic mollicute *Spiroplasma kunkelii*. Mol Genet Genomics 271:325–338

Zhao Y, Hammond RW, Lee IM, Roe BA, Lin S, Davis RE (2004b) Cell division gene cluster in *Spiroplasma kunkelii*: functional characterization of *ftsZ* and the first report of *ftsA* in mollicutes. DNA Cell Biol 23:127–134

MIX
Papier aus verantwortungsvollen Quellen
Paper from responsible sources
FSC® C105338

Printed by Books on Demand, Germany